AA002292

2007 International Conference on Power Electronics and Drive Systems

Bangkok, Thailand
27-30 November 2007

Pages 949-1422

IEEE Catalog Number:	CFP07PEL-PRT
ISBN 10:	1-4244-0644-7
ISBN 13:	978-1-4244-0644-9

Copyright © 2007 by The Institute of Electrical and Electronics Engineers, Inc.
All Rights Reserved

Copyright and Reprint Permissions: Abstracting is permitted with credit to the source. Libraries are permitted to photocopy beyond the limit of U.S. copyright law for private use of patrons those articles in this volume that carry a code at the bottom of the first page, provided the per-copy fee indicated in the code is paid through Copyright Clearance Center, 222 Rosewood Drive, Danvers, MA 01923.

For other copying, reprint or republications permission, write to IEEE Copyrights Manager, IEEE Operations Center, 445 Hoes Lane, Piscataway, New Jersey USA 08854. All rights reserved.

IEEE Catalog Number:	CFP07PEL-PRT
ISBN 10:	1-4244-0644-7
ISBN 13:	978-1-4244-0644-9
LOC:	2006933010

Additional Copies of This Publication Are Available from:

IEEE Service Center
445 Hoes Lane
Piscataway, NJ 08854

Phone:	(800) 678-IEEE
	(732) 981-1393
Fax:	(732) 981-9667
E-mail:	customer-service@ieee.org

Organizers/Committees

Organizers:

Chulalongkorn Univ.
Center of Excellence in Electrical Power Tech., Chulalongkorn Univ.
King Mongkut's Univ. of Tech. Thonburi
King Mongkut's Inst. of Tech. Ladkrabang
King Mongkut's Inst. of Tech. North Bangkok
IEEE Thailand Section
IEEE IAS/PELS Joint Chapter, Singapore Section

Technical Co-Sponsors:

IEEE Power Electronics Society
IEEE Industry Applications Society
IEEE Industrial Electronics Society
IEEE IAS/PELS/IES Joint Chapter, Thailand Section

Organizing Committees

General Chairman	Doncker, R. D.	RWTH-Aachen Univ.
Advisory Board	Lavansiri, D.	Chulalongkorn Univ.
	Jaovisidha, V.	Chulalongkorn Univ.
	Tandhavatana, S.	IEEE Thailand Section
	Pungprasert, V.	IEEE Thailand Section
	Leelarasmee, E.	Chulalongkorn Univ.
	Pichetchumroen, V.	King Mongkut's Inst. of Tech. Ladkrabang
	Yingwatana, A.	King Mongkut's Inst. of Tech. North Bangkok
	Liang, Y. C.	National Univ. of Singapore
	Panda, S. K.	National Univ. of Singapore
General Co-Chairmen	Karnasuta, K.	IEEE Thailand Section
	Vilathgamuwa, D. M.	Nanyang Tech. Univ.
Organizing Committee **Chairman**	Phoomvuthisarn, S.	Center of Excellence in Electrical Power Tech. Chulalongkorn Univ.
Co-Chairman	Kulvitit, Y.	Chulalongkorn Univ.
	Yungyuen, U.	King Mongkut's Univ. of Tech. Thonburi
	Khan-ngern, W.	King Mongkut's Inst. of Tech. Ladkrabang
	Chunkag, V.	King Mongkut's Inst. of Tech. North Bangkok
Technical Program Committee	Khan-ngern, W.	King Mongkut's Inst. of Tech. Ladkrabang
	King Jet, T.	Nanyang Tech. Univ.
	Sangwongwanich, S.	Chulalongkorn Univ.
	Chunkag, V.	King Mongkut's Inst. of Tech. North Bangkok
	Sirisukprasert, S.	Kasetsart Univ.
	Boonyaroonate, I.	King Mongkut's Univ. of Tech. Thonburi
Treasurers	Bunnagulrote, B.	Center of Excellence in Electrical Power Tech. Chulalongkorn Univ.
	Battul, D.	School of Electrical & Electronic Engineering Singapore Polytechnic
Publications	Tarateeraseth, V.	Srinakharinwirot Univ.
	Jangwanitlert, A.	King Mongkut's Inst. of Tech. Ladkrabang
Tutorials	Chunkag, V.	King Mongkut's Inst. of Tech. North Bangkok
	Liutanakul, P.	King Mongkut's Inst. of Tech. North Bangkok
Local Arrangements	Kinnares, V.	King Mongkut's Inst. of Tech. Ladkrabang
	Polmai, S.	King Mongkut's Inst. of Tech. Ladkrabang
	Yutthagowith, P.	King Mongkut's Inst. of Tech. Ladkrabang
	Kittiratsatcha, S.	King Mongkut's Inst. of Tech. Ladkrabang
	Fuengwarodsakul, N.	King Mongkut's Inst. of Tech. North Bangkok
Publicity	Suwankawin, S.	Chulalongkorn Univ
Exhibition	Jangwanitlert, A.	King Mongkut's Inst. of Tech. Ladkrabang
Secretariat	Suwankawin, S.	Chulalongkorn Univ.

International Steering Committee

Acarnley, P.	Univ. of Newcastle	Kennel, R.	Univ. of Wuppertal
Akagi, H.	Tokyo Inst. of Tech.	Kolar, J. W.	Swiss Federal Inst. of Tech. (ETH) Zürich
Alex, Q. H.	North Carolina State Univ.	Lai, J. S.	Virginia Polytechnic Inst. and State Univ.
Amaratunga, G. A. J.	Univ. of Cambridge	Longya, X.	Ohio State Univ.
Bhat, A. K. S.	Univ. of Victoria	Lorenz, R. D.	Univ. of Wisconsin-Madison
Boroyevich, D.	Virginia Polytechnic Inst. and State Univ.	Matsuse, K.	Meiji Univ.
Bose, B. K.	Univ. of Tennessee	Mohan, N.	Univ. of Minnesota
Chan, C.C.	Univ. of Hong Kong	Nakaoka, M.	Yamaguchi Univ.
Clare, J. C.	Univ. of Nottingham	Ninomiya, T.	Kyushu Univ.
Dehong, X.	Zhejiang Univ.	Okuma, S.	Nagoya Univ.
Divan, D.	Georgia Inst. of Tech.	Qian, Z.	Zhejiang Univ.
Elbuluk, M. E.	Univ. of Akron	Rahman, M. A.	Memorial Univ. of Newfoundland
Enjeti, P.	Texas A&M Univ.	Schroeder, D.	Univ. of Munich
Ertan, B.	Middle East Technical Univ.	Sekiya, H.	Chiba Univ.
Forsyth, A.J.	Univ. of Manchester	Sen, P.C.	Queen's Univ.
Green, T. C.	Imperial College	Shoyama, M.	Kyushu Univ.
Guo, X. D.	Harbin Inst. of Tech.	Suetsugu, T.	Fukuoka Univ.
Holtz, J.	Univ. of Wuppertal	Teck, O. B.	McGill Univ.
Hui, R. S. Y.	City Univ. of Hong Kong	Tenti, P.	Univ. of Padova
Husain, I.	Univ. of Akron	Tsai, P. C.	National Tsing Hua Univ.
Jahns, T. M.	Univ. of Wisconsin–Madison	Undeland, T. M.	Norwegian Univ. of Science and Tech.
Jain, P.	Queen's Univ.	Wu, B.	Ryerson Univ.
Jezernik, K.	Univ. of Maribor	Wyk, J. D. V.	Virginia Polytechnic Inst. and State Univ.
Kazimierczuk, M. K.	Wright State Univ.	Zhengming, Z.	Tsing Hua Univ.

History of PEDS Conference

International Conference on Power Electronics and Drive Systems, PEDS Conference, originated in Singapore and the first PEDS conference was held in Singapore in 1995. The aim of the PEDS Conference is to provide a forum for participants from the industry and academia in the area of power electronics and drives to exchange ideas and have interactions. The conference is biennial and since 1995 the PEDS Central Committee, Singapore in collaboration with overseas organizing committees, have organized PEDS Conference series held in various Asia Pacific (IEEE Region 10) countries. All the PEDS conferences are being held in technical co-sponsorship with the IEEE Power Electronics Society and IEEE Industry Applications Society.

PEDS Conference	Venue
PEDS 1995	Singapore
PEDS 1997	Singapore
PEDS 1999	Hong Kong
PEDS 2001	Bali, Indonesia
PEDS 2003	Singapore
PEDS 2005	Kuala Lumpur, Malaysia
PEDS 2007	Bangkok, Thailand

List of Reviewers

Abbaszadeh, K.
Abe, S.
Acarnley, P. P.
Adnani, M. E.
Afjei, E.
Ahmad, G.
Ahmed, M.
Al-Haddad, K.
Amirifar, R.
Ang, S.
Apte, A. A.
Attaviriyanupap, P.
Awad, H.
Azli, N. A.
Baiju, M. R.
Bakan, A. F.
Beig, A. R.
Bhadra, S. N.
Bharanikumar, R.
Bhat, A. K. S.
Bina, M. T.
Biswas, S. K.
Boonyaroonate, I.
Bunlaksananusorn, C.
Chang, K.-T.
Chen, H.
Chen, J.-J.
Cheng, M.-Y.
Chengfeng, Y.
Cheung, N. C.
Chiba, A.
Chien, F. T.
Chiu, H.-J.
Choi, B.
Chou, J.-H.
Chunkag, V.
Clare, J. C.
Colli, V. D.
Corzine, K. A.
Covic, G. A.
Cruden, A.
Dahono, P. A.
Daming, Z.
Dianguo, X.
Doki, S.
Dong-Hee, L.
Duffy, M.
Dzung, P. Q.
Elbuluk, M. E.
Ertugrul, N.
Eskander, M.
Farhangi, S.
Filho, E. R.
Forsyth, A.

Fujiwara, O
Fukuda, S.
Garvey, S. D.
Grabner, H.
Griva, G.
Gueldner, H.
Guo, Y.
Hagh, M. T.
Hakimie, H.
Hamzah, M. K.
Hamzah, N.
Hanamoto, T.
Hava, A. M.
Hayashi, Y.
Hennen, M. D.
Higuchi, T.
Ho, S.-T.
Hofmann, W.
Hori, Y.
Howe, D.
Hsieh, G.-C.
Hua, S.
Huang, L.
Huang, S.-J.
Hung, J. Y.
Hur, J.
Hussien, Z. F.
Idris, N. R. N.
Jain, P. K.
Janakiraman, P. A.
Jangwanitlert, A.
Jerome, J.
Jianxin, S.
Khan, P. K. S.
Khan-Ngern, W.
Kim, I.-S.
Kim, Y.-H.
Kinnares, K.
Kobayashi, S.
Kolar, J. W.
Komurcugil, H.
Kubota, H.
Kulvitit, Y.
Kurokawa, F.
Lafoz, M.
Lai, C.-K.
Lai, J.-S.
Lecci, A.
Lee, D.-C.
Lee, E.-W.
Lee, S. C.
Lee, Y.-S.
Li, D. D.
Li, G.

Li, H.
Li, J.
Li, W.
Liang, Y. C.
Liaw, C.-M.
Lin, C.-H.
Lin, R.-L.
Liserre, C.
Lo, Y.-K.
Loh, A.
Lorenz, L.
Low, K.-S.
Manmek, T
Markadeh, G. A.
Marques, G. D.
Martins, J. F. A.
Matsui, M.
Matsuo, K.
Mekhilef, S.
Morales-Castorena, A.
Morimoto, M.
Morimoto, S.
Mukerjee, R.
Muni, B. P.
Murthy, S. S.
Mutoh, A.
Muyeen, S. M.
Nagaraju, J.
Narayanan, G.
Nho,N. V.
Noguchi, T.
Nussbaumer, T.
Okou, A. F.
Omar, A. M.
Pai, F.-S.
Palandurkar, M. V.
Pan, C.-T.
Panda, S. K.
Patel, H. K.
Phuong, L. M.
Pichetjamroen, V.
Ping, H. W.
Pires, A.
Pires, V.
Polmai, S.
Ponce, M.
Qian, Z.
Rafael, S.
Rahman, M. A.
Ramasamy, A. K.
Rashad, E. E. M.
Ratanapanachote, S.
Rizk, J.
Ruan, X.

Saied, B. M.
Sangwongwanich, S.
Saudemont, C.
Saxena, T. K.
See, K. Y.
Senthilkumar, R.
Shaojun, X.
Sharma, V. K.
Shieh, H.-J.
Shimizu, T.
Shin, G.-H.
Shing, C. S.
Shinnaka, S.
Shuhua, F.
Singh, B.
Sirisukprasert, S.
Soltani, J.
Sopavanit, C.
Staines, C. S.

Sumedha
Sun, K.
Suwankawin, S.
Tahami, F.
Takahashi, R.
Tanaka, T.
Tarnekar, S. G.
Tenti, P.
Thounthong, P.
Tomita, H.
Tseng, K.-J.
Tsui, M.
Vaclavek, P.
Vaez-Zadeh, S.
Veszpremi, K.
Vijayarajan, K.
Vilathgamuwa, D. M.
Villasenor, A. G.
Wang, C.-M.

Wang, H.-P.
Wang, L.
Weiming, M.
Wen, F.-L.
Wolbank, T. M.
Wu, L.
Wu, T.-F.
Xu, D. (David)
Xu, D. (Dehong)
Xu, L.
You, K.
Yousfi, D.
Zhang, X.
Zhengyu, L.
Zhong , Q.-C.
Zhu, J.
Zirn, O.
Zolghadri, M. R.

This page intentionally left blank.

Table of Contents

Fuel cell systems and applications 1
Bernard Davat

Recent Trends iin Power Qualliity Improvements Techniiques 58
Bhim Singh

Power Electronics for Future Utility Applications 213
Rik W. De Doncker, Christoph Meyer, Robert U. Lenke, Florian Mura

Digital Control Generations -- Digital Controls for Power Electronics through the Third Generation 221
Philip T. Krein

Power Electronics and Control of Renewable Energy Systems 226
F. Iov, M. Ciobotaru, D. Sera, R. Teodorescu, F. Blaabjerg

Design and Evaluation of a 60 000 rpm Permanent Magnet Bearingless High Speed Motor 249
T. Schneider, A. Binder

Performance Investigation of Two-, Three- and Four-Phase Bearingless Slice Motor Configurations 257
M.T. Bartholet, S. Silber, T. Nussbaumer, J.W. Kolar

Compensation of Pole Position Estimation Error for Sensor-less IPMSM Drives with DC Link Current
Detection 265
Hisao Kubota, Yusuke Shibano, Takayuki Kobayashi

A Novel Dual-Stator Hybrid Excited Synchronous Wind Generator 270
Liu Xiping, Lin Heyun, Yang Chengfeng, Fang Shuhua, Guo Jian

Application of Multi-level Multi-domain Modeling in the Design and Analysis of a PM Transverse Flux
Motor with SMC Core.. 275
Youguang Guo, Jianguo Zhu, Dikai Liu, Haiyan Lu, Shuhong Wang

New Approximate 2DOF Digital Controller for DC-DC Converter with Second-Order Differential
Characteristics 280
Eiji Takegami, Kohji Higuchi, Kazushi Nakano, Satoshi Tomioka, Kazushi Watanabe, K.K. Densei-Lambda

High Accuracy CMOS Current Sensing Circuit for Current Mode Control Buck Converter................. 286
Yuang-Shung Lee, Chih-Jen Hsu

Small Signal Analysis of a dual-switch forward Converter with non-ideal transformer in Current-
Programmed Control 291
Weiping Zhang, Yuzhou Lei, Xiaoqiang Zhang, Yuanchao Liu

High Frequency Transformer Designs for Improving Cross Regulation in Multiple-Output Flyback
Converters 295
Kusumal Chalermyanont, Pairote Sangampai, Anuwat Prasertsit, Surapon Theinmontri

Operation of a wye Connected Three- Level Active Power Filter under Nonideal Conditions 299
H.B. Zhang, A.M. Massoud, S.J. Finney, B.W.Williams, T.C. Lim, H. Hotait

Application of GPRS Techniques for Wide-Area Power Quality Monitoring 305
Shun-Yu Chan, Jen-Hao Teng, David Chang, Li-Yuan Chin

Design and Development of Autotransformer Based 24-Pulse AC-DC Converter fed Induction Motor
Drive... 310
Bhim Singh, Vipin Garg, G.Bhuvaneswari

Power Quality Monitoring System Using Real-Time Operating System 318
Krisda Yingkayun, Suttichai Premrudeepreechacharn, Kosol Oranpiroj

Technology Performance Comparison of Triacs Subjected to Fast Transient Voltages 322
L. Gonthier, A. Passal

Table of Contents

On-line Junction Temperature Measurement of CoolMOS Devices..327
Andreas Koenig, Thomas Plum, Peter Fidler, Rik W. De Doncker

Analytical Design of High-Power MTO Thyristors...333
Thomas Plum, Rik W. De Doncker

A Novel Gate Driver with Output Voltage Having Double Source Voltage...................................338
K. I. Hwu, Y. T. Yau

Effects of Internal Feedback and Gate-Drive Signal on the Turn-off Loss of MOSFET ZVS.......342
Youthana Kulvitit, Puckapon Opanuruk, Tanvaa Tansatit

A Novel Bridge Type FCL Based on Single Controllable Switch...350
Wanmin Fei, Yanli Zhang, Qi Wang

A Novel Isolation Power Supply for Gating Multiple Devices in FACTS Equipment...................354
Yanli Zhang, Wanmin Fei, Zhengyu Lu

Voltage and Frequency Controller for Parallel Operated Isolated Asynchronous Generators........357
Bhim Singh, Gaurav Kumar Kasal

DSP controlled Semiconductor based High-Voltage Source...363
F. Martin, T. Leibfried, O. Kerz, K. Mossner

Open Switch Fault Diagnosis for a Doubly-Fed Induction Generator..368
W. Sae-Kok, D M Grant

Rapid Analysis & Design Methodologies of High- Frequency LCLC Resonant Inverter as Electrodeless Fluorescent Lamp Ballast..376
Yong-Ann Ang, David Stone, Chris Bingham, Martin Foster

Analysis and Control of Dual-Output LCLC Resonant Converters, and the Impact of Leakage Inductance..............382
Y. Ang, C. M. Bingham, M. P. Foster, D. A. Stone

A Novel QR ZCS Switched-Capacitor Bidirectional Converter..388
Yuang-Shung Lee, Yi-Pin Ko, Chien-An Chi

Analysis of a Half - Bridge Inverter for a Small- Size Induction Cooker Using Positive-Negative Phase-Shift Control under ZVS and NON-ZVS Operation...394
P. Achara, P. Viriya, K. Matsuse

Adaptive Phase Control Method for Load Variation of Resonant Converter with Piezoelectric Transformer..401
S. T. Yun, J. M. Sim, J. H. Park, S. J. Choi, B. H. Cho

Adaptation of Motor Parameters in Sensorless PMSM Drives..406
Antti Piippo, Marko Hinkkanen, Jorma Luomi

Development of 150000 r/min, 1.5 kW Permanent- Magnet Motor for Automotive Supercharger........414
Toshihiko Noguchi, Masaru Kano

Analysis and Performance Evaluation of Radial Flux Air-Cored Permanent Magnet Machines with Concentrated Coils..420
P.J. Randewijk, M.J. Kamper, R-J. Wang

Analysis and Experimental Investigation for Field-Control Capability of a Novel Hybrid Excitation Claw-Pole Synchronous Machine...427
Yang Chengfeng, Lin Heyun, Liu Xiping, Fang Shuhua, Guo Jian

A single-Capacitor Turn-off Snubber for Interleaved Boost Converter with Coupled Inductor.......433
S.-Y. Tseng, J. Z. Shiang, Y.-H. Su

Buck-Boost Converter Associated with Active Clamp Forward Converter for PV Power System........440
S. Y. Tseng, W. C. Chen, Y. J. Li, J. S. Kuo

Table of Contents

Comparison of Three-Phase DC-DC Converters vs. Single-Phase DC-DC Converters 448
Christian P. Dick, Andreas Konig, Rik W. De Doncker

Applying Modified One-Comparator Counter-Based PWM Control Strategy to Flyback Converter 456
K. I. Hwu,, Y. H. Chen

Analysis of Conducted EMI Reduction on a Boost Converter Using Progressive Inductor Winding Technique 460
Kritsada Saritsiri, Werachet Khan-Ngern

Practical Issues Concerned with Zero sequence component and Harmonic Compensation in Four-Wire systems 465
E. Pashajavid, K. Kanzi, M. Tavakoli Bina

Automated Design and Implementation of Resonant Controllers for Current Control of Shunt Active Filters 470
W. Lenwari, M. Sumner, P. Zanchetta

A Modular Structured Multilevel Inverter Active Power Filter with Unified Constant-Frequency Integration Control for Nonlinear AC Loads 475
P. Y. Lim, N. A. Azli

HCC PWM Control of the Single-Phase Bi- Directional Buck Converter giving IEEE 519 Compliance at any Power Factor 480
A. N. Arvindan, V. K. Sharma

Passive EMI Filter Performance Improvements with Common Mode Voltage Cancellation Technique for PWM Inverter 488
C. Khun, W. Khan-Ngern, M. Kando

Novel Auxiliary Diagnosis Method for State-of-Health of Lead-Acid Battery 493
Yu-Hua Sun, Hurng-Liahng Jou, Jinn-Chang Wu

Electromechanical Model of a Longitudinal Mode Piezoelectric Transformer 498
Shine-Tzong Ho

Latest Development of Transformer Parasitic Inductive Components and Lossless Inductive Snubber-Assisted Series Resonant High-Frequency ZCS-PFM DC-DC Converter for RF Generator 504
Hisayuki Sugimura, Manabu Ishitobi, Bishwajit Saha, Sang Pil Mun, Soon Kurl Kwon, Mutsuo Nakaoka

A General Method for Deciding the Input Filter Capacitance of Flyback Switching AC-DC Converter with Peak Current-Controlled Mode 510
Jiaxin Chen, Jianguo Zhu, Youguang Guo

Design of High Performance and Low Cost Line Impedance Stabilization Network for University Power Electronics and EMC Laboratories 515
D. Sakulhirirak, V. Tarateeraseth, W. Khan-Ngern, N. Yoothanom

A Robust Output Current Control Method with Disturbance Observer for Matrix Converter under Unbalanced Input Voltage 521
Kazuo Oka, Kouki Matsuse

FPGA Design of Single-phase Matrix Converter Operating as Cycloconverter 527
Z. Idris, M.K. Hamzah, A. Saparon, N.R.Hamzah, N.Y. Dahlan

Input and Output Ripple Analysis of AC Chopper 534
Arwindra Rizqiawan, Dessy Amirudin, Deni, Pekik Argo Dahono

A Three-level 4 × 3 Conventional Matrix Converter 541
Runjie Rong, Poh Chiang Loh, Peng Wang, Frede Blaabjerg

A novel primary-side controlled contactless battery charger 546
Yi-Hwa Liu, Shun-Chung Wang, Rong Ceng Leou

Table of Contents

Research on Digital Soft-switch Welding/Cutting Inverter Power Source 551
G.R. Zhu, Z. Liu, X. Li, B.Y. Liu, S.X. Duan, Y. Kang

Design of an Adjustable High Output Voltage Asymmetrically Switched Class D Converter 556
M. Rentzsch, H. Guldner, C. Ditmanson

New Direct High Frequency Soft-Switching Inverter-Fed AC-DC Converter with Voltage Doubler for Consumer Magnetron Drive 563
Hisayuki Sugimura, Bishwajit Saha, Hidekazu Muraoka, Sang Pil Mun, Tomokazu Mishima, Hideki Omori, Mutsuo Nakaoka

Complete loading Characteristics Modeling of an Axial Flux Permanent Magnet Synchronous Machine Using Ck Spline Functions 569
Z. Lakhdari, F. Amrane, L. Adélaide, Ph. Makany

The Bearingless 2-Level Motor 574
P. Karutz, T. Nussbaumer, W. Gruber , J.W. Kolar

Analysis and Design of a Sliding Mode Controller for Buck Converters Operating in DCM with Adaptive Hysteresis Band Control Scheme 581
Hung-Chih Lin, Tsin-Yuan Chang

Buck Converter Simulation Technique Based on the Fourier Transform 587
Acacio M. R. Amaral, A. J. Marques Cardoso

ANALYSIS OF HOPF BIFURCATION IN DC-DC LUO CONVERTER USING CONTINUOUS TIME MODEL 595
A.Kavitha, G.Uma

Analysis of a Mixed-Signal Control for DC-DC Converters based on Hysteresis Modulation And Estimated Inductor Current 600
D. Trevisan, S. Saggini, P. Mattavelli, L. Corradini, P. Tenti

Power Quality Study in Macao 607
Sio-Un Tai, Man-Chung Wong, Ming-Chui Dong, Ying-Duo Han

Some Findings on Harmonic Measurement in Macao 614
Sio-Un Tai, Man-Chung Wong, Ming-Chui Dong, Ying-Duo Han

Coordinated design of PSS and TCSC dynamics model for power system network oscillations 620
M. Tarafdar Haque, A. Roshan Milani, A. Lafzi

An Analytic Approach To Harmonic Analysis of 48-Pulse Voltage Source Inverter 626
B. Geethalakshmi, P. Dananjayan

Detailed losses Analysis of High-Frequency Planar Power Transformer 632
Yu Ma, Peipei Meng, Junming Zhang, Zhaoming Qian

Design of a Nuclear Magnetic Resonance Fast Field Cycling Air Cored Magnet 636
Duarte M. Sousa, Gil D. Marques, Pedro J. Sebastiao,, Antonio C. Ribeiro

Using DFT to Obtain the Equivalent Circuit of Aluminum Electrolytic Capacitors 643
Acácio M. R. Amaral, Gustavo M. Buatti, Hugo Ribeiro, A.J. Marques Cardoso

A Mathematical Analysis on Vector Inversion Generators 648
D. J. Thrimawithana, U. K. Madawala

Novel Multi-Level High Voltage Pulsed Power Generator 654
D. J. Thrimawithana, U. K. Madawala

Potential and Electric Field Distribution Analysis of Field Limiting Ring and Field Plate by Device Simulator 660
C.N. Liao, F.T. Chien, Y.T. Tsai

Table of Contents

Wire and Wireless Linked Remote Control for the Group Lighting System Using Induction Lamps 665
Kyu Min Cho, Jae Eul Yeon, Ma Xian Chao, Hee Jun Kim

Induction Heating with Traveling Magnetic Field for Uniform Heating to Flat Metal 671
T. Sekine, H. Tomita, Y. Saito, S. Obata, S. Yoshimura

Three-Phase (LC)(L)-Type Series-Resonant Converter with Capacitive Output Filter 677
M. Almardy, A.K.S. Bhat

Analysis of a Full-Bridge Inverter for Induction Heating Using Asymmetrical Phase-Shift Control under ZVS and NON-ZVS Operation 685
N. Yongyuth, P. Viriya, K. Matsuse

FPGA-Based Phase-Shift ZVS Full-Bridge DC-DC Converter Using One-Comparator Counter-Based PWM Control Strategy 692
K. I. Hwu, Y. T. Yau

A Simplified Power Control Scheme for Resonant Inverter with Purely Resistive Load 697
Pramoch Dorkmai, Youthana Kulvitit, Tanvaa Tansatit

Voltage Injection Based Initial Rotor Position Estimation Method for Three-Phase Star- Connected Switched Reluctance Machines 703
P. Somsiri, P. Champa, P. Wipasuramonton, K. Tungpimonrut, P. Aree

Control Scheme for Switched Reluctance Drives with Minimized DC-Link Capacitance 710
Christoph R. Neuhaus, Rik W. De Doncker, Nisai H. Fuengwarodsakul

Multiphase Torque-Sharing Concepts of Predictive PWM-DITC for SRM 716
Helge J. Brauer, Martin D. Hennen, Rik W. De Doncker

A New Two Phase Configuration for Switched Reluctance Motor with High Starting Torque 722
E. Afjei, K. Navi, S. Ataei

Application of Power Electronics for Damping of Torsional Vibrations 726
T. Zoller, T. Leibfried, A. M. Miri

Application of Battery Energy Operated System to Isolated Power Distribution Systems 731
Bhim Singh, A. Adya, A.P. Mittal, J.R.P Gupta

Pulse Doubling in 18-Pulse AC-DC Converters 738
Bhim Singh, Sanjay Gairola

Magnetic Field Analysis and Control Strategy of Permanent Magnet Actuator for Low Voltage Vacuum Circuit Breaker 745
Fang Shuhua, Lin Heyun, Yang Chenfeng, Liu Xiping, Guo Jian

Analysis of Transformer Inrush Current under Harmonic Source 749
Chien-Lung Cheng, Jim-Chwen Yeh, Shyi-Ching Chern, Yi-Hung Lan

Voltage Sag Compensation Performance by DSTATCOM with Series Inductor and Energy Storage 755
Sumate Naetiladdanon

Cooperative Operation of Active Power Filters by Instantaneous Complex Power Control 760
Elisabetta Tedeschi, Paolo Tenti, Paolo Mattavelli

Impact of Adjustable Speed PWM drives on Operation and Harmonic Losses of Nonlinear Three Phase Transformers 767
M.A.S. Masoum, Paul S. Moses, Amir S. Masoum

Real-Time Implementation of Voltage Dip Mitigation using D-STATCOM with Fast Extraction of Instantaneous Symmetrical Components 773
Thip Manmek, Chathura P. Mudannayake

Table of Contents

Combined System of Static Synchronous Series Compensation and Passive Filter applied to Wind Energy Conversion System 781
A. Singer, W. Hofmann

Control of active injector for multi-pulse rectifiers operating on variable frequency supplies 788
Ismael Araujo-Vargas, Andrew J. Forsyth, F. Javier Chivite-Zabalza

36-pulse hybrid ripple injection for high performance aerospace rectifiers 796
F. Javier Chivite-Zabalza, Andrew J. Forsyth, Ismael Araujo-Vargas

A 48-pulse converter using dc-ripple injection 804
F. Javier Chivite-Zabalza, Andrew J. Forsyth

A Study of Different Possible Switched Mode Chopper Circuits for Multi-Magnet Based DC Electromagnetic Levitation System 812
Subrata Banerjee, Dinkar Prasad, Jayanta Pal

Power Supply with Potential Use in Magnetic Stimulation 817
Duarte M. Sousa, Antonio Ferraz

A Novel Maximum Power Point Tracking Method for the Photovoltaic System 824
Hurng-Liahng Jou, Wen-Jung Chiang, Jinn-Chang Wu

Maximum Power Point Algorithm in PV Generation: An Overview 829
Hardik P. Desai, H. K. Patel

A DC-Module-Based Power Configuration for Residential Photovoltaic Power Application 836
Bangyin Liu, Shanxu Duan, Yong Kang

Analysis and Improvement of Maximum Power Point Tracking Algorithm Based on Incremental Conductance Method for Photovoltaic Array 842
Bangyin Liu, Shanxu Duan, Fei Liu, Pengwei Xu

Application of Maximum Power Point Tracker with Self-organizing Fuzzy Logic Controller for Solar-powered Traffic Lights 847
Noppadol Khaehintung, Phaophak Sirisuk

Supply-side Current Harmonics Control of Three Phase PWM Boost Rectifiers Under Distorted and Unbalanced Supply Voltage Conditions 852
Xinhui Wu,, Sanjib K. Panda, Jianxin Xu

A Two-stage Converter with a Coupled-Inductor 858
Hirotaka Nakanishi, Yoshihiro Tomihisa, Terukazu Sato, Takashi Nabeshima, Kimihiro Nishijima, Tadao Nakano

Three-Phase AC to DC Converter with Minimized DC Bus Capacitor and Fast Dynamic Response 863
U. Kamnarn, Y. Kanthaphayao, V. Chunkag

A Simple Effective Duty Cycle Controller for High Power Factor Boost Rectifier 869
Hussain S. Athab, P. K. Shadhu Khan

A Cost Effective Method of Reducing Total Harmonic Distortion (THD) in Single-Phase Boost Rectifier 874
Hussain S. Athab, P. K. Shadhu Khan

Comparison of Different Methods to Detect Static Air Gap Asymmetry in Inverter Fed Induction Machines 880
T.M. Wolbank, P. Macheiner

Analysis of the Synchronous Torques in a Split Phase Induction Motor 886
P. Scavenius Andersen, D. G. Dorrell, N. C. Weihrauch, P. E. Hansen

On-Line Diagnosis of Three-Phase Closed Loop Induction Motor Drives Using an Eigenvalue aß-Vector Approach 894
J. F. Martins, V. Fernao Pires, A. J. Pires

xiv

Table of Contents

Design and Development of a 36-Pulse AC-DC Converter for Vector Controlled Induction Motor Drive 899
Bhim Singh, Sanjay Gairola

Comparison of Outer- and Inner-Rotor Switched Reluctance Machines 907
Martin D. Hennen, Rik W. De Doncker

Optimization of Predesign of Switched Reluctance Machines Cross Section Using Genetic Algorithms 912
Satit Owatchaiphong, Christian Carstensen, Rik W. De Doncker

Shaft Position for an 8/6 Switched Reluctance Machine: Theoretical concept, FEM analysis and Experimental results. 917
Silviano Rafael, P.J. Costa Branco, A.J. Pires

Sensorless Control of Brushless Doubly-Fed Reluctance Machines using an Angular Velocity Observer 922
Milutin G Jovanovic, David G Dorrell

A Half-Bridge PV System with Bi-direction Power Flow Controlling and Power Quality Improvement 930
C.L. Shen, S.T. Peng

Response of DSTATCOM under Voltage Flicker In Farm Wind 937
K. Aodsup, P. N. Boonchiam, A. Sode-Yome, P. Kongsuk, N. Mithulananthan

A Comparative Study of Fixed Speed and Variable Speed Wind Energy Conversion Systems Feeding the Grid 941
S.S. Murthy, Bhim Sing, P.K. Goel, S.K. Tiwari

Prediction of Wind Power Generation based on Chaotic Phase Space Reconstruction Models 949
Dong Lei, Wang Lijie, Hu Shi, Gao Shuang, Liao Xiaozhong

Power Flow Control for Efficiency Improvement in a Forward-Flyback Mixed Converter 954
Yoshito Kusuhara, Asahi Nakayama, Tamotsu Ninomiya, Shin Nakagawa

Hammerstein Model-Based Robust Control of DC/DC Converters 959
F. Alonge, F. D'ippolito, T. Cangemi

A New Model Control DC-DC Converter to Improve Dynamic Characteristics 968
F. Kurokawa, S. Sukita

Fuzzy Incremental Controller for the 3rd Order Buck Converter 973
M. Veerachary, Deepen Sharma

Design of a Single-Stage Single-Switch Power- Factor-Corrected (S4-PFC) AC/DC Converter 977
P. Kongthawornwattana, C. Bunlaksananusorn, S. Kittiratsatcha

A DSP-Based Unified Three Phase/Switch/Level Unity Power Factor Rectifier Using Feedback Linearization for DC-Bus Voltage Control 983
Ali Moallem, Hesameddin Mirzaee Teshnizi, Mohammadreza Zolghadri

A Soft-Switched AC-DC Symmetrical Boost Converter with Power Factor Correction 989
A. Jangwanitlert, J. Songboonkaew

Education Reforming for Power Electronics 994
Weiping Zhang, Xiaohan Guan, Dongyan Zhang

A Novel Current Control System for PMSM Considering Effects from Inverter in Overmodulation Range 999
Smith Lerdudomsak, Shinji Doki, Shigeru Okuma

Modelling of the Feeding Network of a Linear Synchronous Machine and Estimation of Model Parameters 1006
J. Rost, H. Gueldner, R. Hellinger, A. Weller

Analysis of Losses in Inverter Fed Large Scale Synchronous Machines using 2D FEM Software 1012
Samer Shisha, Chandur Sadarangani

xv

Table of Contents

Position sensorless control of the Reluctance Synchronous Machine considering High Frequency inductances 1017
H.W. De Kock, M.J. Kamper, O.C. Ferreira, R.M. Kennel

Carrier PWM algorithm in overmodulation range for Multileg Multilevel Inverter 1027
Nguyen Van Nho, Hong Hee Lee

Carrier Based Single-state PWM Technique In multilevel Inverter 1033
Nguyen Van Nho, Quach Thanh Hai, Hong Hee Lee

Implementation of a Single-carrier Multilevel PWM Technique Using Field Programmable Gate Array (FPGA) 1041
N. A. Azli, L. Y. Teng, P. Y. Lim

SPACE VECTOR PWM FOR MULTILEVEL INVERTERS - A FRACTAL APPROACH 1047
Anish Gopinath, M.R. Baiju

Elimination of Harmonics in a Five-Level Diode-Clamped Multilevel Inverter Using Fundamental Modulation 1055
Sule Ozdemir, Engin Ozdemir, Leon M. Tolbert, Surin Khomfoi

Compensation of DC-Link Oscillations of Cascaded H-Bridge Converters 1060
M. Tavakoli Bina, B. Eskandari

Combined DC-Filter and optimized Modulation to Absorb DC-Link Oscillations of Cascaded H-Bridge Converters 1065
M. Tavakoli Bina, B. Eskandari

Control Strategies of a Hybrid Multilevel Converter for Expanding Adjustable Output Voltage Range 1070
Shoji Fukuda, Takatsugu Yoshida, Shigeta Ueda

High Efficiency Single Phase Multi-level Inverter by New Controlled Switch Signal 1078
Ruthapong Kumchaiyo, Itsda Boonyaroonate

FPGA Implementation of Quasi-BLDC Drive 1082
C.S. Soh, C. Bi, K.K. Teo

A Practical Method to Eliminate the Conduction Torque Ripple in BLDCM Using Cascade Topology 1088
Xiaofeng Zhang, Zhengyu Lu, Yu Ma, Zhaoming Qian

Program Architecture for Realizing Design Optimization of a BLDC Motor 1092
Dong-Hun Kim, Giwoo Jeung, Heung-Geun Kim, In Dong Kim

Stable Operation of the Brushless Doubly-Fed Machine (BDFM) 1096
Shiyi Shao, Ehsan Abdi, Richard Mcmahon

Sail Generator Feasibility Study 1102
Ha Pham Ngoc, Yasuaki Matsui, Pathom Attaviriyanupap, Osamu Iso

Braking Circuit of Small Wind Turbine Using NTC Thermistor under Natural Wind Condition 1109
Y. Matsui, A. Sugawara, S. Sato, T. Takeda, K.Ogura

Flywheel Energy Storage Drive for Wind Turbines 1115
K. Veszpremi, I. Schmidt

Theory, Simulation and Experimental Verification of a New Integral Cycle Robust Control Strategy for Self Excited Induction Generators 1123
S.S. Murthy, A.J.P. Pinto

Performance Comparison of DC Link Voltage Controllers in Vector Controlled Boost Type PWM Converter for Wind Turbine System 1129
W. Sudmee, B. Neammanee

Analysis and Design of Class DE Amplifier with Nonlinear Shunt Capacitance 1136
Hiroo Sekiya, Takayuki Watanabe, Tadashi Suetsugu, Marian K. Kazimierczuk

xvi

Table of Contents

A Novel Control Strategy of the Class-D Stereo Audio Amplifier......................1142
Kyu Min Cho, Won Seok Oh, Hai Xu, Hee Jun Kim

Robust H_infinity Control Design for PFC Rectifiers......................1147
F. Tahami, H. Molla Ahmadian, A. Moallem

Parallel Operation of Power Factor Corrected AC-DC Converter Modules With Two Power Stages......................1152
Aravind Pothana, Krishna Vasudevan

Noise Radiation of Switched Reluctance Drives......................1160
K. A. Kasper, M. Bosing, R. W. De Doncker, S. Fingerhuth, M. Vorlander

Iron Losses in Electrical Machines Due to Non Sinusoidal Alternating Fluxes......................1167
J. A. Walker, D. G. Dorrell, E. Ritchie

Design Requirements for Doubly-Fed Reluctance Generators......................1174
D. G. Dorrell

A Magnetic Gear Box for application with a Contra-rotating Tidal Turbine......................1182
Laxman Shah, A. Cruden, Barry W. Williams

Mechatronic . Advanced Computational Intelligence......................1187
D. Schroder, H. Schuster, C. Westermaier

New Space Vector Control Approach for Four Switch Three Phase Inverter (FSTPI)......................1195
Phan Quoc Dzung, Le Minh Phuong, Pham Quang Vinh, Nguyen Minh Hoang, Tran Cong Binh

The Development of Artificial Neural Network Space Vector PWM for Four-Switch Three- Phase Inverter......................1202
Phan Quoc Dzung, Le Minh Phuong, Pham Quang Vinh

Voltage Losses Compensation Using Artificial Neural Network for Estimation Nonlinear Characteristic of Switches......................1208
N. Pothi, S. Premrudeepreechacharn, C. Rakpenthai

A Simple Carrier-Based PWM Method For Three-Phase Four-Leg Inverters Considering All Four Pole Voltages Simultaneously......................1213
Nakharet Chudoung, Somboon Sangwongwanich

Inverted Sine Carrier Pulse Width Modulation for Fundamental Fortification in DC-AC Converters......................1221
R.Nandhakumar, S.Jeevananthan

Fault Detection and Reconfiguration Technique for Cascaded H-bridge 11-level Inverter Drives Operating under Faulty Condition......................1228
Surin Khomfoi, Leon M. Tolbert

Investigation into Harmonic Losses in a PWM Multilevel cascaded H-Bridge Inverter Fed Induction Motor......................1236
Prasopchok Hothongkham, Vijit Kinnares

Extend the Use of Auxiliary Circuit to Start up, Shut down, and Balance of the Modified Diode Clamped Multilevel Inverter......................1242
Ahmed Ali Ashaibi, S.J. Finney, B.W. Williams, Ahmed Massoud

Five-Level Z-Source Neutral-Point-Clamped Inverter......................1247
F. Gao, P. C. Loh, F. Blaabjerg, R. Teodorescu, D. M. Vilathgamuwa

Capacitor Voltage Balancing Using Redundant States for Five-Level Multilevel Inverter......................1255
Hadi A Hotait, Ahmed M Massoud, Steve J. Finney, Barry W. Williams

Sliding Mode Repetitive Control of PWM Voltage Source Inverter......................1262
Sufen Chen, Y. M. Lai, Siew-Chong Tan, Chi K. Tse

Output Current Ripple Analysis of Five-Phase PWM Inverters......................1267
Deni, E. G. Supriatna, P. A. Dahono

xvii

Table of Contents

An Improved 'DC-DC Type' High Frequency Transformer-Link Inverter by Employing Regenerative Snubber Circuit 1274
Z. Salam, S. M. Ayob, M. Z. Ramli, N. A. Azli

A Novel Dimming Technique for Cold Cathode Fluorescent Lamp 1278
K. I. Hwu, Y. H. Chen

Time Delay Compensation For A DSP-Based Current-Source Converter Using Observer-Predictor Controller 1284
Huu-Phuc To, Muhammed Fazlur Rahman, Colin Grantham

Implementation of Hysteresis Current Control for Single-Phase Grid Connected Inverter 1290
Krismadinata, Nasrudin Abd Rahim, Jeyraj Selvaraj

Use of Air-Cored Axial Flux Permanent Magnet Generator in Direct Battery Charging Wind Energy Systems 1295
F.G. Rossouw, M.J. Kamper

Transverse Flux Machines for Sustainable Development - Road Transportation and Power Generation 1301
D. Svechkarenko, A. Cosic, J. Soulard, C. Sadarangani

Low Voltage Ride-Through Capability for Wind Turbines based on Current Source Inverter Topologies 1308
Pierluigi Tenca, Andrew A. Rockhill, Thomas A. Lipo

Optimal Control of Direct Driven Feed Axes with Flexible Structural Components 1316
Ekkehard Batzies, Tobias Scholler, Volkmar Welker, Oliver Zirn

Leakage Energy Recovered Narrow Pulsed Voltage Generator Associated with Ultrasound Generator for Liquid Food Sterilization 1321
S. Y. Tseng, Y. D. Chang, P. L. Huang, T. F Wu, Y. M. Chen

Energy Harvesting from Exercise Bicycle 1327
Suchart Janjornmanit, Samart Yachiangkam, Aswin Kaewsingha

Modeling and Analysis of Igniter for HID Lamps 1330
Weiping Zhang, Qiang Cheng

Design of a Single Bi-directional DC-DC Converter for Onboard Energy Improving of Zero Emission Electric Vehicles 1335
Werachet Khan-Ngern

Speed Sensorless Control with Neuron MRAS Estimator of an Induction Machine 1340
Dong Lei, Yang Dong, Liao Xiaozhong

Adaptive Flux model for commissioning of signal injection based zero speed sensorless flux control of induction machines 1346
T.M. Wolbank, M.A. Vogelsberger, R.H. Stumberger

Design and Performance of a Single Stator, Dual Rotor Induction Motor 1352
S. Sinha, N. K. Deb, N. Mondal, S. K. Biswas

Investigation of skew effect on the Performance of Self - Excited Induction Generators 1356
B. Sawetsakulanond, V. Kinnares

Analysis of Double Loops Discrete Single Input PI Fuzzy for Single phase Inverter 1363
S.M. Ayob, Z. Salam, N.A. Azli

A new three-phase varying-band hysteresis current controller for voltage-source inverters 1368
Vinciane Chereau, Francois Auger, Luc Loron

Diode-Assisted Buck-Boost Current Source Inverters 1376
F. Gao, C. Liang, P. C. Loh, F. Blaabjerg

Table of Contents

Single-Stage Fluorescent Lamps Electronic Ballast Using Class-DE Low dv/dt Rectifier for Power-Factor Correction........1383
Chainarin Ekkaravarodome, Adisak Nathakaranakule, Itsda Boonyaroonate

Output Impedance Design Consideration of Three Control Schemes for Bus Converter in On-Board Distributed Power System........1388
Seiya Abe, Masahiko Hirokawa, Tamotsu Ninomiya

Optimal Generation Rescheduling for Security Operation of Power Systems Using Optimal Control Theory........1394
J. Q. Sun, K. W. Chan, D. Z. Fang

Improvement of Transient Response of Thermal Power Plant Using VVVF Inverter........1398
N. Matsui, F. Kurokawa

A Novel Circuit Topology for Three-Phase Four-Wire Distribution Electronic Power Transformer........1404
H.Mirmousa, M.R.Zolghadri

A Half-Bridge DC/DC Converter for Plasma Cutting Machine........1412
N. Sanajit, A. Jangwanitlert

Ripple Estimation for Paralleled Converter System with Automatic Interleaving Function........1417
Teruhiko Kohama, Ryota Tsunesada, Tamotsu Ninomiya

Design of a New Hysteretic PWM Controller for All Types of DC-to-DC Converters........1423
Min Lin, Takashi Nabeshima, Terukazu Sato, Kimihiro Nishijima

Implementation of Fuzzy Logic Controller with Bifurcation Control of a Current-mode Boost Converter........1429
Noppadol Khaehintung, Phaophak Sirisuk, Anantawat Kunakorn

Phase Advance Approach to Expand the Speed Range of Brushless DC Motor........1434
Binhminh Nguyen, Minh C. Ta

Nonlinear Decoupled Control for a Six-Phase Series-Connected Two Induction Motor Drive Using the Sliding-Mode Technique........1442
J. Soltani, N. R. Abjadi, Gh. R. Arab Markadeh

The Decoupled Stator Flux and Torque Sliding-Mode Control of Induction Motor Drive Taking the Iron Losses into Account........1449
M.Hajian, J.Soltani, S.Hosein Nia, G.R.Arab

AN EFFICIENT DIRECT TORQUE CONTROL SCHEME FOR SPLIT PHASE INDUCTION MOTOR........1455
A. Khajeh, J. S. Moghani, M. Shahbazi

A Method of Speed Sensorless Vector Control Parallel -Connected Dual Induction Motors Fed by One Inverter in a Rotor Flux Feedback Control........1460
Jun Nishimura, Kazuo Oka, Kouki Matsuse

A Combined Model Flux Observer for Vector Control of Traction Asynchronous Motors........1465
F. Tahami, S. Chini Foroosh

Torque Ripple Elimination for Doubly-Fed Induction Motors under Unbalanced Source Voltage........1471
Hong-Geuk Park, Ahmed G. Abo-Khalil, Dong-Choon Lee, Kwang-Myoung Son

Online H8 Speed Control of Sensorless Induction Motors with Rotor Resistance Estimation........1477
Peda V Medagam, Farzad Pourboghrat

Analysis and Comparative Study on the Performance between Standard and High Efficiency Induction Machines operating as Self - Excited Induction Generators........1483
B. Sawetsakulanond, V. Kinnaraes

A simple Approach to Capacitance Determination of Self - Exited Induction Generators for Terminal Voltage Regulation........1489
B. Sawetsakulanond, V. Kinnares

xix

Table of Contents

Symmetrical Components-Based Control Technique of Doubly Fed Induction Generators under Unbalanced Voltages for Reduction of Torque and Reactive Power Pulsations................................1495
S. Wangsathitwong, S. Sirisumrannukul, S. Chatratana, W. Deleroi

A New Switching Technique for Direct Torque Control of Induction Motor using Four-Switch Three-Phase Inverter 1501
Phan Quoc Dzung, Le Minh Phuong, Pham Quang Vinh, Nguyen Minh Hoang, Nguyen Xuan Bac

Detection of Some Parameters of Induction Motors a Proposal and Its Verification................................1507
H. Bulent Ertan, Volkan Sezgin, Baris Colak

Comparison of Basic Direct Torque Control Designs for Permanent Magnet Synchronous Motor................................1514
M. N. Abdul Kadir, S. Mekhilef, W.P. Hew

Improved DSVM-DTC Based Current Sensorless Permanent Magnet Synchronous Motor Drive................................1520
Bhim Singh, Devendra Goyal

A High Performance Direct Torque Control Scheme of Permanent Magnet Synchronous Motor................................1527
Dong-Hee Lee, Young-Joo An, Eui-Chel Nho

Low Cost Position Sensor for Permanent Magnet Linear Drive................................1533
Ralf Wegener, Florian Senicar, Christian Junge, Stefan Soter

Design of One Rotary-linear Permanent Magnet Motor with Two Independently Energized Three Phase Windings................................1538
L. Chen, W. Hofmann

Position Estimation of Permanent Magnet Synchronous Motor Using Un-known Input Observer 1543
Masaru Hasegawa, Satoshi Yoshioka, Keiju Matsui

Switched Reluctance Motor Drive for Electric Motorcycle Using HFNN Controller 1549
Chih-Hong Lin

STATE - SPACE AVERAGING, SIMULATION, STABILITY STUDIES FOR STEP UP POSITIVE OUTPUT SWITCHED CAPACITOR DC-DC CONVERTER................................1555
E. Jayashree, G. Uma, M. Vaigundamoorthi

Active Clamp Interleaved Boost Converter with Coupled Inductor for High Step-up Ratio Application 1560
S. Y. Tseng, J. Z. Shiang, W. S. Jwo, C. M. Yang

Active Clamp Interleaved Flyback Converter with Single-Capacitor Turn-off Snubber for Stunning Poultry Applications................................1567
S. Y. Tseng, C. T. Hsieh, H. C. Lin

Novel Current Feedforward Average Current Mode Control Technique to Improve Output Dynamic Performance of DC-DC Converters................................1575
P. Chrin, C. Bunlaksananusorn

Stability Analysis of Cascaded DC-DC Power Electronic System................................1581
M. Veerachary, S. Bala Sudhakar

Averaged Switch Modeling of DC/DC Converters using New Switch Network................................1586
Chien-Min Lee, Yen-Shin Lai

Soft Transition Operation of UPS in High- Power-Factor Mode of Three-Phase Front- End Rectifier................................1590
G. A. Dhomane, H. M. Suryawanshi

Specific Harmonic Power Suppression of Direct- Power-Controlled Current-Source PWM Rectifier................................1595
Toshihiko Noguchi, Kohji Sano

Frequency-Controlled LCC Resonant Converter with Synchronous Rectifier 1601
Yu Ma, Xiaogao Xie, Zhaoming Qian

Selection of the Filter Capacitor for Power Supplies using 1-Phase Diode Rectifier................................1605
N. Mondal, S. K. Biswas, S. Sinha, N. K. Deb

Table of Contents

High Performance Single-Phase Voltage Regulator with a Simple Circuit Topology 1610
Chien-Ming Wang, Ching-Hung Su, Chang-Hua Lin, Maw-Yang Liu, Kuo-Lun Fang

Small-Signal Modeling of Series Resonant Converter 1615
Weiping Zhang, Peng Mao, Yuanchao Liu

Modelling of Three phase Z-Source Boost Buck Rectifiers 1620
D M Vilathgamuwa, P C Loh, K Karunakar

A NEW SINGLE-PHASE CONTROLLED RECTIFIER USING SINGLE-PHASE MATRIX CONVERTER WITH REGENERATIVE CAPABILITIES 1626
R. Baharom, M.K. Hamzah, A. Saparon, S.Z. Mohammad Noor, N.R.Hamzah

Implementation of Space Vector Modulated 3. to 3 . Matrix Converter Fed Induction Motor 1632
S. Ganesh Kumar, S. Siva Sankar, S. Krishna Kumar, G. Uma

A Single-Phase High-Power-Factor Neutral-pointer Clamped Multilevel Rectifier 1636
Yun Xu, Yunping Zou, Chengzhi Wang, Wei Chen, Bangyin Liu

Two Phase Inverter Drive of Three Phase Motor 1641
Saksit Jangjaempradit, Masayuki Morimoto

Predictive Current Controller for Inverter Fed Medium Voltage Drives with LC Filter 1645
T. Laczynski, A. Mertens

Novel Control Strategy of Instantaneous Power Based CVCF Inverter 1651
Akira Sato, Toshihiko Noguchi

An Improved Parallel Processing UPS Using a Voltage-Controlled Voltage Source Inverter 1657
S.W. Lee, H. Dehbonei, S.H. Ko, S.R. Lee, B.H. Jang, Y.H. Moon, T.K. Ko

A PEMFC/Battery Hybrid UPS System for Backup and Emergency Power Applications 1662
Yuedong Zhan, Jianguo Zhu, Youguang Guo, Hua Wang

Design of the Two Parallel Inverter Modules by Circular Chain Control Technique 1667
K. Piboonwattanakit, W. Khan-Ngern

Investigation of Topologies of Low Voltage Multilevel Inverters 1672
Yanli Zhang, Wanmin Fei, Shoufang Wang

Solution for PWM converter switching for Voltage Source Inverter using Non- Traditional Method 1677
V. Jegathesan, Jovitha Jerome

Piecewise Linear Control Surface for Single Input Nonlinear PI-Fuzzy Controller 1682
S. M. Ayob, Z. Salam, N. A. Azli

Open-Loop Control of a Stepping Motor through IP Network 1686
K. Matsuo, T. Miura, T. Taniguchi

Fuzzy Logic Controller for Electric Vehicle Braking Strategy.....**Fig 4. adjusted due to text re-flow 1691
Xixi. Wang, K.W.Eric Cheng, Xiaozhong Liao, Norbert C. Cheung, Lei Dong

Skid Steering in 4-Wheel-Drive Electric Vehicle 1697
Gao Shuang, Norbert C. Cheung, K. W. Eric Cheng, Dong Lei, Liao Xiaozhong

A Flexible Multi-Pulse Control Strategy for Universal Nail Collator 1703
Chien-Lung Cheng, Shyi-Ching Chern, Jim-Chwen Yeh, Ming-Yi Wu

Cycloconverter Based Three Phase Induction Motor to Replace Flywheel of the Process Machine 1708
M.V. Palandurkar, M. A. Chaudhari, J. P. Modak, S. G. Tarnekar

A Novel Zero-Voltage-Switching Single-Stage High-Power-Factor Electronic Ballast 1712
Chien-Ming Wang, Ching-Hung Su, Chang-Hua Lin, Maw-Yang Liu, Kuo-Lun Fang

Opto-Mechatronic System Design of the LED Projector by Using Brushless DC Motor 1717
Jian-Long Kuo, Tzu-Hsuan Fang

xxi

Table of Contents

The Color Measurement System of PWM-Controlled LCD by Using Back-Propagation Neural Network................1722
Jian-Long Kuo, Xian-Lin Liu

Gapped Air-cored Power Converter for Intelligent Clothing Power Transfer..1727
Y. Lu, K.W.E.Cheng, Y. L. Kwok, K. W. Kwok, K.W. Chan, N.C.Cheung

Simulation Program for Switching Converters Using Numerical Fourier Transform..............................1734
Yoshihiro Tomihisa, Hirotaka Nakanishi, Terukazu Sato, Takashi Nabeshima, Kimihiro Nishijima, Tadao Nakano

The Most Suitable Application of SiC Diode...1740
Tomoaki Makino, Atsushi Hirota, Satoshi Nagai

Multi-Domain System Simulation and Rapid Prototyping of Digital Control Algorithms using VHDL-AMS...1744
P.J. Randewijk

Reforming Power Electronics Laboratory...1752
Xiaohan Guan, Weiping Zhang, Xusen Zhao, Yuanchao Liu

Online performance monitoring and testing of electrical equipment using Virtual Instrumentation........1757
S.S. Murthy, Raghu K. Mittal, Avneesh Dwivedi, G. Pavitra, Sonika Choudhary

A Balancing Strategy and Implementation of Current Equalizer for High Power LED Backlighting.......1762
Chang-Hua Lin, Tsung-You Hung, Chien-Ming Wang, Kai-Jun Pai

Modeling of the Parasitical Capacitance Effect in LCD Panel and Corresponding Elimination Strategy.................1767
Chang-Hua Lin, Tsung-You Hung, Chien-Ming Wang, Kai-Jun Pai

On-line SOC Estimation of Battery for Wireless Tram Car...1773
Hiroyuki Miyamoto, Masayuki Morimoto, Katsuaki Morita

Narrow- control-bandwidth Operation of Piezoelectric-transformer Converter....................................1777
Weiping Zhang, Xiaoqiang Zhang, Yuzhou Lei, Yuanchao Liu

Modified Map of Variable Active Passive Reactance for Stability Evaluation with Consideration of Capacitor Mode...1782
S. Mohammad Shariatmadar, Jalal Nazarzadeh

Design of the Longitudinal Mode Piezoelectric Transformer..1788
Shine-Tzong Ho

The Comparison of Conducted EMI Emission and Electrical Performances of Lamps..........................1794
C. Uyaisom, W. Khan-Ngern

Neural Identification of Average Model of STATCOM using DNN and MLP...1799
M. Tavakoli Bina, S. Rahimzadeh

Hybrid Simulation of Power Systems with Dynamic Phasor SVC Transient Model..............................1804
E. Zhijun, K. W. Chan, D. Z. Fang

CONTROL OF CURRENT- SOURCE ACTIVE POWER FILTER USING UNIT VECTOR TEMPLATE IN THREE PHASE FOUR WIRE UNBALNCED SYSTEM...1810
K. Vadirajacharya, Pramod Agarwal, H.O. Gupta

Improved Control of Three Phase Active Filters Using Genetic Algorithms...1816
Bhim Singh, Varun Singhal

A Fuzzy Adaptive Detecting Approach of Harmonic Currents for Active Power Filter.........................1822
Yilong Qu, Weipu Tan, Yihan Yang

Comparative Evaluation of Harmonic Extraction Techniques for Three-Phase Three-Wire Active Power Filter...1827
R. Chudamani, Krishna Vasudevan, C.S. Ramalingam

xxii

Table of Contents

Hybrid Passive Filter Design for Distribution Systems with Adjustable Speed Drives 1834
M.A.S. Masoum, A. Ulinuha, S. Islam, K. Tan

A Graphic User Interface-based Program for Voltage Sag Calculation 1840
T. Tayjasanant, K. Yossombut, P. Sawatpipat

Operational Characteristics of Fault Current Limiting Reactor Combined with Multi- Functional Inverter 1846
S. H. Ko, S. H. Lim, S. R. Lee, S. W. Lee, I. C. Kim, S. H. Ko, H. S. Kim

Low Cost AC Solid State Circuit Breaker 1851
W. Pusorn, W. Srisongkram, W. Subsingha, S. Deng-Em, P. N. Boonchiam

A Variable Gain Control Scheme of Digital Automatic Voltage Regulator for AC Generator 1857
Dong-Hee Lee, Jin-Woo Ahn, Tae-Won Chun

A Graphic User Interface-based Program for Harmonic Impedance Calculation 1862
T. Tayjasanant

The analysis and simulation of power circuits for AC high-voltage converters 1868
Y.Y. Skorokhod, S.I. Volskiy

A Single Stage Flyback PFC Converter for Testing Distance Relay Systems 1875
V. Fernao Pires, J. F. Martins, J. Fernando Silva

H-Infinity Control Theory Apply to New Type Arc-suppression Coil System 1880
Yilong Qu, Weipu Tan, Yihan Yang

Characteristics of a novel topology of a DC-AC Converter for Fuel Cells 1885
K. Fukushima, T. Ninomiya, I. Norigoe, Y. Harada, K. Tsukakoshi, Z. Dai

A Comparative Study of PWM Schemes for Grid Connected PV Cell 1891
Vineeta Agarwal, Alok Vishwakarma

This page intentionally left blank.

Prediction of Wind Power Generation based on Chaotic Phase Space Reconstruction Models

Dong Lei[1,2], Wang Lijie [1], Hu Shi[3], Gao Shuang [1], Liao Xiaozhong[1,2]

[1]Department of Automatic Control, Beijing Institute of Technology, China

[2]Key Laboratory of Complex System Intelligent Control and Decision, Ministry of Education, China

[3]Chu Kochen Honors College, Zhejiang University, China

Abstract–The development of wind generation has rapidly progressed over the last decade, but it must be integrated into power grids and electric utility systems. However, it cannot be dispatched like conventional generators because the power generated by the wind changes rapidly because of the continuous fluctuation of wind speed and direction. So it is very important to predict the wind power generation. This paper discusses why the wind power generation can be predicted in short-term, and how to setup the construction of an ANN (Artificial Neural Network) prediction model of wind power based on chaotic time series. The analysis of modeling with low dimensions nonlinear dynamics indicates that time series of wind power generation have chaotic characteristics, and wind power can be predicted in short-term. Phase space reconstruction method can be used for ANN model design. The data from the wind farm located in the Saihanba China are used for this study.

Index Terms-- chaotic dynamic system, forecast, neural network and wind power prediction.

I. INTRODUCTION

The wind power generation has rapidly increased over the last decade, and by the year 2020 wind power generation will supply about 12% of the total world electricity demands [8]. The wind power installations of China are 1336MW in 2005. The wind power installations of the largest wind farm in China, Saihanba Wind Farm, will exceed 1000MW by the year 2008. The wind energy is free, so all wind-generated electric energy is accepted as it comes. However, its availability is not known in advance. Because the wind power generation is fluctuating due to variations in wind direction and speed,

wind power generation integrated in electrical power system may cause several problems, including power quality, stability and especially power dispatching. These problems increase as the penetration of wind energy increases[1]-[4].

Prediction of wind power, along with load forecasting, permits scheduling the connection or disconnection of wind turbines or conventional generators, thus achieving low spinning reserve and optimal operating cost. The horizon of prediction may vary: long-term horizon, a year; short-term horizon, 36hours, 10 hours, and 2hours. Different horizons correspond to different objectives. There are two main state-of-art approaches, one based on physical deterministic modeling[3][5]-[7] and a second one based on statistical or time series modeling[1]-[2][4][8]. Since wind power is a function of wind speed, predictions of wind power are generally derived from predictions of wind speed[1][2]. Unfortunately, the wind turbines are spread out over a significant geographical area, and the hub height speed of each turbine cannot be estimated accurately because the wind speed is related with the terrain shape, the number and site of anemometer towers, and pressure profiles. Furthermore, the output power of the wind turbines is different during the dynamic period of wind speed because of the different behavior of wind generation system controllers.

This paper will outline the methods using the total wind power generation time series of a wind farm as input to predict the wind power generation for short-term. The wind power generation time series are analyzed in this paper and it shows that time series of wind power generation have a finite correlation dimension, the maximum Lyapunov exponent is positive, and wind power generation is predictable in short term. The ANN

978-1-4244-0644-9/07/$25.00 ©2007 IEEE

model of wind power generation based on chaotic time series is achieved, and the construction of ANN model is discussed in the second part of this paper.

II. RECONSTRUCTION OF DYNAMICAL PHASE SPACE

The wind power generation system is a complex dynamical system. The reconstruction of dynamics from the time series of wind power generation should be done before discussing the chaotic behavior of time series of wind power generation, which is built around Takens' embedding theorem. This theorem provides the mathematical basis of the dynamic reconstruction problem. It states that model reconstruction of a nonlinear dynamical system using just one observation of the system should succeed to a certain extent, and the reconstruction is independent of which signal component is used[12].

With a chaotic time series, consider a D-dimensional compact manifold M, Takens' embedding theorem says with $D_E \geq 2D+1$, the map $\Phi : M \to R^{D_E}$ defines a corresponding trajectory, the trajectory of the reconstructed space is diffeomorphism with original dynamical system, where D_E is the embedding dimension[12].

Typically, a wind power time series x_1, x_2, x_3, x_4, x_5, $\cdots$, x_{n-1}, x_n is a series of record values of observations of a dynamical system at regular, discrete time intervals (Δt). Assume the embedding dimension is D_E, and the time delay is $\tau = k\Delta t$, then the reconstructed phase space becomes

$$Y(i) = [x(i), x(i+\tau), x(i+2\tau), ..., x(i+(m-1)\tau)]$$

$$i = 1, 2, ..., N , \quad N = n - (m-1)\tau \qquad (1)$$

Embedding dimension D_E and time delay τ are two important parameters to perform the dynamic reconstruction. They will impact the aspect of the optimal prediction of a reconstructed dynamical model.

A. Determine The Time Delay

A nonlinear approach is used to measure time delay τ called the mutual information I. As far as the quality of prediction with neural networks is concerned, it is best to choose the value of τ somewhere in between 1 and the first local minimum of the mutual information

function $I(\tau)$[13].

The mutual information of a time series as follows

$$
\begin{aligned}
I(Q,S) &= H(Q) - H(Q|S) = H(Q) + H(S) - H(S,Q) \\
&= -\sum_i P(q_i)\log P(q_i) - \sum_j P(s_j)\log P(s_j) \\
&\quad + \sum_i \sum_j P(s_j, q_i)\log P(s_j, q_i)
\end{aligned}
\qquad (2)
$$

where, Q, S are two discrete variables, $H(*)$: information entropy, $H(S,Q)$: mutual information entropy, $H(Q|S)$: conditional entropy, $P(q_i)$: appearance probability of q_i event, $P(q_i, s_j)$: simultaneous appearance probability.

According wind power time series shows in Fig.1, which are from the wind farm located in the Saihanba China. Assume $S = x(k)$ and $Q = x(k+\tau)$, the $I(Q,S)$ shows as Fig.2. When the time delay τ changes from 1 to 50, the first local minimum of the mutual information function $I(\tau)$ is 12, so we choose 12 as time delay of the wind power time series.

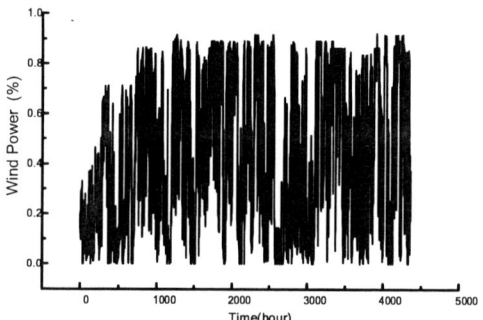

Fig. 1.　Wind power generating capacity time series

(Oct. 2005 ~ Mar. 2006)

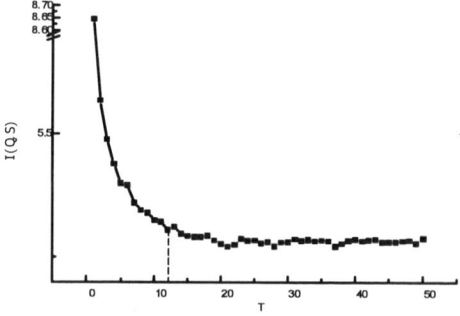

Fig. 2.　Mutual information versus time delay

B. Determine The Embedding Dimension

Paper [14] presents a method to determine the

minimum embedding dimension. Consider a wind power time series x_1, x_2, x_3, x_4, x_5, $\cdots$, x_{n-1}, x_n. The time delay has been determined above, so the reconstructed phase space is shown as (1). Define

$$a(i,m) = \frac{\| Y_{m+1}(i) - Y_{m+1}(n(i,m)) \|}{\| Y_m(i) - Y_m(n(i,m)) \|} \tag{3}$$

$$E(m) = \frac{1}{N-m\tau} \sum_{i=1}^{N-m\tau} a(i,m) \tag{4}$$

$$E1(m) = \frac{E(m+1)}{E(m)} \tag{5}$$

where, $i = 1,2,\cdots,N-m\tau$. $Y_{m+1}(n(i,m))$ is the nearest neighbour of $Y_{m+1}(i)$, and $Y_m(n(i,m))$ is the nearest neighbour of $Y_m(i)$. If m is qualified as an embedding dimension by the embedding theorem, then any two points which stay close in the m-dimensional reconstructed space will be still close in the $(m+1)$-dimensional reconstructed space[14].

Suppose the time series comes from an attractor, $E1(m)$ will stop changing when m is greater than some value m_0. Then the m_0+1 is the minimum embedding dimension. To investigate the wind power time series of Fig. 1, $E1(m)$ is calculated and is shown in Fig. 3. When m is greater than 7 the $E1(m)$ stops changing, and then the embedding dimension of wind power generation time series in Fig. 1 is 8.

III. LYAPUNOV CHARACTERISTIC EXPONENTS

It has to confirm that the wind power generation system is chaotic before applying chaos theory to predict the wind power. To satisfy this need, chaotic characteristics of wind power must be determined. In practice, it can be defined that a bounded deterministic dynamical system with at least one positive Lyapunov exponent is a chaotic system[12]. The deterministic and bounded properties are obvious of the wind power generation system in terms of finite installation power capacity and finite attractor dimension. Then the criterion of at least one positive Lyapunov exponent plays a key role in the characterization of a process as chaotic,

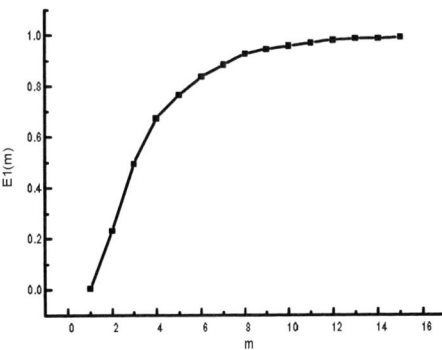

Fig. 3. The values E1(m) for wind power time series

because the Lyapnov exponents not only show qualitatively the sensitive dependence on initial conditions but also give a quantitative measure of the average rate of separation or attraction of nearby trajectories on the attractor.

An algorithm developed by Wolf et al [15], which implements the theory in a very simple and direct fashion, can be used to calculate the largest Lyapunov exponent. As mentioned above, the initial point of the reconstructed phase space of wind power time series (1) is $Y(t_0)$. Assume the nearest neighbour of the initial point is $Y_0(t_0)$. Let L_0 denote the Euclidean distance between them. Next, we have to iterate both points forward for a fixed evolution time, which should be of the same order of magnitude as the time delay τ, and finally they evolve to two points, $Y(t_1)$ and $Y_0(t_1)$, along with their trajectories separately. The distance of the two new points is $L_0' = |Y(t_1) - Y_0(t_1)| > \varepsilon$, where $\varepsilon > 0$. Then another nearest neighbour $Y_1(t_1)$ of the point $Y(t_1)$ should be found, which has the minimum distance between the pair of points, $L_1 = |Y(t_1) - Y_1(t_1)| < \varepsilon$. This procedure is repeated until the initial point $Y(t_0)$ reaches the end of the time series, shown as Fig. 4. Finally, the largest Lyapunov exponent is calculated according to the equation

$$\sigma_{\max} = \frac{1}{t_M - t_0} \sum_{i=0}^{M} \ln \frac{L'_i}{L_i} \tag{6}$$

Where M is the total number of replacement steps.

By using equation (6), the largest Lyapunov exponent for the attractor presented in Fig. 1 can be calculated. The largest Lyapunov exponent converges very well to

$\sigma_{\max} = 0.1578$. This is a firm proof for the chaotic behaviour of the wind power generation system.

IV. PREDCITON OF WIND POWER GENERATION WITH ANN MODEL

ANN model is suitable for wind power prediction. However, the ANN model is difficult to be configured. There are 3 steps for ANN model design, 1) identify the appropriate input parameters, 2) set up appropriate network structure, and 3) design a training algorithm that yields the best training performance. Shuhui Li, et al [5] presents an ANN model with 4 input data, which are wind speed and direction from two meteorological towers.Rohrig, K., et al [16] presents another ANN model with 8 input data in terms of 3 wind speed values, 3 direction values, and 2values related to the time series. Unfortunately, these patterns are limited by the meteorological towers and cannot give out the hub height wind speed for each turbine.

This issue can be considered with the application of the chaos-based ANN model. An ANN serves as the reconstructed phase space model of the wind power generation system. The input data are the

$$Y(i) = [x(i), x(i+\tau), x(i+2\tau), ..., x(i+(m-1)\tau)]$$,

$i = 1, 2, ..., N$, $N = n - (m-1)\tau$, and the number of input parameters is the embedding dimension m. The predicted value of the time series point $i+(m-1)\tau+1$, i. e. $x(i+(m-1)\tau+1)$, is set as the output of the chaos-based ANN model. The nonlinear function

$$\hat{\Phi} : M \to R^{D_E}$$ will be obtained by training the ANN.

This ANN model is diffeomorphism with original dynamical system of wind power generation system, and has locally (i.e. short-term) predictable ability.

The configuration of ANN for wind power prediction shows in the Fig. 5. And Fig. 6 shows the comparison of measured and one-step predicted wind power generation 1-houre ahead over a selected period of 200 continuous time steps of Jan. 2006 on the Saihanba wind farm China. The results show that the ANN model presented in this paper performs well.

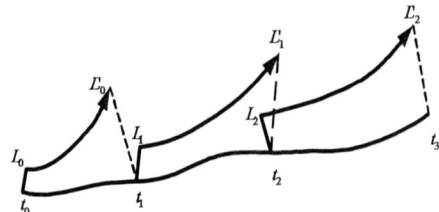

Fig. 4.   Wolf method for largest Lyapunov exponent calculation

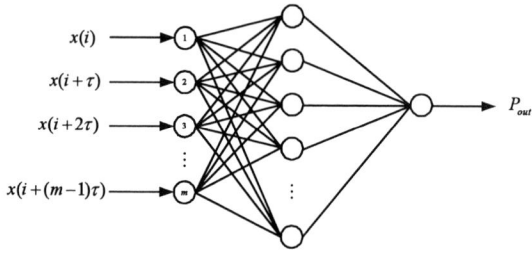

Fig. 5.   Multi-layer neural network for wind power generation prediction

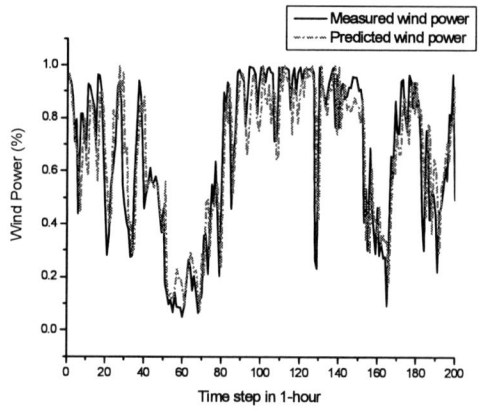

Fig. 6.   Predicted and measured wind power generation

V. CONCLUSIONS

The time series of wind power generation are analyzed in this paper. The wind power system indicates typical chaotic characteristics. According to the chaotic behaviors of wind power generation, the ANN model based on phase space reconstruction is used for wind power prediction. The design method of ANN model is discussed in this paper. And the results show that the chaotic model is a good choice for the prediction of wind power generation.

ACKNOWLEDGEMENT

The authors are grateful to Mr. Hu Guodong and Mr. Liu Guozhong of the Datang ChiFeng Saihanba Wind-Powered Electricity Co., Ltd, and to Yang Ge of the Vestas for their help in collecting the experimental data used in this study.

REFERENCES

[1] Alexiadis, M.C., Dokopoulos, P.S. and Sahsamanoglou, H.S., "Wind speed and power forecasting based on spatial correlation models," Energy Conversion, IEEE Transactions on, Volume: 14, Issue: 3 , Sept. 1999, pp. 836 - 842.

[2] Ioannis G. Damousis, Minas C. Alexiadis, John B. Theocharis, Petros S. Dokopoulos, "A Fuzzy Model for Wind Speed Prediction and Power Generation in Wind Parks Using Spatial Correlation," IEEE TRANSACTIONS ON ENERGY CONVERSION, VOL. 19, NO. 2, JUNE 2004, pp. 352-361.

[3] Ferreira, L.A.F.M., "Evaluation of short-term wind predictability, Energy Conversion," IEEE Transactions on , Volume: 7 , Issue: 3 , Sept. 1992 pp. 409 – 417.

[4] Li, S., "Wind power prediction using recurrent multilayer perceptron neural networks," Power Engineering Society General Meeting, 2003, IEEE, Volume: 4, 13-17 July 2003, pp. 2325-2330.

[5] Shuhui Li, Wunsch, D.C. and O'Hair, E.A., Giesselmann, M.G., "Using neural networks to estimate wind turbine power generation," Energy Conversion, IEEE Transactions on, Volume: 16 , Issue: 3 , Sept. 2001, pp. 276 – 282.

[6] Watson, S.J., Landberg, L. and Halliday, J.A., "Application of wind speed forecasting to the integration of wind energy into a large scale power system, Generation, Transmission and Distribution," IEE Proceedings-, Volume: 141, Issue: 4, July 1994, pp. 357 – 362.

[7] P. Pison and G.N.Kariniotakis, "Wind Power Forecasting using Fuzzy Neural Networks Enhanced with On-line Prediction Risk Assessment," 2003 IEEE Bologna PowerTech Conference, June 23-26, Bologna, Italy.

[8] El-Fouly, T.H.M., El-Saadany, E.F. and Salama, M.M.A., "One day ahead prediction of wind speed using annual trends," Power Engineering Society General Meeting, 2006. IEEE, 18-22 June 2006 Page(s):7 pp. 1-7.

[9] Barbounis, T.G., Theocharis, J.B., Alexiadis, M.C., and Dokopoulos, P.S., "Long-term wind speed and power forecasting using local recurrent neural network models," Energy Conversion, IEEE Transactions on Volume 21, Issue 1,   March 2006 pp. 273 – 284.

[10] Karki, R., Po Hu and Billinton, R., "A simplified wind power generation model for reliability evaluation," Energy Conversion, IEEE Transactions on Volume 21, Issue 2, June 2006 pp. 533 – 540.

[11] Ahlstrom, M.L. and Zavadil, R.M., "The Role of Wind Forecasting in Grid Operations & Reliability," ransmission and Distribution Conference and Exhibition: Asia and Pacific, 2005 IEEE/PES 15-18 Aug. 2005 Page(s):1 – 5.

[12] Simon Haykin and Xiao Bo Li, "Detection of signals in chaos", PROCEEDINGS OF THE IEEE, VOL. 83, NO. 1, JANUARY 1995, pp. 95-122.

[13] Rape, R., Fefer, D. and Drnovsek, J., "Time series prediction with neural networks: a case study of two examples," Instrumentation and Measurement Technology Conference, 1994. IMTC/94. Conference roceedings. 10th Anniversary. Advanced Technologies in I & M., 1994 IEEE 10-12 May 1994 pp. 145 - 148 vol.1.

[14] Cao,L. (1997), "Practical method for determining the minimum embedding dimension of a scalar time series." Physica D, 110 (1997), pp.43 – 50.

[15] Wolf A, Swift J B, Swinney H L and Vastano J A, "Determining Lyapunov exponents from a time series," Physica D 16, 1985, pp. 285–317.

[16] Rohrig, K. and Lange, B., "Application of wind power prediction tools for power system operations," Power Engineering Society General Meeting, 2006. IEEE 18-22 June 2006 pp. 5.

Power Flow Control for Efficiency Improvement in a Forward-Flyback Mixed Converter

Yoshito Kusuhara*, Asahi Nakayama**, Tamotsu Ninomiya**, and Shin Nakagawa***

*Kyushu Polytechnic College Dept.of Product. Electro. Syst.Engrg.Shii 1665-1,Kokuraminami-ku,Kitakyushu,Japan
** Kyushu University Dept. of Elec. & Electro. Syst. Engrg. Motooka 744, Nishi-ku,Fukuoka,Japan
***FIDELIX Co.,Matuyama 2-15-14, Kiyose, Tokyo ,Japan

Abstract— A novel DC-DC converter have been proposed in the previous paper where both the forward and the flyback actions are mixed. This novel converter has some prominent features of the voltage reduction for the main switch, the current-ripple reduction of output inductor, the size reduction, and the power loss reduction. This converter is named "Forward-Flyback-Mixed converter" and is abbreviated as "FFB converter". And, this paper is clarified that the efficiency improvement is achieved with power flow control for efficiency improvement, moreover it is experimentally analyzed that the efficiency of FFB changes compared with the third winding.

Index Terms–A novel DC-DC converter, Forward-Flyback-Mixed converter, a higher efficiency, Power flow control for efficiency improvement

I. INTRODUCTION

Recently, the demand for the switching converter energy saving have been demanded from the viewpoint of the energy protection and the standby power requirement at a high level. The following items are enumerated as an element of the technology demanded as a converter, namely, the higher efficiency, the higher stability, the higher reliability. We have been proposed and analyzed the FFB converter aiming at making to highly effective in these item.

This paper presents a novel DC-DC converter where both Forward and Flyback converter action are mixed. Previously, we study steady state characteristics and analysis of the FFB Converter, and as a result, confirm that this novel converter have four operation mode. And, for investigate design method of the FFB converter that get higher efficiency, this paper presents that the characteristics and current distribution ratio of the FFB converter are confirmed in the case that third windings number. As a result, we are clarified point that show maximum efficiency in with several values of transformer turns ratio.

Furthermore, this paper presents the detailed power flow control for efficiency improvement of load characteristics with another winding turns ratio in the FFB converter by analysis equations and experiment. And its highly efficient is clarified by the experiment, and it is performed that the efficiency improvement is achieved and the FFB converter has an experimental

result of a high efficiency of 94% was obtained for the input voltage of 140V and the output load of DC 18V /4A.

II. CIRCUIT CONFIGURATION AND ITS OPERATION.

The circuit configuration of the FFB converter is shown in Fig.1. This converter is composed of the main switch Q_1 of MOS-FET, three schottky barrier diodes of RD_1, RD_2 and RD_3 , high-frequency transformer T_1 of PQ-core, reactor L_o. The basic composition is a mixed circuit of forward and Flyback converters. In this topology, both diodes of RD_2 and RD_3 hold on during a certain interval in a switching period, and then the output reactor current is constant. This results in the ripple current reduction and power loss reduction in the output reactor. The waveforms of the magnetizing current i_L and the output reactor current i_{Lo} are shown in Fig.2 and the switching state of each element is shown in Table 1. For the interval of State 1, Q_1 and RD_1 are kept on, and RD_2 and RD_3 are kept off. During this interval, i_L increases, and the energy is stored in the transformer. At the same time, the primary power is supplied to the secondary side. When Q_1 and RD_1 are turned off, the state changes into state 2, and RD_2 and RD_3 turn on. At this time, the current i_{Lo} maintains constant. At the time when both the currents i_L and i_{Lo} become equal to each other, diode RD_3 turns off, and the state changes into state 3.

Table1 Switching sequence of the FFB converter

State	Q_1	RD_1	RD_2	RD_3
State1	ON	ON	OFF	OFF
State2	OFF	OFF	ON	ON
State3	OFF	OFF	ON	OFF

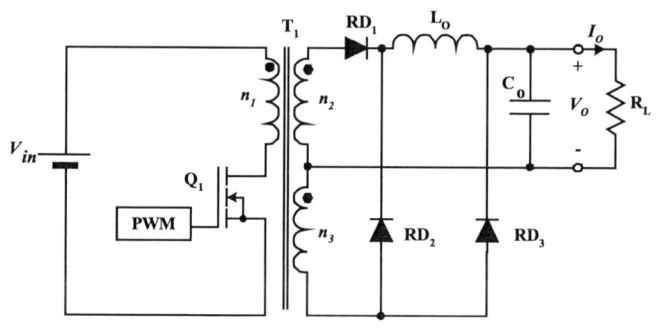

Fig.1 Configuration of Forward-Flyback-Mixed converter.

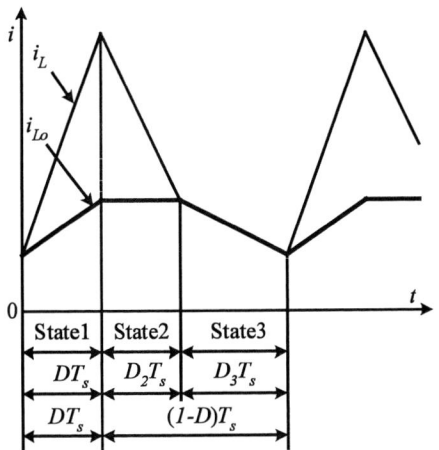

Fig.2 Waveform of magnetizing and output reactor current.

III. ANALYSIS OF CURRENT EQUATION OF EACH ELEMENT FOR THE FFB CONVERTER.

To analyze the operation of the proposed converter, waveforms equation is derived. The waveforms of current of each elements Q_1,RD_1,RD_2,RD_3,L_o of the FFB converter is analyzed. Where, it is D, D_2, D_3 for the duty ratio of each state, and Ts denotes the one cycle ,DT_s denotes interval of state1, D_2T_s denotes interval of state2, V'_{in} ($=n_2V_{in}/n_1$)denotes the secondary side conversion voltage , V_o denotes output voltage , L'_p denotes the secondary side conversion inductance of the excitation inductance n_1 denotes the turn number of first winding, n_2 denotes the turn number of secondary and n_3 denotes the turn number of third winding, the analytical waveform are shown in Fig.3,and the experimental waveform are shown in Fig. 4. $\bar{i}_Q, \bar{i}_{RD1}, \bar{i}_{RD2}, \bar{i}_{RD3}, \bar{i}_{Lo}$ show the average current respectively.

[1] Current equation of switch Q_1
(i) $0<t<DT_s$

$$i_Q(t) = i_Q(0) + \left(\frac{V_{in}}{L_p} + \frac{n_2}{n_1} \frac{V'_{in} - V_o}{L_o} \right) \cdot t \qquad (1)$$

(ii) $DT_s<t<T_s$
$$i_Q(t)=0 \qquad (2)$$

(iii) Mean value

$$\bar{i}_Q = \left[i_Q(0) + \left\{ i_Q(0) + \left(\frac{V_{in}}{L_p} + \frac{n_2}{n_1} \frac{V_{in}' - V_o}{L_o} \right) DT_s \right\} \right] \frac{D}{2} \qquad (3)$$

[2]Current equation of diode RD_1
(i) $0<t<DT_s$

$$i_{RD1}(t) = \frac{n_1}{n_2 + n_3} i_Q(0) + \left(\frac{V'_{in} - V_o}{L_o} \right) \cdot t \qquad (4)$$

(ii) $DT_s<t<T_s$
$$i_{RD1}(t)=0 \qquad (5)$$

(iii) Mean value

$$I_{RD1} = \left\{ \frac{n_1}{n_2 + n_3} i_Q(0) + \left(\frac{n_1}{n_2 + n_3} i_Q(0) + \frac{V'_{in} - V_o}{L_o} DT_s \right) \right\} \frac{D}{2} \qquad (6)$$

[3] Current equation of diode RD_2
(i) $0<t<DT_s$
$$i_{RD2}(t)=0 \qquad (7)$$

(ii) $DT_s<t<D_2T_s$

$$i_{RD2}(t) = \frac{n_1}{n_2 + n_3} i_Q(0) + \left(\frac{V'_{in} - V_o}{L_o} DT_s \right) \qquad (8)$$

(iii) $D_2T_s<t<T_s$

$$i_{RD2}(t) = \left\{ \frac{n_1}{n_2 + n_3} i_Q(0) + \left(\frac{V'_{in} - V_o}{L_o} DT_s \right) \right\} - \left(\frac{V_o}{L'_p + L_o} \right) \left(t - \overline{(D + D_2)}T_s \right) \qquad (9)$$

(iv) Mean value

$$\bar{i}_{RD2} = (D_2 + D_3) \left(\frac{n_1}{n_2 + n_3} i_Q(0) + \frac{V'_{in} - V_o}{L_o} DT_s \right) - \frac{V_o}{L'_p + L_o} \cdot \frac{D_3^2 T_s}{2} \qquad (10)$$

[4] Current equation of diode RD_3
(i) $0<t<DT_s$
$$i_{RD3}(t)=0 \qquad (11)$$

(ii) $DT_s<t<D_2T_s$

$$i_{RD3}(t) = \left(\frac{n_1}{n_3} \cdot \frac{V_{in}}{L_p} - \frac{V'_{in} - V_o}{L_o} \right) DT_s - \frac{V_o}{L'_p} (t - DT_s) \qquad (12)$$

(iii) $D_2T_s<t<T_s$
$$i_{RD3}(t)=0 \qquad (13)$$

(iv) Mean value

$$\bar{i}_{RD3} = \left(\frac{n_1}{n_3} \cdot \frac{V_{in}}{L_p} - \frac{V'_{in} - V_o}{L_o} \right) \frac{D \cdot D_2}{2} T_s \qquad (14)$$

[5] Current equation of output reactor L_O
(i) $0<t<DT_s$

$$i_{Lo}(t) = \frac{n_1}{n_2 + n_3} i_Q(0) + \left(\frac{V'_{in} - V_o}{L_2} \right) \cdot t \qquad (15)$$

(ii) $DT_s<t<D_2T_s$

$$i_{Lo}(t) = \frac{n_1}{n_2 + n_3} i_Q(0) + \left(\frac{V'_{in} - V_o}{L_o} DT_s \right) \qquad (16)$$

(iii) $D_2T_s<t<T_s$

$$i_{Lo}(t) = \left\{ \frac{n_1}{n_2 + n_3} i_Q(0) + \left(\frac{V'_{in} - V_o}{L_o} DT_s \right) \right\} - \left(\frac{V_o}{L'_p + L_o} \right) \left(t - \overline{(D + D_2)}T_s \right) \qquad (17)$$

(iv) Mean value

$$\bar{i}_{LO} = \frac{1}{2} \left((1 + D_2) \frac{V'_{in} - V_o}{L_o} DT_s \right) + \frac{n_1}{n_2 + n_3} i_Q(0) \qquad (18)$$

In the above mentioned equation, $i_Q(0)$ is shown as follows by the initial current value of the switch Q_1. When I_o is a output current, it is shown as equation(19).

$$i_Q(0) = \frac{n_2 + n_3}{n_1} \left[I_o - \frac{DT_s}{2} \left[\left\{ (D_2 + 1) \frac{V'_{in} - V_o}{L_o} \right\} + D_2 \left(\frac{n_1}{n_3} \frac{V_{in}}{L_p} - \frac{V'_{in} - V_o}{L_o} \right) \right] \right] \qquad (19)$$

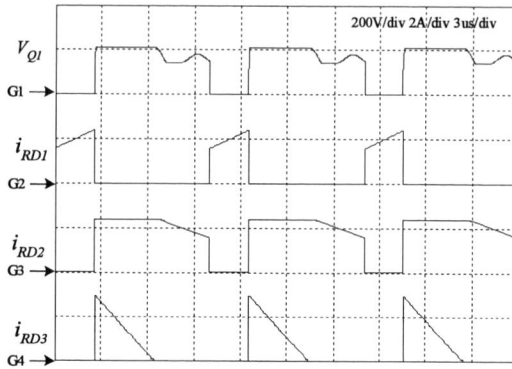

Fig.3 Each device waveform(analysis).

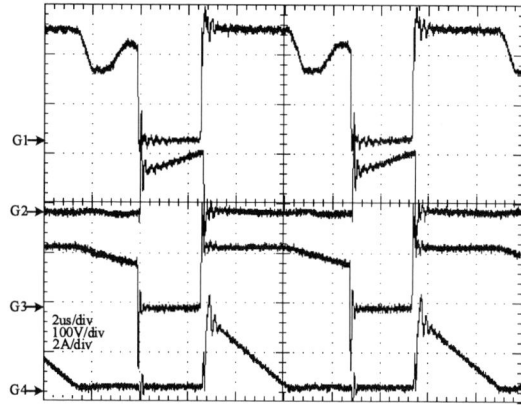

Fig.4 Each device waveform(experiment).

IV. THE MECHANISM OF OUTPUT POWER FLOW AND CURRENT DISTRIBUTION.

Fig.5 shows the mechanism of the power flow for the FFB converter, it operates as a forward converter when Q_1 is on, and as a Flyback converter when Q_1 is off, the power is supplied to the secondary side by Q_1 on and off. Thus, the FFB does the combined Forward and Flyback converter operation.

Where, the output current distribution by the forward side average current $\bar{i}_{Lo}$ flow output reactor L_o and flyback side average current $\bar{i}_{RD3}$ flow diode RD_3 are considered. The current of each element has the relation of equation (20). Where, I_{Lo} denotes r.m.s current of L_o, I_{RD3} denotes r.m.s current of RD_3. If the ripple current is very small, it is not issue for $\bar{i}_{Lo} = I_{Lo}$. When I_{RD3p} denotes peak current of RD_3, the equation (20) becomes equation (21) –(22)from equation (11)-(19) .And ,I_{RD3} become equation(23) from equation (14). Moreover, The current distribution rate from the forward side and the flyback side is defined as $I_{Lo\%}$ and $I_{RD3\%}$, it is shown equation (24). Furthermore, η denotes efficiency ,P_{cLoss} denotes fixed loss, r denotes loss resistance, P_{rLoss} denotes resistance

loss. These are equation (25)-(26). P_{rloss} finally becomes equation(28) by equation(27). To analyze the efficiency characteristic, when the power loss rate $Pl_{oss\%}$ is defined as $P_{loss\%} = P_{loss}/V_oI_o$, it is shown in equation (29).

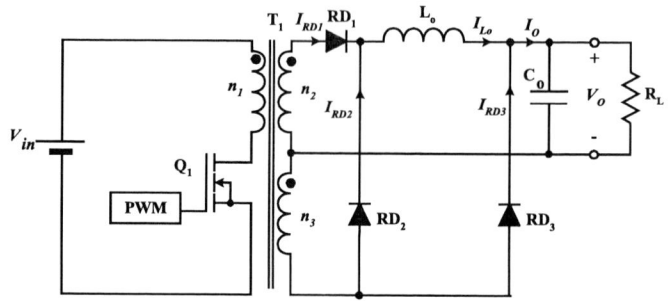

Fig.5 The mechanism of the power flow for the FFB converter.

$$\bar{i}_{Lo} = \bar{i}_{RD1} + \bar{i}_{RD2} \qquad I_o = \bar{i}_{Lo} + \bar{i}_{RD3} \qquad (20)$$

$$\bar{i}_{Lo} = I_{Lo} = I_o - \bar{i}_{RD3} = I - \frac{D_2}{2}I_{RD3p} \qquad (21)$$

$$I_{RD3p} = \left(\frac{n_1}{n_3} \frac{V_{in}}{L_p} - \frac{V_{in}' - V_o}{L_2} \right) DT_s \qquad (22)$$

$$I_{RD3} = \sqrt{\frac{D_2}{3}} I_{RD3p} \qquad (23)$$

$$I_{Lo\%} = \frac{I_{Lo}}{I_{Lo} + I_{RD3}} \qquad I_{RD3\%} = \frac{I_{RD3}}{I_{Lo} + I_{RD3}} \qquad (24)$$

$$\eta = \frac{V_oI_o}{V_oI_o + P_{loss}} \times 100 = \frac{V_oI_o}{V_oI_o + (P_{rLoss} + P_{cLoss})} \times 100 [\%] \qquad (25)$$

$$P_{rLoss} = r\left(I_{Lo}^2 + I_{RD3}^2 \right) = r\left(\overline{i_{Lo}^2} + \overline{i_{RD3}^2} \right) = r\left\{ I_o^2 - 2\bar{i}_{RD3}I_o\left(\bar{i}_{RD3} \right)^2 + \overline{i_{RD3}^2} \right\} \quad (26)$$

$$\bar{i}_{RD3} = \frac{D_2}{2}I_{RD3p} \qquad \overline{i_{RD3}^2} = \frac{D_2}{3}I_{RD3p}^2 \qquad (27)$$

$$P_{rloss} = r\left(I_o^2 - D_2 I_{RD3p}I_o + \left(\frac{D_2^2}{4} + \frac{D_2}{3} \right)I_{RD3p}^2 \right) \qquad (28)$$

$$P_{loss\%} = \frac{r}{V_o}\left(I_o - D_2 i_{RD3p} + \left(\frac{D_2^2}{4} + \frac{D_2}{3} \right)\frac{i_{RD3p}^2}{I_o} \right) + \frac{P_{closs}}{V_oI_o} \qquad (29)$$

V. EXPERIMENTAL EFFICIENCY WITH CURRENT DISTRIBUTION.

The above is declared by the experiment. In the experimental parameters of the FFB converter are chosen as shown in Table 2.The efficiency of the transformer winding ratio n_1:n_2:n_3=22:3:3,22:3:4, 22:3:5 are shown in Fig. 6. It is understood that the maximum efficiency has changed compared with the winding ratio. The experimental results of the power loss with the same winding ratio are shown in Fig.7. It is inefficient that there are a lot of numbers of n_3 winding when the output current is large. This is because bias of the current distribution grow so that average current of RD_3 may become small when the n_3 winding increases. The current distribution rate and the power efficiency for the same transformer winding ratio are compared and examined by the experiment.

956

Table 2 Experimental parameter.

Parameter	Symbol	Constant
Input voltage	V_{in}	140[V]
Exiting inductance	L_p	390[μH]
Output inductance	L_2	68[μH]
Primary winding	n_1	22
Secondarily winding	n_2	3
Third winding	n_3	3,4,5
Switching frequency	f	95[kHz]
Output voltage	V_o	16[V]

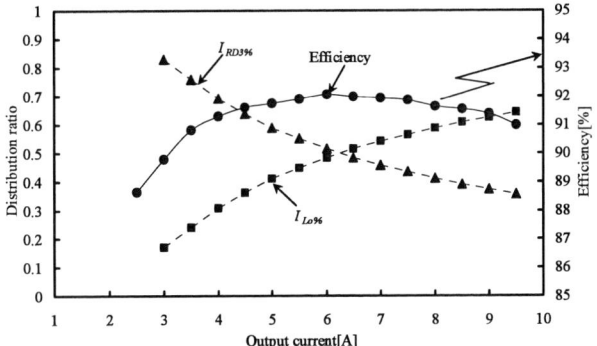

Fig.8 Current distribution and efficiency characteristics (n_1:n_2:n_3=22:3:3.)

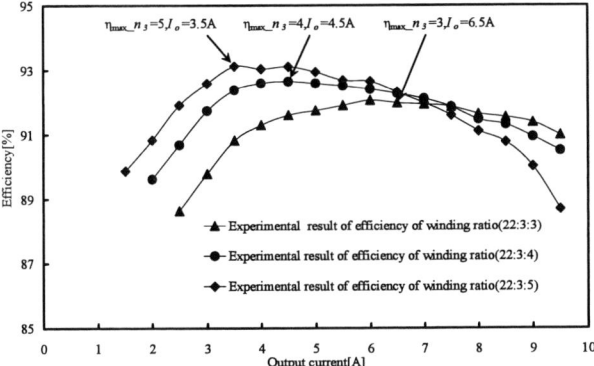

Fig.6 Efficiency characteristics of the FFB converter(22:3:3,4,5).

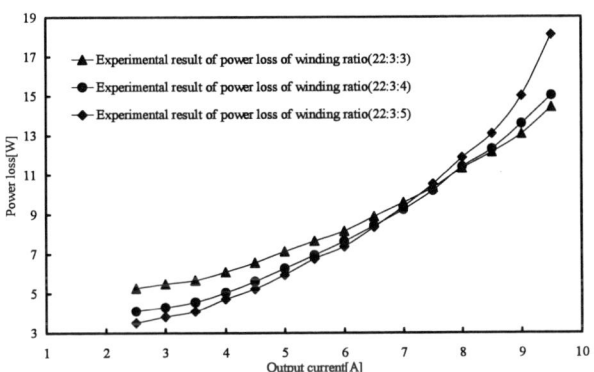

Fig.7 Power loss characteristics of the FFB converter(22:3:3,4,5).

The output current distribution and the efficiency characteristics of the winding ratio n_1:n_2:n_3=22:3:3 are shown in Fig.8. It is understood to show the maximum efficiency from this experimental results when the output current distribution is $I_{Lo\%}=I_{RD3\%}$. The experimental efficiency of the each winding ratio and the experimental effective current of RD$_3$ and L$_o$ are shown in Fig. 9– Fig. 11. It is understood to show the maximum efficiency from these characteristics when I_{Lo} is equal to I_{RD3}. If the output current I_{omax} becomes a maximum efficiency as a result based on equation (21) is requested, it becomes equation(30).

$$I_{o\max} = \left(\sqrt{\frac{D_2}{3}} + \frac{D_2}{2} \right) I_{RD3p} \tag{30}$$

Fig.9 Experimental results of efficiency and effective current for RD$_3$ and L$_o$ (n_1:n_2:n_3=22:3:3).

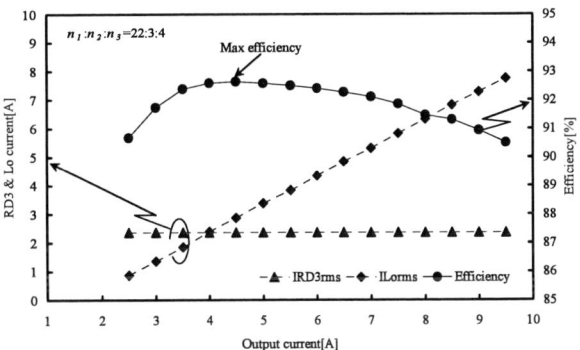

Fig.10 Experimental results of efficiency and effective current for RD$_3$ and L$_o$ (n_1:n_2:n_3=22:3:4).

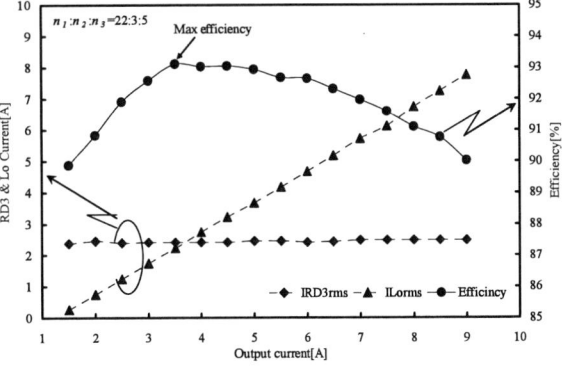

Fig.11 Experimental results of efficiency and effective current for RD$_3$ and L$_o$ (n_1:n_2:n_3=22:3:5).

From the above mention, the calculation value and the experiment value in which the maximum efficiency are shown become Table 3.

Table 3 Analytical and experimental value of maximum efficiency

$n_1:n_2:n_3$	Duty ratio	Analysis[A]	Experiment[A]	Max efficiency[%]	Output power[W]
22:03:03	0.42	7.1	6.5	92	94
22:03:04	0.36	5.1	4.5	92.6	70
22:03:05	0.31	3.9	3.5	93.2	55

Finally, the result of analysis and experiment on the power loss ratio $P_{loss\%}$ are shown in Fig.12-Fig.14. The efficiency is the highest when the power loss rate is minimum, an analytical result by equation (29) is shown by $P_{loss\%}$(Ana), and the experimental result is shown by $P_{loss\%}$(Exp) .From this, the analysis and the experiment agreed almost, and confirmed the validity of the analysis.

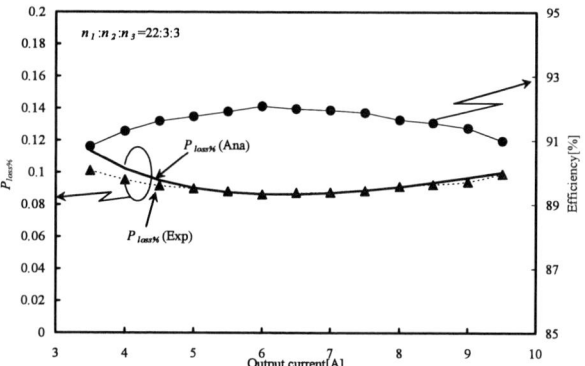

Fig.12 Results of analysis and experiment for the power loss rate
$(n_1:n_2:n_3=22:3:3)$

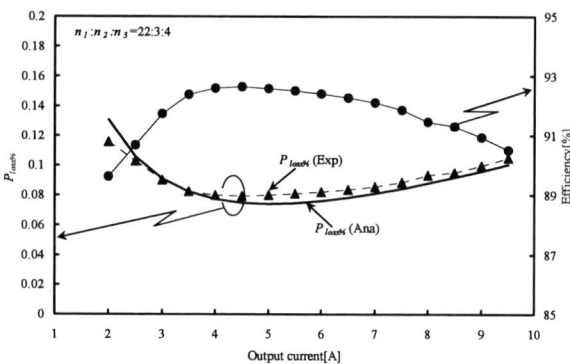

Fig.13 Results of analysis and experiment for the power loss rate
$(n_1:n_2:n_3=22:3:4)$

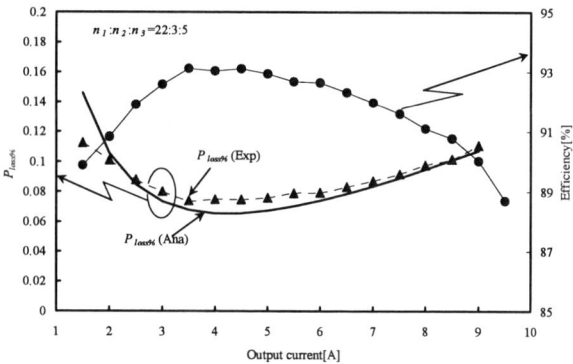

Fig.14 Results of analysis and experiment for the power loss rate
$(n_1:n_2:n_3=22:3:5)$

VI. CONCLUSION.

A novel circuit topology of the FFB converter has been proposed, and its equation of static characteristics have been analyzed by each waveform. And, the change in the efficiency characteristic by the current distribution was confirmed by the experiment. It defined with the power loss, and the efficiency characteristic was analyzed.As the result, the following issues have been clarified.

(1) It has been understood that the maximum efficiency changes with the winding ratio.

(2) The maximum efficiency is obtained by assuming output reactor current I_{Lo} and diode RD$_3$ current $I_{RD3}$to be 50%.

(3) The maximum efficiency is controlled compared with the winding ratio, and the minimum point of the power loss rate becomes the maximum efficiency.

(4) The validity of the analysis was confirmed from power loss rate $P_{\%loss}$ analysis agreement with the experiment result.

Lastly, the future work is to clarify the mechanism of efficiency improvement with energy distribution, and it is necessary to examine the efficiency characteristic of other operation modes.

REFERENCES

[1] S.Nakagawa,Y.Kusuhara, T.Ninomiya , "Steady state Characteristics of Mixed Forward-Flyback DC-DC Converter",IEICE Technical Report,Vol.103,No.38,E2003-5,May.2003.

[2] Y.Kusuhara, S.Nakagawa ,T.Ninomiya , "Steady state Analysis of Mixed Forward-FlybackConverter", IEICE,Technical Report, Vol.103No.303, E2003-38, September 2003.

[3] Y.Kusuhara, T.Ninomiya, S.Nakagawa,"Steady-State Characteristics of a Novel Forward-Flyback-Mixed Converter" IEEE 37th Power Electronics Specialists Conference Record,pp.3018-3023, June 2006.

[4] Y. Kusuhara, T. Ninomiya, S. Nakagawa,"Steady-State Analysis of a Novel Forward-Flyback-Mixed Converter" Proceedings of EPE-PEMC06, pp.3018-3023, September 2006.

[5] Y.Kusuhara, T.Ninomiya, S.Nakagawa,"Steady State characteristics of a Novel Forward-Flyback Converter" IEICE Transactions on Electronics, Vol.J89-B, No.7, pp. 1307-1314,July 2006.

[6] Y.Kusuhara,A.Nakayama,T.Ninomiya,S.Nakagawa "Static and Dynamic Characteristics of a Forward-Flyback-Mixed Converter", APEC'06, Record, pp.768-773, Feburary 2007.

[7] Y.Kusuhara, S.Nakagawa ,T.Ninomiya , "Comparison of Power Loss Distribution between Forward-Flyback Type and Conventional converter",IEICE, Technical Rport,Vol.104, No.406, EE2004-40, Novenber.2004.

[8] Y.Kusuhara,A.Nakayama,T.Ninomiya,S.Nakagawa,"Efficiency Characteristics of Forward-Flyback Mixed coverter   (Seond)", IEICE,Technical Report, Vol.106, No.494, EE2006-43, January 2006.

Hammerstein Model-Based Robust Control of DC/DC Converters

F. Alonge*, *IEEE Member*, F. D'Ippolito*, *IEEE Member*, and T. Cangemi*

*University of Palermo, Dipartimento di Ingegneria dell'Automazione e dei Sistemi, Viale delle Scienze, 90128 PALERMO

(*Italy*)

Abstract– This paper deals with model-based robust control of DC/DC power electronic converters. The converter is modelled by means of its static characteristic and a few continuous-time linear and time-invariant (LTI) models corresponding to contiguous ranges of duty-cycle. The model appears as a Hammerstein model in which the values of the parameters of the LTI part depend on the actual duty-cycle operating range. This suggests to describe the converter as an uncertain system to be controlled using robust control techniques. Frequency domain approach is used for describing the nominal model and the uncertainty. In view of applying robust control, identification of the LTI models is performed by means of simulation experiments carried out on a converter switched model implemented on MATLAB-SIMULINK environment. Internal Model Control (IMC) structure is employed for the controller design, but its implementation is performed using the equivalent classical feedback control structure. Comparison with a PI controller, designed by means of phase margin assignment, shows the peculiarities of the proposed approach.

Index Terms-- Power converters, Hammerstein model, Model identification, Robust control.

I. INTRODUCTION

The aim of this paper is that of giving a systematic procedure for designing a controller for switching DC/DC converters operating in CCM. The peculiarities of this procedure are: a) control-oriented modelling of the converter has to be effected so that both the steady state and transient behaviours be accurately described; b) modelling has to be performed as quickly as possible without the need of carrying out experiments on a prototype of the converter; c) only a software package should be used for modelling, simulation and controller design; d) despite of the high nonlinearities inherent in the converter functioning, the controller has to be able to work in all the desired operating ranges.

Requirement a) can be satisfied using a Hammerstein model which, as is well known, consists of a nonlinear static characteristic followed by a linear model [1]. Modelling of converters by means of Hammerstein models has been already proposed in the literature for Buck [2] and Boost [3] converters; identification is a suitable approach for obtaining this model.

Requirement b) implies that modelling has to be performed using simulation. In order to avoid tedious and consuming time simulation experiments in PSpice environment [4], identification of the model is performed using data obtained by means of implementation of a switched model on a PC.

Requirement c) brings to the use of a software package which allows to perform simulation, data processing for identification, controller design techniques and suitable graphical interface. In this paper, MATLAB/SIMULINK is extensively used for satisfying this requirement.

Requirement d) needs some comments. Due to the high nonlinearities inherent in the converter operations, it is not possible to describe the converter by means of only one LTI model; instead, it has been shown that using only one static characteristic and few LTI models it is possible to describe the converter in its contiguous operating ranges [3]. According to the requirement d), it is then convenient to employ the following approach for controller design: 1) the converter is described as an uncertain process using the above LTI models; 2) a robust controller is designed for this uncertain process. Obviously, the use of robust control techniques allows to cope also with uncertainties arising from model identification carried out by means of simulation.

In this paper, the described procedure is validated by designing a robust controller for a Boost converter modelled according to the Hammerstein structure. For identification and validation purposes, a switched model [5]-[6] is implemented in MATLAB/SIMULINK environment. The steady state characteristic of the converter, i.e. the steady state output vs. duty cycle is then constructed and assumed as the static nonlinearity of the Hammerstein model. Few linear, time-invariant, discrete-time models are identified with the aim of describing the converter in all the contiguous regions of interest. The corresponding continuous-time models are computed using the step-response invariance method and, then, among them a nominal model is chosen. This nominal model and the remaining models are employed for defining a continuous-time uncertain process. In order to describe the above uncertain process, the frequency domain approach is employed. Continuous-time robust control techniques are used for the controller design. Simulation results are shown with the aim of validating the described design procedure. Comparison with PI control, designed by means of phase margin assignment, shows the peculiarities of the approach described in the paper.

This work was supported by University of Palermo (ex 60%).

II. MODELLING OF THE DC/DC BOOST CONVERTER USING THE HAMMERSTEIN APPROACH

The circuit model of the DC/DC Boost converter is given in Fig. 1, where v_i and v_o are the input and the output, respectively. The static gain, i.e. the output to input voltage ratio for a constant value of v_i, depends on the duty cycle of the PWM signal supplying the MOSFET. It is assumed that the values of the circuit parameters are chosen so that it operates in Continuous Conduction Mode (CCM).

As is well known, a Hammerstein model [1] is a nonlinear model consisting of a nonlinear static characteristic, $f(\cdot)$, followed by a LTI model, as shown in Fig. 2, where x is an intermediate and inaccessible variable. In the paper it is assumed that LTI be a continuous-time model given by:

$$\alpha_c(p)y(t) = \beta_c(p)x(t), \tag{1}$$

where:

$$
\begin{aligned}
x(t) &= f(u(t)), \\
\alpha_c(p) &= p^{n_\alpha} + a_{c,n_\alpha-1}p^{n_\alpha-1} + \ldots + a_{c,1}p + a_{c,0} \\
\beta_c(p) &= b_{c,n_\beta}p^{n_\beta} + b_{c,n_\beta-1}p^{n_\beta-1} + \ldots + b_{c,1}p^1 + b_{c,0},
\end{aligned} \tag{2}
$$

where p and $u(t)$ are the derivative operator and the duty-cycle, respectively.

The transfer functions corresponding to (1) is given by:

$$G_p(s) = \frac{b_{c,n_\beta}s^{n_\beta} + b_{c,n_\beta-1}s^{n_\beta-1} + \ldots + b_{c,1}s^1 + b_{c,0}}{s^{n_\alpha} + a_{c,n_\alpha-1}s^{n_\alpha-1} + \ldots + a_{c,1}s + a_{c,0}}, \tag{3}$$

Assuming that the gain of the linear part of the Hammerstein model be equal to 1, i.e.:

$$\lim_{s=0} G_p(s) = \frac{b_{c,0}}{a_{c,0}} = 1, \tag{4}$$

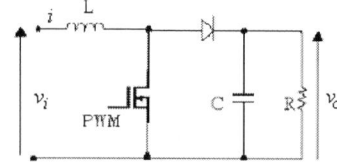

Fig. 1 Circuit model of the Boost converter.

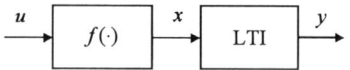

Fig. 2 Hammerstein model.

function $f(\cdot)$ represent the static characteristic of the converter.

The problem now is that of identification of $G_p(s)$. It is important to note that the data useful for identification are samples of the measurable variables acquired with sample time T_s. Consequently, the identification of $G_p(s)$ can be performed by identifying the corresponding discrete-time transfer function $G_{d,p}(z)$ and then computing the parameters of $G_p(s)$.

A. Modelling the linear part of the Boost converter

The Boost converter can be described by a Hammerstein model whose linear part consists of a second order model because it contains two components able to store energy.

Consequently, the continuous-time transfer function of the linear part is given by:

$$G_p(s) = \frac{b_{c,2}s^2 + b_{c,1}s + b_{c,0}}{s^2 + a_{c,1}s + a_{c,0}}, \tag{5}$$

to which the following discrete-time transfer function corresponds:

$$G_{d,p}(z) = \frac{b_0 + b_1 z^{-1} + b_2 z^{-2}}{1 + a_1 z^{-1} + a_2 z^{-2}}. \tag{6}$$

Since the duty cycle is constant within a sampling period T_s, the discrete-time model (6) can be derived using the step response invariance method. According to this method, $G_{d,p}(z)$ is obtained by means of discretization of the model depicted in Fig. 3, where:

$$G_{zoh}(s) = \frac{1 - e^{-sT_s}}{s}.$$

Putting:

$$G_p(s) = b_{c,2}\frac{(s - z_1)(s - z_2)}{(s - p_1)(s - p_2)},$$

the following expression is obtained:

$$G_{d,p}(z) = (1 - z^{-1})Z\{W_{-1}(s)\},$$

where:

$$W_{-1}(s) = b_{c,2}\frac{(s - z_1)(s - z_2)}{s(s - p_1)(s - p_2)},$$

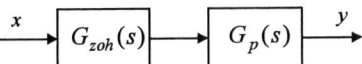

Fig. 3 Model to be discretized.

is the Laplace transform of the step response, $z = e^{-sT_s}$ and $Z\{W_{-1}(s)\}$ denotes the z-transform of $W_{-1}(s)$.

As is easy to verify, the following expression of $G_{d,p}(z)$ is obtained:

$$G_{d,p}(z) = \frac{N(z)}{D(z)}, \tag{7}$$

where:

$$N(z) = (R_0 + R_1 + R_2) - [R_0(p_{d1} + p_{d2}) + R_1(1 + p_{d2}) + R_2(1 + p_{d1})]z^{-1} + (R_0 p_{d1} p_{d2} + R_1 p_{d2} + R_2 p_{d1})z^{-2}, \tag{8}$$

$$D(z) = 1 - (p_{d1} + p_{d2})z^{-1} + p_{d1} p_{d2} z^{-2}, \tag{9}$$

$$p_{d1} = e^{p_1 T_s},$$
$$p_{d2} = e^{p_2 T_s}, \tag{10}$$

where R_0, R_1 and R_2 are, respectively, the residues of the poles 0, p_1 and p_2 of $W_{-1}(s)$ given by:

$$R_0 = b_{c,2} \frac{(-z_1)(-z_2)}{(-p_1)(-p_2)},$$

$$R_1 = b_{c,2} \frac{(p_1 - z_1)(p_1 - z_2)}{p_1(p_1 - p_2)}, \tag{11}$$

$$R_2 = b_{c,2} \frac{(p_2 - z_1)(p_2 - z_2)}{p_2(p_2 - p_1)}.$$

As is easy to verify, the following equation holds:

$$R_0 + R_1 + R_2 = b_{c,2}. \tag{12}$$

Then, the following procedure is suggested for the determination of the parameters of $G_p(s)$.

a) Identify the parameters b_0, b_1, b_2, a_1 and a_2 of $G_{d,p}(z)$ (cf. (6)) with the constraint:

$$\lim_{z \to 1} G_{d,p}(z) = \frac{b_0 + b_1 + b_2}{1 + a_1 + a_2} = 1. \tag{13}$$

b) Compute the poles, p_{d1} and p_{d2}, of $G_{d,p}(z)$ and, then, the residues R_0, R_1 and R_2 by solving the equation (cf. (6), and (8)):

$$\begin{bmatrix} 1 & 1 & 1 \\ p_{d1} + p_{d2} & 1 + p_{d2} & 1 + p_{d1} \\ p_{d1} p_{d2} & p_{d2} & p_{d1} \end{bmatrix} \begin{bmatrix} R_0 \\ R_1 \\ R_2 \end{bmatrix} = \begin{bmatrix} b_0 \\ -b_1 \\ b_2 \end{bmatrix}. \tag{14}$$

Note that the determinant of the coefficient matrix, $(p_{d1} - p_{d2})(1 + a_1 + a_2)$, is different from zero for $p_{d1} \neq p_{d2}$, because $(1 + a_1 + a_2) \neq 0$ (cf. (13)).

c) Compute p_1 and p_2 as follows (cf. (10)):

$$p_1 = \frac{1}{T_s} \ln(p_{d1}),$$
$$p_2 = \frac{1}{T_s} \ln(p_{d2}). \tag{15}$$

d) Compute $z_1 z_2$ as follows (cf. the first equation of (11)):

$$z_1 z_2 = \frac{R_0 p_1 p_2}{b_{c,2}} = \frac{R_0 p_1 p_2}{b_0}. \tag{16}$$

e) Compute $(z_1 + z_2)$ from the second and third equations of (11), as follows:

$$z_1 + z_2 = \frac{R_1(p_2^2 + z_1 z_2) + R_2(p_1^2 + z_1 z_2)}{R_1 p_2 + R_2 p_1}. \tag{17}$$

f) Compute z_1 and z_2 by solving the second order equation:

$$s^2 - (z_1 + z_2)s + z_1 z_2 = 0.$$

B. Switched model of the Boost converter

The switched model of the Boost converter is given by:

$$\dot{x} = A_{on} x + b_{on} v_i, \tag{18}$$

when the MOSFET is ON and:

$$\dot{x} = A_{off} x + b_{off} v_i, \tag{19}$$

when the MOSFET is OFF, where:

$$x = \begin{bmatrix} i & v_o \end{bmatrix}^T,$$

$$A_{on} = \begin{bmatrix} -\dfrac{R_L}{L} & 0 \\ 0 & -\dfrac{1}{RC} \end{bmatrix}, \quad b_{on} = \begin{bmatrix} \dfrac{1}{L} \\ 0 \end{bmatrix},$$

$$A_{off} = \begin{bmatrix} -\dfrac{R_L}{L} & -\dfrac{1}{L} \\ \dfrac{1}{C} & -\dfrac{1}{RC} \end{bmatrix}, \quad b_{off} = b_{on}.$$

and R_L takes into account the resistance of the inductor.

C. Identification of the Hammerstein Model.

The first step for the identification of a Hammerstein model of the Boost DC/DC converter is the determination of the nonlinear static characteristics. Due to the stability properties of the Boost converter and condition (13), which assures that the linear part of the Hammerstein model have unitary gain, the nonlinear static characteristic can be obtained starting from open loop simulation experiments carried out at constant duty cycle values, for a given input voltage v_i.

The second step is to identify the linear part of the Hammerstein model. To this end, the procedure described in [3] can be used; this procedure is also described here for the sake of completeness.

a) The operating input range is subdivided into m intervals; each interval is delimited by two values of duty cycle, d_i and d_{i+1}, $i = 1, \cdots m$.

b) For each duty-cycle couple (d_i, d_{i+1}), the couple $[x(d_i), x(d_{i+1})]$ of voltages is obtained from the above static characteristic.

c) A PRBS sequence is generated with each couple $[x(d_i), x(d_{i+1})]$ and, then, applied to the switched model of the converter in order to obtain data useful for identification. The PRBS sequence is generated at a frequency $f_{PRBS} = \dfrac{1}{\Delta} f_d$ where f_d is the frequency of the duty-cycle and Δ is a positive integer number. The data useful for identification are couples $(x(k), v_o(k))$ where $x(k) \triangleq x(kT_s)$, $v_o(k) \triangleq v_o(kT_s)$, in which $f_s = \Gamma f_d$ is the sampling frequency, where Γ is a positive integer number, and T_s is the corresponding sampling period.

d) The regression model is obtained as follows:

d.1) putting $y(k) \triangleq v_o(k)$, $n_\alpha = n_\beta = 2$, model (1) becomes:

$$y(k) + a_1 y(k-1) + a_2 y(k-2) = b_0 x(k) + \\ + b_1 x(k-1) + b_2 x(k-2) ; \quad (20)$$

d.2) in order to avoid the solution of a constrained identification problem, b_0 is obtained from (13) and substituted in (20); moreover, a system noise $e(k)$ is added in order to take into account for modelling errors, measurement errors and so on, thus obtaining:

$$y(k) - x(k) = a_1 [x(k) - y(k-1)] + a_2 [x(k) + \\ - y(k-2)] + b_1 [x(k-1) - x(k)] + \\ + b_2 [x(k-2) - x(k)] + e(k) , \quad (21)$$

d. 3) putting:

$$\tilde{y}(k) = y(k) - x(k),$$
$$r_{aj}(k) = x(k) - y(k-j),$$
$$r_{bj}(k) = x(k-j) - x(k),$$

and repeating N times equation (21), for $k = k_i, \cdots k_{i+N-1}$, where k_i is the initial data acquisition time, the following regression model is obtained:

$$\tilde{y} = Fp + e , \quad (22)$$

where:

$$\tilde{y} = [\tilde{y}(k_i) \quad \tilde{y}(k_i+1) \quad \cdots \quad \tilde{y}(k_i+N-1)]^T ,$$

$$F = \begin{bmatrix} r_{a1}(k_i) & r_{a2}(k_i) & r_{b1}(k_i) & r_{b2}(k_i) \\ r_{a1}(k_i+1) & r_{a2}(k_i+1) & r_{b1}(k_i+1) & r_{b2}(k_i+1) \\ \cdot & \cdot & \cdot & \cdot \\ r_{a1}(k_i+N-1) & r_{a2}(k_i+N-1) & r_{b1}(k_i+N-1) & r_{b2}(k_i+N-1) \end{bmatrix}$$

$$p = [a_1 \quad a_2 \quad b_1 \quad b_2]^T ,$$

$$e = [e(k_i) \quad e(k_i+1) \quad \cdots \quad e(k_i+N-1)]^T .$$

e) The parameter vector p is computed so as to minimize the cost function:

$$J = \sum_{k=k_i}^{k_i+N-1} e^2(k) .$$

The solution of this problem, obtained by means of the least square algorithm, is given by [7]:

$$p = (F^T F)^{-1} F^T \tilde{y} . \quad (23)$$

D. Application of the described identification procedure

The identification procedure described in the previous Subsection is applied for modelling a Boost converter having the following parameters:

$$R = 10 \ \Omega, \ L = 67 \ \mu H, \ C = 200 \ \mu F, \ R_L = 0.095 \ \Omega .$$

1. Using the open loop switched model implemented in Matlab-Simulink environment, the static nonlinear characteristic of Fig. 4 is obtained corresponding to an input voltage $v_i = 10$ V.

2. A PRBS sequence is generated in Matlab for $f_{PRBS} = f_d = 100$ kHz (cf. steps C. b) and c)). The sampling frequency for simulation is $f_s = 10$ MHz.

3. The models identified following steps C. d)-e) are shown in Table I.

4. According to the procedure A. b)-f) the continuous – time models shown in Table II are obtained.

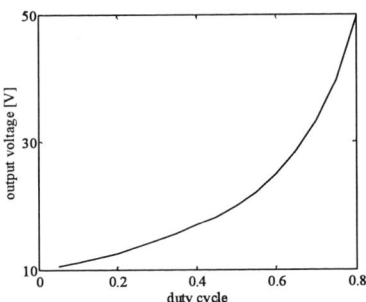

Fig. 4 Static characteristic for $v_i = 10$ V.

TABLE I

DISCRETE-TIME TRANSFER FUNCTIONS

Duty-cycle couples	Transfer function $G_{d,p}(z)$
$[0.3, 0.35]$	$\dfrac{-6.21310 \times 10^{-7} z^2 + 0.0003929 z + 0.001637}{z^2 - 1.207s + 0.209}$
$[0.35, 0.4]$	$\dfrac{8.197 \times 10^{-6} z^2 + 0.0002422\, z + 0.001511}{z^2 - 1.249s + 0.2511}$
$[0.4, 0.45]$	$\dfrac{-1.676 \times 10^{-5} z^2 + 9.657 \times 10^{-5} z + 0.00164}{z^2 - 1.425s + 0.4272}$
$[0.45, 0.5]$	$\dfrac{-2.887 \times 10^{-5} z^2 - 2.376 \times 10^{-5} z + 0.001244}{z^2 - 1.663z + 0.664}$
$[0.5, 0.55]$	$\dfrac{-1.263 \times 10^{-6} z^2 - 6.152 \times 10^{-5} z + 0.001105}{z^2 - 1.844z + 0.8449}$
$[0.55, 0.6]$	$\dfrac{3.828 \times 10^{-6} z^2 - 9.779 \times 10^{-5} z + 0.0008107}{z^2 - 1.927z + 0.9274}$

TABLE II

CONTINUOUS-TIME TRANSFER FUNCTIONS

Duty-cycle couples	Transfer function $G_p(s)$
$[0.3, 0.35]$	$\dfrac{-6.21310 \times 10^{-7} s^2 + 256.2s + 4.017 \times 10^7}{s^2 + 1.565 \times 10^5 s + 4.017 \times 10^7}$
$[0.35, 0.4]$	$\dfrac{8.197 \times 10^{-6} s^2 + 236.1\,s + 3.252 \times 10^7}{s^2 + 1.382 \times 10^5 s + 3.252 \times 10^7}$
$[0.4, 0.45]$	$\dfrac{-1.676 \times 10^{-5} s^2 + 298.5s + 2.555 \times 10^7}{s^2 + 8.506 \times 10^4 s + 2.555 \times 10^7}$
$[0.45, 0.5]$	$\dfrac{-2.887 \times 10^{-5} s^2 + 352.8s + 1.453 \times 10^7}{s^2 + 4.095 \times 10^4 s + 1.453 \times 10^7}$
$[0.5, 0.55]$	$\dfrac{-1.263 \times 10^{-6} s^2 + 669.8s + 1.133 \times 10^7}{s^2 + 1.685 \times 10^4 s + 1.133 \times 10^7}$
$[0.55, 0.6]$	$\dfrac{3.828 \times 10^{-6} s^2 - 980.9s + 7.441 \times 10^6}{s^2 + 7539s + 7.441 \times 10^6}$

III. HAMMERSTEIN MODEL-BASED ROBUST CONTROL

The approach followed for designing the controller is the model-based robust control in the frequency domain for continuous-time models [8].

The closed loop control system to be designed is shown in Fig. 5; this scheme is equivalent to that of Fig. 6 because of the plant is that depicted in Fig. 2, where $G_p(s)$ is the transfer function of the linear part of the Hammerstein model. As already said, in view of

application of robust control techniques, it is convenient to describe the linear part as an uncertain process consisting of a nominal model, $\tilde{G}_p(s)$, and a set of transfer functions (cf. Table II) which satisfy the following inequality:

$$P = \left\{ G_p(s) : \left| \frac{G_p(j\omega) - \tilde{G}_p(j\omega)}{\tilde{G}_p(j\omega)} \right| < \Delta_m(\omega) \right\}, \qquad (24)$$

where $\Delta_m(\omega)$ is the multiplicative uncertainty bound.

A. Modelling of P

For modelling the process P, it is sufficient o define the nominal transfer function and the uncertainty bound. The nominal transfer function chosen is that corresponding to the duty-cycle range $[0.4, 0.45]$. This transfer function is given by:

$$\tilde{G}_{p,1}(s) = \tilde{K}_{p,1} \frac{s^2 - 1.781 \times 10^7 s - 1.525 \times 10^{12}}{s^2 + 8.506 \times 10^4 s + 2.555 \times 10^7},$$

where $\tilde{K}_{p,1}(s) = -1.676 \times 10^{-5}$. It displays a gain equal to 1 and the following pole-zero location:

poles: $p_1 = -301.425, p_2 = -8.475 \times 10^4$,

zeros: $z_1 = 1.789 \times 10^7, z_2 = -8.519 \times 10^4$.

This location suggests the following remarks.

Remark 1- The pole p_2 and the zero z_2 are near each other and, consequently, they can be cancelled from the nominal transfer function $\tilde{G}_{p,1}(s)$. It can be easily verified that the transfer function with or without the above zero-pole couple are practically superimposed if the gain of the new nominal transfer function remains equal to 1. The new candidate nominal transfer function, $\tilde{G}_{p,2}(s)$, is given by:

$$\tilde{G}_{p,2}(s) = \tilde{K}_{p,2} \frac{(s - z_1)}{(s - p_1)}, \quad \tilde{K}_{p,2} = \tilde{K}_{p,1} \frac{(-z_2)}{(-p_2)}.$$

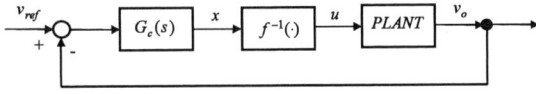

Fig. 5 Basic control scheme.

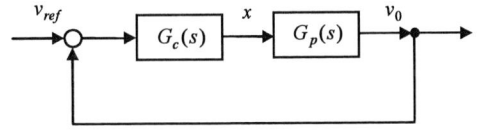

Fig. 6 Basic equivalent control scheme.

Remark 2- The zero z_1 is far from the pole p_1 and, consequently, it can be considered as a zero at the infinity. This allows to choose the nominal transfer function, $\tilde{G}_p(s)$, as follows:

$$\tilde{G}_p(s) = \frac{\tilde{K}_p}{(s-p_1)} = \frac{K}{1+s\tau},\qquad(25)$$

where:

$$\tilde{K}_p = \tilde{K}_{p,2}(-z_1), \ K = \frac{\tilde{K}_p}{-p_1}, \ \tau = \frac{1}{-p_1}.$$

The Bode diagrams of the transfer functions $\tilde{G}_{p,2}(s)$ and $\tilde{G}_p(s)$ are given in Fig. 7. These diagrams show that the two functions differ in the phase for high values of the angular frequency, when the amplitudes of the two transfer functions are less than about 70 dB.

In order to complete the modelling process of P it is necessary to compute the uncertainty bound $\Delta_m(\omega)$. This is accomplished starting from the nominal transfer function $\tilde{G}_p(s)$ (cf. (25)) and the remaining transfer functions displayed in Table II. The resulting uncertainty bound is given in Fig. 8.

B. Controller design via IMC control scheme

As is well known, the controller of the feedback control structure of fig. 6 is conveniently designed using the IMC control structure of Fig. 9, where ξ is an equivalent disturbance and $G_q(s)$ is the transfer function

of the IMC controller. As is easy to verify, the controller of the feedback structure of Fig. 6 is given by:

$$G_c(s) = \frac{G_q(s)}{1 - \tilde{G}_p(s)G_q(s)},\qquad(26)$$

The IMC controller is given by $G_q(s) = \tilde{G}_q(s)G_f(s)$, where $G_f(s)$ is the transfer function of the IMC filter and $\tilde{G}_q(s)$ is obtained by minimizing the cost function:

$$J = \|e(\cdot)\|_2^2 = \int_0^\infty e^2(t)dt,\qquad(27)$$

where $e(t) = v_{ref} - v_0$ (cf. Fig. 7). In order to cope with overshoot problems during transients arising when the input v_{ref} supplies the whole control system, it is convenient to design $\tilde{G}_q(s)$ for a smooth input given by:

$$V(s) = \frac{\beta\gamma}{s(s+\beta)(s+\gamma)},\qquad(28)$$

which in the time domain has the properties:

$$v(0) = \dot{v}(0) = 0, \ \lim_{t\to\infty} v(t) = 1.$$

For design purposes, it is convenient to consider this input generated at the output of a system having transfer function given by:

$$W(s) = \gamma\sqrt{\frac{\beta}{2}}\frac{1}{s+\gamma},\qquad(29)$$

supplied by the input:

$$V'(s) = \frac{\sqrt{2}\beta}{s+\beta}.$$

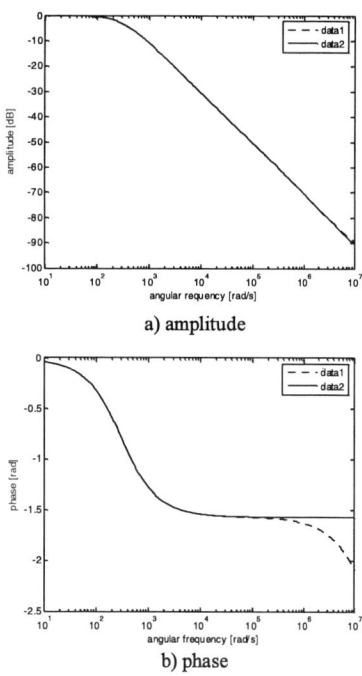

a) amplitude

b) phase

Fig. 7 Bode diagrams of $\tilde{G}_{p,2}(s)$ and $\tilde{G}_p(s)$; data 1: $\tilde{G}_{p,2}(s)$,

data 2: $\tilde{G}_p(s)$.

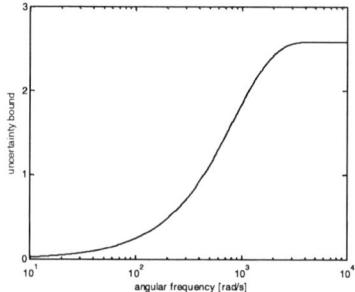

Fig. 8 Uncertainty bound $\Delta_m(\omega)$.

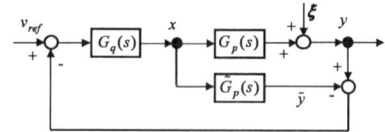

Fig. 9 IMC control scheme.

Note that this input is bounded, i.e. it satisfies the condition:

$$\|v'(t)\|_2^2 = \int_0^\infty |v'(t)|^2 dt \le 1 .$$

Since both $\tilde{G}_p(s)$ and $V(s)$ are minimum-phase, the controller $\tilde{G}_q(s)$ which minimizes (27) is given by [8]:

$$\tilde{G}_q(s) = \tilde{G}_p^{-1}(s) = \frac{1}{K}(1 + s\tau) . \qquad (30)$$

The IMC filter $G_f(s)$ is chosen so that the IMC controller $G_q(s)$ be proper and the whole system be type 1. The following $G_f(s)$ satisfies the above requirements:

$$G_f(s) = \frac{1}{1 + s\lambda} , \qquad (31)$$

independently of the value of λ that will be chosen later in order to satisfy the robust behaviour condition.

From (25), (30) and (31), the structure of the feedback controller $G_c(s)$ is given by (cf. (26)):

$$G_c(s) = K_p(1 + \frac{1}{sT_i}) , \quad K_p = \frac{\tau}{K\lambda} , \quad T_i = \tau \qquad (32)$$

It follows that the robust control procedure employed in this paper brings to a PI-type controller able to assure the desired robustness properties to the whole system.

To this end, only the parameter λ has to be chosen. This parameter is chosen so that the following robust behaviour condition be satisfied:

$$|\tilde{G}_p(j\omega)\tilde{G}_q(j\omega)G_f(j\omega)\Delta_m(\omega)| +$$
$$+ |[1 - \tilde{G}_p(j\omega)\tilde{G}_q(j\omega)G_f(j\omega)]W(j\omega)| < 1, \ \forall \omega, \qquad (33)$$

Obviously, if this condition is satisfied, the first term of (33) is less than 1 and, consequently, it is also satisfied the robust stability condition.

As is easy to verify, choosing $\beta = 5000$ and $\gamma = 10000$ (cf. (29)), the value of λ that satisfies (33) is given by $\lambda = 0.0017$.

Remark 3- Note that the choice of an input whose Laplace transform is minimum-phase affects the design of the robust controller only in the computation of λ, because the structure of the input affects that of $W(s)$.

IV VALIDATION OF THE CONTROLLER

The designed controller has been implemented in MATLAB-SIMULINK environment together with the switched model (18)-(19) with the aim of validating it. To cope with windup problems, the implementation of the PI

robust controller has been carried out using the antiwindup scheme described in [9] and shown in Fig. 10, where SAT denotes saturation of the duty cycle [0.05,0.95], LUT denotes a look-up table that stores the nonlinear characteristic of the Hammerstein model, and:

$$G_{ca}(s) = K_p , \quad G_{ch}(s) = \frac{1}{1 + sT_i} ,$$
$$v_{ref} = \frac{V_r}{s} , \quad G_i(s) = \frac{\beta\gamma}{(s + \beta)(s + \gamma)} .$$

Note that $G_i(s)v_{ref}$ generates the input $V(s)$ of the amplitude V_r.

A. Robust Controller

The responses of the closed loop control system of Fig. 10 are shown in Fig. 11 and 12, for two reference voltages belonging to two different ranges of duty cycle.

B. Comparison with a classical PI controller

The robust controller is now compared with a PI controller designed using the classical method of phase margin assignment starting from the nominal model (25). More precisely, the PI controller is designed so as to obtain a phase margin of 45 degrees in correspondence to a crossover frequency of 800 rad/s. The implementation of this controller, whose parameters are given by $K_p = 1.1695$, $T_i = 5.658 \times 10^{-4}$, is effected according to the scheme of Fig. 10. The corresponding results are shown in Figs. 13 and 14.

Comparison of Figs. 11 and 13 shows that the robust controller allows to obtain a smaller band of oscillation in the current and a smaller ripple in the output voltage error at the steady-state. The same comments can be made by comparing Figs. 12 and 14.

Looking at Figs. 11-14 it appears that the increment of the ripple in the output voltage error caused by the robust PI controller is less than that caused by the classical PI controller.

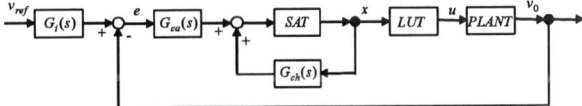

Fig. 10 Closed loop control scheme.

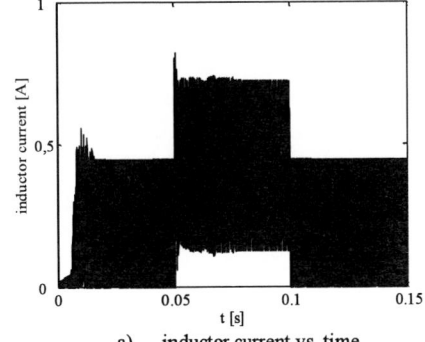

a)   inductor current vs. time.

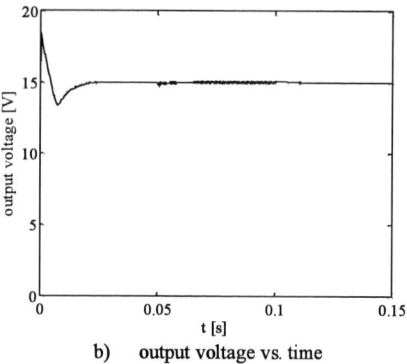

b)   output voltage vs. time

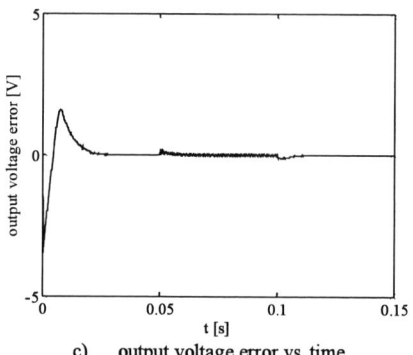

c)   output voltage error vs. time

Fig. 11 Responses of the closed loop system: reference voltage 15 V,
$R = 5 \ \Omega$ for $t \in [0.05, 0.1)$, $R = 10 \ \Omega$ for $t \in [0, 0.05)$ and $t \in [0.1, 0.15]$.

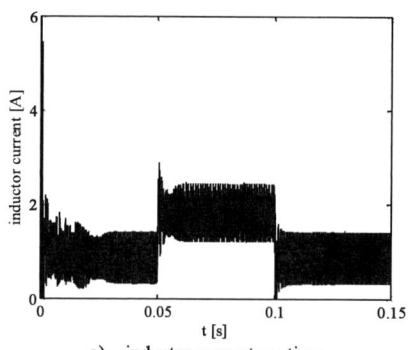

a)   inductor current vs. time.

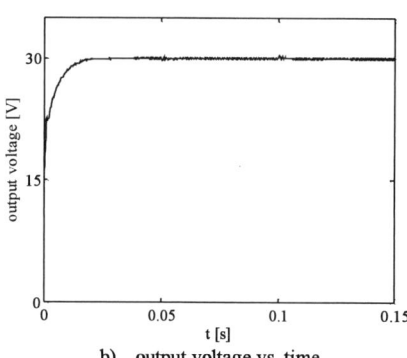

b)   output voltage vs. time.

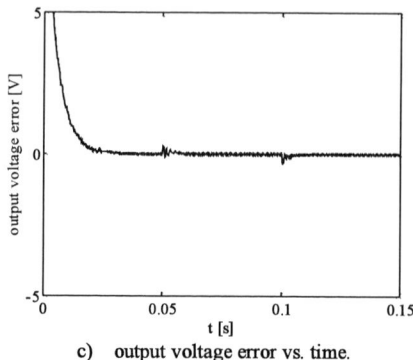

c)   output voltage error vs. time.

Fig. 12 Responses of the closed loop system: reference voltage 30 V,
$R = 5 \ \Omega$ for $t \in [0.05, 0.1)$, $R = 10 \ \Omega$ for $t \in [0, 0.05)$ and $t \in [0.1, 0.15]$.

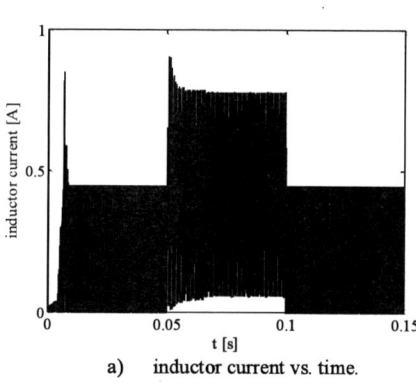

a)   inductor current vs. time.

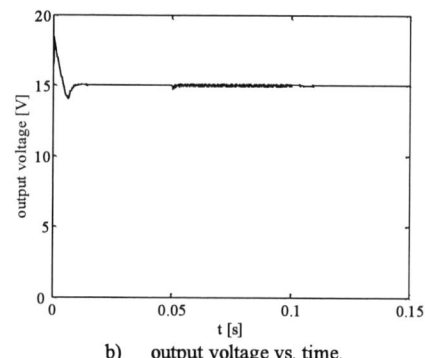

b)   output voltage vs. time.

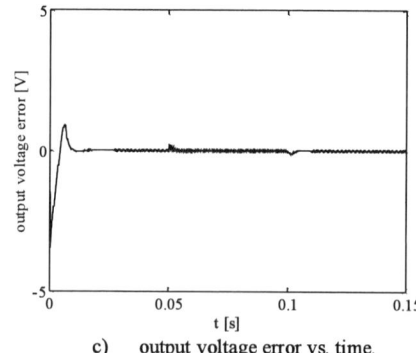

c)   output voltage error vs. time.

Fig. 13 Responses of the closed loop system: reference voltage 15 V,
$R = 5 \ \Omega$ for $t \in [0.05, 0.1)$, $R = 10 \ \Omega$ for $t \in [0, 0.05)$ and $t \in [0.1, 0.15]$,
classical PI.

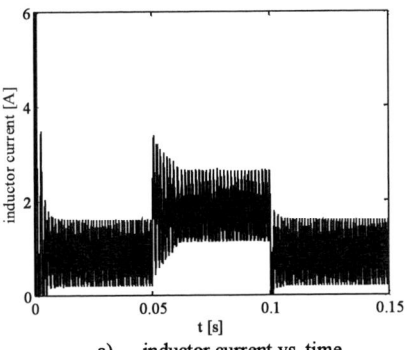

a)   inductor current vs. time

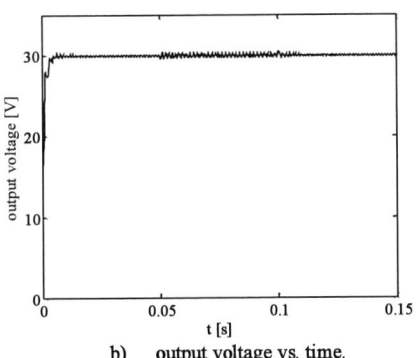

b)   output voltage vs. time.

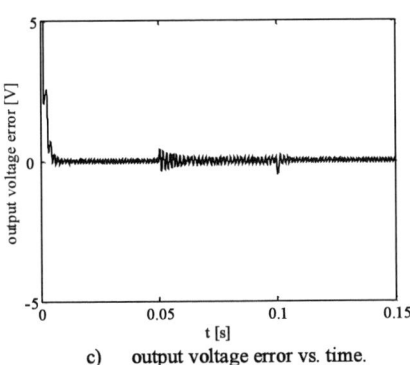

c)   output voltage error vs. time.

Fig. 14 Responses of the closed loop system: reference voltage 30 V, $R = 5\ \Omega$ for $t \in [0.05, 0.1)$, $R = 10\ \Omega$ for $t \in [0, 0.05)$ and $t \in [0.1, 0.15]$, classical PI.

III. Conclusions

In the paper, a method is illustrated for identifying and designing a robust controller for a Boost DC/DC converter described by means of a Hammerstein model. Identification is carried out by means of simulation performed on a switched model of the converter. The robust controller in question is designed using an IMC control structure which allows to obtain the set of all the stabilizing feedback controllers for the system in question. The design of the controller is done choosing a nominal model of the converter and considering the other models, describing the converter in contiguous intervals of the duty-cycle, as a model family describing an uncertain process. The design in question allows to incorporate the reference input into the design requirements. This input can be chosen so as to obtain a desired behaviour during the transients occurring just after the application of the reference voltage. A classical PI controller is also designed with the aim of comparing both the controllers. Several simulation experiments, show that the procedure based on the simulation is suitable for modelling purposes of DC/DC converters by means of Hammerstein models. Moreover, the use of robust control techniques is suitable for designing controllers based on the above models. Finally, it is shown that these control techniques allows to obtain better results compared to classical control techniques.

References

[1] R. Haber, and L. Keviczky, "Non linear System Identification-Input-Output Modelling Approach", *Vol. 1: Nonlinear System Parameter Identification*, Kluwer Academic Publishers, 1999.

[2] A. Balestrino, A. Landi, M. Ould-Zmirli, and L. Sani , "Automatic nonlinear auto-tuning method for Hammerstein modelling of electrical drives," *IEEE Trans. Ind. Electron.*, Vol. 48, No. 3, pp. 645-655, June 2001.

[3] F. Alonge, F. D'Ippolito, F.M. Raimondi, and S. Tumminaro, "Nonlinear modelling of DC/DC converters using the Hammerstein's approach," *IEEE Trans. Power Electron.*, vol. 22, no. 4, pp. 1210-1221, July 2007.

[4] M.H. Rashid, *SPICE for Circuits and Electronics Using PSice*, Prentice Hall, Englewood Cliffs, New Jersey, 1990.

[5] D. Marksimovic, A.M. Stankovic, V.J. Thottuvelil, and G.C. Verghese, "Modelling and simulation of power electronic converters," *Proc. of IEEE*, vol. 89, no. 6, pp. 898-912, June 2001.

[6] A. Borisaljevic, M.M.R. Iravani, and S.B. Dewan, "Modelling and analysis of a digitally controlled high power switch-mode rectifier," *IEEE Trans. Power Electron.*, vol. 20, no. 2, pp. 378-394, March 2005.

[7] J.L. Cassidys and J.L. Junkins, *Optimal Estimation of Dynamics Systems*, Chapman & Hall, 2004.

[8] M. Morari, and E. Zafiriou, *Robust Process Control*, Prentice-Hall International, Inc., 1989.

[9] K. Astrom, and T. Hagglund, *PID Controllers: Theory, Design, and Tuning*, 2nd ed., Instrument Society of America, 1995, pp. 80-92.

A New Model Control DC-DC Converter to Improve Dynamic Characteristics

F. Kurokawa*,** and S. Sukita**

* Department of Electrical and Electronic Engineering
** Graduate School of Science and Technology
Nagasaki University
1-14 Bunkyo-machi, Nagasaki, 852-8521 Japan

Abstract– **We have already reported the novel digital control method "Static Model Reference" which is the improved smart digital reference of output voltage. In this case, the VCO (Voltage controlled oscillator) is used as A/D signal converter because it has superior response and is low cost. However, prior attempts to improve the transient response have been inconclusive because this model depends on the steady-state analysis. This paper presents a new model control method of switching dc-dc converter to improve the dynamic characteristics. Furthermore, the relationship between the parameters of control circuit and the dynamic characteristics is discussed. As a result, it is seen that the dc-dc converter with a new model control has a superior transient response compared with that of the conventional control. Especially, the overshoot of the reactor current is suppressed to within 34% and this result is approximately 30% smaller than that of the conventional "static model reference" control.**

Index Terms—**DC-DC Converter, Digital Control, Model, Voltage Controlled Oscillator.**

I. INTRODUCTION

The concern with digital control for switching power supply has been growing for the last several years [1]-[5] because there has been a renewal of interest in a new approach to improve the control method among the engineer of the power supply for CPU. The central problem of digital control circuit is the design of the A-D converter and digital PWM signal generator. The problem which we have to consider next is to present the typical digital control method. We have already reported the novel digital control method "Static Model Reference" [6] which is the smart digital reference of output voltage. In this case, the VCO (Voltage controlled oscillator) is used as both A-D signal converter and digital PWM generator [7], [8]. The VCO has superior response and low cost performance. However, prior attempts to improve the transient response have been inconclusive because this model depends on the steady-state analysis.

This paper presents a new model control method of a switching dc-dc converter using both the static and dynamic functions to improve the dynamic characteristics in the switching power supply. Furthermore, the relationship between the parameters of control circuit and the dynamic characteristics is discussed.

This work is supported in part by the Grant-in-Aid for Scientific Research (No.18310117) of JSPS (Japan Society for the Promotion of Science) and the Ministry of Education, Science, Sports and Culture.

II. TECHNICAL WORK PREPARATION

Figure 1 shows the digitally controlled dc-dc converter using a new model control circuit. . In this figure E_i, R and e_o are the dc input voltage, load resistance and output voltage, respectively. L and C are the reactor and output smoothing capacitor. The model senses the output voltage E_o, output current i_o and input voltage E_i. The output current i_o is detected as the voltage e_s by a sensing resistor R_s.

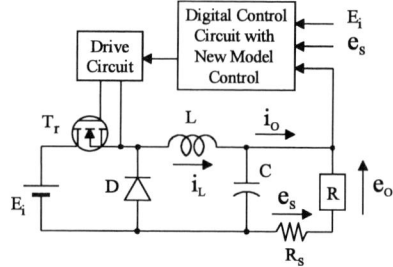

Fig. 1. Basic configuration of the digitally controlled
dc-dc converter.

Figure 2 shows the configuration of the digital control circuit using a new model controller. This controller is divided into the P-I-D controller and the model reference portions.

In the P, I and D controllers, the output voltage e_o of the dc-dc converter is input to the voltage controlled oscillator VCO_{EO} as E_{1EO} through a pre-amplifier circuit, and converted to the frequency f_{EO} as shown in Fig.3. f_{EO}^* and E_{1EO} are the frequency and the voltage at the equilibrium operating point of the dc-dc converter. f_{EO} and E_{1EO} are represented as follows;

$$E_{1EO} = A_{EO}(E_o - E_o^*) + E_{BEO} \qquad (1)$$

$$f_{EO} = G_{EO}E_{1EO} + B_{EO} \qquad (2)$$

where A_{EO} and E_{BEO} are the gain and bias voltage of the pre-amplifier of VCO_{EO} and G_{EO} and B_{EO} are the slope and intercept of VCO_{EO}, respectively.

f_{EO} is sent to the comparator in the P-controller, the up-counter in the D-controller and a new model controller. In I and D controllers, f_{EO} is counted during the pulse counting interval βT_S and the number N_n of pulses is generated [6]-[8]. In this case, the suffix n denotes the n-th period of the switching period T_S. $N_{D,n}$ is multiplied by the differential coefficient K_D and $K_D N_{D,n}$

978-1-4244-0644-9/07/$25.00 ©2007 IEEE

is generated at the multiplier. $\Sigma N_{I,n}$ is generated by the latch and adder. In this case, N_{INT} is the predetermined reference value in the I-controller and corresponds to the desired output voltage of the dc-dc converter. Lastly, the modified reference value $N_{RM,n}$ is generated by the adder, subtracter and static model reference as follows:

$$N_{RM,n} = N_{R,n-1} - (K_D N_{D,n} + K_I \Sigma N_{I,n}) \qquad (3)$$

where $N_{R,n-1}$ is the predetermined reference value of (n-1)-th switching period in a new model controller. In this case, the model reference $N_{R,n-1}$ is used considering delay of one switching cycle when the model reference is calculated by the digital signal processor (DSP). The frequency f_{EO} of the voltage controlled oscillator VCO_{EO} is sent to the comparator as a digital PWM signal generator and compared with the modified reference value $N_{RM,n}$. Therefore, the VCO is performed as both the A-D signal converter and the digital PWM generator.

The on-time $T_{on,n+1}$ of (n+1)-th switching period in the drive circuit is represented as follows:

$$T_{on,n+1} = N_{RM,n} / f_{EO,n+1} \qquad (4)$$

Usually, in equation (4), $f_{EO,n+1}$ is fixed at f_{EO}^* and $N_{RM,n}$ is varied to control $T_{on,n+1}$ in the drive circuit because the steady-state output voltage error is zero in the integral operation mode of the digitally controlled dc-dc converter. Conventionally, though $N_{RM,n}$ is varied to control $T_{on,n+1}$, $N_{R,n-1}$ is fixed in equations (3) and (4). In this paper, however, $N_{R,n-1}$ is intentionally varied for the variations of the input voltage, load current and input voltage of the dc-dc converter.

Figure 4 shows the circuit configuration of a new model controller. The output voltage is converted to f_{EO} through the pre-amplifier and VCO_{EO} in Fig.2. The f_{EO} is counted during βT_S and then $N_{EO,n-2}$ is generated. The simplified voltage versus frequency characteristics of the voltage controlled oscillators VCO_C and VCO_{EI} are similar to that of VCO_{EO} in Fig.3. In Fig.4, $N_{C,n-2}$ and $N_{EI,n-2}$ are the digital values corresponding to the analog values e_S and E_i, respectively. The signal S_{CM} is the information of continuous current and discontinuous current modes, which is determined from the value of e_S in the pre-amplifier circuit.

$N_{EO,n-2}$, $N_{C,n-2}$ and $N_{EI,n-2}$ are represented as follows;

$$N_{EO,n-1} = \beta T_S \left\{ G_{EO} \left(A_{EO} \left(E_o - E_o^* \right) + E_{BEO} \right) + B_{EO} \right\} \qquad (5)$$

$$N_{C,n-1} = \beta T_S \left\{ G_C \left(A_C R_S I_o + E_{BC} \right) + B_C \right\} \qquad (6)$$

$$N_{EI,n-1} = \beta T_S \left\{ G_{EI} \left(A_{EI} \left(E_i - E_i^* \right) + E_{BEI} \right) + B_{EI} \right\} \qquad (7)$$

where A_C and E_{BC} are the gain and bias voltage of the pre-amplifier of VCO_C. G_C and B_C are the slope and intercept of VCO_C. A_{EI} and E_{BEI} are the gain and bias voltage of the pre-amplifier of VCO_{EI} and G_{EI} and B_{EI} are the slope and intercept of VCO_{EI}, respectively.

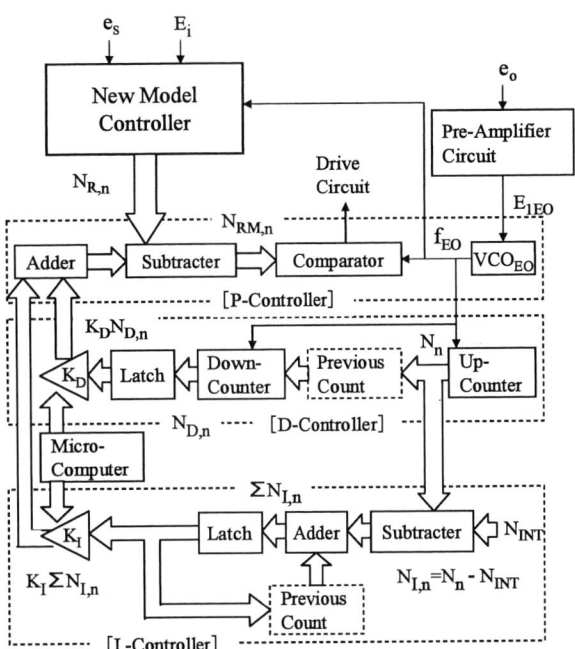

Fig. 2. Circuit configuration of the digital control circuit with a new model controller.

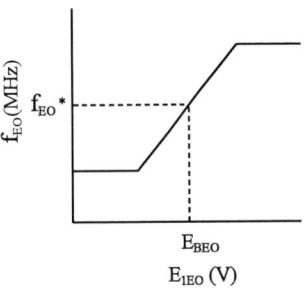

Fig. 3. Simplified voltage versus frequency characteristics of VCO_{EO}.

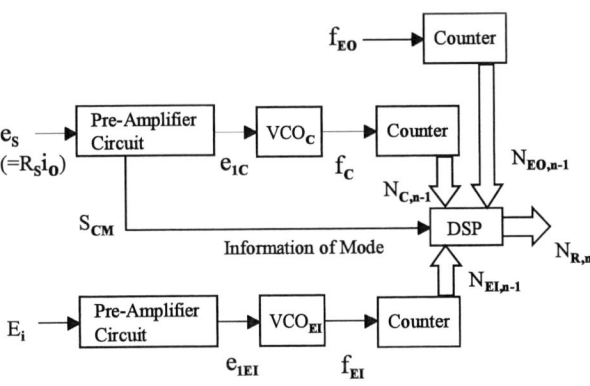

Fig. 4. Circuit configuration of a new model controller

In the buck type dc-dc converter shown in Fig.1, the reference value $N_{R,n}$ is derived as follows:

For the continuous reactor current mode:

$$N_{R,n} = f_{EO}{}^*T_S(E_o{}^* + rN_{I,n})/N_{E,n} \qquad (8)$$

For the discontinuous reactor current mode:

$$N_{R,n} = f_{EO}{}^*\sqrt{2E_o{}^*T_SLN_{I,n}/\{N_{E,n}(N_{E,n} - E_o{}^*)\}} \qquad (9)$$

where N_E is given by the next equation.

$$N_{E,n} = E_i{}^* + \frac{N_{EI,n-1} - \beta T_S(G_{EI}E_{BEI} + B_{EI})}{\beta T_S G_{EI} A_{EI}} \qquad (10)$$

These equations (8), (9) and (10) have been previously presented in [6].

In a new model controller, the following equation N_I is proposed by parameters of digital control circuit.

$$N_{I,n} = E_o{}^* \Bigg/ \left(\frac{1}{M}\sum_{m=n-1-M}^{n-1} R_m\right) \qquad (11)$$

m is the number of the sampling point of the moving average. In the proposed method, the resistor R of output load is calculated as follows;

$$R_{n-1} = \frac{E_{o,n-1}}{I_{o,n-1}}$$

$$= \frac{G_C A_C R_S}{G_{EO} A_{EO}} \cdot \frac{N_{EO,n-1} - \beta T_S(G_{EO}E_{BEO} + B_{EO}) + E_o{}^*\beta T_S G_{EO} A_{EO}}{N_{C,n-1} - \beta T_S(B_C + G_C E_{BC})} \qquad (12)$$

The moving average of the R in Eq. (11) is obtained from Eq. (12). N_I is substituted for Eqs. (8) and (9) corresponding to the static model control [6] and then $N_{R,n}$ is transferred to the subtracter in Fig. 2.

Furthermore, in Fig. 4, the gain of preamplifier for e_S is switched against two currents modes because $N_{C,n-2}$ is smoothly changed in the discontinuous current mode when the gain of discontinuous current mode [9] is larger than that of the continuous mode. The current gain A_C is small in the continuous reactor current mode and is large in the discontinuous one, respectively.

Figure 5 shows the regulation characteristics against the changes of input voltage E_i and output current I_O. The solid line denotes the simulated results of the digitally P-I-D controlled dc-dc converter with a new model controller and the broken and dash-dotted lines denote those of the conventional digitally P-I-D controlled dc-dc converter without a model controller. It is often observed that the dynamic characteristics of the digitally controlled dc-dc converter are deteriorated when the relatively large integral coefficient is used to extend the regulation range of the output voltage for the variations of the input voltage and load current. So, in the digitally P-I-D controlled dc-dc converter with a new model controller, the integral coefficient K_I is very small and is equal to 0.003 because the novel digital controller with a new

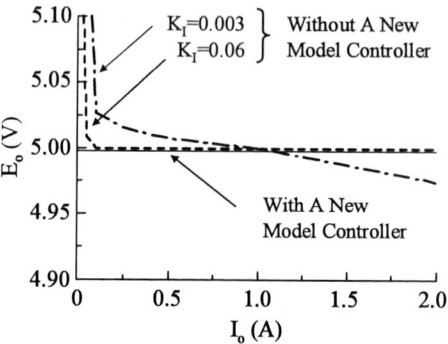

(a) Against the change of input voltage E_i.

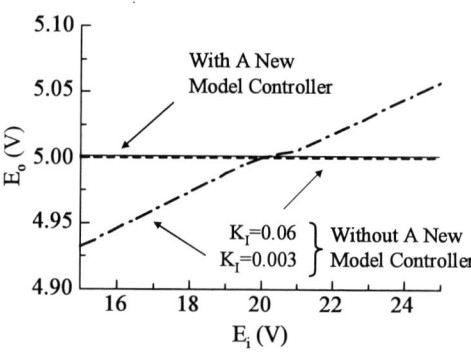

(b) Against the change of output current I_o.

Fig. 5. Regulation characteristics.

model controller is presented to extend the regulation range of the output voltage for the variations of the load current both in the continuous reactor current mode and the discontinuous reactor current mode without using the large integral coefficient K_I in the digital P-I-D controller.

In the conventional digitally P-I-D controlled dc-dc converter, it is clarified in Fig. 5 that the regulation range is very small when K_I is equal to 0.003 corresponding to that with a new model controller. The integral coefficient K_I over 0.05 is necessary to regulate the output voltage in this case.

III. TRANSIENT RESPONSE

Figure 6 through 8 show the simulated transient response of the dc-dc converter with a new model controller in step change of the load resistor R from 100Ω (discontinuous reactor current mode) to 2.5Ω (continuous reactor current mode). The simulator is PSIM. The switching frequency is 100kHz.

Figure 6 shows the transient response, taking the differential coefficient K_D as a parameter. The circuit parameters are K_I=0.01, M=1, E_i=20V, $E_o{}^*$=5V, L=0.18mH, C=500μF, r=0.6Ω, $f_{EO}{}^*$=63MHz, G=22MHz/V and A_{EO}=10. In this case, the circuit performance is same to the previous presented "static model reference" control because Eqs. (3) and (4) are equal to the reference values $N_{R,n-2}$ in [6] when M is equal

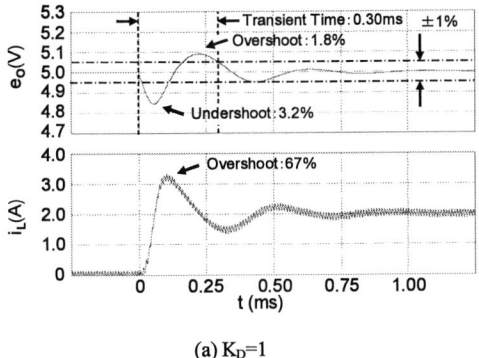

(a) $K_D=1$

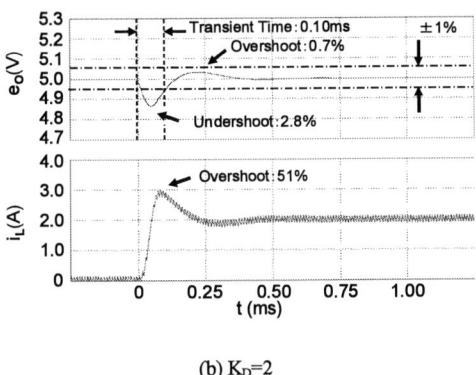

(b) $K_D=2$

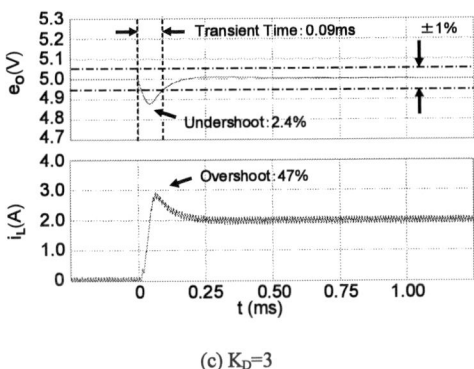

(c) $K_D=3$

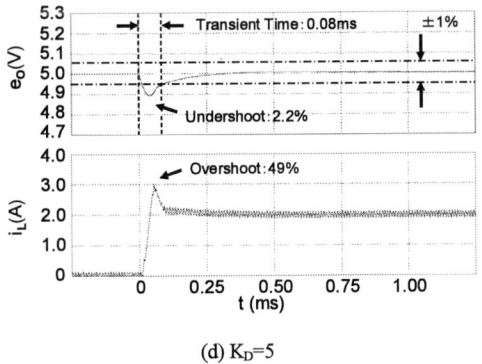

(d) $K_D=5$

Fig. 6. Transient responses in step change of the load resistor R, taking K_D as a parameter.

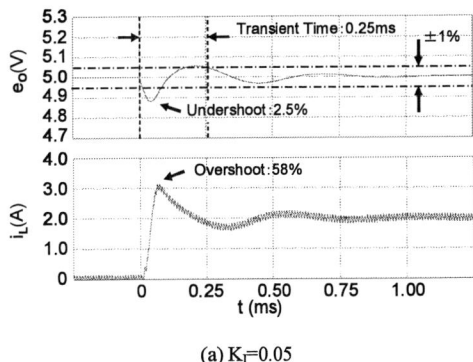

(a) $K_I=0.05$

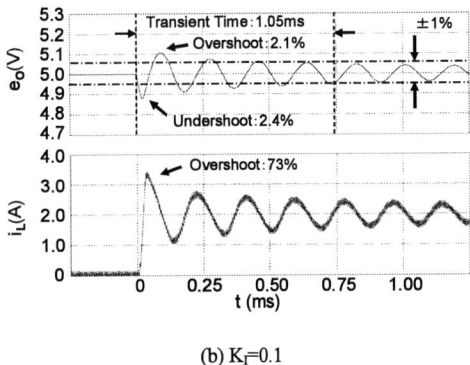

(b) $K_I=0.1$

Fig. 7. Transient responses in step change of the load resistor R, taking K_I as a parameter.

to unity. From this figure, it is seen that the circuit performance can be improved by making the differential coefficient K_D as large in the stable operation. So, $K_D = 3$ is selected in this paper. On the other hand, the transient response of the output voltage and the reactor current are not settling when K_I is large as shown in Figs. 7(a) and (b). In these figures, K_D is equal to 3 and the other parameters are the same as those of Fig. 6 except K_I.

Figure 8 shows the transient response, taking the number M of the sampling points as a parameter. In this figure, K_D is equal to 3 and the other parameters are the same as those of Fig. 6 except M. The superior transient response is obtained in case of M=5 as shown in Fig. 8(a). The undershoot, overshoot and transient time of the output voltage are 2.5%, 0% and 0.11ms, respectively. The transient time becomes long when M is too large. It is revealed that the proposed control method has a superior dynamic response when M is selected to the optimum value. Especially, the overshoot of the reactor current is 34% and is suppressed compare with 47% of the conventional "static model reference" control as shown in Fig. 6(c). This result is approximately 30% smaller than the conventional static model controlled dc-dc converter. The superior characteristics are caused by Eqs. (3) and (4) using the moving average of the load resistance R.

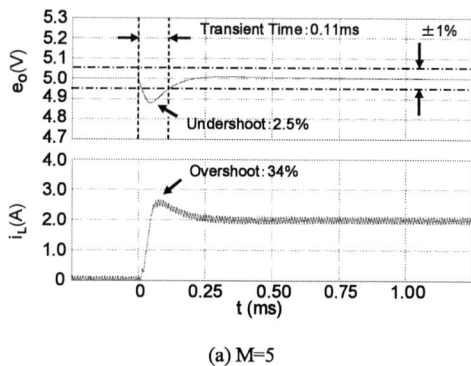

(a) M=5

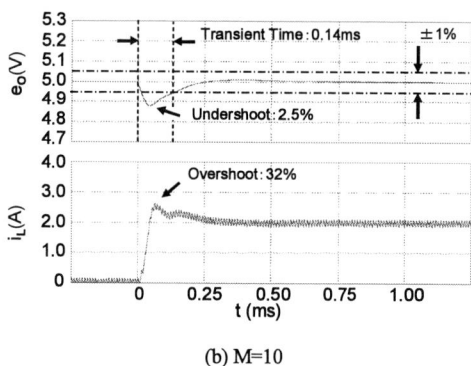

(b) M=10

Fig. 8. Transient responses in step change of the load resistor R, taking M as a parameter.

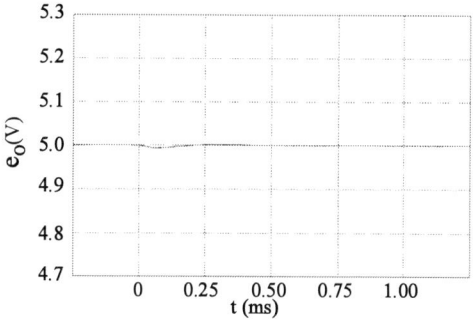

Fig. 9. Transient responses in step change of the input voltage E_i.

Figure 9 shows the simulated transient response of the dc-dc converter with a new model controller in step change of the input voltage E_i from 20V to 16V in the continuous reactor current mode(R=2.5Ω). The circuit parameters are the same as those of Fig. 8(a). It is seen in this figure that the undershoot against the transient response is very short in step change of the input voltage in step change of the input voltage.

IV. CONCLUSION

A new model control method of switching dc-dc converter to improve the dynamic characteristics are discussed.

As a result, it is seen that the dc-dc converter with a new model control has a superior transient response compared with that of the conventional control. Especially, the overshoot of the reactor current is suppressed to within 34% and this result is approximately 30% smaller than that of the conventional "static model reference" control. Therefore, it is confirmed that the moving average of the load resistance is useful.

These results suggest the next generation model of the switching power supply. The proposed smart digital control method, A-D converter and digital PWM methods are useful to realize the digitally controlled switching power supply.

REFERENCES

[1] L. Guo, J. Y. Hung and R. M. Nelms: "PID controller modifications to improve steady-state performance of digital controllers for buck and boost converters", Proceedings of Seventeenth Annual IEEE Applied Power Electronics Conference, no.9.3, pp. 381-388, March 2002.

[2] W. E. Bury, D. Czarkowski, J. Dzieza, S. Lewis and S. Ramamurthy: "DSP H∞ controller for a variable output dc-dc converter: design and implementation", Proceedings of IEEE International Telecommu-nications and Energy Conference, pp. 415-420, Oct. 2002.

[3] D. Maksimovic, R. Zane and R. Erickson: "Impact of digital control in power electronics", Proceedings of 2004 International Symposium on Power Semi-conductor Devices & ICs, Kitakyushu, pp. 13-22, May 2004.

[4] O. Garcia, P. Zumel, A. de Castro, J. A. Cobos and J. Uceda: "An automotive 16 phases dc-dc converter ", 2004 35th Annual IEEE Power Electronics Specialists Conference Record, pp.350-355, June 2004.

[5] J. Li, F. C. Lee and Y. Qiu:"New digital control architecture eliminating the need for high resolution DPWM", IEEE Power Electronics Specialists Conference Record, pp.814-819, June 2007.

[6] F. Kurokawa, K. Tanaka and H. Eto: "Performance characteristics of switching dc-dc power converter with static model reference", Proceedings of the ICEMS '06, pp.1-5, Nov. 2006.

[7] F. Kurokawa, M. Sasaki, S. Hiura and H. Matsuo: "1 MHz high-speed digitally controlled dc-dc converter", Invited Paper, IEICE Trans. on Commun., Vol. E87-B, No. 12, pp. 3437-3442, Dec. 2004.

[8] F. Kurokawa, T. Nakashima, K. Tanaka and W. Okamoto: "A new fast digitally controlled dc-dc converter", IEEE Power Electronics Specialists Conference Record, pp.798-802, June 2007.

[9] H. Matsuo and K. Harada: "Characteristics of dc-dc converter in the discontinuous mode of the reactor current," Trans. IEICE of Japan, vol. J93-C, no.6, pp.123-129, June 1973.

Fuzzy Incremental Controller for the 3rd Order Buck Converter

M. Veerachary, Deepen Sharma

Department of Electrical Engineering
Indian Institute of Technology Delhi
New Delhi, India

Abstract--In this paper a fuzzy incremental controller is proposed for higher order converter topology, 3rd order buck converter. Main advantage of the proposed converter is the input source current is smooth and exhibits lower ripple current/ EMI. As the converter is of higher order nature controller parameters tuning is a difficult task in conventional design and hence an adaptive controller, fuzzy PI type controller, is proposed in this paper. Fuzzy PI tuning factors selection is discussed and then total closed-loop regulated converter system is simulated in SIMULINK using fuzzy logic tool box. Experimental results, obtained using dSPACE, are given for validating the proposed concept and simulations.

Index Terms—Fuzzy logic controller, 3rd order buck converter, Incremental controller, Load voltage regulation.

I. INTRODUCTION

Switch mode power supplies (SMPS) gaining increased importance in the dc-dc power distribution systems. There are several different topologies were evolved, buck, boost and buck-boost type, and working satisfactorily. Each of these topologies has their own application depending on the type voltage conversion requirement. In step-down conversion applications buck topology is most widely used. However, there are certain issues involved with the conventional buck topology such as high ripple content in the source current, increased EMI filtering requirements, etc. To overcome some of these problems a modified buck topology, called 3rd order buck converter (TOBC), is proposed in this paper. Although this converter is single switch topology, but it consists of two capacitors, one inductor and hence the converter belongs to higher order family.

There are various control algorithms like PID control, current mode programmed control, sliding mode control, etc widely utilized in the control of switch-mode dc-dc converters. These control techniques are further implemented with varying control tools such as analog or microprocessor, etc. Conventional solutions for controller requirements were based on classical control theory or modern control theory such as PID family controllers, state-feedback controllers, self-tuning controllers, model reference adaptive controllers, etc [1]. Owing to the need of accurate mathematical models for these controllers, they are sensitive to parameter variation. However, all

these schemes needs accurate model and controller accuracy is also depends on the mathematical formulations. To eliminate the problems related to conventional controller parameter dependency/ tuning, an adoptive fuzzy logic controller (FLC) is proposed in this paper for voltage regulation purpose. This paper focuses on using real time hardware in-loop control tool called dSPACE for controlling the proposed converter, where the control circuit can be easily modeled in MATLAB/Simulink. The reason behind choosing FLC is that, it is an efficient control and simple to implement in MATLAB/Simulink using FLC toolbox. The paper shows that the control algorithm represented by FLC in MATLAB/Simulink and using the dSPACE as control toolbox is able to control the proposed converter model.

II. MODELING OF THE PROPOSED CONVERTER

The proposed converter topology, shown in Fig. 1, can be operated in several operating modes depending on the load, switching frequency and supply voltage. However, inductor current must be continuous; otherwise the benefits obtained from this converter are lost. As a result the converter is analyzed here for continuous inductor current mode (CICM) only. In CICM operation the circuit has two operating modes; Mode-1: S-ON (0<t<dT); Mode-2: S-OFF (dT<t<T). In each mode of operation the circuit is linear and its behaviour can easily be described by the state-space model [2] given by

$$\dot{x} = [A_K][x] + [B_K][u] \; ; v_o = [C_K][x] \qquad (1)$$

where $\quad x = \begin{bmatrix} i_L & v_{C1} & v_{C2} \end{bmatrix}^T, u = \begin{bmatrix} v_g \end{bmatrix}, k = 1,2.$

These equations are transformed into SIMULINK for closed-loop converter performance evaluation.

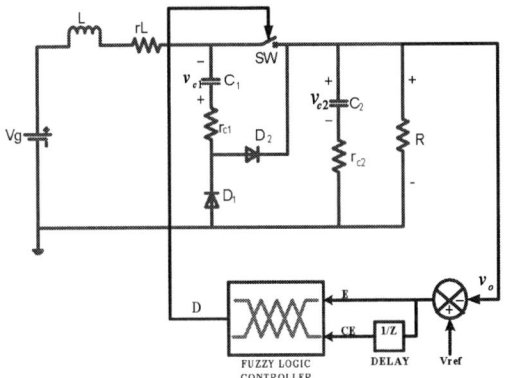

Fig.1. Block diagram of closed-loop 3rd order buck converter.

This work was supported by MHRD, Govt. of India under the R & D Project: Design and Development of Fuzzy Controllers for High Frequency DC-DC Conversion Systems.

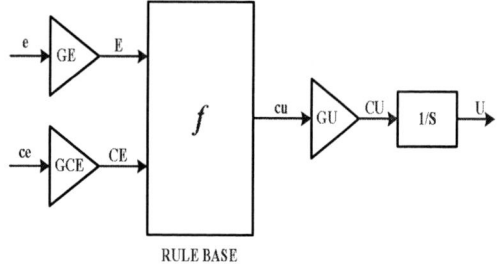

Fig. 2. Block representation of fuzzy incremental control logic.

III. CONTROLLER DEVELOPMENT

Recent trend is towards the fuzzy logic controllers [3]-[4]. They have generated a good deal of interest in certain applications where the inputs are imprecise or the converter is of a higher order resulting into complex mathematical modeling. Therefore in such cases FLC's have certain advantages over the conventional controllers such as: (i) no requirement of accurate mathematical model, (ii) inputs needn't be necessarily precise, (iii) inherent nonlinearity handling capability and (iv) more robust than conventional nonlinear controllers. The causes of nonlinearity in the power converters include a variable structure within a single switching period, saturating inductances, voltage clamping, etc. With the advent of higher order converters it is resulting in more complex mathematical models. Thus FLC seems to be a viable option as a controller for such a situation.

Owing to the non-linearity associated with fuzzy controllers, it is more difficult to set the controller gains compared to conventional proportional plus integral (PI) controllers. Thus certain basic tuning procedure need to be followed that carries tuning rules from the PI domain over to fuzzy domain. In general the procedure initially includes the tuning of conventional PI controller, followed by replacing it with an equivalent non-linear fuzzy controller and eventually fine-tuning of the nonlinear fuzzy controller is done as per the desired system response. The step-by-step FLC tuning process is given below.

Step-1: Tuning of Conventional PI controller

In general the derivation of an equivalent proportional and integral gain for the fuzzy incremental PI controller is obtained by first finding out the approximate K_P and K_I values from the conventional PI Controller. The most widely used *Ziegler-Nichols* tuning method has been followed here. Based on the final gain values obtained for the conventional PI controller, using the Ziegler-Nichols tuning method, the equivalent PI gains for the FLC is obtained. It is possible to obtain an equivalent fuzzy PI controller using only error and change in error as inputs to the rule base. However, it is generally reported in various literature that it is difficult to write rules for the integral action. Problems with *integrator windup* also come at times, which affects the control performance and it needs to be handled carefully. So certain modified approach can be used of which

incremental control [5] is one of them as is used in this paper.

Step-2: Transferring Conventional PI gains to Equivalent Fuzzy gains

The discrete approximation of the standard time domain PI controller's equation is given as

$$ u_n = K_P \left[e_n + \frac{1}{T_i} \sum_{j=1}^{n} e_j T_s \right] \tag{2} $$

where:
n = Time instant, T_S = Sampling time.
u = Controller Output, K_P = Proportional Gain
T_I = Integral Time, e = Error between the reference & the process output

The fuzzy incremental controller (FINC) given in the Fig. 2 is almost the same configuration as the PI controller except for the integrator on the output side. The output from the rule base is therefore the change in output (CU), and GU is the gain on the output. The control signal U is the sum of all previous increments, i.e.

$$ u_n = \sum_i (c u_i * G U * T_s) \tag{3} $$

An incremental controller actually adds a change in control signal 'Δu' to the current control signal, i.e.

$$ u_n = u_{n-1} + \Delta u_n \tag{4} $$

where $\Delta u_n = K_P (e_n - e_{n-1} + \frac{1}{T_i} e_n T_s)$,

so the controller can be represented in general as:

$$ u_n = \sum_{i=1}^{n} (E_i + C E_i) * G C U * T_s \tag{5} $$

This on simplification gives:

$$ u_n = GCE * GCU * \left[\frac{GE}{GCE} \sum_{i=1}^{n} e_i * T_s + e_n \right] \tag{6} $$

Step-3: Fuzzy Incremental Controller Gains Calculation

With the above equivalent relationship developed between the conventional and incremental fuzzy PI gains, the gains for the incremental FLC are being calculated in the 3rd step as given below. For the conventional PI controller the gains obtained by Ziegler-Nichols tuning were: K_P=0.1 and K_I=19.5. So a PI controller with these gains can be conventionally represented as:

$$ K_P + \frac{K_I}{s} = 0.1 + \frac{19.5}{s} \tag{7} $$

Now using the above equivalences obtained in Step-2, the equivalent gains for the incremental FLC is calculated as GCE =0.005 and GU= 10 assuming GE =1. These gains are more or less approximate and are further finely tuned online for the desired response.

974

TABLE I

FUZZY RULE BASE

CE\E	NB	NS	ZE	PS	PB
NB	(PB)	(PB)	(PS)	(ZE)	(NS)
NS	(PB)	(PS)	(PS)	(NS)	(NS)
ZE	(PB)	(PS)	(ZE)	(NS)	(NB)
PS	(PS)	(PS)	(NS)	(NS)	(NB)
PB	(PS)	(ZE)	(NS)	(NS)	(NB)

TABLE II

FUZZY GAINS PERFORMANCE COMPARISON

FUZZY GAINS	K_P	K_I
GCE (Increased)	Increases	Decreases
GCU (Increased)	Increases	No Effect
GE (Increased)	Decreases	Increases

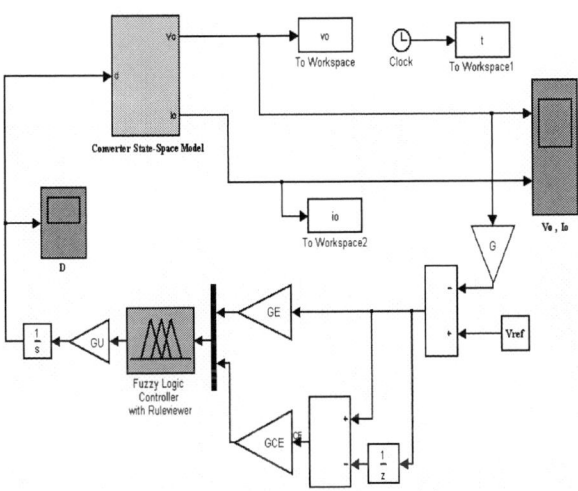

Fig. 3. MATLAB simulation model.

IV. RESULTS AND DISCUSSIONS

Having derived the equivalent gains for the *FINC* as given in step 3 above, it is simulated in MATLAB/Simulink platform where the converter state-space averaged model represented by its state space matrices given in [1] are used. The converter specifications considered for both simulation and experimental purpose are: $V_g = 24$ V, $V_o = 15$ V, $P_0 = 7.5$ W and the parameter values used in these studies are: L = 400 μH, $C_1 = 10$ μF, $C_2 = 220$ μF, $R_{load} = 30$ Ω ±50%. The SIMULINK simulation diagram is shown in Fig. 3. The FLC is implemented using MATLAB/Simulink Fuzzy Logic tool box, where the membership function taken for both the input variables, error (E) & change in

error (CE) of load voltage, as well as the output variable is triangular, ranging from -1 to +1. Five fuzzy levels are used for these triangular membership functions, i.e. positive big (PB), positive small (PS), zero (ZE), negative small (NS), negative big (NB) resulting into 25 linguistic rules as specified in the rule base, Table I. The weights for the rule base are specified in accordance to how E and CE are varying. These weights are repeatedly tuned after every simulation until the desired response is obtained. To verify the closed-loop converter regulation capability several simulations have been performed and demonstration purpose the following cases are given: (i) load disturbance 30 → 15 Ω and then back to 30 Ω, (ii)

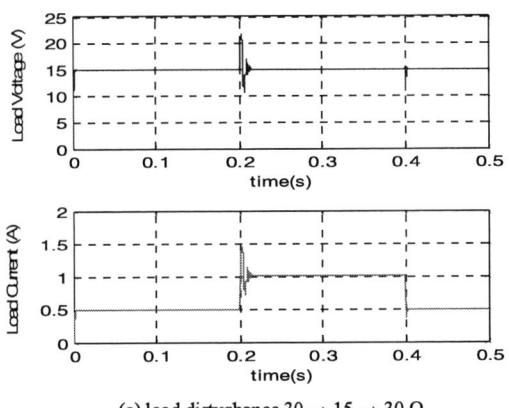

(a) load disturbance 30 → 15 → 30 Ω

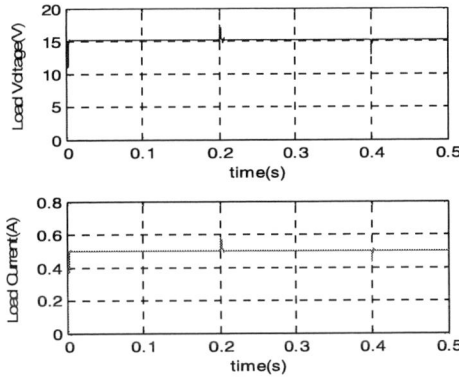

(b) source disturbance 24 → 28 → 24 V

Fig. 4. Dynamic response characteristics of load voltage regulation (simulation).

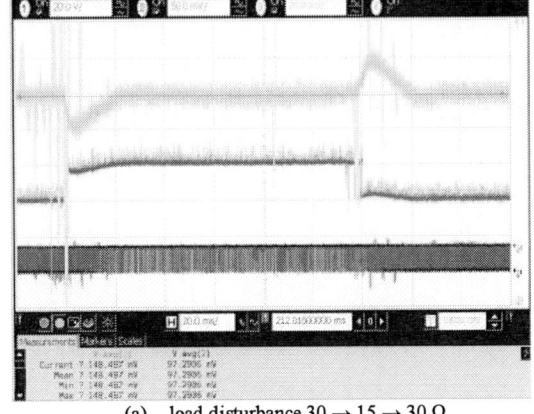

(a)　load disturbance 30 → 15 → 30 Ω

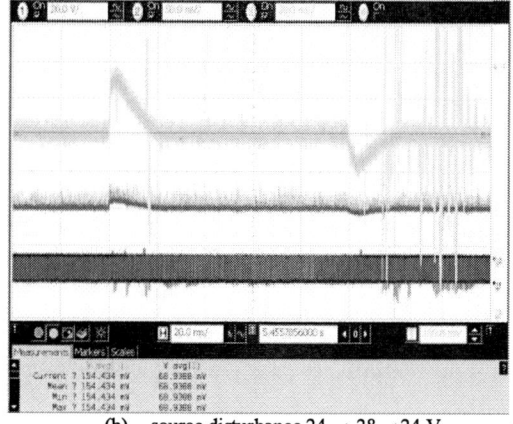

(b)   source disturbance 24 → 28 →24 V

Fig. 5. Dynamic response characteristics of load voltage regulation (experimental).

source disturbance 24 → 28 V and then back to 24 V. These simulation results are given in Fig. 4. Note that in all these cases load voltage regulation was achieved.

To verify the proposed theory as well as the simulation results experimental prototype converter system has been built and tested. For controller implementation the dSPACE (ds1104) board [6], which is real-time hardware in a loop simulation platform, is used. The experimental results were presented here, for the cases as was used in simulations, are shown in Fig. 5. From the simulation and experimental results presented in Figs. 4 and 5 it is seen that the load voltage is getting regulated for both load as well as source perturbation cases. The results given are for the load perturbation of 50% of the nominal load and ±4 V of source perturbation is considered. Also during the course of experimentation fuzzy gains were seen to have effect on the controller performance during the disturbances. The experimental observation regarding the dependency of the dynamic response of the converter under study can be obtained from the Table II.

V. CONCLUSIONS

Eventually it can be concluded that the proposed 3rd order buck converter with the incremental fuzzy logic controller is easily regulated under both load and source perturbations. Further, the dependency of controller performance on the equivalent fuzzy gains indicates that properly tuned gains would lead to a much improved dynamic response of the higher order converters.

ACKNOWLEDGEMENT

The authors would like to thank the MHRD, Govt. of India for supporting this research through R & D Project entitled Design and Development of Fuzzy Controllers for High Frequency DC-DC Conversion Systems.

REFERENCES

[1] Veerachary, M, Deepen Sharma, "Adaptive Hysteretic Control of 3rd Order Buck Converter", *IEEE International Conference, PEDES*-2006, pp.

[2] Robert W. Erickson, Dragan Maksimovic, "Fundamentals of Power Electronics", *Springer International Edition, Second Edition, 2001.*

[3] Y. F. Liu and P. C. Sen, "A general unified large-signal model for current programmed dc-to-dc converters," *IEEE Trans. Power Electron*, 1994, Vol. 9, pp. 414-424.

[4] W. C. So, C. K. Tse, and Y. S. Lee, "A fuzzy controller for dc-dc converters," *IEEE PESC Conf. Rec.*, 1994, pp. 315–320.

[5] C. C. Lee, "Fuzzy logic in control systems: Fuzzy logic controller—Parts I and II," *IEEE Trans. Syst., Man, Cybern*, 1990, Vol. 20, pp. 404–435.

[6] dSPACE user manual, 2006.

Design of a Single-Stage Single-Switch Power–Factor-Corrected (S^4-PFC) AC/DC Converter

P. Kongthawornwattana*, C. Bunlaksananusorn*, and S. Kittiratsatcha*

* Faculty Engineering, King Mongkut's Institute of Technology Ladkrabang (KMITL), Bangkok 10520, Thailand

*Abstract–*This paper presents the design of a Single-Stage Single-Switch Power-Factor-Corrected (S^{4}PFC) AC/DC converter. The converter under study is an integration of boost and flyback converters. The converter operation is equivalent to its two-stage counterpart with switches in the PFC and DC/DC converter stages turned on and off at the same time. Based on this observation, power circuit and control loop design of the converter can be carried out using standard design equations and methods, after the bulk capacitor voltage has been determined. This design concept is applied to design a 120W (12V, 10A) S^{4}PFC AC/DC converter. Experimental results are given to confirm validity of the proposed design method.

*Index Terms--*Single-Stage Single-Switch Power-Factor-Corrected AC/DC converter

I. INTRODUCTION

Modern electronic equipment and systems invariably use a Switched Mode Power Supply (SMPS) as a power source. The SMPS is an AC/DC converter that converts a voltage from the AC mains into a DC voltage at the level required by the load. Traditionally, the SMPS uses a diode bridge rectifier with a bulk output capacitor at its input stage. This input configuration is infamous for poor power utilization, i.e. low power factor, and being a source of harmonic currents. Since the enforcement of harmonics standards such as IEC1000-3-2, it becomes mandatory that a Power Factor Correction (PFC) circuit be incorporated into the SMPS. The common PFC solution is to connect a boost converter in between the bridge rectifier and the DC-DC converter. The boost converter is controlled to draw the averaged sinusoidal input current from the AC mains using the average current control technique [1]. Although this approach gives a near unity power factor and very low harmonic currents, it increases size and cost of the SMPS due to the additional PFC stage. To reduce the size and cost, several circuit topologies that integrate the PFC stage and DC-DC converter stage into a single circuit have recently been proposed [2]. Among them, a Single-Stage Single-Switch Power Factor Corrected (S^4-PFC) AC/DC converter proposed by Redl et. al. [3] has gained wide acceptance due to its good potential for practical usage. The converter uses only one power switch and single control circuit to achieve high power factor and fast output regulation, as compared with the two-stage approach which requires two sets of power switches and control circuits. The reduced part count greatly contributes to the size and cost reduction of the converter.

The existing papers on S^4-PFC AC/DC converters have mainly focused on circuit improvement to reduce the voltage across the bulk capacitor [4, 5]. There is yet to be a publication that describes the design of this type of converter. Therefore, the purpose of this paper is to present the design of the S^4-PFC AC/DC converter.

The paper is organized as follows. The principle of operation of the S^4-PFC AC/DC converter is concisely described in Section II. Section III explains the design concept. The prototype circuit design is undertaken in Section IV. Section V presents experimental results. Section VI gives a conclusion of the paper.

II. DESCRIPTION OF S^4-PFC AC/DC CONVERTER

The S^4-PFC boost-flyback AC/DC converter is shown in Fig. 1. V_{in} is a full-wave sinusoidal voltage rectified from the AC mains. L_1 and L_2 are an input inductance and magnetizing inductance of the flyback transformer respectively. C_1 is a bulk energy storage capacitor which helps balance power between the input and output. Because the switching frequency of the power switch, M, is much higher than the frequency of V_{in}, it can be assumed that V_{in} is constant during one switching period of M.

L_1 is operated in discontinuous conduction mode (DCM) throughout the converter's operating range as this mode of operation yields good power factor. L_2 is operated in DCM at light loads and in continuous conduction mode (CCM) at heavy loads. Fig. 2 depicts possible circuit configurations of the converter within one switching period. When M is turned on (Fig. 2(a)), L_1 is charged by V_{in} through M, L_2 is charged by C_1 also through M, and the load R_L receives the power discharged from C_L. When M is turned off (Fig. 2(b)), L_1 is discharged to C_1 through D_1 and L_2 discharged to C_L and R_L, through D_3. When i_{L1} reaches zero (Fig. 2(c)), D_1 ceases to conduct, but i_{L2} continues to flow since L_2 is larger than L_1. At light loads, i_{L2} will reach zero before the end of a switching period and thus D_3 cease to conduct as shown in Fig. 2(d). At heavy loads, i_{L2} will continue to flow until the end of a switching period, and thus the circuit in Fig. 2(d) will not occur. Fig. 3 shows a sketch of i_{L1} and i_{L2} over one switching period. Note that i_{L1} is always in DCM, whereas i_{L1} is in DCM at light loads and CCM at heavy loads. Fig. 4 shows a sketch of V_{in} and i_{L1} over two periods of V_{in}. It is an inherent property of the boost inductance L_1 that, when operated in DCM, will draw the averaged current waveform, I_{L1} close to a sinusoid, which produces high power factor.

978-1-4244-0644-9/07/$25.00 ©2007 IEEE

Referring back to Fig. 1, the converter's output voltage is regulated by a simple PWM control loop. The error amplifier amplifies the difference between the reference voltage and the feedback output voltage. The resulting control signal, V_c, is then compared with the sawtooth signal, V_s, producing a duty cycle signal to drive the power switch, M, to maintain a constant output voltage. For the converter to have good regulation characteristics, the error amplifier must be properly compensated.

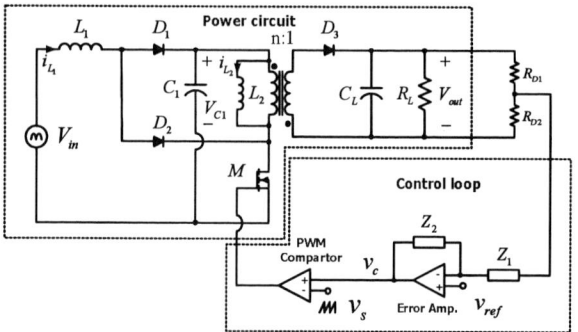

Fig. 1. S⁴-PFC boost-flyback AC/DC converter

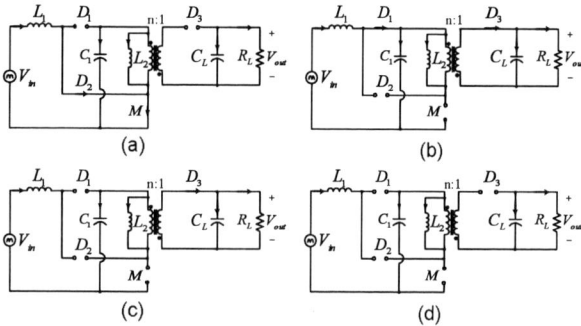

Fig. 2. Circuit topologies in one switching period

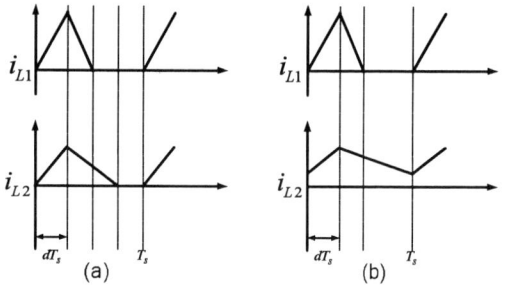

Fig. 3. Waveforms of i_{L1} and i_{L2}: (a) light loads (b) heavy loads

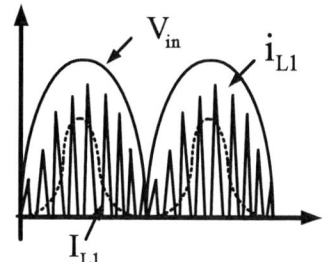

Fig. 4. Input voltage and current waveforms

III. DESIGN CONCEPT

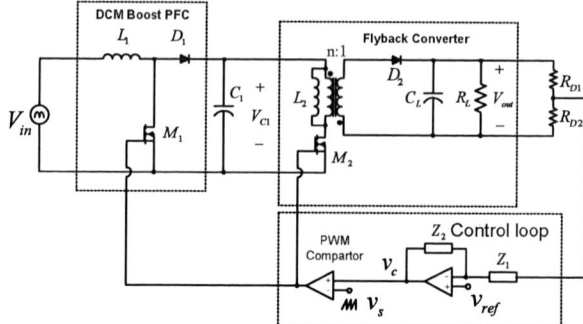

Fig. 5. DCM boost PFC cascaded with flyback converter

The operation of converter in Fig. 1 is the same as that of a DCM boost PFC cascaded with a flyback converter in Fig. 5, given that the power switches M_1 and M_2 are driven by the same duty cycle signal, and turned on and off at the same time. Therefore, the converter in Fig. 5 can be considered to be an equivalent circuit of the converter in Fig. 1. Thus, it will be used to develop the design equations for the S⁴-PFC converter.

A. Power Circuit Design

In Fig. 5, V_{C1} is the output of the boost converter and the input of the flyback converter. It is normally an unknown and depends on the line and load conditions. The expression for V_{C1} can be determined from the equivalent circuit in Fig. 5 by equating the input power to the output power. At light loads, Fig. 5 will be a combination of a DCM boost PFC and DCM flyback converter, where application of power balance yields

$$\left(\frac{1}{T_{in}}\right)\left(\frac{V_{in,pk}^2}{V_{C1}}\right)\left(\frac{L_2}{L_1}\right)\int_0^{T_{in}}\left(\frac{\left(\sin^2(\omega t)\right)dt}{V_{C1}-V_{in,pk}\sin(\omega t)}\right) = 1 \quad (1)$$

where $V_{in,pk}$ is a peak rectified input voltage and T_{in} a time period of the rectified input voltage. At heavy loads, Fig. 5 will be a combination of a DCM boost PFC and CCM flyback converter and the power balance gives

$$\left(\frac{T_S}{T_{in}}\right)\left(\frac{\left(\frac{(nV_{out})}{V_{C1}+(n\cdot V_{out})}\right)^2 V_{in,pk}^2 V_{C1}}{2L_1}\right)\int_0^{T_{in}}\left(\frac{\left(\sin^2(\omega t)\right)dt}{V_{C1}-V_{in,pk}\sin(\omega t)}\right) - P_{out} = 0 \quad (2)$$

where n is a primary-to-secondary turn ratio of the flyback transformer, T_S is a switching period, and P_{out} is an output power. Equations (1) and (2) are an implicit function and V_{C1} can be solved by a numerical method. A sample plot of V_{C1} versus P_{out} is given Fig. 6. The point, where (1) and (2) intersect, defines a DCM/CCM boundary of L_2. V_{C1} is largest and independent of P_{out}, when L_2 is in DCM. V_{C1} becomes lower as L_2 enters CCM.

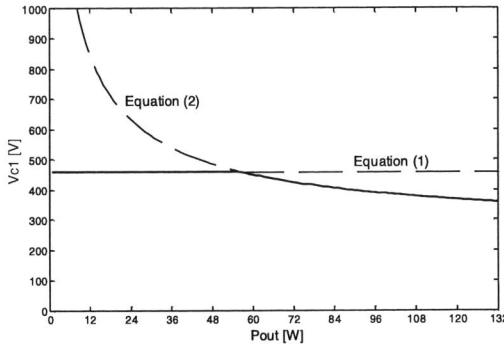

Fig. 6. V_{C1} versus P_{out} curve

B. Control Loop Design

From Fig. 5, it can be observed that control loop design of the S^4-PFC AC/DC converter is equivalent to that of the flyback converter with V_{C1} as the input voltage. The control block diagram of the converter, hence, can be drawn as shown in Fig. 7.

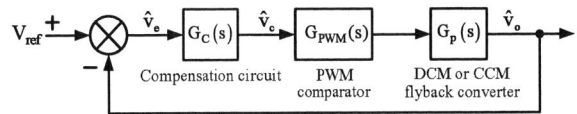

Fig. 7. Control block diagram

From (1) and (2) it can be seen that V_{in}, P_{out}, L_1, L_1/L_2, and n have to be specified before V_{C1} can be computed. The ranges of V_{in} and P_{out} are normally specified in a specification. The inductance L_1/L_2 affects V_{C1} only when L_2 is in DCM; the larger is L_1/L_2, the lower is V_{C1}. By contrast, the turn ratio n affects V_{C1} only when L_2 is in CCM; the larger is n, the lower is V_{C1}. The input inductance L_1 should be small to ensure its operation in DCM. Given V_{in}, P_{out} and the arbitrarily selected L_1, L_1/L_2, and n, (1) and (2) can be solved for V_{C1}. Many rounds of iterations may be required before the right values of L_1, L_1/L_2, and n that yield the satisfied value of V_{C1} are found. Next, it is essential to check if the selected L_1 will be operated in DCM throughout the converter's operating range. This can be verified by the inductance equation of the DCM boost PFC [6]

$$ L_1 < \frac{\eta V_{in,min}^2 T_S \left(V_{C1,min} - \sqrt{2} V_{in,min} \right)}{2 P_{out,max} V_{C1,min}} \quad (3) $$

where $V_{in,min}$ is a minimum RMS input voltage, $V_{C1,min}$ is a minimum bulk capacitor voltage, η is an efficiency of the boost PFC (≈ 0.8), and $P_{out,max}$ is a maximum output power.

The bulk capacitor, C_1, can be chosen based on its hold-up time, as given by (4)

$$ C_1 > \frac{2 P_{out,max} t_H}{V_{C1,min}^2 - \left(V_{C1,min} - 20 \right)^2} \quad (4) $$

where t_H is a hold-up time. The output capacitor, C_L, can be selected as in a standard flyback converter [7]

$$ C_L > \frac{D_{max} T_S}{R_{min}} \frac{V_{out}}{\Delta V_{out}} \quad (5) $$

where $D_{max} = n V_{out}/(V_{C1,min} + n V_{out})$ is a maximum duty cycle and R_{min} a minimum load resistance.

To specify ratings of the components and switching devices, voltage and current stresses that these devices have to endure in the actual circuit must be determined. For this purpose, simulation of Fig. 1 can be performed to estimate the values.

Transfer functions of each block are given below:

- Flyback converter in DCM [8]:

$$ G_p(s) = K_p \frac{\left(1 + \dfrac{s}{\omega_{ZC}}\right)\left(1 - \dfrac{s}{\omega_{zRHP}}\right)}{\left(1 + \dfrac{s}{\omega_{P1}}\right)\left(1 + \dfrac{s}{\omega_{P2}}\right)} \quad (6) $$

where $K_p = \dfrac{V_{C1}}{n\sqrt{K}}$, $K = \dfrac{2 L_2 f_S}{n^2 R_L}$, $M = \dfrac{n V_{out}}{V_{C1}}$, $\omega_{ZC} = \dfrac{1}{r_{C_L} C_L}$,

$\omega_{zRHP} = \dfrac{n^2 R_L}{L_2} \dfrac{1}{M(M+1)}$, $\omega_{P1} = \dfrac{2}{R_L C_L}$, $\omega_{P2} = \dfrac{n^2 R_L}{L_2} \dfrac{1}{(M+1)^2}$. r_{CL} is

an equivalent series resistance (ESR) of C_L.

- Flyback converter in CCM [8]:

$$ G_p(s) = K_p \frac{\left(1 + \dfrac{s}{\omega_{ZC}}\right)\left(1 - \dfrac{s}{\omega_{zRHP}}\right)}{\left(1 + \dfrac{s}{Q\omega_o} + \dfrac{s^2}{\omega_o^2}\right)} \quad (7) $$

where $K_p = \dfrac{V_{C1}}{n D'^2}$, $D' = 1 - D$, $\omega_{ZC} = \dfrac{1}{r_{C_L} C_L}$, $\omega_{zRHP} = \dfrac{n^2 D'^2 R_L}{D L_2}$,

$\omega_o = \dfrac{n}{\sqrt{L_2 C_L}} \sqrt{\dfrac{D'^2 R_L}{R_L + r_{C_L}}}$, $Q = \dfrac{1}{\omega_o} \dfrac{1}{\dfrac{L_2}{n^2 D'^2 R_L} + r_{C_L} C_L}$.

- PWM comparator:

$$ G_{pwm}(s) = \frac{1}{V_s} \quad (8) $$

- Error amplifier's compensation circuit

$$ G_C(s) = K_D \frac{\left(1 + \dfrac{s}{\omega_{Z1}}\right)\left(1 + \dfrac{s}{\omega_{Z2}}\right)}{\left(\dfrac{s}{\omega_I}\right)\left(1 + \dfrac{s}{\omega_{P1}}\right)\left(1 + \dfrac{s}{\omega_{P2}}\right)} \quad (9) $$

where $K_D = \dfrac{R_{D2}}{R_{D1} + R_{D2}}$, $\omega_{Z1} = \dfrac{1}{R_{C3} C_{C2}}$, $\omega_{Z2} = \dfrac{1}{(R_{C2} + R_{C1}) C_{C1}}$

$\omega_I = \dfrac{1}{R_{C1}(C_{C2} + C_{C3})}$, $\omega_{P1} = \dfrac{1}{R_{C2} C_{C1}}$, $\omega_{P2} = \dfrac{1}{R_F \left(\dfrac{C_{C2} C_{C3}}{C_{C2} + C_{C3}}\right)}$.

Design of the error amplifier's compensation circuit involves positioning of poles and zeros of $G_c(s)$ to give the open-loop transfer function, $G_c(s)G_{pwm}(s)G_p(s)$: (1) a high DC gain for good output voltage regulation, (2) a crossover frequency between one tenth and one fourth of the switching frequency for fast response, and (3) a phase margin of at least $45°$ for an adequate stability margin.

IV. PROTOTYPE CIRCUIT DESIGN

In this section the design of a S^4-PFC AC/DC converter is illustrated. The converter has the following specifications: $V_{in,AC} = 220V_{rms}\pm10\%$, $f_{in,AC} = 50Hz$, $V_o=12V$, $\Delta V_o = 5\%$ of V_o, $I_o = 1\text{-}10A$, and switching frequency ($f_s = 1/T_s$) = 100kHz. The 50Hz AC voltage, $V_{in,AC}$, is rectified by an input bridge rectifier, giving the full-wave sinusoidal voltage, V_{in}, as an input to the converter. V_{in} is ranged from 198V to 242V in terms of an RMS value, or 280V to 342V in terms of a peak value. The time period of V_{in} (T_{in}) is 10ms. The converter's output power (P_{out}) is ranged from 12W to 120W.

In the design, L_1/L_2, n, L_1, and L_2 are selected to be 0.4, 5, 100μH and 250μH respectively. Given these values, the minimum and maximum bulk capacitor voltage can be computed from (1) and (2), yielding $V_{C1,min} = 368V$ (at $V_{in,min} = 198V_{rms}$, $P_{out,max} = 120W$) and $V_{C1,max} = 561V$ (at $V_{in,max} = 242V_{rms}$, $P_{o,min}=12W$). These two extreme operating points are marked on the V_{C1}-P_{out} plot shown in Fig. 8. Next, it is essential to check if the selected L_1 will be operating in DCM throughout the converter's operating range. Substitution of $V_{in,min} = 198V_{rms}$, $V_{C1,min} = 368V$, $P_{out,max} = 120W$, $T_S = 10μs$ into (3) found $L_1 < 312μH$. Thus, the selected L_1 of 100μH is a valid value. Given the hold-up time, t_H, of 5ms, C_1 is selected according to (4), which gives $C_1 > 84μF$. Here, C_1 of 110μF is chosen. Since this capacitor must be able to withstand $V_{C1,max} = 561V$, it is assembled from two 220μF, 450V electrolytic capacitors connected in series. From (5), C_L of 990μF is chosen. It is assembled from three 330μF, 50V, ESR= 0.13Ω, electrolytic capacitors connected in parallel.

The ratings of devices or components in the circuit can be determined with an aid of simulation. Fig.9 shows the PSPICE simulation result of the power switch current in Fig. 1, at he full-load condition ($V_{in,min} = 198V_{rms}$, $I_{o,max} = 10A$). Only five cycles of the current is shown here for clarity. The RMS value of this current over one period of

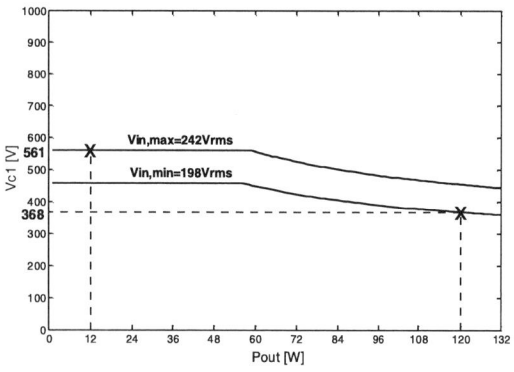

Fig. 8. V_{C1} versus P_{out} curve of the designed converter

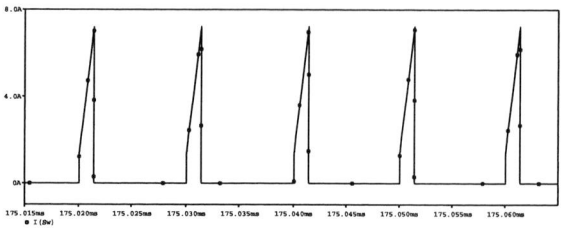

Fig. 9. Simulated waveform of the power switch current

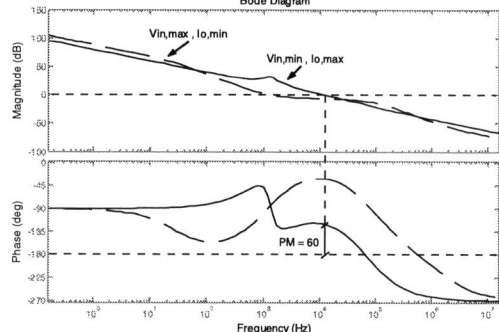

Fig. 10. Open loop frequency responses of the designed converter

V_{in} can be computed using a built-in mathematical function in PSPICE program, which yields the value of 1.5A. Thus, the power switch current handling capability must be rated higher than 1.5A. Ratings of other devices can be quantified in a similar manner.

The error amplifier's compensation circuit in (9) is designed assumed that the converter is operated at the full-load condition ($V_{in,min} = 198V_{rms}$, $I_{o,max} = 10A$). Under this condition, L_2 is in CCM and thus the transfer function of a flyback converter in CCM in (7) is used in the design. Substituting the relevant parameters into (7) and the peak-to-peak sawtooth voltage, V_s, of 1.8V into (8), the transfer function $G_{pwm}(s)G_p(s)$ can be determined

$$G_P(s)G_{PWM}(s) = 63.14\frac{\left(1+42.57\times10^{-6}s\right)\left(1-2.41\times10^{-6}s\right)}{\left(1+60.09\times10^{-6}s+20.16\times10^{-9}s^2\right)} \quad (10)$$

Given $G_{pwm}(s)G_p(s)$ in (10), the error amplifier's compensation circuit can be designed to give the open-loop transfer function the desired frequency response. The crossover frequency, f_c, is chosen at 10kHz. The two zeros of $G_c(s)$ are placed at $\omega_{z1}= \omega_{z2} = 1kHz$. The first pole of $G_c(s)$ is at origin, the second pole is placed at $\omega_{p1}= 3kHz$, and the third pole placed at $\omega_{p21}= 80kHz$. Based on the selected f_c and pole and zero location of $G_c(s)$, the compensation circuit component values are calculated and rounded off to the nearest standard values, obtaining $R_{C1} = 400kΩ$, $R_{C2} = 240kΩ$, $R_{C3} = 150kΩ$, $C_{C1} = 220pF$, $C_{C2} = 1nF$, and $C_{C3} = 20pF$. With the designed compensation circuit, the open-loop frequency response (i.e. the frequency response of $G_c(s)G_{pwm}(s)G_p(s)$) is plotted and shown in Fig. 10 by the solid line. The plot indicates the high DC gain, the phase margin of $60°$ and the crossover frequency of about 10kHz. It predicts that the converter will be stable and exhibit good output voltage regulation.

980

The dashed line in Fig. 10 shows the open-loop frequency response at the light-load condition ($V_{in,\,max}$ = 242V_{rms}, $I_{o,max}$ = 1A). At this condition, L_2 is in DCM and hence the converter's transfer function has changed to (6). In spite of this, the designed compensation circuit is still able to maintain the converter's stability and high DC gain. However, the deterioration in the output voltage response should be expected, as the crossover frequency has reduced to about 1kHz. Fig. 11 shows a circuit schematic of the designed prototype converter.

V. EXPERIMENTAL RESULTS

Tables I and II show the output voltage and input power factor measured from the prototype converter. The output voltage is well regulated at around 12V throughout the converter's operating range. This result agrees with the prediction in Fig. 10 that the high DC loop gain would yield the converter good output voltage regulation. The input power factor is above 0.9 in most cases, except some conditions at high lines and light loads that the power factor falls under 0.9. The low power factor is caused by the much distorted input current occurred at those conditions. The AC input voltage and current waveforms at the nominal input voltage and maximum load current (i.e. V_{in} = 220V_{rms}, I_o = 10A) are shown in Fig. 12. It should be noted that these waveforms were measured at the input of an EMI filter located on an AC side of the input bridge rectifier. The current waveform was analyzed and its harmonics were found to be below the limits set by the IEC1000-3-2 Class D harmonic standards.

Fig. 13 shows the secondary diode current waveform, i_{D3}. At I_o = 4A (P_o = 48W), i_{D3} is in DCM (Fig. 13(a)). At I_o = 5A (P_o = 60W), i_{D3} is in CCM (Fig. 13(b)). Therefore, transition of L_2 from DCM to CCM has taken place at some points between these two loading conditions. This result corresponds with the prediction in Fig. 8, in which the operational mode transition of L_2 was seen to have occurred when the output power is approximately 56W. The output voltage transient responses of the converter due to the load application/rejection are shown in Fig. 14. In Fig. 14(b), the load application/rejection causes L_2 to change between DCM and CCM, while L_2 remains in DCM in Fig. 14(a) and CCM in Fig. 14(c) after the load change. Of the three cases, the output response when L_2 is in CCM (Fig 14(c)) is the most satisfactory because the

error amplifier's compensation circuit has been designed for this mode of operation. The output response is sluggish when L_2 is in DCM (Fig. 14(a)), due to the dwindling crossover frequency in this mode of operation (see the dashed line in Fig. 10).

TABLE I
MEASURED OUTPUT VOLTAGE

Vin	Vout					
	1A	2A	4A	6A	8A	10A
198V	12.00	12.00	11.99	11.99	11.98	11.97
242V	12.00	12.00	11.99	11.99	11.98	11.97

TABLE II
MEASURED INPUT POWER FACTOR

Vin	PF					
	1A	2A	4A	6A	8A	10A
198V	0.803	0.913	0.960	0.961	0.955	0.948
242V	0.718	0.846	0.939	0.950	0.950	0.944

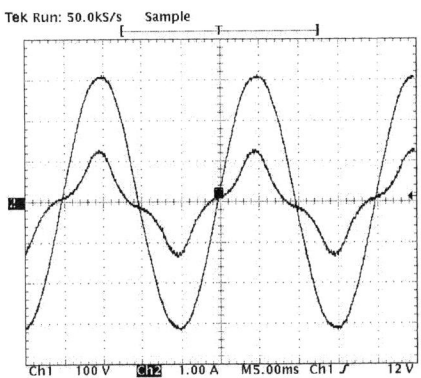

Fig. 12. Waveforms of the AC input voltage and current

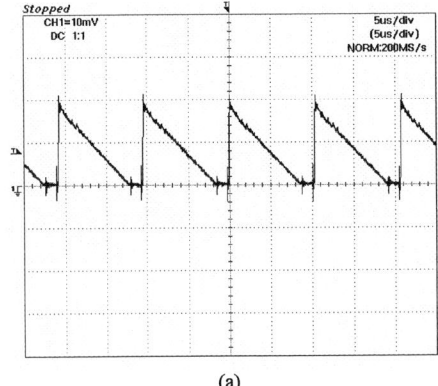

(a)

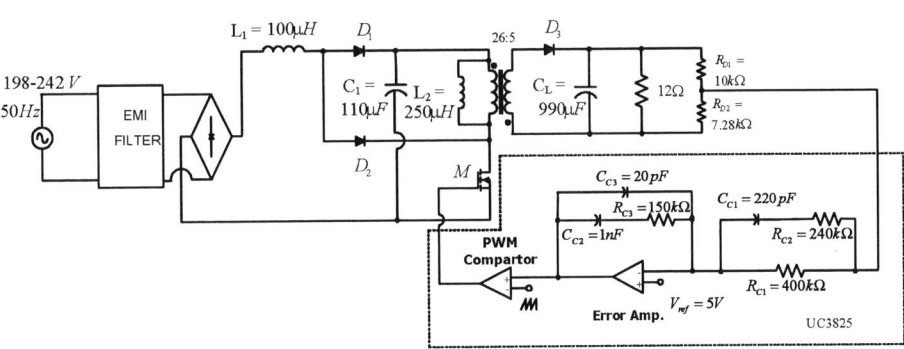

Fig. 11. Circuit schematic of the prototype S^4-PFC AC/DC converter.

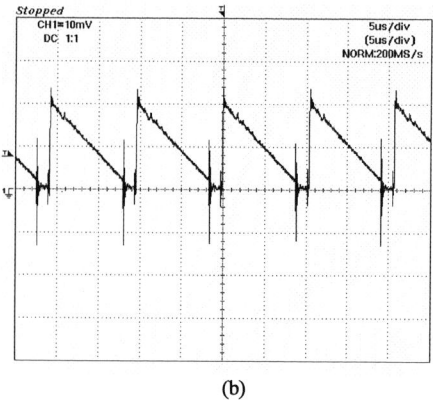

Fig. 13. Waveforms of i_{D3} when: (a) $I_o = 4A$ (b) $I_o = 5A$

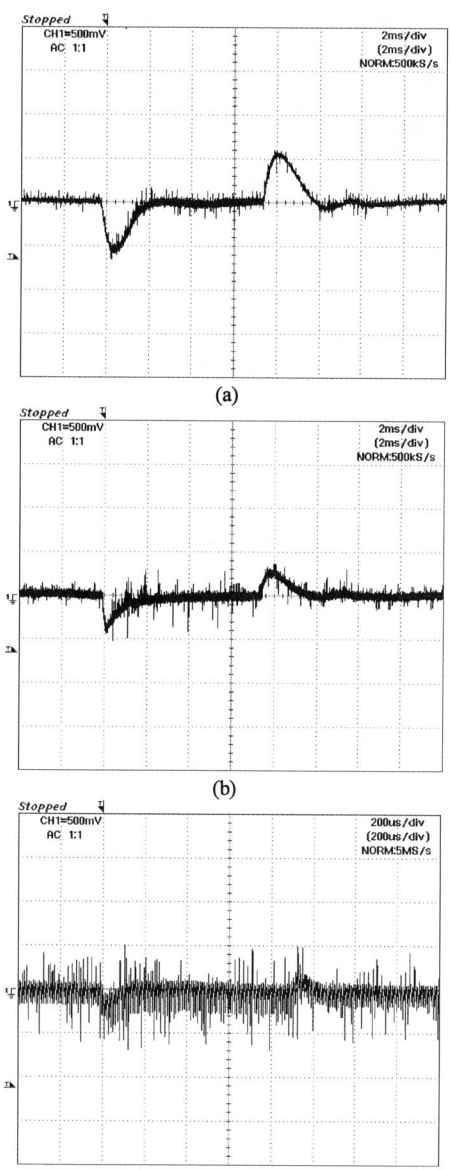

(a)

(b)

(c)

Fig. 14. Output voltage transient response, when: (a) I_o is stepped from 1A to 3A (b) I_o is stepped from 3A to 6A (c) I_o is stepped from 6A to 10A

VI. CONCLUSION

Design of a Single-Stage Single-Switch Power-Factor-Corrected (S⁴PFC) boost-flyback AC/DC converter has been described in this paper. The design is based on the equivalent circuit in Fig. 5 which is a DCM boost PFC cascaded with a flyback converter, where the flyback converter is operated in DCM at light loads and CCM at high loads. From the equivalent circuit, it is seen that the bulk capacitor voltage, V_{C1}, is the output of the DCM boost PFC and the input of the flyback converter. The expression for V_{C1} can be derived from the power balance principle. As given in (1) and (2), V_{C1} is a nonlinear function of the input voltage, V_{in}, the output power, P_{out}, the input inductance L_1, the inductance ratio L_1/L_2 and the flyback transformer's turn ratio, n. Given V_{in}, P_{out} and the arbitrarily selected L_1, L_1/L_2, and n, (1) and (2) can be solved for V_{C1}. It is an iterative process to find L_1, L_1/L_2, and n that gives the satisfied value of V_{C1}. From the equivalent circuit, it is observed that control loop design of the S⁴-PFC AC/DC converter is the same as that of the flyback converter with V_{C1} as the input voltage. Therefore, once the value of V_{C1} has been determined, design of the error amplifier compensation circuit can be proceeded using the standard frequency response design method. In the paper, various experimental results are presented to support the design validity.

REFERENCES

[1] L. H. Dixon, "Average current mode control of switching power supplies", Unitrode Power Supply Design Seminar, SEM-700, 1990.

[2] C. Qian and K. M. Smedly, "A topology survey of single-stage power factor correction with a boost type input-current shaper", *IEEE Applied Power Electronics Conference*, pp. 460-467, 2000.

[3] R. Redl, L. Balogh, and N.O Sokal, "A new family of single-stage isolated power-factor correctors with fast regulation of the output voltage", *Power Electronics Specialists Conference*, pp. 1137-1144, 1994.

[4] F. Tsai, P. Markowski, and E. Whitecomb, "Off-line flyback converter with input harmonic current correction", *IEEE International Telecommunications Energy Conference*, pp. 120-124, 1996.

[5] J. Qian, Q. Zhao, and F.C. Lee, "Single-stage single-switch power-factor-correction ac/dc converters with DC-bus voltage feedback for universal line applications", *IEEE Transactions on Power Electronics*, Vol. 13, No. 6, pp. 1079-1088, 1998.

[6] F. K. Siu, "Analysis and Measurement of DCM Power Factor Correctors", M.S. thesis, The Hong Kong University of Science and Technology, 1998.

[7] V. Quercioli, "Pulse Width Modulated (PWM) Power Supplies", Elsevier, Amsterdam, 1993.

[8] W. Kleebchampee, "Modeling of a Current Mode Controlled Flyback Converter with optocoupler Feedback", M.S. thesis, King Mongkut's Institute of Technology Landkrabang, 2004.

A DSP-Based Unified Three Phase/Switch/Level Unity Power Factor Rectifier Using Feedback Linearization for DC-Bus Voltage Control

Ali Moallem   Hesameddin Mirzaee Teshnizi   MohammadReza Zolghadri

School of Electrical Engineering
Sharif University of Technology
Tehran, Iran 11365–9363
Email: alimoallem@ee.sharif.edu, hesam_mt@ee.sharif.edu, zolghadr@sharif.edu

Abstract—In this paper, a unified hybrid vector control of vector modulated VIENNA I rectifier is presented. Feedback linearization technique is used for the DC-link voltage control. For this purpose, at first a mathematical model of the VIENNA I rectifier is derived. Using this model, it is shown that the dynamic of the rectifier falls into two categories; linear inner dynamic which consists of inner current control and neutral-point potential control and nonlinear outer dynamic which consists of the DC-link voltage control. Based on these facts, a hybrid vector control scheme meaning linear vector control of inner dynamics and nonlinear vector control of outer dynamic is devised. Simulations carried out demonstrate the validity of the proposed control scheme. Simulation results are validated via experimental setup results.

Index Terms—VIENNA rectifier, Space vector modulation, Vector control, PFC, Feedback linearization.

I. INTRODUCTION

AC/DC power rectifiers which convert three-phase AC input to DC output is widely used in telecommunication power supplies, welding equipments, etc. In the years ago, diode rectifiers and/or phase controlled rectifiers have extensively been used to meet these demands. But with the strict standards of power quality in recent years which are also going in a more strict direction for the near future, superior functionality in operations like power factor correction (PFC) and total harmonic distortion (THD) reduction have become mandatory. Three-phase rectifier systems have highly versatile topology and complexity, depending on the application area and the required operational behavior. For the determination of the circuit concept, the following basic requirements are given: [1].

- approximately sinusoidal current consumption
- resistive mains behavior
- controlled output voltage
- low-blocking voltage stress on the power transistors
- high power density
- high efficiency
- low complexity of the power and control circuits
- high reliability

Considering these, three-phase/level/switch neutral-point clamped Vienna I rectifier shown in Fig. 1 is a very suitable

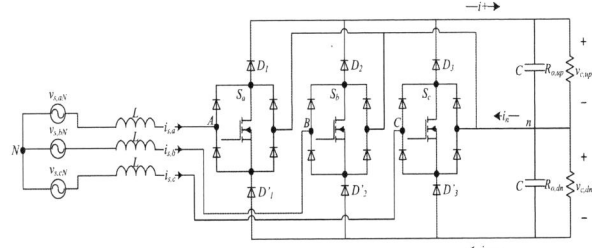

Fig. 1.   Three-phase/level/switch Vienna I rectifier

topology for three-phase unidirectional rectification. It is especially advantageous regarding the stresses on the switching devices. Being compared to the conventional six-switch boost rectifier, it has proven to possess better efficiency, higher switch utilization, smaller AC filter size and only three active switches with half the blocking voltage capability. A vector modulated vector controlled system for VIENNA I rectifier is proposed in [2]. In [1], [3] experimental results for this three-level vector modulated vector controlled system are presented; however, in both, the following drawbacks can be seen: First, assuming operation in continuous conduction mode, the vector control system becomes unstable in low load conditions because the DC-Link voltage dynamic is nonlinear but a linear controller is applied. Second, using a simple PLL for synchronization, the vector control system fails to adapt itself to a variable frequency and distorted mains.

In this paper, a hybrid vector control system of the three-level space vector modulated VIENNA I rectifier using an enhanced phase-locked loop (EPLL) is being designed, simulated and experimentally verified. The hybrid vector control scheme and enhanced PLL has made the system unified in the regard that it can better cope with low load condition, variable frequency and distorted mains. In section II the mathematical modeling of the rectifier is presented. Development of the whole vector control system and principle of operation of enhanced PLL are presented in section III. Finally, in section IV and section V simulation and experimental results are shown respectively.

978-1-4244-0644-9/07/$25.00 ©2007 IEEE

II. VIENNA I MATHEMATICAL MODELING

The power stage of the VIENNA I rectifier is shown in Fig. 1. Writing down KVL equations at the utility side yields:

$$v_{s,aN} = Ri_{sa} + L\frac{di_{sa}}{dt} + v_{r,An} + v_{nN} \tag{1}$$

$$v_{s,bN} = Ri_{sb} + L\frac{di_{sb}}{dt} + v_{r,Bn} + v_{nN} \tag{2}$$

$$v_{s,cN} = Ri_{sc} + L\frac{di_{sc}}{dt} + v_{r,Cn} + v_{nN} \tag{3}$$

Because the input source is balanced and there is no connection between the source side and DC-link side neutrals, the following relations also hold:

$$v_{s,aN} + v_{s,bN} + v_{s,cN} = 0 \tag{4}$$

$$i_{sa} + i_{sb} + i_{sc} = 0 \tag{5}$$

From (1) to (5) it can be deduced that:

$$v_{nN} = -\frac{1}{3}(v_{r,An} + v_{r,Bn} + v_{r,Cn}) \tag{6}$$

Writing KCL equations at the DC side yields:

$$
\begin{aligned}
C\frac{dv_{o,up}}{dt} &= i^+ - i_{o,up} \\
&= \sum_{k=a}^{c}(1-s_k)\frac{[sign(i_{sk})+1]}{2}i_{sk} - i_{o,up}
\end{aligned} \tag{7}
$$

$$
\begin{aligned}
C\frac{dv_{o,dn}}{dt} &= i^- - i_{o,dn} \\
&= \sum_{k=a}^{c}(1-s_k)\frac{[sign(i_{sk})-1]}{2}i_{sk} - i_{o,dn}
\end{aligned} \tag{8}
$$

It is known that by applying the cosine-based park transformation "P", three-phase quantities in a three-wire system can be transferred into two-phase quantities in source voltage positive sequence synchronous d-q frame:

$$P = \frac{2}{3}\begin{bmatrix} \cos(\omega t) & \cos(\omega t - \frac{2\pi}{3}) & \cos(\omega t + \frac{2\pi}{3}) \\ -\sin(\omega t) & -\sin(\omega t - \frac{2\pi}{3}) & -\sin(\omega t + \frac{2\pi}{3}) \end{bmatrix} \tag{9}$$

Applying the above transformation to (1)-(3), the ac side dynamic in source voltage synchronous d-q frame is obtained:

$$\frac{di_{s,d}}{dt} = \frac{v_{s,d}}{L} - \frac{R}{L}i_{s,d} + \omega i_{s,q} - \frac{v_{r,d}}{L} \tag{10}$$

$$\frac{di_{s,q}}{dt} = \frac{v_{s,q}}{L} - \frac{R}{L}i_{s,q} - \omega i_{s,d} - \frac{v_{r,q}}{L} \tag{11}$$

Where $v_{s,dq} = P_{2\times3} \cdot v_{s,abcN}$, $v_{r,dq} = P_{2\times3} \cdot v_{r,ABCN}$ and $i_{s,dq} = P_{2\times3} \cdot i_{s,abc}$. To obtain the equation that governs the neutral-point potential (NPP) dynamic control, relations (7) and (8) are subtracted respectively and to obtain the equation that governs the DC-link voltage control, relations (7) and (8) are added accordingly, so the following equations are resulted:

$$
\begin{aligned}
C\frac{d\Delta v_o}{dt} &= i_n - (i_{o,up} - i_{o,dn}) \\
&= \sum_{k=a}^{c}(1-s_k)\cdot i_{sk} - (i_{o,up} - i_{o,dn})
\end{aligned} \tag{12}
$$

$$
\begin{aligned}
\frac{C}{2}\frac{dv_o}{dt} &= i_{in} - \frac{1}{2}(i_{o,up} + i_{o,dn}) \\
&= \frac{1}{2}\sum_{k=a}^{c}(1-s_k)\cdot sign(i_{sk})\cdot i_{sk} \\
&\quad - \frac{1}{2}(i_{o,up} + i_{o,dn})
\end{aligned} \tag{13}
$$

Where $\Delta v_o = v_{o,up} - v_{o,dn}$ is the neutral-point potential, $v_o = v_{o,up} + v_{o,dn}$ is the output DC-link voltage, i_n is the neutral-point current and $i_{in} = \frac{1}{2}\sum_{k=a}^{c}(1-s_k)\cdot sign(i_{sk})\cdot i_{sk}$. Multiplying both sides of (13) by v_o and writing it in terms of output power, the following equation is obtained:

$$\frac{C}{2}\frac{dv_o}{dt} = \frac{P_{out}}{v_o} - \frac{1}{2}(i_{o,up} + i_{o,dn}) \tag{14}$$

Knowing that P_{out} serves as the control input in the hybrid vector control structure, it can be seen that DC-link dynamic is nonlinear while load currents are considered as a disturbance in the model.

Another important set of relations are those which constitute power balance of the rectifier. Knowing this and writing complex power at input terminals of the rectifier in synchronous d-q frame, it follows that:

$$S_{in} = P_{in} + jQ_{in} = \frac{3}{2}V_{s,dq}\cdot I_{s,dq}^* \tag{15}$$

Expanding (14) in terms of its components and equating real and imaginary parts, it results in:

$$P_{in} = \frac{3}{2}(V_{s,d}\cdot I_{s,d} + V_{s,q}\cdot I_{s,q}) \tag{16}$$

$$Q_{in} = \frac{3}{2}(V_{s,q}\cdot I_{s,d} - V_{s,d}\cdot I_{s,q}) \tag{17}$$

Using the fact that active power at input and output terminal of the rectifier must be equal, the following power balance equation is resulted:

$$P_{out} = \frac{3}{2}(V_{r,d}\cdot i_{s,d} + V_{r,q}\cdot i_{s,q}) = v_o \cdot i_{in} \tag{18}$$

Equations (10) and (11) describe AC-side dynamic, equations (12) and (14) describe neutral-point potential and DC-link voltage dynamics respectively, and equation (18) describes power balance at the input port of the VIENNA I rectifier. Using these relations, a hybrid vector control scheme is introduced in section III.

III. VIENNA I HYBRID VECTOR CONTROL SCHEME

A. Hybrid Vector Control Scheme

Under balanced condition the rectifier model is presented using equations (10), (11), (12), (14) and (18). To control inner current dynamics, PI controllers in synchronous d-q frame have been used as follows:

$$v_{r,d}^* = -PI(i_{s,d,ref}^* - i_{s,d}) + v_{s,d} + \omega L i_{s,q} \tag{19}$$

$$v_{r,q}^* = -PI(i_{s,q,ref}^* - i_{s,q}) + v_{s,q} - \omega L i_{s,d} \tag{20}$$

$$PI = k_p\left(\frac{s + \frac{k_i}{k_p}}{s}\right) \tag{21}$$

The inner current dynamics have been decoupled and linearized using compensating terms. This can be seen by inserting (19) and (20) into (10) and (11). Doing so, the current control closed loop transfer function can be written as follows:

$$\frac{I_{s,d}}{I^*_{s,d}} = \frac{I_{s,q}}{I^*_{s,q}} = \frac{(\frac{k_p}{L}) \cdot (\frac{s+\frac{k_i}{k_p}}{s+\frac{R}{L}})}{s + (\frac{k_p}{L}) \cdot (\frac{s+\frac{k_i}{k_p}}{s+\frac{R}{L}})} \quad (22)$$

The PI controller gains are chosen according to pole placement rule such that the resulted inner current dynamic is a linear first-order one with a valid bandwidth; so, it follows that:

$$k_p = \frac{L}{\tau_i} = \frac{L\omega_i}{2\pi} \quad (23)$$

$$k_i = \frac{R}{\tau_i} = \frac{R\omega_i}{2\pi} \quad (24)$$

Where L and R are input inductance and resistance, respectively, τ_i is the current control loop time constant and ω_i is its respective angular frequency. The current references for current control loops are being generated based on requirements enforced by control actions such as regulating DC-link voltage to its reference value and power factor correction. The former means that output average active power must be equal to its reference value (P_{in}) and the latter requires input reactive power to be zero ($Q_{in} = 0$). Therefore from equations (16) and (17), current references are being produced as (25):

$$\begin{bmatrix} I^*_{s,d} \\ I^*_{s,q} \end{bmatrix} = \frac{2}{3(V^2_{s,d} + V^2_{s,q})} \begin{bmatrix} V_{s,q} & -V_{s,d} \\ V_{s,d} & V_{s,q} \end{bmatrix} \cdot \begin{bmatrix} P_{in} \\ Q_{in} \end{bmatrix} \quad (25)$$

As it can be seen from equation (25), P_{in} serves as the control input in the vector control scheme and is supplied by DC-link voltage controller. Taking P_{in} as the control input, it is shown in (14) that DC-link dynamic is nonlinear. To compensate this nonlinearity, feedback linearization technique as described in [4] is being used. The aim of the this nonlinear control is to define a new control input named "u" such that the previously nonlinear output voltage dynamic reduces to a linear one on which well-known linear control techniques can be applied. For this purpose, an outer transformation law f is being designed as follows:

$$P_{out} = f(v_o, u, i_{o,up}, i_{o,dn}) = v_o \cdot (u + \frac{1}{2}(i_{o,up} + i_{o,dn})) \quad (26)$$

Where P_{out} is the old control input, u is the new control input, v_o is the DC-link voltage and $i_{o,up}$ and $i_{o,dn}$ are upper and lower DC load currents. Performing the above transformation on (14) leads to the following linear dynamic:

$$\frac{C}{2}\frac{dv_o}{dt} = u \quad (27)$$

To practically implement the outer voltage transformation law, the values for output voltage, v_o, and sum of output load currents, ($i_{o,up} + i_{o,dn}$) are needed. Output voltages are being sensed for the purpose of the control system, so v_o is easily computed. To calculate the sum of output load currents,

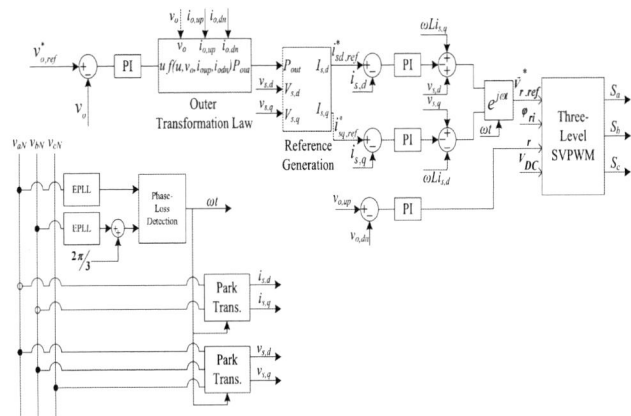

Fig. 2. VIENNA I Hybrid Vector Control Scheme

($i_{o,up} + i_{o,dn}$), an online load current estimator using (13) has been implemented digitally as follows:

$$(i_{o,up}[n] + i_{o,dn}[n]) = \frac{C}{T_s}(v_o[n] - v_o[n-1]) +$$

$$\sum_{k=a}^{c}(1 - s_k) \cdot sign(i_{sk}[n]) \cdot i_{sk}[n] \quad (28)$$

Which n denotes the nth sample of each electrical quantity and s_k is each phase's switching state that is determined from vector control system's three-level space vector modulator.

Designing a linear PI controller for (27) according to the pole placement rule to obtain a linear first-order system results in the following controller gains:

$$k_p = \frac{C_t}{\tau_v} = \frac{C_t\omega_v}{2\pi} \quad (29)$$

$$k_i = \frac{1}{R_l\tau_v} = \frac{\omega_v}{2\pi R_l} \quad (30)$$

Where C_t and R_l are output total capacitance and total load resistance, respectively, τ_v is the voltage control loop time constant and ω_v is its respective angular frequency. The whole hybrid vector control scheme is based on a novel three-level space vector modulator introduced in [2]. It means that the vector control commanded rectifier input terminal PWM voltages are developed using this special modulator. The three-level space vector modulator has five inputs: magnitude and phase of vector control forced rectifier input terminal voltage, $|\vec{V}_{ref}|$, φ_{vref}, input current phase, φ_{ri}, overall DC-link voltage, V_{dc}, and partial DC-link voltages unbalance parameter, r. The latter is used to balance the output capacitor voltages meaning that instead of using (12) in which i_n serves as control input, a new control input r is introduced which is related to neutral-point potential as follows:

$$r = \frac{V_{o,up} - V_{o,dn}}{V_{o,up} + V_{o,dn}} = \frac{\Delta v_o}{V_{dc}} \qquad -1 \le r \le 1 \quad (31)$$

Where r is the DC-link unbalance parameter which can vary between +1 and -1, Δv_o is the neutral-point potential and V_{dc} is the overall DC-link voltage. Based on (31), a PI controller

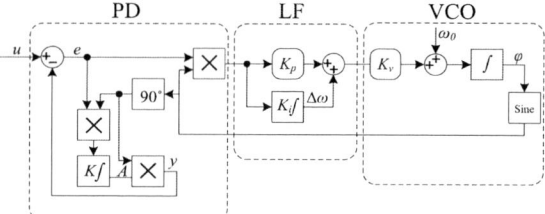

Fig. 3.   EPLL block diagram

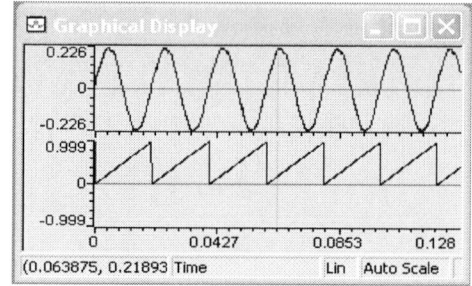

Fig. 4.   Phase (a) source voltage and its extracted phase

is designed such that the output capacitor voltages are kept balanced and the closed loop balancer exhibits appropriate dynamic. The entire vector control scheme is shown in Fig. 2.

B. Enhanced Phase-Locked Loop (EPLL)

Because the whole hybrid vector control scheme is in the synchronous d-q frame, it is required to have an exact and synchronized knowledge of the input voltage angular frequency, ωt. This is very important regarding superior functionality of vector control system in which park transformations have been used. [3], [5] have used only one hardware phase-locked loop (PLL) on phase (a) to obtain input voltage angular phase. This solitary PLL on phase (a) has two major disadvantages besides its extra hardware cost and labor: First, it compromises phase extraction under distorted and variable mains frequency situations and Second, it prevents rectifier from working in case of phase (a) disconnection which is a possible scenario in real world. To overcome this deficiency, in this paper a special software-based PLL is employed. In this regard, for synchronization of the vector control system to the utility line voltage, the system benefits from an adaptive nonlinear notch filter also named as extended phase-locked loop (EPLL) [6]. This unique PLL as shown in Fig. 3 synchronizes to its input signal fundamental component attributes in fixed/variable mains center frequency. This feature makes the vector control system unified regarding variable frequency situations. In addition, by using two of these EPLLs on two different phases for phase extraction, the chance of malfunction under any phase loss is zeroed. As Fig. 3 shows, its overall structure is in accordance with a conventional PLL. This basic structure has three independent internal parameters: K, K_pK_v, and K_iK_v. Parameter K dominantly controls the speed of convergence of amplitude A. Parameters K_pK_v and K_iK_v control the rate of convergence of the phase and frequency. The convergence rate increases with the increase in the values of the parameters; so, appropriate bandwidth and transient response can be obtained. Moreover, since EPLL is actually a bandpass filter, by correctly setting its parameters the functionality of the vector control system is not endangered. The continuous-time differential equations governing an EPLL are derived from the block diagram of Fig. 3 as [6]:

$$\begin{cases} \dot{A}(t) = K \cdot e(t) \cdot \sin(\varphi(t)) \\ \dot{\omega}(t) = K_i \cdot e(t) \cdot \cos(\varphi(t)) \\ \dot{\varphi}(t) = \omega(t) + \frac{K_pK_v}{K_i} \cdot \dot{\omega}(t) \end{cases} \quad (32)$$

To implement (32) digitally a discrete time version of it is obtained using Backward Euler method. Assuming a sampling time of T_s the discrete-time version of (32) is obtained as follows [6]:

$$\begin{cases} A[n] = A[n-1] + \mu_1 \cdot e[n-1] \cdot \sin(\varphi[n-1]) \\ \omega[n] = \omega[n-1] + \mu_2 \cdot e[n-1] \cdot \cos(\varphi[n-1]) \\ \varphi[n] = \varphi[n-1] + T_s \cdot \omega[n-1] \\ \qquad\quad + \mu_3 \cdot e[n-1] \cdot \cos(\varphi[n-1]) \end{cases} \quad (33)$$

Where $\mu_1 = KT_s$, $\mu_2 = K_iT_s$ and $\mu_3 = K_pK_vT_s$ are called step sizes. Equations in (33) are implemented using TMS320F2812 DSP. Fig. 4 demonstrates phase (a) source voltage and its extracted phase using EPLL.

IV. SIMULATION RESULTS

Simulations have been carried out in the MATLAB/SIMULINK environment to verify the proposed method. The rectifier output power is 200 W, DC link voltage is 100 V and switching frequency is 8 kHz. The designed inductor is 2.25 mH and output capacitors are each 100μF. Fig. 5 and Fig. 6 show the simulation results of vector control system introduced in [5] for a load variation from full load to 5% of full load. As it can be seen, the input current shape is far from sinusoidal and output capacitor voltages are on the edge of instability. Simulation results for feedback linearization of DC-Bus voltage dynamic are shown in Fig. 7 and Fig. 8 for a load variation from full load to 5% of full load. It can be seen that even at this very low load condition, the sinusoidal shape of input current is preserved and output capacitor voltages are sustained at reference value with acceptable ripple.

V. EXPERIMENTAL RESULTS

The rectifier prototype has the same specifications used in simulation and is shown in Fig. 9. It consists of a six-pulse diode bridge and three current bidirectional switches each made up of two IRF530N MOSFETs. Because the system has a three-wire structure, input currents are measured using two LTS25-NP *LEM* sensors and are then transferred to a signal conditioning board to prepare them for the DSP. Input voltages are measured using a three-phase Y-Y transformer and output voltages are measured using two differential voltage isolators. Then, the resulting signals are transferred to interface board to prepare them for DSP analog input. The digital control

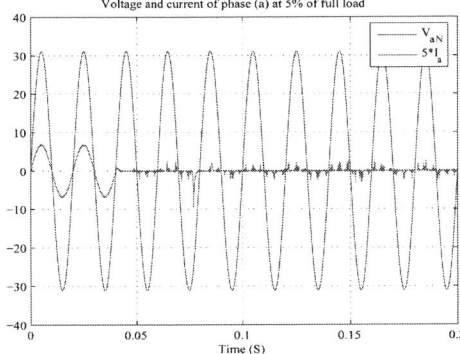

Fig. 5. Phase (a) source voltage and input current under linear vector control after severe load reduction

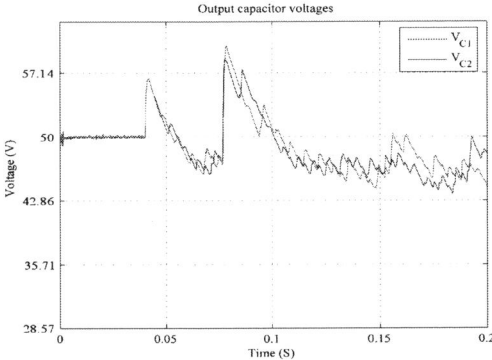

Fig. 6. Output DC-link voltage under linear vector control after severe load reduction

system and space vector modulator are implemented with TMS320F2812 DSP board. The sampling period T_{samp} and switching period T_{sw} are both 125 μs. Switching pulses are symmetrical with respect to half the switching period. Input over-current protection and output over-voltage protection are performed using dedicated software polling routines. Table I shows the conditions under which the setup is tested.

Based on the conditions mentioned in Table I, the following

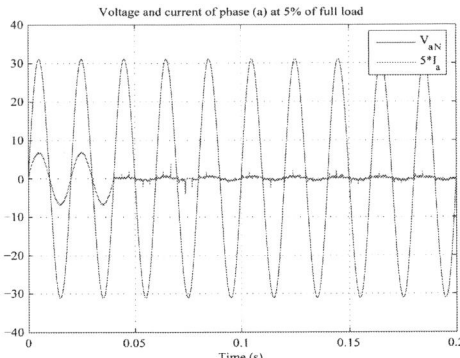

Fig. 7. Phase (a) source voltage and input current under hybrid vector control after severe load reduction

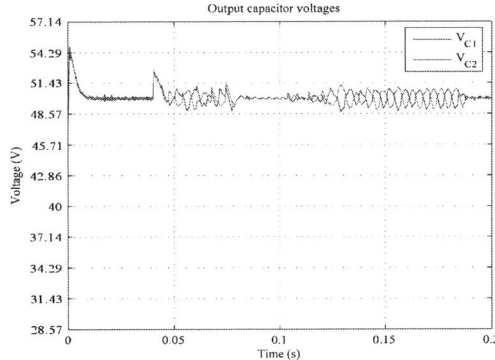

Fig. 8. Output DC-link voltage under hybrid vector control after severe load reduction

Fig. 9. VIENNA I experimental setup

figures are resulted. Fig. 10 shows input phase (a) voltage and its corresponding current and Fig. 11 shows output DC-link voltage experimental results for the linear vector control scheme when a transition of 100% to 15% of full load operation is applied. As it can be seen, although the PFC operation is somehow preserved, the input current is far from sinusoidal with great harmonic content, and the DC-link voltage has become unstable. In Fig. 12 and Fig. 13 exactly the same results for the feedback-linearized hybrid vector control scheme are illustrated. From these figures, it is clear that besides keeping the output DC-link voltage stable, the hybrid vector control scheme has caused input current to maintain a higher sinusoidal quality.

TABLE I
RECTIFIER PARAMETERS

Parameter	Value
$V_{s,abc}$	$20\ V_{rms}$
P_{out}	$200\ W$
V_{DC}	$100\ V$
f_{sw}	$8\ kHz$
$(THD\%)_{Vin}$	3%
$(Unbalance\%)_{Vin}$	5%

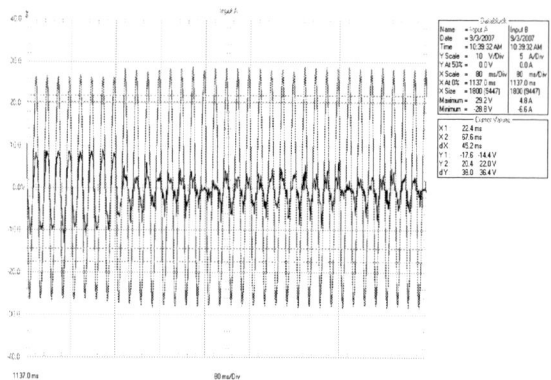

Fig. 10. Phase (a) source voltage and input current under linear vector control after severe load reduction

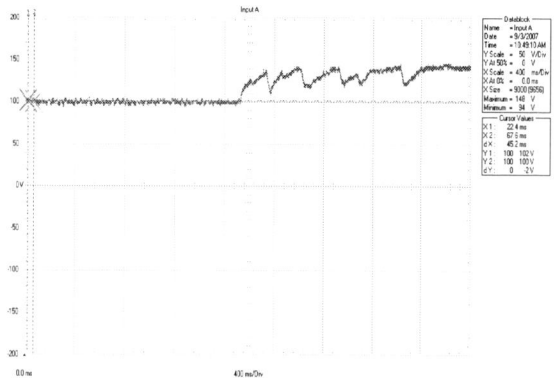

Fig. 11. Output DC-link voltage under linear vector control after severe load reduction

VI. CONCLUSION

In this paper, feedback linearization technique is applied to DC-link voltage control of the vector controlled vector modulated VIENNA I based on a novel PLL and online load current estimator. As a result, besides enabling the system to cope with severe load reduction, the low load operation region has been extended without forcing the system into instability. Implementing the whole vector control system digitally, the

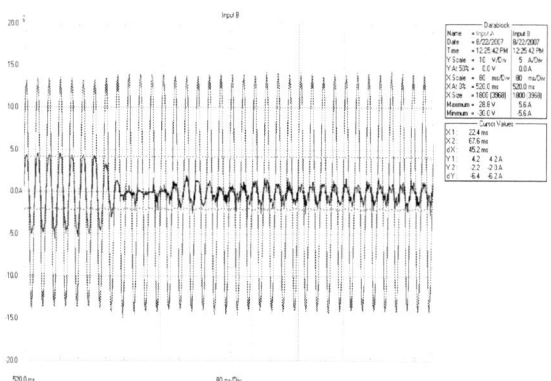

Fig. 12. Phase (a) source voltage and input current under hybrid vector control after severe load reduction

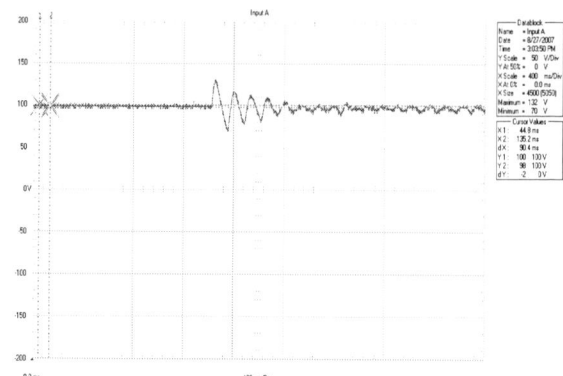

Fig. 13. Output DC-link voltage under hybrid vector control after severe load reduction

extra hardware cost for analog implementation is avoided. This together with the EPLL has made the system unified.

REFERENCES

[1] J. W. Kolar and F. C. Zach, "A novel three-phase utility interface minimizing line current harmonics of high-power telecommunications rectifier modules," *IEEE Trans. Ind. Electron.*, vol. 44, no. 4, pp. 456–467, Aug. 1997.

[2] T. Viitanen and H. Tussa, "Three-level space vector modulation - an application to a space vector controlled unidirectional three-phase/level/switch vienna i rectifier," in *Proc. EPE'03*, Toulous, France, 2003.

[3] ——, "Space vector modulation and control of unidirectional three-phase/level/switch vienna i rectifier with lcl ac filter," in *Proc. IEEE PESC'03*, vol. 3, June 2003, pp. 1066–1068.

[4] J.-J. Slotine and W. Li, *Applied nonlinear control.* Prentice-Hall, 1991.

[5] T. Viitanen and H. Tuusa, "Experimental results of vector controlled and vector modulated vienna i rectifier," in *Proc. IEEE PESC'04*, vol. 6, June 2004, pp. 4637–4643.

[6] M. Karimi-Ghartemani and M. R. Iravani, "A method for synchronization of power electronic converters in polluted and variable-frequency environments," *IEEE Trans. Power Syst.*, vol. 19, pp. 1263–1270, Aug. 2004.

A Soft-Switched AC-DC Symmetrical Boost Converter with Power Factor Correction

A. Jangwanitlert[*], J. Songboonkaew[**]

[*] Faculty of Engineering, King Mongkut's Institute of Technology Ladkrabang, Bangkok Thailand 10520
[**] Faculty of Engineering, Thonburi University, Bangkok Thailand, 10160
Email: kjanuwat@kmitl.ac.th, sojirasak@hotmail.com

Abstract – **This paper presents a soft-switched ac-dc symmetrical boost converter with power factor correction. The rectifier is modified boost voltage that is well suited for low line input applications and operated with half of the switch voltage stress than those found in standard boost converter. Soft-switching in the boost converter is achieved under Zero-Voltage Switching (ZVS) turn on and Zero-Current Switching (ZCS) at turn on, quasi resonant technique. In this paper, the operation principle and operation mode are demonstrated. The experimental results obtained from a prototype are agreement with the simulation ones.**

Index Terms – **ac-dc symmetrical boost converter, power factor correction, soft switching**

I. INTRODUCTION

Power factor correction (PFC) has become an increasingly demanded feature in ac-dc power supplies in recent years. Standards such as EN61000-3-2 have imposed restriction on line current harmonic pollution. It has become standard practice to implement an ac- dc converter by placing a boost converter between diode bridge rectifier and a dc bus capacitor as shown in Fig. 1 [1]. The boost converter provides the shapes of the input current to be appropriate and ac-dc converter can operate with a near unity power factor.

Although the topology in Fig. 1 is the standard one for single phase ac-dc PFC, it is not the best one for all cases and applications. Therefore, different variations on this converter have been proposed [2-5] and many of these have been attempted to reduce conduction losses and component numbers.

Modified, reduced component boost PFC converters are generally suited for applications. A low input line voltage (100V $\pm$ 15%) should be used better than a high input line voltage (220V $\pm$ 15%). This is because more current needs to flow through the converter to deliver the same amount of power to the output than when input line voltage is high. Consequently, the conduction losses are decreased. Also, reduction of the number of components is more significant in conduction losses. The symmetrical boost converter shown in Fig. 2(a) is proposed as a converter that can be used in applications where a high output voltage is required to be obtained from a low input line such as the front-end converter in a two-stage ac-dc converter where a 300V output is typically needed to be fed into the input of a dc-dc full-bridge stage [3]. In this particular case, the input voltage of dc-dc full-bridge stage is kept high although the front-end converter is

limited to low input line operation which limits the current circulating in the dc-dc full-bridge stage, and thus reducing conduction losses.

Modified, reduced component boost PFC converters generally achieve their component reduction by replacing the diode bridge rectifier and other converter diodes with an additional switch. This makes it difficult to implement converters with the same soft-switching techniques used in standard boost PFC converter (i.e. [2] – [5]) because most of these techniques involve the addition of auxiliary circuits with active switches to the main power circuit. In case of modified boost PFC converters, this would mean either implementing an active auxiliary circuit for each switch, which would be impractical and expensive, or implementing a single switch as an auxiliary circuit that would assist the soft-switching of both switches. An alternative approach for soft-switching in modified boost converters is to use quasi-resonant techniques [3]-[4] that can be cheaper to implement the auxiliary circuit using merely a few passive components.

The modified, reduced component boost PFC converter, "a symmetrical boost converter", has zero-voltage switching (ZVS) quasi resonant boost converter as shown in Fig. 2(b), which depends on the L_{in} and C_r in the proposed paper. This paper can operate with soft-switching technique and PFC in low line applications and is economical because it requires only two active switches and four diodes for main power circuit and a few passive components to obtain soft-switching. The turn-off losses of a switch in this converter are approximately 25% of those of a switch in a standard boost converter because turn-off losses are proportional to the square of current across that switch before turning off and voltage stress of a switch in the proposed converter are half those of that switch in standard boost converter. Therefore, the characteristic makes the use of the zero-current switching (ZCS) and zero-voltage switching (ZVS) techniques simpler for implementing soft switching in the converter.

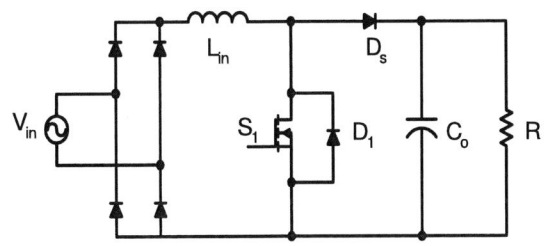

Fig.1 Standard boost ac-dc PFC converter.

978-1-4244-0644-9/07/$25.00 ©2007 IEEE

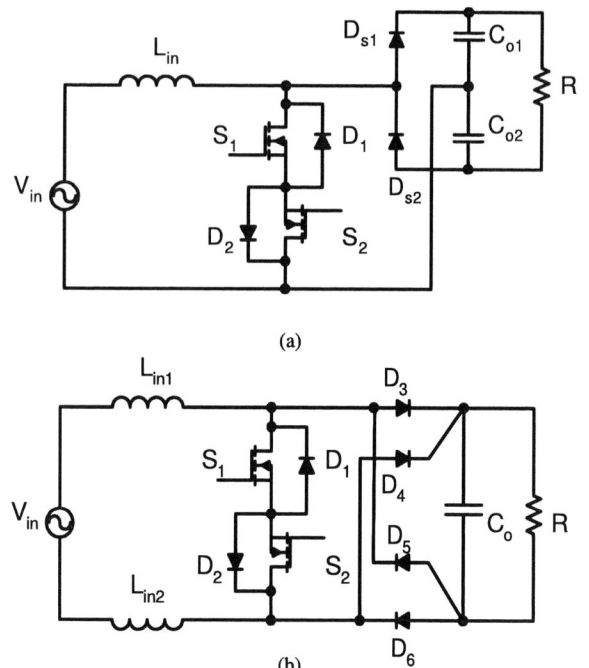

(a)

(b)

Fig. 2(a) Boost PFC ac-dc converter with reduced component and switch voltage stress [2], (b) Proposed soft-switched boost PFC ac-dc converter.

In this paper, the converter and the operation modes are discussed. The equations are also analyzed. The control strategies are explained and feasibility of the converter is shown with the experimental results.

II. OPERATION PRINCIPLE

For simplicity of analysis, the following assumptions are made:

Input inductors are replaced as constant current sources, I_{in}

Voltage across the output capacitor C_o is constant

L_{in1} and L_{in2} are identical

Parasitic capacitances C_{r1} and C_{r2} are identical

There are 3 operation modes in a half cycle. The operation main waveforms are shown in Fig. 3.

Mode 1 (t_0-t_1): At time t_0, Switch S_1 is turned. The current flowing from source V_{in} through L_{in1}, S_1, D_2 and L_{in2}. The current rises linearly as the voltage across L_{in1}, L_{in2} is equal to one half of V_{in}. The other part of current is from discharging current of capacitor C_o to R load.

$$V_{Lin1} = V_{Lin2} = \frac{V_{in}}{2} \qquad (1)$$

Mode 2 (t_1-t_2): Switch S_1 is turned off. The current from V_{in} flowing through Diode D_3, load, D_6, and L_{in2}, respectively. This current is charging to load. The input current decreases linearly. The voltage across L_{in1} and L_{in2} is depicted in (2), where $L_{in} = L_{in1} + L_{in2}$.

$$V_{Lin} = V_{in} - V_o \qquad (2)$$

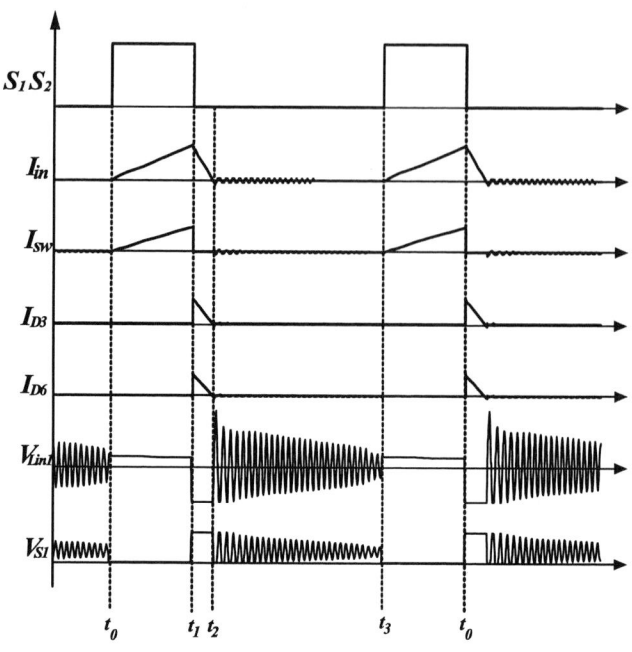

Fig. 3 Main waveforms.

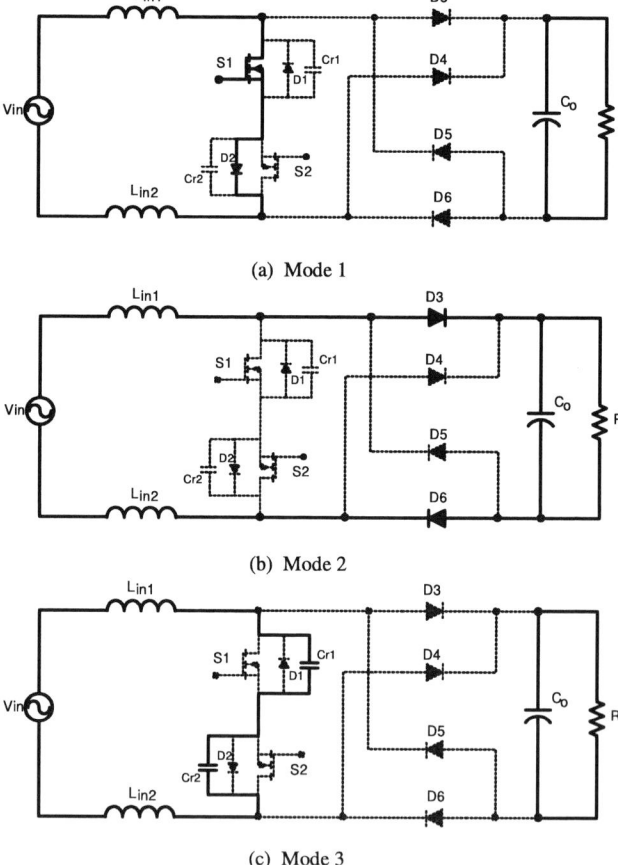

(a) Mode 1

(b) Mode 2

(c) Mode 3

Fig. 4 Operation Modes

Mode 3 (t_2-t_3): The current from source V_{in} flowing through L_{in1}, capacitors C_{r1}, C_{r2}, and L_{in2}, respectively. At this mode, it is resonant circuit between L_{in1}, L_{in2}, C_{r1}, and

C_{r2}. The equation is given by (3). In addition, the output capacitor is discharging through load R.

$$f_r = \frac{1}{2\pi \sqrt{(L_{in1} + L_{in2})\dfrac{C_r}{2}}} \quad (3)$$

where $C_{r1}=C_{r2}=C_r$.

At, $T/2+t_0$, Switch S_2 begins to conduct. The current flowing through L_{in2}, S_2, D_1 and L_{in1}, respectively. That is a second half cycle starting. The second half cycle works in the same manner as a first one except the switch S_2 is turned on instead of switch S_1.

From (1) and (2), the output voltage can be solved in (4).

$$V_o = \left(\frac{\Delta_1 + D}{\Delta_1}\right)V_{in} \quad (4)$$

where Δ_1 is the t_{off} of switch (not including t_{off} circuit) and D is the duty cycle.

However, the eq. (4) must be multiplied by two due to the fact that the second half cycle has the same output equation. Therefore, the equation is rewritten by (5)

$$V_o = 2\left(\frac{\Delta_1 + D}{\Delta_1}\right)V_{in} \quad (5)$$

III. ZCS, ZVS CONDITION AT TURN ON

When Mode 1 is starting, the current is from source to switch. That means the current of switch is from zero at turn on. That provides the ZCS at turn on. Moreover, the oscillation voltage from the resonant circuit in Mode 3 is occurred. It is underdamped oscillation before Mode 1 returns. That means the voltage across the switch is going to be zero. The switch is turned on under the ZVS.

IV. EXPERIMENTAL RESULTS

To verify the operation principle, the practical aspects of an ac-dc symmetrical boost converter are implemented. The experimental results are demonstrated under the specifications as follows:

- Output power P_o =300 W,
- Output voltage V_o= 480 V,
- Input voltage V_{in} = 100 V,
- Switching frequency f_s = 50 kHz,
- Duty cycle D = 0.3

Therefore, a prototype has been assembled and tested with the devices summarized in table I.

Due to the parasitic capacitors from the MOSFETs being selected to meet the required switching frequency, the resonance condition (3), which is 2.5 MHz, is verified to achieve ZVS turned-on as shown in Fig. 5(a). In addition, Fig. 5(a) shows ZCS of switch S_1 at turn on.

Fig. 5(b) is the simulation result that is compared with Fig. 5(a) in order to confirm the idea.

Moreover, Fig. 6 shows the input voltage and input current. The voltage is ac voltage at 100 V and input current is like triangle and chopped by switch S1 and S2. There is a high frequency of 50 kHz added in 50 Hz fundamental. The experimental result shows in Fig. 6(a) and the simulation result shows in Fig 6(b) to emphasize the main idea.

TABLE I. DEVICE SPECIFICATIONS

Switch S_1, S_2	IRFP460
Diode D_1-D_4	HFA25TB60
Lin_1,Lin_2	10 uH
Ro	800 Ω
C_{r1},C_{r2}	440 pF
C_o	50 uF

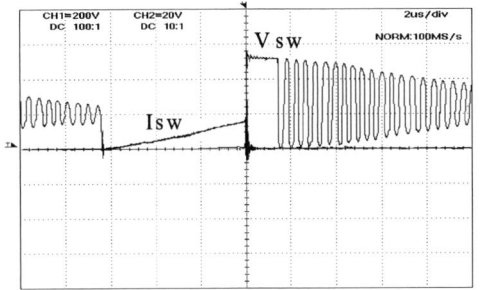

(a) experimental result

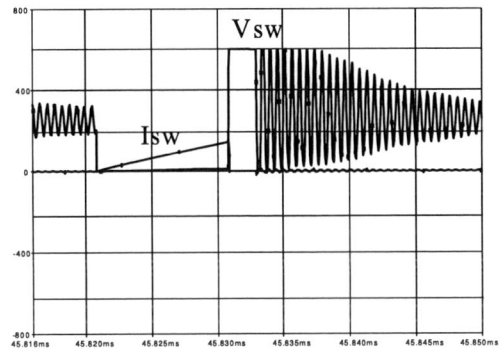

(b) simulation result

Fig. 5 Zero-current switching at turn on and off and Zero-voltage switching condition at turn-on.

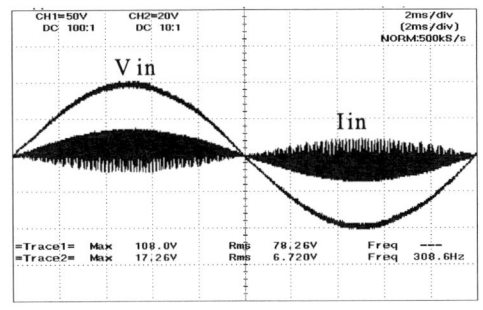

(a)

Fig. 6 Input current (I_{in}) and voltage (V_{in}): experiment.

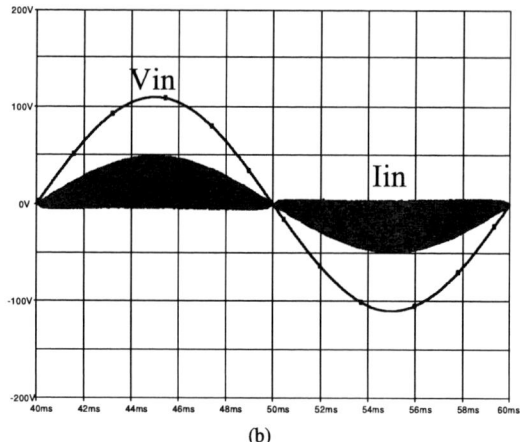

(b)

Fig. 6 Input current (I_{in}) and voltage (V_{in}): simulation.

Furthermore, when duty cycle is 0.3 and input voltage 100 V following to (5), the output voltage and output current that provide 480 V, 0.75 A average. The average value is close to Fig. 7(a) and (b). Fig. 8 shows the efficiency vs. duty cycle. When duty cycle of switch is increased, the efficiency goes down. The efficiency is from 88 % -97 %. Additionally, if the switching frequency is changed from 20-80 kHz, it affects to the efficiency not much that is shown in Fig. 9.

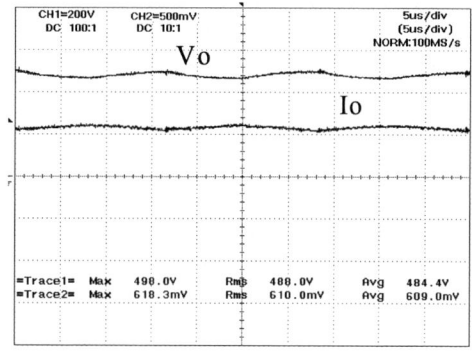

(a) experimental result

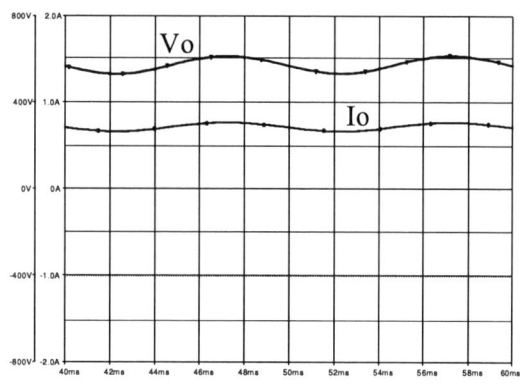

(b) simulation result

Fig.7 DC output voltage (V_o) and current (I_o).

Moreover, following to the EN61000-3-2 standard, the Total Harmonic Distortion current (THD_i) is 13.46 % [6]. The experimental result from Fig. 10 has 6.5 % of THD_i that is below the standard. That means it is agreement with the regulations.

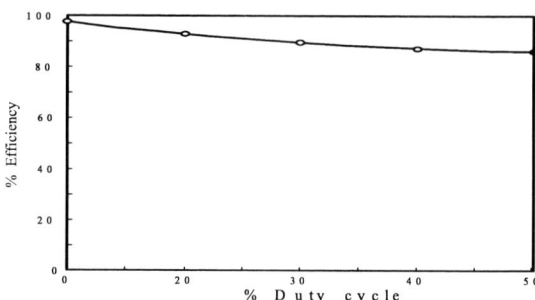

Fig. 8 % Efficiency vs. % duty cycle.

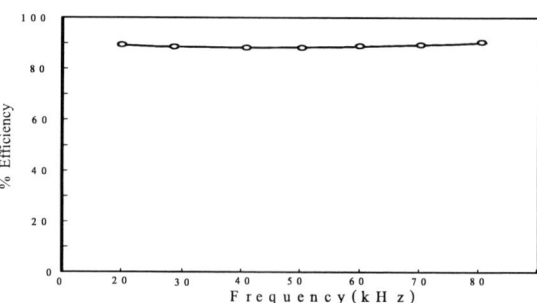

Fig. 9 Efficiency vs. frequency.

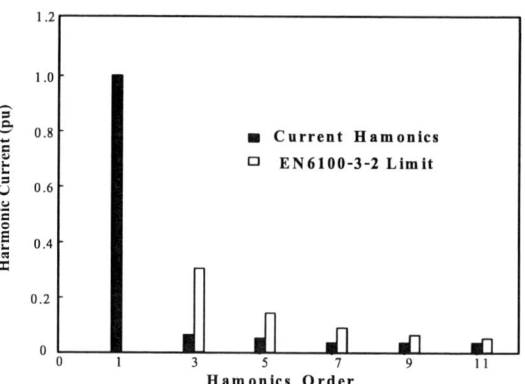

Fig. 10 % THD_i vs. harmonics order.

V. CONCLUSION

An ac-dc symmetrical boost converter has been presented in this paper. Its features include near unity power factor operation, ZVS turned-on, and ZCS turned on and low cost. Moreover, the operation modes of converter have been demonstrated along with the control method. The feasibility of this converter has been shown with results obtained from the simulation and experimental prototype.

A prototype has been constructed and designed at rated output power of 300W, 50 kHz, and input voltage of 100 V. The experimentally measured efficiency at full load and frequency of 20 kHz is 95 %. Also, the THD$_i$ is 6.5 % that can be accepted following the EN61000-3-2 standard.

REFERENCES

[1] F. Liccardo, P. Marino, and M. Triggianese, "High power three phase four wires synchronous active front-end," *Proc. in IEEE ISIE*, Vol.2, 2006, pp.1172-1177.

[2] S. Maniktala, "Things to try," *Switching power supply design & optimization*, McGraw-Hill, 2005, pp. 341-342.

[3] E. P. Nowicki, "A comparison of single-phase transformer-less PWM frequency changers with unity input power factor," *Proc. in IEEE APEC*, Vol.2, 1994, pp.879-885.

[4] L. S. Barreto, E. A. Coelho, V. J. Farias, L. C. Freitas, and J. B. Vieira, "The bang-bang hysteresis current waveshaping control technique used to implement a high power factor power supply," *IEEE trans. Power Electronics*, Vol. 19, No. 1, 2004. pp. 160-168.

[5] K-S. Park; Y-H. Kim, "A new high-efficiency zero-voltage-switching ac-dc boost converter using energy recovery circuits," *Proc. in Industrial Application Conference*, Vol. 4, 2001, pp. 2466 – 2472.

[6] "Limit for harmonic current emissions," IEC International Standard, Edition 2.1, 2001, pp.13-37.

Education Reforming for Power Electronics

Weiping Zhang, Xiaohan Guan, and Dongyan Zhang

Green Power & Energy System Laboratory, North China University of Technology, Beijing 100041, P.R China
Tel. (Fax): 86-010-88802880, Email: gxh@ncut.edu.cn

Abstract– A set of updated course contents of power electronics is published in this paper. The updated course content is arranged around the controllable switching devices and emphasizes the analysis of the converter system. Digital analysis and design method is based on simulation. The analysis of some typical power electronic installations helps students understand the application of this course in practice and grasp some practical analysis methods. The main contents include: (1) Power electronic devices (give priority to controllable devices), (2) Analysis and design of main circuits, (3) Operational principle of control circuits, (4) Digital analysis and design method based on simulation, (5) Introduction and study of power electronic installations. The updated course combines the research fruits with course contents tightly.

Index Terms-- Education Reforming, Power electronics.

I. INTRODUCTION

At present, a great influence has been made to various fields by the development of computer, digital communication and internet technology. This situation has caused some new requirements to power electronics technology such as high frequency, high efficiency, small volume and less weight. Simultaneously because the traditional energy sources have been in short supply, more and more attention has been paid to the renewable energy, such as solar energy, wind power, water power, etc. Power electronics is the key technology to utilize these new kinds of energy sources. In these new fields, the traditional technology using thyristors as the switching components can not meet the new requirements, so the study on the controllable switching devices and the corresponding topologies as well as the relative technologies have become a new important trend of today's power electronics.

The new situation mentioned above has caused the education reforms on power electronics in universities all over the world. Different approaches have been taken and good results obtained. For example, the University of Minnesota has restructured the relative courses, such as power electronics and electric machines/drives, since 1994 through NSF funding. These courses are redesigned carefully using a top-down approach where the topology and control are described in the context of applications. A building block (power pole) methodology has been developed, which provides a common basis for describing all practical converter topologies and combines PFC (Power Factor Correction) and the design of magnetic components in it. The application in industry

is emphasized, which prepares students for industry as well as advanced courses and research [1].

In Japan, Japanese Electrotechnical Committee has organized an expert committee to study the education reform on power electronics and the following conclusions are obtained. (1) The control technology in energy transformation should be emphasized. (2) In undergraduate courses the theory should be centered on and the practical application be introduced. (3) Both the hardware and the software should be cared for. (4) Simulation plays an important role [2].

European reformation centers on Aalborg University in Denmark, which takes an approach where project-oriented and problem-based learning is organized. One project is carried out at each semester. Normally 6 students work together in a project group, except for their Bachelor or Master project, under the supervision of a supervisor. The students are cultivated the ability to learn and study when given a problem. As well the abilities concerned communication, cooperation, organization and management are also required to some extent. Almost all the necessary skills for future engineers are exercised during the projects [3].

In China, Huazhong University of Science and Technology has put forward a new series of power electronics courses. They divide the whole contents into a few parts, centering on the power devices, based on the basic four types of converters and two kinds of control methods, and aiming at two kinds of applications. Some thinking mode peculiar to power electronics are concluded and run through the analysis of various circuits [4].

In addition, many web-based interactive power electronic courses are developed by universities all over the world, such as iPES established by Swiss Federal Institute of Technology Zurich [5], the homepages by Fachhochschule Darmstadt in German [6], Utsunomiya University in Japan [7] and Huazhong University of Science and Technology in China [8], etc. Usually the operation states of the converters are displayed in animation, while the current routes and waveforms are given synchronously. Students can simulate the adjusting of the converters in computer, which stimulates their interests in power electronics greatly.

Based on the reformation carried out all over the world, we can summarize that the novel ideas are focusing on the power electronics systems instead of on the various main circuits. Practical applications are emphasized, which is helpful for students' abilities to solve practical problems and to design circuits, paving the way to their future carriers and research. In this paper, based on the development of power electronics industry

---

This work was supported by 2005 Beijing Municipal College Education & Teaching Reform Project: Reform and Exploration of Course System for Power Electronics

978-1-4244-0644-9/07/$25.00 ©2007 IEEE

in China and consulting the education reforms results mentioned above, we have put forward a set of updated course system of power electronics. Not only will the new course system be introduced in detail, but also its characteristics will be summarized afterwards.

II. A Set of Updated Course System for Power Electronics

Nowadays, controllable switching devices are popularly employed as the main switching components in the high frequency power conversion equipments in the fields of information, aviation, new energy and so on. Taking these factors into account, we put forward some updated course contents which arrange the course around the controllable switching devices and its application in power conversion. Simulation and digital analysis approaches are added, and practical applications emphasized. Theory, simulation and practical applications are paid equal attention to establish foundation for students' future engagement in relative industries and further study. The thinking of "Devices serve circuits, circuits serve system and system serves application." and "digital analysis and design approaches" have been taken as the main clues for this reforming. This set of updated course contents include the following five parts [9][10][11]: (1) Power electronic devices (introducing the outer characteristics and how to use it in the power conversion), (2) Analysis and design of main circuits (Taking controllable switching devices as switching components in power conversion), (3) Operational principle of control circuits, (4) Digital analysis and design technology, (5) Introduction and study of power electronic installations. The above five parts will be introduced in detail as the following.

A. Power electronic devices

In this part, the stress is put on the controllable switching devices such as GTR, MOSFET and IGBT. The theory of semiconductor is weakened, while the outer characteristics and application backgrounds of the devices are reinforced. The whole contents are divided into two levels: an elementary one and an advanced one. The elementary level is suitable for the undergraduates. The students are required to understand the devices' structures and know the devices' outer characteristics (paying more attention to steady state characteristics and less attention to dynamic characteristics) and limiting parameters as well. Via the comparison of the devices parameters, the main performance and appropriate application of each device will be highlighted. Ultimately the students should have the ability to select a suitable switching device for a certain application. The advanced level is for the graduates. The dynamics characteristics and the influence of application environments are the emphases. Students are asked to build up suitable devices' dynamic models for analyzing the states of the devices in the context of the circuit.

Driving and protecting circuits are important contents as well. Undergraduates are asked to grasp some typical driving circuits and chips. Several typical circuits will be introduced in the class for the students to learn some basic analysis methods, and the other circuits are left for study on their own. The graduates are asked, first, to grasp the matching principles both between the circuits and the devices and between the driving and main circuits; second, to grasp the method to design driving circuits; third, to have the ability to modify the typical driving circuits and chips on the background of a practical application. At the same time, the latest development in power electronic devices, such as power integration, PT (Piezoelectric transformer), intelligent power module etc., will be introduced precisely.

B. Analysis and design of main circuits

In this paper the operational principles of four types of basic power electronic converters are introduced. Some laws to analyze the main circuits are put forward: Ideal circuits first and practical circuits later; Simulation first and theoretical analysis later; Imitative design first and original design later; Typical circuits first and general circuits later. The whole contents are divided into an undergraduate level and a graduate level as well.

The course contents for undergraduates include: (a) Operational principles, steady state characteristics and selection of devices for the ideal DC/DC, DC/AC, AC/DC and AC/AC converters, consulting the devices knowledge introduced in the part (1); (b) Introduction to low power typical circuits such as PFC, rectifier, forward and flyback converters; (c) Simulation technique for main circuits (by the general purpose simulation software of PSpice); (d) Design and selection of power magnetic devices, for example, consulting the table offered in reference [12]; (e) The blocks of the control circuits suitable for the main circuits.

The graduates are asked to study further the following contents: (a) Models for circuit analysis and dynamics of practical converters; (b) Theoretical analysis and simulation of the parasitic components' affection via ideal models, which requires the knowledge on the switching devices' parasitic components introduced in the part (1); (c) The analysis of typical from media to high power level systems, (d) Introduction to new technologies, taking circuits used in present industry as examples, such as soft switching, full-bridge phase shift, active clamped converters and so on.

The objects of the above arrangement is to have the undergraduates grasp some elementary knowledge of power electronics and do some experiments through imitating the typical systems, and to have the graduates design some reasonable converters according to the given specifications via simulation and theoretical analysis. Moreover the devices' performances learned in the part (1) are verified in the context of main circuits.

C. Operational principle of control circuits

For a practical power electronic converter, only a main circuit is not enough, and the control and auxiliary circuits must be integrated together. The switching devices in the main circuits must be controlled according to some control strategy, so that a complete closed-loop converter system will be formed. Two basic control ways, current-control-mode (including average and peak current-programmed control) and voltage-control- mode, are the main study objectives. The analysis of the closed-loop converter systems' characteristics, such as transient, frequency, AC stability and large signal characteristics, are the emphases.

The contents for undergraduates are as the followings: (a) Operational principles of typical PWM-controlling chips; (b) Typical control techniques and their merits and demerits; (c) Introduction to some typical compensation networks; (d) Method to analyze the closed-loop characteristics of typical circuits based on Mathcad and Matlab program; (e) Measuring technique in time domain.

The contents for graduates are: (a) Performance analysis and parameters design of some typical compensation networks; (b) The matching between the main circuits and control chips or techniques, and the matching between the main circuits and compensation circuits; (c) Small signal-AC models and dynamics analysis of power electronics systems; (d) Method to analyze the closed-loop characteristics based on the waveforms averaging models—large signal and small signal simulation approaches; (e) Analysis of the closed-loop characteristics of typical power electronic equipments; (f) Measuring technique in frequency domain and finding out the relationship between frequency characteristics and time-domain specifications; (g) Matching technology about installations and loads, line and environments.

A complete power electronic conversion system should be made up of the control circuits and the main circuits, so the analysis of the main circuits in the part (2) forms the basis of the study of the closed-loop system. By the comparison of the open-loop with closed-loop characteristics, the students are asked to deeply understand the importance of the feedback and have the ability to improve the dynamics and frequency characteristics.

D. Digital analysis and design technology

At present, almost all the experiments and original ideas, from the design of integrated circuits to that of the practical electronics systems, electrical installation and testing equipment, can be verified by the computer simulation via general or special purpose simulation software. In some universities and research centers, digital design method based on the simulation results has become the dominant design method for graduates and researchers; furthermore, it is being introduced into undergraduates' course. Compared with the traditional theoretical design, digital design has many merits such as saving time, saving efforts and easy to be understood by the designers.

As general purpose simulation software, PSpice has been utilized widely in power electronic circuit simulation. Therefore we introduce PSpice in both undergraduates' and graduates' courses to assit the theoretical study. These are some typical applications: (a) When the dynamic models of the power switching devices been input, computers can simulate the influence of the dynamic and parasitic parameters on the devices' performance, which will help to understand the qualitative analysis in part (1). (b) Input the model of ideal converters, their steady state performance will be simulated promptly. Applying parametric sweep approach and combining with the data processing programs by Mathcad and Matlab, we can draw some figures about the designed parameters vs. the specifications, based on which the parameters can be optimized. (c) Input the small signal model of main circuit and control circuit, various dynamics and closed-loop characteristics can be analyzed, which will verify the theoretical analysis results given in part (2) and (3).

The simulation helps students find out the relationships and differences between the theory and the practical circuits and deepen the understanding of the theory. Consequently, digital analysis and design technology can offers a valuable method for students to improve the ability to solve practical problems in the future.

E. Introduction and study of power electronics installations

It is well known that power electronics installations applied in industry are the crystallization of human being's intelligence on power electronics. If a student can analyze thoroughly several typical installations and grasp their designing ideas, methods and adjusting techniques, he or she will get a deep understand of the application of power electronics in practice. Therefore, we have proposed some practical analysis methodologies to students in our course.

For example: (a) Qualitative and semi-quantitative analysis method. According to the systems requirements and the loads' characteristics, we can infer some essential functions of the system. (b) Tearing up method—to divide a system into several functional modules. The circuit would be divided into many function modules according to the system's function. Each module expresses clearly one of the functions. (c) Model analysis method. According to its functions, a simplified model of the practical circuit can be built up. Based on this simplified model, the performance will be easily analyzed by computer or in theory. (d) Macro models method. A macro model can be established to describe the main functions of a certain module, based on which the whole system's performance will be analyzed by circuit theory, control theory or mathematical methods.

At present, we have already had some installations that can be provided to our students to study. They are (a) Switching power supply designed by TOPSwitch-II, (b) 300W switching power supply for computer, (c) Testing platform for power magnetic elements, (d) Demonstrators of electronic ballast for fluorescent lamp, (e) Demonstrators of electronic ballast for 1k~12kW HID (High Intensity Discharge) lamp, (f) Demonstrators of solar energy/wind power, (g) Measuring system of small signal characteristics, (h) Experimental system of 2500W full-bridge phase shift converters, (i) Power converter of 50~150W PT, (j) Electromagnetic shield laboratory—testing platform for active/passive EMI

(Electromagnetic Interruption) filters, (h) DSP (Digital Signal Processor) controller for power converters. Through the study of several typical installations, not only are the students given some intuitional impression on the application of power electronics in industry, but also the ahead 4 parts are put into practice comprehensively.

III. THE NOVEL FEATURES OF THE SET OF UPDATED COURSE SYSTEM

The proposed course contents have the following distinctive features.

A. Arranging the course contents around the controllable switching devices and its applications

In order to meet the requirements to power electronics of high frequency, high efficiency and high performances in the field of information, aviation and renewable energy utilization, etc, the new course contents are focusing on the controllable switching devices and their application in power conversion.

B. Emphasis on the complete power electronics systems instead of on the main circuits

For power electronics, our opinion is that "Devices serve circuits, circuits serve system and system serves application." Through the reforming, we want that teacher and students should pay more attention to the whole converter as a complete system, including not only the main circuit, but also the relative circuits. Through analyzing some classical power converters, undergraduate students can learn their designing ideas, methods and adjusting techniques, they will get a deep understand of the application of power electronics in practice. These typical application equipments can serve as a platform for the graduate students to improve their properties further or verify their new ideas.

C. Digital analysis and design technology plays important role.

Digital analysis and design technology has become a more and more powerful tool to assist theory analysis. At present, if a suitable circuit model can be established to describe a special phenomenon, we would always find a digital analysis program for computer to analyze its performance, whose important role is obvious. In our course theory, simulation and experiments promote each other and theoretical, simulation and experimental results confirm each other. Putting emphasis on digital analysis and design technology is a farsighted practice.

D. Through analyzing some power electronics installations some analysis methodologies are provided

These methodologies are not aiming at a particular converter, but can be used to analyze almost all the power electronics circuits, simple or complex, open-loop or closed-loop, taught in the course or to be met in the future … There is an old Chinese saying: Give a man a fish and you will feed him for a day. Teach him how to fish and you will feed him for a lifetime. Even though not for a lifetime, but for a long time these methodologies will help students solve various problems and even some new methods will be put forward by themselves under the elicitation of these methodologies. Moreover in other disciplines besides power electronics, these methodologies are helpful.

E. The layers are perfectly clear

Another contribution of this paper is that the power electronics knowledge system is divided into two levels, the undergraduate level and the graduate level. Almost all the five parts have been divided. The undergraduate course is the base. After this course the students are asked to know about all the aspects of power electronics and to establish the foundation for the graduate course. The graduate course is in higher level. The two courses supplement each other, but each forms a complete system in oneself.

F. The new research results have been introduced into the course

The new research results have been turned into course contents. The course contents are enriched as well the research is promoted.

IV. CONCLUSIONS

A set of updated course contents of power electronics is published in this paper. The updated course content is arranged around the controllable switching devices and its application in power conversion. It emphasizes the complete power electronics systems, including main circuits, control circuits as well as auxiliary circuits. Simulation results have been used as the basis of digital analysis and design approaches and assist the theoretical analysis. By the analysis of some classical power electronics installations the students are taught some practical analysis methods and get a deep understanding of the power electronics' application. The new research results have been introduced in the course.

REFERENCES

[1] Ned Mohan, William P. Robbins, Paul Imbertson, et.al., "Restructuring of First Courses in Power Electronics and Electric Drives That Integrates Digital Control", *IEEE Transactions on Power Electronics, Vol.18, No.1,* Jan. 2003, pp.429-437.

[2] Xu Dehong, "Status of the Reforming for Power Electronics Courses in Universities at Home and Abroad", *2003 National Power Electronics and Drives Teaching Seminar,* Hangzhou, China, pp.1-7.

[3] Frede Blaabjerg, Remus Teodorescu, Zhe Chen, "Problem-based and project-oriented learning. An other way of implementing research based teaching in power electronics", *The 2005 International Power Electronics Conference,* pp.948-953.

[4] Huazhong University of Science and Technology, http://202.114.4.28/2005/C47/kcms-2.htm.

[5] http://www.ipes.ethz.ch.

[6] http://schmidt-walter.fbe.fh-darmstadt.de/smps-e/smps_e.html.

[7] Janos Hamar, Hirohito Funato, Satoshi Ogasawara, "E-Learning in Power Electronics: The State of the Art", *The*

2005 International Power Electronics Conference, pp.751-756.

[8] Zhang Rong, Zhou Yunbin, Yang Lisha, Yang Yinfu, "A New Approach of Education in Power Electronics Employing Flash", *The 2005 International Power Electronics Conference*, pp.954-957.

[9] Ying Jianping, "Fundamentals of Power Electronic Technology", *China Machine Press*, Beijing, 2003.

[10] Lin Weixun, "Modern Power Electronic Circuits", *Zhejiang University Press*, Hangzhou, 2002.7.

[11] Zhang Weiping, "Modeling and Control for DC-DC Converters", *China Electric Power Press*, Beijing, 2006.1.

[12] Ned Mohan, Tore M. Undeland, William P. Robbins. "Power Electronics: Converters, Applications, and Design", 3rd ed. (the gravure). *High Education Press*, Beijing, 2004.1.

A Novel Current Control System for PMSM Considering Effects from Inverter in Overmodulation Range

Smith Lerdudomsak, Shinji Doki and Shigeru Okuma
Department of Electrical Engineering and Computer Science
Nagoya University, Japan

Abstract—**In this paper, we propose a novel current control system for PMSM when an inverter operates in the overmodulation range. Because of the effect from harmonic components generated from an inverter in this range, an unstable problem in current control system will occur. For solving this problem, the method of harmonic currents compensation is proposed. However, the difficulty in this method is that how much we can estimate harmonic current accurately.**

There are two new topics in harmonic components estimation those we use in our proposed current control system. First, the harmonic voltages estimation method based on the conventional Sine-PWM modulation method of an inverter. Second, the harmonic currents estimation method that not consider only a model of PMSM, but also an effect from current controllers. The effectiveness of our proposed system is confirmed by computer simulation and experimental results.

I. INTRODUCTION

Recently, control of permanent magnet synchronous motor (PMSM) in the overmodulation range of an inverter becomes important research topic due to the needs for expanding operating range and achieving faster response [1]~[4]. Waveform of inverter output voltage and its frequency spectrum in the overmodulation range based on the conventional Sine-PWM modulation method are shown together in Fig.1. From this figure, we can find that in this overmodulation range, fundamental component of inverter output voltage (dashed line) can be much increased compared with the case of conventional linear range (theoretically, as high as 127%) [2].

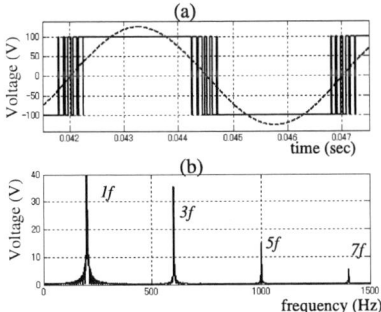

Fig. 1.   (a)Wavefrom of voltage in the overmodulation range (b)Power spectrum of volatge in the overmodulation range

However, there are two main problems occur when using this overmodulation range. First problem is the nonlinear voltage relation and second problem is the large amount of harmonic components in inverter output voltages.

For solving the nonlinear voltage relation problem, the nonlinear amplitude compensation method was proposed [2], by using this compensation method, we can use overmodulation range in the opened loop control system, such as V/f control, of PMSM without any problem.

However in closed loop current control system of PMSM, only using this nonlinear amplitude compensation is still not enough. Because the harmonic currents that generated from inverter harmonic voltages are also fed-back to the closed loop current control system and will cause the problem as discribed in the belowing.

(1) Current controllers will try to suppress these harmonic currents by generating extra reference voltages to an inverter.

(2) Becuase in the overmodulation range, inverter is already operating near voltage saturation limit. Then, an inverter will become saturated easily by these extra reference voltages.

(3) When output voltage of an inverter is saturated, there is not enough voltage for controlling fundamental component of currents, then current responses will be deteriorated or eventually become unstable. And this problem will be more serious when gains of controller are higher or the control period are longer.

For solving the problem from harmonic currents, the current control system with harmonic currents compensation was proposed [3][4]. In this method, first, harmonic voltages are estimated from inverter switching pattern. Next, by using PMSM harmonic model, harmonic currents are estimated and then compensated from real currents before being fed-back to current control loop, hence harmonic currents will not affect to current control system.

However, in the above method, an inverter is controlled by space-vector modulation[3][5] or special modulation method[4][6], then the conventional carrier based Sine-PWM modulation method cannot be applied directly. In our proposed system, we pay attention on the carrier based Sine-PWM modulation method, which is very easy to use in real applications. Proposed harmonic voltages estimation method (Inverter Model) will be discribed detailly in the next section.

Moreover, harmonic currents estimation method in [3] and [4] considered only a harmonic model of PMSM. This method

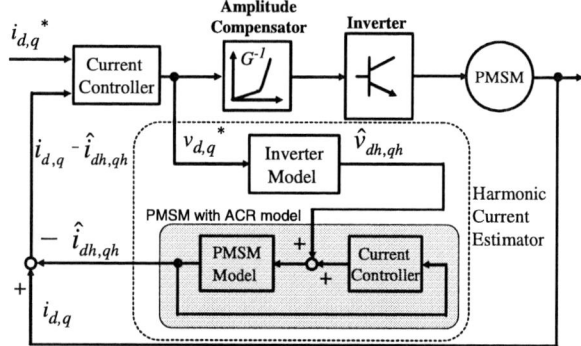

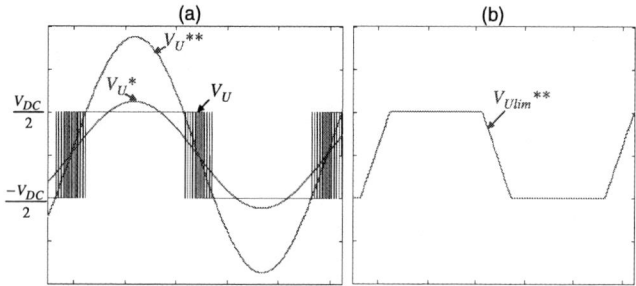

Fig. 2. Proposed currents control system with harmonic currents compensation

Fig. 3. Waveform of inverter reference and output voltage in overmodulation range

will give accurate estimation results in steady state condition but will cause an unstable problem in transient condition due to the difference of system poles. For solving this problem, in our proposed method, we also consider both current controllers as well as PMSM model (PMSM with ACR model) in estimation of harmonic currents. The method for achieving this PMSM with ACR model are discribed in section III and our proposed current control system is shown in Fig.2.

II. INVERTER MODEL

For estimating harmonic voltages by using only reference voltages data, an inverter model is necessary. In this time, we pay attention on conventional and simple Sine-PWM modulation method, however an inverter model that propose here can also be applied to other modulation method too.

In Sine-PWM modulation method, waveforms of u-phase inverter reference voltages and output voltage are shown together in Fig.3(a), when V_U^* is u-phase reference voltage calculated from current controllers, V_U^{**} is reference voltage after amplitude compensation and V_U is inverter output voltage. If the reference voltage V_U^{**} is limitted by $\pm V_{DC}/2$ limiter, it becomes V_{Ulim}^{**} as showm in Fig3(b). This V_{Ulim}^{**} voltage can be used to represent the output voltage V_U in harmonic voltage estimation.

Hence, u-phase harmonic voltage V_{Uh} can be estimated as shown in (1) by the difference between V_{Ulim}^{**} and reference voltage V_U^* which represents fundamental component of inverter output voltage. The harmonic voltages in v-phase and verter output voltage. The harmonic voltages in v-phase and

w-phase V_{Vh}, V_{Wh} can also be estimated by the same method as shown in (2) and (3) respectively.

$$V_{Uh} = V_{Ulim}^{**} - V_U^* \qquad (1)$$

$$V_{Vh} = V_{Vlim}^{**} - V_V^* \qquad (2)$$

$$V_{Wh} = V_{Wlim}^{**} - V_W^* \qquad (3)$$

Next, we can estimate the harmonic voltages in d-q axis rotational frame v_{dh}, v_{qh} by using the conventional Park transformation with harmonic voltages in three phases system as shown in (4) and (5) respectively (when θ_{re} is rotor angle in electrical degree). These harmonic voltages will be used in harmonic currents estimation as will be discribed later in next section.

$$v_{dh} = \sqrt{\frac{2}{3}}(cos(\theta_{re})V_{Uh} + cos(\theta_{re} - \frac{2\pi}{3})V_{Vh} \qquad (4)$$
$$+ cos(\theta_{re} + \frac{2\pi}{3})V_{Wh})$$

$$v_{qh} = \sqrt{\frac{2}{3}}(-sin(\theta_{re})V_{Uh} - sin(\theta_{re} - \frac{2\pi}{3})V_{Vh} \qquad (5)$$
$$- sin(\theta_{re} + \frac{2\pi}{3})V_{Wh})$$

III. PMSM WITH ACR MODEL

In this section, we will discribe about the method of harmonic currents estimation. Based on state equations of PMSM as shown in (6) the harmonic model of PMSM can be written as shown in (7) and this harmonic model was used in harmonic currents estimation [4]. When $[i_d \ i_q]^T$ are d and q axis currents, $[v_d \ v_q]^T$ is d and q axis voltages, $[i_{dh} \ i_{qh}]^T$ are d and q axis harmonic currents, $[v_{dh} \ v_{qh}]^T$ are d and q axis harmonic voltages, ω_{re} is rotational speed, R, L_d, L_q, k_e are resistance, d and q axis inductances, and back emf. constant respectively.

$$\begin{bmatrix} \dot{i_d} \\ \dot{i_q} \end{bmatrix} = \begin{bmatrix} -\dfrac{R}{L_d} & \omega_{re}\dfrac{L_q}{L_d} \\ -\omega_{re}\dfrac{L_d}{L_q} & -\dfrac{R}{L_q} \end{bmatrix} \begin{bmatrix} i_d \\ i_q \end{bmatrix} \qquad (6)$$
$$+ \begin{bmatrix} \dfrac{1}{L_d} & 0 \\ 0 & \dfrac{1}{L_q} \end{bmatrix} \begin{bmatrix} v_d \\ v_q - \omega_{re}k_e \end{bmatrix}$$

$$\begin{bmatrix} \dot{i_{dh}} \\ \dot{i_{qh}} \end{bmatrix} = \begin{bmatrix} -\dfrac{R}{L_d} & \omega_{re}\dfrac{L_q}{L_d} \\ -\omega_{re}\dfrac{L_d}{L_q} & -\dfrac{R}{L_q} \end{bmatrix} \begin{bmatrix} i_{dh} \\ i_{qh} \end{bmatrix} \qquad (7)$$
$$+ \begin{bmatrix} \dfrac{1}{L_d} & 0 \\ 0 & \dfrac{1}{L_q} \end{bmatrix} \begin{bmatrix} v_{dh} \\ v_{qh} \end{bmatrix}$$

The method in (7) can estimate harmonic currents accurately in steady state condition, however in transient condition, this estimation method will give the unstable results due to the mismatch between poles of PMSM and current control system,

1000

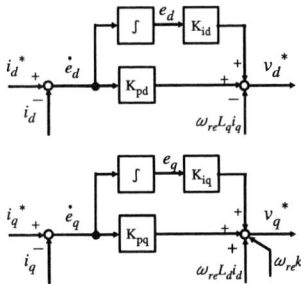

Fig. 4. d-axis and q-axis current controllers

the simulation results of this method in transient condition will be shown in next section.

Because when PMSM is composed in currents control system, characteristic of the system will depend on both PMSM and current controllers. Then, for estimating of harmonic currents, we should make the model by considering both model of PMSM and current controllers (PMSM with ACR model as shown in Fig.2). The method for achieving this model is discribed as the belowing.

First, we will start from the method of generating state equations for current control system of PMSM in linear range of inverter. And we will use the same method to expand for state equations in overmodulation range of an inverter. The Block diagram of PI current controllers with feedforward decoupling controllers are shown in Fig.4 and d-q axis reference voltages can be written in (8) and (9)！％When $[i_d^* \ i_q^*]^T$ are d and q axis reference currents, $[v_d^* \ v_q^*]^T$ are d and q axis reference voltages calculated by current controllers, $[\dot{e}_d \ \dot{e}_q]^T$ are d and q axis current errors, $[e_d \ e_q]^T$ are integration values of d and q axis current errors, $K_{pd}, K_{pq}, K_{id}, K_{iq}$ are d and q axis proportional and integral gains respectively.

$$v_d^* = K_{pd}\dot{e}_d + K_{id}e_d - \omega_{re}L_q i_q \qquad (8)$$

$$v_q^* = K_{pq}\dot{e}_q + K_{iq}e_q + \omega_{re}L_d i_d + \omega_{re}k_e \qquad (9)$$

In linear range we can assume that inverter reference voltages are same as inverter output voltages, then we can substitute d-q axis voltages v_d, v_q in (6) with reference voltages v_d^*, v_q^* from the above equations. And after some rearrangment, the differentiate of d and q axis currents will become as shown in (10) and (11)！％

$$\ddot{i}_d = -\frac{R}{L_d}\dot{i}_d + \frac{K_{pd}}{L_d}\dot{e}_d + \frac{K_{id}}{L_d}e_d \qquad (10)$$

$$\ddot{i}_q = -\frac{R}{L_q}\dot{i}_q + \frac{K_{pq}}{L_q}\dot{e}_q + \frac{K_{iq}}{L_q}e_q \qquad (11)$$

From Fig.4, $\dot{e}_d$ and $\dot{e}_q$ can be calculated as shown in (12) and (13)！％

$$\dot{e}_d = -i_d + i_d^* \qquad (12)$$

$$\dot{e}_q = -i_q + i_q^* \qquad (13)$$

After substitution of (12) and (13) into (10) and (11), the differentiate of of d and q axis currents will become as shown

in (14) and (15)！％

$$\ddot{i}_d = -(\frac{R}{L_d} + \frac{K_{pd}}{L_d})\dot{i}_d + \frac{K_{id}}{L_d}e_d + \frac{K_{pd}}{L_d}i_d^* \qquad (14)$$

$$\ddot{i}_q = -(\frac{R}{L_q} + \frac{K_{pq}}{L_q})\dot{i}_q + \frac{K_{iq}}{L_q}e_q + \frac{K_{pq}}{L_q}i_q^* \qquad (15)$$

By collecting (12)~(15), the state equations of PMSM current control system in linear range can be written in matrix form as shown in (16). And we can use the same method for achieving the state equations in the overmodulation range.

$$\begin{bmatrix} \ddot{i}_d \\ \ddot{i}_q \\ \dot{e}_d \\ \dot{e}_q \end{bmatrix} = \begin{bmatrix} -(\frac{R}{L_d} + \frac{K_{pd}}{L_d}) & 0 & \frac{K_{id}}{L_d} & 0 \\ 0 & -(\frac{R}{L_q} + \frac{K_{pq}}{L_q}) & 0 & \frac{K_{iq}}{L_q} \\ -1 & 0 & 0 & 0 \\ 0 & -1 & 0 & 0 \end{bmatrix}$$

$$\times \begin{bmatrix} i_d \\ i_q \\ e_d \\ e_q \end{bmatrix} + \begin{bmatrix} \frac{K_{pd}}{L_d} & 0 \\ 0 & \frac{K_{pq}}{L_q} \\ 1 & 0 \\ 0 & 1 \end{bmatrix} \begin{bmatrix} i_d^* \\ i_q^* \end{bmatrix} \qquad (16)$$

In the overmodulation range, inverter output voltages are the sum of reference voltages and harmonic voltages generated by inverter itself as shown in (17) and (18), when reference voltages v_d^*, v_q^* are obtained from (8) and (9) and harmonic voltages v_{dh}, v_{qh} from (4) and (5) in previos section.

$$v_d = v_d^* + v_{dh} \qquad (17)$$

$$v_q = v_q^* + v_{qh} \qquad (18)$$

After substitution of (17) and (18) into (6)！ａnd with some rearrangement as before, we can obtain the state equations of current control system in the overmodulation range as shown from (19) ~ (22)！％

$$\ddot{i}_d + \ddot{i}_{dh} = -(\frac{R}{L_d} + \frac{K_{pd}}{L_d})(\dot{i}_d + \dot{i}_{dh}) + \frac{K_{id}}{L_d}(e_d + e_{dh}) \quad (19)$$
$$+ \frac{K_{pd}}{L_d}i_d^* + \frac{1}{L_d}v_{dh}$$

$$\ddot{i}_q + \ddot{i}_{qh} = -(\frac{R}{L_q} + \frac{K_{pq}}{L_q})(\dot{i}_q + \dot{i}_{qh}) + \frac{K_{iq}}{L_q}(e_q + e_{qh}) \quad (20)$$
$$+ \frac{K_{pq}}{L_q}i_q^* + \frac{1}{L_q}v_{qh}$$

$$\dot{e}_d + \dot{e}_{dh} = -(i_d + i_{dh}) + i_d^* \qquad (21)$$

$$\dot{e}_q + \dot{e}_{qh} = -(i_q + i_{qh}) + i_q^* \qquad (22)$$

From the above equations, if we consider only harmonic components, the state equations will become as (23), these state equations are the PMSM with ACR model and we can use this model in estimation for harmonic currents.

$$
\begin{bmatrix} \dot{i}_{dh} \\ \dot{i}_{qh} \\ \dot{e}_{dh} \\ \dot{e}_{qh} \end{bmatrix} = \begin{bmatrix} -(\dfrac{R}{L_d} + \dfrac{K_{pd}}{L_d}) & 0 & \dfrac{K_{id}}{L_d} & 0 \\ 0 & -(\dfrac{R}{L_q} + \dfrac{K_{pq}}{L_q}) & 0 & \dfrac{K_{iq}}{L_q} \\ -1 & 0 & 0 & 0 \\ 0 & -1 & 0 & 0 \end{bmatrix}
$$

$$
\times \begin{bmatrix} i_{dh} \\ i_{qh} \\ e_{dh} \\ e_{qh} \end{bmatrix} + \begin{bmatrix} \dfrac{1}{L_d} & 0 \\ 0 & \dfrac{1}{L_q} \\ 1 & 0 \\ 0 & 1 \end{bmatrix} \begin{bmatrix} v_{dh} \\ v_{qh} \end{bmatrix} \quad (23)
$$

IV. SIMULATION AND EXPERIMENTAL RESULTS

In this section we will show the effectiveness of our proposed method by computer simulation and experimental results. There are two topics that we will evaluate, first the accuracy of estimated harmonic currents by proposed method, and second the improve results after we using estimated harmonic currents in harmonic currents compensation process. The details of each topic will be discribed detailly in the following subsection. Parameters of PMSM that we used in computer simulation and experiment are shown together in TABLE I.

TABLE I
PMSM PARAMETERS

pole pairs P_n	2
EMF constant k_e	0.104V/(rad/s)
Resistance R	0.45Ω
d-axis inductance L_d	4.15mH
q-axis inductance L_q	16.74mH

A. Harmonic Currents Estimation Results

The setting parameters of current controllers and inverter are shown together in TABLE II. The operating condition is that when PMSM running at 1,600 rpm, references currents are changed in step from 0 to 3A 120deg., then inverter will change from linear range to overmodulation range automatically. And we also assume that an inertia of the system is very large, then speed remains constant after the load change.

TABLE II
CURRENT CONTROL SYSTEM PARAMETERS (A)

PI-Gain of current controllers	2,000rad/s
Current control period	100μsec
Inverter carrier frequency	10kHz
DC link voltage V_{DC}	50V

First, we used the proposed harmonic voltages estimation method discribed in sectionII, however for harmonic currents estimation method, we used (7) in sectionIII that considered only the harmonic model of PMSM. The simulation results of d-axis current by this method are shown in Fig.5. From these results, we can find that when using only harmonic

model of PMSM in harmonic currents estimation, the dynamic mismatch between real current and estimated current occurs (Fig.5(a) and (b)). And if we use this estimated harmonic current in compensation process, current control system may be unstable easily (Fig.5(c)). The results of q-axis current are not different from the results of d-axis current, then we show here only the d-axis current results.

Next, we use our proposed harmonic currents estimation method in (23) and its simulation results are shown in Fig.6. From these results, we can find that, with the proposed method, the dynamic of estimated harmonic current matches with the dynamic of real current, and after compensation, only fundamental component can be achieved. If we focus on the results in steady state condition which are shown in Fig.7, we also find that with the proposed method, harmonic current can be correctly estimated both in amplitude and phase angle, and after compensation from real current, harmonic components due to an inverter in overmodulation range will be completely canceled out and only high carrier-frequency component remains.

Experimental estimation results by using our proposed method in transient and steady state condition are also shown in Fig.8 and Fig.9. The experimental results also match with the simulation results, however because of the change in DC link voltage in transient condition and the effect from harmonic currents due to non-sinusoidal back emf. of PMSM itself, some difference occurs.

B. Harmonic Currents Compensation Results

Next, we will use the estimated harmonic currents in the compensation process and show its simulation and experimental results. The setting parameters of current controllers and inverter in this time are shown together in TABLE III. The larger in PI-controller gains and the longer in control period compared with previous section is for making the condition that system will be unstable easily in overmodulation range. The operating condition is that when PMSM running at 1,600 rpm constant speed, references currents are changed in step from 0 to 3.5A 120deg. and inverter will change from linear to overmodulation range (almost six-step condition).

TABLE III
CURRENT CONTROL SYSTEM PARAMETERS (B)

PI-Gain of current controllers	3,500rad/s (Simulation)
PI-Gain of current controllers	2,500rad/s (Experiment)
Current control period	400μsec
Inverter carrier frequency	10kHz
DC link voltage V_{DC}	50V

First, we will consider the case that there is not harmonic currents compensation. The results those will be shown here are amplitude of inverter reference voltage (before amplitude compensation) V^*, voltage limiter V_{lim}, reference torque T^* and real torque response T. Each of value can be calculated as shown in (24) ~ (27) respectively.

$$V^* = \sqrt{v_d^{*2} + v_q^{*2}} \qquad (24)$$

$$V_{lim} = \sqrt{\frac{3}{2}} \frac{4}{\pi} \frac{V_{DC}}{2} \qquad (25)$$

$$T^* = P_n(k_e i_q^* + (L_d - L_q)i_d^* i_q^*) \qquad (26)$$

$$T = P_n(k_e i_q + (L_d - L_q)i_d i_q) \qquad (27)$$

The simulation results when there is not harmonic currents compensation are shown in Fig.10. From these results we can find that without harmonic currents compensation, the effect from harmonic currents will cause inverter saturation, hence there is not enough voltage for controlling torque (currents), and then unstable in torque response occurs.

The simulation results in case that there is harmonic currents compensation are shown in Fig.11. From these results, we can find that with harmonic currents compensation, inverter output voltage is saturated only in transient condition and become less that limitted voltage in steady state condition, the torque of PMSM can be stable controlled even in the overmodulation range.

The experimental results in steady state and transient condition when there is not harmonic currents compensation are shown in Fig.12 and Fig.13 respectively. From these results, we can find that unstable problem occurs in both steady state and transient condition.

The experimental results in steady state and transient condition when there is harmonic currents compensation are shown in Fig.14 and Fig.15 respectively. These results match with simulation results as shown before, and we can find that with our proposed current control system, torque (currents) can be stable controlled even in the overmodulation range.

REFERENCES

[1] H.Nakai, H.Ohtani, E.Satoh and Y.Inaguma "Development and Testing of the Torque Control for the Permanent-Magnet Synchronous Motor," *IEEE Trans. Industrial Electronics*, vol. 52,no.3,Jun. 2005 pp. 800–806.

[2] A.H.Hava,R.J.Kerkman and T.A.Lipo "Carrier-Based PWM-VSI Overmodulation Strategies:Analysis, Comparison, and Design," *IEEE Trans. Power Electronics*, vol. 13,no.4,Jul. 1998 pp. 674–689.

[3] A.M.Khambadkone,J.Holtz "Compensated Synchronous PI Current Controller in Overmodulation Range and Six-Step Operation of Space-Vector-Modulation-Based Vector-Controlled Drives" *IEEE Trans. Industrial Electronics*, vol. 49,no.3,June. 2002 pp. 574–580.

[4] H.W.Kim,N.V Nho and M.J.Youn "Current Control of PM Synchronous Motor in Overmodulation Range," *IECON 2004*, pp. 896-901, 2004.

[5] J.Holtz ,W.Lotzkat and A.M.Khambadkone "On Continuous Control of PWM Inverters in Overmodulation Range Including the Six-Step" *IEEE Trans. Power Electronics*, vol.8, July. 1993 pp. 546–553.

[6] N.V Nho and M.J.Youn "Two-Mode Overmodulation in Two-Level VSI," *Proc. Conf. Rec. IEEE PEDS Conference 2003*, pp. 1241-1244, 2003.

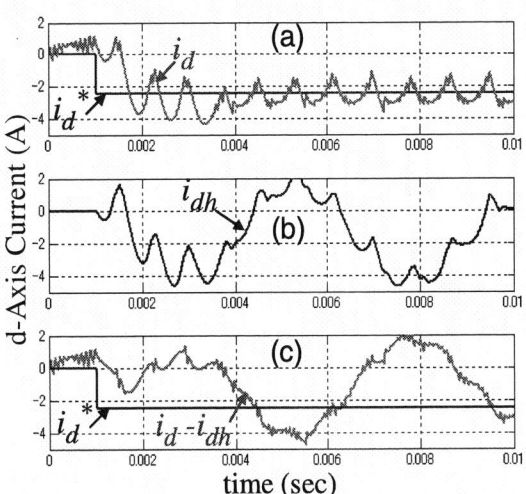

Fig. 5. Simulation Results (transient condition): Harmonic current estimation when using only model of PMSM (a)reference and real currents (b)estimated harmonic current (c)reference and compensated currents

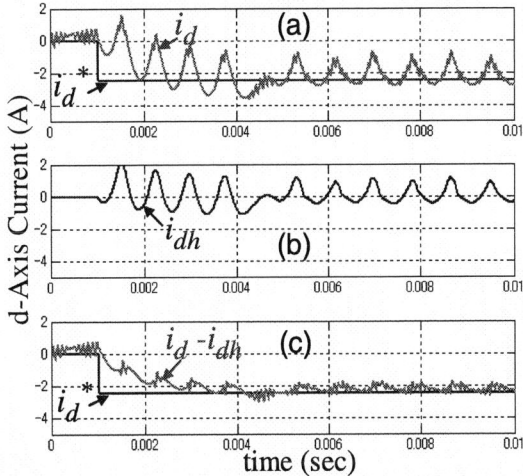

Fig. 6. Simulation Results (transient condition): Harmonic current estimation when using proposed PMSM+ACR model (a)reference and real currents (b)estimated harmonic current (c)reference and compensated currents

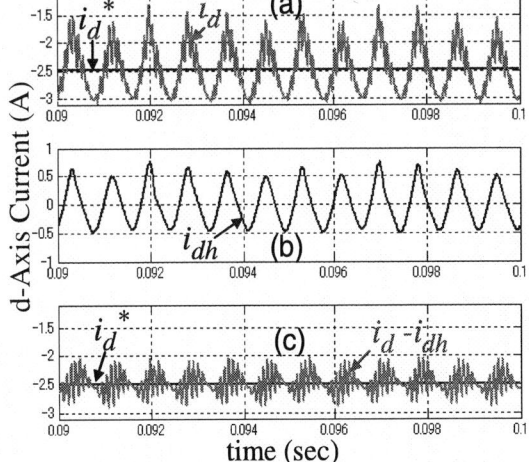

Fig. 7. Simulation Results (steady state condition): Harmonic current estimation when using proposed PMSM+ACR model (a)reference and real currents (b)estimated harmonic current (c)reference and compensated currents

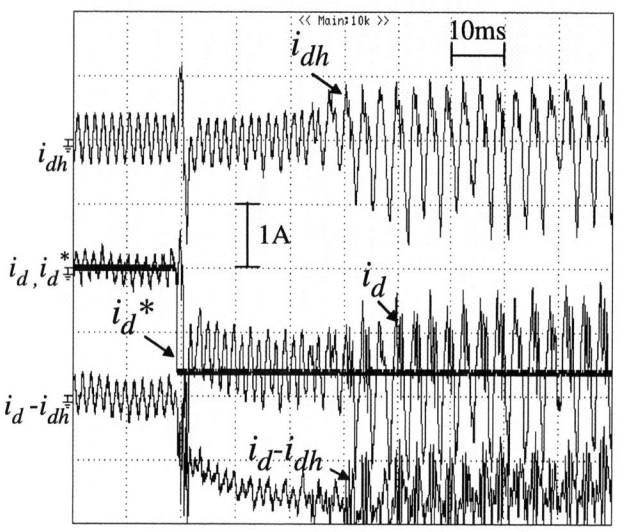

Fig. 8. Experimental Results (transient condition): Harmonic current estimation when using proposed PMSM+ACR model

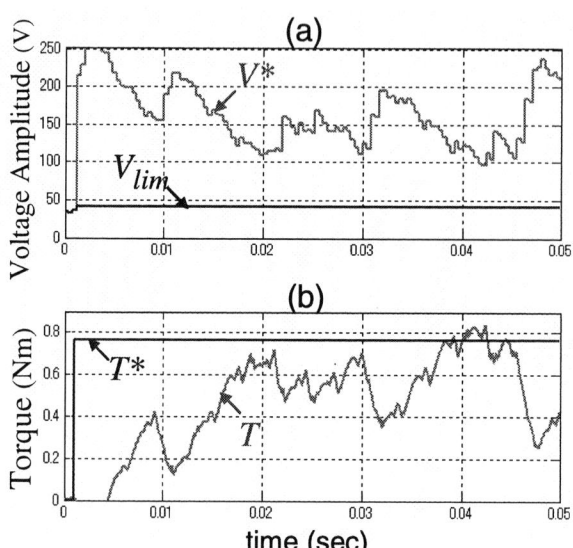

Fig. 10. Simulation Results : Without harmonic currents compensation (a)referecne voltage amplitude and voltage limitter (b)reference and real torque response

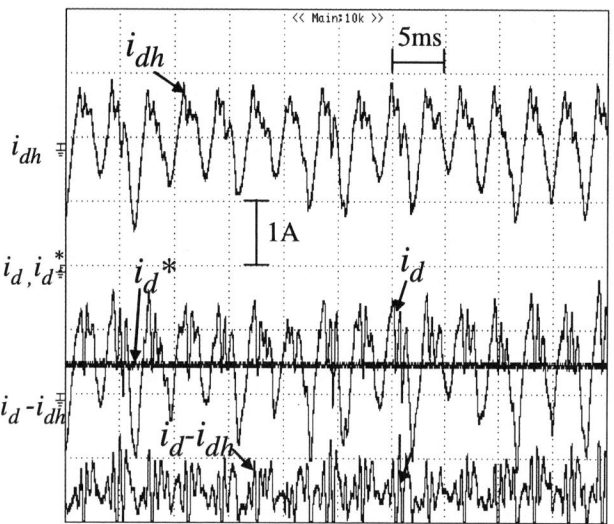

Fig. 9. Experimental Results (steady state condition): Harmonic current estimation when using proposed PMSM+ACR model

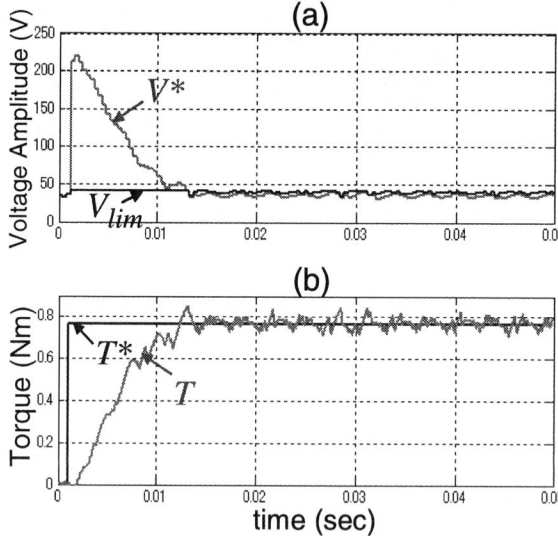

Fig. 11. Simulation Results : With harmonic currents compensation (a)referecne voltage amplitude and voltage limitter (b)reference and real torque response

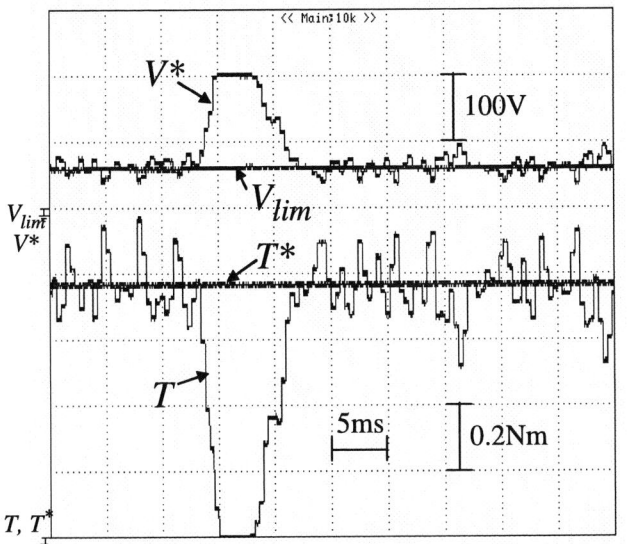

Fig. 12. Experimental Results (steady state condition): Without harmonic currents compensation

Fig. 14. Experimental Results (steady state condition): With harmonic currents compensation

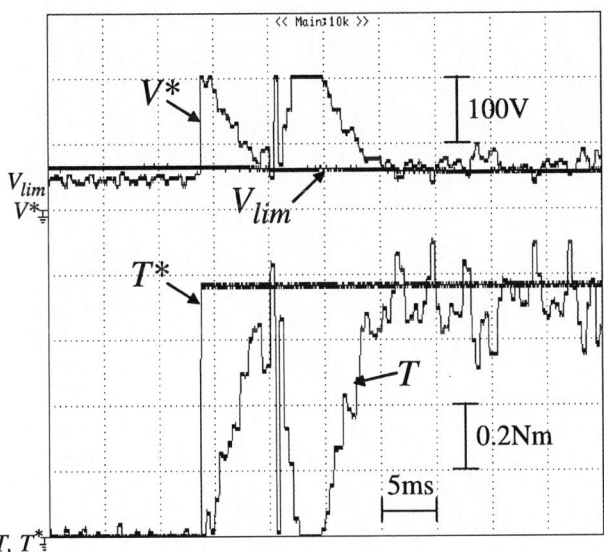
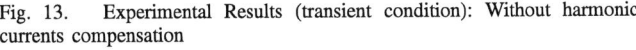

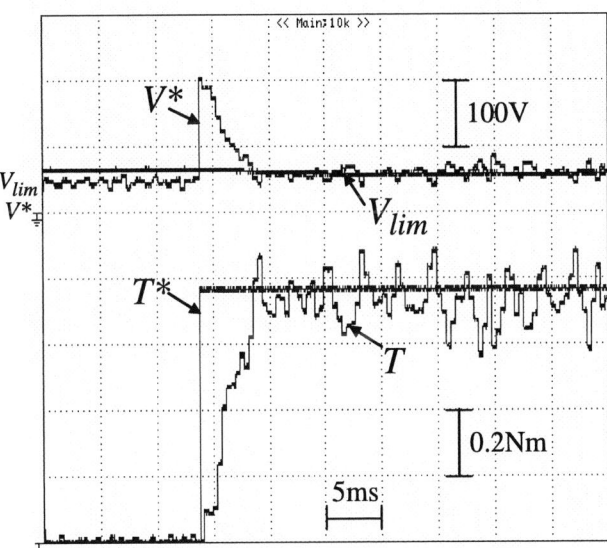

Fig. 13. Experimental Results (transient condition): Without harmonic currents compensation

Fig. 15. Experimental Results (transient condition): With harmonic currents compensation

Modelling of the Feeding Network of a Linear Synchronous Machine and Estimation of Model Parameters

J. Rost*, H. Gueldner*, R. Hellinger**, A. Weller**

* Technische Universität Dresden, Lehrstuhl Leistungselektronik, Dresden, Germany

** Siemens AG, Erlangen, Germany

Abstract—A method for modelling a linear synchronous drive-based transportation system is presented. The linear drive consists of power converter units, a cable network and long stator windings. These transmission elements can be modelled using two-port theory with frequency-dependent parameters. For characterisation of the frequency dependence of these parameters, polynomial functions, rational functions or equivalent circuits can be used. These characterisation techniques are compared with regard to their relative error. Furthermore, a method for transmission line parameter estimation during commissioning of the system is proposed.

Index Terms—modelling, multiport circuits, estimation

I. INTRODUCTION

In the considered transportation system, a linear long stator synchronous machine is fed via a cable network. This cable system has a maximum length of 50 km. Due to the distributed nature of the system and the significant influence of the cable, the transfer characteristic can no longer be neglected, and the behaviour of the cable must be modelled. The system operates at frequencies from 0 to 300 Hz. Because of the lax dynamic requirements, only the steady-state behaviour of the cable system must be modelled. In literature, several models in the time or frequency domain are proposed, e.g., in [1] and [2], time domain models of the power supply system are detailed. These models are used to examine and simulate the complete network under operational conditions. Another application area can be seen in the drive systems of deep-sea oil pumps as described in [3] and [4]. In these papers, π-equivalent circuits with frequency-dependent two-poles are used to model the transfer characteristic of the feeder cable at high frequencies to avoid overvoltages. The control algorithms will run on a digital platform. Thus, the transfer behaviour of the cable can be modelled by an n-port description with frequency-dependent parameters. Alternatively, a π-equivalent circuit similar to the model in [3] and [4] could be used.

Another necessary task is to determine model parameters. Due to the influence of, for example, ambient temperature, installation quality and cable parameter accuracy, the theoretical calculated start values never meet the accuracy requirements of the system. Hence, the parameters must be estimated during commissioning of the system and during its operation.

In this paper, several steady-state models of the cable based on n-port theory are examined and compared. The second part describes a procedure to estimate the parameters during commissioning of the system.

II. SYSTEM DESCRIPTION

The considered system consists of a three-phase voltage source inverter (VSI), a cable network and a linear synchronous machine. Because of the physical length of the track (up to 50 km), the long stator winding is divided into several parts with a maximum length of 1 km, each of which is connected to the cable network by contactors, as shown in Fig 1. During operation, only one section of the long stator winding is connected to the cable system. Thus, each section can be modelled separately. The three-phase drive system is considered to be balanced, i.e. it can be assumed that the source, the cable system, and the long stator winding can be described as single-phase systems. This property yields the black-box model shown in Fig 2.

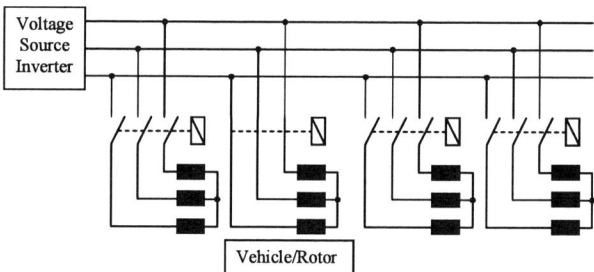

Fig. 1. Schematic illustration of the linear synchronous drive with one VSI

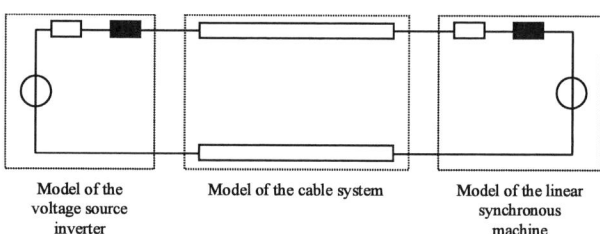

Fig. 2. Black-box model of the linear drive with one VSI during operation

Alternatively, it is possible to feed the linear synchronous machine by two voltage source inverters, as shown in Fig. 3. Thereby, both inverters must be synchronised to avoid current flow from one inverter to the other. This

978-1-4244-0644-9/07/$25.00 ©2007 IEEE

part of the control system is outside the scope of this paper.

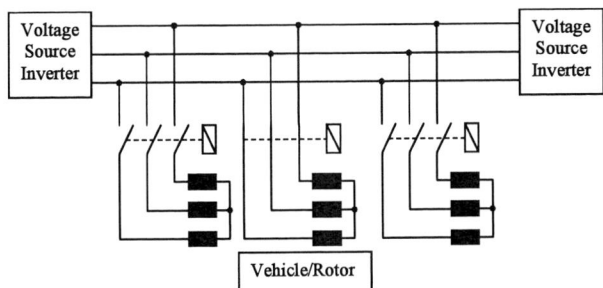

Fig. 3. Schematic illustration of the linear synchronous drive with two VSIs

Analogue to the system with one supply, the above-mentioned drive can be modelled by a reduced system consisting of models of the voltage source inverter, the cable system and the linear synchronous machine, as shown in Fig. 4.

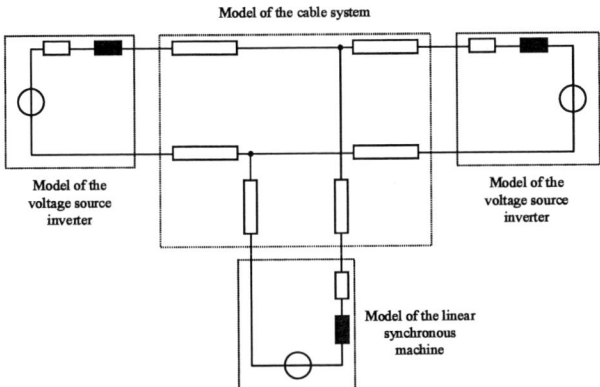

Fig. 4. Black-box model of the linear drive with two VSIs during operation

III. MODELLING OF THE DRIVE SYSTEM WITH A SINGLE SUPPLY IN STEADY STATE

In the initial considerations, it is assumed that the cable system consists only of one cable; hence, the system can be described by transmission line theory. As described in [5] and [6], the following system of differential equations can be used to derive the transmission line model in steady state. Thereby, values R', L', G', and C' represent the parameters of the cable related to the length of the line.

$$\frac{\partial v}{\partial z} = -\left(R' + L'\frac{\partial}{\partial t}\right)i$$
$$\frac{\partial i}{\partial z} = -\left(G' + C'\frac{\partial}{\partial t}\right)v \tag{1}$$

In case the cable system is supplied by a sinusoidal voltage source with constant frequency and amplitude, the steady-state behaviour of a cable with length l can be described by the transmission line equations in (2). Thereby, $\underline{V}_i$ and $\underline{I}_i$ represent the voltage and the current at

the input and $\underline{V}_o$ and $\underline{I}_o$ at the output of the cable in the frequency domain.

$$\underline{V}_o = \cosh(\gamma l)\cdot\underline{V}_i - Z_0\sinh(\gamma l)\cdot\underline{I}_i$$
$$\underline{I}_o = -\frac{1}{Z_0}\sinh(\gamma l)\cdot\underline{V}_i + \cosh(\gamma l)\cdot\underline{I}_i \tag{2}$$

with:

$$\gamma = \sqrt{(R' + j\omega L')\cdot(G' + j\omega C')},$$
$$Z_0 = \sqrt{\frac{(R' + j\omega L')}{(G' + j\omega C')}},$$

For a given constant frequency f_o, the transmission line can also be described by two-port theory [7]. If admittance parameters $\underline{Y}_{11}$, $\underline{Y}_{12}$ and $\underline{Y}_{22}$ are used, the cable behaviour can be characterised by (3), with parameters as per Fig. 5.

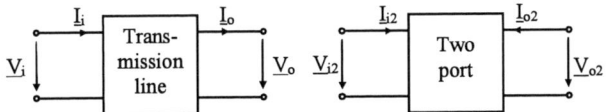

Fig. 5. Definition of the input and output quantities in transmission line and two-port theory

$$\begin{pmatrix} \underline{I}_{i2} \\ \underline{I}_{o2} \end{pmatrix} = \begin{pmatrix} \underline{Y}_{11} & \underline{Y}_{12} \\ \underline{Y}_{12} & \underline{Y}_{22} \end{pmatrix}\begin{pmatrix} \underline{V}_{i2} \\ \underline{V}_{o2} \end{pmatrix} \tag{3}$$

From the comparison of (2) and (3), equations for the two-port parameters can be derived. The results are given in (4) to (6).

$$\underline{Y}_{11} = \frac{\cosh(\gamma l)}{Z_0\sinh(\gamma l)} \tag{4}$$

$$\underline{Y}_{12} = -\frac{1}{Z_0\sinh(\gamma l)} \tag{5}$$

$$\underline{Y}_{22} = \frac{\cosh(\gamma l)}{Z_0\sinh(\gamma l)} \tag{6}$$

In the above-mentioned model, it is assumed that the transmission line parameters are frequency-independent and thus the frequency-dependent influences such as skin and proximity effect are neglected. In the actual system, however, these effects are significant, as shown in Fig. 6. Nonetheless, the steady-state behaviour of the cable for a given constant frequency f_0 can be described by two-port theory, as in the case of frequency-independent parameters [9]. Contrary to equations (4) to (6), the frequency dependence of the two-port parameters is no longer defined by the transmission line equation with constant parameters. To avoid measuring the frequency dependence of any cable type as shown in Fig. 6 and to consider the effects of ambient conditions, it is preferred to measure parameters $\underline{Y}_{11}$, $\underline{Y}_{12}$ and $\underline{Y}_{22}$ directly and to approximate frequency dependence by a mathematical function. The first possibility is the use of predetermined sample points and a linear interpolation to calculate interim values. Thus, the two-port parameters can be determined by (7), where $\underline{Y}_{na}$ and $\underline{Y}_{nb}$ represent two neighbouring sample points.

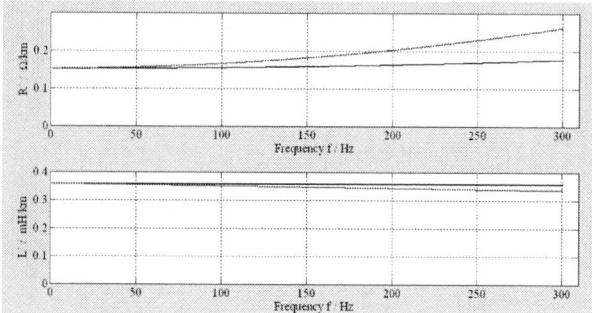

Fig. 6.　Frequency dependence of cable parameters R' and L'
　　　red line: without cross-bonding;, blue line: with cross-bonding

$$\underline{Y}_n(f) = \underline{Y}_{na} \cdot \left(1 - \frac{f - f_a}{f_b - f_a}\right) + \underline{Y}_{nb} \cdot \frac{f - f_a}{f_b - f_a} \qquad (7)$$

The accuracy of this method can only be influenced by the distance between the neighbouring sample points. Hence, this method has been examined for two different cable types in several configurations in order to find an appropriate number of sample points. The parameters of these cable types are given in Table I.

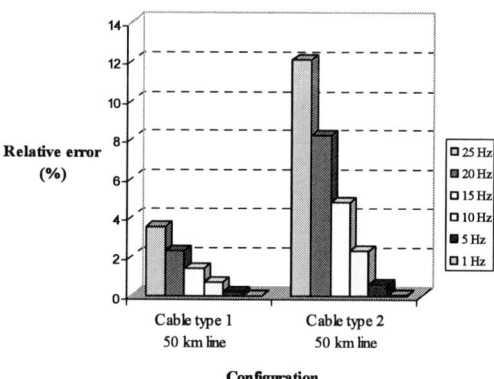

Fig. 7. Relative approximation error of the real part of $\underline{Y}_{11}$ (linear)

Fig. 7 shows the results of a simulation for the real part of admittance parameter $\underline{Y}_{11}$. It can be seen that the difference between two sample points has to be smaller than 6 Hz to achieve a relative approximation error of less than 1%. Another method to describe the frequency dependence of the transmission line parameters is quadratic interpolation. The real and imaginary part of parameter $\underline{Y}_n$ can be determined by (8) and (9), with use of the three sample points $\underline{Y}_{na}$, $\underline{Y}_{nb}$ and $\underline{Y}_{nc}$.

$$\text{Re}\{\underline{Y}_n\} = K_{r2}(f - f_a)^2 + K_{r1}(f - f_a) + K_{r0} \qquad (8)$$

$$\text{Im}\{\underline{Y}_n\} = K_{j2}(f - f_a)^2 + K_{j1}(f - f_a) + K_{j0} \qquad (9)$$

with:

$$K_{2r} = \frac{\text{Re}\{\underline{Y}_{nb}\} - \text{Re}\{\underline{Y}_{nc}\}\left(\dfrac{f_b - f_a}{f_c - f_a}\right) - \text{Re}\{\underline{Y}_{na}\}\left(1 - \dfrac{f_b - f_a}{f_c - f_a}\right)}{(f_b - f_a)^2 - (f_b - f_a)(f_c - f_a)},$$

$$K_{1r} = \frac{\text{Re}\{\underline{Y}_{nb}\} - \text{Re}\{\underline{Y}_{na}\} - K_{2r}(f_b - f_a)^2}{(f_b - f_a)},$$

$$K_{0r} = \text{Re}\{\underline{Y}_{na}\},$$

$$K_{2j} = \frac{\text{Im}\{\underline{Y}_{nb}\} - \text{Im}\{\underline{Y}_{nc}\}\left(\dfrac{f_b - f_a}{f_c - f_a}\right) - \text{Im}\{\underline{Y}_{na}\}\left(1 - \dfrac{f_b - f_a}{f_c - f_a}\right)}{(f_b - f_a)^2 - (f_b - f_a)(f_c - f_a)},$$

$$K_{1j} = \frac{\text{Im}\{\underline{Y}_{nb}\} - \text{Im}\{\underline{Y}_{na}\} - K_{2j}(f_b - f_a)^2}{(f_b - f_a)}, \text{ and}$$

$$K_{0j} = \text{Im}\{\underline{Y}_{na}\}.$$

Evaluation of the two cable types using quadratic interpolation showed only a small improvement of the relative error. For example, the approximation error of cable type 2 at a sample distance of 10 Hz could be reduced from 2.3% to 1.8%. The complete results are shown in Fig. 8.

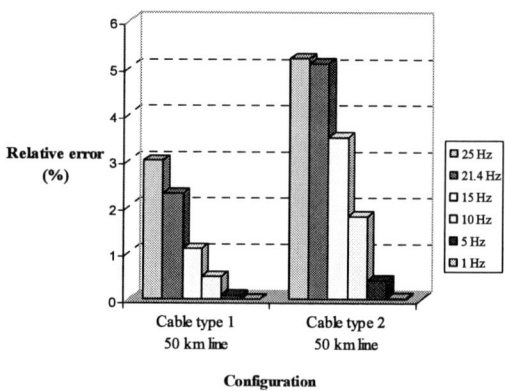

Fig. 8. Relative approximation error of the real part of $\underline{Y}_{11}$ (quadratic)

As an alternative to the interpolation methods, frequency dependence can be described by a polynomial function. The accuracy of this method can be influenced by the degree of the polynomial. The common descriptions of the real and imaginary part are given in (10) and (11).

$$\text{Re}(\underline{Y}_n) = k_{r0} + k_{r1} \cdot f + k_{r2} \cdot f^2 + \ldots + k_{rn} \cdot f^n \quad (10)$$

$$\text{Im}(\underline{Y}_n) = k_{j0} + k_{j1} \cdot f + k_{j2} \cdot f^2 + \ldots + k_{jn} \cdot f^n \quad (11)$$

There are several optimisation algorithms for finding the polynomial coefficients. Fig 9 shows the results for polynomials of the 7th, 9th and 11th degree. Thereby, the three different optimisation functions, absolute error, relative error and maximum error, were used. The coefficients were determined with the Matlab function "fminsearch()". It can be seen that a relative error of less than 1% can be attained, if at least a 9th-degree polynomial and the maximum error function are used. The disadvantage of this approach is an increase in the total error. It must be considered that the maximum relative approximation error of the imaginary part of parameters $\underline{Y}_{11}$ and $\underline{Y}_{12}$ is encountered at small frequencies due to the very low value of these parameters. Another approach is the reduction of the modelled frequency range, e.g. from 10 to 300 Hz. Thus, the relative approximation error is reduced. Fig. 10 shows the results of this method. In addition, the frequency range from 1 and 10 Hz can be modelled with an interpolation approach.

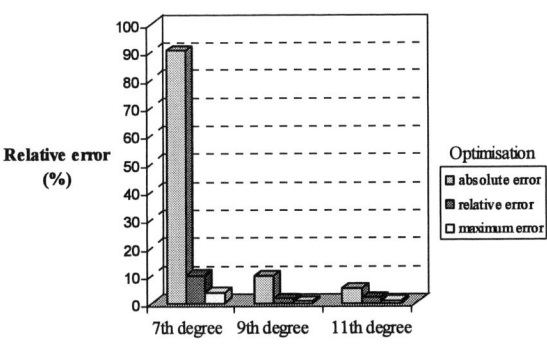

Fig. 9.   Relative approximation error of the real part of $\underline{Y}_{11}$ (50 km line, cable type 2, polynomial, 1 to 300 Hz)

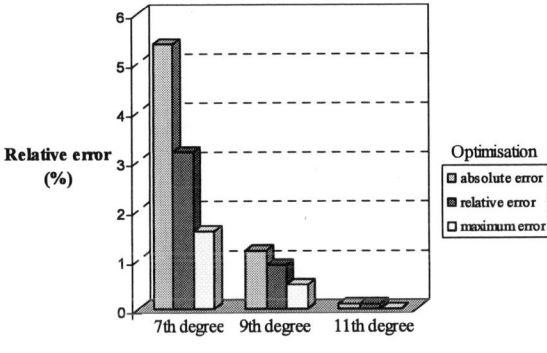

Fig. 10. Relative approximation error of the real part of $\underline{Y}_{11}$ (50 km line, cable type 2, polynomial, 10 to 300 Hz)

The final modelling approach to be considered is a rational function. This function can be derived from the transmission line equations in (6) to (8), with approximations of the frequency dependence of the cable parameters as shown in Fig. 6. In [8], the frequency dependence of R' was approximated by a quadratic function and that of L' by a linear function in the range of 0 to 300 Hz. The hyperbolic functions in (6) and (8) can be calculated by the following Taylor series.

$$\cosh(\gamma l) = 1 + \frac{1}{2!}(\gamma l)^2 + \ldots + \frac{(\gamma l)^{2n}}{(2n)!} + \ldots \qquad (12)$$

$$\sinh(\gamma l) = (\gamma l) + \frac{1}{3!}(\gamma l)^3 + \ldots + \frac{(\gamma l)^{2n+1}}{(2n+1)!} + \ldots \qquad (13)$$

Fig. 11 shows the simulation results using the above-mentioned approximations.

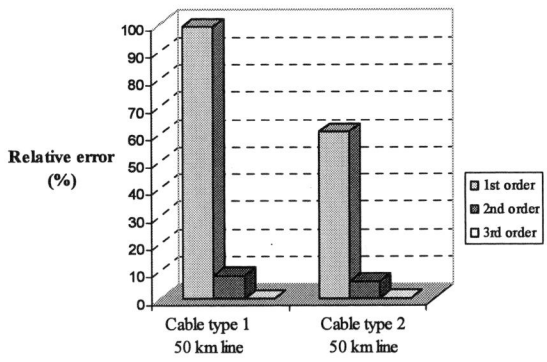

Fig. 11. Relative approximation error of the real part of $\underline{Y}_{11}$ (Taylor series)

TABLE I
CABLE PARAMETERS OF THE EXAMINED LINES

Property	Cable type 1	Cable type 2
R' / Ω/km	0.287	0.152
L' / mH/km	0.285	0.358
C' / μF/km	0.365	0.304
Cross-bonding	none	None

A comparison of the above-mentioned methods shows that the best approximation of the frequency dependence of the two-port parameters is a combination of a 9th-degree polynomial and a linear interpolation. The 9th-degree polynomial describes the frequency range from 10 to 300 Hz and the interpolation method below 10 Hz. Thereby, it was required that the relative approximation error be less than 1% and the number of sample points be reduced.

IV. MODELLING OF THE DRIVE SYSTEM WITH DOUBLE SUPPLY IN STEADY STATE

In the previous paragraph, the cable system was described using two-port theory. As shown in Fig. 4, the system with a double supply can be represented by a three-port parameter and hence, the steady-state behaviour of the cable can be modelled by n-port theory. In (14), the system equations are given for an analogue definition of the input and output quantities as in Fig. 5.

$$\begin{pmatrix} \underline{I}_{i1} \\ \underline{I}_{i2} \\ \underline{I}_o \end{pmatrix} = \begin{pmatrix} \underline{Y}_{11} & \underline{Y}_{12} & \underline{Y}_{13} \\ \underline{Y}_{12} & \underline{Y}_{22} & \underline{Y}_{23} \\ \underline{Y}_{13} & \underline{Y}_{23} & \underline{Y}_{33} \end{pmatrix} \begin{pmatrix} \underline{V}_{i1} \\ \underline{V}_{i2} \\ \underline{V}_o \end{pmatrix} \qquad (14)$$

The frequency dependence of the three-port parameters can be described with the methods discussed above.

V. IDENTIFICATION OF STEADY-STATE PARAMETERS DURING COMMISSIONING

Estimation of the cable system parameters in (14) is independent of the operational parameters (frequency, drive configuration, etc.); this means that appropriate switch states, as shown in Fig. 12, can be used to estimate the three-port parameters for a given number of frequencies. In a second step, one of the described frequency dependence models can be derived using the least-squares method. Only the estimation procedure of the three-port parameters is described in this paper.

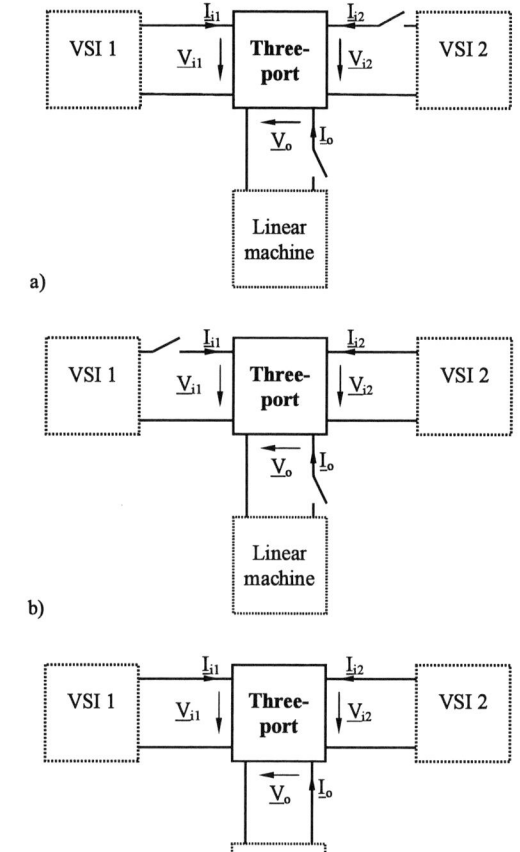

a)

b)

c)

Fig. 11.   a) Supply of the open cable network from VSI 1
        b) Supply of the open cable network from VSI 2
        c) Supply of the long stator winding without vehicle from VSI1 and 2

For the complete identification of model parameters, three different states are necessary. When the voltage and the current at the terminals of the linear synchronous machine can be measured, the following equations will describe the system behaviour of the above-mentioned switch states for a fixed frequency f_0. Instead of the admittance parameters in (15) to (17), the impedances of the three-port parameter are used. Thereby, for example, $\underline{V}_{i13k}$ represents the voltage at input 1, switch state 3 and the k-th measurement of this switch state.

State a)

$$\begin{pmatrix} \underline{V}_{i11k} \\ \underline{V}_{i21k} \\ \underline{V}_{o1k} \end{pmatrix} = \begin{pmatrix} \hat{\underline{Z}}_{11k} & \hat{\underline{Z}}_{12k} & \hat{\underline{Z}}_{13k} \\ \hat{\underline{Z}}_{12k} & \hat{\underline{Z}}_{22k} & \hat{\underline{Z}}_{23k} \\ \hat{\underline{Z}}_{13k} & \hat{\underline{Z}}_{23k} & \hat{\underline{Z}}_{33k} \end{pmatrix} \cdot \begin{pmatrix} \underline{I}_{i11k} \\ 0 \\ 0 \end{pmatrix} = \begin{pmatrix} \hat{\underline{Z}}_{11k}\underline{I}_{i11k} \\ \hat{\underline{Z}}_{12k}\underline{I}_{i11k} \\ \hat{\underline{Z}}_{13k}\underline{I}_{i11k} \end{pmatrix}$$

(15)

State b)

$$\begin{pmatrix} \underline{V}_{i12k} \\ \underline{V}_{i22k} \\ \underline{V}_{o2k} \end{pmatrix} = \begin{pmatrix} \hat{\underline{Z}}_{11k} & \hat{\underline{Z}}_{12k} & \hat{\underline{Z}}_{13k} \\ \hat{\underline{Z}}_{12k} & \hat{\underline{Z}}_{22k} & \hat{\underline{Z}}_{23k} \\ \hat{\underline{Z}}_{13k} & \hat{\underline{Z}}_{23k} & \hat{\underline{Z}}_{33k} \end{pmatrix} \cdot \begin{pmatrix} 0 \\ \underline{I}_{i22k} \\ 0 \end{pmatrix} = \begin{pmatrix} \hat{\underline{Z}}_{12k}\underline{I}_{i22k} \\ \hat{\underline{Z}}_{22k}\underline{I}_{i22k} \\ \hat{\underline{Z}}_{23k}\underline{I}_{i22k} \end{pmatrix}$$

(16)

State c)

$$\begin{pmatrix} \underline{V}_{i13k} \\ \underline{V}_{i23k} \\ \underline{V}_{o3k} \end{pmatrix} = \begin{pmatrix} \hat{\underline{Z}}_{11k} & \hat{\underline{Z}}_{12k} & \hat{\underline{Z}}_{13k} \\ \hat{\underline{Z}}_{12k} & \hat{\underline{Z}}_{22k} & \hat{\underline{Z}}_{23k} \\ \hat{\underline{Z}}_{13k} & \hat{\underline{Z}}_{23k} & \hat{\underline{Z}}_{33k} \end{pmatrix} \cdot \begin{pmatrix} \underline{I}_{i13k} \\ \underline{I}_{i23k} \\ \underline{I}_{o2k} \end{pmatrix}$$

(17)

The identification has an explicit solution if (18) is true. If the measurements can be assumed to be error-free, the parameters can be estimated by (19) to (24).

$$\det \mathbf{I_k} = \begin{vmatrix} \underline{I}_{i11k} & 0 & 0 & 0 & 0 & 0 \\ 0 & \underline{I}_{i11k} & 0 & 0 & 0 & 0 \\ 0 & 0 & \underline{I}_{i11k} & 0 & 0 & 0 \\ 0 & 0 & 0 & \underline{I}_{i22k} & 0 & 0 \\ 0 & 0 & 0 & 0 & \underline{I}_{i22k} & 0 \\ 0 & 0 & \underline{I}_{i13k} & 0 & \underline{I}_{i23k} & \underline{I}_{o3k} \end{vmatrix} = \underline{I}_{i11k}^3 \underline{I}_{i22k}^2 \underline{I}_{o3k} \neq 0$$

(18)

Solution for error-free parameters:

$$\hat{\underline{Z}}_{11k} = \frac{\underline{V}_{i11k}}{\underline{I}_{i11k}}$$

(19)

$$\hat{\underline{Z}}_{12k} = \frac{\underline{V}_{i21k}}{\underline{I}_{i11k}}$$

(20)

$$\hat{\underline{Z}}_{13k} = \frac{\underline{V}_{o1k}}{\underline{I}_{i11k}}$$

(21)

$$\hat{\underline{Z}}_{22k} = \frac{\underline{V}_{i22k}}{\underline{I}_{i22k}}$$

(22)

$$\hat{\underline{Z}}_{23k} = \frac{\underline{V}_{o2k}}{\underline{I}_{i22k}}$$

(23)

$$\hat{\underline{Z}}_{33k} = \frac{\underline{V}_{o3k} - \hat{\underline{Z}}_{13k}\underline{I}_{i13k} - \hat{\underline{Z}}_{23k}\underline{I}_{i23k}}{\underline{I}_{o3k}}$$

(24)

In reality, the measured data will contain errors and thus the above-mentioned equations will provide an estimation which is subject to error. To reduce measurement error, several measurements are carried out and the obtained parameters are averaged as described in (25) to (30).

$$\hat{\underline{Z}}_{11} = \frac{1}{n} \cdot \sum_{k=1}^{n} \hat{\underline{Z}}_{11k} \qquad (25)$$

$$\hat{\underline{Z}}_{12} = \frac{1}{n} \cdot \sum_{k=1}^{n} \hat{\underline{Z}}_{12k} \qquad (26)$$

$$\hat{\underline{Z}}_{13} = \frac{1}{n} \cdot \sum_{k=1}^{n} \hat{\underline{Z}}_{13k} \qquad (27)$$

$$\hat{\underline{Z}}_{22} = \frac{1}{n} \cdot \sum_{k=1}^{n} \hat{\underline{Z}}_{22k} \qquad (28)$$

$$\hat{\underline{Z}}_{23} = \frac{1}{n} \cdot \sum_{k=1}^{n} \hat{\underline{Z}}_{23k} \qquad (29)$$

$$\hat{\underline{Z}}_{33} = \frac{1}{n} \cdot \sum_{k=1}^{n} \hat{\underline{Z}}_{33k} \qquad (30)$$

Alternatively, the three-port parameters can be estimated by the least-squares method. Instead of the open network configurations mentioned in Fig. 11, variants with shorted terminals of the second voltage source inverter as shown in Fig. 12 can be used. These configurations reduce the measurement error due to the higher current amplitudes.

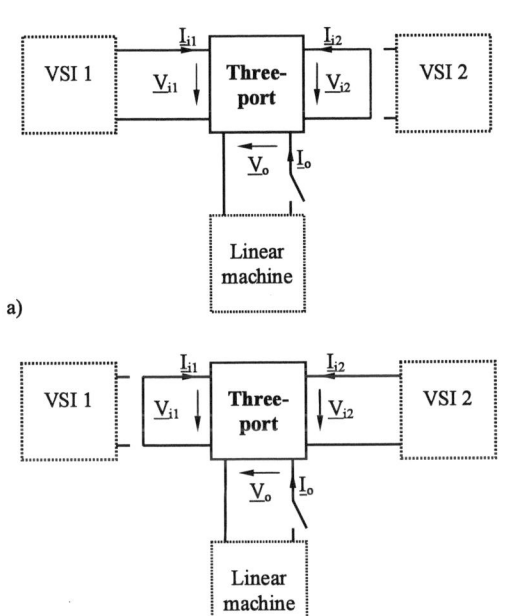

Fig. 12. a) Supply of the partly shorted cable network from VSI 1
b) Supply of the partly shorted cable network from VSI 2

VI. CONCLUSION

In the presented paper, a modelling procedure for the feeding network of a linear long stator synchronous machine has been developed. The steady-state behaviour of the cable network can be described by n-port theory in combination with a frequency-dependent model of the n-port parameters. A polynomial function has been found to be optimal in the range from 10 to 300 Hz and linear interpolation below 10 Hz. Furthermore, a strategy to determine the three-port parameters of the cable system during its commissioning has been explained.

REFERENCES

[1] H.W. Dommel, "Digital Computer Solution of Electromagnetic Transients in Single- and Multiphase Networks", IEEE Trans. Power Apparatus and Systems, Vol. PAS-88, pp. 388 – 399, April 1969

[2] J.R. Marti, "Accurate Modelling of Frequency-dependent Transmission Lines in Electromagnetic Transient Simulation", IEEE Trans. Power Apparatus and Systems, vol. PAS-101, No. 1, pp. 147 – 157, Jan. 1982

[3] M. Grotzbach, J. Schorner, "Harmonics Study of a CSI-Fed Subsea Cable Transmission", Harmonics and Quality of Power, 2000 Proceedings, Ninth International Conference on Volume 3, 1 – 4 Oct. 2000, pp. 948 – 954, Vol. 3

[4] R.O. Raad, T. Henriksen, H.B. Raphael, A. Hadler-Jacobsen, "Converter-fed Sub Sea Motor Drives", Industry Applications, IEEE Transactions on Volume 32, Issue 5, Sept – Oct. 1996, pp. 1069 – 1079

[5] H.-G. Unger, "Elektromagnetische Wellen auf Leitungen", Heidelberg: Hüthig, 1996

[6] K. Küpfmüller, G. Kohn, "Theoretische Elektrotechnik und Elektronik – Eine Einführung", Berlin: Springer Verlag, 1993

[7] P. Vielhauer, "Lineare Netzwerke – Operatorenrechnung, Leitungstheorie, n-Tor-Theorie, Netzwerksynthese", Berlin: Verlag Technik: 1982

[8] J. Leonhardt, "Entwurf eines Streckenkabelsystems für einen linearen Synchronantrieb", unpublished

[9] J.R. Marti, L. Marti, H.W. Dommel, "Transmission Line Models for Steady-State and Transients Analysis", Athens Power Tech, 1993. APT 93. Proceedings. Joint International Power Conference, Volume 2, September 5 – 8, 1993, pp. 744 – 750

Analysis of Losses in Inverter Fed Large Scale Synchronous Machines using 2D FEM Software

Samer Shisha* and Chandur Sadarangani**
School of Electrical Engineering, Electrical Machines and Power Electronics
KTH (Royal Institute of Technology), Teknikringen 33, SE-100 44 Stockholm, Sweden
*samer.shisha@ee.kth.se, **chandur.sadarangani@ee.kth.se

Abstract--This paper presents a study of eddy current rotor losses in solid pole synchronous machines, fed by a voltage source inverter. The main emphasis is on DTC (Direct Torque Control) application. The DTC operation loss analysis is carried out using FEM (Finite Element Method)-software.

The article considers a novel modelling technique that can be used when simulating inverter-fed machines in 2D-FEM application. In addition this article includes a brief introduction into how DTC operates.

Index Terms--Direct Torque Control, FEM Software Utilisation, Losses, Synchronous Machines.

I. INTRODUCTION

The machine that is analysed is a field wound solid pole synchronous machine in the kW range.

Since the stator lamination losses can be analysed using different analytical methods and relatively simple simulation models, the main focus is placed on the solid rotor.

II. CONSIDERATION

DTC [1] is a control scheme that is stochastic in nature, where the switching is non-periodic, yielding a wide spectrum of harmonics, which in turn induce currents in the solid rotor, causing losses and heating in the latter. This heating effect is similar to that found in direct on line start-up of solid pole synchronous machines [2]. The heating effect is one of the reasons why the focus is placed on the rotor.

However, unlike direct online start, the currents induced in the rotor will be of different frequencies and the frequency spectrum will change with different switching patterns. This unpredictability of the DTC spectrum makes it difficult to model.

Apparently there is a need to emulate the DTC system itself and as no method was found to be representative enough of the DTC system, it was found necessary to develop such a model.

III. FEM MODELS

The stochastic nature of DTC makes it difficult to use a constant switching frequency in the inverter circuit coupled to the FEM model. Therefore a novel method was developed in which DTC signals are obtained from an independent DTC simulation.

Values are derived from the DTC simulation and then fed into the FEM programme. This way, a wide frequency spectrum of the DTC will be covered, thus the losses can be obtained for the specific operating condition.

IV. DTC CONTROL SYSTEM

DTC is based on control of the magnitudes of the flux linkage and the torque directly. This is done by choosing an optimal voltage space vector (out of six available voltage vectors) that will maintain the controlled parameters within the required hysteresis band.

The torque is maintained constant with an imposed ripple as shown in Fig. 1. The airgap flux linkage can be controlled in two different manners. It can either be held constant where it will follow a circular path around the airgap, or it can be allowed to utilise the full voltage magnitude in the machine. The latter yields a hexagonal airgap mmf path.

Fig. 2 shows the possible mmf paths, along with the six possible voltage vectors, where Ψ_S is the flux linkage vector and U1-U6 are the switching vectors. The sectors are also shown in red crested arrows indicating the sector in the vicinity of each switching vector [3].

The optimal voltage vector is chosen from a table which is based on two widely accepted notions:

1- The instantaneous torque is proportional to the flux magnitudes, of the rotor and stator, and it is also proportional to the sine of the angle between them [3].

2- The flux linkage of the stator is proportional to the integral of the applied voltage reduced by the armature resistance voltage drop. The latter can in most cases be neglected for simplicity.

978-1-4244-0644-9/07/$25.00 ©2007 IEEE

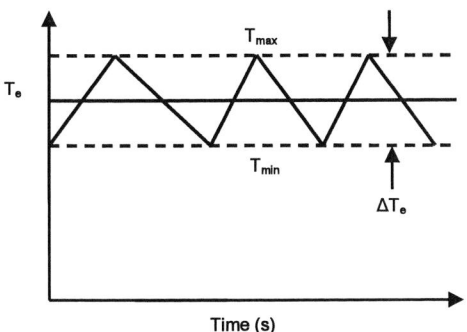

Fig. 1. Illustrated torque ripple in a DTC controlled machine.

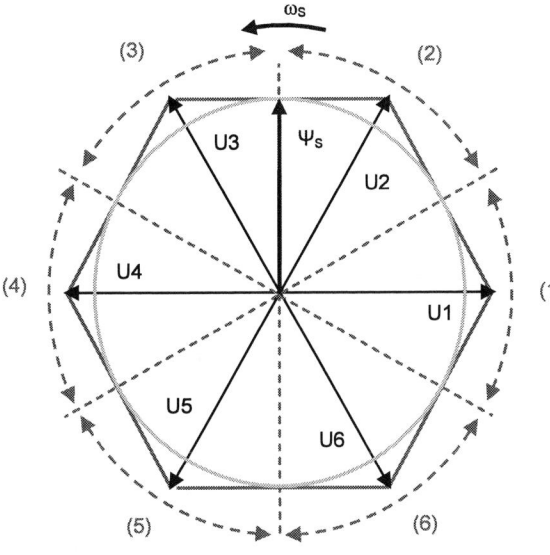

Fig. 2. MMF paths in the airgap.

Fig. 3 shows a simplified block diagram of the DTC system used to obtain the results presented here. The block named optimal switching vector is the main control block where the optimal voltage vector is chosen.

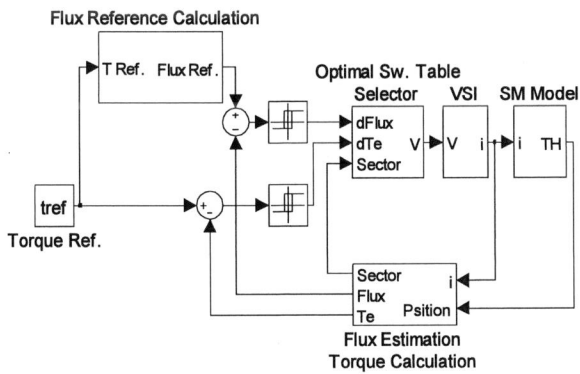

Fig. 3. Simplified DTC Block Diagram.

The choice of the optimal voltage vector is made using a table that, depending on the flux linkage error and the torque error, will find the voltage vector that would yield the necessary change in the torque and/or flux magnitude(s), which would correct these latter back to

their reference values. The table is given in TABLE 1 and is obtained from [3].

TABLE 1
OPTIMAL SWITCHING TABLE

Error Signal		Sector					
DΨ	DT	1	2	3	4	5	6
1	1	U2	U3	U4	U5	U6	U1
	-1	U6	U1	U2	U3	U4	U5
-1	1	U3	U4	U5	U6	U1	U2
	-1	U5	U6	U1	U2	U3	U4

V. RESULTS OF ANALYTICAL DTC MODEL

The armature currents obtained from the analytical DTC simulation for the three phases are given in Fig. 4. The fundamental of the current is found to be 0.69 p.u. of the rated current. It is possible to note the superimposed harmonics on the fundamental of the current waveforms of the three phase currents.

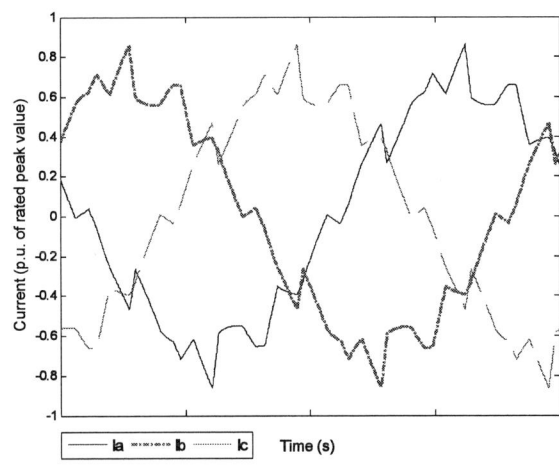

Fig. 4. Armature currents in the three phases of the machine.

The torque is shown in Fig. 5, and the airgap flux linkage magnitude vs. the rotor angle is given in Fig. 6.

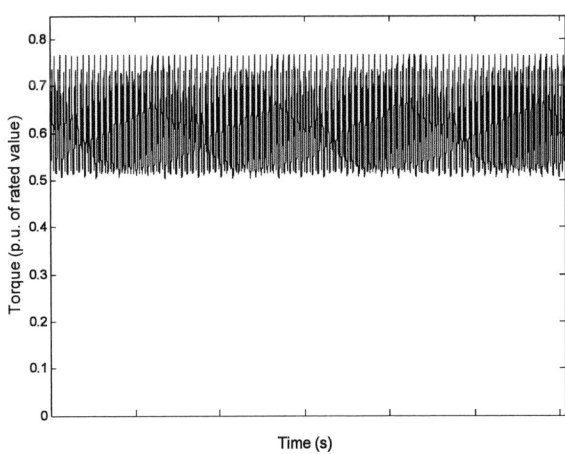

Fig. 5. Torque from DTC simulation.

The flux linkage clearly follows the circular path required with an error determined by the error margin, marked by the red and black circles. The same follows for the torque where the torque level is maintained and the ripple is determined by the error band, which is evident in Fig. 5.

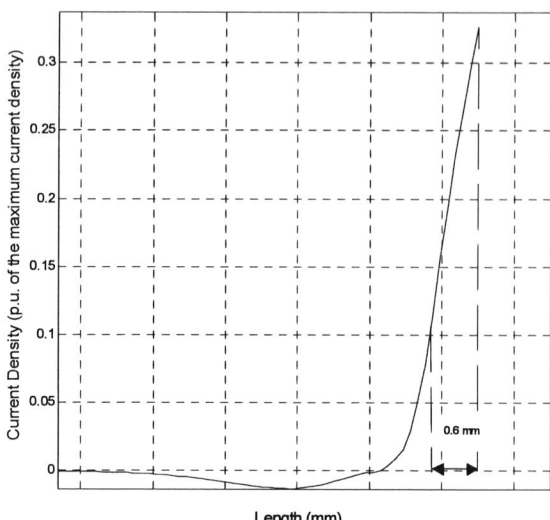

Fig. 7. Evaluation line of the pole-plate current distribution.

Fig. 8b. shows the power density distribution on the rotor pole-plate surface obtained from FEM simulations. These losses are as expected located at the face of the pole-plate, where the loss distribution follows the current distribution.

This is for one instant of time in a DTC cycle, the shape of the induced currents and hence the loss distribution will change with the switching pattern at each instant of time.

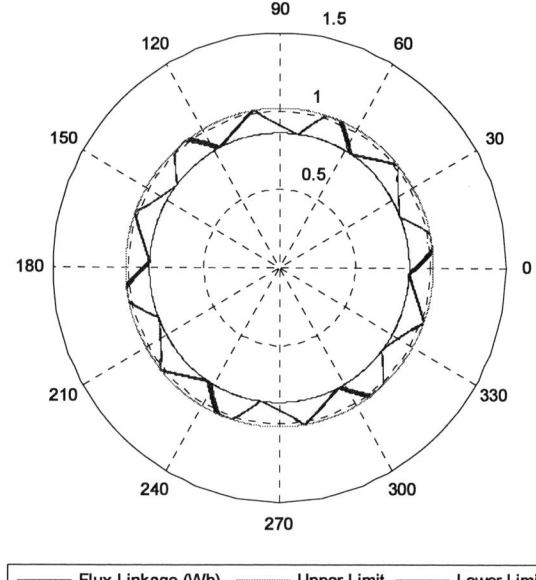

Fig. 6. Flux linkage vs. rotor angle.

VI. RESULTS OF FEM SIMULATION

The results obtained from the analytical model simulation, are used in a dynamic FEM simulation software. As the machine is the same in both models the results should not differ fundamentally.

Fig. 8a shows the current induced in the rotor pole-plate. As is expected the high frequency components of the airgap flux linkage induce high frequency currents in the pole-plate. As a consequence of the skin effect, the high frequency currents will be located at the top layer of the plate in the vicinity of the airgap, where the skin depth is given by:

$$\delta = \sqrt{\frac{\rho}{\pi \cdot \mu \cdot f}} \tag{1}$$

As can be seen in Fig. 7, the current density distribution - along an evaluation line through the midpoint of the pole-plate - does not follow a simple exponential function, as is expected. This is due to the effects of saturation and the non-linear behaviour of the pole-plates. However, assuming that the skin-depth is defined as the depth where the amplitude falls to e^{-1}, its value can be determined to 0.6mm.

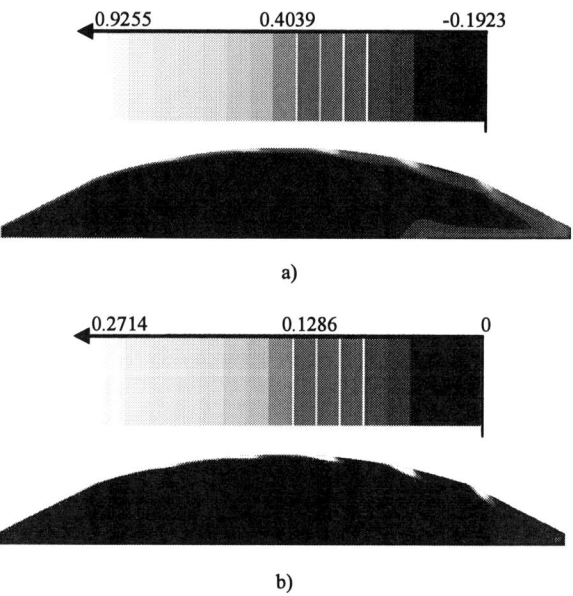

Fig. 8. Induced current- and power density in the rotor pole-plate at one instant of time. a) Shows current density in p.u. of the maximum current density level in the pole-plate and b) shows power density distribution on the rotor pole-plate in p.u. of the maximum value of power density in the plate.

Fig. 9 shows the armature currents, where these differ from those obtained through analytical simulation due to the fact that the interaction of the field induced by the pole-plate currents is not accounted for in the analytical model. The average torque is found to be at approximately the same level as that in the analytical simulations, where these are measured at 0.6131 p.u. c.f. 0.6077 p.u. respectively. The torque is shown in Fig. 10 where the ripples are seen to be significantly greater than those of the analytical simulation, which is due to the induced currents in the rotor.

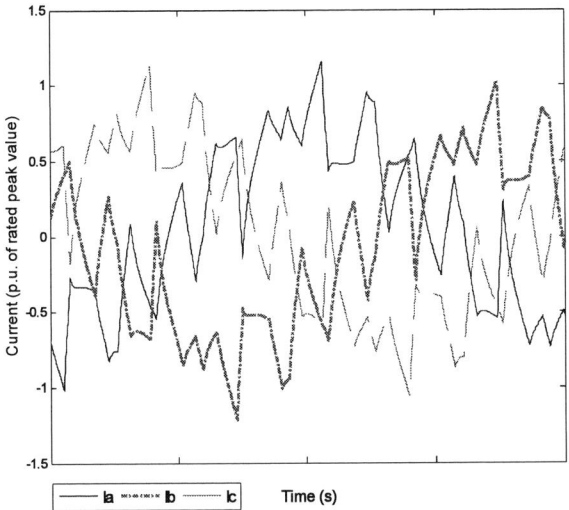

Fig. 9. FEM armature current.

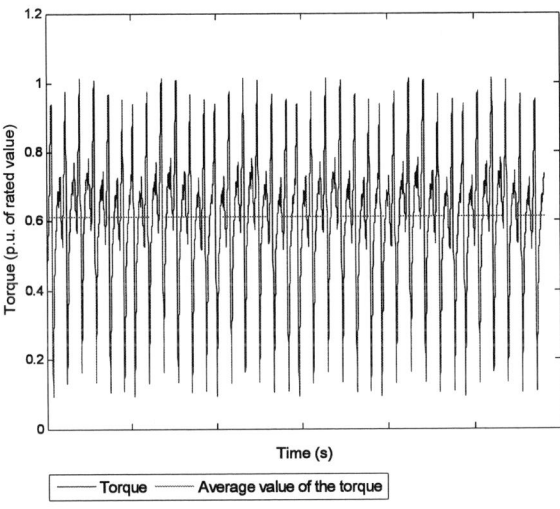

Fig. 10. Torque obtained from FEM simulation.

VII. CURRENT AMPLIFICATION AND ATTENUATION

By performing a frequency analysis of the current in the armature and in the rotor pole-plate, it is possible to get an idea of the harmonics that are the main source of losses in the rotor and how these can be disposed to changes when seen from the rotor side.

A frequency analysis clearly shows that the harmonics prevailing in the armature current are the 5th, 7th, 11th, 13th, 17th and 19th, see Fig. 11. Taking into consideration the direction of rotation of these latter harmonics, they

will generate the 6th, 12th and 18th order harmonics in the rotor.

If the losses in the pole-plates are to be calculated for each harmonic separately and using the method of superposition, the amplitude of each airgap mmf harmonic (as seen by the rotor) can be calculated using addition of power components of each pair of stator components, as given below:

$$I_{R6} = \sqrt{I_{S5}^2 + I_{S7}^2}$$
$$I_{R12} = \sqrt{I_{S11}^2 + I_{S13}^2} \quad (2)$$
$$I_{R18} = \sqrt{I_{S17}^2 + I_{S19}^2}$$

Where I_{R6}, I_{R12} and I_{R18} are the combined harmonic magnitudes in the airgap mmf seen by the rotor and I_{S5} - I_{S19} are the magnitudes of the stator current harmonics. These are shown in their p.u. equivalents in Fig. 11, where the armature components and calculated rotor components are plotted in the same chart.

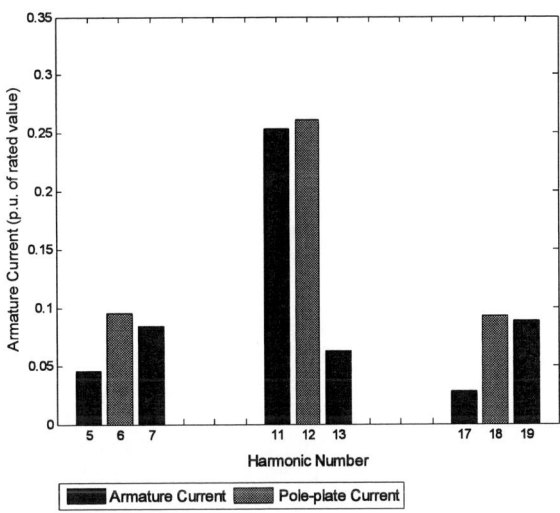

Fig. 11. Current components of the stator and the expected pole-plate components.

Using the method given above, it is expected that the currents induced in the rotor pole-plate, and hence the losses, would be proportional to the magnitudes of the values calculated for the rotor current components. However a careful analysis of the actual current in the rotor pole-plate, evaluated directly from FEM results, reveals a different picture.

By conducting a frequency analysis of the current in the pole-plate, it is possible to note that the amplitude of the loss producing current components are not identical to those obtained from the method of superposition described above. Rather, some of the given harmonic components seem to be amplified while others are attenuated.

Considering the significant harmonic components, it is obvious from Fig. 12 that the pole-plate current spectrum - obtained directly from the FEM - do not match the values calculated using the method of superposition, also shown in the same figure. In the actual pole-plate current harmonic results it is obvious that the 12th harmonic is attenuated while the two other components are amplified.

This is due to the fact that the phase of the current components is such as to partially cancel in the case of the 12th harmonic while the 6th and 18th harmonics partially augment.

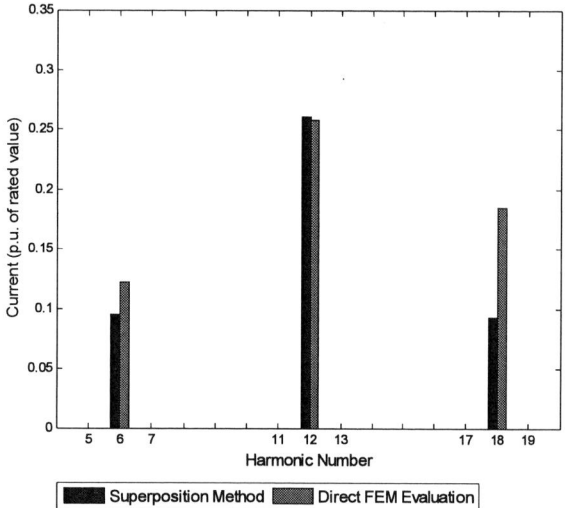

Fig. 12. Current harmonic amplitude in the rotor pole-plate: calculated using superposition method vs. directly obtained from FEM simulation results.

VIII. CONCLUSION

It has been shown that using FEM software, it is possible to find losses induced by harmonics due to inverter control of machines. Despite the complicated control strategy, which is almost impossible to implement directly in FEM, it was possible to develop a method that yielded the required results.

By using a combined model where the results of analytical simulation are used in FEM analysis, it has been found that it is possible to implement such a model to a wide range of applications concerning machine analysis and design.

A revelation which has been made is that the mmf components found on the stator side do not contribute proportionally to the losses on the rotor side. Instead some components are amplified while others are attenuated. This is due to the phase difference between the components when these are transformed to the rotor side and depends on how these superpose in the rotor pole-plate. This fact shows that the superposition method is not adequate for evaluation of rotor losses, hence making dynamic FEM modelling essential when conducting such analysis.

REFERENCES

[1] M. Depenbrock, "Direct Self-Control (DSC) of Inverter-Fed Induction Machines", *IEEE Transactions on Power Electronics.* Vol. 3. No 4, October 1988.

[2] Yujing Liu and Holger Persson, "Surface Currents in Solid Poles of Large Synchronous Machines at Direct Start", *International Conference on Electrical Machines (ICEM September 2006)*, Paper ID: No 639.

[3] Peter Vas, "Sensorless Vector and Direct Torque Control", Oxford University Press 1998.

Position sensorless control of the Reluctance Synchronous Machine considering High Frequency inductances

H.W. de Kock and M.J. Kamper
Electrical Engineering Department
University of Stellenbosch
Banghoek Street, Stellenbosch, 7600,
South Africa
Email: hugodekock@ieee.org, kamper@sun.ac.za

O.C. Ferreira and R.M. Kennel
Electrical Machines and Drives
University of Wuppertal
Rainer-Gruenter Street, Wuppertal, 42119,
Germany
Email: oferreira@ieee.org, kennel@ieee.org

Abstract—**Position sensorless control of the Reluctance Synchronous Machine (RSM) at zero and low speed is possible using High Frequency (HF) carrier injection. The standard *d-q* model of the RSM reveals greatly varying machine parameters for different working points of fundamental current: this fact needs to be taken into account in the design of the fundamental current controller to ensure stability. Similarly, the HF machine parameters vary greatly and should be considered in the demodulation scheme for the estimated rotor position. This paper shows that the effective HF inductances are different to the standard fundamental model inductances, how they are influenced by saturation and how this knowledge can be used to ensure stable and reliable sensorless control.**

I. INTRODUCTION

The RSM has, among other advantages, no windings on the rotor and therefore no copper losses on the rotor, i.e. high efficiency may be obtained using the correct control algorithm. The RSM may substitute the industry standard Induction machine for applications that require only a limited constant power region, however it has to be controlled without a position sensor, to comply with cost and reliability requirements. The large saliency of the rotor structure 1) is the source of reluctance torque, 2) causes $\frac{di_d}{dt} \neq \frac{di_q}{dt}$ and 3) is an essential criterion to be able to detect the rotor position at zero and low speed. The saliency may be expressed as a difference in *d*-axis and *q*-axis inductance, however a clear distinction should be made between 1) fundamental model secant inductance, 2) fundamental model tangential inductance and 3) effective HF model tangential inductance, as applicable to the three cases above respectively.

There are different classes with different methods of position sensorless control, which also apply to the RSM [1]–[9]. It is possible to combine different methods from different classes, as shown in [6], [8], [9]. Here, a separate excitation is used in the low speed region and is then dropped out in the higher speed region where only the fundamental excitation is used to obtain the rotor position. The algorithm presented in this paper also supports this idea, however it

focuses on sensorless control in the low speed region using HF carrier injection. More specifically, an alternating HF voltage is applied and the resulting HF currents are used to feed a Phase Locked Loop (PLL). The carrier injection and demodulation are done within an estimated synchronous reference frame, similarly to [7].

In this paper, a distinct difference is made between the fundamental model and the HF model of the machine, which is supported by FE results and practical measurements. It is shown how the fundamental excitation influences the HF model parameters and the implication thereof on position estimation using the PLL is highlighted. Based on these measurements, necessary improvements on the PLL structure are suggested. It is shown that implementation of these improvements leads to a stable and reliable position estimator for zero and low speed, without any restrictions on the optimal torque producing current vector.

II. FUNDAMENTAL MACHINE MODEL

Neglecting hysteresis and eddy-current losses, the model for the RSM in the synchronously rotating *d-q* reference frame is given by (1) through (4). The voltage equations may be rewritten in complex vector notation by assigning the $\Re$-axis to the *d*-axis, and assigning the $\Im$-axis to the *q*-axis. The torque equation may also be rewritten in polar form as the cross product of the flux linkage vector with the current vector, as in (7).

$$u_d = R \cdot i_d + \frac{d\psi_d}{dt} - \omega_e \cdot \psi_q \tag{1}$$

$$u_q = R \cdot i_q + \frac{d\psi_q}{dt} + \omega_e \cdot \psi_d \tag{2}$$

$$T_m = \frac{3p}{2} \left(i_q \cdot \psi_d - i_d \cdot \psi_q \right) = T_L + J_{eq} \frac{d\omega_m}{dt} + B_{eq}\omega_m \tag{3}$$

$$\omega_e = p \cdot \omega_m = \frac{d\theta_e}{dt} \tag{4}$$

978-1-4244-0644-9/07/$25.00 ©2007 IEEE

$$\vec{i_r} = i_d + ji_q = i_r \angle \phi_r \tag{5}$$

$$\vec{\psi_r} = \psi_d + j\psi_q = \psi_r \angle \delta_r \tag{6}$$

$$T_m = \frac{3p}{2}\vec{\psi_r} \times \vec{i_r} = \frac{3p}{2}\psi_r i_r \sin(\phi_r - \delta_r) \tag{7}$$

$$\vec{u_r} = u_d + ju_q = u_r \angle \alpha_r = R \cdot \vec{i_r} + \frac{d\vec{\psi_r}}{dt} + j\omega_e\vec{\psi_r} \tag{8}$$

For a 1.5 kW RSM, the FE model is shown in Fig. 1(a) and the FE results, Figs. 1(b), 1(c), 1(d) and 1(e), describe this RSM in polar coordinates. The input to the FE program is $\vec{i_r}$ and the electro-static solution gives $\vec{\psi_r}$. T_m is solved using (7), and $\vec{u_r}$ is solved in the steady state using (8) with ω_e as an independent variable. Fig. 1(b) shows the input current $\vec{i_r}$ to the FE program, the outer circle corresponding to the rated current. Fig. 1(c) shows the corresponding flux linkage $\vec{\psi_r}$, where it can be noted that saturation occurs in the d-axis since the ellipses tend closer in this direction. Fig. 1(d) shows the torque magnitude $|T_m|$ with current angle ϕ_r. Filled circles indicate maximum torque per ampere (MTPA) with corresponding MPTA points shown on Figs. 1(b), 1(c) and 1(e). Fig. 1(e) shows the voltage evaluated for rated speed using (8). Considering the voltage limitation $|\vec{u_r}| < \frac{Vdc}{2}$ for sinusoidal PWM, as indicated by the dotted circle, achievable operating conditions are indicated on Figs. 1(b) through 1(e) by solid lines and unachievable operating conditions by dotted lines.

Focusing on the speed region below base speed, the curve for the MTPA points on the current plane (i.e. the optimal torque producing locus for the current vector) may be approximated by a straight line, i.e. constant current angle current control may be used. Choosing a constant current angle corresponding to the MTPA point for the rated current ($\phi_r = 60°$), the loss in torque for non-rated conditions is small, as can be seen in Fig. 1(d). Therefore, the FE results of interest, with respect to the control algorithm below base speed, is only those relating to the current angle of $\phi_r = 60°$.

Even though the transformation to the synchronously rotating d-q reference frame should remove dependencies on θ_e, the flux linkages nevertheless remain functions of θ_e, e.g. due to stator slot openings. Fig. 1(f) shows the ripple components of the torque, ψ_d and ψ_q, expressed in percentage. Also, even though the d-axis and q-axis are perpendicular to each other, cross coupling and cross saturation may not be ignored. Therefore $\psi_d = \psi_d(i_d, i_q, \theta_e)$ and $\psi_q = \psi_q(i_d, i_q, \theta_e)$.

The relationship between $\vec{\psi_r}$ and $\vec{i_r}$ may be expressed in terms of inductances. Now, the tangential inductance L_t is defined as the partial derivative of flux linkage with respect to current $\frac{\partial\psi}{\partial i}$, while secant inductance L_s is defined as flux linkage divided by current $\frac{\psi}{i}$ [10]. The tangential inductance is useful as in (9) through (11). It is common that the secant inductances are used to write the torque equation as $T_m = \frac{3p}{2}(L_{ds} - L_{qs})i_d i_q$, indicating that the source of

the torque is the saliency $L_{ds} > L_{qs}$. Taking saturation into account $L_{ds} \neq L_{dt}$ and $L_{qs} \neq L_{qt}$.

More FE results for the fundamental model are shown in Fig. 2, with reference to the inductances defined above. Flux linkages, differential inductances and secant inductances are shown as functions of i_r, evaluated along three different axes, namely $\phi_r = 89.9°$, $\phi_r = 0.1°$ and $\phi_r = 60°$ (zero current avoided for secant inductance). Here, the differences between the secant and tangential inductances and also the non-linearity due to saturation are visible. The fundamental model tangential inductances shown here can be compared to the effective HF tangential inductances, shown in the next section.

$$\frac{d\psi_d}{dt} = \frac{\partial\psi_d}{\partial i_d} \cdot \frac{di_d}{dt} + \frac{\partial\psi_d}{\partial i_q} \cdot \frac{di_q}{dt} + \frac{\partial\psi_d}{\partial\theta_e} \cdot \frac{d\theta_e}{dt}$$
$$= L_{dt} \cdot \frac{di_d}{dt} + M_{dt} \cdot \frac{di_q}{dt} + \frac{\partial\psi_d}{\partial\theta_e} \cdot \omega_e \tag{9}$$

$$\frac{d\psi_q}{dt} = \frac{\partial\psi_q}{\partial i_d} \cdot \frac{di_d}{dt} + \frac{\partial\psi_q}{\partial i_q} \cdot \frac{di_q}{dt} + \frac{\partial\psi_q}{\partial\theta_e} \cdot \frac{d\theta_e}{dt}$$
$$= M_{qt} \cdot \frac{di_d}{dt} + L_{qt} \cdot \frac{di_q}{dt} + \frac{\partial\psi_q}{\partial\theta_e} \cdot \omega_e \tag{10}$$

$$M_t = M_{qt} = M_{dt} \tag{11}$$

A. HF machine model

The HF model for the machine may be given by (12) through (14). In these equations, cross-saturation and cross-coupling are taken into account, trans-versatility [9] (expressed by (14)) is assumed, and it is ensured that the frequency of the applied voltage is much higher than the rated fundamental. It is common to ignore the mutual inductances, however as pointed out in [9], care has to be taken. In [9], the authors take $L_{dtHF} = L_{dt}$, $L_{qtHF} = L_{qt}$, and $M_{tHF} = M_t$. However it seems that for this case, the rated fundamental frequency being 50 Hz and the HF of interest being 1000 Hz, this assumption is not valid and therefore a distinction is made.

$$u_d \approx L_{dtHF}\frac{di_d}{dt} + M_{dtHF} \cdot \frac{di_q}{dt} \tag{12}$$

$$u_q \approx L_{qtHF}\frac{di_q}{dt} + M_{qtHF} \cdot \frac{di_d}{dt} \tag{13}$$

$$M_{tHF} = M_{qtHF} = M_{dtHF} \tag{14}$$

The values for L_{dtHF}, L_{qtHF} and M_{tHF} are also dependent on the working point, and have been calculated in an experimental setup: using a RSM with a locked rotor, the fundamental current was controlled to specified values, an additional HF voltage command was added to the current controller output (indicated by the $+ =$ sign) and the measured HF currents were used to calculate L_{dtHF}, L_{qtHF} and M_{tHF}. In Fig. 3 the amplitudes of the HF currents, $|i_{dHF}|$ and $|i_{qHF}|$, are shown as functions of the fundamental current i_r. Three cases for fundamental current are evaluated: along the q-axis, along the d-axis and along the $\phi = 60°$ axis; the first two begin only of theoretical interest and the last being of practical

1018

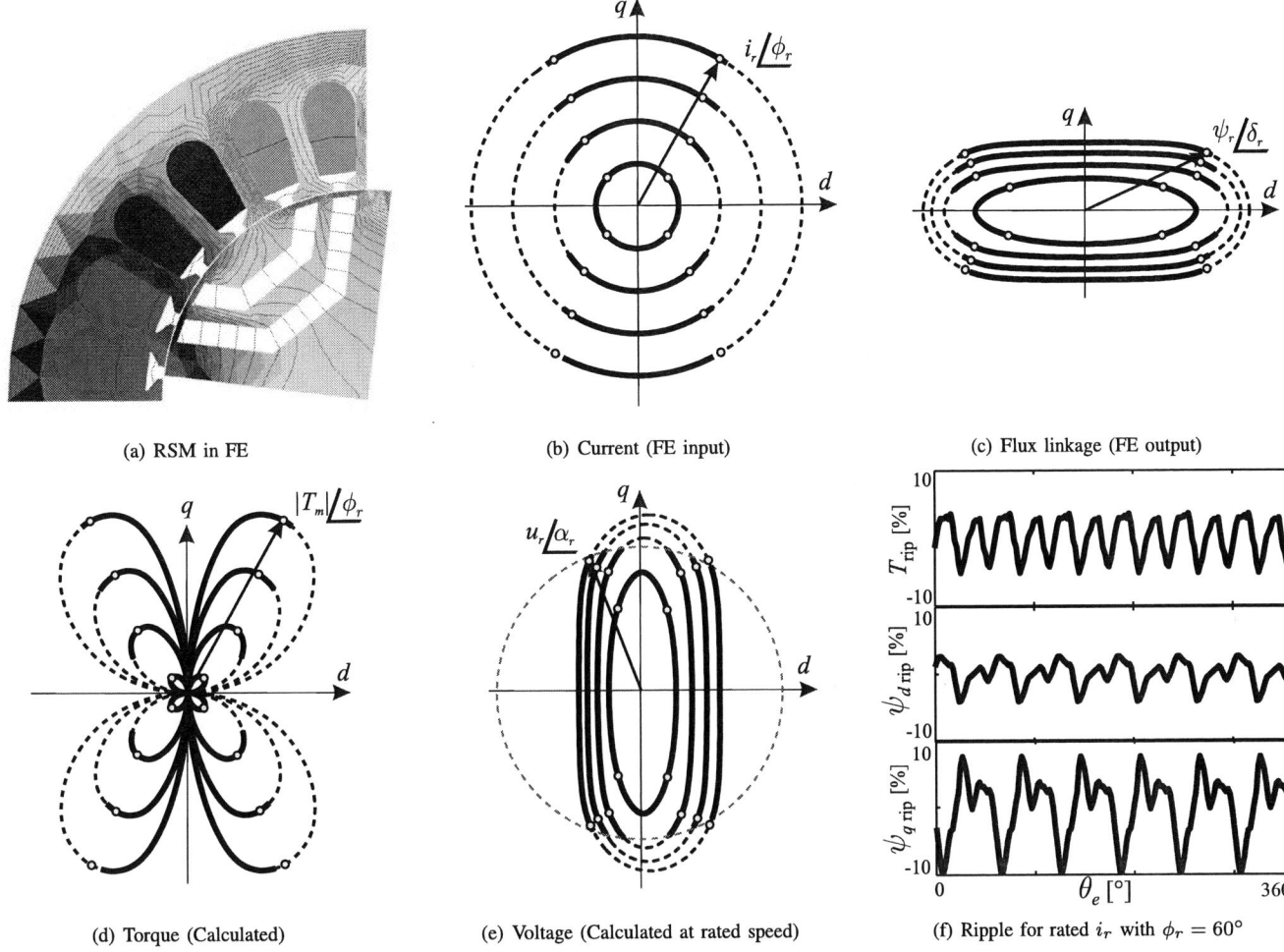

(a) RSM in FE   (b) Current (FE input)   (c) Flux linkage (FE output)

(d) Torque (Calculated)   (e) Voltage (Calculated at rated speed)   (f) Ripple for rated i_r with $\phi_r = 60°$

Fig. 1. Fundamental model

interest. Also note that pulsating HF voltages are applied, either in the d-axis or the q-axis.

Fig. 3(a) shows that increasing i_q causes saturation, such that L_{qtHF} decreases and therefore $|i_{qHF}|$ increases. There is a small amount of cross-coupling since there is measurable $|i_{dHF}|$, however it is close to the ideal zero. Fig. 3(d) shows that increased saturation in the q-axis does not affect $|i_{dHF}|$ much. There is a small amount of cross-coupling since there is measurable $|i_{qHF}|$, however it is close to the ideal zero. Fig. 3(b) shows that increased saturation in the d-axis affects $|i_{qHF}|$, i.e. cross-saturation causes decreased L_{qtHF} and therefore increased $|i_{qHF}|$. There is a small amount of cross-coupling since there is measurable $|i_{dHF}|$, which increases as i_d increases. Fig. 3(e) shows that increasing i_d causes saturation, such that L_{dtHF} decreases and therefore $|i_{dHF}|$ increases. There is a small amount of cross-coupling since there is measurable $|i_{qHF}|$, however it is close to the ideal zero. Although these results are interesting, it is not of much practical use, since the RSM will never be controlled along these axes.

Fig. 3(c) shows that increasing i_r with $\phi = 60°$ causes saturation in the q-axis, such that L_{qtHF} decreases and therefore $|i_{qHF}|$ increases. There is a small amount of cross-coupling since there is measurable $|i_{dHF}|$, however it is close to the ideal zero. Fig. 3(f) shows that increasing i_r with $\phi = 60°$ causes saturation in the d-axis, such that L_{dtHF} decreases and therefore $|i_{dHF}|$ increases. There is a large amount of cross-coupling since there is comparable $|i_{qHF}|$. Here it is also noted that as i_r increases, L_{qtHF} decreases and therefore $|i_{qHF}|$ increases. The significance of this result is clear for sensorless control: by injecting HF voltage in the d-axis and expecting only to find $|i_{dHF}|$, one would normally drive $|i_{qHF}|$ to zero in order to align the estimated and actual d-q reference frames [7]; in this case it is predicted that this method will result in an estimation error, since a large part of $|i_{qHF}|$ is due to cross-coupling and not due to misalignment of the reference frames. One can easily minimize the error caused by cross-coupling by refering to Fig. 3(c), where the effect of cross-coupling is smaller. Therefore, injection of the HF voltage on the q-axis and use of $|i_{dHF}|$ as an indication of misalignment is suggested.

1019

Using the measured data of the HF currents, the effective HF inductances may be calculated with very little approximation. For example, as noted in Fig. 3(c), the effect of mutual inductance is minimal and therefore using the data represented by this figure with (13), $|u_{qHF}| \approx L_{qtHF}\omega_{HF}|i_{qHF}|$. Then, with L_{qtHF} known, the data represented by Fig. 3(f) may be used to find the mutual inductance with (13), $0 \approx L_{qtHF}|i_{qHF}| + M_{tHF}|i_{dHF}|$. Then finally, still using the data represented by Fig. 3(f), now with (12) and with L_{qtHF} and M_{tHF} known, L_{dtHF} may be calculated. These are the inductances that come into play for the sensorless position estimation using HF carrier injection. Comparing these effective HF tangential inductances shown in Fig. 4, to the fundamental model tangential inductances in Fig. 2, insight into the machine behaviour is obtained. The significance of these results for sensorless control in the low speed region is explained next.

III. POSITION ESTIMATION AT ZERO AND LOW SPEED

For low and zero speed, position estimation using HF carrier injection relies on the presense of an anisotropy. The anisotropy manifests itself as a difference in inductances, i.e. $L_{dA} \neq L_{qA}$. The anisotropy might be there due to the physical structure and/or saturation. In the case that the anisotropy is aligned with the actual dq-axes, then $L_{dA} = L_{dtHF}$ and $L_{qA} = L_{qtHF}$. The anisotropy, located at position θ_A is what is "seen" and locked onto by the PLL.

The HF voltage injection and position estimation take place within an estimated reference frame that is aligned with the anisotropy, when the PLL is active, as shown in Fig. 5. In this frame, applying a voltage of (15) will result in the current given by (16). In this derivation, a slow rate of change of $\tilde{\theta}_A = \theta_A - \hat{\theta}_A$ is assumed. Misalignment is indicated by $\hat{i}_{dA}$ and this signal is used to track θ_A, however it is plauged by a large variable DC offset (filtered out by the BPF) and the amplitude of the HF component $\hat{i}_{dA}$ is dependent on L_{dA} and L_{qA} and therefore changes according to the working point. Careful consideration is neccesary when setting up the parameters for the PLL. The required signal $\tilde{\theta}_A$ may be obtained as in (17) and (18), assuming that $\sin 2\tilde{\theta}_A \approx 2\tilde{\theta}_A$, i.e. assuming that the PLL works properly. To avoid aliasing, ω_{HF} has to be chosen so that (19) hold, where ω_0 is the maximum fundamental frequency that will be used in this sensorless scheme (the switch-over frequency to a back-EMF based sensorless algorithm) and $\omega_s = 2\pi f_s$ is the sampling frequency.

$$\hat{\tilde{u}}_{rA} = j \cdot U_{HF} \cos\left(\omega_{HF}t\right) \tag{15}$$

$$\text{BPF}\left\{\hat{\tilde{i}}_{rA}\right\} = \frac{U_{HF}}{2\omega_{HF}} \frac{\sin\left(\omega_{HF}t\right)}{L_{dA} \cdot L_{qA}} \cdot [\ldots$$
$$\ldots (L_{qA} - L_{dA}) \sin 2\tilde{\theta}_A + \ldots \tag{16}$$
$$\ldots + j \cdot (L_{qA} + L_{dA}) - j \cdot (L_{qA} - L_{dA}) \cos 2\tilde{\theta}_A]$$

$$\tilde{\theta}_A \approx \text{LPF}\left\{K_\theta \cdot \hat{i}_{dA} \cdot \sin\left(\omega_{HF}t\right)\right\} \tag{17}$$

$$K_\theta = \frac{2\omega_{HF}}{U_{HF}} \frac{L_{dA} \cdot L_{qA}}{L_{qA} - L_{dA}} \tag{18}$$

$$2\omega_0 < \omega_{HF} < \frac{\omega_s - 2\omega_0}{2} \tag{19}$$

The first tests with the PLL is to look at the relationship between θ_e and θ_A. The fundamental current is controlled using the measured θ_e and the PLL is activated. Two cases are studied: firstly injection of the HF voltage on the $q\hat{A}$-axis (using $\hat{i}_{dA}$ for demodulation) and secondly on the $d\hat{A}$-axis (using $\hat{i}_{qA}$ for demodulation). From these results, shown in Fig. 6, many interesting things can be seen, and it is supported by the FE results in Fig. 2(f) and the experimentally obtained results in Fig. 4. As expected, applying fundamental i_q, the HF inductance difference only becomes greater and it is no problem for the PLL to find the anisotropy. Furthermore, the anisotropy seems to be aligned with the actual dq reference frame. Applying fundamental i_d, we expect the inductance difference to be become smaller and then to change sign. This is clearly visible in the case where HF voltage on the $q\hat{A}$-axis is applied: here the phase shift is eventually $90°$, since the PLL parameters are constant. In the case where HF voltage on the $d\hat{A}$-axis is applied, the estimation becomes unstable: this might be due to HF mutual coupling as indicated before. Of most interest is that the estimation should work when fundamental current with $\phi_r = 60°$ is applied. From the results, it seems that it is indeed no problem to estimate the rotor position, using injection in either direction.

The next step then, is to use the estimated angle given by the PLL for the field oriented control (FOC), i.e. completely sensorless. The test results are shown in Fig. 7. Some of these results are rather surprising and do not support the theory presented. In all the cases however, the system is rather unstable, causing jerk reactions on the rotor. The important results with $\phi_r = 60°$, especially concerning injection on the $q\hat{A}$-axis, are not pleasing at all. To find a solution, it was decided to take one step back, and closely evaluate θ_e vs. θ_A in the steady state for the same conditions as in Fig. 6(e).

The obtained results, as shown in Fig. 8, are very interesting. Here it is firstly clear that the estimation does not work so well under no-load conditions (the anisotropy due to physical structure might be very small), and secondly, there exists a relationship between i_r and the steady state error between θ_e and $\hat{\theta}_A$. It seems like the anisotropy is shifting away from the dq-axes with increasing load (more saturation).

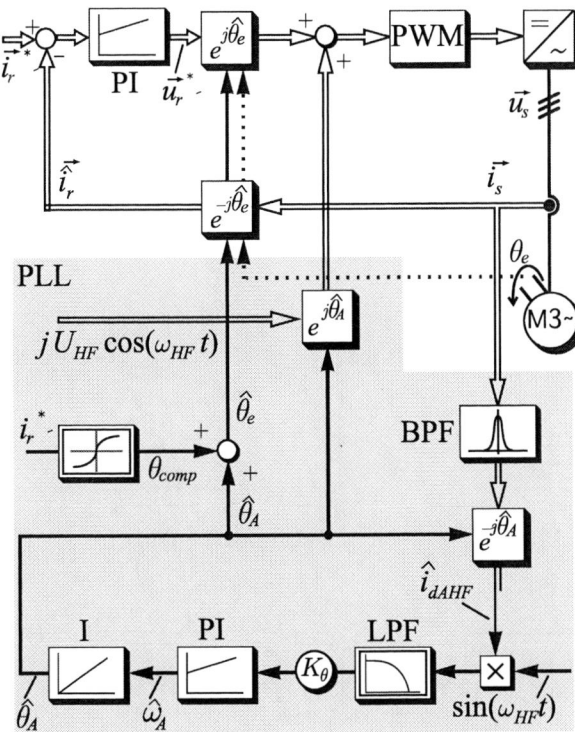

Fig. 5. θ_e estimation PLL

conditions. The estimated rotor position may be obtained as the sum of the estimated anisotropy position and a load dependent compensation function. Using this estimated rotor position, field oriented control with the optimal torque producing current vector may be performed at zero and low speed. The dynamic performance of this scheme needs critical examination.

For full speed-range sensorless control, this scheme needs to be combined with a back-EMF method. Most importantly, the groundwork has been laid for such a combined scheme, and a better understanding of the machine behaviour has been obtained.

REFERENCES

[1] M. Arefeen, M. Ehsani, and A. Lipo, "Sensorless position measurement in synchronous reluctance motor," *Power Electronics, IEEE Transactions on*, vol. 9, no. 6, pp. 624–630, 1994.
[2] R. Lagerquist, I. Boldea, and T. Miller, "Sensorless-control of the synchronous reluctance motor," *Industry Applications, IEEE Transactions on*, vol. 30, no. 3, pp. 673–682, 1994.
[3] M. Jovanovic, R. Betz, and D. Platt, "Sensorless vector controller for a synchronous reluctance motor," *Industry Applications, IEEE Transactions on*, vol. 34, no. 2, pp. 346–354, 1998.
[4] A. Consoli, F. Russo, G. Scarcella, and A. Testa, "Low- and zero-speed sensorless control of synchronous reluctance motors," *Industry Applications, IEEE Transactions on*, vol. 35, no. 5, pp. 1050–1057, 1999.
[5] S.-J. Kang, J.-M. Kim, and S.-K. Sul, "Position sensorless control of synchronous reluctance motor using high frequency current injection," *Energy Conversion, IEEE Transaction on*, vol. 14, no. 4, pp. 1271–1275, 1999.
[6] A. Vagati, M. Pastorelli, F. Scapino, and G. Franceschini, "Impact of cross saturation in synchronous reluctance motors of the transverse-laminated type," *Industry Applications, IEEE Transactions on*, vol. 36, no. 4, pp. 1039–1046, 2000.
[7] M. Linke, R. Kennel, and J. Holtz, "Sensorless speed and position control of synchronous machines using alternating carrier injection," in *Electric Machines and Drives Conference, 2003. IEMDC'03. IEEE International*, vol. 2, 2003, pp. 1211–1217 vol.2.
[8] P. Guglielmi, M. Pastorelli, G. Pellegrino, and A. Vagati, "Position-sensorless control of permanent-magnet-assisted synchronous reluctance motor," *Industry Applications, IEEE Transactions on*, vol. 40, no. 2, pp. 615–622, 2004.
[9] P. Guglielmi, M. Pastorelli, and A. Vagati, "Impact of cross-saturation in sensorless control of transverse-laminated synchronous reluctance motors," *Industrial Electronics, IEEE Transactions on*, vol. 53, no. 2, pp. 429–439, 2006.
[10] M. Gyimesi and D. Ostergaard, "Inductance computation by incremental finite element analysis," *Magnetics, IEEE Transactions on*, vol. 35, no. 3, pp. 1119–1122, 1999.

According to these measurements, a compensation function may be added to the scheme, i.e. $\hat{\theta}_e = \hat{\theta}_A + \theta_{comp}$. Fig. 8(f) shows that such a compensation may give positive results. Indeed, using this compensation it is possible to control the machine sensorless with no restriction on the optimal torque producing current at low speed, as shown in Fig. 8(g), and at zero speed, as shown in Fig. 8(h). Note that in this scheme, the anisotropy tracking is done in the estimated reference frame with $\hat{\theta}_A$, and the FOC is done in the estimated reference frame with $\hat{\theta}_e$, where $\hat{\theta}_A \neq \hat{\theta}_e$ under loaded conditions.

As a final remark, the waveform of the error $\theta_e - \hat{\theta}_e$ shown in Fig. 8(g), has remarkable similarity to the FE results for the θ_e dependent flux-ripple $\psi_{q\ rip}$, shown in Fig. 1(f). Since the $\frac{\partial \psi}{\partial \theta_e}$ terms are not included in the HF model, and are by no means compensated, it may come as no surprise that they are present in the error function $\theta_e - \hat{\theta}_e$.

IV. CONCLUSION

This paper has presented an in-depth view of the RSM, with reference to the fundamental and HF models as applicable to sensorless control. It has suggested that the fundamental model is different to the HF model: future work needs to confirm these results with a dynamic field solution FE program.

Using alternating HF carrier injection and a PLL, an ansitropy is tracked, which is aligned with the actual dq reference frame, but moves away under loaded (saturated)

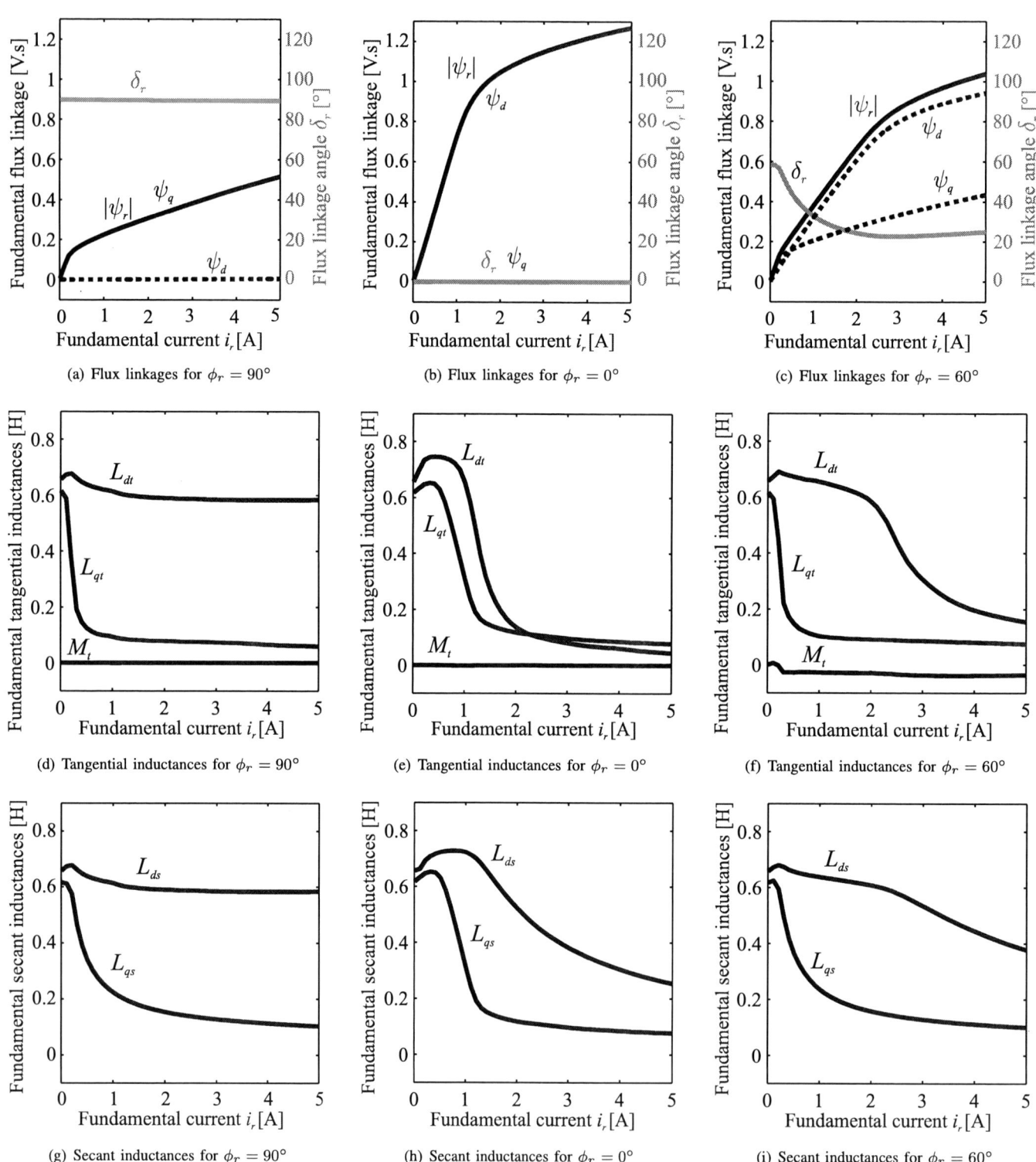

(a) Flux linkages for $\phi_r = 90°$

(b) Flux linkages for $\phi_r = 0°$

(c) Flux linkages for $\phi_r = 60°$

(d) Tangential inductances for $\phi_r = 90°$

(e) Tangential inductances for $\phi_r = 0°$

(f) Tangential inductances for $\phi_r = 60°$

(g) Secant inductances for $\phi_r = 90°$

(h) Secant inductances for $\phi_r = 0°$

(i) Secant inductances for $\phi_r = 60°$

Fig. 2. FE results: fundamental model along $\phi_r = 89.9°$, $\phi_r = 0.1°$ and $\phi_r = 60°$

1022

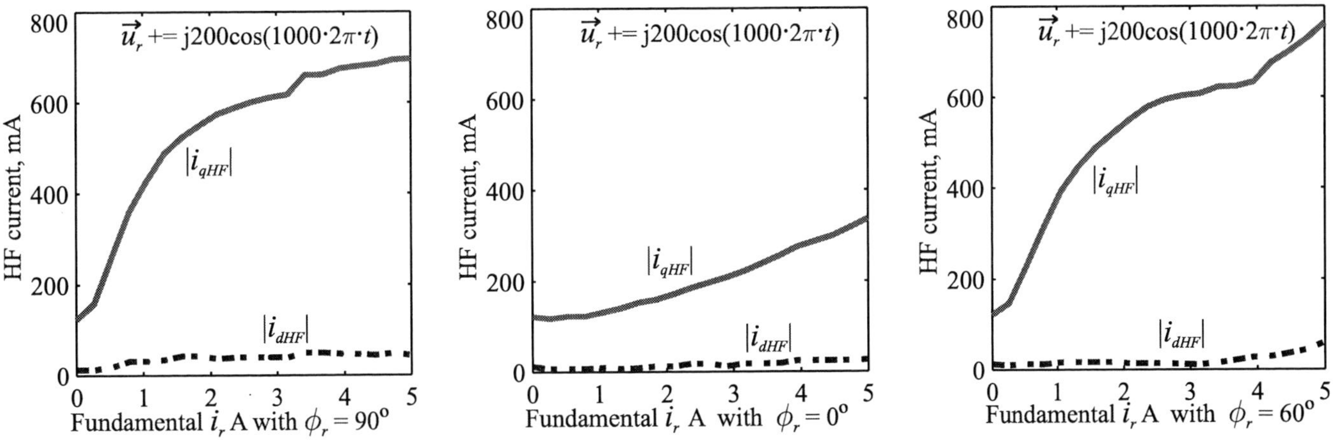

Fig. 3. Practical experiment to study the influence of saturation on the HF parameters

(a) u_{qHF} and i_q
(b) u_{qHF} and i_d
(c) u_{qHF} and i_r with $\phi_r = 60$
(d) u_{dHF} and i_q 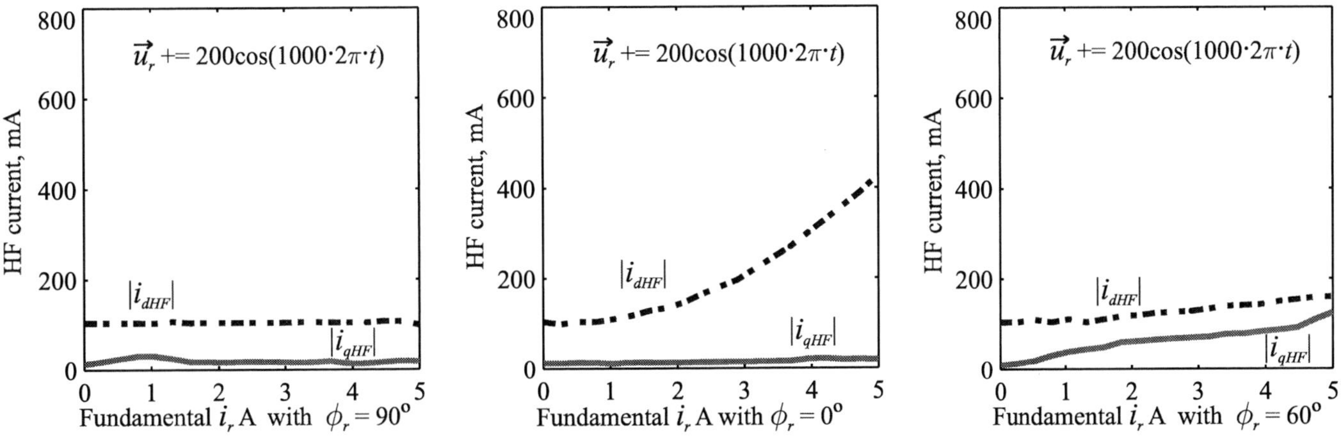
(e) u_{dHF} and i_d
(f) u_{dHF} and i_r with $\phi_r = 60$

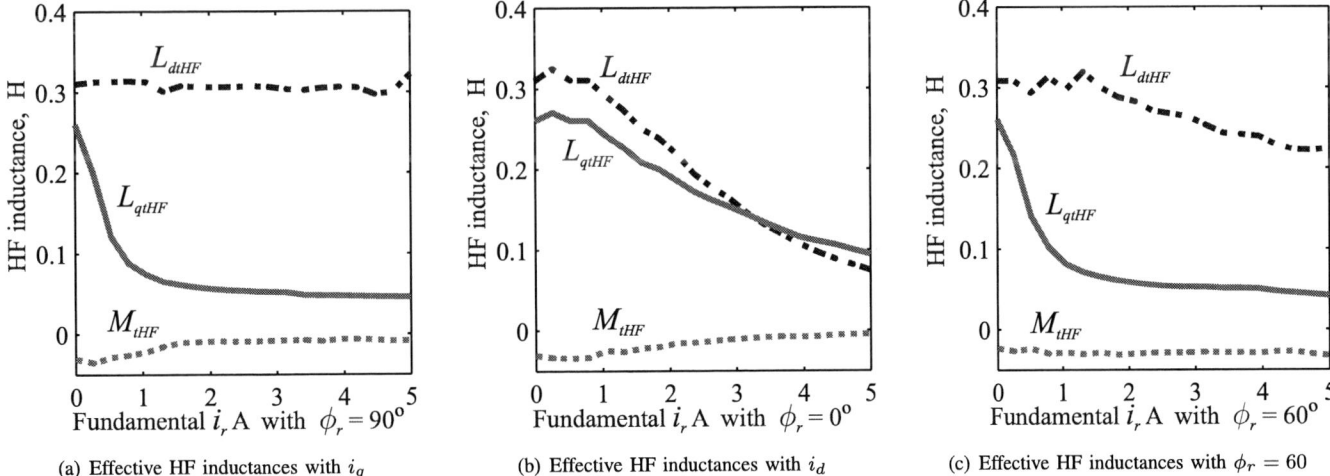

(a) Effective HF inductances with i_q
(b) Effective HF inductances with i_d
(c) Effective HF inductances with $\phi_r = 60$

Fig. 4. Calculated HF inductances from measured data

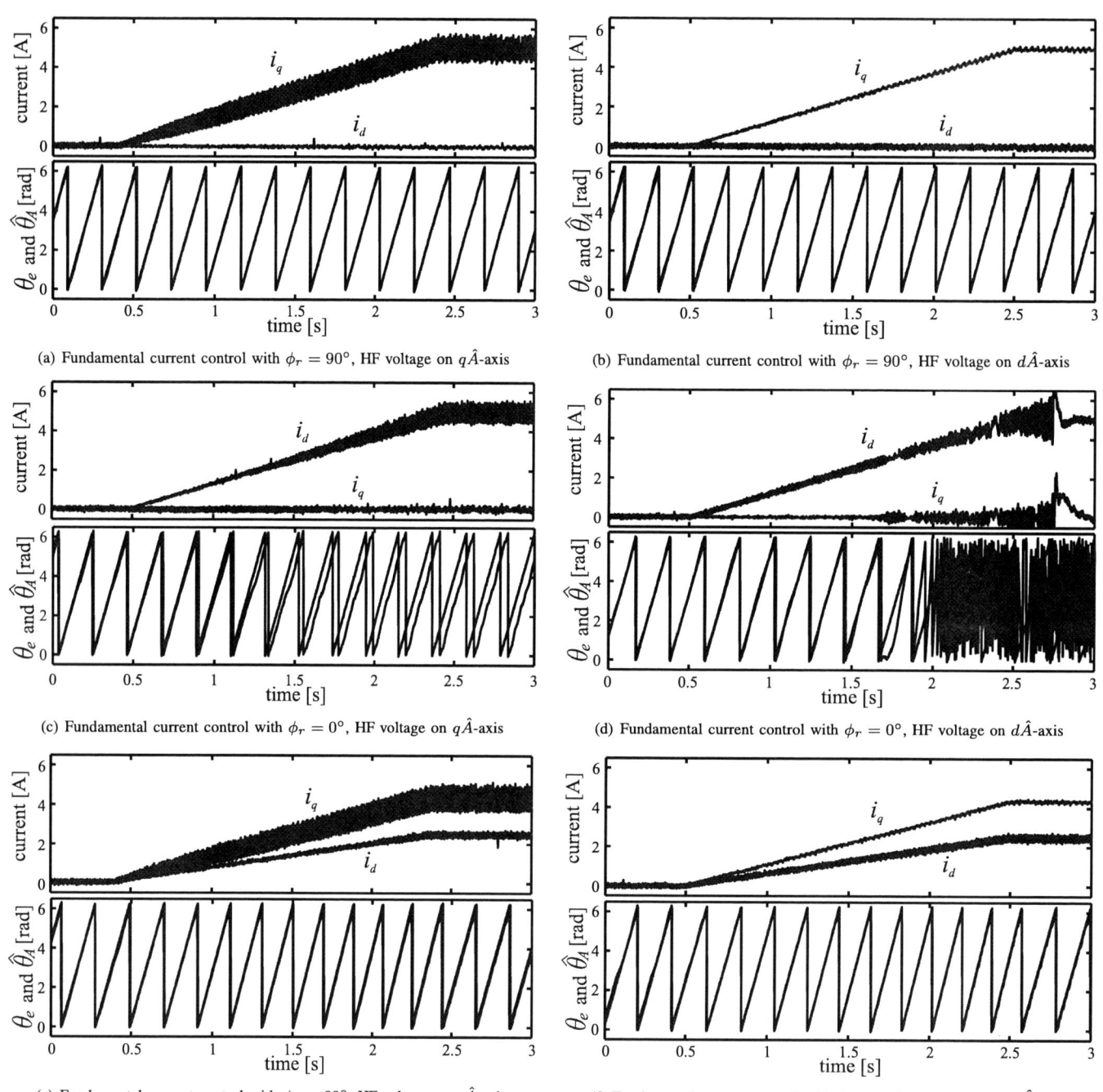

(a) Fundamental current control with $\phi_r = 90°$, HF voltage on $q\hat{A}$-axis

(b) Fundamental current control with $\phi_r = 90°$, HF voltage on $d\hat{A}$-axis

(c) Fundamental current control with $\phi_r = 0°$, HF voltage on $q\hat{A}$-axis

(d) Fundamental current control with $\phi_r = 0°$, HF voltage on $d\hat{A}$-axis

(e) Fundamental current control with $\phi_r = 60°$, HF voltage on $q\hat{A}$-axis

(f) Fundamental current control with $\phi_r = 60°$, HF voltage on $d\hat{A}$-axis

Fig. 6. Field orientated control using measured θ_e, comparison with estimated $\hat{\theta}_A$

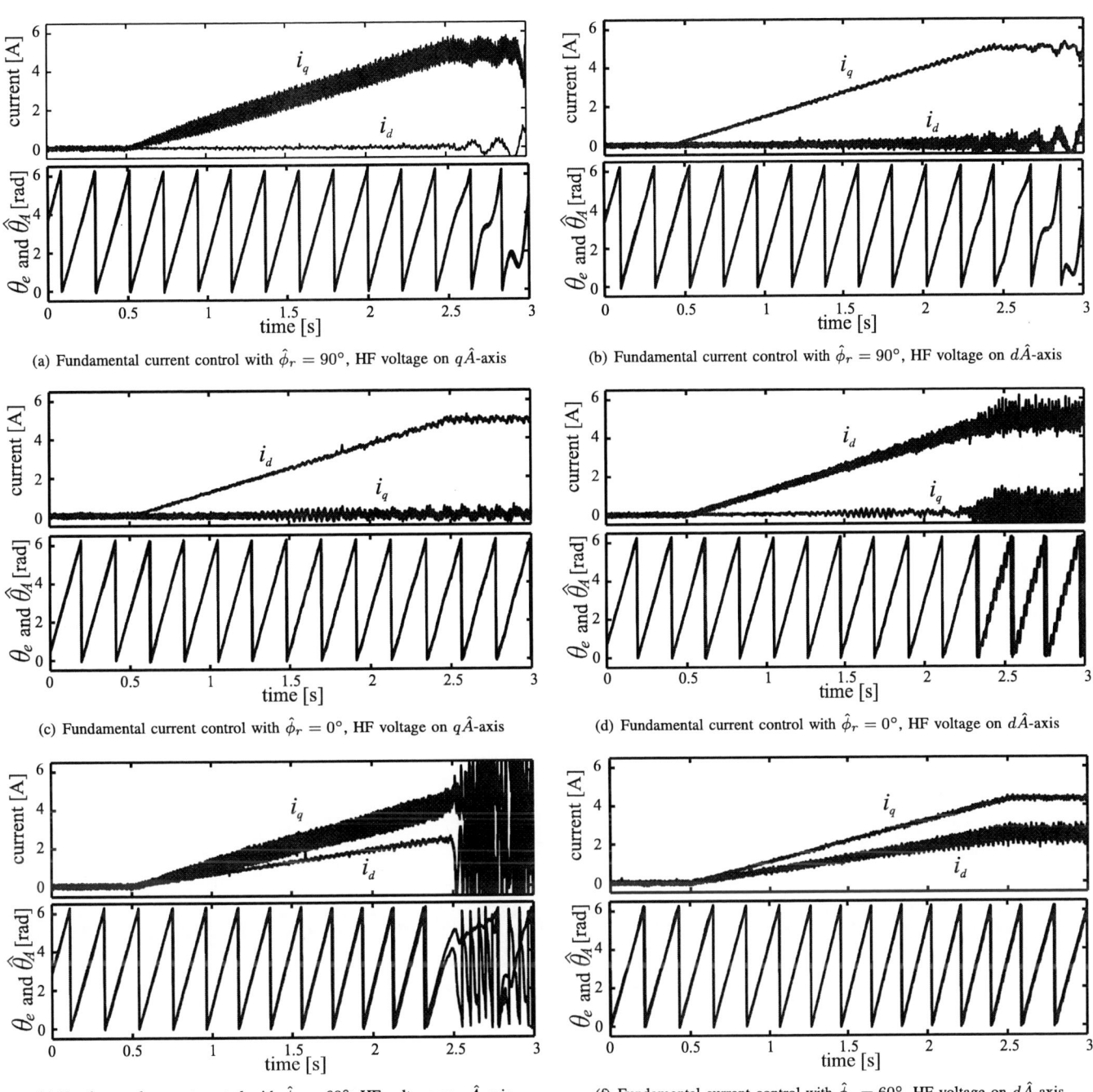

(a) Fundamental current control with $\hat{\phi}_r = 90°$, HF voltage on $q\hat{A}$-axis

(b) Fundamental current control with $\hat{\phi}_r = 90°$, HF voltage on $d\hat{A}$-axis

(c) Fundamental current control with $\hat{\phi}_r = 0°$, HF voltage on $q\hat{A}$-axis

(d) Fundamental current control with $\hat{\phi}_r = 0°$, HF voltage on $d\hat{A}$-axis

(e) Fundamental current control with $\hat{\phi}_r = 60°$, HF voltage on $q\hat{A}$-axis

(f) Fundamental current control with $\hat{\phi}_r = 60°$, HF voltage on $d\hat{A}$-axis

Fig. 7. Field orientated control using estimated $\hat{\theta}_A$, comparison with measured θ_e

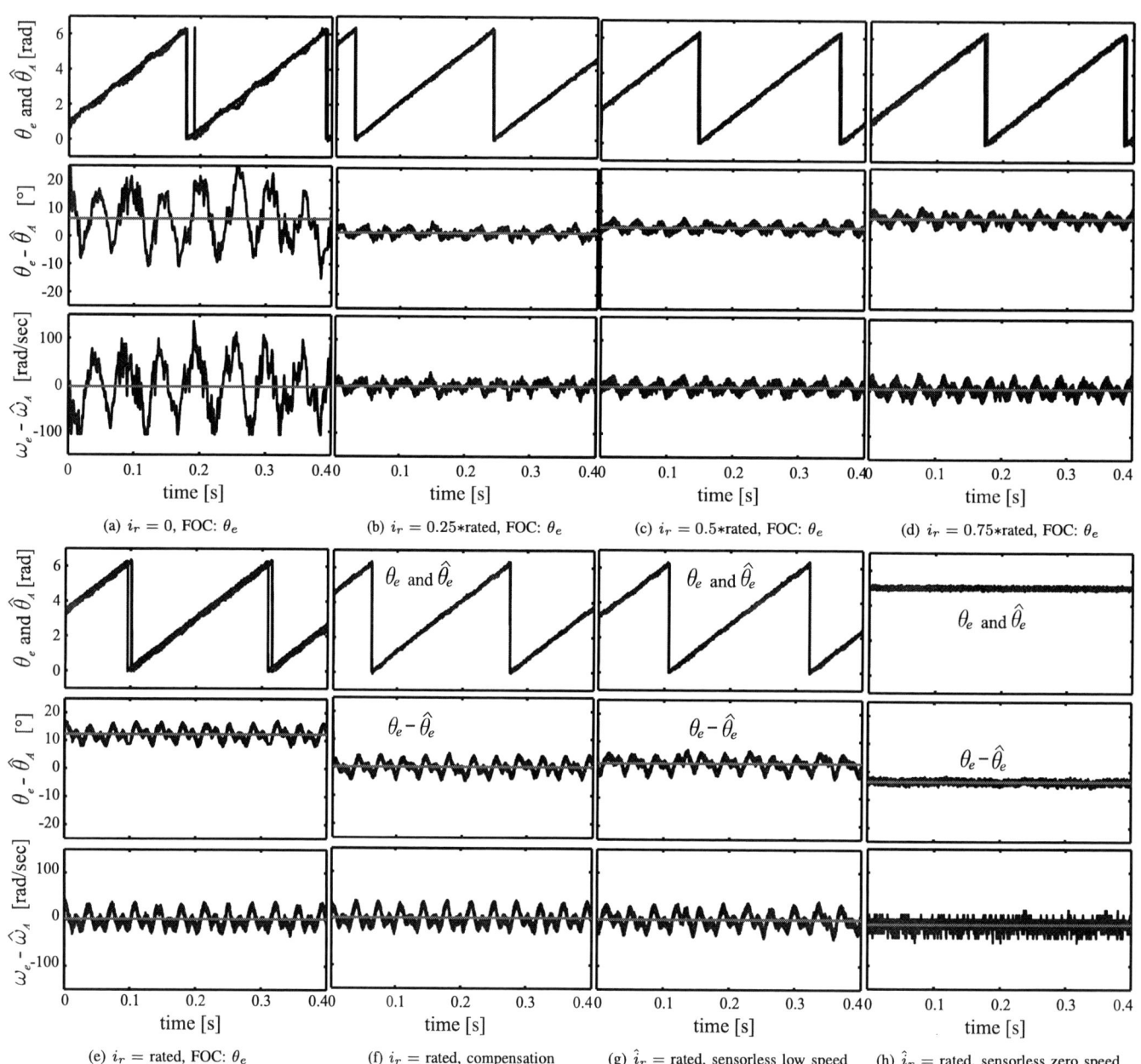

(a) $i_r = 0$, FOC: θ_e (b) $i_r = 0.25*$rated, FOC: θ_e (c) $i_r = 0.5*$rated, FOC: θ_e (d) $i_r = 0.75*$rated, FOC: θ_e

(e) $i_r = $ rated, FOC: θ_e (f) $i_r = $ rated, compensation   (g) $\hat{i}_r = $ rated, sensorless low speed   (h) $\hat{i}_r = $ rated, sensorless zero speed

Fig. 8. Steady state tests with $\phi_r = 60°$, compensation and sensorless control

Carrier PWM algorithm in overmodulation range for Multileg Multilevel Inverter

Nguyen Van Nho
HochiminhCity University Of Technology
Department of Electrical Engineering
268 LyThuongKiet, District 10, Hochiminh City
Email: nvnho@hcmut.edu.vn

Hong Hee Lee
University Of Ulsan
Department of Electrical Engineering
680-749 San 29, Muger 2-dong, Ulsan, Korea
Email: hhlee@mail.ulsan.ac.kr

Abstract—This paper presents a new carrier based overmodulation method in multiphase multilevel inverters for unbalanced dc voltage sources. Two-mode overmodulation in multiphase multilevel inverters is deduced in a similar way as in three-phase two-level inverter. It is different from 3-phase inverter, the overmodulation range in multiphase inverters is larger and therefore it is efficient to utilise this range for applications. The offset voltage can be properly designed for achieving required performance. The method will be demonstrated by simulation results.

I. INTRODUCTION

Recently, multiphase inverters have drawn a great attention of specialists in several applications [1]. Two typical multileg and multilevel inverters are NPC and cascaded types (Fig.1 and 2). The multiphase motor drives are advantageous to three phase motor drive in some applications for reducing power demanding on each phase and having a better reliability. In literatures, multiphase motors are mainly investigated from two-level voltage source inverters. Three-phase four wire/leg inverter, which has achieved a significant role in cleaning power utility, introduces an another application of multiphase inverter. Their control theory and practical applications have been investigated for many years.

Multiphase (>3) multilevel inverters have still not been used in practice in a large scale, however developing a generalised PWM theory for multiphase multilevel inverter can be very useful for applications since

- it helps systematically and comfortably to deduce a proper PWM method for individual inverter topology.
- A generalised PWM approach can save time for solving PWM control of novel or hybrid inverter topologies.

Space vector PWM methods have been developed by many researchers. The vectorial descriptions can advantageously give a simple graphical explanation for understanding close relationship between the parameters of the considered electrical system. Space vector diagram can present well the voltage conditions on the load side. However, so far space vector PWM methods have shown to be disadvantageous or even becoming sophisticated in the following issues as [2]:

- Control of four-leg inverter
- Control of multiphase multilevel inverter

- control of common mode and corresponding PWM performance in multilevel inverter
- control of multilevel inverter on condition of unbalance dc voltage sources.
- formulation of a unified and generalised PWM theory, applicable to inverters, not depending on number of levels, number of phases, level of dc voltage unbalance and modulation modes (linear under-, over-modulation).

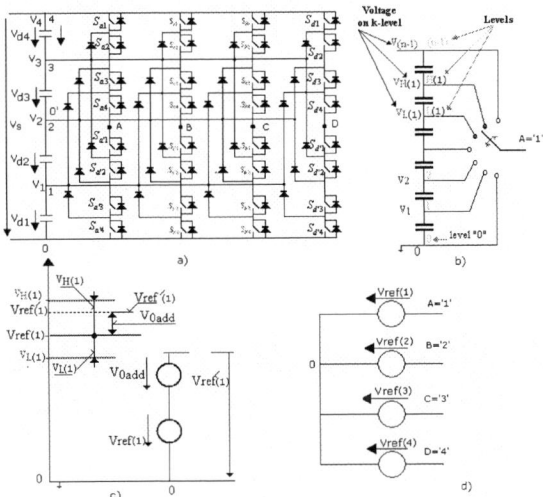

Fig. 1: a) 4-leg 5-level NPC inverter; b) c) and d) Analysis of inverter voltages

For 3-phase two-level inverters, overmodulation has been mostly solved by space vector PWM methods [3]-[5]. For these, the problem as linear control characteristic is often realised using look-up tables or on-line approximate functions. The principle control between two-limit trajectories has introduced a simple overmodulation technique, which can be properly applied to other converter topologies [6]. Similar approaches have been applied for multilevel inverters. For high number of levels and particulaly, if the dc voltage sources vary, the space vector overmodulation will become less flexible to adopt various PWM requirements for being lack of the offset control. For variable dc sources, determining space vector diagram and following switching state sequence for given reference voltage vector will be very complicated. Even in this case, the

978-1-4244-0644-9/07/$25.00 ©2007 IEEE   1027

carrier based PWM technique will present an effcient technique for controlling power converters. Overmodulation in multiphase inverter can play a more significant role comparing to three phase inverter because of its larger operating range. In the paper, carrier based overmodulation will be presented. First, some basic analysis of multiphase multilevel inverter circuits, which is applicable to both under- and over-modulation range, will be mentioned briefly [2]. Then, the proposed overmodulation will be described. Finally, simulation will be implemented to demonstrate the proposed method .

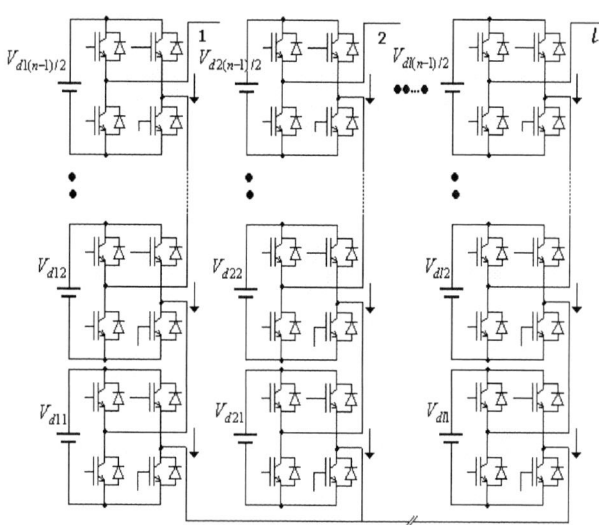

Fig. 2: Circuit diagrams of Multiphase Multilevel cascaded inverter

II. DESCRIPTION OF INVERTER-LEG VOLTAGE

For l-leg n-level inverter, each relevant vector can be described in the l-dimension coordinates. Then, any voltage vector can be expressed as:

$$\vec{V} = [V_1, V_2, ..., V_l]^T \qquad (1)$$

Reference voltage in single-leg inverter: To get convenience for making PWM algorithms, the dc neutral point is supposed to be shifted to the lowest level (0) as shown in Fig.1, 2. Define $V_{ref(x)}$, $x = 1, 2, ..., l$ reference leg voltages between inverter outputs and dc-neutral point "0".

For optimising switching losses in a sampling time period, the reference voltage $V_{ref(x)}$ can be implemented by alternating between two closest levels termed as active low $V_{L(x)}$ and high voltages $V_{H(x)}$, which correspond to levels of $L_{(x)}$ and $H_{(x)}$ on dc side (Fig.1b) and satisfy the conditions as:

$$0 \le V_{L(x)} \le V_{ref(x)} \le V_{H(x)} \le V_{(n-1)} \qquad . \qquad (2)$$

$$\begin{cases} V_{L(x)} \le V_{ref(x)} < V_{H(x)} & if \quad V_{ref(x)} < V_{n-1} \\ V_{H(x)} = V_{n-1} & if \quad V_{ref(x)} = V_{n-1}. \end{cases} \qquad (3)$$

$$H_{(x)} = L_{(x)} + 1; \ x = 1, 2, ..., l . \qquad (4)$$

The phase leg voltage can be analysed as sum of the active voltage and the relevant offset as follows:

$$\vec{V}_{ref} = \vec{V}_{12} + V_{0ref}\vec{I} ; \qquad (5)$$

where $\vec{V}_{ref} = [V_{ref(1)}, V_{ref(2)}, ..., V_{ref(l)}]^T$;

$\vec{I} = [1, 1, ..., 1]^T$ and $\vec{V}_{12} = [V_{12(1)}, V_{12(2)}, ..., V_{12(l)}]^T$.(6)

For symetrical l −phase inverter, the phase components of active voltage vector can be described as:

$$V_{12(k)} = V_m \cos(\theta - 2\pi k / l + 2\pi / l) . \qquad (7)$$

The offset voltage can be any value within the limits V_{0Min} and V_{0Max}, described as:

$$- Min = V_{0Min} \le V_{0ref} \le V_{0Max} = (V_S - Max) \ (8)$$

where

$$\begin{aligned} Min &= Min \ [V_{12\,(1)}, V_{12\,(2)}, ..., V_{12\,(l)}]; \\ Max &= Max \ [V_{12\,(1)}, V_{12\,(2)}, ..., V_{12\,(l)}] \end{aligned} \qquad (9)$$

To get a proper balancing of switching losses between switching devices, the offset voltage can be selected for minimum common mode voltage [1].

III. CARRIER BASED PWM METHODS FOR MULTI-LEG MULTILEVEL INVERTER

For carrier PWM implementing, components of reference modulating vector $\vec{v}_{ref} = [v_{ref(1)}, v_{ref(2)}, ..., v_{ref(l)}]^T$, will be compared with PD multicarrier waveforms for producing the required voltage vector $\vec{V}_{ref}$. The reference modulating vector can be determined from vector of active low levels as $\vec{L} = [L_{(1)}, L_{(2)}, ..., L_{(l)}]^T$ - and nominal modulating signals $\vec{\xi}_{ref}$ as follows [2]:

$$\vec{v}_{ref} = \vec{L} + \vec{\xi}_{ref} \qquad (10)$$

where $\vec{\xi}_{ref} = [\xi_{ref(1)}, \xi_{ref(2)}, ..., \xi_{ref(l)}]^T$ is determined as

$$\vec{\xi}_{ref} = [V_{Ad}]^{-1}(\vec{V}_{ref} - \vec{V}_L). \qquad (11)$$

1028

The parameter $V_{Ad(x)}$- active dc voltage source of x. phase as the difference between the two active voltage levels $V_{H(x)}$ and $V_{L(x)}$ as

$$V_{Ad(x)} = V_{H(x)} - V_{L(x)} \; ; \; x = 1,2,...,l \qquad (12)$$

Corresponding matrix of active dc voltage sources $V_{Ad(x)}$ is of $l \times l$ -dimension and described as:

$$[V_{Ad}] = \begin{bmatrix} V_{Ad(1)} & 0 & ... & 0 \\ 0 & V_{Ad(2)} & ... & 0 \\ ... & ... & ... & 0 \\ 0 & 0 & 0 & V_{Ad(l)} \end{bmatrix} \qquad (13)$$

Modified carrier based PWM For multileg multilevel inverter
The principle of modified carrier based PWM can be applied to both undermodulation and overmodulation ranges. The modified PWM method can be obtained by adding an extra common mode voltage V_{0add} to all phase leg voltages. Then, the resulted voltages and modulating signals will be modified (Fig.3) as

$$\vec{V}'_{ref} = \vec{V}_{ref} + V_{0add}\vec{I} \qquad (14)$$

$$\vec{v}'_{ref} = \vec{L} + [V_{Ad}]^{-1}(\vec{V}_{ref} + V_{0add}\vec{I} - \vec{V}_L) \qquad (15)$$

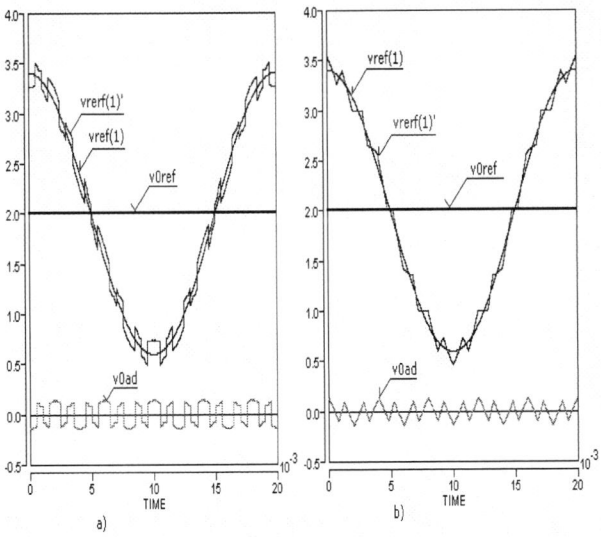

Fig. 3: Five-phase five-level inverter. Modified carrier PWM with minimum common mode and balance dc sources: Diagrams of control signals for a) SVPWM and b) DPWM

IV. PROPOSED   LINEAR OVERMODULATION

Let's suppose sinusoidal waveforms of required phase voltages, which have phase displacement of π / l [rad]. Undermodulation and overmodulation range: the maximum amplitude of phase voltage in undermodulation can be deduced equal to (l is number of phases):

$$V_{(1)mM} = V_S \Big/ \left(2\cos\frac{\pi}{2l} \right) \qquad (16)$$

For five-phase inverter, this limit is $0.527V_S$. Over this limit, overmodulation occurs and its maximum voltage at the six step mode attains a value of $V_{(1)m} = 0.6366V_S$. Let's define the unit modulation index of $m = 1$ corresponding to six step mode. Any deduced   fundamental voltage $V_{(1)m}$ corresponds to modulation index of:

$$m = \frac{V_{(1)m}}{2V_S / \pi} \qquad (17)$$

From the previous consideration, undermodulation and overmodulation control in the 5-phase inverter correspond to the   ranges of $0 \leq m \leq 0.8278$ and $0.8278 < m \leq 1$, respectively. Compared with 3-phase inverter, the lower limit of overmodulation for 5-phase inverter is reduced   more about 10%.

Proposed overmodulation: A linear PWM control including overmodulation range, can be simply implemented using the principle control between two limit trajectories [6],[7]: for a given modulation index m,   the active voltages can be deduced from the defined active voltages $V_{x12,mA}, V_{x12,mB}$ of corresponding two limit modulation indexes of m_A and m_B by:

$$V_{x12,m} = (1-\eta)V_{x12,mA} + \eta V_{x12,mB} \qquad (18)$$

$$\eta = \frac{m - m_A}{m_B - m_A}$$

For undermodulation, the lower limit modulation index   and related voltage amplitude are zero $m_A = 0$, $V_{x12,mA} = 0$. For two-mode overmodulation, three limit trajectories are needed, corresponding to: 1) upper limit of linear modulation range, 2) mediate limit and 3) six-step mode. The first trajectory   presents   a sinusoidal waveform , whose amplitude is defined in (16). The third one corresponding to

1029

fundamental voltage of $\dfrac{2V_S}{\pi}$, has a square waveform. Its active voltage (the 1st phase) can be easily deduced and drawn as shown as $V_{12,3(1)}$ in Fig.5c.

The second trajectory devides overmodulation into two separate ranges: mode 1 for lower range characterised with negligble low-order harmonics and mode 2 for higher range with an increasing of harmonic content before attaining six-step mode. One of simple leg voltages for deducing this trajectory is described as follows:

$$\text{Define } V_{x0} = \begin{cases} V_S & for & f_x > V_S \\ f_x & for & 0 \le f_x < V_S \\ 0 & for & f_x < 0 \end{cases} \quad (19)$$

Where $f_x = 0.5V_S + V_S \cos(\theta - \dfrac{2\pi x}{l} + \dfrac{2\pi}{l})$;

$x = 1,2,3,..,l$ (20)

Overmodulation under oscilating dc sources: Let's assume

that $V_S(t) \ge V_{SMin}$ in considered period, then the previously overmodulation can generate linearly a voltage up to the amplitude of $2V_{SMin}/\pi$.

Modified overmodulation
Limitation range of PWM modulation for unbalanced and oscilating dc voltage sources: The maximum fundamental output voltage is limited by total dc voltage V_S, obtained in the six-step mode can be set equal to:

$V_{(1)mMAX} = \dfrac{2V_{SMIN}}{\pi}$, where V_{SMIN} is a selected constant

value, which does not exceed the total dc source value V_S in the considered fundamental period, i.e. $V_{SMIN} \le V_S(t)$, $0 \le t \le T$. The same results can be concluded for amplitudes of low-order harmonics in overmodulation range. The unbalance between dc sources causes a reduction of PWM quality during each sampling time period. As a result, it influences significantly on the PWM performance of high switching frequency. The flexible control of offset function can be used to produce reference leg voltages in modified PWM techniques. Because of reduction of offset voltage range in overmodulation, it can be seen a small difference between reference voltages in modified PWM methods.

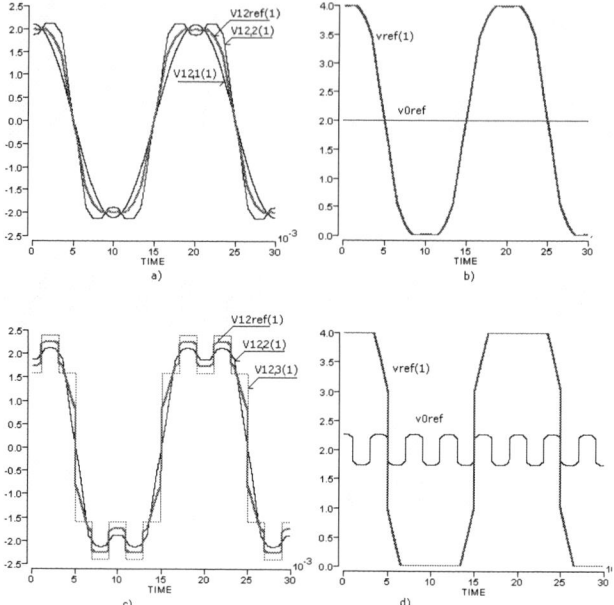

Figure 4: Derivation of reference active voltage $V_{12ref(1)}$ from two limit trajectories for a) mode 1 and c)mode 2. Diagrams of orresponding reference modulating signals $V_{ref(1)}$ and offset for b) mode 1 and d) mode 2.

V. SIMULATION RESULTS

Diagrams of simulation results for undermodulation in 5-phase 5-level inverter on balanced dc sources using software DYNAST were shown in Fig.5. There were drawn the diagrams of active voltage, corresponding reference leg voltage for medium common mode, phase load voltage and currents for m=0.5. For existence of 3-harmonic in overmodulation mode 1, the waveforms of load currents are distorted as shown in Fig.6. In the diagrams of Fig.7, the overmodulation mode 2 presents a high influence of harmonics of low order. For mode 1, the third harmonic amplitude appears strongly. It amplitude increases influence in mode 2 and attains a value of more than 30% of fundamental voltage. Influence of other harmonics are negligible in mode 1, but they increase significantly to about 10% of the fundamental value in mode 2 (Fig.8). For unbalanced dc sources, the diagrams of modified modulating signals were calculated and drawn as shown in Fig.9. Finally, similar diagrams of two-mode overmodulation for unbalanced and oscilating dc sources were demonstrated in Fig.11 and 12.

CONCLUSIONS
The paper has presented a simple carrier overmodulation method for multileg multilevel inverters. The control in overmodulation can be realised in a similar way as for three-phase inverter. Carrier based PWM approach has shown to be simple for controlling power converters on various complicated conditions.

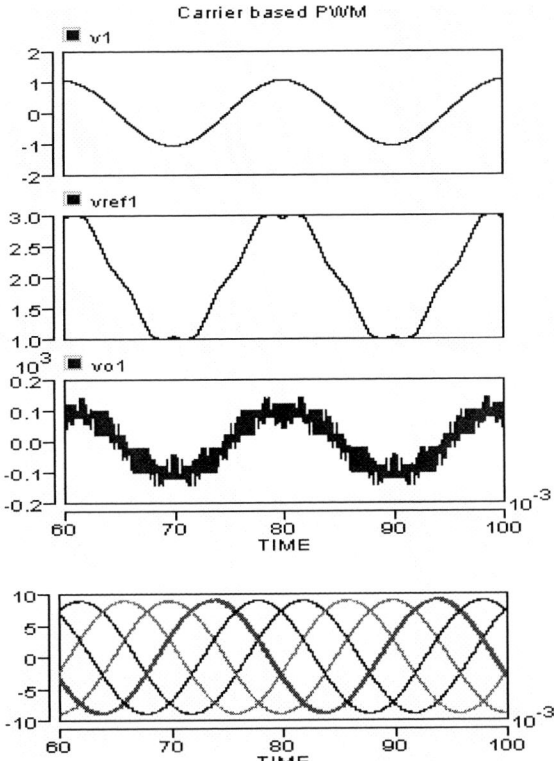

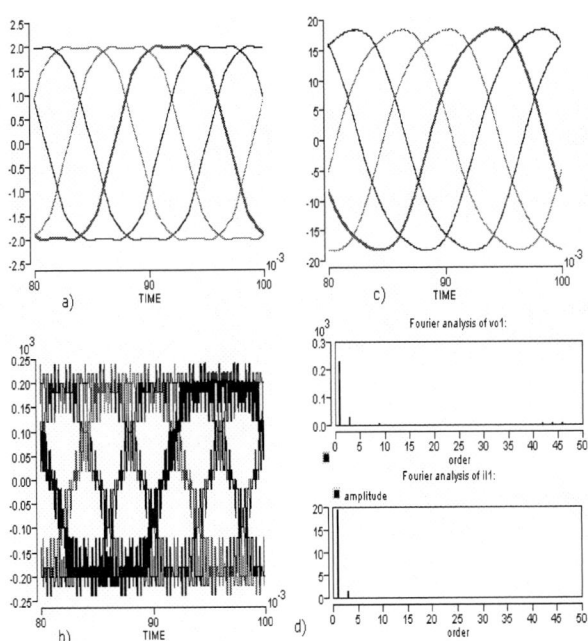

Figure 5: Simulation results. 5-phase 5-level inverter. Undermodulation for balanced dc sources: $V_{d1}=V_{d2}=V_{d3}=V_{d4}=100V$, m=0.5. PWM control with medium common mode voltage. From top to bottom: Diagrams of active voltage, reference voltage, phase load voltage and currents. f_{sw}=3KHz; R=10 Ω, L=20mH.

Figure 6: Simulation results. 5-phase 5-level inverter for balanced dc sources, R=10 Ω, L=0.02H, f_{sw}=3kHz. Overmodulation mode 1, $\eta = 0.5$. Diagrams of a) active voltages, b) load voltages, c) load curents and d) Fourier analysis of harmonic components of phase load voltage and current

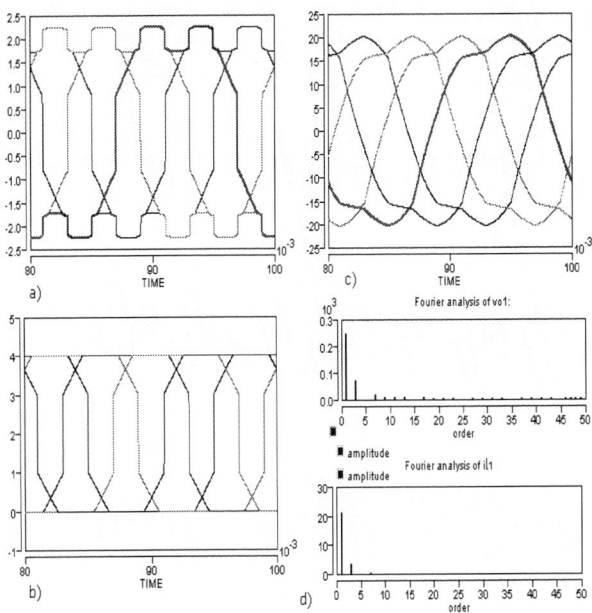

Figure 7: Simulation results. 5-phase 5-level inverter for balanced dc sources, R=10 Ω, L=0.02H, f_{sw}=3kHz. Overmodulation mode 2, $\eta = 0.5$. Diagrams of a) active voltages, b) modified modulating signals in DPWM, c) load curents and d) Fourier analysis of harmonic components of load voltages and currents.

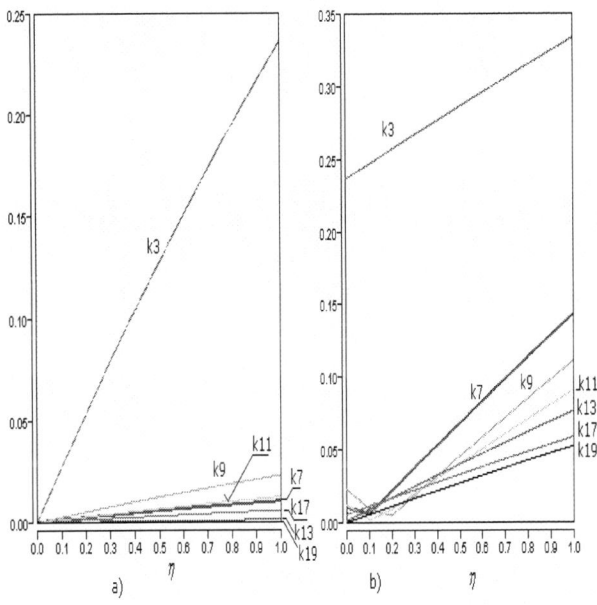

Figure 8: 5-phase inverter. Diagrams of ratio functions $k_n = V_{n(m)} / V_{1(m)}$ in relation to modulation index in a) mode 1 and b) mode 2.

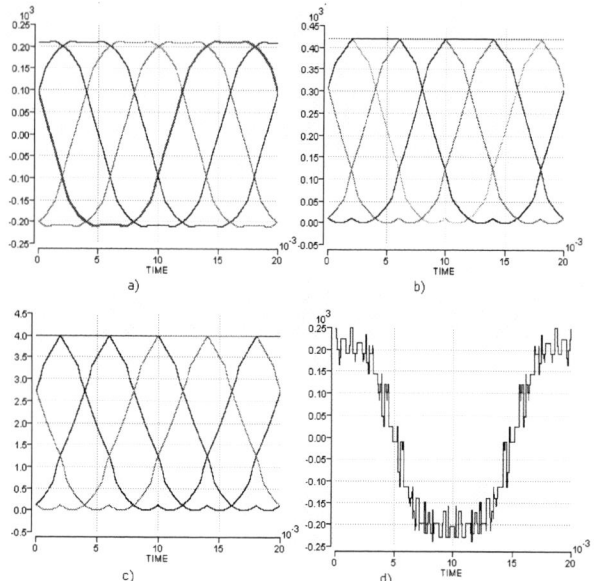

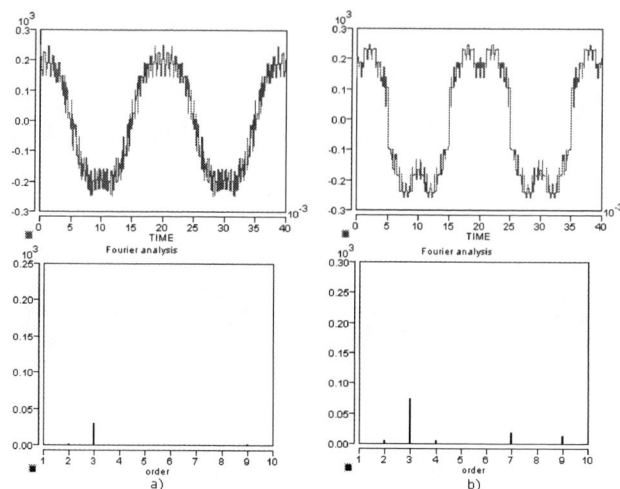

Figure 11: Simulation results. 5-phase 5-level inverter with unbalanced dc sources. Diagrams of phase voltage and harmonic analysis deduced from Fig.11 for a) mode 1 and b) mode 2.

Figure 9: 5-phase 5-level NPC inverter. Dc sources: V_{d1}=90[V]; V_{d2}=130[V]; V_{d3}=120[V]; V_{d4}=80[V]. Selected Min. total dc voltage V_{sMin}=420[V]. Minimum common mode against dc midpoint voltage of $V_{mid}=V_{d1}+V_{d2}$. Overmodulation mode 1 with $\eta = 0.5$. Carrier frequency 1500Hz, output frequency 50Hz. Diagrams of active voltages, modified leg voltages, reference modulating signals and A-phase output voltage.

ACKNOWLEDGMENT

We would like to thank Vietnam National University-HCM, Vietnam for partial support and Ministry of Commerce, Industry and Energy and Ulsan Metropolitan City, which partly supported my research through the Network-based Automation Research Center (NARC) at University of Ulsan.

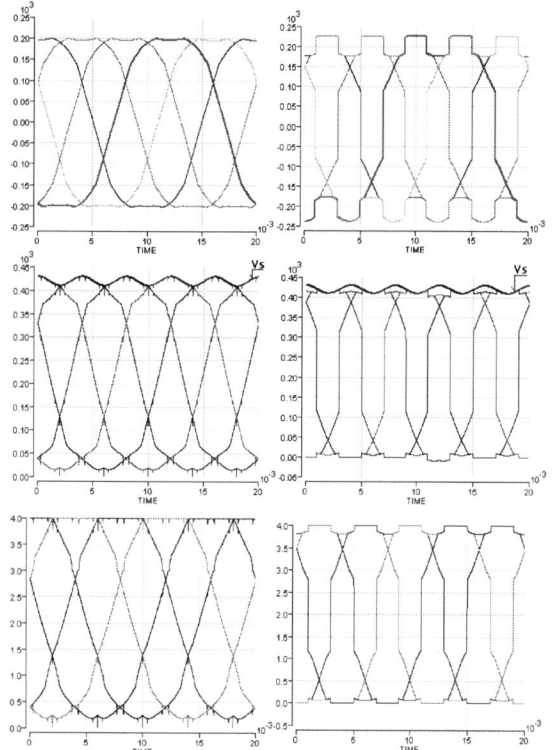

Fig. 10: 5-phase 5-level NPC inverter with variable dc source:. Dc sources V_{d1}=90+10cos(5wt)[V]; V_{d2}=130[V]; V_{d3}=120[V]; V_{d4}=80[V]; V_{sMin}=400[V]; $V_{mid}=V_{d1}+V_{d2}$; a) Overmodulation mode 1 with $\eta = 0.5$ and b) overmodulation mode 2 with $\eta = 0.5$.

From top to bottom, diagrams of active voltages, modified leg voltages and reference modulating signals.

REFERENCES

[1] John W. Kelly, Elias G. Strangas, and John M. Miller, "Multiphase Space Vector Pulse Width Modulation", IEEE Transactions on Energy Conversion,, Vol. 18, No. 2, June 2003

[2] N.V.Nho,H.H.Lee,"Carrier PWM Algorithm For Multi-leg Multilevel Inverters', EPE 2007 - 12th European Conference on Power Electronics and Applications 2 - 5 September 2007, Aalborg, Denmark

[3] J.Holtz, W. Lotzkat and Khambadkone, "On continuous control of PWM inverters in the overmodulation range including the six step mode", IECON, 18th ann. Conf, IEEE Indus. Electronics San Diego, 1992, p.307

[4] *Hava, A.M.; Kerkman, R.J.; Lipo, T.A., "Carrier-based PWM-VSI overmodulation strategies: analysis, comparison, and design "*, Power Electronics, IEEE Transactions on , Volume:13 Issue: 4, July 1998, Page(s): 674 -689

[5] S. Bolognani, M. Zigliotto, "Novel digital control of SVM inverters in the overmodulation range", *IEEE Trans. Industry Applications, vol.33, no.2, pp. 525-530,1997*

[6] N.V.Nho, M.J. Youn " Two-mode overmodulation in two-level VSI using principle control between limit trajectories", CD-ROM Proceedings PEDS 2003, pp.1274-1279

[7] N.V.Nho,M.J. Youn, Comprehensive Study On SVPWM and Carrier Based PWM Correlation In Multilevel Inverters, IEE-Proceedings Electric Power Applications, Jan. 2006, Vol.153, No.1,pp.149-158

1032

Carrier Based Single-state PWM Technique In multilevel Inverter

Nguyen Van Nho[1], Quach Thanh Hai[2]
HochiminhCity University Of Technology
Department of Electrical Engineering
268 LyThuongKiet, District 10, Hochiminh City
Email: nvnho@hcmut.edu.vn

Hong Hee Lee[3]
University Of Ulsan
Department of Electrical Engineering
680-749 San 29, Muger 2-dong, Ulsan, Korea
Email: hhlee@mail.ulsan.ac.kr

Abstract— In the paper, a novel analysis of carrier based PWM methods for multilevel inverters is presented. The space vector PWM and carrier based PWM correlations are investigated in a nominal two-level switching diagram. The obtained results will be applied to design various carrier PWM techniques. In this paper, a carrier based single-state PWM technique, which reduces number of switchings and optimizes active voltage errors will be presented. This technique can be advantageous in high level inverters. The carrier base PWM approach shows a flexible offset control. The proposed method is mathematically formulated and demonstrated by simulation and experimental results.

I. INTRODUCTION

Nowadays, for increasing use in practice and fast developing of high power devices and related control techniques, multilevel inverters have become more attractive to researchers and industrial companies. Two common inverter topologies are NPC and cascaded multilevel inverters (Fig.1 and 2). In recent days, for reducing hardware construction cost, it has been shown a try to develop prospective hybrid multilevel inverters. There are basically three PWM schemes for controlling multilevel inverters as: Carrier based PWM, space vector PWM and selective harmonics elimination PWM methods.

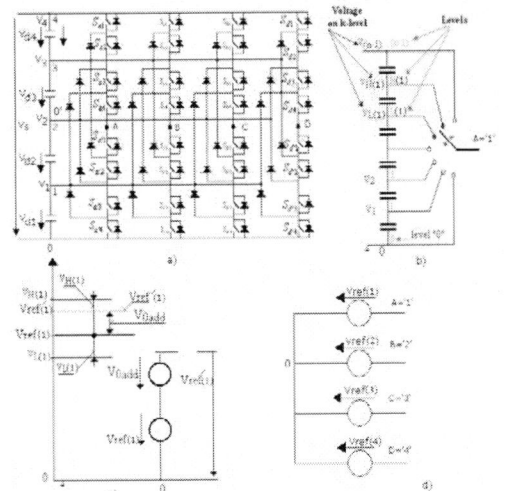

Fig. 1: a) 4-leg 5-level NPC inverter; b) c) and d) Analysis of inverter voltages

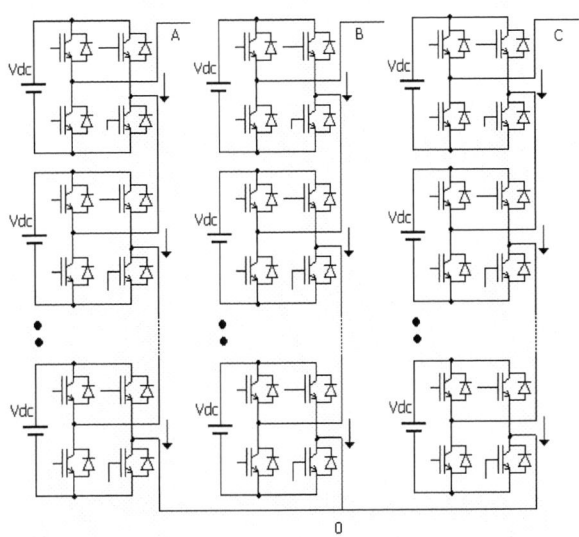

Figure 2: 3-phase Cascaded multilevel inverter

The comprehensive correlation between carrier based PWM and SVPWM have been derived in the recent work [1],[2]. Compared to the space vector PWM methods, the carrier based PWM methods can be advantageously utilised in: 1) controlling common mode voltage, 2) controlling of complicated inverter topologies as 4-leg, 5-leg-,... multilevel inverters, 3) compensation of unbalanced dc sources. It will be shown that the carrier PWM technique can become a possible solution for some approximate PWM methods, which use one or two switching states in a switching state sequence and produce reference voltages with certain voltage error. The drawbacks as nonlinear control characteristics and existence of low-order harmonics of output voltages will be compensated by reduced number of switchings in each sampling time period. A common characteristic of carrier based approximate PWM methods is that the offset function can be properly designed to control the PWM performance. Single state space vector PWM method has been described in some recent paper [3], one of its drawback is the limitation of output voltage range. The methods of selecting the voltage vector in Direct torque control and hysteresis current loop control for AC motor drive systems are some typical and well-known applications of single state PWM technique. In the paper, the carrier based single state PWM will be proposed for minimum voltage error. The only controllable parameter of this method

978-1-4244-0644-9/07/$25.00 ©2007 IEEE

is the offset voltage, which does not influence on the active voltage but able to set approximately common mode voltage and balance switching losses. In the paper, it will be shown that any PWM scheme of multilevel inverter can be centered in a nominal two-level switching state diagram. This makes the PWM study to become more advantageous and comfortable. The proposed method is explained for NPC inverters, its proper modifying can be also applied to cascade topologies.

II. CIRCUIT DESCRIPTION AND NOMINAL SWITCHING DIAGRAM IN MULTILEVEL INVERTER FOR BALANCE DC VOLTAGE SOURCES

Assumption: each dc voltage cell is constant and equal to a unit. Define reference leg voltages between output and dc-neutral point "0", consisting of active voltages $v_{x12}, x = a, b, c$ and reference common mode v_{0ref} (Fig.4) as:

$$v_{xref} = v_{x12} + v_{0ref} . \qquad (1)$$

Or in the vector form as:

$$\vec{v}_{ref} = \vec{v}_{12} + v_{0ref}\vec{I}$$

Active voltages, which exist at the three phase load voltages can be determined from the amplitude and phase angle of voltage vector as follows:

$$v_{a12} = v_{ref}\cos\theta$$
$$v_{b12} = v_{ref}\cos(\theta - 2\pi/3) ; \qquad (2)$$
$$v_{c12} = v_{ref}\cos(\theta - 4\pi/3)$$

Define Max and Min as maximum and minimum values from three phase active voltages as:

$$Max = Max(v_{a12}, v_{b12}, v_{c12}) \qquad (3)$$
$$Min = Min(v_{a12}, v_{b12}, v_{c12})$$

Reference common mode voltage can be proposed of any value, varying within the limits of v_{0Max} and v_{0Min}:

$$
\begin{aligned}
v_{0Max} &= (n-1) - Max \\
v_{0Min} &= -Min
\end{aligned} \qquad (4)
$$

where $\vec{v}_{ref} = \left[v_{aref}, v_{bref}, v_{cref}\right]^T$ and

$$\vec{v}_{12} = \left[v_{a12}, v_{b12}, v_{c12}\right]^T . \qquad (5)$$

Active Low/or High level: each reference leg voltage v_{xref} is produced by subsequent alternating between the two lower and higher active levels as $L_{(x)}$ and $H_{(x)}$, for which the following conditions will be satisfied as:

$$
L_{(x)} = \begin{cases} n_{(x)} & if \quad 0 \le v_{xref} < (n-1) \\ n_{(x)} - 1 & if \quad v_{xref} = (n-1) \end{cases} \qquad (6)
$$

$$H_{(x)} = L_{(x)} + 1$$

where $n_{(x)} = Int(v_{xref}); x = a, b, c$. $\qquad (7)$

Each components of vector $\vec{L} = [L_a, L_b, L_c]^T$ presents possibly a lower level of phase leg voltage in a switching state sequence.

Nominal switching time diagram: To investigate commutation process in a triangle period, a vertical shift of coordinates by the vector $\vec{L} = [L_a, L_b, L_c]^T$ can be implemented as shown in Fig.4. Three-phase active PD carrier bands are overlapped. The diagram is redrawn in a *nominal two-level switching time diagram* as shown in Fig.5. In this nominal diagram, the commutation instants occur depending on the relative voltage level termed nominal modulating signals ξ_x, $x = a, b, c$ defined as:

$$\xi_x = v_{xref} - L_{(x)}; 0 \le \xi_x \le 1; \qquad (8).$$

Or

$$\vec{\xi} = \vec{v}_{ref} - \vec{L}$$

Nominal switching states sequence: The nominal two-level switching diagram in Fig.5 shown that switching time digram in multilevel inverters can be explained using that of two-level inverter. Therefore, in nominal switching diagram, let's define nominal switching states as:

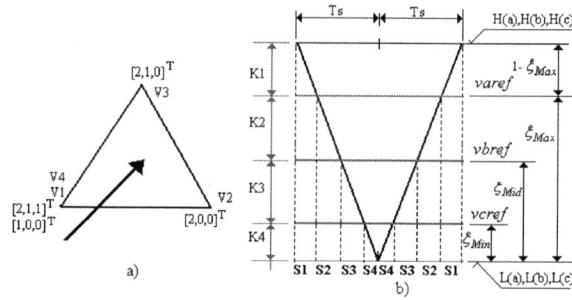

Figure 3: a) reference vector in vector triangle, b) Switching time diagrams in multicarrier PWM.

$$
\begin{aligned}
\vec{s}_1 &= [0,0,0]^T \\
\vec{s}_2 &= [s_{2a}, s_{2b}, s_{2c}]^T \\
\vec{s}_3 &= [s_{3a}, s_{3b}, s_{3c}]^T \\
\vec{s}_4 &= [1,1,1]^T
\end{aligned} \qquad (9)
$$

The first and last states as $\vec{s}_1; \vec{s}_4$ remind two active zero redundant states in switching state sequence of two-level inverter. For the remaining two states as $\vec{s}_2; \vec{s}_3$, the vector components can be determined from relative positions of nominal modulating signals ξ_x.

Define maximum, medium and minimum values of the three phase nominal *modulating signals* as:

1034

$$\xi_{Max} = Max(\xi_a, \xi_b, \xi_c)$$
$$\xi_{Mid} = Mid(\xi_a, \xi_b, \xi_c) \quad (10)$$
$$\xi_{Min} = Min(\xi_a, \xi_b, \xi_c)$$

The components of vectors $\vec{s}_2 ; \vec{s}_3$ can be derived as follows:

$$s_{2x} = \begin{cases} 1 & if & \xi_x \geq \xi_{Max} \\ 0 & else \end{cases}$$

$$s_{3x} = \begin{cases} 1 & if & \xi_x \geq \xi_{Mid} \\ 0 & else \end{cases} \quad (11)$$

Switching state sequence in multilevel inverter

$\vec{S}_1, \vec{S}_2, \vec{S}_3, \vec{S}_4$: can be easily deduced from active low voltage level $\vec{L}$ and nominal switching states as

$$\vec{S}_j = \vec{L} + \vec{s}_j \quad (12)$$

Switching time duties: Reference voltage vector can be described as:

$$\vec{v}_{ref} = K_1 \vec{S}_1 + K_2 \vec{S}_2 + K_3 \vec{S}_3 + K_4 \vec{S}_4 \quad (13)$$

The switching time duties as K_1, K_2, K_3, K_4 can be determined easily from nominal switching diagram as:

$$K_1 = 1 - \xi_{Max};$$
$$K_2 = \xi_{Max} - \xi_{Mid}; K_3 = \xi_{Mid} - \xi_{Min}; K_4 = \xi_{Min}$$
$$K_{14} = 1 - \xi_{Max} + \xi_{Min}$$
$$K_1 + K_2 + K_3 + K_4 = 1 \quad (14)$$

Conventional PWM techniques attain zero active voltage error. For improving output quality, the offset can be regulated within the range of (v_{0Min}, v_{0MAX}). An extra adjustment of the offset within the range defined in (15) can be implemented in various modified PWM methods.

$$-K_4 \leq v_{0add} = e_0 = \xi_{0add} \leq K_1 \quad (15)$$

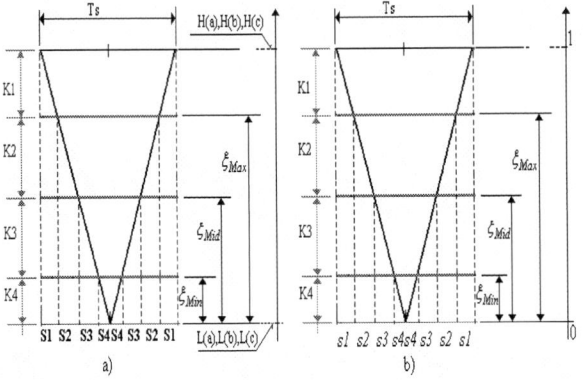

Figure 4: a) Switching time diagram deduced in a) new defined coordinates and b) Nominal switching time diagram.

III. PROPOSED SINGLE STATE PWM METHOD

In single-state PWM technique, there is no commutation in a sampling period. The reference vector $\vec{v}_{ref}$ will be modified and attain one from 4 relevant vectors $\vec{S}_1, \vec{S}_2, \vec{S}_3$ and $\vec{S}_4$, defined as (Fig.5b):

$$\vec{v}_{ref}^{'} = \vec{S}_j ; j \in \{1,2,3,4\} \quad (16)$$

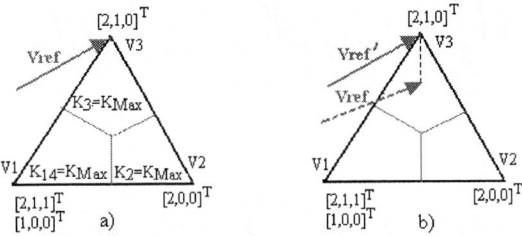

Figure 5: a) Reference vector in single-state switching and function K_{max} in a triangle, and b) Principle of single-state PWM method.

Define function K_{Max} as the largest from three time duties of K_{14}, K_2 and K_3:

$$K_{Max} = Max(K_{14}, K_2, K_3) \quad (17)$$

The value of K_{Max} in a triangle area is described in Fig.5a. In sub-area, where the reference vector $\vec{V}_{ref}$ is located and the condition $K_j = K_{Max}, j \in \{14,2,3\}$ is satisfied, the active error will be minimized if $\vec{v}_{ref}^{'} = \vec{S}_j$. Particularly, if $K_{Max} = K_{14}$, both vectors $\vec{S}_1$ and $\vec{S}_4$ have the same active error $|\vec{e}_{12}|$ and minimizing the offset error e_0 can be considered as condition for selecting the vector $\vec{v}_{ref}^{'}$. For example, the vector $\vec{S}_1$ is select if

$$\left| Offset(\vec{S}_1 - \vec{v}_{ref}) \right| < \left| Offset(\vec{S}_4 - \vec{v}_{ref}) \right|. \quad (18)$$

Or $\dfrac{K_2 + 2K_3 + 3K_4}{3} < 1 - \dfrac{(K_2 + 2K_3 + 3K_4)}{3}$

Table 1: Algorithm for single-state PWM method

Conditions	Selected vectors
$K_{14} > K_2 ; K_{14} > K_3$ $K_2 + 2K_3 + 3K_4 < 1.5$	$\vec{v}_{ref}^{'} = \vec{S}_1$
$K_2 > K_3 ; K_2 > K_{14}$	$\vec{v}_{ref}^{'} = \vec{S}_2$
$K_3 > K_2 ; K_3 > K_{14}$	$\vec{v}_{ref}^{'} = \vec{S}_3$
$K_{14} > K_2 ; K_{14} > K_3$ $K_2 + 2K_3 + 3K_4 > 1.5$	$\vec{v}_{ref}^{'} = \vec{S}_4$

Table 2: Relation between active errors and corresponding selected vectors

Active voltage error e_{12}	$\vec{v}_{ref}'$
$\dfrac{2}{3}\sqrt{K_2^2 + K_3^2 + K_2 K_3}$	$\vec{S}_1$
$\dfrac{2}{3}\sqrt{K_{14}^2 + K_3^2 + K_{14} K_3}$	$\vec{S}_2$
$\dfrac{2}{3}\sqrt{K_{14}^2 + K_2^2 + K_{14} K_2}$	$\vec{S}_3$
$\dfrac{2}{3}\sqrt{K_2^2 + K_3^2 + K_2 K_3}$	$\vec{S}_4$

The corresponding active errors are deduced in Table 2 and drawn in Fig.6.

If reference vector is located at the center of triangle, i.e. $K_{14} = K_2 = K_3 = 1/3$, active error can achieve a maximum value of

$$e_{12Max} = 2/(3\sqrt{3}). \qquad (19)$$

The influence of offset control is demonstrated by corresponding reference modulating signals for 5-level inverter in Fig.7. There have been calculated reference modulating signals for cases of minimum common mode (Fig.7a, c and e) and medium common mode (Fig.7b, d and f). From the study, single state PWM with minimum common mode offset presents advantageous compared to that of medium common mode for reduced number of extra commutations in a large modulation index range.

IV. OVERMODULATION IN SINGLE-STATE PWM METHODS

Because of producing output voltage with nonzero error, the single-state PWM method has a non-linear control characteristic and generates low-order harmonic voltages for the whole modulation range. Compared to conventional PWM methods, overmodulation in single-state PWM method losses its original meaning. However, overmodulation can be supposed to be an approach to extend the reference fundamental voltage $V_{(1)mref}$ to a maximum value in six-step mode, i.e., attaining a value of $2\dfrac{V_S}{\pi}$. That is, the overmodulation happens if the reference fundamental voltage $V_{(1)mref}$ exceeds the value of $V_s/\sqrt{3}$.

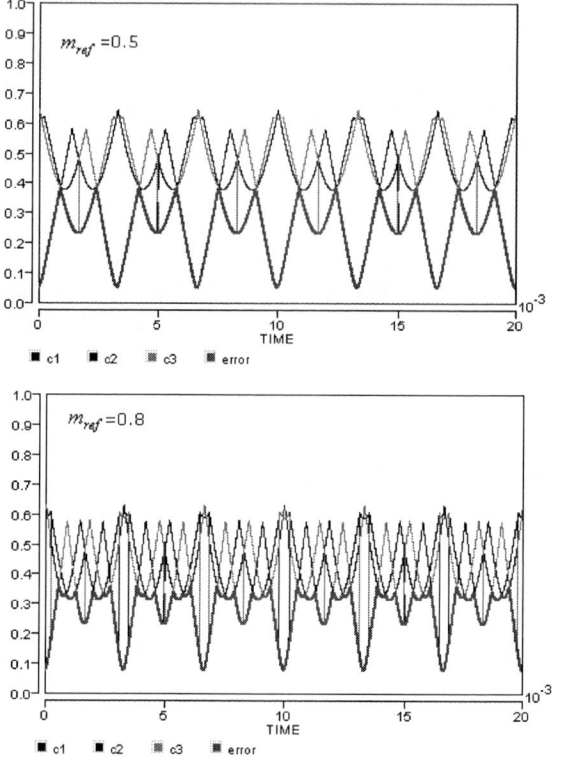

Figure 6: 5-level inverter. Active errors a c1,c2,c3 and "error" in single-state PWM for using corresponding vectors $\vec{V}_1, \vec{V}_2, \vec{V}_3$ and $\vec{V}_{proposed}$.

The active voltages in single-state carrier overmodulation can be deduced from the principle control between two-limit trajectories [1], for which the active modulating signals corresponding modulating index of m, $m_A \leq m \leq m_B$ can be deduced from the active signals of the corresponding limit modulation indexes of m_A, m_B as:

$$v_{x12,m} = (1 - \eta) v_{x12,A} + \eta v_{x12,B}$$

Where $\eta = (m - m_A)/(m_B - m_A)$.

Compared to limit deduced in [3], the carrier based single-state PWM approach can help to get a maximum modulation index up to that of six-step method. Three limit active modulating signals corresponding to modulation indexes of 1,1.055 and 1.1 are proposed (Fig.8). In Fig. 8d, reference modulating signal is deduced for $m_{ref} = 1.03$.

V. SIMULATION AND EXPERIMENTAL RESULTS

Diagrams of the phase leg voltage and line-line voltage in five-level inverter for several modulation indexes in under- and over-modulation have been calculated and drawn in Fig.9,10 and 11. The voltage quality can be more precisely evaluated through the following diagrams of Fourier analysis of low harmonic components and THD factor as shown in Fig.12 and Fig.13. For comparison, corresponding diagrams for 7-level inverter are also included in Fig.14 and 16. The single-state PWM method can be advantageously operated at medium and high modulation index, where harmonic amplitude can be reduced to about 5% to index of 0.45 for 7 level inverter, while this happens around index of 0.8 for 5-level inverter.

Control characteristics: The diagrams of nonlinear control characteristics of five- and seven-level inverters are calculated and drawn in Fig.15 and Fig.16. A better linearity characteristic is obviously obtained for higher level inverters.

Experimental results: A harware set - Five-level cascaded inverter has been built to validate the theoretical analysis. The hardware parameters are as following: IGBT IRG450UD, three-phase load R=20 Ω ; each dc source Vdc=35.6Vdc; the measuring osciloscope TDS2012 and Control kit eZdsp TMS320F2812. The diagrams of leg voltage, line-line voltage and its FFT analysis for modulation indexes of 0.4, 0.85 and 1.05 are measured and shown in Fig.17,18 and 19. These diagrams are similar to the simulated waveforms in Fig.9-11.

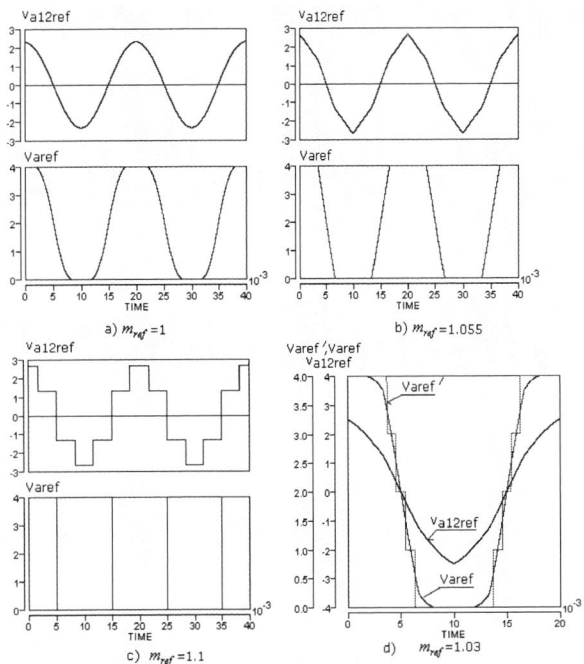

Figure 8: a),b) and c) Diagrams of A-phase active voltage and leg voltage for limit modulation indexes of 1,1.055 and 1.1 with minimum common mode and d) derivation of reference modulating signals for m=1.03.

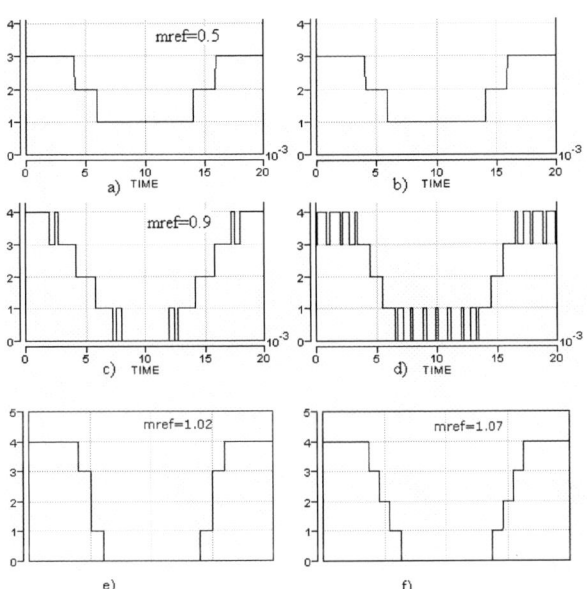

Figure 7: 5-level inverter: Diagrams of leg-pole voltage for single-state PWM method with minimum CM offset for a) $m_{ref} = 0.5$, c) $m_{ref} = 0.9$, and with medium CM offset for and for b) $m_{ref} = 0.5$, d) $m_{ref} = 0.9$ and e) f) overmodulation.

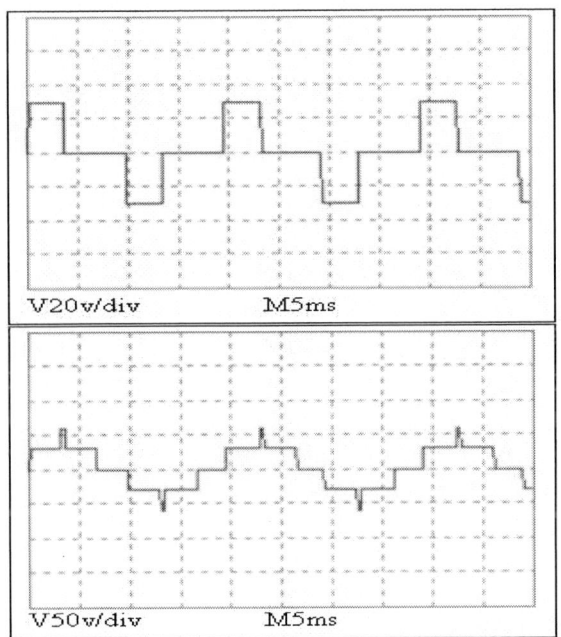

Figure 9: Five-level inverter. Simulation results. Diagrams of leg voltage and line-line voltage for m$_{ref}$=0.4.

1037

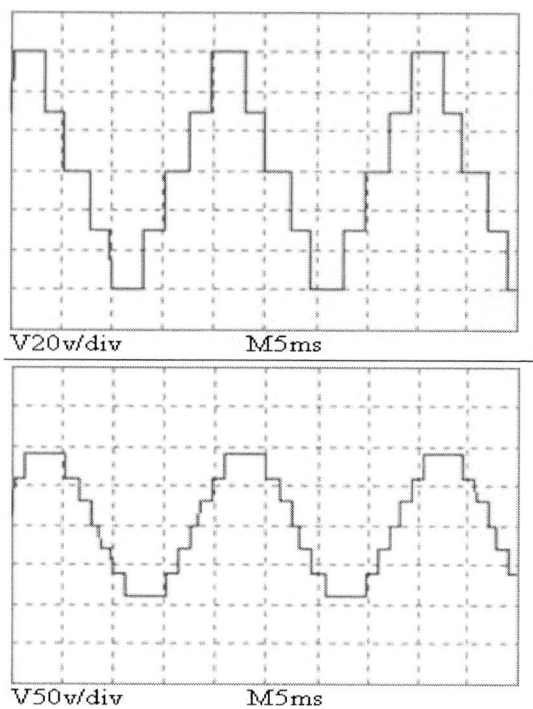

V20v/div   M5ms

V50v/div   M5ms

Figure 10: Five-level inverter. Simulation results. Diagrams of leg voltage and line-line voltage for m_{ref}=0.85.

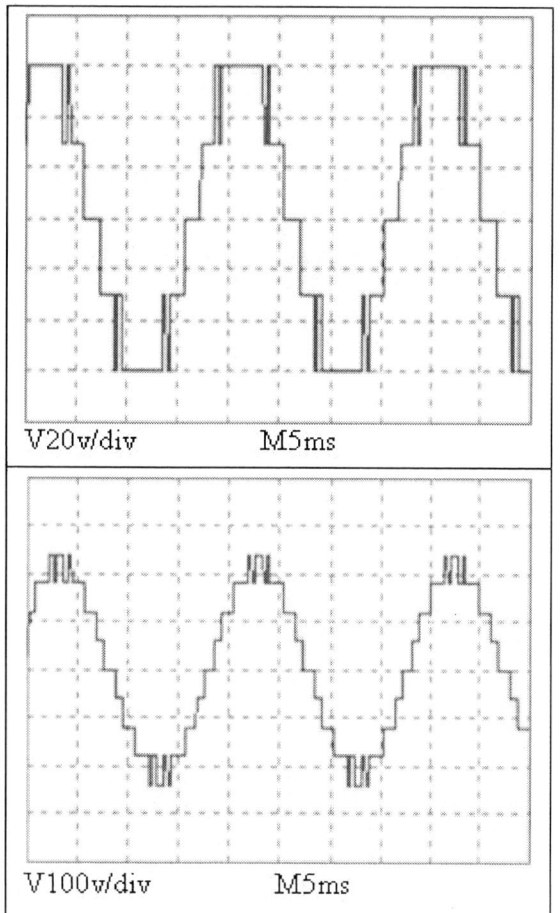

V20v/div   M5ms

V100v/div   M5ms

Figure 11 : Five-level inverter. Simulation results. Diagrams of leg voltage and line-line voltage for m_{ref}=1.05.

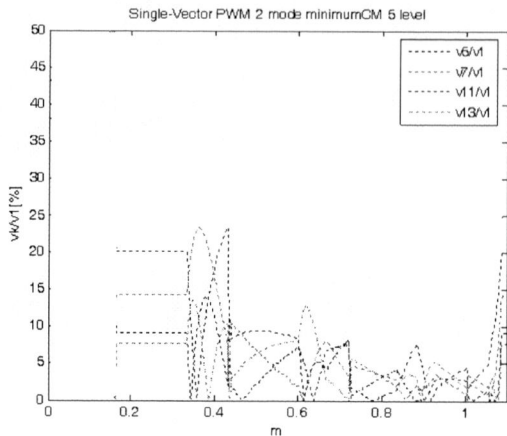

Figure 12: Single-state PWM for 5-level inverter. Harmonic voltage diagrams

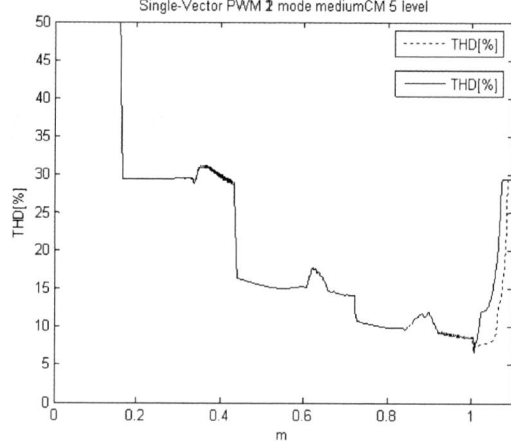

Figure 13: 5-level inverter. Diagram of voltage THD factor

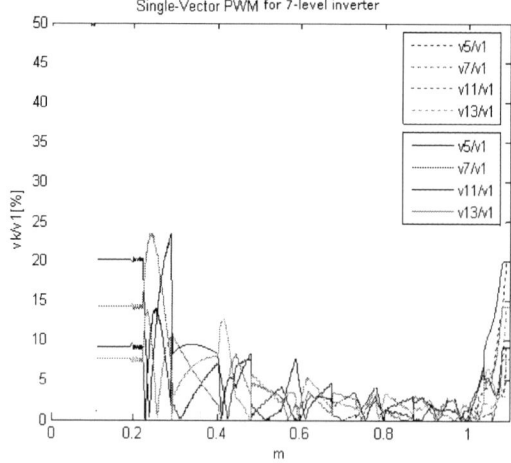

Figure 14: 7-level inverter. Single state PWM method: Fourier analysis diagram of harmonic voltage components .

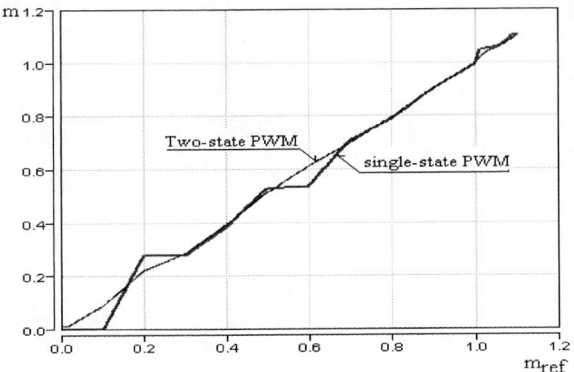

Figure 15: 5-level inverter. Diagrams of control characteristics of fundamental voltages for entire modulation index range in single-state and two-state PWM methods.

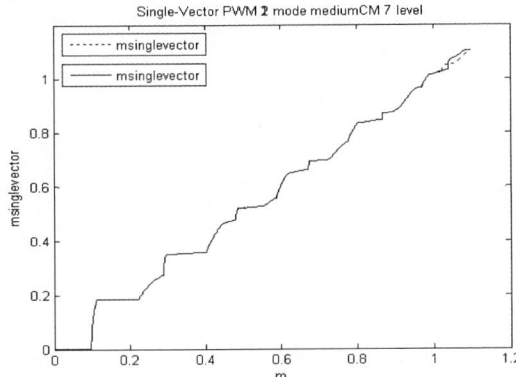

Figure 16: Single state PWM method: Control characteristics for 7-level inverter

VI. CONCLUSIONS

In the paper, the carrier based single-state PWM method for reducing the switching losses for multilevel inverter has been proposed. It has been introduced a proper tranformation of switching state diagram of multicarrier modulation into a nominal switching time diagram. The proposed PWM method can be extended to full voltage range to six-step mode. The PWM performance as switching losses and its distribution on switching devices and common mode voltage can be improved by corresponding selection of common mode voltage.

REFERENCES

[1]   N.V.Nho, M.J.Youn," Comprehensive study on Space vector PWM and carrier based PWM correlation in multilevel invertors", IEE Proceedings Electric Power Applications, Vol.153, No.1, pp.149-158, Jan. 2006

[2]   N.V.Nho,H.H.Lee," Optimised Discontinuous PWM for multilevel inverter with variable load power factor", PESC 2006

[3]   Jose Rodríguez, Luis Morán, Pablo Correa and Cesar Silva,"A Vector Control Technique for Medium-Voltage Multilevel Inverters', IEEE TRANSACTIONS ON INDUSTRIAL ELECTRONICS, VOL. 49, NO. 4, AUGUST 2002

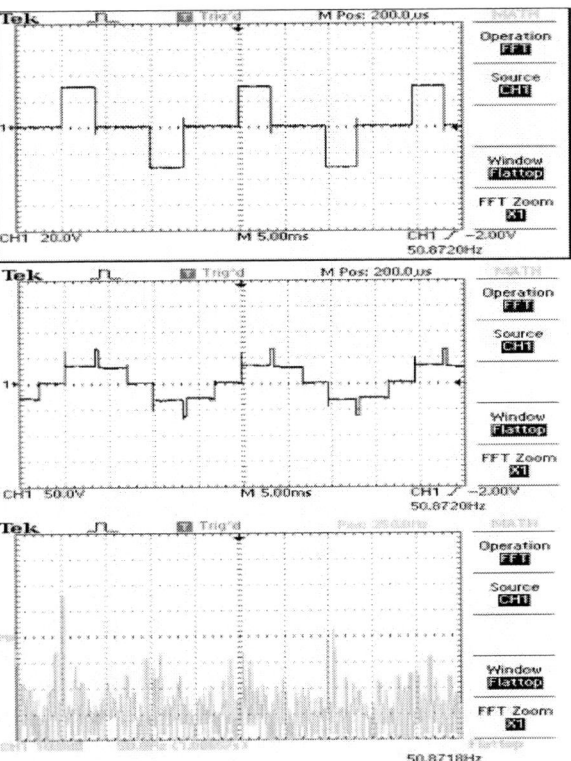

Figure 17: Five-level inverter. Experimental results. Diagrams of leg voltage, line-line voltage and Fourier analysis of line-line voltage for mref=0.4.

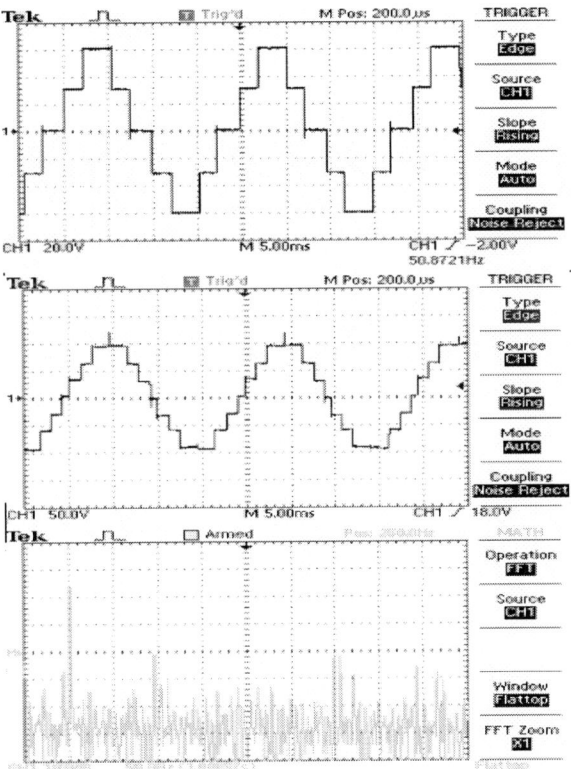

Figure 18: Five-level inverter. Experimental results. Diagrams of leg voltage, line-line voltage and Fourier analysis of line-line voltage for mref=0.85.

1039

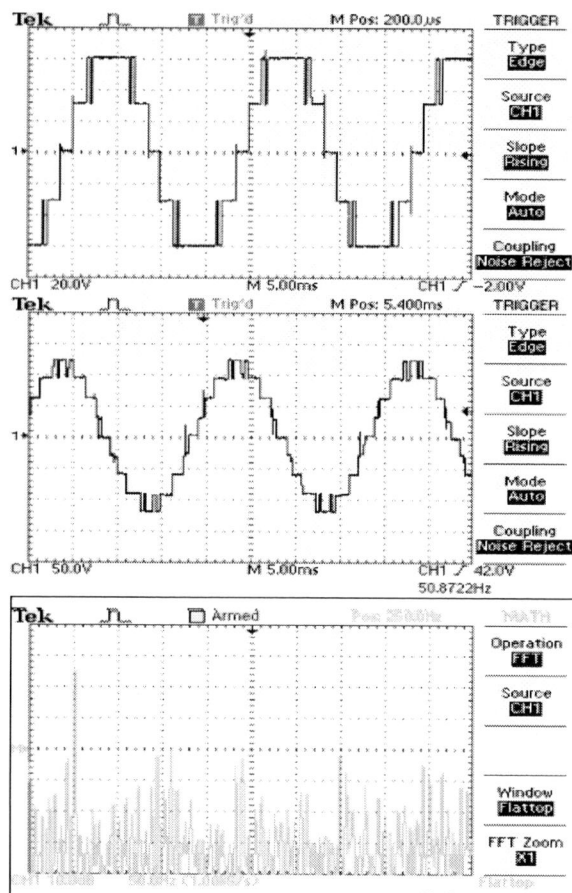

Figure 19 : Five-level inverter. Experimental results. Diagrams of leg voltage, line-line voltage and Fourier analysis of line-line voltage for m_{ref}=1.05

Implementation of a Single-carrier Multilevel PWM Technique Using Field Programmable Gate Array (FPGA)

N. A. Azli and L. Y. Teng
Energy Conversion Department
Faculty of Electrical Engineering
Universiti Teknologi Malaysia
Johor, Malaysia
naziha@ieee.org

P. Y. Lim
School of Engineering and Information Technology
Universiti Malaysia Sabah
Sabah, Malaysia
lpy@ums.edu

Abstract—**The application of Field Programmable Gate Array (FPGA) in the development of power electronics circuits control scheme has drawn much attention due to its high computation speed, shorter design cycle and higher density. A single-carrier multilevel PWM technique has been recognized for a modular structured multilevel inverter (MSMI). This paper presents the implementation of the PWM technique using FPGA. A simulation study has been conducted on the operation of the MSMI based on this technique using MATLAB/Simulink. For hardware implementation, the gate signals of the MSMI power devices have been generated using MAX+PLUS II software and downloaded into an FPGA device (FLEX10K20) from Altera. The gate signals generated by the FLEX10K20 have been verified as similar to that obtained from the simulation study.**

Keywords—FPGA, modular structured multilevel inverter, single-carrier multilevel PWM technique

I. INTRODUCTION

Multilevel inverters have drawn much of attention in recent years particularly in high voltage and high power applications. This is due to several of its advantages compared to the conventional two-level output inverters in handling high power conversion. The main feature of a multilevel inverter is its ability to operate at high DC-bus voltages when using series connections of power devices and reduced output voltage harmonics by switching between multiple voltage levels. Generally, the development of a multilevel inverter system can be broadly divided into two issues namely, power circuit topology and switching technique. For circuit topology, three main types of development have been reported [1]: (i) diode-clamped multilevel inverters (DCMI), (ii) flying capacitor multilevel inverters (FCMI) and (iii) modular structured multilevel inverters (MSMI) or typically known as the cascaded H-bridge multilevel inverters.

The second aspect that defines the multilevel inverter performance is the switching technique. This is closely related to the harmonic profile of the multilevel inverter output voltage waveform. Work on extending various switching techniques that have been applied on the conventional inverter topology such as sinusoidal Pulse Width Modulation (SPWM), selective harmonic elimination PWM (SHEPWM), optimized PWM

(OPWM) and space vector PWM (SVPWM) to the multilevel inverter topology has been reported in literature [2]-[4].

A multicarrier multilevel PWM technique with three carrier disposition schemes was suggested in [2] as Phase Opposition Disposition (POD), Phase Disposition (PD) and Alternative Phase Opposition Disposition (APOD). A single-carrier multilevel PWM technique has also been proposed in [5] for an MSMI for fuel cell applications. The technique has also been highlighted in [6] and [9] as a basis to a regular sampled PWM switching strategy for the MSMI. For practical implementation purposes, the switching techniques for the multilevel inverters have been realized using various tools. The OPWM switching technique for instance has been implemented on an MSMI using a Digital Signal Processor (DSP) as elaborated in [7]. The same switching technique has also been successfully implemented using an FPGA [8]. Microcontrollers have also been used as a gate signal generator based on the sinusoidal PWM switching technique for an MSMI [9]. Furthermore, development of MATLAB/Simulink blocks used in conjunction with MATLAB/Real-Time Workshop and dSPACE/Real-Time Interface to generate the gate signals for an MSMI has been reported in [10].

The previous work have shown that the hardware implementation using the various tools described earlier is capable of producing gate signals that are in good agreement with the simulation results. However, in some implementation, the accuracy of the signals produced is affected due to the sampling rates. Some of the work presented generation of gate signals output by referring to a look-up table. In this case time is consumed due to the offline calculations of the switching angles for the MSMI. In microcontrollers, additional hardware is needed to improve the performance due to the insufficiency of the functions provided [9]. Furthermore, there is a speed restriction due to the limited number of processing units in the microcontroller that causes difficulties in running concurrent tasks.

This paper presents the implementation of a single-carrier multilevel PWM technique for an MSMI using FPGA. It describes the hardware development of the gate signals generator. The performance of the gate signals

978-1-4244-0644-9/07/$25.00 ©2007 IEEE   1041

generator is evaluated in terms of the outputs obtained from the FPGA device in comparison to that obtained from the results of a simulation study using Matlab/Simulink.

II. MODULAR STRUCTURED MULTILEVEL INVERTER TOPOLOGY

Fig. 1 shows the configuration of a single-phase 5-level MSMI. The MSMI is unique when compared to other types of multilevel inverters because it consists of several modules that require separate DC sources. When compared to other types of multilevel inverters, the MSMI requires less number of components with no extra clamping diodes or voltage balancing capacitors. As shown in Fig. 1, each module of the MSMI is formed by a single-phase full-bridge inverter. The number of levels is unlimited by stacking up the modules.

The output voltage of an MSMI is equal to the summation of the output voltage of the respective modules that are connected in series. The number of modules (M), which is equal to the number of DC sources required, depends on the number of levels (N) of the MSMI. The relationship between N and M for an MSMI is described by (1). For example, for an output voltage consisting of five levels which include $+2V_{DC}$, $+V_{DC}$, 0, $-V_{DC}$ and $-2V_{DC}$, the number of modules needed is 2.

$$M = \frac{(N-1)}{2} \qquad (1)$$

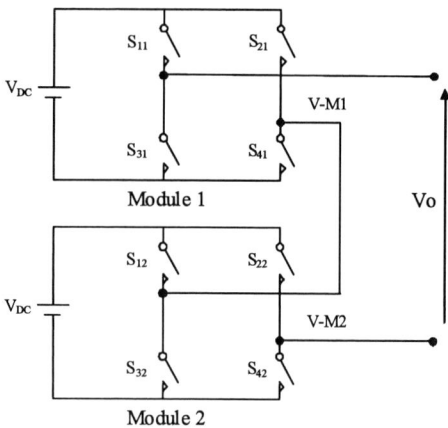

Fig. 1 Single-phase 5-level MSMI configuration

III. SINGLE-CARRIER MULTILEVEL PWM TECHNIQUE

Basically, this switching technique for the MSMI is based on the classical unipolar PWM switching strategy. A significant difference in this switching technique is the use of a single-carrier and two sampled sinusoidal modulation signals. The position of the modulation signals can be controlled by changing the value of the modulation index (m_a). The MSMI will produce a 5-level output only

when m_a is greater than 0.5. Otherwise, the MSMI will give a 3-level output. In this case, the modulation scheme works exactly as the classical unipolar PWM switching technique by giving the 3-level output $+V_{DC}$, 0 and $-V_{DC}$ for the MSMI. Apart from that, the harmonics performance of the switching technique is very identical to the multicarrier multilevel PWM technique with POD scheme.

Fig. 2 illustrates the single-carrier multilevel modulation strategy. The modulation signals have the same frequency (f_o) and amplitude (A_m). The sinusoidal signals are sampled by a triangular carrier signal with frequency f_c and amplitude A_c once in every cycle. Intersection between the sampled modulation signals and the carrier signal defines the switching instant of the PWM pulses. In order to ensure quarter wave symmetric properties of the PWM output waveform, the starting point of the modulation signals is phase shifted by one period of the carrier wave. In addition, the frequency modulation ratio (m_f) must also be an even number [8].

For an N-level inverter, m_a and m_f for the single-carrier multilevel modulation strategy are defined as:

$$m_a = \frac{A_m}{\frac{(N-1)}{2} A_c} \qquad (2)$$

$$m_f = \frac{f_c}{f_o} \qquad (3)$$

Further details on the single-carrier multilevel PWM technique can be obtained from [5].

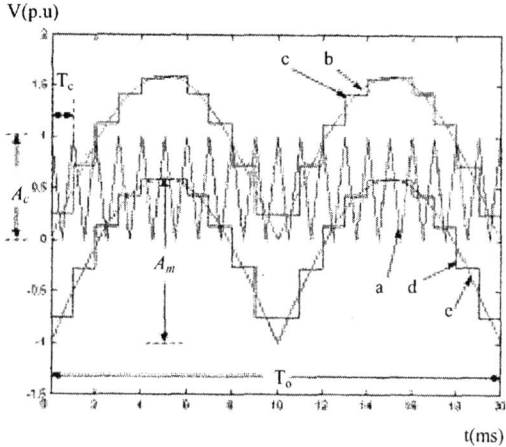

Fig. 2 Single-carrier multilevel PWM technique
a - the carrier signal
b - absolute sinusoidal modulation signal
c - sampled sinusoidal modulation signal of b
d - shifted absolute sinusoidal modulation signal
e - sampled sinusoidal modulation signal of d

IV. HARDWARE IMPLEMENTATION USING FPGA

Fig. 3 shows a basic block diagram on how the hardware implementation of the single-carrier multilevel PWM technique for the MSMI is designed. Before creating the block diagram, tools and materials provided by the Altera's UP1 board has been recognized. Referring to Fig. 3 the inputs to the circuit are m_a and m_f. The parameter m_a is controlled externally by an 8 bits ADC because of the absent of an analog to digital conversion function on the UP1 board. The parameter m_f is controlled by a 25.175 MHz on-board oscillator. The generation of the modulating signals and triangular carrier are configured in the FLEX10K FPGA device. Output of the I/O pins of the FLEX10K are the gate signals for each of the MSMI power devices.

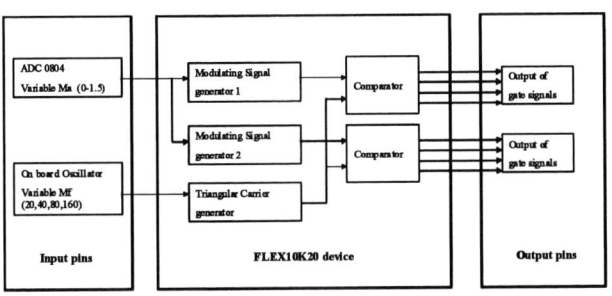

Fig. 3 Block diagram of the hardware environment

Fig. 4 illustrates the FPGA circuit design blocks for gate signals generation based on the single-carrier multilevel PWM technique. In general, it consists of 4 main blocks which are the modulating signal generator 1, modulating signal generator 2, triangular carrier generator and the low frequency signal generator. The modulating signal generators are responsible for generating 2 sampled modulating signals while the triangular carrier generator will produce a sampled triangular output. Both of the sampled modulating signals are compared separately with the sampled triangular carrier.

Comparators are labeled as LPM_COMPARE with 16 bits input. Subsequently, output from the comparators are ex-ORed with the low frequency signal to generate the gate signals for the power devices of module 1 and module 2 of the MSMI. Low frequency gate signals are generated directly from the low frequency generator block labeled as LOW_FRE.

Basically, an LPM_COUNTER acts as a clock frequency divider to supply clock signals to other block sets. The input of this counter is assigned to the on-board oscillator. In addition, a multiplexer is used to select the clock input to the triangular carrier generator. The parameter m_f is controlled by selecting different types of clock to the triangular carrier generator. In this work, the available m_f are limited to 20, 40, 80 and 160 only. This is mainly due to the clock input of the design that is provided by the fixed on-board 25.175 MHz oscillator.

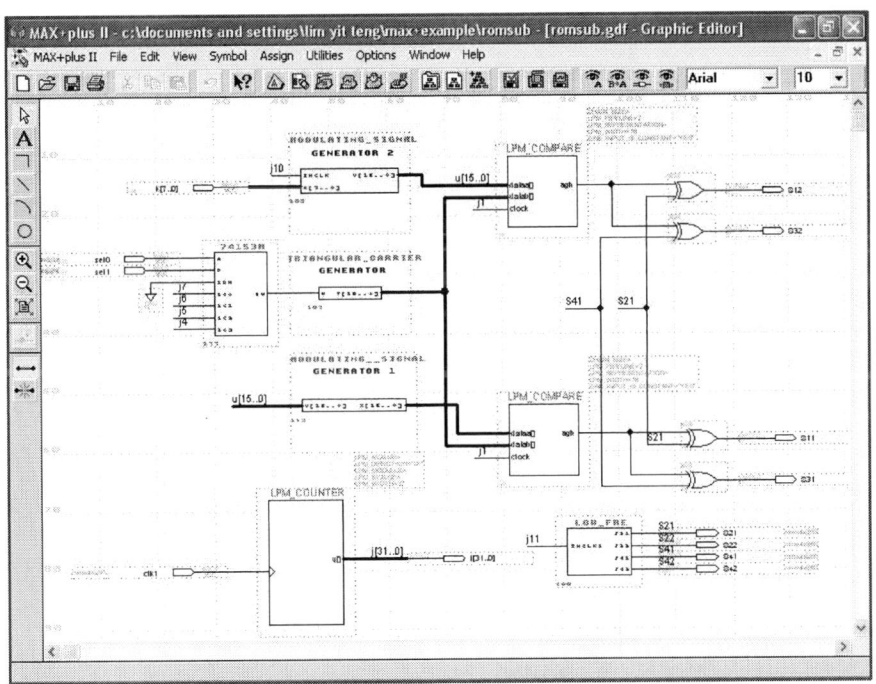

Fig. 4 FPGA circuit design blocks for gate signals generation

Fig. 5 shows the subsystem of the MODULATING_SIGNAL GENERATOR 2 block set. It contains 3 function blocks labeled by 7bit, LPM_ROM and LPM_MULT. Basically, the FPGA is a digital device, so all the design is based on digital representation. In this work, the digital representation scheme is the same as the representation used in an ADC whereby 1V is represented by 85 in decimal value. Therefore, 1 decimal value represents 11.8mV in the actual voltage value. In this subsystem, the modulating signal is sampled by 244 steps per cycle. As stated before, the modulating signal generated is an absolute signal whereby the signal repeats after half cycle. As a result, 122 steps value representation are required. Offline calculations are made based on (4) to obtain all the step values. The step values are calculated by increasing u from 0 to 121.

$$f(u)=(\sin(((180/488)+(360/488) \times u) \times \pi/180)) \times 85$$
(4)

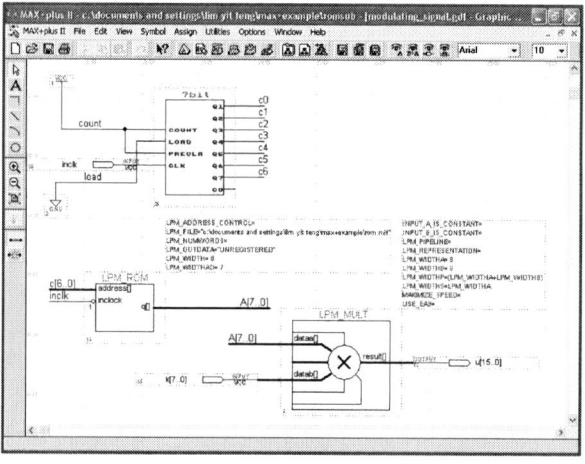

Fig. 5 Subsystem of the *MODULATING_SIGNAL GENERATOR 2*

All the step values are then stored into the LPM_ROM by creating an MIF file. The 7bit counter is designed to count up from 0 to 121. The counter reads the sampled modulating signal value which is stored in the ROM to generate a sampled modulating signal. The modulating signal output from the LPM_ROM block is transferred to the LPM_MULT multiplier. The sampled modulating signal is multiplied by an amplitude value from the ADC. As a result, a modulating signal that represents different m_a values can be generated.

Fig. 6 shows the subsystem of the MODULATING_SIGNAL GENERATOR 1. Basically, it consists of an LPM_ADD_SUB function block and an LPM_CONSTANT block. The LPM_ADD_SUB is set to operate as a 16 bits adder. This block function is simple whereby the input of the adder labeled as u[15..0] is fed from the MODULATING_SIGNAL

GENERATOR 2. This modulating signal input is added to a constant 1. The adder produces a sampled modulating signal with the same pattern as the signal from MODULATING_SIGNAL GENERATOR 2 but it is shifted up by magnitude 1. This way, components being used to generate the modulating signal for module 1 of the MSMI are reduced. It also minimizes the space used for the design so that it can be fit into the EPF10K20 device. The output signal from this block set will be compared with the triangular carrier to generate the gate signals for module 1 of the MSMI.

Fig. 7 illustrates the block circuit development of the TRIANGULAR_CARRIER GENERATOR. The basic idea of the design is similar to the modulating signal generator. A 7bit counter labeled as tria7bit is used together with an LPM_ROM to generate the triangular carrier signal. The triangular signal is sampled by 98 steps per cycle. This indicates that the counter should be able to count up from 0 to 97. Values for every step sample were calculated offline. All the values calculated are stored into an MIF file and kept in the LPM_ROM. The frequency of the clock input can be selected using the switches on the UP1 board. The m_f value can be determined by selecting the clock input to the counter.

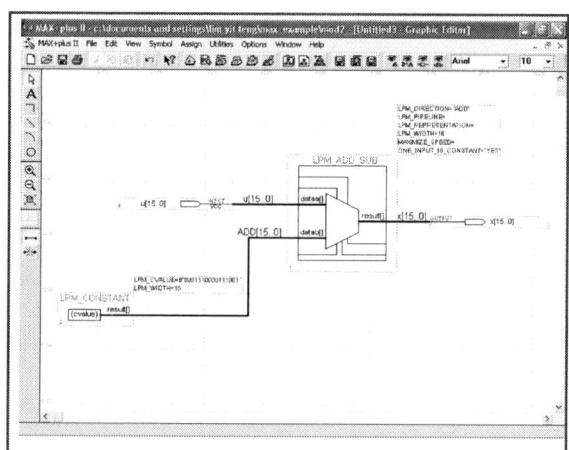

Fig. 6 Subsystem of the *MODULATING_SIGNAL GENERATOR 1*

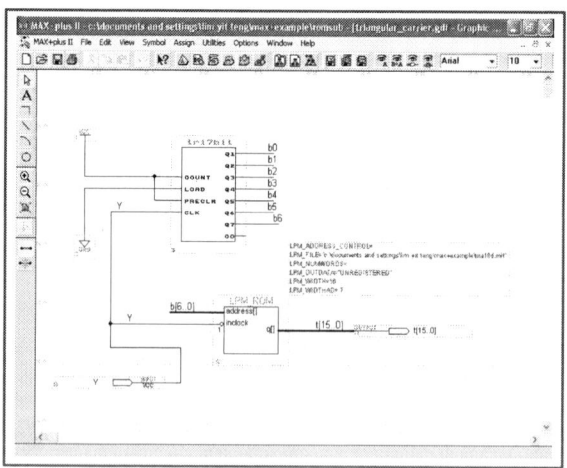

Fig. 7 Subsystem of *TRIANGULAR_CARRIER GENERATOR*

Fig. 8 shows the developed LOW_FRE block. This subsystem is responsible to generate the 4 low frequency gate signals for the MSMI power devices. These 4 signals always remain as a low frequency signal without being affected by m_a. A 7bit counter and an LPM_ROM are used to fulfill this purpose. The LPM_ROM will generate logic low output for the first half cycle of a sinusoidal signal and logic high output for the second half cycle.

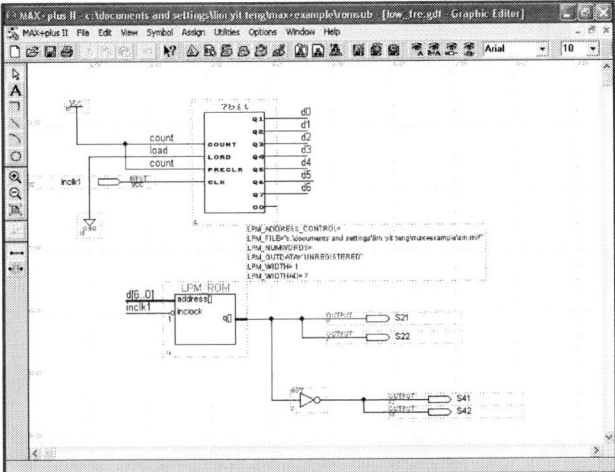

Fig. 8 Subsystem of the *LOW_FRE*

V. RESULTS AND ANALYSIS

Before comparing the experimental results and the simulation results, the performance of the gate signal generator should be stated. Normally, the sampling rate, timer interrupts and switching instant resolution are taken into account for the performance evaluation. In this work, the sampling time achieved is 81.96μs. Thus the switching instant resolution obtained is 1.48°.

The gate signals waveforms obtained from the simulation results using MATLAB/Simulink are compared with that generated by the EPF10K20 device. Comparison are made for the gate signals waveforms based on variations of parameters m_a and $m_f = 20$ and 40 respectively. Fig. 9 shows the gate signals obtained for the power devices in module 1 of the MSMI for $m_a= 0.4$ and $m_f=20$ based on simulation and experimental results. With m_a less than 0.5, all four gate signals for the power devices in module 2 of the MSMI are at low frequency and do not contribute to its output voltage. Fig. 10 shows the gate signals for the power devices in each of the MSMI module when m_a is set at 0.8 while maintaining the same value of m_f. Based on the results obtained, the gate signals generated by the EPF10K20 device are found to be in good agreement with that generated through simulation. The number of pulses in each of the gate signals are the same for both the simulation and experimental results.

With $m_f = 20$, comparison is also made by measuring the pulse width of the gate signals from the simulation output and the EPF10K20 device. In this work, pulse width for the simulation signals are viewed and measured within the MATLAB/Simulink environment while for the experimental results, a measurement function in the digital oscilloscope has been used to measure every pulse width.

Table 1 shows the comparison among the pulse width measured for $m_a = 0.4$ while Table 2 shows the comparison for $m_a = 0.8$. Referring to Table 1, the highest percentage differences of pulse width are given by pulses 1 and 10 which are 3% and 3.38% respectively. For the other pulses, the percentage difference is around 1%. The average percentage difference is 1.278%. Referring to Table 2, 5 pulses are generated in half a period for the MSMI module 1. The average of percentage difference in pulse width is quite low whereby for module 1 is around 0.92% and for module 2 is 1.79%. For module 2, 6 pulses are generated in a half period. The highest percentage difference is 3%.

Overall, the pulse widths percentage difference between the simulation and experimental results for both cases described above are still within the acceptable level with the average of between 0.92% and 1.278%. It must also be noted that the measurement made within the MATLAB/Simulink environment is subject to a maximum of 4 decimal points only. Further work is required to evaluate the performance of the FPGA based gate signal generator based on the single-carrier multilevel PWM technique for the MSMI. Testing the FPGA based gate signal generator on the actual MSMI circuit may provide further verification on its feasibility based on the MSMI output voltage harmonic spectrums for various m_a and m_f values.

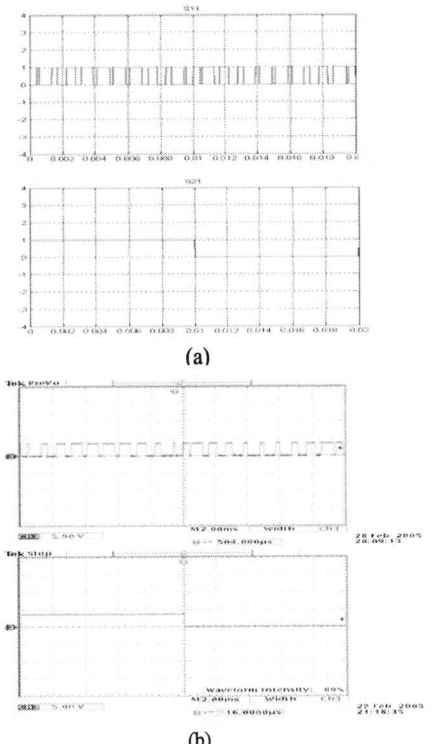

(a)

(b)

Fig. 9 Gate signals for module 1 of the MSMI for $m_a = 0.4$ and $m_f = 20$ (a) simulation (b) EPF10K20 device

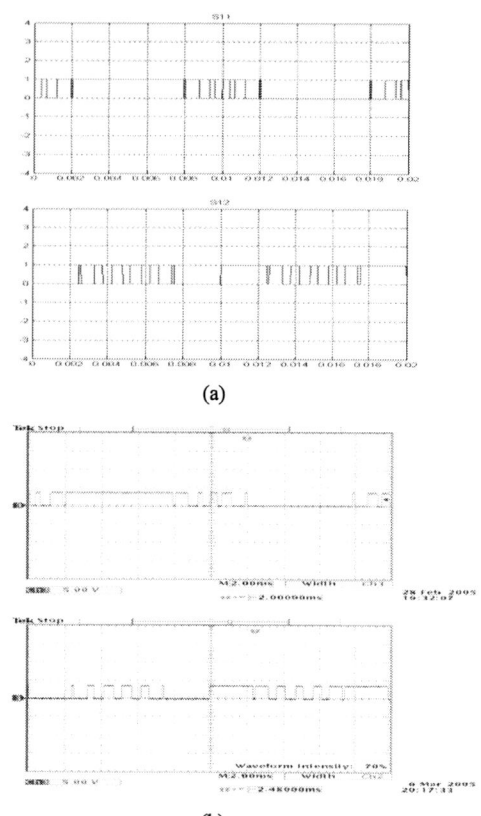

(a)

(b)

Fig. 10 Gate signals for module 1 (upper) and module 2 (lower) of the MSMI for $m_a = 0.8$ and $m_f = 20$ (a) simulation ((b) EPF10K20 device

Table 1 Comparison on pulse width of gate signals for $m_a = 0.4$, $m_f = 20$

Pulse	MATLAB/Simulink output tm (ms)	EPF10K20 device output tf (ms)	$\frac{tm - tf}{tm} \times 100\%$
1	0.1266	0.1228	3.00
2	0.3644	0.3686	1.15
3	0.5603	0.5632	0.52
4	0.7123	0.7168	0.63
5	0.7834	0.7782	0.66
6	0.7834	0.7782	0.66
7	0.7118	0.7168	0.70
8	0.5594	0.5632	0.68
9	0.3635	0.3686	1.40
10	0.1271	0.1228	3.38

Table 2 Comparison on pulse width of gate signals for $m_a = 0.8$, $m_f = 20$

Pulse	MATLAB/Simulink output tm (ms)	EPF10K20 device output tf (ms)	$\frac{tm - tf}{tm} \times 100\%$
Module 1			
1	0.2622	0.2662	1.53
2	0.7531	0.7475	0.74
3	5.9371	5.9392	0.04
4	0.7532	0.7475	0.76
5	0.2622	0.2662	1.53
Module 2			
1	0.1262	0.1228	2.69
2	0.4271	0.4198	1.71
3	0.5783	0.5734	0.85
4	0.5784	0.5734	0.86
5	0.4269	0.4198	1.66
6	0.1266	0.1228	3.00

VI. CONCLUSIONS

The work presented has shown that the implementation of the single-carrier multilevel PWM technique for the MSMI using FPGA is able to produce proper gate signals for the MSMI power devices with high computation speed and accuracy on the output generated. Outputs from the FPGA FLEX10K20 have been verified by the results obtained from the MATLAB/Simulink simulation. A sampling time of 81μs corresponding to 1.48° resolution for the gate signals pulse widths has been achieved.

VII. REFERENCES

[1] J. Lai and F. Z. Peng, "Multilevel Converters-A New Breed of Power Converters." *IEEE Transactions on Industry Applications*, Vol. 32, No.3, pp.509-517,1996.

[2] Carrara, G., Gardella, S. and Marchesoni, M., "A New Multilevel PWM Method: A Theoretical Analysis", *IEEE Transactions on Power Electronics*, 7 No. 3, pp. 497-505, 1992.

[3] Azli, N.A. and Yatim, A. H. M., "Optimal Pulsewidth Modulation (PWM) Online Control of a Modular Structured Multilevel Inverter (MSMI)", *Proceedings of the 4th IEEE International Conference on Power Electronics and Drive Systems PEDS'01*, 22-25 October 2001, Bali, Indonesia.

[4] Schibli, N.P., Nguyen, T. and Rufer, A. C. , "A Three Phase Multilevel Converters for High Power Induction Motors." *IEEE Transactions on Power Electronics*, 13 No. 5, pp. 978-986, 1998.

[5] Naziha Ahmad Azli and Abdul Halim Mohd Yatim,"A Modular Structured Multilevel Inverter for Fuel Cell Applications.", *Jurnal Teknologi D. 32.*, 2000.

[6] Naziha Ahmad Azli, Abdul Halim Mohd Yatim and Faridah Mohd Taha, "Regular Sampled Pulsewidth Modulation (PWM) Switching Strategies for a Modular Structured Multilevel Voltage Source Inverter (VSI)", *Proceedings of World Engineering Congress WEC '99*, 19-22 July 1999, Kuala Lumpur.

[7] Naziha Ahmad Azli and Abdul Halim Mohd Yatim, "DSP-based Online Optimal PWM Multilevel Control for Fuel Cells Power Conditioning System." *Proceedings of the 27th Annual Conference of the IEEE Industrial Electronics Society*, 29 Nov. – 2 Dec. 2001, Denver, USA.

[8] F. Salim and N.A. Azli, "Development of an FPGA Based Gate Signal Generator for a Multilevel Inverter." *Proceedings of 2003 International Conference on Power Electronics and Drive Systems PED 2003*, Singapore. 17-20 November 2003.

[9] J. A. Aziz and Z. Salam, "A PWM Strategy For The Modular Structured Multilevel Inverter Suitable For Digital Implementation" *Technical Proceedings Power Electronics Congress*, pp. 160- 164, 2002.

[10] N. A. Azli and M. S. Bakar, "A DSP-based Regular Sampled Pulsewidth Modulation (PWM) Technique for a Multilevel Inverter", *Proceedings of 2004 International Conference on Power System Technology*, Singapore, 21-24 November 2004.

SPACE VECTOR PWM FOR MULTILEVEL INVERTERS
- A FRACTAL APPROACH

Anish Gopinath   M.R. Baiju
Department of Electronics & Communication, College of Engineering Trivandrum, Kerala, India
Email: mrbaiju@ece.cet.ac.in

Abstract– **This paper proposes a space vector PWM (SVPWM) technique based on fractal approach for multilevel inverters. The inherent fractal structure in the space vector representation of multilevel inverters is brought out in this paper. The proposed method utilizes the inherent fractal structure of the space vector representation of multilevel inverters. Sector identification and switching vector determination in SVPWM through this fractal approach reduces the computational complexity. The proposed method does not use any look up tables for sector identification. The switching space vectors are also directly determined without using any look up tables. The proposed scheme can be extended to an n-level inverter without increase in complexity. Simulation results are presented for a 5-level inverter.**

Index terms:Multilevel inverter, Space Vector PWM, fractal

I. INTRODUCTION

Multilevel Inverter technology finds significant applications in the area of high-power medium-voltage energy control [1]. Multilevel Inverters generate sinusoidal voltages from discrete voltage levels and Pulse Width Modulation (PWM) strategies accomplish this task of generating sinusoids of variable voltage and frequencies. Several techniques for implementation of PWM for inverters have been developed [2]. The two main techniques of PWM generation for multilevel inverters are Sine-Triangle PWM (SPWM) and Space Vector PWM (SVPWM) [2]. Multilevel Sine-Triangle PWM involves comparison of a reference signal with a number of level shifted carriers to generate the PWM signal [4]. Space Vector PWM involves synthesizing the reference voltage space vector by switching among the three nearest voltage space vectors [2 - 10]. Space Vector PWM is considered a better technique of PWM implementation owing to its associated advantages of (i) better fundamental output voltage (ii) better harmonic performance (iii) Easier implementation in Digital Signal Processor and Microcontrollers. Techniques utilizing the equivalence of SVPWM with SPWM can also automatically generate the SVPWM signals from the instantaneous reference phase voltages [11, 12]. The implementation of SVPWM involves (i) Identification of the sector in which the tip of the reference vector lies (ii) Determination of the three nearest voltage space vectors (iii) Determination of the duration of each of these switching voltage space vectors and (iv) Choosing an optimized switching sequence. The sector identification can be done by using coordinate transformation of the

reference vector into a two dimensional coordinate system [8]. The sector can also be determined by resolving the reference phase vector along a, b and c axii and by repeated comparison with discrete phase voltages [9,10]. After identifying the sector, the voltage vectors at the vertices of the sector are to be determined. Once the switching voltage space vectors are determined the switching sequences can be identified using lookup tables [5-10]. The calculation of the duration of the voltage vectors can be simplified by mapping the identified sector to correspond to a sector of 2-level inverter [7]. To obtain optimum switching, the voltage vectors are to be switched for their respective durations, in a sequence such that only one switching occurs as the inverter moves from one switching state to another [3]. Conventional techniques involve look up tables for achieving this optimum switching sequence [5-10].

In the present paper, a novel approach for generation of space vector PWM for multilevel inverter based on fractals is presented. It is motivated from the fact that the switching vector representation of any multilevel inverter has an inherent fractal structure, with the basic unit of this structure being the triangle made of the vertices of three adjacent inverter voltage space vectors. As the number of levels increases, it can be viewed that, each sector gets further divided into smaller triangular regions or sectors. The present work is pivoted on this idea, and an algorithm is also proposed for generating the sectors of higher level inverter from the triangular regions of an equivalent 2-level inverter. The proposed method uses simple arithmetic for determining the sector and does not require look up tables. The switching vectors are also directly determined using simple arithmetic and hence does not require look up tables. The paper explains the proposed method for a 5-level inverter and presents the simulation results.

II. INHERENT FRACTAL STRUCTURE IN THE SPACE VECTOR REPRESENTATION OF MULTILEVEL INVERTER

The voltage space vector represents the combined effect of the three reference phase voltages at a particular instant. The voltage space vectors are represented in the (α, β) plane. The generation of PWM using space vector, SVPWM, involves approximating the instantaneous reference voltage space vector by switching the three nearest inverter voltage space vectors. The equilateral triangle formed by the three nearest vectors as vertices is

978-1-4244-0644-9/07/$25.00 ©2007 IEEE   1047

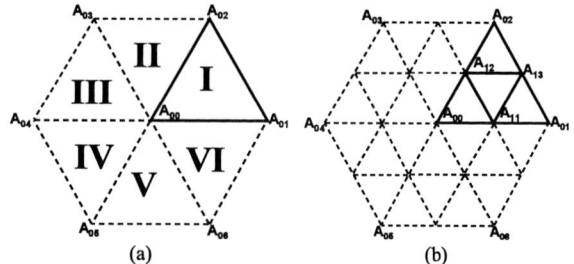

Fig. 1 (a) Voltage Space vectors of 2-level inverter (b) Voltage Space vectors of 3-level inverter

referred as a sector. Fig. 1(a) shows the voltage space vectors of 2-level inverter. The voltage space vector locations for a 3-level inverter are shown in Fig.1 (b) where A_{00}, A_{01}, A_{02}, A_{03}, A_{04}, A_{05} and A_{06} are same as locations of voltage space vectors of 2-level inverter. Consider the region marked I in the case of 2-level inverter (Fig. 1(a)), formed by the vectors located at A_{00}, A_{01}, and A_{02}. In the case of 3-level inverter, as in Fig.1 (b), this region has three additional voltage space vectors. It may be noted that the three additional voltage space vectors are located at the midpoints of each side of the sector of equivalent 2-level inverter. The three additional switching voltage space vectors together with switching voltage space vectors located at A_{00}, A_{01} and A_{02} results in four sectors within $\Delta A_{00}A_{01}A_{02}$ of 3-level inverter.

Fig. 2 and Fig. 3 show the locations of voltage space vectors and the corresponding inverter states for a 5-level inverter. In Fig.2, if we consider the triangular region formed by the space vectors located at A_{00}, A_{01} and A_{02}, besides the voltage space vectors of 3-level inverter, nine additional voltage space vectors are present. The nine additional vectors are located at A_{21}, A_{22}, A_{23}, A_{24}, A_{25}, A_{26}, A_{27}, A_{28}, and A_{29}. It may be noted that the nine additional vectors are located at the midpoints of the sides of sectors of 3-level inverter. The nine additional vectors together with the voltage space vectors of 3-level inverter results in 16 sectors within $\Delta A_{00}A_{01}A_{02}$ of 5-level inverter Fig. 2. In this manner each sector in the voltage space vector representation of an equivalent 2-level inverter is divided into four smaller sectors, resulting in voltage space vector locations of 3-level inverter. Each of the sectors of 3-level inverter is further divided into four smaller sectors resulting in switching space vectors of 5-level inverter. This process gets repeated for generation of space vectors of higher level inverters.

The space vector representation of a higher level inverter can be conceived as generated from the space vector representation of 2-level inverter, wherein the sectors of 2-level inverter get progressively divided and subdivided. The basic structure, a triangle (sector) is transformed by further dividing itself into smaller triangles. A basic structure that evolves by dividing itself into structures similar to it has an associated fractal [13]. The switching voltage space vector representation of multilevel inverters also has an associated fractal. In fractal theory, the fractal structure with triangle as basic structure and that which gets divided into four smaller triangular regions, joined by the midpoints of the sides of the triangle is called the Sierpinski triangle [13].

III. TRIANGULARISATION

As explained in the previous section, due to the inherent fractal structure associated with multilevel inverter, the voltage space vector representation of 2-level inverter grows to that of higher level inverters by repeated division of each sector. At every stage, the triangular region is divided into four smaller triangular regions, due to the presence of the additional voltage vectors. The three additional voltage vectors are located at the mid point of each side of sector. The sectors of higher level inverter can therefore be generated by such repeated *triangularisation*. Let us consider region I of the 2-level inverter formed by vertices A_{00}, A_{01} and A_{02}. The coordinates of three vertices are (α_{00},β_{00}), (α_{01},β_{01}) and (α_{02},β_{02}) respectively, Fig. 4(a). The coordinates of the three new voltage space vectors located at A_{11}, A_{12} and A_{13} can be obtained from the coordinates of A_{00}, A_{01} and A_{02} as

Coordinates of A_{11} are

$$\alpha_{11} = \frac{1}{2}(\alpha_{00} + \alpha_{01}) \qquad (1)$$
$$\beta_{11} = \frac{1}{2}(\beta_{00} + \beta_{01})$$

Coordinates of A_{12} are

$$\alpha_{12} = \frac{1}{2}(\alpha_{00} + \alpha_{02}) \qquad (2)$$
$$\beta_{12} = \frac{1}{2}(\beta_{00} + \beta_{02})$$

Coordinates of A_{13} are

$$\alpha_{13} = \frac{1}{2}(\alpha_{01} + \alpha_{02}) \qquad (3)$$
$$\beta_{13} = \frac{1}{2}(\beta_{01} + \beta_{02})$$

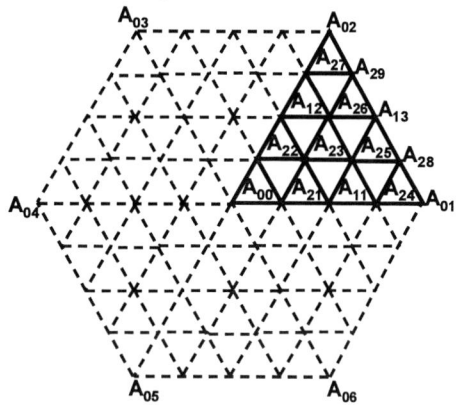

Fig. 2. Switching space vectors of 5-level inverter

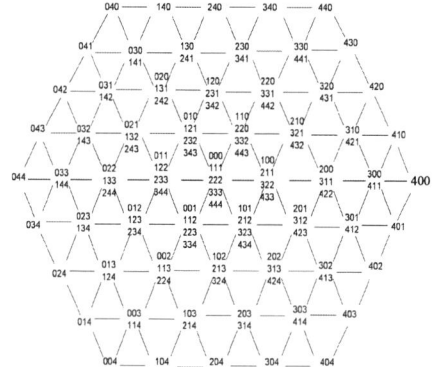

Fig. 3. Inverter states of 5-level inverter.

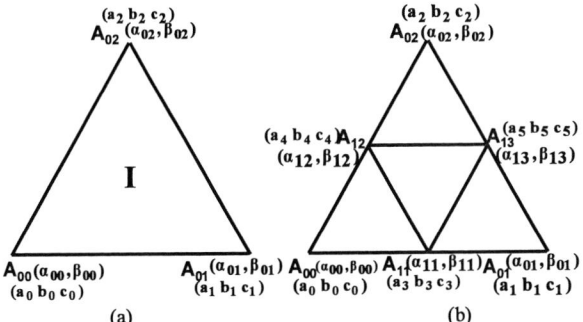

(a) (b)

Fig. 4. Triangularisation of region I of equivalent 2-level inverter.

The associated switching vector states can also be determined in a similar manner. The inverter switching states corresponding to the voltage space vector located at A_{00}, A_{01} and A_{02} are $(a_0\ b_0\ c_0)$, $(a_1\ b_1\ c_1)$ and $(a_2\ b_2\ c_2)$ respectively. The switching states of the new voltage space vectors at A_{11} $(a_3\ b_3\ c_3)$, A_{12} $(a_4\ b_4\ c_4)$ and A_{13} $(a_5\ b_5\ c_5)$ can be obtained as

$$x_3 = \frac{1}{2}(x_0 + x_1) \qquad (4)$$

$$x_4 = \frac{1}{2}(x_0 + x_2) \qquad (5)$$

$$x_5 = \frac{1}{2}(x_1 + x_2) \qquad (6)$$

Where x takes a, b and c for the respective phases. The equations (1) - (6) represent the arithmetic procedure used in the proposed method for dividing a triangular region (sector) into four similar regions, by generating three additional vectors situated at the mid points of the sides forming the original triangular region. This is referred as the *triangularisation algorithm* in the proposed work. This algorithm can be summarized as:-

a. Determine the sector (triangular region) on which the triangularisation is to be performed.

b. Determine the midpoints of each side of the sector by equations (1) - (3). These are the coordinates of the three new vectors which will divide the sector into four smaller, but similar triangular regions.

c. Determine the inverter states corresponding to these vectors using equations (4) - (6).

In order to generate sectors of higher level inverter by progressively dividing sectors of 2-level inverter, the

triangularisation algorithm is repeatedly applied. In fractal theory, algorithms whose repeated iterations would results in the pattern to grow or evolve, is referred as Iterated Function System [13]. In the present work, the repeated iteration of triangularisation algorithm will grow the inherent fractal structure in space vector representation of multilevel inverters. Therefore, the triangularisation algorithm can be viewed as the Iterated Function System for this fractal structure.

IV. GENERATION OF PWM FOR MULTILEVEL INVERTERS THROUGH FRACTALS

This section explains the proposed method for generation of SVPWM for a 5-level inverter using the inherent fractal structure associated with the switching space vector representation of multilevel inverter. The implementation of SVPWM involves four stages ; (i) Sector Identification (ii) Determination of switching voltage space vectors (iii) Determination of the duration of each of the determined switching voltage space vectors and (iv) Determination of an optimum switching sequence.

A. Sector Identification and Switching vector determination.

Sector identification determines the triangle that encloses the tip of the reference space vector. The vertices of the triangle represent the locations of switching voltage space vectors used to synthesize the reference space vector. If V_a, V_b, V_c denote the instantaneous values of the three reference phase voltage, then the (α, β) coordinates of the corresponding space vector can be obtained as

$$V_\alpha = \frac{3}{2} V_a$$
$$V_\beta = \frac{\sqrt{3}}{2}(V_b - V_c) \qquad (7)$$

The (α, β) components of the space vector of an n-level inverter can be normalized through division by V_{dc}/n-1, where V_{dc} is the dc link voltage. In the case of a 5-level inverter I, the voltage V_{dc} in the normalized space vector representation is therefore represented by a vector of length 4, Fig 6.

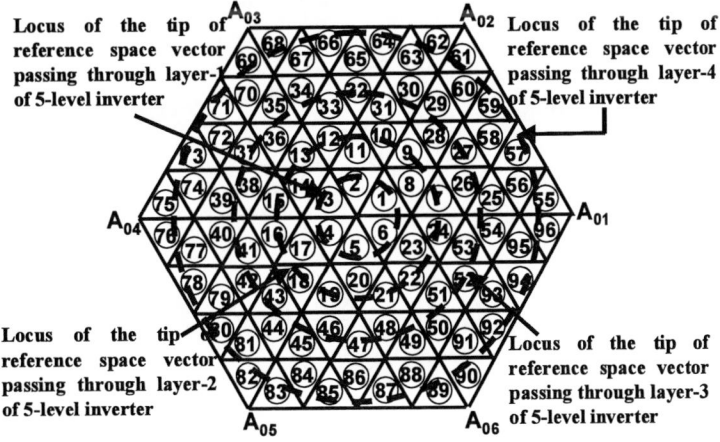

Fig. 5. The loci of tip of a reference space vector when passing through each of the four layers of a 5-level inverter are shown

1049

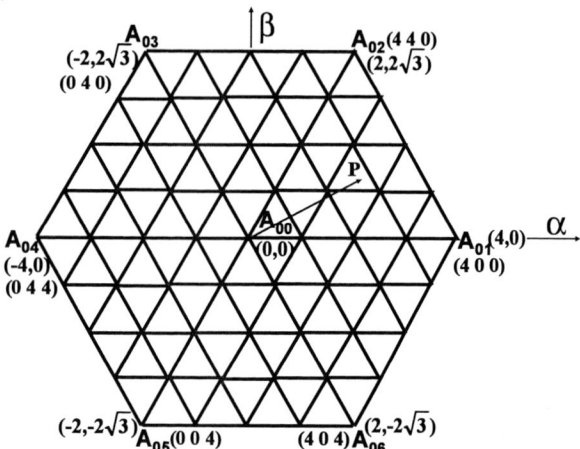

Fig. 6. Normalized space vector representation of 5-level inverter (with coordinates of the vectors of equivalent 2-level inverter) and the instantaneous position of reference voltage space vector at an instant.

For a 5-level inverter, the switching vectors located at the six vertices of the hexagon forming the periphery are same as the vectors of equivalent 2-level inverter, but they have the switching states as 4 0 0, 4 4 0, 0 4 0, 0 4 4, 0 0 4, 4 0 4.

Let us consider a reference space vector $\overrightarrow{A_{00}P}$ for explaining the proposed method of sector identification for a 5-level inverter Fig (Figs 6 & 7). The first step in the proposed sector identification method is to determine the location of the tip of the reference space vector $\overrightarrow{A_{00}P}$ from among the six regions of the equivalent 2-level inverter. The step is implemented by comparison of instantaneous reference phase voltages [9]. In this case the reference space vector $\overrightarrow{A_{00}P}$ is located in region I of equivalent 2-level inverter. The region I, from Fig 1(a), is formed by the vertices A_{00}, A_{01} and A_{02}. The coordinates of the vertices are $(0,0)$, $(4,0)$ and $(2,2\sqrt{3})$ respectively, as shown in Fig. 7(a). The switching states corresponding to the vectors located at A_{00}, A_{01} and A_{02} are also shown in Fig. 7(a). The switching states of the vector located at A_{00}, A_{01} and A_{02} are (0 0 0, 1 1 1, 2 2 2, 3 3 3, 4 4 4), (4 0 0) and (4 4 0) respectively.

The next step is to divide region I into four smaller triangular regions by applying the *triangularisation algorithm* in section III. The equations (1) - (3) will

generate the coordinates of the new voltage space vectors and equations (4) - (6) generate the inverter states corresponding to these new switching vectors. The three new voltage space vectors divide region I into four smaller triangular regions marked R_1, R_2, R_3, R_4 in Fig. 7(b). The location of the tip of the reference voltage space vector $\overrightarrow{A_{00}P}$ from among these four triangular regions is found by determining the region whose centroid is closest to the tip of reference space vector. The coordinates of the centroid of an equilateral triangle can be determined as the average of coordinates of the 3 vertices. For an equilateral triangle with the coordinates of the vertices as $(\alpha_1,\beta_1),(\alpha_2,\beta_2),(\alpha_3,\beta_3)$, the coordinates of the centroid $(\alpha_{cent},\beta_{cent})$ is given by,

$$\alpha_{cent} = \frac{1}{3}(\alpha_1 + \alpha_2 + \alpha_3) \qquad (8)$$

$$\beta_{cent} = \frac{1}{3}(\beta_1 + \beta_2 + \beta_3) \qquad (9)$$

The triangle with centroid closest to tip of reference space vector is $\Delta A_{11}A_{12}A_{13}$.

For a 5-level inverter the *triangularisation algorithm* has to be applied once more. The application of equations (1) - (6) to $\Delta A_{11}A_{12}A_{13}$ will generate further three new voltage space vectors and also the inverter states corresponding to these new voltage vectors, thus dividing it into four smaller triangles, Fig. 7(c). From among these four triangles, the triangle enclosing the reference space vector $\overrightarrow{A_{00}P}$ is $\Delta A_{23}A_{25}A_{26}$ as its centroid is closest to tip of reference space vector. The $\Delta A_{23}A_{25}A_{26}$ corresponds to sector 27, from Fig. 4, of the 5-level inverter. The sector is identified and the inverter states corresponding to the switching vectors located at the vertices of the identified sector are also generated simultaneously.

B. Calculation of switching time for the vectors.

The switching voltage vectors approximate the volt-second of the reference space vector by operating for specific durations. The determination of duration of operation of the switching voltage space vectors is simplified by mapping the sector that is identified to enclose the reference space vector to a sector of 2-level inverter [7].

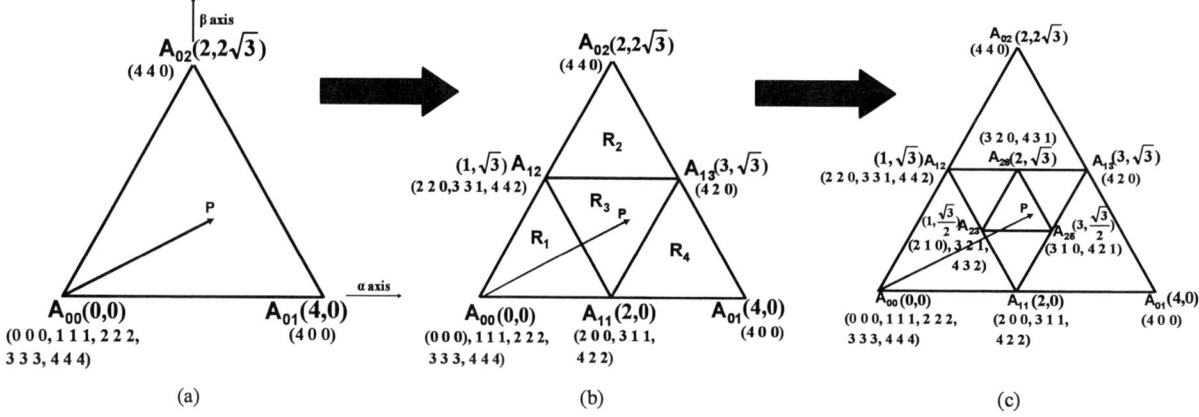

Fig. 7. Sector Identification and switching vector determination when tip of reference vector lies in region I of equivalent 2-level inverter

The mapping is done by choosing one of the three vectors of the identified sector to coincide with the actual zero vector in the voltage space vector representation of the inverter. In the present work the vector selected to coincide with the actual zero vector is referred as *virtual zero vector*. The vector with minimum value for the sum of magnitudes of α and β coordinates is chosen to be the *virtual zero vector* for a particular sector. The sum of magnitudes of α and β coordinates represent the total offset of the vector from actual zero vector. Therefore, in the present work the *virtual zero vector* chosen is at minimum offset from zero vector.

For the voltage reference vector $\overline{A_{00}P}$, the sector identified is sector 27. The voltage space vector with tip located at A_{23} becomes the virtual zero vector for sector 27. The sector 27 thus gets mapped to sector 1 (refer Fig. 5) of the 5-level inverter. The determination of duration now reduces to that of a 2-level inverter since after mapping one of the vectors of the identified sector coincides with the zero vector. The durations of the vectors can be determined using the conventional equations for 2-level inverter [3].

C. Determination of optimized switching sequence s

Once the switching vectors are determined and their respective durations are calculated, then vectors are to be switched in an optimum sequence such that only one switching occurs when the inverter changes its state. In the space phasor PWM technique, the optimum switching is achieved by using the redundant states of the zero vector for alternate switching cycles. In the proposed work, the optimum switching is achieved by using two redundant switching states of the respective *virtual zero vector* in the alternate cycles.

The tip of the reference vector $\overline{A_{00}P}$ is located in sector 27. The switching states corresponding to the switching vectors at the vertices of sector 27 with redundancies are A_{23} (2 1 0, 3 2 1, 4 3 2), A_{25} (3 1 0, 4 2 1) and A_{26} (3 2 0, 4 3 1). For sector 27, the virtual zero vector at A_{23} has three redundant switching states while the voltage vectors at A_{25} and A_{26} has two redundancies each. In the present work, if the redundancy of the virtual zero vector is greater than two, the last two redundant switching states are selected for the virtual zero vector. For the other two vectors, if redundant states are more than one, the last redundant state is selected. This strategy of choosing the last two redundant states for the *virtual zero vector* and the last redundant state for the other vectors will achieve optimum switching sequence. The selection will result in switching states [(3 2 1),(4 3 2)] for the *virtual zero vector*. The other vectors has switching states [(4 2 1)] and [(4 3 1)]. With these states, switching sequence of inverter for the sector number 27 is 3 2 1→4 2 1→4 3 1→4 3 2 during a sampling interval and 4 3 2→4 3 1→4 2 1→3 2 1 for the subsequent sampling interval. It may be noted that only one switching occurs as the inverter changes state.

To demonstrate this strategy further, let us consider another sector, sector 7 of layer2 of Fig. 5. It may be noted from Fig. 2 that sector 7 is formed by vertices A_{21}, A_{11} and A_{23}. From Fig. 3, the switching inverter states corresponding to the vectors at these vertices are A_{21}(1 0 0, 2 1 1, 3 2 2, 4 3 3), A_{11}(2 0 0, 3 1 1, 4 2 2) and A_{23}(2 1 0, 3 2 1, 4 3 2).

The switching vector located at A_{21} is the *virtual zero vector* for sector 7. In this case, the virtual zero has four redundant states (1 0 0, 2 1 1, 3 2 2, 4 3 3) and the other two vectors have three redundant states each. As per the strategy adopted in the proposed work, the last two redundant states are selected for the virtual zero vector i.e. (3 2 2, 4 3 3). For the other two vectors, among their redundancies the last redundant state is selected i.e. for vector located at A_{11} state (4 2 2) and for the vector at A_{23} state (4 3 2) are selected. With these states the sequence of inverter switching will be 3 2 2→4 2 2→4 3 2→4 3 3 for one sampling interval and 4 3 3→4 3 2→4 2 2→3 2 2 for the subsequent sampling interval. It may be noted that this is optimum switching where only one switching is involved as the inverter changes state.

Therefore, in the proposed work, optimum switching sequence is achieved by adopting a strategy of selecting the switching state of the vectors from the available redundancies. This simple strategy of choosing the switching states from the available redundant states achieves optimum switching in every sector and hence does not require look up tables.

V. SIMULATION RESULTS

The proposed method of generation of SVPWM has been simulated for a 5-level inverter with a DC link voltage of 300V. Fig. 8 shows the locations of switching vectors of a 5-level inverter generated through the proposed method. It can be seen from the figure that all the 61 space vector locations of the 5-level inverter has been generated. Fig. 9 shows the sectors identified by the proposed technique when the tip of the reference vector moves through sectors 7 to 24 (layer 2 operation, refer Fig. 5). The figure shows that the sector numbers are identified properly in the correct sequence. Fig. 10 shows the sectors identified when the tip of the reference space vector passes through sectors 25 to 54, corresponding to layer 3 operation, refer Fig. 5. Similarly, Fig. 11 shows the sectors identified for layer 4 operation, when the tip of reference space vector moves through sectors 55 to 96. It can be seen from the figures that the sectors are identified properly for operation in each of the three layers.

The Fig. 12 and Fig. 13 show the switching pattern for two successive sampling instants when the tip of reference space vector lies in sectors 80 and 83 respectively. For sector 80, the switching sequence is 0 1 3→0 1 4 →0 2 4 → 1 2 4. For sector 83, the switching sequence achieved is 0 0 3→ 0 0 4→1 0 4→1 1 4. These figures show that optimum switching is achieved in the proposed method.

Fig. 14 shows the plot of pole voltage for layer 2 operation. The inverter switches between the levels of

1051

300V, 225V and 150V. Fig. 15 shows the phase voltage corresponding to layer 2 operation simulated with an induction motor model. Fig. 16 shows the pole voltages for layer3 operation. The inverter switches between levels of 300V, 225V, 150V and 75V. The layer 3 operation thus corresponds to the 4-level operation of the inverter. The plot of phase voltage for layer 3 operation is shown in Fig. 17.

Fig. 18 and Fig. 19 show the plot of pole voltages and phase voltages of the inverter for layer 4 operation. It can be seen that the levels of the pole voltages for layer4 operation are 300V, 225V, 150V, 75V and 0V.

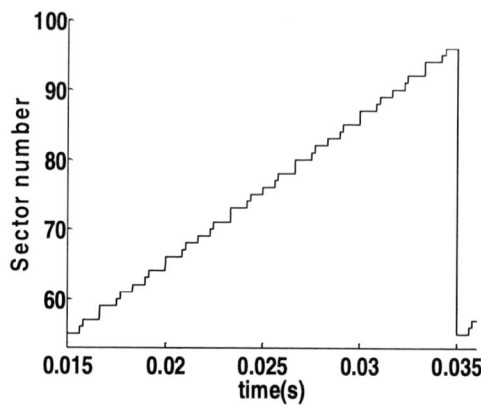

Fig. 11. Sectors identified for layer 4 operation

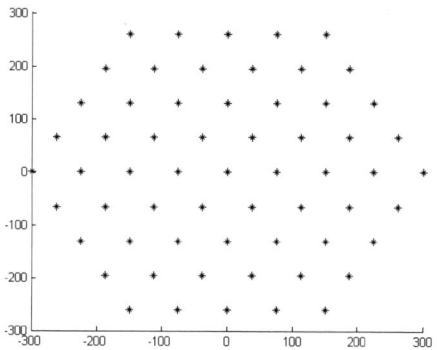

Fig. 8. Switching vectors of 5-level inverter generated through the proposed method

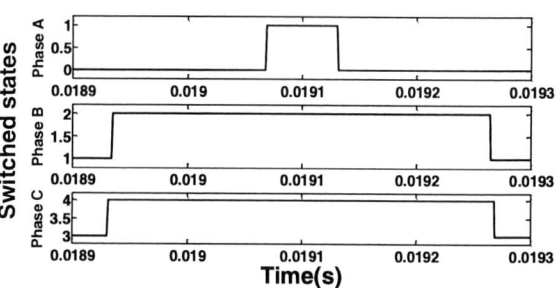

Fig. 12. Switching sequence for sector 80 for two Successive sampling instants (simulation results).

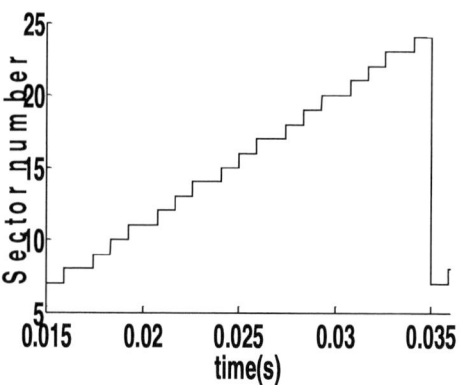

Fig. 9 Sectors identified for layer 2 operation

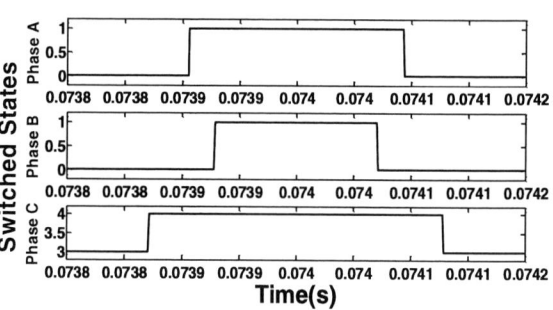

Fig. 13. Switching sequence for sector 83 for two Successive sampling instants (simulation results).

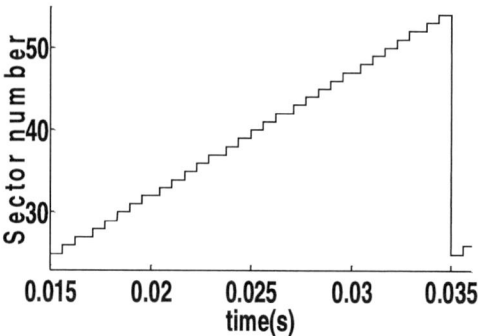

Fig. 10 Sectors identified for layer 3 operation

Fig. 14. Pole voltage for layer 2 operation

1052

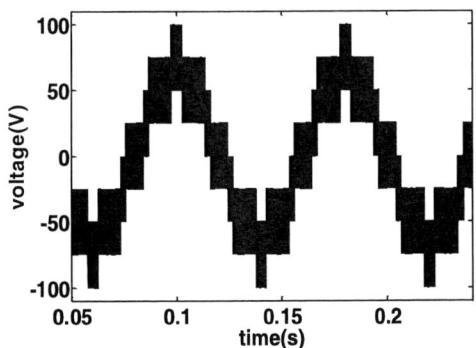

Fig. 15. Phase voltage for layer 2 operation

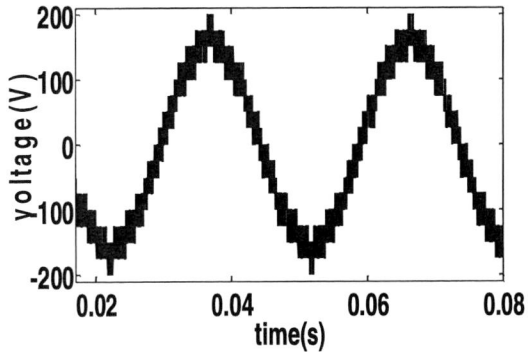

Fig. 19. Phase voltage for layer 4 operation

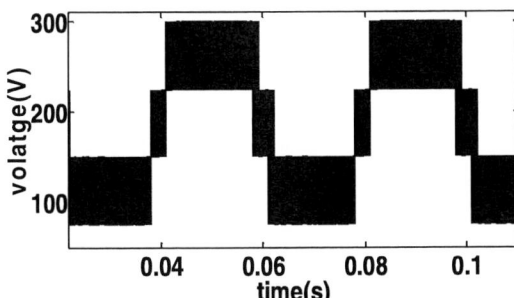

Fig. 16. Pole voltage for layer 3 operation

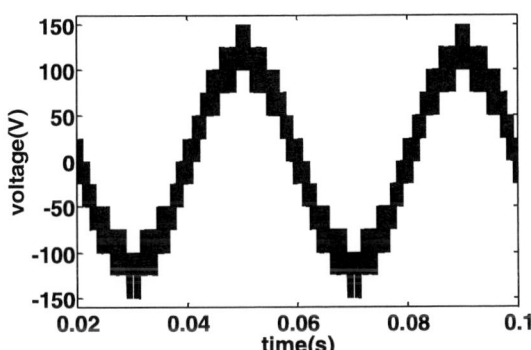

Fig. 17. Phase voltage for layer 3 operation

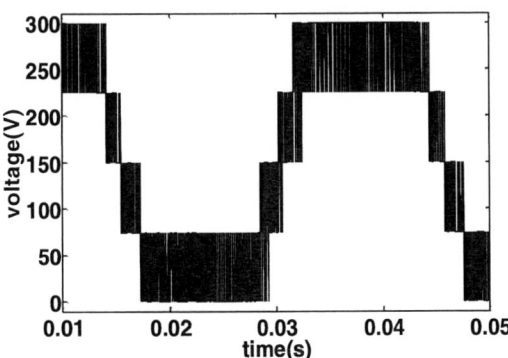

Fig. 18. Pole voltage for layer 4 operation

VI. CONCLUSION

This paper establishes the inherent fractal structure in the voltage space vector representation of multilevel inverter and proposes a method of generating SVPWM for multilevel inverter using this fractal structure. The sector identification algorithm proposed does not need look up tables and the inverter switching states corresponding to the sector are generated simultaneously along with sector identification. Optimum switching of the vectors is also achieved without using look-up tables. The simulation results are presented for a 5-level inverter, and the method can be extended for higher level inverters.

REFERENCES

[1] Jose Rodriguez, Jih-Sheng Lai and Fang Zheng Peng," Multilevel Inverters: A Survey of Topologies, Controls and Applications ", IEEE Transactions on Industrial Electronics, vol.49, No.4, August 2002, pp 724-738.

[2] Joachim Holtz, "Pulse width Modulation- A Survey",IEEE Transactions on Industrial Electronics, vol.39,No.5, December 1992,pp 410-420.

[3] Heinz Willi Van Der Broeck, Hans-Christoph Skudenly, and Georg Viktor Stanke, " Analysis and Realization of Pulsewidth Modulator Based on Voltage Space Vectors", IEEE Transactions on Industry Applications, vol.24, No.1, January/February 1988.

[4] Keliang Zhou and Danwei Wang, "Relationship Between Space-Vector Modulation and Three-Phase Carrier-Based PWM: A Comprehensive Analysis", IEEE Transactions on Industrial Electronics, vol.49, No.1, February 2002, pp 186-196.

[5] B.P. McGrath, Donald.G.Holmes, Thomas Lipo, " Optimized Space Vector switching sequences for Multilevel inverters", IEEE Transactions on Power Electronics, vol.18,No.6, November 2003, pp 1293-1301.

[6] Yo-Han Lee,Bum-Seok Suh and Dong-Seok Hyun, " A Novel PWM Scheme for a Three-Level Voltage Source Inverter with GTO Thyristors",IEEE Transactions on Industrial Applications, vol.32,No.2, March/April 1996, pp 260-268.

[7] A.K.Gupta and A.M Khambadkone " A General Space Vector PWM Algorithm for Multilevel Inverters, Including Operation in Overmodulation Range", IEEE Transactions on Power Electronics, vol.22,No.2, March 2007,pp 517-526.

[8] N.Celanovic and Dushan Boroyevich, " A Fast Space-Vector Modulation Algorithm for Multilevel Three-Phase Converters",IEEE Transactions on Industry Applications, vol.37,No.2, March/April 2001,pp 637-641.

[9] P.F Seixas, M.A. Severo Mendes, P.Donoso Garcia, A.M.N. Lima, " A Space Vector PWM Method for Three-Level Voltage Source Inverters", Proc. APEC, vol.1, 2000,pp 549-555.

[10] V.T Somasekhar and K.Gopakumar, "Three-level inverter configuration cascading two-level inverters", Proc. IEE- Electr. Power Appl, vol.150, No.3, May 2003, pp 245-254.

[11] R.S.Kanchan, M.R.Baiju, K.K.Mohapatra, P.P.Ouseph and K.Gopakumar, " Space vector PWM signal generation for Multilevel Inverters using only sampled amplitudes of reference phase voltages ",Proc. IEE, vol.152,No.2,March 2005,pp 297-309.

[12] Baiju M.R, Gopakumar. K, Somasekhar V.T, Mohapatra K.K and Umanand L, " A space vector based PWM method using only the instantaneous amplitudes of reference phase voltages for three-level inverters", EPE J., 2003, 13,(2), pp. 35-45.

[13] Heinz – Otto Peitgen, Hartmut Jurgens, Dietmar Soupe, "Chaos and Fractals – New Frontiers of Science", Second Edition, Springer.

Elimination of Harmonics in a Five-Level Diode-Clamped Multilevel Inverter Using Fundamental Modulation

Sule Ozdemir[1], Engin Ozdemir[1], Leon M. Tolbert[2], and Surin Khomfoi[3]

[1] Kocaeli University, Faculty of Technical Education, Department of Electrical Education, 41380, Umuttepe, Kocaeli, Turkey
[2] The University of Tennessee, Electrical and Computer Engineering Department, 311 Ferris Hall, Knoxville, TN 37996, USA
[3] King Mongkut's Institute of Technology, Faculty of Engineering, Department of Electrical Engineering, Ladkrabang, Bangkok 10530 Thailand

Abstract — In this study, elimination of harmonics in a five-level diode-clamped multilevel inverter (DCMLI) has been implemented by using fundamental modulation switching. The proposed method eliminates harmonics by generating negative harmonics with switching angles calculated for selective harmonic elimination method. In order to confirm the proposed method, first Matlab/Simulink and PSIM simulation results are given. Then the proposed method is also validated by experiments with Opal-RT controller and a 10 kW three-phase, five-level DCMLI prototype.

Index Terms — Fundamental switching, harmonic elimination, and multilevel inverter.

I. INTRODUCTION

The problem of eliminating harmonics in switching inverters has been the focus of research for many years. The current trend of modulation control for multilevel inverters is to output high quality power with high efficiency. For this reason, popular traditional PWM modulation methods are not the best solution for multilevel inverter control due to their high switching frequency. The selective harmonic elimination method has emerged as a promising modulation control method for multilevel inverters. The major difficulty for the selective harmonic elimination method is to solve the equations characterizing harmonics; however, the solutions are not available for the whole modulation index range, and it does not eliminate any number of specified harmonics to satisfy the application requirements. The proposed harmonic elimination method is used to eliminate any number of harmonics and can be applied to DCMLI application requirements.

The diode clamped inverter, particularly the three-level one, has drawn much interest in motor drive applications because it needs only one common voltage source. Also, simple and efficient PWM algorithms have been developed for it, even if it has inherent unbalanced dc-link capacitor voltage problem [1]. However, it would be a limitation to applications beyond four-level diode clamped inverters for the reason of reliability and complexity considering dc-link balancing and the prohibitively high number of clamping diodes [2]. Multilevel PWM has lower *dV/dt* than that experienced in some two-level PWM drives because switching is between several smaller voltage levels [3].

In this study, Engin Ozdemir was supported and sponsored by TUBITAK postdoctoral research fund R-2219 for his studies in USA.

In this paper, a five-level DCMLI switching at the fundamental frequency is proposed. However, many interesting PWM techniques have been proposed for controlling these inverters. The lower order harmonics are eliminated by choosing the switching angles that do not generate specifically chosen harmonics. In this study, an harmonic elimination technique is presented that allows one to control a multilevel inverter in such a way that it is an efficient low total harmonic distortion (THD) inverter that can be used to interface distributed dc energy sources to a main ac grid or as an interface to a motor drive powered by fuel cells, batteries, or ultra-capacitors.

II. DIODE-CLAMPED MULTILEVEL INVERTER

An *m*-level diode-clamped multilevel inverter typically consists of *m* – 1 capacitors on the dc bus and produces *m* levels of the phase voltage [4]. A three-phase five-level structure of a DCMLI is shown in Fig. 1. Each of the three phases of the inverter shares a common dc bus, which has been subdivided by four capacitors into five levels. The voltage across each capacitor is V_{dc}, and the voltage stress across each switching device is limited to V_{dc} through the clamping diodes.

Table 1 lists the output voltage levels possible for one phase of the inverter with the negative dc rail voltage V_0 as a reference. State condition 1 means the switch is on, and 0 means the switch is off. Each phase has five complementary switch pairs such that turning on one of the switches of the pair requires that the other complementary switch be turned off. The complementary switch pairs for phase leg *a* are (S_{a1}, $S_{a'1}$), (S_{a2}, $S_{a'2}$), (S_{a3}, $S_{a'3}$), and (S_{a4}, $S_{a'4}$). Table 1 also shows that in a diode-clamped inverter, the switches that are on for a particular phase leg are always adjacent and in series.

TABLE I
DCMLI VOLTAGE LEVELS AND SWITCHING STATES

Voltage V_a	SWITCH STATE							
	S_{a1}	S_{a2}	S_{a3}	S_{a4}	$S_{a'1}$	$S_{a'2}$	$S_{a'3}$	$S_{a'4}$
$V_4 = 4Vdc$	1	1	1	1	0	0	0	0
$V_3 = 3Vdc$	0	1	1	1	1	0	0	0
$V_2 = 2Vdc$	0	0	1	1	1	1	0	0
$V_1 = Vdc$	0	0	0	1	1	1	1	0
$V_0 = 0$	0	0	0	0	1	1	1	1

978-1-4244-0644-9/07/$25.00 ©2007 IEEE

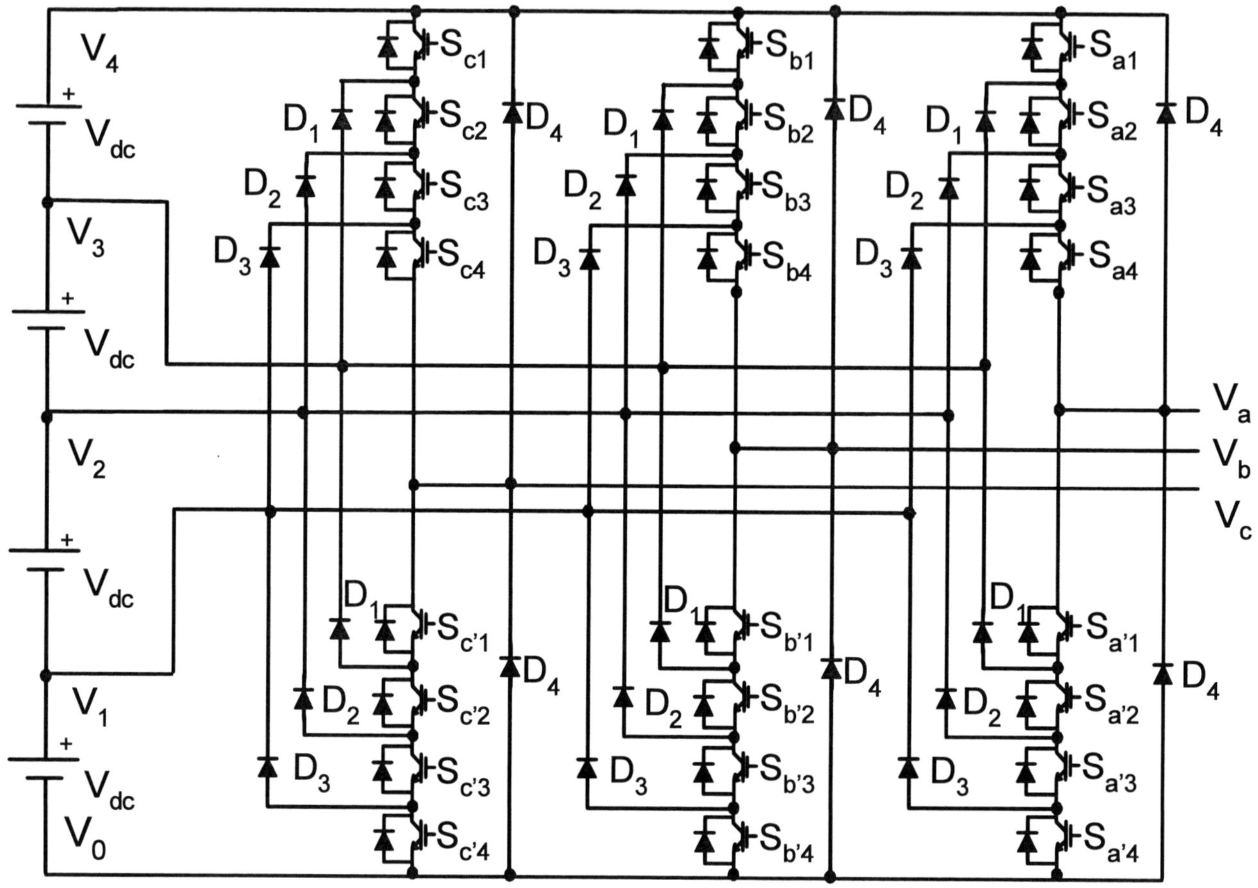

Fig. 1. A three-phase five-level diode-clamped multilevel inverter schematic.

The following are the some advantages and disadvantages of the DCMLI:

Advantages:
- As the number of levels increases the harmonic content of the output waveform decreases the filter size.
- Lower switching losses due to the devices being switched at the fundamental frequency without increasing the harmonic content in the output.
- Reactive power flow can be controlled, as this does not cause unbalance in the capacitor voltages.
- Fast dynamic response.
- Back to back operation is possible.

Disadvantages:
- High number of clamping diodes is required as the number of levels increase.
- Active power transfer causes unbalance in the DC-bus capacitors, this complicates the control of the system.

Fig. 2 shows one of the three line-line voltage waveforms for a five-level DCMLI. The line voltage V_{ab} consists of a phase-leg a voltage and a phase-leg b voltage. The resulting line voltage is a 9-level staircase waveform. This means that an m-level diode-clamped inverter has an m-level output phase voltage and a $(2m-1)$-level output line voltage.

The simplest way to control a multilevel converter is to use a fundamental frequency switching control where the switching devices generate an m-level staircase waveform that tracks a sinusoidal waveform. In this control, each switching device only needs to switch one time per fundamental cycle, which results in low switching losses and low electromagnetic interference. Considering the symmetry of the waveform, there are only two switching angles (θ_1 and θ_2) that need to be determined in this control strategy, as shown in Fig. 2. Note that the angles θ_1 and θ_2 are measured with respect to the reference angle $\pi/4$.

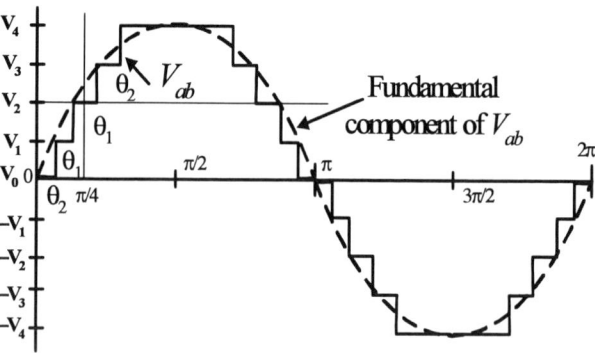

Fig. 2. Line voltage waveform for a five-level DCMLI.

III. PROPOSED METHOD FOR FIVE-LEVEL DCMLI

An important issue in multilevel inverter design is that the voltage waveform is near sinusoidal and the lower order harmonics are eliminated. A key concern in the fundamental switching scheme is to determine the switching angles in order to produce the fundamental voltage and not generate specific higher order harmonics. Often iterative techniques are used to calculate the switching angles [5], though such an approach does not guarantee finding all the possible solutions. Some other fundamental modulation techniques were presented in [6-8].

Previous work in [9] has shown that the transcendental equations characterizing the harmonic content can be converted into polynomial equations, which are then solved using the method of resultants from elimination theory. The work presented here is based on the previous work in [10]. However in this paper, the proposed calculation method results are used for switching of the DCMLI instead of a cascaded H-bridges multilevel inverter.

As shown in Fig. 2, a multilevel inverter can produce a quarter-wave symmetric voltage waveform synthesized by several DC voltages [10]. By applying Fourier series analysis, the output voltage can be expressed as

$$V(t) = \sum_{n=1,3,5...}^{\infty} \frac{4V_{DC}}{n\pi}(\cos(n\theta_1) + (\cos(n\theta_2) + ... \tag{1}$$
$$+ (\cos(n\theta_s)) \sin(n\omega t)$$

where s is the number of DC sources, and V_{DC} is the voltage of each DC level. For the DCMLI, the switching angles must satisfy the condition $0 < \theta_1 < \theta_2 < ...... \theta_s < \pi/4$. In order to minimize harmonic distortion and to achieve adjustable amplitude of the fundamental component, up to s-1 harmonic contents can be removed from the voltage waveform.

In general, the most significant low-frequency harmonics are chosen for elimination by properly selecting angles among different level inverters, and high-frequency harmonic components can be readily removed by using additional filter circuits. To keep the number of eliminated harmonics at a constant level, all switching angles must satisfy the condition $0 < \theta_1 < \theta_2 < ...... \theta_s < \pi/4$, or the total harmonic distortion (THD) increases dramatically. Considering the symmetry of the waveform, only two switching angles need to be determined in this strategy, which are θ_1 and θ_2 shown in Fig. 2. For a five-level multilevel inverter, the harmonic equations are given as

$$\cos(\theta_1) + \cos(\theta_2) = m_a \tag{2}$$
$$\cos(5\theta_1) + \cos(5\theta_2) = 0$$

where modulation index is $m_a = \pi V_1/(4V_{DC})$. By defining $x_i = \cos(\theta_i)$ and then using the trigonometric identity given below

$$\cos(5\theta) = 5\cos(\theta) - 20\cos^3(\theta) + \tag{3}$$
$$16\cos^3(\theta) + 16\cos^5(\theta)$$

The polynomial equations given below are used for calculation of the two switching angles (θ_1 and θ_2).

$$p_1(x_1, x_2) = \sum_{n=1}^{2} x_n - m = 0$$
$$p_5(x_1, x_2) = \sum_{n=1}^{2} (5x_n - 20x_n^3 + 16x_n^5) = 0 \tag{4}$$

The switching angles of the DCMLI were determined such that the 5th order harmonic was eliminated while at the same time controlling the value of the fundamental. The appropriate polynomial harmonic equations were first derived as given above, and the equations are solved and modulation index versus switching angles (θ_1 and θ_2) calculated. The proposed fundamental modulation features low switching losses because all the switches operate at the fundamental frequency. In addition, the converter can operate over a fairly wide range of modulation indices based on selective harmonic elimination.

An m-file in Matlab was used to perform all of the above calculations. The switching angles versus the modulation index m_a are shown in Fig. 3. Also, some modulation indices have more than one set of solutions with different values for their residual harmonics and thus THD. For practical applications, the set of switching angles with the lowest THD is used for simulation study and real time implementation.

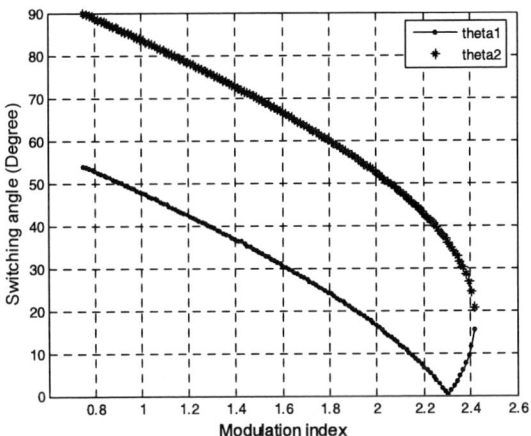

Fig. 3. Switching angles (θ_1 and θ_2) solutions versus modulation index.

IV. SIMULATION RESULTS

Matlab/Simulink and PSIM software packages were linked and run concurrently to perform this simulation implementation. Blocks in Matlab/Simulink generate the proposed fundamental switching pattern. SimCoupler module, an add-on module to the PSIM software, provides an interface between Matlab/Simulink and PSIM software packages for co-simulation [11]. The SimCoupler module enables Matlab/Simulink users to implement and simulate power circuits in their original circuit form, thus greatly shortening the time to set up and simulate a system, which includes electric circuits and motor drives. First, the DCMLI power circuit is simulated in PSIM, and the switching control in Matlab/Simulink. The line voltage waveform and harmonic spectrum of the output voltage that revealed elimination of fifth harmonic are shown in Fig. 4 and Fig. 5, respectively, for 60 Hz.

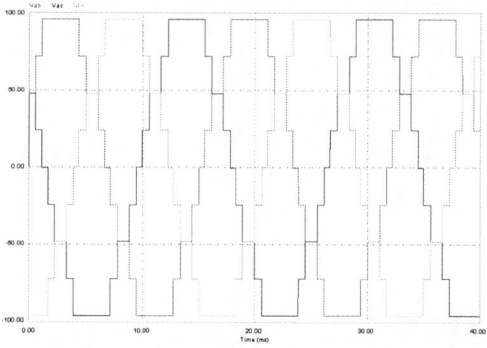

Fig. 4. Line voltage waveforms of the DCMLI output for 60 Hz.

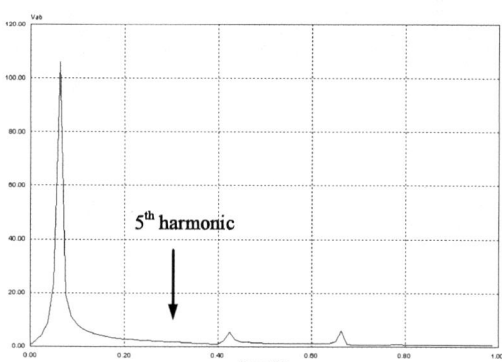

Fig. 5. The harmonic spectrum of line voltage for 60 Hz.

V. EXPERIMENTAL RESULTS

A prototype was built to verify the operation and efficiency of the proposed method. A picture of the laboratory experimental setup is represented in Fig. 6 that includes power supply, five-level DCMLI circuit, three-phase induction motor, and measurement equipment. The DCMLI supplies a three-phase induction motor rated at 250 W, 1725 rpm, 1.5 A, and 208 V coupled with a dc generator (250 W) as a load of the induction motor.

In this work, the RT-Lab real-time computing platform from Opal-RT-Technologies Inc. [12] was used to interface the computer to the five-level DCMLI. The Opal RT-Lab system is utilized to generate gate drive signals and interfaces with the gate drive board. This system allows one to implement the switching algorithm in Simulink that is then converted to *C* code using real-time workshop from Mathworks. The RT-Lab software provides icons to interface the Simulink model to the digital I/O board and converts the *C* code into executables.

The step size for the real-time implementation was 10 microseconds, which was used to obtain an accurate resolution for implementing the switching times. The real-time implementation is accomplished by placing the data in a lookup table and therefore does not require high computational power for implementation.

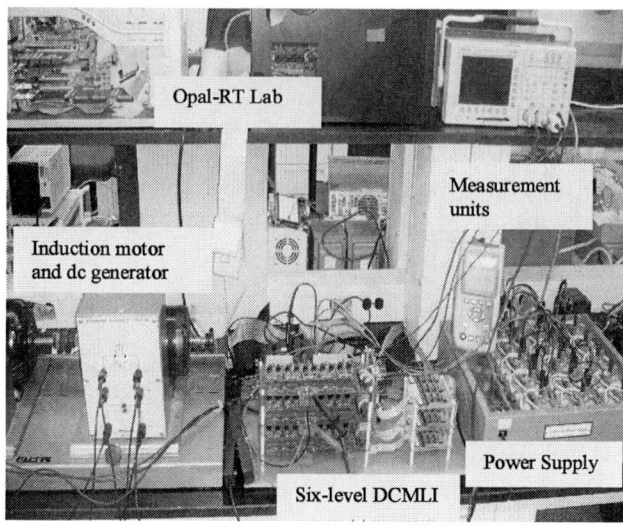

Fig. 6. The five-level DCMLI experimental setup.

Fig. 7(a) shows the three-phase line-to-line voltages and motor current waveforms, and Fig. 7(b) illustrates its corresponding fast Fourier transform (FFT) spectrum that shows the 5th, 13th and triplen harmonics are absent from waveform. The output voltage has nine levels, and the wave shape is near sinusoidal. Using the data in Fig. 7(b), the total harmonic distortion (THD) of the line-line voltage was computed to be 10 %. Since the magnitude of the lower order harmonics is very low, there is no need for a large filter circuit.

Experimental verification that the low order harmonics are indeed eliminated is also presented by driving a three-phase induction motor from a five-level DCMLI. Since five-level DCMLI can generate a nine-level line-to-line staircase waveform, the generated voltage and current waveforms are sinusoidal shape even at fundamental switching frequency. The THD level of output current of the inverter was as low as 1.9% at full load with fundamental frequency switching.

The conversion efficiency of the inverter, which was measured with a Yokogawa PZ 4000 power analyzer, at low power levels is more than 95%, and the maximum efficiency is about 98.5%, which is comparatively higher than conventional PWM inverters. Negligible switching losses occur for this system due to the low switching frequency employed. In the present case, the main causes of inverter losses are related to semiconductor conduction losses. Each switch in the inverter switches only once per cycle when performing fundamental frequency switching; this results in high efficiency.

Experimental waveforms, efficiency and harmonic measurements, obtained from a small-scale lab prototype, have been used to validate the proposed fundamental switching modulation method and the feasibility of the application of fundamental frequency modulated the three-phase five-level DCMLI. The proposed technique, while maintaining switching at fundamental frequency, is simple and requires relatively simple control circuitry and the output quality becomes better as the number of levels increases. The proposed fundamental frequency operated DCMLI based power system presented a good performance concerning efficiency and power quality.

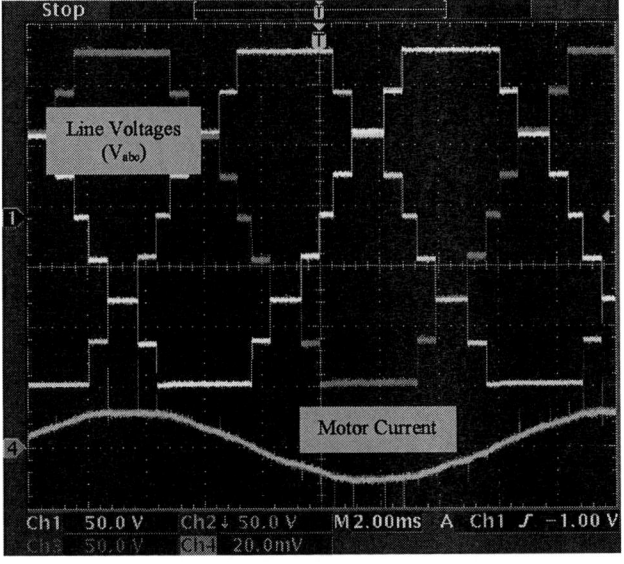

(a)

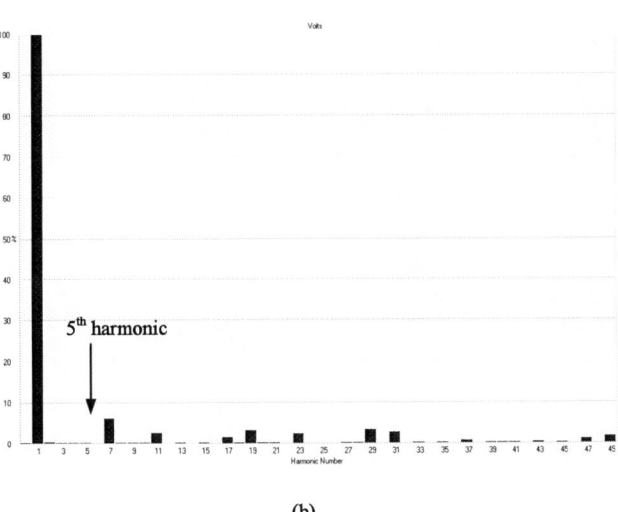

(b)

Fig. 7. (a) Three-phase line voltages and motor current waveforms (b) FFT spectrum of output voltage in experiment.

VI. CONCLUSION

The selected harmonic elimination is a popular issue in multilevel inverter design. The proposed selective harmonic elimination method for DCMLI has been validated in both simulation and experiment. The simulation and experimental results show that the proposed algorithm can be used to eliminate any number of specific lower order harmonics effectively and results in a dramatic decrease in the output voltage THD. In the proposed harmonic elimination method, the lower order harmonic distortion is largely reduced in fundamental switching. Furthermore, in the experiments reported here, an induction motor load is connected to the three-phase five-level DCMLI, and the current as well as the voltage waveforms are collected for analysis. The FFTs of these waveforms show that their harmonic content is close to the simulated values.

REFERENCES

[1] J. Rodríguez, J. S. Lai, F. Z. Peng, "Multilevel inverters: a survey of topologies, controls, and applications," *IEEE Transactions on Industrial Electronics*, vol.49, no.4, pp. 724-738, 2002.

[2] J. S. Lai, F. Z. Peng, "Multilevel converters-a new breed of power converters," *IEEE Transactions on Industry Applications*, vol. 32, no.3, pp. 509-517, 1996.

[3] L. M. Tolbert, F. Z. Peng, T. G. Habetler, "Multilevel converters for large electric drives," *IEEE Transactions on Industry Applications*, vol. 35, no. 1, pp. 36–44, Jan./Feb. 1999.

[4] S. Khomfoi, L. M. Tolbert, "Multilevel Power Converters," Chapter 17, *Power Electronics Handbook,* 2nd Edition, Elsevier, ISBN 978-0-12-088479-7, pp. 451-482, 2007.

[5] J. N. Chiasson, L. M. Tolbert, K. J. McKenzie, Z. Du, "A complete solution to the harmonic elimination problem," *IEEE Transactions on Power Electronics*, vol. 19, no. 2, pp. 491-499, 2004.

[6] J. Wang, R. Wei, Q. Ge, Y. Li, "The harmonic selection elimination of multilevel converters" *IEEE International Conference on Electrical Machines and Systems*, vol.1, pp. 419-422, 2003.

[7] E.Y. Guan,, et. al. "Fundamental modulation: multi-pattern scheme with an entire range of modulation indices for multilevel cascaded converters," *IEEE Industrial Electronics and Application Conference*, pp. 1-6, 2006.

[8] Z. Pan, F. Z. Peng, et. al., "Voltage balancing control of diode-clamped multilevel rectifier/inverter systems," *IEEE Transactions on Industry Applications*, vol. 41, pp. 1698–1706, 2005.

[9] J. Chiasson, L. M. Tolbert, K. J. McKenzie, Z. Du, "Control of a multilevel converter using resultant theory," *IEEE Transactions on Control Systems Technology*, vol. 11, no. 3, pp. 345-354, 2003.

[10] Z. Du, L. M. Tolbert, J. N. Chiasson, "Active harmonic elimination for multilevel converters," *IEEE Transactions on Power Electronics*, vol. 21, no. 2, pp. 459-469, 2006.

[11] Powersim Inc, *PSIM User's Guide Version 6,* Powersim Inc, 2003, http://www.powersimtech.com.

[12] Opal-RT technology Inc, *RT-LAB User's manual*, Opal-RT technology Inc, Version 6, 2001, http://www.opal-rt.com.

Compensation of DC–Link Oscillations of Cascaded H-Bridge Converters

M. Tavakoli Bina* and B. Eskandari*

* Faculty of Electrical Engineering, K. N. Toosi University of Technology, P. O. Box 16315–1355, Tehran 16314, Iran,
E-mail: tavakoli@ieee.org

Abstract—Single-phase AC applied voltage of an H–bridge converter produces second harmonic on top of the DC-link voltage. Three-phase unbalanced voltages make similar effects on the DC-link voltages, as well. Nevertheless, for applications that need higher voltages, series connection of H-bridges could lower the amplitude of the oscillations. Low-frequency oscillations are considerable when the number of cascaded H-bridges is less than four, introducing the worst case oscillations for a single H-bridge converter. This paper proposes various external DC active filter circuits, aiming at cancelling these oscillations of cascaded H-bridges up to three. Proposed circuits are simulated, and their performances on compensation of oscillations are compared to select the best choice. Simulation results confirm that the proposed methods limit the DC–link oscillations on DC-link of H-bridges. Also, the presented methods are compared in terms of both their advantages and disadvantages.

Index Terms—Active-filtering, auxiliary compensation, DC-link Oscillations, H-bridge, S-bridge.

I. INTRODUCTION

CASCADED H-bridge multilevel converters can potentially be used as an alternative to the series connection of semiconductor switches to increase the system voltage [1]-[4]. Figure 1 shows a typical cascade H-bridge converter in which the harmonic performance is expected to be improved compared to the converters with series connected switches. This topology has found high-voltage high-power applications such as modular multilevel AC–AC converters (M^2LC) [5]-[6]. The M^2LC includes four modules of cascaded H-bridge converters of type shown by Fig. 1. However, each H-bridge sub-module of Fig. 1 exchanges active power between the electrical network and the load through the other H-bridge converters. This power exchange depends on the magnitude of the fundamental voltage of each H-bridge converter as well as the magnitudes of low order harmonics. The exchanged power would influence considerably on the DC–link voltage of the H-bridge converter, causing low frequency oscillations. Further, when H-bridge converters introduce different power exchanges from each other, then balance of capacitor voltages is a major concern that could possibly lead to instability. Figure 1(b) depicts the SPWM technique that is used to modulate two cascaded H-bridge converters [7]. It can be seen that the difference in the output voltage

magnitudes of the two H-bridge converters can affect the balancing of the two capacitor voltages.

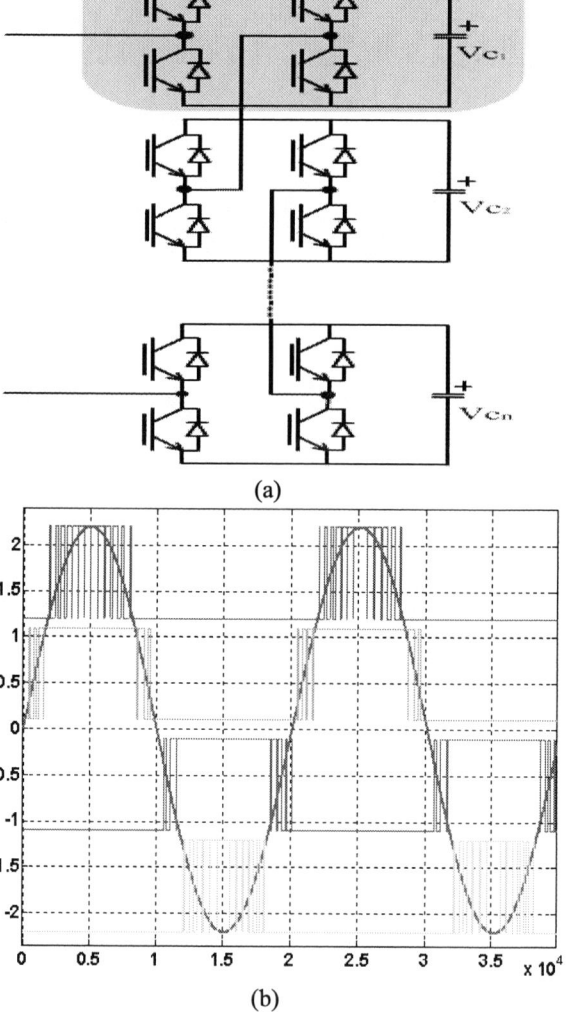

(a)

(b)

Figure 1: (a) General topology of Cascade H-bridge converters, and (b) the effect of difference in fundamental voltage magnitudes of two H-bridge converters on the DC-link voltage balancing.

This paper examines various methods to compensate the low-frequency oscillations that appear on top of the DC-link voltages of H-bridge multilevel converters. Different passive/active filtering circuits are proposed that is connected to the DC–link capacitor, resulting in

---

This work was performed in the Research Laboratory of K. N. Toosi University of Technology.

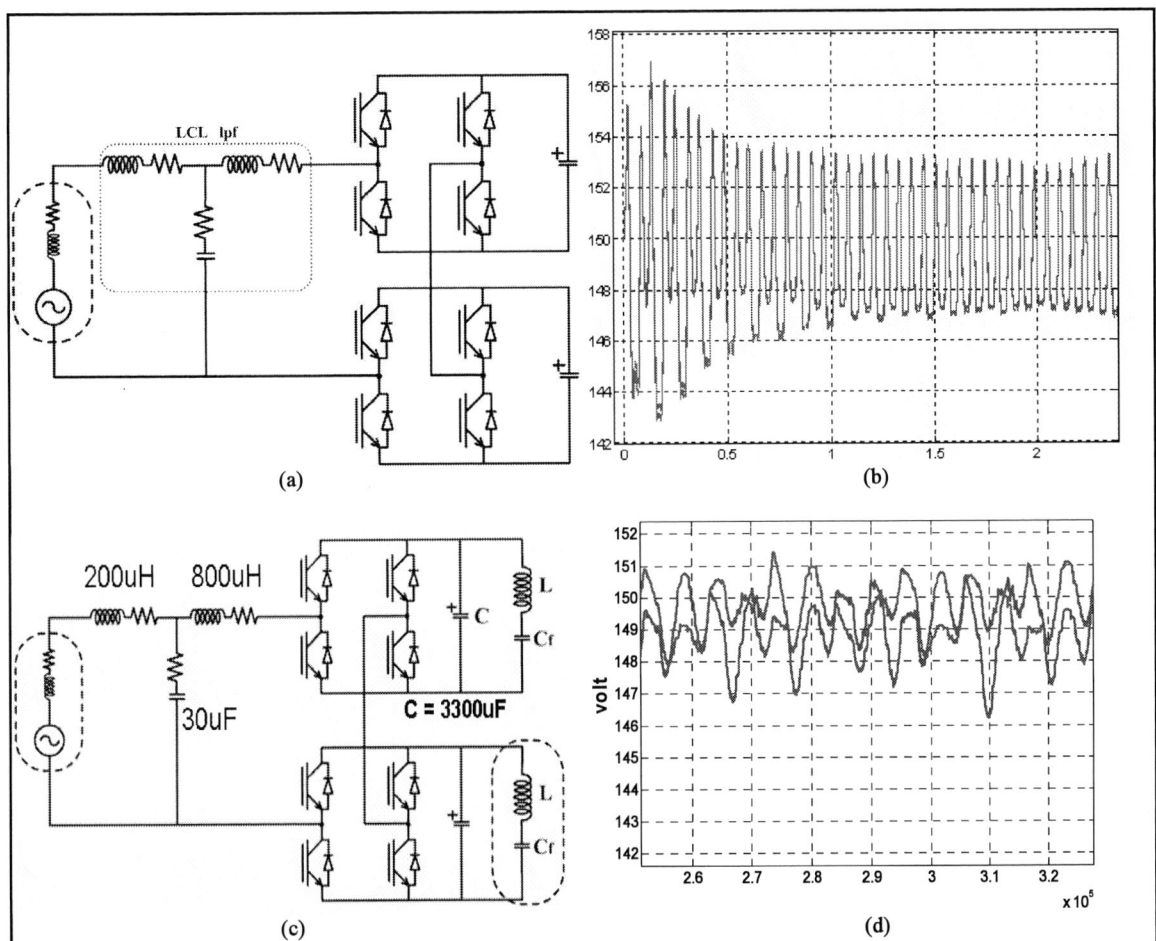

Figure 2: (a) Cascaded two H-brige converter without any compensators, (b) DC-link oscillations excluding any DC-link filters, (c) a tuned passive LC filter compensates the DC-link oscillations, and (d) simulation results when DC capacitor oscillations are absorbed by the passive filter.

compensation of the voltage oscillations. All DC-filters are examined and simulated with MATLAB to compare their performances along with suitability for the cascade H-bridge converters. Simulation results confirm that the DC active-filtering proposals compensate the DC–link oscillations much effective than other introduced methods.

II. DC ACTIVE -FILTERING OF OSCILLATIONS

Here we examine connection of various DC passive/active filtering circuits across the DC capacitor to compensate the low-frequency oscillations imposed by the exchange of active power. Predominant frequency of these oscillations is 100/120 Hz when power system operates at 50/60 Hz. Figure 2(a) shows a cascaded two H-bridge converter, excluding any DC-link compensator, which is simulated with SIMULINK. A PI controller is employed to control the phase of the H-bridges output voltages such that the average DC-voltage remains fixed at 150 V [8]-[9]. Switching pulses are swapped between the two H-bridges to force both capacitor voltages vary

similarly. Figure 2(b) illustrates the DC capacitor variations when no filter is designed to absorb the oscillations. It is clear that the oscillations cannot be damped (peak-to-peak variation is about 8V or 5.33%). Hence, the following sub-sections suggest and examine various topologies that use passive/active filters to damp more efficiently the oscillations.

A. Passive LC-filter compensation (PLC): proposition 1

The low-frequency oscillations of the DC-link can be filtered using a tuned LC passive filter that its natural frequency is 100 Hz. Figure 2(c) shows a cascade converter consisting of two H-bridges as well as two LC passive filters, which is simulated by SIMULINK. This would effectively reduce the oscillations, needs no extra semiconductor switches and control circuits. Nevertheless, the disadvantage of the PLC method is that the 100 Hz passive LC filter needs big parameters as well as occupying huge space [10].

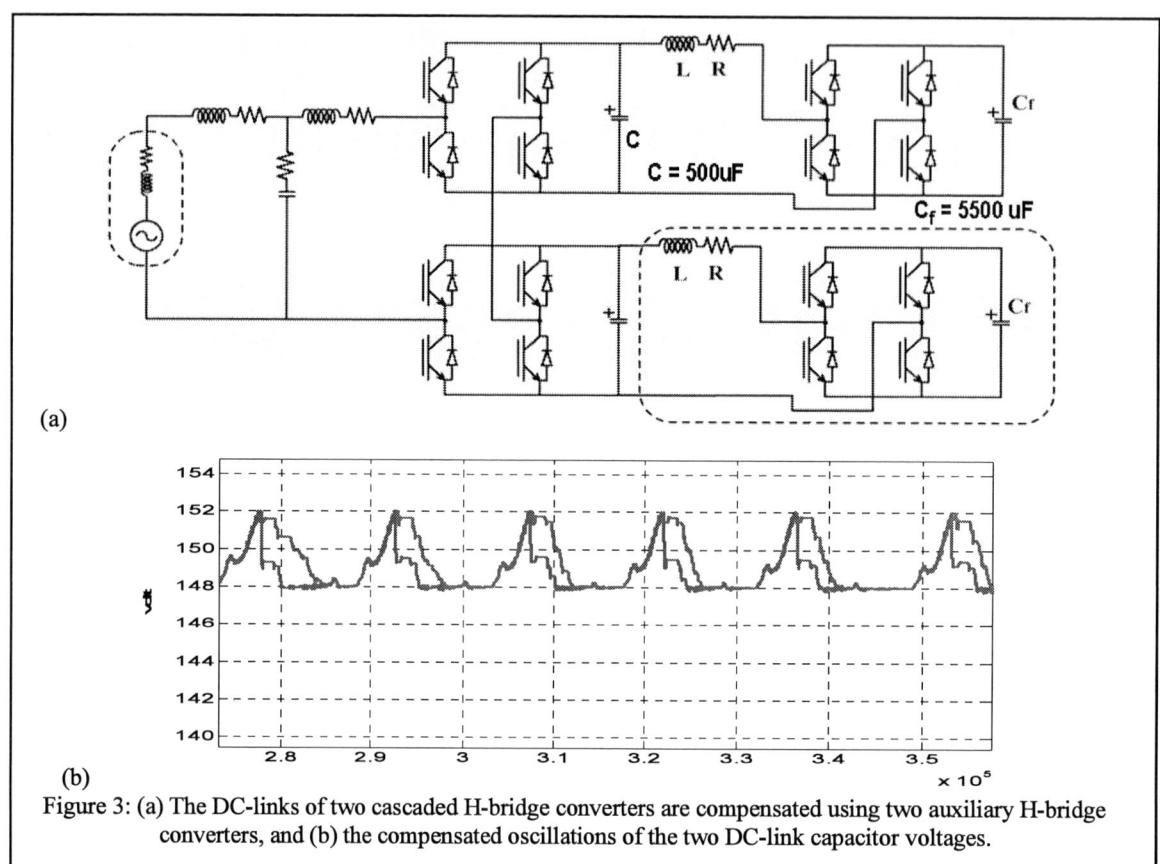

(a)

(b)

Figure 3: (a) The DC-links of two cascaded H-bridge converters are compensated using two auxiliary H-bridge converters, and (b) the compensated oscillations of the two DC-link capacitor voltages.

B. Auxiliary H-brige compensation (AHC): proposition 2

Here an H-bridge converter (with DC capacitor C_f) is connected across the DC-link capacitor C of each main H-bridge converter through the inductance L like it is shown by Fig. 3(a) [11]. Capacitance C_f is much bigger than that of C, while the average voltage of both capacitors is considered identical. The two H-bridges operate in a way that when the voltage of capacitor C starts rising, the capacitor C_f is charged by its H-bridge converter. When voltage of capacitor C starts getting lower, then the capacitor C_f is discharged to prevent the decrease in voltage of capacitor C. It is noticeable that the voltage variations in capacitor C_f is lower than capacitor C because C_f is bigger than C.

Any oscillations on the capacitor C is transferred to the capacitor Thus, the low-frequency oscillations on the capacitor voltage V_C are sampled from the capacitor voltage V_{C_f}. A PI controller is used to regulate the average voltage of the capacitor C_f. Also, the capacitor voltage is considered to be slightly bigger than V_C. This would prevent the flow of current from C_f through the parallel diodes to the capacitor C.

It should also be mentioned that the switching signals are swapped between the two H-bridge converters of Fig.

3(a) symmetrically. This implies that the two capacitors (C_f) have similar variations. Hence, taking sample from one of the capacitors is compared to a reference voltage, and the resultant error is sent to a PI controller.

One disadvantage over the *AHC* is the big capacitance C_f, which causes dynamically slow tracking of the DC-link voltage. Also, since the average DC voltages of both C and C_f are identical, a voltage difference between these two capacitors is needed to control the current exchange through the inductance L. Therefore, the *AHC* is unable to damp completely the oscillations of the DC-links. Figure 3(b) provides simulated oscillations of the two DC-links using the *AHC*, which is much lower than those of the uncompensated case.

C. Auxiliary S-bridge compensation (ASC): proposition 3

This proposal uses two identical capacitors (C_{f1} and C_{f2}) along with three switches for each DC-link like it is illustrated by Fig. 4(a). Capacitances C_{f1} and C_{f2} do not have to be big, while their average DC voltages are equal to three quarters of the DC-link voltage of capacitor C. Three switches are operated in a way that when the DC-link voltage is increased beyond than a positive band (ΔV), the two vertical switches are turned on and the horizontal switch is turned off. This makes both capacitances C_{f1} and C_{f2} to operate in

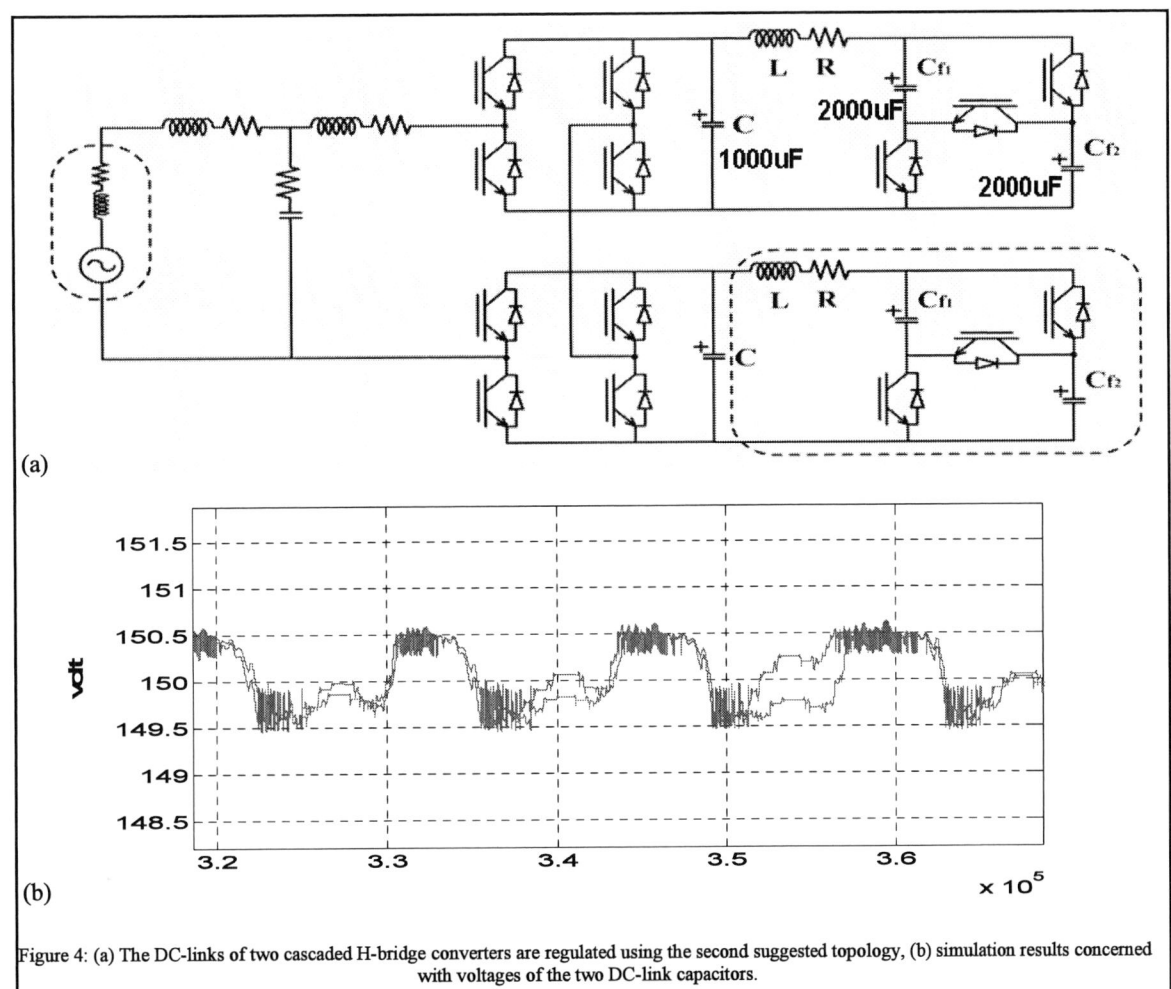

(a)

(b)

Figure 4: (a) The DC-links of two cascaded H-bridge converters are regulated using the second suggested topology, (b) simulation results concerned with voltages of the two DC-link capacitors.

parallel, absorbing charging currents through the inductance L from the DC-link C by the following slope:

$$\frac{di_L}{dt} = \frac{V_L}{L} = \frac{V_{DC} - \frac{3}{4}V_{DC}}{L} = \frac{V_{DC}}{4L} \qquad (1)$$

Where the voltage V_L is the drop on inductance L and i_L is the absorbed current. This relatively big slope forces the main DC-link voltage to come down more rapidly. Accordingly, when the DC-link voltage is dropped below a negative band $(-\Delta V)$, the two vertical switches are turned off and the horizontal switch is turned on. The two capacitors C_{f1} and C_{f2} operate in series, injecting current to the DC-link C through the inductance L as follows:

$$\frac{di_L}{dt} = \frac{V_L}{L} = \frac{V_{DC} - \frac{6}{4}V_{DC}}{L} = -\frac{V_{DC}}{2L} \qquad (2)$$

Here the negative slope reverses the current from

C_{f1} and C_{f2} to charge the DC-link capacitor C. Note that here the positive slope is different from the negative one as it is illustrated by Fig. 4(b). Simulation results show that peak-to-peak of the oscillations is smaller than 1V (or 0.6%). In practice, the hysteresis band could limit the performance of the proposal because the switching frequency could be very high when the chosen ΔV is small.

Like the PI controller of the AHC, the controller of the ASC takes DC-voltage sample from either C_{f1} or C_{f2}; because voltage variations of the two capacitors are identical. Table I summarizes the simulation results obtained by all suggested methods, including the AHC, the ASC, the PLC, and the uncompensated case.

It can be seen from the results gathered in Table I that while the uncompensated case contains considerable low-order oscillations, other suggested methods lower the oscillations from 11.1% well below 3.5%. Amongst the analyzed methods, the ASC introduces the lowest level of oscillations (smaller than 0.6%). Then, the AHC achieves 2.6% maximum oscillations, and the PLC up to 3.3%. Since the proposed circuits use low-power elements, total cost of the added device could be reasonable enough to apply the suggested methods to the H-bridge converters.

TABLE I: SIMULATION RESULTS CONCERNED WITH THE PEAK-TO-PEAK OSCILLATIONS OF THE DC-LINK OF H-BRIDGE CONVERTERS ALONG WITH THE ADVANTAGES AND DISADVANTAGES OF ALL SUGGESTED METHODS

Compensation Method	Oscillations (maximum Peak-to-Peak)	Oscillations (average Peak-to-Peak)
AHC	2.6%	2.6%
ASC	0.6%	0.6%
PLC	3.13%	1.82%
Uncompensated	11.1%	8.9%

III. CONCLUSIONS

The DC-link oscillations related to the H-bridge cascade converters is significant when the number of H-bridges is smaller than four. This situation is worse when the three-phase applied voltages are unbalanced. The oscillations on the DC-link are modulated through the H-bridge converter, and enter the AC system. This has considerable impacts on harmonic performance as wellas the efficiency of the converter. While an uncompensated simulation with two H-bridges show considerable oscillations on the DC-links, three methods are proposed and examined to lower the peak-to peak oscillations. These methods are the passive LC-filter compensation (PLC), the auxiliary H-bridge compensation (AHC) and the auxiliary S-bridge compensation (ASC). These methods are also simulated with SIMULINK to compare their effectiveness on lowering the DC-link oscillations. Simulation results show that these three methods successfully control the oscillations, amongst them the ASC performs as the best solution. Nevertheless, there are both advantages and disadvantages for each method that are needed to be taken into account in practical implementations.

ACKNOWLEDGEMENT

The authors would like to thank the support of the research Laboratory of power quality and reactive power control in K. N. Toosi University of Technology.

REFERENCES

[1] J. S. Lai, F. Z. Peng "Multilevel Inverters: A survey of topologies, controls, and applications", IEEE Transactions on Industrial Electronics, vol. 49, August 2002, pp. 724-738.

[2] J. Rodriguez, J. S. Lai, F. Z. Peng "Multilevel Converters- A new breed of power converters", IEEE Transactions on Industry Applications, vol. 32, no. 3, May/June 1996, pp. 509-517.

[3] J. Rodriguez, J. S. Lai, F. Z. Peng, "Multilevel Converters- A new breed of power converters", IEEE Transactions on Industry Applications, vol. 32, No. 3, pp. 509–517, May/June 1996.

[4] D. Soto, T. C. Green, "A Comparison of High Power Converter Topologies for the implementation of FACTS Controllers", IEEE Transactions on Industrial Electronics, vol. 49, No. 5, pp. 1072–1080, October 2002.

[5] A. Lesnicar, and R. Marquardt "An Innovative Modular Multilevel Converter Topology Suitable for a Wide Power Range", ETG-Fachtagung, Bad Nauheim, Germany, 2003.

[6] R. Marquardt and A. Lesnicar "New Concept for High Voltage – Modular Multilevel Converter" IEEE PESC 2004.

[7] Bai Hua, Zhao Zhengming, Meng Shuo,Liu Jianzheng, Sun Xiaoying, "Comparison of Three PWM Strategies, SPWM,SVPWM & One-cycle Control", IEEE PEDS 2003 Singapore.

[8] Rajesh Gupta, Arindam Ghoshb, Avinash Joshi, "Control of cascaded transformer multilevel inverter based DSTATCOM", ELSEVIER- Electric Power System Research, October 2006.

[9] Feel-soon Kang, Su Eog Cho, Sung-Jun Park, Cheul-U Kim, Toshifumi Ise, "A new control scheme of a cascaded transformer type multilevel PWM inverter for a residential photovoltaic power conditioning system", ELSEVIER-2004.

[10]] Wang Yi, Li Heming, Shi Xinchun, Zhu, "Harmonic Analysis and Filter Design for Medium-Voltage Multilevel PWM Inverters", IEEE PEDS 2003, Singapore.

[11] Toshihisa shimizu, "Dc Active Filter Concept and Single-phase p-q Theory for Single-phase Power Converters", IEEE PEDS2003, Tutorial A, Singapore.

Combined DC–Filter and optimized Modulation to Absorb DC–Link Oscillations of Cascaded H-Bridge Converters

M. Tavakoli Bina* and B. Eskandari*

* Faculty of Electrical Engineering, K. N. Toosi University of Technology, P. O. Box 16315–1355, Tehran 16314, Iran,
E-mail: tavakoli@ieee.org

Abstract--Cascaded H–bridge converters can be employed suitably for applications that need higher voltages, avoiding series connection of semiconductor switches. However, DC-link oscillations are a practical issue when converters operate under three-phase unbalance condition and/or single phase operation. Resultant low frequency oscillations would affect the harmonic performance as well as efficiency of the converter. This paper proposes four various external DC active filter circuits, aiming at cancelling these oscillations. Proposed circuits are simulated, and their performances on compensation of oscillations are compared to select the best choice. Furthermore, an optimized switching pulse train is developed to attenuate the harmonics objectively. The solution is then combined with the selected DC active filter circuit to limit the peak of the oscillations. The combined method can be applied to modular AC–AC multilevel converter as an application of the suggested technique. Simulation results confirm that the combined method limits the DC–link oscillations of cascaded H-bridges more suitably compared to other proposed circuits, and can be employed for single-phase applications as well as unbalanced three–phase systems.

Index Terms— Active DC-filter, cascaded converters, DC-link oscillations, combinatorial minimization, optimized modulation.

I. INTRODUCTION

MULTILEVEL converters can potentially overcome the practical issues concerned with the series connection of semiconductor switches to increase the system voltage [1]-[3]. Figure 1(a) shows a typical cascade H-bridge converter in which the harmonic performance is expected to be improved in addition to raising the output voltage. However, each H-bridge sub-module exchanges active power between the electrical network and the load through the other H-bridge converters. This power exchange depends on the magnitude of the fundamental voltage of each H-bridge converter, as well as the magnitudes of low order harmonics.

Furthermore, the exchanged power would affect considerably on the DC–link voltage of the H-

bridge converter, causing low frequency oscillations. Further, when H-bridge converters introduce different power exchanges from each other, then balance of capacitor voltages is a major concern that could possibly lead to instability.

This paper proposes a DC-filter to be connected across the DC–link capacitor, resulting in compensation of the voltage oscillations. The suggested method is then examined and simulated with MATLAB to assess its performance along with suitability for the cascaded H-bridge converters. Moreover, the method is modified to get better performance on attenuating the DC–link oscillations. Then a conventional optimized switching technique is combined with this modified circuit to damp significantly the DC–link oscillations. Simulation results confirm that this latter combined proposal presents considerable improvement in adsorption of the DC–link oscillations.

II. PROPOSED DC-FILTER

Here we examine connection of various DC passive/active filtering circuits [8] across the DC capacitor to compensate its low-frequency oscillations imposed by the exchange of active power. Predominant frequency of these oscillations is 100/120 Hz when power system operates at 50/60 Hz. Figure 1(a) shows an uncompensated cascade converter (two H-bridges), which is simulated with SIMULINK. A PI controller is employed to control the phase of the H-bridges output voltages such that the average DC-voltage remains fixed at 150 V. Switching pulses are swapped between the two H-bridges to force both capacitor voltages vary similarly. Figure 1(b) illustrates the DC capacitor variations when no filter is designed to absorb the oscillations.

It is clear from Fig. 1 that the oscillations cannot be significantly damped (peak-to-peak variation is about 16V or 10.67%). Hence, the following sub-sections suggest and examine new designs that use active filters to damp the oscillations effectively.

This work was performed in the Research Laboratory of K. N. Toosi University of Technology.

978-1-4244-0644-9/07/$25.00 ©2007 IEEE

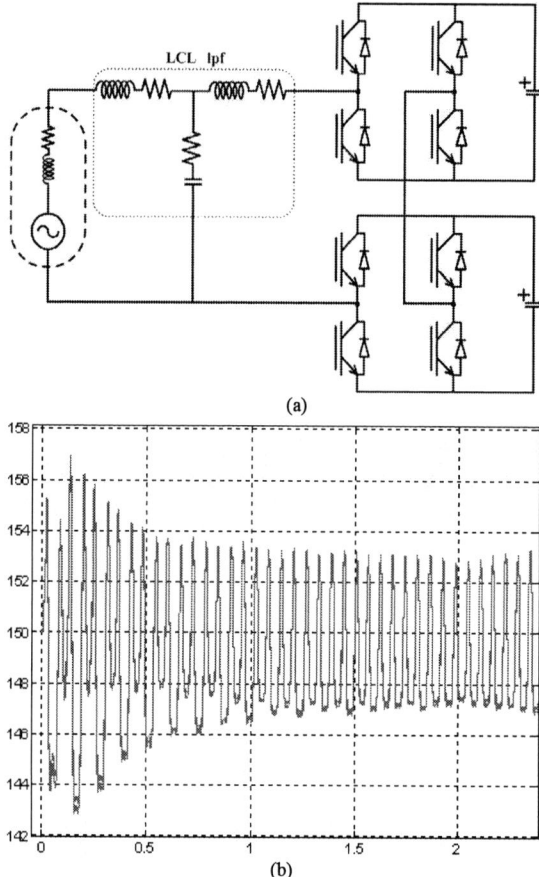

(a)

(b)

Figure 1. (a) Cascaded two H-brige converter without any compensators, (b) DC-link oscillations excluding any DC-link filters.

The low-frequency oscillations of the DC-link can also be filtered using a tuned LC passive filter that its natural frequency is 100 Hz. This would effectively reduce the oscillations. Nevertheless, the disadvantage of the PLC method is that the 100 Hz passive LC filter has big parameters, and occupies huge space.

A. Independent DC source (IDC): Proposition 1

Design of Fig. 2(a) uses the idea of passive LC-filter to damp the oscillations out; but, the inductance of the passive filter is replaced with the primary winding of a transformer. The secondary of transformer is connected to a low-power low-voltage H-bridge converter through a passive low-pass *LCL* filter. The suggested design provides much more satisfactory attenuation of DC-link oscillations compared to the PLC. Thus, the control and operation of this proposal is examined in detail.

First, the oscillations of DC-link voltage (V_f) is extracted by subtracting the average value from the exact value of the DC-link voltage like it is depicted by Fig. 2(b). Using the zero-order hold (ZOH)

function of SIMULINK, the 10 kHz sampled signals are converted to continuous-time signals for analyzing the sampled continuous-time system. Then, the volt-second balance law is applied to each switching period (100 μs) to find the duty ratio of each switch as follows:

$$V_{dc} \times t_{on} = V_f \times 100(\mu s) \qquad (1)$$

Where the source V_{dc} is the DC voltage of the compensating H-bridge converter, and t_{on} is the on-time duration of each switch. Computing t_{on} of the switches, they are applied to the extra H-bridge converter. The whole design is simulated with MATLAB, where Fig. 3(c) shows both voltages of primary and secondary windings of the transformer with a turn ratio slightly bigger than one. Also, Fig. 3(d) illustrates the compensated DC-link oscillations, which is considerably attenuated up to about 2.5 V peak-to-peaks (or 1.67%).

It is noticeable that the compensating elements including the DC source, transformer, low-pass filter and switches operate under low-voltage low-power conditions. This also makes low extra-cost to the cascaded converter. However, in practice, as the magnitudes of oscillations are being lowered, using (1), duty ratios of switches are being decreased as well. This implies a limit on reduction of magnitudes of oscillations as a drawback of the method. To remedy this issue, the voltage V_{dc} is needed to be decreased when the oscillations are going to be considerably attenuated. This also will add extra-cost to the converter for developing regulated controllable DC voltage (V_{dc}).

B. Variable DC source (VDC): improvement of proposition 1

One disadvantage over the proposition 1 is that presence of a variable DC source would allow damping of the oscillations down to a desirable level. This issue can be overcome by replacing the independent DC source V_{dc} with a full-bridge rectifier that is supplied through the capacitor C_5 in Fig. 2(d). Capacitor C_5 absorbs the DC value of the DC-link voltage, where the oscillations are transferred to the rectifier. Thus, the average DC value of capacitor C_7 depends directly to the magnitude of DC-link oscillations. This way provides a controllable DC source, which in turn can compensate even low magnitude oscillations by having the possibility of proper switching pulse width modulation according to (1).

Also, the control algorithm is like that of the IDC method stated by (1) as shown by Fig. 3. However,

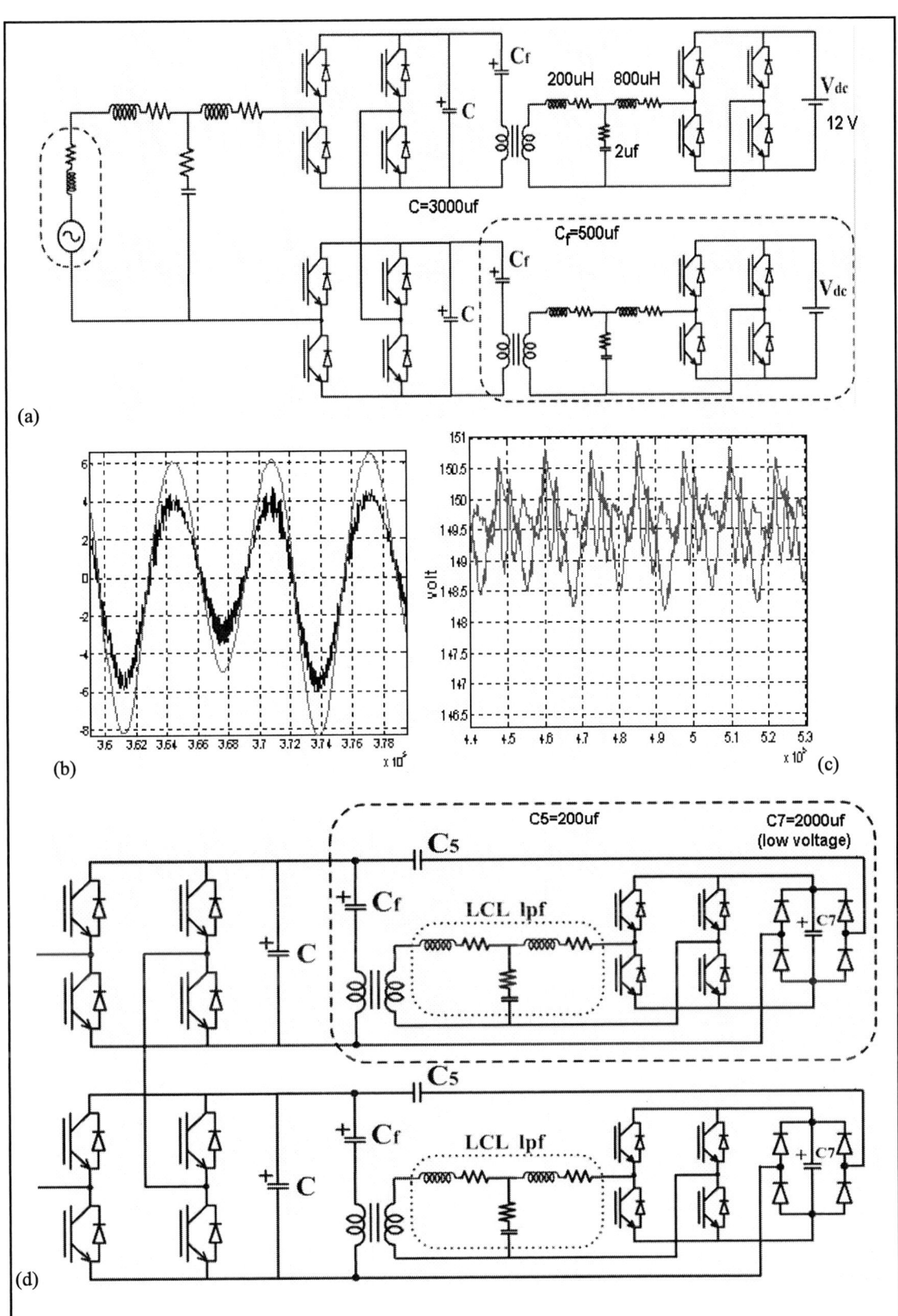

Figure 2: (a) The DC-links of two cascaded H-bridge converters are regulated by the suggested topology of IDC, (b)–(c) simulation results concerned with voltages of the two DC-link capacitors, and (d) the modified proposed method of the VDC.

the only modification is that V_{C7} is used for sampling instead of V_{dc}. Figure 3 illustrates how the DC-link oscillations are extracted out of the capacitor voltage V_{C7}. It also calculates the on-time durations t_{on}, and eventually generates the switching pulses required for the auxiliary H-bridge converter.

III. OPTIMAL SWITCHING TECHNIQUE

Conventional multilevel PWM schemes produce fundamental voltages for all H-bridges within the cascaded converter, which are different. This would cause real power exchange between the H-bridges themselves as well as the cascade converter and the network, making voltages of the H-bridges unbalanced. To remedy this issue, an optimization problem is arranged to find the best switching instants [4]-[7].

Assume N cascade H-bridge converters in which each H-bridge is switched such that its output voltage introduces both half-wave and quarter-wave symmetry. Thus, the Fourier series include only odd sinusoidal terms ($\sin(n\omega t)$). Figure 4(a) presents an example for $N = 4$, where determination of five switching instants is enough to recognize the whole period. Also, Fig. 4(b) gives the resultant output voltage of the whole cascade converter, where the general description of the nth harmonic voltage using the Fourier series for N H-bridges and K switching instants within each quarter-wave is calculated as below:

$$V_{a-n} = \frac{4}{n\pi} \sum_{i=1}^{N} \sum_{j=1}^{K} (-1)^{k+1} \cos(n\alpha_{ij}) \qquad (2)$$

Where the angle α_{ij} is the jth switching instant of the ith H-bridge converter, and V_{a-n} is the nth harmonic of the cascade converter voltage.

It should be noted that the Fourier series (2) indicates half-wave symmetry in the voltage waveform. Now we can arrange an optimization problem with an objective of minimizing the remaining odd harmonics starting from the third up to $2NK-1$ [5]. This also is subjected to several constraints. For example, one necessary condition is needed to be satisfied. The fundamental components of the AC output voltages for all the H-bridges of the cascade converter have to be identical. Also, obvious conditions on angles α_{ij} have to be maintained satisfy as follows:

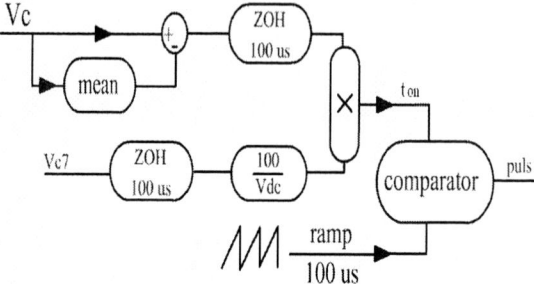

Figure 3. Control of on-duration pulses for the auxiliary H-bridge converter based on the extracted oscillations.

$$\text{Minimize} \quad \sum_{\substack{n=3 \\ (n \text{ is odd})}}^{2NK-1} V_{a-n}^2$$

$$\text{Subject to}: \begin{cases} \dfrac{4}{\pi} \sum_{j=1}^{K} (-1)^{k+1} \cos(\alpha_{ij}) = \dfrac{|V_{a-1}|}{N}, & \text{for } i=1,2,\ldots,N \\ 0 < a_{i1} < a_{i2} < \cdots < a_{iK}, & \text{for } i=1,2,\ldots,N \end{cases}$$

$$(3)$$

The above optimization problem is considered to be combined with the variable DC source compensation proposal to damp the DC-link oscillations. Nevertheless, we added other conditions to the optimization problem to make sure that the *pulse widths are bigger than a certain value* (e.g. 10 μs) for implementation purposes. This value depends on the switch specifications, including the on and off times that limits the switching frequency. Moreover, two adjacent switching instants need to be bigger than a certain value as well. Table I lists the resultant switching instants described by angles in degrees. These switching angles are repeated for other parts of the waveform because of the assumed quarter-symmetry as well as half-wave symmetry of the voltage waveform.

TABLE I
CALCULATED SWITCHING INSTANTS OF A CASCADED CASE FOR $N = 4$.

Cascaded H-Bridge No.	1	2	3	4
Switching Instant 1 in DEG	5	25	27	10
Switching Instant 2 in DEG	7	27	38	44
Switching Instant 3 in DEG	20	35	45	55
Switching Instant 4 in DEG	23	64	72	56
Switching Instant 5 in DEG	40	67	73	60

Table II summarizes the simulation results obtained by the suggested methods, including the uncompensated case, the IDC, the VDC and combination of optimal switching plus the VDC.

1068

TABLE II
SIMULATION RESULTS CONCERNED WITH THE PEAK-TO-PEAK
OSCILLATIONS OF THE DC-LINK OF H-BRIDGE CONVERTERS.

Compensation Method	Oscillations (maximum Peak-to-Peak)	Oscillations (average Peak-to-Peak)
IDC	2.0%	1.5%
VDC	2.0%	1.5%
VDC + optimized modulation	1.5%	1.2%
Uncompensated	10.67%	8.9%

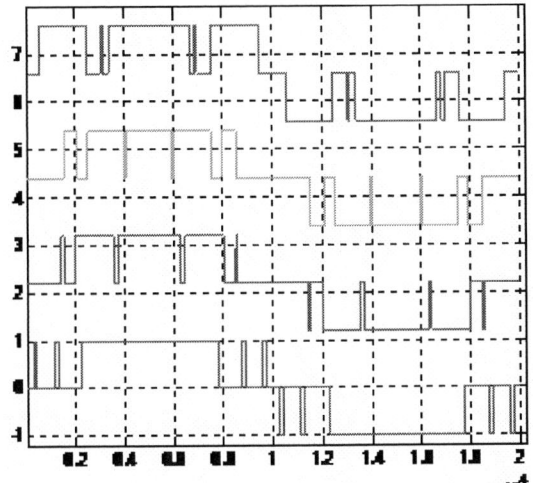

(a)

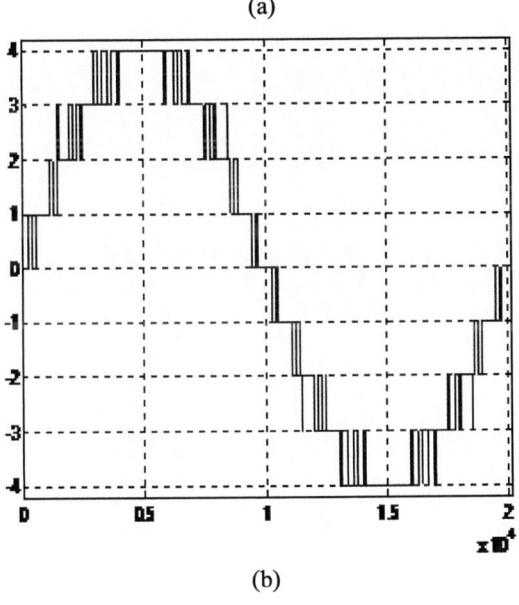

(b)

Figure 4. (a) Optimized solutions for a cascaded converter including four H-bridges, and (b) resultant output voltage of the cascade converter.

IV. CONCLUSIONS

This paper describes the DC-link oscillations concerned with H-bridge cascade converters when the number of H-bridges is smaller than four. While an uncompensated simulation with two H-bridges show considerable oscillations on the DC-link, an active DC-filter method is proposed and examined to lower peak-to peak of the oscillations. This method is called independent DC-source (IDC) in which there would be a limit in reduction of the magnitude of the oscillations. Thus, the method is improved by proposing variable DC source (VDC) method. Furthermore, an available optimized switching technique is programmed, and combined with the improved proposal VDC to manage harmonic cancellation along with filtering the DC–link oscillations. These methods are all simulated with SIMULINK to compare their performance on lowering the DC-link oscillations. Simulation results show that these methods successfully control the oscillations; among them the combined optimized modulation along with the VDC performs as the best solution.

ACKNOWLEDGEMENT

The authors would like to thank the support of the research Laboratory of power quality and reactive power control in K. N. Toosi University of Technology.

REFERENCES

[1] J. S. Lai, F. Z. Peng, "Multilevel Inverters: A survey of topologies, controls, and applications'", *IEEE Transactions on Industrial Electronics*, vol. 49, pp. 724–738, August 2002.

[2] M. marchesoni and P. Tenca, "Theoretical and Practical Limits in Multilevel MPC Inverters with Passive Front Ends", *European Power Electronics (EPE'01)*, CD Record, pp. p.1–p.12, September 2001.

[3] J. Rodriguez, J. S. Lai, F. Z. Peng, "Multilevel Converters- A new breed of power converters", *IEEE Transactions on Industry Applications*, vol. 32, No. 3, pp. 509–517, May/June 1996.

[4] N. A. Azli and W. S. Ning, "Application of Fuzzy Logic in an Optimal PWM Based Control Scheme for a Multilevel Inverter", IEEE PEDS'03, November 2003, Singapore.

[5] J. Aziz, Z. Salam, S. I. Safie, "Analytical Approach to Obtain the Harmonics Spectra on a Five Level Cascaded Inverter Subjected to a New Modulation Scheme", IEEE PEDS 2003, Singapore.

[6] Zhong Du, Leon M. Tolbert, John N. Chiasson, "Active Harmonic Elimination for Multilevel Converters", IEEE TRANSACTIONS ON POWER ELECTRONICS, vol. 21, no. 2, March 2006.

[7] Mohamed S.A. Dahidah, Vassilios G. Agelidis, Machavaram V. Rao, "Hybrid genetic algorithm approach for selective harmonic control", Electric Power System Research ELSEVIER, in Press 2007.

[8] Toshihisa shimizu, "Dc Active Filter Concept and Single-phase p-q Theory for Single-phase Power Converters", IEEE PEDS2003, Tutorial A, November 2003, Singapore.

Control Strategies of a Hybrid Multilevel Converter for Expanding Adjustable Output Voltage Range

Shoji Fukuda*, *Senior Member, IEEE*, Takatsugu Yoshida*, and Shigeta Ueda**, *Member, IEEE*

* Hokkaido University, Japan

** Tomakomai National College of Technology, Japan

Abstract--This paper proposes a new arrangement and new control strategies for a hybrid multilevel converter aiming at a high power high efficiency converter system with being free from harmonic distortion of the output. The strategies enable one to expand the adjustable output voltage range up to 100% with keeping very low harmonic distortion of the output voltages. The triangular-carrier based PWM and space-vector based PWM strategies are applied to the voltage control. Effectiveness is verified by simulation and experimental results. The proposed hybrid multilevel converter is applicable to medium voltage networks as a high power grid-interface converter.

Index Terms--harmonic distortion, high power converter, hybrid converter, multilevel converter

I. INTRODUCTION

Requirement for high power converters, which reach the megawatt level and are connected to the medium voltage network, have been growing in utility and drive applications [1]-[6]. Multilevel power converters have been receiving increasing attention for those purposes [7][8].

Conventionally, a multi-pulse converter system consisting of several six-pulse converter units producing a square-wave voltage was well-known [1]-[3]. The output voltages of the units were phase-shifted from each other required for the final output and the phase-shift transformers align the fundamental components. Recently many multilevel topologies and modulation strategies have been introduced and studied extensively for utility applications [9]-[11] and drive applications [12]-[16]. These converters are suitable in high voltage and high power applications due to their ability to synthesize waveforms with better harmonic spectrum and attain higher voltages with a limited device rating. In the family of multilevel converters, topologies based on a diode-clamped converter [17][18], a capacitor-clamped converter [7][18][20] and a cascaded H-bridge converter [7][9][21] are particularly attractive because of the capability of a direct link to the grid without transformers. However, they have some drawbacks: the diode-clamped and capacitor-clamped converters have a difficulty in balancing the dc voltage levels, the cascaded H-bridge converter requires many isolated dc voltage supplies, and they include lower order harmonics in the output voltages because of their staircase waveforms.

The power semiconductor technology suggests a trade-off in the selection of power devices in terms of switching frequency and voltage sustaining capability. Normally, the voltage blocking capability of faster devices such as IGBTs and the switching speed of high voltage devices like GTOs and IGCTs are found to be limited. With an H-bridge topology, realization of multilevel inverters using a hybrid approach involving IGCTs and IGBTs in co-operation was proposed [22].

The authors proposed a hybrid multilevel converter system [23][24][1] aiming at grid interface converters for medium voltage (6.6 kV) and high power (up to several MVA) applications. It consisted of a twelve-pulse converter using GTOs[2] and a diode-clamped converter using IGBTs connected in series by transformers. The topology therefore requires transformers but has several advantages: the control is easy, the output voltage has eleven-levels and is free from lower order harmonics, and the total device losses can be minimized. By employing different devices under different operating conditions, the power conversion capability of the whole system can be optimized. However it had the following drawbacks:

- adjustable output voltage range was narrow, namely, 32%,
- transformer required special arrangement.

In order to eliminate these drawbacks this paper proposes the phase difference control between the two GTO converters. As a result, the adjustable ac output voltage range is expanded to 100% with keeping almost free from harmonic distortion in the output voltage, and a costly phase-shifted transformer is eliminated.

The effectiveness of the new control strategy and new arrangement is verified by a 2kW prototype. Experiment results demonstrate that the output voltage and current are almost free from harmonic distortion, and the responses are quick and stable with closed-loop control.

---

[1] With the original hybrid converter system [23-24], the GTO inverter-1 and GTO inverter-2 in Fig. 1 operate with a fixed phase difference, 30°. The transformer-2 has two secondary windings per phase and the output voltage is phase-shifted by 30° from that of the transformer-1 to restore the phase difference between the two GTO inverters. Thus, the $6m\pm1$ (*m*: integer) harmonics in the output voltage produced by the GTO inverters are eliminated.

[2] In this paper GTOs represent high power devices. High power devices such as IGCTs and IEGTs can be also employed instead of GTOs.

II. HYBRID MULTILEVEL CONVERTER SYSTEM

Fig. 1 shows the proposed hybrid multilevel converter system. It consists of two 2-level GTO converters and a 3-level IGBT converter connected in series by three output transformers. The neutral-point-clamped (NPC) topology enables the IGBT converter to share a common high DC voltage, V_{DC}, with the GTO converters. Hereafter these converters are assumed to operate in an inverter mode for the sake of simplicity. Let the resultant reference output voltage per phase be sinusoidal as shown in Fig. 2(a),

$$v^* = V_1^* \sin \omega t \qquad (1)$$

The two GTO inverters operate in square wave switching at the same switching rate as a reference to minimize the switching losses of GTOs, and operate with a phase difference 2α for adjusting the output voltage of the GTO inverters. The output voltages, v_{GTO1} and v_{GTO2}, are shown in Fig. 2(b) and (c), respectively. The combined voltage of the two GTO inverters is

$$v_{GTO} = v_{GTO1} + v_{GTO2} \qquad (2)$$

and is shown in Fig. 2(d). The GTO inverters produce a main part of the resultant output voltage. The IGBT inverter operates in PWM switching at a higher switching rate than a reference and is assigned to generate the voltage difference between a reference and GTO inverters:

$$v_{IGBT}^* = v^* - v_{GTO}. \qquad (3)$$

Its waveform is shown in Fig. 2(e). From (3), the IGBT inverter has two functions.

  i) It offsets harmonic components of the output voltage generated by the GTO inverters.
  ii) It generates an additional fundamental output voltage.

The resultant output voltage therefore contains only a fundamental component. It is important to note that the above harmonic compensation principle holds only if the IGBT inverter operates in a linear modulation mode. Thus, a reference of the IGBT inverter should be confined to

$$\left| v_{IGBT}^* \right| \leq V_{DC} / 2. \qquad (4)$$

If Eq. (4) does not hold for some periods, the IGBT inverter cannot follow the reference for those periods. Distorted resultant output voltage would result for the periods. Refer to carrier-based PWM in Ref. [25].

III. CARRIER BASED PWM

Double-triangular carrier based PWM [25] is suitable for analog implementation of PWM control for NPC inverters. Here carrier based PWM is discussed. The fundamental component of the combined GTO inverter output voltage will be

$$v_{1GTO} = \left(\frac{4\sqrt{3}}{\pi} V_{DC} \cos \alpha \right) \sin \omega t \qquad (5)$$

and it reaches to more than 80% of the resultant output voltage. Therefore the output voltage is controlled mainly by α. However straightforward α-control causes a problem: *the IGBT inverter operates in an over-modulation mode because* **(4)** *does not hold in some periods*. Distorted output voltage results for the periods. To solve the problem this paper proposes to inject a third harmonic component to an original reference of the IGBT inverter. The third-harmonic injection strategy is well-known for expanding the output voltages of three-phase sinusoidal inverters, but here it is used for a different purpose.

A. Third Harmonic Injection

Define a modified reference v^{**}_{IGBT} by adding a third harmonic component as

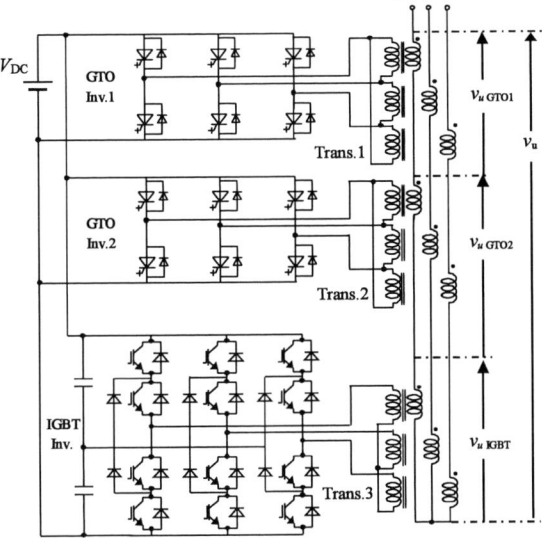

Fig. 1. Arrangement of hybrid multilevel converter system.

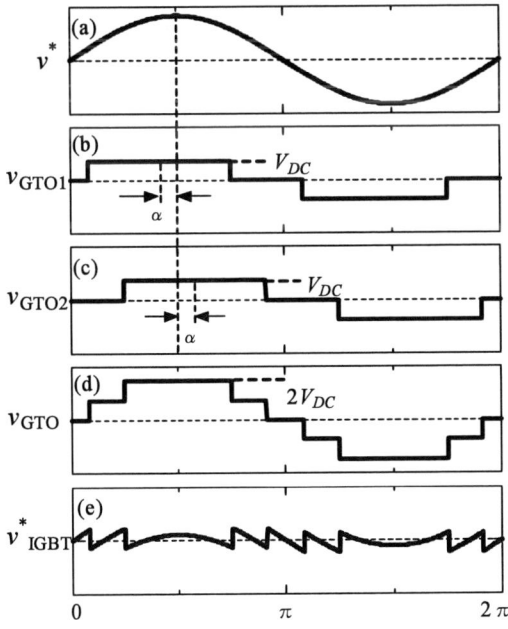

Fig. 2. Output voltages of respective inverters.

$$v_{IGBT}^{**} = v_{IGBT}^{*} + V_3 \sin 3\omega t$$
$$= MV_{DC} \sin \omega t - v_{GTO} + A_3 V_{DC} \sin 3\omega t \qquad (6)$$
$$M \equiv V_1^* / V_{DC}, \qquad A_3 \equiv V_{3ad} / V_{DC}$$

where M denotes the output voltage ratio and A_3 represents the amplitude of an added third harmonic component.

An example of an original reference voltage, v_{IGBT}^*, is shown in the middle figure of Fig. 3. One can observe that (4) does not hold for some periods, during which the IGBT inverter ceases switching. If a third harmonic voltage v_3 is added to v_{IGBT}^* it lowers the part of the original reference voltage above $V_{DC}/2$ and raises the voltage below $V_{DC}/2$. The bottom figure of Fig. 3 shows the modified reference v_{IGBT}^{**}. One can observe that it is confined within $\pm V_{DC}/2$. In this manner over-modulation operation can be eliminated and harmonic distortion of the output voltage can be suppressed.

B. Adjustment of α and A_3 as a function of M

The aim of the proposed converter is to expand the adjustable output voltage range without increasing harmonic distortion of the output voltage. Therefore the amplitude A_3 must be carefully adjusted in association with α so that the IGBT inverter can operate in a linear modulation mode. Since the shape of v_{IGBT}^* changes remarkably at some special values of α the whole adjustable range is divided into five ranges in terms of α.

(i) α < π/6

Let the third-harmonic component to be injected be
$$v_3 = A_3 V_{DC} \sin 3\theta \qquad (7)$$
Fig. 3 shows typical voltage waveforms. One can observe that the original reference v_{IGBT}^* of the IGBT inverter exceeds $V_{DC}/2$ around at $\theta=\pi/6-\alpha$ and $\theta=\pi/6+\alpha$. If the maximum values of v_{IGBT}^* at $\theta=\pi/6-\alpha$ and $\pi/6+\alpha$ take a value $V_{DC}/2$ the reference is confined within $V_{DC}/2$. These conditions are described by

$$v_{IGBT}^{**}\big|_{\theta=\pi/6-\alpha} = MV_{DC} \sin(\pi/6-\alpha) + A_3 V_{DC} \cos 3\alpha = V_{DC}/2 \qquad (8)$$

and

$$v_{IGBT}^{**}\big|_{\theta=\pi/6+\alpha} = MV_{DC} \sin(\pi/6+\alpha) - V_{DC} + A_3 V_{DC} \cos 3\alpha = V_{DC}/2. \qquad (9)$$

From (8) and (9) α and A_3 in terms of M are given by

$$\sin \alpha = 1/\sqrt{3}M \qquad (10)$$

$$A_3 = \frac{M \cdot \sin\left\{-\dfrac{\pi}{6}+\sin^{-1}\left(\dfrac{1}{\sqrt{3}M}\right)\right\}+\dfrac{1}{2}}{\cos\left\{3 \cdot \sin^{-1}\left(\dfrac{1}{\sqrt{3}M}\right)\right\}} \qquad (11)$$

If α and A_3 are set as indicated in (10) and (11) the over-modulation is eliminated as the modified reference v_{IGBT}^{**} indicates in the bottom trace of Fig. 3. The valid boundaries of (10) and (11) are defined as follows. An

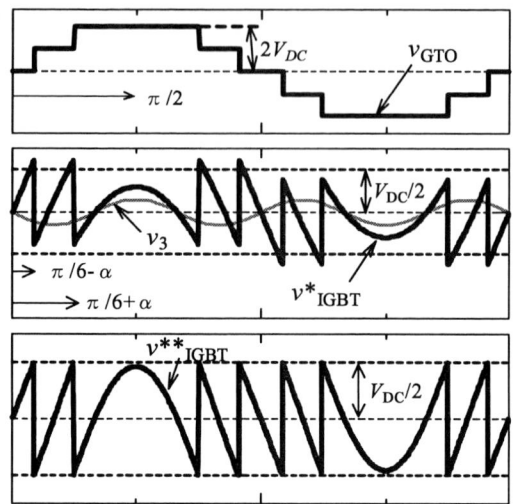

Fig. 3. Voltage waveforms for α ≤ π/6. Upper: GTO inverters, middle: original reference of IGBT inverter and third-harmonic to be added, lower: modified reference of IGBT inverter.

extreme value of v_{IGBT}^* occurs at $\theta=\pi/2$ and it must be lower than or equal to $V_{DC}/2$. Thus, one has a boundary,
$$A_3 \geq M - 5/2. \qquad (12)$$
Also the extreme value must be higher than or equal to $-V_{DC}/2$. Thus, one has another boundary,
$$A_3 \leq M - 3/2. \qquad (13)$$
By combining (11)-(13) the valid region for (10) and (11) is obtained as
$$2.33 \geq M > 1.79 \qquad (14)$$
Because of limited paper length the deduction process is omitted here but there are additional four ranges as follows. Details are given in Ref. [26].

(ii) π/6 ≤ α < π/3

$$\sin \alpha = 2/\sqrt{3}M \qquad (15).$$

$$A_3 = \frac{M \cdot \sin\left\{-\dfrac{\pi}{6}+\sin^{-1}\left(\dfrac{2}{\sqrt{3}M}\right)\right\}-\dfrac{1}{2}}{\cos\left\{3 \cdot \sin^{-1}\left(\dfrac{2}{\sqrt{3}M}\right)\right\}} \qquad (16)$$

The above equations (15) and (16) are valid for
$$1.7 \geq M > 4/3 \qquad (17)$$

(iii) α = π/3

$$\alpha = \pi/3, \quad A_3 = (1-M)/2 \qquad (18)$$
The above equations (18) are valid for 2/3 ≤ M < 4/3.

(iv) π/3 < α ≤ π/2

$$\alpha = \sin^{-1}(1/\sqrt{3}M),$$

$$A_3 = \frac{M \cdot \sin\left\{-\dfrac{\pi}{6}+\sin^{-1}\left(\dfrac{2}{\sqrt{3}M}\right)\right\}-\dfrac{1}{2}}{\cos\left\{3 \cdot \sin^{-1}\left(\dfrac{2}{\sqrt{3}M}\right)\right\}} \qquad (19)$$

The above equations (19) are valid for $1/\sqrt{3} \leq M < 2/3$.

(v) α=π/2

$$A_3 = M / 6 \quad (20)$$

The above equation (20) is valid for $0 \leq M < 1/\sqrt{3}$.

C. Voltage Control

It is important that the output voltage can be controlled in a wide range with keeping the IGBT inverter operating in a linear modulation range. It can be realized by adjusting α and A_3 simultaneously according to Figs. 4 and 5 as a function of a reference voltage M.

However, there are two singular ranges where third harmonic injection cannot exclude the over modulation operation for the IGBT inverter. With these ranges a simulation based trial-and-error approach is conducted, and α and A_3 are decided as follows.

vi) $2.33 < M \leq 2.5$　The authors suggest to apply (10) for α, and to assume a linear change for A_3 between A_3 at M=2.33 in (11) and A_3=0 at M=2.5.

vii) $1.7 < M \leq 1.79$　The authors suggest to extend the valid ranges for (10) and (15) up to and down to M=1.75 respectively for α, and to assume a linear change for A_3 between $A_3(M$=1.7) in (11) and $A_3(M$=1.79) in (16).

Simulation results verify that a linear relation holds between V_1 and M, and THD is kept low in both ranges vi) and vii) as indicated in Figs. 8 and 9.

IV.　SPACE VECTOR BASED PWM

Space-vector-based PWM is suitable for computer based implementation of PWM control for NPC inverters. Here space vector based PWM is discussed.

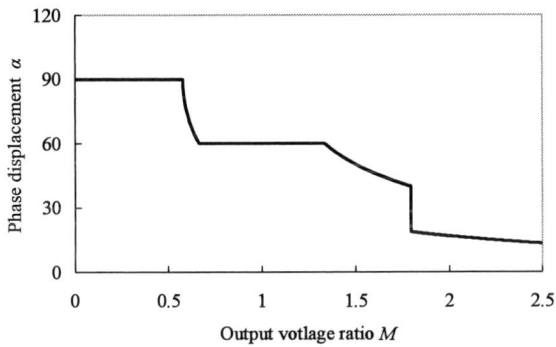

Fig. 4. Phase-difference α as a function of output voltage M.

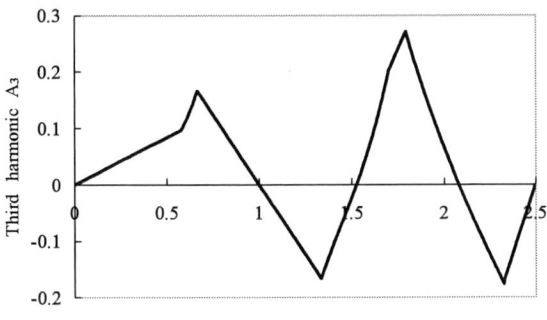

Fig. 5. Amplitude of third-harmonic voltage A_3
as a function of output voltage M.

Another method that enables one to fully adjust the output voltage is to employ space vector based PWM [27]. It automatically introduces an appropriate phase difference α and third harmonic voltage A_3 as a function of a reference voltage M without taking account of Figs. 4 and 5.

A. Space Vectors Produced by Hybrid Multilevel Inverter

The definition of the space vector is given by

$$v = \sqrt{\frac{2}{3}}\left(v_u \cdot e^0 + v_v \cdot e^{j2\pi/3} + v_w \cdot e^{j4\pi/3}\right) \quad (21)$$

where v_u, v_v and v_w are three-phase voltages. The IGBT inverter has three-level output voltages, $V_{DC}/2$, 0. $-V_{DC}/2$. Substituting these voltages into (21) one has 27 vectors shown in Fig. 6(a). Each apex of the triangles in the hexagon denotes the 27 vectors [28]. The IGBT inverter can therefore produce arbitrary vectors inside the blue hexagon by means of PWM. The side length of the blue hexagon is $\sqrt{2/3}V_{DC}$.

The same procedure is applied to the GTO inverters. However it is important to note that the GTO inverters supply voltages to a load through Δ-Y connection transformers. The output phase voltage on the load side is therefore $\sqrt{3}$ times higher than and leading by 30° from the corresponding output phase voltage on the inverter side. Considering these facts one has the hexagons indicating 8 vectors in Fig. 6(b) and (c). The side length of the red hexagon is $\sqrt{2}V_{DC}$.

Fig. 6. Region in which space vectors generated by the hybrid multilevel inverter system can be located

The apexes of the six triangles in the hexagon indicate the vectors the GTO inverter can produce.

Consequently, the space vectors that the proposed inverter system can produce are summarized in Fig. 6(d) combining the hexagons in Fig. 6(a)-(c). In the figure, as the IGBT inverter operates in PWM switching, the hybrid inverter system can produce any vectors inside the whole region given in Fig. 6(d). Since the inverter system can produce 259 vectors the state-vector based modulation seems to be very complicated. However it is not so difficult if a required reference vector is resolved into appropriate vectors that the respective three inverters can produce.

B. Specification of Reference Vectors to Each Inverter

Define three-phase reference voltages as

$$v*_u = V_1^* \sin \omega t = MV_{DC} \sin \omega t$$
$$v*_v = MV_{DC} \sin(\omega t - 2\pi/3) \qquad (22)$$
$$v*_w = MV_{DC} \sin(\omega t - 4\pi/3)$$

Substituting (22) into (21) one has a reference vector $v*$ as

$$v* = \sqrt{\frac{3}{2}} MV_{DC} e^{j(\omega t - \pi/2)} = \frac{3}{2} MV e^{j\theta}$$
$$V \equiv \sqrt{2/3} V_{DC}, \quad \theta \equiv \omega t - \pi/2 \qquad (23)$$

A reference vector rotates in the counterclockwise direction with an angular speed ω. Consider how to specify three vectors that each of the three inverters have to produce. With the symmetrical nature it is sufficient to treat a 60°-sector in Fig. 6(d). The sector corresponding to the period $0 \leq \theta < \pi/3$ is enlarged in Fig. 7.

The first step to be taken is to identify in which blue hexagon a reference vector $v*$ is located. This is done by resolving $v*$ into two components: $v*_0$ in the real axis and $v*_{60}$ in the direction of 60° from the real axis as

$$v* = v*_0 + v*_{60}.$$

The absolute values of $v*_0$ and $v*_{60}$ are obtained as

$$|v*_0| = \frac{\sqrt{3}}{2} MV \sin\left(\frac{\pi}{3} - \theta\right)$$
$$|v*_{60}| = 3\sqrt{3} MV \sin\theta .$$

Using $|v*_0|$ and $|v*_{60}|$ the hexagon in which a reference vector exists is identified according to Table I.

The second step is to assign a vector to each of the GTO inverters. The GTO inverters must produce the vector located at the center of the blue hexagon in which a reference vector exists. In the case of Fig. 7 the reference vector $v*$ exists in the #4 hexagon. Therefore the GTO inverters are assigned to produce the vector **OB**, or v_{GTO}. How to assign a vector to each GTO inverter is discussed in the following section. The IGBT inverter is required to produce the vectors $v*_{IGBT}$.

The third step is to assign vectors to the IGBT inverter so that it can produce $v*_{IGBT}$ by PWM. This is done using the information of the amplitude $|v*_{IGBT}|$ and angle Φ. Details including how to control the neutral point potential are given in Ref. [28].

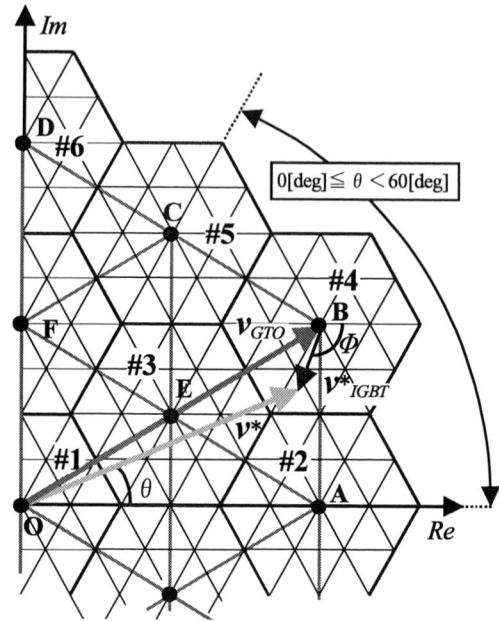

Fig. 7. Space vectors generated by the hybrid multilevel inverter system for 0° ≤θ<60°.

TABLE I
IDENTIFICATION OF HEXAGON WHERE REFERENCE VECTOR IS LOCATED.

Hexagons	Conditions of identification
#1	$\left\|v*_0\right\| + \left\|v*_{60}\right\| < V$
#2	$\left\|v*_0\right\| \geq 2V$ & $\left\|v*_{60}\right\| < V$
#3	$V \leq \left\|v*_0\right\| + \left\|v*_{60}\right\| < 3V$ & $\left\|v*_0\right\| < 2V$ & $\left\|v*_{60}\right\| < 2V$
#4	$\left\|v*_0\right\| + \left\|v*_{60}\right\| \geq 3V$ & $\left\|v*_0\right\| \geq V$ & $\left\|v*_{60}\right\| \geq V$
#5	$\left\|v*_0\right\| < V$ & $\left\|v*_{60}\right\| \geq 2V$

C. Selection and Sequence of Space Vectors

There are several combinations for the GTO inverters to create a vector v_{GTO}: two combinations if the vector is located in the #2, #4 and #5 hexagons, four in the #3 hexagon, and thirteen in the #1 hexagon. Furthermore it is noted that each GTO inverter is allowed to make only one switch transition per every 60°-period because of the square-wave switching. Therefore, it is crucial to select two vectors and decide the sequence to provide them to a load.

Every time when a reference vector moves across the boundary between the adjacent blue hexagons switching occurs at either of the GTO inverters. The time duration when $v*$ crosses the one boundary to another boundary varies depending on the amplitude M of the reference vector $v*$. This time duration corresponds to the phase difference 2α between the two GTO inverters in (5). Therefore the space vector modulation automatically adjusts α so that the hybrid inverter system can produce a required output voltage. It also automatically injects an appropriate zero-sequence voltage[3] to a reference so that the IGBT inverter can operate in a linear modulation range in order to fully offset the harmonic voltages generated by the GTO inverters. This is a significant advantage over the carrier based PWM in Chapter II.

In Fig. 7 the reference vector v^* rotates in the counter clockwise direction with the angular speed ω. Assume that at first v^* is located in the #2 hexagon with M=2.3. Then the GTO inverters generate the vector **OA**. Also assume that the #1 and #2 GTO inverters generate the vectors **OE** and **EA**, respectively. At the instant when v^* crosses the boundary between the #2 and #4 hexagons the #2 GTO inverter makes a switch transition to the vector **EB**. As a result, the GTO inverters switch the vector to **OB** and keep generating it until v^* crosses the next boundary. At the instant when v^* crosses the boundary between the #4 and #5 hexagons the #1 GTO inverter makes a switch transition to the vector **OF**. As a result, the GTO inverters switch the vector to **OC** because the vectors **EB** and **FC** are identical, and keep generating it until v^* crosses the boundary between the #5 and #6 hexagons.

There is a small singular voltage range where the GTO inverter cannot operate in square-wave switching, namely, $\sqrt{3}(=1.732) \leq M \leq 2\sqrt{7}/3(=1.764)$. In order for the GTO inverters to maintain square-wave switching, the author permits the IGBT inverter to conduct over-modulation operation only in this special range. As the singular range is very narrow no harmful influence is observed on the output voltages.

V.   SIMULATION RESULTS

A. Basic Characteristics

The output characteristics when using the three control methods are compared by simulation. They are:
- Method-1: with α and A_3 control,
- Method-2: with only α control,
- Original hybrid multilevel converter[1]

The state-vector based modulation is not included because it exhibits the identical characteristics to those with the method-1. The parameters used for the simulation are listed in Table II.

Fig. 8 shows the fundamental output voltage V_1 versus M characteristics. The methods 1 and 2 exhibit the identical characteristic: the full adjustable output voltage range of M=0 to 2.7 is obtained, and the linear relation between V_1 and M holds for $0 \leq M \leq 2.5$, or 100%. The original converter exhibits the adjustable output voltage range of M=1.6 to 2.8, and the linear relation holds for $1.7 \leq M \leq 2.5$, or 32%. Therefore, the α-control is quite useful for expanding the adjustable output voltage range.

Fig. 9 compares THD of the output current. If only the α-control is employed some high THD areas exist because of over-modulation operation of the IGBT inverter, but the A_3-control eliminates them.

---

[3] Space vector modulation automatically injects zero-sequence voltage including third harmonic component. This is the reason why singular voltage range is so narrow compared to that of the carrier based PWM.

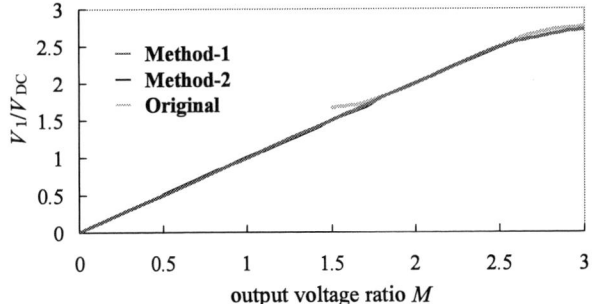

Fig. 8. Comparison of fundamental output voltage under three control strategies.

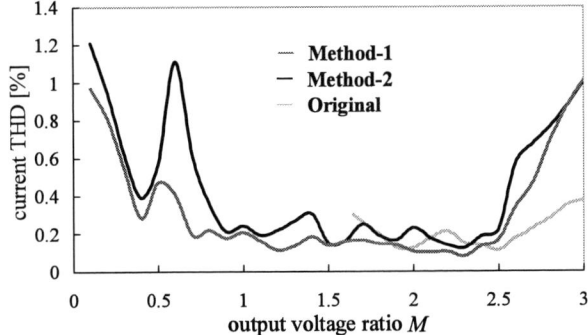

Fig. 9. Comparison of current THD under three control strategies.

TABLE II
PARAMETERS USED FOR SIMULATION AND EXPERIMENT

DC Source voltage V_{DC}	50[V]
Carrier frequency of IGBT inverter	2.4[kHz]
Sampling frequency	4.8[kHz]
DC link capacitor of IGBT inverter C	1000[μF]
AC load R_L	10.3[Ω]
AC load L_L	30.9[mH]
Output frequency f	50[Hz]

Thus, the method-1 exhibits low current THD over the entire adjustable output voltage range.

B. Output Estimation and Expected Applications

Here an attainable output kVA and device losses of a proposed hybrid inverter system are estimated [26]. It is noted that IGCTs are used instead of GTOs for the 2-level inverters taking account of current situations on the devices. The following devices are selected:
- IGCTs: V_{DRM}=6000V, I_{TGQM}=3000A [29],
- IGBTs: V_{DRM}=3300V, I_C=1200A [30],
- FRDs: V_{RRM}=6000V, I_{FRMS}=1700A [31].

Those devices establish the following voltage and current ratings of the hybrid inverter system:

DC supply voltage: V_{DC}=3000V,
Peak output AC voltage: V_1=7500V,
Peak output AC current: I_L=600A.

The maximum attainable output is estimated as 6.75MVA at M=2.5. The IGCT inverters produce 86%, while the IGBT inverter produces 14% of the output. The total device losses are 0.7% of the output assuming a load with an 80% power factor, where the IGBT inverter produces 80% of the total device losses at 2.4kHz switching, while the IGCT inverters produce 20% at 50Hz switching. THD of the resultant output voltage is as low as 3.2% when the harmonics up to the 60th (3kHz) are taking account of.

The above estimation indicates that the proposed hybrid converter is suitable for high power grid-interface converters [32] at medium voltage [33] because of a stringent restriction [34] on the voltage harmonic distortion of the converters. One of the prospective applications of the proposed hybrid multilevel converter is to a line side inverter of a wind-farm that connects wind turbine generators to the grid through the DC link. Wind farms are generally located in remote areas and, thus, voltage variations at the point of common coupling are quite large. The line side inverters are therefore required to have an adjustable ac voltage range of ±20% on the nominal value. Besides the inverters for wind-farms, application to a line interactive UPS system for crucial loads and an active power line conditioner such as STATCOM for a medium voltage grid of 6.6kV-class is expected.

VI. EXPERIMENTAL RESULTS

A small size prototype of 2kVA was built, and was used for a stand alone inverter system feeding an *L-R* load as well as a grid-interface inverter.

Figs. 10 (a) and (b) show voltage and current waveforms when the space-vector modulation was used at *M*=2.3 and 1.3, respectively. The parameters are the same as in Table II. The first trace from the top shows the GTO inverter output voltage v_{GTO}. The GTO inverters produce 85% and 93% of the resultant output voltage at *M*=2.3 and 1.3, respectively. The second trace shows the IGBT inverter output voltage v_{IGBT}. The IGBT inverter offsets the harmonic components generated by the GTO inverters and simultaneously produces 15% and 7% of the resultant output voltage, respectively. The third trace shows a resultant output voltage v and it is almost sinusoidal. It is not shown in the figures but the line-to-line output voltage has 21 levels at its maximum. The fourth trace shows the output current. It is free from harmonics.

Grid-interface inverter operation with closed-loop current control was performed employing the carrier based modulation. The parameters are listed in Table III.

TABLE III
PARAMETERS USED FOR EXPERIMENT

Output	2kW
DC source voltage V_{DC}	65[V]
Line-to-line AC mains voltage	150[V]
Carrier frequency of IGBT inverter	2.4[kHz]
Sampling frequency	4.8[kHz]
DC link capacitance of IGBT inverter C	1000[μF]
AC coupling reactor L_C	2.5[mH] (7.25%)

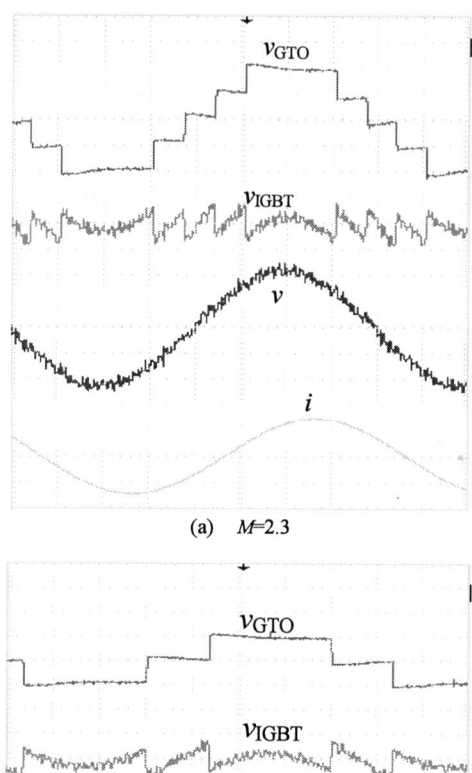

(a) *M*=2.3

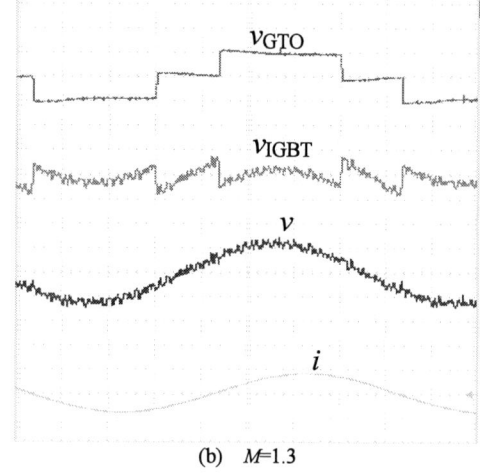

(b) *M*=1.3

Fig. 10. Experimental voltage and current waveforms. Upper: GTO inverter voltage v_{GTO}, upper middle: IGBT inverter voltage v_{IGBT}, lower middle: resultant output voltage v_u, lower: output current i_u. 50Hz, 50V/div., 5A/div., and 5ms/div.

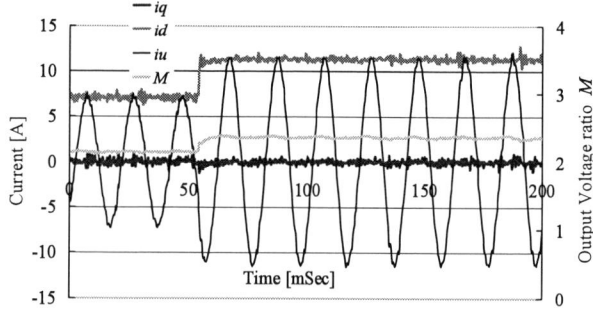

Fig. 11. Step responses of output current i_d.

Fig. 11 shows step responses: a *d*-axis reference $i_d{}^*$ was increased stepwise from 7.2A to 12A at 53ms, while a *q*-axis reference $i_q{}^*$ was kept at 0A constant. Quick and stable responses with unity PF operation are observed.

VII. CONCLUSIONS

This paper proposes a new arrangement and new control strategies for a hybrid multilevel converter system. The adjustable output voltage range is expanded to 100%, while the total harmonic distortion of the output voltage is suppressed below 5% over a wide adjustable output voltage range. It is suitable for high power grid-interface converters at medium voltage because a stringent restriction on the voltage harmonic distortion of the converters is required. A 2kVA prototype was build to verify the effectiveness of the proposed arrangement and control strategies. Experimental and simulation results demonstrate that a full adjustable voltage range is accomplished, voltage and current are almost free from harmonic distortion over a wide output voltage range, and closed loop control exhibits stable and rapid current responses.

REFERENCES

[1] S. Mori, K. Matsuo, M. Takeda, and M. Seto, "Development of a large static var generator using self-commutated inverters for improving power system stability," IEEE PES 1992 Winter Meeting, 92 WN 165-1.

[2] F. Ichikawa, K. Suzuki, T. Nakajima, S. Irokawa, and T. Kitahara, "Development of self-commutated SVC for power systems," PCC-Yokohama, pp.609-614, 1993.

[3] C. Schauder,M. Gernhardt, E. Stacey, T. Lemak, L. Gyugyi, T.W. Cease, and A. Edris, "Development of a ±100 MVAR static condenser for voltage control of transmission systems," IEEE Trans. Power Delivery, vol.10, pp. 1486–1493, 1995.

[4] K. Suzuki, M. Yajima, M. Nohara, S. Ueda, H. Satou, and Y. Eguchi, "Control method for 50MVA self-commutated static var compensator," Trans. of IEEJ, vol.117-B, no.7. pp.953-959. 1997. (in Japanese)

[5] M. Tobita and R. Kushibiki, "Development of new high power converter using IEGT," IPEC-Tokyo, pp.970-975, 2000.

[6] M. Prodanovic, T.C. Green, "Control and filter design of three-phase inverters for high power quality grid connection," Trans. on IEEE Power Elect., vol. 18, no.1, 2003.

[7] J.S. Lai and F.Z. Peng, "Multilevel Converters - A New Breed of Power Converters," IEEE IAS, pp.2348-2356, 1995.

[8] J. Rodriguez, J. Lai, and F. Peng, "Multilevel inverters: a survey to topologies, controls, and applications," IEEE Trans. Industrial Electronics. vol.49, No.4, pp.724-738, 2002.

[9] F.Z. Peng, J.S. Lai, J.W. Mckeever, and J. VanCoevering, "A multilevel voltage-source inverter with separate DC source for static var generation," IEEE Trans. Ind. Applicat., vol. 32, pp.1130-1138, 1996.

[10] F.Z. Peng, J.W. Mckeever, and D.J. Adams, "A power line conditioner using cascade multilevel inverters for distribution systems," IEEE IAS, pp.1316-1321, 1997.

[11] B.A. Renz et al, "AEP unified power flow controller performance,"IEEE PES Winter Meeting, 1998, Paper PE-042-PWPD-0-12.

[12] P. Hammond, "A new approach to enhance power quality for medium voltage ac drives," IEEE Trans. Ind. Applicat., vol. 33, pp. 202-208, 1997.

[13] J. Rodriguez, L. Moran, A. Gonzalez, and C. Silva, "High voltage multilevel converter with regeneration capability," IEEE PESC, pp. 1077-1082, 1999.

[14] M. Koyama, Y. Shimomura, H. Yamaguchi, M. Mukunoki, H. Okayama, and S. Mizoguchi, "Large capacity high efficiency three-level GCT inverter system for steel rolling mill drives," EPE, 2001.

[15] Madhav D. Manjrekar, Peter Steimer, Thomas A. Lipo, "Hybrid multilevel power conversion system: a competitive solution for high power applications," IEEE IAS, 1999.

[16] J. Rodriguez, J. Pontt, G. Alzamora, N. Becker, O. Einenkel, J.L. Cornet, "Novel 20MW downhill converter system using three-level converters," IEEE IAS, 2001.

[17] A. Nabae, I. Takahashi and H. Akagi," A new neutral-point-clamped PWM inverter," Trans. on IEEE Ind. Applicat. vol. IA-17, no.5, pp.518-523, 1981

[18] N. S. Choi, J. G. Cho, and G. H. Cho, "A general circuit topology of multilevel inverter," IEEE PESC, 1991.

[19] T. A. Meynard and H. Foch, "Multi-level conversion: High voltage choppers and voltage-source inverters," IEEE PESC, pp. 397–403, 1992.

[20] C. Hochgraf, F. Lasserter, D. Divan, and T.A. Lipo, "Comparison of multilevel inverters for static var compensation," IEEE IAS, pp. 921-928, 1994.

[21] M. Marchesoni, M. Mazzucchelli, S. Tenconi, "A Non conventional Power Converter for Plasma Stabilization," IEEE PESC, pp. 122-129, 1988.

[22] M.D. Manjrekar and T.A. Lipo, "A Generalized Structure of Multilevel Power Converter," IEEE PEDES'98, pp. 62-67.

[23] S. Fukuda and M. Kitano, ''Control strategy for a three phase series-connected hybrid inverter system," IEEE PESC, 2001.

[24] S. Fukuda, Y. Kubo and D. Li, "Introduction of a series-connected hybrid multi-converter system," IEEE IAS, 2002.

[25] B. Velaerts, P. Mathys, E. Takakis, and G. Bingen, "A novel approach to the generation and optimization of three-level PWM waveforms," IEEE PESC, 1255-1262, 1988.

[26] S. Fukuda, T. Yoshida and S. Ueda, "A hybrid multi converter system having full adjustable output voltage range," IEEJ, PCC-Nagoya, 2007.

[27] S. Fukuda, T. Yoshida and S. Ueda, "A hybrid multilevel converter system with extended adjustable output voltage range," IEEE PESC, 2007.

[28] S. Fukuda, Y. Matsumoto and A. Sugawa, "Optimal regulator-based control of NPC boost rectifiers for unity power factor and reduced neutral-point-potential variations," IEEE Trans. Ind. Electronics, vol. 46, no. 3, 1999.

[29] ABB Semiconductors AG data sheets on Integrated Gate-Commutated Thyristor, 5SHY 30L6010.

[30] E. Kraft, A. Steimel and J. Stenke, "Three-level high-power inverters with IGCT and IGBT elements compared on the basic of measurements of the device loss," EPE 1999.

[31] ABB Semiconductors AG data sheets on Fast Recovery Diode, 5SDF 10H6004.

[32] R.W. De Doncker, "Medium-voltage power electronics technologies for future decentralized power systems," PCC-Osaka, pp.927-932, 2002.

[33] M. Crappe, L. Gertmar, A. Habock, W. Leonhard, and D. Povh, "Power electronics and control by microelectronics in future energy systems," EPE Journal, vol. 10, no. 1, 2000.

[34] IEEE Industry Applications Society/Power Engineering Society, *IEEE Recommended Practice and Requirements for Harmonic Control in Electrical Power Systems*. Piscataway, NY: IEEE 1992.

High Efficiency Single Phase Multi-level Inverter by New Controlled Switch Signal

Ruthapong Kumchaiyo and Itsda Boonyaroonate

Electrical Engineering Dept.,King Mongkut University of Technology Thonburi.

126 Prachautid Rd. Bangmood Tungkru Bangkok Thailand 10140

e-mail mutdking@hotmail.com Fax +6624709033

Abstract- **This research reports a high efficiency single phase multi-level inverter by new controlled switch signal which is simply and not complexity that is generated by analog control circuit. The circuit topology of the proposed inverter is based on a conventional full-bridge inverter with one bidirectional auxiliary. The experimental results show the proposed controlled switch signal can increase the efficiency of a single phase multi-level inverter which is 2% higher than the old controlled switch signal[2] at 220 V_{AC} output ,92 watts output and 10 kHz switching frequency.**

Keywords-**New controlled switch signal.**

I. INTRODUCTION

Multi-level inverters are now becoming an established topology for use in higher power application, where they offer the advantage of substantially lower harmonic content in the output voltage for a given switching frequency, together with significantly reduce switching stresses [1]-[12].

The power loss of multi-level inverter is relative to the efficiency which is consist of conduction loss and switching [12]. We can increase the efficiency by decrease the power loss in multi-level inverter so in this research, we present the new controlled switch signal to increase the efficiency of multi-level inverter that is reduce the switching loss by some power switch is controlled by the PWM signal and the other power switch is controlled by square wave signal. The proposed controlled switch signal is generate from very simply analog control circuit.

The main advantages of this paper are simply analog control circuit and the multi-level inverter will be small and cheap because the size of heatsink is decrease when the power loss of power circuit is reduced.

Figure 1. show the power circuit of multi-level that is using a full-bridge inverter and a bidirectional auxiliary switch.

II. CIRCIUT OPERATION

This multi-level inverter has five level of voltage output. There are $+V_S/2$, $+V_S$, 0, $-V_S$ and $-V_S/2$ (V_S=155.5 V_{DC})

Level 1, $+V_S/2$: The auxiliary switch, S5 is ON, connecting the load positive terminal to point A, through diodes D5 and D8, and Disp4 is ON, connecting the load negative terminal to ground. All other controlled switches are OFF; the voltage applied to the load terminal is $+V_S/2$. Fig. 2 shows the current paths that are active at this level.

Level 2, $+V_S$: S1 is ON, connecting the load positive terminal to, and S4 is ON, connecting the load negative terminal to ground. All other controlled switches are OFF; the voltage applied to the load terminal is $+V_S$. Figure 3 shows the current paths that are active at this level.

Level 3, 0: This level has two operation of circuit

Case I, +0: S4 and D3 are ON, short-circuit the load. All other controlled switches are OFF; the voltage applied to the load terminals is +0.Figure 4 shows the current paths that are active at this level.

Case II:-0: S3 and D4 are ON, short-circuit the load. All other controlled switches are OFF; the voltage applied to the load terminals is -0.Figure 5 shows the current paths that are active at this level.

Level 4, $-V_S$: S2 is ON, connecting the load negative terminal to V_S, and D3 is ON, connecting the load positive terminal to ground. All other controlled switches are OFF; the voltage applied to the load terminals is $-V_S$. Figure 6 shows the current paths that are active at this level.

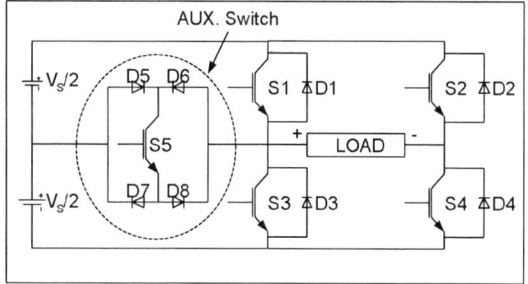

Figure 1. Proposed single phase multi-level inverter

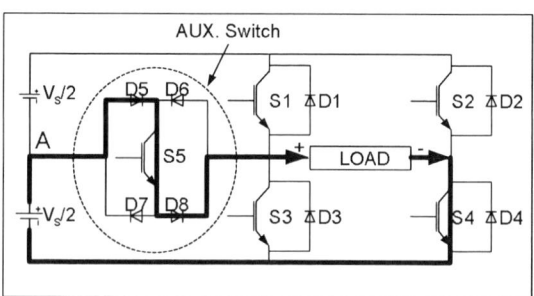

Figure 2. Switching combination required to generate output voltage level $+V_S/2$.

978-1-4244-0644-9/07/$25.00 ©2007 IEEE

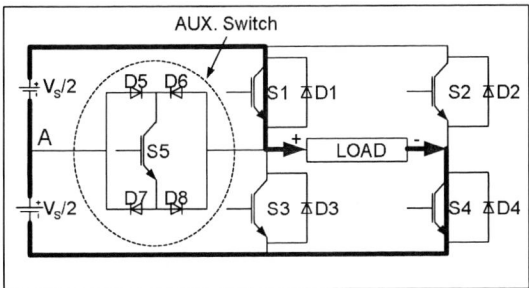

Figure 3. Switching combination required to generate output voltage level +V$_S$.

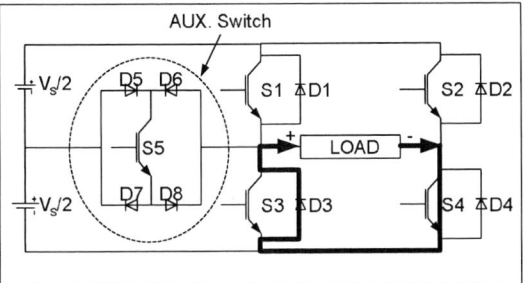

Figure 4. Switching combination required to generate output voltage level +0.

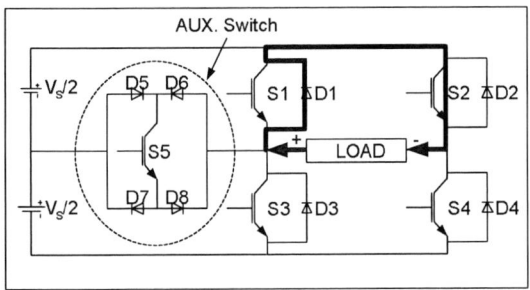

Figure 5. Switching combination required to generate output voltage level -0.

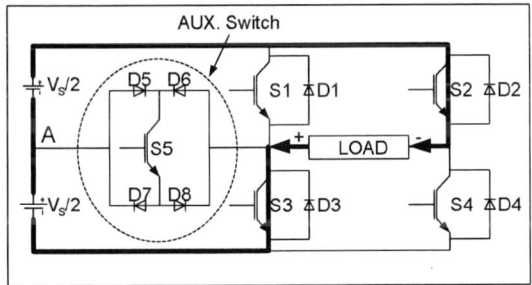

Figure 6. Switching combination required to generate output voltage level −V$_S$.

Level 5, -V$_S$/2: The auxiliary switch, S5 is ON, connecting the load positive terminal to point A, through diodes D6 and D7, and S2 is ON, connecting the load negative terminal to V$_S$. All other controlled switches are OFF; the voltage applied to the load terminals is −V$_S$/2. Figure 7 shows the current paths that are active at this level.

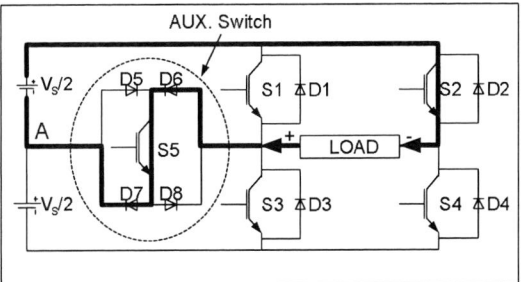

Figure 7. Switching combination required to generate output voltage level-V$_S$/2.

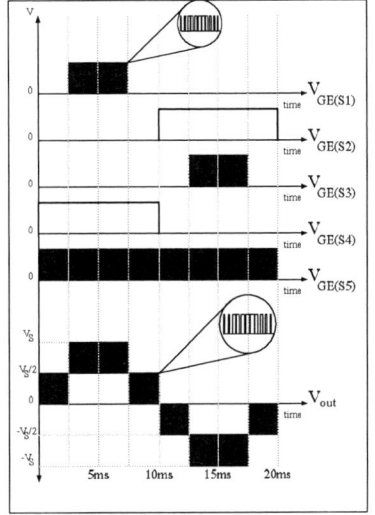

Figure 8. New controlled switches signal wave form and output voltage wave form.

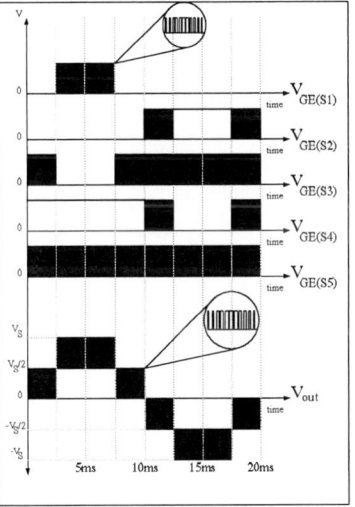

Figure 9. Old controlled switches signal wave form and output voltage wave form. [2]

From figure 2 to figure 7, Figure 8 shows the new controlled switches signal wave form and output voltage wave form and

Figure 9 shows the old controlled switches signal wave form and output voltage wave form.[2]

III. EXPERIMENTAL RESULTS

In experimentation, a 220 V output prototype single phase multi-level inverter was built and tested. We use the IGBT IRG4BC30U as S1-S4, IRG4BC30UD as S5, DIODE MUR8100E as D5-D8,$V_S/2$ is about 155.5 V and load is 100 W incandescent lamp. The switching frequency is 10 kHz. Both controlled switch signal are tested at the same power circuit. Table 1 shows experimental voltage, current and power of single phase multi-level inverter with the old controlled switches signal. Table 2 shows experimental voltage, current and power of single phase multi-level inverter with the new controlled switches signal. Figure 10 shows the experimental voltage and current wave form of single phase multi-level inverter with the old controlled switches signal. Figure 11 shows the experimental voltage and current wave form of single phase multi-level inverter with the new controlled switches signal. Figure 12 shows the Experimental voltage THD of single phase multi-level inverter with the old controlled switches signal. Figure 13 shows the Experimental voltage THD of single phase multi-level inverter with the new controlled switches signal.

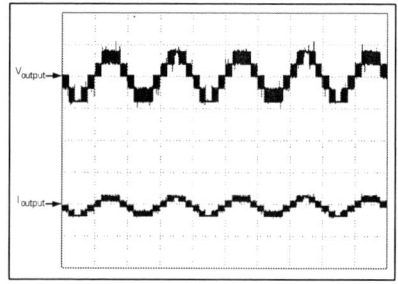

(Vertical V_{OUTPUT} 400 V/Div, I_{OUTPUT} 2 A/Div Horizontal 10 ms/Div)
Figure 10 Output voltage and output current wave form of single phase multi-level inverter with the old controlled switches signal.[2]
(V_{OUTPUT}= 228V, I_{OUTPUT}= 462 mA)

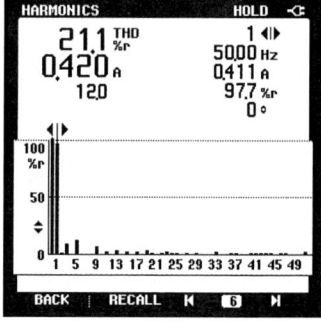

(Vertical V_{OUTPUT} 400 V/Div, I_{OUTPUT} 2 A/Div Horizontal 10 ms/Div)
Figure11. Output voltage and output current wave form of single phase multi-level inverter with the new controlled switches signal.
(V_{OUTPUT}= 227 V, I_{OUTPUT}= 458 mA)

Table 1 Experimental results of Single phase multi-level inverter with the old controlled switches signal.[2]

INPUT		OUTPUT	Efficiency
Source 1	Source 2		
155.8 V_{DC}	155.4 V_{DC}	221.9 V_{AC}	
0.397 A	0.399 A	1.936 A	91.36 %
49.4 W	50.2 W	91 W	

Table 2 Experimental results of Single phase multi-level inverter with the new controlled switches signal.

INPUT		OUTPUT	Efficiency
Source 1	Source 2		
155.8 V_{DC}	155.4 V_{DC}	221.0 V_{AC}	
0.396 A	0.398 A	2.146 A	93.30 %
49.2 W	49.4 W	92 W	

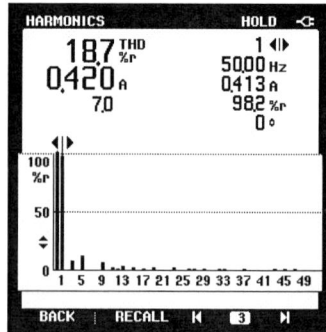

Figure 12. Total harmonics distortion of single phase multi-level inverter with the old controlled switches signal.[2]

Figure13. Total harmonics distortion of single phase multi-level inverter with the new controlled switches signal.

IV. CONCLUSIONS

From the experimental results, we can see that the wave form of output voltage and output current from both controlled switch signals are not different but the efficiency of single phase multi-level inverter with new controlled switch signal is 93.30 % at 92 W and 10 kHz switching frequency which is 2% higher than controlled by old controlled switch signal and lower percentage of harmonic content in the output voltage.

REFERENCES

[1] David M Baker, Vassilios G Agelidis and Jian Yi Dhen, "A Five-Level Zero Average Current Error Controlled Single Phase Grid-Interactive Inverter" Power Electronic Drives and Energy Systems for Industrail Growth, 1998. Proceedings. 1998 International conference on Volume 1, 1-3 Dec. 1998 Page(s):50-55 Vol.1

[2] Gerardo Ceglia, Víctor Guzmán, *Member, IEEE*, Carlos Sánchez, Fernando Ibáñez, Julio Walter, and María I. Giménez "A New Simplified Multilevel Inverter Topology for DC-AC Conversion" Power Electronics, IEEE Transactions on Volume 21, Issue 5, Sept. 2006 Page(s):1311-1319

[3] Aguillon Garcia, J. Fernandez-Nava, J.M. Banuelos-Sanchez, P, "Unbalanced voltage effects on a single phase multilevel inverter due to control strategies" Telecommunications Energy Conference, 2004. INTELEC 2004. 26th Annual International 19-23 Sept. 2004 Page(s):140 - 145 Digital Object Identifier 10.1109/ INTLEC.2004.1401457

[4] Lin, B.-R.; Hung, T.-L, "Analysis and implementation of a single-phase multilevel inverter for power quality improvement" Industrial Electronics, 2002. ISIE 2002. Proceedings of the 2002 IEEE International Symposium on Volume 4, 8-11 July 2002 Page(s):1235 - 1240 vol.4 Digital Object Identifier 10.1109/ISIE.2002.102596

[5] Poh Chiang Loh; Bode, G.H.; Holmes, D.G.; Lipo, T.A, "A time-based double-band hysteresis current regulation strategy for single-phase multilevel inverters" Industry Applications, IEEE Transactions on Volume 39, Issue 3, May-June 2003 Page(s):883 - 892 Digital Object Identifier 10.1109/TIA.2003.810667

[6] Diong, B.; Corzine, K, "WTHD-optimal staircase modulation of single-phase multilevel inverters" Electric Machines and Drives, 2005 IEEE International Conference on 15-18 May 2005 Page(s):1341 – 1344

[7] Bouhali, O.; Berkouk, E.M.; Francois, B. Saudemont, C, "Direct generalized modulation of electrical conversions including self stabilization of the DC-link for a single phase multilevel inverter based AC grid interface" Power Electronics Specialists Conference, 2004. PESC 04. 2004 IEEE 35th Annual Volume 2, 20-25 June 2004 Page(s):1385 - 1391 Vol.2

[8] Bode, G.H.; Holmes, D.G, "Hysteresis current regulation for single-phase multilevel inverters using asynchronous state machines" Industrial Electronics Society, 2003. IECON '03. The 29th Annual Conference of the IEEE Volume 2, 2-6 Nov. 2003 Page(s):1203 - 1208 Vol.2 Digital Object Identifier 10.1109/IECON.2003.1280224

[9] Loh, P.C.; Bode, G.H.; Holmes, D.G.; Lipo, T.A, "A time-based double band hysteresis current regulation strategy for single-phase multilevel inverters" Industry Applications Conference, 2002. 37th IAS Annual Meeting. Conference Record of the Volume 3, 13-18 Oct. 2002 Page(s):1994 - 2001 vol.3 Digital Object Identifier 10.1109/ IAS.2002.1043806

[10] Beristain, J.; Bordonau, J.; Raventos, O.; Rocabert, J.; Busquets, S.; Mata, M, "A New Single-Phase HF-Link Multilevel Inverter" Power Electronics Specialists Conference, 2005. PESC '05. IEEE 36th 2005 Page(s):237 – 243 Digital Object Identifier 10.1109/ PESC.2005.1581630

[11] Baker, D.M.; Agelidis, V.G.; Chen, J.Y., "A five-level zero average current error controlled single-phase grid-interactive inverter" Power Electronic Drives and Energy Systems for Industrial Growth, 1998. Proceedings. 1998 International Conference on Volume 1, 1-3 Dec. 1998 Page(s):50 - 55 Vol.1 Digital Object Identifier 10.1109/PEDES.1998.1329989

[12] Radermacher, H.; Schmidt, B.D.; De Doncker, R.W.; "Determination and comparison of losses of single phase multi-level inverters with symmetric supply" Power Electronics Specialists Conference, 2004. PESC 04. 2004 IEEE 35th Annual Volume 6, 20-25 June 2004 Page(s):4428 - 4433 Vol.6 Digital Object Identifier 10.1109/PESC.2004.1354783

FPGA Implementation of Quasi-BLDC Drive

C.S. Soh[1,2], C. Bi[1,2], and K.K. Teo[1]

[1]Data Storage Insitute, 5 Engineering Drive 1, Singapore 117608
[2]National University of Singapore, ECE, 4 Engineering Drive 3, Singapore 117576

Abstract—For the reduction of acoustic noise and vibration generated in self-sensing brushless dc motor operation, a self-sensing quasi brushless DC drive (QBLDC) mode is presented. The structure of the drive and the related algorithms are introduced in detail in the paper. The drive can be implemented on field programmable gate arrays (FPGA). The performance of the drive presented has been confirmed by the testing results of the spindle motor used in hard disk drive.

Index Terms—Self-sensing, BLDC, Spindle Motor, Acoustics and FPGA.

I. INTRODUCTION

The introduction of high energy permanent magnet materials coupled with the increasing concerns for power efficiency has opened the gateway for Permanent Magnet Synchronous Motor (PMSM). The benefits of using PMSM include power density and high efficiency and low acoustics noise. As such, it has become an attractive option for industrial applications, such as Hard Disk Drives (HDD). The motor deployed HDD are a sub category belonging to PMSM, commonly known as Brushless DC Motor (BLDCM). Compared to PMSMs, this kind of BLDCM has several unique features. The rotor of BLDCMs has got surface-mounted permanent magnet constructing a smooth-air-gap machine. As such, reluctance torque contributed by inductance variations can be neglected. In addition, the rotor utilizes fractional-slots which in turn make the cogging torque negligible. These, together with other features, such as sinusoidal/trapezoidal back-emfs and a well symmetrized three-phase structure, create an unique PMSM or a BLDCM. [1]

In PMSM drives, the motor usually has a rotor position sensor, such as encoder or resolver. In BLDCM drive, usually three hall effect sensors are used as rotor position sensors. However, these sensors are undesirable as they incur additional cost and space. As such, self-sensing, or sensorless control, is often being deployed for the BLDCM drives. There are many categories of self-sensing solutions [1]-[3], such as the back-EMF voltage sensing, back-EMF integration, flux estimation and detection of freewheeling diodes conduction. In HDD, the self-sensing operation is accomplished by the utilizing the back-EMF zero crossing points (ZCPs) as rotor position information. In a BLDC drive, each gate turns on for 120° and for each phase, there will be two silent periods, each of 60°, where the terminals are floating. It is during the silent phase that the ZCPs will

occur and to detect these ZCPs, a common method is to create a virtual neutral point and compare it to the voltage terminal. And the resulting signal will have ZCPs equivalent to the back-emf ZCPs. However, due to gate commutation, spikes occur and these will result in false ZCPs.

Back-emf based methods, however, fail during starting where the back-emf are zero or small. As such, during starting, the motor is driven open loop with six-stepping on a skewed frequency to a speed, ω_{co}. At ω_{co}, the back-emf is sufficiently large to be detected and all voltage terminals are floated for back-emf detection. However, due to the removal of gating signals, the motor will be decelerating and this will result in declining back-emfs. Thus, in the determination of ω_{co}, it is tuned higher to take into account of this decline. Beyond ω_{co}, the system advances to the self-sensing BLDC drive.

In BLDC drive, during the 120° conduction segment, the goal is to inject rectangular stator phasor currents during that period. However, the motor being inductive, voltage spikes will occur during commutation. These spikes are undesirable as they are a source of both electrical and mechanical noise, and vibration.

In recent years, owing to the progress of VLSI technology, the field programmable gate array (FPGA) has gained world wide acceptance. It has traditionally been perceived as a essential platform for Application Specific Integrated Circuit (ASIC) prototyping. However, in recent years, it has gained significant market share in end-product solutions as fundamentally, FPGA offers fast time to market, low design/manufacturing cost and risk, extremely high processing performance, and programmability.

The research goal of this paper is to design a Self-Sensing Quasi-BLDC (QBLDC) Drive on FPGA. The FPGA design comprises of

(1) an innovative back-emf based method for rotor position detection with zero delay;

(2) a six step δ° masked open loop starting for bumpless crossover to BDLC/QBLDC drive, and

(3) a QBLDC drive for reduction of acoustics and vibration generated in the motor operation.

In (1), the proposed method takes an integrative approach in back-emf zero-crossing-points (ZCP) detection and BLDC/QBLDC drive implementation. Heuristic logic is adopted in the implementation and accurate BDLC/QBLDC drive with zero delay ZCP detection for wide speed range is achieved. In (2), the proposed methodology proposes a novel gate signal masking on a six step open loop self starting enabling

---

This work was supported by Data Storage Institute, Singapore

978-1-4244-0644-9/07/$25.00 ©2007 IEEE

back-emf detection possible without a complete removal of gating signals. The strategy offers the advantages of i. an earliest possible crossover while making no assumption on the crossover frequency, ii. smooth crossover as the motor rotation is continued iii. continuance of frequency skewing during detection. In (3), the proposed drive aims to reduce the drive acoustics and runout by the reduction of the drive current spikes. It uses a trapezoidal drive instead of rectangular drive for gating. By utilizing such an approach, the current spikes are reduced and its effect on acoustics and vibration are reduced.

II. BRUSHLESS DC (BLDC) OPERATION

In a BLDC drive, the motor is typically driven by a three-phase inverter circuit as shown in Fig. 1. It consists of six power semiconductor transistors with a protection diode connected in parallel to each of these transistors.

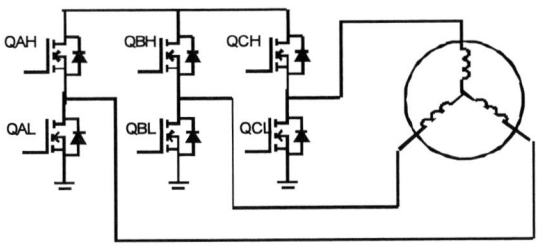

Fig. 1. Bridge Circuit for BLDC Drive

Each transistor is gated by a 120°-conduction drive, in which each gate turns on for 120 electrical degrees in each cycle. Commutation occurs at every 60 electrical degrees of rotation in the sequence QAH, QCL, QBH, QAL, QCH and QBL. For maximum torque production, the gating with respect to the back-emf is given in Fig. 2. For each phase, there will be two unexcited 60° periods, where the voltage terminals are floating or unexcited. During this unexcited phase, the phase voltage gives the phase back-emf. Optimally, the zero crossing points (ZCPs) of the phase back-emfs should occur mid-way in the silent period. These ZCPs represent position information and it's based on this that self-sensing operation using back-emf ZCP detection is established.

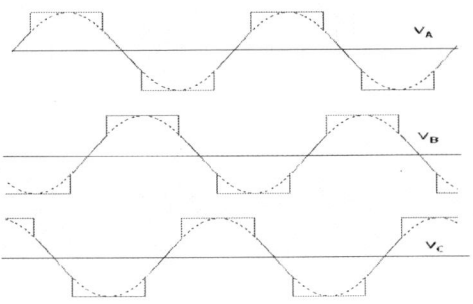

Fig. 2. Terminal Voltage waveforms

III. SELF-SENSING OPERATION USING BACK-EMF ZCP DETECTION

As highlighted, during the unexcited phase, the phase voltage gives the phase back-emf which would encompass the ZCP. In the detection of ZCP, however, the entire phase voltage is not required as at the instants of ZCPs, the terminal voltages would be equal to the neutral voltage. Consequently, the detection of ZCPs is equivalent to the detection of the instants whereby VAN or VBN or VCN equals to zero. Whilst the terminal voltages are available, the neutral voltage might not be available. A common method is to reconstruct a "virtual" neutral which provides a equivalence to the actual neutral. To derive the ZCP signals, the terminal voltage is thus compared to the virtual neutral voltage and the comparison will provide signals with ZCPs corresponding to the back-emf ZCPs. Fig. 3 illustrates the notion pictorially depicting the terminal voltages comparison with the virtual neutral constructed.

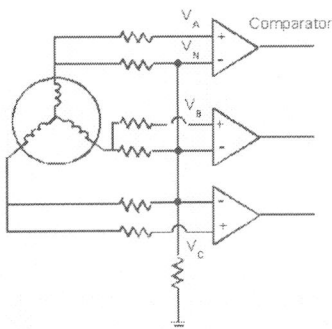

Fig. 3. ZCP Detection Circuit

Practically, however, the simplistic topology fail as voltage spikes will occur for every commutation. As a result, false ZCPs are immersed amongst true ZCPs. This problem has been well documented and researched, among which many methods center on the usage of filters which inevitably result in phase delays.

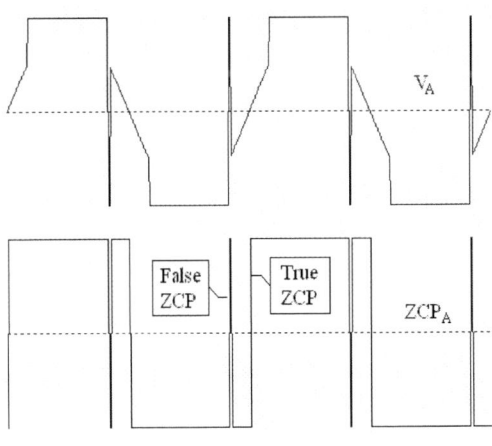

Fig. 4. Comparator Waveforms

1083

Most of the methods focus on the removal or exclusion of these false ZCPs when they occur. In this proposed method, an adoption of an integrative Zero Delay ZCP Detection - BLDC drive is utilized.

In a BLDC drive, a commutation occurs 30 electrical degrees (30°) after every true ZCP is detected. To estimate this 30°, the following algorithm is proposed.

a. A global free running counter is implemented in the design and latches are used to store the counter counts for all the ZCPs.

b. Positive edge transitions will be used to latch the 0°, 120° and 240° ZCPs counter counts whereas negative edge transitions will be used to latch the 60°, 180° and 300° ZCPs counter counts.

c. 30° time lag is to be estimated using the 60° time lapse from the last ZCP to latest ZCP detected. For example, to estimate the 30° time lag after the occurrence of 0° ZCP, the 30° time lag would be equivalent to the point whereby the counter has further increased by ½ × (Latch0° - Latch300°) since the 0° ZCP. The factor of ½ is easily accomplished by a single bit right shift of (Latch0° - Latch300°).

At this juncture, the algorithm, however, too suffers from the presence of false ZCPs, resulting in incorrect counter counts being latched. However, achieving zero delay ZCP detection is only possible if all ZCPs are taken as true as when it occurs. This constraint implies that the latching must be done for all ZCPs, true or false.

This constraint can be elegantly respected by the inclusion of false ZCP avoidance within the BLDC drive algorithm. An unique property derived from the signals is during the 30° time lag for commutation, after an active ZCP detection, inactive ZCPs should never occur during the 30° time lag. For example, 0° ZCP counter count will be latched by positive or rising edge ZCP transitions. However, in the 30° time lag, inactive negative or falling edge ZCP will not occur if the preceding active ZCP is true. Hence, in the presence of an inactive ZCP during the 30° time lag interval, the interval will be made inactive, latch will be restored to its previous count and active edge ZCP transition is awaited. As an additional level of heuristic control, it can be observed that apart from satisfying the 30° time lag, commutation should occur only at the active level. Thus, the integrated algorithm becomes

a. A global free running counter is implemented in the design and latches are used to store the counter counts for all the ZCPs.

b. Positive edge transitions will be used to latch the 0°, 120° and 240° ZCPs counter counts whereas negative edge transitions will be used to latch the 60°, 180° and 300° ZCPs counter counts. 30° time lag is to be estimated using the 60° time lapse from the last ZCP to latest ZCP detected.

c. Occurrence of inactive edge transitions will be reset ZCP counter counts to its previous counter values. Negative edge transitions will trigger a reset of the 0°, 120° and 240° ZCPs counter counts whereas positive edge transitions will trigger a reset of the 60°, 180° and 300° ZCPs counter counts.

d. If the current count minus the ZCP latched count equals to the estimated 30° time lag and current ZCP signal level remains active, commutation occurs. Waiting for 60° ZCP counter counts commences.

e. Pictorially, the algorithm is provided in Fig. 5 and its schematic provided in Fig. 6.

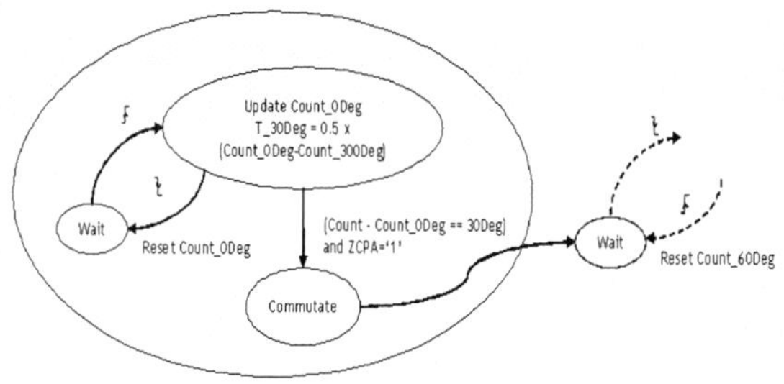

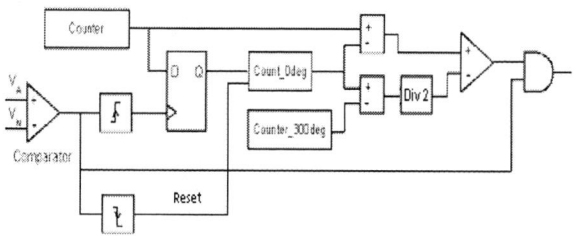

Fig. 5. Flowchart of proposed algorithm

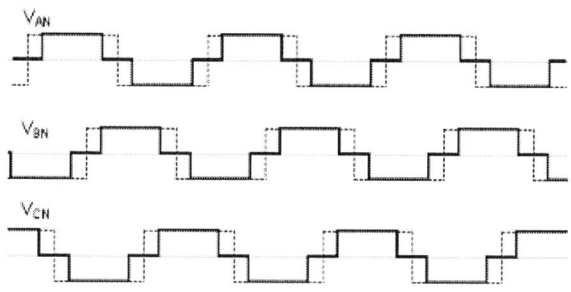

Fig. 6. Schematic of integrated Zero Delay ZCP Detection

IV. CROSSOVER FROM SIX STEPPING TO SELF-SENSING BLDC OPERATION

In self-sensing bldc operation, position feedback from back-emf ZCPs is not available at starting or low speeds. The motor is typically open loop started from standstill in a six stepping mode with a skewed frequency to ω_{co} speed. At ω_{co}, two conditions must be met; (i) back-emf is sufficiently large to be detected (ii) ZCPs occur during the unexcited phase of gating. Fig. 7 gives the gating signals with respect to the ZCP signals in an ideal crossover.

Fig. 7. Gating & Back-Emf Waveforms

1084

However, practically, it is tuned to achieve the gating and ZCP phasor relationship as in Fig. 7. Such an approach, apart from requiring time and effort, is system dependent. Another approach is to switch off all the gates and detect the ZCPs as soon as the phase currents decrease to zero. This method, though simplistic, causes the motor to decelerate and its corresponding back-emf to diminish which inevitably requires a higher ω_{co}. Furthermore, the deceleration of the motor will cause a "bump" in the crossover. In addition, in the event of a failure, restarting is difficult as it is to be restarted on a decelerating motor.

In this proposed methodology, a novel gate signal masking technique is utilized, making back-emf detection possible without a complete removal of gating signals. To provide insights into this, it would be beneficial if the phenomenon at six stepping is better understood. During open loop six stepping, generally, the wave will not be that as intended. Typically, the drive is over driven to provide a higher starting torque for acceleration as well as overcoming frictional, viscous torque. The phasor relationship between the gating and backemf will differ by a larger angle than that seen in Fig. 7. Generally, in an extreme case, the phasor relation is similar to that illustrated in Fig. 8.

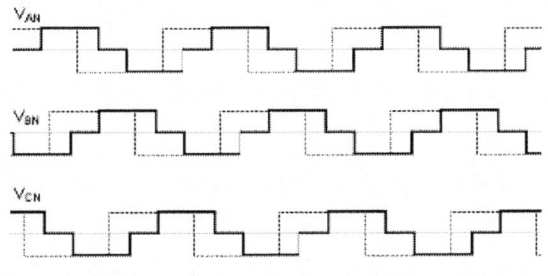

Fig. 8. Gating & Back-Emf Waveforms

As mentioned, to extract the ZCPs, the system can be tuned or all the gates turned off. In this method, this matter can be elegantly rectified by performing gate masking.

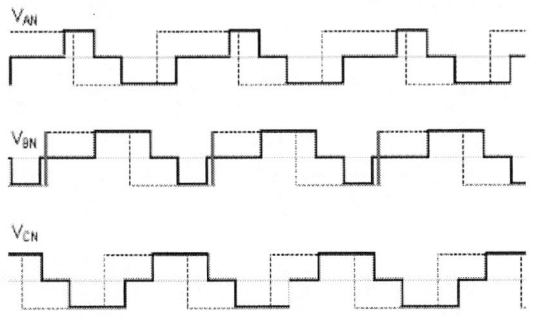

Fig. 9. Gating & Back-Emf Waveforms with δ masking, δ = 60°

The motor is similarly started on a six-stepping mode. However, the gating signals are masked out an arbitrary angle, δ, as soon as the motor rotates. This will increase the silent phase window and for any $\delta \geq 60°$ this will guarantee at least one ZCP detection. Fig. 9 illustrates the possible detection of a ZCP with gate signals masking.

In doing so, since only 60° of the gating signals are masked out, the inertia of the motor will keep rotation going in contrast with the complete removal of gating signals. This means even if the window/back-emf is missed, the same back-emf amplitude is available for detection after a 300° angle rotation. Thus, since the motor is not made to decelerate, a smooth and more robust transfer is possible. In addition, the assumption that the back-emf is large enough for detection is no longer necessary as such strategy allows the skewing of frequency to be continued till back-emf is large enough for ZCP to be detected.

An extension of the method is to mask the gating signals by δ=120°. In doing so, an additional 60° window will make two ZCPs available for detection. This will greatly reduce the possibility of erroneous detection since dual detection is provided. However, it must be noted that, by setting up gate masking, the starting torque will be reduced. Nevertheless, it should not be a concern as the motor is commonly overdriven at starting.

V. SELF-SENSING QUASI-BLDC (QBLDC) OPERATION

In BLDC operation, for torque production, the transistors are commutationally switched as shown in Fig. 10 to provide rectangular stator currents. While the objective is to inject phasor rectangular currents, in reality, current waveforms as shown in Fig. 11 are injected. The reason being, the motor is inductive in nature will accumulate energy when driven. During commutation, the interruption of inductive currents will result in current spikes. There are two types of spikes belonging to two different sources. For those spikes at the side (in-phase spikes), they are to the turning on and off of its corresponding gates, whereas for the spikes at the middle of the rectangular currents (adjacent-phase spikes), they are due to the turning on and off of its adjacent gates. Nevertheless, both of these spikes are undesirable as they are a source of noise and vibration.

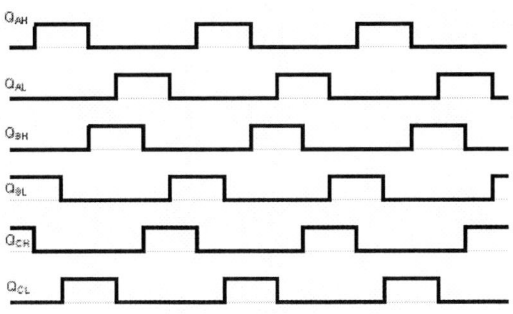

Fig. 10. BLDC Gating Signals

1085

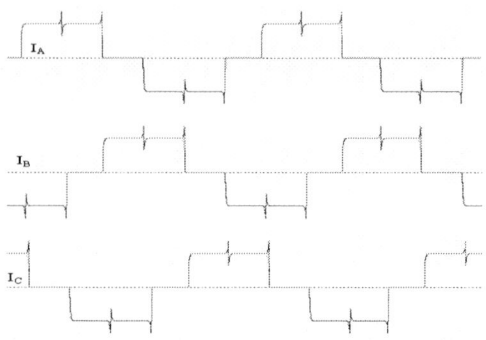

Fig. 11. BLDC Phase Currents

To reduce these spikes a quasi-BLDC drive is proposed. In this drive, trapezoidal phase voltages as shown in Fig. 12 are to be achieved instead of rectangular phase voltages. Execution wise, it means to compare a trapezoidal reference to a triangular signal.

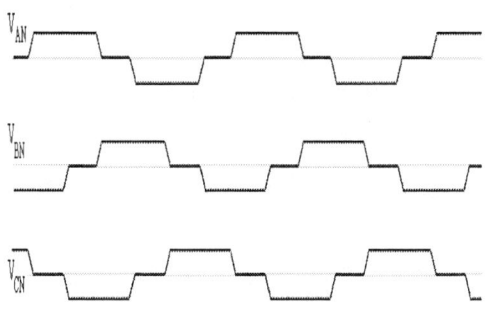

Fig. 12. QBLDC Phase Voltages

The purpose of such a drive is to control the injection and withdrawal of phase currents to reduce the in-phase spikes. On a single phasor topology, it works fine and these spikes can be significantly reduced. However, in a three phase topology, such a method gives rise to a peculiar development. The in-phase spikes will reduce but adjacent-phase spikes will increase. In-phase spikes can be reduced as voltages are slowly injected. However, due to a three phase topology, the overlap in gating between the commutating phases will result in an increased current in the non-commutating phase. This non-commutating phase, then, suffers from adjacent-phase spikes. This is illustrated in

Fig. 13 under (I). In the turning off of phase A and turning on of phase B, there will be instants where both QAH and QBH are turned on. This implies a spike in current in phase C.

To deal with this problem, the PWM trapezoidal switching in (II) is proposed. In this proposed switching, to turn on a gate, the trapezoidal reference and the normal PWM is used. However, to turn it off, the logic inverse of the adjacent sequential gating signal is used. For example, QBH is turned on following QAH is turned off.

Therefore, to turn off QAH, the NOT of QBH is applied. In doing so, at any one time, only one gate from the upper leg is turned on. This will ensure that the current following through QCL remains relatively constant and also the currents through QAH and QBH are respectively decreasing and increasing slowly. In self-sensing bldc operation, position feedback from back-emf ZCPs is not available at starting or low speeds. The motor is typically open loop started.

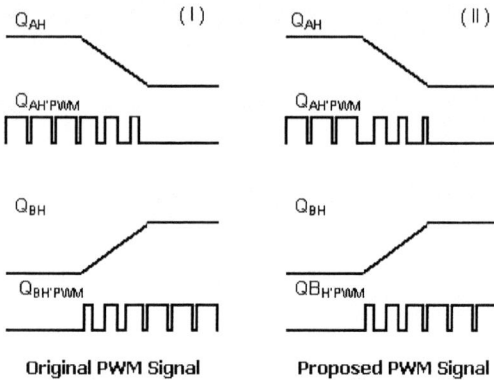

Fig. 13. PWM Signals for QBLDC Drive

VI. FPGA IMPLEMENTATION AND EXPERIMENTAL RESULTS

The overview of the FPGA implementation is provided in Fig. 14. Each of the modules have been coded with VHDL and is implemented on a XC4VFX12, a Xilinx Virtex-4 FX FPGA Device. The design is synthesized, implemented and simulated entirely on Xilinx ISE 8.1. The design was then used to drive a Hard Disk Spindle.

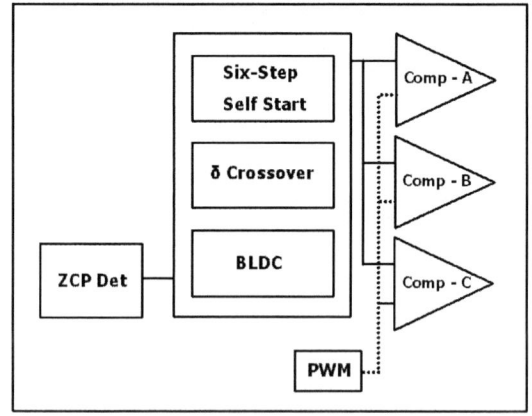

Fig. 14. Block Diagram of FPGA Design

Fig. 15 gives the terminal voltage waveform VA, the virtual neutral VN, the ZCP waveform from a comparator and IA, the phase current. It can be observed that false ZCP are not treated as true and the true ZCP are in fact used for commutation.

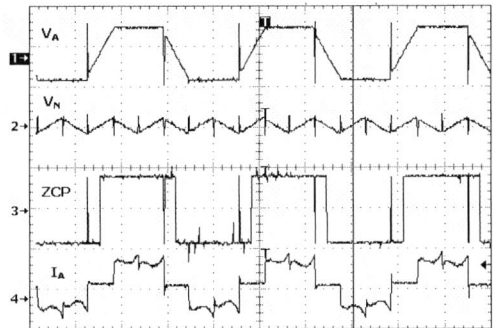

Fig. 15. Waveforms for BLDC Drive

In the implementation of QBLDC, the BLDC entity is swapped with QBLDC. The current waveform comparison between the two drives are given in Fig. 16.

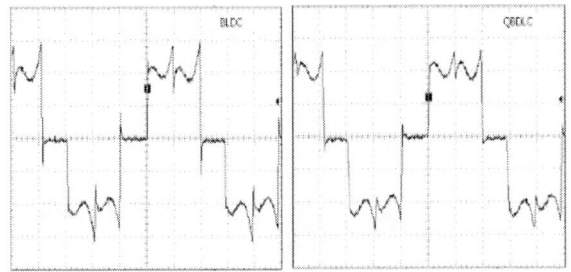

Fig. 16. Current Waveforms for BLDC & QBLDC Drive

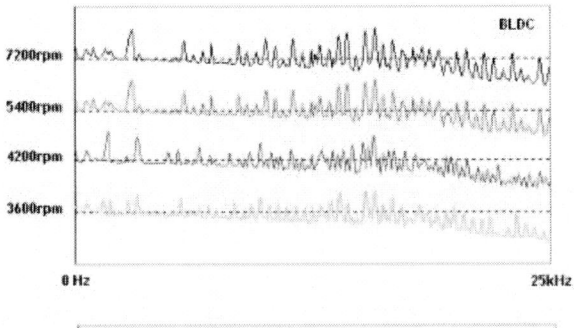

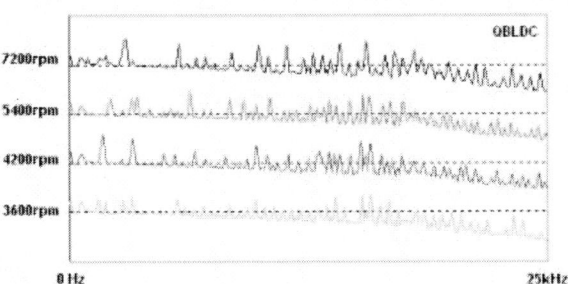

Fig. 17. Noise Waveforms for BLDC & QBDLC Drive

As mentioned, BLDC drive given gives as spiky current and QBDLC drive has reduced the spikes by 40%. Fig. 17 shows the acoustics noise captured for four different operating speeds with Acoustics Chamber. It can be seen that QBLDC gives a lower noise floor compared to BLDC. This is intuitive as a spike is considered to be rich in frequencies and a reduction will result in a reduction of noise frequency across a wide range.

VII. CONCLUSIONS

In this paper, a self-sensing quasi brushless DC drive mode for brushless dc motors is presented. This drive comprises of

(1) an innovative zero delay back-emf based method for rotor position detection;
(2) a six step $\delta°$ masked open loop starting for bumpless crossover to BDLC/QBLDC drive;
(3) a QBLDC drive for reduction of acoustics and vibration.

By integrating all the schemes put forward, the drive is implemented on field programmable gate arrays. Tests results showed that, when the presented self-sensing drive is used, the voltage spikes induced by the commutation can be reduced significantly, and the acoustic noise and vibration of the BLDCM can thus be observably reduced. These testing results prove the effectiveness of the proposed drive.

ACKNOWLEDGEMENT

The authors would like to thank the staffs at DSI for its help and support.

REFERENCES

[1] Q. Jiang, C. Bi, R.Y. Huang, "A New Phase-Delay-Free Method to Detect Back EMF Zero-Crossing Points for Sensorless Control of Spindle Motors", *IEEE Transactions on Magn*, vol 25, no 6, pp 4358-4363, Nov 1989.
[2] J.P. Johnson, M. Ehsani, and Y. Guzelgunler, "Review of sensorless methods for brushless DC," *34th IAS Annual Meething Conf. Rec. 1999 IEEE*, vol. 1, 1999, pp. 143-150
[3] K. Iizuka, H. Uzuhashi, and M. Kano, "Microcomputer control for sensorless brushless motor," *IEEE Trans. Ind. Appl.* vol IA-21, no. 4, pp.595-601, Jul./Aug. 1985

A Practical Method to Eliminate the Conduction Torque Ripple in BLDCM Using Cascade Topology

Xiaofeng Zhang, Zhengyu Lu, Yu Ma, Zhaoming Qian

Department of Electrical Engineering

Zhejiang University, Hangzhou,310027, P.R.China

Abstract-**This paper presents a comprehensive research on torque ripples of brushless dc motor drive in conduction region based on the mathematical models.The phenomenon of torque ripples of conduction exists because of diode freewheeling happened under the conventional PWM schemes.Through theoretical and mathematical analysis, A new method for reducing the torque ripple in brushless dc motors with a single current sensor has been proposed by adding the DC-DC converter that regulates the amplitude of the current. So in such drives, torque ripples are theoretically eliminated in conduction region. Effectiveness and feasibility of the proposed control method is verified by experiments. Torque ripples due to diode freewheeling current and conventional high-frequency PWM modulation have vanished. So it is preferable for the drive system of high accuracy-controlling BLDCM.**

Keywords- **Diode freewheeling, Conduction region, Torque ripple**

I. INTRODUCTION

Brushless dc motor(BLDCM) with trapezoidal back-EMF make it possible to use a single dc-link current sensor to regulate the phase current flowing through two motor phases.it has linear torque to phase current and speed to voltage,low acoustic noise and fast dynamic response.But torque ripple generated is the main drawback of BLDCM, however, which deteriorates the precision of BLDCM. Much work have been focused on the commutation torque ripples in BLDCM[1][3].However,the torque in the conduction regions is also influenced by PWM methods owing to the fly-wheel diode ripple current in the nonconducted phase.

In this paper, the phenomenon and reason of torque ripple are investigated in the conduction region. Based on the theoretic analysis,A new front_PWM-end_ON modulation scheme and corresponding circuit topology are proposed.The diode freewheeling in the nonconducted phase is eliminated completely by the new control strategy.

II. MATHEMATICAL MODEL OF THE BRUSHLESS DC MOTOR

The conventional BLDCM drive diagram is shown in Fig.1.The general driving method,unipolar PWM of

120° conduction makes fewer switching loss and current ripple, so H_PWM-L_ON modulation (shown in Fig.2) are selected for BLDC speed regulation.

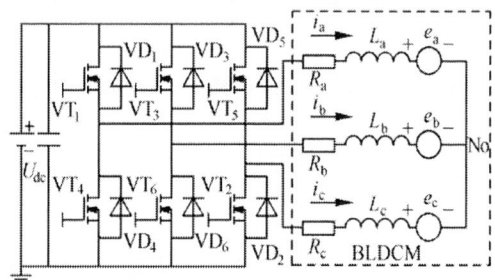

Fig.1 The diagram of the driving system for BLDCM

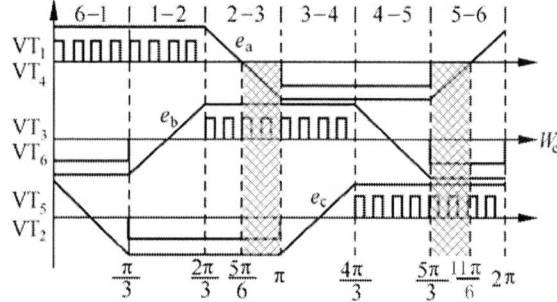

Fig.2 The sequence chart of conventional
H_PWM-L_ON modulation

The BLDCM has three stator windings and permanent magnets on the rotor.Since both the magnet and the stainless steel retaining sleeves have high resistivity, rotor-induced currents can be neglected and no damper windings are modeled[2]. Hence the circuit equations of the three windings in phase variables are described as Eq.(1).

$$
\begin{bmatrix} U_{ka} \\ U_{kb} \\ U_{kc} \end{bmatrix} = \begin{bmatrix} R & 0 & 0 \\ 0 & R & 0 \\ 0 & 0 & R \end{bmatrix} \bullet \begin{bmatrix} i_a \\ i_b \\ i_c \end{bmatrix} + \frac{d}{dt} \begin{bmatrix} L & 0 & 0 \\ 0 & L & 0 \\ 0 & 0 & L \end{bmatrix} \bullet \begin{bmatrix} i_a \\ i_b \\ i_c \end{bmatrix} + \begin{bmatrix} e_a \\ e_b \\ e_c \end{bmatrix} + \begin{bmatrix} U_{N_0} \\ U_{N_0} \\ U_{N_0} \end{bmatrix}
$$

(1)

Where, U_{ka}、 U_{kb}、 U_{kc}:Phase terminal voltage with reference to ground. i_a、i_b、i_c:Phase current. e_a、e_b、e_c:Back EMF of 3-phase. U_{N0}:The voltage on the neutral node of the motor with reference to ground. L_s:Magnetizing inductance. M:Mutual inductance($L=L_s-M$). R: Phase resistance.

The electromagnetic torque is

$$T_e = \frac{e_a i_a + e_b i_b + e_c i_c}{\Omega} \qquad (2)$$

Where T_e is the belectromagnetic torque of motor, Ω is mechanical angular velocity of rotor.

III. ANALYSIS OF TORQUE RIPPLES IN THE CONDUCTION REGION

The BLDC motor runs in $120°$ conduction mode. In the ideal condition, two of the three phases are active and the idle phase is nonconducted at all times. There is no current in the nonconducted phase winding in principle. The parasitic diode in the inverter, which is connected with the nonconducted phase winding, will turn on once the terminal voltage of this phase that is over the DC link voltage or under the zero voltage. Thus, a freewheeling current flows in the nonconducted phase caused the torque ripple in the conduction region(two dashed area exploded in Fig.2). This ripple current is caused inevitably using the tranditional PWM methods. The reason of generating the diode freewheeling currents in the nonconducted phase will be analized in this section.

There are some generally used PWM schemes in the practical application such as unipolar PWM method including PWM_ON, ON_PWM, H_PWM-L_ON, H_ON-L_PWM, and bipolar PWM method including H_PWM-L_PWM. The following analysis is based on H_PWM-L_ON scheme that exploded in Fig.2.

Phase A is assumed to be idle while analying, the waveform of e_a is shown in Fig.2 and four analized steps are : Step a：$2\pi/3-5\pi/6$; Step b：$5\pi/6-\pi$; Step c：$5\pi/3-11\pi/6$; Step d：$11\pi/6-2\pi$.

During these four steps, the terminal voltages of three phase A, B, C can be deduced from(1) as follows.

$$\begin{cases} U_{ka} = e_a + U_{N0} \\ U_{kb} = S_b \cdot U_{dc} \\ U_{kc} = S_c \cdot U_{dc} \end{cases} \qquad (3)$$

Where U_{dc} is the dc-link voltage, S is the switching function(1 denotes up_branch on and down_ branch off, 0 denotes up_branch off and down_branch on).

Assumed $i_b = -i_c$ =const; $e_b = -e_c$ =const during these four step, U_{N0} can be deduced from(1) and (3).

$$U_{N0} = \frac{1}{2}U_{dc}(S_b + S_c) \qquad (4)$$

According to the conditions of S_b and S_c, the simplified representation is as follows.

$$U_{N0} = \begin{cases} 0 & S_b = S_c = 0 \\ \frac{1}{2}U_{dc} & S_b = 1, S_c = 0; S_b = 0, S_c = 1 \\ U_{dc} & S_b = S_c = 1 \end{cases} \qquad (5)$$

Step a: the value of phase A's back EMF is $0 \le e_a \le U_{dc}/2$, according to (3) and (5), the following results can be derived: $U_{N_0} = 0$ ($S_b = S_c = 0$, dotted lines describe this closed circuit shown in Fig.3), $U_{N_0} = U_{dc}/2$ ($S_b = 1$ and $S_c = 0$, solid arrow lines describe this closed circuit shown in Fig.3). Thus, $U_{ka} = e_a$ or $e_a + U_{dc}/2$, the terminal voltage of phase A is $0 \le U_{ka} \le U_{dc}$, at this moment, There are not diode freewheeling currents in the nonconducted phase A.

Step b: the value of phase A's back EMF is $-U_{dc}/2 \le e_a \le 0$, according to (3) and (5), the following results can be derived: $U_{N_0} = 0$ ($S_b = S_c = 0$, solid arrow lines *2* describe this closed circuit shown in Fig.4), $U_{N_0} = U_{dc}/2$ ($S_b = 1$ and $S_c = 0$, solid arrow lines *1* describe this closed circuit shown in Fig.4). Thus, $U_{ka} = e_a$ or $e_a + U_{dc}/2$, the terminal voltage of phase A is $-U_{dc}/2 \le U_{ka} \le U_{dc}/2$, at that moment, The diode freewheeling currents occur in the nonconducted phase A(dotted lines describe this closed circuit shown in Fig.4).

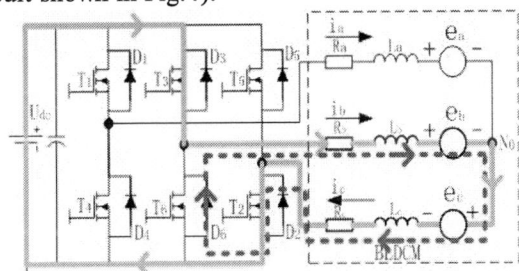

Fig.3　Equivalent circuit of phase A in step a

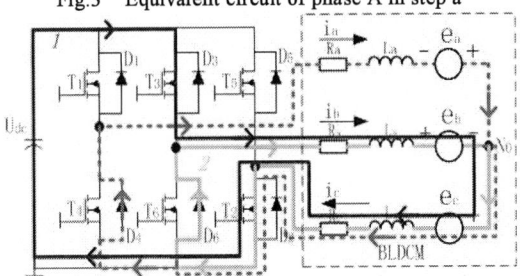

Fig.4　Equivalent circuit of phase A in step b

Step c: the condition are similar to step b. ***Step d:*** the condition are similar to step a.

Assumed that $i_a = 0$, $i_b = -i_c > 0$, before the diode freewheeling currents existing, the electro-magnetic torque can be derived from(2):

$$T_{e\text{-}bc} = -\frac{2E_m i_c}{\Omega} \qquad (6)$$

During current ripples appearing, the currents of 3-phase are $i'_a > 0$, $i'_b > 0$, $i'_c < 0$, and $i'_a + i'_b + i'_c = 0$, the back EMFs are $e_a < 0, e_b = -e_c = E_m$. The electromagnetic torque can be derived from(2).

$$T_{e\text{-}abc} = \frac{e_a i'_a + E_m i'_b - E_m i'_c}{\Omega} = -\frac{2E_m i'_c}{\Omega} + \frac{(e_a - E_m)i'_a}{\Omega} \quad (7)$$

Because of $i'_a > 0$, $i'_c = i_c$, and $e_a < 0$, hence $T_{e-abc} < T_{e-bc}$.

From the above theoretic analysis, we can find when the terminal voltage of phase A is below the zero voltage, the branch diode connected with phase A is in freewheeling satate and the freewheeling current flows in this phase. Accordingly, the electromagnetic torque of BLDC motor trails off under these conditions. The torque ripples also exist in the conduction region under the other unipolar PWM methods. If selecting bipolar PWM method H_PWM-L_PWM, there will be not conduction torque ripples, but this method has more switching loss and high-frequency switching current ripples than unipolar PWM method. So, it is not the best modulation for BLDC motor.

IV. APPROACH FOR THE PROPOSED METHOD

Although a new PWM scheme, PWM_ON_PWM is designed for eliminating the freewheeling current[4]. But the point of back EMF crossing zero voltage level must be estimated precisely, so it will loss of veracity in the high speed region. To resolve the problem, a practical and simple control technique (front_PWM-end_ON) and corresponding driving topology(shown in Fig.5) that can eliminate the conduction torque ripples is introduced in this section.

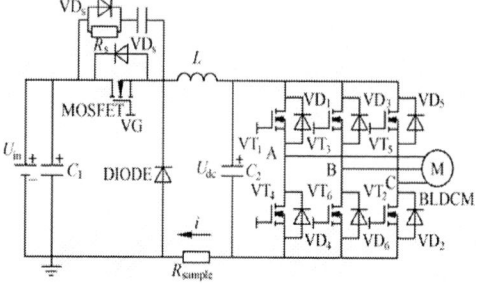

Fig.5　The proposed circuit configuration of BLDCM

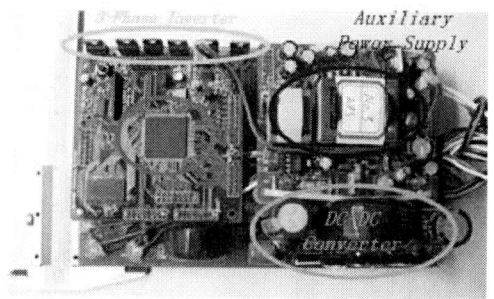

Fig.6 The photo of controller prototype

The proposed cascade structure adopts BUCK converter and 3-phase inverter. BUCK converter can regulate the amplitude of the dc-link current by changing the duty ratio of switching device, and the inverter which is controlled in such a way as to supply 3-phase rectangular current with a pulse width of 120°conduction instead of conventional PWM modulation.The proposed control strategy which can eliminate the freewheeling current over entire speed range is analized by taking nonconducted phase A for example as follows.

Step a and step b: the value of phase A's back EMF is $-U_{dc}/2 \le e_a \le U_{dc}/2$, and $U_{N_0} = U_{dc}/2$ ($S_b = 1$, $S_c = 0$), so $0 \le U_{ka} \le U_{dc}$.During this interval, There are not diode freewheeling currents in the nonconducted phase A.***Step c and step d***($S_b = 0, S_c = 1$) are similar to ***Step a and step b.***Therefore, the conduction torque ripples and high-frequency switching torque ripples are eliminated absolutely.

V. EXPERIMENTAL RESULTS

Experiments are carried out to verify the effectiveness of the proposed method. The parameters of BLDCM prototype are shown in table.I, accordingly the controller prototype is shown in Fig.6.

TABLE I: PARAMETERS OF BLDCM PROTOTYPE

Parameter	Value
U_{rated} (V)	24
P_{rated} (W)	240
Ω_{rated} (rad/m)	248
Poles	9
R_{phase} (Ω)	0.4
L_{phase} (mH)	0.18
f_{buck} (KHz)	50

To implement the new control method, the TMS320LF2407A DSP is employed in the prototype. Digtal PID speed control, constant current control and constant voltage control operate every $20\ \mu s$ sampling time. Fig.7 and Fig.8 show the experiment results of the phase current adopting the conventional unipolar H_PWM-L_ON and the new front_PWM-end_ON modulation respectively. The comparision indicates torque ripples during conduction region eliminated effectively by the new proposed control technique.

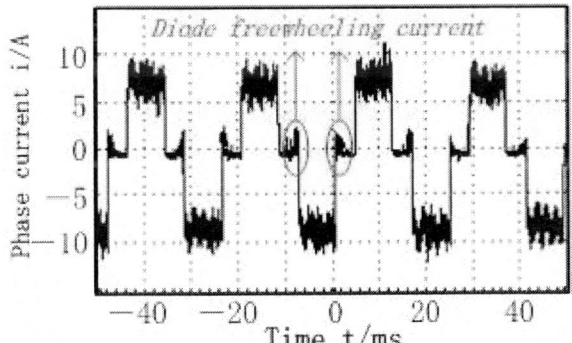

Fig.7   The current waveforms of a phase at conventional
H_PWM-L_ON modulation

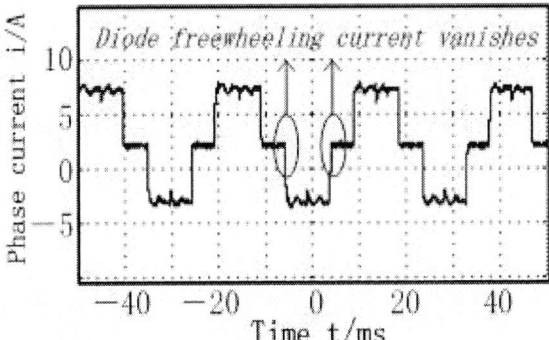

Fig.8   The current waveforms of a phase at new modulation

VI. CONCLUSIONS

In this paper, the characteristics of diode freewheeling in the nonconduction phase which cuased torque ripples are investigated, based on the analysis, a new torque ripple reduction method based on buck converter has been proposed for BLDC motor drives using a single dc current sensor.In such control method, the conduction torque ripples are eliminated completely, meanwhile,lower torque ripples are obtained as well in the commutation region.Subsequently effectiveness and feasibility of the proposed control method is verified through experiments.

REFERENCES

[1] Joong-Ho Song and Ick Choy,"Commutation torque ripple reduction in brushless DC motor drives using a single DC current sensor," *IEEE. Trans. Power Electronics,* vol.19, March 2004, pp. 312-319.

[2] Luk P.C.K and Lee C.K,"Efficient modeling for a brushless DC motor drive,"*International Conference on Industrial Electronics,Control and Instrumentation,* vol.1,September 1994, pp. 188-191.

[3] Carlson.R,Lajoie-Mazenc.M and Fagundes J.C.d.S, "Analysis of torque ripple due to phase communtation in brushless DC machines," *IEEE. Trans. Industry Applications,* vol.28, May-June 1992, pp. 632-638.

[4] Wei Kun,Ren Junjun,Teng Fanghua,"A Novel PWM Scheme to Eliminate the Diode Freewheeling In the Inactive Phase in BLDC Motor,"*in Record of the 35th IEEE Power Electronics Specialists Conference,* vol.3,2004, pp. 2282-2285.

Program Architecture for Realizing Design Optimization of a BLDC Motor

Dong-Hun Kim*, Giwoo Jeung*, Heung-Geun Kim*, and In Dong Kim**

* School of Electrical Eng. & Computer Science, Kyungpook National University, Daegu, South Korea
** School of Electrical and Control Engineering, Pukyong National University, Busan, South Korea

Abstract--The paper presents on new developments in the application of Continuum Design Sensitivity Analysis to design optimisation. Fast convergence, the ability to use existing electromagnetic software without the need to access source codes and independence of computing times on the number of design variables are the distinctive features of the proposed implementation. The design procedure of shape optimization has been integrated into MS Excel spreadsheets and Visual Basic editor so that the methodology allows designers to incorporate their own EM packages as a design tool. The computationally challenging problem of reducing cogging torque in a brushless direct current motor has been selected to illustrate the advantages of the approach in 2D and 3D optimization.

Index Terms—BLDC Motor, Cogging Torque, Design Sensitivity Analysis, Optimization.

I. INTRODUCTION

Continuum Design Sensitivity Analysis (CDSA) has recently gained momentum as an alternative optimization technique. The physical meaning of pseudo sources of an adjoint system in a CDSA when applied to shape optimization was explored in [1] and the approach was reported to avoid the need to access the source codes of commercial programs. Moreover, the computing times required to find an optimal solution are not affected by the number of design variables. The initial, very encouraging, results have prompted the researchers to pursue this technique further as it appears to be very competitive compared, for example, with stochastic methods [2].

Brushless Direct Current (BLDC) motors are rapidly gaining popularity. They are used in industries such as Appliances, Automotive, Aerospace, Consumer, Medical, Industrial Automation Equipment and Instrumentation. The exterior BLDC motor considered here has the same shape as a permanent magnet DC motor without brushes and a commutator and has the permanent magnets rotating outside.

In a permanent magnet motor the cogging torque (a term used to describe non-uniform angular velocity) appears as motion "jerkiness", especially at low speeds, and is undesirable because it introduces vibration and noise. It is thus of great practical importance to gain a deeper understanding of the phenomenon and find ways to reduce it [3].

In this paper further advances are reported on the application of CDSA, in combination with commercial electromagnetic software (on this occasion a program called MagNet [4, 5] was used), with the aim of aiding the efficient design optimization of electromagnetic devices. In our previous publication [1] we focused on showing a way of obtaining the first order sensitivity information using another general purpose FEM package called OPERA and utilizing only the post-processing data. Although such an implementation is fairly straightforward, an attempt has now been made to provide a higher level interface to make the incorporation of CDSA into any electromagnetic software even easier. Thus, with the view of facilitating understanding of the method itself, a more general and very common program architecture consisting of MS Excel spreadsheets and the Visual Basic (VB) editor has been proposed, taking full advantage of the powerful application programming interface (API) provided by MagNet. As Excel and VB are familiar tools for designers, the whole process is more transparent and easier to follow.

A BLDC motor was selected as an example due to the increasing range of applications and because reduction of cogging torque is a challenging (and thus 'real life') problem for electromagnetic simulation and modelling. It was decided to introduce cubic spline interpolation functions to prevent impractical saw-toothed pole shapes resulting from the optimization. Both 2D and 3D simulations have been attempted and comparisons are made.

II. PROGRAM ARCHITECTURE FOR DESIGN OPTIMIZATION

It is not difficult to set up a routine to optimise the pole shape of a BLDC motor, with the objective of minimising the cogging torque, by implementing the CDSA because the adjoint system of equations does not need to be solved [1]. However, the purpose of this work was to develop a more general approach and provide a framework for design optimization using CDSA with the level of flexibility required to tackle other design problems.

The design details necessary for shape optimization are contained in the MS Excel spreadsheets and VB editor communicating with a standard FE analysis package as shown by the left diagram of Fig. 1. The VB script file, containing an optimization algorithm and command language used in the FE software, controls the overall design procedure as depicted in the flow chart of

---

This work has been supported by EIRC (I-2007-0-261-01), which is funded by MOCIE (Ministry of commerce, industry and energy) and ETEP (Electric Power Industry Technology Evaluation & Planning).

978-1-4244-0644-9/07/$25.00 ©2007 IEEE

Fig. 1. At each stage of the iterative design process, information about changes to geometric parameters and performance data are stored and graphically visualized on the Excel spreadsheets which are divided into two parts: the pre- and post-design stage, respectively.

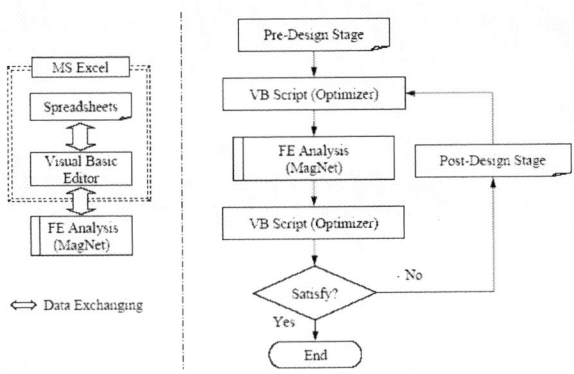

Fig. 1. Program architecture and flow chart of design optimization.

There are some precautions that need to be taken prior to executing the main optimization program for minimizing the cogging torque of BLDC motors as explained below.

A. Pre-Design Stage

1) FE Modelling

First, a conventional FE geometrical model of the initial structure, material properties, boundary conditions, symmetries, etc., relevant to the motor of interest has to be set up. In addition, a pole shape for the motor must be specified using carefully defined points (vertices) – which will act as design variables – using user-defined parameters provided by the FE software. This allows continuing shape changes of the design to be automatically reflected in the FE model based on the parameter values assigned through sensitivity information computed during the optimization process.

2) Definition of the Design Problem

To search for a solution in the possible design space effectively, the design problem has to be cast into a mathematical form, typically by formulating an objective function. Since it has been shown that the reduction of the cogging torque is successfully accomplished by minimizing the variation of the co-energy stored in a magnetic system versus the rotor position [3], the objective function may be written as

$$F = \sum_{i=1}^{nr} (W_i - W_o)^2 \qquad (1)$$

where nr is the number of rotor positions considered, W_i the stored co-energy computed at the i-th position and W_o the constant target value.

Before running the main optimization routine, the designer needs to construct a database containing all the information about the model (e.g. number of design variables, constraint conditions, and initial coordinates for each design variable, etc.). The database is stored in the Excel spreadsheet and is referred to by the main program if and when necessary. This is helpful in easy

identification of the problem as well as for the purpose of modification or further expansion of the database. Fig. 2 shows an example of such a design database embedded in the spreadsheet.

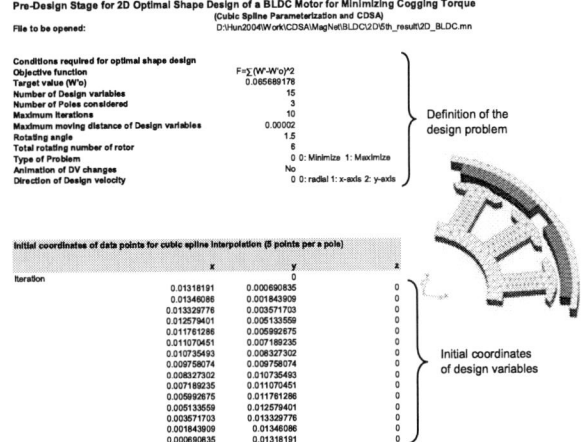

Fig. 2. An example of a spreadsheet for pre-design stage.

B. VB Script File

As already mentioned, the flow chart shown in Fig. 1 has been programmed into a VB script file utilizing the VB editor and the command languages available within the FE software. The VB script file, as the main optimization program, causes the Excel spreadsheets and the FE analysis to continually communicate with each other and reads/writes information necessary to perform optimization into the spreadsheets.

As mentioned earlier, an optimiser based on the first derivative information about the design variables **p** can be easily constructed using CDSA. After solving only one FE model, the design sensitivity may be obtained using the post-processing data and the formula given in [3]

$$\frac{dF}{d\mathbf{p}} = 2\sum_{i=1}^{nr} (W_i - W_o) \cdot \int_{\gamma} [(v_1 - v_2)\mathbf{B}_1 \cdot \mathbf{B}_2]V_n \, d\Gamma \qquad (2)$$

where γ is the movable boundary interface on which design variables are designated, v the magnetic reluctivity, **B** the magnetic flux density and V_n an inner product of a normal vector outward to γ and a directional vector imposed on each design variable. The subscripts, 1 and 2, in (2) denote either side of the interface between air and iron regions, respectively. This formula quantitatively represents the variation of the stored magnetic energy when γ is deformed.

C. Post-Design Stage

During the iterative design process, the results are stored and graphically illustrated to allow easy visual checks on whether the solution to the design problem defined is progressing satisfactorily. This has been achieved by utilizing Excel spreadsheet and its 'chart wizard' functions. The main characteristics of the design process – such as the convergence of the objective function, the variation of a pole face shape, node-based

1093

sensitivity and cogging torque waveform – are automatically displayed after each iteration is compared with the values for the initial model, as shown in Fig. 3.

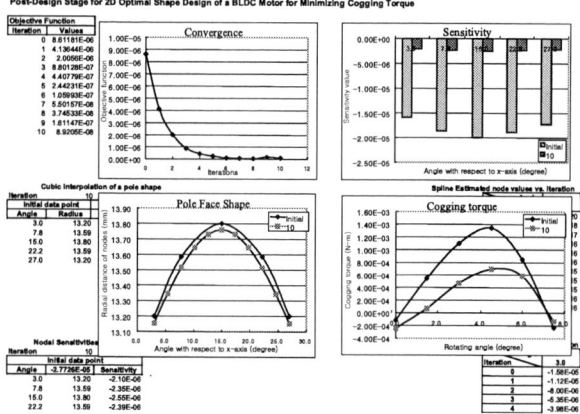

Fig. 3. An example of a spreadsheet for post-design stage.

III. EXAMPLE

To illustrate the performance of the proposed system, a BLDC motor with 8 permanent magnets and 12 salient stator poles has been considered and optimization carried out with the aim of minimizing the cogging torque. The outer radii of stator teeth, magnet and rotor yoke are 13.8mm, 15.3mm and 16mm, respectively. The depth of the teeth is 2.5mm, whereas that of the magnet and the yoke is 3.8mm. Only one-eighth of the problem needs to be modelled owing to symmetry, as shown in Fig. 4.

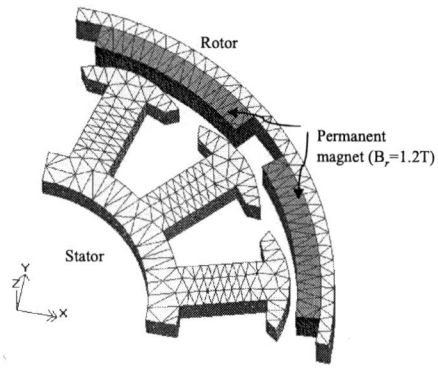

Fig. 4. 3D model used for optimization.

To investigate the fringing effect on the cogging torque, an optimization of the pole face shape has been performed using CDSA in conjunction with commercial software (MagNet 6 [7]). Each pole face is described by 11 finite element vertices forming the outline of the stator pole, as in Fig. 5. To prevent a saw-toothed pole face shape, a cubic spline interpolation curve has been introduced [6]. Thus the movement of the 5 control points marked with dotted circles on the pole face in Fig. 5 constrains their corresponding 11 vertices to be positioned on a smooth curve. In order to take into account manufacturing limitations, a geometrical constraint that the pole face shapes should be the same and symmetry is imposed on the design variables when

moving points in the radial direction. The reduction of cogging torque is accomplished by (1) expressing the variation of the co-energy stored in a magnetic system versus the rotor positions. Due to the $15°$ periodicity of the cogging torque, the objective function is calculated every $1.5°$ from $0°$ to $15°$ in both 2D and 3D nonlinear FE analyses

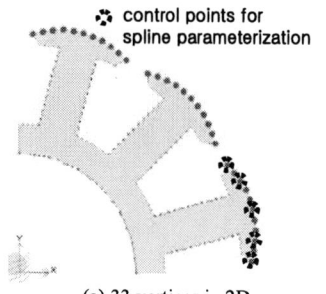

(a) 33 vertices in 2D

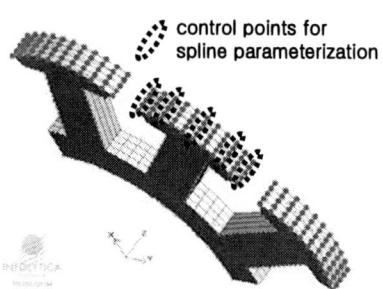

(b) 165 vertices in 3D

Fig. 5. Design variables in 2D and 3D optimization.

A. 2D Optimization

To examine the effectiveness of the spline curve constraint, 11 vertices per pole face are selected as design variables and allowed to move independently according to their own sensitivity values. After 10 iterations, optimum pole shapes with and without the spline parameterization were obtained as shown in Fig. 6. Even though the maximum variation of design variables with respect to the initial shape is less than 0.1 mm, the cogging torque of the optimized pole using the spline parameterization in Fig. 7 has been reduced by nearly 50% of its initial value, based on this 2D FE analysis and the spline constraint produces a larger reduction than allowing a free movement of all vertices.

B. 3D Optimization

In order to accomplish a 3D shape optimization with spline parameterization, the stator has been decomposed into four independent layers with a thickness of 0.3125 mm and the common surface of adjacent layers is allowed to be deformable in order to facilitate the conformity of the FE mesh with the continuing shape changes. After 11 iterations, the optimal pole face shape was achieved. Fig. 8 shows the difference between the pole face shapes optimised using 2D and 3D analyses on a cutting plane parallel to the z-axis and located at $10.2°$ from the x-axis. The variation of the 3D optimised pole face divided into the four layers with respect to the shaft center is presented in Fig. 9. After creating a 3D model by the extrusion technique according to the 2D optimised pole shape, a comparison of the cogging torque waveforms obtained from the 2D and 3D optimised pole shapes is

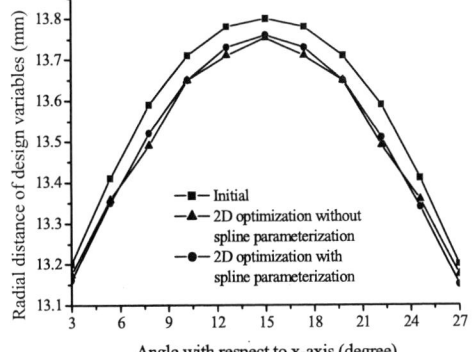

Fig. 6. Variation of pole shapes with respect to shaft center.

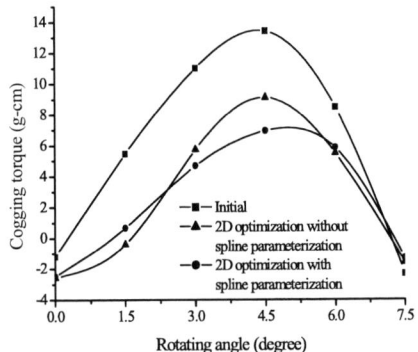

Fig. 7. Cogging torque waveform before and after 2D optimization.

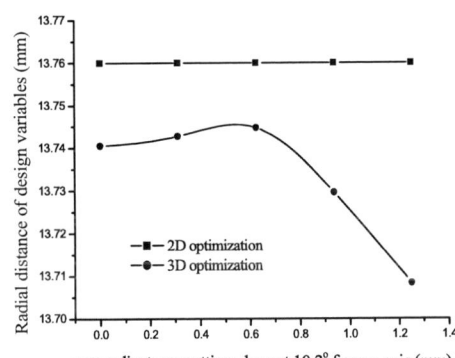

Fig. 8. Comparison of optimised shapes from 2D and 3D analyses.

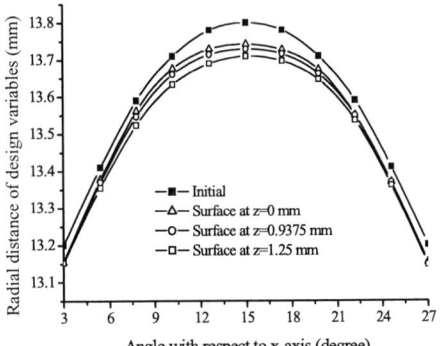

Fig. 9. Comparison Variation of 3D optimised pole face with respect to shaft center.

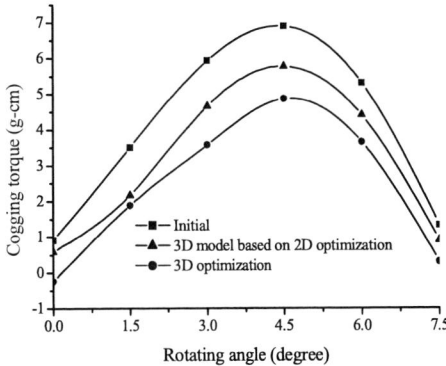

Fig. 10. Cogging torque waveforms for the initial, 2D, and 3D optimal shapes.

shown in Fig. 10. It is revealed that the 3D optimised pole reduces the cogging torque by 30% of the initial value whereas the 2D analysis suggests only a 16% reduction due to neglecting the fringing effect.

IV. CONCLUSIONS

A flexible program architecture combining MS Excel spreadsheets and Virtual Basic script with a standard FE software package is proposed with the aim of aiding the efficient design optimization for minimising the cogging torque of a BLDC motor. The results show that the Continuum Design Sensitivity Analysis is a very efficient optimization technique offering much reduced computational effort due to the fact that computing times do not depend on the number of design variables. Reduction of cogging torque in a brushless dc motor by shaping a pole face has been accomplished and importance of 3D modelling emphasized.

REFERENCES

[1] Dong-Hun Kim, K. S. Ship and J. K. Sykulski, "Applying continuum design sensitivity analysis combined with standard EM software to shape optimisation in magnetostatic problems", *IEEE Trans. Magn.*, vol 40, pp. 1156-1159, 2004.

[2] M. Farina and J. K. Sykulski, "Comparative study of evolution strategies combined with approximation techniques for practical electromagnetic problems", *IEEE Trans. Magn.*, vol 37, pp. 3216-3220, 2001.

[3] Dong-Hun Kim, Il-Han Park, Joon-Ho Lee and Chang-Eob Kim, "Optimal shape design of iron core to reduce cogging torque of IPM motor", *IEEE Trans. Magn.*, vol 39, pp. 1456-1459, 2003.

[4] A. Stochniol, E. M. Freeman and D. A. Lowther, "A user oriented shell for electromagnetic CAD of axisymmetric devices", *IEEE Trans. Magn.*, vol 28, pp. 1782-1784, 1992.

[5] Quang Vo and David A. Lowther, "A paradigm for the non-routine design of electromagnetic devices using a case based reasoning approach", *IEEE Trans. Magn.*, vol 36, pp. 1669-1672, 2000.

[6] Chank-hyun Kim, Hyang-beom Lee and Il-Han Park, "B-spline parameterization of finite element models for optimal design of electromagnetic devices", *IEEE Trans. Magn.*, vol 35, pp. 3763-3765, 1999.

[7] Infolytica Corporation, *MagNet 6 User's Guide*, 2005.

Stable Operation of the Brushless Doubly-Fed Machine (BDFM)

Shiyi Shao, Ehsan Abdi, and Richard McMahon

Engineering Department, Cambridge University, Cambridge, CB3 0FA, UK

Abstract—A simple method of controlling the Brushless Doubly-Fed Machine (BDFM) is presented. The controller comprises two Proportional-Integral (PI) modules and requires only the rotor speed feedback. The machine model and the control system are developed in MATLAB. Both simulation and experimental results are presented. The performance of the system is presented in the motoring and generating operations. The experimental tests included in this paper were carried out on a 180 frame size BDFM with a nested-loop rotor.

Index Terms—Brushless Doubly-Fed Machine, Closed-loop control, Proportional-Integral (PI) controller, Stable operation.

I. INTRODUCTION

The Brushless Doubly-Fed Machine (BDFM) shows commercial promise as a variable speed drive and generator. The technology has significant advantages as a generator in wind turbines as the brushless operation of the machine reduces maintenance and increases reliability. This is particularly important as more and more installations are being constructed offshore and in difficult-to-reach places. The BDFM also promises significant benefits in the form of motor in water pumps and gas compressors. The motivation is the reduced inverter rating; the typically low starting torque required for fluid loads means that this advantage can be maximised.

In order to progress the BDFM towards commercial service, not only must the machine be designed to give best possible performance in terms of high efficiency, output power, and other steady-state measures, but also the machine must be fully controllable so that it can always maintain these desired operating conditions. For example, as a wind turbine generator, the machine must operate at a specific shaft speed set by wind conditions to gain the maximum power output. Similar requirements apply for variable speed drive applications, such as pump drives [1].

The contemporary BDFM is a single frame induction machine with two 3-phase stator windings of different pole numbers, and a special rotor design. Typically one stator winding is connected to the mains or grid, and hence has a fixed frequency, and the other is supplied with variable voltage at variable frequency from a converter.

The modelling, design and operation of the BDFM have been studied comprehensively, for example in [2-6].

However less attention has been given to the control and stability of the machine. A 'direct torque control' method was proposed for the BDFM, although never implemented [7,8], and a number of variations of a field oriented control scheme were proposed and some of these were implemented [9,10].

Nevertheless, the stability of the machine over wide range of operating speeds has not been demonstrated using these control schemes. The open-loop instability of the BDFM was investigated in [11], but it was shown in [12] that a closed-loop control system is essential for stable operation over wide speed range. Roberts et al. in [13] presented two stabilizing control algorithms for the BDFM, but results were shown only in simulation. The stability of the BDFM under closed-loop current control was investigated recently in [14] and stability in response to a relatively small and slow torque perturbation was achieved.

In applications where high dynamic performance is not required, such as water pumps, a simple and practical controller may be of significant benefit, decreasing the cost and complexity of the system. This paper presents a simple controller designed for the BDFM operating in a variable speed drive and generator. The control system comprises phase angle and speed controllers for the stator control winding. A coupled-circuit model for the BDFM is used to predict the machine performance under the proposed control algorithm and the results are validated experimentally.

II. BDFM DYNAMIC MODEL

The coupled-circuit model is one of several methods to analyse electrical machines. The main advantage of this model is that it is a dynamic model and hence enables the observation of transient behaviour of the machine. A generalised coupled-circuit model for the BDFM is presented in [15] and a method to calculate the model parameters is given. The model is based on the general electrical machine coupled-circuit equation:

$$v = Ri + \omega_r \frac{dM}{d\theta_r} i + M \frac{di}{dt} \qquad (1)$$

where v and i are the voltage and current vectors, and R and M are the resistance and inductance matrices. ω_r and θ_r are respectively the rotor angular speed and position. The torque generated by an electrical machine can be

978-1-4244-0644-9/07/$25.00 ©2007 IEEE

determined by considering instantaneous power transfer in the system [16] and is given by

$$T_e = \frac{1}{2} i^T \frac{dM}{d\theta_r} i \qquad (2)$$

In the BDFM, it is convenient to partition v and i into stator 1, stator 2, and rotor quantities, noting that the rotor voltage will always be zero. For the BDFM, the mutual inductance matrix between two stator windings is zero. This is because two stator pole numbers are chosen so that the stator windings in the BDFM do not couple to each other [2]. The BDFM coupled-circuit equations can be therefore written as

$$\frac{d}{dt} \begin{bmatrix} i_{s1} \\ i_{s2} \\ i_r \end{bmatrix} = \begin{bmatrix} M_{s1} & 0 & M_{s1r} \\ 0 & M_{s2} & M_{s2r} \\ M_{s1r}^T & M_{s2r}^T & M_r \end{bmatrix}^{-1} \left\{ - \left(\begin{bmatrix} R_{s1} & 0 & 0 \\ 0 & R_{s2} & 0 \\ 0 & 0 & R_r \end{bmatrix} + \omega_r \begin{bmatrix} 0 & 0 & \frac{dM_{s1r}}{d\theta_r} \\ 0 & 0 & \frac{dM_{s2r}}{d\theta_r} \\ \frac{dM_{s1r}^T}{d\theta_r} & \frac{dM_{s2r}^T}{d\theta_r} & 0 \end{bmatrix} \right) \begin{bmatrix} i_{s1} \\ i_{s2} \\ i_r \end{bmatrix} + \begin{bmatrix} v_{s1} \\ v_{s2} \\ 0 \end{bmatrix} \right\} \qquad (3)$$

The torque equation from (2) and (3) can be shown to be:

$$T_e = \begin{bmatrix} i_{s1}^T & i_{s2}^T \end{bmatrix} \begin{bmatrix} \frac{dM_{s1r}}{d\theta_r} \\ \frac{dM_{s2r}}{d\theta_r} \end{bmatrix} \begin{bmatrix} i_r \end{bmatrix} \qquad (4)$$

The resistance and inductance matrices in (3) and (4) can be calculated from the machine geometrical dimensions as described in [15]. The BDFM employed in this paper has the resistance and inductance parameters given in [5]. Using (3) and (4), a dynamic model was implemented in *Simulink* [17]. A block diagram of the model is shown in figure 1.

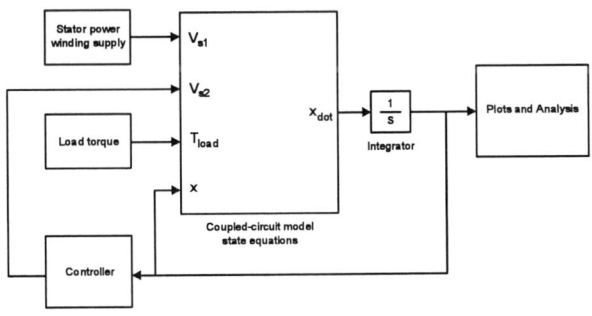

Fig. 1. The block diagram of the coupled-circuit model implemented in Simulink

III. CONTROLLER DESIGN

The control winding is supplied with a converter and therefore the amplitude, frequency and phase of the voltage can be controlled. The proposed controller comprises a phase angle controller and a speed controller. Both controllers are based on the PI (Proportional-Integral) control algorithm. The block diagram of the control system is shown in figure 2.

A. Phase Angle Control

The torque produced by the BDFM comprises synchronous and induction torques [4]. However, in general, the induction torque will be relatively small and therefore the torque can be expressed as [4]:

$$T = \frac{3(p_1 + p_2)}{\omega_1} \frac{\left| V_{s1} V_{s2}'' s_2 \right|}{\omega_1 L_r' s_1} \sin(\delta) \qquad (5)$$

where V_{s1} and V_{s2} are the magnitude of the voltages applied to stator winding 1 and 2 respectively, ω_1 is stator 1 angular frequency, s_1 and s_2 are the slips, p_1 and p_2 are stator 1, stator 2 number of pole pairs, and L_r is the rotor inductance. ' ' ' denotes that the quantity or parameter is referred. δ is the so called load angle and is described in detail in [4]. Changing the phase of V_{s2} on a real machine changes δ causing the rotor to advance or retard in the synchronous reference frame.

A PI controller has been developed to adjust the phase angle of the control winding voltage, as shown in figure 2. The controller parameters were determined experimentally and are shown in table I.

B. Speed Control

The frequency of the control winding is adjusted using a PI controller as shown in figure 2. The magnitude of the control winding voltage is then determined using a conventional V/f control algorithm. Through this method, the flux of the control winding is maintained at an appropriate level. The pattern that is used for the controller is shown in figure 3. The corresponding parameters are tabulated in table I.

TABLE I
THE CONTROLLER PARAMETERS

Controller parameters		value
Phase angle controller	P_{Phase}	0.025-0.05s
	I_{Phase}	0.5s
Frequency controller	P_{Speed}	0.005
	I_{Speed}	0.5
V/f controller	F_1	±2Hz
	F_2	±42Hz
	V_1	19V
	V_2	153V

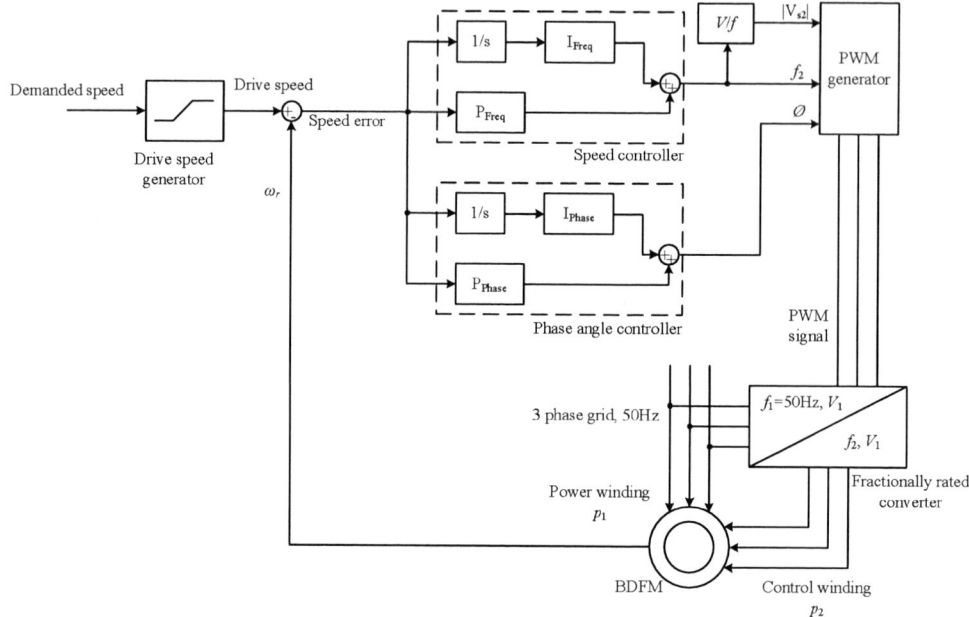

Fig. 2. Block diagram of the control system comprising phase angle and speed controllers

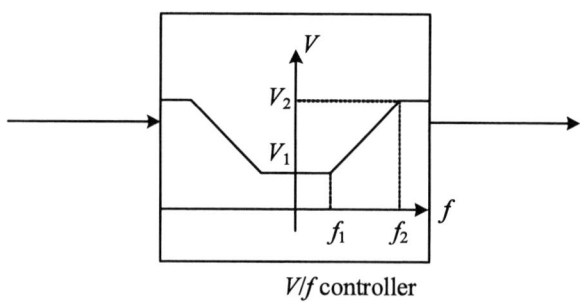

Fig. 3. The *V/f* (voltage-versus-frequency) pattern used in the control system

IV. EXPERIMENTAL AND SIMULATION RESULTS

A. BDFM Operation

The BDFM can be operated in several modes for which a brief description is given here. Details can be found in [2, 15]. The *synchronous*, or *doubly fed*, mode of operation of the BDFM is the desirable operating mode. The shaft speed is independent of the machine torque and is given by:

$$\omega_r = \frac{\omega_1 \pm \omega_2}{p_1 \pm p_2} \quad (6)$$

where ω_1 and ω_2 are angular frequencies of the supplies to the stator 1 and 2 respectively. A further relationship for the so-called *natural speed* ω_n, that is the synchronous speed when the control winding is fed with dc, is given by:

$$\omega_n = \frac{\omega_1}{p_1 + p_2} \quad (7)$$

Maintaining operation of the BDFM in the synchronous mode over a wide speed range is the major challenge for controller design.

The BDFM has also asynchronous modes of operation where the shaft speed is dependent on the loading of the machine, as well as the supply frequency. The machine can be operated as a cascaded induction machine by exciting stator 1 or stator 2 and leaving the other winding shorted in each case. A cascade induction machine formed from p_1 and p_2 pole pair induction machines has characteristics which resemble an induction machine with $p_1 + p_2$ pole pairs. In the sequel, this mode of operation will be referred to as *cascade* mode. This mode of operation can be used for starting to avoid the need to use a bi-directional converter for the control winding [18]. Once the natural speed is approached, operation can transfer to the synchronous mode.

B. Experimental Rig

Table II gives the physical data for the BDFM used throughout this and the work described in [3-5, 13, 15. 18]. The machine is shown in figure 4 in the experimental rig.

A DC machine is mechanically coupled to the BDFM in order to provide required test conditions for operation in motoring and generating modes. Since the DC machine is not equipped with a suitable control system, it was not possible to carry out specific tests such as the response to arbitrary steps in torque. The torque step response shown in figure 8 was performed by switching off the armature and field windings supplies.

The stator power (4-pole) winding is supplied from a constant voltage and frequency supply at 120V and 50Hz. The stator control (8-pole) winding is fed with a variable voltage, variable frequency inverter.

Fig. 4. Prototype BDFM machine (left) on test rig with torque transducer and DC load machine (right)

TABLE II
PROTOTYPE MACHINE SPECIFICATIONS

Parameter	Value
Frame size	D180
Stator 1 pole-pairs	2
Stator 2 pole-pairs	4
Stator slots 48	48
Rotor slots 36	36
Rotor design	'Nested-loop' design consisting of 6 'nests' of 3 concentric loops of pitch 5/36, 3/36 and 1/36 of the rotor circumference. Each nest offset by 1/6 of the circumference, for the details see [15].

C. Variable Speed Drive Operation

The machine is started in the cascade mode and is then switched into the closed-loop synchronous mode at around 500rpm (natural speed). A starting procedure is shown in figure 5. The machine is driving a fan type load with the following characteristics:

$$T = KI_f^2\omega_r \qquad (8)$$

The constant K of the DC machine is 1.80Nm/A. The field current I_f was set so that the load torque is 10Nm at 500rpm.

The speed of the drive can be controlled in the synchronous mode over a wide range, below, at and above the natural speed. Figure 6 shows the response to a change in demanded speed from 0rpm to 1000rpm. The BDFM is driving a fan type load described by equation (8). The shaft torque is also shown in figure 6.

In order to make a comparison of the performance of the proposed control method to the vector control algorithm implemented on a similar size BDFM by Poza et al., a test was carried out at the same condition as presented in [10]. The results are shown in figure 7. The machine is driving a constant load torque of 7Nm. A speed change of 400rpm is applied to the machine. The shaft speed settles in less than 2s. This is almost the same as was achieved in [10] with a field oriented control scheme.

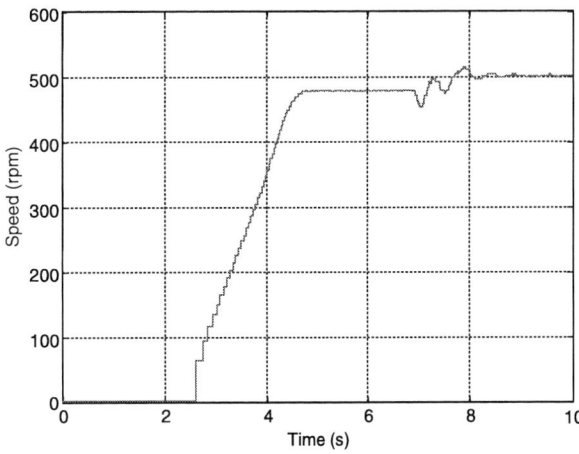

(a) Shaft speed

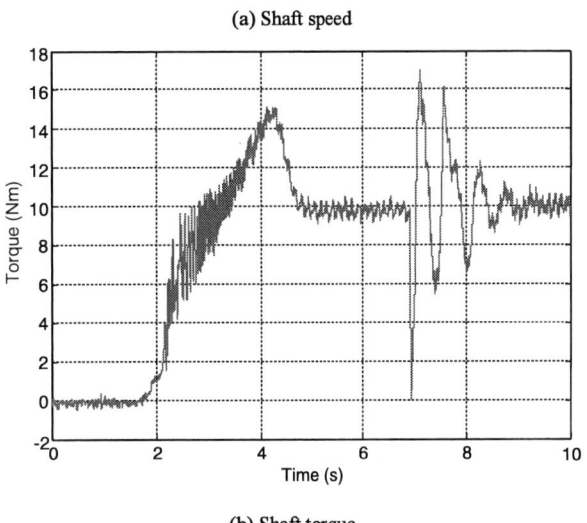

(b) Shaft torque

Fig. 5. Startup procedure for the BDFM driving a fan type load: the machine is started in a cascade mode; then at t=6.7s, the machine is switched into the synchronous mode. The results are from experimental tests.

D. Variable Speed Generator Operation

As mentioned earlier, the experimental setup is not able to provide an arbitrary torque perturbation for test purposes. However, a torque perturbation was applied to the BDFM operating as a generator by switching off the supplies to the DC, giving an effective step in torque of 42Nm. The results are shown in figure 8.

As the perturbation in speed is small, it is partly masked by the digitization error evident in figure 8-b.

V. CONCLUSIONS

The experimental results presented in this paper show that the closed loop controller described is able to achieve stable operation of the BDFM over wide speed range in the synchronous mode. The control algorithm is a simple and practical method and is easy to implement. Further, it only requires the measurement of rotor speed. Although the dynamic response of the system may not be as good as that achievable with a more complex controller based on, for example, a vector control algorithm, it is suitable for low demand applications such as a water pump where

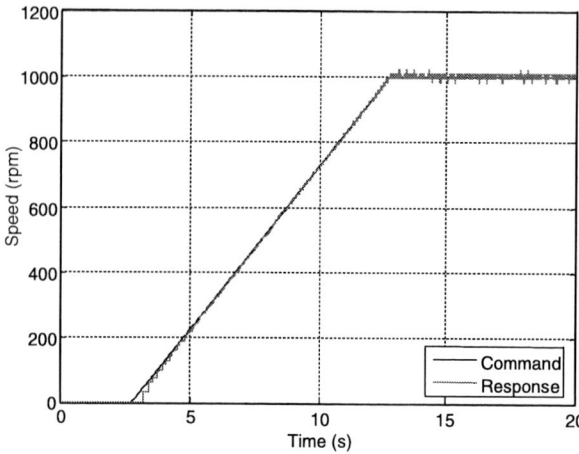

(a) Shaft speed

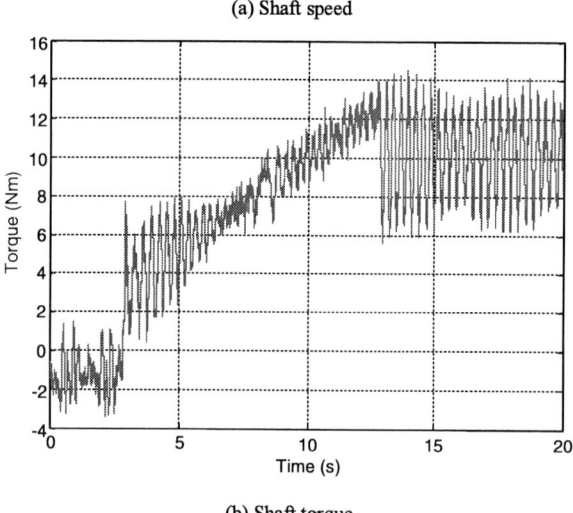

(b) Shaft torque

Fig. 6. Dynamic response of the BDFM to a 1000rpm speed change. The machine is driving a fan type load provided by a DC machine. The results are from experimental tests.

a fast dynamic response is not needed. In such applications, low system cost and complexity are important issues.

In addition, the starting of a BDFM drive in the cascade mode, against a load, and the subsequent transition to the synchronous mode has been demonstrated. This approach is of practical value as the converter feeding the control winding need only handle power flow into the BDFM above natural speed, a mode in which the BDFM will give highest efficiency.

Two important directions for future work have emerged. Work will be undertaken to improve further the dynamic response of the controller. In addition, at any particular operating point, the control winding voltage can be varied to alter the power factor of the power winding. Previous work has shown that the efficiency of the BDFM can be optimised by control of the flow of reactive power through the machine. The challenge is to incorporate this feature in a practical controller.

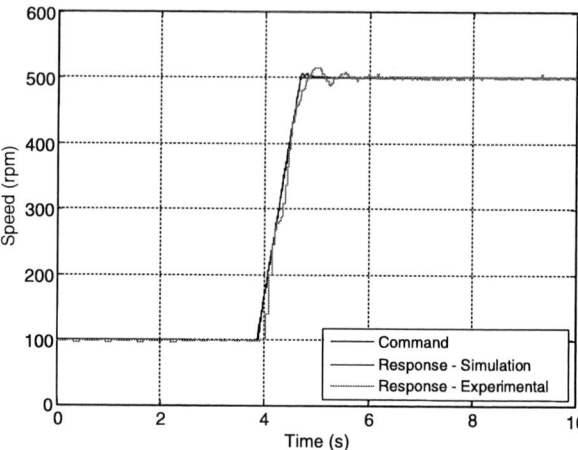

Fig. 7. Dynamic response of the BDFM to a 400rpm speed change when driving a constant load torque of 7Nm.

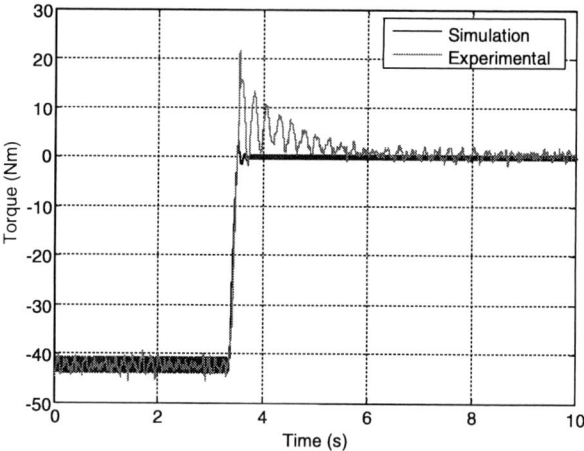

(a) Shaft torque

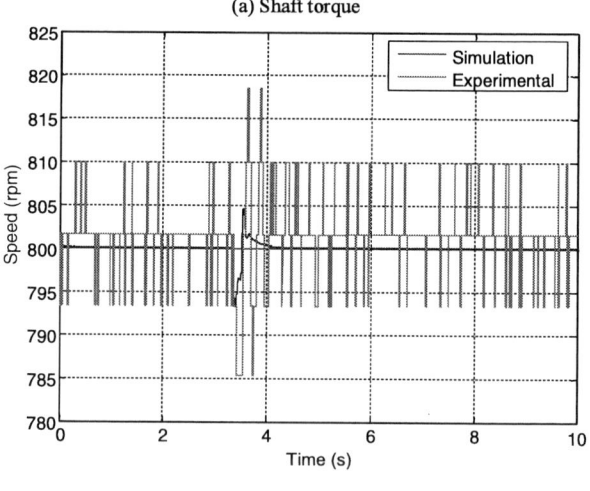

(b) Shaft speed

Fig. 8. Dynamic response of the BDFM to a torque change of 42Nm.

REFERENCES

[1] M. Boger, A. Wallace, and R. Spee; "Investigation of appropriate pole number combinations for brushless doubly fed machines as applied to pump drives", *IEEE Transactions on Industry Applications*, 31(5):1022-1028, 1996.

[2] S. Williamson, A. C. Ferreira, and A. K. Wallace; "Generalised theory of the brushless doubly-fed machine. part I: analysis",

Electrical Power Applications, IEE Proceedings, 144:111-122, 1997.

[3] P. C. Roberts, R. A. McMahon, P. J. Tavner, J. M. Maciejowski, and T. J. Flack; "Equivalent circuit for the brushless doubly fed machine (BDFM) including parameter estimation and experimental verification", *Electrical Power Applications, IEE Proceedings*, 152(4):933-942, July 2005.

[4] R. A. McMahon, P. C. Roberts, X. Wang, and P. J. Tavner; "Performance pf BDFM as generator and motor", *Electrical Power Applications, IEE Proceedings*, 153(2):289-299, March 2006.

[5] Ehsan Abdi-Jalebi; "Modelling and Instrumentation of Brushless Doubly-Fed (Induction) Machines", PhD Thesis, University of Cambridge, September 2006.

[6] J. Poza, E. Oyarbide, D. Roye, and M. Rodriguez; "Unified reference frame dq model of the brushless doubly fed machine", *Electrical Power Applications, IEE Proceedings*, 153(5):726-734, September 2006.

[7] W. R. Brassfield, R. Spee, and T. G. Habetler; "Direct torque control for brushless doubly fed machines" *IEEE Transactions on Industry Applications*, 32(5):1098.1103, 1996.

[8] I. Sarasola, J. Poza, M. Rodriguez, and G. Abad; "Direct torque control for brushless doubly fed induction machine", *IEEE International Electric Machines & Drives Conference IEMDC2007*, Antalya, 2007.

[9] D. Zhou, R. Spee, and G. C. Alexander; "Experimental evaluation of a rotor flux oriented control algorithm for brushless doubly-fed machines", *IEEE Transactions on Power Electronics*, 12(1):72-78, 1997.

[10] J. Poza, E. Oyarbide, and D. Roye, 'New vector control algorithm for brushless doubly-fed machines'. Proc. 2002 IEEE IECON Conf., Seville, Spain, November 2002

[11] D. Zhou and R. Spee; "Experimental evaluation of a rotor flux orientated control algorithm for brushless doubly-fed machines", *25th IEEE Power Electronics Specialists Conference PESC1994*, vol. 2, pp. 1229-1236, Taipei, Taiwan, June 1994.

[12] A. Kemp, M. Boger, E. Wiedenbrug, and A. Wallace; "Investigation of rotor current distributions in brushless doubly-fed machines", *IEEE Industry Application Society Annual Meeting*, pp. 638-643, San Diego, CA, October 1996.

[13] P. C. Roberts, T. J. Flack, J. M. Maciejowski, and R. A. McMahon; "Two stabilising control strategies for the brushless doubly-fed machine (BDFM)", *International Conference on Power Electronics, Machines and Drives, PEMD2002*, pp. 341-346, Bath, UK, June 2002.

[14] I. Sarasola, J. Poza, E. Oyarbide, and M. A. Rodriguez; "Stability analysis of a brushless doubly-fed machine under closed loop scalar current control", 32nd Annual Conference of the IEEE Industrial Electronics Society, IECON2006, pp. 1527-1532, Paris, France, 2006.

[15] Paul Roberts, "A study of Brushless Doubly-Fed (Induction) Machines", PhD Thesis, University of Cambridge, September 2004.

[16] P. C. Krause, O. Wasynczuk, and S. D. Sudho, *"Analysis of Electric Machinery and Drive Systems"* IEEE Press Wiley, New York, 2nd edition, 2002.

[17] The Mathworks Inc. *Using Simulink*, 2005. www.mathworks.com.

[18] D. Feng, P. C. Roberts, and R. A. McMahon, *"Control Study on Starting of BDFM,"* 41st International Proceedings of Universities Power Engineering Conference (UPEC), Vol. 2, pp. 660-664, 6-8 Sept. 2006.

Sail Generator Feasibility Study

Ha Pham Ngoc*, Yasuaki Matsui*, Pathom Attaviriyanupap* and Osamu Iso**

* Tokyo Institute of Technology, Japan
** Kansai Electric Power CO., INC., Japan

Abstract-- In this paper, the authors propose a new small size vertical axis wind power generation system for operating in standalone mode at low speed under weak wind conditions. The proposed system consists of a Sail Turbine and an Outer Zipper Rotor Generator. The Sail Turbine is designed so that it can catch energy from weak wind and also can protect itself from damage under strong wind. The generator is designed for standalone operation without brush and gear. The feasibility study of the proposed generation system is conducted using Fourier Expansion. Analytical results show that the target output can be achieved.

Index Terms—Fourier Expansion, Outer Zipper Rotor Generator, Sail Turbine, Wind Power Generation

I. INTRODUCTION

According to the Kyoto Protocol, agreed to by 171 participating countries, greenhouse gas emissions in Japan will be reduced by 6 percent over the years 2008-2012 compared with emission levels in 1990 [1]. The Japanese government has encouraged an increase in the contribution of low-emission clean energy resources such as wind power.

Usages of clean renewable sources such as wind and solar power have been vigorously developed over recent years, particularly wind energy [2-7]. At present, more than 40 GW wind generators have been installed worldwide and the number will increase in the near future [8]. It is expected that 3 GW wind generators will be connected to the Japanese power system in 2010 [9].

However, large size conventional wind power generation systems with propellers are not suitable for Japan, where useable space is limited. Moreover, a high tower and propeller, part of conventional wind power generation systems, are at risk to be damaged due to lightning and storms. Hence, maintenance cost is high.

The authors have developed a smaller wind power generation system (sail generator) for use in Japan. The proposed wind power generation system uses a sail to catch wind energy and convert it to electricity through a multi-pole generator. The proposed wind generation system is designed to operate in standalone mode at low speed under weak wind conditions in urban areas. It is a brushless generator and is connected directly to the load without a gearbox. As a result, it has many advantages over the conventional propeller type wind power generation system including (1) low maintenance cost; (2) less damage risk due to lightning and strong wind.

This work was supported by Kansai Electric Power CO., INC., Japan

The target specifications of the generator are as follows:

1. Small size with a diameter of about 2 meters.
2. Operates under weak wind (around 6 m/s) at low speed (30 rpm)
3. DC excitation, multi-pole, gearless and brushless
4. Single phase standalone at 300 W, 100 V, 60 Hz
5. Small copper loss

The objectives of this paper are to (1) develop a sail generation model and study the operation principle; (2) conduct a feasibility study of the generator.

II. SAIL TURBINE

A vertical axis 2-meter diameter wind turbine has been designed for its low maintenance cost. We named it "Sail Turbine".

A. Design and Principle

A model Sail Turbine is shown in Fig. 1. The Sail Turbine has one mast in the center holding a sail made of cloth or plastic. There is a string attached to the end of the sail in order to adjust the direction of the sail when it is rotating.

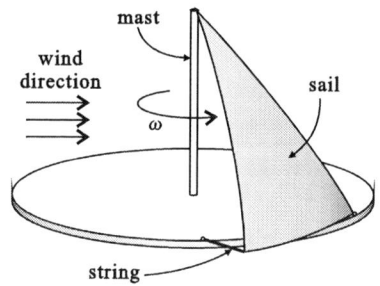

Fig. 1. Sail Turbine Model (one sail)

When the wind speed exceeds the safe limit on the system, the mast is designed to automatically fall off along with the sail to avoid damages from strong wind. The turbine will be restarted after the gust is over.

Figure 2 shows the principle of the Sail Turbine. The mechanism of the Sail Turbine can be explained through 4 states as follows:

a) The sail catches the wind and produces a torque to rotate the turbine counterclockwise.

b) When the sail is nearly parallel to the direction of the wind, it moves into a new position which is decided by the length of the string.

c) In the new position, the torque produced still keeps the turbine rotating counterclockwise.

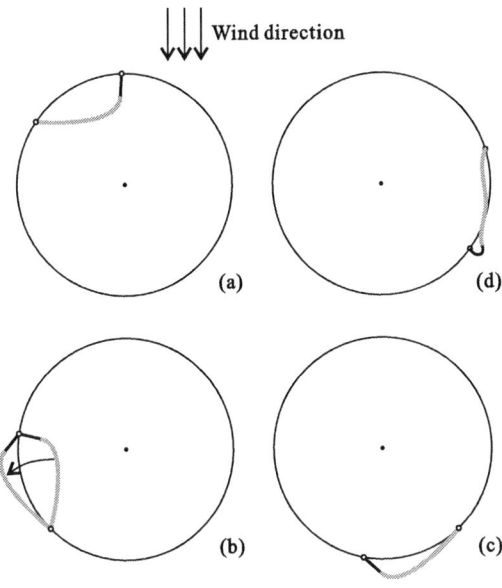

Fig. 2. Sail Turbine Principle

d) When the sail is nearly parallel to the direction of the wind, the string goes loose and the sail moves freely with the wind. In this state, the sail does not catch any wind and the turbine's moment of inertia keeps the rotation counterclockwise. This state continues until the string reaches its limit and pulls the sail back into the original position of state (a).

B. Windmill Test

Figure 3 shows a windmill model of a sail generator using one sail. The windmill was tested using a fan which generates weak wind. Experiments showed that the windmill can catch the wind and rotate well.

Fig. 3 Windmill Test

III. OUTER ZIPPER ROTOR GENERATOR

A vertical axis 2-meter diameter generator is designed to generate power from the Sail Turbine. Excitation is conduced by a DC current instead of a permanent magnet in order to reduce the cost. We named the generator "Outer Zipper Rotor Generator" or OZR generator according to its shape. Because the rotor is outside, the Sail Turbine can be installed directly on it.

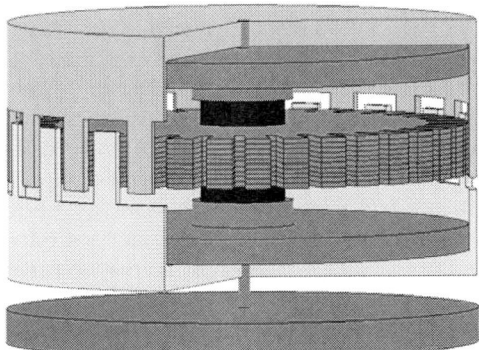

Fig. 4. Three Dimension Shape of OZR Generator

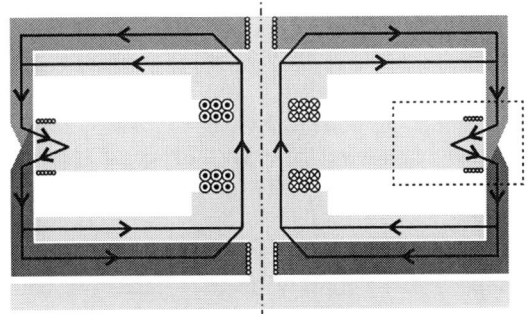

Fig. 5. Flux Path of OZR Generator
(Vertical Cross Section)

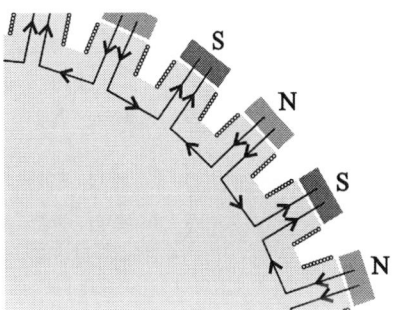

Fig. 6. Flux Path of OZR Generator
(Central Cross Section)

A. Principle

The shape of an OZR generator is shown three dimensionally in Fig. 4. The stator is grey while the rotor is colored in blue and pink. The DC excitation coil (black) is wound around the vertical axis of the stator. The armature pole is made from a stack of laminated iron plates and is also wound on the stator side. The laminated iron plates are used to reduce the loss on the iron surface. Because there is no winding on the rotor, a brush is not necessary. The rotor is composed of the upper part (blue) and the lower part (pink) aligned like a zipper.

The OZR generator is spurred into action by a DC current supported from an external voltage source. Instead of using one excitation coil for each pair of poles,

1103

the generator uses only one excitation coil in the center so that the copper loss can be reduced. Many turns of the excitation coil are used to make the excitation current as near constant as possible. The excitation current is injected so that the linkage flux between two windings flows from the center of the stator to the upper part, the lower part of rotor and back to the center of the armature, as seen in Figs.5 and 6.

Because the reluctance between stator and rotor varies as the rotor rotates, the linkage flux between two coils changes resulting in induced voltage [10, 11]. The OZR generator functions similar to a synchronous generator.

B. Model

Figures 7 and 8 show a model of an OZR Generator. The outer diameter is designed to be around 2 meters so that the Sail Turbine can fit right on it. The thickness of the generator is around 0.1 meters. The characteristics of OZR Generator depend on width w, gap d, and height h of armature pole.

The number of poles is set to be 240 in order to achieve the output of 60 Hz when the rotation speed is at 30 rpm.

IV. FEASIBILITY STUDY

Assume that the input voltage of excitation coil is constant at V_1, we can find out the theoretical output of the OZR Generator as seen in the following calculation.

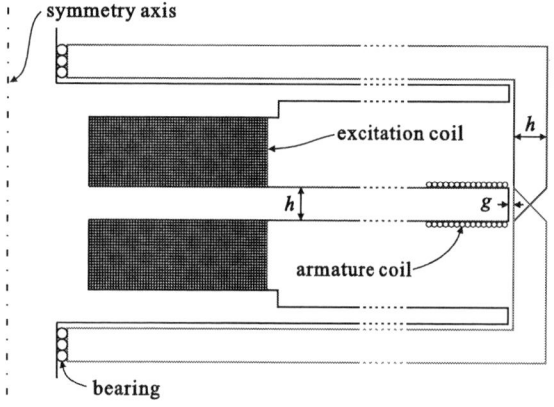

Fig. 7. OZR Generator model
(Vertical Cross Section)

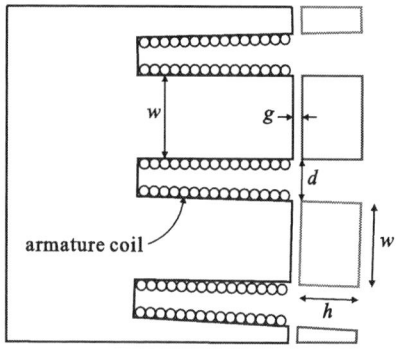

Fig. 8. OZR Generator model
(Central Cross Section)

Let coil 1 and coil 2 represent the excitation and armature coil, respectively. The flux linkage of coil 1 (λ_1) is derived from two components: the flux linkage of coil 1 due to current i_1 acting alone and the flux linkage of coil 1 due to current i_2 acting alone. Flux linkage in coil 2 is considered similarly. Flux linkage can be represented mathematically as:

$$\lambda_1 = L_1 i_1 + M i_2 \tag{1}$$
$$\lambda_2 = M i_1 + L_2 i_2 \tag{2}$$

It should be noted that i_1, i_2, L_1 and L_2 in (1) and (2) are functions of rotor position.

In this research, the reactance variation coefficient (h_s and h_m) are used to represent the changes of self and mutual reactance (inductance) in term of rotor location or electrical angle θ. We can rewrite (1) and (2) in term of h_s and h_m as follows:

$$\lambda_1 = h_s L_1 i_1 + h_m M i_2 \tag{3}$$
$$\lambda_2 = h_m M i_1 + h_s L_2 i_2 \tag{4}$$

where L_1, L_2 and M are the maximum self inductance of coil 1, coil 2 and mutual inductance between two coils, respectively. It should be noted that L_1, L_2 and M in (3) and (4) are constant and can be computed from (5) to (7).

$$L_1 = \frac{n_1^2}{\Re_0} \tag{5}$$

$$L_2 = \frac{n_2^2}{\Re_0} \tag{6}$$

$$M = \frac{\beta n_1 n_2}{\Re_0} \tag{7}$$

Parameters n_1 and n_2 are the number of turns of excitation and armature coils, $\Re_0$ is the minimum value of air gap reluctance, β is the magnetic coupling coefficient between excitation and armature coils.

Assuming that the system's moment of inertia is large enough that the rotation speed ω are nearly constant, multiply (3) and (4) by ω we have

$$\omega\lambda_1 = x_1 h_s i_1 + x_m h_m i_2 \tag{8}$$
$$\omega\lambda_2 = x_m h_m i_1 + x_2 h_s i_2 \tag{9}$$

where $x_1 = \omega L_1$, $x_2 = \omega L_2$, and $x_m = \omega M$.

The voltage equations can be formulated as:

$$V_1 = r_1 i_1 + \frac{d\lambda_1}{dt} \tag{10}$$

$$v_2 = r_2 i_2 = -\frac{d\lambda_2}{dt} \tag{11}$$

where V_1 is the DC excitation voltage, r_1 is the resistance in excitation coil, i_1 is the excitation current, v_2 is the armature induced voltage, r_2 is the total resistance on the armature side (= armature coil resistance (r_{2w}) + load resistance (r_{2L})), i_2 is the armature current.

Replace (8) and (9) into (10) and (11), we get

$$V_1 - r_1 i_1 = \frac{dh_s}{d\theta} x_1 i_1 + \frac{dh_m}{d\theta} x_m i_2 + h_s x_1 \frac{di_1}{d\theta} + h_m x_m \frac{di_2}{d\theta} \quad (12)$$

$$-r_2 i_2 = \frac{dh_m}{d\theta} x_m i_1 + \frac{dh_s}{d\theta} x_2 i_2 + h_m x_m \frac{di_1}{d\theta} + h_s x_2 \frac{di_2}{d\theta}. \quad (13)$$

Currents i_1 and i_2 are determined by comparing frequency of i_1, i_2, h_s and h_m after employing Fourier Expansion. Detail of calculation is shown in the Appendix.

A. Analysis Result of OZR Generator

Assuming that h_s and h_m can be approximately calculated through the change of air gap reluctance, the output of the OZR Generator can be solved as in the previous section. Two designs for the OZR Generator have been studied and compared.

1) Design 1

To achieve the target output 300 W, 100 V, 60 Hz, the parameters of OZR generator were calculated and shown in Table 1.

Table 1 Parameters of Design 1

Parameter		Value
Open circuit output voltage		100 Vrms
Rated output		300 W, 60 Hz
Rated speed		30 rpm
Number of poles		240 poles
Rotor outer diameter		around 2 m
Armature pole width, gap, height		12 mm, 12 mm, 10 mm
Air gap length		2 mm
Excitation	Number of turns	7000 turns
	Winding diameter, length	1 mm, 4400 m
	Winding resistance (r_1)	94 Ω
	Winding reactance (x_1)	84 kΩ
	Input DC voltage	43.2 V
Armature	Number of turns/pole	24 turns/pole
	Winding diameter/length	2 mm, 300 m
	Winding resistance (r_{2w})	1.6 Ω
	Winding reactance (x_2)	3.9 Ω
Load resistance (r_{2L})		30.0 Ω

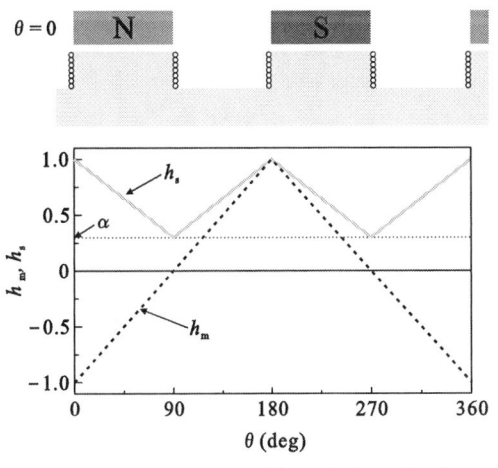

Fig. 9. The change of h_s and h_m in Design 1

Relations between h_s, h_m and θ are shown in Fig. 9. The self reactance reaches maximum at $\theta = 0$ and 180 degree and minimum at $\theta = 90$ and 270 degree. The mutual reactance is minimum at $\theta = 0$ degree and maximum at $\theta = 180$ degree. At $\theta = 90$ and 270 degree, the total flux from excitation coil into excitation coil is 0 so the mutual reactance is also 0. Parameter α is the minimum self reactance coefficient where each rotor pole is in between two armature poles (the magnetic reluctance of the air gap is maximum). The reactance coefficient α and β are 0.3 and 0.8 (estimated).

Figure 10 shows the analysis result of Design 1. To achieve 300 W with maximum of flux density of 0.6 T, the DC excitation voltage must be 43.2 V. Copper loss in excitation and armature coil is 23 W and 16 W respectively. The total copper loss is 39 W (about 10 % of the output).

According to Fig. 10, the excitation current contains DC component around 0.46 A, and ripples when θ changes. The output voltage and current contain second harmonic component. It can be seen that the linkage fluxes in the excitation coil λ_1 is nearly constant because the winding reactance in the excitation coil is large (84 kΩ). This means there is not much eddy current loss in the iron core of the excitation coil. In contrast, the linkage flux in the armature coil λ_2 varies with θ. Hence laminated iron plates are needed for armature coils.

2) Design 2

In Design 2, we changed the armature pole width, gap and height to 16 mm, 8 mm and 12 mm, respectively in order to increase α to 0.5, but kept β and the pole number at 0.8 and 240. All parameters are shown in Table 2.

Relations between h_s, h_m and θ are shown in Fig. 11. It should be noted that the shape of h_s and h_m are changed due to changes of pole size. The self reactance still reaches maximum at $\theta = 0$ and 180 degrees. However, from $\theta = 60$ degrees to 120 degrees the area of air gap that the flux from the excitation coil travels through does not change and remains minimum (parameter α is now 0.5). The mutual reactance change is also different from Design 1 but the position of maximum, minimum and 0 crossing all remain the same.

Figure. 12 shows the analysis result of Design 2. To achieve 300 W with maximum of flux density of 0.6 T, the DC excitation voltage must be 41.0 V. Copper loss in excitation and armature coil is 19 W and 9 W respectively. The total copper loss is 28 W (better than in Design 1).

According to Fig.12, the performance of OZR is improved because the AC component in the excitation coil and harmonic component in the armature coil are reduced. The excitation current contains a DC component around 0.44 A. The ripple of it is smaller than in Design 1. The linkage flux in the excitation coil λ_1 is bigger because the pole size is now bigger.

1105

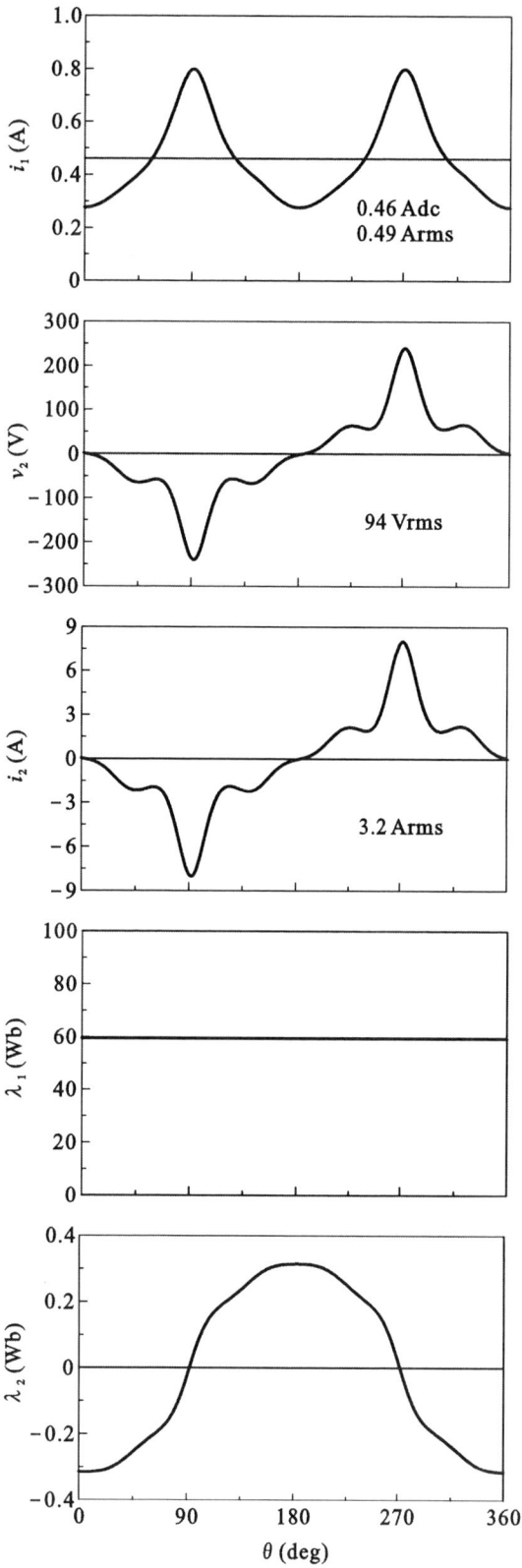

Fig. 10. Analysis result of OZR Generator Design 1

Table 2 Parameters of Design 2

Parameter		Value
Open circuit output voltage		100 Vrms
Rated output		300 W, 60 Hz
Rated speed		30 rpm
Number of poles		240 poles
Rotor outer diameter		around 2 m
Armature pole width, gap, height		16 mm, 8 mm, 12 mm
Air gap length		2 mm
Excitation	Number of turns	7000 turns
	Winding diameter, length	1 mm, 4400 m
	Winding resistance (r_1)	94 Ω
	Winding reactance (x_1)	134 kΩ
	Input DC voltage	41.0 V
Armature	Number of turns/pole	11 turns/pole
	Winding diameter/length	2 mm, 180 m
	Winding resistance (r_{2w})	1.0 Ω
	Winding reactance (x_2)	1.3 Ω
Load resistance (r_{2L})		31.3 Ω

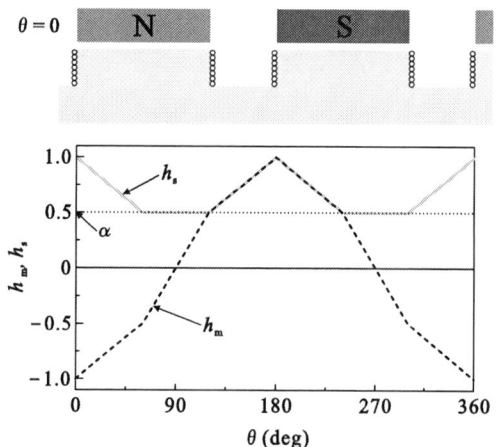

Fig. 11. The change of h_s and h_m in Design 2

V. Conclusions

The authors have proposed a new vertical axis 2-meter diameter wind generator system as a combination of a Sail Turbine and an Outer Zipper Rotor Generator (OZR Generator).

A model of the Sail Turbine was made and tested under weak wind. The experiment showed that the Sail Turbine can catch weak wind and rotate well.

On the other hand, the feasibility of a multi-poles, brushless, gearless, single excitation OZR generator was studied. Two designs for the generator were analyzed using Fourier Expansion. Analytical results show that the output target of 300 W, 100 V, 60 Hz can be achieved. The performance of the OZR generator can be improved by adjusting the armature pole size.

The proposed generation system is expected to have low investment and maintenance cost. It is designed to operate under weak wind in urban areas.

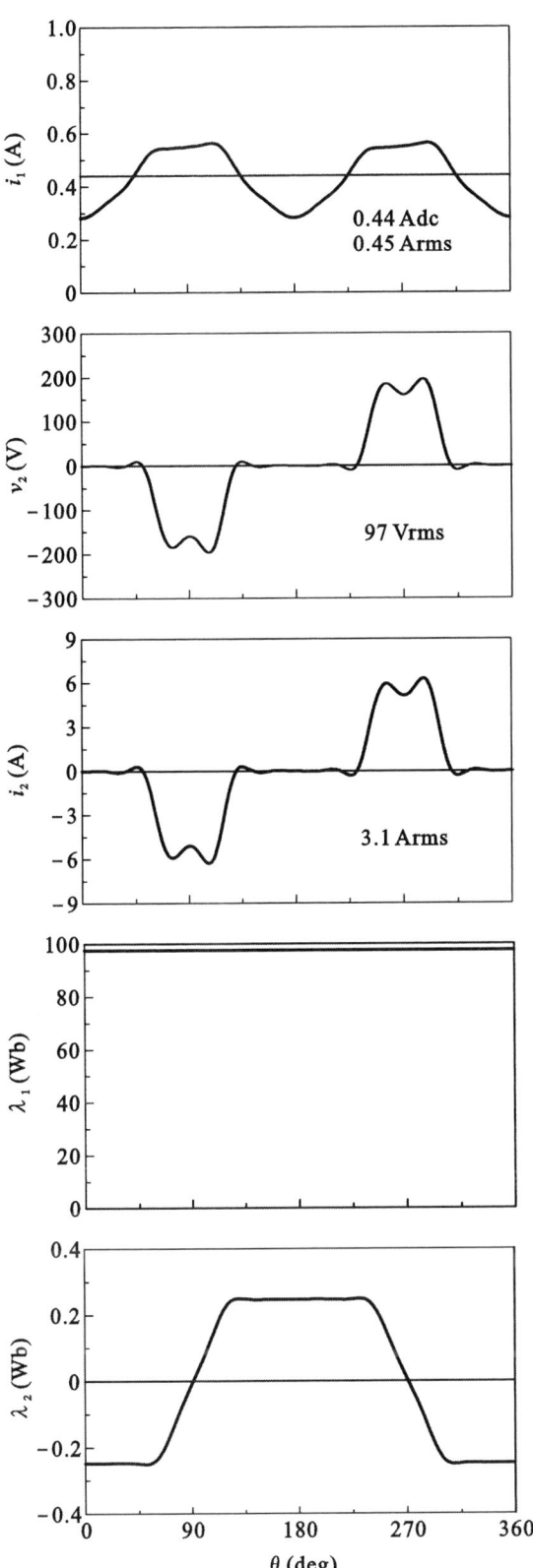

Fig. 12. Analysis result of OZR Generator Design 2

APPENDIX

The variable i_1, i_2 and coefficient h_s, h_m can be written using Fourier Expansion as

$$i_1(\theta) = \frac{1}{N}\left(y_1 + y_2 e^{i\theta} + \ldots + y_{N-1}e^{-2i\theta} + y_N e^{-i\theta}\right) \quad (A1)$$

$$i_2(\theta) = \frac{1}{N}\left(z_1 + z_2 e^{i\theta} + \ldots + z_{N-1}e^{-2i\theta} + z_N e^{-i\theta}\right) \quad (A2)$$

$$h_s(\theta) = \frac{1}{N}\left(A_1 + A_2 e^{i\theta} + \ldots + A_{N-1}e^{-2i\theta} + A_N e^{-i\theta}\right) \quad (A3)$$

$$h_m(\theta) = \frac{1}{N}\left(B_1 + B_2 e^{i\theta} + \ldots + B_{N-1}e^{-2i\theta} + B_N e^{-i\theta}\right) \quad (A4)$$

If we replace (A1) to (A4) into (12) and (13) and compare the frequency we have the following simultaneous equation

$$\begin{pmatrix} x_1\tilde{A}^* + r_1 D^* & x_m \tilde{B_1}^* \\ x_m \tilde{B_2}^* & x_2\tilde{A} + r_2 D \end{pmatrix}\begin{pmatrix} Y^* \\ Z \end{pmatrix} = \begin{pmatrix} C_1 \\ C_2 \end{pmatrix} \quad (A5)$$

where

$$\tilde{A} = \frac{1}{N^2}\begin{bmatrix} A_1 & A_N & A_{N-1} & \cdots & A_3 & A_2 \\ A_2 & A_1 & A_N & \cdots & A_4 & A_3 \\ \vdots & \vdots & \vdots & & \vdots & \vdots \\ A_{N-1} & A_{N-2} & A_{N-3} & \cdots & A_1 & A_N \\ A_N & A_{N-1} & A_{N-2} & \cdots & A_2 & A_1 \end{bmatrix} \quad (A6)$$

$$\tilde{A}^* = \frac{1}{N^2}\begin{bmatrix} A_1 & A_N & A_{N-1} & \cdots & A_3 & A_2 \\ A_2 & A_1 & A_N & \cdots & A_4 & A_3 \\ \vdots & \vdots & \vdots & & \vdots & \vdots \\ A_{N-2} & A_{N-3} & A_{N-4} & \cdots & A_N & A_{N-1} \\ A_{N-1} & A_{N-2} & A_{N-3} & \cdots & A_1 & A_N \end{bmatrix} \quad (A7)$$

$$\tilde{B_1}^* = \frac{1}{N^2}\begin{bmatrix} B_2 & B_1 & B_N & \cdots & B_4 & B_3 \\ B_3 & B_2 & B_1 & \cdots & B_5 & B_4 \\ \vdots & \vdots & \vdots & & \vdots & \vdots \\ B_{N-1} & B_{N-2} & B_{N-3} & \cdots & B_1 & B_N \\ B_N & B_{N-1} & B_{N-2} & \cdots & B_2 & B_1 \end{bmatrix} \quad (A8)$$

$$\tilde{B_2}^* = \frac{1}{N^2}\begin{bmatrix} B_N & B_{N-1} & B_{N-2} & \cdots & B_3 & B_2 \\ B_1 & B_N & B_{N-1} & \cdots & B_4 & B_3 \\ \vdots & \vdots & \vdots & & \vdots & \vdots \\ B_{N-2} & B_{N-3} & B_{N-4} & \cdots & B_1 & B_N \\ B_{N-1} & B_{N-2} & B_{N-3} & \cdots & B_2 & B_1 \end{bmatrix} \quad (A9)$$

$$D = \frac{-i}{N}\begin{bmatrix} 0 & 0 & 0 & \cdots & 0 & 0 \\ 0 & 1 & & \cdots & 0 & 0 \\ & & 1/2 & \cdots & 0 & 0 \\ \vdots & \vdots & \vdots & & \vdots & \vdots \\ 0 & 0 & 0 & \cdots & -1/2 & 0 \\ 0 & 0 & 0 & \cdots & 0 & -1 \end{bmatrix} \quad (A10)$$

$$D^* = \frac{-i}{N}\begin{bmatrix} 1 & 0 & 0 & \cdots & 0 & 0 \\ 0 & 1/2 & 0 & \cdots & 0 & 0 \\ 0 & 0 & 1/3 & \cdots & 0 & 0 \\ \vdots & \vdots & \vdots & & \vdots & \vdots \\ 0 & 0 & 0 & & -1/2 & 0 \\ 0 & 0 & 0 & & 0 & -1 \end{bmatrix} \qquad (A11)$$

$$C_1 = \frac{-x_1}{N^2}\begin{bmatrix} A_2 y_1 & A_3 y_1 & \cdots & A_{N-1} y_1 & A_N y_1 \end{bmatrix}^T \qquad (A12)$$

$$C_2 = \frac{-x_m}{N^2}\begin{bmatrix} B_1 y_1 & B_2 y_1 & \cdots & B_{N-1} y_1 & B_N y_1 \end{bmatrix}^T \qquad (A13)$$

$$Y^* = \begin{bmatrix} y_2 & y_3 & \cdots & y_{N-1} & y_N \end{bmatrix}^T \qquad (A14)$$

$$Z = \begin{bmatrix} z_1 & z_2 & \cdots & z_{N-1} & z_N \end{bmatrix}^T \qquad (A15)$$

It should be noted that the DC component y_1 is computed directly from

$$y_1 = \frac{N V_1}{r_1} \qquad (A16)$$

Assume that the changes of h_s and h_m are known, the parameters A_1, A_2, ... A_N and B_1, B_2, ... B_N can be calculated using Fourier Expansion. Therefore, Y^* and Z can be calculated by solving simultaneous equation (A5), then i_1 and i_2 can be solved.

ACKNOWLEDGEMENT

The authors wish to thank Dr. Satoru Ihara, a retired Professor of Tokyo Institute of Technology, for his valuable suggestions.

REFERENCES

[1] UNITED NATION: Kyoto Protocol to The United Nations Framework Convention on Climate Change (1998)

[2] S. Bhowmik, R. Spee and Johan. H. R. Enslin, "Performance Optimization of Double Fed Wind Power Generation Systems", *IEEE Trans. Industry Applications*, vol. 35, no. 4, pp. 949-958, Jul/Aug. 1999.

[3] M. A. Mueller, "Design of Low Speed Switched Reluctance Machines for Wind Energy Converters", *9th International Conference on Electrical Machines and Drives*, 1999.

[4] E. Muljadi, and C.P. Butterfield, "Pitch-Controlled Variable-Speed Wind Turbine Generation", *IEEE Trans. Industry Applications*, vol. 37, no. 1, pp. 240-246, Jan/Feb. 2001.

[5] T. Kinjo, T. Senjyu, N. Urasaki and H. Fujita, "Output Levelling of Wind Power Generation System by EDLC Energy Storage", *The 30th Annual Conference of the IEEE Industrial Electronics Society*, 2004.

[6] Choom poo-inwai, WJ Lee, P. Fuangfoo, M. Williams, and J. R. Liao, "System Impact Study for the Interconnection of Wind Generation and Utility System" *IEEE Trans. Industry Applications*, vol. 41, no. 1, pp. 163-168, Jan/Feb. 2005

[7] M. Karrari, W. Rosehart, and O. P. Malik, "Comprehensive Control Strategy for a Variable Speed Cage Machine Wind Generation Unit", *IEEE Trans. Energy Conversion*, vol. 20, no. 2, pp. 415-423, Jun. 2005.

[8] G. Lalor, A. Mullane, and M. O'Malley, "Frequency Control and Wind Turbine Technologies", *IEEE Trans. Power Systems*, vol. 20, no. 4, pp. 1905-1913, Nov. 2005.

[9] World Wide Fund for Nature Japan http://www.wwf.or.jp

[10] S. J. Chapman, *Electric Machinery Fundamentals*, McGRAW Hill International Edition, 2005.

[11] A. E. Fitzgerald, C. Kingsley Jr., S. D. Umans, *Electric Machinery*, McGRAW Hill, 2003.

Braking Circuit of Small Wind Turbine Using NTC Thermistor under Natural Wind Condition

Y. Matsui*, A. Sugawara*, S. Sato*, T. Takeda*, and K.Ogura*

* Niigata University/ Department of Electrical and Electronic Engineering, 8050 Ikarashi-2, Niigata, Japan

Abstract–An electric brake by a three-phase short-circuit system used as the brake equipment of a small wind turbine damages the rotor blades by a rapid revolution stopping of the generator. Moreover, the generators windings may also be damaged by a large short-circuit current. In this paper, the electric braking circuit using the NTC thermistors (negative temperature coefficient resistors) is proposed as a braking system for a cheaper and safe stop of the small wind turbine. The effect under the natural wind condition is examined by the field test using a 500W small wind turbine. AS a result, it is clarified that the generator can decelerates gradually and reduce the burst current by using the braking circuit that composed of the NTC thermisitors.

Index Terms—electric brake, NTC thermistors, small wind turbine, wind power

I. INTRODUCTION

Recently, constructions of wind power generation and photovoltaic energy system against environmental problems such as global warming are advanced to the use of natural energy in Japan [1]. We can see small wind turbines of about 300-500 W for environmental consciousness and stand-alone power supply systems at some parks, schools, and on streets. Small wind turbines are defined as the swept area by the rotor blades with less than 200 m^2 (the rotor diameter of 16 m) by IEC (International Electrotechnical Commission) [2,3]. A basic configuration of the small wind turbine as the block chart is shown in Fig. 1. The components are rotor blades, a generator, a control unit with a battery, and so on. Large or middle scale wind turbines with more than 20 m in height are connected with power line to provide electricity infrastructure. On the other hand, small wind turbines are aimed for the security of the stand alone power supply, and the electricity generated by them is stored in the battery and then supplied generally as shown in Fig. 1. references. The small wind turbine should have a brake system to prevent the overspeed so that an extremely large centrifugal force may work by the strong wind. In addition, the brake system is also necessary for the overcharge prevention of the battery or the maintenance.

The method of controlling or stopping the rotor revolution is divided into a mechanical brake, an aero brake, or an electric brake [4].

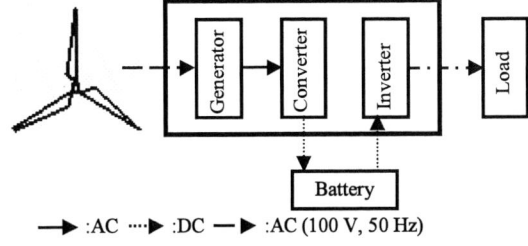

Fig. 1. Basic structure of small wind turbine

Disk brakes are generally used for the mechanical brake. The brake pad stops the revolution of the rotor blades by friction. The aero brake is divided into a pitch control to which the angle of blades is changed in the direction passed through a wind by sensing the wind velocity or the generation output, a stall control which makes a stalling state for the flow of the wind behind the blades, or a yaw control which diverts the direction of the blade revolution. The electric brake decreases the speed of the rotor revolution because of the magnetic force between the permanent magnet and the generator windings with the 3-phase short circuit. Furthermore, the control and/or braking methods combine them. However, the electric braking method causes some damages that are for the rotor blades by a rapid revolution stopping of the generator and the burnt wire of the generator windings by the large short circuit current. The conventional braking methods also make a rapid stopping of the rotor blades. In addition, the cost of establishment and maintenance is expensive. To solve these problems, it is necessary that a cheap and safe brake system stops the revolution of the generator gradually and reduces the burst current.

In this report, the electric braking circuit using NTC thermistors (negative temperature coefficient resistors) [5] is proposed as a new stop technology for the small wind turbine [6]. The effect in natural wind condition is examined by the field test. A discharge equipment that consists of three-phase resistors decreases the speed of rotor revolution more gradually than the three-phase short-circuit system. However, the rotor blades are rotating slowly, and cannot be stopped completely because the current keeps flowing. In that respect, NTC thermistor shows the initial large resistance of several ohms and then shows about zero ohms by the self Joule heating. At first, a little current flows into the generator windings, the braking torque is small, and then the

978-1-4244-0644-9/07/\$25.00 ©2007 IEEE

braking torque grows large when the resistance value of the NTC thermistors become low. They return to the original state when the current flowing into the NTC thermistor is lost and the temperature fallen. The NTC thermistors can be used repeatedly about 700 times [5]. Therefore, the number of revolution of the rotor blades decreases gradually by using NTC thermisitors, and they stop safely after a sufficient time progress.

If simple resistances are used without changing the value instead of NTCs, they work as a gently braking circuit and a sudden decrease in the revolution of the blades can be controlled. But the rotor blades continue rotating slowly and cannot be stopped completely since the resistance value does not change and the windings of the generator do not become state of the short circuit. To solve this problem, an attached control circuit that changes resistance small gradually is needed. However the attached control circuit makes the braking circuits complex, and cause the increase of power consumption by the control circuit. On the other hand, the braking circuit using NTCs proposed in this paper doesn't need the attached control circuit because the value of resistance decreases automatically by self-Joule-heating. Therefore, the braking circuit using NTCs can be composed simpler than using resistances.

II. ELECTRICAL BRAKING CIRCUIT USING NTC THERMISTORS

A. NTC thermistor

NTC thermisitor is a semiconductor device that has the initial resistance of several ohms, and the value of resistance of several ohms, and the value of resistance decreases according to the temperature rise exponentially. In this study, NTC thermistor "1R30A" is used from the advantages that the maximum current is large and the initial resistance of the braking circuit is able to be changed easily by their series connection. The specification of 1R30A is shown in Table I. Figure 2 shows the temperature characteristic.

TABLE I
SPECIFICATION OF 1R30A

Registance [Ω] at 25°C	1
Maximum current [A]	30
Registance [Ω] at Max current	0.03

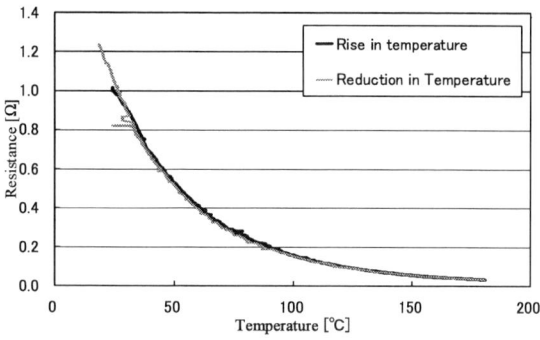

Fig. 2. Temperature characteristic of 1R30A

In Fig. 2, the value of resistance of 1R30A decreases from 1 ohm in initial to about 0.02 ohms with rising the temperature. And it shows that 1R30A returns to the original state when its temperature falls to the previous temperature. In addition, the unit price of 1R30A is cheap with 2.1 US$.

B. The structure of Electric braking circuit using NTC thermistors

The short circuit current at the rated speed of the 500 W small wind turbine is 24 A. Among the available NTC thermistors, only 1R30A (Resistance at 25 °C: 1Ω, Rated current: 30 A) stands a maximum current more than 24 A. The braking circuit is comprised by a series-connected some 1R30A to each phase of a three-phase with Y-connection. The rotor blades can be decelerated more gradually for an advantage to connect the larger number of NTCs in series because the initial resistance value of each phase increases and then the electric current flowing into the circuit is reduced. Moreover, it is also the advantage that we can produce the braking circuit with a different initial resistance value as it has the large maximum current of 30 A. The name of the braking circuit is written "1R30A×X" in this report. The X is the quantity of 1R30A which is connected to each phase in series.

III. EXPERIMENTAL SETUP

In this research, the electric braking circuit using NTC thermistors as a cheap and safe stop technology for a small wind turbine is proposed, and the effects using the 500 W small wind turbine are examined by the field test in Niigata city, Japan.

A. Specification of the small wind turbine

The small wind turbine is shown in Fig. 3. Table II shows the specification. The inertia moment of the rotor blades is 2.88 kg · m². A synchronous generator, an induction generator, and a permanent-magnetic generator are usually used for a wind power generation. The rate of three phase's permanent-magnetic generator with exciting arrangement which can generate electricity independently to small wind turbines is increasing. Therefore, the 500 W

Fig. 3. Small wind turbine

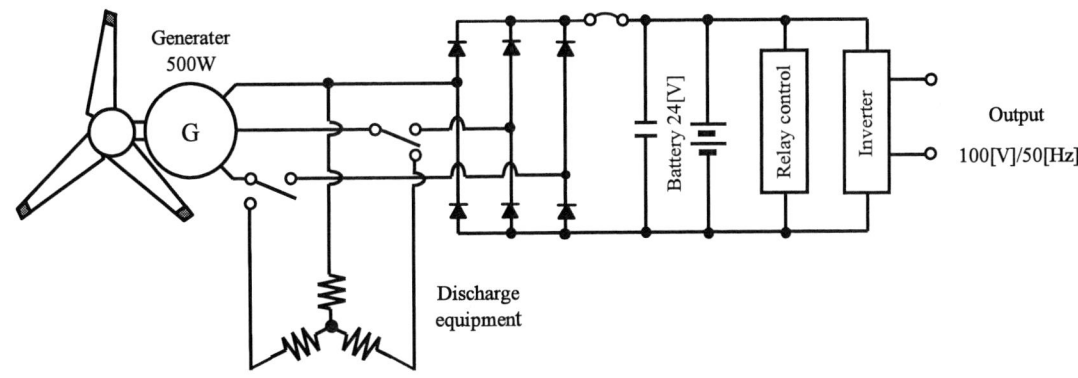

Fig. 4. Control system of the small wind turbine

TABLE II
SPECIFICATION OF THE SMALL WIND TURBINE

Model	FD2.5-500
Type of generator	Permanent-magnetic Synchronous generator
Rotor diameter [m]	2.5
Rated power [W]	500
Rated wind velocity [m/s]	8
Cutin wind velocity [m/s]	3
Cutout wind velocity [m/s]	25
Height of tower [m]	5.5

propeller shape small wind turbine which has this three phase's permanent-magnetic generator is targeted in the field test in this paper.

B. Control system of the small wind turbine

Figure 4 shows the controller of the small wind turbine used by this research. The 500W generator G generates electric power by the revolution of the rotor blades as a 3-phase alternating output power. The output power is rectified and charged a battery [7]. The electric power stored with the battery temporarily is transformed into the commercial power supply of 100 V/50 Hz with a single phase in Japan by the inverter. Figure 5 shows the operation model of the controller. Calm, the battery voltage of 20 V, and the electric supply stop state to the load are assumed as an initial state. If a wind blows over the cutin speed, the generator G generates. The battery charge starts, when the voltage exceeds the battery charge start voltage of 24 V. When the battery voltage exceeds the electric supply voltage of 28 V, the inverter works and the system will begin to supply to the load. If the battery charge continues and the battery voltage reaches up to 30 V, the discharge equipment is connected to protect the battery by the relay. After the battery is consumed until the voltage of 28 V, the discharge equipment turns off and the battery is recharged. And the flow chart is repeated. On the other hand, if the wind stops, the electric power of the battery is consumed by the load and the battery voltage falls. If the battery voltage is less than the battery protection voltage of 20 V, the electric power stops to supply to the load.

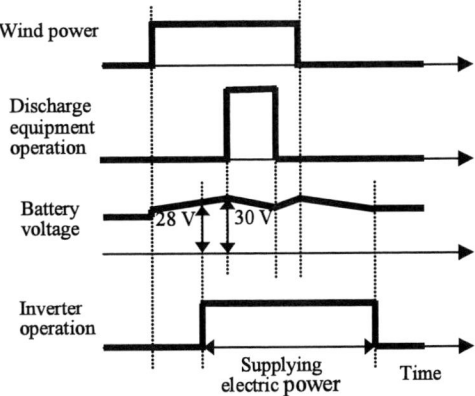

Fig. 5. Operation model of control system

For the control of small wind turbines, when the battery voltage exceeds the battery protection voltage and/or a wind velocity exceeds the cutout wind velocity, it is necessary to stop the revolution of the rotor blades safely.

In this research, the number of revolution of the rotor blades is decreased by connecting the braking circuit using NTC thermistors to the generator output terminal when the wind velocity exceeds the cutout wind velocity of 25 m/s or the battery voltage exceeds 30 V.

C. Experiment on braking characteristic for small wind turbine

A schematic diagram of the experimental device is shown in Fig. 6. In Fig. 6, the generator output of the small wind turbine is opened and the rotor blades are rotated. When the number of revolution of the rotor blades has reached to arbitrary number, the switch SW is connected to the A-circuit formed with the NTC thermistors or the B-circuit. The line current, the number of revolution, and the variation of the wind velocity are measured. In this experiment, the quantity of 1R30A which is connected to each phase in series is changed up to 5, and the braking circuit with the different resistance is examined. In addition, the SW is connected with B-circuit as a comparison to the conventional 3-phase short-circuit brake.

The definition of the relaxation time is shown in Fig. 7.

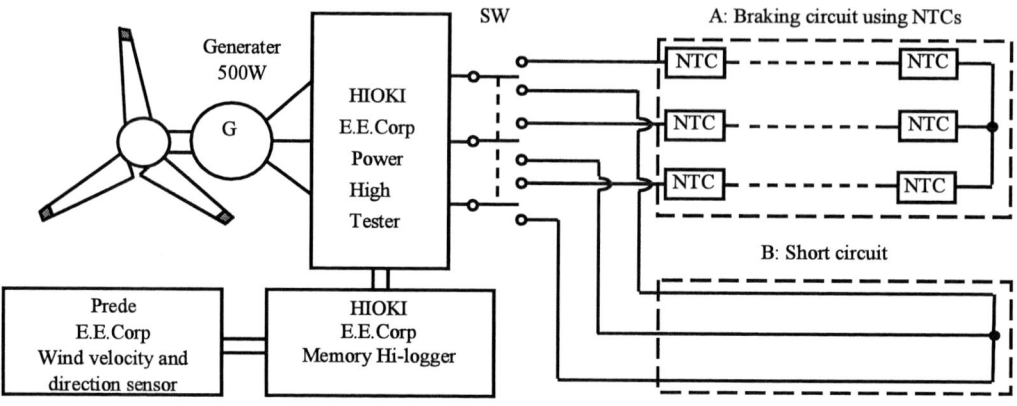

Fig. 6. Schematic diagram of experimental device

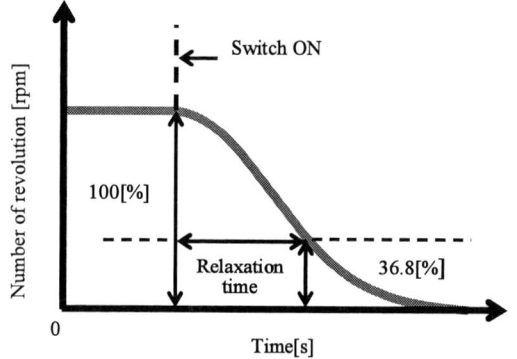

Fig. 7. Relaxation Time

In this paper, the relaxation time for braking of the generator is defined as follows, and it is measured. After the switch SW turns ON, the number of revolution decreases exponentially. The relaxation time is defined as time until it decreases to 1/e =36.8 % for the initial number of revolution of the generator defined as 100 %.

IV. EXPERIMENT RESULT AND DISCUSSIONS

The waveforms of the number of revolution and the line current for the short circuit and the braking circuit using NTCs (1R30A×3) are shown in Figs. 8 and 9, respectively. The initial number of revolution of the generator is 360 rpm. Figure 8 shows that the number of revolution decreases rapidly and the large line current flows suddenly soon after the short circuit brake is used. It is the reason that the large short circuit current flows to the generator windings and the load to the generator has entered an overloaded state.

On the other hand, figure 9 shows that the number of revolution decreases gradually for the braking circuit using NTCs as compared with the short circuit brake and the maximum value of the line current can be reduced. It is reason that the larger line current doesn't flow to the generator windings rapidly and the load to the generator does not affect the revolution of the generator suddenly

The waveforms of the number of revolution and the line current using the discharge equipment connected with the generator output are shown in Fig. 10.because the initial resistance of NTC thermistors works as a load. The number of revolution and the line current rise and fall with the variation of the wind velocity. The rotor blades cannot be decelerated enough under the natural wind condition with the discharge equipment. Therefore, it is effective that the braking circuit using NTC thermistors is used for the controller to prevent the overcharge of the battery

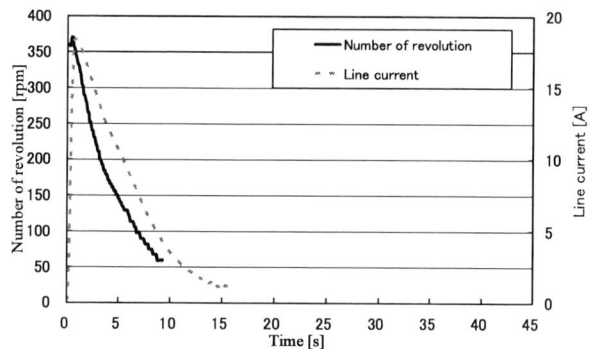

Fig. 8. Number of revolution and line current
(Short circuit, average wind velocity: 5m/s)

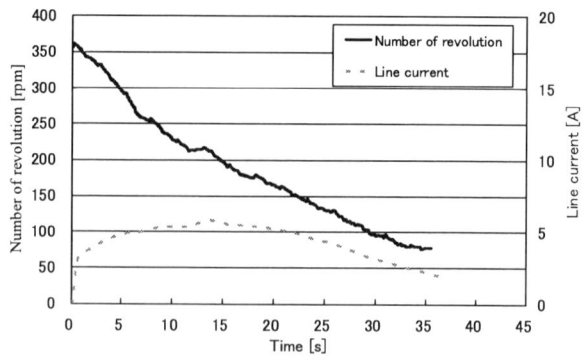

Fig. 9. Number of revolution and line current
(1R30A×3, average wind velocity: 5m/s)

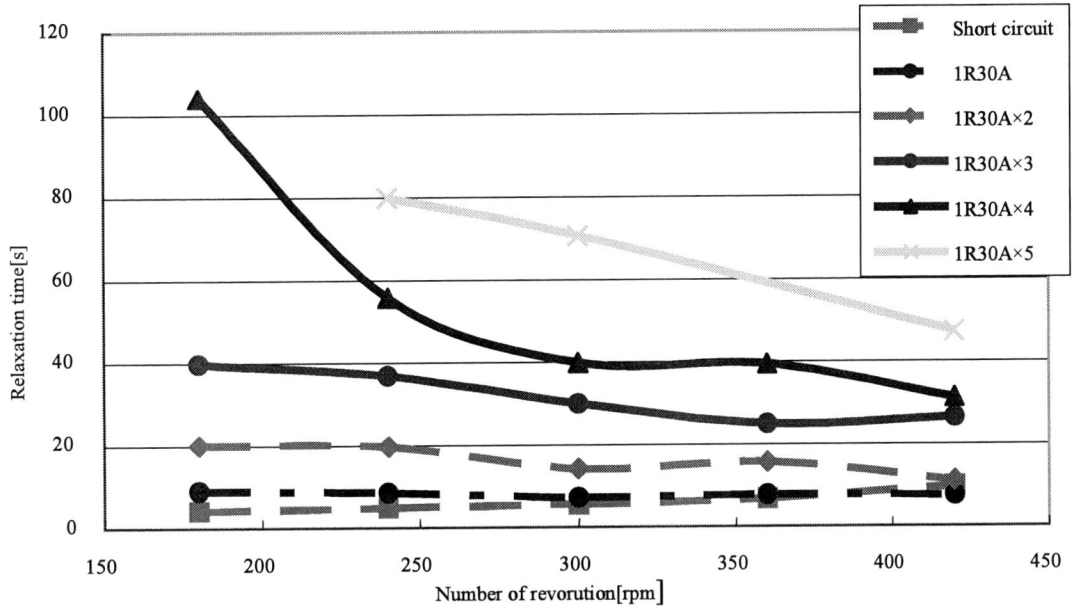

Fig. 12. Schematic diagram of experimental device

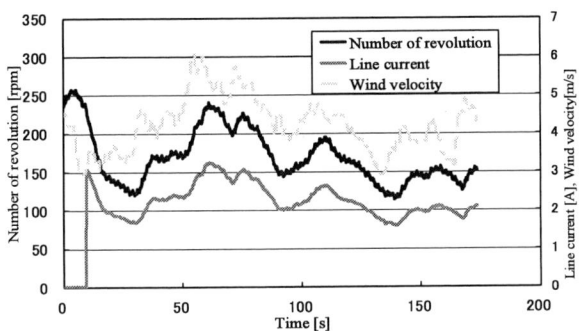

Fig. 10. Number of revolution and line currrent.
(Discharge equipment, average wind velocity: 4.2m/s)

The maximum current flowing into the generator windings after braking at the arbitrary number of revolution is shown in Fig.11. It is possible that the current value flowing into the generator windings is reduced by the braking circuit using NTCs compared with the short circuit brake.

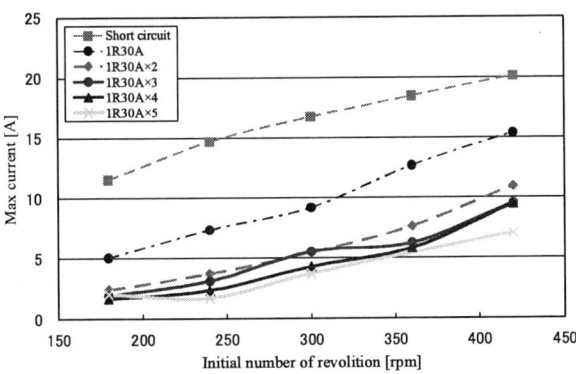

Fig. 11. Maximum current.

The relaxation time in the range of 0-420 rpm is shown in Fig.12. The relaxation time becomes long by the braking circuit using NTCs compared with the short-circuit brake. In other words the revolution of the generator has decelerated gradually.

V. CONCLUSIONS

Characteristics on the electric braking system for the 500W small wind turbine are measured by the field test. By inserting the braking circuit comprised by the Y-connection NTC thermistors into the generator output terminal, the gently braking of the rotor blades is possible. In addition, the braking circuit can reduce the current value flowing into the generator windings. Therefore, the braking circuit using NTCs can brake the rotor blades safely and is cheap compared with the conventional braking method of the small wind turbine.

ACKNOWLEDGEMENT

This study was supported by Foundation of The Institute of Environmental Geological Sciences. Authors thank to Mr. Teruya Nakayama (the chairman of the board of directors) and the laboratory all staffs.

1113

REFERENCES

[1] G. Boyle: "Renewable Energy Power for a Sustainable Future," *Oxford University Press*, pp. 19-21, 1996, the United Kingdom.

[2] P. Gipe: "WIND POWER FOR HOME & BUSINESS," *Chelsea Green Publishing Vermont*, pp. 133-146, 1993, America

[3] W. Kellogg, M. H. Nehrir, G. Venkataramanan, V. Gerez, "Optimal unit sizing for a hybrid wind/photovoltaic generating syatem," *Electric Power System Research*, vol. 39, 1996, pp. 35-38.

[4] T. Uie, A. Sugawara, K. Itagaki, H. Kitamura, H. Kaizu, "High Efficiency Operating of Small Wind Turbine Controlled by Resistance Load," *JAPAN WIND ENERGY ASSOCIATION*, vol. 21, No.1, 1997, pp. 45-50.

[5] O. Mrooz, A. Kovalski, J. Pogorzelska, O. shpotyuk, M. Vakiv, B. Butkiewicz, J. Maciak,"Thermoelectrical degradation processes in NTC thermistors for in-rush current protection of electronic circuits," *Microelectronics Reliability*, vol. 41, 2001, pp. 773-777.

[6] Ziyad M.salamah, I. Safari, "The Effect of the Windmill's Parameters on the Capacity Factor" *IEEE Transactions on Energy Conversion. Japan*, vol. 10, pp. 747-751, December 1995.

[7] T. Wakui, K. Yamaguchi, T. Hashizume, E.Outa, Y.Tanzawa, "Effect of Operating Methods of Wind Turbine Generator System on Net Power Extration under Wind Velocity Fluctuations in Fields." *Microelectronics Reliability*, vol. 16, 1999, pp. 843-846.

Flywheel Energy Storage Drive for Wind Turbines

K. Veszprémi and I. Schmidt
Budapest University of Technology and Economics
Department of Electric Power Engineering
Egry J. u. 18. Budapest H–1111 Hungary
E–mail: veszpremi@vet.bme.hu; URL: http://www.vet.bme.hu

Abstract – **The main problem of the wind power is its stochastic availability. The pulsation of the wind speed causes power pulsation, resulting in deterioration of the power quality. To compensate it, energy storage is necessary. Considering the wind spectrum, different storage systems can be used for the different frequencies of the wind speed variation. The short time turbulent power pulsation can be balanced by flywheel energy storage drive. This paper deals with this method. After the theoretical considerations of the system the possible power circuits and control methods are discussed. The operation of the system is investigated by simulations, using acceptable simplifications. The results help to design the overall system and its elements.**

Index Terms – **Energy storage, Flywheel, IGBT converter, Wind turbine.**

I. INTRODUCTION

The electric power provided by a wind turbine is highly dependent on the wind speed, approximately proportional to its cube: $p=Cv^3$. Relatively small variation of the wind speed may produce power fluctuation, resulting in voltage pulsation, when the electrical energy is supplied to a stand–alone load or weak grid [4]. Van der Hoven determined the wind spectrum (Fig.1.) by a model taking into account the stochastic property of the wind speed [1]. It is applied widely. The few days variation of the wind speed (synoptic peak) caused by the weather fronts and the daily variation depending on the location (diurnal peak) can be forecasted by the meteorology. The few minutes variations (turbulent peak) caused by the wind turbulence are also important. These affect significantly the quality of the power provided by the wind turbine. The few day power variation e.g. can be balanced by hydrogen based energy storage, while the short time turbulent power pulsation e.g. can be balanced by flywheel energy storage drive [2]. This paper deals with the later one.

The flywheel energy storage is rediscovered nowadays for wide range of powers due to its advantages over the other energy storage systems [2], [4], [5]: reliability, long–life, cost, fast response.

The flywheels can be classified in many respects:

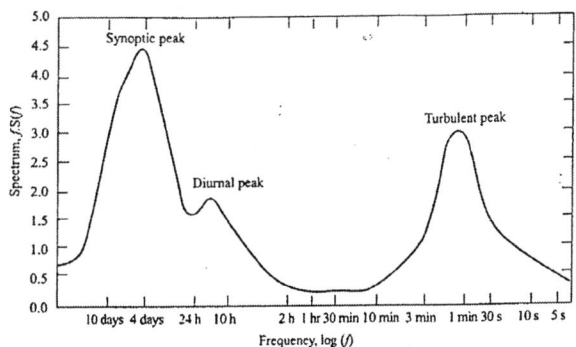

Fig.1. The wind spectrum by Van der Hoven [1].

- Considering the maximal speed, low–speed ($\omega_{FWmax}<10000$rpm) and high–speed ($\omega_{FWmax}>10000$rpm) flywheels can be distinguished. High–speed flywheels are mainly used in embedded applications (transport, satellites) because of their much lower weight [5].
- The material of the flywheel traditionally is steel or nowadays is steel–plastic composite for the high–speed applications.
- The gearless direct driven flywheel systems are preferred considering the losses.
- The type of the bearing is mainly determined by the maximal speed.
- The driving electrical machine can be: squirrel–cage induction machine, brushless excitation synchronous machine, permanent magnet synchronous machine or switched reluctance machine.

The electrical part of the flywheel drive is investigated mainly in this paper. First the aim, the mechanics and the operating range of the flywheel drives are investigated on system level. Then the possible power circuits and control methods are discussed. The operation of the proposed circuit and method is investigated by simulations using appropriate simplifications. The results help to design the overall system and its elements.

II. FLYWHEEL ENERGY STORAGE DRIVE

The flywheel energy storage system applies the kinetic energy (E_{FW}) of a ω_{FW} angular speed rotating mass with J_{FW} inertia. The maximal kinetic energy corresponds to the maximal angular speed:

This work was supported by the Hungarian N.Sc. Found (OTKA No. T046916) for which the authors express their sincere gratitude.

978-1-4244-0644-9/07/$25.00 ©2007 IEEE

$$E_{FW} = \frac{1}{2} J_{FW} \omega_{FW}^2 , \quad E_{FW\,max} = \frac{1}{2} J_{FW} \omega_{FW\,max}^2 \qquad (1)$$

To utilize the kth part of the E_{FWmax} energy, the following expressions can be written:

$$\Delta E_{FW\,max} = E_{FW\,max} - E_{FW\,min} = k E_{FW\,max} , \qquad (2)$$

$$E_{FW\,min} = \frac{1}{2} J_{FW} \omega_{FW\,min}^2 = (1-k) E_{FW\,max} , \qquad (3)$$

$$\omega_{FW\,min} = \sqrt{1-k}\, \omega_{FW\,max} . \qquad (4)$$

Example usual practical values are: $k=0.9$, $\omega_{FWmin}=0.316\omega_{FWmax}$, when the 90% of the stored energy can be utilized.

The kinetic energy can be controlled by the m_{FW} torque (in fact by the p_{FW} power) of the electrical drive of the flywheel:

$$p_{FW} = -\frac{dE_{FW}}{dt} = -m_{FW}\omega_{FW} . \qquad (5)$$

The power supplied to the lines (generator power) is considered to be positive (see. Fig.4. and Fig.5. for the positive directions of the powers). At deceleration, decreasing the ω_{FW} speed (discharging) energy is taken out, at acceleration, increasing the ω_{FW} speed (charging) energy is supplied to the flywheel. An advanced flywheel drive system (Fig.2a.) consists of a flywheel (FW), an electrical machine (EM), a power electronics unit (PE) and the electrical supply lines (L). J_{FW} is the resultant inertia. The EM, PE and L units must be capable of bidirectional power flow. The electrical machine (EM) is in motor mode at charging ($p_{FW}<0$) and in generator mode at discharging ($p_{FW}>0$).

The usual operating range of the PE+EM electrical drive is presented in Fig.2b. on the ω_{FW}–m_{FW} plane. In the $\omega_{FWmin}\leq\omega_{FW}\leq\omega_{FWmax}$ normal operating range the maximal power demand is $-P_{FWmax}$ at charging and $+P_{FWmax}$ at discharging. The maximal driving torque of the drive is $M_{FWmax}=P_{FWmax}/\omega_{FWmin}$ and the maximal braking torque is $-M_{FWmax}$. It can be established that the flywheel drive is a monodirectional two quadrant drive and its normal operating range is the field weakening. This also has the advantage of reducing iron losses and improving efficiency. The nominal motor mode operating point of the drive should be set to point 2, where: $M_{FWn}=M_{FWmax}$, $\omega_{FWn}=\omega_{FWmin}$ and $P_{FWn}=-M_{FWn}\omega_{FWn}=-P_{FWmax}$.

Assuming lossless conditions, the drive can accelerate the flywheel from standstill to ω_{FWmax} (charge it by E_{FWmax} kinetic energy through points 1–2–3) in time:

$$T = T_{FWstn} \frac{2-k}{2-2k} . \qquad (6)$$

The $T_{FWstn} = \dfrac{J_{FW}\omega_{FWn}}{M_{FWn}} = C T_{EMstn}$ is the nominal starting time of the drive, which is $C=J_{FW}/J_{EM}$ (it is the ratio of the inertias) times grater than the T_{EMstn} nominal starting time of the electrical machine.

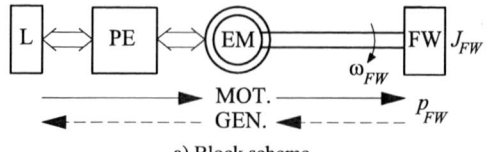

a) Block scheme.

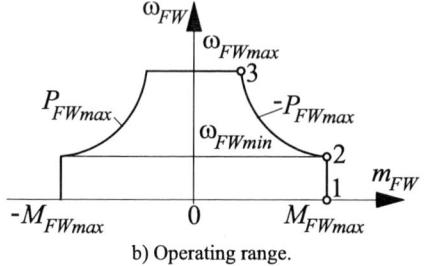

b) Operating range.

Fig.2. Advanced flywheel drive system.

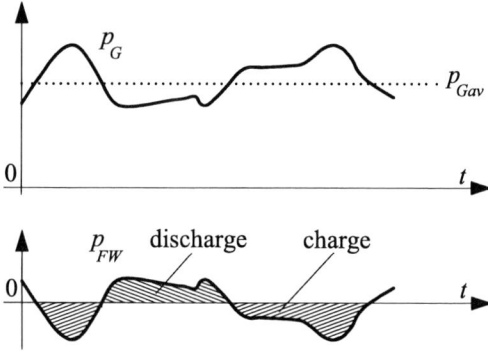

Fig.3. The wind turbine (generator) power (p_G) and the flywheel power (p_{FW}).

The T_{FWstn} part of T is the acceleration time from zero to $\omega_{FWn}=\omega_{FWmin}$ angular speed (from point 1 to 2), the $T-T_{FWstn}$ part is the acceleration time from ω_{FWmin} to ω_{FWmax} angular speed (from point 2 to 3). E.g. if $k=0.9$, $T=5.5T_{FWstn}$. The maximal time interval to input or output $\pm P_{FWn}=\pm P_{FW\,max}$ power by the flywheel drive during its operation is:

$$\Delta T = T - T_{FWstn} = \frac{k}{2-2k} C T_{EMstn} . \qquad (7)$$

E.g. if $T_{EMstn}=1$s, $k=0.9$ and $C=20$, then $\Delta T=90$s. Accordingly the flywheel drive can balance only the short time power pulsation of the wind turbine.

The task of the flywheel drive is to balance the pulsation of the wind turbine power (p_G) caused by the turbulence. According to Fig.3. the necessary flywheel power is $p_{FW}=p_{Gav}-p_G$. The p_{FW} power must remain in the $\pm P_{FWmax}$ interval, while the energy fluctuation got by the integral of the power ($\Delta E_{FW} = \int p_{FW} dt$) must remain in the $\pm\Delta E_{FWmax}$ interval. Satisfying these two conditions, the flywheel operates in the $\omega_{FWmin}<\omega_{FW}<\omega_{FWmax}$ angular speed range and capable of balancing the pulsation of the p_G wind turbine power.

The practically frequently applied voltage source inverter–fed squirrel–cage induction machine flywheel (IFW) drive is investigated in the following [3], [4].

III. THE POWER CIRCUITS

The application of the flywheel drive is demonstrated by assuming a wind turbine with permanent magnet synchronous generator (SG in Fig.4. and Fig.5.). Both the machine side (CFWM) and line side (CFWL) converters are two–level voltage source converters. Applying IGBT switches and pulse width modulation (PWM), $f_{FWmax} \approx 1000$Hz is accessible (as an example: with a two–pole machine it means that $n_{FWmax} \approx 60000$rpm speed is reachable).

In the first version (Fig.4.) the power circuit of the flywheel drive is totally independent of the power circuit of the wind turbine generator drive, since it is connected directly to the $f_L = 50$Hz frequency three–phase lines. In this case the CFWM and CFWL converters must be designed to the P_{FWmax} power of the flywheel.

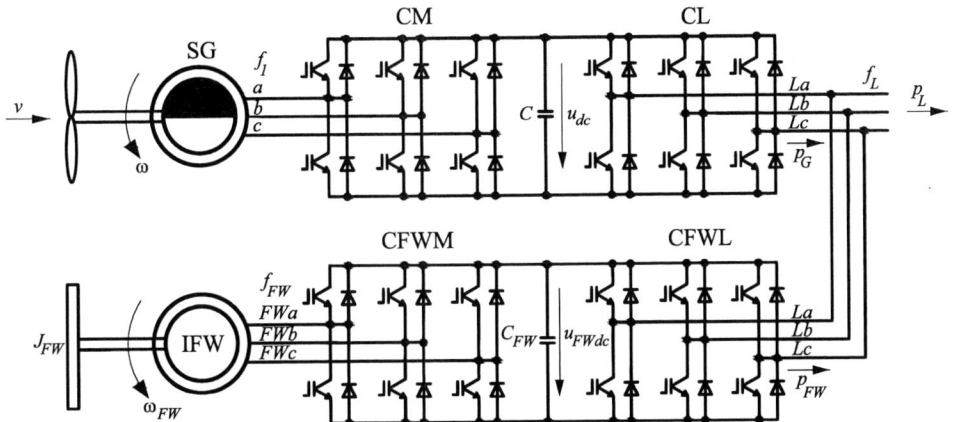

Fig.4. Permanent magnet synchronous wind turbine generator with flywheel energy storage connected to the lines.

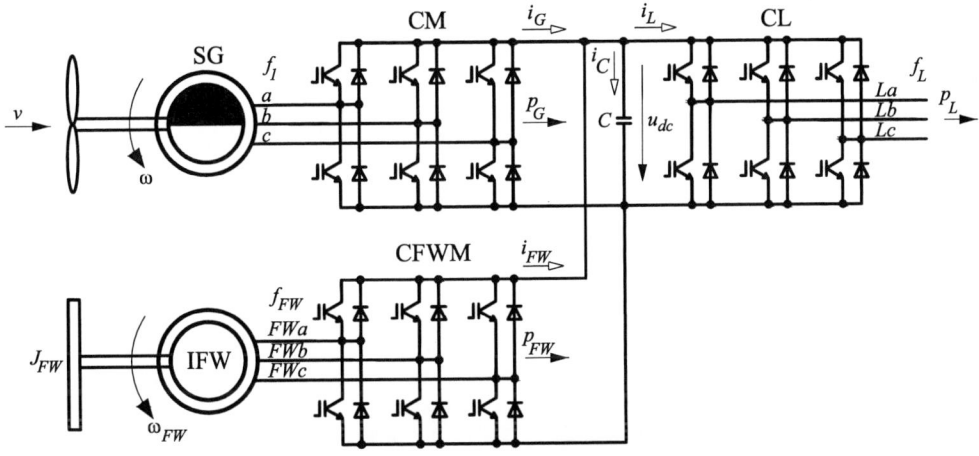

Fig.5. Permanent magnet synchronous wind turbine generator with flywheel energy storage connected to the *dc* link.

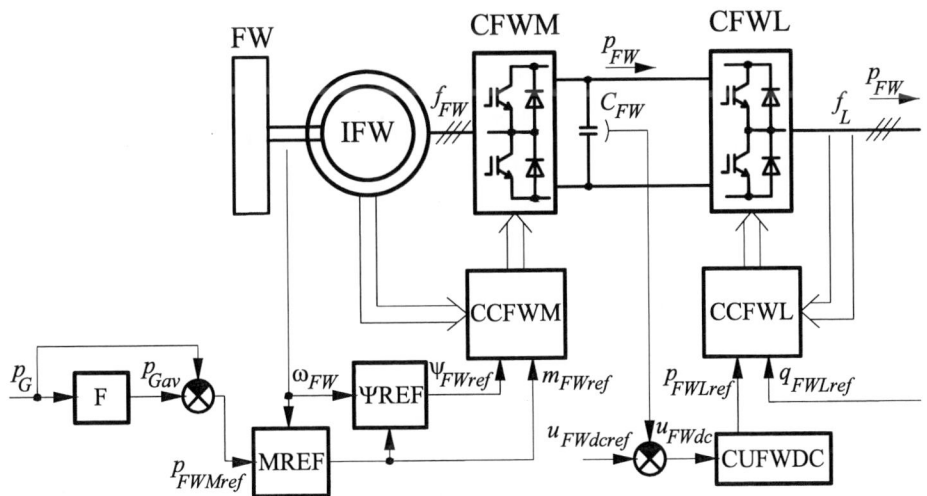

Fig.6. The control circuit of the flywheel energy storage drive connected to the three–phase lines.

1117

As can be seen, the power circuit of a variable frequency advanced wind turbine generator drive is the same as the power circuit of the variable frequency advanced flywheel drive. As a result, if the same dc link voltage is selected ($U_{dc}=U_{FWdc}$) then the CL and CFWL converters can be merged. Such a connection is presented in Fig.5. The fact remains that the CM converter must be designed to the $\max(p_G)$ generator power, while the CFWM converter to the P_{FWmax} flywheel power.

The advantages of the second version (Fig.5.) are:
- The resultant power rating of the power electronics of the wind turbine – flywheel system is less (there are only three converters and CL can be designed to P_{Gav}).
- The efficiency of the power circuit is higher.

Its disadvantages are:
- The flywheel drive is not independent, limiting the applicability.
- Sensing the p_G generator power in the case of a double–fed wind turbine generator is more complicated.

The advantages and disadvantages have approximately the same importance. Both versions can be applied practically.

In a wind farm more wind turbine generators can be served by one flywheel energy storage drive with proper power rating. In such a case, if the second version of the power circuit is applied (Fig.5.), the CL converter is also common for more generators.

IV. CONTROL CIRCUIT

The control circuit of a flywheel energy storage drive connected to the three–phase lines is presented in Fig.6. The p_{Gav} average value is produced by the F filter from the measured p_G wind turbine generator electric power. The difference of these two powers provides the reference value of the electric power of the flywheel drive:

$$p_{FWMref} = p_{Gav} - p_G \qquad (8)$$

The filter must be designed according to the pulsation caused by the turbulence to filter out the turbulent peaks. From the p_{FWMref} reference value and the ω_{FW} flywheel speed the MREF unit creates the torque reference in the following way:

$$m_{FWref} = -\frac{p_{FWMref} - p_{FW\ell}}{\omega_{FW}}, \qquad (9a)$$

$$\text{if } \omega_{FW\,min} < \omega_{FW} < \omega_{FW\,max},$$

and

$$m_{FWref} = m_{FW\ell}, \qquad (9b)$$

$$\text{if } \omega_{FW} \leq \omega_{FW\,min} \text{ or } \omega_{FW} \geq \omega_{FW\,max}.$$

Where $-(p_{FWMref} - p_{FW\ell})$ is the effective mechanical power, $p_{FW\ell}$ is the ω_{FW} speed dependent loss of the drive and $m_{FW\ell} = p_{FW\ell} / \omega_{FW}$ is the corresponding torque. The $m_{FW\ell}$ motor mode torque is necessary to maintain ω_{FW}

speed constant. If the ω_{FW} speed exceeds the ω_{FWmax} speed by more value than allowed ($\Delta\omega_{FWmax}$) for any reason, then the drive independently of anything gets to safety braking operation until reaching $\omega_{FW}<\omega_{FWmax}$. (Instead of the MREF torque reference unit, power controller also can be applied, but the p_{FW} flywheel drive power also must be measured in this case.)

From the ω_{FW} and m_{FWref} signals the ΨREF flux reference unit determines the ψ_{FWref} flux reference value of the IFW squirrel–cage induction machine. The machine side CCFWM controller controls the torque and the flux of the IFW induction machine by the CFWM converter.

This control can be implemented in different ways. If field oriented current vector control is applied, ψ_{FWref} is the reference value of the rotor flux of IFW. If direct torque and flux control is used [5], the ψ_{FWref} is the reference value of the stator flux of IFW.

The line side CUFWDC voltage controller controls the u_{FWdc} dc voltage by its prescribed p_{FWLref} active power reference. The reference value of the reactive power (q_{FWLref}) is determined by external demands. The CCFWL controller controls the electrical active and reactive power of the flywheel drive by the CFWL converter. The control can be implemented by line oriented current vector control or by direct active and reactive power control. The CUFWDC voltage controller provides constant u_{FWdc} voltage and $C_{FW}u_{FWdc}^2/2$ energy. This is satisfied, if on the machine and line sides connected to the C_{FW} capacitor the active powers are the same. There is a slight difference between the p_{FWLref} and p_{FWMref} power references caused by the losses. The flywheel energy storage drive with such a power and control circuits can be applied to any kind of wind turbine generator.

The control circuit of the flywheel energy storage drive connected to the dc link (Fig.5.) is similar, but the dc voltage control is done by the common CL line side converter.

The presented control circuit does not contain that part which is necessary to speed–up the flywheel drive from standstill (initial charge).

V. SIMULATION RESULTS

The system in Fig.6. with the proposed control is simulated by Matlab Simulink. The CCFWM is implemented as field oriented current vector control.

To simplify and speed–up the simulation, the following simplifications are done:
- The wind generator is represented by its power p_G only.
- The converters are implemented by controllable ideal sinusoidal voltage sources.

The induction machine (IFW), the lines (L), the mechanical part (FW) and the dc link are modelled accurately by their equations.

Per–unit system is used, the values are referred to the nominal values of the flywheel drive.

The modelled process is the following: into the p_G power of the wind generator a sinusoidal turbulent part is injected with constant amplitude (ΔP_G) and period (T_p):

$$\Delta p_G = \Delta P_G \, sin\!\left(2\pi t / T_p\right) \qquad (10)$$

The parameters are: T_p=1min=18850pu (corresponding to Fig.1.); k=0.75 (ω_{FWmin}=1pu; ω_{FWmax}=2pu); P_{Gav}=1pu mostly; T_{EMstn}=1s=314pu; C=7; T_{FWstn}=7*314=2198pu; (ΔT=10.5s which is less than 1min, but the p_{FW} is mostly far from P_{FWmax}=1pu during the simulations).

Initially the flywheel is accelerated to a speed corresponding to the mean value of the maximal and minimal flywheel energy (with the above parameters it is ω_{FW}=1.581pu).

The following cases are investigated:

A. Case I.: Perfect Compensation

If the limits of the operation are not reached, the compensation is perfect. Such a case is demonstrated in Fig.7. The data are: ΔP_G=0.4pu; calculation time is 40000pu.

The mean value of the flywheel speed during the compensation depends on the phase of the Δp_G at the time instant when the compensation is started. Here it is started at the negative maximum of Δp_G (14137pu time instant). In this way, the first decelerating (discharging) quarter period (decreasing the speed to its minimum) is followed by two accelerating (charging) quarter periods with the same energy each (increasing the speed to its maximum). Then the discharging and charging are repeated periodically, resulting in the same speed mean value as the initial (Fig.7b), which is in equal distance from the maximum and minimum flywheel energy.

The process can be followed in Fig.7. The speed limits are not reached (Fig.7b). The line power is constant during the compensation (Fig.7c). By the speed changing, the rotor flux magnitude of the IFW induction machine is also modified in the field weakening range (Fig.7d). It is controlled by the d component of the $\bar{i}_{FW}$ current vector of the IFW induction machine (Fig.7a).

B. Case II.: The Effect of the Δp_G Phase

To demonstrate the effect of phase of Δp_G on the mean value of the flywheel speed Fig.8. is drawn. Here at 9425pu time instant (where the phase of Δp_G is at the starting of its negative period) the compensation is started (ΔP_G=0.4pu). The negative period of Δp_G decreases the speed to its lower limit (1pu), so the limit of the compensation is reached (Fig.8a). But as can be seen, the next maximal speed is below its upper limit (2pu) and the next speed minimum (caused by the periodic disturbance) is exactly at the lower limit, the limit is only osculated (Fig.8a). The compensation is perfect from now (Fig.8b), only the mean value of the speed is smaller than in case I., resulting in less reserve in the stored energy. There are methods, which can ensure not to reach the limits [5] and to have the same reserve always, but for this the constant line power control (p_L=const.) must be given up sometimes.

At ω_{FWmin}=1pu speed the rotor flux magnitude also reaches its 1pu maximum (Fig.8c).

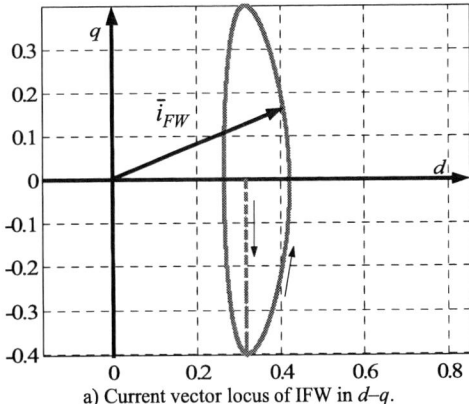

a) Current vector locus of IFW in d–q.

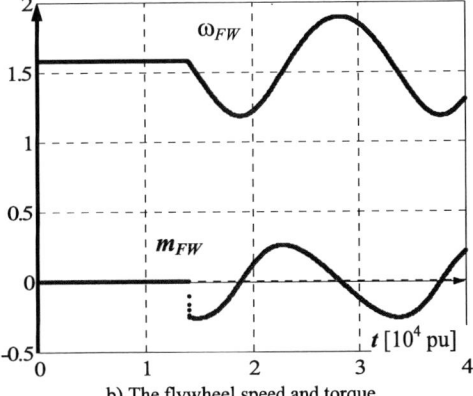

b) The flywheel speed and torque.

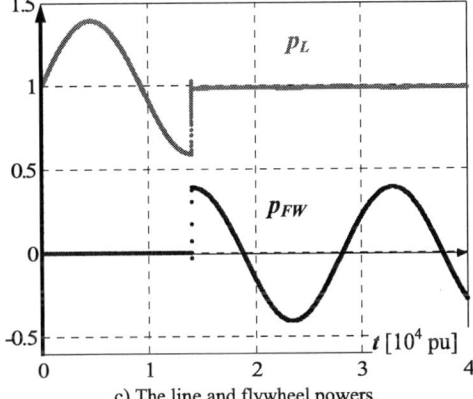

c) The line and flywheel powers.

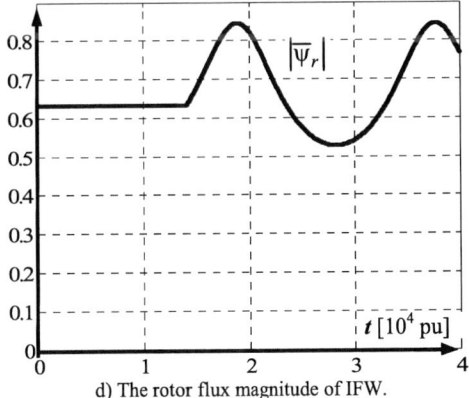

d) The rotor flux magnitude of IFW.

Fig.7. Perfect compensation.

1119

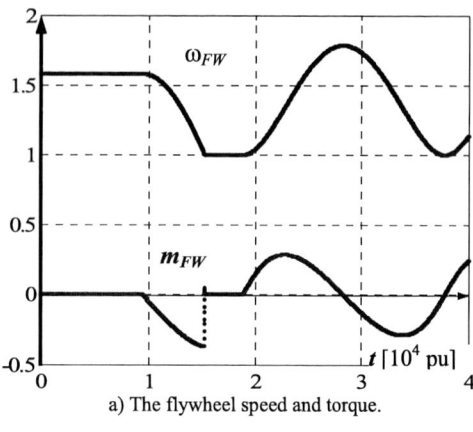

a) The flywheel speed and torque.

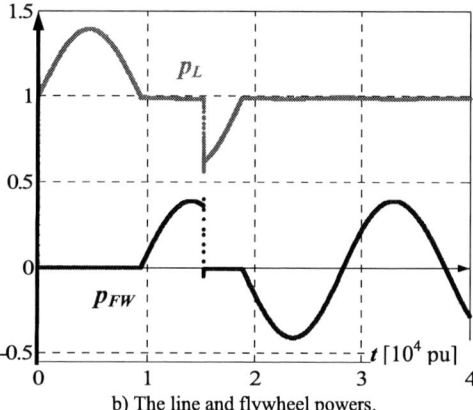

b) The line and flywheel powers.

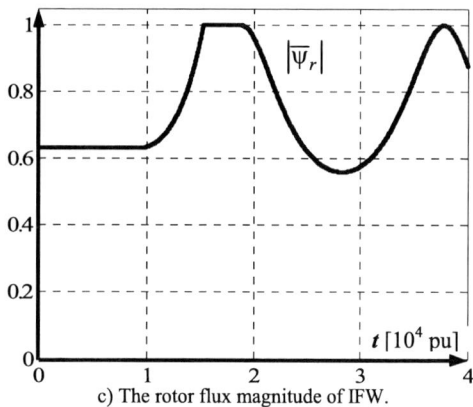

c) The rotor flux magnitude of IFW.

Fig.8. The effect of the phase of Δp_G on the mean value of the flywheel speed.

C. Case III.: Reaching the Speed Limits Periodically

If the ΔP_G is higher, the speed limits are reached. Fig.9. is drawn for ΔP_G=0.6pu. In this case the compensation is not perfect (Fig.9c.). Reaching the speed limits (time intervals 3, 5, 7) can be identified in the current vector locus (Fig.9a) and rotor flux magnitude time function (Fig.9d) also, where IFW is in the minimum (time interval 5) and maximum (time intervals 3, 7) flux operation (minimum and maximum i_d current).

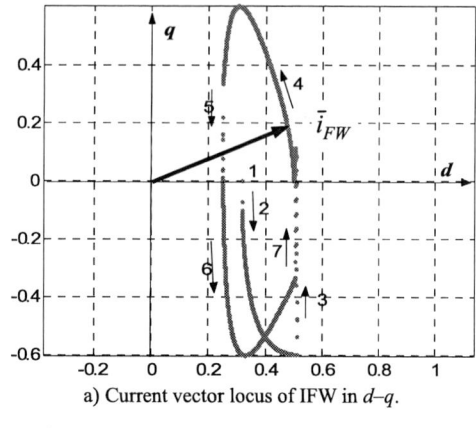

a) Current vector locus of IFW in d–q.

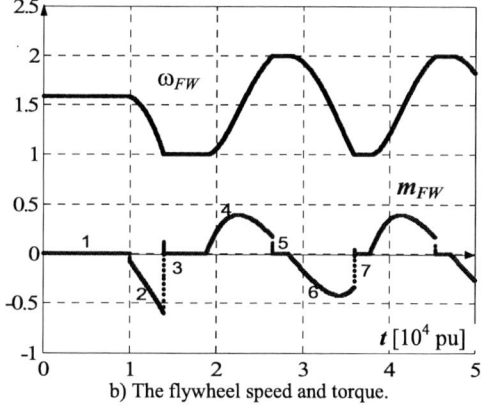

b) The flywheel speed and torque.

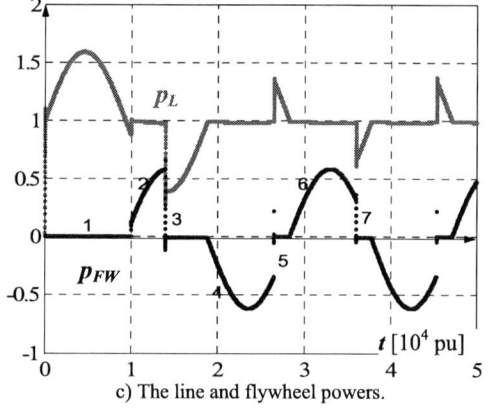

c) The line and flywheel powers.

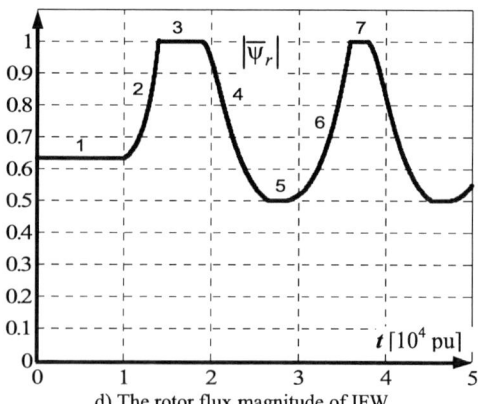

d) The rotor flux magnitude of IFW.

Fig.9. Reaching the speed limits periodically.

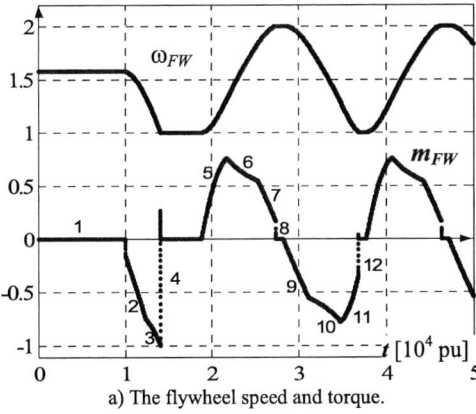

a) The flywheel speed and torque.

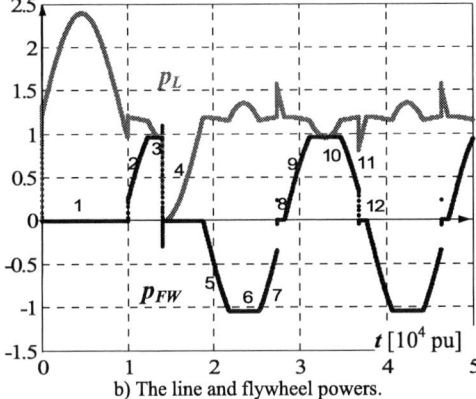

b) The line and flywheel powers.

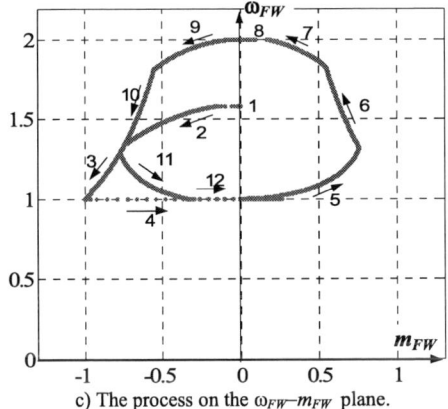

c) The process on the ω_{FW}–m_{FW} plane.

Fig.10. Reaching all limits.

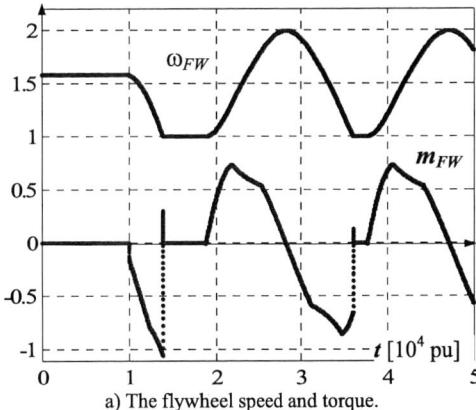

a) The flywheel speed and torque.

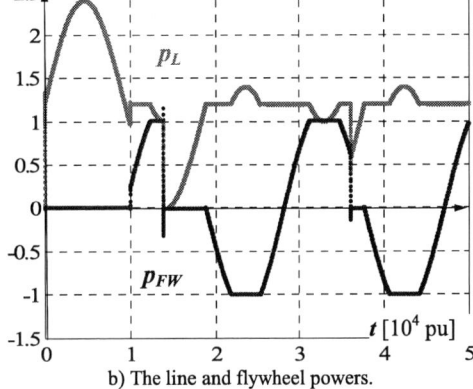

b) The line and flywheel powers.

Fig. 11. Compensating the copper losses of IFW.

D. Case IV.: Reaching All Limits

If the ΔP_G is much more high, the P_{Lmax} limit also can be reached. To simulate this, two times larger T_{FWstn} is assumed (not to be always at speed limit), $\Delta P_G=1.2$pu is injected and the average generator power is set to 1.2pu (not to get negative p_G). Fig.10. shows the process. Considering the limits, the process is most demonstrative on the ω_{FW}–m_{FW} plane.

As can be seen, at time intervals 3, 6 and 10 (at the $\pm P_{FWmax}$ limits) p_{FW} is slightly bellow the limits. It is caused by the electrical losses in the machine IFW, since by (9) the mechanical power is prescribed. This is true not only in these intervals and in this case, but only here it can be identified clearly.

As a result, the compensation is not totally perfect anywhere, p_L is not totally smooth where it is expected (also in the other cases).

To compensate the electrical losses of the driving system, more methods can be used:

– The more accurate but more complicated method is to use a power controller (e.g. PI) to set m_{FWref} instead of using (9ab). The feedback power must be the terminal electrical power of the flywheel system. It compensates all the losses.

– Implementing a loss model to calculate the electrical losses of IFW, it can be used to modify the reference value of the flywheel power P_{FWMref}.

To demonstrate the loss compensation, the second method is implemented for the copper losses of the induction machine. Case IV. is recalculated, the results are given in Fig.11.

The p_{FW} flywheel power reaches exactly its limits (Fig.11b). The process is slightly modified: the upper speed limit is not reached (the time interval 8 is missing). It is because in time interval 6 the absolute value of the p_{FW} accelerating power is less (as the result of the loss compensation).

VI. CONCLUSIONS

Using the acceptable simplifications, system–level investigation is possible. The most important conclusions of the results for periodic turbulent wind are the followings:

- The mean value of the flywheel speed during the compensation depends on the phase of the Δp_G at the time instant when the compensation is started.
- Coming out the speed limit, perfect compensation is found (if ΔP_G is small), the limit is only osculated. Not necessary to ensure the mean value of the speed to be at a value corresponding to the mean value of the maximal and minimal flywheel energy, but the compensation reserve must be considered. To ensure symmetrical, constant compensation reserve, the constant line power control (p_L=const.) must be given up sometimes.
- The effect of the losses of the flywheel induction machine IFW can be compensated.

REFERENCES

[1] Van der Hoven, I., "Power spectrum of horizontal wind speed in the frequency range from 0.0007 to 900 cycles per hour", *Journal of Meteorology*, vol.14. pp.160-164. 1957.

[2] James A. McDowall, "Opportunities for electricity storage in distributed generation and renewables", *in Proc. of IEEE Saft America Inc. Transmission and Distribution Conf. and Exposition, 2001 IEEE/PES*, vol.2. pp.1165-1168.

[3] Cárnedas R., Pena R., Asher G. and Clare J., "Control strategies for enhanced power smoothing in wind energy systems using a flywheel driven by a vector–controlled induction machine", *IEEE Trans. Ind. Electronics*, Vol.48, No.3, pp.625-635. 2001.

[4] Cárnedas R., Pena R., Asher G. and Clare J., "Power smoothing in wind generation systems using a sensorless vector controlled induction machine driving a flywheel", *IEEE Trans. Energy Conv.*, Vol.19, No.1. pp.206-216, 2004.

[5] Cimuca G., Breban S., Radulescu M., Saudemont C. and Robyns B., "Control strategy for an induction machine–based flywheel energy storage system associated to a variable–speed wind generator", *in Proceedings of OPTIM. 2006.*

Theory, Simulation and Experimental Verification of a New Integral Cycle Robust Control Strategy for Self Excited Induction Generators

S.S. Murthy* and A.J.P. Pinto **

* Professor, Department of Electrical Engineering, Indian Institute of Technology, 110011 Delhi. India
** Research Scholar, National Institute of Technology Karnataka, Surathkal, 575025, India

Abstract–This paper presents a new control strategy based on integral cycle control of the excitation capacitors to achieve excellent voltage regulation for all types of loads including direct on line starting of induction motors. The integral control strategy not only eliminates switching harmonics but also the requirement of exactly sized capacitors, which are hard to procure. In addition the implemented system is self starting without the need of any external energy source/storage. The paper presents simulation and experimental results both of which show excellent voltage regulation that can be obtained at any desired operating point on the magnetization curve of the core material including the linear region, a region where no operation was possible using the previously available control methods.

Index Terms-- Induction generators, Standby generators, Integral cycle control

I. INTRODUCTION

The induction motor operated as an induction generator with terminal capacitors in stand alone mode specially for renewable energy applications, offers considerable advantages due to its ruggedness, low cost, brush-less squirrel cage rotor, manufacturing simplicity, low maintenance and wide off-the-shelf range as compared to the synchronous machine, but has a major drawback of poor voltage regulation. The rigid frequency and voltage of the grid makes the equivalent circuit model suitable for steady state analyses of induction motor in generating mode. However its analysis and operation as a stand-alone power source is complicated, since now both voltage and frequency are variables, and involve solving non linear equations of higher order [1]. Under these conditions proper selection of equipment and the prediction of the system performance are essential for successful implementation of the scheme. The operation of such a machine in any remote or stand alone condition therefore requires a suitable voltage regulator. Considerable literature exists describing various arrangement such as, contactor or thyristor switched capacitors, thyristor controlled inductors, saturable reactors, etc [2,3]. Devices like STATCOM in effect provide a virtual bus for the induction machine to operate. All these systems introduce a lot of harmonics or are expensive to build and complicated to program. The excitation capacitors too need to be sized properly for acceptable operation. The STATCOM based controller overcomes these problems but is not self-starting. The greatest drawback of most of the systems is their inability to handle unintentional capacitive loads as in the case of power factor improvement capacitors left on line, when the motor is disconnected by some fault, resulting in dangerous over voltages. To overcome these shortcomings which prevent wide acceptance of induction generators in stand-alone engine-driven applications, the present controller was developed. The machine and load were modeled for dynamic analysis and simulation studies were carried out using MATLAB-SIMULINK. The encouraging simulation results were instrumental in building the hardware and technology demonstration reported here.

II. NEW THEORY

An SEIG system consists of an induction machine driven by a prime mover. A three-phase capacitor bank provides self-excitation and load VAR requirements. The machine is driven at constant speed and capacitor is required to achieve self excitation. The machine will build up voltage and operate at a constant voltage level, which is a function of the magnetizing characteristics and the excitation capacitance. Extensive literature is available on the selection of minimum capacitance for self excitation and determining the operating voltage. Synchronous speed test or other methods are used to determine the non linear relationship between the magnetizing reactance and the terminal voltage. The terminal voltage of the machine can be said to depend on the machine's magnetic part's capacity to absorb the capacitive volt amperes at that voltage. Better insight of the machine is obtained from the real and reactive power flow diagrams as shown in Fig. 1 from which,

$$P_{in} = P_{Load} + P_{loss(Machine)} + P_{loss(cap)} \tag{1}$$

$$Q_{Cap} = Q_{Machine} + Q_{Load} \tag{2}$$

Therefore considering the purpose of magnetic saturation to be that of absorbing the excess capacitive volt amperes and thereby arriving at stable operating point, should means be provided to absorb the excess capacitive vars the machine will operate stably at any desired voltage level even if it be in the linear region.

978-1-4244-0644-9/07/$25.00 ©2007 IEEE

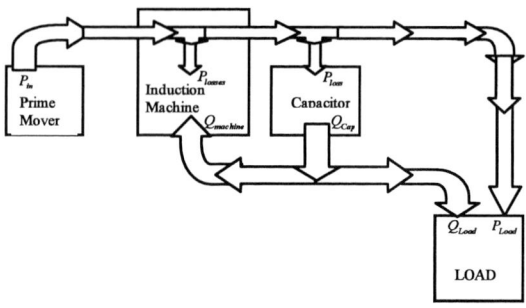

Fig 1. P-Q flow chart for self excited induction generator.

III. DYNAMIC MODELING AND SIMULATION

Connection of excitation capacitors across the terminal can be represented in the d-q model of the machine as two additional equations with the capacitor voltages as state variables. Similarly the connection of the load across the terminal can be represented in four more equations shown here. The complete arrangement is shown in Fig. 2.

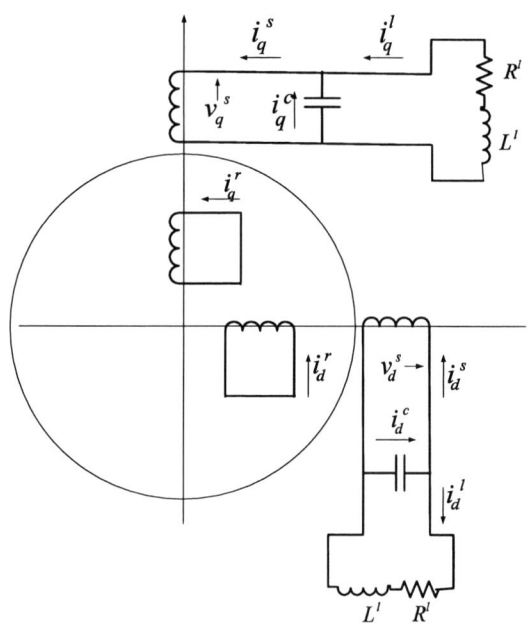

Fig. 2. d-q model of the induction machine complete with excitation capacitors and load.

Referring to Fig. 2, it is seen that

$$i_q^l = -i_q^s - Cpv_q^s \tag{3}$$

$$i_d^l = -i_d^s - Cpv_d^s \tag{4}$$

and

$$v_q^s = i_q^l \left(R^l + L^l p \right) \tag{5}$$

$$v_d^s = i_d^l \left(R^l + L^l p \right) \tag{6}$$

Combining the above four equations in the standard d-q model[4] of the induction machine we get the dynamic model of a self excited induction generator complete with excitation capacitor and load, giving due representation to

the capacitive voltages and inductor currents as state variable is in (7)

$$
\begin{pmatrix} v_q^s \\ v_d^s \\ 0 \\ 0 \\ i_q^l \\ i_d^l \\ 0 \\ 0 \end{pmatrix} =
\begin{pmatrix}
(R+Lp) & 0 & Mp & 0 & 0 & 0 & 0 & 0 \\
0 & (R+Lp) & 0 & Mp & 0 & 0 & 0 & 0 \\
Mp & k\omega M & (R+Lp) & k\omega L & 0 & 0 & 0 & 0 \\
-k\omega M & Mp & -k\omega L & (R+Lp) & 0 & 0 & 0 & 0 \\
-1 & 0 & 0 & 0 & Cp & 0 & 0 & 0 \\
0 & -1 & 0 & 0 & 0 & Cp & 0 & 0 \\
0 & 0 & 0 & 0 & 1 & 0 & -(R^l+L^l p) & 0 \\
0 & 0 & 0 & 0 & 0 & 1 & 0 & -(R^l+L^l p)
\end{pmatrix}
\begin{pmatrix} i_q^s \\ i_d^s \\ i_q^r \\ i_d^r \\ v_q^c \\ v_d^c \\ i_q^l \\ i_d^l \end{pmatrix}
$$
....(7)

Simulation setup carried out using MATLAB-Simulink is in Fig 3. In order to investigate the acceptability of the proposed integral cycle control under extreme conditions, magnetic saturation is not represented. This approach will then also permit operation in the linear region of the saturation curve. Capacitance value Cp is selected to be greater than that required to achieve self excitation; Loading is controlled by varying R^l and L^l, other variables in (7) being machine parameters. Since the capacitive vars absorbed by the machine is a function of the loading on the machine too, for low values of capacitance just above minimum required for self excitation, loading of the machine with resistive load only is sufficient to absorb the excess capacitive vars. By monitoring the line voltages and switching the terminal capacitors- switching operation being carried out only at current zero crossings- the machine model can operate in any average condition, i.e., of excess capacitive vars, deficient capacitive vars or just balanced condition corresponding to voltage build-up, voltage decay or stable voltage operation.

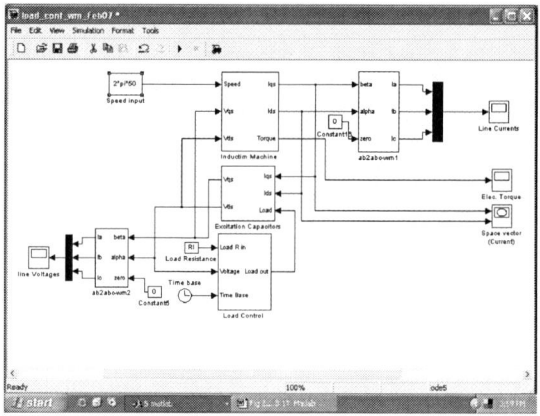

Fig. 3. MATLAB simulation: Simulink setup

A. Simulation Results

Results of the simulation are shown in Fig 4. It shows the variation in line voltage over time while the capacitive vars are varied in a controlled fashion. The machine is driven at constant speed with no load and minimum required capacitance for self excitation. The exponential build up of terminal voltage on achieving self excitation is seen.

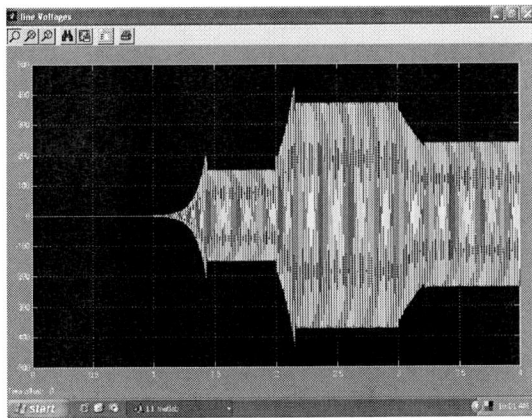

Fig. 4. Simulation results showing terminal voltage being controlled at desired values.

Referring to Fig. 4, at 1.4s when the line voltage exceeds the set-point, control system is activated and the capacitors are switched off and on for integral number of cycle in the manner of 'duty cycle'. Upon change of the set-point, at 2s, the capacitors are fully on line and the voltage again rises until the new set-point is reached and the capacitors are again on integral cycle based 'duty cycle'. At 3s, when the line voltage set-point is reduced, the capacitors are switched off at their respective current zeros, until the voltage decays to the new reduced set-point, thus activating the control scheme and stabilizing the voltage.

The flow chart representing the operating logic of the integral cycle and duty cycle based controller is in Fig 5.

IV. EXPERIMENTAL SETUP

The block diagram of the experimental setup is shown in Fig 6 and the Schematic diagram of the implemented hardware set up is shown in Fig. 7 and Fig 8. The capacitors are connected to the 10-hp, 400/440-V, delta connected induction machine's terminals through Samikron SKM75GB123 IGBTs which are in turn connected as shown in Fig 9 to form an AC switch. For cold starting, i.e., without the use of an external power source, a normally closed contactor (not shown) bypasses the IGBTs, connecting the capacitors directly across the terminal of the induction machine and providing a low resistance path for the extremely small excitation currents resulting from residual magnetism of the rotor and thus permitting voltage buildup. The contactor opens when the terminal voltage reaches 50% of rated voltage of machine which is sufficient to power up the control circuits. Hall effect current and voltage sensors are used as isolation devices between power and control circuits. Fig 10 is a photograph of the complete hardware setup.

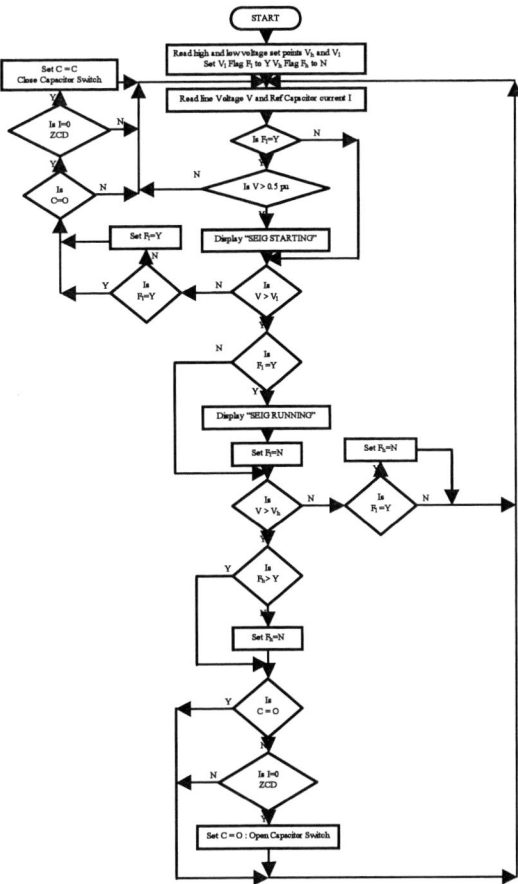

Fig. 5. Integral cycle & duty cycle control flow chart.

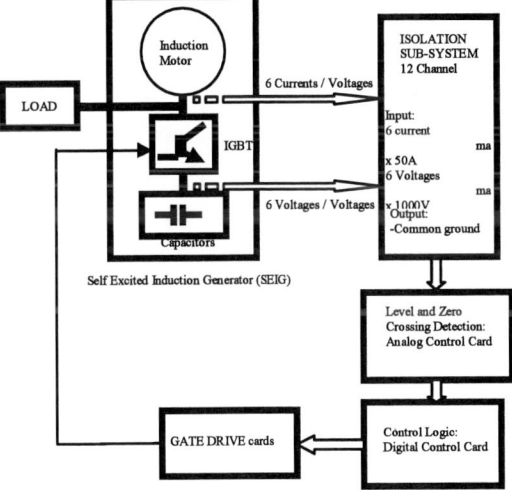

Fig 6. Block diagram

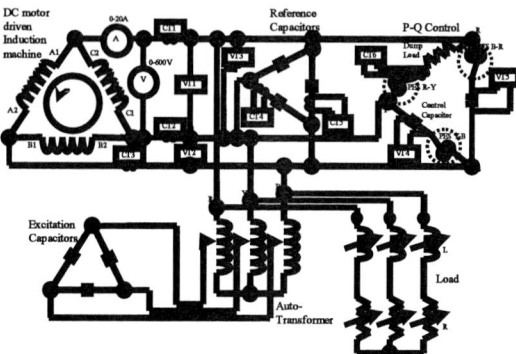

Fig. 7. Circuit diagram of experimental setup

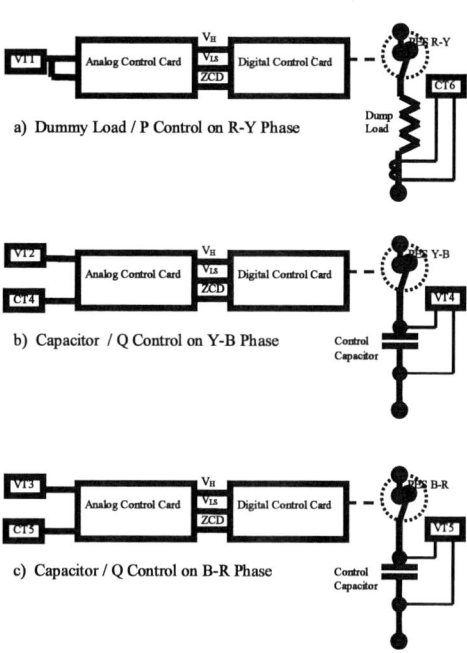

a) Dummy Load / P Control on R-Y Phase

b) Capacitor / Q Control on Y-B Phase

c) Capacitor / Q Control on B-R Phase

Fig 8. Experimental setup: Control circuit diagram

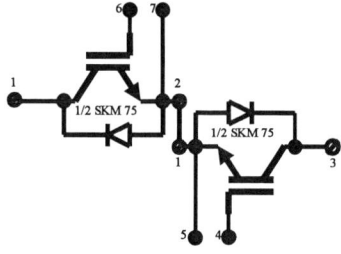

Fig. 9. Bidirectional power electronic switch: Circuit diagram

The Control logic shown in flow chart, Fig 5, is implemented using CMOS logic ICs. The control circuit monitors and compares machine terminal voltage with the set point. Switching of the capacitors on high or low voltage in each phase is implemented only at the zero crossing of currents, irrespective of the point at which the decision to switch takes place. The dead band between high and low voltage set-point is continually adjustable.

Since switching is at zero currents, the device losses are considerably reduced, in addition the integral cycle switching scheme eliminates the need for snubber circuits, and accordingly none have been provided, resulting in reduced component count and complexity.

Fig. 10. Hardware setup

A. *Experimental Results*

The operation of the control scheme may be observed from Figs. 11, 12 and 13 which are Oscilloscope prints taken while the control system was in operation and the machine was loaded with R-L load. The top, middle and bottom traces respectively represent the line voltage the line current and the capacitor voltage. Fig. 11 shows the switching in and out of one bank of capacitors in a 'duty cycle' manner. The capacitor bank is 'on' for an integral number of cycles and 'off' for an integral no of cycles. The ratio between the numbers of 'on' to 'off' cycles depends on the loading of the machine and the dead-band between the high and low voltage set-points. The rise in line voltage while the capacitors are on line and fall in line voltage when the capacitors are disconnected may be observed in all there prints.

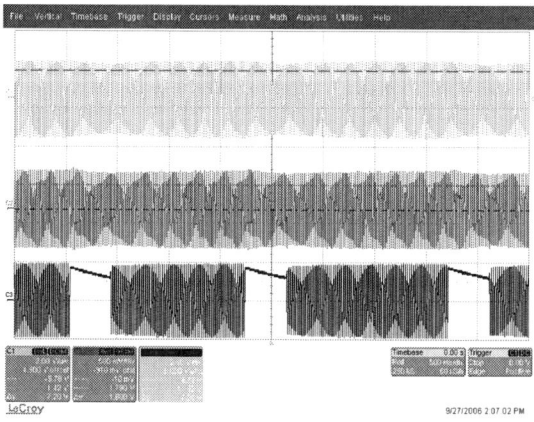

Fig. 11. Oscilloscope print showing 'duty cycle' base integral control action.

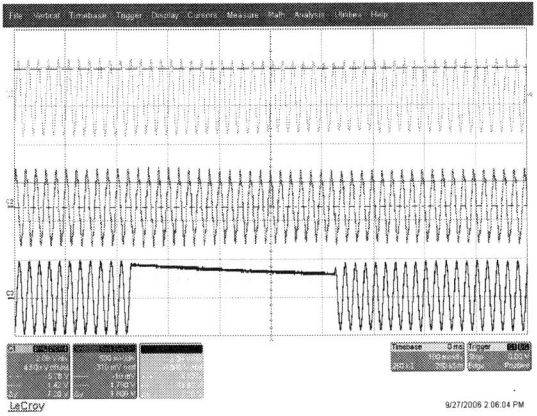

Fig. 12. Oscilloscope print showing Capacitor voltage during off state.

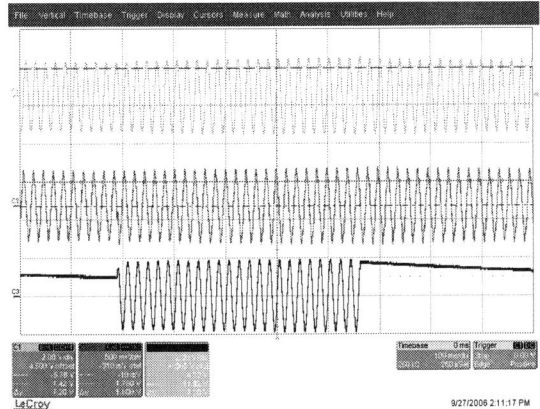

Fig. 13. Oscilloscope print showing capacitor voltage during on state.

Capacitor voltage during the 'off' time is seen in more detail in Fig. 12, the droop in the capacitor voltage during this time is due to the discharge resistors connected across the terminals of the capacitors which should not be removed for safety reasons. The voltage across the capacitor during the on period may be observed in Fig. 13. The absence of switching surges in the traces of the line voltages in the three prints may also be observed. In Figs. 14 and 15, the Oscilloscope prints show further detail at different operating conditions corresponding to different duty cycles.

Fig 16 shows the variation in duty cycle during 'direct on line (DOL) starting' of an induction motor.

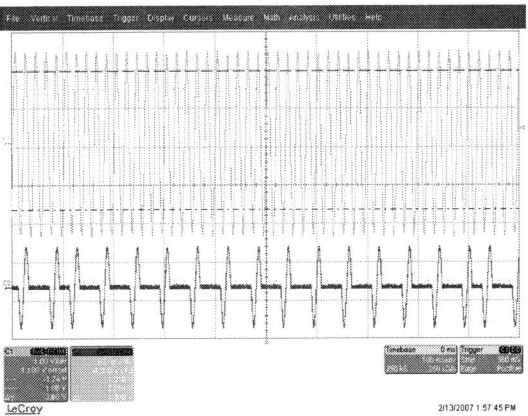

Fig. 14. Operation with R control: Duty Cycle 1/2 and 1/3

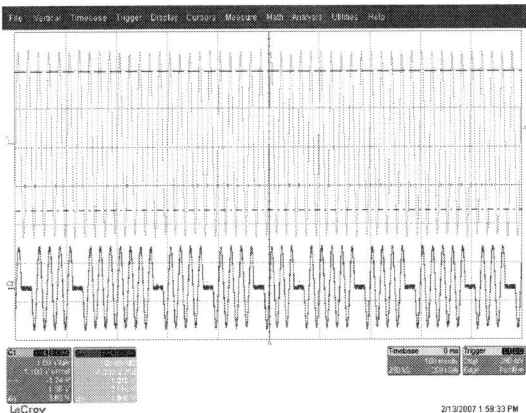

Fig. 15. Operation with R control: Duty Cycle 3/4 and 4/5

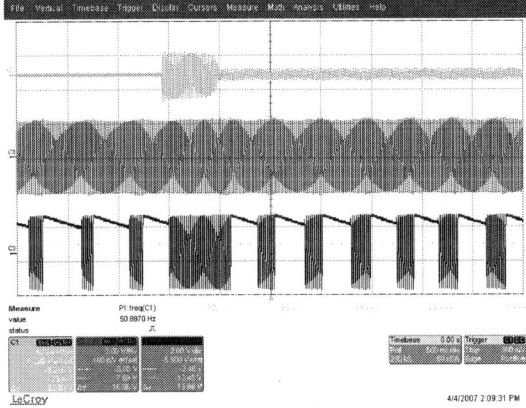

Fig. 16   DOL Starting of an induction motor

1127

V. CONCLUSION

A practically working scheme using SEIG to maintain required power quality for consumers under different types of prime movers for stand alone power generation was found wanting for a long time ever since the SEIG was found to be an alternative for the conventional synchronous generator. This paper effectively bridges that gap as a good voltage and frequency regulating scheme using integral cycle control and has been designed, implemented and successfully tested under different types of loads as reflected by encouraging results presented here. It is hoped that this would be the breakthrough in commercializing SEIG for decentralized power generation using engine, small hydro and other non-conventional energy systems.

REFERENCES

[1] S.S. Murthy, O.P. Malik and A.K. Tandon, "Analysis of self-excited induction generator", IEE Proc. C. Gener. Transm. Distrib., vol. 129, (6), pp 260-265. 1982.

[2] S.S. Murthy and B. Singh, "Capacitive VAR controllers for induction generators for autonomous power" proc. IEEE Power Electronics, Drives and Energy Systems Conf., 1996, pp 679-686.

[3] B. Singh, S.S. Murthy and S. Gupta, "Analysis and design of STATCOM based Voltage Regulator for self-Excited Induction Generators" IEEE Transactions on Energy conversion, Vol 19, No 4, Dec 2004.

[4] Howard E. Jordan, "Digital Computer Analysis of Induction Machines in Dynamic systems" IEEE Trans. on Power Apparatus and systems, Vol. 86, No 6, pp. 722-728. June 1967.

Performance Comparison of DC Link Voltage Controllers in Vector Controlled Boost Type PWM Converter for Wind Turbine System

W. Sudmee, B. Neammanee

Department of Electrical Engineering, King Mongkut's Institute of Technology North Bangkok
1518 Bangsue, Bangkok 10800, Thailand Phone: (+66) 2913-2500-24 Ext: 8420, Fax: (+ 66)2585-7350

Abstract–This paper presents control algorithms to regulate the DC link voltage in a vector controlled boost type PWM converter with particular applications to a wind energy conversion system. The algorithms provide near-sinusoidal input currents and unity power factor and low output voltage ripple. Moreover, the designed PWM exhibits fast transient response to load variations, and is capable of regenerative operation. An experiment with the 7.5 kW converter is set up to investigate and compare the control performances of three designed DC link voltage controllers comprising PI-fuzzy, PI-fuzzy with measured load current feed-forward and PI-fuzzy with observed load current feed-forward.

Index Terms–Line side converter, load current feed-forward, fuzzy logic control, current observer

Nomenclature

v_{sa}, v_{sb}, v_{sc}	=	phase voltage in stationary reference frame
i_{sa}, i_{sb}, i_{sc}	=	phase current in stationary reference frame
v_{sd}, v_{sq}	=	voltage in d, q-axis
i_{sd}, i_{sq}	=	current in d, q-axis
I_{conv}	=	converter current
I_{load}	=	load current
I_{cap}	=	capacitor current
V_{dc}	=	DC link voltage
V_m	=	phase peak
P_{conv}	=	converter power
K_p	=	proportional gain
K_i	=	integral gain
θ	=	angle of phase voltage
P_{loss}	=	power loss
p_2, q_2, r_1	=	constant gain
W_1, W_2	=	gain weight
$\hat{x}(k)$	=	estimate state
K	=	observer gain
Φ	=	characteristic polynomial
R	=	resistance
L	=	inductance
C	=	capacitance

This work is supported by Faculty of Engineering, King Mongkut's Institute of Technology North Bangkok (KMITNB).

I. Introduction

A boost type PWM converter provides a bi-directional between DC-AC interface with high quality currents on the AC side and controllable power factor [1]. The converter uses as the front end (line side) in AC drive applications where it allows regeneration capability and nearly sinusoidal input currents. The boost converter of interest in this paper is used in a wind energy conversion system as front end interface to the AC grid system. When induction machines are used for grid-connected variable speed wind turbine, two back to back converters are required for squirrel cage machines or doubly fed induction machines [2]. Control of the AC side currents is usually carried out in a d-, q-axis synchronous reference frame. High bandwidth current response is then achieved and furthermore decoupled control of the active and reactive power flow between the converter and the grid is possible. In a wind energy system, there may be a number of generation sources/loads connected to the DC link. Power flow of the converter must be controlled to regulate the DC link voltage according to a reference value [3, 4].

This paper focuses on control strategies to regulate the DC link voltage. The mathematical models of line side converter, voltage decoupling control, phase locked loop (PLL) and control block diagrams are presented. The converter uses decoupling control to regulate the DC link voltage and drive the power converter by space vector PWM. The control problems created by non-linear impedance on the DC side are addressed particularly when the characteristics of the DC side generation/load are unknown. Three designed DC link voltage controllers comprising PI-fuzzy, PI-fuzzy with measured load current feed-forward and PI-fuzzy with observed load current feed-forward are developed to improve the performance of the controller to load disturbance rejection. The last controller uses a load current observer needed for feed-forward compensation. The three controllers are implemented on designed DSC boards and tested with a 7.5kW converter to compare their control performance.

II. Mathematical Model of Line Side Converter

Figure 1 shows a three phase line side converter that converts AC-DC voltage or DC-AC. The per-phase voltage equations can be written as in (1)-(3).

978-1-4244-0644-9/07/$25.00 ©2007 IEEE

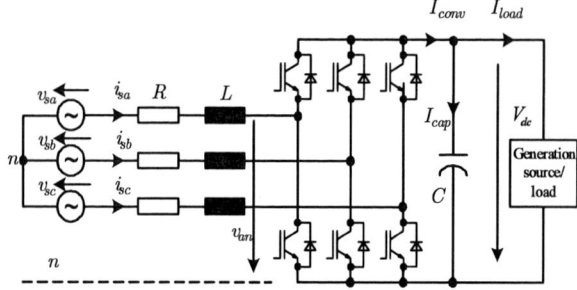

Fig. 1. Line side converter.

$$v_{sa} = Ri_{sa} + L\frac{di_{sa}}{dt} + v_{an} \quad (1)$$

$$v_{sb} = Ri_{sb} + L\frac{di_{sb}}{dt} + v_{bn} \quad (2)$$

$$v_{sc} = Ri_{sc} + L\frac{di_{sc}}{dt} + v_{cn} \quad (3)$$

Transforming (1)-(3) in three phases in stationary reference frame to two phase synchronous reference frame (d-, q-axis) gives

$$v_{sd} = (Ls + R)i_{sd} - \omega Li_{sq} + v_{dn} \quad (4)$$

$$v_{sq} = (Ls + R)i_{sq} + \omega Li_{sd} + v_{qn} \quad (5)$$

For balance three phase, $v_{sd} = V_m$ and $v_{sq} = 0$, the converter power, P_{conv}, and current, i_{conv}, can be calculated by (6) and (7) [5].

$$P_{conv} = \frac{3}{2}\left(v_{sd}i_{sd} + v_{sq}i_{sq}\right) \quad (6)$$

$$I_{conv} = \frac{P_{conv}}{V_{dc}^*} = \frac{3}{2}\frac{\left(v_{sd}i_{sd} + v_{sq}i_{sq}\right)}{V_{dc}^*} \quad (7)$$

The capacitor current, I_{cap}, and DC link voltage, V_{dc}, can be calculated by (8) and (9).

$$I_{cap} = I_{conv} - I_{load} \quad (8)$$

$$V_{dc} = \frac{1}{C}\int I_{cap}dt \quad (9)$$

With the transformation of (4)-(9) into s-domain, a block diagram in the d-, q-axis reference frame can be constructed as shown in the right hand side of Fig. 2 (Rectifying model and its block diagram in right hand side). It can be seen from (4) and (5) that voltage ωLi_{sq} is cross-coupled with v_{sd} and ωLi_{sd} with v_{sq}. To have independent control of V_{dc}, the coupling voltage should be compensated by a controller [5].

III. CONTROL OF LINE SIDE CONVERTER

One of the control objectives is to regulate the DC link voltage without the voltage cross coupling. To achieve this objective, it is necessary to compensate the coupling components ωLi_{sq} and ωLi_{sd} with a PI controller. The control equations are as follows.

$$v_{dn}^* = \left(K_{pi} + \frac{K_{ii}}{s}\right)\left(i_{sd}^* - i_{sd}\right) + \omega Li_{sq} + V_m \quad (10)$$

$$v_{qn}^* = \left(K_{pi} + \frac{K_{ii}}{s}\right)\left(i_{sq}^* - i_{sq}\right) - \omega Li_{sd} \quad (11)$$

If the voltage couplings are compensated, the DC link voltage depends only on the current i_{sd}. The PI controller is then used to control the bus voltage by the current i_{sd}. The control equation is shown in (12).

$$i_{sd}^* = \left(K_{pi} + \frac{K_{ii}}{s}\right)\left(V_{dc}^* - V_{dc}\right) \quad (12)$$

With (10)-(12), the line side converter control with a voltage decoupling block diagram can be constructed as in the left hand side of Fig. 2. In the block diagram, the voltage cross coupling is compensated by feeding i_{sd} and i_{sq} with gain ωL. After the voltage is decoupled, the DC link voltage is controlled by two loops: current and voltage loops. The current i_{sq} does not affect the bus voltage and therefore it is not used. The value of i_{sq} is set to zero.

The controller of the line side converter is controlled in the d-, q-axis reference frame. The transformation between three phases and d-, q-axis requires an angle of phase voltage. A PLL is used to calculate the angle of phase voltage, θ^*, and its control block diagrams are shown in Fig. 3 [6]. The parameter θ^* in the transformation is obtained by integrating frequency command ω^*. If the frequency is identical to the grid frequency, the voltages v_{sd} and v_{sq} become DC values, depending on the angle θ^*. A PI regulator is used to obtain that value of ω^*, which drives the feedback voltage v_{sq} to the command value v_{sq}^*. The magnitude of the controlled quantity v_{sq} determines the phase difference between the utility voltages and $\sin(\theta^*)$ or $\cos(\theta^*)$. This system sets $v_{sq}^* = 0$.

IV. PROPOSED CONTROLLER STRUCTURE

A. Small Signal Model

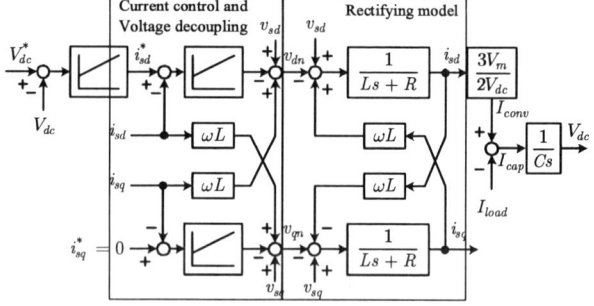

Fig. 2. Line side converter with voltage decoupling.

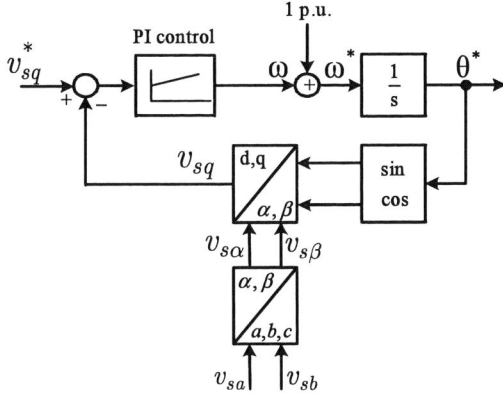

Fig. 3. Phase locked loop.

To obtain the relationship between the DC link voltage and the current i_{sd}, the power balance equation is used. The power balance between the DC link side and the induction machine side is (see Fig. 1)

$$V_{dc}i_{load} + \frac{1}{2}CV_{dc}^2 + P_{loss} = V_{dc}i_{conv}$$

$$= k\left[v_{sd}i_{sd} - R(i_{sd}^2 + i_{sq}^2) - \frac{1}{2}L(i_{sd}^2 + i_{sq}^2)\right] \quad (13)$$

To regulate V_{dc}, it is necessary to control the real power flow by i_{sd}. The quadrature current i_{sq} is normally close to zero implying close-to-unity power factor operation. From (13), the converter losses can be neglected and $i_{sq} = 0$, a small signal transfer function for an operating points V_{dco}, i_{sdo} and i_{loado} can be obtained from

$$\frac{\partial V_{dc}}{\partial i_{sd}} = \frac{k(v_d - 2Ri_{sdo} - i_{sdo}Ls)}{V_{dco}\left(\dfrac{i_{loado}}{V_{dco}} + \dfrac{\partial i_{load}}{\partial V_{dc}}\bigg|_{V_{dco}i_{sdo}} + Cs\right)} \quad (14)$$

Equation (14) is used to design a linearized controller. However, the transfer function is dependent on the operating points. Furthermore, the pole of (14) is strongly dependent on i_{load}. In general, i_{load} may be a function of V_{dc}. For example, a static load could be placed across the DC link and or i_{cap} may be supplied from a generator operating with a constant power. As the transfer function can be varied over operational range due to unknown load and generating conditions, a controller should be, therefore, appropriately designed.

B. PI-Fuzzy Control

A fuzzy logic control (FLC) is characterized by "IF-THEN" rules. It is suitable for complex nonlinear models and parameter variations. There are two inputs of FLC: error $e(k) = V_{dc}^* - V_{dc}$ and change of the error $\Delta e(k) = e(k) - e(k-1)$. Indexes k and $k-1$ indicate the present state and the previous state of the system, respectively. One of the membership functions of the FLC inputs, $e(k)$, is defined on common normalized

domain. Triangular membership functions are chosen for NM, NS, ZE, PS, PM fuzzy sets and trapezoidal membership functions are chosen for fuzzy sets NB and PB as shown in Fig. 4 a). The second input is $\Delta e(k)$, which have small change, then there are three memberships function as shown in Fig. 4 b).

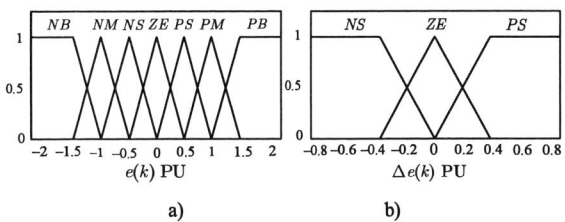

a)   b)

Fig. 4. Input membership function a) voltage error, $e(k)$ and b) change of voltage error, $\Delta e(k)$.

where

NB	= negative big,	NM	= negative medium,
NS	= negative small,	ZE	= zero,
PS	= positive small,	PM	= positive Medium,
PB	= positive big		

The output of the system is an incremental change in the control signal Δi_{sd}. The control signal is obtained by $i_{sd}(k) = i_{sd}(k-1) + \Delta i_{sd}(k)$. There are in total 21 rules base implemented for computing the output built based on the characteristics of the step response as shown in Table I. For example, if the output falls far away from the set point, a large control signal that pulls the output toward the set point is expected, whereas a small control signal is required when the output is near or approaches the set point [7].

TABLE I FUZZY RULES BASE

		Voltage error, $e(k)$						
		NB	NM	NS	ZE	PS	PM	PB
Change of voltage error, $\Delta e(k)$	NS	NB	NM	NS	NS	ZE	PS	PM
	ZE	NB	NM	NS	ZE	PS	PM	PB
	PS	NM	NS	ZE	PS	PS	PM	PB

Takagi-Sugeno's fuzzy model [8] is employed in this paper because its output calculation is straightforward. The fuzzy model is represented in form of "If x is A and y is B then $z = f(x, y)$" and its reasoning process is shown in Fig. 5

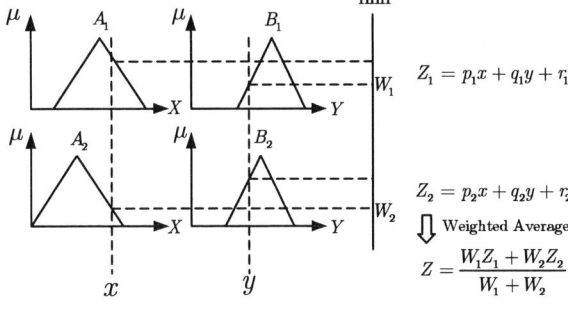

Fig. 5. Reasoning process of Takagi-Sugeno fuzzy model.

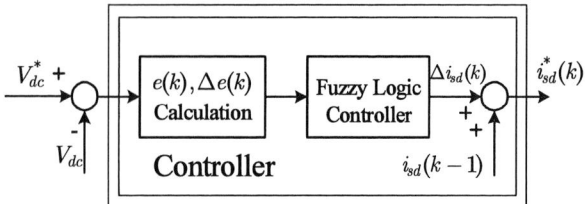

Fig. 6. PI-fuzzy control block diagram.

C. PI-fuzzy with Feed-forward Compensation

To improve DC link voltage regulation, feed-forward compensation is used. The compensation is based on the parametric model of the system and relates the disturbance current i_{load} to i_{sd}. Neglecting the energy variation in L because the input filter inductance is small and considering that the energy variation in C is also negligible because the voltage is kept relatively constant by controller action then the steady state equation may be used to derive the feed-forward compensation term. From Fig. 1, the steady state power balance for unity-power factor operation, $i_{sq}^* = 0$, is given by [9, 10]

$$V_{dc}i_{load} + P_{loss} = k(v_{sd}i_{sd} - Ri_{sd}^2) \quad (15)$$

where the feed-forward term (i_{sd}) is given by

$$i_{sd} = \frac{v_{sd} - \sqrt{v_{sd}^2 - 4R\frac{1}{k}(V_{dc}i_{load} + P_{loss})}}{2R} \quad (16)$$

Equation (16) is the exact compensation but it is not easy to solve in real time, although a look-up table implementation is straightforward. For a simpler expression, i_{sd} can be obtained by neglecting the inductor copper losses ($R = 0$), which yields

$$i_{sd} = (V_{dc}i_{load} + P_{loss})/(kv_{sd}) \quad (17)$$

This term is employed to implement the DC link current (i_{load}) for feed-forward compensation. The feed-forward control block diagram is shown in Fig. 3. In this structure, the value of (i_{load}) is fed forward to the output of the fuzzy controller which itself provides only the dynamic adjustment due to the steady state nature. The d-axis reference current i_{sd}^{ref} including the feed-forward compensation term is

$$i_{sd}^{ref} = i_{sd}^* + i_{sd}^{comp} \quad (18)$$

The load current (i_{load}) can be obtained by 1) direct measurement or 2) DC link current observer. The DC link current observer is a preferable method to avoid a current transducer and the associated electronic circuits.

D. Dc Link Current Observer

Feed-forward control of the DC link load current can provide good dynamics in three-phase active rectifier control. To obtain the load current for the feed-forward signal without additional current sensors, a load current

observer obtains a second-order discrete state equation with a sampling rate of T_s for the three phase PWM rectifier as in

$$\begin{bmatrix} V_{dc}(k+1) \\ i_{load}(k+1) \end{bmatrix} = \begin{bmatrix} 1 & -\dfrac{T_s}{C} \\ 0 & 1 \end{bmatrix} \begin{bmatrix} V_{dc}(k) \\ i_{load}(k) \end{bmatrix} + \begin{bmatrix} \dfrac{T_s}{C} \\ 0 \end{bmatrix} i_{conv} \quad (19)$$

$$V_{dc}(k) = \begin{bmatrix} 1 & 0 \end{bmatrix} \begin{bmatrix} V_{dc}(k) \\ i_{load}(k) \end{bmatrix} \quad (20)$$

$$\hat{x}(k+1) = (\Phi - KC)\hat{x}(k) + \Gamma u(k) + Ky(k) \quad (21)$$

In this paper, we use $C = 925\mu F$, $T = 0.1ms$ and $T_s = 1ms$. Substituting these into (21) gives the value of gain K. The load current observer equation is expressed in (22) and the observer block diagram is shown in Fig. 7.

$$\begin{bmatrix} \hat{V}_{dc}(k+1) \\ \hat{I}_{load}(k+1) \end{bmatrix} = \left[\begin{bmatrix} 1 & -0.1081 \\ 0 & 1 \end{bmatrix} - \begin{bmatrix} 0.7027 \\ -1.3632 \end{bmatrix} [1\ 0] \right] \begin{bmatrix} \hat{V}_{dc}(k) \\ \hat{I}_{load}(k) \end{bmatrix}$$

$$+ \begin{bmatrix} 0.1081 \\ 0 \end{bmatrix} I_{conv}(k) + \begin{bmatrix} 0.7027 \\ -1.3632 \end{bmatrix} V_{dc}(k) \quad (22)$$

The DC link current observer served as an input of the feed-forward control can be implemented by C language on a DSC controller board. The feed-forward of DC link current observed with PI-fuzzy is shown in Fig. 8. The observer receives the DC link voltage V_{dc} to estimate I_{load}, which will then be transformed to $i_{sd}^{comp.}$ by a gain of $2V_{dc}/3V_m$ and fed to the output of PI-fuzzy to generate i_{sd}^{ref}.

E. Overall Control Block Diagram

The overall control block diagram shown in Fig. 9 has two main parts. The first part is a power unit consisting of

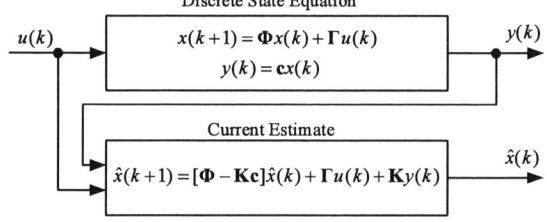

Fig. 7. Block diagram of load current observer.

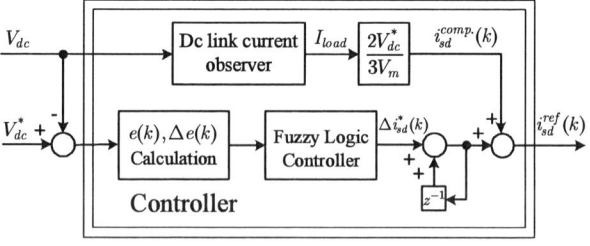

Fig. 8. PI-fuzzy with load current feed-forward.

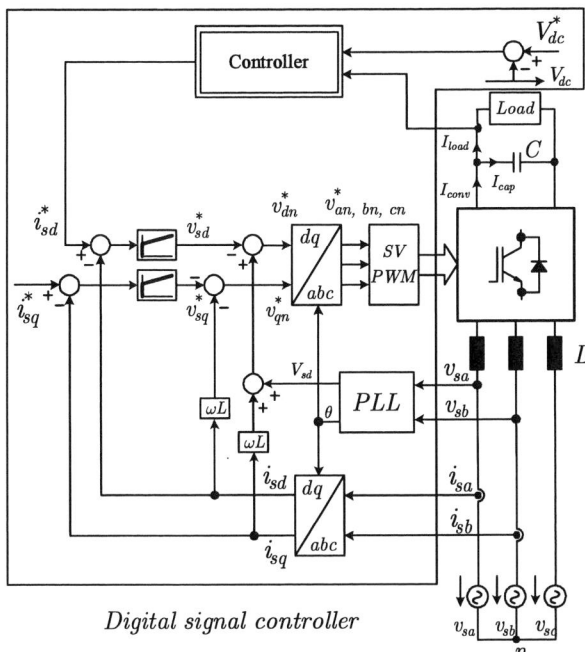

Fig. 9. Control block diagram.

three single phase inductors, a power converter, DC link capacitors and loads (right hand side). The second part is a control unit consisting of a PLL, an axis transformation (three phase to d-, q-axis), an axis transformation from d-, q-axis to a three phase, space vector PWM (left hand side), two PI controllers and proposed controllers (on the top). The digital signal controller controls the system with two references. One is i_{sq}^* for reactive power control (zero reactive power) and the other is V_{dc}^* for DC link voltage control (550V). Both references are the inputs to the proposed controllers. The output of the controllers will be changed to d-, q-axis voltage which is then transformed to three phase voltage. This voltage is converted to drive signal by the SVPWM to control the converters. All the control schemes described are implemented in the controller block labeled "Proposed Controllers" (on the top of the Fig. 9).

V. CASE STUDIES

A. Experimental Systems

A test system is shown in Fig. 10. The upper right corner of the figure is composed of a three phase 7.5 kW converter, DC link capacitors (925 μF), three single phase inductors (4.5mH), three phase variable voltage transformer, loads, voltage and current sensors. The lower part is composed of a data acquisition system connected to a DSC board, an oscilloscope and a personal computer. The controller board uses a high performance 16 bits dsPIC30f6010. The upper left corner is a variable DC voltage supply for tests in the inverting mode. This part can increase the DC voltage higher than the bus voltage to supply the active power in inverting operation. This system is designed to regulate the DC link voltage at 550 V and unity power factor with bi-directional power

flow. The controller is implemented in the DSC and linked with a personal computer via RS232 port for transferring wind speed data to the DSC board and the other for sending parameters to the computer for control verification.

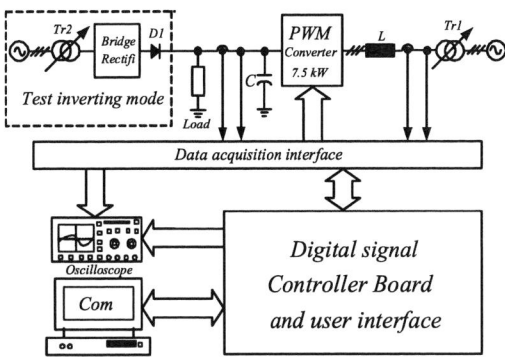

Fig. 10. The test system.

B. Step Load Current Rejection

The control performance is tested in two modes: 1) rectifying mode where the active power flows from the grid to the DC link and load respectively and 2) inverting mode where the active power flows from the DC link to grid. In both operations, the power factor is controlled closed to unity. A conventional PI controller is also implemented for comparison. Figures 11 (a)-(d) show the DC link voltage and currents i_{sd} and i_{sq} response for load disturbance rejection in the rectifying mode for PI (controller A), PI-fuzzy (controller B), PI-fuzzy with measured load current feed-forward (controller C) and PI-fuzzy with observed load current feed-forward (controller D), respectively. A step load of 1.8kW is connected at the DC link at $t = 0.1s$ and removed at $t = 1.2ms$.

Figure 12 shows the same condition as in Fig. 11, but the four controllers are operated in inverting mode. A step current of 3A (approximately 1.65kW) is connected at the DC link at $t = 0.1s$ and removed at $t = 1.2ms$. The performances of the controllers for the two operation modes are summarized in Table II. As can be seen from Table II, the performance of controller B can reduce the percentage of overshoot and undershoot to 10.91% of load current rejection, compared with those of controller A. When the current is measured directly for controller C, the percentage of overshoot and undershoot are further reduced by 1.82 % and 0.91 %, respectively. With a current observer to estimate i_{load}, the output response still yields significant improvement, with a reduction to 3.64% and 5.45% in the overshoot and undershoot. Although controller D is not as good as controller C, there is still a considerable improvement over controllers A and B. The increase/decrease of setting time has the same direction with the percentage of overshoot and undershoot. For the inverting mode, the percentage of overshoot and undershoot have the same trend as shown in the lower of Table II.

Figure 13 shows the measurement, I_L and observed load current, $\hat{I}_L$. When the load current is stepped at

1133

$t = 80\ ms$, the load current observer takes $16\ ms$ to track the load current. The delay time is caused by the low pass filter. Figures 14 and 15 show the AC side converter current and the associated phase voltage for steady state operation in rectifying and inverting modes. When i_{sq} is set at zero, the phase displacement between the phase voltage and current become $0°$ and $180°$, respectively.

VI. CONCLUSION

This paper has implemented three control algorithms for regulating the DC link voltage of a boost type front end converter in wind energy conversion systems. The three algorithms are PI-fuzzy, PI-fuzzy with measured load current feed-forward and PI-fuzzy with observed load current feed-forward. A conventional PI controller is also implemented for performance comparison. It can be shown from the results that the control performance of a PI controller can be improved by using a PI-fuzzy, the developed rules base which can be designed to separate PI gain for rectifying and inverting modes. Further improvement is obtained by load current feed-forward, which gives the best output response for load current rejection and nonlinear load injection. However, it requires a load current sensor. Alternatively, a load current observer can be used to obtain the feed-forward compensation term with a slightly lower performance.

ACKNOWLEDGEMENT

The authors would like to give a special thank to Dr. Somporn Sirisumrannukul for valuable comments.

REFERENCES

[1] N. R. Zargary and G. Joos, "Performance Investigation of a Current-controlled Voltage-regulated PWM Rectifier Rotating and Stationary Frames," *IEEE transactions on Industry Electronics,* Vol. 42, No. 4, Aug. 1995, pp. 396-401.

[2] R. Pena, J.C. Clare, G.M. Asher, "A doubly fed induction generator using back to back PWM converters and its application to variable speed wind energy generation," *IEE-Proceeding part B* (Electric Power and Applications), May 1996, pp. 231-241.

[3] C.T. Rim, N.S. Choi, G.C. Cho and G.H. Cho, "A complete DC and AC Analysis of Three-phase Current PWM Rectifier Using d-q Transformation," *IEEE Transactions on Power Electronics,* Vol. 9, Issue. 4, July 1994, pp. 390-396.

[4] R. Jones and I. Gilmore, "Benefits of Sinusoidal Rectifiers on Variable Speed Wind Turbines High Quality Mains Power from Variable Speed Wind Turbines," *17th British Wind Energy Association Conference,* 1995, pp.339-345.

[5] Jinhwan Jung, Sun-Kyoung Lim and Kwanghee Nam, "A feedback linearizing control scheme for PWM converter-inverter having a very small DC-link capacitor," IEEE Transactions On Industry Application, Vol 35, No 5, 12-15 Oct. 1998, pp.1497-1503.

[6] Kaura, V. and Blasko, V., *Operation of a Phase Locked Loop System Under Distorted Utility Condition,* IEEE on Industry Application, Vol. 33, Issue 1, Jan.-Feb1997, pp. 58-63.

[7] D. Driankov, H. Hellendoorn, M. Reinsrank, "An Introduction to Fuzzy Control," *Springer-Verlag,* 1993.

[8] T.Takagi and M. Sugeno, "Fuzzy Identification of Systems and Its Applications to Modeling and Control," *IEEE Transactions on System,* Vol 15, 1985, pp.116-132.

[9] R. Cardenas, R. Pena, G.M. Asher and J.C. Clare, "Control Strategies for Energy Recovery from Flywheel Using a Vector Controlled Induction Machine," *IEEE Power Electronics Specialist Conf.,* Vol.1, 18-23 June 2000, pp. 454-459.

[10] Zhingfu Zhou, Yanzhen Liu, and P.J. Unsworth, "Design of DC Link Current Observer for a 3-phase Active Rectifier with Feed-forward Control," *IEEE Industrial Applications Conf.,* Vol.1, 3-7 Oct. 2004, pp. 461-468.

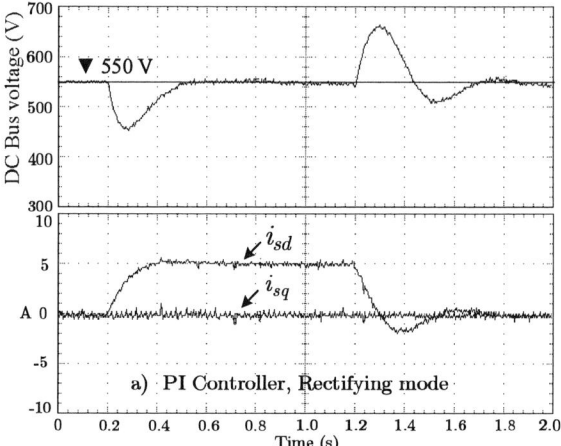

a) PI Controller, Rectifying mode

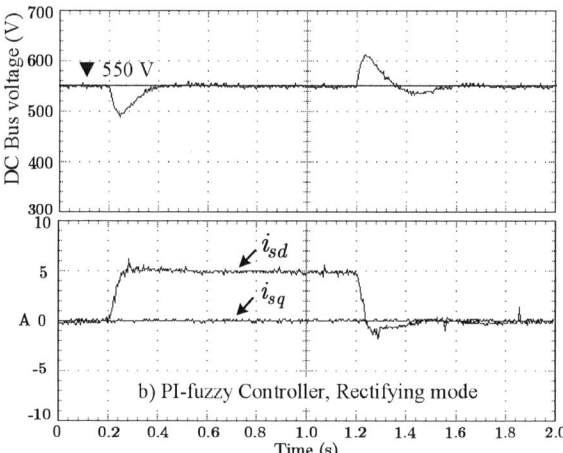

b) PI-fuzzy Controller, Rectifying mode

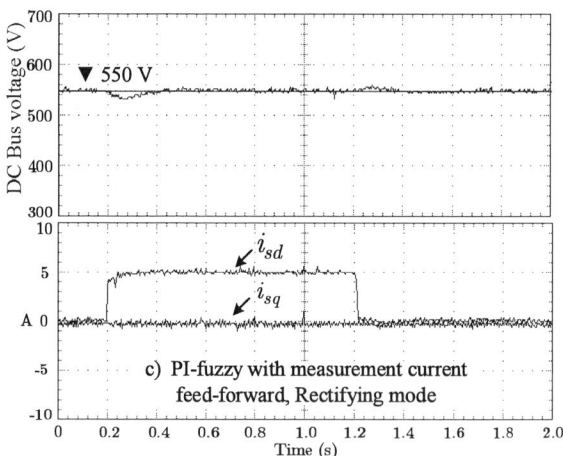

c) PI-fuzzy with measurement current feed-forward, Rectifying mode

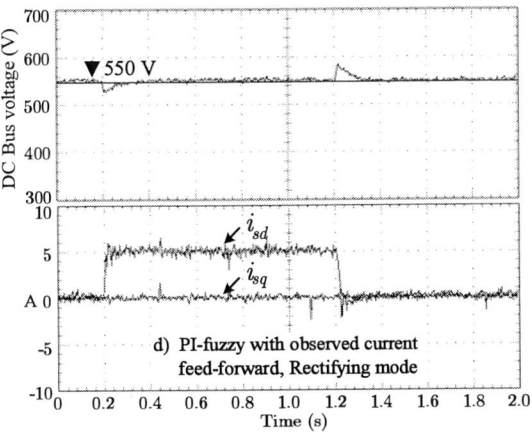

Fig. 11. Control performance of each controller with step load (rectifying mode).

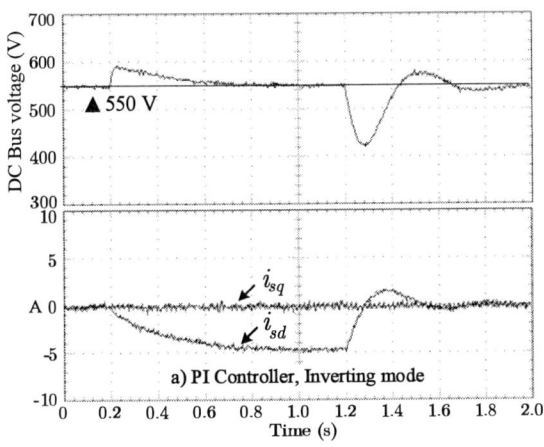

a) PI Controller, Inverting mode

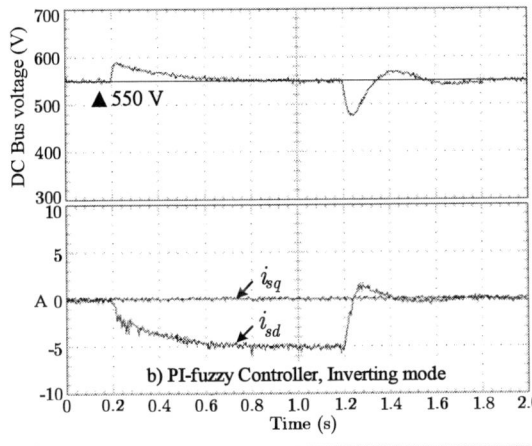

b) PI-fuzzy Controller, Inverting mode

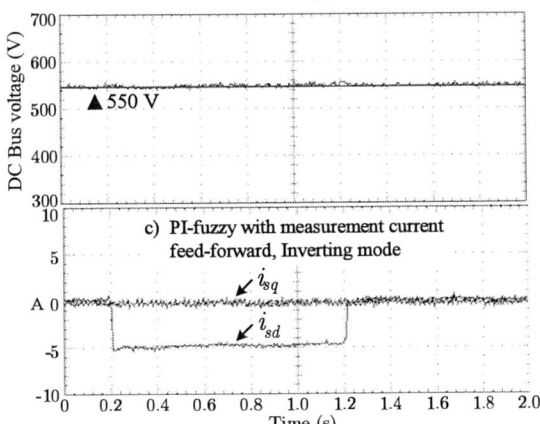

c) PI-fuzzy with measurement current feed-forward, Inverting mode

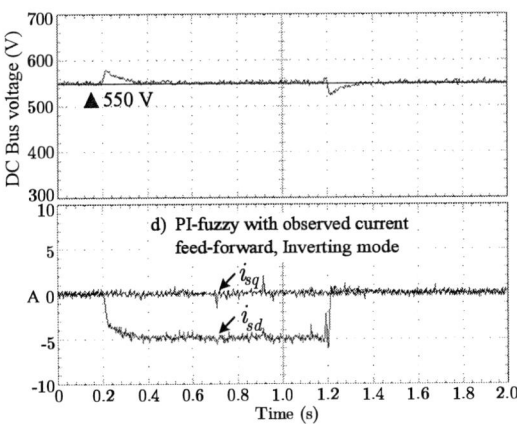

d) PI-fuzzy with observed current feed-forward, Inverting mode

Fig. 12. Control performance of each controller with step load (inverting mode).

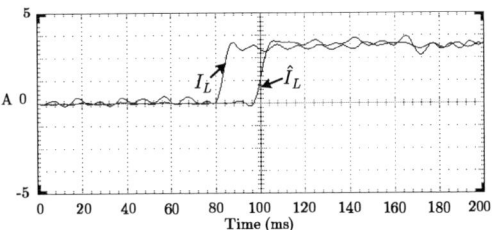

Fig. 13. Measurement, I_L and observed load current, $\hat{I}_L$.

TABLE II COMPARISON OF CONTROL PERFORMANCE MODE: A) PI, B) PI-FUZZY, C) PI-FUZZY WITH MEASUREMENT LOAD CURRENT FEED-FORWARD AND D) PI-FUZZY WITH OBSERVED LOAD CURRENT FEED-FORWARD

Mode	Controller type	Undershoot (%)	Overshoot (%)	Settling time step up/down
Rectifying	A	16.36	18.18	300/650 ms
	B	10.91	10.91	200/350 ms
	C	1.82	0.91	100/100 ms
	D	3.64	5.45	100/120 ms
Inverting	A	21.82	7.27	500/500 ms
	B	14.55	7.27	400/400 ms
	C	0.36	0.36	5/5 ms
	D	3.64	5.45	150/120 ms

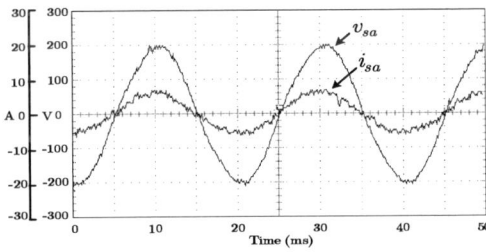

Fig. 14. Steady state performance for rectifying mode.

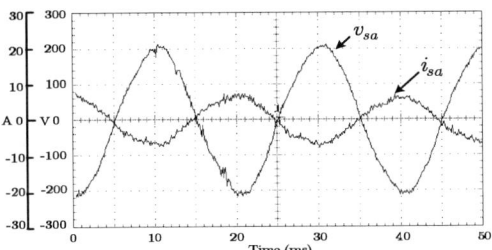

Fig. 15. Steady state performance for inverting mode.

1135

Analysis and Design of Class DE Amplifier with Nonlinear Shunt Capacitance

Hiroo Sekiya , Takayuki Watanabe , Tadashi Suetsugu[†] and Marian K. Kazimierczuk[‡]

Graduate School of Advanced Integration Science, Chiba Univ., Chiba, 263-8522 JAPAN

Email: sekiya@faculty.chiba-u.jp

[†] Dept. of Electronics Engineering, Fukuoka Univ., Fukuoka, 814-0180 JAPAN

Email: suetsugu@fukuoka-u.ac.jp

[‡]Dept. of Electrical Engineering, Wright State Univ., Dayton, Ohaio, 45435-0001 USA

Email: marian.kazimierczuk@wright.edu

Abstract—This paper presents the analysis and design of class DE amplifier with nonlinear shunt capacitance. In the past, the researchers clarified class DE amplifier with linear shunt capacitance analytically. For higher frequency operation, however, lower shunt capacitance is required for the design. In this case, shunt capacitance has nonlinearity because of output capacitance of MOSFET. Therefore, it is important to clarify the effect of this nonlinearity to the operation of class DE amplifier analytically. In this paper, analytical expressions are derived as waveforms and design equations. From this analysis, it is clarified that the nonlinearities of shunt capacitances affect the waveforms in the dead-time intervals. The validity of our analysis is confirmed by PSpice simulations and experiments.

I. INTRODUCTION

Class DE amplifier [1]–[5] is realized by adding shunt capacitance to class D amplifier and achieving class E zero-voltage switching (ZVS) and zero-derivative switching (ZDS) at the transistor turn-on [6].

Since the class E switching should be satisfied with two conditions simultaneously, it is difficult to determine the elemental values of class DE amplifier. Therefore, several analyses were carried out to design it [1]–[5]. In all analyses, however, it is assumed that shunt capacitances are linear. In the real circuits, the shunt capacitance is realized by the sum of an external capacitance and the MOSFET shunt capacitances. The problem is that the parasitic capacitance is nonlinear, [7]–[14], but the external capacitance is linear. The external capacitance is much higher than the transistor output capacitance for frequencies in the hundred kHz or a few MHz range. Therefore, the all analyses of the class DE amplifier until now have assumed that the shunt capacitances are linear. However, the shunt capacitances are lower and lower as the operating frequency becomes high. Therefore, the parasitic capacitances of MOSFETs are dominant under high frequency operation so their nonlinearities cannot be neglected. For above reason, it is quite important to clarify the effects of the nonlinearities of the shunt capacitances to the operation and design of the class DE amplifier analytically.

This paper presents the analysis and design of the class DE amplifier with nonlinear shunt capacitances. Analytical expressions are derived as waveforms and design equations. From this analysis, it is clarified that the nonlinearities of shunt

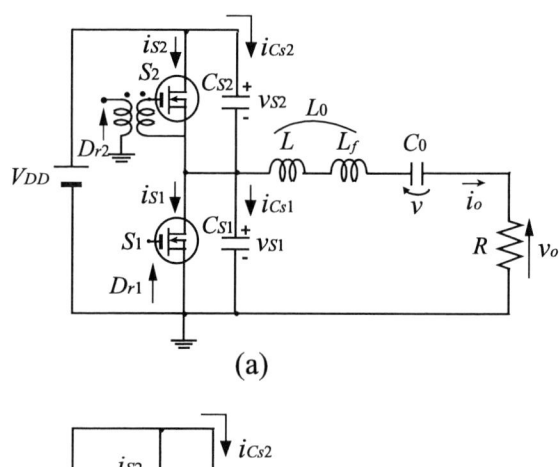

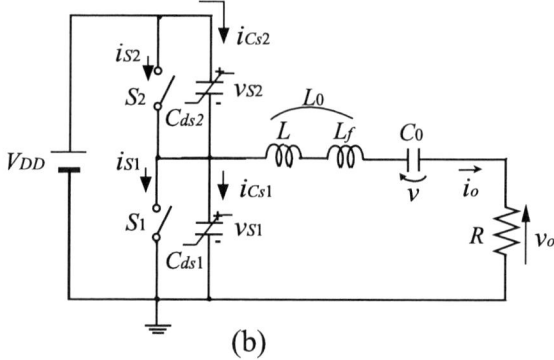

Fig. 1. Class DE amplifier. (a) Circuit topology. (b) Equivalent circuit.

capacitances affect the waveforms in the dead-time intervals and the ratio of the supply voltage to the built-in potential of the MOSFET body diode is a very important parameter. The parameters of the nonlinear shunt capacitances are determined from the operating frequency, the load resistance, and the ratio of the supply voltage, and the built-in potential. This result means that the operating frequency is restricted by the nonlinearity of the shunt capacitance if the load resistance and the supply voltage are given as design specifications. By carrying out PSpice simulation, it is confirmed that simulated waveforms are similar to the analytical ones quantitatively, which indicates the validity of our analysis.

978-1-4244-0644-9/07/$25.00 ©2007 IEEE

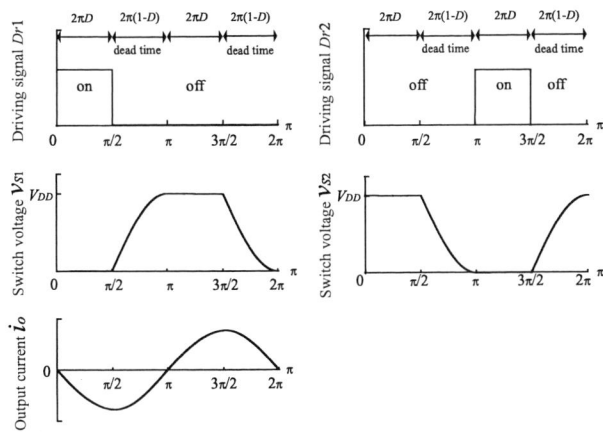

Fig. 2. Nominal waveform of class DE amplifier for D=0.25.

II. CIRCUIT DESCRIPTION

Figure 1 depicts the circuit topology of the class DE amplifier[1]-[5]. The example waveforms of the class DE amplifier are shown in Fig. 2 for the switch on duty ratio is 0.25. The switches S_1 and S_2 are driven by a driving pattern of D_1 and D_2 shown in Fig. 2, respectively. The driving pattern generates a dead time during the period when one switch has turned off and before the other switch has turned on. During the dead time, the output current i_o charges one shunt capacitance and discharges the other. The midpoint voltage between two switches v_{S1} becomes equal to V_{DD} or zero at the end of the dead time, allowing the class E switching conditions to be satisfied [6]. Class E switching conditions means that both the switch voltage and the slope of the switch voltage are zero at the switching instant, namely,

$$v_{S1}(2\pi) = 0, \qquad \frac{dv_{S1}(\theta)}{d\theta}\bigg|_{\theta=2\pi} = 0, \qquad (1)$$

$$v_{S2}(\pi) = 0, \qquad \frac{dv_{S2}(\theta)}{d\theta}\bigg|_{\theta=\pi} = 0. \qquad (2)$$

Because of the class E switching, the class DE amplifier achieves high power conversion efficiency for high-frequency operation.

III. THE NONLINEARITY OF SHUNT CAPACITANCE

Generally, shunt capacitances C_{S1} and C_{S2} are realized by the sum of the external capacitances and the MOSFET output capacitances. The problem is that the transistor output capacitance is nonlinear, but the external capacitance is linear. The external capacitance is much higher than the transistor output capacitance at low operating frequencies. Therefore, the all analyzes of the class DE amplifier until now have assumptions that the shunt capacitances are linear elements. However, the shunt capacitances C_{S1} and C_{S2} are lower and lower as the operating frequency becomes high. Therefore, the MOSFET output parasitic capacitances are dominant for high-frequency operations so the designers cannot eliminate their nonlinearities. For the above reason, it is quite important to

TABLE I
SWITCHING PATTERN

	$0 \leq \theta < \frac{\pi}{2}$	$\frac{\pi}{2} \leq \theta < \pi$	$\pi \leq \theta < \frac{3\pi}{2}$	$\frac{3\pi}{2} \leq \theta < 2\pi$
D_{r2}	OFF	OFF	ON	OFF
D_{r1}	ON	OFF	OFF	OFF

clarify the effects of the nonlinearities of the shunt capacitances on the operation and design of the class DE amplifier analytically.

Since the power MOSFET contains a p-n junction body diode, the parasitic capacitance C_{ds} can be expressed by

$$C_{ds} = \frac{C_{j0}}{\left(1 + \frac{v_S}{V_{bi}}\right)^m}, \qquad (3)$$

where V_{bi} is the built-in potential which typically ranges from 0.5 to 0.9 V, v_S is the drain-to-source voltage, C_{j0} is the capacitance at $v_S = 0$, and m is the grading coefficient of the diode junction [7]–[14].

IV. ANALYSIS OF CLASS DE AMPLIFIER WITH NONLINEAR SHUNT CAPACITANCE

A. Assumptions

The analysis in this paper is based on the following assumptions.

a. Both shunt capacitances are only parasitic capacitances of MOSFETs whose nonlinearities are given by (3) for $m = 0.5$.

b. All passive elements except shunt capacitances are linear elements and do not have parasitic resistances.

c. The quality factor of the resonant filter $Q = \omega L0/R$ is high enough to generate pure sinusoidal output voltage. The current through the L_0 C_0 circuit and the load resistance is sinusoidal at the operating frequency.

d. The duty ratios of the two switches are fixed at $D = 0.25$.

e. It is assumed that the switching pattern is the same as that in Fig. 2 and in Table I.

From the above assumptions, the equivalent model of the Class DE amplifier is given as shown in Fig. 1(b).

B. Waveform Equations

The analysis is performed in the interval $0 \leq \theta < 2\pi$, where, $\theta = \omega t$ represents the angular time. The output current is assumed to be sinusoidal

$$i_o(\theta) = I_o \sin(\theta + \varphi). \qquad (4)$$

Time Interval: $0 \leq \theta < \pi/2$

The switch S_1 is turned on at $\theta = 0$. In this time interval, the switch S_1 is ON and S_2 is OFF as shown in Table I. Therefore, the switch voltages are given by

$$v_{S1}(\theta) = 0, \qquad v_{S2}(\theta) = V_{DD}. \qquad (5)$$

Since they are constant, the currents through the shunt capacitance C_{S1} and C_{S2} are given as

$$i_{CS1}(\theta) = i_{CS2}(\theta) = 0. \qquad (6)$$

Because of the bottom switch is open, its current is

$$i_{S2}(\theta) = 0. \tag{7}$$

From (4), (6) and (7), the current through the switch S_1 is obtained as

$$i_{S1}(\theta) = -i_o(\theta) = -I_o \sin(\theta + \varphi). \tag{8}$$

Time Interval: $\pi/2 \le \theta < \pi$

The switch S_1 is turned off at $\theta = \pi/2$. In this time interval, both switches S_1 and S_2 are OFF as shown in Table I. Therefore,

$$i_{S1}(\theta) = i_{S2}(\theta) = 0. \tag{9}$$

From KCL, the relationship for the current is

$$\omega C_{ds1} \frac{d v_{S1}(\theta)}{d\theta} + \omega C_{ds2} \frac{d v_{S2}(\theta)}{d\theta} = -I_o \sin(\theta + \varphi). \tag{10}$$

Moreover, when we consider the relation of the switch voltages, namely, $v_{S1} = V_{DD} - v_{S2}(\theta)$, the following equation is obtained.

$$\frac{d v_{S2}(\theta)}{d\theta} = \frac{I_o}{\omega(C_{ds1} + C_{ds2})} \sin(\theta + \varphi). \tag{11}$$

Since $v_{S2}(\pi/2) = V_{DD}$, we have

$$\omega \int_{V_{DD}}^{v_{S2}} \left(\frac{C_{j0}}{\sqrt{1 + \frac{V_{DD} - v_{S2}}{V_{bi}}}} + \frac{C_{j0}}{\sqrt{1 + \frac{v_{S2}}{V_{bi}}}} \right) d v_{S2} = \int_{\frac{\pi}{2}}^{\theta} I_o \sin(\theta' + \varphi) d\theta'. \tag{12}$$

The voltage v_{S2} is obtained in the analytical form by performing the integration of (12)

$$v_{S2}(\theta) = \frac{V_{DD}}{2} - \frac{1}{2} \left[V_{DD}^2 - 4V_{bi}^2 \left(-1 + \frac{V_{DD}}{V_{bi}} + \left\{ 1 + \frac{V_{DD}}{2V_{bi}} - \frac{1}{2} \left[\frac{I_o(\cos(\theta + \varphi) + \sin\varphi)}{2\omega V_{bi} C_{j0}} - 1 + \sqrt{1 + \frac{V_{DD}}{V_{bi}}} \right]^2 \right\}^2 \right) \right]^{\frac{1}{2}}. \tag{13}$$

In (13), the sign "$\pm$" of the second term in the right-hand side changes at the boundary of $v_{S2} = V_{DD}/2$. Namely, "+" is valid in the interval $\pi/2 \le \theta < 2\pi/3$ and "$-$" is valid in the interval $2\pi/3 \le \theta < \pi$. Now, the class E switching conditions are considered. From (2) and (10), we have

$$I_o \sin(\pi + \varphi) = 0 \tag{14}$$
$$\varphi = 0, \pi. \tag{15}$$

The phase shift of the output current $\varphi = \pi$ since it means that the amplitude of the output I_o is positive. From (12) and $v_{S2}(\pi) = 0$, we get

$$\cos\varphi = -\frac{4\omega V_{bi} C_{j0}}{I_o} \left(\sqrt{1 + \frac{V_{DD}}{V_{bi}}} - 1 \right). \tag{16}$$

The amplitude of the output current is given as

$$I_o = 4\omega C_{j0} V_{bi} \left(\sqrt{1 + \frac{V_{DD}}{V_{bi}}} - 1 \right). \tag{17}$$

By substituting (17) into (13), the switch voltage v_{S2} becomes

$$v_{S2}(\theta) = \frac{V_{DD}}{2} - \frac{1}{2} \left[V_{DD}^2 - 4V_{bi}^2 \left(-1 + \frac{V_{DD}}{V_{bi}} + \left\{ 1 + \frac{V_{DD}}{2V_{bi}} - \frac{1}{2} \left[2 \left(1 - \sqrt{1 + \frac{V_{DD}}{V_{bi}}} \right) \cos\theta - 1 + \sqrt{1 + \frac{V_{DD}}{V_{bi}}} \right]^2 \right\}^2 \right) \right]^{\frac{1}{2}}. \tag{18}$$

Time Interval: $\pi \le \theta < 3\pi/2$

The switch S_2 is turned on at $\theta = \pi$. In this interval, S_1 is OFF and S_2 is ON as shown in Table I. Since the switch voltages are constant

$$v_{S1}(\theta) = V_{DD}, \qquad v_{S2}(\theta) = 0, \tag{19}$$

the currents through the shunt capacitances C_{S1} and C_{S2} are

$$i_{CS1}(\theta) = i_{CS2}(\theta) = 0. \tag{20}$$

Because S_1 is open, the current through the switch S_1 is

$$i_{S1} = 0. \tag{21}$$

From (4), (20) and (21), the current through the switch S_1 is derived as

$$i_{S1}(\theta) = -i_o(\theta) = I_o \sin\theta. \tag{22}$$

Time Interval: $3\pi/2 \le \theta < 2\pi$

The switch S_2 is turned off at $\theta = 3\pi/2$. In this interval, both switches S_1 and S_2 are OFF as shown in Table I, namely,

$$i_{S1}(\theta) = i_{S2}(\theta) = 0. \tag{23}$$

From

$$\omega C_{ds1} \frac{d v_{S1}(\theta)}{d\theta} + \omega C_{ds2} \frac{d v_{S2}(\theta)}{d\theta} = -I_o \sin\theta \tag{24}$$

and $v_{S2} = V_{DD} - v_{S1}(\theta)$, the differential equation for the switch voltage v_{S1} is obtained

$$\frac{d v_{S1}(\theta)}{d\theta} = \frac{I_o}{\omega(C_{ds1} + C_{ds2})} \sin\theta. \tag{25}$$

Because $v_{S1}(3\pi/2) = V_{DD}$, we can write

$$\omega \int_{V_{DD}}^{v_{S1}} \left(\frac{C_{j0}}{\sqrt{1 + \frac{V_{DD} - v_{S1}}{V_{bi}}}} + \frac{C_{j0}}{\sqrt{1 + \frac{v_{S1}}{V_{bi}}}} \right) d v_{S1} = \int_{\frac{3\pi}{2}}^{\theta} I_o \sin\theta \, d\theta. \tag{26}$$

1138

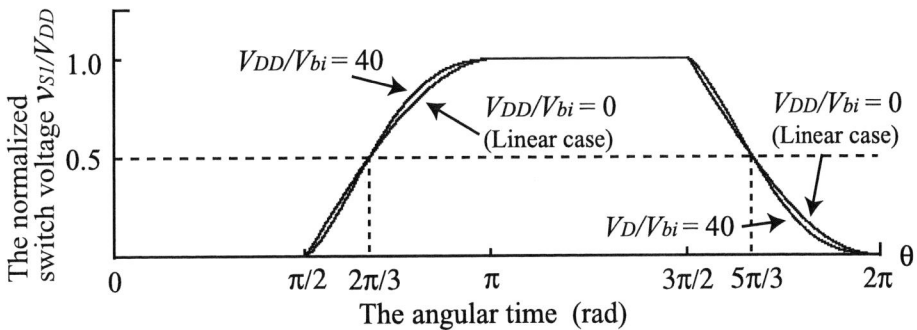

Fig. 3. Normalized voltage of the bottom switch v_{S1}/V_{DD}.

By calculating the integration of (26) and substituting (17), the switch voltage v_{S1} is derived analytically as

$$
\begin{aligned}
v_{S1}(\theta) &= \frac{V_{DD}}{2} - \frac{1}{2}\left[V_{DD}^2 - 4V_{bi}^2\left(1 - \frac{V_{DD}}{V_{bi}}\right) \right.\\
&+ \left\{ 1 + \frac{V_{DD}}{2V_{bi}} - \frac{1}{2}\left(2 - 1 - \sqrt{1 + \frac{V_{DD}}{V_{bi}}}\right)\cos\theta \right.\\
&\left.\left. \left(1 + \sqrt{1 + \frac{V_{DD}}{V_{bi}}}\right)^2 \right\}^2 \right]^{\frac{1}{2}}.
\end{aligned}
\tag{27}
$$

In (27), the sign "$\mp$" of the second term in the right-hand side changes at the boundary of $v_{S1} = V_{DD}/2$. Namely, the sign "+" is valid in the interval $3\pi/2 < \theta < 5\pi/3$ and "$-$" is valid in the interval $5\pi/3 < \theta < \pi$.

When both side of (18) are divided by V_{DD}, the following equation;

$$
\begin{aligned}
\frac{v_{S2}(\theta)}{V_{DD}} &= \frac{1}{2} - \frac{1}{2}\left[1 - 4\frac{V_{bi}^2}{V_{DD}^2}\left(1 - \frac{V_{DD}}{V_{bi}}\right) \right.\\
&+ \left\{ 1 + \frac{V_{DD}}{2V_{bi}} - \frac{1}{2}\left(2 - 1 - \sqrt{1 + \frac{V_{DD}}{V_{bi}}}\right)\cos\theta \right.\\
&\left.\left. \left(1 + \sqrt{1 + \frac{V_{DD}}{V_{bi}}}\right)^2 \right\}^2 \right]^{\frac{1}{2}}.
\end{aligned}
\tag{28}
$$

is given. The resulting waveform equations indicate that the difference between amplifiers with the nonlinear shunt capacitances and linear ones appear only during the dead time intervals. Figure 3 shows the normalized voltage waveforms for the bottom switch v_{S1}/V_{DD}. From this figure, it is found that the slope of the switch voltage increases when V_{DD}/V_{bi} increases. Namely, the maximum currents through the shunt capacitances are larger due to the nonlinearities of the shunt capacitances.

C. Power Relations

The input power is given by

$$
P_{in} = V_{DD}I_D.
\tag{29}
$$

In (29), the dc input current I_D is expressed as the average of the supply current from the dc voltage source V_{DD}, which is given by

$$
\begin{aligned}
I_D &= \frac{1}{2\pi}\int_0^{2\pi}\{i_{S2}(\theta) + i_{CS2}(\theta)\}d\theta \\
&= \frac{1}{2\pi}\int_{\frac{3\pi}{2}}^{\pi} I_o\sin\theta\, d\theta \\
&= 2\frac{\omega C_{j0}V_{bi}}{\pi}\left(\sqrt{1 + \frac{V_{DD}}{V_{bi}}} - 1\right).
\end{aligned}
\tag{30}
$$

The output power P_o is derived from (17) as

$$
\begin{aligned}
P_o &= \frac{I_oV_o}{2} = \frac{RI_o^2}{2} \\
&= 8\omega^2 C_{j0}^2 RV_{bi}^2\left(\sqrt{1 + \frac{V_{DD}}{V_{bi}}} - 1\right)^2.
\end{aligned}
\tag{31}
$$

On the other hand, P_o is also derived from (17) as

$$
P_o = \frac{V_o^2}{2R} = \frac{V_{DD}^2}{2R\pi^2}.
\tag{32}
$$

If the power losses on the parasitic resistance are neglected and the identical operation is assumed, the drain efficiency achieves 100% conversion, Namely,

$$
P_{in} = P_o
\tag{33}
$$

From (29) – (31), the following equation is obtained:

$$
\omega C_{j0}R = \frac{V_{DD}/V_{bi}}{4\left(\sqrt{1 + \frac{V_{DD}}{V_{bi}}} - 1\right)}
\tag{34}
$$

D. Fourier Analysis

The output voltage $v_o(\theta)$ and the voltage across the resonant inductance $v_L(\theta)$ are expressed as

$$
v_o(\theta) = V_o(-\sin\theta),
\tag{35}
$$

$$
v_L(\theta) = V_L(-\cos\theta),
\tag{36}
$$

where the amplitudes V_o and V_L are

$$
V_o = RI_o,
\tag{37}
$$

$$
V_L = \omega LI_o.
\tag{38}
$$

1139

From (37) and (38), we obtain

$$\frac{V_L}{V_o} = \frac{\omega L}{R}. \tag{39}$$

The amplitudes V_o and V_L can be calculated from the Fourier integrals

$$V_o = \frac{1}{\pi} \int_0^{2\pi} s_1(\theta) \sin(\theta) d\theta = \frac{V_{DD}}{\pi}, \tag{40}$$

$$V_L = \frac{1}{\pi} \int_0^{2\pi} s_1(\theta) \cos(\theta) d\theta = \frac{V_{DD}}{2}, \tag{41}$$

Eqs. (40) and (41) show same expressions as those in [1]. From this result, the output voltage is only determined by supply voltage and the nonlinearity of the shunt capacitance never affect to the relation of supply and output voltage if class E switching conditions are satisfied.

E. Load Network Elements

Finally, we derive the design equations of the load network. From (40), the output power is

$$P_o = \frac{V_o^2}{2R} = \frac{V_{DD}^2}{2R\pi^2}. \tag{42}$$

Moreover, from (29), (30) and 100% drain efficiency, the output power is given as

$$P_o = P_{in} = \frac{2\omega V_{bi} C_{j0} V_{DD} \left(\sqrt{1 + \frac{V_{DD}}{V_{bi}}} - 1\right)}{\pi}. \tag{43}$$

By substituting (43) to (42), the load resistance R is given as

$$R = \frac{V_{DD}^2}{2\pi^2 P_o} = \frac{V_{DD}/V_{bi}}{4\pi\omega C_{j0}\left(\sqrt{1 + \frac{V_{DD}}{V_{bi}}} - 1\right)}. \tag{44}$$

From (39), (40) and (41), the inductance L is given as

$$L = \frac{R}{2\omega}. \tag{45}$$

Because of the definition of the loaded quality factor Q, the inductance L_0 is given as

$$L_0 = \frac{QR}{\omega} \tag{46}$$

From (45) and (46), the inductance L' is obtained as

$$L' = L_0 - L = \frac{R(2Q - \pi)}{2\omega}. \tag{47}$$

The identical resonant filter with resonant frequency $f = 2\pi\omega$ is realized by L' and C_0. From $\omega = 1/\sqrt{L'C_0}$ and (47), the resonant capacitance C_0 is expressed analytically as;

$$C_0 = \frac{1}{\omega R(Q - \frac{\pi}{2})} \tag{48}$$

V. SIMULATION VERIFICATION

In this section, the design example is shown for class DE amplifier with nonlinear shunt capacitance. It is assumed that the shunt capacitance are composed of only parasitic capacitance on MOSFETs and it can be varied arbitrarily.

A. Design Example

At first, the design specifications are given as; the operating frequency $f = 20$MHz, the supply voltage $V_D = 10$V, the output power $P_o = 1$ W, the built-in potential of MOSFET $V_{bi} = 0.7$ V and the loaded quality factor $Q = 10$.

From (44), the load resistance R is calculated $R = 5.07 \Omega$. The inductance of the resonant circuit L_0 is given from (46), that is, $L_0 = 0.404 \mu$H. The capacitance of the resonant circuit C_0 is obtained from (48), that is, $C_0 = 0.186$nF. Finally, C_{j0} in the nonlinear shunt capacitance is derived from (34), namely, $C_{j0} = 0.613$nF.

B. PSpice Simulation

In this paper, the simulations are performed using the circuit simulator PSpice. We use a power MOSFET Spice model Level 3.

Figure 4 shows the waveforms from analysis and simulations. Figures 4 (a) and (b) are the waveforms from analytical waveform equations, simulated waveforms for $f = 20$MHz, respectively. The analytical equations denote that the waveforms of class DE amplifier with nonlinear shunt capacitance independent on the operating frequency. Both waveforms of switch voltage achieve class E switching conditions and the waveforms in Fig. 4 (b) are quite similar to those in Fig. 4 (a). Zero slope of voltage switching can be confirmed since the output current i_o is zero at the turn on instant. These results indicate the validity of our analysis.

VI. CONCLUSION

This paper has presented the analysis and design of the class DE amplifier with nonlinear shunt capacitances. Analytical expressions are derived as waveforms and design equations. From this analysis, it is clarified that the nonlinearities of shunt capacitances affect the waveforms in the dead-time intervals and the ratio of the supply voltage to the built-in potential of the MOSFET body diode is a very important parameter. The parameters of the nonlinear shunt capacitances are determined from the operating frequency, the load resistance, and the ratio of the supply voltage, and the built-in potential. This result means that the operating frequency is restricted by the nonlinearity of the shunt capacitance if the load resistance and the supply voltage are given as design specifications. By carrying out PSpice simulation, it is confirmed that simulated waveforms are similar to the analytical ones quantitatively, which indicates the validity of our analysis.

ACKNOWLEDGEMENT

This research was partially supported by Saneyoshi Scholarship Foundation and Grant-in-Aid for scientific research (No. 17760296 and 18560269) of JSPS.

REFERENCES

[1] H. Koizumi, T. Suetsugu, M. Fujii, K. Shinoda, S. Mori, and K. Ikeda, "Class DE high-efficiency tuned power amplifier , " *IEEE Trans. Circuits Syst.*, vol.43, no.1, pp.51-60, Jan. 1996.
[2] K. Shinoda, T. Suetsugu, M. Matsuo and S. Mori, "Idealized operation of the class DE amplifier and frequency multipliers , " *IEEE Trans. Circuits Syst.*, vol.45, no.1, pp.34-40, Jan. 1998.

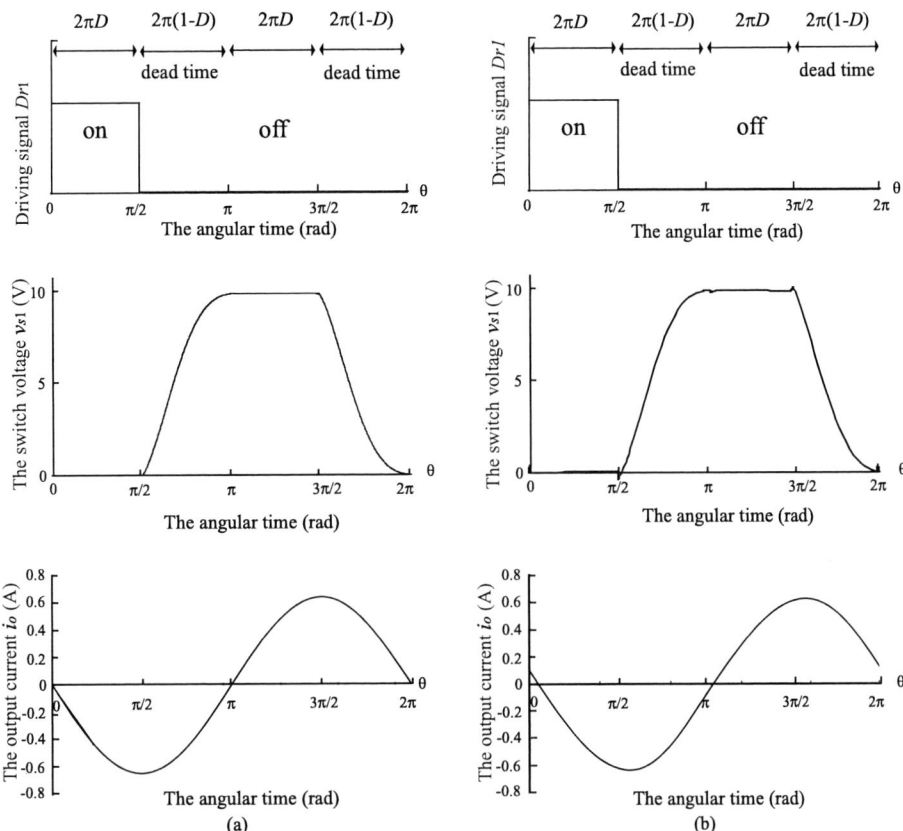

Fig. 4. Example waveforms of class DE amplifier with nonlinear shunt capacitance. (a)Waveforms from analytical equations, (b)Simulated waveforms for f=20MHz.

[3] M. Albulet, "An exact analysis of class-DE amplifier at any output Q, " *IEEE Trans. Circuits Syst.*, vol.CAS-46, no.10, pp.242-248, Apr. 1999.

[4] H. Sekiya, S. Mori and I. Sasase, "Exact analysis of class DE amplifier with FM and PWM control, " *IEICE Trans. on Communications*, vol.E86-B, no.10, pp.3082–3093, Oct. 2003.

[5] H. Sekiya, H. Koizumi, S. Mori, I. Sasase, J. Lu, and T. Yahagi, "FM/PWM control scheme in class DE inverter, " *IEEE Trans. Circuits and Systems I*, vol. 51, no. 7, pp. 1250 - 1260, July 2004.

[6] N.O. Sokal and A.D. Sokal, "Class E - A new class of high-efficiency tuned single-ended switching power amplifiers," *IEEE J. Solid-State Circuits*, vol.SC-10, no.3, pp.168-176, June 1975.

[7] M. J. Chudobiak, "The Use of Parasitic Nonlinear Capacitors in Class E Amplifiers," *IEEE Trans.Circuits Syst-I*, vol.41, no.12, pp.941-944, Dec. 1994.

[8] P. Alinikula, K. Choi, and S. I. Long, "Design of class E power amplifier with nonlinear parastic output capacitance," *IEEE Trans. Circuits and Systems II*, vol. 46, no. 2, pp. 114 – 119, Feb. 1999.

[9] C. K. T. Chan and C. Toumazou, "Design of a class E power amplifier with non-linear transistor output capacitance and finite DC-feed inductance," *The 2001 IEEE International Symposium on Circuits and Systems*, vol. 1, pp. 129 – 132, May 2001.

[10] S. W. Ma, H. Wong, and C. K. Ho, "Piece-wise linear approximation of MOS nonlinear junction capacitance in high-frequency class E amplifier design," *2003 IEEE Conference on Electron Devices and Solid-State Circuits*, no. 223 – 236, Dec. 2003.

[11] T. Suetsugu and M. K. Kazimierczuk, "Comparison of Class-E Amplifier With Nonlinear and Linear Shunt Capacitance,"*IEEE Trans.Circuits Syst-I*, vol.50, no.8, pp.1089-1097, Aug 2003.

[12] P. M. Gaudo, C. Bernal, and A. Mediano, "Output power capability of class-E amplifiers with nonlinear shunt capacitance," *2004 IEEE MTT-S International*, vol. 2, pp.891-894, June 2004.

[13] T. Suetsugu and M. K. Kazimierczuk, "Analysis and Design of Class E Amplifier With Shunt Capacitance Composed of Nonlinear and Linear Capacitance", *IEEE Trans.Circuits Syst-I*, vol.51, no.7, pp.1261-1268, July 2004.

[14] H. Sekiya, Y. Arifuku, H. Hase, J. Lu and T. Yahagi, "Investigation of class E amplifier with nonlinear capacitance for any output Q and finite dc-feed inductance," *IEICE Trans. on Fundamentals*. vol. E89-A, no. 4, pp. 873 - 881, Apr. 2006.

A Novel Control Strategy
of the Class-D Stereo Audio Amplifier

Kyu Min Cho*, Won Seok Oh**, Hai Xu***, and Hee Jun Kim***

* Dept. of Information and Communications, Yuhan College, Korea
** Dept. of Electrical Engineering, Yuhan College, Korea
*** Div. of Electrical and Computer Engineering, Hanyang University, Korea

Abstract-- **This paper presents a class-D stereo audio amplifier that controls directly the current of speakers using three phase full bridge circuit. And a new switching method for the current control of the amplifier is proposed. The proposed switching strategy is an improved version of the dead time minimization method, which can reduce current intermittence that commonly occurs with the conventional dead time minimization method whenever the current changes polarity. Dead time is not always adapted in the proposed method, which makes it different from the conventional practice of applying dead time in every current polarity changing point. In this paper, the specific strategy for the stereo operation is discussed. With the experimental results, usefulness of the control strategy is confirmed**

Index Terms— **class-D, amplifier, three phase full bridge circuit, current control.**

I. INTRODUCTION

In the most cases of the audio amplifier that needs wide bandwidth, the linear amplifier is used for the main power amplification. The linear amplifier, however, has a very low efficiency. Nowadays, the class-D amplifier is widely adopted in some audio amplifier applications in order to overcome low efficiency. As the class-D audio amplifier has a discrete switching operation, the high-dynamic characteristics of the switching method are required.

For high performance, high order passive filter, multiple control loops and modern control theory are adopted in some cases, such as a sliding mode control. Sometimes, the small linear amplifier is used to fill up the dynamic characteristics added on the class-D amplifier that is a main power amplifier. For some applications of audio amplifier, however, which does not need high fidelity, it is desirable to be able to construct by low cost, compact size and low power consumption.

In this paper, a new switching method for the current control of the class-D amplifier is proposed, and a scheme of stereo operation using three-phase full-bridge circuit configuration is presented, which has a DC source and six switching devices. The proposed amplifier is controlled with only a current control loop and the switching harmonics included in the output current are filtered by a series filter inductance. Since the main circuit configuration and the control circuit are very simple, the proposed amplifier can be constructed in a compact size.

The detailed algorithm of the proposed switching method is discussed in this paper, including the

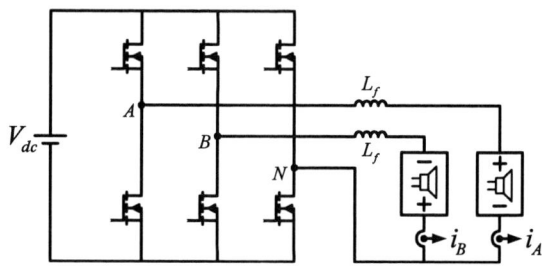

Fig. 1. Configuration of the proposed amplifier.

independent actions of the switching devices of the upper arm and lower arm according to the current polarity and the strategy of driving with the three-phase full-bridge circuit as a stereo amplifier. Finally, experimental results of the proposed class-D stereo audio amplifier are presented and discussed.

II. A CURRENT CONTROLLED CLASS-D STEREO AUDIO AMPLIFIER USING THREE PHASE PULL BRIDGE CIRCUIT

A. Configuration of the Main Circuit

Fig. 1 shows the main circuit of the proposed amplifier. It is a conventional three phase full bridge circuit. However, the load configuration is not a normal three phase configuration. Two speakers are connected at channel A-N and channel B-N through the inductors to filter the switching harmonics. Therefore, two switching devices can be reduced in comparison with conventional two channel stereo class-D amplifier, which uses full bridge circuit. Even though two channels produce stereo effects, the sound level of each channel is alike. Therefore, the speakers are connected with the reverse polarity each other to reduce the load of common phase N. In this case, of course, the reference of each channel has each other polarity.

B. Direct Current-Control Method of Speaker

In most audio amplifiers, the voltage of speaker is controlled. In the proposed amplifier, however, the current which flows through speaker is controlled. If the speaker is equivalently presented by a series RL circuit with variable back electromotive force, which is generated by the elasticity of speaker, we expect that the proposed method of direct current control of the speaker could amplify the original sound more correctly than the voltage control method. As the main circuit of the

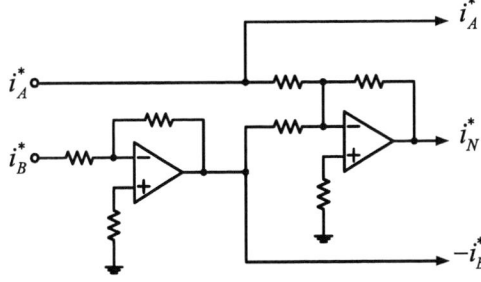

Fig. 2. Reference signal generator.

proposed amplifier uses three phase full bridge circuit, the reference current of the phase N must be built up and it can be easily generated by a reference generator as shown in Fig. 2. Since one of speaker is connected with reverse polarity, the reference signal is also to be inverted.

III. A NEW SWITCHING STRATEGY FOR THE CURRENT CONTROL

A. Proposed Switching Strategy

Fig. 3 shows the basic diagrams of the proposed switching strategy, where I_p and I_n represent current polarities, and G_p and G_n are switching signals of the upper arm and lower arm, respectively. Basically, the gate signal G_p for the upper arm is applied when the current is positive and the gate signal G_n for the lower arm is applied when the current is negative. Thus, we have to know the polarity information of the current. Therefore, the dead time is not needed except instant of current polarity change. In order to prevent the arm short at the instant of current polarity change, dead time is needed. Since the dead time is prepared from the negative edge of the last gate signal G_n instead of the negative edge of the flag signal I_p that means the current is positive, the applied dead time duration should be either less than the prepared dead time or zero. It can be seen that the applied dead time is zero in the case of Fig. 3. Of course, sometimes the dead time will be applied at once in every polarity changing instant of the current. However, at that time the current is almost zero and the dead time effect is negligible. As a result, dead time compensation is no longer required while reducing the possibility of current intermittence and arm short in the proposed switching strategy. And it is very useful for the high frequency application as like class-D amplifiers.

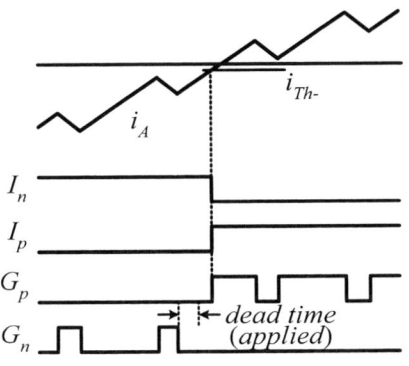

Fig. 3. Diagrams of the proposed method.

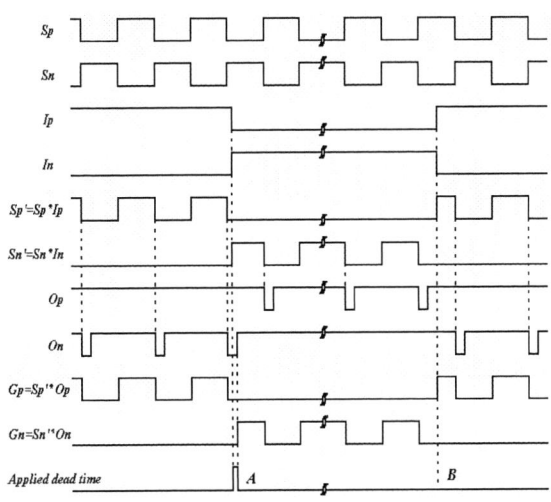

Fig. 4. Specific timing diagrams for the proposed switching strategy.

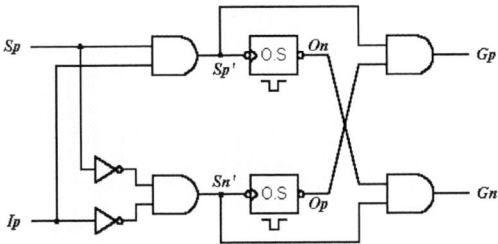

Fig. 5. An example circuit of the signal generation.

B. Detailed Scheme of the Gating Signal Generation

The specific timing diagram for the generation of the proposed switching signal is presented in Fig. 4. After taking logic AND of $S_{p,n}$ and $I_{p,n}$, $S_p{}'$ and $S_n{}'$ can be derived. At every falling edge of these signals, the appropriate dead time O_p and O_n are applied. The gating signals of the upper and lower arms are then given by (1) and (2)

$$G_p = S_p * I_p * O_p \tag{1}$$
$$G_n = S_n * I_n * O_n . \tag{2}$$

In Fig. 4, it can be seen that the actual dead time is applied in Point A. The applied dead time is either the same as the prepared dead time or smaller. In Point B's case, it can be seen that the actual dead time is not applied. In many cases, the dead time has not been applied.

Fig. 5 shows an example circuit generating the gating signals. It can be constructed simply with some additional ICs on the typical gating signal generator of conventional switching method with dead time.

C. Characteristics of the Proposed Method

The characteristics of the proposed switching strategy is analyzed for the passive RL loads.

If the increasing ratio and decreasing ratio of the current are similar in the short duration near the polarity changing instant, (3) is satisfied, as shown in Fig. 6.

1143

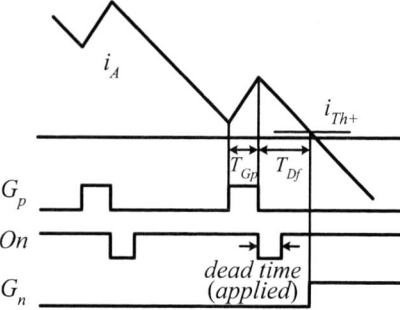

Fig. 6. Diagrams at the polarity changing instant.

$$T_{Df} > T_{Gp}, \tag{3}$$

where T_{Gp} is the duration of last gating signal for the upper arm and T_{Df} is the free-wheeling duration. At the end of free-wheeling duration, gating signal for the lower arm has to be applied for supplying negative current.

Thus, the sufficient condition that does not require dead time is given by

$$T_{Gp} \geq T_d, \tag{4}$$

where T_d denotes the prepared dead time.

When the triangular PWM is adopted, assume that the displacement angle is α, T_{Gp} can be represented by

$$T_{Gp} = \frac{1 - a \sin \alpha}{2} T_s \tag{5}$$

where T_s denotes the sampling period, and a represent the modulation index of the inverter output. Equation (5) is not affected from the back EMF of active load. From (4) and (5) we get

$$a \sin \alpha \leq 1 - \frac{2T_d}{T_s}. \tag{6}$$

Thus, the sufficient condition that satisfy (6) for all range of $0 < a \leq 1$, is given by

$$\alpha \leq \sin^{-1}\left(1 - \frac{2T_d}{T_s}\right). \tag{7}$$

Finally, (7) is represented using the displacement power factor as follows:

$$\cos \alpha \geq 2\sqrt{\frac{T_d}{T_s}\left(1 - \frac{T_d}{T_s}\right)} \tag{8}$$

Fig. 7 shows the operating region without the adapted dead time, which is evaluated according to the modulation index and load displacement power factor in the case of the single phase inverter, which adopts

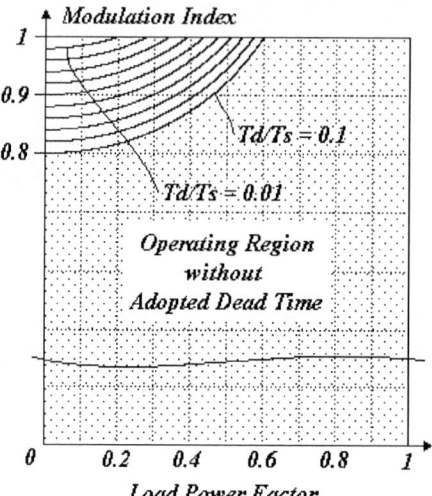

Fig. 7. Operating region without applied dead time.

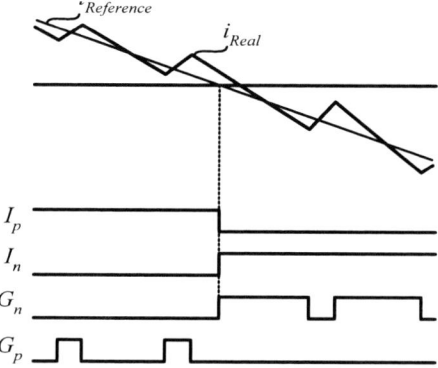

Fig. 8. The reference and real currents with the proposed gating signals in the current control system.

triangular modulation. It can be seen that the dead time is not applied over the wide region at the instant of current polarity change. Therefore, the dead time compensation is no more needed and the possibility of arm short is almost eliminated.

D. Usefulness of the Proposed Switching Strategy for the Current Control System

Current control is required in many applications of PWM power converters. Especially in PWM power converters, current control systems are essential for situations that demand the highest performance, such as in the vector control system of ac motor drives, the ac/dc PWM converter system, and in the active power filter system for high-quality source line condition. Existing current control systems can easily adopt the proposed switching strategy that uses the polarity information of the reference current instead of the real current. Fig. 8 shows the reference current and real current with gating signals in the case of the proposed switching strategy. It can be seen that current control should be performed correctly and efficiently to eliminate problems arising from real current detection.

IV. EXPERIMENTAL RESULTS

At first, simulation results for the proposed class-D amplifier, which is controlled by using the proposed switching strategy, are shown in Fig. 9. Fig. 9 (a) shows 5-kHz sine wave response. The current control is very well done by the proposed switching strategy although the reference frequency is very high. It is confirmed that the output current has only 0.1756-% of THD by the harmonic analysis of the 500-kHz range. And the simulated music reproduction is also well done with the proposed class-D amplifier by using three phase full bridge circuit.

In order to verify the effectiveness of the proposed switching strategy in the application of a current-controlled class-D stereo amplifier using a three-phase full bridge circuit was built and tested.

Fig. 10 shows output waveforms of the reference signal generator presented in Fig 2. It can be seen that the signal level of the common phase N is a relatively small to other main two channels A and B.

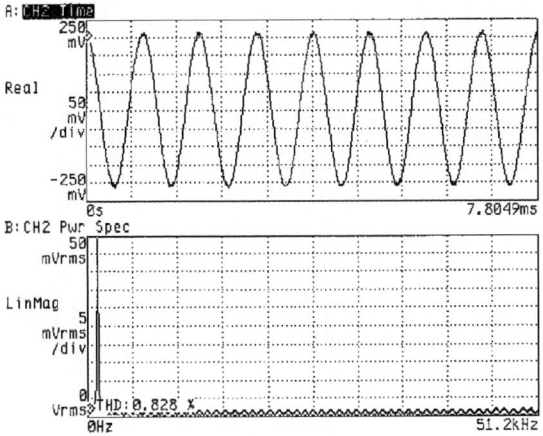

Fig. 11. Measuring result of total harmonic distortion for the 1-kHz sine wave.

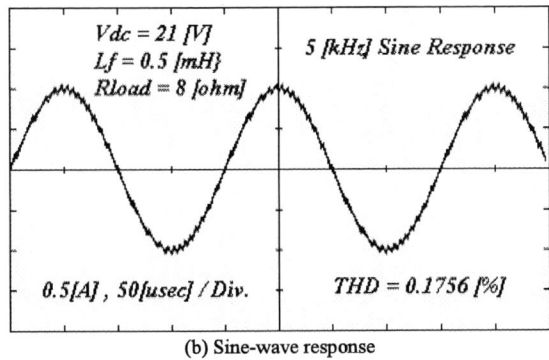

(b) Sine-wave response

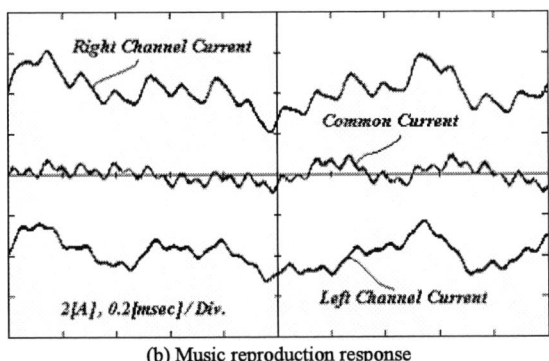

(b) Music reproduction response

Fig. 9. Simulation results for the proposed class-D amplifier using a novel switching strategy.

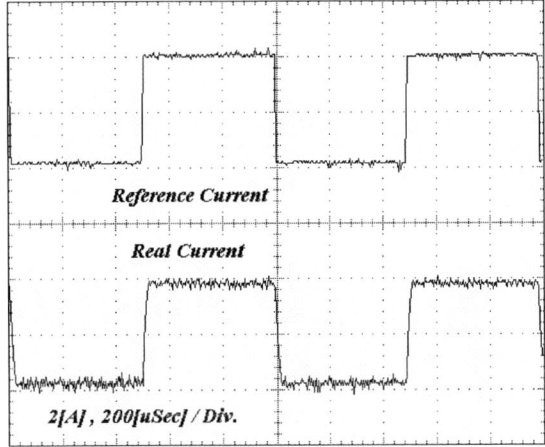

(a) Waveforms of the reference and real currents.

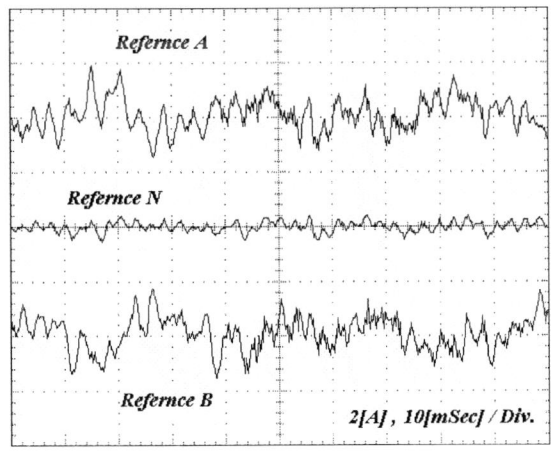

Fig. 10. Three phase reference signal waveforms.

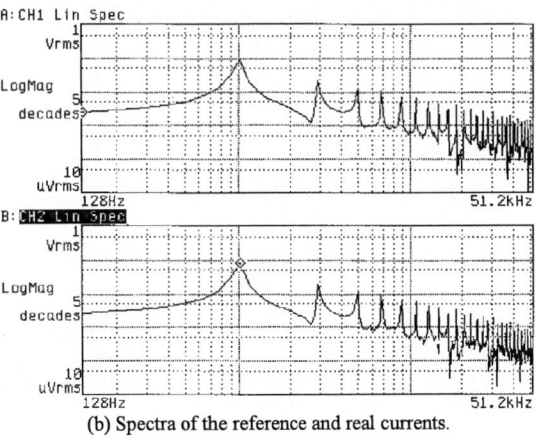

(b) Spectra of the reference and real currents.

Fig. 12. Waveforms and spectra of the 1-kHz rectangular response.

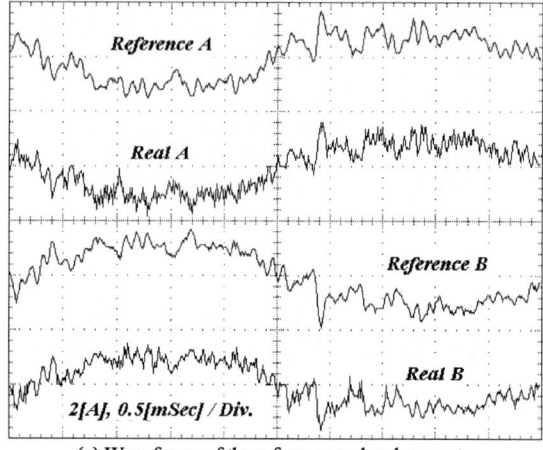

(a) Waveforms of the reference and real currents.

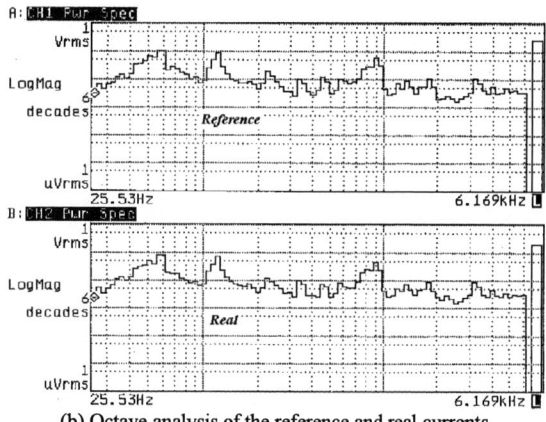

(b) Octave analysis of the reference and real currents.

Fig. 13. Waveforms and octave analysis of music sound reproduction.

Fig. 11 shows measuring result of total harmonic distortion for the 1-kHz sine wave outputs. 0.828-% of THD is measured in the frequency range to 51.2-kHz.

Fig. 12(a) shows the 1-kHz rectangular response and Fig. 12(b) shows the spectra of the reference and real currents, showing an excellent response over the audible frequency band.

Fig. 13(a) represents the music sound reproduction response. The octave analysis of the music sound signals shown in Fig. 13(b) confirmed that original sound was reproduced well, with the reference sound wave matching the real sound wave.

V. CONCLUSIONS

In this paper, a novel class-D stereo audio amplifier using three phase full bridge circuit is proposed and a new switching strategy to control directly the output current which flows through the speaker is also proposed. The proposed class-D stereo amplifier can reduce two switching devices compare with the conventional class-D stereo audio amplifier, which is constructed by two full bridge circuits. And the proposed switching strategy is especially efficient in the area of current control system, which uses the polarity information of the reference current without the problems entailed by real current

detection. Moreover, even in cases of high frequency switching operation, good switching performance is still possible, underlining the viability of the proposed switching strategy for not only class-D amplifier but also the other wide practical applications. From the experimental results, the validity and viability of the proposed class-D stereo audio amplifier and its switching strategy were confirmed.

REFERENCES

[1] H.R. Camenzind, "Modulated Pulse Width Audio Power Amplifier", *IEEE Trans. AE*, vol. AU-14, no. 3, pp. 136-140, 1996.

[2] Ronaldo C. Oliveira, Ernane A.A. Coelho and Joao B. Vieira Jr., Luiz C. Freitas and Valdeir J. Farias, "Switching Power Amplifiers with Soft Commutation for Audio Applications", *IEEE ISCS Conf. Rec.*, pp.557-560, 1996.

[3] Karsten Nielsen, "High-fidelity PWM-based amplifier concept for active loudspeaker systems with very low energy consumption" *J. Audio Eng. Soc.*, vol. 45, No. 7/8, pp. 1329-1333, 1998.

[4] F.A. Himmelstoss, K.H. Edelmser , "High Dynamic Classs-D Power Amplifier", *IEEE Trans. CE*, Vol. 44, No. 4, pp. 1329-1333, Nov. 1988.

[5] J.F. Silva, "PWM Audio Power Amplifiers: Sigma Delta Versus Sliding Mode Control", *IEEE Int. Conference on Electronics, Circuits and Systems*, Vol. 1, pp.359-362, 1998.

[6] David Leggate and Russel J. Kerman, "Pulse based dead time compensator for PWM voltage inverters" *IEEE IECON Conf. Rec.*, vol. 1, pp.474-481, 1995.

[7] S.-G. Jeong and M.-H. Park, "The analisis and compensation of dead time effects in PWM inverters," *IEEE Trans. Ind. Electron.*, vol. 38, no. 2, pp. 108-114, 1991.

[8] J.-W. Choi and S.-K. Sul, "A new compensation strategy reducing voltage/current distortion in PWM VSI systems operating with low output voltages," *IEEE Trans. Ind. Appl.*, vol. 31, no. 5, pp. 1001-1008, 1995.

[9] D. Leggate and R.J. Kerman, "Pulse-based dead time compensator for PWM voltage inverters," in *Conf. Rec. IEEE IECON*, 1995, vol. 1, pp. 474-481.

[10] W. S. Oh, Y. T. Kim, and H. J. Kim, "Dead time compensation of a current-controlled inverter using the space vector modulation method," *Int. J. Electron.*, vol. 80, no. 2, pp. 277-289, 1996.

[11] R. P. Joshi and B. K. Bose, "Base/gate drive suppression of inactive power devices of a voltage-fed inverter and precision synthesis of AC voltage and DC link current waves," in *Conf. Rec. IEEE IECON*, 1990, pp. 1034-1040.

[12] J.-S, Choi, J.-Y. Yoo, S.-W. Lim, and Y.-S. Kim," in *Conf. Rec. IEEE IAS*, 1999, vol. 4. pp. 2188-2193.

Robust H∞ Control Design for PFC Rectifiers

F. Tahami, H. Molla Ahmadian, A. Moallem

Department of Electrical Engineering, Sharif University of Technology, Tehran, Iran

Abstract– **Power factor correction (PFC) converter circuits are non-linear system due to the contribution of their multiplier. This non-linearity reflects the difficulty of analysis and design. Models that reduce the system to a linear system involve considerable approximation, and produce results that are susceptible to instability problems. Our goal is to design a controller that achieves some performance specifications despite these uncertainties in modeling. This paper addresses a robust H∞ control problem for switched circuits with parameter uncertainties. The necessary and sufficient conditions for existence of strong robust H∞ dynamic compensators and static output feedback controllers are established in terms of linear matrix inequality (LMI) approach. It is shown that the dynamics of the voltage loop can be significantly improved using an H∞ controller.**

Index Terms-- **Power Factor Correction rectifier, circuit averaging, H∞ norm, Linear Matrix Inequality.**

I. INTRODUCTION

The power-factor-correction (PFC) rectifiers are now commonly used in low power supply systems connected to the AC networks providing input-line harmonics in accordance with harmonic distortion standards. In general, near ideal rectifier systems contain two quite different types of control systems; the inner wide-bandwidth controller causes the instantaneous input current waveform to follow the input voltage by emulating a loss-free resistor, and the outer low-bandwidth controller, which varies the emulated resistance as necessary to balance the average ac input and dc load powers [1]. Apart from the boost converter, the nonlinear time-varying nature of the system does not allow linearization the equations needed to design the rectifier averaged current controllers of most converter topologies [2].

Different strategies have been proposed for improving controller design. Nishida [3] proposed a simplified analogue controller-based predictive instantaneous current control scheme for the single-phase voltage-fed rectifier. An on-line neural network is proposed by Insleay, et al., [4] to shape the input line currents of a PWM rectifier and provide unity power factor operation. In [5] an optimal control problem incorporating the appropriate control objectives was formulated for a discrete-time hybrid model of a boost converter. A variable structure control algorithm was proposed by Carrasco, et al., [6] to improve the robustness, stability and dynamic characteristic of the PWM power factor corrector employing a neural network controller. Power factor correction circuits with robust current control techniques were investigated in [7, 8].

Robust control can be used for the analysis and design of control system models including nonlinearities that are either hard to model or too complicated. Robust control system design based on H_∞ control problem for linear systems and nonlinear systems is a popular research area. It has been emphasized that H_∞ control problem in linear systems can be cast into the form of first-order matrix polynomial inequalities called linear matrix inequalities (LMI's), which belong to the group of convex problem [9].

In this paper the robust H_∞ control problem for control of PFC rectifiers is considered in sense of dynamic output feedback. Simulation results verified the experimental ones with a very good matching.

II. MODELING OF PFC RECTIFIERS

The simplest and less expensive approach to realize a near-ideal rectifier is to employ a full-wave rectifier followed by a dc-dc converter as depicted in Fig. 1 [1]. Among the different dc-dc converters, the boost converter is the most suitable for use in implementing PFC rectifiers, because the boost inductor is in series with the line input terminal, the inductor will achieve smaller current ripple and it is easier to implement average current mode control. A conventional PFC rectifier based on a PWM converter controlled by two interconnected feedback loops, a wide bandwidth current loop and a slow voltage loop.

For design of this controller, the rectifier can be modeled using the loss-free resistor (LFR) concept. Perturbation and linearization of the LFR leads to a small-signal equivalent circuit than predicts the relevant small-signal transfer function.

This loop cannot attempt to remove the capacitor voltage ripple that occurs at the second harmonic of ac

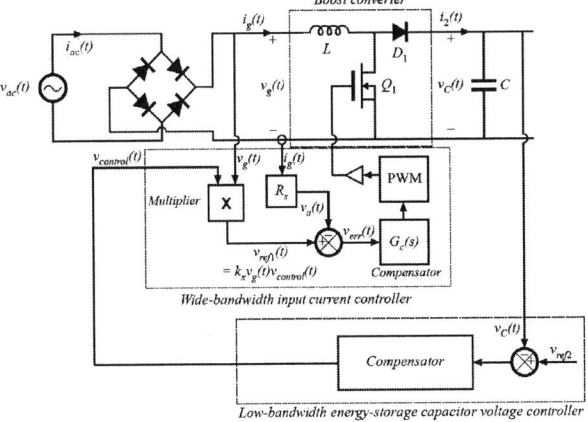

Fig. 1. Realizing of an ideal rectifier by a PWM converter [1]

line frequency. Therefore, for the purpose of designing the low-bandwidth outer control loop, it is unnecessary to model the system high-frequency behavior. It is desired to model only the low-frequency components excited by slow variations in the control, the load and the line voltage amplitude. High-frequency switching harmonics are removed via averaging the waveforms over a switching cycle. Hence, we average over the switching period to remove the switching harmonics, and then we average again over one-half of the ac line period to remove the even harmonics of the ac line frequency. The resulting model is valid for frequencies sufficiently less than the ac line frequency. The equivalent circuit is time-invariant but nonlinear. We can perturb and linearize it to construct a small-signal ac model that describes how slow variations in the control signal, load and input affect the rectifier output. The small-signal equivalent circuit is given in Fig. 2. Expressions for the parameters g_2, j_2, and r_2, are as below [1]:

$$g_2 = \frac{2}{R_e} \frac{V_g}{V} \tag{1}$$

$$r_2 = \frac{V}{I_2} \tag{2}$$

$$j_2 = \frac{V_{g,rms}}{V R_e^2} \frac{dR_e}{dV_{control}} \tag{3}$$

A. Modeling the Inner wide-bandwidth average current controller

To aid in the design of the inner feedback loop that controls the ac line current waveshape, a converter model is needed that describes how the converter average input current depends on the duty cycle. The problem is that the variations in the duty cycle, as well as in the ac input and current are not small. As a result, in general the small signal assumptions are violated. When the rectifier operates near steady-state, the output voltage of a well-designed system exhibits small variations. In the special case of the boost rectifier, this is sufficient to linearize the equations of the average current controller, even though the ac input variations are not small. The averaged control-to-input current transfer function is [1]:

$$\frac{i_g(s)}{d(s)} = \frac{V}{sL} \tag{4}$$

III. OUTPUT FEEDBACK H_∞ CONTROL

The optimal H_∞ control problem is to find a controller K that uses y as input to generate control u so the induced L_2 gain of the closed loop system is less than γ, a specified performance level. When the systems are LTI, then this induced L_2 gain is the H_∞ norm of the closed loop transfer function.

Consider the linear time-invariant augmented plant P and the controller K, shown in Fig. 3 and described by:

$$P : \begin{cases} \dot{x} = Ax + B_w w + Bu \\ z = C_z x + D_{zw} w + D_z u \\ y = Cx + D_w w \end{cases} \tag{5}$$

$$K : \begin{cases} \dot{x}_k = A_k x_k + B_k y \\ u = C_k x_k + D_k y \end{cases} \tag{6}$$

Where y, z, u and w are the output feedback, the augmented output, the control signal and disturbance, respectively. We denote the closed loop transfer matrix from w to z by T_{zw}. Given a scalar $\gamma > 0$, the controller K is said to be an H_∞ controller if the following two conditions hold: 1- the closed loop system is asymptotically stable, 2- $\|T_{zw}\|_\infty < \gamma$.

Where $\|T_{zw}\|_\infty$ is called the H_∞ norm of the transfer matrix T_{zw} and is defined as follows:

$$\|T_{zw}\|_\infty = \sup(\|T_{zw}\| | \mathrm{Re}\, s > 0) \tag{7}$$

After some manipulations it can be shown that the closed loop system can be described by:

$$\begin{aligned} \dot{x}_{cl} &= A_{cl} x_{cl} + B_{cl} w \\ z &= C_{cl} x_{cl} + D_{cl} w \end{aligned} \tag{8}$$

Where the coefficient matrices A_{cl}, B_{cl}, C_{cl} and D_{cl} are [8]:

$$A_{cl} = \begin{bmatrix} A + B_k C & BC_k \\ B_k C & A_k \end{bmatrix}$$

$$B_{cl} = \begin{bmatrix} B_w + BD_k D_w \\ B_k D_w \end{bmatrix}$$

$$C_{cl} = \begin{bmatrix} C_z + D_z D_k C & D_z C_k \end{bmatrix} \tag{9}$$

$$D_{cl} = D_{zw} + D_z D_k D_w$$

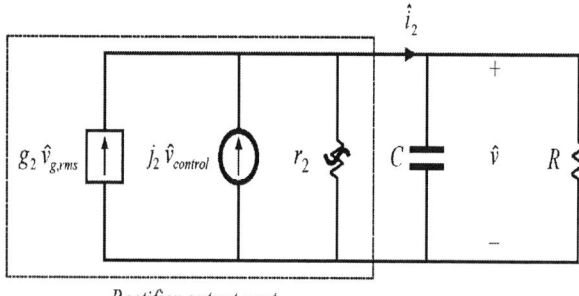

Fig. 2. The small signal equivalent circuit of the voltage loop

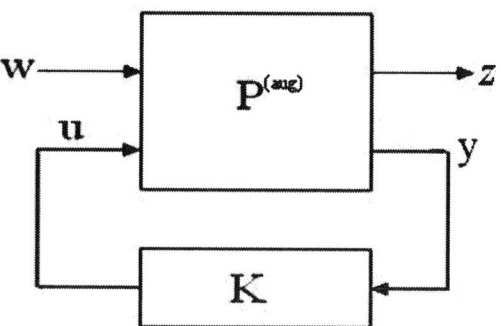

Fig. 3. Augmented Plant with Controller

The H_∞ control problem in linear systems can be cast into the form of first-order matrix polynomial inequalities called linear matrix inequalities (LMI's) which belong to the group of convex problems. A system is said to be quadratically stable if there exists a positive-definite quadratic Lyapunov function that decreases along every trajectory of the system. A necessary and sufficient condition for quadratic stability can be directly formulated in terms of a finite number of linear matrix inequalities [9].

By virtue of the Bounded Real Lemma [10], A_{cl} is stable and the H_∞ norm of T_{zw} is smaller than γ, if and only if a symmetric matrix Q exists that [11]:

$$\begin{bmatrix} A_{cl}^T Q + Q A_{cl} & Q B_{cl} & C_{cl}^T \\ B_{cl}^T Q & -\gamma I & D_{cl}^T \\ C_{cl} & D_{cl} & -\gamma I \end{bmatrix} < 0; \quad Q > 0 \qquad (10)$$

By replacing A_{cl}, B_{cl}, C_{cl} and D_{cl} in (9) from (10), Bilinear Matrix Inequality (BMI) achieved. In order to turn BMI's constraints into LMI's we partition Q such that the following inequalities are held [11]:

$$Q = \begin{bmatrix} Y & N \\ N^T & \Delta \end{bmatrix} > 0, \ Y > 0 \qquad (11)$$

$$Q^{-1} = \begin{bmatrix} X & M \\ M^T & V \end{bmatrix} > 0, \ X > 0 \qquad (12)$$

Where $Y, X \in \mathbf{R}^{m \times m}$, $M, N \in \mathbf{R}^{m \times k}$, $\Delta, V \in \mathbf{R}^{k \times k}$, and m and k are number of system and controller states respectively. X and Y are new LMI optimization variables. Knowing $QQ^{-1} = I$, then:

$$\begin{cases} YX + NM^T = I \\ N^T X + \Delta M^T = 0 \end{cases} \qquad (13)$$

In order to turn all constraints into LMI's we define change of controller variables as follow [11]:

$$\begin{aligned} \hat{A}_k &:= N A_k M^T + N B_k C X + Y B C_k M^T + Y A X + Y B D_k C X \\ \hat{B}_k &:= N B_k + Y B D_k \\ \hat{C}_k &:= C_k M^T + D_k C X \\ \hat{D}_k &:= D_k \end{aligned} \qquad (14)$$

With use of new variables, we can extract the following symmetric matrix:

This equation is LMI related to $\hat{A}_k$, $\hat{B}_k$, $\hat{C}_k$, $\hat{D}_k$, X and Y (LMI variables).

If the M and N matrices are of full row rank and the matrices $\hat{A}_k$, $\hat{B}_k$, $\hat{C}_k$, $\hat{D}_k$, X and Y are available then we can compute controller matrices A_k, B_k, C_k, and D_k. For full order design, one can always assume that M and N have full row ranks. Hence the variables A_k, B_k, C_k, and D_k can be replaced by $\hat{A}_k$, $\hat{B}_k$, $\hat{C}_k$, and $\hat{D}_k$ without loss of generality [11]. The constraint $Q > 0$ is equivalent to:

$$\begin{bmatrix} X & I \\ I & Y \end{bmatrix} > 0 \qquad (16)$$

Hence the LMI optimization problem is to minimize γ with symmetric matrices X and Y subject to the constraints in (15) and (16).

We can solve this optimization problem using MATLAB LMI toolbox [9]. Once the above LMI problem solved, one can find the nonsingular matrices M and N. Then from (13) the controller parameters will be obtained as follow [11]:

$$\begin{aligned} D_k &= \hat{D}_k \\ C_k &= \left(\hat{C}_k - D_k C X \right) M^{-T} \\ B_k &= N^{-1} \left(\hat{B}_k - Y B \hat{D}_k \right) \\ A_k &= N^{-1} \left(\hat{A}_k - N B_k C X - Y B C_k M^T - Y A X - Y B D_k C X \right) M^{-T} \end{aligned} \qquad (17)$$

Note that in this design of controller we supposed that the controller has the same order as augmented plant ($m = k$).

IV. ILLUSTRATIVE EXAMPLE OF AN H_∞ CONTROL DESIGN FOR A PFC RECTIFIER

As an example, we consider a boost rectifier with the features as tabulated in Table I.

TABLE I
THE PARAMETERS OF THE BOOST RECTIFIER CIRCUIT

Parameter	Symbol	Quantity
Energy storage capacitance	C	470μF
Inductance	L	8mH
MOSFET on- resistance	R_{ON}	0.3 Ω
Diode forward voltage drop	V_D	0.7V
Inductor resistance	R_L	1.4 Ω
Input voltage (rms)	v_g	110V
Output voltage	v	200V
Input frequency	f_{ac}	50Hz
Switching frequency	f_s	40kHz
Output Power	Pout	250W

Following the method given in previous section, we can

$$\begin{bmatrix} AX + B\hat{C} + X^T A^T + \hat{C}^T B^T & A + B\hat{D}C + \hat{A}^T & B_w + B\hat{D}B_w & XC_z^T + \hat{C}^T D_z^T \\ \left(A + B\hat{D}C + \hat{A}^T \right)^T & YA + \hat{B}C + A^T Y^T + C^T \hat{B}^T & YB_w + \hat{B}D_w & C_z^T + C^T \hat{D}_z^T \\ B_w^T + B_w^T \hat{D}^T B^T & B_w^T Y^T + D_w^T \hat{B}^T & -\gamma I & D_{zw} + D_z \hat{D}_k D_w \\ C_z X^T + D_z \hat{C} & C_z + \hat{D}_z C & D_{zw}^T + D_w^T \hat{D}_k^T D_z^T & -\gamma I \end{bmatrix} < 0 \qquad (15)$$

now synthesize the H_∞ controllers for the voltage and current control loops.

A. The Voltage Loop Controller

Using loop-shaping ideas the augmented plant of the voltage control loop may be constructed as shown in Fig. 4. The control goal is to minimize the infinity norm of the transfer function from V_{ref} to z.

The design criteria are described by the performance weighting function for modifying the tracking error and the input scaling matrix W. In order to properly attenuate the second line harmonic that appears at the output voltage of the converter, the bandwidth of W must be limited to a fraction of the AC line frequency. We arbitrarily choose it as:

$$W = \frac{100}{10s+1} \tag{21}$$

We used MATLAB to solve the optimization problem with the LMI and equality constraints. The desired controller is obtained as:

$$k_v(s) = \frac{403.8s + 10770}{s^2 + 278.4s + 8.631} \tag{22}$$

It's a second order compensator which can simply be implemented by analog and digital circuits. The bandwidth of controller is $0.031 rad/s$ which is quite smaller than the second harmonic of ac line frequency.

B. The current loop controller

The augmented plant is shown in Fig. 5. The control goal is to minimize the transfer function of I_{ref} to z.

The bandwidth of W is chosen sufficiently large to precisely regulate the input current. For the problem criteria formulation W is chosen as:

$$W = \frac{100}{7.9577 \times 10^{-6} s + 1} \tag{23}$$

The desired controller is obtained as:

$$G_i = \frac{2.749 \times 10^8 s + 3.814 \times 10^{13}}{s^2 + 5.478 \times 10^6 s + 6.351 \times 10^{11}} \tag{24}$$

V. SIMULATION RESULTS

The effectiveness of the proposed method was investigated by conducting a series of computer

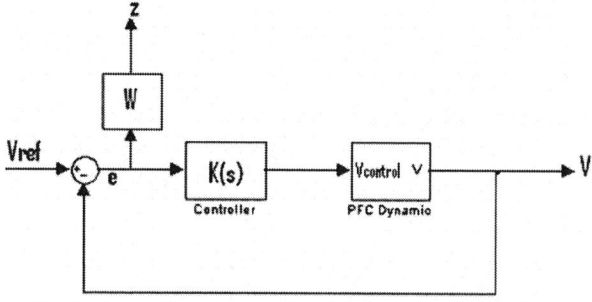

Fig. 4. Construction of augmented plant for designing the voltage controller

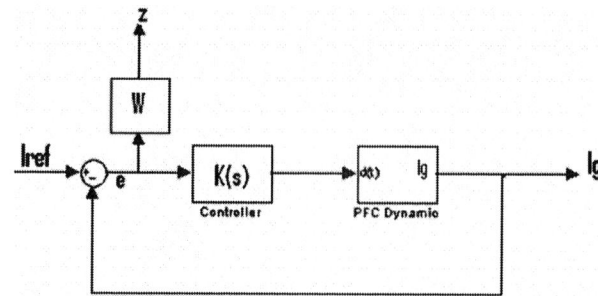

Fig. 5. The augmented plant for current loop control of PFC boost

simulations for the boost rectifier of Fig. 1.

The DC output voltage after turning on under the rated load is shown in Fig. 6. The response of the H_∞ voltage control loop is obviously faster compared with the conventional PID controller. Furthermore, the output voltage undershoot with H_∞ is about half of that of PID controller. Fig. 7 compares the steady state input current for the H_∞ and the PID controllers at the rated load. The steady state responses are rather similar.

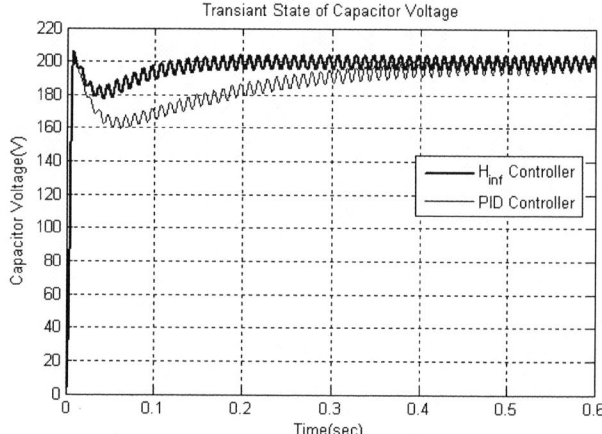

Fig. 6. The output voltage for conventional and the proposed controllers

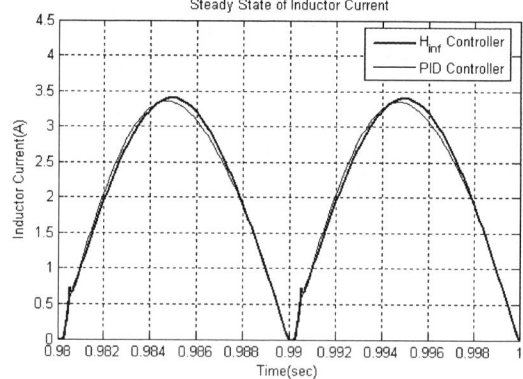

Fig. 7. Comparison between the steady state input currents for the proposed robust control and the traditional PID control.

VI. EXPERIMENTAL RESULTS

To verify the validity of the proposed controller, a boost PFC rectifier with the same values and regulation circuits that were described in table I has been built and tested. The control system was designed around a TMS320F2812 Digital Signal Processor. The

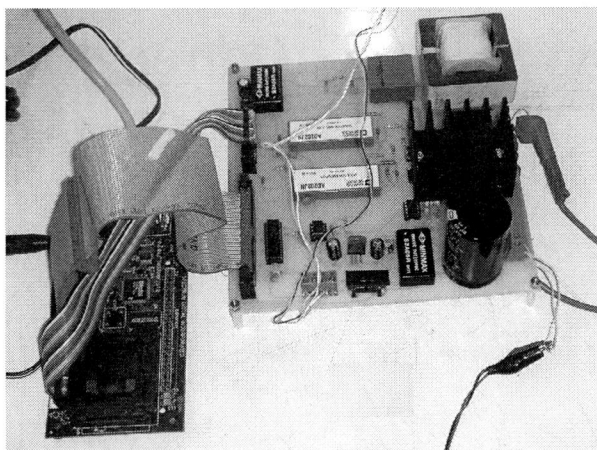

Fig. 8. The experimental setup

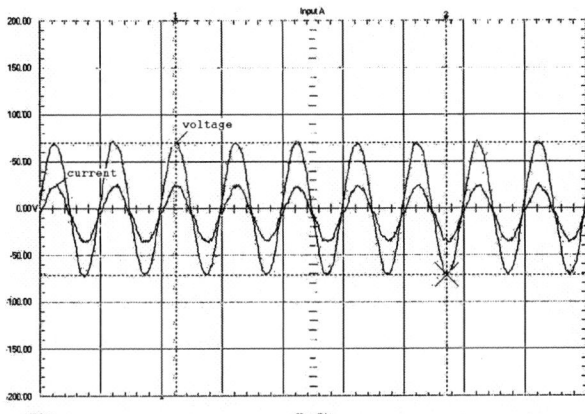

Fig. 9. The input current and voltage into the rectifier circuit

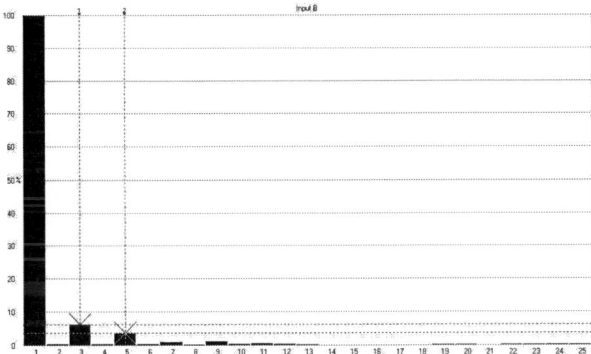

Fig. 10. The spectrum of the line currents

experimental setup together with DSP evaluation board is shown in Fig. 8. Fig. 9 shows the waveforms of the input current and voltage into the rectifier circuit. It is evident that the average inductor current exactly follows the waveform of the reference current.

Fig. 10 shows the spectrum of the line currents. The AC current contains negligible harmonic content. The THD is better than %7.5.

VII. CONCLUSIONS

In this paper, we have considered the robust H_∞ control problem of the PFC rectifier systems. In terms of LMI, we designed the H_∞ dynamic compensators for both voltage and current loops. It is shown by simulation that how the output voltage dynamic response can be significantly improved by H_∞ controllers. Improved voltage loop dynamics potentially enables the use of smaller tank capacitor at the output, operation at a smaller output voltage, and use of smaller, not over-designed components in the following dc/dc converter stage. Low input current harmonics is also experimentally verified on a 250 W boost high power factor rectifier.

REFERENCES

[1] R. W. Erickson and D. Maksimovic, Fundamentals of Power Electronics, 2nd ed., Kluwer Academic Publishers, 2000.

[2] A. Fernandez, J. Sebastian, P. Villegas, M. M. Hernando, and J. Garcia, "Dynamic limits of a power factor pre-regulator," in Proc. IEEE PESC'03, 2003, pp. 1697–1602.

[3] Y. Nishida, M. Nakaoka, "Simplified Predictive Instantaneous current control for single-phase and three-phase voltage-fed PFC converters," IEE Proc.-Electr. Power Appl., Vol. 144, No. I, January 1997.

[4] A. Insleay, N. R. Zargari, G. Joos, "A Neural Network Controlled Unity Power Factor Three Phase Current Source PWM Front-end Rectifier for Adjustable Speed Drives", 'Power Electronics and Variable-Speed Drives', 26 - 28 October 1994, Conference Publication No 399.

[5] A. G. Beccuti, G. Papafotiou and M. Morari, "Optimal Control of the Boost dc-dc Converter," Proceedings of the 44th IEEE Conference on Decision and Control, and the European Control Conference 2005 Seville, Spain, December 12-15, 2005 WeA09.5.

[6] J. M. Carrasco, J. M. Quero, F. P. Ridao, A. Perales, and L. G. Franquelo, "Sliding Mode Control of a DC/DC PWM Converter with PFC Implemented by Neural Networks, " IEEE Trans. Circuits and Systems—I: Fundamental Theory and Applications, vol. 44, no. 8, Aug. 1997.

[7] Z. Yang, P. C. Sen, "Power Factor Correction Circuits with Robust Current Control Technique," IEEE Trans. Aerospace and Electronic Systems vol. 38, no. 4 Oct. 2002.

[8] E. Figueres, J. Benavent, G. Garcerá, and M. Pascual, "Robust Control of Power-Factor-Correction Rectifiers with Fast Dynamic Response", IEEE TRANSACTIONS ON INDUSTRIAL ELECTRONICS, VOL. 52, NO. 1, FEBRUARY 2005.

[9] P. Gahinet, A. Nemirovski, A. Laub, and M. Chilali, The LMI Control Toolbox, The Mathworks Inc., 1995

[10] S. P. Boyd, L. El Ghaoui, E. Feron, and V. Balakrishnam, Linear Matrix Inequalities in System and Control Theory, Philadelphia, PA: SIAM, 1994.

[11] C. Scherer, P. Gahinet, M. Chilali, "Multi-objective Output-Feedback Control via LMI Optimization," IEEE Trans. on Automatic Control, vol. 42, NO. 7, JULY 1997; pp 896-911.

Parallel Operation of Power Factor Corrected AC-DC Converter Modules With Two Power Stages

Aravind Pothana
Department of Electrical Engineering
Indian Institute of Technology Madras
Chennai, India 600036
Email: ee05s026@smail.iitm.ac.in

Krishna Vasudevan
Department of Electrical Engineering
Indian Institute of Technology Madras
Chennai, India 600036
Email: Krishna@ee.iitm.ac.com

Abstract—**This paper presents a new control method for paralleling single-phase power factor corrected ac-dc converter modules with two power conversion stages. The control algorithm varies the number of converter modules in operation and their power sharing in accordance with the load power demand. This control method offers advantages such as tight output dc voltage regulation, very low output voltage ripple, inherent input power factor correction, flexibility to expand power capability, and high overall efficiency. The paper presents the topological composition of the ac-dc converter module. The principle of operation of the new control method is explained. Simulation results are presented. The results demonstrate that the proposed topology and control method are able to produce tight regulation and good transient response.**

I. INTRODUCTION

Switching mode ac-dc converters systems with power factor correction (PFC) are being widely used as power supplies for telecommunications, airport navigation equipment and other electronic systems. A typical dc power supply system with input power factor correction ability, powered by a single-phase ac utility has been realized by three different methods in the literature. The first method is the single stage method, involving a single power conversion stage. Here an ac-dc converter is responsible for both input power factor correction and output voltage regulation [1], [2], [3], [4]. However this method lacks the ability to tightly regulate the dc output voltage [5]. The second approach uses a two-stage power conversion scheme. The first stage is a power factor corrected ac-dc converter stage to attain good input power factor with low total harmonic distortion and to provide a loosely regulated dc link voltage. This stage is cascaded with a dc-dc converter stage to enable fast and fine output voltage regulation, and elimination of low frequency ripple present in the loosely regulated dc link voltage [5], [6]. But the overall efficiency in this method is less [9]. The third approach uses parallel power processing by a PFC ac-dc converter stage and a dc-dc converter stage [7], [8], [9]. It has an integrated control method that facilitates power factor correction at the input and voltage regulation at the output. Significant conversion efficiency improvement can be achieved in this method but, at the expense of complicated power and control stages [10].

A modular development approach where single-phase converter units are operated in parallel to form a high power converter system has several advantages [11]. Some of them are as follows

- Power expandability offers great flexibility in the development of power converter products for different power level.
- Less requirement for maintenance and repair of power converter modules because of the use of standard single-phase converter units
- If the trigger equipment are synchronized and phase shifted, the ripple contents of output voltage can be reduced significantly.
- System reliability and operational redundancy are improved.
- With proper operation, control and management, the overall power conversion efficiency and the life of the converter module can be increased.

Parallel operation of ac-dc converter modules facilitating power expandability has been carried out by the authors of [12] using the two power conversion stage method. However the second power conversion stage in their approach is essentially used to step down the output voltage of the first stage which is a PFC boost topology, rather than for improving the dynamic response of the output. In the scheme of [12] the transient response of the output voltage of each module is sensitive to the transient response of the first stage, which however is inherently slow [5]. Therefore the regulation of the dc output voltage is not tight though a two stage scheme is used. Moreover the control scheme in this method distributes the total load current to each module equally, reducing the efficiency of the system at light loads. A control algorithm that is capable of operating only that many number of converter modules, which are necessary to supply the power demanded actually by the load, and nearly all the converter modules that are in operation are working with nominal power, can greatly improve the overall efficiency of the system irrespective of the load demand. Such an approach has been proposed for dc-dc converters in [13]. A control method on the lines of such a methodology has been attempted in [14]. But this method, apart from being complex, is suitable only for output voltages greater than the input line voltage. It also has a high degree of output voltage ripple.

978-1-4244-0644-9/07/$25.00 ©2007 IEEE

A simple multi-module control mechanism for paralleling standard PFC ac-dc converter modules with two stages has been presented here. Apart form providing electrical isolation, tight dc output voltage regulation, very low output voltage ripple, inherent input power factor correction, flexibility to expand power capability, the control mechanism improves the overall efficiency of the system by selecting the number of converter modules in operation depending on the load demand. The proposed converter topology, along with the control mechanism explained in detail, is presented in Section II. Simulation results are presented and analyzed in Section III. Section IV summarizes the work with concluding remarks.

II. PRINCIPLE OF OPERATION

In the multi-module control mechanism that has been presented here, a single-phase PFC ac-dc flyback converter is cascaded with a dc-dc buck converter to form a single ac-dc converter module. The circuit diagram for the module along with the control structure in block diagrammatic form is shown in Fig. 1.

A. PFC AC-DC Converter

The function of the front-end converter in an ac-dc converter module is to provide power factor correction apart from regulating the voltage across its output. The front-end converter in a module acts as a variable power, constant DC voltage supply to the dc-dc converter. Though the boost converter [2] can also be used for PFC applications, a flyback converter [3] has been used here, because of its simplicity of operation, and the electrical isolation it provides. Further, features like start up and short circuit protection are obtained through a single switch. In addition, unlike the case of a boost converter, the line voltage need not necessarily be lower than the output voltage.

The control structure for the module consists of a current loop and a voltage loop. The current loop facilitates power factor correction, by controlling the mean input current $i_{in-mean}(t)$. It is made proportional to the input voltage $v_{in}(t)$ through the equivalent input conductance G_e [16] as explained below.

The mean input current over a switching cycle is

$$i_{in-mean}(t) = \frac{1}{T_s} \int i_{in}(t)dt \qquad (1)$$

The switch S1 turns on at the beginning of each switching cycle. The mean of the switch current $i_{in}(t)$ is obtained by integrating continuously the sensed current and dividing by the fixed switching time period T_s. The switch is closed as long as the calculated mean input current is less than the product of $v_{in}(t)$ and G_e, and is open after that until new switching cycle begins. Therefore the condition for opening the switch S1 is (2). At the end of each switching period the integrator is reset back to zero. This loop therefore provides amplitude sizing and wave shaping of the ac input current drawn.

$$i_{in-mean}(t) \geq v_{in}(t) \times G_e \qquad (2)$$

The voltage feedback loop stabilizes the output voltage of the front-end converter against variations in load power, ac line voltage, and component characteristics, by controlling the equivalent input conductance G_e. This loop brings the steady state output voltage of the PFC ac-dc converter v_s to the desired voltage reference V_{refc}.

Following the principle outlined in [5], the output voltage to equivalent input conductance transfer function (3) used to design the controller for the voltage loop can be derived as

$$\frac{\hat{v}_s(s)}{\hat{v}_{control}(s)} = \frac{V_{in,rms}^2}{1 + sC(R/2)} \qquad (3)$$

However the voltage feedback loop cannot attempt to remove the storage capacitor voltage ripple that occurs at second harmonic of the AC line frequency, since doing so would require that G_e change significantly at the second harmonic frequency [5]. This would introduce significant distortion, phase shift, and power factor degradation in the ac line current waveform. Consequently, this loop must have sufficiently small gain at ac line frequency, and hence its bandwidth must be low . Therefore the dynamic response of the PFC ac-dc converter at its output is slow.

B. DC-DC Converter

The dc-dc converter basically acts as a current regulated constant output voltage source to the load. The converter control is realized using a modification of the one-cycle control method [15] in order to achieve current regulation, as shown in Fig. 1. In the control structure, the dc-dc converter is presented with a voltage and current reference. The voltage reference is the output voltage to be provided by the multi-module setup with a common outer voltage loop. The current reference is provided to individual converters by this loop after processing the voltage error. The current reference decides the contribution of a particular converter module to the load power in the overall multi-module setup. The inductor current error, after being processed by a proportional controller, is allowed to modulate the voltage reference to the one-cycle control loop. This ensures that the voltage reference is only temporarily raised or lowered in ordered to achieve the required change in current level of the inductor current. This however requires a feedback of the average inductor current. Instead of using a filter to obtain this value, in order to avoid delays, the average inductor current is obtained by integrating over a cycle. The integrator is reset every cycle. In this manner, the delay in obtaining the average value is reduced to one switching cycle. Further, in the proposed scheme, the switch S2 of the dc-dc converter is operated simultaneously with switch S1. It may be noted that during the ON time of the switch S2, the voltage across capacitor C and the diode voltage are equal except for a small ON state switch voltage drop across S2. The voltage across the capacitor C is already being sensed for the front-end converter output voltage regulation. Therefore instead of sensing the voltage across the output diode, the implementation has been slightly modified, and the capacitor

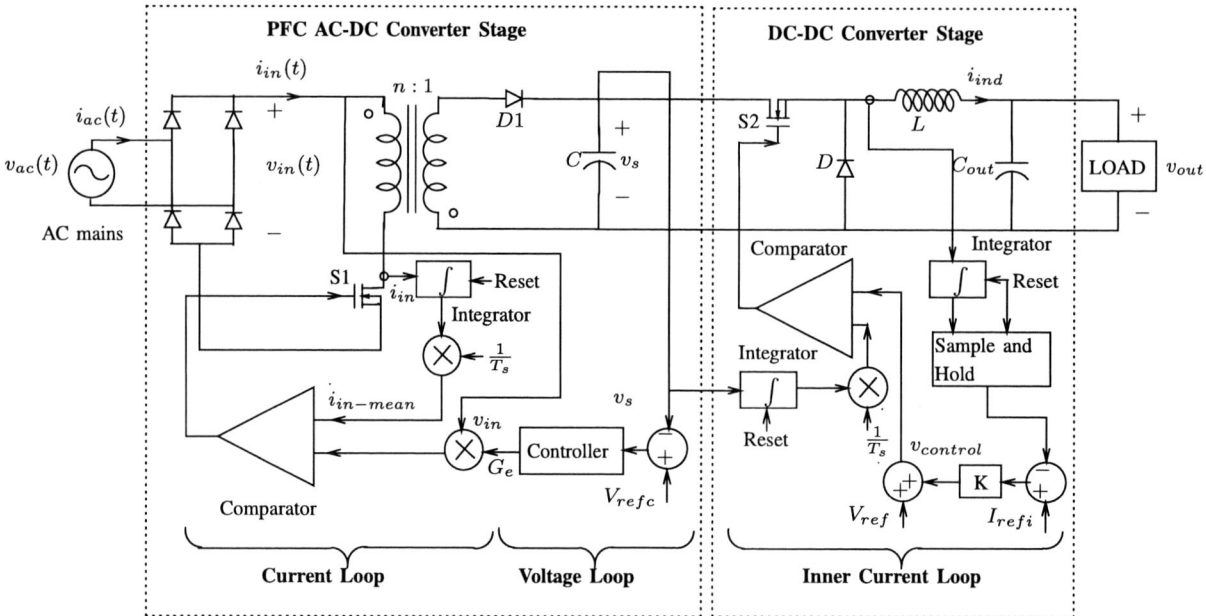

Fig. 1. AC-DC Converter Module

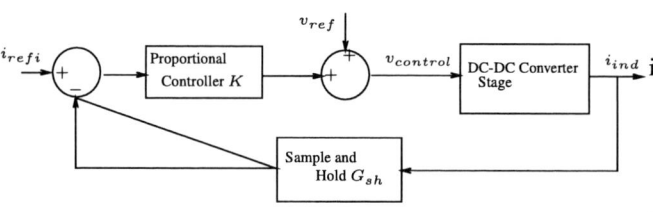

Fig. 2. Block Diagram Representation of The DC-DC Converter Along With The Inner Current Loop

voltage v_s is sensed and used. The error caused is taken care of by the closed loop operation.

1) Modeling The Inner Current Loop Control System: The dc-dc converter along with its inner current loop can be represented by a block diagram as shown in Fig. 2

The inductor current to duty ratio transfer function for the dc-dc buck converter is

$$\frac{\hat{i}_{ind}(s)}{\hat{d}(s)} = \frac{v_s(\frac{1}{R} + C_{out}s)}{1 + \frac{L}{R}s + LC_{out}s^2} \qquad (4)$$

Based on the principle of one cycle control outlined in [15], the transfer function for the duty ratio $d(s)$ of switch S2 to control reference $v_{control}(s)$ (Fig. 1) can be derived as

$$\frac{\hat{d}(s)}{\hat{v}_{control}(s)} = \frac{1}{v_s} \qquad (5)$$

Multiplying eqn(4), and eqn(5)

$$\frac{\hat{i}_{ind}(s)}{\hat{v}_{control}(s)} = \frac{\frac{1}{R} + C_{out}s}{1 + \frac{L}{R}s + LC_{out}s^2} \qquad (6)$$

The first order approximation of a sample and hold circuit is

$$G_{sh} = \frac{1}{1 + \frac{T}{2}s} \qquad (7)$$

where T the sampling period of the sample and hold circuit. In this case however sampling period is equal to the switching period of the switch S2.

Using the block diagram of the current loop in Fig. 2 and equations (6), and (7) the small-signal ac variation in i_{ind} about a quiescent value can be derived as

$$\hat{i}_{ind} = \frac{A}{B}(K\hat{i}_{refi} + \hat{v}_{ref}) \qquad (8)$$

where

$$A = \frac{1}{R} + (\frac{T}{2R} + C_{out})s + \frac{C_{out}T}{2}s^2$$

$$B = 1 + \frac{K}{R} + (\frac{L}{R} + \frac{T}{2} + KC_{out})s + (\frac{LT}{2R} + LC_{out})s^2$$
$$+ \frac{LC_{out}T}{2}s^3$$

In (8) $\hat{v}_{ref}$ and $\hat{i}_{refi}$ are small-signal ac variations in the two reference inputs v_{ref} and i_{refi} about their respective quiescent values. The voltage reference v_{ref} is a constant, therefore $\hat{v}_{ref}$ is always zero. The small signal transfer function for inductor current to the current reference is

$$\frac{\hat{i}_{ind}}{\hat{i}_{refi}} = \frac{KA}{B} \qquad (9)$$

The proportional gain constant K, in the inner current loop of the dc-dc converter can be designed with the help of (6), (7), (9) and the block diagram shown in Fig. 2.

1154

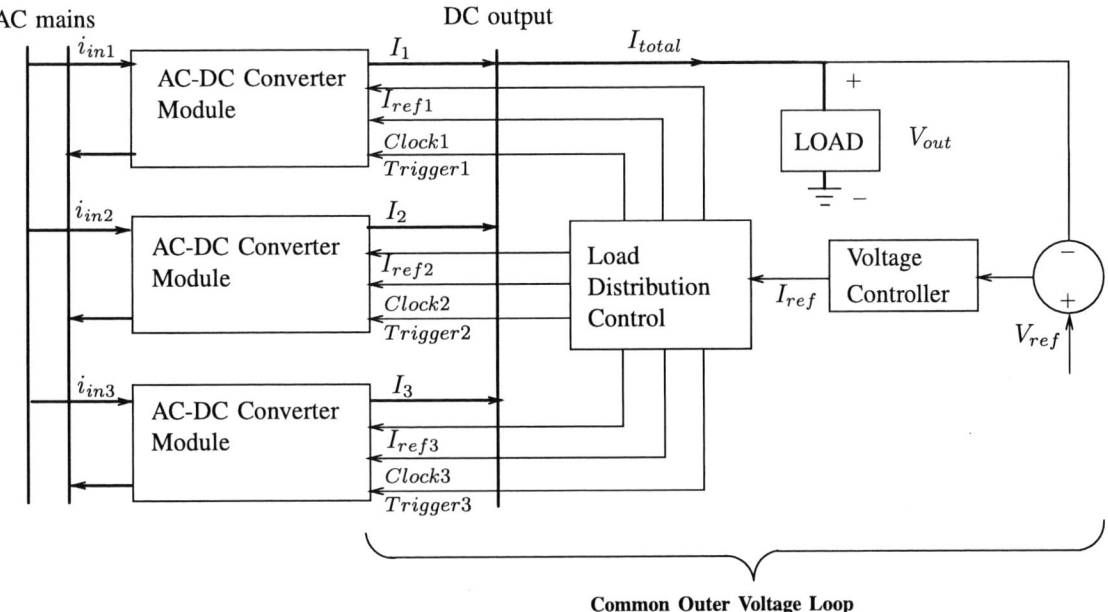

Fig. 3. Block Diagram Representation of Converter Modules in Parallel

C. The Common Output Voltage Feedback Loop

The common outer voltage loop plays a prominent role in the parallel operation of the AC-DC converter modules. The block diagram representation of the whole system along with the common outer voltage feedback loop is shown in Fig. 3. This loop has high bandwidth compared to the voltage loop of the front-end converter, and is the reason for fast dynamic response at the output load voltage. This loop is common to all the converter modules that are present in the system. The output voltage signal that has been sensed from the common output voltage bus is compared with the output reference voltage V_{ref}, and the error is sent to a voltage controller. The output of the voltage controller, which is the total current reference I_{ref} is input to the load distribution control unit. The load distribution control unit decides on the number of converter modules that should be in operation, the extent of phase delay for the clock of each converter module in order to minimize output voltage ripple, and the current that must flow in each converter module. The load distribution control algorithm that is used in the load distribution control unit(LDC) is explained below.

1) Load Distribution Control Algorithm: The load distribution control algorithm that is used here is an adaptation of the method proposed in [13]. The total output current of the system I_{total} is the sum of the inductor currents of the dc-dc converters, in the converter modules that are ON. If N is the number of converter modules that are ON

$$I_1 + I_2 + \cdots + I_N = I_{total} \qquad (10)$$

Hence, calculation of the current reference values I_{refi} of each module by the LDC is based on

$$I_{ref,1} + I_{ref,2} + \cdots + I_{ref,N} = I_{ref} \qquad (11)$$

Fig. (4) shows state graph of the load distribution control algorithm assuming the total number of converter modules to be 3. In the figure, the circles indicate the state of operation of the assembly of paralleled modules. The variable Z indicates the state of the LDC. Its value is equalto the number of converter modules that are in operation in that state. The circles also show the current references of the modules, and the phase angle by which each module switching clock should be shifted in each state. Transiting from one state to another depends on conditions determined in the boxes.

The load distribution control algorithm described by the state graph of Fig. (4) is arrived at from the following requirements. It is assumed that the converters are identically rated.

- The maximum current (i.e. the reference current $I_{ref,i}$) in a single module should not exceed the current rating I_{rat} of each module.
- The minimum current in a single module should not be lower than I_{min} to avoid discontinuous operation.
- Most importantly, the change from $Z = i$ to $i + 1$ or $Z = i$ to $i - 1$ should not affect the current references $I_{ref,j}$ for $j < i - 1$ modules.

The functioning principle may be explained as follows. If one module is operating(that means $Z = 1$) and the total current reference I_{ref} is less than I_{rat} the LDC is held in that state. If I_{ref} increases to a value higher than I_{rat}, the LDC leaves the state 1 and enters states 2 or 3 depending on the value of I_{ref}, and the first module gets a reference value of $I_{rat} - I_{min}$. It could have well been fixed at I_{rat}, but if the I_{ref} increases to just over I_{rat} and as the minimum current

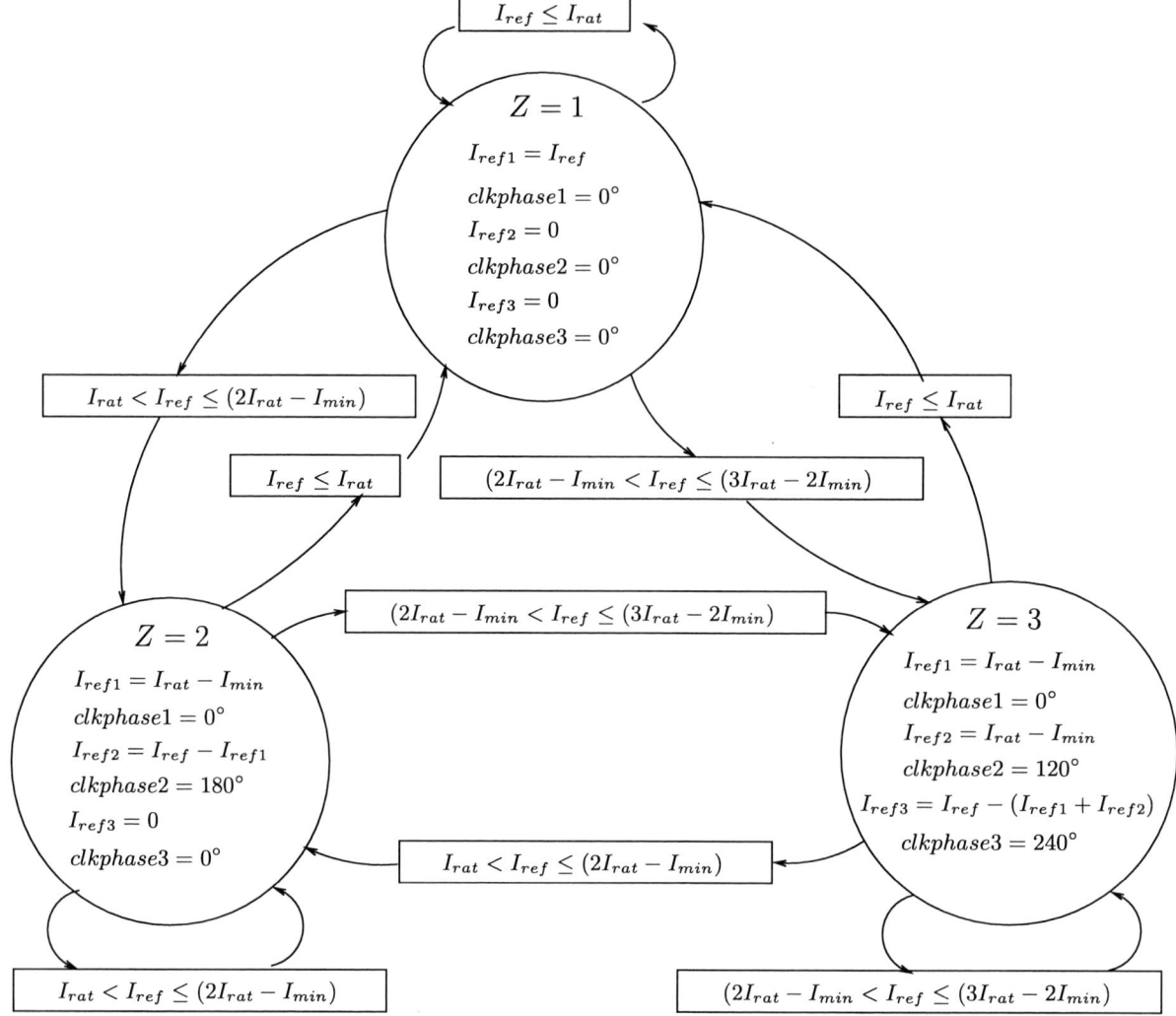

Fig. 4. State Graph of the Load Distribution Control Method.

for a converter module to operate is I_{min}, we can be in a situation where module 1 is fixed at I_{rat} and the next module is not ON because the residual current is not more than I_{min}.

In states of LDC where $Z > 1$, ie, more than one converter module is ON, $Z-1$ modules get a current reference of $I_{rat}-I_{min}$ and the last converter carries the residual current.

It can be observed that current is not being shared equally, but is assigned to $i-1$ converters and the remaining value of the total reference current is given as reference to the ith converter.

2) Modeling The Common Outer Voltage Loop Control System: The common outer voltage loop has been modeled using small-signal analysis, for the case where a small-signal variation in load about its quiescent value does not cause a change in the number of converter modules in operation. In this case the variation in the load is felt only by the ith converter module of the i converter modules in operation. The remaining $i-1$ converter modules act as constant current sources of value $I_{rat}-I_{min}$. In such a case, the common outer

voltage loop design may be carried out with the ith converter module alone. The common outer voltage loop along with the inner current loop and the dc-dc converter of the ith converter module can be represented using block diagrams as shown in Fig. (5).

The load distribution controller in this case can be modeled as total current reference i_{ref} subtracted for an offset value of $(i-1)(i_{rat}-i_{min})$ as shown in Fig. (5). The transfer function for a small signal variation in output voltage $\hat{v}_{out}$ with respect to small signal variation in inductor current $\hat{i}_{ind}$ for a dc-dc buck converter is

$$\frac{\hat{v}_{out}}{\hat{i}_{ind}} = \frac{1}{\frac{1}{R} + C_{out}s} \qquad (12)$$

From (9) and (12), the small-signal ac variation in $\hat{v}_{out}$ about a quiescent value for variations in $\hat{v}_{ref}$ and $\hat{i}_{refi}$ can be derived as

$$\hat{i}_{ind} = \frac{C}{B}(K\hat{i}_{refi} + \hat{v}_{ref}) \qquad (13)$$

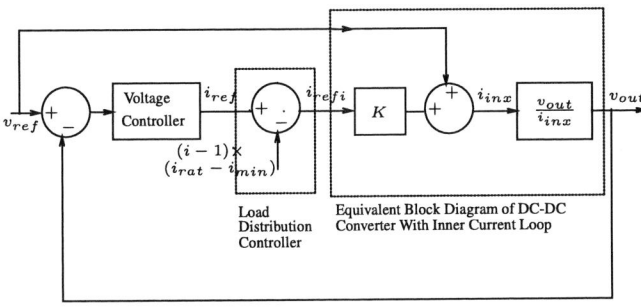

Fig. 5. Block Diagram Representation of The ith DC-DC Converter Along With Its Inner Current Loop And Common Outer Voltage Loop

where

$$C = 1 + \frac{T}{2}s \tag{14}$$

If $K\hat{i}_{refi} + \hat{v}_{ref}$ is represented as $\hat{i}_{inx}$, (13) can be rewritten as

$$\hat{i}_{ind} = \frac{C}{B}\hat{i}_{inx} \tag{15}$$

The voltage controller, in the common outer voltage loop can be designed with the help of (13) and the block diagram shown in Fig. (5).

III. SIMULATION RESULTS

Simulation of the system operation has been carried out in SABER simulation software, with the following design parameters.

Input Voltage of AC-DC Converter Module $V_{in,rms}$ = 230 V rms, 50 Hz
Output Voltage of AC-DC Converter Module V_{out} = 42 V
Output Voltage of the front-end Converter V_s = 50 V
Rated Power of each module = 210 W
Minimum Current each module can supply I_{min} = 1 A
Maximum Current each module can supply I_{rat} = 5 A
Switching Frequency of front-end Converter = 50 KHz
Switching Frequency of DC-DC Converter = 50 KHz
Total number of converter modules = 3

The proportional gain controller K in the inner current loop of the dc-dc converter and the voltage controller in the common outer voltage loop have been designed based on the modeling of both the loops carried out in the previous sections. The bode plots for $\frac{\hat{i}_{ind}}{\hat{v}_{control}}$ (Fig. 2) for load currents of 1A to 5 A are shown in Fig. 6. The value of K has been designed to be 10. The bode plots for $\frac{\hat{v}_{out}}{\hat{i}_{inx}}$ (Fig. 5) for load currents of 1A to 5 A at $K = 10$ are shown in Fig. 7. A PID controller has been used as voltage controller for the common outer voltage loop. The K_p, K_i and K_d values that have been derived at are 200, 25k, and 1m respectively.

Initially the system was simulated with a load of 1 A. The load was changed to cause a step increase in current to 11 A at 150 ms and a step decrease to 7 A at 200 ms. Though

there was a step change in load power of about 420 W at 150 ms and 168 W at 200 ms, it can be observed from the output voltage waveform in Fig. 8 that the maximum deviation in the output voltage during the transients at 150 ms and at 200 ms were a mere 0.67 V and 0.2 V with settling times of 11.3 ms and 6 ms respectively, indicating fast dynamic performance even during large changes in load.

Fig. 9 shows the output filter inductor current waveforms of the first, second and third converter modules. Before 150 ms, the first converter module was supplying a current of 1 A, the second and third converter module were OFF. With sudden increase in load current to 11 A at 150 ms, second and third converter modules turn ON, and the current through the first second and third converter modules can be found to 4 A, 4 A and 3 A respectively. At 200 ms, with the sudden reduction in load to 7 A the currents in the first second and third converter modules change to 4 A, 3 A, and 0 A respectively. Good current sharing performance can be observed from the results. The current sharing in each module before and after the load changes is in accordance with the load distribution control algorithm.

The output voltage waveforms of first, second and third PFC ac-dc converters are shown in Fig. 10. Comparison of Fig 8 with Fig. 10 shows that the output load voltage is immune to the transient and twice the line frequency variation in voltages across the output capacitors of the front-end converters. Instead the dynamics of the output voltage are dependent on the high bandwidth common outer voltage loop of the system.

The input current waveform is shown in Fig. 11. Input power factor has been calculated to be greater than 0.96, under all conditions. This proves that the proposed method achieves high power factor correction irrespective of the load.

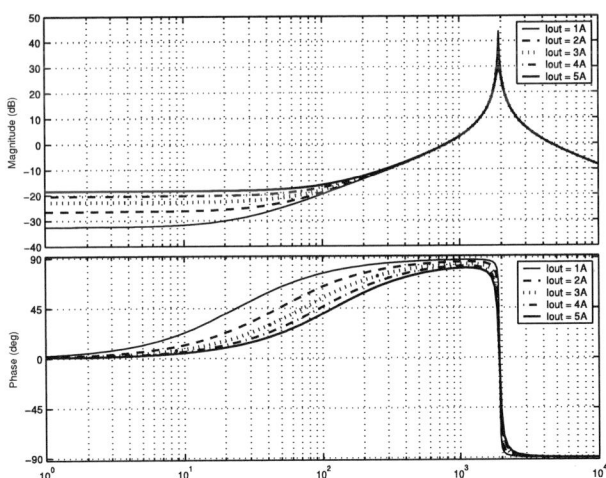

Fig. 6. Bode Plot For $\frac{\hat{i}_{ind}}{\hat{v}_{control}}$ at load currents of 1A, 2A, 3A, 4A and 5A

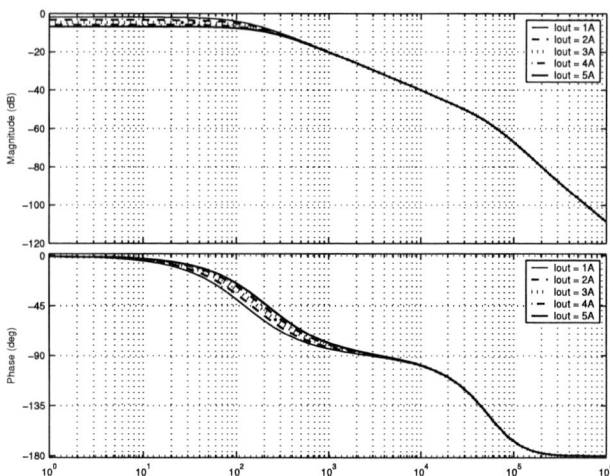

Fig. 7. Bode Plot For $\frac{\hat{v}_{out}}{\hat{i}_{inx}}$ at load currents of 1A, 2A, 3A, 4A and 5A

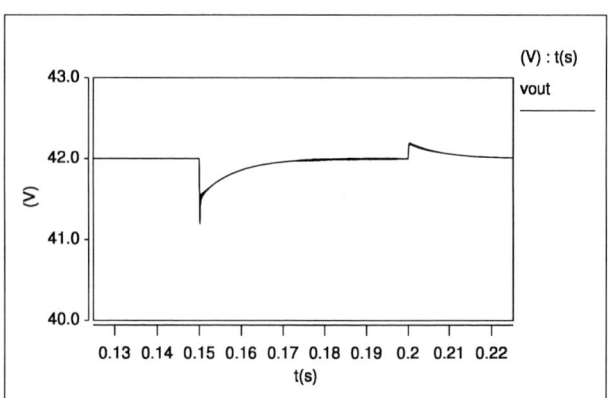

Fig. 8. Output voltage waveform representing step change in load at 150 ms and 200 ms

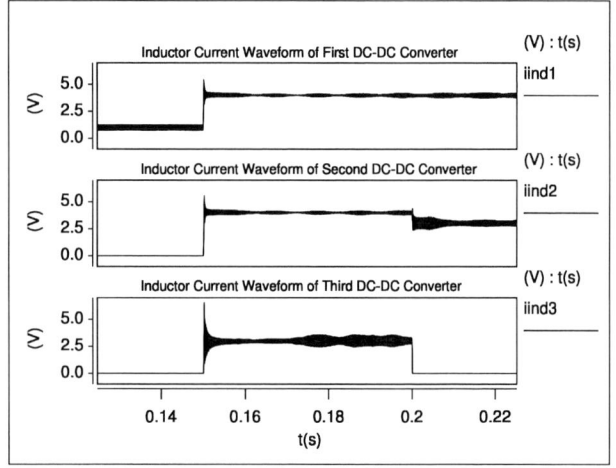

Fig. 9. Output filter inductor current waveforms of the ac-dc converters modules. Current Distribution in each converter module for different loads can be observed

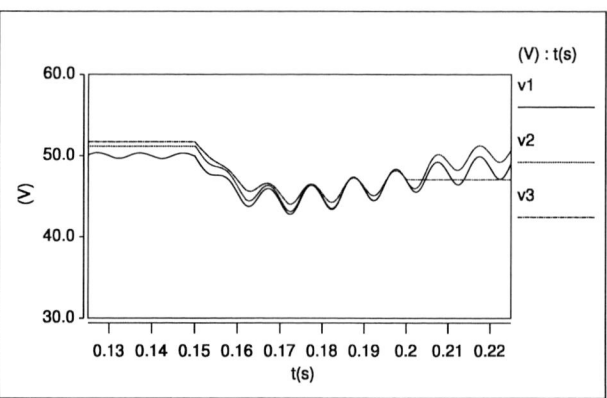

Fig. 10. Output voltage waveforms of front-end converters. The slow dynamic response of the PFC ac-dc converter stage is evident here

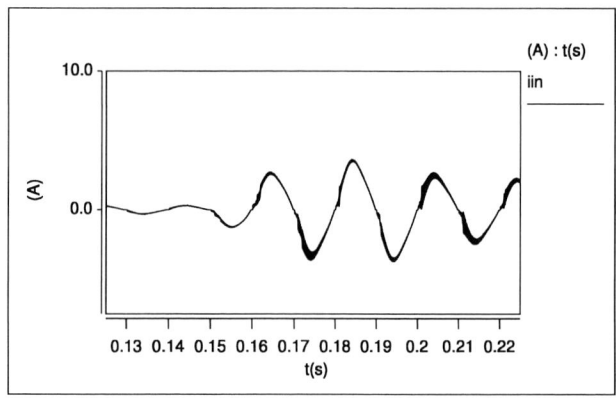

Fig. 11. Input current waveform. High input power factor has been achieved

IV. CONCLUSION

In this paper, a simple multi-module control mechanism for paralleling standard PFC ac-dc converter modules with two stages was presented. It has been demonstrated using simulated results that tight output dc voltage regulation, low output voltage ripple, input power factors above 0.96, can be achieved using the proposed approach. The implementation of the proposed power and control methods is currently in progress.

REFERENCES

[1] N. P. Papanikolaou, E. J. Rikos, E. C. Tatakis," A Novel technique for high power factor correction in flyback converters: Theoretical analysis and design guidelines," Proc. IEE-Elect. Power Applicat., vol. 148, no. 2, March 2001, pp. 177-186.

[2] W. Tang, Y. Jiang G. C. Hua, F. C. Lee, I. Cohen,"Power Factor Correction with flyback converter employing charge control", in Proc. APEC'93, 1993, pp.293-298.

[3] H. Wei, I. Batarseh, "Comparision of Basic Converter Topologies for Power Factor Correction", in Proc. Southeastcon'98, 1998, pp.348-353.

[4] R. Oruganti, M. Palanipan."Inductor Voltage Control of Buck-Type Single-Phase AC-Dc Converter", IEEE Transactions on Power Electronics, vol. 15, No. 2, March 2000, pp.411-417.

[5] Robert W.Erickson, Dragan Maksimovic, "Fundamentals of Power Electronics", Second Edition.

[6] L. Dixon Jr., "High Power Factor Preregulators for off-line supplies", in Proc. Unitrode Power Supply Design Sem., 1991.

[7] Y. Jiang, F. C. Lee, G. Hua, W. Tang, "A novel single-phase power factor correction scheme" in Proc. APEC 1993, March 1993, pp. 287-292.

[8] Sangsun Kim, Prasad N.Enjeti, "A Parallel-Connected Single Phase Power Factor Correction Approach With Improved Efficiency", IEEE Transactions on Power Electronics. Vol.19, No. 1,January 2004, pp.87-93

[9] Aman Kumar Jha, K. Hari Babu, B. M. Karan."Parallel Power Flow AC/DC Converter with High Input Power Factor and Tight Output Voltage Regulation for Universal Voltage Application", in Proc. PEDES, Dec. 2006, pp.1-7.

[10] D. D. C. Lu, D. K. W. Cheng, Y. S. Lee,"A Novel Single-Phase Power-Factor-Corrected Voltage Regulator", in Proc. PESC 2001, Vol. 2, June 2001, pp.936-941.

[11] S.Y.R Hui, H.Chung, "Paralleling Power Converters for AC-DC Step-Down Power Conversion With Inherent Power Factor Correction", IEE proceedings on Electrical Power Applications. Vol.146, No.2, March 1999, pp.247-252.

[12] M. Phattanasak, V. Chunkag,"Paralleling of Single-Phase AC/DC Converter with Power-Factor Correction" in PESC 2004, Vol. 2, June 2004, pp.1576 - 1580.

[13] Stefan Huth, "DC/DC-Converters in Parallel Operation with Digital Load Distribution Control", in Proc. IEEE ISIE'96, vol.2, 1996, pp.808-813.

[14] Sheng-Hua Li, Chang-Ming Liaw,"Paralled DSP-Based Soft Switching-Mode Rectifiers With Robust Voltage Regulation Control", IEEE Transactions on Power Electronics, Vol. 19, No 4, July 2004, pp.937-946.

[15] Keyue M. Smedley, Slobodan Cuk,"One-Cycle Control of Switching Converters ", IEEE Transactions on Power Electronics, vol. 10, No. 6, November 1995, pp.625-633.

[16] A. de Castro, P. Zumel, O. Gaecia, T. Riesgo, and J. Uceda, Concurrent and Simple Digital Controller of an AC-DC Converter With Power Factor Correction Based on FPGA, IEEE Trans. Power Electron., vol. 18, Jan. 2003, pp, 334-343.

Noise Radiation of Switched Reluctance Drives

K. A. Kasper and M. Bösing and R. W. De Doncker
Institute of Power Electronics and Electrical Drives
RWTH Aachen University
Jaegerstrasse 17-19, 52066 Aachen, Germany
Phone: +49 241 8096949
Email: kr@isea.rwth-aachen.de

S. Fingerhuth and M. Vorländer
Institute of Technical Acoustics
RWTH Aachen University
Neustrasse 50, 52066 Aachen, Germany

Abstract— To simulate noise emissions of switched reluctance machines (SRMs), the entire signal transmission chain has to be modeled, starting with the radial force and ending with the radiated sound. Especially the last element of this transmission chain, the radiation characteristic of an SRM, is very rarely addressed in literature. This paper gives a brief introduction to noise development in SRMs and then presents the first phase-exact measurement of the radiation characteristic of an SRM that is known to the authors. Furthermore, methods to simulate the sound radiation are introduced.

I. INTRODUCTION

Switched reluctance machines (SRMs) are an appealing alternative to conventional machines for many applications due to their low manufacturing costs and high reliability. Unfortunately, unpredictable noise emissions of SRMs have so far prevented a wide spread use in industry. For the development of effective noise reduction measures, it is important to understand the entire acoustic signal transmission chain as displayed in Fig. 1, starting with the phase current and ending with the radiated sound. With this knowledge, simulation models for the acoustic behavior of SRMs can be constructed.

However, most publications in this field focus on the excitation (e.g. [1]) or the mechanical transfer function (e.g. [2]). So the radiation characteristic of SRMs has hardly been studied in detail. In [3], [4] some theoretical approximations for an overall radiation efficiency are given, but no experimental verification is provided. Furthermore, the directivity of the sound radiation is not taken into account. The radiated noise is not the same for all directions around the machine. This is important if the acoustic behavior of an SRM within a larger application shall be modeled, e.g. an assembly line or an electric vehicle.

After a brief introduction to the noise development mechanisms in SRMs, this paper presents the first phase-exact measurement of the radiation characteristic of an SRM known to the authors. Furthermore, simulation methods are introduced that can reproduce the measured behavior.

II. NOISE DEVELOPMENT IN SWITCHED RELUCTANCE MACHINES

The noise development in switched reluctance machines can be traced back to electromagnetic and aerodynamic effects. The latter appear primarily in high speed machines, more precisely at high track speeds of the rotor teeth. Rotating pressure fields in front of the rotor teeth radiate a tone that can be calculated as the product of the rotation frequency and the number of rotor teeth. Since the pressure waveform is not sinusoidal, the spectrum also contains the harmonics of the fundamental frequency. Aerodynamic noise cannot be influenced by control measures but it can be avoided by closing the gaps between the rotor teeth with non-magnetic material or by encapsulating the rotor completely [5]. Therefore, aerodynamic effects are not considered in the following. The main cause for electromagnetic noise development in SRMs is the radial force that is caused by the phase currents. It excites vibrations of the stator (also called structure-borne sound) which are then radiated as airborne sound. The signal transmission path is shown in Fig. 1.

The mechanical transfer function and the radiation characteristic are linear systems, thus it is obvious that all frequencies present in the airborne sound have to exist already in the exciting force. Only the amplitude and the phase are altered, e.g. mechanical resonances influence the sound of the machine, of course, but do not add new frequencies. Therefore, the frequency domain can be used for the following analysis. Since the focus of this work is the radiation characteristic, the overview of the noise development mechanisms is kept short. More details can be found in [4].

A. Radial Force Spectral Composition

What does the radial force in a switched reluctance machine look like? At every current pulse at a stator tooth (once per rotor tooth during one machine rotation) a radial force pulse occurs. Thus, the fundamental frequency of the force equals the fundamental frequency of the current, the so-called electrical fundamental frequency (mechanical speed multiplied by rotor tooth number). Since neither current nor force are sinusoidal, their spectra contain harmonics of this fundamental frequency. In signal theory, the force $s(t)$ at a stator tooth can be described as the convolution of a series of Dirac pulses $III(t/T)$ by one force pulse $p(t)$, see equation (1). T is the time interval between the force pulses, $1/T$ is therefore the electrical fundamental frequency. Transformed into the frequency domain this becomes the multiplication of a series of Dirac pulses $|T| \cdot III(Tf)$ with the spectrum of a single force pulse $P(f)$, see equation (2).

978-1-4244-0644-9/07/$25.00 ©2007 IEEE

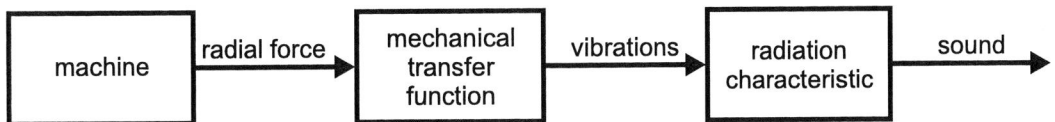

Fig. 1. Acoustic Signal Transmission Chain of Switched Reluctance Machines

$$s(t) = III(\frac{t}{T}) * p(t) \qquad (1)$$

$$S(f) = |T| \cdot III(Tf) \cdot P(f) \qquad (2)$$

Consequently, the spectrum of the radial force $S(f)$ contains only the harmonics of the electrical fundamental frequency. The intensity of these harmonics depends on the spectrum $P(f)$ and therewith on the waveform $p(t)$ of a single force pulse and can thus be influenced via the current waveform. A more detailed analysis of the radial force and methods for an estimation of the spectrum can be found in [6]. Methods to achieve acoustically optimized radial force waveforms are described in [7].

B. Mechanical Transfer Function and Vibrations

The radial forces act on the stator teeth of the SRM and cause surface vibrations via a mechanical transfer function. This function can be described as a superposition of mass-spring-damper systems representing the different vibration modes with their eigenfrequencies. Figure 2 shows four mode shapes of an SRM stator. It is typical for SRMs that certain vibration mode shapes match with the exciting force mode shape and are therefore directly excited. In an 8/6 machine (eight stator teeth, six rotor teeth), for example, every phase will excite a mode 2, because the radial force is always acting on two opposing stator teeth simultaneously. Following that logic, in an 16/12 machine mode 4 is excited whereas mode 2 is not excited. If a harmonic of the radial force comes close to the eigenfrequency of such a directly excited mode, strong vibrations in the shape of the respective mode will be the consequence. Measurements at our institute have shown, that especially if the directly excited mode is a mode 2, it usually dominates the mechanical transfer function and thus the resulting vibrations. Other vibrations are also present but usually do not play a role for the overall sound pressure level. The reason for this behavior is, that for typical stator sizes the mode 2 eigenfrequency falls in the same frequency range as the lower order harmonics of the exciting radial force whereas the other modes' eigenfrequencies are far higher and are therefore only excited by higher force harmonics which contain less energy.

Included in the mechanical transfer function is also the effect that certain harmonics of the radial force cancel out due to symmetric excitation while others are amplified, e.g. in the vibrations of a four phase machine the even harmonics are strongly diminished. A detailed description of the modal superposition can be found in [3].

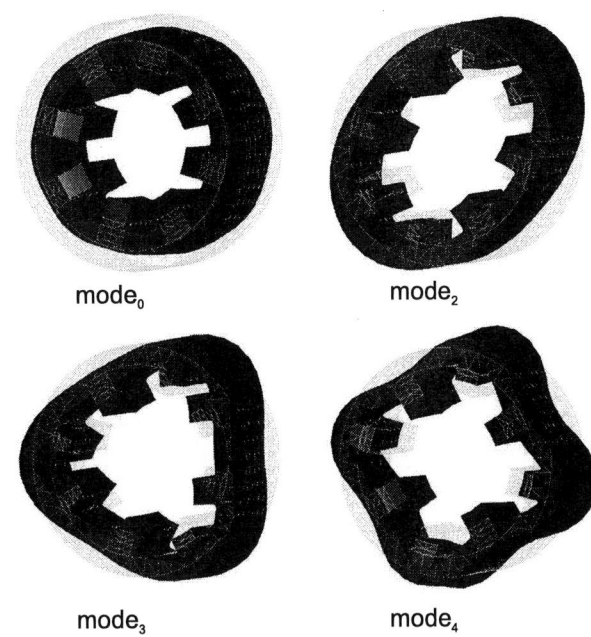

Fig. 2. Vibration Modes of an SRM Stator

Getting back to our initial example, the vibrations of an 8/6 machine can be described as two mode 2 shapes rotated by $45°$ against each other with a phase difference of $\pi/2$. One is caused by the excitation of phase 1 and 3, the other one by phase 2 and 4.

C. Expected Radiation Characteristic

Since the vibrations can be split into a superposition of different vibration modes, it also makes sense to describe the overall radiation characteristic as the superposition of the radiation characteristics of the single vibration modes. For a 2D approximation, the vibration modes can be regarded as acoustic multipoles. Their radiation characteristic can be described analytically. The following equations are based on [8]. A general expression for the sound pressure $\underline{p}$ radiated by an acoustic multipole is:

$$\underline{p}(r, \theta, t) = j\frac{kc\rho_0 \hat{Q}}{4\pi r}e^{j(2\pi ft - kr)}G(\theta) \qquad (3)$$

$k = {}^{2\pi f}/c$ is the wavenumber, c is the speed of sound in the respective medium, Q is the source strength, r is the distance from the center of the source, $G(\theta)$ is the directivity in dependence of the angle θ around the source. For a 2D quadrupole with the diameter d this is:

$$G(\theta) = -\frac{k^2 d^2}{2}\cos(2\theta). \qquad (4)$$

The sound pressure waveform in the time domain $p(t)$ can be described as the real part of $\underline{p}$

$$p(t) = \Re(\underline{p}). \qquad (5)$$

Note that the condition $kd \ll 1$ has to be fulfilled in order to get multipole behavior. A quadrupole that represents a mode 2 vibration radiates four lobes. The two quadrupoles of an 8/6 machine will therefore radiate eight lobes as shown in Fig. 3. The second quadrupole is rotated $45°$ and phase shifted $\pi/2$ compared to the first one. This leads to the following sound pressure distribution:

$$p(t) = \frac{k^3 d^2 c\rho_0 Q}{8\pi r}(\cos(2\theta)\cos(2\pi\frac{t}{T}) + \sin(2\theta)\sin(2\pi\frac{t}{T}))$$

$$(6)$$

The directivity is proportional to the rms value $p_{\mathrm{rms}}(\theta)$ defined as

$$p_{\mathrm{rms}}(\theta) = \sqrt{\frac{1}{T}\int_{t_0}^{t_0+T} p^2(t)\mathrm{d}t} \ . \qquad (7)$$

Inserting (6) in (7) shows that p_{rms} is independent of θ. That means that in the ideal case, the lobes of the two quadrupoles add up to constant sound pressure around the circumference of the machine.

However, although there is no directivity for the sound pressure level, the radiation is not the same as for a mode 0 (breathing mode). The quadrupole source can still be identified because there will be a phase difference of π between the lobes of one quadrupole, marked by '+' and '-' in Fig. 3, and of $\pi/2$ to the respective lobes of the other quadrupole.

So far, only the radiation in the radial direction has been considered. A 3D approximation of the radiation has to take into account the axial length of the machine. A possible model for a mode 2 vibration of a cylinder is a finite length line array quadrupole. In the radial direction, the behavior is identical to the 2D model introduced above. However, additional lobes and phase jumps occur in the polar direction due to the finite length

of the radiator. Figure 4 shows the phase of the radiated sound mapped onto a hemisphere. The phase difference between the dark and light areas is π. The mathematical derivation of this effect can be found in [8]. It is quite lengthy, not the focus of this work and therefore not presented here.

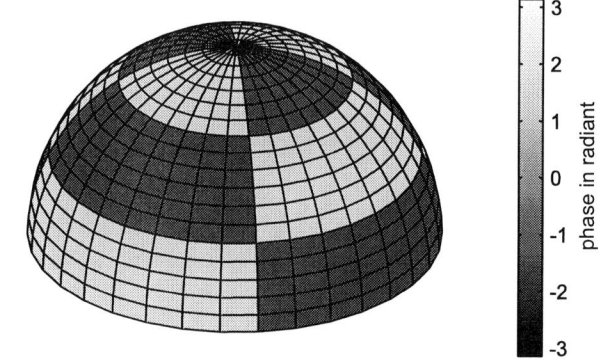

Fig. 4.   Phase of the Radiation of a 180 mm Line Array Multipole at 2350 Hz

This is the radiation characteristic that can be expected according to acoustic theory. But so far, no experimental verification has been performed.

III. EXPERIMENTAL SETUP

A. Drive Under Test

A four phase switched reluctance machine with eight stator teeth and six rotor teeth was chosen for the measurement of the radiation characteristic. Its rated power is $2.7\,\mathrm{kW}$ and its stator diameter is $120\,\mathrm{mm}$, the length of the stator stack is $180\,\mathrm{mm}$. The housing is a simple metal shell, which is especially important for measurements of the radiation characteristics. Cooling fins etc. could distort the results. Figure 5 shows a picture of the machine.

Fig. 5.   Picture of the 8/6 GRM Used for the Measurements

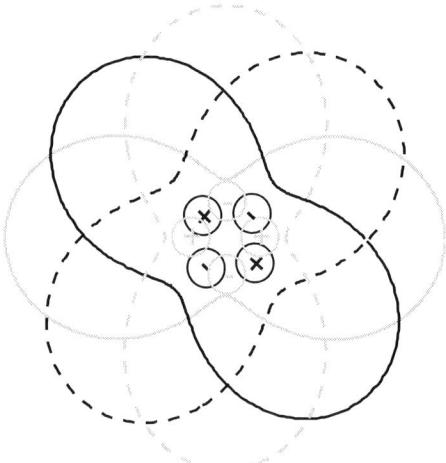

Fig. 3.   2D Radiation Lobes of 8/6 SRM, quadrupole caused by phases 1 and 3 (grey) and quadrupole caused by phases 2 and 4 (black)

1162

B. Eddy Current Test Bench

In order to avoid the noise of a second electrical machine, a special test bench was developed using an eddy current brake as a load. Since the measurements have to be performed in an anechoic chamber, the test bench was built lightweight to be portable. The SRM sets a torque and the brake is operated under speed control. When the test bench speed is below the reference value, the DC current in the excitation coils is reduced, otherwise it is increased. The noise level generated by the test bench is very low, because no alternating magnetic fields are present in the airgap that could cause vibrations. Only bearing noise and other mechanically generated sounds exist. To keep this mechanical noise as low as possible, the entire active braking system comprised of the excitation coils, the braking disc and its bearings is mechanically decoupled from the aluminum frame by rubber buffers.

The electromagnetic design of the eddy current brake, which includes the geometric dimensions and material parameters of the disc and the excitation coils, is based on analytic equations published in [9]. Figure 6 shows the torque speed diagram of the eddy current brake for three excitation currents. All operating points below the curve for the maximum current can be measured.

Since the entire braking power is transformed to heat in the braking disc, it gets extremely hot during operation. Also the excitation coils heat up due to copper losses. In order to keep the setup simple and portable, no complex cooling system was installed. Only two fans are attached to the frame that cool the disc and also the excitation coils. However, these fans create substantial noise and hence are turned off during measurements. So the thermal design limits the operation time of the brake depending on the load. However, since acoustic measurements usually require only a few seconds of operation at a certain operating point, this is acceptable. Between the measurements the fans are turned on and the temperature of the disc and the coils is reduced quickly. For safety reasons, these temperatures and the speed are monitored at all times by the brake control unit. If the maximum values are exceeded, emergency turn-off is activated. In this case, the power supply of the drive under test is turned off by the brake control. That also happens automatically if the brake control has a failure or looses power.

C. Measurement Setup

For the measurements, the test bench was installed in the floor of an anechoic chamber and covered with a wooden board so that only the SRM protruded, see Fig. 7. That way, also the remaining test bench noise could be avoided and the SRM could be measured directly above a plain acoustically reflecting surface. This is useful if the results shall be used for the verification of radiation simulations, because such a surface is a clearly defined boundary condition. A 90° arc with a radius of 2 m holding 19 equally spaced microphones was constructed, starting on the ground and ending above the machine. That way, the entire spatial noise radiation can be measured step by step by turning the SRM. For this purpose, the test bench mounting is specially designed and equipped with a degree scale. A step size of 10° was chosen, leading to 18 measurement positions to cover half a rotation of the machine or a quarter sphere. For symmetry reasons, the noise emissions in the second quarter sphere will be identical and do not have to be measured.

Furthermore, one phase current was recorded to provide a trigger signal to determine the correct phasing of the measurements. Additionally, a reference microphone was placed at ground level and rotated together with the machine so that it always pointed at the same spot on the machine. The sound measured by this microphone has to be the same for all machine positions, otherwise the operating point has changed (e.g. because of thermal effects) and the measurement cannot be used.

The measurements were done at half of the maximum torque for two different speeds, 3385 rpm and 1118 rpm. At these speeds, the machine's mode 2 eigenfrequency of about 2350 Hz is hit by a harmonic of the radial force leading to clear mode 2 vibrations.

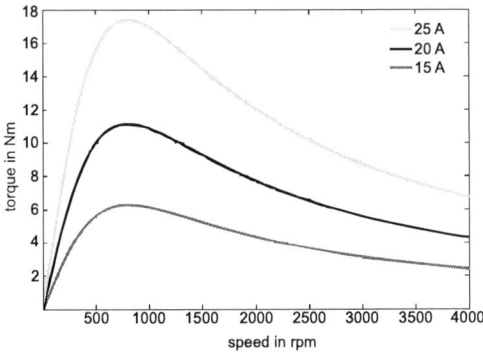

Fig. 6. Torque Speed Diagram of Eddy Current Brake

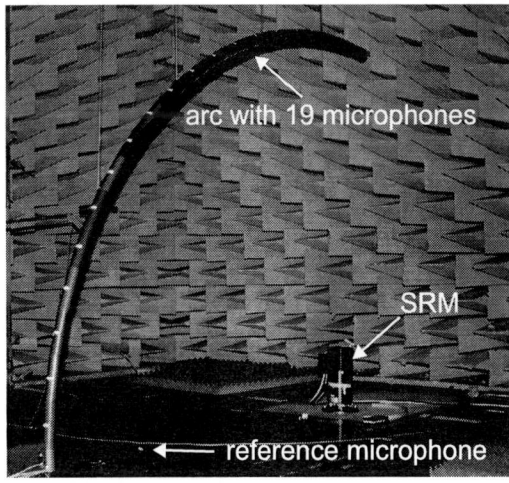

Fig. 7. Measurement Setup in Anechoic Chamber

IV. MEASUREMENT RESULTS

A. Sound Radiation Analysis

Before the evaluation, the distortion caused by the arc itself was compensated in all microphone channels. (It was measured before by creating defined test tones with a loudspeaker.) Thus, the sound pressure is the same as for perfect free-field conditions. All channels of the measurements for the single positions were shifted in time relative to the other positions until the measured phase currents matched. That way, the relative phasing of the different measurement positions can be evaluated.

The spectra of the measured sound pressures at 3385 rpm clearly show a maximum at the mode 2 eigenfrequency 2350 Hz of the machine (which was previously determined via vibration measurements), see Fig. 8 as an example. That means, that the mode 2 eigenfrequency dominates the noise, which is in accordance with the theory presented above. At that frequency, the sound pressure level and the phase are evaluated by mapping the microphone data on a quarter sphere, shown in Fig. 9. The sound pressure in decibel is coded as distance from the center, the phase is color-coded. The machine position is upright in the center of the sphere. It is clearly visible, that the sound pressure level in the radial direction is higher than in the axial direction, as expected. The noise radiation at the bottom, meaning in the radial direction around the machine, has the circumferential shape you would expect from a quadrupole. The lobes opposite of each other on the left and right side have the same phasing, shown by the dark color, the other one in between them is shifted by π, indicated by the light color. The behavior in the third dimension can be explained with the finite length of the machine: The steps of π when going from the bottom toward the 'north pole' are very similar to those in Fig. 4.

It has to be noted that it is not possible to correctly visualize a phase from $-\pi$ to π in greyscale since both ends should have the same color. So at the wrap-around single white spots may appear in a black area. In the following figures, the data has

been prepared in a way that the important results can be seen although greyscale is used.

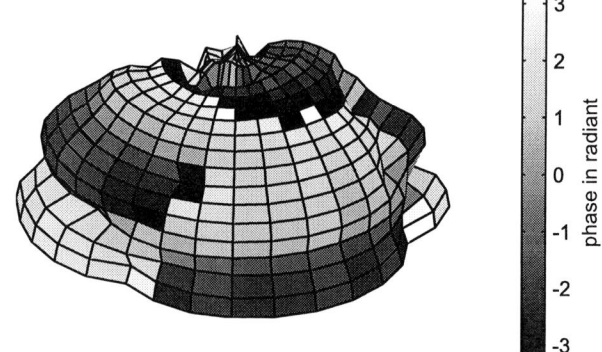

Fig. 9. Measured Radiation Characteristic of 8/6 SRM at 2350 Hz

Interestingly, only one quadrupole seems to exist, whereas two were expected. This is due to asymmetries in the machine geometry. Because of the flange mounting, screws and a slightly different contact between stator stack and housing, the stiffness of the structure is not symmetrical. This leads to several eigenfrequencies with a mode 2 like shape. This phenomenon was further investigated using vibration measurements. The results are presented in the next section. One of these frequencies, 2950 Hz, is also excited by a harmonic of the radial force at 3385 rpm. It can be identified in the airborne sound in Fig. 8. Its effect on the overall noise emissions is quite small because of the lower amplitude. The measured phase of the radiated sound pressure at that frequency is shown in Fig. 14, mapped onto a hemisphere. The quadrupole behavior at the bottom can be recognized, although colored plots can show this better.

B. Analysis of Vibration Behavior

As already mentioned, the difference between the mode 2 shape eigenfrequencies is caused by asymmetries of the structure. An experimental modal analysis was performed in order to determine the vibration behavior of the machine.

In a first measurement, the surface acceleration of the machine was measured on the housing above a stator tooth. Then the rotor was aligned to this phase and a current pulse was applied, leading to a force pulse similar to a dirac pulse, thus exciting a broad frequency range. This was done for all phases. Figure 10 contains the frequency response measured above phases 1 and 2 for the excitation of the respective phase. The main resonance frequency can easily be identified in both plots. Further accelerometers around the circumference of the machine showed that the dominating vibration always has mode 2 shape. Its amplitude is almost identical for both phases, but the frequencies differ. For phase 1 we get the already well known 2350 Hz, but in the direction of phase 2, the mode 2 eigenfrequency is only 2230 Hz. This is due to small holes in the stator at the teeth belonging to phases 2 and 4. They were created as a side effect of the manufacturing process and decrease the stiffness of the structure at that

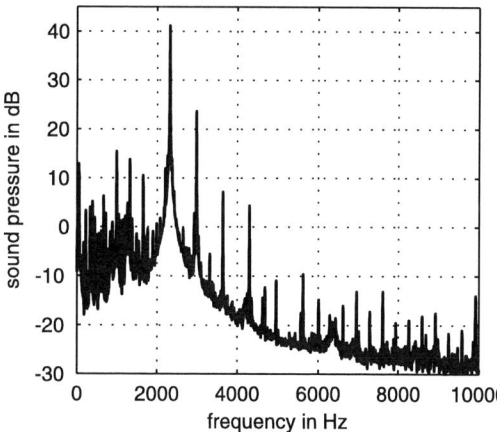

Fig. 8. Spectrum of the Sound Pressure at 3385 rpm

location. Therefore, the eigenfrequency is lower. During the measurement of the radiation characteristic at 3385 rpm, this frequency was not excited by a harmonic of the radial force. Consequently, no sound was measured at that frequency (see Fig. 8).

A second measurement was performed to investigate the influence of asymmetries in the axial direction. Four accelerometers were placed in a row on the housing above one stator tooth, thus covering the entire axial length of the machine. Then the drive was run at the same operating point as during the measurements of the radiation characteristic. The spectra of the vibrations measured by the sensor at the top and the bottom are displayed in Fig. 11 . Both plots show the eigenfrequency at 2350 Hz, but it is more prominent at the top than at the bottom. At the bottom, where the structure is significantly stiffer due to the flange mounting, there is an additional eigenfrequency at 2950 Hz. This one is also present at the top, but with a much lower amplitude. From the analysis of the radiation characteristic at this frequency, we know that it radiates similarly to a quadrupole and therefore has mode 2 shape (Fig. 14).

Summing this up, it could be shown that the asymmetries in the machine structure lead to a set of mode 2 shape eigenfrequencies with different frequencies instead of just one resonance. All these modes can be directly excited by harmonics of the radial force.

V. SIMULATION OF NOISE RADIATION

In order to find simulation methods suitable for the prediction of SRM noise radiation, a combined finite element and boundary element model of the stator of the 8/6 machine was constructed using the software Virtual Lab. The input data is the radial force on the stator teeth. In the first step, the surface vibrations are calculated using the finite element method in the frequency domain. These are then used as input data for the second step, the calculation of the sound pressure at certain points in the environment. The measurement setup

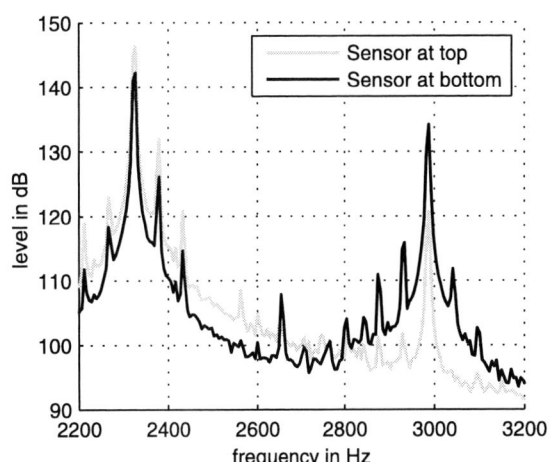

Fig. 11.   Surface Acceleration at 3385 rpm at Two Positions

was reconstructed for this purpose, so the floor was modeled as a sound reflecting surface. The sound pressure was calculated at the microphone positions of the measurements, so the results are directly comparable. Figure 12 shows again the measured phase of the radiation at 2350 Hz mapped onto a quarter sphere. For easier comparison this time without the amplitude. Figure 13 presents the simulated phase. Both plots show the same principal behavior, especially the same phase steps already described above. In the region closer to the ground, meaning in the radial direction around the machine, they are very similar. Also the phenomenon that only one quadrupole dominates the noise emissions is represented because asymmetries of the stator are included in the finite element model. The higher regions are a bit disturbed in the measurements because they measure sound from all multipoles, so the results are not as clear as in the simulations.

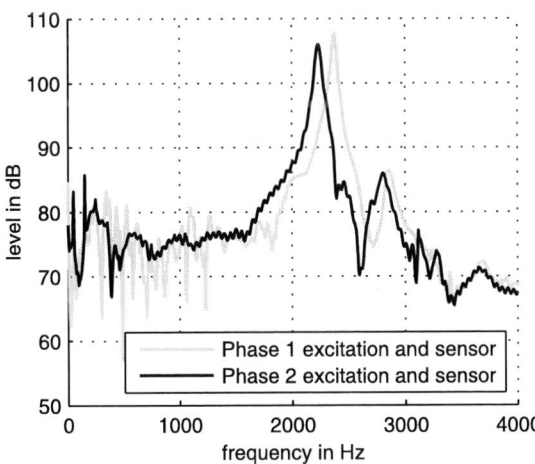

Fig. 10. Frequency Response Above Stator Teeth for Dirac Pulse Excitation

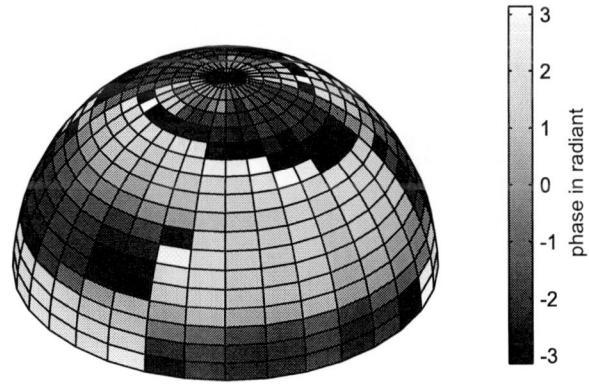

Fig. 12.   Measured Phase of Radiated Sound of 8/6 SRM 2350 Hz

The second mode 2 eigenfrequency at 2950 Hz behaves similarly. Figure 14 shows the measured phase of the sound pressure while Fig. 15 shows the simulated values. At the bottom, simulation and measurement agree very well. But

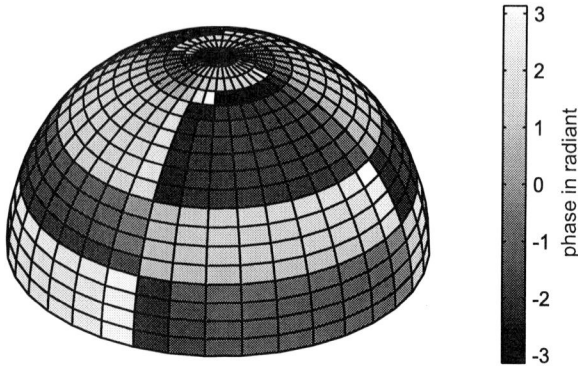

Fig. 13. Simulated Phase of Radiated Sound of 8/6 SRM 2350 Hz

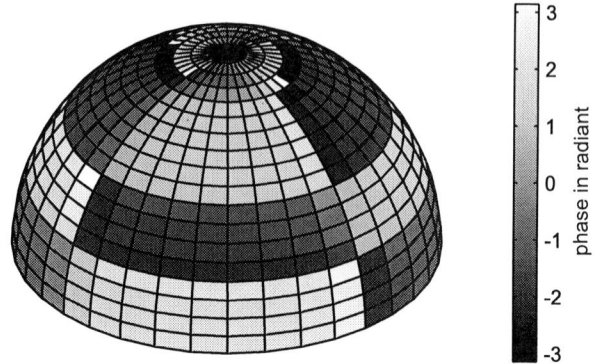

Fig. 15. Simulated Phase of Radiated Sound of 8/6 SRM 2950 Hz

moving upwards, the simulation predicts a number of phase jumps that could not be measured. The difference is larger than for the lower eigenfrequency. The reason for this is probably the inaccuracy of the model. Currently, only the stator is modeled for the simulation of the radiated noise. A more complex model that also includes the housing and the mounting conditions is being developed. It is expected that the results will improve further, especially the amplitudes of the simulated sound pressure. Currently, there is still quite a large error due to the housing that was neglected. However, the similarities between the analytic model using only quadrupoles (Fig. 4) and the complex boundary element model (Fig. 13) are so apparent, that these simple models seem to be an interesting alternative, especially when computation time is an issue. Also other vibration shapes than mode 2 can be simulated that way.

line array multipoles. Hence, the existing theory, e.g. [4], is supported. The influence of asymmetries in the machine structure on the vibration behavior and on the radiation was investigated. Furthermore, a combined finite element and boundary element model was introduced that can reproduce the measured radiation characteristic. If computation time is an issue, simple analytic models using finite length line array multipoles can provide a reasonable good approximation of the vibration behavior of switched reluctance machines.

ACKNOWLEDGMENT

The authors would like to thank the German Research Foundation (DFG) for financing the studies presented in this paper.

REFERENCES

[1] C. Pollock and C.-Y. Wu, "Acoustic noise cancellation techniques for switched reluctance drives," *IEEE Transactions on Industry Applications*, vol. 33, no. 2, pp. 477–484, March 1997.

[2] P. O. Rasmussen, F. Blaabjerg, J. K. Pedersen, P. C. Kjaer, and T. J. E. Miller, "Acoustic noise simulation for switched reluctance motors with audible output." Lausanne: EPE, 1999.

[3] J. O. Fiedler, K. A. Kasper, and R. W. De Doncker, "Spectral composition of stator vibrations resulting from modal superposition in srm." Dublin: 3RD IET International Conference on Power Electronics, Machines and Drives PEMD, 2006.

[4] J. O. Fiedler, "Design of Low-Noise Switched Reluctance Drives," Ph.D. dissertation, ISEA, RWTH Aachen, 2006.

[5] J. O. Fiedler, K. A. Kasper, and R. W. De Doncker, "Acoustic noise in switched reluctance drives: An aerodynamic problem?" in *IEEE International Conference on Electric Machines and Drives IEMDC*, 2005, pp. 1275–1280.

[6] J. O. Fiedler and R. W. De Doncker, "Simplified calculation of radial force spectrum in srm for acoustic noise prediction in preliminary machine design." Dublin: 3RD IET International Conference on Power Electronics, Machines and Drives PEMD, 2006.

[7] K. A. Kasper, J. O. Fiedler, D. Schmitz, and R. W. De Doncker, "Noise Reduction Control Strategies for Switched Reluctance Drives." Windsor: VPPC, 2006.

[8] L. E. Kinsler, A. R. Frey, A. B. Coppens, and J. V. Sanders, *Fundamentals of Acoustics*. Wiley & Sons, 2000.

[9] W. Zimmermann, "Rechnung und Versuch bei der scheibenförmigen Wirbelstrombremse," *Archiv für Elektrotechnik*, vol. 10, pp. 133–156, 1921.

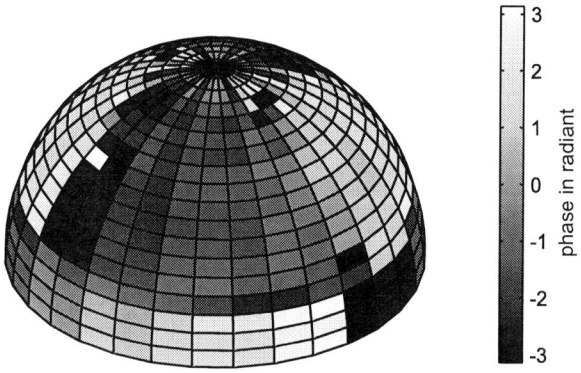

Fig. 14. Measured Phase of Radiated Sound of 8/6 SRM 2950 Hz

VI. CONCLUSION

With the first phase-exact measurement of the radiation characteristic of a switched reluctance machine it could be shown that the noise is radiated very similarly to finite length

Iron Losses in Electrical Machines Due to Non Sinusoidal Alternating Fluxes

J. A. Walker*, D. G. Dorrell*, and E. Ritchie**

* University of Glasgow, Glasgow, UK
** Aalborg University, Aalborg, Denmark

Abstract—**This paper shows how the flux waveform in the core of an electrical machine can be vary non-sinusoidally which complicates the calculation of the iron loss in a machine. A set of tests are conducted on a steel sample using an Epstein square where harmonics are injected into the flux waveform which vary in both magnitude and phase. The measurements are put forward to illustrate the additional iron losses. Further work is suggested so that a set of linear approximations or interpolation of look-up tables can be used to calculate the iron loss from an analytical of finite element analysis of a machine.**

Index Terms—**Iron losses, measurement, Epstein square, harmonics.**

I. INTRODUCTION

Iron losses in electrical steels tend to be calculated using Steinmetz or Generalized Steinmetz equations [1], or from look-up loss tables. The common methods of testing sheet steels, as described in the IEC and ASTM standards [2]-[6], all require sinusoidal flux density waveforms, which are normally enforced through some form of feedback control signal.

When electrical sheet steel is used as the lamination material in an electric motor, the flux density waveforms that occur include higher harmonics and may be strongly non-sinusoidal. If the harmonic content is such that the polarity of the flux density waveform changes more than twice within any one cycle, a minor hysteresis loop will be created within the major dynamic hysteresis loop, as shown in Fig. 1.

This paper investigates the variation of the iron losses due to the effects of non-sinusoidal flux in a motor. The work was carried out as part of a research study into the magnetization and iron losses in switched reluctance and permanent magnet motors [7]. Both of these machines are likely to experience non-sinusoidal flux waves in their cores. An extreme example of a brushless permanent magnet machine operating out of synchronism will be investigated to illustrate the variation of flux waves in different parts of the machine. Then a series of tests will be conducted using an Epstein square with various harmonic injections to assess the affect on iron loss.

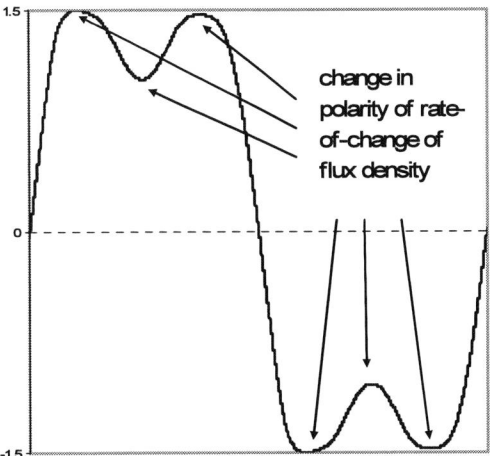

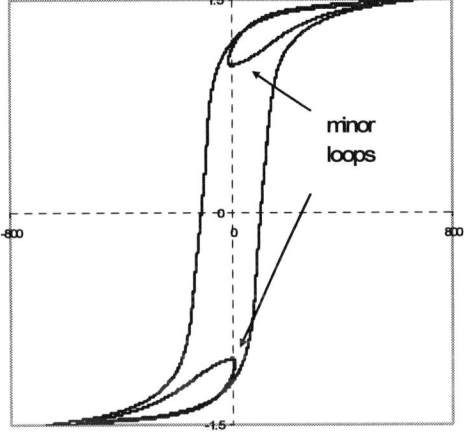

Fig. 1. Criteria for minor hysteresis loops.

These concentrate on low frequency flux harmonics (i.e., 3rd and 5th) but high harmonic losses are also generated in electrical machine due to issues such as slotting and MMF harmonics and, in recent times, due to inverter switching [8].

The waveforms represented in Fig. 1 can be numerically analysed using the methods in [9] while an in-depth analysis of flux waveforms with harmonic content (both low frequency and high frequency) was put forward in [10] which successfully implements a Preisach model for the losses in a non-sinusoidal model (that copes well with the hysteresis, eddy current and excess losses).

This work was carried out while Dr Walker was a PhD student with The University of Glasgow. She is grateful to the EPSRC, UK, Robert Bosch GmbH, Germany and the SPEED laboratory for financial support during her studies.

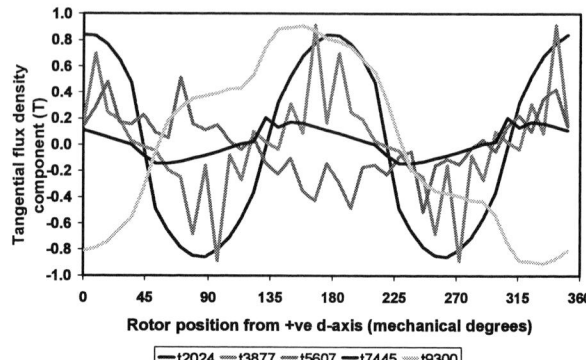

Fig. 5. Tangential flux density components of selected mesh elements.

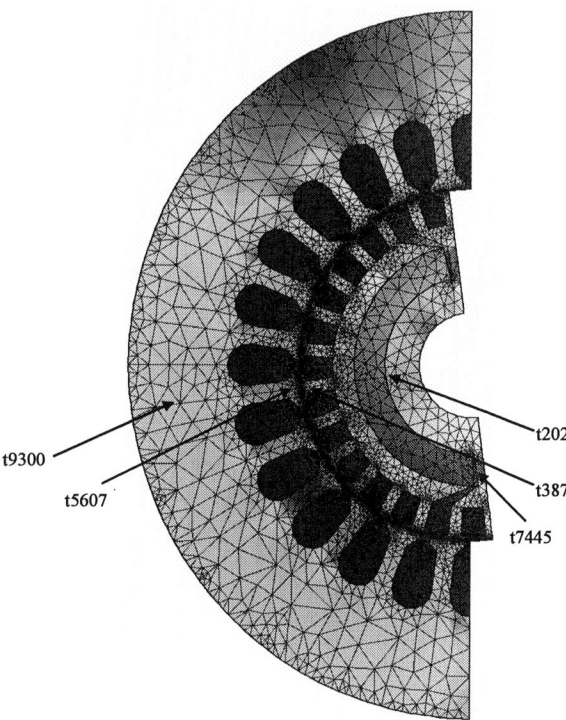

Fig. 2. Motor cross section showing selected finite element mesh numbers.

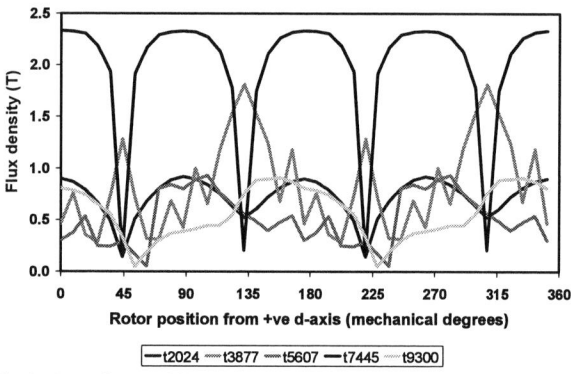

Fig. 3. Waveforms of flux density magnitude of each selected element.

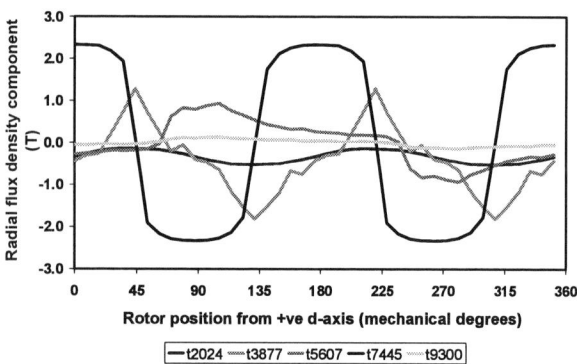

Fig. 4. Radial flux density components of selected mesh elements.

II. MACHINE SIMULATION OF AN UNSYNCHRONIZED BRUSHLESS PERMANENT MAGNET MOTOR

Depending on the excitation conditions of the motor (specifically the phase advance angle when operating under sinusoidal excitation), the flux density waveforms in different areas of the motor cross section will be non-sinusoidal and in some cases will exhibit the minor loop criteria. Fig. 2 shows one half of the motor cross-section of a split-phase line-start internal permanent magnet test motor. This machine has its magnets deeply embedded in the rotor because there is a cage structure on the rotor surface which is used to start the machine. The geometry is therefore quite complex. Five elements from the finite element mesh are illustrated and the waveforms for these elements are studied in detail.

This machine is operating out of synchronism – the rotor is rotating in one direction while the rotational component of the stator MMF is rotating in the opposite direction at the same velocity. The cage is also assumed to be removed. While this is an unrealistic situation designed to highlight the harmonic effects, similar situations will occur in a single phase machine, where there is a component of MMF in the counter-rotating direction. The corresponding flux density (magnitude) waveforms can be seen in Fig. 3. This is magnitude of the flux density so to illustrate the oscillating nature of the flux then the normal components of the flux density are shown in Fig. 4 and the tangential components are shown in Fig. 5. These clearly show the complex nature of the flux in various parts of the machine.

The element with the highest flux density is t7445 which is an element in the radial steel section adjacent to the magnets. This is embedded in the rotor and is not influenced greatly by the slotting of the stator and rotor. This waveform is therefore almost symmetrical and contains some low odd harmonics. It oscillates at twice rotational speed since the rotor is rotating in one direction and the stator MMF is rotating in the other direction; and the stator MMFs dominate.

However the elements in the rotor tooth (t3877) and stator tooth (t5607) contain higher harmonics that will contain slotting effects and these can be seen clearly in Fig. 3. Interestingly it is the tangential component of flux density (Fig. 5) that exhibits the high-frequency flux ripple (particularly the stator tooth element t5607 which is close to the stator surface) which illustrates the effects zigzag flux has on the tooth-tip flux. These slotting effects will cause minor loops in the B/H loop and additional tooth-tip losses.

The stator core-back element t9300 has mostly tangential flux and Fig. 5 shows the waveform to be asymmetrical and liable to some minor loops in the B/H loop hence increasing the stator yoke losses. In the next section the effects of adding harmonics to flux waves is investigated using measurements from an Epstein square tester. These will be in terms of addition of 3rd and 5th harmonics to produce asymmetrical waveforms in a similar fashion to the waveform for t9300 in Fig. 5. This element represents the largest component of steel in the machine and therefore the flux harmonics in this waveform are likely to produce the largest change in total iron loss.

The flux for the element behind the magnet (t2024) in Fig. 5 has a twice rotational speed oscillation since the magnets contribute no tangential flux (it is centred in at the back of the magnet) and most of the flux flowing through the element will be due to the stator MMF flux which is flowing across the back of the magnet.

III. STEEL LOSS MEASUREMENTS

A. Measured losses with harmonic injection

A set of measurements were conducted using an Epstein square arrangement. This was set up and calibrated against a single sheet tester and it included full compensation [7]. A set of 3rd harmonic injections were used to explore the variation of iron losses at different fundamental frequencies. These are listed in Table I. The fundamental magnitude is a nominal 1.5 T. Further 5th harmonic injections were also tested which are illustrated later. Fig. 6 shows the flux waveforms for the waves listed in Table I.

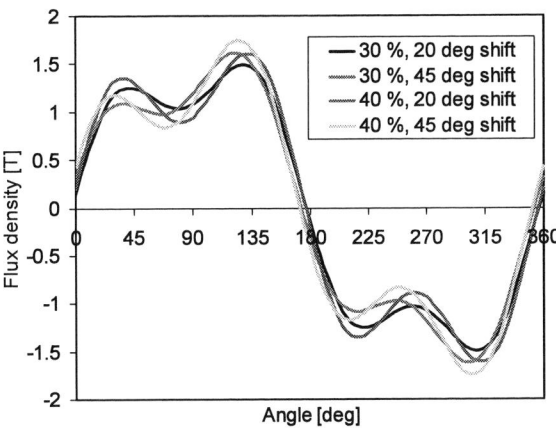

Fig. 6. Nominal flux waves for B/H loops in Figs. 7 to 10.

TABLE I
MINOR LOOP FREQUENCIES FOR DIFFERENT AMPLITUDES AND PHASES OF HIGHER HARMONIC

Major loop frequency [Hz]	Injected harmonic	Amplitude	Phase shift	Minor loop frequency [Hz]
22.32	3rd	30%	20°	116.82
22.32	3rd	30%	45°	103.41
22.32	3rd	40%	20°	93.02
22.32	3rd	40%	45°	86.96
52.08	3rd	30%	20°	245.09
52.08	3rd	30%	45°	284.09
52.08	3rd	40%	20°	206.61
52.08	3rd	40%	45°	257.73

The occurrence of minor loops leads to a marked increase in specific loss for a given peak flux density. The area of the minor hysteresis loops is dependent on the amplitude of the higher harmonics present in the flux density waveform. In Fig. 2, there is a dominant third harmonic with amplitude of 40 % of the fundamental component. With a higher amplitude 3rd harmonic, the area of the minor hysteresis loop increases (although the frequency of the minor loop would be decreased, as it would take longer to traverse the loop).

The area and frequency of the minor hysteresis loop are also dependent on the phase of the higher harmonic content with respect to the fundamental, i.e., there will be a significant difference in the area and frequency of minor loops created by a 40 % amplitude 3rd harmonic at phase shifts of 20° and 45°. Table I gives some examples of the difference in frequency of minor loops when the injected harmonic is of different amplitude and phase shift.

It can be noted from Table I that in the case of the 22.32 Hz tests, the frequency of the minor loops is greatest when the phase shift and amplitude of the injected 3rd harmonic are minimized. In the second set of tests at 52.08 Hz, the opposite is true; the minor loop frequency is greatest when the phase shift and amplitude are greatest. One might expect that the relationship between the minor loop frequency (the inverse of this is the period of the minor loop) and the amplitude and phase shift of the injected harmonics would be the same, regardless of the fundamental frequency of the flux density waveform. However, the frequency of the minor loops is also dependent on the shape of the applied field waveform; the applied field affects the positioning and shape of the minor loop, and hence its frequency. The relationship between applied field and flux density is dependent on frequency.

The dependence on the applied field waveform, and the fundamental frequency, is best illustrated by examining the complete dynamic hysteresis loop, including minor loops at different fundamental frequencies. Fig. 7 shows the dynamic hysteresis loop for a flux density waveform with injected 3rd harmonic of 30 % amplitude, at a phase shift of 20 degrees and a fundamental frequency of 22.32 Hz. This can be compared with Fig. 8, which shows the dynamic hysteresis loop for a flux density waveform with the same additional harmonic component, but at a frequency of 52.08 Hz. The positioning of the minor loops can be seen to be different, and as such the area of the loops also differs. Figs. 9 and 10 further illustrate the change in phase relationship of the harmonic injection with respect to the fundamental.

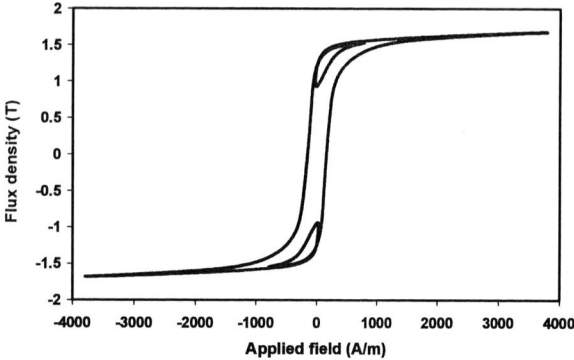

Fig. 7. Torque-speed characteristic of an induction motor of 30 % amplitude and 20° phase shift, at 22.32 Hz.

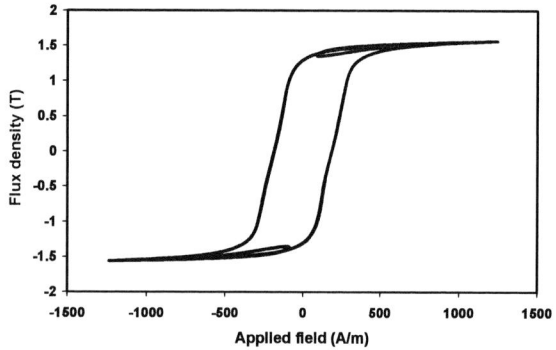

Fig. 8. Hysteresis loop with injected 3rd harmonic of 30 % amplitude and 20° phase shift, at 52.08 Hz.

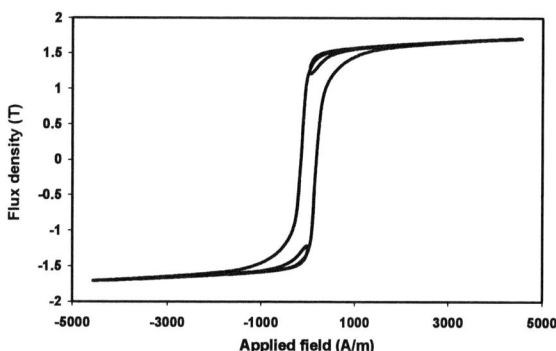

Fig. 9. Hysteresis loop with injected 3rd harmonic of 30 % amplitude and 45° phase shift, at 22.32 Hz.

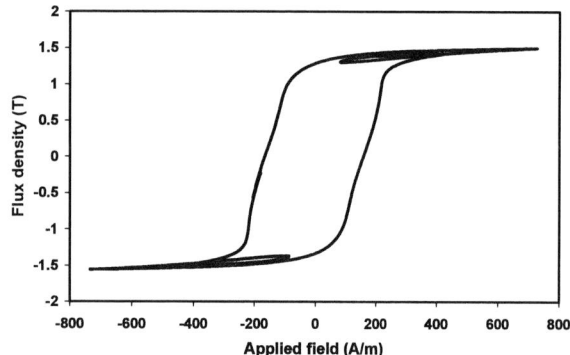

Fig. 10. Hysteresis loop with injected 3rd harmonic of 30 % amplitude and 45° phase shift, at 52.08 Hz.

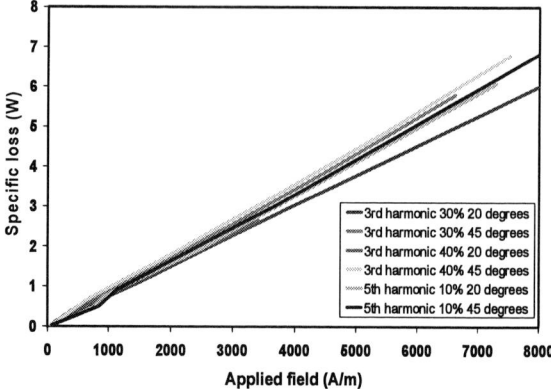

Fig. 11. Comparison of measured losses at 22.32 Hz.

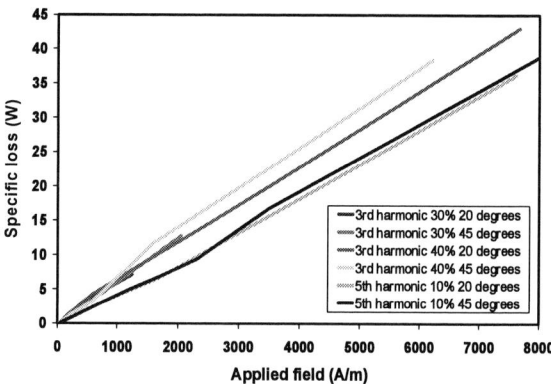

Fig. 12. Comparison of measured losses at 52.08 Hz.

B. Comparison of results

The results for the harmonic loss measurements are shown in Fig. 11 at 22.32 Hz and Fig. 12 for 52.08 Hz. The units are in watts rather than watts per kilogram to illustrate the amount of power dissipated in the steel sample and to highlight the increase in losses when moving from 22.32 Hz to 52.08 Hz. In addition to the losses highlighted in Table I, further harmonics were injected at these two frequencies: there were 5th harmonics with 10 % injection and 20° and 45° phase shift. The lower harmonic content reflects the point that as the harmonic number increases then the magnitude tends to decrease.

The measured results are listed in full in the Appendix for various values of H. These characteristics suggest that, even though the losses exhibit non-linear variation, the loss across a range of H appears to be almost linear, and this is investigated in the next section.

IV. IRON LOSS DATA MODIFICATION

A. Variation of Losses with H

The losses at each harmonic amplitude and phase shift can be compared, as shown in Fig. 11 for a fundamental frequency of 22.32 Hz. A similar graph is shown in Fig. 9 for 52.08 Hz. Fig. 12 shows that the loss is greater when the amplitude of the injected harmonic is greater and when the phase shift is greater. The larger the amplitude of the injected harmonic then the greater the area of the associated minor hysteresis loop, and hence the loss. The losses, as expected, are significantly increased when the frequency of the fundamental component of the flux density is increased. As the frequency of the major hysteresis loop is increased, the ratio of minor loop frequency to major loop frequency increases, so that in addition to the increased losses due the increase in major loop frequency, there is an extra increase in loss due to the increased minor loop relative frequency. As such, the relationship between specific loss and frequency is much more complex for hysteresis loops containing minor loops than for the simple major loop that results from a sinusoidal flux density waveform. However the results in Figs. 11 and 12 show a possibility of linear approximation of the characteristics.

B. Linearised loss characteristics

If the results in the Appendix are linearised then the errors are noted in Tables A.I and A.II. However, the error is a function of the applied H field and this is highlighted in Figs. 13 to 18 for 22.32 Hz and Figs. 19 to 24 for 52.08 Hz. The 22.32 Hz results show excellent correlation. There are errors but these tend to be at very low H.

This illustrates that there may be the possibility of using the approximated linear functions to obtain fast calculations for the iron loss. However this requires extensive testing of the material but this may be possible on an automated test facility that can step through harmonics in order with different values of H, percentage harmonic injection and phase shift. These tests would produce multi-dimensional arrays that can be utilised in conjunction with linear interpolation or to generate the linear (or even higher-power) functions. These can be used with analytical of finite element solutions to give post-processing iron-loss calculations.

V. CONCLUSIONS

This paper has described an investigation into the effects of minor hysteresis loops in the flux of a motor core. An example is shown for the variation of the flux in a motor core using finite element analysis. While this was an extreme example of a rotor turning in a counter-rotating direction to the stator MMF it highlighted that it is possible to observe very non-sinusoidal flux waves in the core of a machine and that there may be substantial harmonics.

Tests were conducted on a set of steel laminations in an Epstein square arrangement to assess the steel loss with different degrees of harmonic injection, both in terms of harmonic magnitude and phase. Some linear relationships were obtained and this suggests that further work should be carried out in this area.

A. Further Work

This project represents the initial investigations into the assessment of losses in an electrical machine and attempt to quantify additional losses due to non-

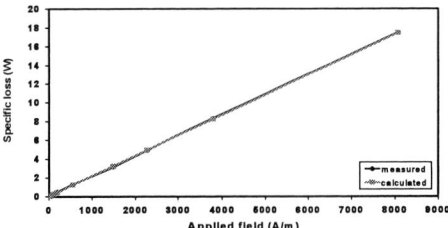

Fig. 13. Specific loss versus applied field for injected 3rd harmonic of 30 % amplitude and 20° phase shift at 22.32 Hz.

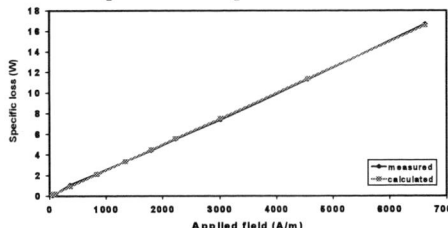

Fig. 14. Specific loss versus applied field for injected 3rd harmonic of 30 % amplitude and 45° phase shift at 22.32 Hz.

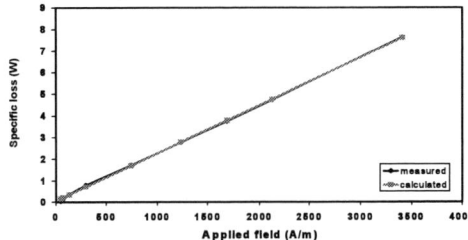

Fig. 15. Specific loss versus applied field for injected 3rd harmonic of 40 % amplitude and 20° phase shift at 22.32 Hz.

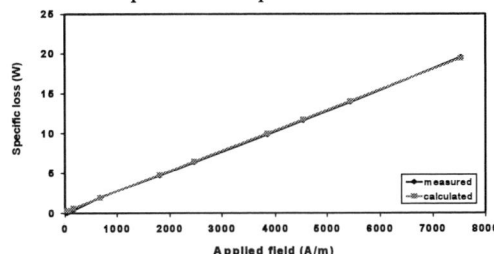

Fig. 16. Specific loss versus applied field for injected 3rd harmonic of 40 % amplitude and 45° phase shift at 22.32 Hz.

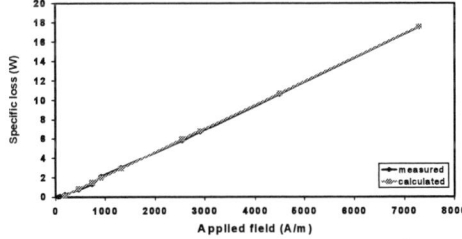

Fig. 17. Specific loss versus applied field for injected 5th harmonic of 10 % amplitude and 20° phase shift at 22.32 Hz.

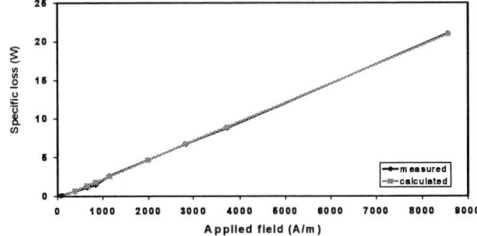

Fig. 18. Specific loss versus applied field for injected 5th harmonic of 10 % amplitude and 45° phase shift at 22.32 Hz.

sinusoidal flux and harmonic flux waves. Further work will be to:

- Investigate the possibility of using linearised empirical formulae using iron loss measurements for material samples as described in Section IV.*B*
- Reference [10] successfully implemented a Preisach calculation and this should be examined
- Develop a "hardware-in-loop" system where the flux waves from a finite element solution are fed into a iron loss tester (either single sheet tester or Epstein square) to obtain the iron losses from measurement rather than calculation. The sum of the element losses gives the overall core loss
- Investigate the effect of cutting and machining of lamination material on the iron losses

REFERENCES

[1] T.J.E. Miller, "SPEED's Electrical Motors", SPEED Laboratory, University of Glasgow, 2006.

[2] ASTM Standard A343-97, "Standard test method for AC magnetic properties of materials at power frequencies using 25 cm Epstein square", ASTM International 1997.

[3] IEC Standard 404 Magnetic Materials Part 2, "Methods of measurement of magnetic, electrical and physical properties of magnetic sheet and strip", IEC 1978

[4] AEG Instruction Manual "Elektroblech-Meßeinrichtung mit 25 cm Epsteinrehmen", AEG GmbH 1962.

[5] ASTM Standard A804/A804M-99, "Standard test methods for alternating-current magnetic properties of materials at power frequencies using sheet-type test specimens", ASTM International 1999.

[6] IEC Standard 60404 Magnetic Materials Part 3, "Methods of measurement of the magnetic properties of magnetic sheet and strip by means of a single sheet tester", IEC 2002.

[7] J. A. Walker, "Aspects of magnetisation and iron loss characteristics in switched-reluctance and permanent-magnet machines", PhD thesis, University of Glasgow, March 2006.

[8] C. A. Hernandez-Aramburo, T. C. Green, and A. C. Smith, "Assessment of Power Losses of an Inverter-Driven Induction Machine with its experimental validation", IEEE Transactions of Industry applications, Vol. 39, No. 4 pp 994-1004, 2003.

[9] E. Cardelli, R. Giannetti and B. Tellini, "Numerical Characterization of Dynamic Hysteresis Loops and Losses in Soft Magnetic Materials", IEEE Transactions of Magnetics, Vol. 41, No. 5, pp 1540-1543, 2005.

[10] E. Barbisio, F. Fiorillo and C. Ragusa, "Predicting Loss in Magnetic Steels Under arbitrary Induction Waveform and with Minor Hysteresis Loops", IEEE Transactions on Magnetics, Vol. 40, No. 4, pp 1810-1819, 2004.

APPENDIX

The measured values of loss with harmonic injection were used to produce the linear characteristics in Figs. 11 and 12 at 22.32 Hz and 52.08 Hz respectively. Tables A.I and A.II compare the linearised values with the measured values over a range of H and gives the % error values when compared to the linearised curves shown in Figs. 13 to 24.

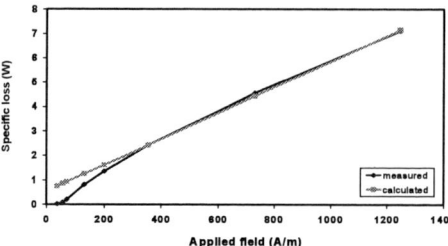

Fig. 19. Specific loss versus applied field for injected 3rd harmonic of 30 % amplitude and 20° phase shift at 52.08 Hz.

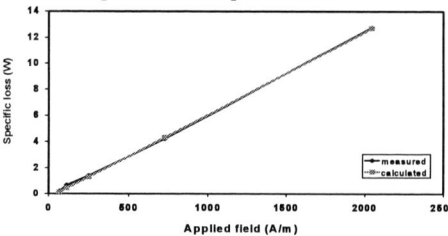

Fig. 20. Specific loss versus applied field for injected 3rd harmonic of 30 % amplitude and 45° phase shift at 52.08 Hz.

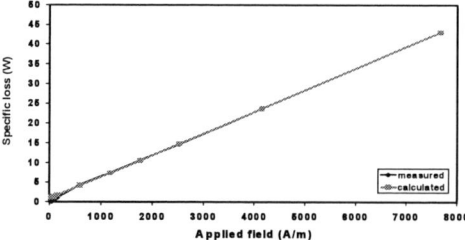

Fig. 21. Specific loss versus applied field for injected 3rd harmonic of 40 % amplitude and 20° phase shift at 52.08 Hz.

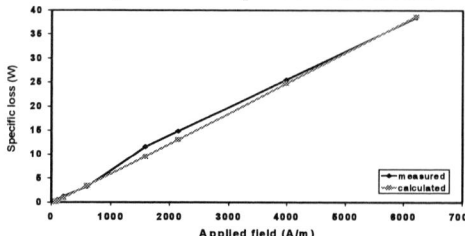

Fig. 22. Specific loss versus applied field for injected 3rd harmonic of 40 % amplitude and 45° phase shift at 52.08 Hz.

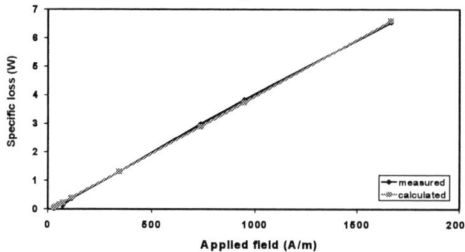

Fig. 23. Specific loss versus applied field for injected 5th harmonic of 10 % amplitude and 20° phase shift at 52.08 Hz.

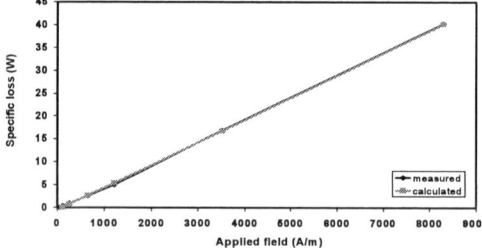

Fig. 24. Specific loss versus applied field for injected 5th harmonic of 10 % amplitude and 45° phase shift at 52.08 Hz.

TABLE A.I
MEASURED AND CALCULATED VALUES FROM LINEAR APPROXIMATIONS WITH ERROR EVALUATION FOR DIFFERENT HARMONIC INJECTIONS AT 22.32 Hz

Test point	Hmax	loss	calculated loss W	% error
3rd harm, 30%, 20°	52.0658	0.0482	0.1324	-174.82
3rd harm, 30%, 20°	65.1635	0.1036	0.1607	-55.17
3rd harm, 30%, 20°	110.3518	0.2924	0.2583	11.64
3rd harm, 30%, 20°	195.6848	0.5081	0.4426	12.88
3rd harm, 30%, 20°	558.2146	1.2621	1.2257	2.88
3rd harm, 30%, 20°	1484.2	3.135	3.2258	-2.89
3rd harm, 30%, 20°	2268.6	4.8976	4.9201	-0.46
3rd harm, 30%, 20°	3791.5	8.3118	8.2096	1.23
3rd harm, 30%, 20°	8072.8	17.4673	17.4572	0.06
3rd harm, 30%, 45°	50.0243	0.041	0.1170	-185.51
3rd harm, 30%, 45°	83.1508	0.1549	0.1998	-29.04
3rd harm, 30%, 45°	367.4011	1.067	0.9105	14.67
3rd harm, 30%, 45°	841.512	2.1675	2.0957	3.31
3rd harm, 30%, 45°	1337.6	3.344	3.336	0.24
3rd harm, 30%, 45°	1795.2	4.4273	4.48	-1.19
3rd harm, 30%, 45°	2234.2	5.517	5.5775	-1.10
3rd harm, 30%, 45°	3015.5	7.4052	7.5307	-1.69
3rd harm, 30%, 45°	4565.3	11.3111	11.405	-0.83
3rd harm, 30%, 45°	6621.2	16.6895	16.545	0.87
3rd harm, 40%, 20°	48.7634	0.0389	0.1442	-270.83
3rd harm, 40%, 20°	72.1868	0.1468	0.1962	-33.69
3rd harm, 40%, 20°	134.172	0.3471	0.3338	3.81
3rd harm, 40%, 20°	299.1017	0.7857	0.7	10.91
3rd harm, 40%, 20°	744.0024	1.72	1.6876	1.88
3rd harm, 40%, 20°	1236.7	2.7687	2.7814	-0.46
3rd harm, 40%, 20°	1687.2	3.7495	3.7815	-0.86
3rd harm, 40%, 20°	2125.6	4.7364	4.7548	-0.39
3rd harm, 40%, 20°	3410.2	7.6272	7.6066	0.27
3rd harm, 40%, 45°	42.5464	0.0251	0.3084	-1129.06
3rd harm, 40%, 45°	62.9219	0.0919	0.3604	-292.22
3rd harm, 40%, 45°	158.5507	0.3632	0.6043	-66.38
3rd harm, 40%, 45°	678.7896	1.9846	1.9309	2.71
3rd harm, 40%, 45°	1804.1	4.71	4.8004	-1.92
3rd harm, 40%, 45°	2472.3	6.3753	6.5043	-2.02
3rd harm, 40%, 45°	3845.1	9.8465	10.005	-1.61
3rd harm, 40%, 45°	4533.1	11.6199	11.7594	-1.20
3rd harm, 40%, 45°	5432.7	13.931	14.0533	-0.88
3rd harm, 40%, 45°	7526	19.5364	19.3913	0.74
5th harm, 10%, 20°	50.5879	0.0274	-0.1760	742.55
5th harm, 10%, 20°	87.0709	0.0965	-0.0866	189.82
5th harm, 10%, 20°	185.9854	0.2661	0.1556	41.50
5th harm, 10%, 20°	465.4995	0.7412	0.8404	-13.39
5th harm, 10%, 20°	735.8366	1.2863	1.5027	-16.83
5th harm, 10%, 20°	919.3351	2.1237	1.9523	8.07
5th harm, 10%, 20°	1311.8	3.0232	2.9139	3.62
5th harm, 10%, 20°	2556.6	5.7518	5.9636	-3.68
5th harm, 10%, 20°	2915.6	6.7018	6.8432	-2.11
5th harm, 10%, 20°	4491.4	10.5898	10.7039	-1.08
5th harm, 10%, 20°	7290.7	17.5133	17.5622	-0.28
5th harm, 10%, 45°	66.5005	0.0553	-0.1350	344.27
5th harm, 10%, 45°	100.4331	0.1186	-0.0509	142.94
5th harm, 10%, 45°	388.127	0.62	0.6625	-6.86
5th harm, 10%, 45°	654.7279	1.0954	1.3237	-20.84
5th harm, 10%, 45°	843.1116	1.4293	1.7909	-25.30
5th harm, 10%, 45°	1143.8	2.6443	2.5366	4.07
5th harm, 10%, 45°	1986.1	4.683	4.6255	1.23
5th harm, 10%, 45°	2836.9	6.6914	6.7355	-0.66
5th harm, 10%, 45°	3723	8.7964	8.9330	-1.55
5th harm, 10%, 45°	8561.1	21.0624	20.9315	0.62

TABLE A.II
MEASURED AND CALCULATED VALUES FROM LINEAR APPROXIMATIONS WITH ERROR EVALUATION FOR DIFFERENT HARMONIC INJECTIONS AT 52.08 Hz

Test point	Hmax	loss	calculated loss W	% error
3rd harm, 30%, 20°	37.9087	0.0248	0.7509	-2927.89
3rd harm, 30%, 20°	57.2259	0.0928	0.8533	-819.50
3rd harm, 30%, 20°	71.4356	0.1997	0.9286	-365.00
3rd harm, 30%, 20°	130.9169	0.805	1.2439	-54.52
3rd harm, 30%, 20°	200.2802	1.3508	1.6115	-19.30
3rd harm, 30%, 20°	354.7596	2.4181	2.4302	-0.50
3rd harm, 30%, 20°	731.5037	4.5577	4.4270	2.87
3rd harm, 30%, 20°	1245.8	7.1129	7.1527	-0.56
3rd harm, 30%, 45°	68.7331	0.205	0.1365	33.44
3rd harm, 30%, 45°	114.1036	0.6617	0.4246	35.84
3rd harm, 30%, 45°	248.9511	1.3837	1.2808	7.43
3rd harm, 30%, 45°	728.119	4.2201	4.3236	-2.45
3rd harm, 30%, 45°	2044.7	12.7547	12.6838	0.56
3rd harm, 40%, 20°	42.5208	0.0381	1.0839	-2744.79
3rd harm, 40%, 20°	61.0703	0.1368	1.1859	-766.88
3rd harm, 40%, 20°	93.858	0.4083	1.3662	-234.61
3rd harm, 40%, 20°	161.5719	1.1349	1.7386	-53.20
3rd harm, 40%, 20°	598.7556	4.2338	4.1432	2.14
3rd harm, 40%, 20°	1171.4	7.3685	7.2927	1.03
3rd harm, 40%, 20°	1756.7	10.5919	10.5119	0.76
3rd harm, 40%, 20°	2523.5	14.6606	14.7293	-0.47
3rd harm, 40%, 20°	4148.4	23.6125	23.6662	-0.23
3rd harm, 40%, 20°	7667.9	42.982	43.0235	-0.10
3rd harm, 40%, 45°	48.79	0.0526	-0.0926	276.09
3rd harm, 40%, 45°	57.7018	0.1152	-0.0365	131.67
3rd harm, 40%, 45°	79.9152	0.299	0.1035	65.40
3rd harm, 40%, 45°	105.2193	0.543	0.2629	51.59
3rd harm, 40%, 45°	210.7015	1.2172	0.9274	23.81
3rd harm, 40%, 45°	608.2667	3.268	3.4321	-5.02
3rd harm, 40%, 45°	1578.4	11.6177	9.5439	17.85
3rd harm, 40%, 45°	2142.8	14.8563	13.0996	11.82
3rd harm, 40%, 45°	3986	25.4928	24.7118	3.06
3rd harm, 40%, 45°	6212.3	38.4302	38.7375	-0.80
5th harm, 10%, 20°	31.0959	0.0217	0.0644	-196.70
5th harm, 10%, 20°	43.5268	0.082	0.1141	-39.16
5th harm, 10%, 20°	55.359	0.1751	0.1614	7.80
5th harm, 10%, 20°	72.8271	0.1243	0.2313	-86.09
5th harm, 10%, 20°	115.8528	0.3623	0.4034	-11.35
5th harm, 10%, 20°	342.6544	1.3093	1.3106	-0.10
5th harm, 10%, 20°	737.8944	2.9883	2.8916	3.24
5th harm, 10%, 20°	947.6126	3.8419	3.7305	2.90
5th harm, 10%, 20°	1666.9	6.5393	6.6076	-1.04
5th harm, 10%, 20°	7613.9	36.1583	30.3956	15.94
5th harm, 10%, 45°	48.6886	0.0316	-0.2614	927.30
5th harm, 10%, 45°	74.2827	0.1338	-0.1360	201.66
5th harm, 10%, 45°	117.354	0.3728	0.0750	79.87
5th harm, 10%, 45°	257.9195	0.9171	0.7638	16.72
5th harm, 10%, 45°	648.2013	2.5702	2.6762	-4.12
5th harm, 10%, 45°	1205.8	4.9547	5.4084	-9.16
5th harm, 10%, 45°	3512.1	16.8592	16.7093	0.89
5th harm, 10%, 45°	8292.8	40.2257	40.1347	0.23

Design Requirements for Doubly-Fed Reluctance Generators

D. G. Dorrell

University of Glasgow, Glasgow, UK

Abstract—**This paper puts forward an analytical model for representing a doubly-fed reluctance machine using rotating field theory and permeance harmonics. The radially-laminated version of the rotor is described then a machine is developed using the theory. The algorithm is implemented and a design for a 7.5 kW machine is developed to illustrate the sizing required. The machine is of a 2/6 pole arrangement. The paper illustrates that the stator should be specially manufactured because the slots need to be able to cope with a much higher electric loading than an equivalent induction machine. The paper also shows that there is still much work to be done in terms of being able to design and specify this type of novel machine.**

Index Terms—**Brushless doubly-fed reluctance generator**

I. INTRODUCTION

Wind turbines use either a cage-rotor induction generator, a variable speed synchronous generator or, more recently, a doubly-fed wound-field induction generator (with the field fed from a converter and main windings fed connected to the grid). These are usually connected to the turbine via a gearbox although large-diameter direct-drive generators do exist. The wound-field induction generator is now very common but it has reliability issues due to the slip rings. Researchers are now investigating brushless doubly-fed generators where there are two sets of 3-phase windings – these will have different pole numbers (say 2 and 6 poles) with one connected to the grid (power winding) and one controlled via a converter (control winding). There are two alternatives for this machine: induction type (with the rotor formed from bars connected in nested loops) and reluctance type (with a salient pole rotor similar to switched reluctance machine or an axially-laminated rotor).

This paper reports on an on-going study of the electromagnetic analysis of the doubly-fed radially-laminated reluctance generator. A recent paper [1] assessed the iron losses generated in the axially-laminated machine and debated whether the correct option should be to use a radially-laminated rotor to reduce the rotor losses. This was expanded upon in [2]. The argument is carried further forward here to assess the magnetizing performance of the machine and develop an equivalent circuit that can be used in a control strategy. The control of these machines has already been studies extensively [3]-[7]. While work on the electromagnetic design was carried out in [8]. The correct design of this type of machine is still not reported upon to any great extent; [9] studied the conversion of an induction machine with a specially constructed rotor. It was found that it was difficult to get the correct electrical loading because there was insufficient slot space.

This paper develops an algorithm for the correct design of the machine. The design will focus on the radially-laminated nominally 7.5 kW machine. Initially the design will take the axially-laminated machine in [1] and redesign the rotor with a radially-laminated rotor. The procedure will then take the stator and redesign the lamination to give the correct slot area. This is done be essentially increasing the slot depth and outer diameter. The air-gap length and average air-gap diameter will be kept constant to simplify the procedure.

II. BASIC ARRANGEMENT

A. Topology

The machine can be modelled using analytical analysis utilizing the conductor density method [9]. This allows the calculation of the equivalent circuit components. The machine that is studied here is a 2/6 pole machine with a radially-laminated rotor as used in [1] and [2]. The initial stator design was from a 4 pole 7.5 kW induction motor while the axially-laminated rotor was initially constructed for a study into synchronous reluctance motor operation. Fig. 1 shows the initial machine cross section. It shows radially-laminated salient-pole rotor similar to a switched-reluctance machine (and the machine that is studied here). For an arrangement with m and n pole windings then the rotor should have $(m \pm n)/2$ poles for correct operation (see below).

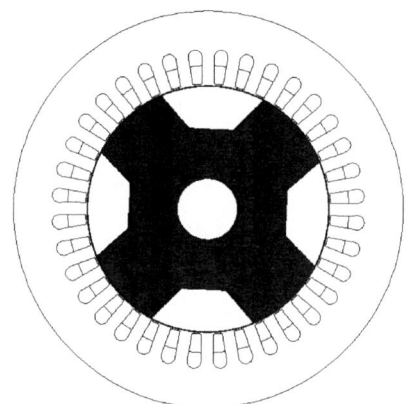

Fig. 1. Rotor arrangement (radial lamination).

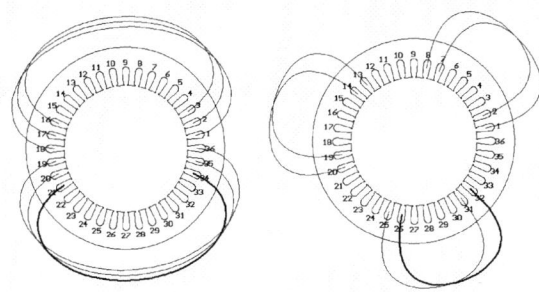

Fig. 2. One phase of the 2-pole and 6-pole windings.

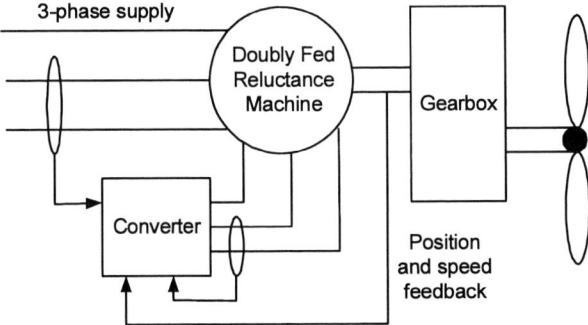

Fig. 3. Doubly Fed Machine Arrangement.

The winding layout, which would be typical for this sort of machine, is shown in Fig. 2. This illustrates one phase winding for each of the two 3-phase windings which are concentric single-layer arrangements, forming a doubly layer arrangement in total. The basic initial parameters are shown in Table I.

B. Basic Operation

The doubly-fed reluctance machine can be considered as being similar to the doubly-fed induction generator in terms of its operation with two windings and the rotor speed being independent of the rotating field velocity. In the induction machine, the windings are coupled directly by a common pole number whereas in the reluctance type of machine the MMFs are modulated by the air-gap permeance to generate flux waves of $p \pm 1$ pole-pairs producing cross coupling. It is necessary to use a rotor with a high d-q reluctance ratio and vector control [5]. One winding is fixed in frequency and volts (grid connection or power winding) and the rotor speed varies according to the mechanical prime mover. There is a synchronizing requirement for the non-grid connected control winding:

$$\omega_c = P\omega_r \pm \omega_p \qquad (1)$$

where the ω is a rotational velocity (rad/sec) and c, r and p represent the control frequency, rotor velocity and grid frequency; P is equal to 4 for the 2/6 machine.

Fig. 3 shows the basic arrangement for the machine if it is used as a generator in a wind turbine. There are two sets of windings, the primary (power) winding and the secondary (control) windings. It is possible to short circuit the secondary winding and run the machine as a motor up to the synchronous speed [3]. This is similar to an induction motor. In (1), the synchronous speed occurs when $\omega_c = 0$ and $\omega_p = 2\pi \times 50$ rad/sec (50 Hz mains

TABLE I
EXPERIMENTAL MACHINE PARAMETERS (INITIAL DESIGN)

Winding	2-pole	6-pole
Winding Connection	star	star
Turns/coil	18	18
Strands in hand	2	2
Series turns per phase	108	108
Wire diameter	0.85 mm	0.85 mm
Phase resistance	1.83 Ω	1.24 Ω
Coils-sides per pole per phase	6	2
Stator Geometry	[mm, p.u. or mm²]	
Stator outside diameter	203.9	
Stator inner diameter	127	
Stack length	202.4	
Slot number	36	
Slot opening	2.92	
Slot depth	18.54	
Tooth width	8.24	
Slot area	106.96	
Rotor geometry	[mm or p.u]	
Air-gap length	0.5	
Pole number	4	
Shaft diameter	40	
Pole arc	117.9	
Shaft diameter	40	

supply), which gives a speed of 750 rpm. Above this speed the machine operated super-synchronously as a generator in a similar manner to a directly-connected induction motor. However, to control the machine correctly via the control winding, position and current feedback is required as illustrated in [3] – [6].

III. ANALYTICAL ANALYSIS OF MACHINE

In this paper an attempt is made to simulate the machine using the conductor density method of analysis as used by several investigators, including the author [9]. This is to test the validity using an air-gap permeance method with such a large air-gap permeance variation. Finite element analysis is used to further test the method.

A. Rotor Current Density

The current density on the stator surface is made up of the 6-pole winding and the 2-pole winding [9]:

$$j_s(y,t) = \text{Re} \sum_{n=-\infty}^{\infty} \left(\bar{J}_6^n e^{j(\omega_6 t - nk3y)} + \bar{J}_2^n e^{j(\omega_2 t - nky)} \right)$$

$$\bar{J}_6^n \big|_{n=1,-5,7,-11,13..} = (1 + a^{n-1} + a^{1-n})\bar{N}_6^n \bar{I}_6 = 3\bar{N}_6^n \bar{I}_6 \qquad (2)$$

$$\bar{J}_2^n \big|_{n=1,-5,7,-11,13..} = (1 + a^{n-1} + a^{1-n})\bar{N}_2^n \bar{I}_2 = 3\bar{N}_2^n \bar{I}_2$$

where y is a linearized distance around the air-gap ($= r\theta$), $a = e^{\frac{j2\pi}{3}}$, r is the mean air-gap radius and $k = 1/r$. This assumes that the frequency of the currents in the windings is constant and balanced. It is also assumed that there are two three-phase sets. Using the equation

$$b(y,t)g(y,t) = \int \mu_0 j_s(y,t)dy + C \qquad (3)$$

expressions for the stator air-gap flux density can be obtained where

$$b_s(y,t) = \mathrm{Re} \sum_{n=-\infty}^{\infty} \frac{j\mu_0}{nkg(y,t)} \left(\bar{J}_6^n e^{j(\omega_6 t - nk3y)} + \bar{J}_2^n e^{j(\omega_2 t - nky)} \right)$$

$$= \mathrm{Re} \sum_{n=-\infty}^{\infty} \frac{j\mu_0 \lambda(y,t)}{nk} \left(\bar{J}_6^n e^{j(\omega_6 t - nk3y)} + \bar{J}_2^n e^{j(\omega_2 t - nky)} \right) \tag{4}$$

This uses an expression for the permeance. However, first it is worth investigating the equivalent circuit for the machine.

B. Equivalent Circuit

The simple equivalent circuit for the machine is shown in Fig. 4 and this shows the cross-coupling of the two windings.

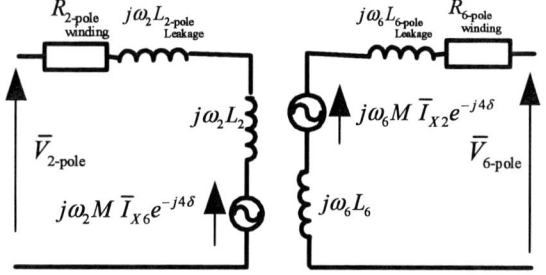

Fig. 4. Equivalent Circuit for radially-laminated rotor machine.

C. Rotor Permeance Model of Salient Pole Rotor

To get the permeance waves for rotor then we can apply the model shown in Fig. 5. There are several models that can be used here to obtain the correct permeance models however we will assume that it varies as a "V" shape across the rotor slot.

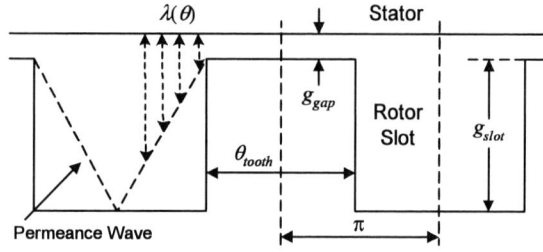

Fig. 5. Rotor permeance representation.

The air-gap can be described using a harmonic permeance wave which is given by

$$g(y,t) = g_o + \sum_{n=1}^{\infty} g_n \cos\left(n\left(w_r t + 4(ky+\delta)\right)\right), \tag{5}$$

However, we can see that we need the permeance wave in (4) rather than the air-gap length. Inverting (5) gives a very complicated series, however, if we take a first approximation then:

$$\lambda(y,t) = \frac{1}{g(y,t)} \approx \frac{1}{g_o'}\left(1 + \frac{g_1}{g_o}\cos\left(4(w_r t + ky + \delta)\right)\right)$$

$$\approx \lambda_0 + \frac{\lambda_1}{2}\left(e^{j4(w_r t + ky + \delta)} + e^{-j4(w_r t + ky + \delta)}\right) \tag{6}$$

Reference [9] reported that the correct value for the constant air-gap permeance is given by

$$\lambda_o = \frac{1}{g_o'} = \frac{\theta_{tooth}}{2\pi g_{gap}} + \frac{2\pi - \theta_{tooth}}{\pi g_{slot}} \tag{7}$$

This gives much better correlation. By changing the rotor

angle then it was also verified that the machine could generate. From the Fourier series for the air-gap:

$$g_o = g_{gap} + \frac{2g_{slot}}{\pi(2\pi - \theta_{tooth})}\left[\frac{\pi^2}{2} - \frac{\pi\theta_{tooth}}{2} + \frac{1}{2}\left(\frac{\theta_{tooth}}{2}\right)^2\right] \tag{8}$$

and

$$g_1 = \frac{2g_{slot}}{\pi\left(\pi - \frac{\theta_{tooth}}{2}\right)}\left[1 - \cos\left(\frac{\theta_{tooth}}{2}\right)\right] \tag{9}$$

(equation (9) is corrected from [9]).

If it is assumed that the air-gap length is 0.5 mm (from Table I) and that the rotor slot depth can go up to between 20 and 30 mm (simply to allow a reasonable rotor core back – in Table I the rotor has a radius of 63 mm while the shaft has a radius of 20 mm which means that rotor yoke and slot must add to 43 mm). The rotor tooth pitch and slot depth can be varied to find optimum values for these.

If the tooth pitch is 157.2 deg (as used in [9]) then the air-gap profile, as obtained from the harmonic series, is shown in Fig. 6. The inverse of this, which is a permeance, is also illustrated together with the first harmonic as obtained from (9).

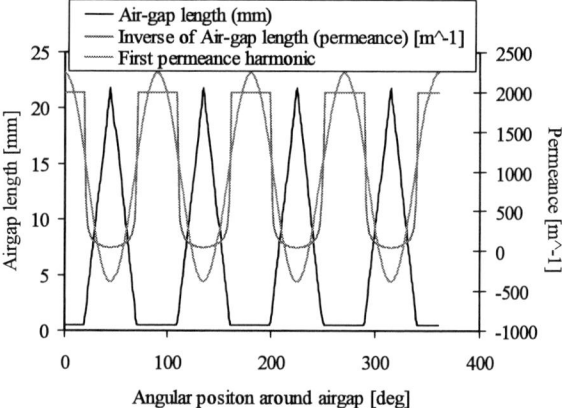

Fig. 6. Air-gap and permeance waves.

If the air-gap length is maintained then the tooth pitch needs to be investigated to obtain the minimum for g_1. This will give the maximum coupling between the different windings. It is possible to obtain expressions from the equations above that would produce the maximum for λ_1. However this is a convoluted exercise and simply ranging the tooth pitch will produce the peak. This is shown in Fig. 7 and the optimum value is found to be (at about 22 mm for the slot depth) 300 deg for the tooth pitch, which is much higher than indicated in [9].

In the equivalent circuit it can be seen that the main inductance associated with the permeance λ_0 is a series inductance so that reducing this would help with the regulation of the machine. The variation of λ_0 is shown in Fig. 8. With such a high tooth pitch the main permeance is high, which is a significant disadvantage. However, this is necessary to get good coupling.

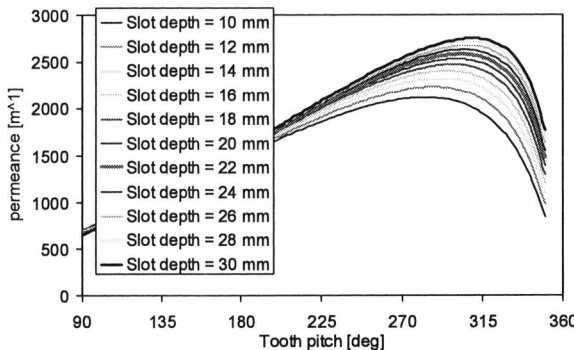

Fig. 7 Variation of λ_1 with rotor tooth pitch and rotor slot depth.

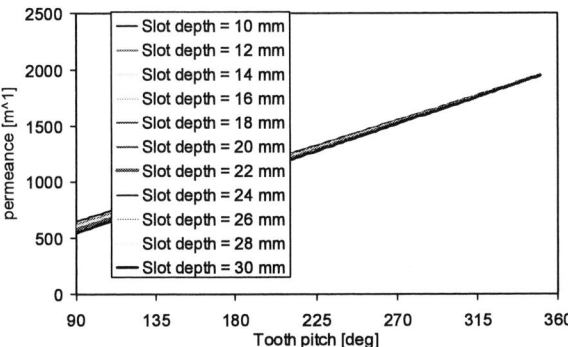

Fig. 7 Variation of λ_0 with rotor tooth pitch and rotor slot depth.

D. Air-gap Flux Waves

Combining (4) and (6) gives

$$b_s(y,t) = \mathrm{Re}\,\frac{j\mu_0\lambda_0}{k}\left(\frac{\bar{J}_6}{3}e^{j(\omega_6 t - k3y)} + \bar{J}_2 e^{j(\omega_2 t - ky)}\right)$$

$$+\mathrm{Re}\,\frac{j\mu_0\lambda_1}{2k}\left(\begin{array}{c}\dfrac{\bar{J}_6}{3}e^{j((\omega_6+4\omega_r)t-7ky-4\delta)} + \bar{J}_2 e^{j((\omega_2+4\omega_r)t-5ky-4\delta)} \\[2ex] +\dfrac{\bar{J}_6}{3}e^{j((4\omega_r-\omega_6)t-ky-4\delta)} + \bar{J}_2 e^{j((4\omega_r-\omega_2)t-3ky-4\delta)}\end{array}\right)$$

$$b_s(y,t) = \mathrm{Re}\left(\bar{B}_6^6 e^{j(\omega_6 t - k3y)} + \bar{B}_2^2 e^{j(\omega_2 t - ky)}\right)$$

$$+\mathrm{Re}\left(\begin{array}{c}\bar{B}_6^{14} e^{j((4\omega_r+\omega_6)t-7ky-4\delta)} + \bar{B}_2^{10} e^{j((4\omega_r+\omega_2)t-5ky-4\delta)} \\[1ex] +\bar{B}_6^2 e^{j((4\omega_r-\omega_6)t-ky-4\delta)} + \bar{B}_2^6 e^{j((4\omega_r-\omega_2)t-3ky-4\delta)}\end{array}\right)$$

$$(10)$$

where

$$\bar{B}_n^n = \frac{j\mu_0\lambda_0\bar{J}_n}{{}^n\!/_2\,k} = \frac{j3\mu_0\lambda_0\bar{N}_n^1\bar{I}_n}{{}^n\!/_2\,k}$$

$$\bar{B}_n^{m\neq n} = \frac{j\mu_0\lambda_1}{2\,{}^n\!/_2\,k}\bar{J}_n = \frac{j3\mu_0\lambda_1\bar{N}_n^1\bar{I}_n}{2\,{}^n\!/_2\,k}$$

$$(11)$$

This illustrates the cross coupling between the 2-pole and 6-pole fields, i.e., $\bar{B}_6^2$ is a 2-pole flux produced by the 6-pole MMF wave modulated by the salient-pole rotor. Interestingly, the 2-pole MMF wave is also modulated to produce a 10-pole flux wave, which will link the 5th winding harmonic of the 2-pole winding. This will induce a voltage set at a different frequency from the supplied frequency unless the rotor is at standstill (producing a backwards-rotating set) or $\omega_r = -2\omega_2$ (which will

produce a forwards-rotating set). The study in [1] found that the 2-pole winding was unbalanced at standstill which illustrated this point. The 6-pole MMF will also produce a 14-pole field which will induce EMFs in the 7th harmonic of the 2-pole winding. This illustrated that care should be taken to ensure the winding has low harmonic content and it may be necessary to skew the rotor.

E. Impedance and Linkage

The electric field is

$$e(y,t) = \int \frac{db(y,t)}{dt}\,dy \tag{13}$$

from which

$$e(y,t) = -\mathrm{Re}\,\frac{1}{k}\left(\frac{\omega_6\bar{B}_6^6 e^{j(\omega_6 t - 3ky)}}{3} + \frac{\omega_2\bar{B}_2^2 e^{j(\omega_2 t - ky)}}{1}\right)$$

$$-\mathrm{Re}\,\frac{1}{k}\left(\begin{array}{c}\dfrac{(\omega_r+\omega_6)\bar{B}_6^{14} e^{j((4\omega_r+\omega_6)t-7ky-4\delta)}}{7} \\[2ex] +\dfrac{(\omega_r+\omega_2)\bar{B}_2^{10} e^{j((4\omega_r+\omega_2)t-5ky-4\delta)}}{5} \\[2ex] +\dfrac{(\omega_r-\omega_6)\bar{B}_6^2 e^{j((4\omega_r-\omega_6)t-ky-4\delta)}}{1} \\[2ex] +\dfrac{(\omega_r-\omega_2)\bar{B}_2^6 e^{j((4\omega_r-\omega_2)t-3ky-4\delta)}}{3}\end{array}\right) \tag{14}$$

And we can further use the equation

$$u(t) = L_{stk}\int_0^{2\pi r} -e(y,t)n(y)dy \tag{15}$$

to obtain the EMF induced into a winding defined by $n(y)$. For the 2- and 6-pole windings and neglecting higher MMF harmonics:

$$n_{2,6}(y) = \bar{N}_{2,6}^1 e^{-j(1,3)ky} + \bar{N}_{2,6}^{-1} e^{j(1,3)ky} \tag{16}$$

By combining (14), (15) and (16) then we obtain expression for the 2- and 6-pole winding EMFs. So that the 2-pole winding has an EMF:

$$u_2(t) = \frac{2\pi L_{stk}}{k^2}\left\{\mathrm{Re}\left[\bar{N}_2^{-1}\left(\left(\omega_2\bar{B}_2^2 e^{j(\omega_2 t)}\right) \\ +\left(\omega_r-\omega_6\right)\bar{B}_6^2 e^{j(\omega_r-\omega_6)t-4\delta)}\right)\right]\right\} \tag{17}$$

and the 6-pole winding has an EMF:

$$u_6(t) = \frac{2\pi L_{stk}}{3k^2}\left\{\mathrm{Re}\left[\bar{N}_6^{-1}\left(\left(\omega_6\bar{B}_6^6 e^{j(\omega_6 t)}\right) \\ +\left(\omega_r-\omega_2\right)\bar{B}_2^6 e^{j(\omega_r-\omega_2)t-4\delta)}\right)\right]\right\} \tag{18}$$

These can be broken down into

$$u_2(t) = \frac{j6\mu_0\pi L_{stk}}{k^3}$$

$$\times\left\{\mathrm{Re}\left[\left(\left(\frac{\omega_2\lambda_0\bar{N}_2^1\bar{N}_2^{-1}\bar{I}_2}{1}e^{j(\omega_2 t)}\right) \\ +\frac{(\omega_r-\omega_6)\lambda_1\bar{N}_6^1\bar{N}_2^{-1}\bar{I}_6}{6}e^{j((4\omega_r-\omega_6)t-4\delta)}\right)\right]\right\} \tag{19}$$

$$\bar{U}_2 = jX_2\bar{I}_2 + j\bar{Z}_M\bar{I}_6 e^{-j4\delta} = j\omega_2 L_2\bar{I}_2 + j\omega_2 M\bar{I}_6 e^{-j4\delta}$$

and

$$u_6(t) = \frac{j2\mu_0\pi L_{stk}}{k^3}$$

$$\times \left\{ \mathrm{Re}\left[\left(\left(\frac{\omega_6\lambda_0\bar{N}_6^1\bar{N}_6^{-1}\bar{I}_6}{3} e^{j(\omega_6 t)} \right) \right. \right. \right.$$
$$\left. \left. \left. + \frac{(\omega_r - \omega_2)\lambda_1\bar{N}_2^1\bar{N}_6^{-1}\bar{I}_2}{2} e^{j((4\omega_r - \omega_2)t - 4\delta)} \right) \right] \right\} \qquad (20)$$

$$\bar{U}_6 = jX_6\bar{I}_6 + j\bar{Z}_M\bar{I}_2 e^{-j4\delta} = j\omega_6 L_6\bar{I}_6 + j\omega_6 M\bar{I}_2 e^{-j4\delta}$$

The assumptions in (19) and (20) are that the rotor is rotating at the correct synchronous speed and the winding coefficients are wholly real.

F. Impedance Matrix

An impedance matrix can be formulated using (19) and (20) where

$$\begin{bmatrix} \bar{V}_2 \\ \bar{V}_6 \end{bmatrix} = \begin{pmatrix} R_2 + j\omega_2(L_2 + L_{2L}) & j\omega_2 Me^{-j4\delta} \\ j\omega_6 Me^{-j4\delta} & R_2 + j\omega_6(L_6 + L_{6L}) \end{pmatrix} \begin{bmatrix} \bar{I}_2 \\ \bar{I}_6 \end{bmatrix} \qquad (21)$$

which is a slight modification to the matrix in [9] which includes the rotor angle δ. The matrix also includes phase winding and leakage inductances. This is a powerful matrix and can be used to simulate the machine under a variety of load and voltage conditions.

G. Winding layout

In the simulation we will assume that the winding layout is fixed and that the wire thickness is varied to match the required current. The coil locations are given in Fig. 2. The 6-pole and 2-pole winding coefficients can be defined by the general equation:

$$\bar{N}_1^n = \frac{1}{2\pi r}\sum_{W=1}^{slot\,no} k_s^n C_w e^{jpnky_W} = \frac{1}{2\pi r}\sum_{W=1}^{slot\,no} k_s^n C_w e^{jpn\theta_W} \qquad (22)$$

where C_W is the number of conductors in the slot, y_W is the linear location of the centre of the slot and θ_W is the angular location of the slot (in mechanical degrees). The slot opening factor is given by

$$k_s^n = \frac{\sin pnk\dfrac{b_s}{2}}{pnk\dfrac{b_s}{2}} \qquad (23)$$

The second and third phase can be defined by

$$\bar{N}_2^n = \bar{N}_1^n e^{jn\frac{2\pi}{3}} = a^n\bar{N}_1^n$$
$$\bar{N}_3^n = \bar{N}_1^n e^{-jn\frac{2\pi}{3}} = a^{-n}\bar{N}_1^n \qquad (24)$$

For the 6-pole and 2-pole windings:

$$\bar{N}_{6-pole}^n = \frac{18k_{s6}^n \times 2 \times \left[\begin{array}{c} -e^{j3n\frac{0.5}{18}\pi} + e^{j3n\frac{6.5}{18}\pi} + e^{j3n\frac{7.5}{18}\pi} \\ -e^{j3n\frac{12.5}{18}\pi} - e^{j3n\frac{13.5}{18}\pi} + e^{j3n\frac{17.5}{18}\pi} \end{array} \right]}{2\pi \times 63.5 \times 10^{-3}}$$

$$k_{s6}^n = \frac{\sin\dfrac{3n}{63.5\times10^{-3}}\times\dfrac{2.92\times10^{-3}}{2}}{\dfrac{3n}{63.5\times10^{-3}}\times\dfrac{2.92\times10^{-3}}{2}}$$

$$\bar{N}_{2-pole}^n = \frac{18k_{s2}^n \times 2 \times \left[\begin{array}{c} -e^{jn\frac{0.5}{18}\pi} - e^{jn\frac{1.5}{18}\pi} - e^{jn\frac{2.5}{18}\pi} \\ +e^{jn\frac{15.5}{18}\pi} + e^{jn\frac{16.5}{18}\pi} + e^{jn\frac{17.5}{18}\pi} \end{array} \right]}{2\pi \times 63.5 \times 10^{-3}}$$

$$k_{s2}^n = \frac{\sin\left(\dfrac{n}{63.5\times10^{-3}}\times\dfrac{2.92\times10^{-3}}{2}\right)}{\dfrac{n}{63.5\times10^{-3}}\times\dfrac{2.92\times10^{-3}}{2}}$$

H. Fluxing levels

Using the equations in (11) current limits can be set for the give flux levels. Ignoring the higher harmonics and just considering the main two flux waves we can rearrange (11) so that

$$\bar{I}_n = \frac{n/_2 k}{j3\mu_0\lambda_0\bar{N}_n^1}\bar{B}_n^n + \frac{2\,n/_2 k}{j3\mu_0\lambda_1\bar{N}_n^1}\bar{B}_n^{m\neq n} \qquad (25)$$

If we assume that the 2- and 6-pole flux waves total a peak air-gap flux of 0.9 T and that the 6-pole winding is the power winding (lower current so that this results in 0.6 T for the 6-pole flux wave and 0.3 T for the 2-pole flux wave) then the rms currents of about 20 A for the 6-pole winding and 10 A for the 2-pole winding. In [9] the 2-pole winding was used as the power winding but as will be seen in the finite element analysis below, it is better to use the 6-pole winding as the power winding.

I. Voltage levels and frequency variation

Using the maximum currents given in the previous section then the cross-coupling voltages can be calculated from equations (19) and (20):

$$\bar{U}_2' = \left| j\omega_2 M\bar{I}_6 e^{-j4\delta} \right| = 126 \text{ V}$$
$$\bar{U}_6' = \left| j\omega_6 M\bar{I}_2 e^{-j4\delta} \right| = 314 \text{ V}$$

These calculations use the windings in Table I and assume the speed is 900 rpm. The winding layouts are given in Fig. 2. At this speed, with the 6-pole winding connected to a 50 Hz supply and the 2-pole winding acting as the control winding, the 2-pole winding is excited at 10 Hz. This assumes that the rotor and main winding phase rotation are in the same direction. The control frequency variation is shown in Fig. 8.

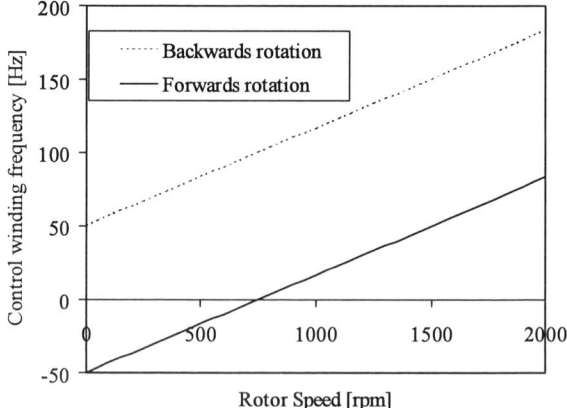

Fig. 8. Variation of control frequency with speed.

J. Winding sizing, slot depth variation and stator yoke width

The number of turns per phase in Table I appears to be approximately correct from the calculation in the previous section if the machine is nominally rated at 400 V line. To simplify the design then the number of conductors (or strands) in hand can be varied to correctly rate the machine winding. One conductor has a cross-section of 0.57 mm². If 6 A/mm² is allowed as a maximum then each conductor can carry a current of 3.4 A. Therefore the 2-pole winding requires 3 strands in hand and the 2-pole power winding requires 6 stands in hand. This gives a 6-pole resistance of 0.41 Ω and a 6-pole winding of 1.22 Ω.

The winding in Table I was fitted into the lamination shown in Fig. 1. The slot fill was 40 % which is reasonable for a hand-wound experimental machine with two sets of windings. It this slot fill it to be maintained with 18×9 conductors rather than 18×4 conductors then the slots depth has to be increased from 20 mm to 32 mm and the outer diameter of the machine has to be increased from 204 mm to 228 mm.

The stator yoke needs to be sized. For any air-gap flux wave, the total flux per length of rotor is given by

$$\Psi_p = \int_{-\pi/pk}^{\pi/pk} \hat{B} \cos pky \, dy = \frac{2\hat{B}}{kp} \tag{26}$$

where p is the pole-pair number and k is the inverse of the average air-gap radius. The yoke will carry half of the flux per pole so that

$$\hat{B}_{yoke} L_{yoke} = \frac{\Psi_p}{2} \tag{27}$$

Therefore

$$L_{yoke} = \frac{\hat{B}_2 + \hat{B}_2/3}{\hat{B}_{yoke} k} = \frac{0.3 + 0.6/3}{1 \times 15.8} = 33 \text{ mm}$$

This was rounded up to 40 mm to reduce the reluctance of the yoke for the 2-pole flux wave.

The modified machine is now shown in Fig. 9 and can be compared to the original machine studied in [9] as illustrated in Fig. 1. This machine is now studied in the next section using an analytical model and finite element simulations.

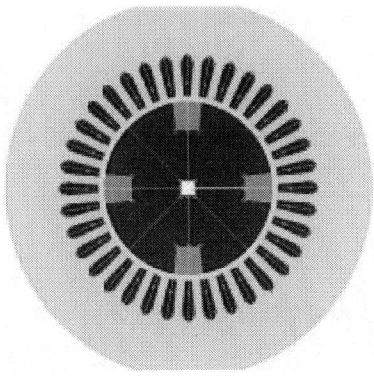

Fig. 9. Modified machine model.

IV. SIMULATIONS

The algorithm described above will be implemented in MATLAB and further design sizing addressed. It is important to size the core correctly. Therefore the machine was first investigated using static finite element analysis.

A. Finite element analysis

The machine was simulated in *SPEED* PC-SREL from the University of Glasgow [10] which is a simulation package for axially-laminated synchronous machines. The finite element analysis bolt-on package for this is PC-FEA and the design was passed through with various modifications in the GDF editor to produce the finite element model shown in Fig. 9. With further modifications to the PC-FEA script the windings were modified so that 3-phase current could be stepped through and the winding flux linkage measured to obtain a terminal voltage in either set of windings. The rotor was static so this represented a locked-rotor simulation. From Fig. 8 it can be seen that the frequency in both sets of windings is 50 Hz.

The first simulation excited the 6-pole winding up to 20 A rms and stepped through one cycle of the current waveform. The voltage waves are shown in Fig. 10 for the 6-pole and Fig. 11 for the 2-pole windings.

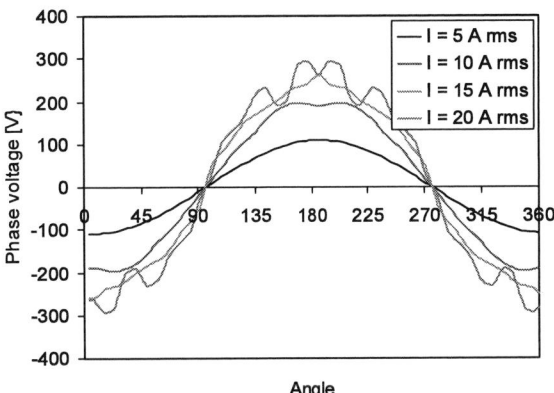

Fig.10. Variation 6-pole voltages for 6-pole winding excitation.

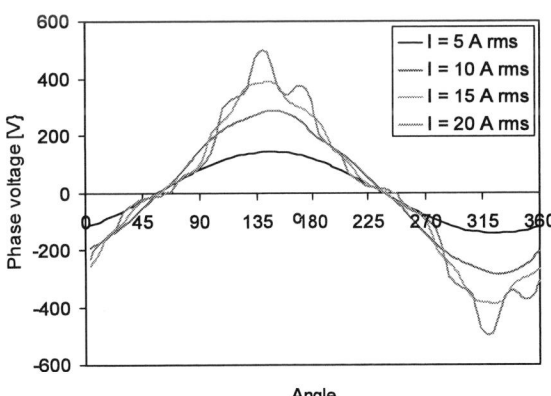

Fig.11. Variation 2-pole voltages for 6-pole winding excitation.

The results in Figs. 10 and 11 show that the machine saturates between 15 and 20 A. The variation of the true rms of the waveforms and the rms of the fundamental are shown in Fig. 12. A flux plot is shown in Fig. 13 for 6-pole excitation at 20 A, and it can be clearly seen that the modulation creates a 2-pole flux distribution. In addition it can be observed that the teeth are very saturated – this leads to the point that the stator teeth are too narrow and further design modifications would be to increase the width of these.

In Fig. 12 it can be observed that the analytical solution overestimated both the self and mutual inductance. To check the non-linear effect of the finite element analysis then a linear solution was run and the results are shown.

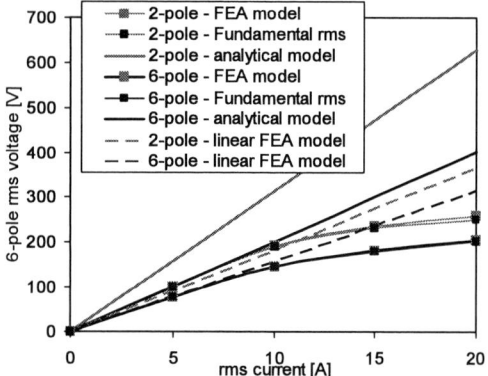

Fig.12. Variation 2-pole and 6-pole voltages with 6-pole excitation.

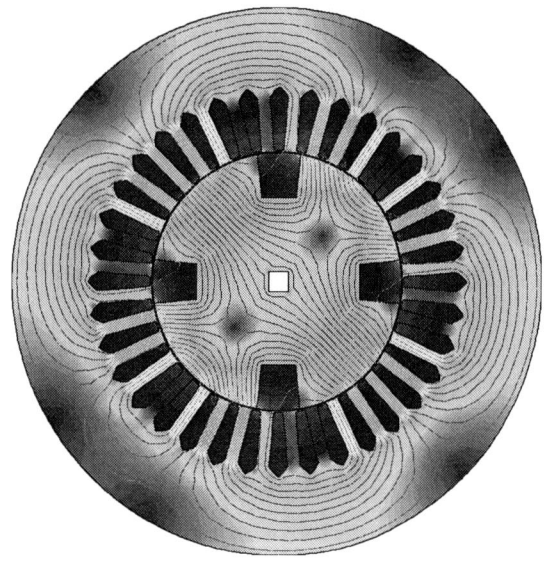

Fig.13. flux plot with 20 A 6-pole excitation.

The simulations were repeated but this time with the 2-pole winding excited. The voltage waveforms are shown in Fig. 14 for the 2-pole winding and Fig. 15 for the 6-pole winding. What can be observed is that when the machine becomes saturated there is a large 3rd order (in both time and space) flux wave that induces a voltage into the 6-pole winding. This illustrates one of the disadvantages of this arrangement – the 6-pole winding is a 3rd spatial harmonic of the 2-pole winding. Therefore a 4/8 pole arrangement may be more practical in terms of spurious and unwanted cross-coupling.

Fig. 16 shows the variation of the voltages when the 2-pole winding is excited and the finite element values are

compared to the analytical model. Again there is heavy saturation and the analytical model greatly overestimates. If the analytical model cross-coupling permeance λ_1 is divided by 2 then it gives an approximation that can be used in load simulations.

In addition, the limit for the 2-pole current appears to be about 3 A and about 15 A for the 6-pole winding.

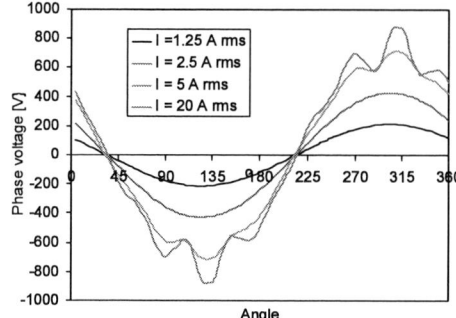

Fig.14. Variation 2-pole voltages for 2-pole winding excitation.

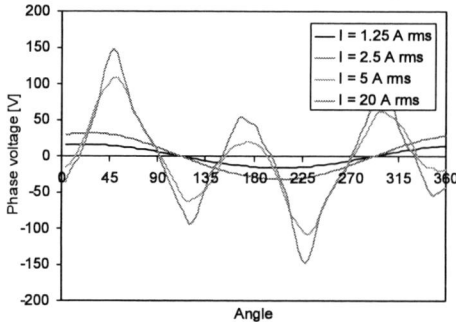

Fig.15. Variation 6-pole voltages for 2-pole winding excitation showing large 3rd harmonic when saturated.

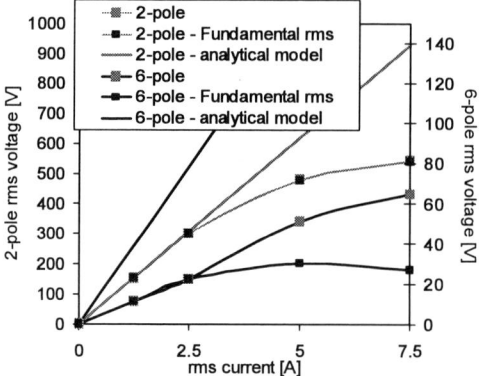

Fig.16. Variation 2-pole and 6-pole voltages with 2-pole excitation.

B. Analytical load simulations

The machine can be operated as an induction machine by simply short-circuiting the 2-pole winding. The synchronous speed is 750 rpm and above this speed the machine should generate. However, here, the 6-pole winding is connected to a 230 V 50 Hz bus-bar and the machine rotor angle δ is varied at 900 rpm with different values of capacitor attached to the 2-pole winding. The output power is shown in Fig. 17 while the 6-pole and 2-pole currents are given in Figs. 18 and 19. The cross coupling permeance λ_1 is halved because of the results found in the FEA section.

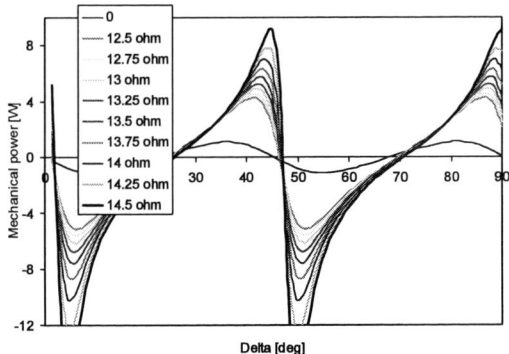

Fig. 17. Output power at 900 rpm and 230 V attached to the 6-pole winding and different capacitive reactances connected to the 2-pole winding (including short-circuit).

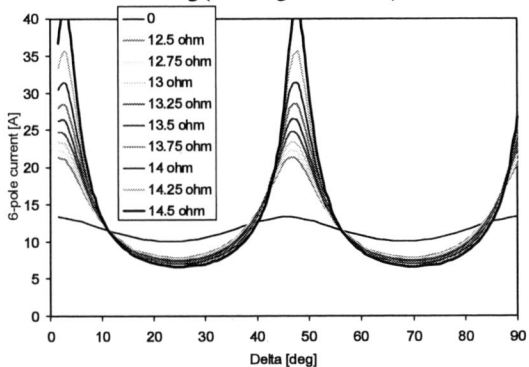

Fig. 18. 6-pole current (rms) at 900 rpm and 230 V attached to the 6-pole winding.

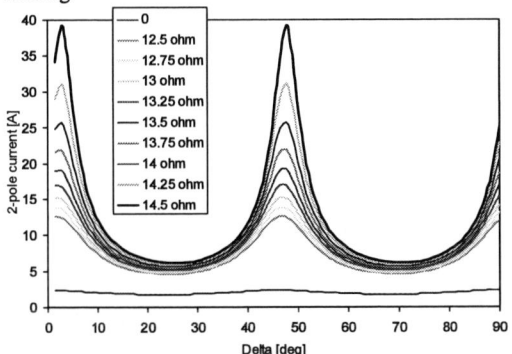

Fig. 19. 2-pole current (rms) at 900 rpm and 230 V attached to the 6-pole winding.

Negative power represents generation. The power factors are shown in Fig. 20. These are a function of the attached capacitor because the 2-pole winding is used to control the magnetization. If more VArs are fed in from the 2-pole winding then the power winding p.f. is improved. While it appreciated that the currents exceed the current levels set in the previous section these simulations are for illustrative purposes. Table II shows a set of results for the cases were the machine is generating 3.5 kW and 7 kW for two different capacitive reactance sets connected to the 2-pole winding.

Further work is required to assess the control needed and to understand the full equivalent circuit of the machine referred to the primary. This machine appears to exhibit characteristics associated with the induction machine and also the synchronous machine.

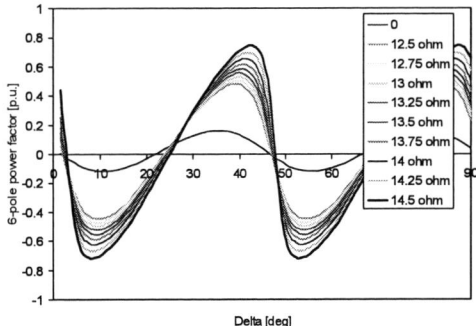

Fig. 20. Power factor for the 6-pole current at 900 rpm and 230 V attached to the 6-pole winding.

TABLE II
SIMULATION RESULTS

P_{out} [kW]	Delta [Deg]	2-pole reactance [ohm]	6-pole current [A]	6-pole p.f.	2-pole current [A]	Mech. P_{in} [kW]
3.4	13.5	13.75	9.8	0.5	7.6	3.7
7.2	6	13.75	20.4	0.51	15.7	8.6
3.2	15	14.5	8.7	0.53	8.1	3.5
7.3	9	14.5	14.9	0.71	13.9	8.3

V. CONCLUSIONS

This paper addresses several issues with regards to the design of a brushless doubly-fed reluctance machine. It is illustrated that this is a complex machine and further work will be to develop further equivalent circuits and design procedures.

REFERENCES

[1] I. Scian, D. G. Dorrell, and P. Holik, "Assessment of Losses in a Brushless Doubly-Fed Reluctance Machine", *IEEE Transactions on Magnetics, INTERMAG special edition*, October 2006.

[2] D. G. Dorrell, I. Scian E. M. Schulz, R. B. Betz and M. Jovanovic, "Electromagnetic Considerations in the Design of Doubly-Fed Reluctance Generators for use in Wind Turbines" *IEEE Industrial Electronics Conference, IECON*, Paris 7-10 Nov. 2006.

[3] R. E. Betz and M. G. Jovanovic, "The Brushless Doubly Fed Reluctance Machine and the Synchronous Reluctance Machine - A Comparison", *IEEE Trans on Industry Applications*, Vol. 36, No. 4, July/August 2000.

[4] R. B. Betz and M. G. Jovanovic, "Theoretical Analysis of Control Properties for the Brushless Doubly Fed Reluctance Machine", *IEEE Trans on Energy Conversion*, Vol 17, No 3, Sept 2002, pp 332-339.

[5] M. G. Jovanovic and R. E. Betz, "The Use of Doubly Fed Reluctance Machines for Large Pumps and Wind Turbines", *IEEE Trans on Industry. Applications*, Vol 38 No 6, Nov 2002, pp 1508-1516.

[6] M. G. Jovanovic and J. Yu, "An Optimal Direct Torque Control Strategy for Brushless Doubly-Fed Reluctance Motors", *IEEE Power Electronics and Drive Systems Conference*, 17-20 Nov 2003, Vol 2, pp 1229-1234.

[7] M. G. Jovanovic, "Control of Brushless Doubly-Fed Reluctance Motors", *IEEE International. Symposium. on Industrial Electronics*, 2005, Dubrovnik, Croatia, pp 1667-1672.

[8] E. M. Shulz, and R. E. Betz, "Optimal Torque per Amp for Brushless Doubly Fed Reluctance Machines", *40th IEEE IAS Meeting*, 2-6 Oct, 2005, Hong Kong, pp 1749-1753.

[9] D. G. Dorrell, I. Scian, and P. J. Holik, "Assessment of the Electromagnetic Performance of Doubly-Fed Reluctance Machines for Use in Variable Speed Generators", *ICEMS*, Nagasaki, Japan, 20-23 Nov 2006.

[10] T.J.E. Miller, "SPEED's Electrical Motors", SPEED Laboratory, University of Glasgow, 2006.

A Magnetic Gear Box for application with a Contra-rotating Tidal Turbine

Laxman Shah, A. Cruden and Barry W. Williams
Department of Electronic and Electrical Engineering
University of Strathclyde
Glasgow, United Kingdom

Abstract: **This paper presents experimental results for the application of a magnetic gear (MG) box to a contra-rotating tidal turbine. The contra-rotating tidal turbine is a novel design comprising two sets of rotors placed axially: the upstream rotor rotates in a clockwise direction whilst the downstream rotor rotates in an anti-clockwise direction. The tidal turbine is a low speed mechanical system similar to a wind turbine, hence the use of a gearbox is considered to permit a standard 4 or 6 pole induction generator to be used. The magnetic gearbox magnetically combines the output of both contra-rotating shafts into a single high speed rotating shaft. A gear box has been designed using 3D Finite Element Analysis and experimentally tested.**

Keywords: **Contra-rotating, Finite-element analysis (FEA), Magnetic gear (MG), Gear ratio (GR).**

1. INTRODUCTION

The contra-rotating tidal turbine system delivers higher efficiency than a conventional single rotor turbine system. Analysis shows that the dual rotor system gives 6-16% more power than a single rotor system and produces less bending stress on the supporting structure due to zero net reaction torque[1,2,3]. The tidal turbine resembles an underwater turbine so its performance is broadly analogous to a wind turbine. A direct coupled generator with a low speed turbine delivers a low machine power density as the mass of the machine depends upon the torque rating. Therefore some form of gear box is advantageous for converting the two low speed rotating shafts into a single high speed shaft at low torque.

NdFeB magnets are common and have been widely used to increase the performance of AC/DC motors, linear motors, actuators, loudspeakers, and magnetic couplings because of its high magnetic strength.

Therefore a magnetic gear (MG) can also be used in place of a conventional mechanical gear box because of its unique features such as, no mechanical contact so less acoustic noise, no lubrication, no frictional heat, less wear and tear, and can facilitate over load protection. In 1987 [4] a magnetic gear using $SmCo_5$ magnets, having a gear ratio of 3:1 with a transmitting torque 5.5 Nm was presented. Similarly in 1991 [5] a magnetic spur gear for micro and miniature electro-mechanical systems application was presented. The maximum transmission torque for the gear was 100 g-cm, using ferrite magnets, although this could be increased by using rare-earth magnets.

In [6] the structure and design of a magnetic worm gear using $SmCo_5$ was presented with a prototype MG of gear ratio 1:33. It was capable of transmitting a torque of 11.5 Nm and confirmed that various types of MG like skew, worm, and bevels gear could be manufactured. The drawback of all these MGs is that only a small portion of the magnets are used to transmit torque.

In 2001 [7,8] a radial flux MG using rare earth magnets was considered good it was argued that its transmission torque density could be increased to more than 100 kNm/m^3. The paper also discussed the operational principle, torque speed characteristics at different gear ratios, and transmission capability of the MG. Similarly in [9], a MG of gear ratio 5.5:1, stack length 26 mm with a 1 mm air gap, and 27 Nm transmission torque having 92 $kN.m/m^3$ torque per volume density, was presented.

The MG presented in this paper has diameter 120 mm with a 15 mm stack length and is capable of a theoretical maximum transmission torque of 12.51 Nm and has a 74 kNm/m^3 torque per volume. The novel feature of the proposed MG, and this papers archival value, is that it has three rotating shafts and is capable of coupling and insulating power from two contra-rotating low speed shafts to a single high-speed output shaft. This paper is divided into three parts: i) Constructions of the MG, ii) Operating principle and iii) Simulation and experimental results.

978-1-4244-0644-9/07/$25.00 ©2007 IEEE

2. CONSTRUCTION OF THE MAGNETIC GEAR BOX

The prototype MG, shown in fig.1, consists of an outer rotor, inner rotor and ferromagnetic pole pieces. The outer rotor has twenty two pairs of NdFeB magnets bonded to its inner circumference whilst the inner rotor has four pairs of the same magnets bonded to its outer circumference.

Fig 1: Prototype magnetic Gear Box.

Twenty six individual pole pieces are located between the outer and inner magnetic rotors to modulate the air gap flux density. The gearbox allows the outer rotor and pole pieces to rotate in opposite directions, driven by the two tidal turbine blades which convert the two input driving shafts into a single high speed output shaft. The direction of rotation of the outer rotor, pole pieces, inner rotor and its arrangements are shown in fig. 2.

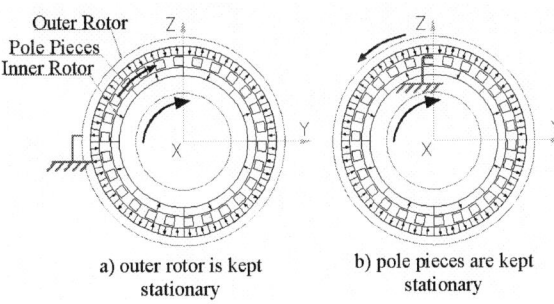

a) outer rotor is kept stationary

b) pole pieces are kept stationary

Fig 2: Rotational direction of the MG at different states.

When the outer rotor is held stationary, the gear ratio (GR) between the pole pieces and inner rotor is 1:6.5 and both rotate in the same direction (fig. 2A). However when the pole pieces are held stationary, the gear ratio between the outer and inner rotors is 1:5.5 and both rotate in opposite directions (fig. 2B).

For the contra-rotating tidal turbine these conditions apply simultaneously; when both the outer rotor and pole pieces rotate with the same speed in the opposite direction then the resulting GR becomes 1:12.

3. OPERATING PRINCIPLE

The flux density distribution with the rotation of the outer rotor is modulated by the stationary ferromagnetic pole pieces, producing a space harmonic flux distribution in the inner air gap of the rotor. The velocity of the space harmonic flux distribution is different from the velocity of the outer rotor. For the inner rotor speed, the number of outer pole pairs and ferromagnetic pole pieces is selected for the specific space harmonic pole pairs. The inner rotor speed will be equal to the velocity of the space harmonic flux distribution.

The GR and the direction of rotation between the outer rotors to the inner rotor, and pole pieces to the inner rotor are expressed with the formulas (1, 2).

$$n_l = \frac{p_h}{p_h - p_p} n_h = -\frac{1}{5.5} n_h \qquad (1)$$

The negative sign shows that the rotational direction of the high speed rotor is opposite to the outer rotor.

$$n_p = \frac{p_h}{P_p} n_h = \frac{1}{6.5} n_h \qquad (2)$$

where n_l, n_h, n_p are the rpm of the outer, inner and pole pieces. In the same way p_h, p_p, are the number of high speed rotor poles and pole pieces.

4. SIMULATION AND EXPERIMENTAL RESULTS

The Integrated Engineering 3D software package 'Faraday' was used to investigate a magneto static field solution. The static torque simulation results on all three rotors over a 90^0 angular displacement of the inner (high speed) rotor, is depicted in fig.3. The same torque characteristics are obtained over a complete period of the torque curve with an angular displacement of the low speed outer rotor of 16.36^o.

The maximum static torque on the low speed rotor (outer rotor and on the pole pieces) was designed to be 10.5 Nm and 12.51 Nm, while at the high speed rotor (inner rotor) 1.92 Nm was achieved.

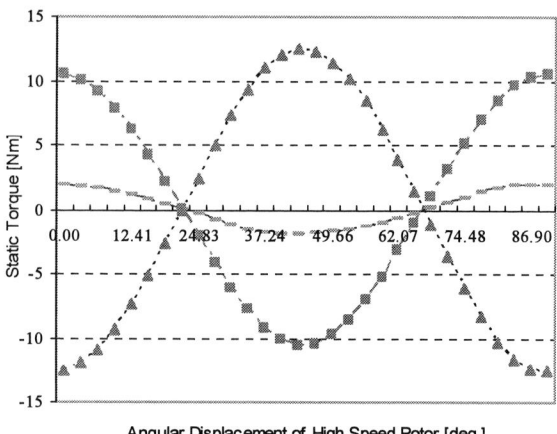

▲ – Pole pieces, — High Speed Rotor, ■ – Outer Rotor

Fig 3: Different rotor torque versus speed.

The wave length of the torque curve is formulated as:

$$\lambda = \frac{360^0}{P/2} \qquad (3)$$

where P is the numbers of poles of the driven rotor.

The simulation results of the torque on the outer rotor versus angular displacement of the outer rotor, and of the inner rotor versus displacement of the inner rotor, are depicted in figs. 4 and 5 respectively. The gear ratio can be verified between any of the rotors using these wavelength torque curves.

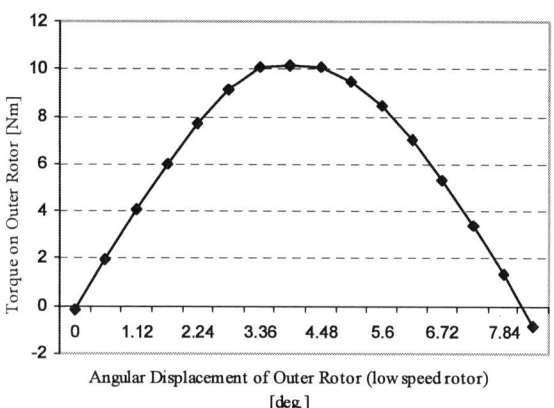

Fig 4: Torque on outer rotor vs. angular displacement of outer rotor.

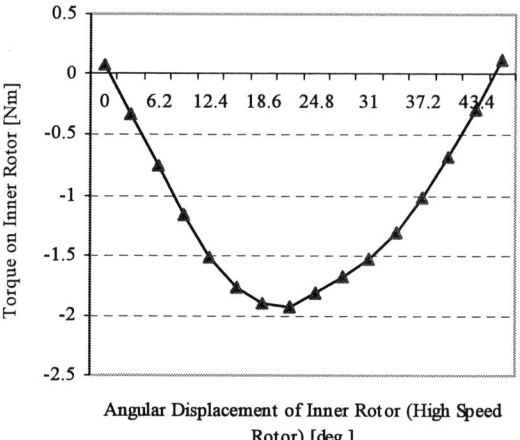

Fig 5: Torque on inner rotor vs. angular displacement of inner rotor.

The half wavelength of the torque curves of the outer and inner rotors are 8.18^0 and 45° respectively, as shown in figs. (4,5). Therefore the gear ratio between the outer to inner rotors is 1:5.5.

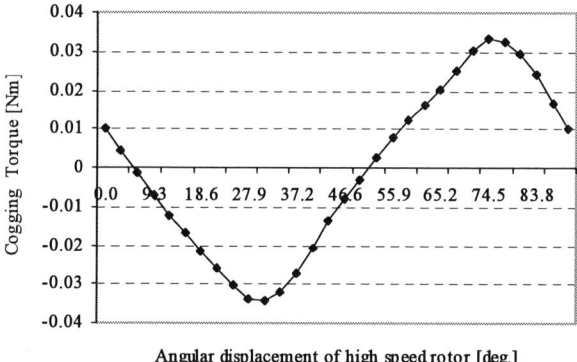

Fig 6: Cogging torque with rotation of the high speed rotor.

The peak cogging torque with the angular displacement of the high speed rotor is 0.034 Nm, as depicted in fig. 6. That is, 1.73% of rated torque of the high speed rotor which is typically less than the permanent magnet machine [10].

A practical test rig was developed as shown in the block diagram fig. 7(a). Due to mechanical constraints a torque transducer could not be connected directly to the outer rotor, therefore it is connected via spur gear. In a contra-rotating application this will not be used. Torque transducers at the pole pieces drive and high speed load end shafts are directly connected to the shafts as shown in fig. 7(b).

1184

The gear ratio has been verified compared to the simulation results by rotating all three rotors one by one with respect to each other.

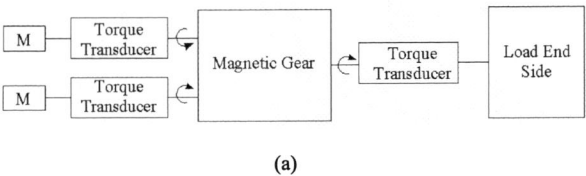

(a)

(b)

Fig 7: Magnetic gear box (a) block diagram and (b) test rig.

Transmission torque capability of the prototype MG from the low speed to high speed rotor is measured by varying the load. The pull-out torque at the low speed rotor (outer rotor) and high speed rotor are measured as 7.96 Nm and 1.01 Nm respectively. The measured outer rotor torque is 25% less than the simulation prediction. The low speed to high speed efficiency with stationary pole pieces at different torque levels at 30 and 40 rpm is depicted in fig. 8. The torques shown in the fig. 8 is measured on the outer rotor shaft. Efficiency at different rpm is different due to cogging torque. The low efficiency is possibly due to the mechanical constrained of using solid (not arch laminated) square steel pole pieces. The other reason for low efficiency is spur gear losses and eddy current losses.

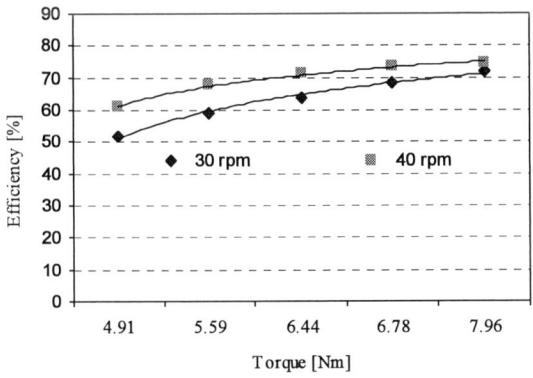

Fig 8: Variation of efficiency at different torques from low speed (outer) to high speed rotor at 30 and 40 rpm.

Gear ratio testing results from the low speed to high speed shafts are depicted in fig 9.

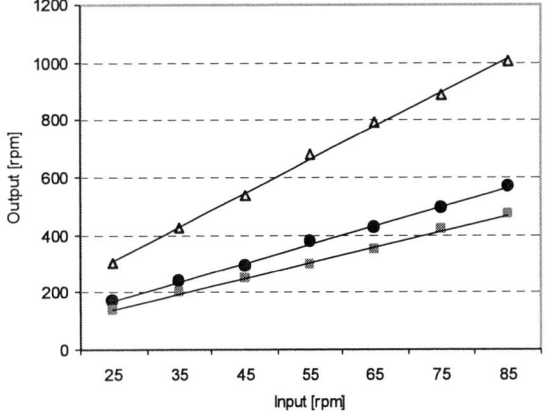

● – Output with only the rotation of the pole pieces.
■ - Output with only the rotation of the outer rotor.
Δ - Output with the combined rotation of both pole pieces and outer rotor.

Fig 9: Output variation speed with the rotation of low speed shafts.

The testing result in fig. 9 verifies the gear ratio of 1:12 to the simulation result when the two input shafts are being driven in opposite direction.

5. CONCLUSION

A prototype MG for contra-rotating tidal turbine application has been successfully made and tested. It is capable of converting two low-speed shafts into one single high-speed shaft. The pull-out torque on the low speed side rotor was found to be 7.96 Nm which is less than that simulated. The reason for this is the geometry of pole pieces does not create a uniform air gap. The maximum efficiency was achieved only 74 % since the torque transducer could not be installed close to the outer coaxial shaft so as to eliminate spur gear losses. Also the pole pieces and yoke should be laminate to reduce losses. Future research will focus on efficiency, analytical analysis of torque transmission capability, and on the variable magnetic gear ratio to control a constant output.

AKNOWLEDGMENT

The author would like to thank Mr. David McCrindle for making the prototype gear and gratefully appreciates to Dr Nand Kishor Singh and Dr Narendra De Silva fruitful discussions.

REFERENCES

[1] Kari Appa, "Counter Rotating Wind Turbine System" *Feasibility Analysis, California Energy Commission, Renewable Energy Technologies*, July 2003, P500-03-055F.

[2] Davenport F.J., Colehour J.L. and Sokhey J.S. "Analysis of Counter-Rotating Propeller Performance" AIAA paper 86-1804, Jan 1985.

[3] J.A. Clarke, G. Connor, A.D. Grant and C.M. Johnston "Design and initial testing of a contra-rotating tidal current turbine" *Energy Systems Research Unit, University of Strathclyde*, Scotland.

[4] K. Tsurumoto and S. Kikuchi "A New Magnetic Gear using Permanent Magnet" *IEEE Transactions on Magnetics*, Vol. Mag.-23, No. 5, 1987, pp. 3622-3624.

[5] Koji Ikuta, Shunichi Makita, Suguru Arimotto "Non-Contact magnetic Gear for Micro Transmission Mechanism" *IEEE CH2957-9/91*, 1991, pp. 125-130.

[6] Shinki Kikuchi "Design and characteristics of a New Magnetic Worm Gear Using Permanent Magnet" *IEEE Transactions on Magnetics*, Vol. 29, No. 6, November 1993, pp. 2923-2925.

[7] K. Atallah and D. Howe "A Novel High-Performance Magnetic Gear" *IEEE Transactions on Magnetics*, Vol.37, No.4, July 2001, pp. 2844-2846.

[8] K. Atallah and D. Howe "Design analysis and realisation of a high performance magnetic gear" *IEEE Proc.-Electr. Power Appl.*, Vol. 151, No. 2, March 2004, pp.135-143.

[9] Peter Omand Rasmussen, Torben Ole Anderson, Frank T. Jorgensen, and Orla Nielsen "Development of a High-Performance Magnetic Gear" *IEEE Transactions on Industry Applications*, Vol. 41, No. 3, May/June 2005, pp. 764-770 .

[10] Ronghai Qu, Thomas A. Lipo "Dual-Rotor, Radial-Flux, Toroidally Wound, Permanent – Magnet Machines" *IEEE Transactions on Industry Applications*, Vol. 39, No.6, Nov/Dec 2003, pp 1665-1673.

Mechatronic – Advanced Computational Intelligence

D. Schröder, Fellow IEEE, H. Schuster, and C. Westermaier

Institute for Electrical Drive Systems
Technische Universität München
Arcisstr. 21, D-80333 Munich, Germany

Abstract— This paper investigates two different strategies to control nonlinear mechatronic systems. The first method is applicable for a plant with known linear part and unknown static nonlinearity. By an intelligent observer the nonlinear characteristic is identified, such that the result of the identification can be used to improve the performance of the controller.
The second approach presented is non-identifier-based. Hence, no identification is necessary if the system possesses some structural attributes. A time-varying control law is employed, which reacts on the present behaviour of the error signal instantaneously. The appealing feature is that bounds for the control error can be fixed a priori.

Index Terms— control systems, identification, mechatronic systems, non-identifier-based control

I. INTRODUCTION

Generally, for the design of a control law in real world applications one of the hardest challenges is that the plant is not exactly known. There may be uncertainties in the structure of the plant and/or in the parameters. Additionally, there might exist one or more (time-varying) nonlinearities with a characteristic not known precisely as well as time-varying parameters.

During the last decades many effort has been spent in control engineering to cope with those systems. A well-known and widespread solution in this situation is the identification of the unknown system-behavior. For further strategies and applications the reader is referred to [9].

Typically, a certain amount of a priori-knowledge about the nonlinear system is available however. For example parts of the structure, the relevant order, or some parameters might be known. Often the position of the nonlinearities is known as well. This motivates the *engineering approach* which uses all existing a priori-knowledge and focusses only on the unknown details of the nonlinear plant during identification.

In contrast to this a non-identifier-based approach is presented in Sec. V. The exact values of parameters are not of interest and also may be time-varying. Structural properties of the plant have to be known only, which offers an enlarged field of applicability.

The comparison of both methods reveals that identification of unknown parameters and/or nonlinearities is superfluous in many cases. This motivates to abandon identification strategies and to use controllers with much less demand on computational power.

II. INTELLIGENT OBSERVER WITH IDENTIFICATION

Throughout this section a plant is considered with a known structure, exactly known and time-invariant linear parameters, and a static but unknown nonlinearity. Typical examples for such systems are mechanical systems with friction [1] or backlash [16]. By an intelligent observer the unknown nonlinear function is identified.

The method is applicable for a class of systems which can be characterized as a linear time-invariant plant with one single static nonlinearity at a known position. (We will only discuss single-input-single-output (SISO) systems, but all derivations are applicable to the MIMO case as well.)

A system with an isolated nonlinearity can be described in state space representation as

$$\begin{aligned}
\dot{\underline{x}} &= A\underline{x} + \underline{b}u + \mathcal{NL}(\underline{x}, u) \\
y &= \underline{c}^T \underline{x} + du
\end{aligned} \tag{1}$$

where $\mathcal{NL}(\underline{x}, u)$ is an isolated nonlinearity. An isolated nonlinearity can be described as a product of a single scalar nonlinear function $\mathcal{NL}(\underline{x}, u)$ with a known input vector $\underline{k}$, i.e: $\mathcal{NL}(\underline{x}, u) = \underline{k} \cdot \mathcal{NL}(\underline{x}, u)$. The nonlinearity is called static if it contains no internal states and is called visible at the output of a SISO-system, if the transfer function $G(s) = \underline{c}^T [sE - A]^{-1} \underline{k}$ is not identically zero.

The design of an observer according to the engineering approach requires the following assumptions:

(i) The linear part of the system (i.e. A, $\underline{b}$, $\underline{c}$, d) is known and time-invariant.
(ii) The point of influence where the unknown nonlinearity acts, i.e. the input vector $\underline{k}$ is known.
(iii) The static nonlinearity is visible at the system output.
(iv) The relevant region of the nonlinearity is excited during the identification process.

A. Approximation of a Static Nonlinearity with a Neural Network

For identification the observer needs an appropriate function approximator. A static nonlinear function can be reproduced with a number of different types of neural networks. For reasons of parameter convergence Radial Basis Function Networks (RBF) and especially General Regression Neural Networks (GRNN) [14] are convenient.

The output of the GRNN is given by the inner dot product of its trainable weights $\widehat{\underline{\Theta}}$ and the network activation

978-1-4244-0644-9/07/$25.00 ©2007 IEEE

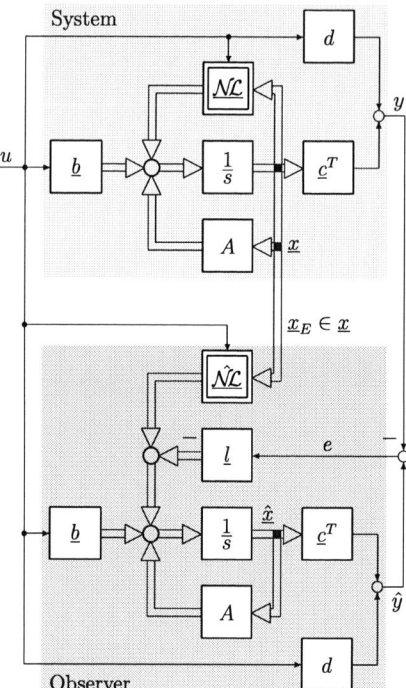

Fig. 1. Block diagram for the system and observer (SISO-case).

function $\underline{\mathcal{A}}(\underline{x}, u)$:

$$\widehat{\mathcal{NL}}(\underline{x}, u) = \widehat{\underline{\Theta}}^T \underline{\mathcal{A}}(\underline{x}, u) \tag{2}$$

This clearly shows the linear influence of the weights $\widehat{\underline{\Theta}}$. If the GRNN is perfectly tuned (i.e. its weights $\widehat{\underline{\Theta}}$ coincide with the best possible weights $\underline{\Theta}$), its output equals the unknown nonlinear function, apart from the inherent approximation error e_A:

$$\mathcal{NL}(\underline{x}, u) = \underline{\Theta}^T \underline{\mathcal{A}}(\underline{x}, u) + e_A(\underline{x}, u) \tag{3}$$

The approximation error decreases with a rising number of neurons and is therefore small, if a GRNN with an appropriate size is used. Note, that the parameters $\widehat{\underline{\Theta}}$ of the GRNN only have impact locally. For this reason possibly existing previous knowledge about the nonlinearity may be incorporated easily.

B. Observer Design

The design of an observer is shown for the case where the argument $\underline{x}$ of the nonlinear function is measurable. For the plant in Eq. (1) the observer is designed according to

$$\begin{aligned} \dot{\hat{\underline{x}}} &= A\hat{\underline{x}} + \underline{b}u + \underline{k} \cdot \widehat{\mathcal{NL}}(\underline{x}, u) - \underline{l}e \\ \hat{y} &= \underline{c}^T \hat{\underline{x}} + du \end{aligned} \tag{4}$$

where $\widehat{\mathcal{NL}}(\underline{x}, u)$ is represented by a neural network according to Eq. (2). The observer error is defined by $e = \hat{y} - y$. The observer vector $\underline{l}$ has to be dimensioned in order to place the observer poles to guarantee stability and desired transient behavior. The combination of the plant with the observer is shown in Fig. 1.

C. Adaptation Law

In order to derive the adaptation law, the observer error e must be calculated, which depends on the network weights $\widehat{\underline{\Theta}}$. We will use a hybrid notation of time-dependent signals and Laplace transfer functions and obtain for the observer error:

$$\begin{aligned} e &= \hat{y} - y = \underline{c}^T (\hat{\underline{x}} - \underline{x}) \\ &= \underline{c}^T \left[sE - A + \underline{l}\,\underline{c}^T \right]^{-1} \underline{k} \cdot (\widehat{\mathcal{NL}} - \mathcal{NL}) \\ &= H(s)(\widehat{\mathcal{NL}} - \mathcal{NL}) \end{aligned} \tag{5}$$

The observer contains a GRNN to approximate the behaviour of the nonlinearity $\mathcal{NL}$. If the adaptable estimate $\widehat{\underline{\Theta}}$ equals the unknown, constant parameter vector $\underline{\Theta}$, the GRNN generates the best possible approximation for the function $\mathcal{NL}$. Assuming $e_A \approx 0$ and substituting (2) and (3) into (5), the the error equation can be written as:

$$e = H(s)\left(\widehat{\mathcal{NL}} - \mathcal{NL}\right) = H(s)\left[\widehat{\underline{\Theta}} - \underline{\Theta}\right]^T \underline{\mathcal{A}}(\underline{x}, u) \tag{6}$$

Defining the parameter error vector by $\underline{\Phi} = \widehat{\underline{\Theta}} - \underline{\Theta}$ we finally arrive at the error equation

$$e = H(s)\underline{\Phi}^T \underline{\mathcal{A}}(\underline{x}_E, u). \tag{7}$$

For this error equation there exists an adaptation law, which forms a globally stable error model [8]. For a properly chosen adaptation law we have to determine whether the transfer function $H(s)$ is *strictly positive real* (SPR). If it fulfills this condition, or can be constructed (with the observer vector $\underline{l}$) in order to fulfill it, the adaptation law is chosen to be

$$\dot{\underline{\Phi}} = \dot{\widehat{\underline{\Theta}}} = -\eta e \underline{\mathcal{A}}(\underline{x}_E, u) \tag{8}$$

with the positive adaptation step size η. If $H(s)$ is not SPR, the delayed activation method is usable [6], [8].

With the stability theory of Lyapunov it can be proven that for such an error model the error asymptotically tends to zero [8]. Additionally, the parameters of the neural network converge to the best possible values [7]

$$\lim_{t \to \infty} \widehat{\underline{\Theta}}(t) = \underline{\Theta} \tag{9}$$

which is a consequence of the characteristics of the GRNN together with the assumptions on the excitation.

D. Simulation Example

We will now demonstrate the concept of learning a static nonlinearity by means of a simulation example. The plant is a simplified model of an electric drive system with nonlinear friction. The state x_1 is the speed of the drive; x_2 is the acceleration torque, which is related to the input u by a first-order dynamics with the time constant $T_2 = 0.005\,\mathrm{s}$. The friction torque depends on the speed x_1 of the rotation mass and is defined by the nonlinear function $\mathcal{NL}(x_1) = (20/\pi) \cdot \arctan(80 \cdot x_1)$. For later purposes let $f(t) = \mathcal{NL}(x_1(t))$. The constant T_1

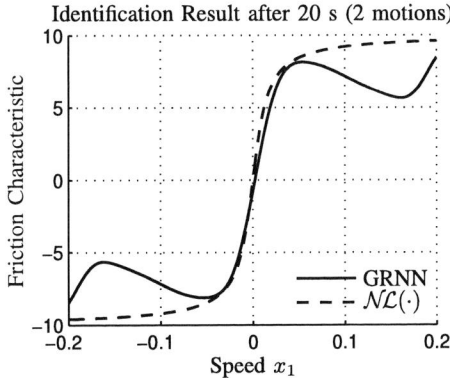
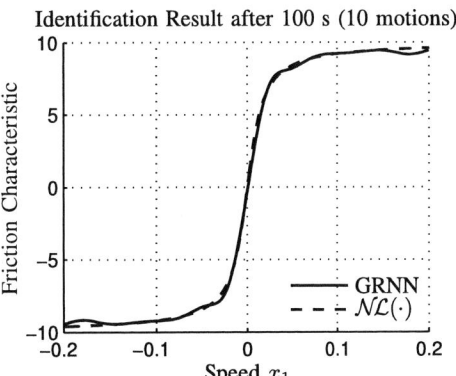

Fig. 2. Identified nonlinear friction characteristic.

is proportionally related to the mass moment of inertia. This plant is described by the equation

$$\begin{bmatrix} \dot{x}_1 \\ \dot{x}_2 \end{bmatrix} = \begin{bmatrix} \frac{1}{T_1}\left(-\mathcal{NL}(x_1) + x_2\right) \\ -\frac{1}{T_2}x_2 \end{bmatrix} + \begin{bmatrix} 0 \\ \frac{1}{T_2} \end{bmatrix} u \quad (10)$$

with the output $y = x_1$. The task is to learn the unknown friction characteristic, such that its influence can be compensated by the control law. During identification, the plant is excited by the signal $u(t) = 9.75\sin(0.2\pi t)$.

Fig. 2 shows the learned characteristic compared to the real characteristic after 20 s and 100 s of learning, which corresponds to 2 motions and 10 motions, respectively. Inside the range $|x_1| \leq 0.15$ a good approximation is given. Because the plant operates too little in the range $|x_1| > 0.15$ there a small approximation error remains. Extending the identification period or changing the input signal, the excitation may be improved in this range, such that a better identification results.

Fig. 3 displays the time diagram of the states x_1 and $\hat{x}_1$ (row 1) and the error $e = \hat{x}_1 - x_1$ (row 2). The deviation is about 0.059 in maximum, which corresponds to 21% of the maximum value of the output $y = x_1$. Within the 100 s of learning the deviation reduces considerably.

To show the influence of the nonlinearity the result is depicted if a Luenberger-observer (without GRNN) is used. Because the nonlinearity is ignored by the observer, a deviation between the observer state $\tilde{x}_1$ and x_1 remains (rows 3 and 4, respectively).

III. CONTROL LAW WITH COMPENSATION

After identification knowledge about the nonlinearity is available, such that feed forward compensation is applicable to cancel out (or to reduce at least) the effect of the nonlinearity. The unobtainable aim is perfect cancellation, which means that the ideally compensated plant is described by Eq. (10) with $\mathcal{NL}(x_1) \equiv 0$. The best we can expect however is approximate compensation only. Hence, the compensated plant behaves not identically but approximately like the linear system

$$F(s) = \frac{y(s)}{u(s)} = \frac{1}{sT_1(T_2s + 1)}, \quad (11)$$

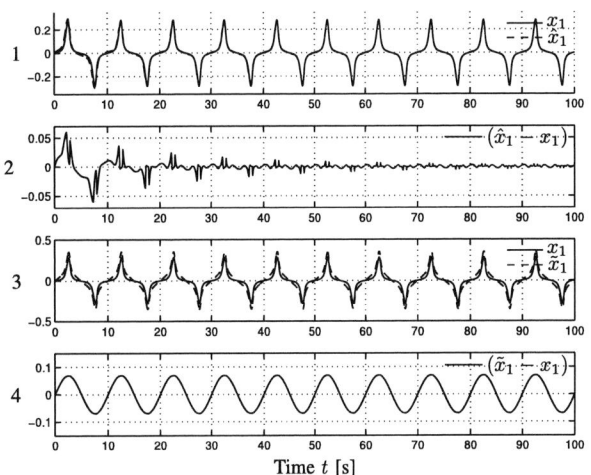

Fig. 3. Error between x_1 and $\hat{x}_1$.

whose dynamics is known due to the assumption of known linear parameters. A standard controller[1] for this system is a proportional output feedback law, tuned according to the classical magnitude optimum [17], [18]. From this well-known optimization criterion the controller

$$u = \frac{1}{2}\frac{T_1}{T_2} \cdot (y^* - y) \quad (12)$$

results. The step response of the closed loop system consisting of controller (12) and plant (11) shows a maximum overshoot of 4% and the rise time $4.7 \cdot T_2$ to arrive at the desired value for the first time. It provides steady state accuracy if no disturbances are applied. A constant disturbance signal z causes for $t \to \infty$ the deviation

$$\Delta = 2\frac{T_2}{T_1} \cdot z . \quad (13)$$

For feed forward compensation, a GRNN uses the identified, constant parameter vector $\widehat{\underline{\Theta}}$ to calculate $\widehat{\mathcal{NL}}(x_1)$ according to Eq. (2). With $\hat{f}(t) = \widehat{\mathcal{NL}}(x_1(t))$ an estimation for the current friction torque $f(t)$ is available.

---

[1]We use a P-Controller for purposes of comparison with Funnel-Control. Commonly for plant (11) a PI-Controller according to the symmetric optimum is used.

1189

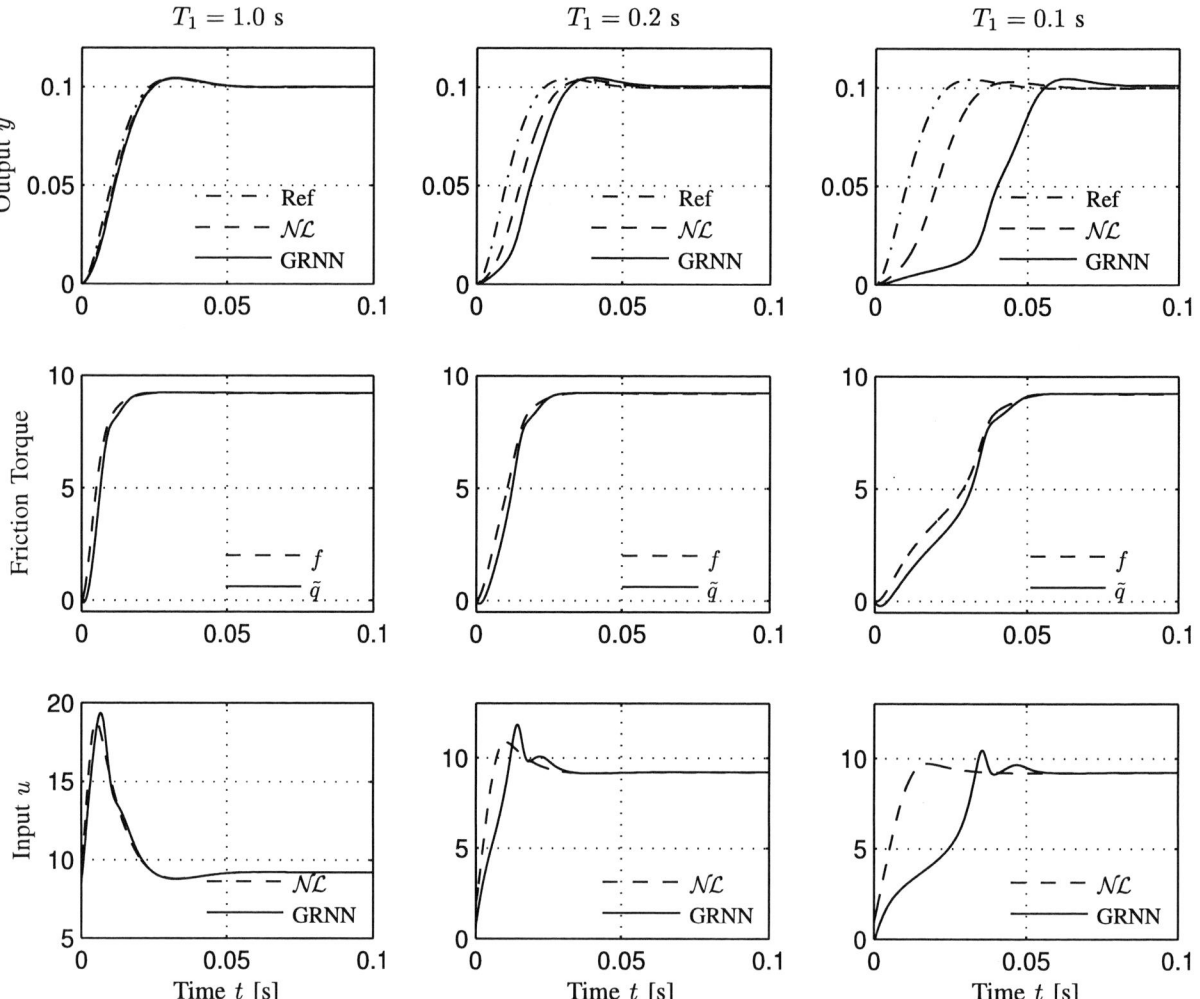

Fig. 4. Step response of the control loop with different compensation of the nonlinear friction characteristic.

Because this estimation is not injectable at the point where the real friction acts but at the plant's input only, the part of the system between input and point of influence must be taken into account. For this the linear filter

$$F_F(s) = \frac{T_2 s + 1}{T_3 s + 1} \quad \bullet\!\!\stackrel{\mathcal{L}}{-}\!\!\circ \quad h_F(t) \qquad (14)$$

is introduced to generate the signal $q(t)$ from the estimated friction via the convolution

$$q(s) = F_F(s) \cdot \hat{f}(s) \quad \bullet\!\!\stackrel{\mathcal{L}}{-}\!\!\circ \quad q(t) = h_F(t) * \hat{f}(t). \quad (15)$$

The polynomial in the numerator cancels the delaying effect of the plant, i.e. the dynamics from the input u to the point of influence of the friction. The denominator is necessary to maintain causality.

The compensation of the nonlinear behaviour is done by adding the signal q to the actuating variable, calculated in Eq. (12), such that the overall control law is given by

$$u = \frac{1}{2}\frac{T_1}{T_2} \cdot (y^* - y) + q. \qquad (16)$$

The addition of q in Eq. (16) has the identical effect as

adding the signal

$$\tilde{q}(s) = \frac{1}{T_2 s + 1} q(s) = \frac{1}{T_3 s + 1}\hat{f}(s) \qquad (17)$$

at the point of influence directly. From Eq. (17) the deteriorating effect of T_3 is evident: even if the GRNN provides a perfect estimation, i.e. $\hat{f}(s) \equiv f(s)$, the compensation signal is delayed due to T_3. This impedes complete suppression of the nonlinear behaviour. To keep the influence of T_3 marginal this constant must be chosen "small enough", i.e. fast in proportion to the dynamics of the control loop. In the simulation the value $T_3 = 1$ ms is used.

Consider the closed loop system – consisting of compensated plant and controller – as a linear system subject to the disturbance $\tilde{q}(t) - f(t)$. With $\hat{f}(s) \equiv f(s)$ (perfect approximation) this perturbation is caused by the delay of the filter exclusively and invoking Eq. (17) we get the following transfer function from the friction to the output:

$$\frac{y(s)}{f(s)} = -\frac{1}{T_1} \cdot \frac{2T_2(T_2 s + 1)}{2T_2 s(T_2 s + 1) + 1} \cdot \frac{T_3 s}{T_3 s + 1} \qquad (18)$$

Apparently the influence of the delaying effect is weighted

TABLE I

VALUES FOR T_1 USED IN THE SIMULATION

T_1	=	1.0 s	0.2 s	0.1 s
$2T_2/T_1$	=	0.01	0.05	0.1

with the factor $1/T_1$. From this, a plant with a large T_1 tolerates more delay. The delay gets more influence if T_1 is small.

Additionally, the inherent approximation error has to be taken into account. Hence, Eq. (2) gives not the exact value of the nonlinearity, but an approximation instead. Because of the inaccuracy $\hat{f}(t) \approx f(t)$ a (small) deviation disturbs the linear control loop.

The situation is analyzed in Fig. 4 for three different values of T_1, given in Tab. I. Pursuant to the magnitude optimum the associated gain of the controller depends on T_1. From Eq. (12) and (13) follows that the amplification of a perturbation is equal to the reciprocal controller gain. Consequently, the resulting deterioration due to an inaccurate compensation highly depends on the ratio T_2/T_1, which is contained in Tab. I as well. The three diagrams in the upper row of Fig. 4 show the output $y = x_1$ of the closed loop system if the constant reference value $y^* = 0.1$ is used. The dash-dot line (labelled with "Ref") marks the behaviour of the linear plant (11) governed by the controller (12) and is according to the magnitude optimum. We regard this as the desired reference behaviour.

The dashed line (labelled with "$\mathcal{NL}$") belongs to the nonlinear plant with friction from Eq. (10), governed by the controller (16). The compensation is not based on the estimation of the friction $\hat{f}(t)$, but on the correct value of the friction $f(t)$ instead. From this, no approximation error is existent in this case. The difference to the reference behaviour is caused by the delaying effect of the filter (14) exclusively.

In contrast to this, the solid line (labelled with "GRNN") gives the case if the compensation is done with the approximation $\hat{f}(t)$ generated by the trained GRNN. Consequently, the time delay of the filtered signal together with the inaccuracy of the estimation degrade the quality of the compensation.

The second row of Fig. 4 shows the development of the real friction torque f (dashed line) as well as the compensating signal $\tilde{q}$ (solid line), computed via Eq. (17), whereas $\hat{f}$ is the output of the GRNN.

In the third row the actuating variable u is depicted, which results if the compensation is based on $f(t)$ ("$\mathcal{NL}$") and on $\hat{f}(t)$ ("GRNN").

The simulation verifies that the quality of the controller is highly related to the mass moment of inertia. A large inertia ($T_1 = 1.0$ s) yields a controller with the ability to suppress perturbations very well. Tab. I points out that only 1% of the inaccuracy deteriorate the output. By virtue of a good identification result the compensation signal and the real friction coincide nearly. In the steady state we have the disturbance $\tilde{q} - f \approx 0.02$, which

causes the negligible deviation $\Delta \approx 0.0002$ (0.2% of the reference value). In this case the estimation error is not of interest. The time delay due to the filter is acceptable, such that the chosen control strategy gives the desired result.

A reduction of the mass moment of inertia or T_1, respectively augments the influence of inaccuracies. The second column in Fig. 4 reveals notable differences between the three curves "Ref", "$\mathcal{NL}$" and "GRNN" if $T_1 = 0.2$ s. The reference behaviour (dash-dot line) remains unchanged. The compensation without approximation error yields steady state accuracy, because for $t \to \infty$ the filter is without influence. But the control loop reacts more sensitive on the delay due to T_3. From Eq. (18) the reduction of T_1 increases the gain of the disturbance transfer function such that the delay of the output is increased also. The case "GRNN" additionally is subject to approximation inaccuracies, which have with 5% more impact (see Tab. I).

The sensitivity to the filter increases again if T_1 is reduced to $T_1 = 0.1$ s. If the estimation is perfectly no steady state error occurs but the output is delayed significantly (curve "$\mathcal{NL}$"). The estimation error additionally deteriorates the output. Tab. I displays that the resulting steady state deviation is given by 10% of the inaccuracy. The latter is about 0.014, such that a steady state error of 1.4% of the reference value occurs. However as the curve "GRNN" in Fig. 4 reveals the arising delay is unacceptable. For this reason the compensation achieved is not satisfying in this case, although the result of the identification seems quite good. But especially for slow speeds the approximation is not perfectly (see Fig. 2). As a result the compensation signal differs to much from the real friction torque and the compensated plant does not react like the linear plant without friction. Because the value of $\tilde{q}$ is not high enough to cancel out the friction totally the mass accelerates not as fast as desired.

IV. APPRAISAL OF IDENTIFICATION AND COMPENSATION

In the previous sections the theoretical approach to identify nonlinearities, to design a nonlinear observer and to control the plant in a classical way is discussed. The described intelligent observer is based on the engineering approach and requires therefore the definite knowledge of the location of the nonlinearity and of the linear parameters. Because the observer employs an identification information about the nonlinearity is available after the learning period. Usually the identification process takes long time, what depends on the nonlinearity and the excitation of the plant.

Together with the time-consuming identification the most severe drawbacks of intelligent control are the known linear parameters and the sensitivity to disturbances. At this point we have to distinguish two different perturbations. On the one hand the perturbations, which act on the controller (after the identification process) and are considered in Fig. 4. On the other hand, perturbations may be present during the identification process itself. (The identification performed in this work is done without external disturbances.) Simulations reveal that even small disturbances may deteriorate the result of the identification seriously. A straightforward solution of this problem is to implement a disturbance model. Such a model is realized by a homogeneous differential equation with an appropriate initial condition, whose output gives an estimate for the disturbance signal. The plant is augmented by the disturbance model, such that the identification algorithm does not tune the unknown parameters of the plant only, but the parameters of the disturbance model as well. In other words, the disturbance signal is represented. For this reason, the identification is able to decide whether the error comes from an unsatisfying estimation of the nonlinearity or is caused by the disturbance signal.

In general, disturbance models are restricted to periodic disturbances, where the order of the signal must be known. However, this prerequisite is a very hard one and cannot be tolerated in practical applications, where for example measurement noise is present.

To circumvent the disadvantages of identification strategies Sec. V presents a more sophisticated method for controlling nonlinear plants with unknown parameters.

V. FUNNEL-CONTROL

A more elegant approach abandons the identification completely. It is a well-known fact that high-gain based controllers go without any knowledge about nonlinearities or the values of the plant parameters. Only structural properties such as known sign of the high frequency gain, relative degree and stability of the zero dynamics must be satisfied. From this point of view high-gain based controllers look like an appealing alternative.

In theory, high gain controllers work very well. The control engineer however finds many reasons why the gain of the controller should be reduced if possible. In [2] funnel-control is presented, which is based on high-gain-feedback control and employs a time-varying proportional gain instead of identified system information. The gain depends on the current behaviour of the closed loop and is not a monotone function. It is possible to control (nonlinear) plants with unknown and/or time-varying parameters. This concept allows to define the transient as well as the asymptotic characteristics of the plant. A decreasing function (funnel) describes the tolerable limit for the absolute value of the control error. The control law forces the control error to evolve inside this funnel. The consideration of the transient behaviour is of high importance in practical applications because constraints for the system outputs must be kept during the transient period as well as asymptotically. The concept is "non-identifier-based", because no identification algorithm is necessary for the calculation of the actuating variable u.

For the definition of the admissible upper limit for the control error e a bounded, continuous and positive function $t \mapsto \psi(t)$ is needed. Its reciprocal value

$$\partial F_\psi(t) := \frac{1}{\psi(t)} \qquad (19)$$

is used to fix the funnel boundary by the functions $\pm \partial F_\psi(t)$. A feasible example is the function

$$\partial F_\psi(t) = a_F \cdot \exp\{-\lambda_F t^2\} + \Delta_F. \qquad (20)$$

The parameters a_F and Δ_F must be chosen such that the initial error $e(t=0)$ is enclosed by the funnel. With λ_F and Δ_F the constriction of the funnel and the maximum for the steady state error are given. The absolute value of the error evolves inside the funnel for all $t \geq 0$, i.e.

$$|e(t)| < \partial F_\psi(t) = 1/\psi(t) \quad \Leftrightarrow \quad \psi(t) \cdot |e(t)| < 1. \quad (21)$$

The funnel itself is defined as the set

$$F_\psi : t \to \{e \in \mathbb{R} \quad | \quad \psi(t) \cdot |e(t)| < 1\} \qquad (22)$$

and is marked by the grey shape in Fig. 5 which illustrates

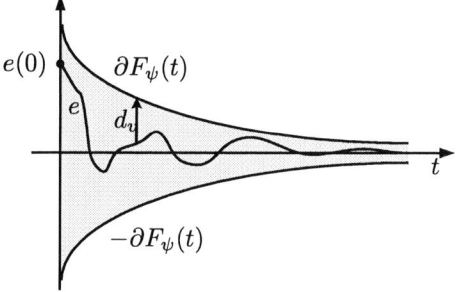

Fig. 5. Constriction of the control error e by the funnel-function $\pm \partial F_\psi(t)$.

the idea of funnel-control. The vertical distance

$$d_v(t) = \partial F_\psi(t) - |e(t)| \qquad (23)$$

between the current value of the control error and the funnel-boundary directly gives the gain $\theta(t)$ by $\theta(t) = 1/d_v(t)$. Using this gain function the time-varying, proportional control law

$$u(t) = \theta(t) \, e(t) \qquad (24)$$

can be derived. In [2] a proof is given that e is bounded by the funnel and that a stable closed-loop system results. The prerequisite for this proof is that the plant belongs to class $\mathcal{P}$, defined in Sec. V-A.

The intuition behind this concept is as follows: If the control error approaches the funnel boundary (i.e. the distance d_v gets small) a large gain results. Together with the underlying high-gain-property of the system class $\mathcal{P}$ the closed loop gets stabilized which in turn prevents the error from leaving the funnel. On the other hand, a small error corresponds to a large distance, such that a reduced gain and a more relaxed control action results.

To keep the gain bounded the vertical distance must never be zero. Hence, the funnel-function clearly must be strictly positive ($\partial F_\psi(t) > 0 \; \forall t > 0$), what is equivalent to the toleration of a (small) steady state error. In theory the funnel may constrict arbitrarily small forcing the remaining control error to be approximately zero. But in applications with measurement noise the funnel must maintain a minimum width to enclose the noisy signal.

1192

From this a (unacceptable) steady state error may result. A remedy for this problem is presented in [12] and [13], respectively: the extension of funnel-control with an integrating control action improves the performance and enables its application therefore. It can be shown that the closed loop stability is maintained and that a funnel with large width is applicable and simultaneously the error converges to zero nevertheless.

A. System Class $\mathcal{P}$

First we define an operator class $\mathcal{Z}$ with the following properties: A operator Z from class $\mathcal{Z}$ maps a continuous function into the space of measurable, locally essentially bounded functions. It describes a dynamical system which is bounded-input-bounded-output stable. The operator class is restricted to systems meeting the condition of causality. Therefore, the output depends on the past input-signal only and is completely independent of the future input-signal. This is fulfilled automatically and is no restriction for "real-world-systems". Z is locally Lipschitz, i.e. the deviation of two output-signals generated by two different input-signals is limited by the deviation of the input-signals. With such an operator we are in a position to define the system class $\mathcal{P}$:

A system with input u and output y, governed by the nonlinear functional differential equation

$$\dot{y} = f((Zy)(t), p(t), u(t)) \qquad (25)$$

belongs to the system class $\mathcal{P}$ – for which a funnel-controller is applicable – if the following properties are fulfilled:

(i) The (nonlinear) function $f(\cdot, \cdot, \cdot)$ is continuous.
(ii) The operator Z belongs to the operator class $\mathcal{Z}$.
(iii) The (external) disturbance $p(t)$ is bounded.
(iv) For nonlinear systems, the generalization[2] of the positive high-frequency-gain concept in linear systems holds true.

The nonlinear operator $(Zy)(t)$ describes the zero dynamics[3] of the plant. Stability of the operator Z is equivalent to the minimum-phase property in the linear case, i.e. to the stability of the zero dynamics in general.

From Eq. (25), the input affects the first derivative of the output directly, which means the relative degree of this system is strictly one. This is a hard restriction for the applicability of funnel-control. The work in [10], [11] and [12] provides a solution for this problem, especially in the field of mechatronics. For a flexible two mass system a simulation study is performed in [13], which is a precursor of [3]. The latter shows the applicability of funnel-control and contains measurement results, whereas this control strategy is used to control a flexible two mass system.

---

[2] It holds for every sequence $(u_n) \subset \mathbb{R}\backslash\{0\}$: $\|u_n\| \to \infty$ as $n \to \infty$
$\Rightarrow \min_{(v,w)} \frac{\langle u_n, f(v,w,u_n) \rangle}{\|u_n\|} \to \infty$ as $n \to \infty$

[3] For a definition of the zero dynamics see Isidori [4].

B. Simulation Example

For purposes of comparison the plant (10) from Example II-D is used again for a simulation. With the bounding function $\partial F_\psi(t) = 11 \cdot \exp\{-3000 \cdot t^2\} + 0.05$ the funnel-controller calculates the time varying gain according to $\theta(t) = 1/(\partial F_\psi(t) - |e(t)|)$. The input is generated by Eq. (24), whereas the control error is calculated from $e = -[100, \ 0.01]\underline{x} + 100y^*$.

In contrast to the engineering approach, which requires perfectly known linear parameters, funnel-control is applicable even if all parameters are unknown. Although the structure of the plant is given as the only information, the controller yields a similar steady state error and an improved transient behaviour compared to the classical approach with the intelligent observer. The rise time depends on the funnel boundary and is chosen similar to the reference behaviour of the magnitude optimum, which again is marked by the dash-dot line ("Ref") in Fig. 6. With funnel-control ("Funnel") no overshoot occurs, the output is well damped. The steady state error is approximately 1.4% of the reference value independent of T_1. This value is situated in the range of the classical method, which shows errors from 0.2% to 1.4%. The sensitivity of the control loop to the parameter T_1 is very low, all three experiments with the three different T_1 from Tab. I give nearly identical results. Hence the range for T_1 with a satisfying disturbance rejection and fast rise time is much improved. The lower row of Fig. 6 contains the input u generated by funnel-control. The comparison with Fig. 4 reveals that both strategies need approximately the same amount of input power.

Because no identification takes place the undesirable condition "persistency of excitation" and the time consuming learning period are not longer necessary. Furthermore, the absence of the learning process enhances the robustness against disturbances. The comparison of both strategies shows clearly that identification is superfluous in this case.

VI. CONCLUSION

This paper presents and compares two different control methods for mechatronic systems. On the one hand, an intelligent observer with a neural network is used to identify the characteristic of an unknown nonlinearity. The disadvantage of this approach is the assumption of a known linear part of the plant. The adaptation law is derived from Lyapunov's stability theory and guarantees stable learning and parameter convergence. Although the learning process is time consuming in general this approach is feasible for some practical applications [1], [5], [15], [16].

To overcome the restriction of known linear parameters a non-identifier-based controller is presented. The main benefit is that structural properties are sufficient to obtain an improved performance during the transient period and the same good steady state error. In other words, knowledge about the parameters or the nonlinearity is not necessary. Consequently, identification strategies are

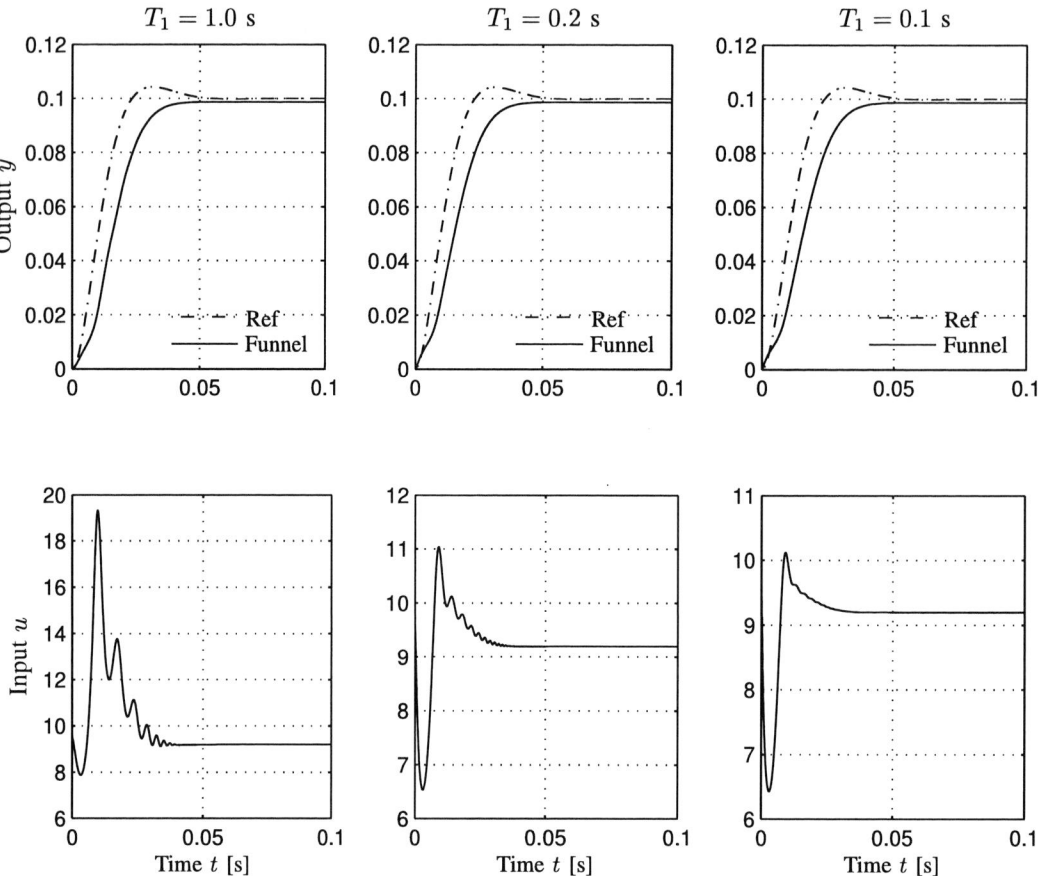

Fig. 6. Step response of the closed loop system governed by funnel control.

not required, which simplifies the controller considerably. The funnel-controller compares favourably, particularly if the robustness to disturbances/ parameters and the long learning period of the neural network is taken into evaluation.

REFERENCES

[1] FRENZ T., SCHRÖDER D.: *On Line Identification and Compensation of Friction Influence of Feed Drives of Machine Tools.* Electronics and Applications (EPE) '97, Trondheim, Norway, Vol. 3, Proceedings pp. 3.927-3.932, 1997.

[2] ILCHMANN, A., RYAN, E.P., SANGWIN, C.J.: *Tracking with Prescribed Transient Behaviour.* ESAIM: Control, Optimisation and Calculus of Variations, Vol. 7, pp. 471-493, 2002.

[3] ILCHMANN, A., SCHUSTER, H.: *PI-funnel control for two mass systems.* unpublished.

[4] ISIDORI, A.: *Nonlinear Control Systems.* Springer–Verlag, London, 2001.

[5] LENZ U., SCHRÖDER D.: *Artificial Intelligence for Combustion Engine Control.* SAE International Congress and Exposition, Detroit, USA, SAE Technical Paper No. 960328, 1996.

[6] MONOPOLI, R.V.: *Model reference adaptive control with an augmented error signal.* IEEE Transactions on Automatic Control, AC-19:474-484, 1974.

[7] MOORE, J. B., HOROWITZ, R., MESSNER, W.: *Functional Persistence of Excitation and Observability for Learning Control Systems.* ASME Journal of Dynamic System Measurements and Control, 114(3), pp. 501-507, 1992.

[8] NARENDRA, K. S., ANNASWAMY, A. M.: *Stable Adaptive Systems.* Prentice Hall, Englewood Cliffs, New Jersey, 1989.

[9] SCHRÖDER, D. (ED.): *Intelligent Observer and Control Design for Nonlinear Systems.* Springer–Verlag, Berlin, 2000.

[10] SCHUSTER H., WESTERMAIER C., SCHRÖDER D.: *High-Gain Control of Systems with Arbitrary Relative Degree: Speed Control for a Two Mass Flexible Servo System.* Proc. of the 8th IEEE Int. Conf. on Intelligent Engineering Systems, INES, Cluj-Napoca, Romania, Sept., 2004.

[11] SCHUSTER H., HACKL, C., WESTERMAIER C., SCHRÖDER D.: *Funnel-Control for Electrical Drives with Uncertain Parameters.* Proc. of the 7th IEEE Int. Power Engineering Conf., IPEC, Singapore, Dec., 2005.

[12] SCHUSTER H., WESTERMAIER C., SCHRÖDER D.: *Non-Identifier-Based Adaptive Speed Control for a Two-Mass Flexible Servo System: Consideration of Stability and Steady State Accuracy.* Proc. of the 14th Mediterranean Conf. on Control and Automation, MED, Ancona, Italy, June, 2006.

[13] SCHUSTER, H., WESTERMAIER, C., SCHRÖDER, D.: *Non-Identifier-Based Adaptive Control for a Mechatronic System Achieving Stability and Steady State Accuracy.* Proceedings of the 2006 IEEE Conf. on Control Applications CCA, Munich, Germany, pp. 1819-1824, Oct. 2006.

[14] SPECHT, D.: *A General Regression Neural Network.* IEEE Transactions on Neural Networks, Vol 2, pp. 568-576, November, 1991.

[15] STRAUB, S.: *Entwurf und Validierung neuronaler Beobachter zur Regelung nichtlinearer dynamischer Systeme im Umfeld antriebsstechnischer Problemstellungen.* Dissertation, Lehrstuhl für Elektrische Antriebssysteme, TU München, 1998.

[16] STROBL D., SCHRÖDER D.: *Neural Observers for the Identification of Backlash in Electromechanical Systems.* IFAC Workshop on Motion Control, Grenoble, France, 1998.

[17] UMLAND, J., SAFIUDDIN, M.: *Magnitude and Symmetric Optimum Criterion for the Design of Linear Control Systems: What Is It and How Does It Compare with the Others?* IEEE Trans. on Industry Applications, Vol. 26, No. 3, pp. 489-497, 1990.

[18] VRANČIĆ, D., STRMČNIK, S., JURIČIĆ, D.: *A magnitude optimum multiple integration tuning method for filtered PID controller.* Automatica, Vol. 37, No. 9, pp. 1473-1479, 2001.

New Space Vector Control Approach for Four Switch Three Phase Inverter (FSTPI)

Phan Quoc Dzung*, Le Minh Phuong*, Pham Quang Vinh**,

Nguyen Minh Hoang***, Tran Cong Binh*

* Faculty of Electrical & Electronic Engineering, HCMC University of Technology, Ho Chi Minh City, Vietnam
** Siemens AG Presentation, Ho Chi Minh City, Vietnam
*** NARC, Ulsan University, Korea

Abstract–This paper is to present a space vector PWM algorithm for four switch three phase inverters (B4, FSTPI) based on the one for six switch three phase inverters (B6, SSTPI) (principle of similarity) where the αβ plan is divided into 6 sectors and the formation of the required reference voltage space vector is done in the same way as for B6 by using effective (mean) vectors. This facilitates the calculation for B4 and some studies on B6 can be applied for B4 as well through this proposed similarity, e.g. the problem with PWM in the overmodulation zone due to the complicated non-linear character there. Matlab/ Simulink is used for the simulation of the proposed SVPWM algorithm for undermodulation, overmodulation mode 1 and 2. This SVPWM approach is also validated experimentally using control board TMS eZDSP2812 (Texas Instrument Co.).

Index Terms-- B4 (Four Switch Three Phase Inverter), B6 (Six Switch Three Phase Inverter), Low-cost inverter, overmodulation, Pulse-Width- Modulation, Space vector, undermodulation.

I. INTRODUCTION

Three phase variable speed drives for asynchronous motors have been used more and more, especially in energy saving drive applications for fans, pumps, air compressors... In many cases, the cost reduction is an important target for the drive. Hence, a reduced number of inverter switches is a promising solution. Several inverter schemes with reduced number of switches have been proposed. Among them the four switch three phase inverter (B4) (fig.1) was introduced with four IGBT switches instead of six (B6) in a conventional three phase inverters [1-8, 10].

In spite of the B4's drawbacks like a higher DC side capacitor voltage and unsymmetrical scheme exposed to the unbalanced capacitor voltage, this inverter has the following advantages over B6 [2]:
- The number of switches is reduced by a third; driving circuits are only two as only two branches are controlled.
- In spite of the switch's higher withstand-able voltage in B4 the cost is still lower thanks to the price ratio of B4 to B6 usually lower than 3/2.
- B4's maximum common mode voltage is just 2/3 of B6's.

This work was supported by Vietnamese Ministry of Science and Technology (MOST).

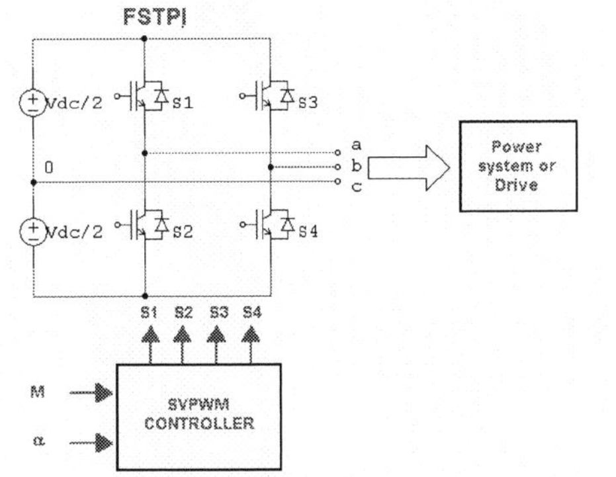

Fig. 1. Four switch three phase inverter (FSTPI)

Despite rather comprehensive study SVPWM's, as proposed by the researchers and based on the formation of the required voltage vectors from the 4 base vectors [2-5, 7, 8, 10], have not yet set up the link with B6's SVPWM, therefore the results well known for B6 are not exploited.

The content of this paper is aimed at presenting a SVPWM for B4 inverter (FSTP) modeled on the basis of a B6 by using the principle of similarity and revealing perspective solution for the PWM in the zone of overmodulation when the PWM is quite complicated due to the nonlinear character of modulation in this extended zone. This issue has not been approached in the above mentioned papers.

II. ANALYSIS OF SPACE VOLTAGE VECTORS AND STATOR FLUX

According to the scheme in fig.1 the switching status is represented by binary variables S1 to S4, which are set to "1" when the switch is closed and "0" when open. In addition the switches in one inverter branch are controlled complementary (1 on, 1 off), therefore:

$S_1 + S_2 = 1$
$S_3 + S_4 = 1$ (1)

Phase to common point voltage depends on the turning off signal for the switch:

$$V_{a0} = (2S_1 - 1) \cdot \frac{V_{dc}}{2}; V_{b0} = (2S_3 - 1) \cdot \frac{V_{dc}}{2}; V_{c0} = 0; \quad (2)$$

Combinations of switching S_1-S_4 result in 4 general space vectors $\vec{V}_1 \rightarrow \vec{V}_4$ (Table 1), components αβ of the voltage vectors are gained from abc voltages by using Clark's transformation:

$$\begin{bmatrix} V_\alpha \\ V_\beta \end{bmatrix} = \frac{2}{3} \begin{bmatrix} 1 & -\frac{1}{2} & -\frac{1}{2} \\ 0 & \frac{\sqrt{3}}{2} & -\frac{\sqrt{3}}{2} \end{bmatrix} \begin{bmatrix} V_a \\ V_b \\ V_c \end{bmatrix} \tag{3}$$

where V_a, V_b, V_c : phase voltages on the load (Y connection), defined by:

$$V_a = \frac{1}{3}\left(2V_{a0} - V_{b0}\right) ;$$

$$V_b = \frac{1}{3}\left(2V_{b0} - V_{a0}\right);$$

$$V_c = -\frac{1}{3}\left(V_{a0} + V_{b0}\right) \tag{4}$$

In order to form the required voltage space vector $\vec{V}_{ref}$, we can use 3 or 4 vectors in one sampling interval T_s. The constant value 0 (zero) vectors can be formed by dividing t_0 (duration of zero vector) among 2 opposite vectors $(\vec{V}_1, \vec{V}_3)$ or $(\vec{V}_2, \vec{V}_4)$ [2, 7].

For three phase induction motors the stator flux linkage vector can be represented as follows [2, 10]:

$$\vec{\Psi} = \int \vec{V}(t)dt \tag{5}$$

In case of the sinusoidal voltage source the standard stator flux linkage vector can be defined:

$$\vec{\Psi}^* = \frac{V_m}{\omega} e^{j\left(\omega t - \frac{\pi}{2}\right)} \tag{6}$$

Thus the flux linkage vector locus will describe a circle with the radius $r = \dfrac{V_m}{\omega}$.

TABLE 1
COMBINATIONS OF SWITCHINGS AND VOLTAGE SPACE VECTORS

S1	S3	$\vec{V} = V_\alpha + jV_\beta$
0	0	$\vec{V}_1 = \dfrac{V_{dc}}{3} e^{-j\frac{2\pi}{3}}$
1	0	$\vec{V}_2 = \dfrac{2V_{dc}}{3} e^{-j\frac{\pi}{6}}$
1	1	$\vec{V}_3 = \dfrac{V_{dc}}{3} e^{j\frac{\pi}{3}}$
0	1	$\vec{V}_4 = \dfrac{2V_{dc}}{3} e^{j\frac{5\pi}{6}}$

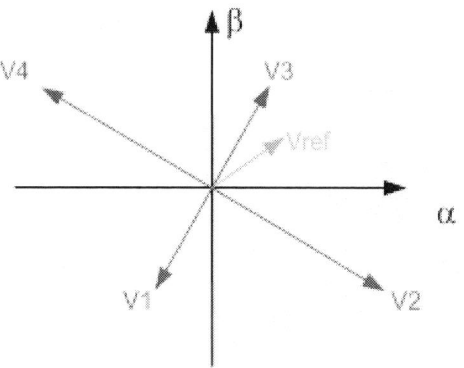

Fig. 2. Voltage space vectors in the plan αβ

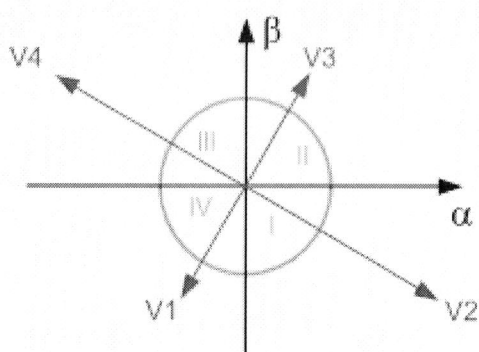

Fig. 3. Sectors used in conventional SVPWM methods for B4

In case the motor is fed from a B4 inverter the flux linkage vector is:

$$\vec{\Psi} = t_n \cdot \vec{V}_n + \vec{\Psi}_0 \tag{7}$$

where n = 1..4 ; t_n : duration of V_n.

If the switching algorithms can ensure the best approximation by minimizing the discrepancy between vector loci $\vec{\Psi}$ and $\vec{\Psi}^*$, the stator voltage performance will be optimized. Thus the design of the algorithm for PWM samples in the inverter phases based on this rule has a very important role. This approach is used with success for B6 inverters.

The performance index used for evaluating the output voltage is the Total Harmonic Distortion factor THD%, defined as follows [2]:

$$THD\% = \frac{\sqrt{\sum_{n=2,3,...} \left(\frac{V_n}{n}\right)^2}}{V_1} \cdot 100\% \tag{8}$$

where n : harmonic number.

III. NEW SVPWM APPROACH FOR B4 INVERTER

SVPWM methods presented in papers [2, 3, 7, 8, 10] are based on the formation of the reference vector on the plan αβ which is divided into four sectors (sector I...IV). The active vectors and their duration in one sampling interval are selected and calculated on the basis of the required $\mathbf{V_{ref}}$ location respective for these sectors (fig.3).

SVPWM method proposed in this paper is based on the principle of similarity of the one for B6 inverters, where plan αβ is divided into 6 sectors and the formation of $\mathbf{V_{ref}}$ is done similarly as for B6. This facilitates the calculation for B4 and some issues for B6 can be applied for B4 thanks to this proposed approach.

To simulate 6 non-zero vectors in B6, in this proposed method, beside the two $\mathbf{V_1}$ and $\mathbf{V_3}$, we use the effective vectors $\mathbf{V_{23M}}$, $\mathbf{V_{34M}}$, $\mathbf{V_{41M}}$ and $\mathbf{V_{12M}}$. These vectors are formed as follows:

$$\vec{V}_{23M} = \frac{1}{2}\left(\vec{V}_2 + \vec{V}_3\right) = \frac{V_{dc}}{3} e^{j0}; \vec{V}_{34M} = \frac{1}{2}\left(\vec{V}_3 + \vec{V}_4\right) = \frac{V_{dc}}{3} e^{j\frac{2\pi}{3}}; \tag{9}$$

$$\vec{V}_{41M} = \frac{1}{2}\left(\vec{V}_4 + \vec{V}_1\right) = \frac{V_{dc}}{3} e^{j\pi}; \vec{V}_{12M} = \frac{1}{2}\left(\vec{V}_1 + \vec{V}_2\right) = \frac{V_{dc}}{3} e^{-j\frac{\pi}{3}}$$

To simulate zero vectors of B6, we use the effective V_{0M}:

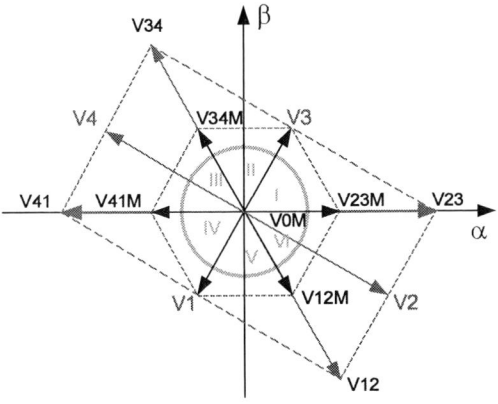

Fig. 4. SVPWM method proposed for B4 on the principle of similarity of B6.

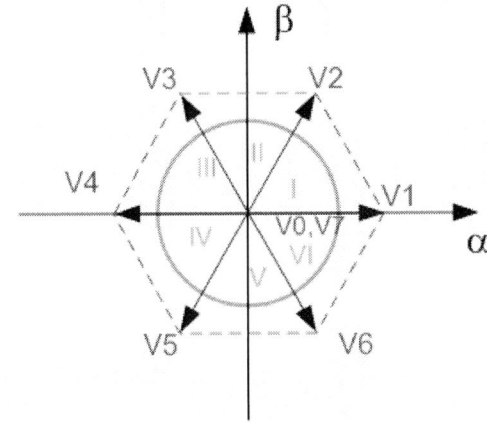

Fig. 5. Basic space vectors in B6 inverter.

TABLE 2
SIMILARITY BETWEEN SPACE VECTORS OF B4 AND B6

Base voltage space vectors of B6	Voltage space vectors of B4
V_1	V_{23M}
V_2	V_3
V_3	V_{34M}
V_4	V_{41M}
V_5	V_1
V_6	V_{12M}
V_0, V_7	V_{0M}

$$\vec{V}_{0M} = \frac{1}{2}\left(\vec{V}_1 + \vec{V}_3\right), \text{ or } \vec{V}_{0M} = \frac{1}{2}\left(\vec{V}_2 + \vec{V}_4\right); \quad (10)$$

The similarity between space vectors of B4 (Fig.4) and B6 (Fig.5) is presented in Table 2. The base vectors in each sector used to form the required space vector V_{ref} is presented in Table 3.

Below we will describe the space vector modulation for B4 inverter based on the modulation for B6 with the principle of similarity.

A. Under modulation (0 < M < 0.907)

In this zone the required voltage space vector rotates in a hexagon. The space vector modulation in this zone is based on the formation of three voltage vectors in sequence in one sampling interval T_s so that the average

output voltage meets the requirement. The calculations of the switching states in B6 and B4 are as follows for ½ T_s [9]:

$$t_x = \frac{\sqrt{3}}{\pi} MT_s \sin(\pi/3 - \alpha);$$

$$t_y = \frac{\sqrt{3}}{\pi} MT_s \sin(\alpha); \quad (11)$$

$$t_z = T_s/2 - t_x - t_y$$

where:

t_x - duration for vector V_x
t_y - duration for vector V_y
t_z - duration for vector V_z

M – the index of modulation $M = V^*/V_{1sw}$ (V^* - amplitude of the required voltage vector, V_{1sw} – peak value of six step voltage).

However in B4 inverter since mean vectors t_{XYM} and zero vectors t_{0M} are formed from the two base vectors the duration of base vectors is equal to ½ as for the above mentioned mean and zero vectors.

In order to ensure the closest following and the minimized discrepancy between $\vec{\Psi}$ and $\vec{\Psi}^*$ loci, the approach shall be done in a symmetrical way from both sides.

It can be used, for example, the effective vectors V_{23M}, V_3, V_{0M} for sector I, where V_{23M}, V_{0M} are defined as (11):

$$t_x = t_{23f} = \frac{\sqrt{3}}{\pi} MT_s \sin(\pi/3 - \alpha); t_y = t_{3f} = \frac{\sqrt{3}}{\pi} MT_s \sin(\alpha);$$

$$t_z = t_{0f} = T_s/2 - t_{23f} - t_{3f};$$

$$t_{2m} = \frac{t_{23f}}{2}; t_{3m} = \frac{t_{23f}}{2}; t_{3z} = \frac{t_{0f}}{2}; t_{1z} = \frac{t_{0f}}{2}$$

(12)

Thus the total durations for base vectors V_1, V_2, V_3 are:

$$t_{V1} = t_{1z};$$

$$t_{V2} = t_{2m}; \quad (13)$$

$$t_{V3} = t_{3f} + t_{3m} + t_{3z}$$

Similarly we can calculate the space vector modulation for the other sectors. The calculation results are shown in Table 4.

B. Overmodulation in mode 1 (0.907 ≤ M ≤ 0.952)

Similarly as for B6, this mode starts when the required V_{ref} goes beyond the circle inscribing the hexagon and reached its sides. When sliding on the hexagon side (M=0.952) the value of t_z is equal to zero:

$$t_x = \frac{\sqrt{3}\cos\alpha - \sin\alpha}{\sqrt{3}\cos\alpha + \sin\alpha} \cdot \frac{T_s}{2}; t_y = \frac{T_s}{2} - t_x; t_z = 0 \quad (14)$$

As for the undermodulation, effective vectors t_{XYM} and

TABLE 3
VECTORS USED IN THE SPACE VECTOR MODULATION B6 & B4

Sector	B6 (V_X, V_Y, V_Z)	B4 (V_X, V_Y, V_Z)
I	V_1, V_2, V_Z	V_{23M}, V_3, V_{0M}
II	V_2, V_3, V_Z	V_3, V_{34M}, V_{0M}
III	V_3, V_4, V_Z	V_{34M}, V_{41M}, V_{0M}
IV	V_4, V_5, V_Z	V_{41M}, V_1, V_{0M}
V	V_5, V_6, V_Z	V_1, V_{12M}, V_{0M}
VI	V_6, V_1, V_Z	V_{12M}, V_{23M}, V_{0M}

1197

TABLE 4
VECTOR DURATIONS IN THE PROPOSED SVPWM METHOD

Sector I	Sector II
$t_x = t_{23f} = \frac{\sqrt{3}}{\pi} MT_s \sin(\pi/3 - \alpha)$;	$t_x = t_{3f} = \frac{\sqrt{3}}{\pi} MT_s \sin(\pi/3 - \alpha)$
$t_y = t_{3f} = \frac{\sqrt{3}}{\pi} MT_s \sin(\alpha)$;	$t_y = t_{34f} = \frac{\sqrt{3}}{\pi} MT_s \sin(\alpha)$;
$t_z = t_{of} = T_s/2 - t_{23f} - t_{3f}$;	$t_z = t_{of} = T_s/2 - t_{3f} - t_{34f}$;
$t_{2m} = \frac{t_{23f}}{2}; t_{3m} = \frac{t_{23f}}{2}$;	$t_{3m} = \frac{t_{34f}}{2}; t_{4m} = \frac{t_{34f}}{2}$;
$t_{3z} = \frac{t_{of}}{2}; t_{1z} = \frac{t_{of}}{2}$	$t_{3z} = \frac{t_{of}}{2}; t_{1z} = \frac{t_{of}}{2}$
$\Rightarrow$ $t_{V1} = t_{1z}$ $t_{V2} = t_{2m}$ $t_{V3} = t_{3f} + t_{3m} + t_{3z}$	$\Rightarrow$ $t_{V1} = t_{1z}$ $t_{V4} = t_{4m}$ $t_{V3} = t_{3f} + t_{3m} + t_{3z}$

Sector III	Sector IV
$t_x = t_{34f} = \frac{\sqrt{3}}{\pi} MT_s \sin(\pi/3 - \alpha)$;	$t_x = t_{41f} = \frac{\sqrt{3}}{\pi} MT_s \sin(\pi/3 - \alpha)$;
$t_y = t_{41f} = \frac{\sqrt{3}}{\pi} MT_s \sin(\alpha)$;	$t_y = t_{1f} = \frac{\sqrt{3}}{\pi} MT_s \sin(\alpha)$;
$t_z = t_{of} = T_s/2 - t_{34f} - t_{41f}$;	$t_z = t_{of} = T_s/2 - t_{41f} - t_{1f}$;
$t_{3m} = \frac{t_{34f}}{2}; t_{4m} = \frac{t_{34f}}{2} + \frac{t_{41f}}{2}$;	$t_{4m} = \frac{t_{41f}}{2}; t_{1m} = \frac{t_{41f}}{2}$;
$t_{1m} = \frac{t_{41f}}{2}$;	$t_{3z} = \frac{t_{of}}{2}; t_{1z} = \frac{t_{of}}{2}$
$t_{3z} = \frac{t_{of}}{2}; t_{1z} = \frac{t_{of}}{2}$	
$\Rightarrow$ $t_{V1} = t_{1m} + t_{1z}$ $t_{V4} = t_{4m}$ $t_{V3} = t_{3m} + t_{3z}$	$\Rightarrow$ $t_{V1} = t_{1f} + t_{1m} + t_{1z}$ $t_{V4} = t_{4m}$ $t_{V3} = t_{3z}$

Sector V	Sector VI
$t_x = t_{1f} = \frac{\sqrt{3}}{\pi} MT_s \sin(\pi/3 - \alpha)$;	$t_x = t_{12f} = \frac{\sqrt{3}}{\pi} MT_s \sin(\pi/3 - \alpha)$;
$t_y = t_{12f} = \frac{\sqrt{3}}{\pi} MT_s \sin(\alpha)$;	$t_y = t_{23f} = \frac{\sqrt{3}}{\pi} MT_s \sin(\alpha)$;
$t_z = t_{of} = T_s/2 - t_{1f} - t_{12f}$;	$t_z = t_{of} = T_s/2 - t_{12f} - t_{23f}$;
$t_{1m} = \frac{t_{12f}}{2}; t_{2m} = \frac{t_{12f}}{2}$;	$t_{1m} = \frac{t_{12f}}{2}; t_{2m} = \frac{t_{12f}}{2} + \frac{t_{23f}}{2}$;
$t_{3z} = \frac{t_{of}}{2}; t_{1z} = \frac{t_{of}}{2}$	$t_{3m} = \frac{t_{23f}}{2}$
	$t_{3z} = \frac{t_{of}}{2}; t_{1z} = \frac{t_{of}}{2}$
$\Rightarrow$ $t_{V1} = t_{1f} + t_{1m} + t_{1z}$ $t_{V2} = t_{2m}$ $t_{V3} = t_{3z}$	$\Rightarrow$ $t_{V1} = t_{1m} + t_{1z}$ $t_{V2} = t_{2m}$ $t_{V3} = t_{3m} + t_{3z}$

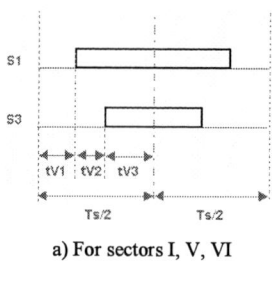

a) For sectors I, V, VI

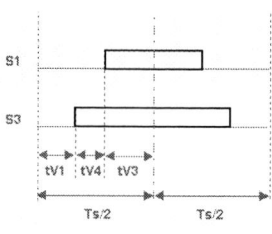

b) For sectors II, III, IV

Fig. 6. Pulse patterns for switching in the proposed method.

C. Overmodulation mode 2 ($0.952 \leq M \leq 1$)

Similarly as for B6, in this overmodulation mode 2, the required $\mathbf{V_{ref}}$ increases up to six step mode.

When M=1, the modulation is done in two cases:

$$t_x = \frac{T_s}{2}; t_y = 0; t_z = 0; \text{for } 0 \leq \alpha \leq \pi/6; \quad (15)$$
$$t_x = 0; t_y = \frac{T_s}{2}; t_z = 0; \text{for } \pi/6 \leq \alpha \leq \pi/3$$

From the above formulas we can induce that the duration of the respective base vectors to form the effective and effective-zero vectors is similar to the two above mentioned modulation cases.

When M = 0.952, values t_x, t_y, t_z are defined as (14). For $0.952 < M < 1$ the linear approximation is used to calculate t_x, t_y, t_z.

IV. SIMULATION OF THE PROPOSED SVPWM FOR B4

Matlab/ Simulink is used for the simulation of the proposed SVPWM for the undermodulation, overmodulation mode 1 and 2. DC voltage V_{dc} = 600V. Output voltage fundamental harmonic f = 50Hz. Switching frequency f_{sw} = 4.8 kHz.

1. Case study 1: For the undermodulation with M = 0.7.
   The phase voltage, line voltage waveforms and trajectory of flux space vector are shown in Fig. 7-9 respectively.

2. Case study 2: For the overmodulation mode 1 and 2.
   The phase voltage, line voltage waveforms, trajectory of flux space vector are shown in Fig. 10-12 for the value of modulation index M = 0.94 and in Fig. 13-15 for the value of modulation index M = 1. The simulation results demonstrate the excellent performance of the proposed SVPWM for B4, while the good responses of the output voltages and flux vector are obtained (fig.7–15, Table 5).

effective -zero vectors t_{0M} are formed from the two base vectors, so the duration of the base vectors is just half of the one for the mentioned vectors.

When M = 0.907, values t_x, t_y, t_z are defined as (12). In case of $0.907 < M < 0.952$ the linear approximation is used to calculate t_x, t_y, t_z.

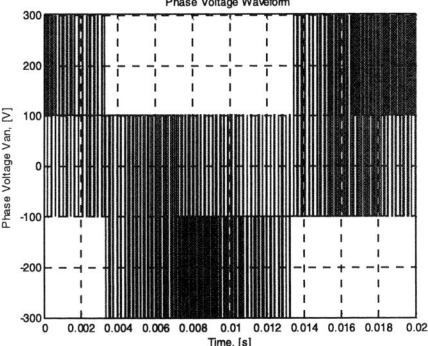

Fig. 7. Phase voltage waveform (M=0.7).

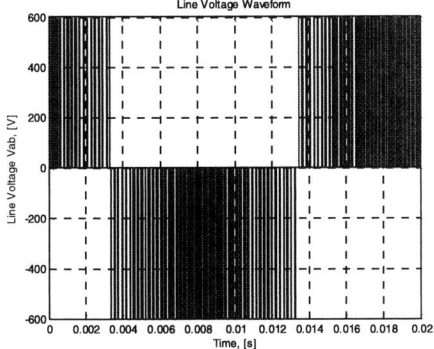

Fig. 8. Line voltage waveform (M=0.7).

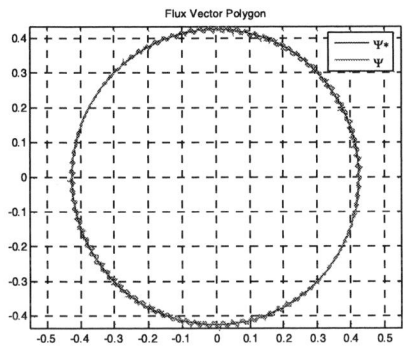

Fig. 9. Trajectory of flux space vector (M=0.7)

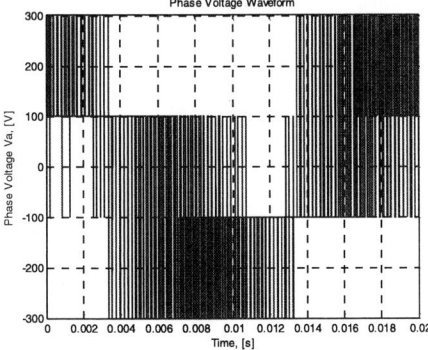

Fig. 10. Phase voltage waveform in case of overmodulation mode 1 (M=0.94).

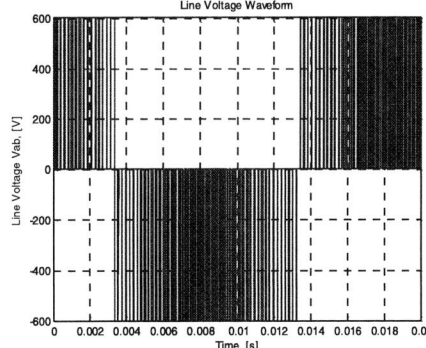

Fig. 11. Line voltage waveform for in case of overmodulation mode 1(M=0.94).

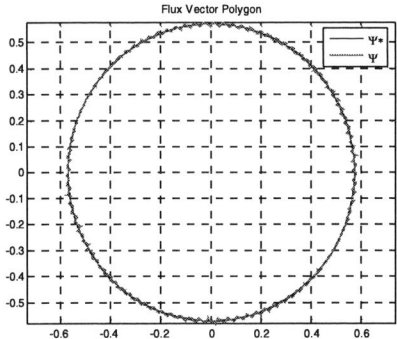

Fig. 12. Trajectory of flux space vector in case of overmodulation mode 1(M=0.94).

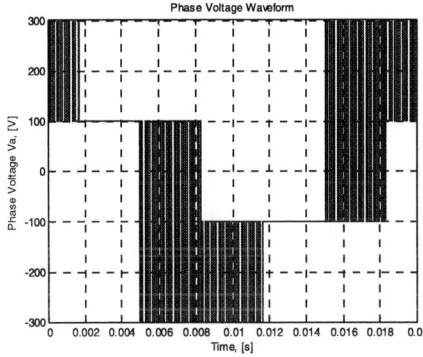

Fig. 13. Phase voltage waveform in case of overmodulation mode 2 (M=1).

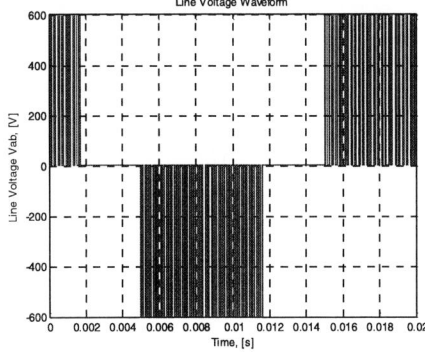

Fig. 14. Line voltage waveform in case of overmodulation mode 2 (M=1).

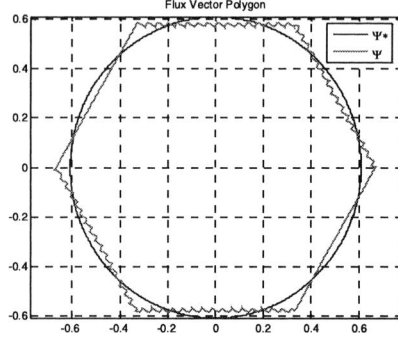

Fig. 15. Trajectory of flux space vector in case of overmodulation mode 2 (M=1).

TABLE 5

TABLE OF SIMULATION RESULTS FOR SVPWM CONTROLLER

Modulation index M	0.7	0.94	1 (six-step)
Reference voltage V_{1ref}, [V]	94.5	126.9	135
Simulated fundamental output phase voltage V_1, [V]	94.5	127.1	135
Distortion factor THD, [%] (for first 35 harmonics)			
for V_{an}	0.0155	0.5349	4.6359
for V_{bn}	0.0146	0.5359	4.6359
for V_{cn}	0.0067	0.5248	4.6353

V. EXPERIMENTS ON THE PROPOSED SVPWM FOR B4

To verify the proposed SVPWM an experiment has been set up. The SVPWM is programmed in the control board TMS eZDSP2812 to generate the command pulses for B4 (IGBT SKM 75GB 123D, Driver SKHI 22A).

The output from FSTPI was connected to a three phase induction motor, which has the follows parameters: f = 50 Hz, 400V,Y, 2HP, 3.4A, cosφ =0.81, 1420 rpm.

The switching frequency of IGBT is 5kHz. The DC link voltage was adjusted at 100V-200V, and the split capacitors are rated at 3300µF.

1. Case study 1: The fundamental harmonic of output voltages is 25Hz. Fig. 16-18 show the command pulse patterns for S_1, S_3; line voltages waveforms of V_{ab}, the phase current waveform of i_a.

2. Case study 2: The fundamental harmonic of output voltages is 50Hz. Fig. 19-21 show line voltages waveforms of V_{ab}, V_{bc}, the phase current waveform of i_a.

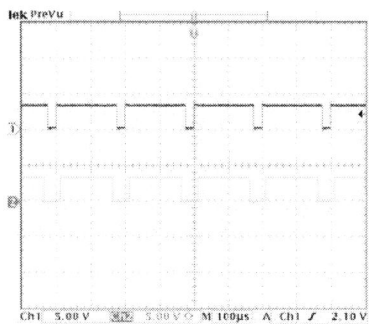

Fig. 16. Command pulse pattern for S1, S3.

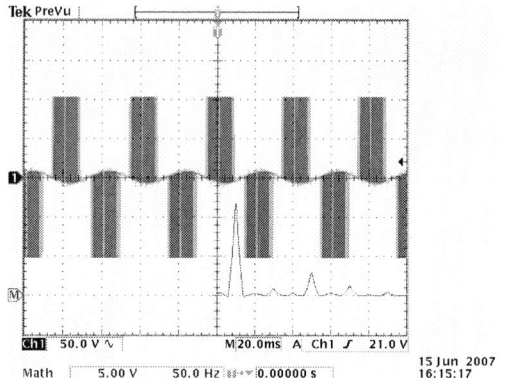

Fig. 17. Line voltage waveform and harmonic spectrum of V_{ab}.

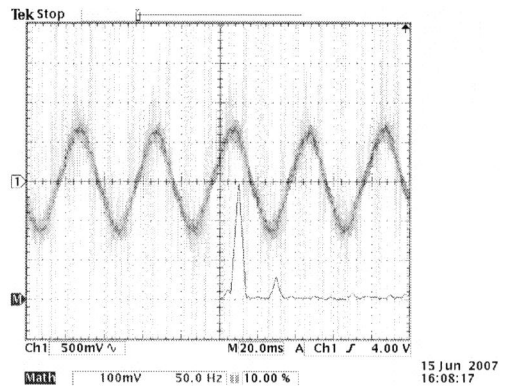

Fig. 18. Current waveform and harmonic spectrum of i_A.

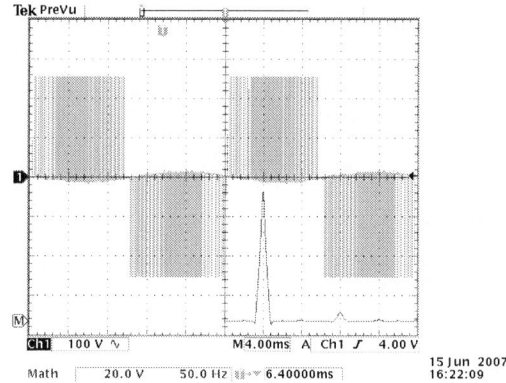

Fig. 19. Line voltage waveform and harmonic spectrum of V_{ab}.

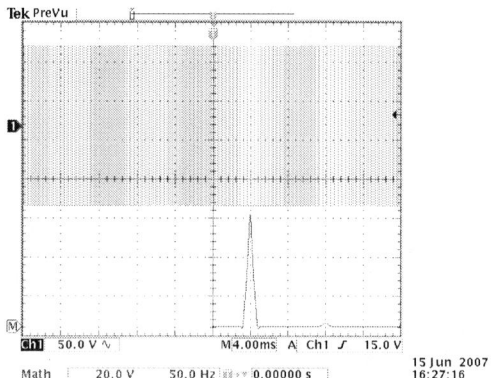

Fig. 20. Line voltage waveform and harmonic spectrum of V_{bc}

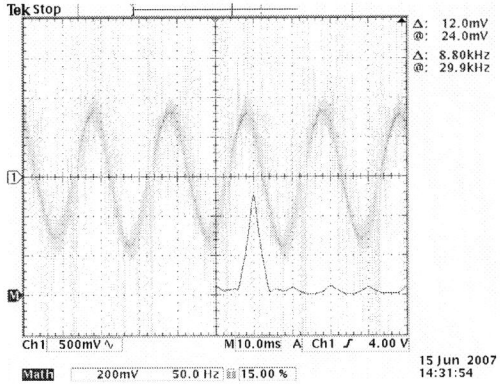

Fig. 21. Current waveform and harmonic spectrum of i_A.

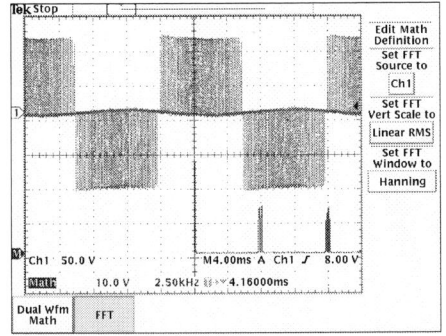

Fig. 22. Line voltage waveform and harmonic spectrum of V_{ab} (M=0.7).

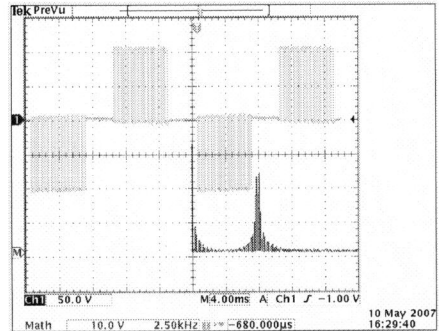

Fig. 23. Line voltage waveform V_{ab} and spectrum analysis (M=1).

3. Case study 3: The fundamental harmonic of output voltages is 50Hz. Fig. 22 shows the line voltage waveform V_{ab} and spectrum analysis for M=0.7 (undermodulation). Fig. 23 shows the line voltage waveform V_{ab} and spectrum analysis for M=1 (six-step mode).

VI. CONCLUSION

The proposed SVPWM in this paper is based on the one for six switch three phase inverters (B6, SSTP) using principle of similarity where the $\alpha\beta$ plan is divided into 6 sectors and the formation of the required reference voltage space vector is done in the same way as for B6 by using the additional effective vectors. This facilitates the SVPWM calculation for B4 and some studies on B6 can be applied for B4 as well through this proposed approach, e.g. SVPWM for the overmodulation. The implementation of the proposed SVPWM is done by simulation and in experiment to serve the practical production of the cost effective inverters in the future.

REFERENCES

[1] H. W. van der Broeck and J. D. van Wyk, "A comparative investigation of a three-phase induction machine drive with a component minimized voltage-fed inverter under different control options," *IEEE Trans. Ind. Appl.*, vol. IA-20, no. 2, pp. 309–320, Mar./Apr. 1984.

[2] Frede Blaabjerg,, Sigurdur Freysson, Hans-Henrik Hansen, and S. Hansen "A New Optimized Space-Vector Modulation Strategy for a Component-Minimized Voltage Source Inverter " *IEEE Trans. on Power Electronics*, Vol. 12, No. 4, July 1997,pp 704-710.

[3] C. B. Jacobina, E. R. C. Da Silva, A. M. N. Lima, and R. L. A Ribeiro. "Vector and scalar control of a four switch three phase inverter". In *Conf. Rec. IAS*, pages 2422-2429, 1995.

[4] F. Blaabjerg, S. Freysson, H. H. Hansen, and S. Hariseri. "Comparison of a space-vector modulation strategy for a three phase standard and a component minimized voltage source inverter". In *Conf. Rec. EPE*, pages 1806-1813, Sevilha - Spain, September 1995.

[5] G. A. Covic, G. L. Peters, and J. T. Boys, "An improved single phase to three phase converter for low cost ac motor drives," in *Proc. PEDS '95*, Singapore, vol. 1, pp. 549–554.

[6] G. T. Kim and T. A. Lipo, "VSI-PWM inverter/rectifier system with a reduced switch count," in *Proc. IAS '95*, pp. 2327–2332.

[7] M. B. R. Correa, C. B. Jacobina, E. R. C. Da Silva, and A. M. N. Lima. "A General PWM Strategy for Four-Switch Three-Phase Inverters" *IEEE Trans. on Power Electronics*, Vol. 21, No. 6, Nov. 2006, pp 1618-1627.

[8] G.I. Peters, G.A.Covic and J.T.Boys," Eliminating output distortion in four-switch inverters with three-phase loads." *IEE Proc.Electr.Power Appl..*vol.IA-34, pp.326-332,1998.

[9] J. O. P. Pinto, B. K. Bose, L. E. B. da Silva, and M. P. Kazmierkowski, "A neural network based space vector PWM controller for voltage-fed inverter induction motor drive," *IEEE Trans. Ind. Applicat.*, vol. 36, pp. 1628–1636, Nov./Dec. 2000.

[10] D. T. W. Liang and J. Li, "Flux vector modulation strategy for a fours witch three-phase inverter for motor drive applications," in *Proc. IEEE PESC*, Jun. 1997, pp. 612–617.

The Development of Artificial Neural Network Space Vector PWM for Four-Switch Three-Phase Inverter

Phan Quoc Dzung*, Le Minh Phuong*, Pham Quang Vinh**,

* Faculty of Electrical & Electronic Engineering, HCMC University of Technology, Ho Chi Minh City, Vietnam
** Siemens AG Presentation, Ho Chi Minh City, Vietnam

Abstract–**This paper presents the development of neural-network-based controller of space vector modulation (ANN-SVPWM) for low-cost voltage-source inverters. This ANN-SVPWM controller completely covers the undermodulation and overmodulation modes with operation extended linearly and smoothly up to square wave (six-step). The ANN controller has the advantage of the very fast implementation of an SVM algorithm that can increase the switching frequency of power switches of the static converter, this point has a very important sense for minimizing DC bus capacitance, unbalanced DC source and the ripple of load current. The ANN controller uses the individual training strategy with the fixed weight and supervised models. A computer simulation program is developed using Matlab/Simulink together with the Neural Network Toolbox for training the ANN-controller.**

Index Terms-- **Artificial Neural Network, Four-Switch Three-Phase Inverter (FSTPI), Low-cost Inverter, overmodulation, Pulse-Width- Modulation, Six-Switch Three-Phase Inverter (SSTPI), Space vector, undermodulation,.**

I. INTRODUCTION

Recently, several scientific researches have been done for Four-Switch Three-Phase Inverters (FSTPI) with the target for reducing the cost of electric drives. Several inverter schemes with reduced number of switches have been proposed [4-8].

However, in comparison with the conventional Six-Switch Three-Phase Inverters (SSTPI), FSTPI has some drawbacks like increased ripples of DC source voltage and load currents, enlarged size of DC-bus capacitor, which reduces the competition of FSTPI with SSTPI. Hence, the solution in improving the switching frequency may be an important meaning for overcoming these above drawbacks.

To enhance the harmonic problem and the implementation of microprocessors in modulating required output voltage for FSTPI, Space Vector PWM methods are used very popular. However, a setback of SVM is that it requires complex online computation that usually limits its operation only up to several kilohertz of switching frequency. Switching frequency can be extended by using a high –speed DSP and simplifying computation with the help of lookup tables which is very large and tends to reduce the pulse width resolution.

This work was supported by Vietnamese Ministry of Science and Technology (MOST).

Power switches recently have been improved in term of switching frequency. Modern ultra-fast IGBTs allow operation at 50 kHz. However, the DSP- based SVM practically fails in this region where artificial- neural – network (ANN) –based SVM would probably take over [1].

The applications of ANN technique have been developed strongly in power electronics for recent years. Several researches of ANN implementation of SVM have been worked out for conventional VSI [1-3].

The proposed back-propagation type feed-forward ANN-SVM for FSTPI (Fig. 1) in this paper has been successfully trained by using two main approaches:

1) SVPWM approach based on the principle of similarity of the one for SSTPI for 3 modes: undermodulation, overmodulation mode 1, overmodulation mode 2) up to six-step mode.
2) Individual training strategy (to overcome the complexity of SVM for three mentioned above modes.

II. SPACE- VECTOR PWM IN UNDERMODULATION AND OVERMODULATION REGIONS USING THE NEW APPROACH

The SVM technique has been well discussed and developed and a number of authors [4-8] have described its operation, but only for the region of undermodulation. For the first time, by using the new approach for SVPWM, similarly to the one of SSTPI, the calculation of duty cycles may be done easily for all modes of

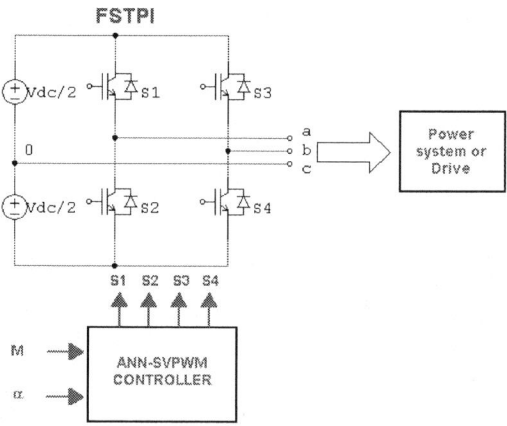

Fig. 1. ANN-SVPWM Controller for FSTPI.

modulation: under-, over-mode-1 , over-mode-2.

To simulate 6 non-zero vectors in SSTPI, using this proposed method, beside the two V_1 and V_3 in conventional SVPWM methods for B4 (Fig.2), it can be used the effective vectors V_{23M}, V_{34M}, V_{41M} and V_{12M} (Fig.3). These vectors are formed as follows:

$$\vec{V}_{23M} = \frac{1}{2}\left(\vec{V}_2 + \vec{V}_3\right) = \frac{V_{dc}}{3}e^{j0}; \vec{V}_{34M} = \frac{1}{2}\left(\vec{V}_3 + \vec{V}_4\right) = \frac{V_{dc}}{3}e^{j\frac{2\pi}{3}};$$

$$\vec{V}_{41M} = \frac{1}{2}\left(\vec{V}_4 + \vec{V}_1\right) = \frac{V_{dc}}{3}e^{j\pi}; \vec{V}_{12M} = \frac{1}{2}\left(\vec{V}_1 + \vec{V}_2\right) = \frac{V_{dc}}{3}e^{-j\frac{\pi}{3}};$$

(1)

To simulate zero vectors of SSTPI, we use the effective V_{0M}:

$$\vec{V}_{0M} = \frac{1}{2}\left(\vec{V}_1 + \vec{V}_3\right),$$

(2)

The base vectors in each sector used to form the required space vector V_{ref} is presented in Table 1.

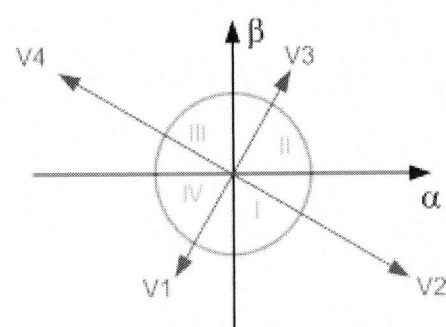

Fig. 2. Sectors used in conventional SVPWM methods for B4.

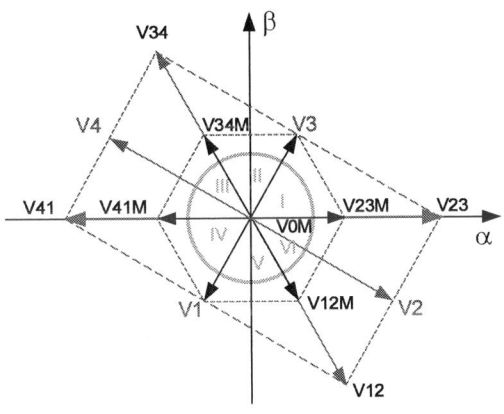

Fig. 3. SVPWM method proposed for FSTPI on the basis of SSTPI.

TABLE 1
VECTORS USED IN THE SPACE VECTOR MODULATION

Sector	FSTPI (Vx, Vy, Vz)
I	V23M, V3, V0M
II	V3, V34M, V0M
III	V34M, V41M, V0M
IV	V41M, V1, V0M
V	V1, V12M, V0M
VI	V12M, V23M, V0M

A. Undermodulation ($0 \le M \le 0.907$)

In the undermodulation mode, the rotating reference voltage remains within the hexagon. The SVM strategy in this region is based on generating three consecutive switching voltage vectors in a sampling period (T_s) so that the average output voltage matches with the reference voltage. The equations for effective duty cycle of the inverter switching states can be described as follows:

$$d_{Vx} = \frac{2\sqrt{3}}{\pi}M\sin(\pi/3 - \alpha);$$

$$d_{Vy} = \frac{2\sqrt{3}}{\pi}M\sin(\alpha);$$

(3)

$$d_{Vz} = 1 - d_{Vx} - d_{Vy}$$

where:

d_{Vx} - duty cycle ($2t_x/T_s$) of switching vector that lags.

d_{Vy} - duty cycle ($2t_y/T_s$) of switching vector that leads.

d_{Vz} - duty cycle ($2t_0/T_s$) of zero-switching vector.

M - modulation factor M = V^*/V_{1sw} (V^* - magnitude of reference voltage vector, V_{1sw} – the peak value of six- step voltage wave).

The timing intervals are obtained by multiplying duty cycles and period $T_s/2$ respectively.

B. Overmodulation mode 1 ($0.907 < M \le 0.952$)

This mode takes place when the reference voltage V^* exceeds the circle inscribed in the hexagon and attains the sides of the hexagon. On the hexagon trajectory M=0.952, duty d_{Vz} equals to 0:

$$d_{Vx} = \frac{\sqrt{3}\cos\alpha - \sin\alpha}{\sqrt{3}\cos\alpha + \sin\alpha}; d_{Vy} = 1 - d_{Vx}; d_{Vz} = 0$$

(4)

In the range of 0.907<M<0.952 the method of linear approximation is used for calculating d_{Vx}, d_{Vy}, d_{Vz}.

C. Overmodulation mode 2 ($0.952 < M \le 1$)

In overmodulation mode 2, the reference vector V* grows further up to six-step mode.

- For $0 \le \alpha \le \pi/6$:

$$d_{Vx} = 1; d_{Vy} = 0; d_{Vz} = 0$$

(5)

- For $\pi/6 \le \alpha \le \pi/3$:

$$d_{Vx} = 0; d_{Vy} = 1; d_{Vz} = 0$$

(6)

In the range of 0.952<M<1 the method of linear approximation is used for calculating d_{Vx}, d_{Vy}, d_{Vz}.

III. NEURAL-NETWORK BASED SPACE VECTOR PWM

The SVM algorithm will be used to obtain training data for the relevant ANN-SVPWM.

A. Undermodulation Region Sub-net:

For any command angle α^*, the duty cycles d_{Vx}, d_{Vy}, d_{Vz} are given by (3) for for all six sectors and the phase turn-on duty cycles can be calculated as:

1) For sector 1: ($0 \leq \alpha^* \leq \pi/3$)

$$d_{A-ON} = \frac{1}{2} + \frac{\sqrt{3}}{\pi} \cdot M \cdot \left(-\sin(\pi/3 - \alpha) - \sin\alpha\right); \quad (7)$$

$$d_{B-ON} = \frac{1}{2} + \frac{\sqrt{3}}{\pi} \cdot M \cdot (-\sin\alpha)$$

2) For sector 2: ($\pi/3 \leq \alpha^* \leq 2\pi/3$)

$$d_{A-ON} = \frac{1}{2} + \frac{\sqrt{3}}{\pi} \cdot M \cdot \left(-\sin(\pi/3 - \alpha)\right); \quad (8)$$

$$d_{B-ON} = \frac{1}{2} + \frac{\sqrt{3}}{\pi} \cdot M \cdot \left(-\sin(\pi/3 - \alpha) - \sin\alpha\right)$$

3) For sector 3: ($2\pi/3 \leq \alpha^* \leq \pi$)

$$d_{A-ON} = \frac{1}{2} + \frac{\sqrt{3}}{\pi} \cdot M \cdot \sin\alpha; \quad (9)$$

$$d_{B-ON} = \frac{1}{2} + \frac{\sqrt{3}}{\pi} \cdot M \cdot \left(-\sin(\pi/3 - \alpha)\right)$$

4) For sector 4: ($\pi \leq \alpha^* \leq 4\pi/3$)

$$d_{A-ON} = \frac{1}{2} + \frac{\sqrt{3}}{\pi} \cdot M \cdot \left(\sin(\pi/3 - \alpha) + \sin\alpha\right); \quad (10)$$

$$d_{B-ON} = \frac{1}{2} + \frac{\sqrt{3}}{\pi} \cdot M \cdot \sin\alpha$$

5) For sector 5: ($4\pi/3 \leq \alpha^* \leq 5\pi/3$)

$$d_{A-ON} = \frac{1}{2} + \frac{\sqrt{3}}{\pi} \cdot M \cdot \sin(\pi/3 - \alpha); \quad (11)$$

$$d_{B-ON} = \frac{1}{2} + \frac{\sqrt{3}}{\pi} \cdot M \cdot \left(\sin(\pi/3 - \alpha) + \sin\alpha\right)$$

6) For sector 6: ($5\pi/3 \leq \alpha^* \leq 2\pi$)

$$d_{A-ON} = \frac{1}{2} + \frac{\sqrt{3}}{\pi} \cdot M \cdot (-\sin\alpha); \quad (12)$$

$$d_{B-ON} = \frac{1}{2} + \frac{\sqrt{3}}{\pi} \cdot M \cdot \sin(\pi/3 - \alpha)$$

Hence, the turn-on and turn-off time interval are determined as follows:

$$T_{A,B-ON} = d_{A,B-ON} \cdot \frac{T_s}{2}; \quad T_{A,B-OFF} = \left(1 - d_{A,B-ON}\right) \cdot \frac{T_s}{2} \quad (13)$$

Equation (7)-(12) can be transformed in general form:

$$d_{A-ON} = \frac{1}{2} + \frac{\sqrt{3}}{\pi} M \cdot h_{10}(\alpha^*); \quad (14)$$

$$d_{B-ON} = \frac{1}{2} + \frac{\sqrt{3}}{\pi} M \cdot h_{20}(\alpha^*)$$

where

$$h_{10}(\alpha^*) = \begin{cases} \left[-\sin(\pi/3 - \alpha) - \sin\alpha\right], & \theta = 1 \\ \left[-\sin(\pi/3 - \alpha)\right], & \theta = 2 \\ \left[\sin\alpha\right], & \theta = 3 \\ \left[+\sin(\pi/3 - \alpha) + \sin\alpha\right], & \theta = 4 \\ \left[\sin(\pi/3 - \alpha)\right], & \theta = 5 \\ \left[-\sin\alpha\right], & \theta = 6 \end{cases} \quad (15)$$

$h_{20}(\alpha^*)$ is obtained similarly. Then, $h_{10,20}(\alpha^*)$ have been used for creating databases which are needed for training undermodulation region Sub-net with one input (α^*) and two outputs (h_{10}, h_{20}). The angle step is 1^0.

A two-layer network is used for implementing this sub-net. The sub-net is obtained by training (supervised) with *trainlm* function – Levenberg –Marquardt algorithm; the squared error acceptable for training is 10^{-4}. The number of neurons of 1st layer is 7 *tansig* neurons, the 2nd layer has 2 *purelin* neurons.

So, the total number of neurons is 9 (convergence obtained for 687 epochs). For implementing undermodulation duty cycles (equations (14)), it can be used one further product net-input-function (inputs: h(α^*), M) and one purelin neuron (weight = $\sqrt{3}/\pi$, bias = 1/2) (Fig.4).

B. Overmodulation model Sub-net

Duty cycles can be determined for all six sectors and the phase turn-on duty cycles can be calculated as follows:

$$K_1 = \sqrt{3}; \quad K_2 = \frac{\sqrt{3}}{\pi} \cdot 0.907$$

$$A = \sin(\pi/3 - \alpha); \quad B = \sin\alpha; \quad C = \frac{K_1 \cos\alpha - \sin\alpha}{K_1 \cos\alpha + \sin\alpha} \quad (16)$$

1) For sector 1: ($0 \leq \alpha^* \leq \pi/3$)

$$d_{A-ON} = \frac{1}{2} - K_2(A + B) + \eta\left(-\frac{1}{2} + K_2(A + B)\right) \quad (17)$$

$$d_{B-ON} = \frac{1}{2} - K_2 B + \eta\left(-\frac{1}{2} + K_2 B + \frac{1}{2}C\right)$$

2) For sector 2: ($\pi/3 \leq \alpha^* \leq 2\pi/3$)

$$d_{A-ON} = \frac{1}{2} - K_2 A + \eta\left(K_2 A - \frac{1}{2}C\right) \quad (18)$$

$$d_{B-ON} = \frac{1}{2} - K_2(A + B) + \eta\left(-\frac{1}{2} + K_2(A + B)\right)$$

3) For sector 3: ($2\pi/3 \leq \alpha^* \leq \pi$)

$$d_{A-ON} = \frac{1}{2} + K_2 B + \eta\left(\frac{1}{2} - K_2 B - \frac{1}{2}C\right) \quad (19)$$

$$d_{B-ON} = \frac{1}{2} - K_2 A + \eta\left(K_2 A - \frac{1}{2}C\right)$$

4) For sector 4: ($\pi \leq \alpha^* \leq 4\pi/3$)

$$d_{A-ON} = \frac{1}{2} + K_2(A + B) + \eta\left(\frac{1}{2} - K_2(A + B)\right) \quad (20)$$

$$d_{B-ON} = \frac{1}{2} + K_2 B + \eta\left(\frac{1}{2} - K_2 B - \frac{1}{2}C\right)$$

5) For sector 5: ($4\pi/3 \leq \alpha^* \leq 5\pi/3$)

$$d_{A-ON} = \frac{1}{2} + K_2 A + \eta\left(-K_2 A + \frac{1}{2}C\right) \quad (21)$$

$$d_{B-ON} = \frac{1}{2} + K_2(A + B) + \eta\left(\frac{1}{2} - K_2(A + B)\right)$$

6) For sector 6: ($5\pi/3 < \alpha^* < 2\pi$)

$$d_{A-ON} = \frac{1}{2} - K_2 B + \eta\left(-\frac{1}{2} + K_2 B + \frac{1}{2}C\right) \quad (22)$$

$$d_{B-ON} = \frac{1}{2} + K_2 A + \eta\left(-K_2 A + \frac{1}{2}C\right)$$

Equations (17)-(22) can be transformed in general form:

$$d_{A-ON} = h_{11a}(\alpha^*) + \eta \cdot h_{21a}(\alpha^*)$$
$$d_{B-ON} = h_{11b}(\alpha^*) + \eta \cdot h_{21b}(\alpha^*) \quad (23)$$

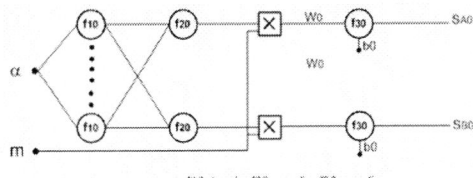

Fig. 4. Undermodulation region subnet.

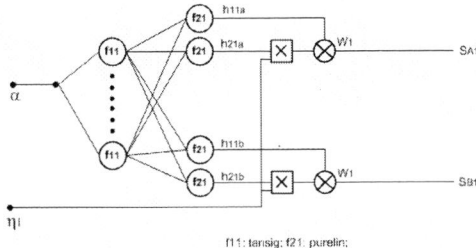

f11: tansig; f21: purelin;

Fig. 5. Overrmodulation mode 1 region subnet.

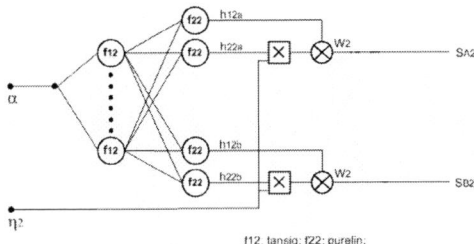

f12: tansig; f22: purelin;

Fig. 6. Overrmodulation mode 2 region subnet.

Then, $h_{11a, 21a, 11b, 21b}$ (α^*) have been used for creating databases which are needed for training overmodulation region sub-net with one input (α^*) and four outputs (h_{11a}, h_{21a}, h_{11b}, h_{21b}). The angle step is 1 degree.

The number of neurons of 1st layer is 10 *tansig* neurons, the 2nd layer has 4 *purelin* neurons. So, the total number of neurons is 14 (convergence obtained for 578 epochs). For implementing complete overmodulation duty cycles in equations (23), it can be used one further product net-input-function (inputs: $h_{21}(\alpha^*)$, η_1), one further sum net-input-function (inputs : $h_{11}(\alpha^*)$, $\eta_1 * h_{21}(\alpha^*)$) (Fig.5).

C. Overmodulation mode 2 Sub-net

Duty cycles can be expressed for all six sectors :

1) For sector 1: (0 ≤ α^* ≤ π/3)
for 0 < α < π/6:

$$d_{A-ON} = 0; d_{B-ON} = \frac{1}{2}C + \eta\left(\frac{1}{2} - \frac{1}{2}C\right); \qquad (26)$$

for π/6 < α < π/3:

$$d_{A-ON} = 0; d_{B-ON} = \frac{1}{2}C + \eta\left(-\frac{1}{2}C\right);$$

2) For sector 2: (π/3 ≤ α^* ≤ 2π/3)

for π/3 < α < π/2:

$$d_{A-ON} = \frac{1}{2}(1-C) + \eta\left(-\frac{1}{2} + \frac{1}{2}C\right); d_{B-ON} = 0; \qquad (27)$$

for π/2 < α < 2π/3:

$$d_{A-ON} = \frac{1}{2}(1-C) + \eta\left(\frac{1}{2}C\right); d_{B-ON} = 0$$

3) For sector 3: (2π/3 ≤ α^* ≤ π)

for 2π/3 ≤ α ≤ 5π/6:

$$d_{A-ON} = (1 - \frac{1}{2}C) + \eta\left(-\frac{1}{2} + \frac{1}{2}C\right);$$

$$d_{B-ON} = \frac{1}{2}(1-C) + \eta\left(-\frac{1}{2} + \frac{1}{2}C\right); \qquad (28)$$

for 5π/6 < α ≤ π:

$$d_{A-ON} = (1 - \frac{1}{2}C) + \eta\left(\frac{1}{2}C\right);$$

$$d_{B-ON} = \frac{1}{2}(1-C) + \eta\left(\frac{1}{2}C\right)$$

4) For sector 4: (π ≤ α^* ≤ 4π/3)
for π < α < 7π/6:

$$d_{A-ON} = 1; d_{B-ON} = \left(1 - \frac{1}{2}C\right) + \eta\left(-\frac{1}{2} + \frac{1}{2}C\right); \qquad (29)$$

for 7π/6 < α < 4π/3:

$$d_{A-ON} = 1; d_{B-ON} = \left(1 - \frac{1}{2}C\right) + \eta\left(\frac{1}{2}C\right)$$

5) For sector 5: (4π/3 ≤ α^* ≤ 5π/3)
for 4π/3 ≤ α ≤ 3π/2:

$$d_{A-ON} = \frac{1}{2}(1+C) + \eta\left(\frac{1}{2} - \frac{1}{2}C\right); d_{B-ON} = 1; \qquad (30)$$

for 3π/2 ≤ α ≤ 5π/3:

$$d_{A-ON} = \frac{1}{2}(1+C) + \eta\left(-\frac{1}{2}C\right); d_{B-ON} = 1;$$

6) For sector 6: (5π/3 ≤ α^* ≤ 2π)
for 5π/3 < α < 11π/6:

$$d_{A-ON} = \frac{1}{2}C + \eta\left(\frac{1}{2} - \frac{1}{2}C\right);$$

$$d_{B-ON} = \frac{1}{2}(1+C) + \eta\left(\frac{1}{2} - \frac{1}{2}C\right); \qquad (31)$$

for 11π/6 < α < 2π:

$$d_{A-ON} = \frac{1}{2}C + \eta\left(-\frac{1}{2}C\right);$$

$$d_{B-ON} = \frac{1}{2}(1+C) + \eta\left(-\frac{1}{2}C\right)$$

Equations (26)-(31) can be written in general form:

$$d_{A-ON} = h_{12a}\left(\alpha^*\right) + \eta \cdot h_{22a}\left(\alpha^*\right)$$
$$d_{B-ON} = h_{12b}\left(\alpha^*\right) + \eta \cdot h_{22b}\left(\alpha^*\right) \qquad (32)$$

Then, $h_{12a, 22a, 12b, 22b}$ (α^*) have been used for creating databases which are needed for training overmodulation mode 2 sub-net with one input (α^*) and four outputs (h_{12a}, h_{22a}, h_{12b}, h_{22b}). The angle step is 1 degree.

The number of neurons of 1st layer is 15 *tansig* neurons, the 2nd layer has 4 *purelin* neurons. So, the total number of neurons is 19 (convergence obtained for 308 epochs).

For implementing complete overmodulation duty cycles in equations (32), it can be designed one further product net-input-function (inputs: h_{22} (α^*),η), one further sum net-input-function (inputs: h_{21} (α^*), ηh_{22} (α^*)) (Fig.6).

D. η_1, η_2 calculation Sub-nets

The coefficient η_1 and η_2 in overmodulation mode 1 and mode 2 respectively are given by method of linear approximation:

$$\eta = \frac{M - M_1}{M_2 - M_1} \tag{33}$$

This equation has been used for generating neural network training data. The input M is varied from 0.907 to 0.952 with step of 0.001, the output is η_1 and varied from 0.952 to 1 with step of 0.001, the output is η_2 (Fig. 7, 8).

E. Code of modulation mode Sub-net

The purpose of this subnet is to define the code of modulation mode: undermodulation : C_m= 3; overmodulation mode 1 : C_m = 1; overmodulation mode 2 : C_m = 2. The input M is varied from 0 to 1 with step of 0.001, the output is C_m.

F. Mode Selection Code Sub-net

This subnet is used for determining which modulation mode will be the choice for generating duty cycles at outputs of ANN-SVM-Controller (S_A, S_B, S_C).

The input of this subnet is C_m; the outputs are e_A, e_B, e_C (Table II) (Fig. 9).

The proposed complete ANN-SVM controller has 2 inputs (α, M) and 2 outputs (WT$_{A-ON}$, WT$_{B-ON}$,) (Fig.10). The outputs are the digital words presenting the turn-on times which are generated by multiplying the duty cycles d_{A-ON}, d_{B-ON} by the value of the sample time $T_s/2$.

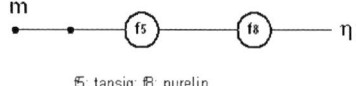

$f5$: tansig; $f8$: purelin

Fig. 7. η_1 calculation subnet.

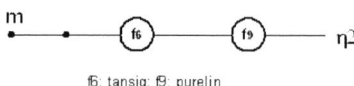

$f6$: tansig; $f9$: purelin

Fig. 8. η_2 calculation subnet.

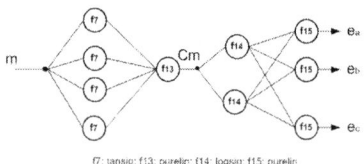

$f7$: tansig; $f13$: purelin; $f14$: logsig; $f15$: purelin

Fig. 9. Code of modulation mode Sub-net.

TABLE II.
MODE SELECTION CODE

C_m	e_A	e_B	e_C
3	1	0	0
1	0	1	0
2	0	0	1

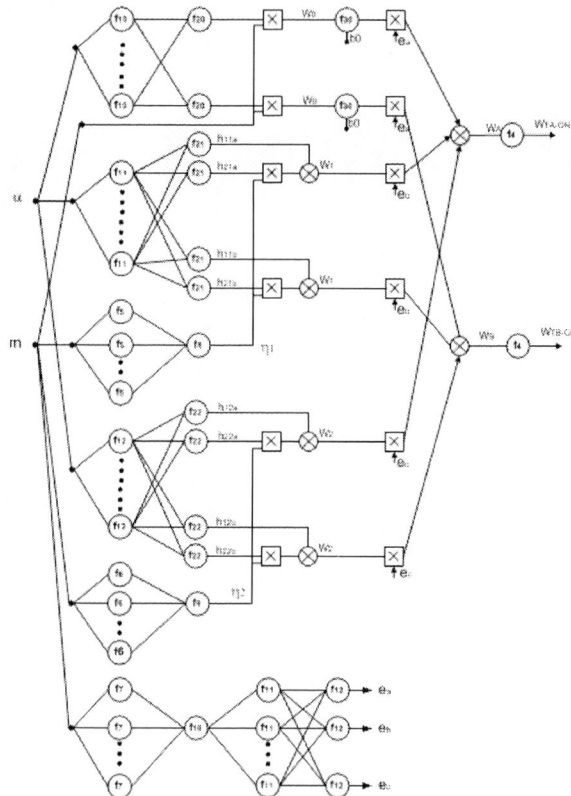

Fig. 10. Complete ANN-SVPWM Controller for FSTPI.

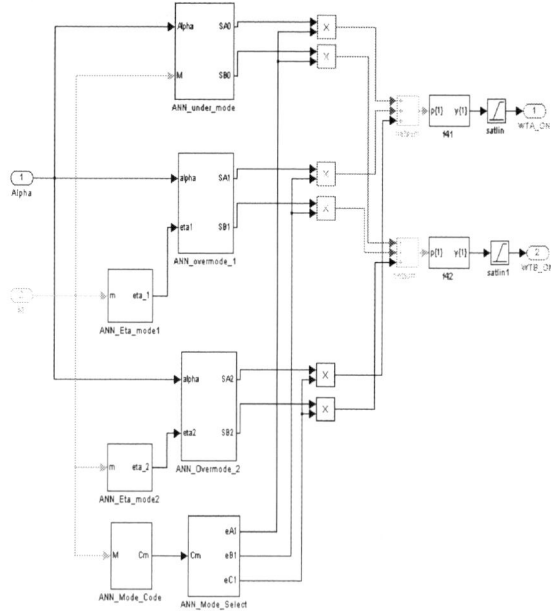

Fig. 12. Simulation model of ANN-SVPWM Controller for FSTPI.

1206

IV. SIMULATION OF ANN-SVPWM CONTROLLER

A Simulink/Matlab program with the toolbox of neural –network is used to train and simulate the complete ANN- SVPWM controller with the above-mentioned subnets for different mode of operation: undermodulation, overmodulation mode 1, 2. Simulation model of VSI is presented in Fig.11, while the simulation model of ANN-SVPWM Controller is shown in Fig.12.

DC source voltage Vd = 400V. The sample time T_S=40µs.

Modulation index : M = 0.5 when 0 ms ≤ t ≤ 40 ms; M = 0.93 when 40 ms ≤ t ≤ 80 ms; M = 0.97 when 80 ms ≤ t ≤ 120 ms; M = 1 when 120 ms ≤ t ≤ 160 ms.

Simulation results demonstrate the excellent performance of the proposed ANN-SVPWM for FSTPI, while the good responses of the output phase and line voltages are obtained (fig.13 – 14, Table III).

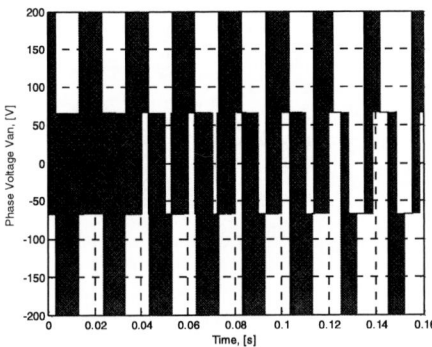

Fig. 13. Phase voltage for undermodulation and overmodulation mode 1, 2 region.

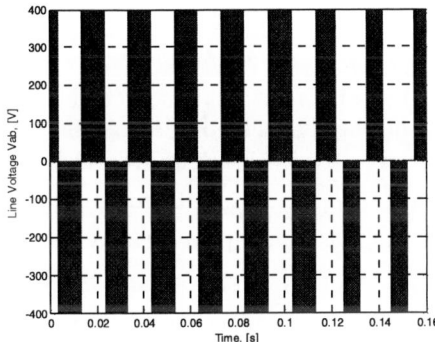

Fig. 14. Line voltage for undermodulation and overmodulation mode 1, 2 regions.

TABLE III
TABLE OF SIMULATION RESULTS FOR ANN-SVPWM CONTROLLER

Modulation index M	0.5	0.7	0.91	1 (six-step)
Reference voltage V_{1ref}, [V]	63.66	89.13	115.9	127.3
Simulated output phase voltage V_{1pANN}, [V]	62.09	86.36	112.2	127.1
Tolerance ε, [%]	2.47	3.11	3.19	0.16

V. CONCLUSION

As above obtainable analysis, the calculation of duty cycles in SVPWM is very complex for the whole range of modulation and requires a large time for conventional DSP to implement, which aims to reduce the switching frequency of power switches. By using ANN for implementing modulator, this limit of frequency may be overcome.

Enlarging the switching frequency by ANN approach also has the sense for minimizing the size and the ripple of DC capacitors, which aims to enhance the performance of FSTPI output voltage.

Finally, the development of a complete artificial-neural-network space- vector- modulation controller (ANN-SVPWM Controller) scheme has been done for a Four-Switch Three-Phase Inverter. Two main approaches have been used successfully in this case: *A new SVPWM algorithm with extension to six-step mode* and *Individual training strategy*.

This ANN-SVPWM-Controller may be implemented in ASIC chip in future.

REFERENCES

[1] J. O. P. Pinto, B. K. Bose, L. E. B. Silva, M. P. Karmierkowski "A Neural Network Based Space Vector PWM Controller for Voltage-Fed Inverter Induction Motor Drive", *IEEE Trans. on Ind. Appl.*, vol.36, no. 6, November/December 2000.

[2] A. Bakhshai, J. Espinoza, G. Joos, H. Jin. "A combined ANN and DSP approach to the implementation of space vector modulation techniques", in *conf. Rec.IEEE –IAS Annu. Meeting*, 1996, pp.934-940.

[3] Phan Quoc Dzung, Le Minh Phuong, Pham Quang Vinh, Nguyen Van Nho, Dao Minh Hien, "The Development of Artificial Neural Network Space Vector PWM and Diagnostic Controller for Voltage Source Inverter", 2006 *IEEE Power India Conference*, New Delhi, India, April 10-12, 2006.

[4] Frede Blaabjerg,, Sigurdur Freysson, Hans-Henrik Hansen, and S. Hansen "A New Optimized Space-Vector Modulation Strategy for a Component-Minimized Voltage Source Inverter " *IEEE Trans. on Power Electronics*, Vol. 12, No. 4, July 1997,pp 704-710

[5] F. Blaabjerg, S. Freysson, H. H. Hansen, and S. Hariseri. "Comparison of a space-vector modulation strategy for a three phase standard and a component minimized voltage source inverter". In *Conf. Rec. EPE*, pages 1806-1813, Sevilha - Spain, September 1995.

[6] M. B. R. Correa, C. B. Jacobina, E. R. C. Da Silva, and A. M. N. Lima. "A General PWM Strategy for Four-Switch Three-Phase Inverters" *IEEE Trans. on Power Electronics*, Vol. 21, No. 6, Nov. 2006, pp 1618-1627.

[7] G. A. Covic, G. L. Peters, and J. T. Boys, "An improved single phase to three phase converter for low cost ac motor drives," in *Proc. PEDS '95*, Singapore, vol. 1, pp. 549–554.

[8] G.-T. Kim and T. Lipo, "VSI-PWM rectifier/inyerter system with a reduced switch count," in *Conf. Rec. IAS*, 1995, pp. 2327 - 2332.

Voltage Losses Compensation Using Artificial Neural Network for Estimation Nonlinear Characteristic of Switches

N. Pothi, S. Premrudeepreechacharn, *Member, IEEE*, and C. Rakpenthai

Abstract—**This paper presents the voltage losses compensation technique for three-phase inverter by considering nonlinear characteristic of switches, dead time, and turn on/off delay. An artificial neural network (ANN) is employed to estimate the nonlinear characteristic of switches. The compensation method improves the modulation index of the open-loop Space Vector Pulse Width Modulation Voltage-Source Inverter (SVPWM-VSI) performing on the constant V/f control technique. DSP TMS320LF2407 was used to control the effect of the voltage losses on the efficiency. The principle is explained qualitatively and extensive experiments have been carried out to verify the propose method.**

Index Terms— **voltage compensation, modulation index, SVPWM, nonlinear characteristic of switches.**

I. INTRODUCTION

THE main problems encountered in open-loop Voltage-Source Inverter (VSI) operating at low frequency are a losses of output voltage and current distortions caused by the nonlinear characteristic of switches, dead time, and turn on/off delay. Normally, nonlinear characteristic of switches vary with the operating condition, such as temperature and magnitude of phase current. Dead time is a small delay times in the gate signal of the turning on device to avoid short-circuit of inverter leg. Turn on/off delay inevitably exists in actual device.

Several compensation methods [1]-[4] have been proposed to compensate voltage losses of drive system at very low frequency. However, the temperature changing of switches is not considered in these papers. In this paper, the artificial neural network (ANN) is used to estimate the nonlinear characteristic of switches. The block diagram of an open-loop V/f controlled VSI incorporating the proposed ANN based SVPWM controller is shown in Fig. 1.

N. Pothi is with Department of Electrical Engineering, School of Engineering, Naresuan University Phayao, Phayao, 56000, Thailand (e-mail: nattapongpo@nu.ac.th).

S. Premrudeepreechacharn is with Department of Electrical Engineering, Faculty of Engineering, Chiang Mai University, Chiang Mai, 50200 Thailand (e-mail: suttic@eng.cmu.ac.th).

C. Rakpenthai is with Department of Electrical Engineering, School of Engineering, Naresuan University Phayao, Phayao, 56000, Thailand (e-mail: chawasak@hotmail.com).

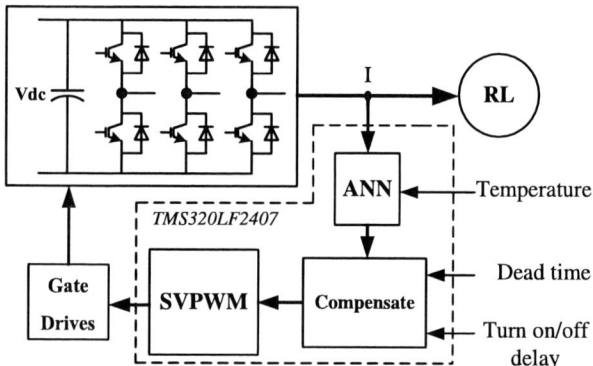

Fig. 1. The block diagram of the proposed compensation method.

II. ANALYSIS OF VOLTAGE LOSSES

This section describes the voltage losses of SVPWM controlled three-phase inverter. The voltage losses in the inverter can be classified in three modes a) nonlinear characteristic of switches, b) dead time, and c) turn on/off delay.

A. Nonlinear characteristic of switches

Fig. 2 shows the relationship between ideal and nonlinear characteristic of switches for one leg of a three-phase inverter, including two active switches, Q_1 and Q_2, and two freewheeling diode when $i_a > 0$ and $i_a < 0$ follows through switching device Q_1 during *on/off* state. Where V_{dc} is the dc bus voltage, V_{igbt} and V_{diode} are voltage losses of active switches (IGBT) and freewheeling diode respectively.

A voltage loss of nonlinear characteristic of switches ($V_{nonlinear}$) increases with magnitude of phase current as following:

$$V_{nonlinear} = V_{th} + \tilde{R} \cdot \left| i_a \right|. \qquad (1)$$

For the power devices, $V_{nonlinear}$ can be divided into

$$V_{igbt} = V_{th,igbt} + \tilde{R}_{igbt} \cdot \left| i_a \right| \qquad (2)$$

$$V_{diode} = V_{th,diode} + \tilde{R}_{diode} \cdot \left| i_a \right|. \qquad (3)$$

978-1-4244-0644-9/07/$25.00 ©2007 IEEE   1208

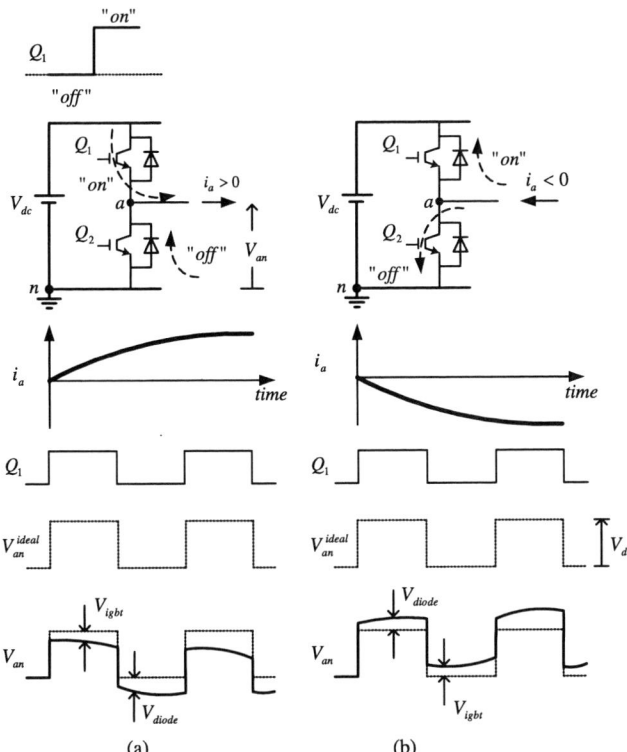

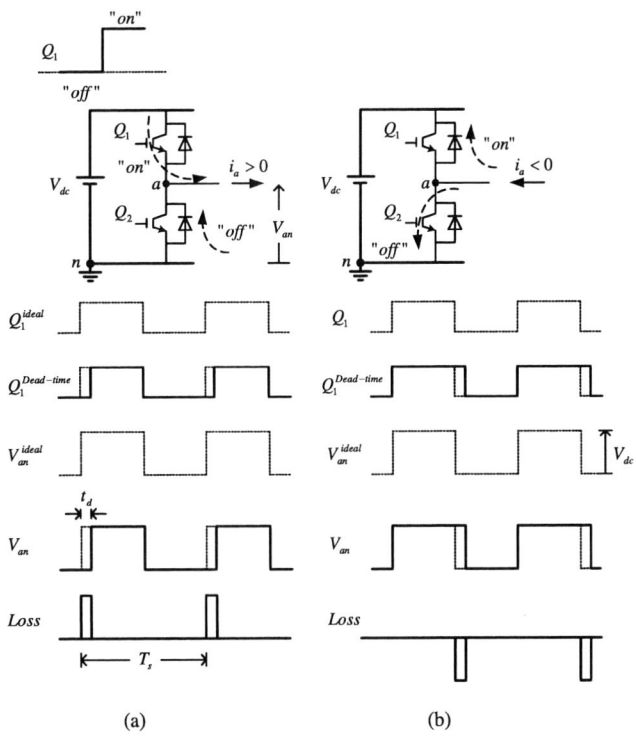

Fig. 2. Relationship between ideal and nonlinear characteristic of switches: (a) For $i_a > 0$ and (b) $i_a < 0$.

Fig. 3. Relationship between ideal and dead time: (a) For $i_a > 0$ and (b) $i_a < 0$.

Where $V_{th,igbt}$ and $V_{th,diode}$ are the threshold voltage of power devices and $\tilde{R}_{igbt}$ and $\tilde{R}_{diode}$ are the nonlinear resistances of power devices at the given temperature. From (2) and (3), $V_{nonlinear}$ over a cycle turn on/off switching ($\Delta V_{nonlinear}$) is derived as

$$\Delta V_{nonlinear} = \frac{t_{on} \cdot V_{igbt} + (T_s - t_{on}) \cdot V_{diode}}{T_s} \cdot sign(i_a) . \quad (4)$$

Base on SVPWM, because $t_{on} = t_{off}$, equation (4) can be explained as

$$\Delta V_{nonlinear} \cong \frac{V_{igbt} + V_{diode}}{2} \cdot sign(i_a) \quad (5)$$

where $sign(i_a)$ is a sign function of phase current, t_{on} is *on* state time, and T_s is the switching period.

B. Dead time

The dead time (t_d) is fixed to prevent short-circuit of a phase-bridge, which is a main cause of a voltage loss and phase current distortion. The relationships between ideal and dead time when $i_a > 0$ and when $i_a < 0$ have been studied on a switching device Q_1 during *on/off* state (Fig. 3).

As a result, the average voltage loss of dead time ($\Delta V_{deadtime}$) is given by

$$\Delta V_{deadtime} = \frac{t_d}{T_s} V_{dc} \cdot sign(i_a) . \quad (6)$$

C. Turn on/off delay

Turn on/off delay inevitably exists in actual device. The compensation time (T_{com}) is estimated from the difference between turn-on delay ($T_{d(on)}$) and turn-off delay ($T_{d(off)}$). That is

$$T_{com} = T_{d(on)} - T_{d(off)} \quad (7)$$

Following (4), the voltage loss with consideration of turn on/off delay of power device (ΔV_{turn}) is given by

$$\Delta V_{turn} = \frac{T_{com}}{T_s} V_{dc} \cdot sign(i_a) . \quad (8)$$

III. ESTIMATION OF NONLINEAR CHARACTERISTIC

In this paper, the nonlinear characteristic of switches is estimated by using an artificial neural network based on back-propagation algorithm. This mean was chosen because it gives a good approximate result and is easy to implement on the DSP. Fig. 4 shows the proposed back-propagation algorithm for estimation of $\Delta V_{nonlinear}$. The structure of ANN is composed of 3 layers i.e. input layer, hidden layer, and output layer. The input layer has 2 units which are average current and switch temperature. Note that the inputs are the average current, obtained from (9), and switch temperature. The average current from (9) is calculated from absolute average three phase current based on Clark's transform. The output is a voltage loss due to nonlinear characteristic of switches.

$$I_{avg} = \frac{2}{3} \cdot \left| \frac{1}{T} \int_0^T (I_a + I_b \cdot e^{j120^\circ} + I_c \cdot e^{j240^\circ}) dt \right| \quad (9)$$

where T is a period of voltage waveform. I_a, I_b and I_c are the magnitude of phase current in phase a, b and c, respectively.

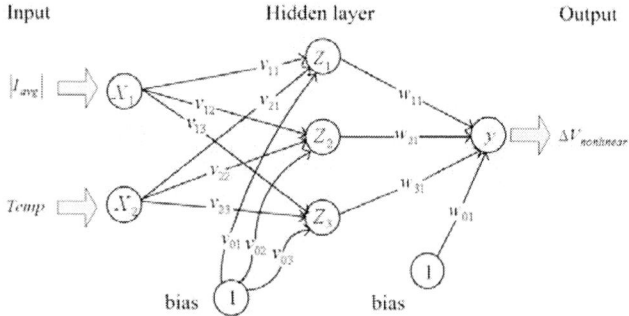

Fig. 4. Back-propagation algorithm for estimate the nonlinear characteristic of switches.

In learning schedule, a leaning rule which tunes the connection weights to reduce the difference between the desired output and the actual output is use to train the network. The back-propagation algorithm developed on MATLAB is used for training. The output layer has 1 unit which is $\Delta V_{nonlinear}$. The hidden layer has 3 units using the bipolar sigmoid for activation function is given as (10)

$$f(x) = \frac{2}{1 + e^{(-2x)}} - 1 \quad (10)$$

The estimation of $\Delta V_{nonlinear}$ by using the back-propagation approach can be written as

$$\Delta V_{nonlinear} \cong f(w_{0k} + \sum_{j=1}^{p} f(v_{0j} + \sum_{i=1}^{n} X_i v_{ij}) \cdot w_{jk}) \quad (11)$$

where v_{ij} and w_{jk} are the weights. The subscripts i, j, k, p and n are the input units, the hidden units, the output unit, the number of hidden units, and the number of input units, respectively.

The advantage of the ANN approach for estimating the nonlinear characteristics of switches is that the voltage loss ($V_{nonlinear}$) estimated by this method is very similar to the actual data. In this study, we tested the ANN to estimate $V_{nonlinear}$ at 25 and $150\,C^\circ$ and found a similar data as those reported in the actual data (Fig. 5). It is likely that the ANN approach can give the best estimation of the $V_{nonlinear}$ at a wide spectrum of temperature ranging between 25 to $150\,C^\circ$.

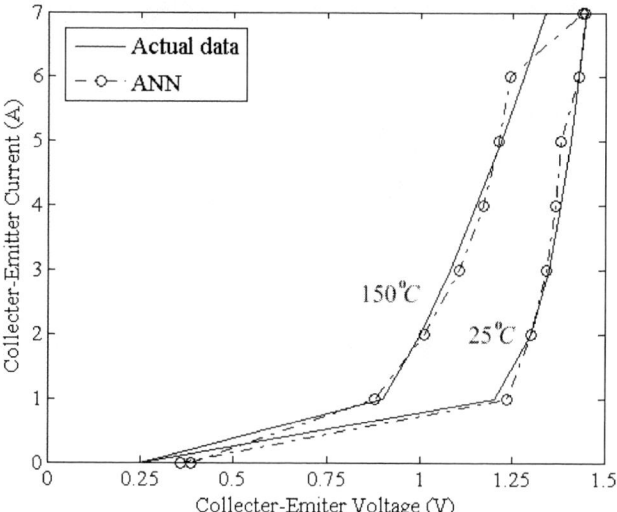

Fig. 5. Comparison of nonlinear characteristic of switches between ANN approach and actual data (HGTG30N60C3D) at 25 and $150\,C^\circ$.

IV. VOLTAGE LOSSES COMPENSATION

Fig. 6 shows the block diagram of the proposed compensation method for improving the modulation index of the SVPWM-VSI performing on the constant V/f control technique.

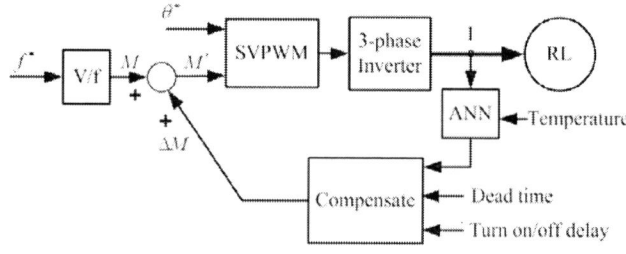

Fig. 6. The block diagram of compensation method

Equation (12) is a calculation of the compensated modulation index (ΔM). It is the sum absolute of a voltage losses from (11), (6), and (8) divided by V_{dc}.

$$\Delta M = \frac{|\Delta V_{nonlinear}| + |\Delta V_{deadtime}| + |\Delta V_{turn}|}{V_{dc}} \quad (12)$$

The modulation index after compensated (M') is the sum of the modulation index (M) and the compensated modulation index (ΔM). That is

$$M' = M + \Delta M \qquad (13)$$

Where M' is the modulation index after compensated.

$\quad M$ is the modulation index.

$\quad \Delta M$ is the compensated modulation index.

V. EXPERIMENTAL RESULTS

This section the proposed algorithm has been implemented in a DSP TMS320LF2407. The three-phase inverter using resistive and inductive (RL) load is used as a test system. The importance parameters are shown in Table I.

TABLE I
PARAMETER FOR EXPERIMENT

Symbol	Quantity	Detail
R_a	2.83 Ω	Resistance load phase a
R_b	2.87 Ω	Resistance load phase b
R_c	2.79 Ω	Resistance load phase c
L_a	12.34 mH	Inductance load phase a
L_b	11.74 mH	Inductance load phase b
L_c	16.31 mH	Inductance load phase c
V_{dc}	100 V	Dc bus voltage
f_s	10 kHz	Switching frequency
t_d	5 μs	Dead time

Fast Fourier Transform (FFT) was employed to evaluate the peak magnitude phase voltage after the steady state has been established. Fig. 7 shows the voltage losses of three-phase inverter. The results show that the proposed method can be compensated voltage losses of three-phase inverter under the test conditions.

Fig. 8 shows the phase current waveform at 1 and 5 Hz operations. The proposed method increases the magnitude but reduce the distortion of the phase current at both frequencies. Without the compensation method, the magnitude of phase current has reduced (Fig. 8b) or totally disappeared at the very low frequency (Fig. 8a). This reduction in the magnitude is caused by the voltage losses of the inverter.

In addition, FFT of the current as shown in Fig. 8 is illustrated in Fig. 9. As the results, it can be seen that the proposed method can reduce the loss of three phase inverter at the low frequencies.

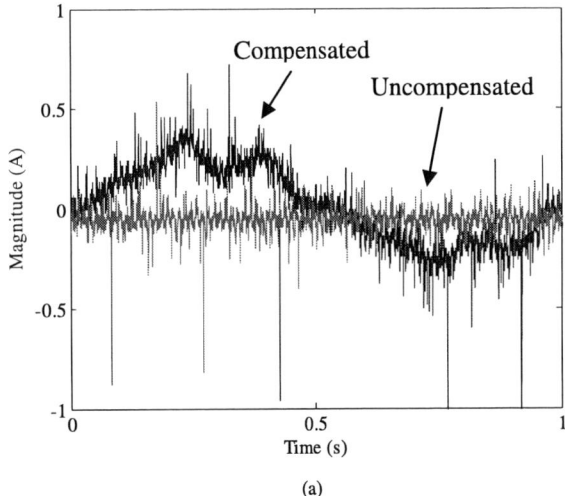

(a)

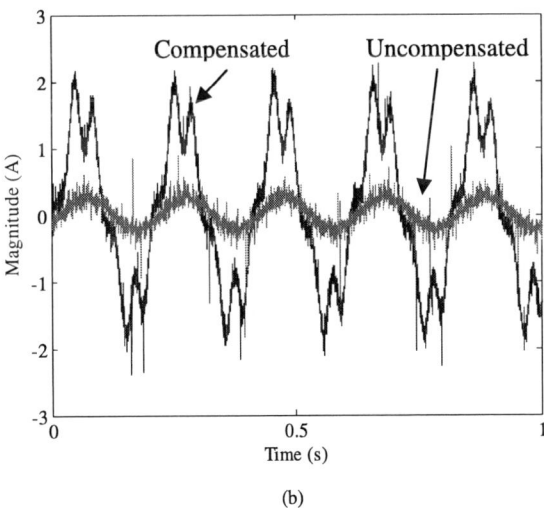

(b)

Fig. 8. The phase current waveform: (a) at 1 Hz, (b) at 5 Hz.

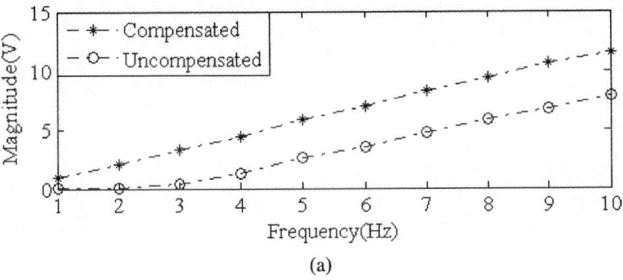

(a)

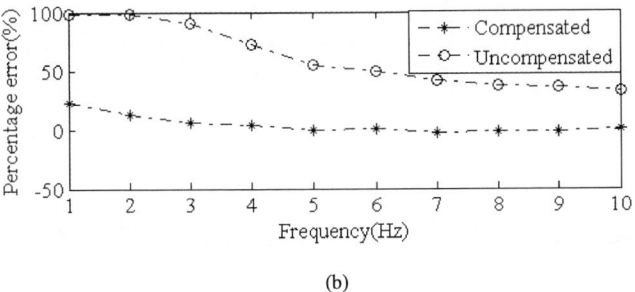

(b)

Fig. 7. The voltage losses of three-phase inverter: (a) peak phase voltage, (b) percentage error.

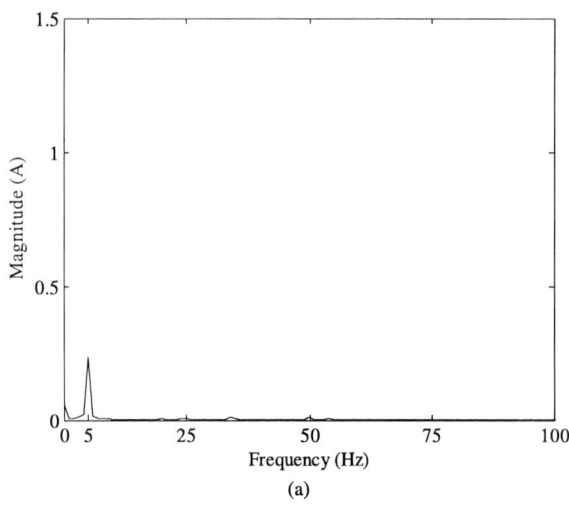

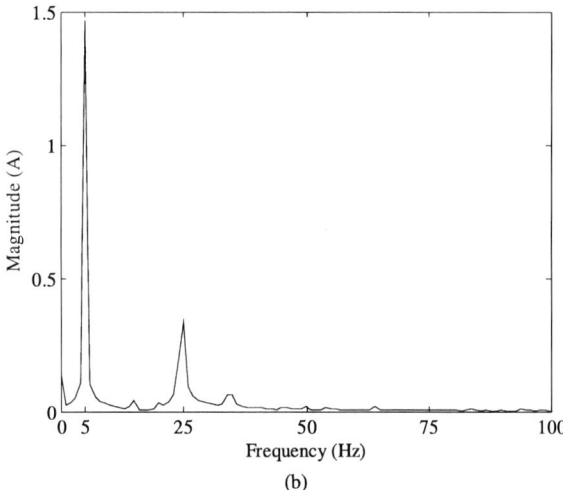

Fig. 9. FFT of phase current at 5 Hz: (a) Uncompensated, (b) Compensated.

VI. CONCLUSION

This paper, the propose algorithms based on SVPWM for voltage compensation is easily implemented on DSP base SVPWM. Here, the back-propagation neural network increases the accuracy for the estimation of a voltage loss of nonlinear characteristic of switches because the effect of temperature on the efficiency is considered. The dead time and turn on/off delay are also considered in the proposed compensation. The experimental results show that the proposed method can reduce the voltage losses and the current distortion, especially for very low frequency operation.

VII. REFERENCES

[1] Blaabjerg, F., Pedersen, J.K. and Thoegersen, P., "Improved Modulation Techniques for PWM-VSI Drives," *IEEE Transactions on Industry Applications,* vol. 44, no. 1, pp. 87-95, Feb. 1997.

[2] Choi, J.W. and Sul, S.K., "Inverter Output Voltage Synthesis Using Novel Dead Time Compensation," *IEEE Transactions on Power Electronics,* vol. 11, no. 2, pp. 221-227, Mar. 1996.

[3] Holtz, J. and Quan, J., "Sensorless Vector Control of Induction Motors at Very Low Speed Using a Nonlinear Inverter Model and Parameter Identification," *IEEE Transactions on Industry Applications,* vol. 38, no. 4, pp. 1087-1095, Jul/Aug. 2002.

[4] Oh, W.S., Kim, Y.T., and Kim, H.J., "Dead time Compensation of Current Controlled Inverter Using Space Vector Modulation Method," *IEEE Transactions on Power Electronics,* vol. 80, no. 2, pp. 277-289, Feb. 1996.

[5] Mondal, S, K., Pinto, J.O.P., and Bose, B.K., "A Neural-Network-Based Space-Vector PWM Controller for a Three-Level Voltage-Fed Inverter Induction Motor Drive," *IEEE Transactions on Industry Applications,* vol. 38, no. 3, pp. 660-669, May/June, 2002.

Nattapong Pothi received the B.Eng. and M.Eng. degree in electrical engineering from Chiang Mai University, Thailand in 2002, and 2006, respectively.

Currently, he is a Lecturer in the Department of Electrical Engineering, Naresuan University Phayao, Thailand. His research interest is the power electronics and drives system.

Suttichai Premrudeepreechacharn received the B.Eng. in electrical engineering from Chiang Mai University, Thailand and M.S. and Ph.D. in electric power engineering from Rensselaer Polytechnic Institute, Troy, NY.

Currently, he is an Associate Professor at the Department of Electrical Engineering, Chiang Mai University, Thailand. His research interests include power quality, high quality utility interface, power electronics and artificial intelligence applied power system.

Chawasak Rakpenthai received the B.Eng., M.Eng. and Ph.D. degrees in electrical engineering from Chiang Mai University, Thailand in 1999, 2003 and 2007, respectively.

Currently, he is a Lecturer in the Department of Electrical Engineering, Naresuan University Phayao, Thailand. His research interests include artificial intelligence in power system, power electronic, power system state estimation, and flexible AC transmission systems (FACTS) devices.

A Simple Carrier-Based PWM Method For Three-Phase Four-Leg Inverters Considering All Four Pole Voltages Simultaneously

Nakharet Chudoung and Somboon Sangwongwanich
Dept of Electrical Eng., Faculty of Eng., Chulalaongkorn Unviersity
Payatai, Bangkok 10330, Thailand somboona@chula.ac.th

Abstract--Complexity of the three-dimensional space vector PWM (3D-SVM) algorithm can be reduced using an equivalent carrier-based continuous PWM method. However, since the neutral-phase voltage was neglected in determining the zero-sequence voltage of the PWM, it introduces an error for an imbalance case, increases complexity of the algorithm, and is inconsistent with the carrier-based PWM for the three-leg inverter. This paper proposes a new and natural standpoint for the carrier-based PWM of the four-leg inverter, by considering all pole voltages including that of the neutral phase simultaneously in the determination of zero-sequence voltage. This new viewpoint renders the full consistency between the PWM methods for the three- and four-leg inverters. It can generate both continuous and discontinuous PWM easily in the same manner as is done for the three-leg inverter. The connection between the discontinuous PWM and the 3D-SVM is also clarified.

Index Terms—**Carrier-based PWM, four-leg inverters, zero-sequence voltage, discontinuous PWM, three-dimensional space vector PWM.**

I. INTRODUCTION

A three-phase voltage source inverter with a neutral leg, i.e. a four-leg inverter as shown in Fig. 1, is capable of providing unbalanced voltages and can handle the neutral current caused by the unbalanced load or source. Since the three-phase four-leg inverter is more advantageous than the three-phase three-leg inverter as compared in Table I, it is generally used for the applications in the three-phase four-wire system where the connection and control of the neutral-point voltage are necessary, e.g. active filters or power converters for renewable energy system [1]-[9]. Several pulse-width-modulation (PWM) methods have also been proposed for the four-leg inverters to generate the required phase voltages which may be balanced or unbalanced.

Table I. Comparison between three- and four-leg inverters in three-phase four-wire system.

	Three-Leg Four-Wire Inverter	Four-Leg Four-Wire Inverter
DC link capacitor	Large due to zero-sequence current	Small (absorbing only switching-ripple current)
PWM algorithm	Simple	Complex
Peak line-to-neutral voltage	0.5 Vg	0.577 Vg

The main concept of the PWM methods presented so far extends the well-known two-dimensional space vector PWM (2D-SVM) to the three-dimensional one (3D-SVM) [3]-[7]. Since, the calculation of the switching timing in the SVM is quite complicated and time-consuming, several attempts have been made to alleviate this problem [7]-[9]. Similar to what is done for the three-leg inverters, [8] presented a carrier-based continuous PWM method which is equivalent to the 3D-SVM, but is much easier to implement. However, in [8] the neutral-pole voltage was neglected and only three phase voltages were considered in the process of determining the zero-sequence (or off-set) voltage. This unnatural viewpoint is fine if the generated voltages are balanced, but it may introduce a pitfall for the imbalance case. Another consequence of such consideration of the zero-sequence voltage is the complexity of the algorithm and the inconsistency with the carrier-based PWM for three-leg inverters. In addition, the connection between the carrier-based discontinuous PWM [9] and the 3D-SVM is not yet clarified.

The objectives of this paper are to propose a new and natural standpoint for the carrier-based PWM of the four-leg inverter, by considering all four pole voltages including that of the neutral phase simultaneously in the determination of zero-sequence voltage. This new viewpoint renders the full consistency between the PWM methods for the three- and four-leg inverters, and is very simple in the implementation. It can generate both continuous and discontinuous PWM easily by adjusting the zero-sequence voltage in the same manner as what is done for the three-phase PWM. In fact, all the PWM methods derived for the three-leg inverter can be applied to the four-leg inverter seamlessly. The connection between the discontinuous PWM and the 3D-SVM is also shown. Simulation and experiment are carried out to verify the correctness of the theoretical results.

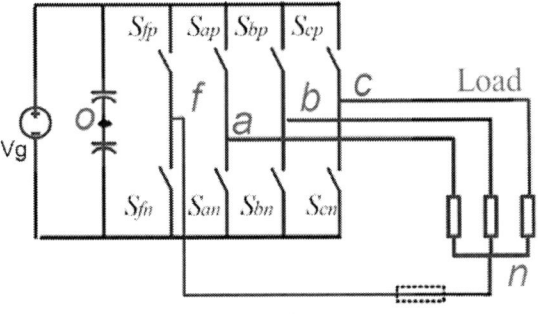

Fig. 1. Four-leg voltage-source inverter.

978-1-4244-0644-9/07/$25.00 ©2007 IEEE

II. CONVENTIONAL AND PROPOSED CARRIER-BASED PWM METHODS

The carrier-based PWM makes use of the fact that as far as the line-to-neutral voltages (1) at the load are kept unchanged, we may add arbitrary zero-sequence voltage v_{fo}^* to the commanded pole voltages. The resultant pole voltages are then given by (2).

Phase voltages:

$$v_{an}^* = v_{af}^* + v_{fn}^* = v_{af}^*, \quad v_{bn}^* = v_{bf}^* + v_{fn}^* = v_{bf}^*,$$
$$v_{cn}^* = v_{cf}^* + v_{fn}^* = v_{cf}^*, \quad v_{fn}^* = 0 \tag{1}$$

where v_{in} is the line-to-neutral voltage, v_{if} is the line-to-f-phase voltage, and '*' denotes the commanded value.

Pole voltages:

$$v_{ao}^* = v_{af}^* + v_{fo}^*, \quad v_{bo}^* = v_{bf}^* + v_{fo}^*, \quad v_{co}^* = v_{cf}^* + v_{fo}^* \tag{2}$$

where v_{io} is the pole voltage referred to the middle point of the dc bus V_g. Since all pole voltages including that of the neutral phase are generated by the inverter, they must satisfy the following conditions:

$$-\frac{V_g}{2} \le v_{ao}^*, v_{bo}^*, v_{co}^* \le +\frac{V_g}{2} \tag{3}$$

$$-\frac{V_g}{2} \le v_{fo}^* \le +\frac{V_g}{2}. \tag{4}$$

A. Conventional carrier-based PWM

In the conventional carrier-based PWM[8], the a,b,c phases are treated differently from the neutral phase f. This standpoint thus emphasizes mainly on the a,b,c phase voltages and views the function of the neutral pole as only to generate the required zero-sequence voltage. As a consequence of this standpoint, the required zero-sequence voltage v_{fo}^* is determined solely from the three phase voltages as shown in (5)-(7), and it is added to the required phase voltages to obtain the commanded pole voltages as previously shown in (2). The PWM signals are then generated using the triangular carrier waveform as illustrated in Fig. 2, where only three phase or pole voltages are taken into consideration.

$$v_{fo}^* = \begin{cases} -v_{max}'/2, & v_{min}' > 0 \\ -v_{min}'/2, & v_{max}' < 0 \\ -(v_{max}' + v_{min}')/2, & \text{Otherwise} \end{cases} \tag{5}$$

where $\quad v_{max}' = \max(v_{an}^*, v_{bn}^*, v_{cn}^*) \tag{6}$

$$v_{min}' = \min(v_{an}^*, v_{bn}^*, v_{cn}^*). \tag{7}$$

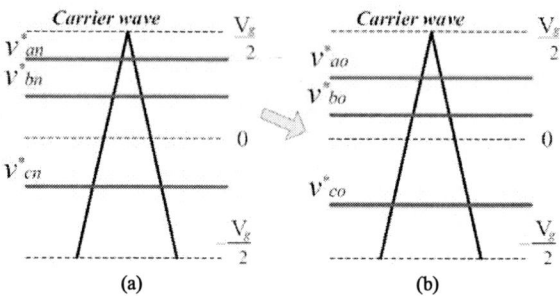

(a)　　　　　　　(b)

Fig. 2. Reference voltages considering only three pole voltages: (a) required phase voltages and (b) commanded pole voltages.

And based on the conceptual diagram in Fig. 2, it is wrongly concluded in [8] that the allowable range of zero-sequence voltage v_{fo}^* is given by:

$$-\frac{V_g}{2} - v_{min}' \le v_{fo}^* \le +\frac{V_g}{2} - v_{max}'. \tag{8}$$

The correct one is shown later in (9). Although, this carrier-based PWM method is shown to be equivalent to the 3D-SVM, the equations for the zero-sequence voltage (5)-(7) are more complicated and are inconsistent with that of the 2D-SVM for the three-leg inverter. This happens because only three of the four pole voltages are considered.

B. Proposed carrier-based PWM

In this paper, the following new standpoint is adopted. Although in most cases the neutral phase voltage v_{fn}^* is zero, the neutral phase f should be treated equally as the other a,b,c phases. We thus propose that all the four pole voltages should be simultaneously considered in the generation of PWM signals as shown in Fig. 3. Following the diagram in Fig. 3, it is clear that the correct allowable range of zero-sequence voltage is

Allowable zero-sequence voltage:

$$-\frac{V_g}{2} - v_{min} \le v_{fo}^* \le +\frac{V_g}{2} - v_{max} \tag{9}$$

where $v_{max} = \max(v_{an}^*, v_{bn}^*, v_{cn}^*, v_{fn}^*) \tag{10}$

$$v_{min} = \min(v_{an}^*, v_{bn}^*, v_{cn}^*, v_{fn}^*). \tag{11}$$

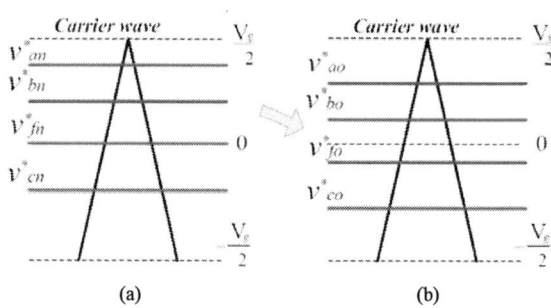

(a)　　　　　　　(b)

Fig. 3. Reference voltages considering four pole voltages: (a) required phase voltages and (b) commanded pole voltages.

Using this new definition of the instantaneous values of the maximum and minimum voltages, the zero-sequence voltage in (5) can be rewritten as:

$$v_{fo}^* = -\frac{v_{max} + v_{min}}{2}. \tag{12}$$

Equation (12) determines the zero-sequence voltage for the 3D-SVM (explained later) and coincides with the same zero-sequence voltage's equation used in the 2D-SVM of the three-leg inverter.. Therefore, the consistency between the PWM methods for the four- and three-leg inverters is rendered by considering all four pole voltages simultaneously. The main difference between considerations of four and three pole voltages exists when the phase voltages are heavily unbalanced and become at some instants all positive or negative simultaneously as shown in Figs. 4 and 5. In such cases, the neutral phase voltage v_{fn}^* (not the phase voltages $v_{an}^*, v_{bn}^*, v_{cn}^*$) becomes the minimum or maximum voltage.

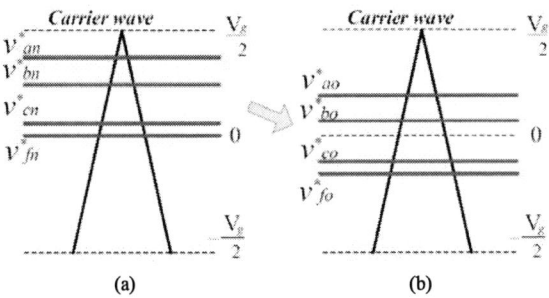

Fig. 4. Reference voltages considering four pole voltages under heavily unbalanced case (I): (a) required phase voltages and (b) commanded pole voltages

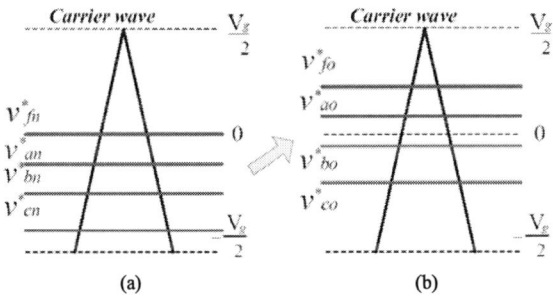

Fig. 5. Reference voltages considering four pole voltages under heavily unbalanced case (II): (a) required phase voltages and (b) commanded pole voltages.

III. CONNECTION BETWEEN CARRIED-BASED PWM AND 3D-SVM

In the SVM, the required voltage vector is synthesized from the switching voltage vectors generated by the inverter. For the four-leg inverter, the vector diagram of the corresponding voltage vectors (14 active vectors and 2 zero vectors) is shown in Fig. 6. This three dimensional polygon generated by the four-leg inverter is divided into 4 prisms, and each prism is divided again

into 4 tetrahedrons. The commanded voltage belonging to one of the tetrahedrons is then synthesized from three adjacent active vectors ($V1, V2, V3$) and one or two zero vectors ($V0p, V0n$) (see Table III for the voltage vectors of each tetrahedron). For each region of the tetrahedron, the polarity of the phase voltages and the corresponding maximum and minimum voltages are given in Table II.

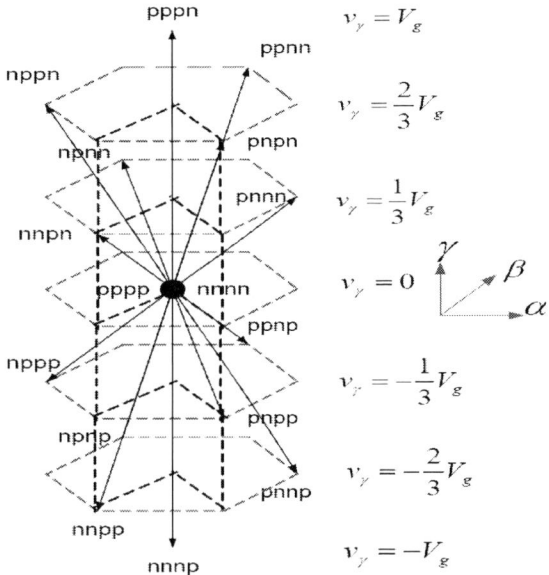

Fig. 6. 3D space vector diagram for the four-leg inverter.

A. Carrier-based continuous PWM and 4-arm switching 3D-SVM

In the 4-arm switching 3D-SVM, the zero voltage vectors $V0p, V0n$ are used equally at the beginning and the end of the switching period as shown in Fig. 7 and Table III. In this case, all the switches of the four legs (arms) will change the status during one switching period. The zero-sequence voltage generated by the 4-arm switching 3D-SVM is calculated to be as shown in Table IV for each area of the tetrahedrons. From Tables II and IV, it is easily found that the zero-sequence voltage generated by the 4-arm switching 3D-SVM is equal to (12) of the carrier-based continuous PWM previously mentioned.

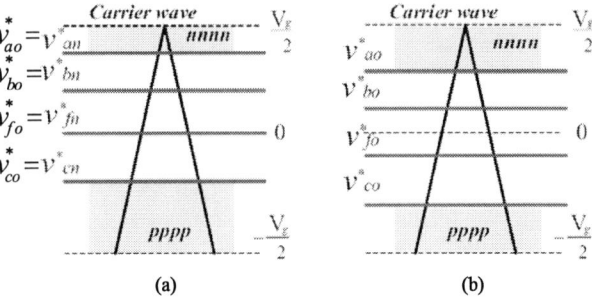

Fig. 7. Distribution of zero vectors in sinusoidal PWM and 4-arm switching 3D-SVM: (a) sinusoidal PWM and (b) 4-arm switching 3D-SVM.

Table II. Polarity and maximum/minimum voltages for each tetrahedron.

		Tetrahedron			
		1	2	3	4
Prism I		$v_{an}^* \geq 0$ (max)	$v_{an}^* \geq 0$ (max)	$v_{an}^* \geq 0$ (max)	$v_{an}^* \leq 0$
		$v_{bn}^* \leq 0$	$v_{bn}^* \geq 0$	$v_{bn}^* \geq 0$	$v_{bn}^* \leq 0$
		$v_{cn}^* \leq 0$ (min)	$v_{cn}^* \leq 0$ (min)	$v_{cn}^* \geq 0$	$v_{cn}^* \leq 0$ (min)
		$v_{fn}^* = 0$	$v_{fn}^* = 0$	$v_{fn}^* = 0$ (min)	$v_{fn}^* = 0$ (max)
II		$v_{an}^* \geq 0$	$v_{an}^* \leq 0$	$v_{an}^* \geq 0$	$v_{an}^* \leq 0$
		$v_{bn}^* \geq 0$ (max)	$v_{bn}^* \geq 0$ (max)	$v_{bn}^* \geq 0$ (max)	$v_{bn}^* \leq 0$
		$v_{cn}^* \leq 0$ (min)	$v_{cn}^* \leq 0$ (min)	$v_{cn}^* \geq 0$	$v_{cn}^* \leq 0$ (min)
		$v_{fn}^* = 0$	$v_{fn}^* = 0$	$v_{fn}^* = 0$ (min)	$v_{fn}^* = 0$ (max)
III		$v_{an}^* \leq 0$ (min)	$v_{an}^* \leq 0$ (min)	$v_{an}^* \geq 0$	$v_{an}^* \leq 0$ (min)
		$v_{bn}^* \geq 0$ (max)	$v_{bn}^* \geq 0$ (max)	$v_{bn}^* \geq 0$ (max)	$v_{bn}^* \leq 0$
		$v_{cn}^* \leq 0$	$v_{cn}^* \geq 0$	$v_{cn}^* \geq 0$	$v_{cn}^* \leq 0$
		$v_{fn}^* = 0$	$v_{fn}^* = 0$	$v_{fn}^* = 0$ (min)	$v_{fn}^* = 0$ (max)
IV		$v_{an}^* \leq 0$ (min)	$v_{an}^* \leq 0$ (min)	$v_{an}^* \geq 0$	$v_{an}^* \leq 0$ (min)
		$v_{bn}^* \geq 0$	$v_{bn}^* \leq 0$	$v_{bn}^* \geq 0$	$v_{bn}^* \leq 0$
		$v_{cn}^* \geq 0$ (max)	$v_{cn}^* \geq 0$ (max)	$v_{cn}^* \geq 0$ (max)	$v_{cn}^* \leq 0$
		$v_{fn}^* = 0$	$v_{fn}^* = 0$	$v_{fn}^* = 0$ (min)	$v_{fn}^* = 0$ (max)
V		$v_{an}^* \leq 0$	$v_{an}^* \geq 0$	$v_{an}^* \geq 0$	$v_{an}^* \leq 0$
		$v_{bn}^* \leq 0$ (min)	$v_{bn}^* \leq 0$ (min)	$v_{bn}^* \geq 0$	$v_{bn}^* \leq 0$ (min)
		$v_{cn}^* \geq 0$ (max)	$v_{cn}^* \geq 0$ (max)	$v_{cn}^* \geq 0$ (max)	$v_{cn}^* \leq 0$
		$v_{fn}^* = 0$	$v_{fn}^* = 0$	$v_{fn}^* = 0$ (min)	$v_{fn}^* = 0$ (max)
VI		$v_{an}^* \geq 0$ (max)	$v_{an}^* \geq 0$ (max)	$v_{an}^* \geq 0$ (max)	$v_{an}^* \leq 0$
		$v_{bn}^* \leq 0$ (min)	$v_{bn}^* \leq 0$ (min)	$v_{bn}^* \geq 0$	$v_{bn}^* \leq 0$ (min)
		$v_{cn}^* \geq 0$	$v_{cn}^* \leq 0$	$v_{cn}^* \geq 0$	$v_{cn}^* \leq 0$
		$v_{fn}^* = 0$	$v_{fn}^* = 0$	$v_{fn}^* = 0$ (min)	$v_{fn}^* = 0$ (max)

Table III. Vector utilization of 4-arm switching 3D-SVM.

Vector Sequence for 4-Arm Switching				
	Tetrahedron			
	1	2	3	4
Prism I	V0n=nnnn	V0n=nnnn	V0n=nnnn	V0n=nnnn
	V1 = pnnn	V1 = pnnn	V1 = pnnn	V1 = nnnp
	V2 = pnnp	V2 = ppnn	V2 = ppnn	V2 = pnnp
	V3 = ppnp	V3 = pppn	V3 = pppn	V3 = ppnp
	V0p=pppp	V0p=pppp	V0p=pppp	V0p=pppp
II	V0n=nnnn	V0n=nnnn	V0n=nnnn	V0n=nnnn
	V1 = pnnn	V1 = pnnn	V1 = npnn	V1 = nnnp
	V2 = ppnn	V2 = npnp	V2 = ppnn	V2 = npnp
	V3 = ppnp	V3 = pppn	V3 = pppn	V3 = ppnp
	V0p=pppp	V0p=pppp	V0p=pppp	V0p=pppp
III	V0n=nnnn	V0n=nnnn	V0n=nnnn	V0n=nnnn
	V1 = npnn	V1 = npnn	V1 = npnn	V1 = nnnp
	V2 = npnp	V2 = nppn	V2 = nppn	V2 = npnp
	V3 = nppp	V3 = nppp	V3 = pppn	V3 = nppp
	V0p=pppp	V0p=pppp	V0p=pppp	V0p=pppp
IV	V0n=nnnn	V0n=nnnn	V0n=nnnn	V0n=nnnn
	V1 = nnpn	V1 = nnpn	V1 = nnpn	V1 = nnnp
	V2 = nppn	V2 = nnpp	V2 = nnpn	V2 = nnpp
	V3 = nppp	V3 = nppp	V3 = pppn	V3 = nppp
	V0p=pppp	V0p=pppp	V0p=pppp	V0p=pppp
V	V0n=nnnn	V0n=nnnn	V0n=nnnn	V0n=nnnn
	V1 = nnpn	V1 = nnpn	V1 = nnpn	V1 = nnnp
	V2 = nnpp	V2 = pnpn	V2 = pnpn	V2 = nnpp
	V3 = pnpp	V3 = pnpp	V3 = pppn	V3 = pnpp
	V0p=pppp	V0p=pppp	V0p=pppp	V0p=pppp
VI	V0n=nnnn	V0n=nnnn	V0n=nnnn	V0n=nnnn
	V1 = pnnn	V1 = pnnn	V1 = pnnn	V1 = nnnp
	V2 = pnpn	V2 = pnnp	V2 = pnpn	V2 = pnnp
	V3 = pnpp	V3 = pnpp	V3 = pppn	V3 = pnpp
	V0p=pppp	V0p=pppp	V0p=pppp	V0p=pppp

As a conclusion, the 4-arm switching 3D-SVM can be realized much easier using the carrier-based continuous PWM with zero-sequence voltage of (12) to place the voltage range covered by the four pole voltages symmetrically balanced round the midpoint of the carrier waveform.

B. Carrier-based discontinuous PWM and 3-arm switching 3D-SVM

In the 3-arm switching 3D-SVM, only either the zero voltage vector $V0n$ or $V0p$ is used during one switching period as shown in Tables V and VII. In this case, the switches of one leg (arm) will not change the status during one switching period. And this leg can be the neutral phase leg if the commanded phase voltages are heavily unbalanced. The zero-sequence voltage generated by the 3-arm switching 3D-SVM is calculated to be as shown in Tables VI and VIII for each tetrahedron. From Tables II, VI and VIII, it is concluded that the zero-sequence voltage generated by the 3-arm switching 3D-SVM is equal to (13) or (14) of the so-called carrier-based discontinuous PWM modes I or II, by which one of the pole voltage (including that of the

Table IV. Zero-sequence voltage generated by 4-arm switching 3D-SVM for each tetrahedron.

Zero-Sequence Voltage for 4-Arm Switching				
	Tetrahedron			
	1	2	3	4
Prism I	$-\dfrac{v_{an}^* + v_{cn}^*}{2}$	$-\dfrac{v_{an}^* + v_{cn}^*}{2}$	$-\dfrac{v_{an}^* + v_{fn}^*}{2}$	$-\dfrac{v_{fn}^* + v_{cn}^*}{2}$
II	$-\dfrac{v_{bn}^* + v_{cn}^*}{2}$	$-\dfrac{v_{bn}^* + v_{cn}^*}{2}$	$-\dfrac{v_{bn}^* + v_{fn}^*}{2}$	$-\dfrac{v_{fn}^* + v_{cn}^*}{2}$
III	$-\dfrac{v_{bn}^* + v_{an}^*}{2}$	$-\dfrac{v_{bn}^* + v_{an}^*}{2}$	$-\dfrac{v_{bn}^* + v_{fn}^*}{2}$	$-\dfrac{v_{fn}^* + v_{an}^*}{2}$
IV	$-\dfrac{v_{cn}^* + v_{an}^*}{2}$	$-\dfrac{v_{cn}^* + v_{an}^*}{2}$	$-\dfrac{v_{cn}^* + v_{fn}^*}{2}$	$-\dfrac{v_{fn}^* + v_{an}^*}{2}$
V	$-\dfrac{v_{cn}^* + v_{bn}^*}{2}$	$-\dfrac{v_{cn}^* + v_{bn}^*}{2}$	$-\dfrac{v_{cn}^* + v_{fn}^*}{2}$	$-\dfrac{v_{fn}^* + v_{bn}^*}{2}$
VI	$-\dfrac{v_{an}^* + v_{bn}^*}{2}$	$-\dfrac{v_{an}^* + v_{bn}^*}{2}$	$-\dfrac{v_{an}^* + v_{fn}^*}{2}$	$-\dfrac{v_{fn}^* + v_{bn}^*}{2}$

neutral phase) will be clamped to either the negative or positive rail of the dc bus (Fig. 8).

$$v_{fo}^* = -\frac{V_g}{2} - v_{min} \quad (13) \quad \text{or} \quad v_{fo}^* = \frac{V_g}{2} - v_{max}. \quad (14)$$

Table V. Vector utilization of 3-arm switching 3D-SVM (mode I).

Vector Sequence for 3-Arm Switching (Mode I)				
	Tetrahedron			
	1	2	3	4
Prism I	V0n=nnnn	V0n=nnnn	V0n=nnnn	V0n=nnnn
	V1 = pnnn	V1 = pnnn	V1 = pnnn	V1 = nnnp
	V2 = pnnp	V2 = ppnn	V2 = ppnn	V2 = pnnp
	V3 = ppnp	V3 = ppnp	V3 = pppn	V3 = ppnp
II	V0n=nnnn	V0n=nnnn	V0n=nnnn	V0n=nnnn
	V1 = npnn	V1 = npnn	V1 = npnn	V1 = nnnp
	V2 = ppnn	V2 = npnp	V2 = ppnn	V2 = npnp
	V3 = ppnp	V3 = ppnp	V3 = pppn	V3 = ppnp
III	V0n=nnnn	V0n=nnnn	V0n=nnnn	V0n=nnnn
	V1 = npnn	V1 = npnn	V1 = npnn	V1 = nnnp
	V2 = npnp	V2 = nppn	V2 = nppn	V2 = npnp
	V3 = nppp	V3 = nppp	V3 = pppn	V3 = nppp
IV	V0n=nnnn	V0n=nnnn	V0n=nnnn	V0n=nnnn
	V1 = nnpn	V1 = nnpn	V1 = nnpn	V1 = nnnp
	V2 = nppn	V2 = nnpp	V2 = nppn	V2 = nnpp
	V3 = nppp	V3 = nppp	V3 = pppn	V3 = nppp
V	V0n=nnnn	V0n=nnnn	V0n=nnnn	V0n=nnnn
	V1 = nnpn	V1 = nnpn	V1 = nnpn	V1 = nnnp
	V2 = nnpp	V2 = pnpn	V2 = pnpn	V2 = nnpp
	V3 = pnpp	V3 = pnpp	V3 = pppn	V3 = pnpp
VI	V0n=nnnn	V0n=nnnn	V0n=nnnn	V0n=nnnn
	V1 = pnnn	V1 = pnnn	V1 = pnnn	V1 = nnnp
	V2 = pnpn	V2 = pnnp	V2 = pnpn	V2 = pnnp
	V3 = pnpp	V3 = pnpp	V3 = pppn	V3 = pnpp

Table VI. Zero-sequence voltage generated by 3-arm switching 3D-SVM (mode I).

Zero-Sequence Voltage for 3-Arm Switching (Mode I)				
	Tetrahedron			
	1	2	3	4
Prism I	$-\frac{V_g}{2} - v_{cn}^*$	$-\frac{V_g}{2} - v_{cn}^*$	$-\frac{V_g}{2} - v_{fn}^*$	$-\frac{V_g}{2} - v_{cn}^*$
II	$-\frac{V_g}{2} - v_{cn}^*$	$-\frac{V_g}{2} - v_{cn}^*$	$-\frac{V_g}{2} - v_{fn}^*$	$-\frac{V_g}{2} - v_{cn}^*$
III	$-\frac{V_g}{2} - v_{an}^*$	$-\frac{V_g}{2} - v_{an}^*$	$-\frac{V_g}{2} - v_{fn}^*$	$-\frac{V_g}{2} - v_{an}^*$
IV	$-\frac{V_g}{2} - v_{an}^*$	$-\frac{V_g}{2} - v_{an}^*$	$-\frac{V_g}{2} - v_{fn}^*$	$-\frac{V_g}{2} - v_{an}^*$
V	$-\frac{V_g}{2} - v_{bn}^*$	$-\frac{V_g}{2} - v_{bn}^*$	$-\frac{V_g}{2} - v_{fn}^*$	$-\frac{V_g}{2} - v_{bn}^*$
VI	$-\frac{V_g}{2} - v_{bn}^*$	$-\frac{V_g}{2} - v_{bn}^*$	$-\frac{V_g}{2} - v_{fn}^*$	$-\frac{V_g}{2} - v_{bn}^*$

Table VII. Vector utilization of 3-arm Switching 3D-SVM (mode II).

Vector Sequence for 3-Arm Switching (Mode II)				
	Tetrahedron			
	1	2	3	4
Prism I	V1 = pnnn	V1 = pnnn	V1 = pnnn	V1 = nnnp
	V2 = pnnp	V2 = ppnn	V2 = ppnn	V2 = pnnp
	V3 = ppnp	V3 = ppnp	V3 = pppn	V3 = ppnp
	V0p=pppp	V0p=pppp	V0p=pppp	V0p=pppp
II	V1 = npnn	V1 = npnn	V1 = npnn	V1 = nnnp
	V2 = ppnn	V2 = npnp	V2 = ppnn	V2 = npnp
	V3 = ppnp	V3 = ppnp	V3 = pppn	V3 = ppnp
	V0p=pppp	V0p=pppp	V0p=pppp	V0p=pppp
III	V1 = npnn	V1 = npnn	V1 = npnn	V1 = nnnp
	V2 = npnp	V2 = nppn	V2 = nppn	V2 = npnp
	V3 = nppp	V3 = nppp	V3 = pppn	V3 = nppp
	V0p=pppp	V0p=pppp	V0p=pppp	V0p=pppp
IV	V1 = nnpn	V1 = nnpn	V1 = nnpn	V1 = nnnp
	V2 = nppn	V2 = nnpp	V2 = nppn	V2 = nnpp
	V3 = nppp	V3 = nppp	V3 = pppn	V3 = nppp
	V0p=pppp	V0p=pppp	V0p=pppp	V0p=pppp
V	V1 = nnpn	V1 = nnpn	V1 = nnpn	V1 = nnnp
	V2 = nnpp	V2 = pnpn	V2 = pnpn	V2 = nnpp
	V3 = pnpp	V3 = pnpp	V3 = pppn	V3 = pnpp
	V0p=pppp	V0p=pppp	V0p=pppp	V0p=pppp
VI	V1 = pnnn	V1 = pnnn	V1 = pnnn	V1 = nnnp
	V2 = pnpn	V2 = pnnp	V2 = pnpn	V2 = pnnp
	V3 = pnpp	V3 = pnpp	V3 = pppn	V3 = pnpp
	V0p=pppp	V0p=pppp	V0p=pppp	V0p=pppp

Table VIII. Zero-sequence voltage generated by 3-arm switching 3D-SVM (mode II).

Zero-Sequence Voltage for 3-Arm Switching (Mode II)				
	Tetrahedron			
	1	2	3	4
Prism I	$\frac{V_g}{2} - v_{an}^*$	$\frac{V_g}{2} - v_{an}^*$	$\frac{V_g}{2} - v_{an}^*$	$\frac{V_g}{2} - v_{fn}^*$
II	$\frac{V_g}{2} - v_{bn}^*$	$\frac{V_g}{2} - v_{bn}^*$	$\frac{V_g}{2} - v_{bn}^*$	$\frac{V_g}{2} - v_{fn}^*$
III	$\frac{V_g}{2} - v_{bn}^*$	$\frac{V_g}{2} - v_{bn}^*$	$\frac{V_g}{2} - v_{bn}^*$	$\frac{V_g}{2} - v_{fn}^*$
IV	$\frac{V_g}{2} - v_{cn}^*$	$\frac{V_g}{2} - v_{cn}^*$	$\frac{V_g}{2} - v_{cn}^*$	$\frac{V_g}{2} - v_{fn}^*$
V	$\frac{V_g}{2} - v_{cn}^*$	$\frac{V_g}{2} - v_{cn}^*$	$\frac{V_g}{2} - v_{cn}^*$	$\frac{V_g}{2} - v_{fn}^*$
VI	$\frac{V_g}{2} - v_{an}^*$	$\frac{V_g}{2} - v_{an}^*$	$\frac{V_g}{2} - v_{an}^*$	$\frac{V_g}{2} - v_{fn}^*$

The relationship between the carrier-based PWM and the 3D-SVM can be then summarized in Table IX, wherein t0 denotes the total time for zero vectors, and tp, tn are the switching times of the zero vectors $V0p, V0n$, respectively. It should be noted that for each tetrahedron there always exists two phases which can be clamped to the negative or positive rail. Any combination of the modes I and II, results also in a discontinuous PWM. This property can be utilized such as to achieve the minimum-loss PWM, by which the phase with the maximum current is clamped to negative or positive rail if allowed; otherwise, the phase with the next large current is clamped.

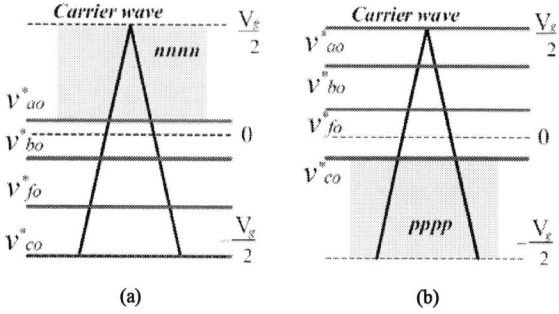

Fig. 8. Distribution of zero vectors in 3-arm switching 3D-SVM modes (I) and (II): (a) clamped to negative rail and (b) clamped to positive rail.

Table IX. Relationship between carrier-based PWM and 3D-SVM

	3D-SVM	Zero-sequence voltage
Continuous PWM	tp=tn	$v^*_{fo} = -(v_{max} + v_{min})/2$
Discontinuous PWM (mode I)	tp=0, tn=t0	$v^*_{fo} = -V_g/2 - v_{min}$
Discontinuous PWM (mode II)	tp=t0, tn=0	$v^*_{fo} = V_g/2 - v_{max}$

IV. EXPERIMENTAL RESULTS

Simulation and experiment are carried out to verify the correctness of the theoretical results, but due to limited space only the experimental results will be given. In the experiment, the dc bus voltage is 540 V, and the switching frequency is 10 KHz. The inverter is driving a load impedance of 50 Ω and 30 mH connected in series. It should be mentioned that for the four-leg inverter, if the current is unbalanced the dead-time effect of the neutral phase may affect each phase voltage unequally. To eliminate this effect, it is compensated in the experiment. Figures 9, 10, 12, and 13 are the waveforms of the commanded pole voltages and the corresponding PWM signals in the case of balanced three-phase voltages. The carrier-based PWM used therein are the sinusoidal PWM, 4-arm switching 3D-SVM, 3-arm switching 3D-SVM (mode I) and (mode II), respectively. These resultant waveforms are similar to those of the three-leg inverter. All the PWM algorithms produce the same phase currents as shown in Fig. 11. The waveforms for the unbalanced voltage commands are shown in Figs. 14, 15, 17, and 18, for the sinusoidal PWM, 4-arm switching 3D-SVM, 3-arm switching 3D-SVM (mode I) and (mode II), respectively. Similarly, all the PWM voltages give the same phase currents as shown in Fig. 16. All the experimental results confirm then that the proposed carrier-based PWM works well as expected in both continuous and discontinuous mode. The proposed

approach, which considers all four pole voltages simultaneously, enables us to develop the PWM theory for the four-leg inverter in the same manner as those of the three-leg inverter, which are well established.

V. CONCLUSIONS

With the standpoint that the neutral phase should be treated in the same way as the other three phases, the carrier-based PWM algorithm for the three-phase four-leg inverter is derived by considering all four pole voltages simultaneously. This new viewpoint renders the similarity of the PWM algorithms for the three- and four-leg inverters. It enables the fruitful research results of the PWM algorithms for the three-leg inverter to be applied to the four-leg inverter seamlessly. As a concrete example, in this paper the equivalence between the continuous/discontinuous carrier-based PWM and the 4-arm and 3-arm switching 3D SVM are shown in exactly the same way as are done for the three-leg inverter. Other interesting algorithms, e.g. the minimum-loss PWM algorithm based on the discontinuous PWM, can also be obtained for the four-leg inverter within this framework. The allowable voltage range for the zero-sequence voltage to be added to the commanded phase voltages is also clarified and unlike the previous research work, it is valid even if the phase voltages are heavily unbalanced. It can be concluded that the proposed framework of considering all four pole voltage simultaneously is more natural than the conventional one.

REFERENCES

[1] Ali. S.M., Kazmierkowski M.P., "Current regulation of four-leg PWM/VSI," Proc. of 24th IEEE IECON, 1998, pp. 1853–1858.

[2] Verdelho. P., Marques. G.D., "A current control system based in αβ0 variables for a four-leg PWM voltage converter," Proc. of IEEE IECON, 1998, pp. 1847-1852.

[3] Zhang. R., Prasad. V.H., Boroyevich. D., F.C. Lee, "Three-dimensional space vector modulation for four-leg voltage-source converters," IEEE Trans. on PE, vol. 17, no. 3, 2002, pp.:314–326.

[4] Prasad.V.H., Borojevic.D., Zhang.R, "Analysis and comparison of space vector modulation schemes for a four-leg voltage source inverter," Proc. of IEEE APEC'97, 1997, pp. 864–871.

[5] Shen. D., Lehn. P.W., "Fixed-frequency space-vector-modulation control for three-phase four-leg active power filters, " IEE Proc. on Electric Power Applications , 149 , 4 , (July 2002) : 268-274

[6] Perales. M.A., Prats. M.M., Portillo. R., Mora. J.L., Leon. J.I., Franquelo. L.G., "Three-dimensional space vector modulation in abc coordinates for four-leg voltage source converters," IEEE Power Electronics Letters, vol. 1, no. 4, 2003, pp. 104–109.

[7] P. Rodríguez, J. Pou, R. Pindado, J. Montanya, R. Burgos and D. Boroyevich, "An Alternative Approach on Three-Dimensional Space-Vector Modulation of Three-Phase Inverters," IEEE Proc. of PESC, 2005, pp. 882-828

[8] Jang-Hwan Kim, Seung-Ki Sul, "A carrier-based PWM method for three-phase four-leg voltage source converters," IEEE Trans. on PE, vol. 19, no. 1, 2004, pp. 66–75.

[9] Ojo. O., Kshirsagar. P., "A carrier-based discontinuous PWM modulation for four-leg inverters and applications," 38th IEEE/ IAS Conf. Rec., 2003., pp. 32 – 39.

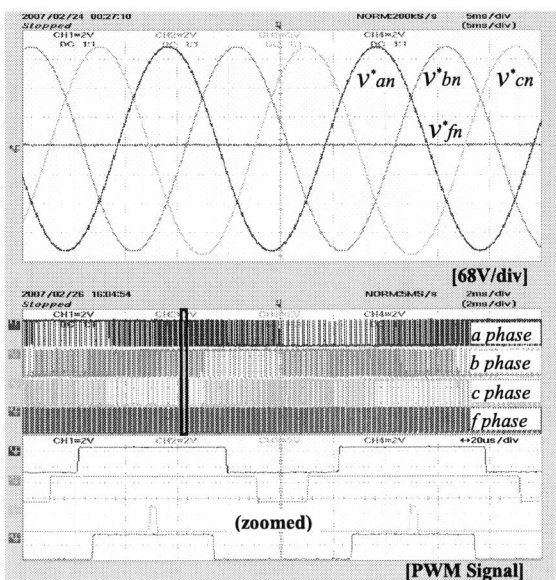

Fig. 9. Commanded pole voltages and PWM signals for sinusoidal PWM ($v_{fo}^* = 0$): balanced case.

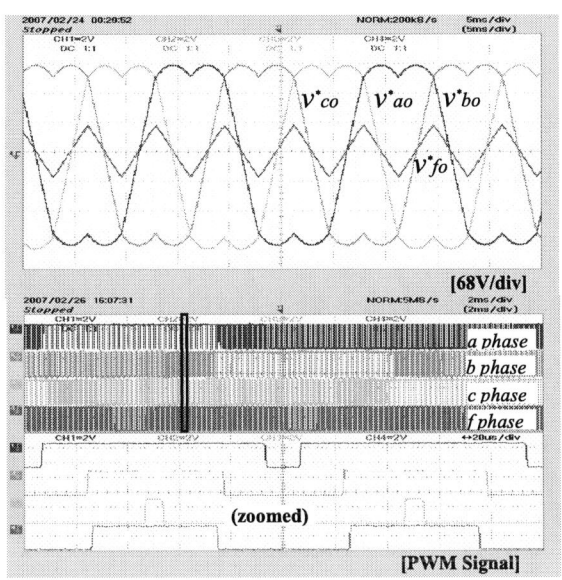

Fig. 10. Commanded pole voltages and PWM signals for 4-arm switching 3D-SVM ($v_{fo}^* = -(v_{max} + v_{min})/2$): balanced case.

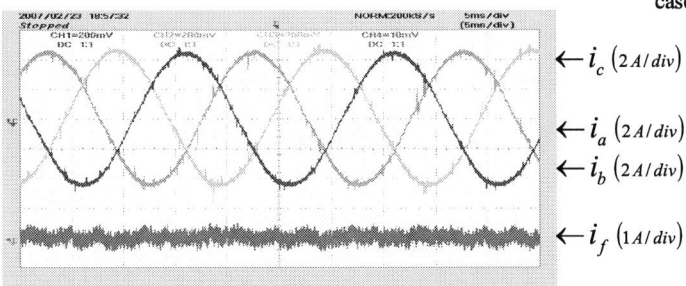

Fig. 11. Phase currents for the case of balanced phase voltages.

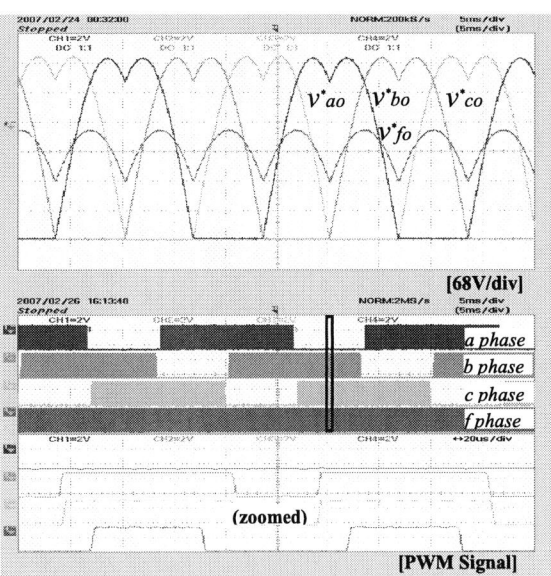

Fig. 12. Commanded pole voltages and PWM signals for 3-arm switching 3D-SVM mode I ($v_{fo}^* = -V_g/2 - v_{min}$): balanced case.

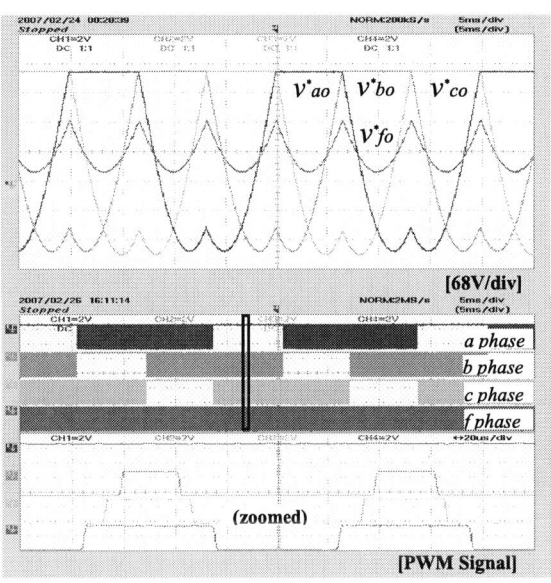

Fig. 13. Commanded pole voltages and PWM signals for 3-arm switching 3D-SVM mode II ($v_{fo}^* = V_g/2 - v_{max}$): balanced case.

1219

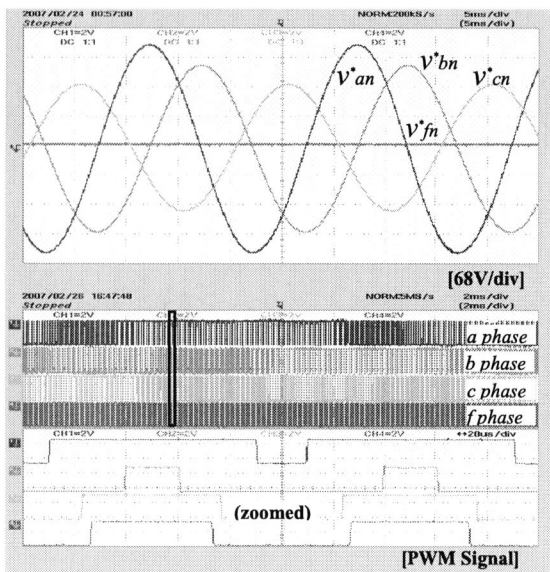

Fig. 14. Commanded pole voltages and PWM signals for sinusoidal PWM ($v_{fo}^* = 0$): unbalanced case.

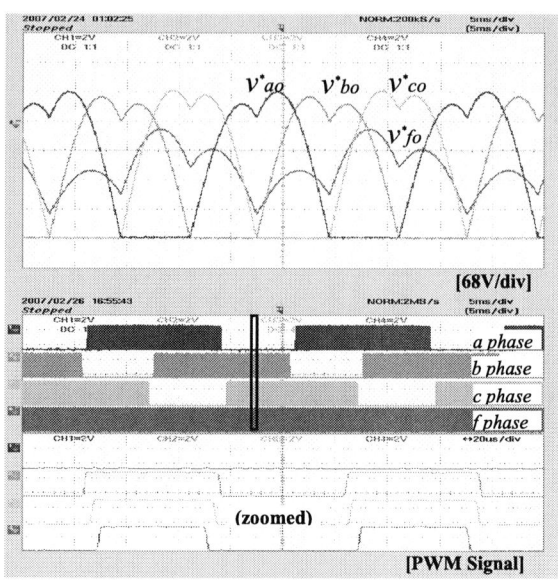

Fig. 17. Commanded pole voltages and PWM signals for 3-arm switching 3D-SVM mode I ($v_{fo}^* = -V_g/2 - v_{min}$): unbalanced case.

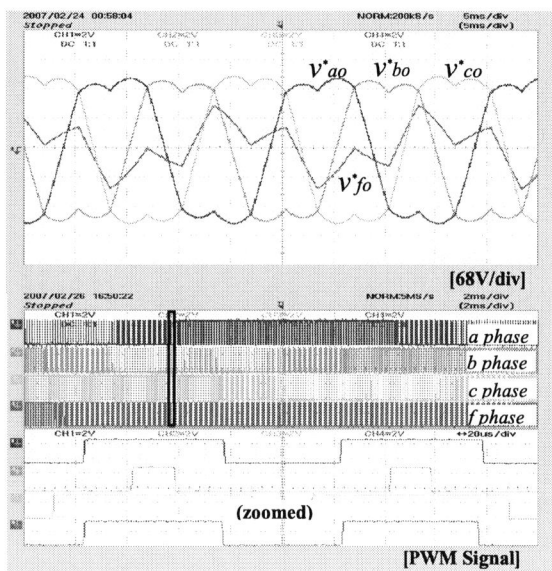

Fig. 15. Commanded pole voltages and PWM signals for 4-arm switching 3D-SVM ($v_{fo}^* = -(v_{max} + v_{min})/2$): unbalanced case.

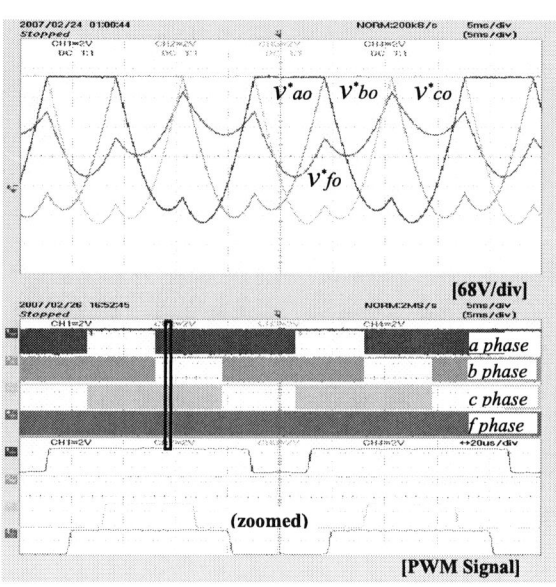

Fig. 18. Commanded pole voltages and PWM signals for 3-arm switching 3D-SVM mode II ($v_{fo}^* = V_g/2 - v_{max}$): unbalanced case.

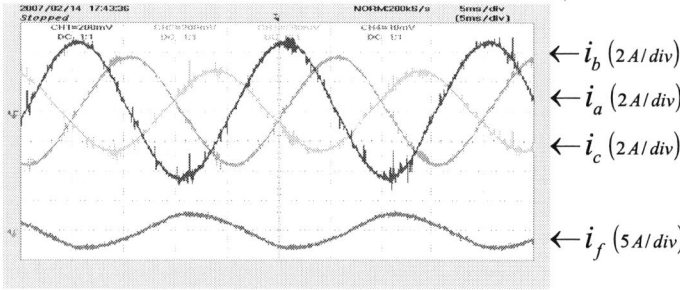

Fig. 16. Phase currents for the case of unbalanced phase voltages.

Inverted Sine Carrier Pulse Width Modulation for Fundamental Fortification in DC-AC Converters

R.Nandhakumar[1]
PTPHC-LV,
ABB Limited,
Bangalore -560058, India.
Email: nandha_raja1982@yahoo.co.in,

S.Jeevananthan[2]
Senior Lecturer, Department of EEE,
Pondicherry Engineering College,
Pondicherry -605014, India.
Email: jeeva_seeni@yahoo.com

Abstract-This paper deals with a novel natural sampled Pulse Width Modulation (PWM) switching strategy for Voltage Source Inverter (VSI) through carrier wave modification. The proposed Inverted Sine Carrier PWM (ISCPWM) method uses a sinusoidal reference signal and an inverted sine carrier which has better spectral quality and higher output limit as compared to the conventional Sinusoidal PWM (SPWM), without any pulse dropping. The ISCPWM strategy reduces the Total Harmonic Distortion (THD) without hampering device switching losses. It also enhances the fundamental output voltage particularly at lower modulation index. The paper also presents mathematical analysis to bring out the salient features of the proposed scheme. The detailed comparison of ISCPWM and SPWM has been also presented.

Index terms - Carrier Modification, Inverted Sine Carrier Pulse Width Modulation (ISCPWM), Total Harmonic Distortion (THD).

I INTRODUCTION

THE Pulse Width Wodulation (PWM) techniques and strategies have been subject of intensive research since 1970's so as get a sinusoidal ac output voltage. Sinusoidal PWM (SPWM) is an effective method to reduce lower order harmonics while varying the output voltage, which has undergone many revisions [1-5]. The SPWM technique has limitations with respect to maximum attainable voltage and power transfer. The maximum attainable fundamental amplitude in the SPWM output waveform is lesser than for the rectangular waveform. In case of the three-phase inverter, the ratio of the fundamental component of the utmost line-to-line voltage to the direct supply voltage is 86.6% and this value indicates poor exploitation of the dc supply.

The reduced circuit complexity delta modulation suitable for half-bridge inverter has been reported for smooth transition between the PWM and single pulse mode (V/f control) in 1981 [3]. The Third Harmonic Injection PWM (THIPWM) method suitable for three-phase inverters has been proposed in which a modulating wave is obtained by adding the third harmonic component to the fundamental sine in right proportion while the carrier is conventional triangular [6]. The triplen harmonic injection PWM (TRIPWM) is a variation of the THIPWM, in which the modulation function is obtained by adding the harmonic components of integer multiples of three to the fundamental sine [7].

In the above mentioned harmonic injection PWM methods, it is possible to increase the fundamental output voltage by about 15% and hence better utilization of the dc power supply and device. Usage of staircase as modulating function with high frequency triangular carrier for three-phase application had demonstrated nearly 10% fundamental improvement in the work reported in 1988 [8]. A modified carrier PWM method was proposed in which any two adjacent cycles of carrier triangular wave are grouped as either "W" shape or "M" shape and then suitable "W" and "M" cycle group conversion are made [9]. This type of carrier requires a digital platform for its implementation and gives about 4% and 19% improvements in fundamental component while working alone and amalgamated with THIPWM reference respectively. All the previous attempts to achieve the same objectives are by either regular sampled or mode-changing methods. However, in regular sampled PWM (digitally based controller), the generation of harmonics is dominated by quantization effects even with frequency ratios as low as 8:1 [10], hence they fail to emulate the properties of (natural) carrier and reference functions. The natural sampled solutions viz. THIPWM relay on mode changing.

The purpose of this paper is to propose a natural sampled single mode solution to fundamental restriction and distortion through the modification of carrier function. The proposed Inverted Sine Carrier PWM (ISCPWM) control scheme for single-phase full-bridge inverter, eliminates some of the limitations of the conventional SPWM viz. poor spectral quality of the output voltage, poor performance with regards to maximum output voltage possible etc. This paper also presents the mathematical analysis of the novel scheme along with the SPWM in addition to computer simulation.

II PWM STRATEGY

A Sinusoidal Pulse Width Modulation

The basic single-phase full-bridge inverter is shown in Fig.1 in which S_1, S_2 will be given PWM pulses for first (positive) output half cycles and S_3, S_4 are gated for the next (negative) half cycle. The unipolar PWM pulse generation with resulting pattern is shown in Fig.2 in which a triangular carrier wave is compared with sinusoidal reference waveform to generate PWM gating pulses. All PWM waveforms presented in this paper are

978-1-4244-0644-9/07/$25.00 ©2007 IEEE

assumed to be synchronous, unipolar PWM voltage switching.

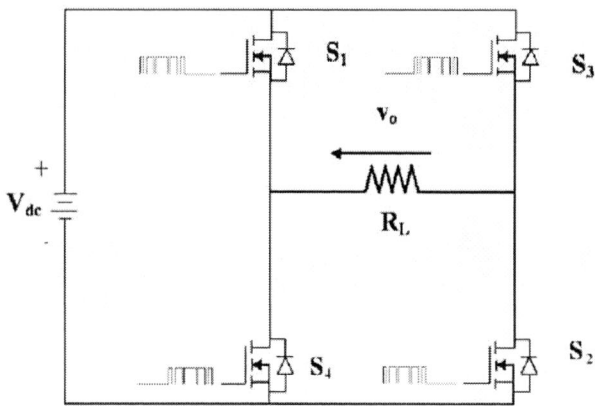

Fig.1 Basic single-phase inverter

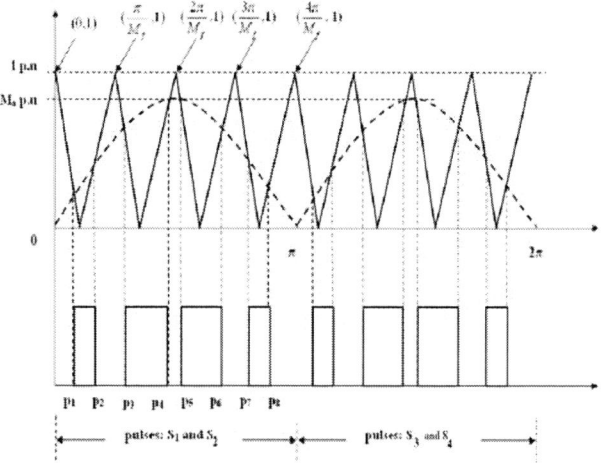

Fig.2 SPWM pulse generation and pattern

B. Mathematical Analysis

The harmonics present in the quasi-square wave and their relative amplitudes always remain the same. With PWM, however, the relative amplitudes of the harmonics change with the modulation index. The use of SPWM, for all its technical benefits, renders most complex calculations relating to inverter behavior. It is generally accepted that the performance of an inverter with any switching strategy can be related to the harmonic content of its output voltage [11].

A precise value of the switching angle and hence duty cycle can be obtained through the triangular (carrier) and the sinusoidal (reference) equations. The modulation pattern of the SPWM control (Fig.2) indicates the switching angles/meeting points (p_1, p_2, p_3...p_i). The PWM control signal is obtained by comparing a high frequency triangular carrier of frequency f_c and amplitude 1 (per unit) and a low frequency sine wave of frequency f_m and amplitude M_a (per unit). Equations for sinusoidal reference and triangular carrier are given by (1) and (2) respectively.

$$Y = M_a \sin x \qquad (1)$$

$$x \pm (\frac{\pi}{2 M_f}) * y = \frac{r}{M_f} \qquad (2)$$

Where, M_a-modulation index, M_f-frequency ratio, and r = for first pair of triangular sections (straight lines), 3 for second pair, 5 for third pair and so on. The '+' sign should be taken for odd numbered line sections and the '-' sign for even number line sections. The equations describing the natural sampled switching angles are transcendental and have the general distinct solutions for odd and even meeting points. The condition for switching angles is given in (3) and (4) respectively for odd and even switching angles.

$$M_a \sin p_i + \frac{2 M_f P_i}{\pi} = i \qquad \text{for i=1, 3, 5... (3)}$$

$$M_a \sin p_i - \frac{2 M_f P_i}{\pi} = 1 - i \qquad \text{for i=2, 4, 6... (4)}$$

Where, 'I' is the number of points and p_i is the ith switching angle. The pattern represented in Fig.2 does have eight switching angles and four PWM pulses. The duty cycle can be calculated by adding the width of the individual pulses. The width of any pulse can be found from subtracting one odd meeting point from immediate even successor. Since the inverter output irrespective of control methods exhibits equal positive and negative half cycles. Which results in zero dc components ($a_0=0$), and also does not posses any even harmonics due to half wave symmetry. Equation (5) gives the generalized Fourier coefficients for the problem considered. In the equation p'_i represents switching angles corresponds to negative half cycle

$$a_n = \frac{V_{dc}}{n\pi\sqrt{2}} \sum_{k=1}^{i-1} \{(\sin np_{k+1} - \sin np_k) - (\sin np'_{k+1} - \sin np'_k)\}$$

$$b_n = \frac{V_{dc}}{n\pi\sqrt{2}} \sum_{k=1}^{i-1} \{(\cos np_{k+1} - \cos np_k) - (\cos np'_{k+1} - \cos np'_k)\} \qquad (5)$$

$$c_n = \sqrt{a_n^2 + b_n^2}$$

III INVERTED SINE CARRIER PWM

The control strategy uses the same reference (synchronized sinusoidal signal), as the conventional SPWM while the carrier triangle is a modified one. The control scheme uses an inverted sine (high frequency) carrier that helps to maximize the output voltage for a given modulation index. Enhanced fundamental component demands greater pulse area. The difference in pulse widths (hence area) resulting from triangle wave and inverted sine wave with the low (output) frequency reference sine wave in different sections can be easily understood. In the gating pulse generation of the proposed ISCPWM scheme shown in Fig.3, the triangular carrier waveform of SPWM is replaced by an inverted sine waveform.

1222

For the ISCPWM pulse pattern, the switching angles may be computed as the same way as SPWM scheme. The equations of inverted sine wave are given by (6) and (7) for its odd and even cycles respectively. The intersections (q_1, q_2, q_3...q_i) between the inverted sine voltage waveform of amplitude 1 p.u and frequency f_c and the sinusoidal reference waveform of amplitude M_a p.u and frequency f_o can be obtained by substituting (1) in both (6) and (7). The switching angles for ISCPWM scheme can be obtained from (8) and (9).

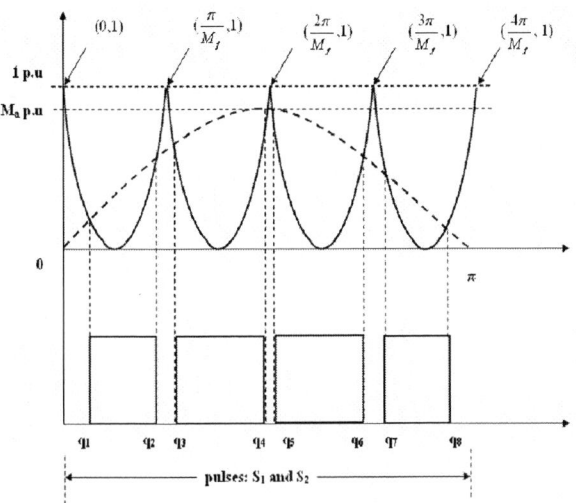

Fig 3 Inverted sine carrier PWM pulse pattern

$$y = 1 - \sin\left(M_f x - \frac{\pi}{2}(i-1)\right) \quad (6)$$

$$y = 1 - \sin\left(M_f x - \frac{\pi}{2}(i-2)\right) \quad (7)$$

$$M_a \sin q_i + \sin\left(M_f q_i - \frac{\pi}{2}(i-1)\right) = 1, \quad for \ i = 1,3,5... \quad (8)$$

$$M_a \sin q_i + \sin\left(M_f q_i - \frac{\pi}{2}(i-2)\right) = 1, \quad for \ i = 2,4,6... \quad (9)$$

It is worthwhile to note that both in SPWM (considered) and ISCPWM schemes, the number of pulses will be equal to M_f and hence the constant switching loss is guaranteed. To have conceptual understanding of wider pulse area and hence the dexterous input dc utilization in the ISCPWM, location of switching angles, duty cycle and their dependence on M_a and M_f are discussed. Fig.4 depicts the influence of M_a on different switching angles (four angles considered in both cases) at a constant M_f of 6. From this figure, it is observed that the odd switching instants vary with negative slope and the even switching instants have positive slope. Variation of all the switching instants against M_a is the straight line and slope of each one is more than it's previous. All the odd switching angles of ISCPWM method happen earlier than similar angles of PWM method, while the situation is reverse in case of even switching angles and hence higher pulse area.

Fig.5 gives the position of first switching angle, p_1/q_1 for various M_f at two M_a values 0.4 and 0.8. Influence of M_f over the switching angles for M_f value above 20 is negligible while for the range below 20 it largely depends on M_f. Both SPWM and ISCPWM upshot nonlinear relationship in the lower M_f ranges. Fig.6 shows the variation of duty cycle for different M_a with constant M_f. The figure demonstrates that duty cycle is higher for ISCPWM throughout the entire range of M_a and the austere liner relationship of duty cycle in SPWM is violated in ISCPWM for lower values of M_a. In addition, in ISCPWM causes M_f dependency of duty. The ISCPWM gives higher duty cycle without any pulse dropping at given modulation index while the relation becomes little non-linear. Fig.7 shows that the dependence of duty cycle on M_f at any M_a value is a constant for even the lowest typical carrier frequency of application.

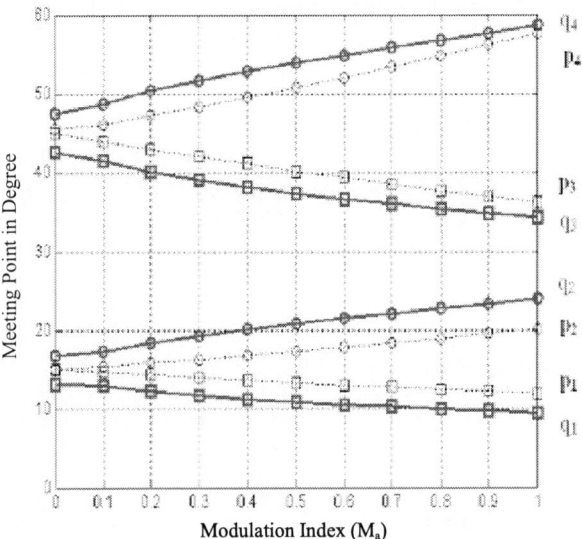

Fig.4 Influence of M_a on Meeting Point

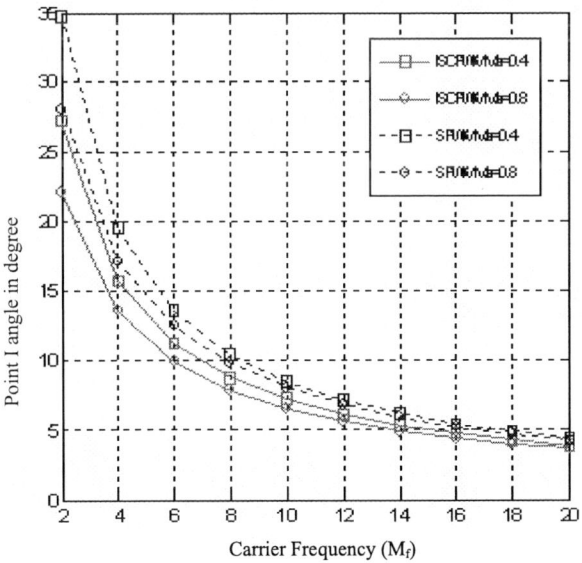

Fig.5 Influence of M_f on Meeting Points

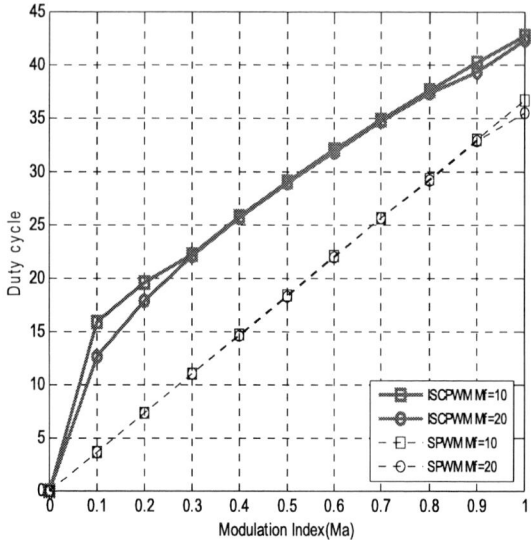

Fig. 6 M_a Vs Duty Cycle

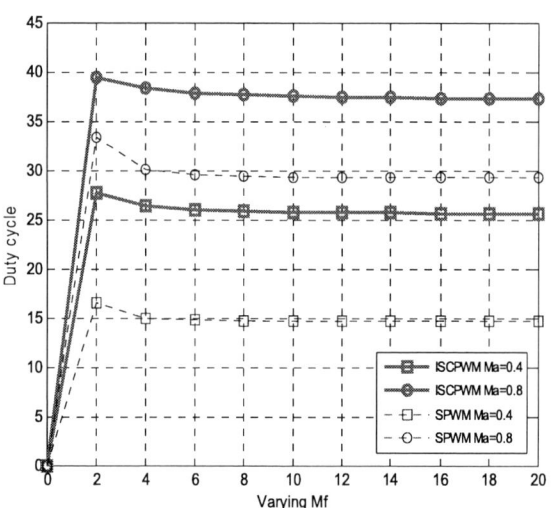

Fig.7 M_f Vs Duty Cycle

IV SIMULATION RESULTS

To show the effectiveness of the proposed modulator, a thorough simulation is performed for different modulation index and carrier frequency values. The ISCPWM scheme achieves fundamental voltage values of range, which can only be obtained by over-modulation, if a conventional SPWM scheme is adopted. Fig.8 shows the output voltage waveforms and harmonic spectrums of SPWM and ISCPWM. Tables 1 and 2 compares the methods for fundamental, harmonics and THD for M_a=0.8, M_f=15 and V_{dc}= 300V. The improved fundamental and reduced THD are evident from the figures, which gives 19.21% fundamental fortification than SPWM. At M_a=1 (verge on linearity), ISCPWM gives 9% higher fundamental than SPWM, while the fortification obtained from the harmonic injection methods with pulse dropping and mode changing is 15%.

The additional advantage in the ISCPWM is it does not require any mode changing like THIPWM. Regrettably, the ISCPWM causes marginal increase in the lower order harmonics, but except third harmonics all other harmonics are in acceptable level (less than 5%). It is worth noting that for three-phase applications, the heightened third harmonics need not be bothered.

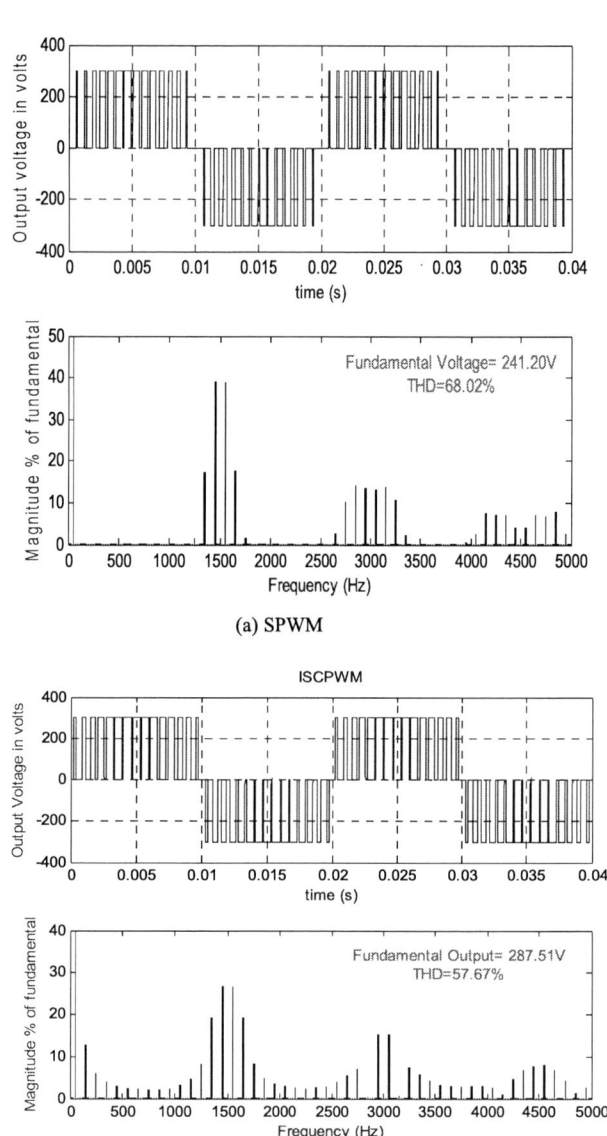

Fig. 8 Output voltage waveforms and their harmonic spectrum

Fig.9 shows the complete fundamental component working range as function of M_a while Fig.10 presents the corresponding THD values. The ISCPWM method gives more fundamental throughout of the inverters working range. Its performance is more appreciable in lower range of modulation index. For instance, at M_a=0.1, ISCPWM gives fundamental component value three times of SPWM at the same time the THD value 40% less. Fig.11 shows the variation of fundamental component with the THD.

Table 1 Comparison of THD, Fundamental and Lower Order Harmonics

Method	THD%	h_1 (V)	h_3 (V)	h_5 (V)	h_7 (V)	h_9 (V)
SPWM	68.02	241.2	0.42	0.28	0.07	0.31
ISCPWM	57.67	287.5	36.75	17.58	11.35	8.21

Table 2 Comparisons of Higher Order Harmonics

Method	$2M_r-3$ h_{27}(V)	$2M_r-1$ h_{29}(V)	$2M_r+1$ h_{31}(V)	$2M_r+3$ H_{33}(V)	$4M_r-3$ h_{57}(V)	$4M_r-1$ h_{59}(V)	$4M_r+1$ H_{61}(V)	$4M_r+3$ h_{63}(V)
SPWM	42.34	93.72	93.72	42.33	34.14	31.78	31.98	33.54
ISCPWM	55.01	76.43	76.84	54.84	1.32	43.69	43.72	2.16

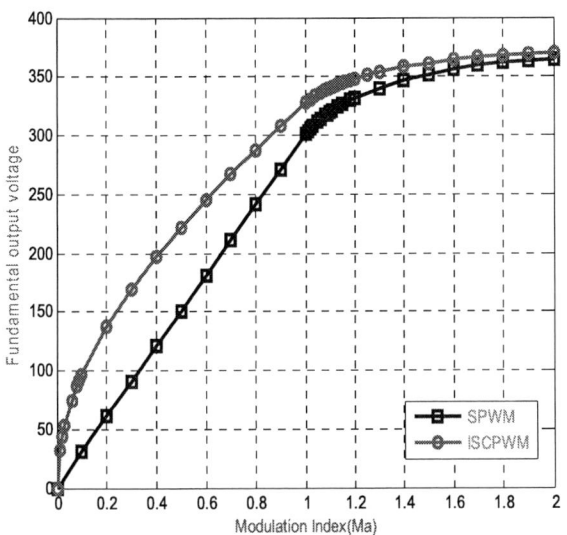

Fig.9 Variation of Fundamental with M_a

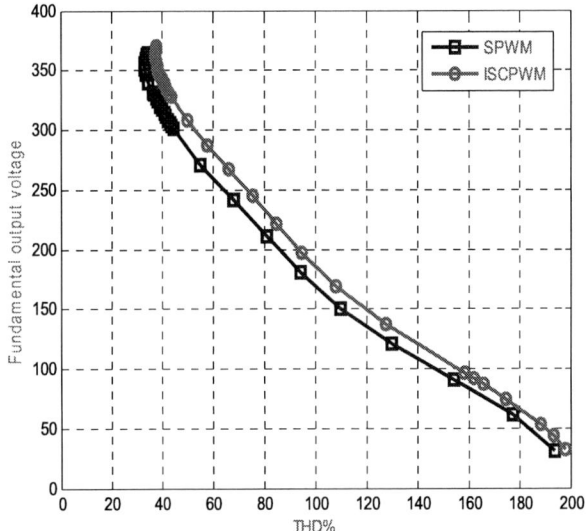

Fig.11 Values of THD for Various Fundamental Output

A Over Modulation

To increase the fundamental amplitude further in the SPWM technique the only way is increasing the M_a beyond 1.0, which is called as an over modulation. Over-modulation causes the output voltage to contain many more low order harmonics (3, 5, 7... etc) and also makes the fundamental component-modulation index relation non-linear. As the proposed ISCPWM gives improved fundamental component, to some extend it replaces the over modulation and avoids pulse dropping. For still higher values of fundamental, ISCPWM also has equally good opportunity to work in the over modulation region. To understand the performance of the schemes in over modulation range, the simulated spectral outputs are presented in Fig.12 for M_a=1.8. The result shows that though the ISCPWM works better than the traditional SPWM in over modulation; its performance cannot be appreciated to the extent as in linear range.

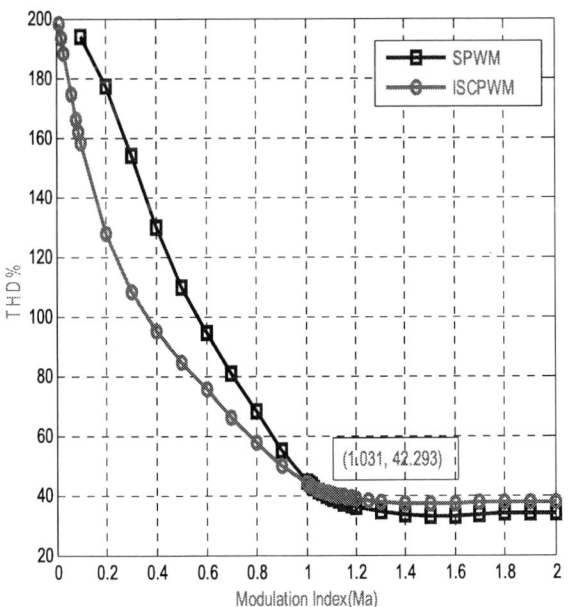

Fig.10 Variation of THD with M_a

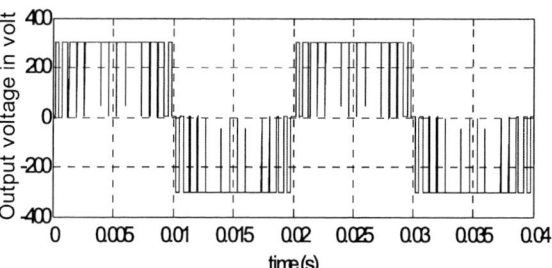

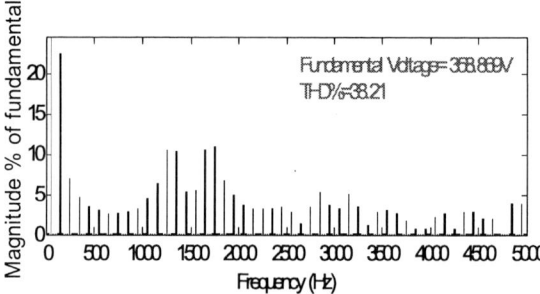

Fig.13 Output Voltage and their Spectrum – Amalgamation

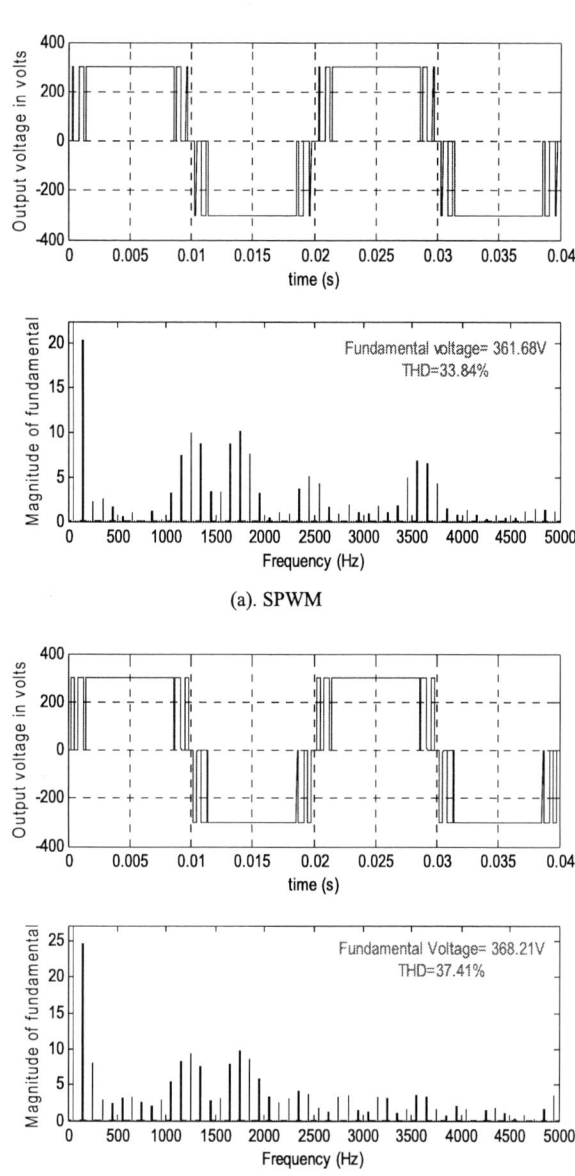

(a). SPWM

(b). ISCPWM

Fig.12 Output voltage and their harmonic spectrum-over modulation

B Amalgamation

The reference modification in harmonic injection PWM methods and carrier modification in the proposed ISCPWM aim at increasing the fundamental through increase in the pulse area. As the aim of both the modifications is same, amalgamation of both reference and carrier modifications will improve the situation further. On the basis of this intuitive notion, it is logical to amalgamate the inverted sine carrier with third harmonic and triplen harmonic injected reference waveforms may be amalgamated in the three-phase system to improve the system further. Fig.13 depicts such results obtained from amalgamated operation with third harmonic injection reference, which results in 19.73% enhancement in fundamental than SPWM, which is greater than the fortification obtained when triplen harmonic injected reference alone is used.

V CONCLUSION

The paper presents a novel PWM scheme for controlling the output of an inverter with improved fundamental component value. The main advantage of this approach is that it adopts a consistent strategy for the entire range of modulation index i.e. it does not require any mode change and also causes exactly same number of switching per cycle. The appreciable improvement in THD in the lower range of modulation depth attracts drive applications where low speed operation is required. The reduced distortions even at low modulation depth provide scope for proposed scheme not only when higher fundamental demanded and also obtaining low fundamental values. The drawbacks of the proposed scheme are marginal boost in the lower harmonics and non-linear fundamental and M_a relation.

REFERENCES

[1] Michael A. Boost, and Phoivos D.Ziogas, "State-of-Art–Carrier PWM Techniques: A Critical Evaluation," *IEEE Transactions Industry Applications*, vol.24, no.2, pp.271-280 March/April 1998.

[2] Joachim Holtz, "Pulse width Modulation-A Survey," *IEEE Transaction Industrial Electronics*, vol.39, no.5, pp.410-420, Dec.1992.

[3] Phoivos D.Ziogas, "The Delta Modulation Technique in Static PWM Inverters" *IEEE Transactions on Industry Application,* vol.1A-17, pp.199-203, March/April 1981.

[4] S Jeevananthan, P.Dananjayan and A. Mohamed Asif Fisal, "A HPWM Method for Thermal Management in a Full-Bridge Inverter with Loss Estimation and Electro-Thermal Simulation", AMSE Periodicals of Modeling, Measurement and Control – Series B: Vol. 73, No.6, pp. 1-20, December 2004.

[5] P.Enjeti, P.D.Ziogas and J.F.Lindsay, "Programmed PWM Techniques to Eliminates Harmonics-A Critical Evaluation," IEEE IAS Conference Record, pp.418-430, 1988.

[6] J.A.Houldsworth and D.A.Grant, "The Use Harmonic Distortion to Increase the Output Voltage of Three-Phase PWM Inverter," *IEEE Transaction on Industry Application*, vol.1 IA-20, pp.1224-1228, Sept/Oct 1984.

[7] K.Taniguchi, Y.ogino and H.Irie, "PWM Technique for Power MOSFET Inverter," *IEEE Transactions on Power Electronics*, vol.3, no.2, p.p328-334, July 1988.

[8] Kjeld Thorborg and Ake Nystrom, "Staircase PWM: An Uncomplicated and Efficient Modulation Techniques for AC Motors Drives", *IEEE Transactions on Power Electronics*, vol.3, no.4, pp.391-398, Oct. 1988.

[9] S.Jeevananthan, P.Dananjayan and S.Venkatesan,"SPWM-An Analytical Characterization, and Performance Appraisal of Power Electronic Simulation Software's", *Proceedings of PEDS2005*, Kulala Lumpur, Malasia, pp.681-686, Nov.28-Dec.1, 2005.

[10] W.G.Dunford and J.D.Van Wyk, "The Calculation of Sub-Harmonics in an Asynchronous PWM Induction Motor Drive," Proceedings of IEEE PESC Conference Record, pp.672-677, 1990.

[11] S Jeevananthan, P Dananjayan and S Venkatesan, "A Novel Modified Carrier PWM Switching Strategy for Single-Phase Full-Bridge Inverter", *Iranian Journal of Electrical and Computer Engineering, Summer Fall-Special Section on Power Engineering,* Vol.4, no.2, pp.101-108, Tehran, Iran, 2005.

BIBLIOGRAPHIES

R.Nandhakumar was born in Pondicherry, India on September 20, 1982. He received his B.Tech degree in Electrical and Electronics Engineering from Regency Institute of Technology, Yanam, India, in 2004 and M.Tech degree in Electrical Drives and Control from Pondicherry Engineering College, Pondicherry, in 2006. He is presently working with ABB Ltd, Bangalore, India as research and development engineer in the field of Real Time Advanced Dynamic Reactive Power Compensation. He is an active member of the team for development and modification of the new generation of technology-STATCON.

S. Jeevananthan was born in Eathamozhy (Nagercoil), India on May 25, 1977. He received his B.E. degree in Electrical and Electronics Engineering from Mepco Schlenk Engineering College, Sivakasi, India, in 1998, and M.E. degree from PSG College of Technology, Coimbatore, India, in 2000. He is currently working towards the Ph.D. degree, focusing on "Performance Evaluation of Innovative Control Strategies for Power Electronic Converters". Since 2001, he has been with the Department of Electrical and Electronics Engineering, Pondicherry Engineering College, Pondicherry, India. He has authored more than 10 papers published in international journals and about 30 papers in proceedings of national/international conference.

Fault Detection and Reconfiguration Technique for Cascaded H-bridge 11-level Inverter Drives Operating under Faulty Condition

Surin Khomfoi[1], Leon M. Tolbert[2]

[1] King Mongkut's Institute of Technology Ladkrabang, Dept. of Electrical Engineering, Faculty of Engineering
Chalongkrung Rd. Ladkrabang Bangkok, 10530 THAILAND
[2] The University of Tennessee, Electrical and Computer Engineering
414 Ferris Hall, Knoxville, TN 37996-2100, USA

Abstract-A fault detection and reconfiguration technique for a cascaded H-bridge 11-level inverter drives during faulty condition is proposed in this paper. The ability of cascaded H-bridge multilevel inverter drives (MLID) to operate under faulty condition is also discussed. Output phase voltages of a MLID can be used as a diagnostic signal to detect faults and their locations. AI-based techniques are used to perform the fault classification. A neural network (NN) classification is applied to the fault diagnosis of a MLID system. Multilayer perceptron (MLP) networks are used to identify the type and location of occurring faults. The principal component analysis (PCA) is utilized in the feature extraction process to reduce the NN input size. The genetic algorithm (GA) is also applied to select the valuable principal components to train the NN.

A reconfiguration technique is also developed. The developed system is validated with simulation and experimental results. The developed fault diagnostic system requires about 6 cycles (~100 *ms* at 60 Hz) to clear an open circuit and about 9 cycles (~150 *ms* at 60 Hz) to clear a short circuit fault. The experiment and simulation results are in good agreement with each other, and the results show that the developed system performs satisfactorily to detect the fault type, fault location, and reconfiguration.

Index Terms—Fault diagnosis, fault tolerance, genetic algorithm, multilevel inverter, neural network, power electronics.

I. INTRODUCTION

For a medium voltage grid, it is troublesome to connect only one power semiconductor switch directly. As a result, a multilevel power converter structure has been introduced as an alternative in high power and medium voltage situations, and also multilevel inverter drive (MLID) systems have become a solution for high power drive applications. A multilevel inverter not only achieves high power ratings, but also enables the use of renewable energy sources. Two topologies of multilevel inverters for electric drive application have been discussed in [1]. The cascaded MLID is a general fit for large automotive all-electric drives because of the high VA rating possible and because it uses several dc voltage sources which would be available from batteries or fuel cells [1].

A possible structure of a three-phase cascaded multilevel inverter drive for an electric vehicle is illustrated in Fig. 1. The series of H-bridges makes for modularized layout and packaging; as a result, this will enable the manufacturing process to be done more quickly and cheaply. Also, the

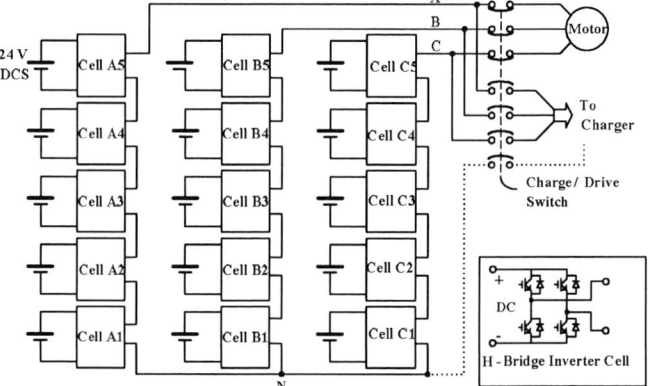

Fig. 1. Three-phase wye-connection structure for electric vehicle motor drive.

reliability analysis reported in [2] indicates that the fault-tolerance of cascaded MLID has the best life cycle cost. However, if a fault (open or short circuit) occurs at a semiconductor power switch in a cell, it will cause an unbalanced output voltage and current, while the traction motor is operating. The unbalanced voltage and current may result in vital damage to the traction motor if the traction motor is run in this state for a long time.

Generally, the passive protection devices will disconnect the power sources or gate drive signals from the multilevel inverter system whenever a fault occurs, stopping the operated process, overlooking the consequence of such accidental shut down. For instance, in the case of a MLID fault such as open or short circuit in a power switch, the fuse in the dc link will blow when the current reaches to the safety limit, disconnecting the dc voltage supply. This may cause vitally consequent damages in the motor if the motor is running at base speed with rated load. Therefore, the passive protection system may not be adequate if the application of a MLID needs a continuous operation or the motor is connected with a large load such as conveyer or hybrid/electric vehicle. It would be better if one can isolate the fault and continue to operate the motor at lower power levels or degraded performance than completely stopping it.

Although a cascaded MLID has the ability to tolerate a fault for some cycles, it would be better if we can detect the fault and its location; then, switching patterns and the modulation

978-1-4244-0644-9/07/$25.00 ©2007 IEEE   1228

index of other active cells of the MLID can be adjusted to maintain the operation under balanced load condition. Of course, the MLID can not be operated at full rated power. The amount of reduction in capacity that can be tolerated depends upon the application; however, in most cases a reduction in capacity is more preferable than a complete shutdown.

A study on fault diagnosis in drives begins with a conventional PWM voltage source inverter (VSI) system [3-5]. Then, artificial intelligent (AI) techniques such as fuzzy-logic (FL) and neural network (NN) have been applied in condition monitoring and diagnosis [6-8]. Furthermore, a new topology with fault-tolerant ability that improves the reliability of multilevel converters is proposed in [9]. A method for operating cascaded multilevel inverters when one or more power H-bridge cells are damaged has been proposed in [2, 10]. The method is based on the use of additional magnetic contactors in each power H-bridge cell to bypass the faulty cell.

One can see from the concise literature survey that the knowledge and information of fault behaviors in the system is important to improve system design, protection, and fault tolerant control. Thus far, limited research has focused on MLID fault diagnosis and reconfiguration. Therefore, a MLID diagnostic system is proposed in this paper that only requires measurement of the MLID's voltage waveforms and does not require measurement of currents.

II. RELIABILITY CONSIDERATIONS OF MLID

Since multilevel inverters contain several semiconductors connected in series to achieve medium voltage and high power demand, one might consider that multilevel inverters are less reliable. In contrast, multilevel cascaded H- bridge inverters using modular series-cells with separated dc sources as depicted in Fig. 1 could improve reliability if the MLID has the ability to detect and bypass the faulty cell. If one of the power cells fails, it can be bypassed and operation can continue at reduced voltage capacity. The definition of reliability given by [11] is "*the probability of a device performing its purpose adequately for the period of time intended under the operating condition encountered*". The word *adequately* permits some application at reduced capacity to be included in the probability calculations [2].

The engineering reliability analysis in a system is usually concerned with the reliability R and/or the probability of failure P. As a system is considered reliable unless it fails, the reliability and probability of failure sum to unity as explained in equation (1) [11].

$$R(t) + P(t) = 1,$$
$$R(t) = 1 - P(t), \quad (1)$$
$$P(t) = 1 - R(t),$$

where $P(t)$ is probability of a system will fail by time t,

$R(t)$ is probability of a system will still be operational by time t. Therefore, (1) can be applied in MLID system reliability analysis. Suppose that the cascaded H-bridge MLID system as shown in Fig. 1 contains N cells and can not tolerate any failures; then, if the probability of a single cell will

function properly during a time interval is R, so that the probability all N cells will function properly during the same time interval is R^N because the MLID system is considered as series system in this case. $P(t)$ and $R(t)$ can be defined as the point density functions; then, $P = \dfrac{dP(t)}{d(t)}$ and $R = \dfrac{dR(t)}{d(t)}$. Next, if the MLID has an extra cell which can tolerate failures, the MLID reliability will become $R^N + [N \times R^{(N-1)} \times (1- R)]$ instead of R^N. It is obvious that the MLID with a tolerated failure cell has a higher reliability than the one without tolerance for failures. A numerical reliability example of a MLID can be illustrated in Table I. Suppose that the MLID in Table I has a cell reliability R of 99% and it contains totally 15 cells. As can be seen, with one tolerated cell in each phase, the reliability of the MLID can increase from 86% to 99.0%; therefore, a fault diagnostic and fault reconfiguration (bypass) system can improve the reliability of the MLID system. In addition, for the case of m tolerated cells, the reliability function can be written as

$$R_m = \sum_{i=0}^{m} \left(\frac{N!}{(N-i)! \times i!} \times R^{(N-i)} \times (1-R)^i \right), \quad (2)$$

where m is number of tolerated cells,
       N is number of cells in MLIDs,
       R_m is total reliability of the system.

TABLE I.
NUMERICAL EXAMPLE OF 15 CELLS MLID WITH 99% RELIABILITY (R) IN EACH POWER CELL.

Number of tolerated cell faults	Reliability Function	Reliability (Percentage)
0	$R_0 = R^N$	86.006%
1	$R_1 = R_0 + [N \times R^{(N-1)} \times (1- R)]$	99.037%
2	$R_2 = R_1 + [(N \times (N-1) \times (R^{(N-2)}) \times (0.5 \times (1-R)^2)]$	99.958%
3	$R_3 = R_2 + [(N \times (N-1) \times (N-2) \times (R^{(N-3)}) \times (0.1667 \times (1-R)^3)]$	99.999%

III. FAULT DIAGNOSTIC METHODOLOGY

Before continuing discussion, it should be emphasized that the multilevel carrier-based sinusoidal PWM is used for controlling gate drive signals for the cascaded MLID. Fig. 2 shows that the output voltages can be controlled by controlling the modulation index (m_a). To expediently understand, the two separate dc sources (SDCS) cascaded MLID structure is used as an example in this section.

A. Diagnostic Signals

The selection of diagnostic signals is very important because the neural network could learn from unrelated data to classify faults which would result in improper classification. Simulation results (using power simulation (PSIM) from Powersim Inc.) of input motor current waveforms during an open circuit fault at different locations of the MLID (shown in Fig. 2 (a)) are illustrated in Fig. 3 and Fig. 4. As can be seen in Fig. 3 and Fig. 4, the input motor currents can classify open

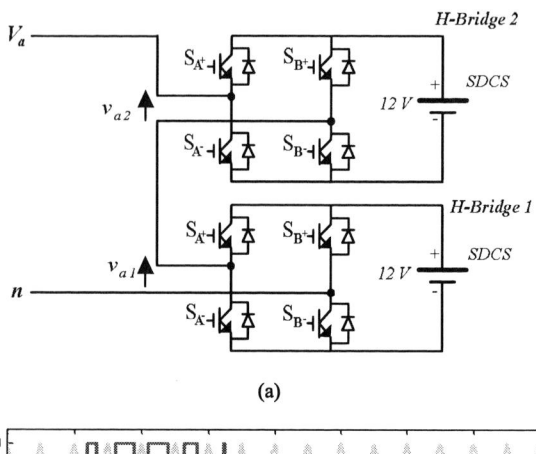

(a)

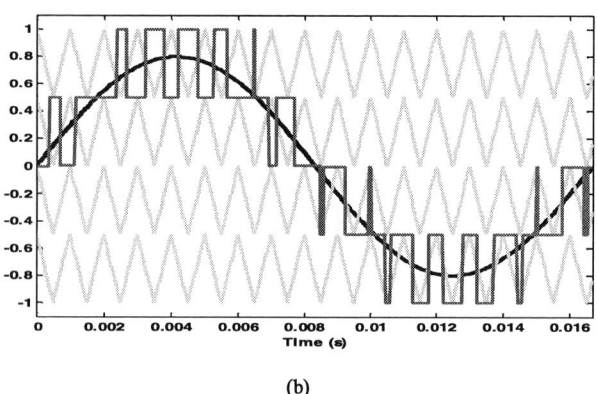

(b)

Fig. 2. (a) Single-phase multilevel-inverter system; (b) Multilevel carrier-based sinusoidal PWM showing carrier bands, modulation waveform, and inverter output waveform ($m_a = 0.8/1.0$).

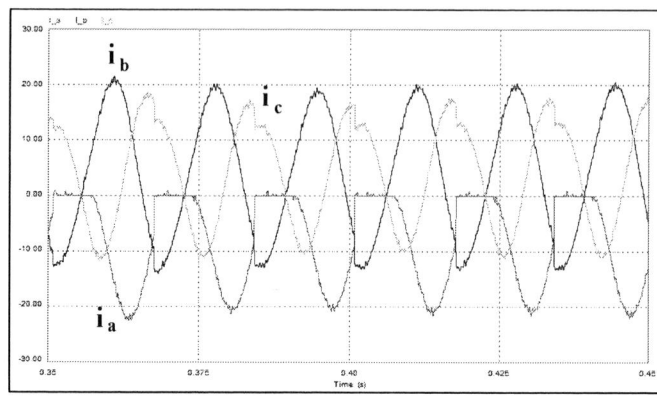

(a)

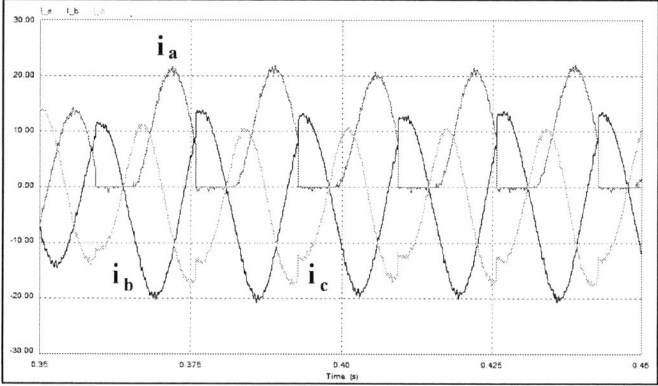

(b)

Fig. 4. Input motor currents during open circuit fault at H-bridge 1: (a) switch S_{A+}, (b) switch S_{B+}.

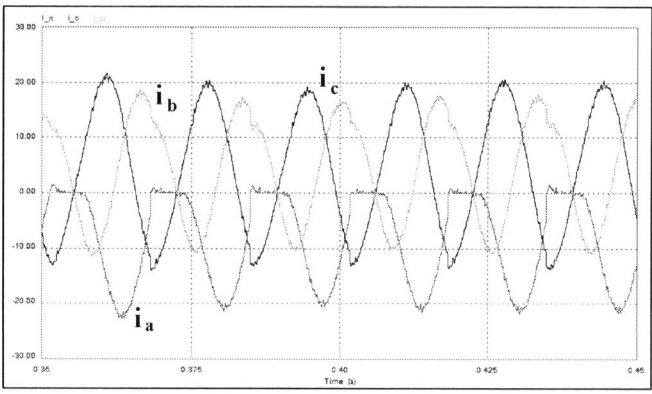

Fig. 3. Input motor currents during open circuit fault at switch S_{A+} of H-bridge 2.

circuit faults at the same power cell by tracking current polarity (see Fig. 4); however, it is difficult to classify the faults at different power cells; the current waveform for a fault of S_{A+} in H-bridge 2 (Fig. 3) looks identical to that for a fault of S_{A+} in H-bridge 1 (Fig. 4 (a)). As a result, the detection of fault locations could not be achieved with only using input motor current signals. Also, the current signal is load dependent: the load variation may lead to misclassification; for instance, light load operation as reported in [12].

Auspiciously, Fig. 2 indicates that an output phase PWM voltage is related to turn-on and turn-off time of associated

switches; hence, a faulty switch can not generate a desired output voltage. The output voltage for a particular switch is zero if the switch has a short circuit fault, whereas the output voltage is about V_{dc} of SDCS if the switch has an open circuit fault. For this reason, the output phase voltage can convey valuable information to diagnose the faults and their locations. The simulation results of output voltages are shown for an MLID with open circuit faults and short circuit faults in Fig. 5. One can see that all fault features in both open circuit and short circuit cases could be visually distinguished.

B. AI-Based Techniques for Fault Diagnosis

It is possible that artificial intelligent (AI) based techniques can be applied in condition monitoring and diagnosis. AI-based condition monitoring and diagnosis have several advantages. For instance, AI-based techniques do not require any mathematical models; therefore, the engineering time and development time could be significantly reduced [13]. The methodology of fault diagnostic system using AI has been reported in [14-16] and will not be repeated here. The discussion of AI presented in this section will be brief, providing only the indispensable notion to elucidate the fundamental AI-based approach applied to a fault diagnosis system in a MLID.

First, the feature extraction of the output voltage signals is performed by using FFT; then, the principal component

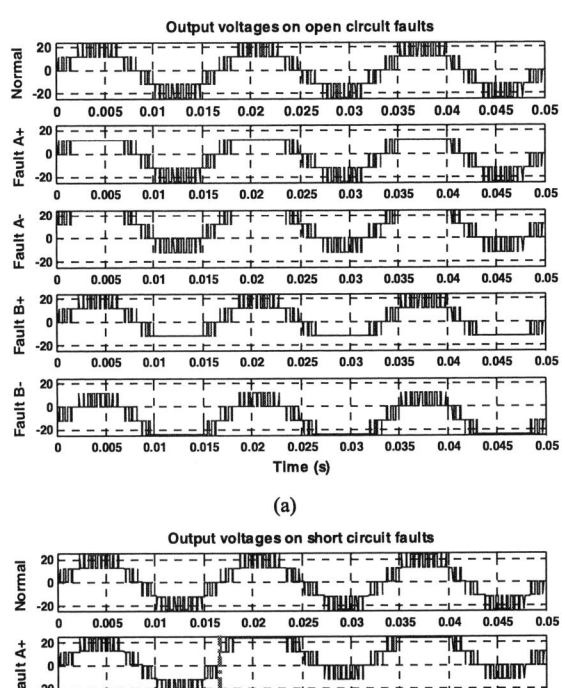

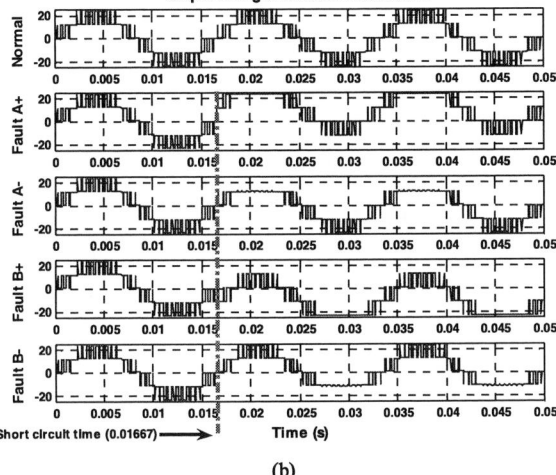

Fig. 5. Simulation of output voltages signals (a) open circuit faults, (b) short circuit faults showing fault features at S_{A+}, S_{A-}, S_{B+}, and S_{B-} of H-bridge 2 with modulation index = 0.8 out of 1.0.

analysis (PCA) is used in the feature extraction process. PCA offers a lower dimensional input space which will also usually reduce the time necessary to train a neural network, and the reduced noise (by keeping only valuable principal components (PCs)) may improve the mapping performance [15]. Next, a genetic algorithm (GA) is applied to search for the best combination of PCs to train the neural network as explained in [16]. The output of the GA is the best combination of PCs which provide the weight and bias matrix of neural networks used for the classification task. After that, the weight and bias matrix of the neural networks will be implemented in Simulink interfacing with FFT and PCA subsystem as shown in Fig. 6. The PCA and GA process will be performed off-line to achieve the best combination of PCs.

Before continuing discussion, it should be mentioned that the methodology of fault diagnosis presented in [15-16] can be applied to any other cascaded H-bridges MLID. However, some minor processes are different such as neural network structure, input/output data set, and principal component (PC) selection. Since the simulation and experiment validation will

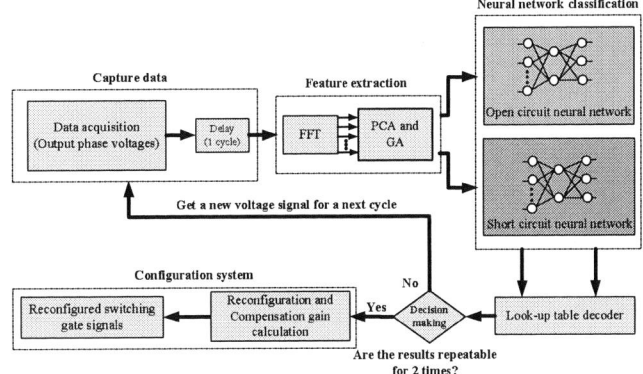

Fig. 6. Fault diagnostic diagram for 11-level MLID with 5 SDCS.

be performed with 11-level MLID, the fault diagnostic processes for the 11-level MLID are explained in the following.

C. Neural Network Structure

The fault diagnostic diagram for an 11-level MLID with 5 SDCS is depicted in Fig. 6. The neural network classification process consists of two networks: open circuit network and short circuit network. The training time and required memory for implementation are reduced with the segregated neural network as reported in [17, 18]. Moreover, in this particular case, the short circuit data set includes the loss of separate dc source (SDCS) condition due to the fuse protection because the fuse may blow before the fault is detected; therefore, the short circuit neural network may contain more complexity than the open circuit neural network. Also, the neural networks may be assigned to have the ability to provide "do not know" conditions. The multilayer feedforward perceptron (MLP) networks are used in both open circuit and short circuit neural networks. The neural network architecture is based upon GA selection as discussed in [16]. The input neurons depend on GA selection; however, for this example, 1 hidden layer with 4 hidden nodes and 6 output nodes are assigned.

D. Input/Output Data

The input/output data set diagram for 11-level MLID is illustrated in Fig. 7. We can see that the set of original input data at each MLID operation point (modulation index) contains five fault classes: normal, Fault A+, A-, B+, and B-. Modulation indices (m_a) are observations changing with desired load. In this particular case, m_a is varied from 0.6 to 1.0 with 0.05 intervals. The original data are divided into two subsets: Open circuit and short circuit. Also, each subset is separated into one training set and two testing sets as shown in Fig. 7. Both open circuit and short circuit neural networks are trained with both open and short circuit training sets. However, the open circuit neural network will be trained with short circuit training set with "do not know" target binary and vice versa with the short circuit neural network as depicted in Fig. 7.

Target binary variables are also illustrated in Table II. Six binary bits are used to code the input/output mapping. The first two bits (counting from the right bits 0 and 1) are utilized to code the faulty switches, the 3rd bit from the right (bit 2) is

1231

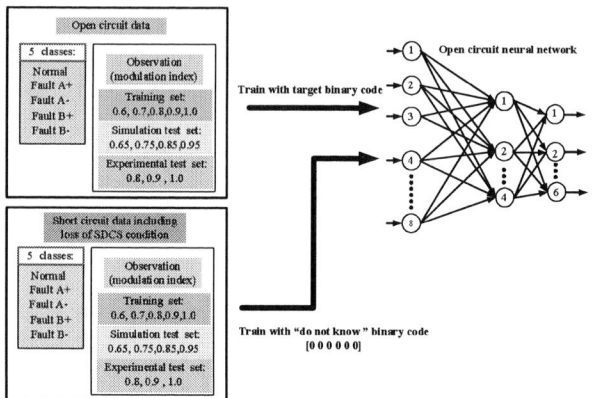

Fig. 7. Training and testing data set diagram.

TABLE II.
TARGET BINARY CODE FOR 11-LEVEL MLID.

Condition		Number of binary bits and their description				
		Faulty cell			Fault type	Faulty switch	
		5	4	3	2	1	0
Normal		1	1	1	1	1	1
Faulty cells	1	0	0	1	-	-	-
	2	0	1	0	-	-	-
	3	0	1	1	-	-	-
	4	1	0	0	-	-	-
	5	1	0	1	-	-	-
Fault types	open	-	-	-	0	-	-
	short	-	-	-	1	-	-
Faulty switches	Fault A+	-	-	-	-	0	0
	Fault A-	-	-	-	-	0	1
	Fault B+	-	-	-	-	1	0
	Fault B-	-	-	-	-	1	1
"Do not know"		0	0	0	0	0	0

used to code the fault type, and the last three bits (bits 3, 4, and 5) are used to code which cell has faulted. Also, the code [1 1 1 1 1 1] is used to represent the normal condition, whereas the code [0 0 0 0 0 0] is used to characterize the "do not know" condition. Therefore, the six output neurons are used for particular 11-level MLID. For instance, if the neural network provides [0 1 1 0 0 1] as the outputs, we can decode the fault type and location as cell 3 is faulty with open circuit fault at switch S_A-. This decoder paradigm can be implemented in Simulink model by using 2-D dimension look-up table as shown in Fig. 6.

E. Principal Component Selection

The selection of principal components (PCs) is significant because input selected PCs can cause uncertainty results: (1) additional unneeded input PCs to the neural network can increase the solution variance; (2) absent necessary input PCs can increase bias. The correlation coefficient (CC) between PCs and target variable is shown in Fig. 8. As can be seen in Fig. 8 (a), 4 PCs have CC higher than 0.4; PC *6, 12, 13,* and *14*. This means that PC *6, 12, 13, 14* are good predictors for the first bit of target variables. Also, PC *15, 18,* and *36* are a good predictors for the second bit of the target variable. We can see that different bits of the target variable have different desired

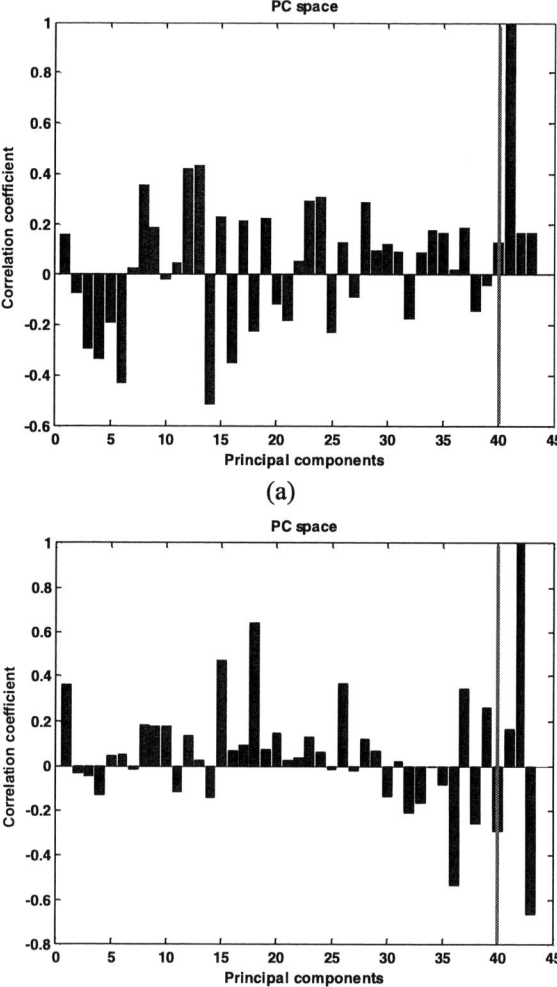

Fig. 8. Correlation coefficient of PCs and target variables: (a) first bit of target variable, (b) Second bit of target variable.

predictors (PCs). Since we have 6 bits of the target variable, it would be better to use a multivariable optimization technique to select the best combination of PCs. Therefore, a genetic algorithm (GA) is used to search for the best combination of PCs to train the neural network as proposed in [16, 18]. By using the methodology proposed in [16], the principal components (PCs) are selected by GA. 8 PCs (*PC 1, 2, 3, 5, 7, 8, 13, and 14*) are selected for an open circuit neural network, whereas 11 PCs (*PC 2, 3, 4, 5, 7, 8, 9, 11, 12, 13, and 14*) are chosen for short circuit neural networks. Therefore, the neural network architecture for open circuit neural network has 8 input neurons, 4 hidden neurons and 6 output neurons, whereas the short circuit neural network architecture has 11 input neurons, 4 hidden neurons and 6 output neurons.

IV. RECONFIGURATION TECHNIQUE

A. Corrective Action Taken

The basic principal of the reconfiguration method is to bypass the faulty cell (H-bridge); then, other cells in the MLID are used to compensate for the faulty cell. For instance, if cell

1232

2 of MLID in Fig. 2 has an open circuit fault at S_{A+}; accordingly, S_{A-} and S_{B-} need to be turned on (1), whereas S_{B+} needs to be turned off (0) to bypass cell 2. The corrective actions taken for other fault locations are shown in Table III. As can be seen, the corrective action would be the same for cases that have similar voltage waveforms during their faulted mode (for instance, see Fig. 5 for a short circuit fault in S_{A+} and open circuit fault in S_{A-}). Therefore, even if the fault may be misclassified (an actual short circuit fault at S_{A+} is misclassified as an open circuit fault at S_{A-} or vice versa), the corrective action taken would still solve the problem.

B. Reconfiguration Method

The reconfiguration diagram for an 11-level MLID with 5 SDCS is illustrated in Fig. 9. The turn-on intervals of each cell are not equal with multilevel carrier-based sinusoidal PWM: cell 1 has the longest turn-on interval, then the turn-on interval decreases from cell 2 to cell 5 as a staircase PWM waveform. The desired output voltage of a MLID can be achieved by controlling modulation index (m_a). For instance, suppose cell 2 has an open circuit fault at S_1 while the MLID operates at $m_a = 0.8/1.0$ (MLID is operated with four cells (cell 1-4)). We can see from Fig. 9 (b) that S_3 and S_4 need to be turned on, then the gate signal of cell 2 will be shifted up to control cell 3, then the gate signal of cell 3 will shift to cell 4, and the gate signal of cell 4 will shift to cell 5 respectively.

This reconfiguration also applies to other phases of MLID in order to maintain balanced output voltage. By using this method, the operation of MLID in a modulation index range of 0.0 to 0.8 (out of 1) can be fully compensated such that the inverter will continue to function like normal operation; however, if MLID operates at $m_a > 0.8$ and has a fault, lower order harmonics will occur in the output voltage since the MLID will operate in the overmodulation region in order to output the full requested voltage as illustrated in Fig. 10.

The compensated gain of the MLID operating at $m_a > 0.8$ is shown in Fig. 11. This compensated gain can also be written as a function of m_a by using polynomial curve fitting. Because the overmodulation region has a nonlinear relationship between modulation index and output fundamental voltage, the compensated gain is calculated in particular modulation indices; then, the polynomial function represents the nonlinear characteristic of this particular application. In addition, this polynomial function can be implemented in a Simulink model. The fitting function can predict the compensated gain with a norm of residuals less than 0.09. The overmodulation region will occur when the MLID operates at $m_a > 0.825$. To relieve this problem, space vector, and third harmonic injection PWM schemes may be used. Also, a redundant cell can be added into the MLID, but the additional part count should be considered. The reconfiguration effect and limitation of this reconfiguration method have been reported in [20].

V. SIMULATION AND EXPERIMENT VALIDATION

A. Simulation Setup

Two simulation programs are used in the simulation setup: Matlab-Simulink and PSIM [21]. Matlab-Simulink is used to

TABLE III.
GATE DRIVE SIGNALS OF CORRECTIVE ACTION TAKEN

Fault types	Locations	Signal S_{A+}	Signal S_{A-}	Signal S_{B+}	Signal S_{B-}
Open circuit	S_{A+}	0	1	0	1
	S_{A-}	1	0	1	0
	S_{B+}	0	1	0	1
	S_{B-}	1	0	1	0
Short circuit	S_{A+}	1	0	1	0
	S_{A-}	0	1	0	1
	S_{B+}	1	0	1	0
	S_{B-}	0	1	0	1

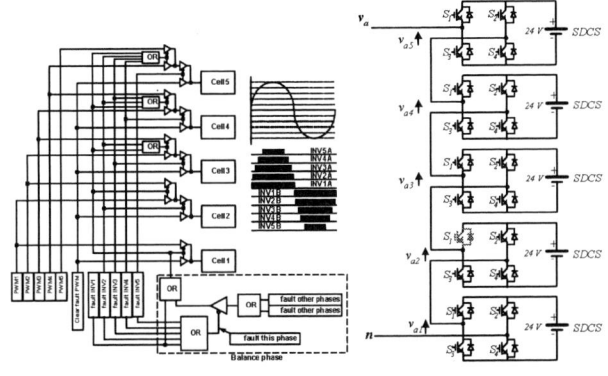

(a)   (b)

Fig. 9. Reconfiguration diagram for MLID with five SDCS: (a) Reconfiguration diagram, (b) H-Bridge 2 Switch S_1 open circuit fault at second level of single-phase multilevel-inverter with 5 SDCS.

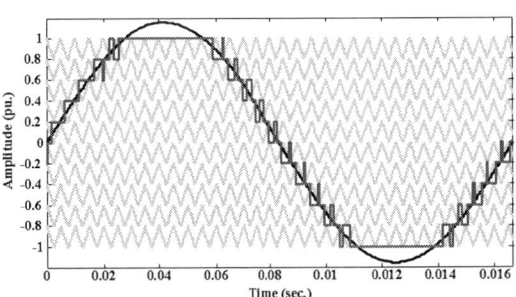

Fig. 10. Multilevel carrier-based sinusoidal PWM with 2 kHz switching frequency for 5 SDCS MLID showing carrier bands, modulation waveform, and inverter output waveform ($m_a = 1.2/1.0$)

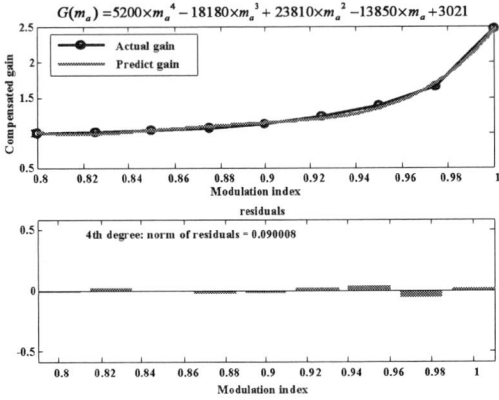

Fig. 11. Compensated gain of the MLID operating at $m_a > 0.8$.

implement feature extraction (FFT and PCA), neural network classification, and reconfiguration. A reconfiguration is corrective method to continuously operate a MLID after the faults are detected. PSIM is used to implement the MLID power circuit. It should be noted that the same Simulink model is used in both simulation and experiment.

B. Experimental Setup

The experimental setup is represented in Fig. 12. A three-phase wye-connected cascaded multilevel inverter using 100 V, 70 A MOSFETs is used. The MLID supplies an induction motor (1/3 hp) coupled with a dc generator load (1/3 hp). The Opal RT-Lab system [22] is utilized to generate gate drive signals and interfaces with the gate drive board. The switching angles are calculated by using a Simulink model based on multilevel carrier-based sinusoidal PWM with 2 kHz switching frequency. A separate individual power supply is supplied to each cell of the MLID, consisting of 5 cells per phase as shown in Fig. 1. Open and short circuit fault occurrences are created by physically controlling the switches in the fault-creating circuit. A Yokogawa DL 1540c is used to measure output voltage signals as ASCII files. The voltage spectrum is calculated and transferred to the Opal-RT target machine.

C. Results

C.1 Open circuit case

The simulation and experimental results are shown in Fig. 13. The faulty power cell (S_{A+}) was placed at cell 2 on phase A (see Fig. 1), and the multilevel inverter drive was operating at 0.8/1.0 modulation index before the fault occurs. We can see that the simulation and experimental results agree with each other. The fault diagnostic system requires about 6 cycles (~100 ms at 60 Hz) to clear the open circuit fault. Obviously, the open circuit fault causes unbalanced output voltage (V_{an}) of the MLID during the fault interval, and the average current on phase A (I_a) has negative polarity during the fault interval.

C.2 Short circuit case

The faulty power cell (S_{A+}) of the short circuit case was placed at power cell 3 on phase A (see Fig. 1), and the multilevel inverter drive was operating at 0.8/1.0 modulation index before the fault occurs. The simulation results of a short circuit fault at cell 3 switch S_{A+} are represented in Fig. 14. The fault diagnostic system also requires about 6 cycles to clear the short circuit fault. Obviously, the peak of the fault current increases about 1.5 times compared with the normal operation. Practically, the fuse protecting the SDCS may blow (disconnect the SDCS from a MLID) before the diagnostic system performs fault clearing so that the output phase-voltage will be zero. The developed diagnostic system can also detect a short fault under the loss of SDCS condition as shown in Fig. 15. The clearing time for this particular case is about 9 cycles.

VI. CONCLUSION

A fault detection and reconfiguration technique for cascaded H-bridge 11-level inverter drives has been developed. The developed fault diagnostic paradigm has been validated in both simulation and experiment. The fault diagnostic system

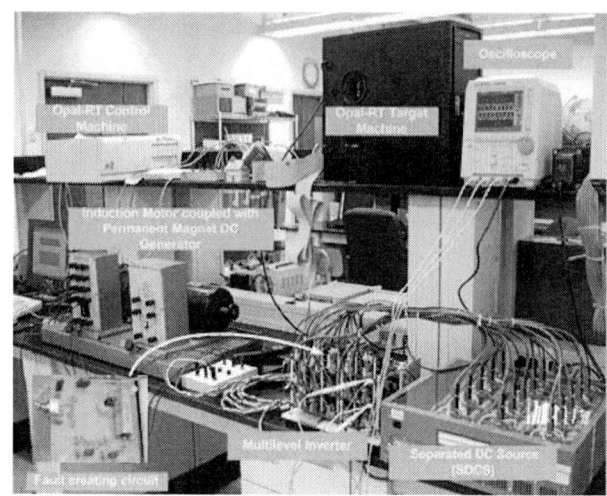

Fig. 12. Experimental setup.

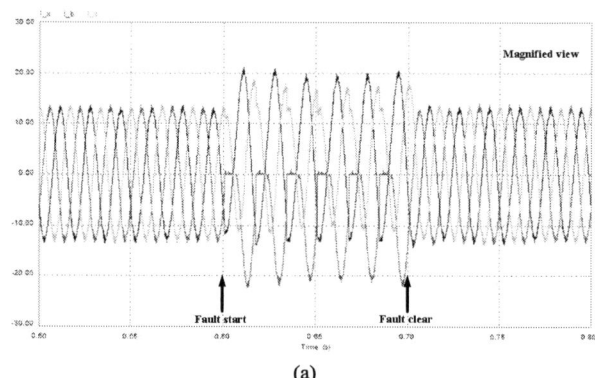

(a)

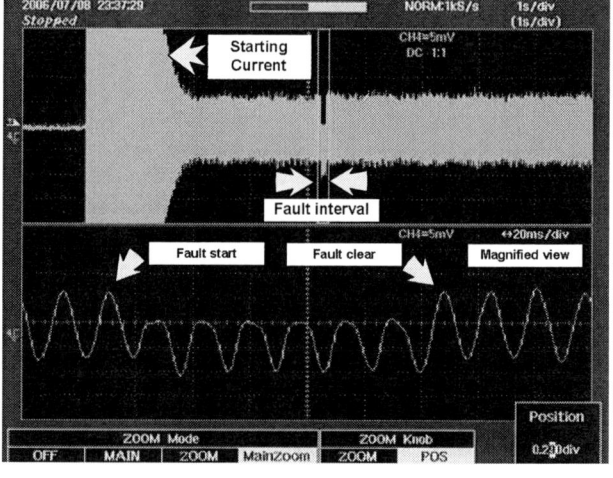

(b)

Fig. 13. Results of the open circuit fault at S_{A+}, cell 2 of the MLID during operation at $m_a = 0.8/1.0$: (a) Simulation of current waveforms, (b) Experimental result showing line current (I_a) at the faulty phase.

requires about 6 cycles (~100 ms at 60 Hz) to clear the open circuit fault and about 9 cycles (~150 ms at 60 Hz) to clear short circuit fault with loss of SDCD. The clearing time of the proposed system can be shorter than this if the proposed system is implemented as a single chip using an FPGA or DSP. The experiment and simulation results in both open circuit fault and short circuit fault with loss of SDCS are in good agreement with each other. The results show that the proposed diagnostic and reconfiguration paradigm can be applied to MLID applications. Therefore, by using the proposed system, the reliability of the MLID system can be increased.

REFERENCES

[1] L. M. Tolbert, F. Z. Peng, T.G. Habetler, "Multilevel Converters for Large Electric Drives," *IEEE Trans. Industry Applications*, vol. 35, no. 1, Jan/Feb. 1999, pp. 36-44.

[2] D. Eaton, J. Rama, and P. W. Hammond, "Neutral Shift," *IEEE Industry Applications Magazine*, Nov./Dec. 2003, pp. 40-49.

[3] D. Kastha, B. K. Bose, "Investigation of Fault Modes of Voltage-fed Inverter System for Induction Motor Drive," *IEEE Trans. Industry Applications*, vol. 30, no. 4, Jul. 1994, pp. 1028-1038.

[4] D. Kastha, B. K. Bose, "On-Line Search Based Pulsating Torque Compensation of a Fault Mode Single-Phase Variable Frequency Induction Motor Drive," *IEEE Trans. Industry Applications*, vol. 31, no. 4, Jul./Aug. 1995, pp. 802-811.

[5] A. M. S. Mendes, A. J. Marques Cardoso, E. S. Saraiva, "Voltage Source Inverter Fault Diagnosis in Variable Speed AC Drives by Park's Vector Approach," in *Proceedings of the 1998 IEE 7th International Conference on Power Electronics and Variable Speed Drives*, pp. 538-543.

[6] S. Hayashi, T. Asakura, S. Zhang, "Study of Machine Fault Diagnosis Using Neural Networks," in *Proceedings of the 2002 Neural Networks, IJCNN '02*, Vol. 1, pp. 956 – 961.

[7] S. Zhang, T. Asakura, X. Xu, B. Xu, "Fault Diagnosis System for Rotary Machines Based on Fuzzy Neural Networks," in *Proceedings of the 2003 IEEE/ASME Advanced Intelligent Mechatronics*, pp. 199-204.

[8] A. Bernieri, M. D'Apuzzo, L. Sansone, M. Savastano, "A Neural Network Approach for Identification and Fault Diagnosis on Dynamic Systems," *IEEE Trans. Instrumentation and Measurement*, vol. 43, no. 6, Dec. 1994, pp. 867-873.

[9] A. Chen, L. Hu, L. Chen, Y. Deng, X. He, "A Multilevel Converter Topology With Fault-Tolerant Ability," *IEEE Trans. on Power Electronics*, vol. 20, no. 2, March. 2005, pp. 405-415.

[10] J. Rodriguez, P. W. Hammond, J. Pontt, R. Musalem, P. Lezana, M. J. Escobar, "Operation of a Medium-Voltage Drive Under Faulty Conditions," *IEEE Trans. on Industrial Electronics*, vol. 52, no. 4, August 2005, pp. 1080-1085.

[11] S. R. Calabro, *Reliability Principals and Practices*, McGraw-Hill, New York, 1962.

[12] K. Rothenhagen, F. W. Fuchs "Performance of Diagnosis Methods for IGBT Open Circuit Faults in Three Phase Voltage Source Inverters for AC Variable Speed Drives," in *Proceedings of the 2005 European Conference on Power Electronic and Applications, Dresden, Germany* pp. P.1-P.10.

[13] P. Vas, *Artificial-Intelligence-Based Electrical Machines and Drives*, Oxford University Press, Inc.,New York, 1999.

[14] S. Khomfoi, L. M. Tolbert, "Fault Diagnostic System for a Multilevel Inverters Using a Neural Network," *IEEE Trans. Power Electronics*, vol. 22, no. 3, May, 2007, pp. 1062-1069.

[15] S. Khomfoi, L. M. Tolbert, "Fault Diagnosis System for a Multilevel Inverters Using a Principal Component Neural Network," 37th IEEE Power Electronic Specialists Conf., June 18-22, 2006, pp. 3121-3127.

[16] S. Khomfoi, L. M. Tolbert, "A Diagnostic Technique for Multilevel Inverters Based on a Genetic-Algorithm to Select a Principal Component Neural Network," *IEEE Applied Power Electronics Conference*, 2007, Anaheim, California, pp. 1497-1503.

[17] S. Mcloone, and G. W. Irwin, "Fast Parallel Off-Line Training of Multilayer Perceptrons," *IEEE Transactions on Neural Networks*, vol. 8, no. 3, May 1997, pp. 646-653.

[18] J. Fieres, A. Grubl, S. Philipp, K. Meier, J. Schemmel, and F. Schurmann, " A Platform for Parallel Operation of VLSI Neural Networks," in *Proceedings of the Brain Inspired Cognitive System (BICS) Conference*, 2004, pp. NC 4.3 1-7.

[19] S. Khomfoi, L. M. Tolbert, B. Ozpineci, "Operation under Faulty Condition of Cascaded H-bridge Multilevel Inverter Drives Including AI-Based Fault Diagnosis and Reconfiguration," *IEEE International Electric Machines and Drives Conference*, May 3-5, 2007, Antalya,

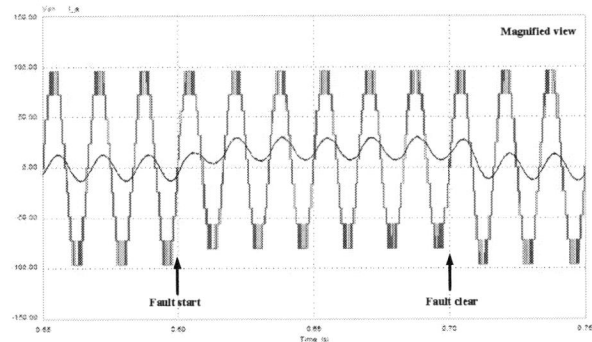

Fig. 14. Simulation results of the short circuit fault at S_{A+}, cell 3 of the MLID during operated at $m_a = 0.8/1.0$.

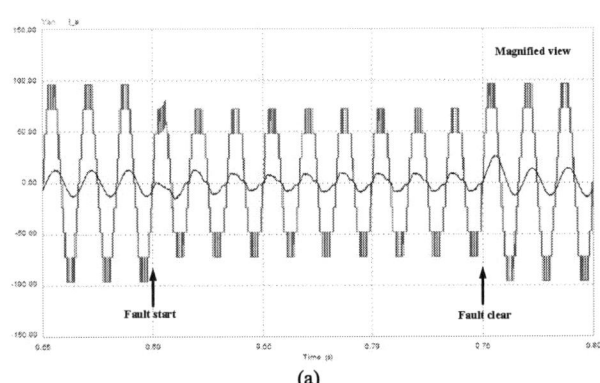

(a)

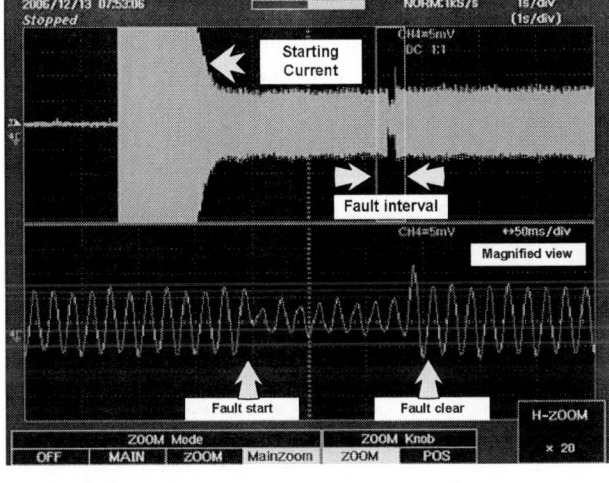

(b)

Fig. 15. Results of the short circuit fault at S_{A+}, cell 3 under loss of SDCS condition at the faulty cell of the MLID during operated at $m_a = 0.8/1.0$: (a) simulation, (b) experiment showing line current (I_a) at the faulty phase.

Turkey, pp. 1649-1656.

[20] S. Khomfoi, L. M. Tolbert, "A Reconfiguration Technique for Multilevel Inverters Incorporating a Diagnostic System Based on Neural Network," *10th IEEE Workshop on Computers in Power Electronics*, July 16-19, 2006, pp. 317-323.

[21] Powersim Inc, *PSIM User's Guide Version 6*, Powersim Inc, 2003, http://www.powersimtech.com.

[22] Opal-RT technology Inc, *RT-LAB User's manual*, Opal-RT technology Inc, Version 6, 2001, http://www. opal-rt.com.

Investigation into Harmonic Losses in a PWM Multilevel cascaded H-Bridge Inverter Fed Induction Motor

Prasopchok Hothongkham, Vijit Kinnares

Dept. of Electrical Engineering, Faculty of Engineering, King Mongkut's Institute of Technology Ladkrabang
Chalongkrung Road, Ladkrabang , Bangkok, Thailand, 10520
Tel.662-3266052-101 Ext.2619 Fax.662-3269902 E-mail:prasopchok_ho@yahoo.com

Abstract--This paper presents harmonic loss considerations in a PWM multilevel cascaded H-bridge inverter fed induction motor. A non-sinusoidal waveform or voltage harmonic has influence on efficiency and performance of a drive system. Harmonic voltages depend upon PWM parameters such as modulation index, switching frequency, PWM schemes, and dc voltage levels etc. The comparison of additional losses between a conventional inverter drive and a multilevel cascaded H-Bridge inverter drive is evaluated through MATLAB/Simulink and experimental results. Harmonic loss evaluation under various PWM strategies such as sinusoidal pulse width modulation (SPWM), third harmonic injection pulse width modulation (THPWM) and space vector pulse width modulation (SVPWM) for multilevel cascaded inverters is given. PWM signals are generated by a dsPIC microprocessor. The harmonic loss calculation is based on frequency dependent loss functions. Moreover experimental results are also given. The results can be guidelines for designing multilevel inverters suitable for induction motor drives.

Index Terms--Core Loss, Copper Loss, Harmonic Loss, Multilevel Inverter

I. INTRODUCTION

PWM inverters are widely used for adjustable speed induction motor drives. Multilevel inverters are becoming an attractive solution for medium and high voltage applications [1]. Harmonic voltages are inevitably generated by those. However, additional losses associated with harmonic voltages have influence on the efficiency and the performance in the induction motor control. Contribution on harmonic loss due to the harmonic voltage of the two-level inverter can be found in [2-5]. However there are a few publications involving harmonic loss associated with multilevel inverter fed induction motors. Therefore this paper is focused on calculation and experiment of harmonic loss in a multilevel inverter fed induction motor. The loss calculation is based on frequency dependent loss factor and individual harmonic voltage squared. The harmonic loss is evaluated through MATLAB/Simulink program. The experimental results are also given in order to verify the proposed loss model.

The obtained results can be reference data for designing multilevel inverters suitable for induction motor drives.

II. MULTILEVEL CASCADED H-BRIDGE INVERTER

A multilevel cascaded H-bridge inverter consists of a series connected single phase full-bridge inverter (H-Bridge inverter) shown in Fig. 1 in order to achieve an increased voltage range and reduced output voltage harmonics [1]. The number of level is related to the number of dc source as

$$n = 2s + 1 \qquad (1)$$

Where n is the number of level

s is the number of dc source

The number of switching devices in the main power circuit of the inverter is

$$N_{sw} = 6(n-1) \qquad (2)$$

Output fundamental voltage of the multilevel cascaded H-bridge inverter is linearly related to the modulation index (m_a) and dc voltage (V_{dc}) as

$$V_{1peak} = m_a V_{dc} \; ; \; \left(0 \leq m_a \leq 1\right) \qquad (3)$$

Where $V_{dc} = sE$

E is the dc voltage input for each full-bridge inverter

m_a is the modulation index

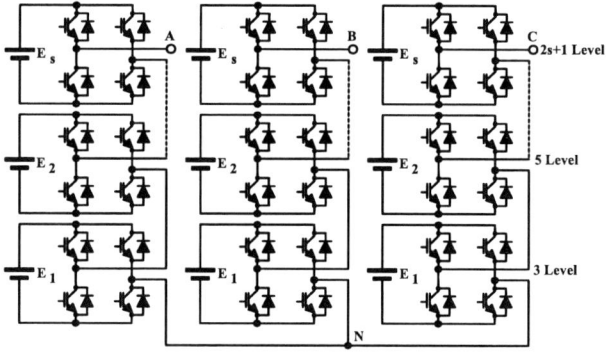

Fig. 1. Main power circuit of Multilevel Cascaded H-bridge Inverter.

For this work, the investigation of harmonic motor losses for 2, 3, 5, 7 and 9 level cascaded inverters is given under various operating conditions.

III. HARMONIC LOSS FUNCTION

Harmonic loss functions are obtained by measuring input power of the induction motor under test supplied by the six step inverter with constant dc link voltage and variable frequency. The harmonic losses can be achieved by subtracting the fundamental power (P_{Fund}) from the input power (P_{in}) by assuming that harmonics input power is almost loss. Thus

$$P_h = P_{in} - P_{Fund} \qquad (4)$$

and

$$P_h = P_{cu1} + P_{cu2} + P_c + P_{stray\text{-}loss} \qquad (5)$$

where

P_h	is harmonic input power
P_{in}	is input power
P_{Fund}	is fundamental power
P_{cu1}	is stator harmonic copper losses
P_{cu2}	is rotor harmonic copper losses
P_c	is harmonic core losses
$P_{stray\text{-}loss}$	is stray-load losses

Total harmonic loss can be expressed as:

$$P_h = \sum_{n \neq 1} \left(\frac{A}{f_n^{\alpha}} + \frac{B}{f_n^{\beta}} \right) V_n^2 \qquad (6)$$

where

P_h is total harmonic loss (Watts)

V_n is the nth harmonic voltage, (V_{rms})

f_n is harmonic frequency at the nth order, (kHz)

A is copper loss coefficient

α is frequency exponent of copper loss

B is core loss coefficient

β is frequency exponent of core loss

The first term represents the harmonic copper loss and the second term represents the harmonic core loss. Basically, the exponent α is 1.5 which takes skin effect into account and β varies between 0.3-0.5 which depends on material type and lamination thickness. A and B are inherent constants for each machine and will depend on machine size and design. The six-step voltage waveform with a range of frequency between at 300 Hz and 30 kHz, which corresponding harmonic voltages are exactly known, is applied to the motor under test. The measured harmonic power is normalized to loss factor (mW/V^2). The total harmonic loss factor is plotted against with frequency. The constants can be achieved by

curve fitting [3-5]. In this work, $A = 0.89$, $B = 0.762$, $\alpha = 1.5$, and $\beta = 0.5$ are found from the curve fitting. As a consequence, frequency dependent loss characteristics obtained from the curve fitting for loss segregation are shown in Fig. 2.

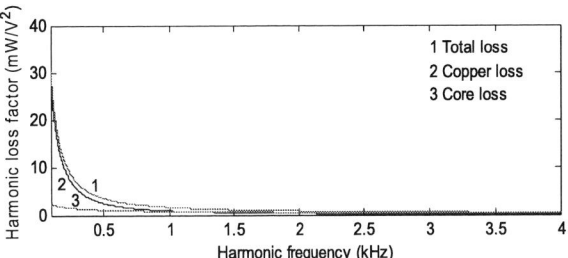

Fig. 2. Harmonic loss factor curves for each loss component.

Figure 2 shows the frequency dependent loss factor (harmonic copper loss (line 2), harmonic core loss (line 3) and total harmonic loss (line 1)) for calculating individual harmonic losses. Individual harmonic losses are calculated by multiplication of harmonic loss factor and harmonic voltage squared. From the loss factor curves implies that harmonic copper loss decreases rapidly whilst harmonic core loss decreases gradually as harmonic frequency increases. Total harmonic loss can be obtained by summing individual harmonic losses as equation (6).

Figs.3-5 show modulation pattern, output voltage waveforms, corresponding harmonic voltage spectra, and calculated individual harmonic losses using the proposed model of the 7-level cascaded H-bridge inverter for carrier-based SPWM, THPWM and SVPWM schemes.

Using the proposed loss factor, it is useful to show individual harmonic losses. Apparently, the individual harmonic losses are influenced by individual harmonic voltages and harmonic frequencies. Dominant harmonic losses are at about switching frequency. The SVPWM scheme give slightly better performance than THPWM scheme interms of harmonic spectra and harmonic loss. From the above example of the presented condition of the various PWM schemes one can see that the amount of harmonic losses could be dependent on PWM parameters such as the modulation index, switching frequency and types of PWM strategies. However, surprisingly, at this condition, the SPWM technique seem to give better performance than others.

IV. RESULTS AND DISCUSSION

MATLAB\Simulink program for generating PWM output voltage of the inverter and harmonic loss calculation is used. Additional loss calculations use loss functions in Fig. 2 in conjunction with equation (6). A comparison of calculated harmonic loss results between various multilevel inverters for the SPWM scheme supplying a 3HP, 380V, 50Hz, squirrel cage induction motor under rated flux level with varying inverter frequency is given in Fig.6. Clearly, the higher number of level, the lower

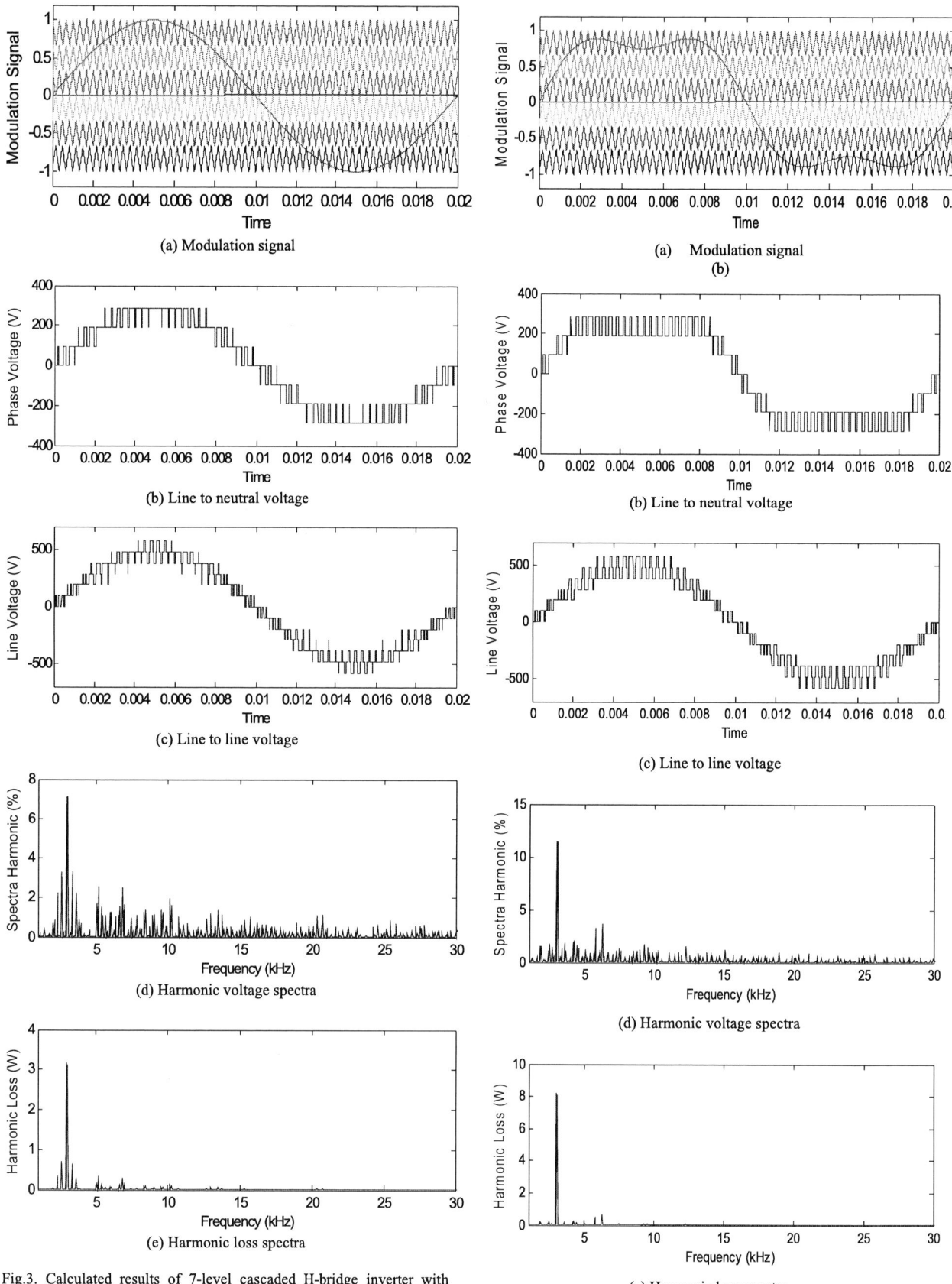

(a) Modulation signal

(b) Line to neutral voltage

(c) Line to line voltage

(d) Harmonic voltage spectra

(e) Harmonic loss spectra

Fig.3. Calculated results of 7-level cascaded H-bridge inverter with carrier-based SPWM at $m_a = 1.0$, fundamental frequency = 50 Hz, switching frequency = 3 kHz.

(a) Modulation signal
(b)

(b) Line to neutral voltage

(c) Line to line voltage

(d) Harmonic voltage spectra

(e) Harmonic loss spectra

Fig.4. Calculated results of 7-level cascaded H-bridge inverter with carrier-based THPWM at $m_a = 1.0$, fundamental frequency = 50 Hz, switching frequency = 3 kHz.

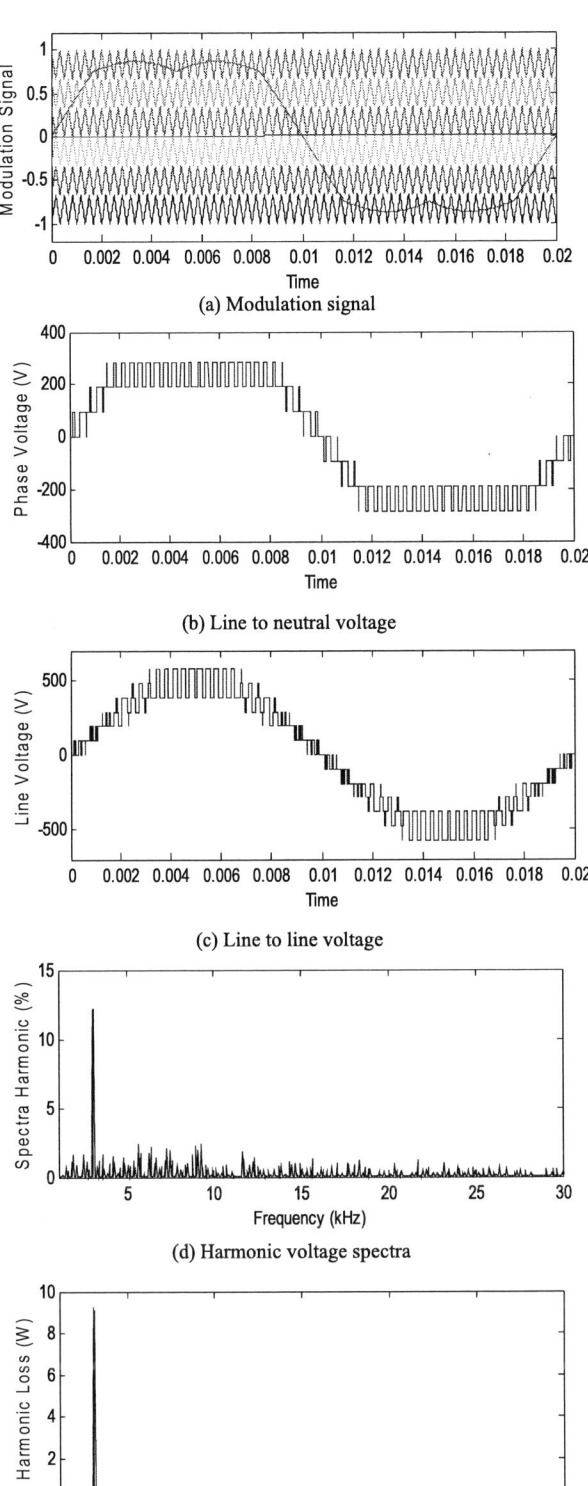

(a) Modulation signal

(b) Line to neutral voltage

(c) Line to line voltage

(d) Harmonic voltage spectra

(e) Harmonic loss spectra

Fig.5. Calculated results of 7-level cascaded H-bridge inverter with carrier-based SVPWM at m_a = 1.0, fundamental frequency = 50 Hz, switching frequency = 3 kHz.

harmonic loss is achieved. However, the two-level inverter produces lower harmonic loss than the 3-level inverter at a low frequency range but at a high frequency range this event is inverted. Increased level number of the

inverter causes a reduction in harmonic distortion of both voltage and current. There is no noticeable difference in harmonic losses between 7 and 9 level. The fact that higher number of level improves the quality of waveforms is also confirmed by Fig.7. Nearly sinusoidal waveform quality of both voltage and current is obtained. Measured results of output voltage and current waveforms as shown in Fig.7 have proved correctness and effectiveness of the implemented circuit which is controlled by a low cost dsPIC microprocessor.

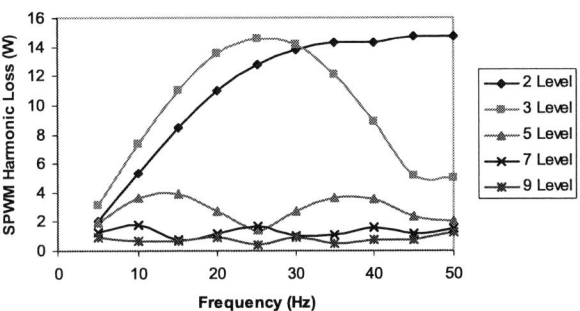

Fig.6. Variation of the harmonic loss with inverter frequency and level number of the inverter.

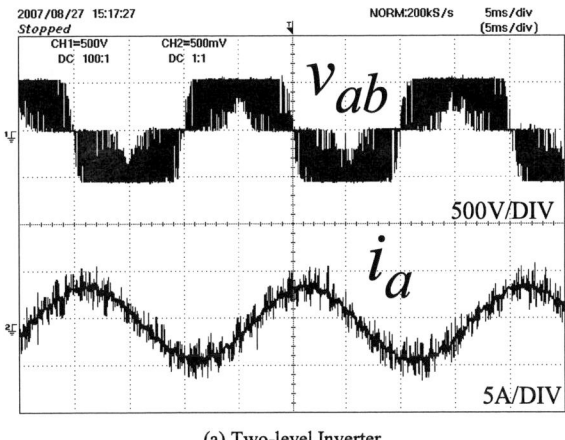

(a) Two-level Inverter

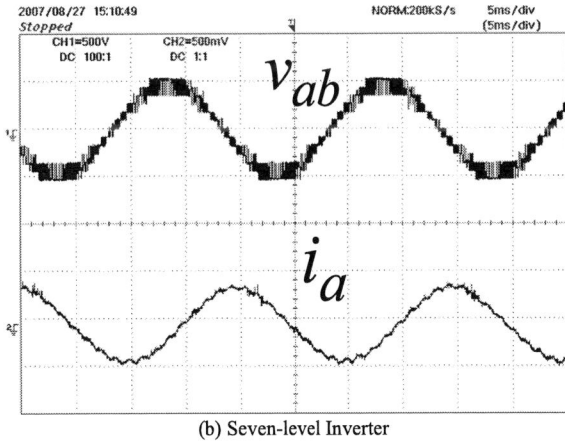

(b) Seven-level Inverter

Fig.7. Measured results of line to line voltage and line current between 2-level inverter and 7-level multilevel cascaded H-bridge inverter with SPWM at inverter frequency = 50 Hz, switching frequency = 4.8 kHz, m_a =1.0.

1239

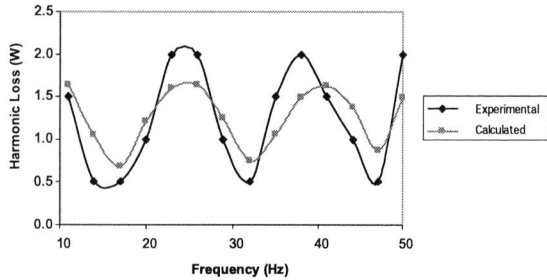

(a) SPWM Technique.

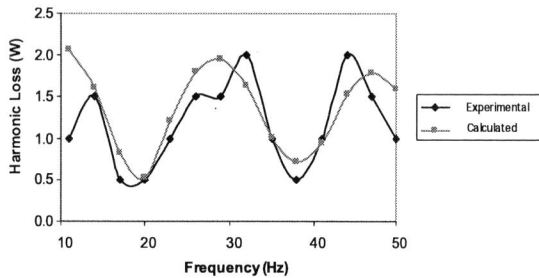

(b) THPWM Technique.

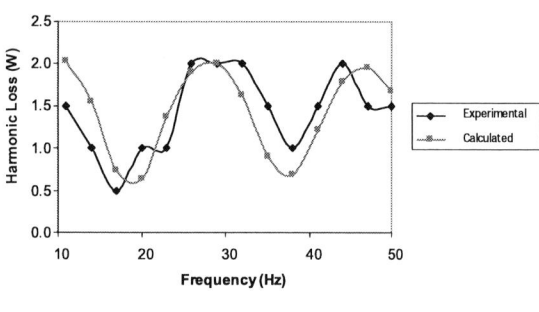

(c) SVPWM Technique.

Fig. 8. Harmonic Losses of the 7-level Inverter for SPWM, THPWM and SVPWM techniques.

In order to confirm the validity of the proposed loss model, power harmonic analyzer, YOKOGAWA, PZ4000 is used. Since the impact of loads on harmonic losses is insignificant, only the no load condition is investigated for the motor under test. Fig. 8 shows considerable agreement between the proposed model and experimental results for SPWM, THPWM and SVPWM schemes. The results show that the measured harmonic losses are in consistent with the proposed model over a wide range of operating inverter frequencies. However, the small amount of harmonic losses may cause fluctuation display of results. As a consequence, it is difficult to conclude the trend of harmonic loss characteristics under inverter frequency variation. It is noted that harmonic loss is influenced by the amplitude of the harmonic voltage. The modulation index of the PWM strategy affects the amplitude of the dominant sidebands. For constant volt/hertz operation the modulation index is proportional to the inverter frequency. Therefore, harmonic losses could change with inverter frequency variation.

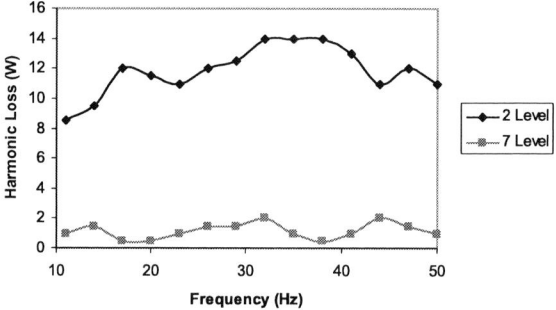

(a) SPWM Technique.

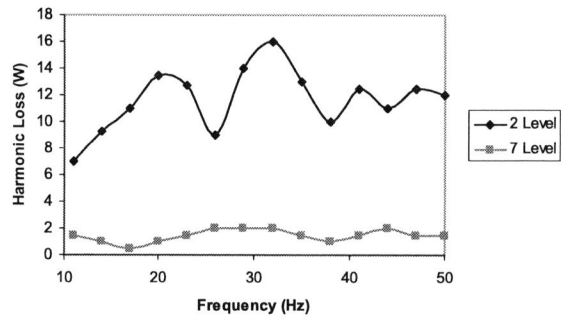

(b) THPWM Technique.

(c) SVPWM Technique.

Fig.9. Comparison of harmonic losses between 2-level and 7-level inverter using by SPWM, THPWM and SVPWM techniques at switching frequency = 4.8 kHz.

Fig.10. Prototype and measurement set-up.

Fig. 9 shows the experimentally comparative harmonic losses between 2-level and 7-level for the various PWM strategies under inverter frequency variations with rated constant volt/hertz. Again apparently, the 7-level cascaded inverter gives lower harmonic losses for all PWM types. As inverter frequency increases, the trend of harmonic losses increases. The behavior of this trend is influenced by the modulation index which is proportionally varied with inverter frequency.

Fig. 10 shows the prototype of the multilevel inverter and measurement set-up. Due to the limitation of the harmonic amount, the larger size of machines will be investigated to confirm the accurate loss model. It is noted that the increased number of level results in higher cost. However, in this work we have found that there is no noticeable difference in harmonic losses between the 7 and 9-level inverters. Therefore, when considering the number of switching devices, the 7-level cascaded inverter is more interested. When comparing with the 2-level PWM inverter, the 7-level PWM inverter offers more advantages in terms of higher output voltage, harmonic reduction, nearly sinusoidal waveform for both voltage and current. However, the number of devices is a major drawback. For suitable drive applications, careful attention must be paid for considerations of energy savings, output voltage level and available dc source.

V. CONCLUSIONS

This work has dealt with calculation and experiment of harmonic losses for the PWM multilevel cascaded H-bridge inverter. The method of harmonic loss calculation is described. The modulation index is influenced on harmonic losses. There is no noticeable between PWM strategies for multilevel cascaded H-bridge inverter, particularly for higher level. Main results are obtained from the comparison between 2 and 7-level cascaded inverter since the 7-level cascaded inverter is a reasonable choice compared to higher level inverter.

ACKNOWLEDGMENT

The authors would like to give a special recognition to thanks Dr.Surin Khomfoi for valuable suggestion and discussion. Many thanks must also be made to Mr.Chatrchai Aimsa-ard for help in the design and the implementation of the prototype and SIAM University for financial support.

REFERENCES

[1] B. P. McGrath, D. G. Holmes, "Multicarrier PWM Strategies for Multilevel Inverters", *IEEE Trans. On Industrial Electronics*, vol.. 49, No. 4, August 2002, pp.858-867

[2] A. Boglietti, P. Ferraris, M. Lazzari and F. Profumo, "Influence of the Inverter Characteristics on Iron Losses in PWM Inverter-Fed Induction Motors", *IEEE Trans. Magnetic*, vol. 2, No. 5, pp.1190-1194, September 1996.

[3] V. Kinnares, J. C. Clare and K. J. Bradly, "A New Technique for Determining & Predicting Harmonic Power Loss in PWM Fed Induction Machine", *ICEM96 Vigo/Spain*, pp.327-331, 1996.

[4] S. Khomfoi, V. Kinnares and P. Viriya, "Investigation into Core Losses due to Harmonic Voltages in PWM Fed Induction Motors", *IEEE 1999 International Conference on Power Electronic and Drive Systems, PEDS'99 Hong Kong*, 27-29 July 1999, pp. 104-109.

[5] T. M. Undeland and N. Mohan, "Overmodulation and Loss Considerations in High-Frequency Modulated Transistorized Induction Motor Drives", *IEEE Trans. on Power Electronics*, vol. 3, No. 4, October 1988, pp.447-452.

[6] N. Mohan., T. M. Undeland and W. P. Robbins, *Power Electronics Converters, Applications, and Design*, 2nd Edition, John Wiley & Sons Inc, Canada, 1995.

[7] S.Khomfoi, "Measurement and Analysis of Inverter Characteristics Based-on Losses in PWM Inverter Fed Induction Motors", *Master Thesis*, King Mongkut's Institute of Technology Ladkrabang, Thailand, May, 2001.

Extend the Use of Auxiliary Circuit to Start up, Shut down, and Balance of the Modified Diode Clamped Multilevel Inverter

Ahmed Ali Ashaibi *
Ahmed.Ashaibi@eee.strath.ac.uk
0044 141 5482443

Dr. S.J. Finney **
S.Finney@eee.strath.ac.uk
0044 141 548 2516

Prof. B.W. Williams **
B.W.Williams@eee.strath.ac.uk
0044 141 548 2386

Dr Ahmed Massoud**
Ahmed.Massoud@eee.strath.ac.uk
0044 141 548 2124

Abstract-- The diode clamped multilevel inverter (DCMLI) is an attractive high voltage multilevel inverter due to its robustness. The main draw back of the DCMLI is dc link capacitor voltage imbalance. A diode clamped inverter has been used in this research. The method used to balance the dc link capacitors is based on auxiliary switch-mode circuits, which are operated in a discontinuous inductor current mode. This research is focused on how to use auxiliary smps circuits to start up, balance, and shutdown the inverter. These functions are an issue in high voltage applications of the multilevel inverter. Simulation and experimental results are presented for the proposed auxiliary circuits operational technique.

Index Terms-- capacitor unbalance, charging the capacitor, Discharging the capacitor, multilevel inverter.

I. Introduction

The multilevel inverter attracted researchers due to its advantages of offering better switch utilization and higher voltage and power ratings [1]. Multilevel inverters produce a stepped AC output by connecting the load to one of the capacitor nodes using electronic switches (IGBTs). Utilization of the Modified Diode Clamped Multilevel Inverter (MDCMLI) shown in Figure1 is the subject of this paper.

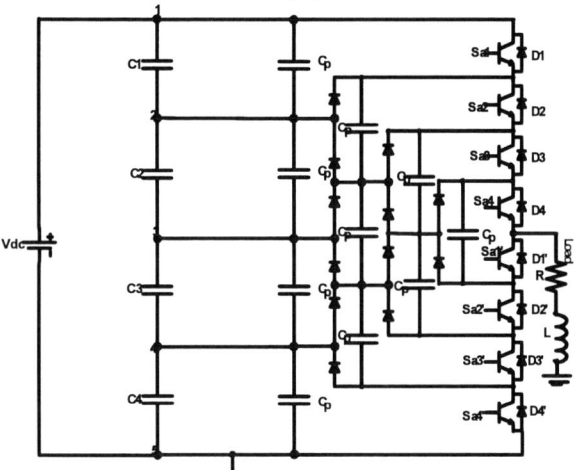

Figure 1 Single leg five- level MDCMLI.

The main drawback of this inverter is the capacitor voltage unbalance caused by different link capacitor loadings. Many balancing techniques have been published. These techniques can be divided into two types, hardware techniques[2-5] and software

techniques[6, 7]. The objective of balancing techniques is to equalize the voltage cross the switches during normal operation, and to purify the output voltage, that is reduce the total harmonic distortion (THD) [8].

The factors that affect capacitor unbalance have been assessed and investigated [9]. The two factors that affect unbalance can be summarised as follows [9], the power factor and the modulation index. Without capacitor balancing techniques, all odd level inverters develop three level output voltages, especially at high modulation indices.

The technique used in this work is an extension of the auxiliary circuits method to balance the inverter capacitors proposed in [2, 5, 9]. The control algorithm coordinates the energy transfer between the capacitors via a switched mode power supply (SMPS) inductor. A discontinuous inductor current mode is used to transfer energy between adjacent capacitors. Figure 2 shows the energy transfer sequence in both directions between the inner and the outer capacitors. This bidirectional energy transfer flexibility can be used during inverter startup and shutdown.

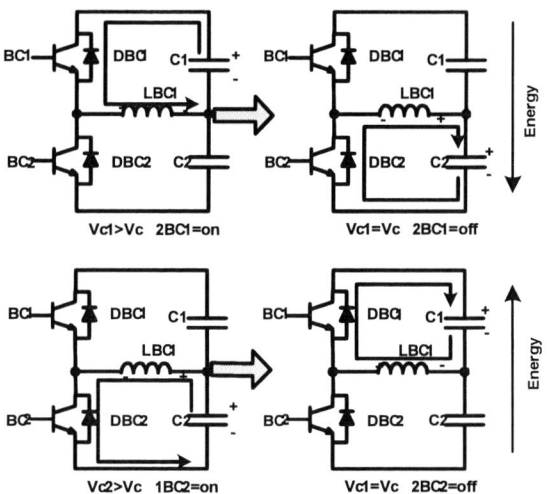

Figure2 Energy transfer between the capacitors and inductor in the balancer

The main objective of this paper is to implement the charging and discharging of the inverter capacitors at inverter start up and shut-down. Direct on-line starting-up and shut-down is an operational issue in high voltage applications. Auxiliary SMPS are key to achieving these goals, and as implemented in [2, 5, 9] were not capable of performing the necessary tasks. Two single low voltage circuits are added and in conjunction with the smps to implement the tasks.

978-1-4244-0644-9/07/$25.00 ©2007 IEEE

1- A boost converter is used to charge the lowest level capacitor. The boost switch is rated at the cell level voltage. Boosting generates the necessary high-voltage from a lower-voltage.

2- High power external resistance circuit is used to discharge the lowest level capacitor. Thus external circuitry is used only to charge and discharge the lowest level multilevel dc link capacitor. The balancing circuits already in the system used to balance the capacitors also transfer the charge to higher level capacitors in the charging mode and transfer the charge from higher level capacitors to lower level capacitors in the discharging mode.

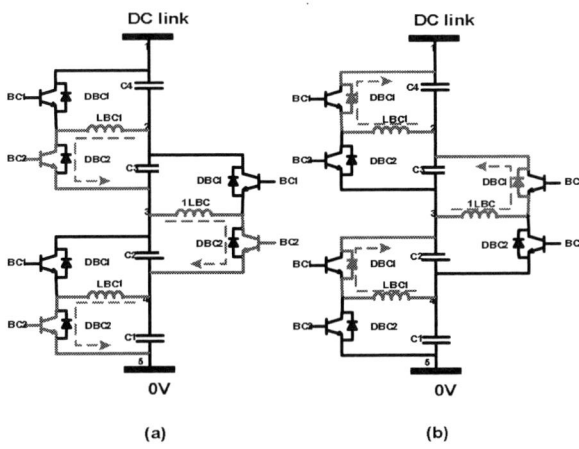

(a)   (b)

Figure 3 Energy transfer up through the auxiliary circuit during charging up
  a)   Dashed path when the lower switch is on.
  b)   Dashed path when lower switch is off and diodes are conducting.

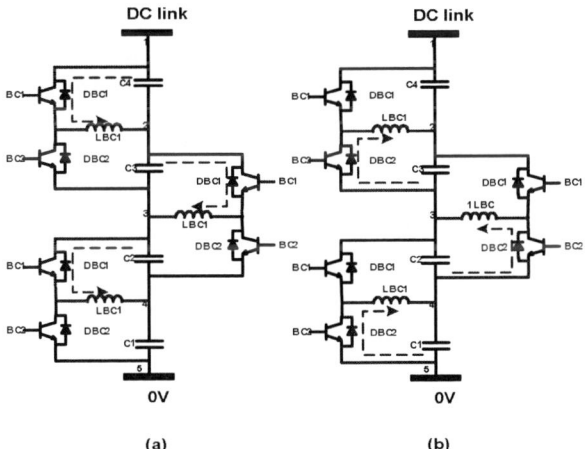

(a)   (b)

Figure 4 Energy transfer down through auxiliary circuit during discharge mode
  c)   Dashed path when the upper switch is on.
  d)   Dashed path when upper switch is off and diodes are conducting.

Figures 3 and 4 show the energy transfer paths between link capacitors during the charging and discharging modes, respectively.

II. Simulation results

Pspice is used to model the inverter system, and sinusoidal pulse width modulation (SPWM) is the modulation technique used. Four carriers are used to produce the five level output voltage. Figures 3 a, b show the output voltage and currents with and without auxiliary circuit balancing respectively.

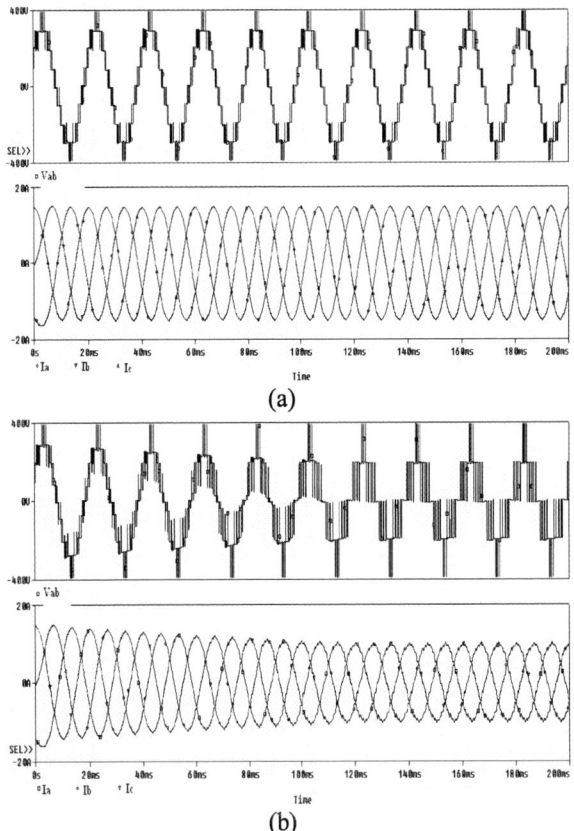

(a)

(b)

Figure 3 Pspice simulation results for MDCMLI. a) V_{ab} voltage, I_a, I_b, I_c, without Auxiliary circuit. b) V_{ab} voltage, I_a, I_b, I_c with Auxiliary circuit.

The auxiliary circuit switching frequency is 2k Hz with a vary on-state duty cycle. A dead-band has been implemented to avoid unnecessary energy transfer between the capacitors during the balancing mode. The results in figure 3 show the affect of the auxiliary circuits on multilevel inverter operation, especially when supplying a high power factor load.

The auxiliary circuits are used for start up and shut down of the inverter. Figure 4 shows the circuit diagram for the proposed dc link capacitor voltage control system, with a boost converter and discharge circuit shown.

1243

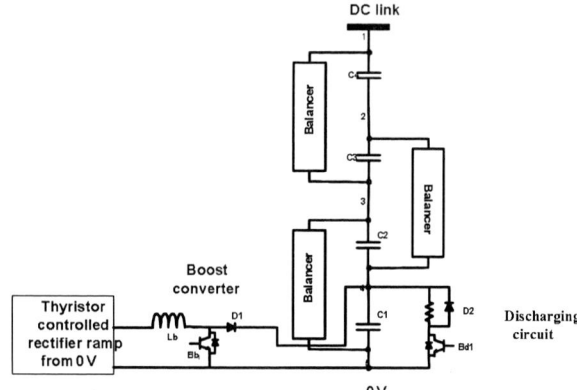

Figure 4 Proposed auxiliary circuit charging and discharging scheme

Figure 5 shows the Pspice simulation results of the dc link capacitor voltages during the three modes of operation, capacitor charging, and normal operation, then shut down

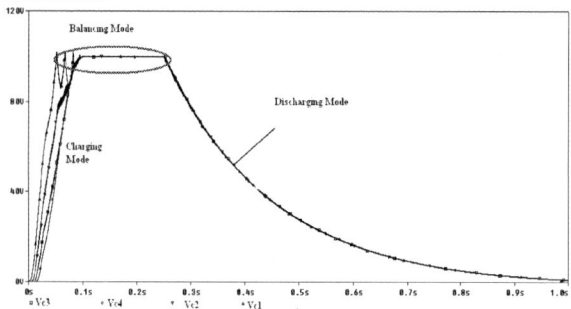

Figure 5 Pspice simulation result for three mode operation (charging, balancing and discharging)

III. Practical results

A modified five level-inverter is used to validate the proposal practically. Each of the three auxiliary smps circuits is controlled by a PIC16F73 microcontroller that generates the two PWM signals needed to operate the IGBTs in the auxiliary circuit. Two analogue to digital converters (ADC) are used to measure the lower capacitor voltage and the voltage difference between the two capacitors. Voltage difference (the voltage difference error) is implemented using an operational amplifier and then offset. The auxiliary circuits are programmed to operate in four modes: stop, charge, discharge, and balance modes. The microcontroller in each SMPS communicates with the main controller using asynchronous communications (USART) with a baud rate of 250kbit/s. The main controller, Infineon TC1796, controls the five-level inverter and the start up to shut down sequencing. The auxiliary circuit duty cycle and operating mode are specified by the main controller. A controlled rectifier is used as the input to the boost converter. The thyristor controlled rectifier implements a soft start to reduce capacitor inrush current. Practically

the soft start is controlled from the main controller and serial controlled digital to analogue converter (DAC). The DAC generates the ramp analogue output voltage which is used to phase angle control the output voltage of the rectifier. The rectifier output is filtered with a smoothing capacitor. The controlled rectifier output is reduced to zero when the dc link capacitor charging mode is finished, as shown in   figure 6.

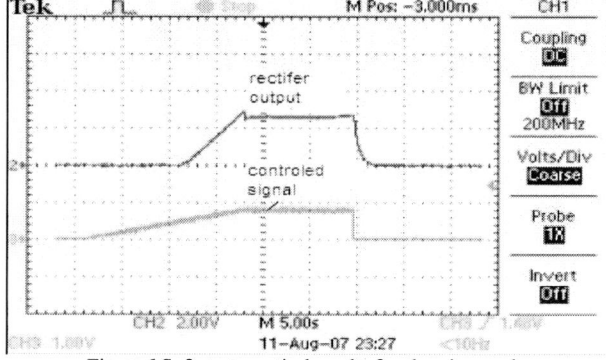

Figure 6 Soft start practical results for charging mode.

The effect of power factor on the auxiliary circuit has been tested and investigated practically. Figures 7(a) and (b) show the line-line voltage V_{ab} and three line currents respectively.

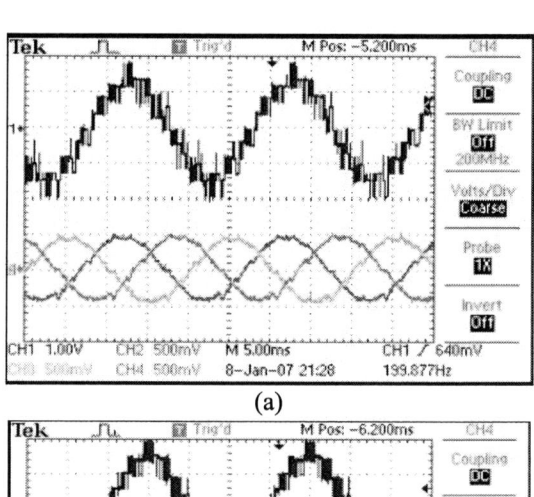

(a)

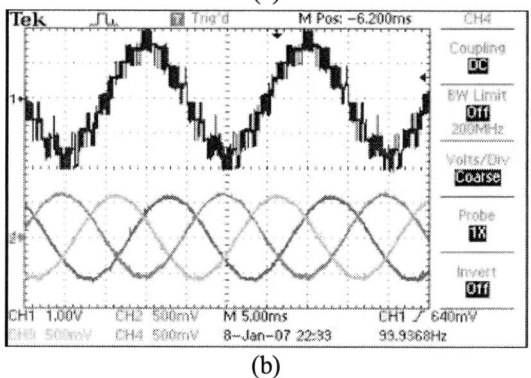

(b)

Figure 7 Practical inverter measurements (a) 100V Line-line voltage and three phase line load currents pf=.96 lagging. Scales (1V=100V) and (100mV=1A), (b) Line-line voltage and three phase line load currents pf=.1 lagging.

Also the energy transferred through the auxiliary circuits has been assessed and measured under various load conditions. Figures 8(a) and (b) show the auxiliary circuit inductor current during high and low power factor loading conditions respectively. The energy transfers from an outer capacitor to an inner capacitor, that is, from C_1 to C_2 and from C_4 to C_3. More energy is transferred in the high power factor case, as shown in figure 8(a). A discontinuous inductor current mode is used to eliminate auxiliary circuit switch turn on and diode recovery losses.

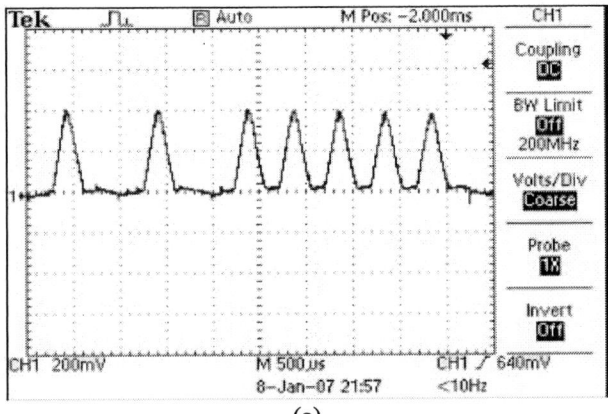

(a)

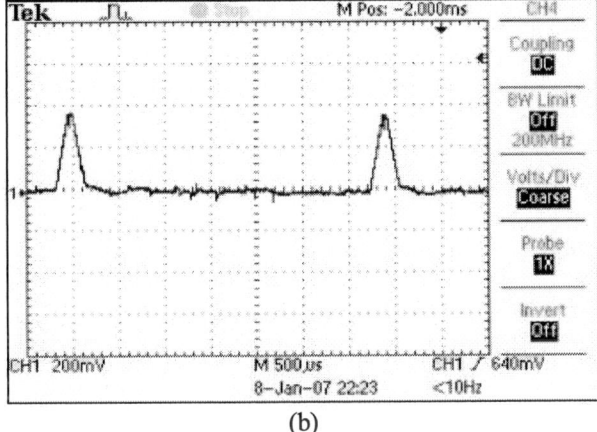

(b)

Figure8 Practical inverter measurements of auxiliary circuit current, (a) pf=.96 lagging, and (b) pf=.1 lagging.

The control algorithm for complete operation is summarised in the following points:-

- The AC input voltage is sampled.
- The DC voltage for each link capacitor and converter boost voltage are specified.

$$V_{c1}=V_{c2}=V_{c3}=V_{c4}=\frac{1}{4}V_{dc}$$

$$V_{boost}=\frac{1}{2}V_{c1}$$

- The main controller transmits the duty cycle to the each auxiliary circuit and specifies the duty cycle for the boost converter.

- The Capacitor voltages are transmitted from each auxiliary circuit to the main controller.
- The main controller ramps the boost converter input voltage to the required voltage using a DAC controlling a phase controlled rectifier.
- The boost converter charges the lower capacitor to the required voltage and simultaneously the auxiliary circuits transfer energy to higher capacitor levels.
- The total link capacitor voltage is increased to greater than the expected DC voltage after rectification of the ac input supply so that the three phase rectifier diodes will block the three phase.
- The AC main contactor is closed by the main controller. No inrush current occurs and SPWM can start to generate the output voltage. Also the main controller changes the auxiliary circuit from a charge up mode to the balancing mode.
- For shut down, PWM operation is stopped and the main AC contactor isolates the ac supply under zero current conditions. The discharge circuit is operated to discharge the lower capacitor. Simultaneously the auxiliary circuits start to transfer the capacitor energy from higher level capacitors to lower level capacitors.
- When all capacitor voltages reduce to zero, the main controller commands the auxiliary circuits to cease operation.

The algorithm has been assessed and tested under different operating conditions for both the boost and auxiliary circuits.
The boost converter switching frequency is 20 kHz and the switching frequency is 2 KHz for the auxiliary circuits.

Case 1- boost duty cycle is 12µs (24% duty cycle) and the duty cycle for the auxiliary circuit is 30µs (6% duty cycle). Figure 9 shows the DC link voltage, rectifier output, and control signal to the phase controlled rectifier. Since no resistive dumping is used across the lowest capacitor at shut down, capacitors discharge slowly through the auxiliary circuit voltage transducer resistance (25kΩ) .

Case 2- duty cycles are the same as for case 1, only the resistive discharge circuit discharges the lower level capacitor after inverter operation has ceased. As shown in figure 10 the discharging time has been dramatically reduced when the resistive discharge circuit is used.

1245

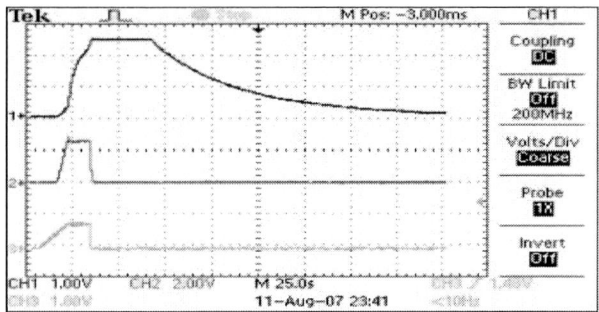

Figure 9 Practical measurements of complete inverter operation. 1) DC link voltage scales (1V=100V). 2) Controlled rectifier output voltage scales (1V=20V). 3) Control signal for rectifier (1V=1V).

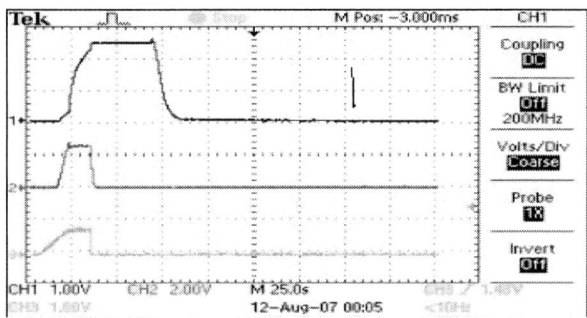

Figure 10 Practical measurements of complete inverter operation with resistive discharge circuit.

Case 3- boost and auxiliary circuit duty cycle have increased to 24% and 16% respectively. Figure 11 shows the affect of increasing the duty cycle for both boost and auxiliary circuit. Faster charging and discharging results as the duty cycle is increased.

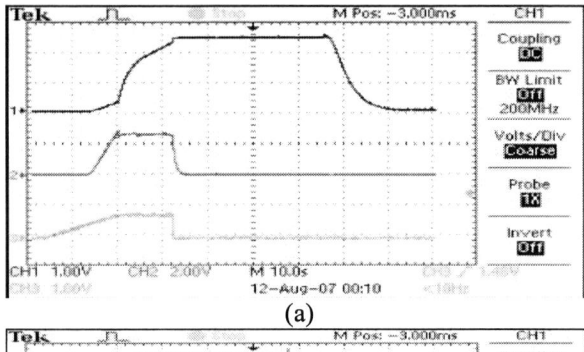

(a)

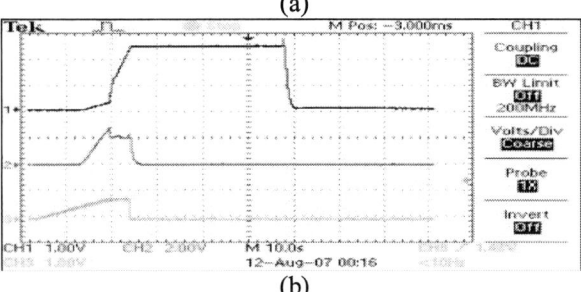

(b)

Figure 11 Practical measurements of complete inverter operation a) boost duty cycle 24% and auxiliary duty cycle 6%. b) Boost duty cycle 36% and auxiliary duty cycle 16%.

IV. Discussion

The resistive discharge circuit reduces the capacitor dc link discharge time from 155 s to 10 s. Also due to a duty cycle increase, the capacitor dc link charge up time is significantly reduced as shown in figure11. These figures show the capacitor charge up time is reduced from 12 s to 3 s and the discharge time is reduced from 10 s to 2 s.

IIV. Conclusion

The auxiliary capacitor balancing circuits not only balance the link capacitor voltages, but also facilitate inverter startup and shut down by charging and discharging the link capacitors. The charging up time and discharge time can be controlled by control the duty cycle of the auxiliary smps.

References

[1] M. Akira Nabae, IEEE, Isao Takahashi, IEEE, and Hirofuni Akagi, Member, IEEE, "A New Neutral-Point-Clamped PWM Inverter," *IEEE Transaction on Industry Application*, Vol. IA-17, No pp 518, 1981.

[2] C.Yiqiang and O. Boon- Teck, "Multimodular multilevel rectifier/inverter link with independent reactive power control," *Power Delivery, IEEE Transactions*, Vol. 13, pp. 902-908, 1998.

[3] A. von Jouanne, S. Dai, and H. Zhang, "A multilevel inverter approach providing DC-link balancing, ride-through enhancement, and common- mode voltage elimination," *Industrial Electronics, IEEE Transactions*, Vol. 49, pp. 739-745, 2002.

[4] S. D. Annette Von Jouanne, Haoran Zhang, "A Simple Method for Balancing the DC-Link Voltage of Three-Level Inverters," *IEEE*, Vol 3 No pp.1341, 2001.

[5] M. S. Newton C., "Novel Technique for maintaining balanced internal DC link voltage in diode clamped five-level inverters," *IEE Proc.-Electr. Power Appl.*, Vol. 146, No pp. 341, 1999.

[6] M. S. Newton C., T. Alexander, "The Investigation and Development of A Multi- Level Voltage Source Inverter," presented at Power Electronics and Variable Speed Drives, No pp 317 ,1996.

[7] M. S. Newton C., "Neutral Point Control for Multi-Level Inverters : theory , design and operational limitations," *IEEE Industry Application Society*, Vol 2 No pp. 1336, 1997.

[8] C. Yiqiang, B. Mwinyiwiwa, Z. Wolanski, and O. Boon-Teck, "Regulating and equalizing DC capacitance voltages in multilevel STATCOM, " *Power Delivery, IEEETransactions*, Vol. 12, No pp. 901-907, 1997.

[9] A. A.-M. A.-W. Ibrahim, "A Practical Method for Capacitor Voltage Balancing of Diode Clamped Multilevel Inverter, PhD. Heriot- watt University , Edinburgh, 2004.

Five-Level Z-Source Neutral-Point-Clamped Inverter

F. Gao[1], P. C. Loh[1], F. Blaabjerg[2], R. Teodorescu[2] and D. M. Vilathgamuwa[1]

[1]School of Electrical and Electronic Engineering
Nanyang Technological University
Nanyang Avenue, Singapore 639798
Email: gaof0001@ntu.edu.sg

[2]Institute of Energy Technology
Aalborg University
DK-9220 Aalborg East, Denmark
Email: fbl@iet.aau.dk ; ret@iet.aau.dk

*Abstract--*This paper proposes a five-level Z-source neutral-point-clamped (NPC) inverter with two Z-source networks functioning as intermediate energy storages coupled between dc sources and NPC inverter circuitry. Analyzing the operational principles of Z-source network with partial dc-link shoot-through scheme reveals the hidden theories in the five-level Z-source NPC inverter unlike the operational principle appeared in the general two-level Z-source inverter, so that the five-level Z-source NPC inverter can be designed with the modulation of carrier-based phase disposition (PD) or alternative phase opposite disposition (APOD) technique. To verify the theoretical findings and practical issues, a scaled down laboratory prototype was constructed and tested with a 1.3 kw induction motor load.

Index Terms--Z-source inverter; five-level NPC inverter; buck-boost.

I. INTRODUCTION

For high voltage inversion, multilevel inverters, especially the diode-clamped inverters and the cascaded inverters, are preferably applied with the significant advantages of reduced voltage stress on semiconductors, improved harmonic performance and decreased electromagnetic interference (EMI) when compared with two-level inverters. In particular, neutral-point-clamped (NPC) inverters are well known for their widely use in industry applications [1]. However, as all traditional inverters demonstrated, the peak ac output voltage of NPC inverter is always constrained by the total dc voltage supplied with individual dc source or controlled dc source, such as the controllable rectifier. To overcome this limitation, a number of voltage boost techniques are adopted, the most straightforward of which is to couple a simple dc-dc boost converter between dc source and inverter circuitry forming two-stage power conversion, which unavoidably increases the control complexity and reduces system reliability. On the other hand, the Z-source topology with the capability of single-stage power conversion was proposed in [2], which uses an X-shape network with the symmetrical placement of passive components to store inductive energy and then release it to the following load achieved by using special shoot-through

techniques [2, 3]. Doing so, the dc-link voltage can be boosted to any desired value theoretically or maintain the original dc value unchanged unlike the topology with boost converter added which always boost dc-link voltage being higher than the input dc voltage. With the consideration of above mentioned characteristics in mind, this paper would, therefore, focus on the possible high voltage development of the advantageous Z-source technique.

The Z-source inverters have to date been developed in voltage source and current source topologies [4]. For voltage-type Z-source inverters, besides the simple two-level structure, three-level topologies have been proposed in NPC-based or dual connection topologies with the control of phase disposition (PD), alternative phase opposite disposition (APOD) [5] and reduced common mode (RCM) modulations [6-8], respectively, for the preferable selection depending on the specific applications and system requirements. For higher level development, however, neither the topology nor the modulation methods of Z-source inverter have been reported in literatures. Aiming for the multilevel development of Z-source inverter, this paper proposes a five-level Z-source NPC inverter firstly, which is controlled using appropriate PD and APOD modulations.

By analyzing the operational principles, the unique features of Z-source network appeared in multilevel Z-source NPC inverter are explored step by step from the full dc-link shoot-through to the unobvious partial dc-link shoot-through for illustrating the necessary processes of power inversion, which is important for deriving five-level Z-source inverter. Depending on the partial dc-link shoot-through technique, the proposed five-level topology can boost its dc-link voltage properly with the expected output waveform quality when using specifically designed modulation schemes. Finally, the experimental verifications were performed by constructing a laboratory prototype and using 1.3kW induction motor as load to test practical issues, whose captured experimental results are presented in Section V.

II. OPERATIONAL ANALYSIS OF Z-SOURCE NETWORK

978-1-4244-0644-9/07/$25.00 ©2007 IEEE

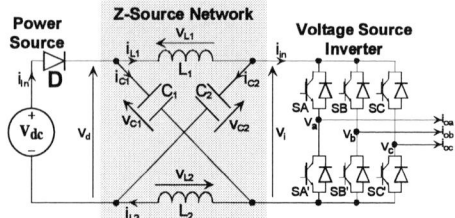

Fig. 1. General illustration of the Z-source inverter.

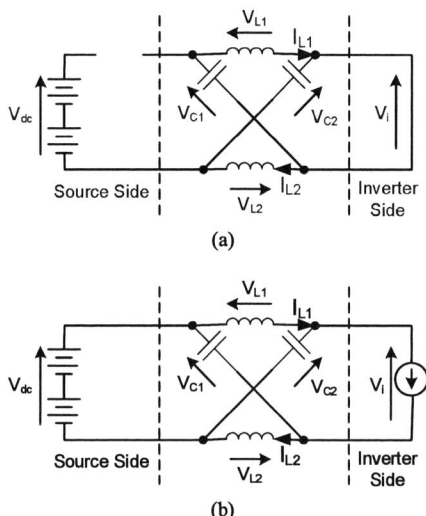

Fig. 2. Equivalent circuits of Z-source inverter at (a) shoot-through states and (b) non-shoot-through states.

Fig. 1 shows the general topology of two-level Z-source inverter, where the only difference compared with the traditional inverters is the inserted Z-source network and a passive diode D between dc supply and inverter circuitry. The shaded two-port network is built using a split inductor and two shunt capacitors, whose X-shape structure allows switches from any phase-leg to turn ON simultaneously with the input diode D naturally reverse-biased to create a

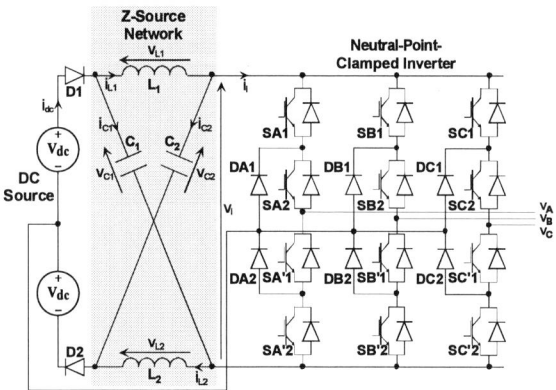

Fig. 3. Topology of Z-source NPC inverter using a single LC impedance network.

shoot-through state without damaging semiconductor devices due to the natural property of current protection by inductors. In order to completely analyze the operational principle of Z-source network, two totally different voltage boost theories would be illustrated in the following subsections that are classified as full dc-link shoot-through and partial dc-link shoot-through.

A. Full DC-Link Shoot-Through

As introduced in [2, 3], the shoot-through of any phase-leg creates a disconnection between input dc source and its following components by naturally reverse-biasing input diode D, whose equivalent circuit is drawn in Fig. 2 (a). During shoot-through interval, the inductive energy stored in the Z-source network is boosted by inductor current circulation between capacitors and inductors. During non-shoot-through interval (traditional active and null states), the Z-source inverter switches as traditional inverters as shown in Fig. 2 (b) with $V_i \geq V_{dc}$, where the inverter circuitry and load are represented by a current source. By averaging the inductor voltage over a switching period to be zero, the peak dc voltage $\hat{v}_i$ and the peak ac output voltage $\hat{v}_x$ ($x = a$, b or c) can be written as [2]:

$$
\begin{cases}
\hat{v}_i = 2V_C - V_{dc} = \dfrac{V_{dc}}{1 - 2T_0/T} = BV_{dc} \\[2mm]
\hat{v}_x = \dfrac{M}{2}\hat{v}_i = \dfrac{MBV_{dc}}{2}
\end{cases}
\tag{1}
$$

Where, M refers to the modulation ratio and T_0 represents the shoot-through duration during the switching period T. $B \geq 1$ is the gain factor used for representing voltage boost capability.

B. Partial DC-Link Shoot-Through

Among multilevel implementations, three-level Z-source

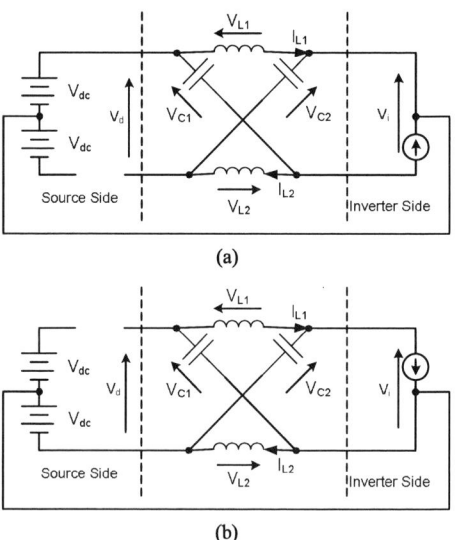

Fig. 4. Equivalent circuit of three-level Z-source NPC inverter at shoot-through states with (a) upper shoot-through and (b) lower shoot-through.

NPC inverter with single Z-source network, as shown in Fig. 3, is an interesting alternative for buck-boost operation, where two input dc sources are necessary to power the inverter in order to form a distinct neutral point so that achieving three-level phase voltage switching whose equivalent circuits under shoot-through and non-shoot-through states can also be drawn as Fig. 2 when designed APOD modulation is assumed [9]. Note that APOD modulation provides the short-circuit path crossing through two phase-legs to short upper and lower dc-links completely during shoot-through interval, which allows the inverter side can still be represented by a short circuitry as Fig. 2(a). Analyzing the equivalent circuits using the same principles as that for the two-level Z-source inverter, the inverter peak dc link and ac output voltages can still be derived as (1).

Besides the shoot-through path linking upper and lower dc-rails, another unobvious shoot-circuit path is available by partially shooting-through dc-links, as the resulted equivalent circuits shown in Fig. 4, which corresponds to the operation of the particular topology of three-level Z-source NPC inverter with single Z-source network, where one dc source is always inserted in the short-circuit path in series, resulting in the total different shoot-through operations. For example, upper dc source is inserted in the shoot-through path formed by inductor L_1, ON switches of SX1, SX2 and SX'1 and the diode DX1 (X = A, B or C) with respect to the denominated upper shoot-through. Alternatively, the lower shoot-through is generated by turning ON SX2, SX'1 and SX'2 simultaneously involving the lower dc source into the shoot-through path. Noting that during these specific shoot-through intervals, only partial dc-links are short-circuit with no previous full dc-links short-circuit generated. Mathematical analysis reveals that the partial dc-link shoot-through can also effectively boost dc-link voltage, which would be mathematically analyzed below.

During shoot-through intervals, the inductor voltage can be expressed as:

$$v_{L1} = v_{L2} = V_{dc} \qquad (2)$$

Alternatively, when in the non-shoot-through states, the inductor voltage can be derived as:

$$v_{L1} = v_{L2} = -\frac{T_0'}{T/2 - T_0'} V_{dc} \qquad (3)$$

Where, T_0' refers to the shoot-through duration in half switching cycle and V_{dc} represents single dc source voltage in Fig. 3. Further, capacitor voltage can be derived as:

$$V_C = 2V_{dc} - v_L = \frac{2T - 2T_0'}{T - 2T_0'} V_{dc} \qquad (4)$$

Therefore, the peak dc-link voltage $\hat{v}_i$ can be expressed as:

$$\hat{v}_i = V_C - v_L = \frac{2T}{T - 2T_0'} V_{dc} \qquad (5)$$

The dc-link voltage v_i would then alternate between $2T \cdot V_{dc}/(T - 2T_0')$ and $T \cdot V_{dc}/(T - 2T_0')$ during non-shoot-through and shoot-through intervals, respectively, unlike full dc-link shoot-through with the value of zero appearing. Assuming only the peak value of v_i will be seen by the inverter, therefore, the peak ac output voltage $\hat{v}_x$ can be derived as:

$$\hat{v}_x = M\frac{\hat{v}_i}{2} = \frac{M}{2}\frac{2T}{T - 2T_0'}V_{dc} = B'\left\{\frac{M}{2}2V_{dc}\right\} = B\{MV_{dc}\} \quad (6)$$

Obviously, full dc-link shoot-through and partial dc-link shoot-through provide the same voltage buck-boost capability for the Z-source inverters since the newly derived boost factor B' is equal to B.

III. OPERATIONAL PRINCIPLES OF FIVE-LEVEL Z-SOURCE NPC INVERTER

Reviewing the existing three-level Z-source NPC inverters shown in Fig. 3 and Fig. 5, obviously, the topology with single Z-source network is cost-efficient due to less passive components used [9]. But integrating the concept developed in the Z-source NPC inverters using either two Z-source networks [6] or single Z-source network, the Z-source three-level NPC inverter can be extended to a newly developed five-level topology with two Z-source networks used, as shown in Fig. 6.

In Fig. 6, the upper clamping leg is connected to the neutral point N1 between voltage sources V_{dc1} and V_{dc2} and the lower clamping leg is linked to the another neutral point N2 formed between V_{dc3} and V_{dc4}, respectively. Besides these two connections, the only remaining middle clamping leg is intentionally connected to the neutral potential N which is formed by cascading two Z-source networks. Once the inverter operates properly, two Z-source networks would boost dc-link voltage equally, therefore, N is a real neutral potential in the inverter operation which can be illustrated by the simplified equivalent circuits shown in Fig. 7. When in a non-shoot-through state, the operation of the proposed inverter is similar to the combination of two three-level Z-source inverters just by cascading two Z-source networks at the neutral potential N as illustrated in

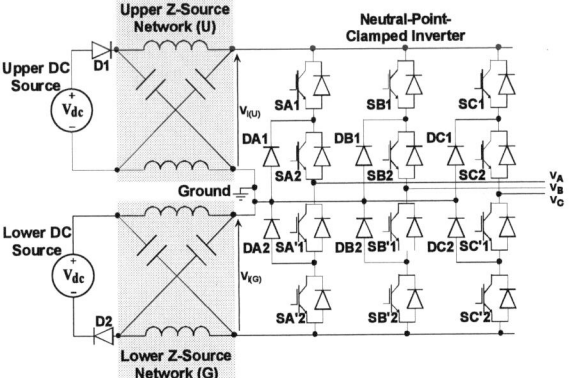

Fig. 5. Topology of three-level Z-source inverters with two Z-source networks.

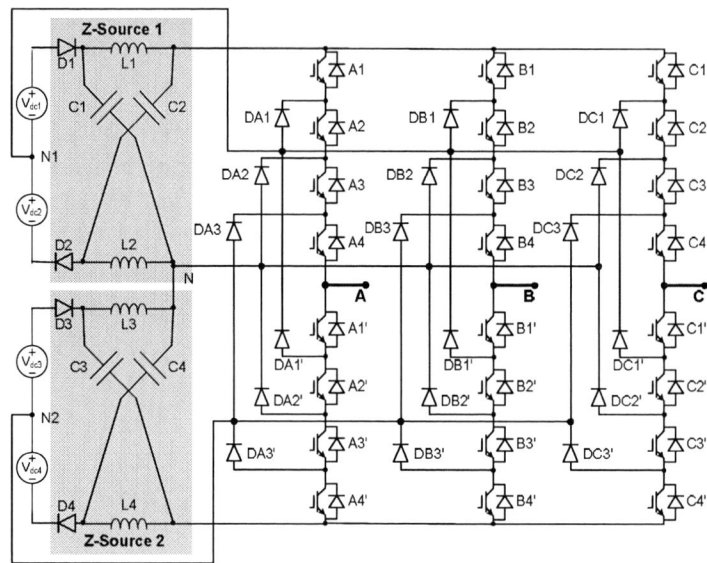

Fig. 6. Topology of five-level Z-source NPC inverter.

Fig. 7(a). When in a shoot-through state, either upper Z-source or lower Z-source network is short-circuit, as shown in Fig. 7 (b) and (c), respectively, where partial dc-link shoot-through is assumed resulting that only the outer equivalent current source is shorted effectively and the remaining Z-source network still works as normal operation whose operational principle is similar to that analyzed in Section II. For example, switching ON A1, A2, A3, A4 and A1' can generate the operation state like that shown in Fig. 7(b). Being similarly, turning ON A4, A1', A2', A3' and A4' can induce the state shown in Fig. 7(c). The additional ON switching, being different from the traditional five-level NPC inverter which always has four switches ON per phase at any moment, can be introduced by changing the switching time of the corresponding switch so as to not increase commutation count. This advantageous behavior has been reported in literatures for modulating the Z-source inverters [3, 6, 9].

The Z-source network, however, would not suffer alternating partial dc-link shoot-through unlike three-level Z-source NPC inverter because when the inner equivalent current source is intended to be short-circuit, any additional switching would cause another unwanted dc-link shoot-through resulting in the full dc-link shoot-through for a single Z-source network, which further complicates the modulation schemes for achieving proper five-level phase switching. Therefore, only outer partial dc-link shoot-through performs the voltage boost operation of five-level Z-source NPC inverter, whose dc-link voltage and peak ac output voltage can be expressed as (5) and (6) again. It is noted that the partial dc-link shoot-through states are intentionally generated alternatively to fully utilize the shoot-through capability for minimizing voltage stress [10]. But the alternating shoot-through operations for single Z-

source network do not appear in the presented five-level Z-source inverter, replacing by the alternating shoot-through operation between two Z-source networks. Besides the partial dc-link shoot-through technique, a simple full dc-

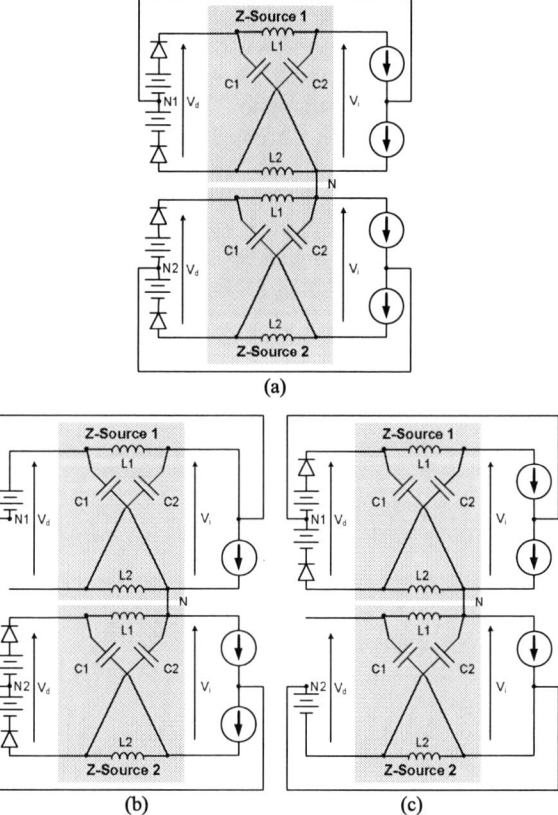

Fig.7. Equivalent circuits of five-level Z-source NPC inverter when in (a) non-shoot-through and (b), (c) shoot-through states.

1250

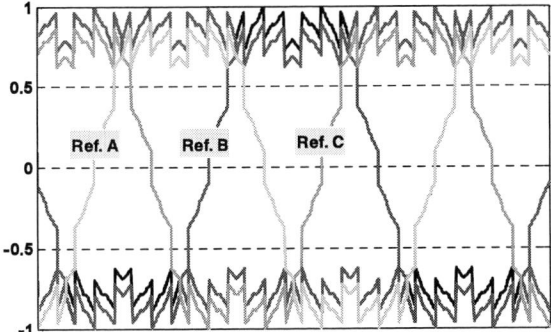

Fig. 8. The modified references for modulating five-level Z-source NPC inverter.

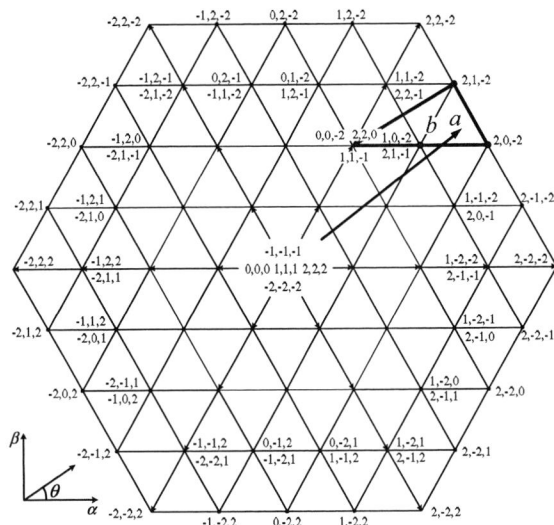

Fig. 10. Space vector diagram of five-level inverter.

link shoot-through can also be assumed by just shorting one phase-leg completely when in shoot-through intervals. Doing so, however, leads to the inferior output waveform quality and increased commutation count.

IV. MODULATION SCHEMES OF FIVE-LEVEL Z-SOURCE INVERTER

In this section, two different carrier-based modulation schemes are presented, both of which can control the five-level inverter effectively for buck and boost operations each with the advantageous characteristic, such as optimized harmonic performance or reduced common mode voltage.

A. Phase Disposition Modulation

Comparing three-phase sinusoidal references with a proper triplen harmonic offset added to a set of disposed triangular carriers with 0° phase shift, the multilevel NPC inverter can achieve optimized harmonic performance [11]. Aiming for realizing the optimized harmonic switching, meanwhile modulating the proposed five-level Z-source inverter properly, this subsection presents the detailed illustration for PD modulation technique with the modified references assumed. With a proper offset added, the exampled references suitable for the five-level Z-source inverter modulation can be drawn as Fig. 8, whose mathematical expressions can be written as:

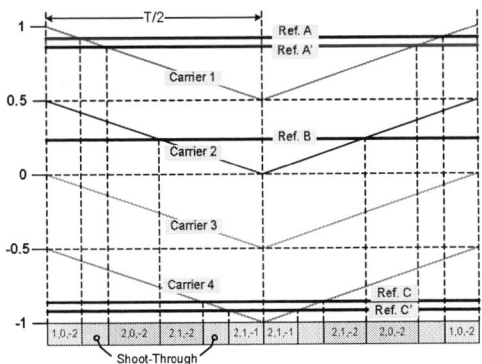

Fig. 9. Exampled switching sequence of five-level Z-source NPC inverter using PD modulation.

$$
\begin{cases}
V_{\max(SX)} = V_{\max} + V_{off(SVM)} + T_1 \\
V_{\max(SX')} = V_{\max} + V_{off(SVM)}
\end{cases}
$$

$$
\begin{cases}
V_{\mathrm{mid}(SX)} = V_{\mathrm{mid}} + V_{off(SVM)} \\
V_{\mathrm{mid}(SX')} = V_{\mathrm{mid}} + V_{off(SVM)}
\end{cases}
$$

$$
\begin{cases}
V_{\min(SX)} = V_{\min} + V_{off(SVM)} \\
V_{\min(SX')} = V_{\min} + V_{off(SVM)} - T_1
\end{cases}
$$

$$
T_1 = \frac{T_0}{T} , \quad X = A, B \text{ or } C \tag{7}
$$

Where, $V_{off(SVM)}$ is a triplen harmonic offset added to force two equivalent null intervals in each switching sequence to be equal.

To clearly illustrate the modulation scheme, an exampled switching sequence for PD modulation is shown in Fig. 9, where four carriers are disposed with 0° phase shift. At the first switching transit from equivalent null state {1, 0, -2} to active state {2, 0, -2}, a shoot-through state is inserted by turning ON switches A1, A2, A3, A4 and A1' simultaneously with respect to the shoot-through state shown in Fig. 7(b). In order to maintain volt-sec average unchanged, shoot-through state should be inserted in the equivalent null interval which can be realized easily by switching ON A1 early with time T_0. This shoot-through duration in carrier-based modulation can be defined by the comparison of two references A and A' of phase A with the particular Carrier 1. On the other hand, the second shoot-through state is inserted at the third switching transit from active state {2, 1, -2} to another equivalent null state {2, 1, -1} by delaying the OFF time of C4' with T_0. In sine-triangle comparison method, this can also be realized by using two references for phase C to compare with Carrier 4 which corresponds to the equivalent circuit shown in Fig. 7(c). It is noted that the shoot-through state cannot be inserted at the second transit in this switching sequence by

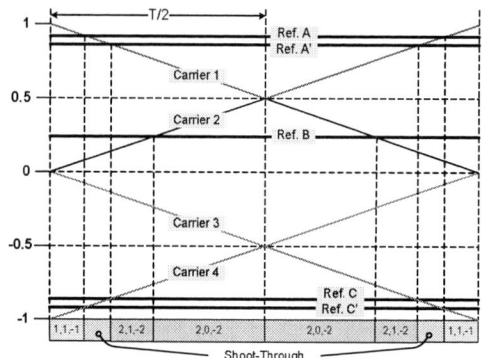

Fig. 11. Exampled switching sequence of five-level Z-source NPC inverter using APOD modulation.

using two references for phase B because the resulted shoot-through state would boost Z-Source 1 again leading to unbalanced dc-link voltage boosting seen from the inverter.

Besides the analyzed switching sequence for the reference phasor locating in the triangle a of five-level space vector diagram as indicated in Fig. 10, other switching sequences appeared in a fundamental cycle can be analyzed using the same method. It is important that two Z-source networks should be boosted equally in each switching period in order to minimize the current ripple and voltage stress [12]. Calculating the commutation counts of the traditional five-level NPC inverter using PD modulation and the five-level Z-source NPC inverter reveals that no additional device commutation is introduced in the Z-source inverter modulation because shoot-through states are inserted by only changing the original switching time of corresponding devices to satisfy voltage boost requirements.

B. Alternative Phase Opposition Disposition Modulation

Additionally, another carrier-based modulation scheme with inferior harmonic performance, named as alternative phase opposition disposition modulation, can be designed for controlling five-level Z-source NPC inverter. Although the APOD modulation does not provide optimized harmonic spectral, it can effectively reduce the common mode voltage comparing to the PD modulation. Again, an exampled switching sequence is shown in Fig. 11 for illustrating the modulation principle. Obviously, except of the shoot-through states, total three switching states, alternating along with the trajectory of bold triangle b in Fig. 10, appear in the exampled switching cycle.

Following the criteria for equally boosting two Z-source networks, a shoot-through state is intentionally inserted at the transit from {1, 1, -1} to {2, 1, -2} because the short-circuit path can be created with no additional switching commutation introduced. For the Z-source 1, switches A1, A2, A3, A4 and A1' are turned on to enter the boost mode and for the Z-source 2, switches C4, C1', C2', C3' and C4' are commanded ON to enter the boost mode simultaneously,

which are realized by comparing the interleaved references of phase A and C to Carrier 1 and 4, respectively. The interleaved references in a fundamental cycle can also be expressed as equation (7) except that the offset $V_{off(SVM)}$ should be replaced by $-0.5(\max(V_a, V_b, V_c)+\min(V_a, V_b, V_c))$.

V. EXPERIMENTAL VERIFICATION

To verify all theoretical findings developed in the proposed five-level Z-source inverter, a scale down experimental setup was constructed which allowed the inverter to drive an induction motor using PD and APOD modulation techniques presented in Section IV, respectively.

The system consisted of a five-level NPC inverter and two Z-source networks which were powered by four dc sources. The Z-source network was assembled using capacitors with C1 = C2 = C3 = C4 = 1000 μF and inductors with L1 = L2 = L3 = L4 = 3.3 mH, implying that the Z-source network was symmetrically constructed. The five-level NPC inverter was built using insulated gate bipolar transistor (IGBT) modules and diodes. At the inverter output side, a 1.3 kW induction motor was connected as load to verify all presented findings and test the inverter's capability for driving motor load.

In Fig. 12, the experimental waveforms with $M = 1.15$ and $T_0/T = 0$ under PD modulation are presented, where $V_{dc1} = V_{dc2} = V_{dc3} = V_{dc4}$ were tuned to be 80V. Obviously, distinct nine-level line voltage switching and five-level phase voltage switching were captured. Moreover, a sinusoidal current of induction motor is observed, both of which illustrate the operational validation of the proposed five-level Z-source NPC inverter under normal operations without voltage boost needed. For boost operation, Fig. 13 shows the captured waveforms under PD modulation with $M = 0.7*1.15$ and $T_0/T = 0.3$, and $V_{dc1} = V_{dc2} = V_{dc3} = V_{dc4} = 80V$ either, in which the amplitude of line voltage, phase voltage and common mode voltage are effectively boosted to desired value. Since the induction motor outputs constant power, therefore, the line current is reduced consequently. A little distortion observed in captured line current was induced by the saturation of induction motor.

Comparatively, Fig. 14 and Fig. 15 show the captured waveforms under APOD modulation with the same operation conditions as those used in Fig. 12 and 13, respectively. It is noted that the common mode voltage appeared in Fig. 14 and 15 are obviously less than the common mode voltage in Fig. 12 and 13 as the inherent advantage of APOD modulation. The current waveforms, however, show inferior harmonic performance since less switching states were used as analyzed above. The proposed inverter can also work well under both normal operation and voltage boost operation conditions.

VI. CONCLUSIONS

This paper presents the design of a five-level Z-source NPC inverter that can perform buck-boost dc-ac energy

1252

inversion. Under special operations with respect to the partial dc-link shoot-through, which profitably appear in the operation of five-level Z-source NPC inverter, the Z-source network shows much different operational principles and boost characteristics. Modulation wise, the proposed inverter can be controlled using PD and APOD modulation techniques with a slight modification of three-phase references. The theoretical findings, together with the practicality of the inverter, have been tested experimentally using a laboratory prototype under both normal and voltage boost conditions.

REFERENCES

[1] M. Fracchia, T. Ghiara, M. Marchesoni, and M. Mazzucchelli, "Optimized modulation techniques for the generalized N-level converter," in *Proc. IEEE PESC'92*, 1992, pp. 1205-1213.

[2] F. Z. Peng, "Z-source inverter," *Industry Applications, IEEE Transactions on*, vol. 39, pp. 504-510, Mar./Apr. 2003.

[3] L. Poh Chiang, D. M. Vilathgamuwa, Y. S. Lai, C. Geok Tin, and Y. Li, "Pulse-width modulation of Z-source inverters," *Power Electronics, IEEE Transactions on*, vol. 20, pp. 1346-1355, Nov. 2005.

[4] P. C. Loh, D. M. Vilathgamuwa, C. J. Gajanayake, L. T. Wong, and C. P. Ang, "Z-Source Current-Type Inverters: Digital Modulation and Logic Implementation," *Power Electronics, IEEE Transactions on*, vol. 22, pp. 169-177, Jan. 2007.

[5] B. P. McGrath and D. G. Holmes, "Multicarrier PWM strategies for multilevel inverters," *Industrial Electronics, IEEE Transactions on*, vol. 49, pp. 858-867, Aug. 2002.

[6] P. C. Loh, F. Blaabjerg, S. Y. Feng, and K. N. Soon, "Pulse-Width Modulated Z-Source Neutral-Point-Clamped Inverter," in *Proc. IEEE APEC '06*, 2006, pp. 431-437.

[7] P. C. Loh, F. Gao, and F. Blaabjerg, "Topological and Modulation Design of Three-Level Z-Source Inverters," in *Proc. IEEE IPEMC'06*, 2006, pp. 1-5.

[8] F. Gao, P. C. Loh, F. Blaabjerg, and D. M. Vilathgamuwa, "Dual Z-source Inverter with Three-Level Reduced Common Mode Switching," in *Proc. IEEE IAS'06*, 2006, pp. 619-626.

[9] L. Poh Chiang, L. Sok Wei, G. Feng, and B. Frede, "Three-Level Z-Source Inverters Using a Single LC Impedance Network," *Power Electronics, IEEE Transactions on*, vol. 22, pp. 706-711, Mar. 2007.

[10] S. Miaosen, W. Jin, A. Joseph, P. Fang Zheng, L. M. Tolbert, and D. J. Adams, "Constant boost control of the Z-source inverter to minimize current ripple and voltage stress,"

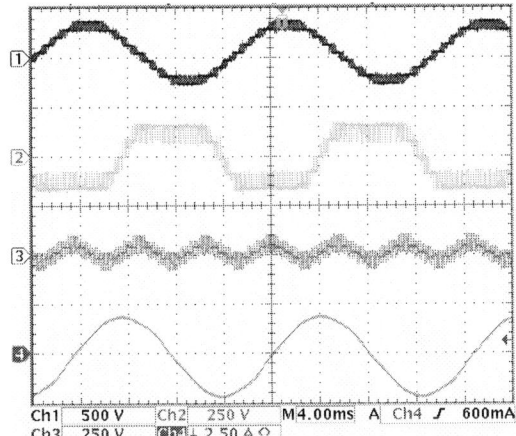

Fig. 12. Experimental results of five-level Z-source NPC inverter using PD modulation with $M = 1$ and $T_0/T = 0$.

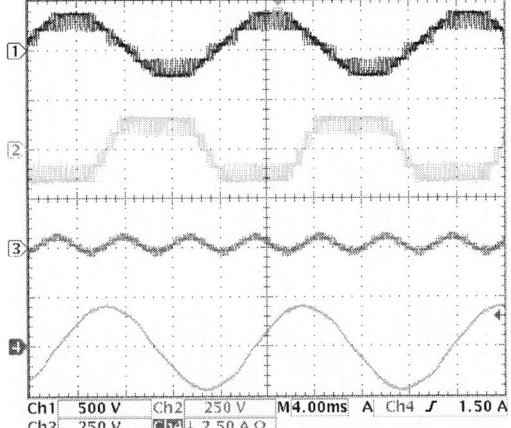

Fig. 14. Experimental results of five-level Z-source NPC inverter using APOD modulation with $M = 1$ and $T_0/T = 0$.

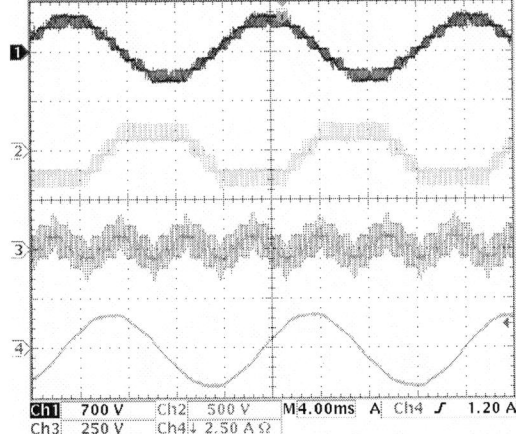

Fig. 13. Experimental results of five-level Z-source NPC inverter using PD modulation with $M = 0.7$ and $T_0/T = 0.3$.

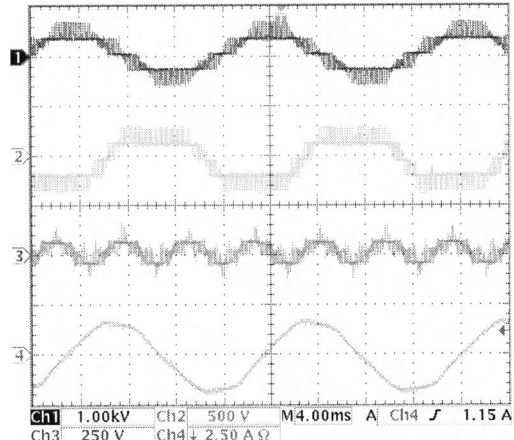

Fig. 15. Experimental results of five-level Z-source NPC inverter using APOD modulation with $M = 0.7$ and $T_0/T = 0.3$.

Industry Applications, IEEE Transactions on, vol. 42, pp. 770-778, May/Jun. 2006.

[11] B. P. McGrath, D. G. Holmes, and T. Lipo, "Optimized space vector switching sequences for multilevel inverters," *Power Electronics, IEEE Transactions on*, vol. 18, pp. 1293-1301, Nov. 2003.

[12] P. C. Loh, F. Blaabjerg, and C. P. Wong, "Comparative Evaluation of Pulsewidth Modulation Strategies for Z-Source Neutral-Point-Clamped Inverter," *Power Electronics, IEEE Transactions on*, vol. 22, pp. 1005-1013, May 2007.

Capacitor Voltage Balancing Using Redundant States for Five-Level Multilevel Inverter

Hadi A Hotait, Ahmed M Massoud, Steve J. Finney, Barry W. Williams

Department of Electronic and Electrical Engineering
University of Strathclyde
Glasgow, United Kingdom
Tel: 0044 141 548 2350

Abstract-- **In this paper a new five-level inverter capacitor voltage balancing technique for high modulation index and high power factor operation is proposed. The proposed redundant state technique is based on dividing the space vector diagram of the five-level inverter into six three-level space vector diagrams. The original centre of the space vector diagram is shifted. New states of the space vector are determined, the dwelling time is calculated as for conventional two-level modulation, and the switching sequence is determined depending on the voltage level of the four dc link capacitors, using a redundant state method.**

*Index Terms—***Multilevel Inverter, Space Vector Modulation, Balancing Technique.**

I. INTRODUCTION

Multilevel level inverters have been advocated as a suitable choice in several medium and high voltage applications because of the difficulty in connecting a single power semiconductor switch directly to the medium voltage grid (3.3 to 13kV) [1]. The purpose of multilevel inverters is to generate a high-voltage waveform using lower voltage switching devices.

The three-level inverter was the first diode clamped multilevel inverter [2]. By increasing the number of inverter levels, the output voltage has steps and a staircase waveform is generated, which produces reduced harmonic distortion. However, a high level number increases control complexity and develops voltage imbalance problems.

The insulated gate bipolar transistor (IGBT) is capable of controlling considerable power at a high switching frequency [3],[4]. Multilevel inverters are being considered for applications such as active filtering, reactive power compensation [5] and large induction motor drives [6].

A disadvantage of the diode clamped multilevel inverter is voltage imbalance of the dc link capacitors. Voltage balancing of the two dc link capacitors of the three-level or "neutral-point clamped inverter [7] is not a problem.
However, when the number of levels is increased above three, capacitor voltage balancing becomes a problem. It is recognized that capacitor voltage balancing of the five

level inverter can only be achieved if the modulation index and power factor are restricted [8]. If certain limits are exceeded, the central capacitors discharge and the outer capacitors over charge. The five-level inverter then produces a three-level output voltage and the outer switches are voltage stressed.

Several techniques have been proposed to over come this capacitor imbalance problem:

- Isolated dc sources [6];
- Auxiliary dc/dc smps circuits [9];
- Active rectifier as a front-end added to a multilevel inverter system [10]; or
- A redundant state method; this technique has modulation index and power factor limitations for five-level inverters and higher levels [11].

A new capacitor voltage balancing technique for high modulation index, high power factor, five-level inverters using a redundant state method, is proposed.

II. PROPOSED TECHNIQUE

The proposed balancing technique modifies the vector space of the five-inverter and converts it into six two-level space vector spaces where the new state spaces are determined and the dwell time is calculated as for two-level inverters. The selection of the state sequence depends on the four capacitor voltages and uses a redundant state method.

The algorithm can be summarized as follows:
 i. Determine the space vector location;
 ii. Determine in which hexagon the space vector lies;
 iii. Determine in which sector the space vector lies;
 iv. Determine the distribution of the sample times using conventional five-level space vector modulation equations;
 v. If the vector lies in hexagons four or five, determine the sector where the space vector lies;
 vi. Move the centre of the five-level space vector to an apex of the inner hexagon, so six new centres are possible for the six new two-level spaces;
 vii. Generate the new space vector states, depending on which sector the space vector lies within, so a new angle and a new reference voltage is generated;

This work was supported by the University of Strathclyde.

978-1-4244-0644-9/07/$25.00 ©2007 IEEE

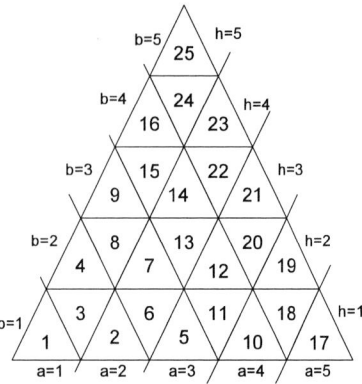

Fig. 1. The region number according to the three parameters a, b and h

viii. Determine the distribution of the sample times from the new modulation index and new angle, using conventional two-level space vector modulation equations;

ix. Measure the voltage of the four capacitors.

x. Determine the sequence of the states using a redundancy method.

A. Determination of the space vector location

The space vector location is determined in two steps. The first step is to determine the sector number where the vector lies. The second step is to determine in which region the vector lies.

Step 1: Sector numbers

The voltage magnitude V_m is obtained by transforming from the a-b-c frame to the $\alpha - \beta$ frame using Park's transformation.

$$\begin{pmatrix} V_\alpha \\ V_\beta \end{pmatrix} = \begin{pmatrix} 1 & -\dfrac{1}{2} & -\dfrac{1}{2} \\ 0 & \dfrac{\sqrt{3}}{2} & -\dfrac{\sqrt{3}}{3} \end{pmatrix} \begin{pmatrix} V_a \\ V_b \\ V_c \end{pmatrix} \qquad (1)$$

Then the vector magnitude and the vector angle can be determined by

$$V_m = \sqrt{V_\alpha^2 + V_\beta^2} \qquad (2)$$

$$\theta = \tan^{-1}\left(\frac{V_\beta}{V_\alpha}\right) \qquad (3)$$

The sector number is given by

$$s = floor\left(\frac{\theta}{\pi/3}\right) + 1 \qquad (4)$$

where **floor** is the function which corrects any real positive number to the nearest, but lower, integer number (e.g. floor (3.2) = 3).

Step 2: Region number

Three parameters identify the vector location hence determine the region number. The first parameter `h` divides the sector horizontally as shown in Fig. 1. The

second parameter `a` divides the sector with 60-degree inclined lines. The third parameter `b` divides the sector with 120-degree inclined lines. The maximum value that a, b and h can obtain for m-levels is m-1.

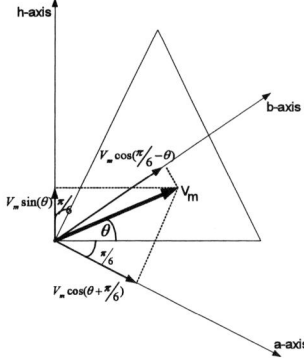

Fig. 2. the three axes a, b, and h for region determination

Using numerical analysis, the region number can be expressed as a function of the three parameters [12]:

$$reg = -a + (b^2 - b + 1) + h \qquad (5)$$

The three parameters are determined as follows:

$$a = floor\left(\frac{V_m \cos\left(\theta + \dfrac{\pi}{6}\right)}{E \cos\dfrac{\pi}{6}}\right) + 1 = floor\left(m_a(m-1)\cos\left(\theta + \frac{\pi}{6}\right)\right) + 1$$

$$b = floor\left(\frac{V_m \cos\left(\dfrac{\pi}{6} - \theta\right)}{E \cos\dfrac{\pi}{6}}\right) + 1 = floor\left(m_a(m-1)\cos\left(\frac{\pi}{6} - \theta\right)\right) + 1 \qquad (6)$$

$$h = floor\left(\frac{V_m \sin\theta}{E \cos\dfrac{\pi}{6}}\right) + 1 = floor\left(m_a(m-1)\sin\theta\right) + 1$$

where $0 \le \theta \le \pi/3$

The modulation index for the multilevel inverter is calculated from

$$m_a = \frac{V_m}{(m-1)E\cos\dfrac{\pi}{6}} \qquad (7)$$

where E is the cell voltage and m is the number of levels per phase voltage.

Fig. 2 indicates the axes a, b, and h where the value of `a` is the projection of vector V_m on to the a-axis. Similarly for b and h.

B. Division of the vector space

The first step of the proposed technique is to locate in which sector and in which sub-sector the space vector lies. Then the reference centre of the five-level space is moved, depending on the location of the space vector, to an apex of the inner hexagon. The vector space of the five-level inverter, shown in Fig. 3a, is simplified into six two-level vector space regions as shown in Fig. 3b.

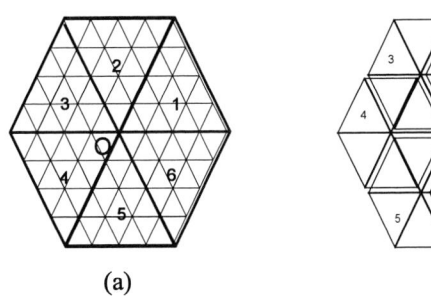

(a)   (b)

Fig. 3. Five level inverter: (a) space vector modulation and (b) simplified space vector modulation

Sector one in hexagon one (Fig. 4b) and sector one in hexagon two (Fig. 4c) in the simplified SVM covers sector one in conventional five-level SVM, Fig. 4a. Table I summaries the sector divisions.

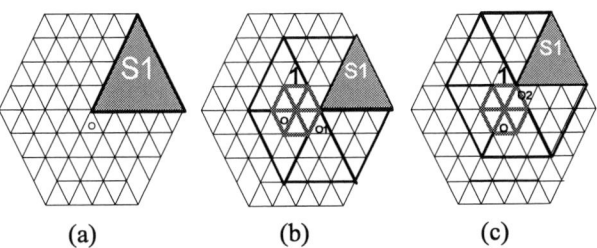

(a)   (b)   (c)

Fig. 4. sections: (a) sector one in conventional SVM, (b) sector one in hexagon one and (c) sector two in hexagon two

TABLE I
SECTORS DIVISIONS

Sector (five-level inverter)	Hex (Shifted Hex)	Sector in Shifted Hex
1	1,2	1
2	2,3	2
3	3,4	3
4	4,5	4
5	5,6	5
6	6,1	6

C. Choosing between the new sectors

As mentioned, all the inner triangles in the last two hexagons in one sector of the five level space are divided into 2 sectors in two different two level spaces with two different centres.

As shown in Fig. 5, triangles numbered (2,5,6,10,11) are located in sector one in hexagon one, so if the space vector is located in any of these triangles, hexagon one with centre O_1, is selected. Triangles numbered (4,8,9,15,16) are located in sector one in hexagon two. So if the space vector is located in any of these triangles, hexagon two, with centre O_2, is selected. Triangle 12 is nearer to centre O_1 than O_2, thus is in sector one in

hexagon one with centre O_1. Triangle 14 is nearer to centre O_2 than O_1 thus triangle 14 is in sector one in hexagon two with centre O_2.

Triangles (7, 13) are equi-distance between the two centres. So as to achieve a symmetrical equal weighting between the two centres, a straight line with angle 30^0 is drawn from the main centre of the space vector O. If the space vector is located below this line, hexagon one with centre O_1 is chosen. If the space vector is located beyond this line, hexagon two with centre O_2 is chosen.

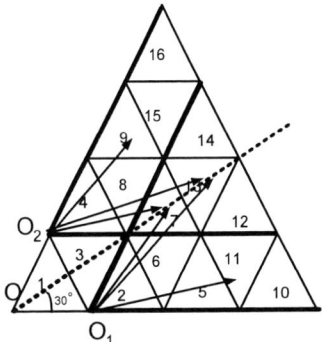

Fig. 5. Division of one sector of conventional SVM into two sectors

D. Moving the Voltage Reference

After dividing the vector space for the five-level inverter into six two-level vector spaces, the centre of these hexagons will be the apexes of the inner hexagon since these states are the redundant states used to balance the four voltage capacitors.

Moving the original centre to another centre depends on the location of the space vector, (in which sector it lies) and its angle (as mentioned).

The first step in moving the centre, is to decompose the voltage reference vector into V_α, V_β.

$$V_\alpha = V_m \cos\theta$$
$$V_\beta = V_m \sin\theta \qquad (8)$$

Then $V_{\alpha 1}, V_{\beta 1}$ (the decomposed vectors of the new centre) are calculated depending on the selected sector.

- Centre O_1

$$V_{\alpha 1} = V_\alpha - V_{dc}; \quad V_{\beta 1} = V_\beta \qquad (9)$$

- Centre O_2

$$V_{\alpha 1} = V_\alpha - \tfrac{1}{2}V_{dc}; \quad V_{\beta 1} = V_\beta - \frac{\sqrt{3}}{2}V_{dc} \qquad (10)$$

- Centre O_3

$$V_{\alpha 1} = V_\alpha + \tfrac{1}{2}V_{dc}; \quad V_{\beta 1} = V_\beta - \frac{\sqrt{3}}{2}V_{dc} \qquad (11)$$

- Centre O_4

$$V_{\alpha 1} = V_{\alpha}; \quad V_{\beta 1} = V_{\beta} \qquad (12)$$

- Centre O_5

$$V_{\alpha 1} = V_{\alpha} + \tfrac{1}{2}V_{dc}; \quad V_{\beta 1} = V_{\beta} + \frac{\sqrt{3}}{2}V_{dc} \qquad (13)$$

- Centre O_6

$$V_{\alpha 1} = V_{\alpha} - \tfrac{1}{2}V_{dc}; \quad V_{\beta 1} = V_{\beta} - \frac{\sqrt{3}}{2}V_{dc} \qquad (14)$$

After determining $V_{\alpha 1}, V_{\beta 1}$, the new voltage vector V_{m1} is

$$V_{m1} = \sqrt{V_{\alpha 1}^2 + V_{\beta 1}^2}$$

and the angle of the new voltage sector is determined depending on the position of the space vector.

if $(V_{\alpha 1} \geq 0 \ \& \ V_{\beta 1} \geq 0)$ then $\theta_1 = \tan^{-1}\left(\dfrac{V_{\beta 1}}{V_{\alpha 1}}\right)$

if $(V_{\alpha 1} \leq 0 \ \& \ V_{\beta 1} \geq 0) \| ((V_{\alpha 1} \leq 0 \ \& \ V_{\beta 1} \leq 0))$;

then $\theta_1 = \pi + \tan^{-1}\left(\dfrac{V_{\beta 1}}{V_{\alpha 1}}\right)$

if $(V_{\alpha 1} \geq 0 \ \& \ V_{\beta 1} \leq 0)$ then $\theta_1 = 2\pi + \tan^{-1}\left(\dfrac{V_{\beta 1}}{V_{\alpha 1}}\right)$

The new voltage vector and angle are determined from the six two-level spaces, with the centre moved as shown in Fig 6.

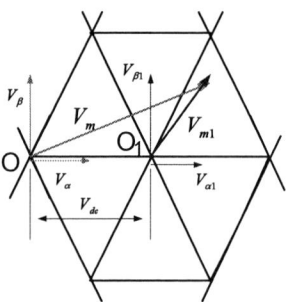

Fig. 6. Moving the original centre to a new centre

E. Time Distribution Determination

The dwell times for the proposed space vector modulation are calculated in the same manner as for the conventional two level inverter case.

The sampling time is distributed between the three states based on equating the volt-second integral. The following equation complies with this principle.

$$\int_0^{T_s} \overline{V}_{ref} = \int_0^{t_1} \overline{V}_1 dt + \int_{t_1}^{t_1+t_2} \overline{V}_2 dt + \int_{t_1+t_2}^{T_s} \overline{V}_0 \qquad (15)$$

Since $\overline{V}_0$ is a zero vector, equation (15) can be re-write as

$$T_s \overline{V}_{ref} = (t_1 \overline{V}_1 + t_2 \overline{V}_2) \qquad (16)$$

Projecting V_{ref} along the sector axis, the equation becomes

$$T_s |\overline{V}_{ref}|\begin{pmatrix}\cos\alpha \\ \sin\alpha\end{pmatrix} = t_1 \frac{2}{3}V_{dc}\begin{pmatrix}1 \\ 0\end{pmatrix} + t_2 \frac{2}{3}V_{dc}\begin{pmatrix}\cos\pi/3 \\ \sin\pi/3\end{pmatrix} \qquad (17)$$

where $0 \leq \alpha \leq 60^0$
From equation (17)

$$t_1 = T_s a \frac{\sin(\pi/3 - \alpha)}{\sin\pi/3}; \quad t_2 = T_s a \frac{\sin\alpha}{\sin\pi/3} \qquad (18)$$

where

$$a = \frac{|\overline{V}_{ref}|}{\dfrac{2}{3}V_{dc}} \qquad (19)$$

To reduce DSP execution time, t_1 and t_2 are calculated from equation 4.21 while t_0 is calculated from

$$t_0 = T_s - t_1 - t_2 \qquad (20)$$

T_s is the sampling time, and V_0, V_1 and V_2 are the voltage vectors corresponding to the first, second, third state.

F. Selecting the Switching States

The switching state sequence for the proposed space vector modulation is slightly different from the switching states sequence for the conventional two-level inverter or five-level inverter.

In conventional five-level SVM, when the vector lies in a particular region (inner triangle) the movement between the nearest three states is chosen, but in the proposed space vector modulation, the sequence of the switching states depends on the sector in which vector lies. For example, if the modulation index is above 0.75 (outer hexagon of the conventional five level space) and the vector lies in hexagon one, sector one, the sequence of the rotation between the states is shown is Fig. 7 (b). The 4 sequences are (100 400 430 211), (211 400 430 322), (322 400 430 433) and (433 400 430 433). These are for the first half of the switching cycle. For the second half of the switching cycle, the same sequence is reversed.

To improve the inverter output voltage quality, certain conditions should be considered when choosing the switching sequence. Namely minimizing switching during the sampling period and reducing dv/dt. These conditions yield low harmonics in the output voltage and current, and reduce capacitor ripple current.

(100 400 430 211) is a switching sequence for the first hexagon first sector where $100(\tfrac{1}{2}t_0)$, $400(t_1)$, $430(t_2)$ and $211(\tfrac{1}{2}t_0)$. Moving this sequence from (100) to (400) results in 3 switch changes at the same time, jumping from V_{dc} to $4V_{dc}$. The worst case from (430) to (211), needs a change of 5 switches in the same sampling period. To overcome this excessive switching problem, inner states are used for a short time. The sequence will then be (**100** *200 300* **400** *410 420* **430** *431 432* **433**). The

1258

sequence dwells for $2\mu s$ at every inner state. The $12\mu s$ dwell is distributed between (t_1, t_2) as shown in fig 7(c). By this method, output dv/dt is decreased.

The same dwell method applies if the modulation index is between 0.5 and 0.75 (third hexagon in conventional five-level SVM). If the vector lies in hexagon one, sector one, the sequence of the rotation between the states is shown in Fig.7 (a). (**100** *200* **300** **310** **320** *321* **322**) is one switching sequence. The sequence dwells for $2\mu s$ at the inner states (200 310 and 321) and the $6\mu s$ dwell is distributed between (t_1, t_2).

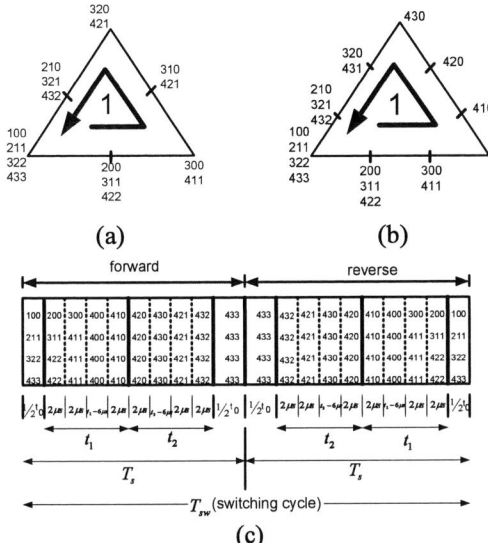

(a)   (b)

(c)

Fig. 7. (a) rotation sequence in sector one hexagon 3, (b) rotation sequence in sector one hexagon 4, and (c) the distribution of the states within the sampling period

G. Redundant States

Redundant switching states are those states for which a particular output phase voltage or line-to-line voltage can be generated by more than one switch combination [13]. Based on the voltage space vector for a five level inverter represented by one sector Fig. 7 (a) and (b), its found that four states that give the same vector have different capacitor charging paths. So there are several combinations that produce the same output voltage but with different capacitor loadings, which is important to the proposed balancing technique.

Among the possible redundant state selection goals are:

- Capacitor voltage balancing
- DC source current control
- Reduction of switching stresses

The proposed technique requires the four capacitor voltages, three phase currents, the sector where the space vector lies, and the centre of the new space vector.

There are 13 possible capacitor voltage imbalance conditions as shown in Table II.

Table II

CAPACITOR VOLTAGE CONDITIONS IN THE FIVE-LEVEL INVERTER

$> \frac{1}{4}V_{dc}$	$< \frac{1}{4}V_{dc}$
C_1	C_2, C_3, C_4
C_2	C_1, C_3, C_4
C_3	C_1, C_2, C_4
C_4	C_1, C_2, C_3
C_1, C_2	C_3, C_4
C_1, C_3	C_2, C_4
C_1, C_4	C_2, C_3
C_2, C_3	C_1, C_4
C_2, C_4	C_1, C_3
C_3, C_4	C_1, C_2
C_1, C_2, C_3	C_3
C_1, C_2, C_4	C_4
C_1, C_3, C_4	C_2
C_2, C_3, C_4	C_1

After measuring the capacitor voltages, the three phase currents are measured since charging and discharging of the capacitor voltages depends on the direction of the output current. If the capacitor is loaded and its current is positive the capacitor is discharging.

As mentioned, the sector in which the space vector lies and the new space vector origin are required to indicate which current should be taken into consideration.

Each centre of the six new sectors has 4 redundancy states as shown in Fig. 8.

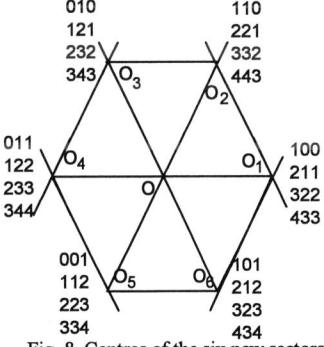

Fig. 8. Centres of the six new sectors

Consider centre O_1 as an example. The four redundant states are (100 211 322 433), which give the same output voltage but different capacitor loadings.

100 means that C_1 is loaded

211 means that C_2 is loaded

322 means that C_3 is loaded

1259

433 means that C_4 is loaded

The phase current and its direction are required so as to determine which capacitor should be loaded. If the space vector uses centre O_1, (taking 433 as an example) this means that phase (a) is on line 4 and phases (b) and (c) are on line 3 as shown in Fig 9(a). So in this case phase current (a) is taken in consideration. If ($i_a > 0$ and $C_4 > 0$) this means that loading C_4 will discharge the capacitor and if ($i_a < 0$ and $C_4 > 0$) loading C_4 will charge the capacitor. The possible cases are shown in TABLE III.

TABLE III

CHARGING AND DISCHARGING OF C$_4$ IN THE SIX SECTORS, TAKING THE CURRENT INTO ACCOUT

Sector	Centre	Current	Current direction	Capacitor condition
1	O_1	i_a	+	Discharge
	O_2	i_c	+	Charge
2	O_2	i_c	+	Charge
	O_3	i_b	+	Discharge
3	O_3	i_b	+	Discharge
	O_4	i_a	+	Charge
4	O_4	i_a	+	Charge
	O_5	i_c	+	Discharge
5	O_5	i_c	+	Discharge
	O_6	i_b	+	Charge
6	O_6	i_b	+	Charge
	O_1	i_a	+	Discharge

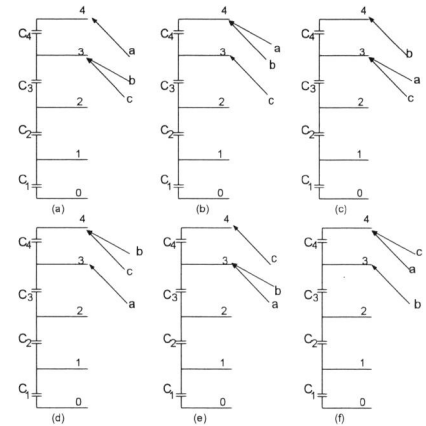

Fig. 9. Current conditions at each centre (a) 433 centre O_1, (b) 443 centre O_2, (c) 343 centre O_3, (d) 344 centre O_4, (e) 334 centre O_5, and (f) 434 centre O_6

Balancing the four capacitor voltages by redundancy states using current direction is complex. At high power factor, the current direction can be ignored since thecurrent and voltage are virtually in phase. The balancing technique is simplified and the power system costs are reduced. Table IV shows the states to charge specific capacitors, in every condition.
(Sector 1, inner triangles (10, 11, 12, 13, 14, 15, 16)).

TABLE IV
SELECTION OF THE STATES FOR CHARGING THE CAPACITORS (SECTOR 1)

To charge:	States
C_1	(**100** 200 300 **400** 410 420 **430** 320 210 **100**)
C_2	(**211** 311 411 **400** 410 420 **430** 320 210 **211**)
C_3	(**322** 422 411 **400** 410 420 **430** 320 321 **322**)
C_4	(**433** 422 411 **400** 410 420 **430** 431 432 **433**)
C_1, C_2	(**100** 200 300 **400** 410 420 **430** 320 210 **211**)
C_1, C_3	(**100** 200 300 **400** 410 420 **430** 320 321 **322**)
C_1, C_4	(**100** 200 300 **400** 410 420 **430** 320 <u>321 432 433</u>)
C_2, C_3	(**211** 311 411 **400** 410 420 **430** 320 321 **322**)
C_2, C_4	(**211** 311 411 **400** 410 420 **430** 431 432 **433**)
C_3, C_4	(**322** 422 411 **400** 410 420 **430** 431 432 433)

H. Practical Results
The practical results when utilizing the proposed balancing technique are shown in Fig. 10. The results are for a 200V dc link, 0.9 modulation index, 0.9 lagging power factor, and a 2.5 kHz switching frequency.

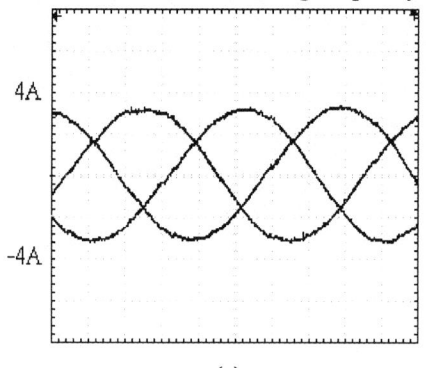

(a)

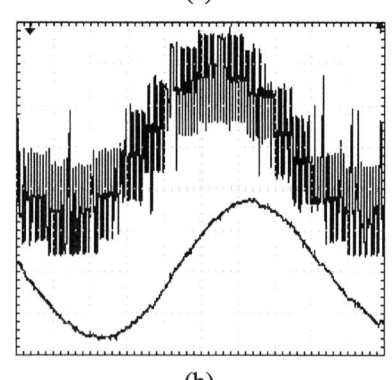

(b)

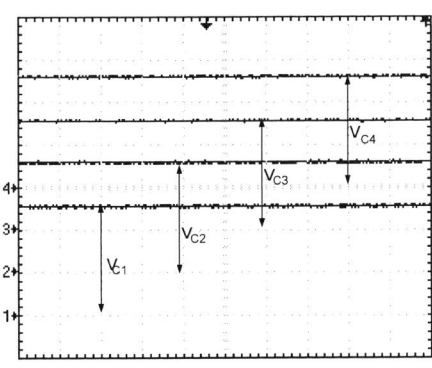

(c)

Fig.10. (a) three phase currents, (b) phase current and its phase voltage, and (c) capacitor voltages

As shown in Fig. 10, the proposed technique produces a sine wave space vector modulation current at a high power factor, high modulation index, and the four capacitors voltages are each balanced at 50V.

III. CONCLUSIONS

Multilevel inverters suffer from capacitor voltage imbalance problems at high modulation indices. Many techniques had been proposed to solve this problem; one of these techniques is the redundant state method. This method has power factor and modulation index limitations when used on a five-level inverter. The proposed technique offers capacitor voltage balance at high power factor and high modulation index by using a redundancy method. Space vector modulation of the five level inverter is simplified to six two-level vector spaces. Each of the six hexagon centres has associated redundant states. The advantage of this technique is balancing of the five level inverter at high modulation index and high power factor without any auxiliary circuits, therefore energy losses are reduced, and hardware costs are decreased. At high power factor, current transducers are not necessary, since the current and voltage are in phase. The disadvantages of this technique are its control complexity and increased device switching frequency.

ACKNOWLEDGEMENT

The authors greatly appreciate the assistance of Dr. T.C. lim.

References

[1] W. Hill and C. Harbourt, 'Performance of Medium Voltage Multi-Level Inverters', IEEE IAS Conference, Vol. 2, 1999, pp. 1186-1192.

[2] Y. Shakweh, 'New Breed of Medium Voltage Converters', Power Engineering Journal, Vol. 14, No. 1, 2000, pp. 12-20.

[3] Y. Baba *et al.,* 'A 1000A, 2500 V pressure mount RC-IGBT', Proceedings of sixth European Conf. on Power Elect.and Appl. (EPE'95), Sevilla, Spain, September 1995, pp.1051-1054.

[4] H. Brunner *et al.,* 'Improved 3.5 kV IGBT-diode chipset and 800 A module applications'
Power Electronics Specialists Conference, 1996. PESC '96, Baveno, Italy, 27th Annual IEEE, Vol. 2, pp. 1748-1753.

[5] N. S. Choi, G. C. Cho, and G. H. Cho, 'Modeling and analysis of a static VAR compensator using multilevel voltage source inverter', IAS Conf., 1993 IEEE, Vol. 2, pp. 901-908.

[6] R. W. Menzies, P. Steimer, and J. K. Steinke, 'Five-level GTO inverters for large induction motor drives', Ind. Appl., IEEE Trans., Vol. 30, July-Aug. 1994, pp. 938-944.

[7] F. Peng and J. Lai, 'Multilevel Converters- A New Breed of Power Converters', IEEE Trans. on Industry Applications, Vol. 32, No. 3, 1996, pp. 509-517.

[8] F. Peng *et al.,* ' Harmonics Optimizations of the Voltage Balancing control for Multilevel Converter/Inverter systems', IEEE Trans. on Industry Application, Vol. 4, 3-7 Oct. 2004, pp. 2194-2201.

[9] R.Rojas, *et al,.* 'PWM Control Method for a Four-Level inverter', IEEE Trans. on Industry Applications, Vol. 142, No. 6, Nov. 1995, pp. 390-396.

[10]L.M.Tolbert, *et al,.* Multilevel Converter For Large Electric drives," IEEE Trans. on Industry Applications,Vol. 35, No. 1, Jan/Feb. 1999, pp. 36-44.

[11] M. Botao *et al.,* 'New SVPWM Control Scheme for Three-Phase Diode Clamping Multilevel Inverter with Balanced *DC* Voltages', IECON [IEEE 28th Annual Conference of the Industrial Electronics Society], Vol. 2, 2002, pp. 903-907.

[12] A. M. Massoud *et al.,* ' Systematic Analytical-Based Generalized Algorithm for Multilevel Space Vector Modulation with a Fixed Execution Time ' 2007, IET, Electric Power Applications.

[13] F. Peng, L. Tolbert, and T. Habetler, 'Multilevel *PWM* Methods at Low Modulation Indices', IEEE Trans. on Power Electronics, Vol. 15, No. 4, 2000, pp. 719.

Sliding Mode Repetitive Control of PWM Voltage Source Inverter

Sufen Chen, Y. M. Lai, Siew-Chong Tan, and Chi K. Tse

Department of Electronic and Information Engineering, The Hong Kong Polytechnic University, Hong Kong, China

Abstract—**This paper proposes a hybrid sliding mode repetitive control scheme that combines both the features of the sliding mode control and the repetitive control to achieve excellent transient and steady-state system performances in voltage source inverters. The principle of equivalent control is adopted to integrate the two methodologies and to facilitate the design and analysis of the proposed scheme. A specific low pass filter is also introduced to improve the transient response of the regulation. Experimental results show that fast dynamical responses are achieved through the sliding mode control, while low harmonic distortions are achieved through the repetitive control.**

Index Terms— **Repetitive control, sliding mode control, THD, voltage source inverter.**

I. INTRODUCTION

The pulse-width modulation (PWM) voltage source inverter (VSI) is extensively used in AC power conditioning systems. Many of such systems require the VSI to have a fast dynamical response, an accurate output regulation, and a low total harmonic distortion (THD), even when the operating load is highly nonlinear and/or distorted. Currently, most of these systems in the industry use the conventional type of linear PWM controllers, which are simple and cost effective for common applications that do not require a tight regulation standard. However, for applications requiring more stringent regulation, nonlinear control methodologies have to be employed. One such methodology is the sliding mode control (SMC) [1]–[3]. It has been shown that the adoption of SMC in the VSI can significantly improve the transient characteristic as well as the robustness of the VSI. However, in terms of the steady-state performance, the use of SMC in VSI cannot satisfactorily alleviate the THD. The presence of harmonic distortion is solely attributed to the imperfect tracking of the sinusoidal varying reference signal. This is a result of having only limited DC gain in the SM controller. Thus, by similar reasoning, most other control methodologies are incapable of fully containing this THD issue.

Theoretically, the alleviation of the THD would mean that the steady-state tracking of the periodic signal must be very accurate. To do this, control approaches specialized for achieving precise tracking of periodic signal have to be incorporated. One approach commonly chosen for this purpose is the repetitive control (RC) [4], [5], which is based on the internal model principle [6]. The basic operating principle of

the RC is to observe the system's external periodic signals for one cycle period, and then to generate in the next cycle period a corresponding compensating signal that ensures precise tracking of the external signals [5]. The recent applications of the RC in the VSI to alleviate the THD are reported in [7], [8]. It has been demonstrated that a very low THD can be obtained using this control methodology. However, the works also reveal that due to its one-cycle-delay requirement, the RC is a relatively slow-responding control technique when dealing with instantaneous transient changes. This makes it inappropriate for VSI that requires a fast dynamical response, unless the RC is incorporated along with a fast responding nonlinear control scheme. This implies a hybrid controller that simultaneously achieves the various objectives of having fast response, accurate regulation, and very low THD.

Following this direction, we propose in this paper a hybrid sliding mode repetitive control (SMRC) scheme that combines the features of SMC and RC to offer excellent transient and steady-state performances to the VSI. Note that previous works on the SMC of power converters are mainly performed in the time domain [9], [10], whereas the work on RC is performed in the frequency domain [8], [11]. Thus, this paper serves to clarify the various issues concerning the combined application of both control methodologies, to provide guidelines towards their theoretical design, and to give experimental verification and evaluation of the proposed controller.

II. STATE-SPACE MODEL OF VOLTAGE SOURCE INVERTER

Fig. 1 shows a typical single phase inverter system. The load to the inverter can either be resistive or wave-rectified depending on the nature of the application. Using the notation described in the figure, the state-space representation of the

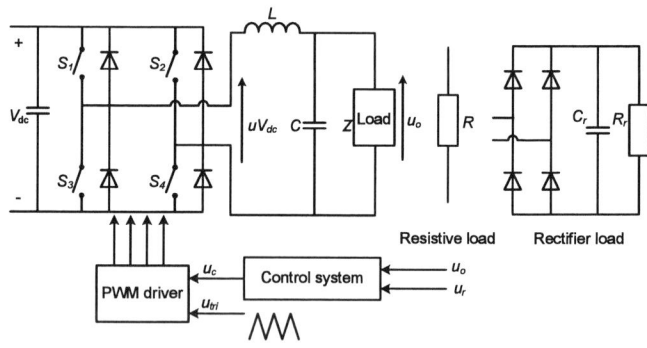

Fig. 1. Single-phase inverter system.

978-1-4244-0644-9/07/$25.00 ©2007 IEEE   1262

inverter dynamics in terms of its PID voltage errors can be written as

$$\dot{\boldsymbol{x}} = A\boldsymbol{x} + Bu + D, \qquad (1)$$

where

$$\boldsymbol{x} = \begin{bmatrix} x_1 \\ x_2 \\ x_3 \end{bmatrix} = \begin{bmatrix} u_r - u_o \\ \dot{u}_r - \dot{u}_o \\ \int (u_r - u_o)dt \end{bmatrix},$$

$$A = \begin{bmatrix} 0 & 1 & 0 \\ -\frac{1}{LC} & -\frac{1}{ZC} & 0 \\ 1 & 0 & 0 \end{bmatrix},$$

$$B = \begin{bmatrix} 0 \\ -\frac{V_{dc}}{LC} \\ 0 \end{bmatrix},$$

$$D = \begin{bmatrix} 0 \\ \ddot{u}_r + \frac{\dot{u}_r}{ZC} + \frac{u_r}{LC} \\ 0 \end{bmatrix};$$

and u_r, V_{dc}, and u_o are the sinusoidal reference voltage, the input voltage, and the output voltage, respectively; C, L, and Z are the capacitance, inductance, and load impedance, respectively; and $u \in \{-1, +1\}$ is the control signal, which respectively generates an input voltage of $-V_{dc}$ and $+V_{dc}$ at the input of the LC filter. If we adopt an amplitude modulation ratio of $m_a = \hat{u}_c(t)/\hat{u}_{tri} < 1$, where u_c is the compensated control signal with amplitude $\hat{u}_c(t)$, and $\hat{u}_{tri}$ is the amplitude of the PWM triangular carrier, then the averaged time-continuous value of the discrete control signal u will be

$$\bar{u} = \frac{u_c}{\hat{u}_{tri}}, \qquad (2)$$

where $\bar{u}$ is the averaged duty cycle bounded by positive and negative unity. Choosing $\hat{u}_{tri} = 1$, the averaged duty cycle will be simplified to $-1 < \bar{u} = u_c < +1$.

III. PROPOSED CONTROL METHODOLOGY

A. Sliding Mode Control

The principle of SMC and its application to power electronics can be found in [1]–[3], [9] and [10]. In this work, the commonly used PID SM voltage control structure is adopted for investigation. The switching control law for this structure is given as

$$\begin{cases} u = +1, & \text{for } S > 0 \\ u = -1, & \text{for } S < 0 \end{cases}, \qquad (3)$$

where sliding function

$$S = \alpha_1 x_1 + \alpha_2 x_2 + \alpha_3 x_3 = \boldsymbol{J}^{\mathrm{T}} \boldsymbol{x} \qquad (4)$$

with sliding coefficients $\boldsymbol{J}^{\mathrm{T}} = [\alpha_1 \ \alpha_2 \ \alpha_3]$. According to SMC theory, the existence condition of a system working in sliding mode operation can be found by using Lyapunov's direct method to solve $S\dot{S} < 0$. By substituting (1), (3) and (4), the existence condition is given by

$$\begin{cases} (-\frac{1}{LC} + \frac{\alpha_3}{\alpha_2})x_1 + (\frac{\alpha_1}{\alpha_2} - \frac{1}{ZC})x_2 - \frac{V_{dc}}{LC} + \ddot{u}_r + \frac{\dot{u}_r}{ZC} + \frac{u_r}{LC} < 0 \\ (-\frac{1}{LC} + \frac{\alpha_3}{\alpha_2})x_1 + (\frac{\alpha_1}{\alpha_2} - \frac{1}{ZC})x_2 + \frac{V_{dc}}{LC} + \ddot{u}_r + \frac{\dot{u}_r}{ZC} + \frac{u_r}{LC} > 0 \end{cases}. \qquad (5)$$

In sliding mode operation, the dynamics of the system can be characterized using the invariance conditions $S = 0$ and

$\dot{S} = 0$. With the condition $S = 0$, the system is represented as a second order homogeneous differential equation. A set of design equations for the under-studied inverter control system can be obtained using the approach described in [10]. Note that the selection of the sliding coefficients should comply with the stability condition of the second order system and the existence condition (5) in the sliding mode operation. With the condition $\dot{S} = 0$, i.e., $\boldsymbol{J}^{\mathrm{T}}A\boldsymbol{x} + \boldsymbol{J}^{\mathrm{T}}Bu_{eq} + \boldsymbol{J}^{\mathrm{T}}D = 0$, the equivalent control signal, which is the averaged value of the control signal u, can be obtained as

$$\begin{aligned} u_{eq} &= -[\boldsymbol{J}^{\mathrm{T}}B]^{-1}\boldsymbol{J}^{\mathrm{T}}(A\boldsymbol{x} + D) \\ &= \frac{LC}{V_{dc}}(\frac{\alpha_3}{\alpha_2} - \frac{1}{LC})(u_r - u_o) + \frac{LC}{V_{dc}}(\frac{\alpha_1}{\alpha_2} - \frac{1}{ZC})(\dot{u}_r \\ &\quad -\dot{u}_o) + \frac{LC}{V_{dc}}(\ddot{u}_r + \frac{\dot{u}_r}{ZC} + \frac{u_r}{LC}) \end{aligned} \qquad (6)$$

where $-1 < u_{eq} < +1$. Since u_{eq} is the equivalent control of sliding mode, which directly controls the inverter, it is reasonable to relate u_{eq} with the averaged duty cycle $\bar{u}$ in (2) to establish the relationship $u_{eq} = \bar{u} = u_c$. For inverter system working in steady state, both the input voltage V_{dc} and the load impedance Z are static, so equation (6) is linear, and it can be written in Laplace form as

$$\begin{aligned} U_{eq}(s) &= U_c(s) \\ &= \left[\frac{LC}{V_{dc}}(\frac{\alpha_1}{\alpha_2} - \frac{1}{ZC})s + \frac{LC}{V_{dc}}(\frac{\alpha_3}{\alpha_2} - \frac{1}{LC})\right][U_r(s) \\ &\quad -U_o(s)] + \left(\frac{LC}{V_{dc}}s^2 + \frac{L}{V_{dc}Z}s + \frac{1}{V_{dc}}\right)U_r(s) \end{aligned} \qquad (7)$$

under the assumption that the initial condition is nulled. This transformation to the s domain is critical for combining SMC with RC. Fig. 2(a) illustrates the block diagram of an inverter with the derived sliding mode controller given in (7). Here, the tracking error is given by

$$E_0(s) = \frac{U_r(s) - D(s) - U_r(s)G_{s2}(s)G_p(s)}{1 + G_{s1}(s)G_p(s)}, \qquad (8)$$

where $G_{s1}(s) = \frac{LC}{V_{dc}}(\frac{\alpha_1}{\alpha_2} - \frac{1}{ZC})s + \frac{LC}{V_{dc}}(\frac{\alpha_3}{\alpha_2} - \frac{1}{LC})$ and $G_{s2}(s) = \frac{LC}{V_{dc}}s^2 + \frac{L}{V_{dc}Z}s + \frac{1}{V_{dc}}$; $G_p(s)$ denotes the inverter plant; and $D(s)$ denotes the external disturbances.

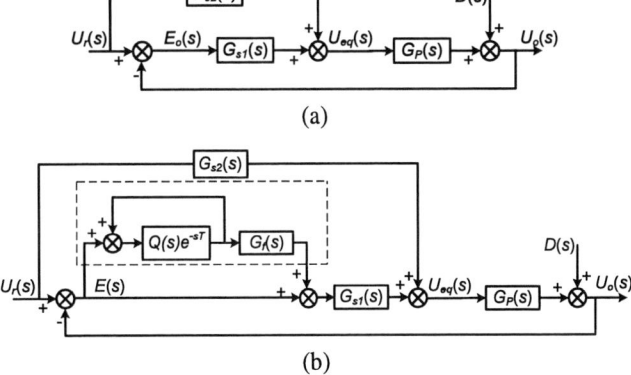

(a)

(b)

Fig. 2. Block diagram of: (a) the SMC, (b) the proposed SMRC.

B. Repetitive Control

The detailed description of RC and its applications to power electronics can be found in [4]–[8], [11] and [12]. Fig. 2(b) shows the block diagram of the proposed SMRC scheme with the inverter plant. It is obtained from the SMC system described in Fig. 2(a) by incorporating a plug-in RC component into the system.

1) Periodic Error Elimination: The tracking error of the SMRC system shown in Fig. 2(b), in terms of $E_0(s)$ of the SMC system, is given by:

$$E(s) = E_0(s) \times \frac{1 - Q(s)e^{-sT}}{1 - [1 - G_f(s)H(s)]Q(s)e^{-sT}}, \quad (9)$$

where $H(s) = \frac{G_{s1}(s)G_p(s)}{1+G_{s1}(s)G_p(s)}$ and T is the period of the reference signal.

Assuming that the original SMC system is asymptotically stable and that

$$\left| [1 - G_f(s)H(s)]Q(s) \right| < 1, \quad (10)$$

where $s = j\omega$, for all ω, the stability condition of SMRC system is satisfied (small gain theorem [13]). Equation (9) can be expressed as $E(s) = E_0(s)\{1 - G_f(s)H(s)Q(s)e^{-sT} - G_f(s)H(s)[1 - G_f(s)H(s)]Q^2(s)e^{-2sT} - G_f(s)H(s)[1 - G_f(s)H(s)]^2Q^3(s)e^{-3sT} + \cdots\}$. Theoretically, by setting $1 - G_f(s)H(s) = 0$ and $1 - Q(jk\omega_0)e^{-jk\omega_0 T} = 0, k = 0, 1, 2, \cdots$, where $\omega_0 = 2\pi/T$, we can ensure that the steady-state tracking error converges to zero at the occurrence of every harmonic. However, due to the non-idealities of the system, such an ideal design condition is practically impossible. Instead, it is more appropriate to consider the design condition:

$$1 - G_f(s)H(s) \approx 0 \text{ and } 1 - Q(jk\omega_0)e^{-jk\omega_0 T} \approx 0, \quad (11)$$

and for harmonics limited to a certain frequency bandwidth.

2) Selection of Repetitive Control Parameters: It is easy to understand from (11) that the parameter $G_f(s)$ should be a zero-magnitude-and-phase compensator for $H(s)$. An intuitive way to design $G_f(s)$ is to make $G_f(s) = 1/H(s)$. However, due to system parameters' drifts (e.g. capacitor's ESR variation), the use of this design will result in a system which is stable at low frequencies, but unstable at high frequencies. Hence, to tackle this issue, we propose to include an additional high-frequency pole to the compensator $G_f(s)$ to remove the high frequency signals that will cause the instability. This makes

$$G_f(s) = \frac{1}{H(s)} \times \frac{p_1}{(s + p_1)}, \quad (12)$$

where p_1 is a high frequency pole. This will not affect the low frequency characteristic of $G_f(s)$. Additionally, careful observation of (11) reveals that $Q(s)$ should be unity while (10) indicates that $Q(s)$ should be less than unity at high frequencies because $1 - G_f(s)H(s)$ tends to unity as $G_f(s)H(s) \to 0$ at high frequencies. Thus, $Q(s)$ should be a low pass filter with unity magnitude and zero phase shift at low frequencies, and slightly lower than "1" at high frequencies to ensure system stability. To satisfy the imposing requirements, a low pass filter:

$$Q(s) = \frac{e^{\tau s}}{\frac{s^2}{\omega_q^2} + \frac{2\zeta s}{\omega_q} + 1}, \quad (13)$$

as given in [12] is adopted. Here, the damping ratio is chosen as $\zeta = 0.707$ to give a flat magnitude with quasi-linear phase shift characteristic within its bandwidth. The time advance τ is selected to equal the effective time delay of the denominator term, i.e., $2\zeta/\omega_q$. This ensures that the low pass filter achieves a zero phase shift at low frequencies. The selection of bandwidth ω_q is influenced by the pre-designed system and $G_f(s)$. A larger value of ω_q improves tracking accuracy, but degrades the system stability. The appropriate ω_q can be determined through simulation and experimental tuning.

C. Reduction of the Dynamic Sensitivity of RC

As mentioned before, the operating characteristic of RC is that it uses information in the current period to generate a corresponding control signal for the next cycle period. This results in a one cycle-period time delay in the control action, which in the event of a disturbance, will introduce a distortion to the transient behavior of the system. This is because the disturbance observed by the RC will be applied to influence the system performance only one cycle-period after its occurrence. However, at that point in time, the disturbance would have already been corrected by the SMC component in the controller. Hence, the correction enforced by the RC can be regarded as a form of parasitic distortion. Note that the RC does not hinder the fast response of the SMC component since there is a direct channel from the feedback to the input of $G_{s1}(s)$ without passing through the RC (see Fig. 2(b)). To resolve the aforementioned problem, we propose to incorporate a specific low pass filter (SLPF):

$$L(s) = \frac{p_2}{s + p_2}, \quad (14)$$

with bandwidth p_2, to the RC component to filter out the high frequency disturbances of the system, i.e., to reduce its sensitivity to dynamics introduced by abrupt changes. Recall that there is already an existing frequency pole p_1 in $G_f(s)$ (see (12)). The inclusion of the SLPF means that there will be an overlap between p_1 and p_2. Hence, it is sufficient to consider only the filter with the lower cutoff frequency for the design. Since the proposed SLPF is typically of a much lower bandwidth, i.e., $p_2 < p_1$, it is sufficient to consider only the SLPF for the design of the RC component. With the filter related to p_1 removed, the SLPF filter assumes both the roles of ensuring system robustness and reducing the sensitivity of RC.

IV. RESULTS AND DISCUSSIONS

An experimental system has been setup to verify the proposed SMRC using dSPACE DS1103 [14]-[16] and Matlab/Simulink to control the IGBT switches in the VSI illustrated in Fig. 1. The calculation step size of DS1103 is set at 15 μs. The sampling rate of the A/D converter is $f_{sp} = 66.7$

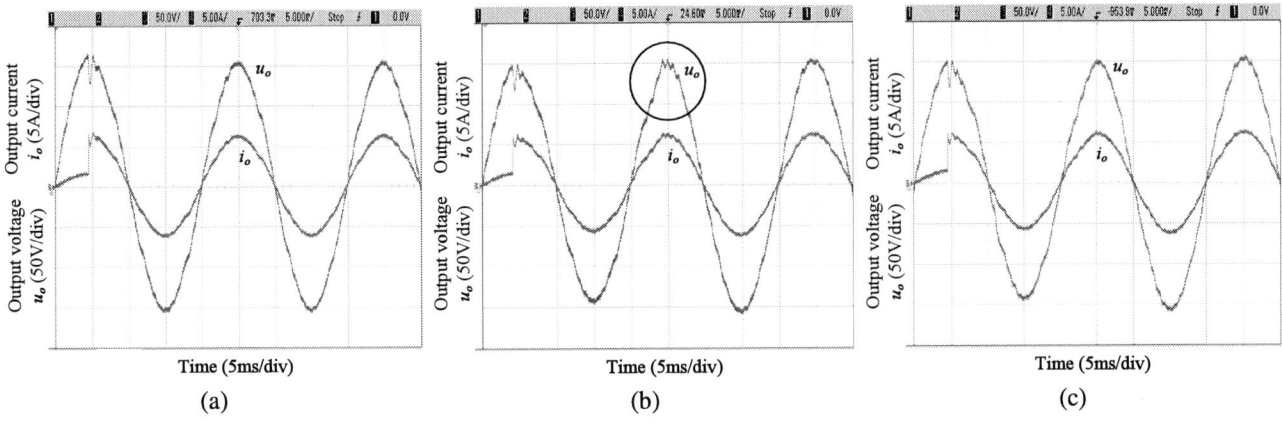

Fig. 3. The output voltage u_o (50V/div) and output current i_o (5A/div) at step load change under (a) SMC; (b) SMRC; and (c) SMRC with SLPF.

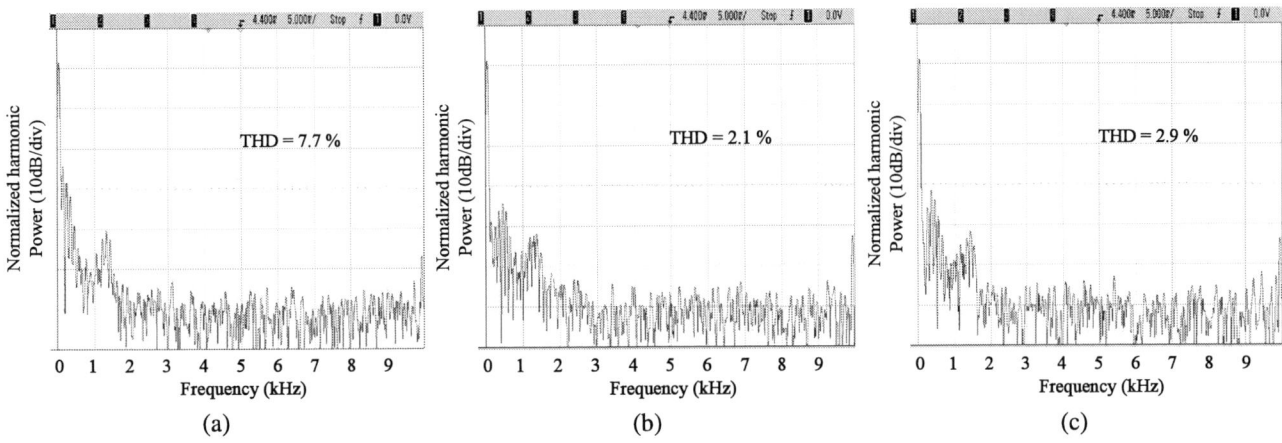

Fig. 4. The frequency spectrum of inverter output voltage under rectifier load with (a) SMC; (b) SMRC; and (c) SMRC with SLPF.

kHz and the PWM switching frequency is $f_{sw} = 10$ kHz, so the oversampling rate of the system is $f_{sp}/f_{sw} = 6.67$. The system parameters are given as $V_{dc} = 250$ Vdc; $U_r = 110$ V$_{(rms)}$, 50 Hz; $L = 1.5$ mH and $C = 20$ μF. The controller is designed with $\frac{\alpha_1}{\alpha_2} = 5.33 \times 10^3$, $\frac{\alpha_3}{\alpha_2} = 4.267 \times 10^7$; $p_1 = 100$ krad/s with ω_q=2500 rad/s; and $p_2 = 10$ krad/s with $\omega_q = 1500$ rad/s.

A. Dynamic Performance

The dynamic performance of the system is tested using a resistive load with step change from 1/4 load (121 W) to full load (484 W). Fig. 3 shows the corresponding waveforms of the output voltage u_o and the output current i_o under SMC, SMRC, and SMRC with SLPF. From the figure, it can be seen that the dynamical responses of all three controllers are similarly optimal at the instance of the step load change. This is credited to the excellent dynamic property of the SMC. On the other hand, the excellent consistency of the dynamical responses also verifies that the RC does not alter the dynamical response capability of the SMC towards handling the transient disturbance. This confirms the discussion of the previous section. However, in the cycle-period following the disturbance, it is observed that while the regulations of the inverter with SMC (Fig. 3(a)) and SMRC with SLPF (Fig. 3(c)) are returned to normal, the regulation with SMRC (Fig.

3(b)) is distorted. As previously discussed, this is caused by the one cycle delay of the RC action. The results concurrently verify that the inclusion of SLPF can alleviate the parasitic distortion from the regulation.

B. Steady-State Performance

The steady-state performance of the proposed control scheme is tested using a rectifier load with $C_r = 330$ μF and $R_r = 50$ Ω. Fig. 4 shows the harmonic contents in the output voltage of the inverter under SMC, SMRC, and SMRC with SLPF. A comparison of the three graphs shows that the THD of SMRC is the lowest (2.1%), followed by SMRC with SLPF (2.9%), and then SMC (7.7%). This shows that the use of SLPF can affect the tracking accuracy of the RC component. Yet, without RC, the use of SMC alone will give a high THD level which may not meet the required regulation standard.

V. CONCLUSION

A sliding mode repetitive control scheme is proposed for the PWM voltage source inverter to achieve excellent dynamic and steady-state performances. The sliding mode and repetitive control methodologies are combined into a single control scheme using the equivalent control principle. The design of the control scheme for both the sliding mode and the repetitive control components have been discussed in the paper. It is demonstrated that a specific low pass filter can be incorporated into the scheme to reduce the system's dynamical sensitivity towards abrupt changes. This can improve the overall transient response of the system. Finally, it can be concluded from the experimental results that both the sliding mode control and the repetitive control method can be combined into a hybrid control scheme that inherits the respective advantages of the individual schemes to achieve excellent control performances in the inverter system.

ACKNOWLEDGMENT

The authors would like to thank Mr. S. Y. Lam for help in developing the experimental prototype.

REFERENCES

[1] J. J. E. Slotine and W. Li, "Sliding control," in *Applied Nonlinear Control.* Englewood Cliffs, NJ: Prentice-Hall, 1991, ch. 7.

[2] M. Carpita and M. Marchesoni, "Experimental study of a power conditioning system using sliding mode control," *IEEE Trans. Power Electron.,* vol. 11, no. 5, pp. 731–742, Sep. 1996.

[3] R. Gupta and A. Ghosh, "Frequency-domain characterization of sliding mode control of an inverter used in DSTATCOM application," *IEEE Trans. Circuits Syst. I,* vol. 53, no. 3, pp. 662–676, Mar. 2006.

[4] M. Nakano and S. Hara, "Microprocessor-based repetitive control," in *Microprocessor-Based Control Systems,* N. K. Sinha, Ed. Dordrecht: Reidel Publ. Co., 1986.

[5] W. S. Yao and M. C. Tsai, "Analysis and estimation of tracking errors of plug-in type repetitive control systems," *IEEE Trans. Automat. Contr.,* vol. 50, no. 8, pp. 1190–1195, Aug. 2005.

[6] B. A. Francis and W. M. Wonham, "The internal model principle for linear multivariable regulators," *Appl. Math. Opt.,* vol. 2, no. 2, pp. 170–194, 1975.

[7] Y. Q. Ye, K. L. Zhou, B. Zhang, D. W. Wang, and J. C. Wang, "High-performance repetitive control of PWM DC–AC converters with real-time phase-lead FIR filter," *IEEE Trans. Circuits Syst. II,* vol. 53, no. 8, pp. 768–772, Aug. 2006.

[8] A. García-Cerrada, O. Pinzón-Ardila, V. Feliu-Batlle, P. Roncero-Sánchez, and P. García-González, "Application of a repetitive controller for a three-phase active power filter," *IEEE Trans. Power Electron.,* vol. 22, no. 1, pp. 237–246, Jan. 2007.

[9] R. R. Ramos, D. Biel, E. Fossas, and F. Guinjoan, "A fixed-frequency quasi-sliding control algorithm: application to power inverters design by means of FPGA implementation," *IEEE Trans. Power Electron.,* vol. 18, no. 1, pp. 344–355, Jan. 2003.

[10] S. C. Tan, Y. M. Lai, C. K. Tse, and M. K. H. Cheung, "A fixed-frequency pulsewidth modulation based quasi-sliding-mode controller for buck converters," *IEEE Trans. Power Electron.,* vol. 20, no. 6, pp. 1379–1392, Nov. 2005.

[11] K. Zhang, Y. Kang, J. Xiong, and J. Chen, "Direct repetitive control of SPWM inverter for UPS purpose," *IEEE Trans. Power Electron.,* vol. 18, no. 3, pp. 784–792, May. 2003.

[12] K. Srinivasan and F. R. Shaw, "Analysis and design of repetitive control systems using the regeneration spectrum," *ASME J. Dynam. Syst. Meas. Control,* vol. 113, no. 2, pp. 216–222, Jun. 1991.

[13] C. A. Desoer and M. Vidyasagar, *Feedback Systems: Input-Output Properties.* New York: Academic Press, 1975.

[14] *DS1103 Hardware Installation and Configuration,* dSPACE GmbH., Paderborn, 2005.

[15] *ControlDesk Experiment Guide,* dSPACE GmbH., Paderborn, 2006.

[16] *RTI and RTI-MP Implementation Guide,* dSPACE GmbH., Paderborn, 2006.

Output Current Ripple Analysis
of Five-Phase PWM Inverters

Deni*, E. G. Supriatna, and P. A. Dahono

School of Electrical Engineering and Informatics, Institute of Technology Bandung,
Jl Ganesha No 10, Bandung 40132, Indonesia
Tel, 62-22-2503315, fax. 62-22-2508132
Email: deni | erik | pekik @konversi.ee.itb.ac.id

Abstract-- **An output current ripple analysis of five phase PWM inverters under sinusoidal and nonsinusoidal operations is presented in this paper. A general expression of output current ripple of five-phase PWM inverters under nonsinusoidal operation is first derived. Under sinusoidal operation it is found that the optimum reference signal that results in minimum output current ripple is a sinusoidal reference signal. Under nonsinusoidal operation, however, it is found that fifth harmonic injection can be used to reduce the output current ripple. Moreover, fifth harmonic injection is useful to increase the maximum modulation index of five-phase PWM inverter. Simulation and experimental results are included to show the validity of the derived expressions.**

Index Terms— **Current Ripple, Five-Phase Inverters**

I. INTRODUCTION

A multiphase motor has reduced torque pulsation, rotor harmonic currents, stator current per phase, dc link current ripple, and higher reliability compare to a three-phase motor [1]-[3]. The smallest multiphase motor which higher than three-phase is five-phase motor. A lot of works on design and control of five-phase ac motor drives were published [4]-[5]. Various modulation techniques for five-phase PWM inverters were also proposed [6]-[8].

In the case of three-phase PWM inverters, it was found that a sinusoidal plus twenty-five percent third harmonic is the optimum reference signal that results in minimum output current ripple [9]. Moreover, the maximum modulation index is increased by using this third harmonic injection. In [10], it was found that the optimum reference signal for five-phase PWM inverter under sinusoidal operation is a pure sinusoidal signal. Harmonic injection cannot be used to reduce the output current ripple.

In [6], it was shown that fifth harmonic injection can be used to increase the maximum modulation index of five-phase inverter. Kelly et al [7] developed SVPWM for multiphase inverter result in better DC bus utilization. Discontinues PWM can also be used to reduce switching frequency loses in five-phase inverter [8]. However, the influence of these modulation signals on the output current ripple has not been investigated.

In most application, five-phase motor is designed to have the non-sinusoidal emf voltage. It was shown that third harmonic injection can be used to increase the produced torque [11]-[12]. Another advantage of using third harmonic injection is higher maximum modulation index comparing to sinusoidal PWM inverter.

In this paper, an output current ripple analysis of five-phase PWM inverters under nonsinusoidal operation is presented. It is assumed that the inverter has to produce third harmonic current to increase the produce torque of five-phase motor. Analysis results show that under this condition, fifth harmonic signal can be injected into the reference signal to reduce the switching-frequency output current ripple. The optimum magnitude of the fifth harmonic signal is determined by the magnitude of the desired third harmonic output voltage. It is also found that the optimum current ripple under nonsinusoidal operation is smaller than under sinusoidal operation. Simulated and experimental results are included to show the validity of the proposed analysis method.

II. GENERAL EXPRESSION

The scheme of five phase inverter that is used in the analysis is shown in Fig. 1. The load is assumed to be connected in polygon connection. If the actual load is connected in star connection, a polygon equivalent connection can be found easily. Each phase of load is represented as a series connection of a resistance, an inductance and an emf. If the actual load emf is not sinusoidal, the derived expression must be modified accordingly. The dc voltage source of the inverter is assumed to be a ripple free constant dc voltage source. The inverter switching devices are assumed as ideal switches.

The ON-OFF signals for the switching devices are obtained by comparing five-phase reference or modulation signals to a high-frequency triangular carrier signal. In this paper, it assumed that the carrier frequency is much higher than the reference signal. Fig. 2 shows the five-phase sinusoidal reference signal with third harmonic injection. If the instantaneous value of the reference signal (phase 1, for example) is higher (lower) than the carrier signal then the upper (lower) switching device of phase 1 receives an ON (OFF) signal.

978-1-4244-0644-9/07/$25.00 ©2007 IEEE   1267

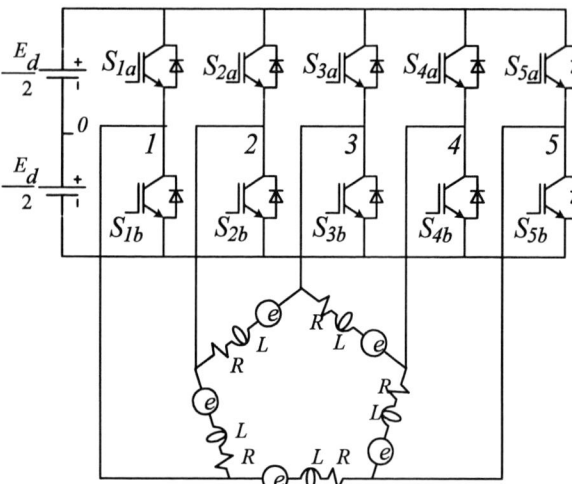

Fig.1 Scheme of five-phase inverter and load connection.

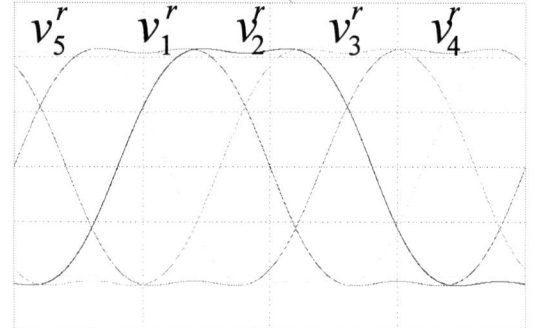

Fig. 2. Five-phase sinusoidal plus third harmonic injection reference signal.

By using this method, the average value (average over one carrier period) of phase-to-zero voltage will proportional to the reference signal.

Let us assume that the reference signal can be written as follow:

$$v_1^r = k\left[\sin\theta + b\sin 3(\theta - \alpha)\right] + s_0 \tag{1}$$

$$v_2^r = k\left[\sin\left(\theta - \frac{2\pi}{5}\right) + b\sin 3\left(\theta - \frac{2\pi}{5} - \alpha\right)\right] + s_0 \tag{2}$$

$$v_3^r = k\left[\sin\left(\theta - \frac{4\pi}{5}\right) + b\sin 3\left(\theta - \frac{4\pi}{5} - \alpha\right)\right] + s_0 \tag{3}$$

$$v_4^r = k\left[\sin\left(\theta - \frac{6\pi}{5}\right) + b\sin 3\left(\theta - \frac{6\pi}{5} - \alpha\right)\right] + s_0 \tag{4}$$

$$v_5^r = k\left[\sin\left(\theta - \frac{8\pi}{5}\right) + b\sin 3\left(\theta - \frac{8\pi}{5} - \alpha\right)\right] + s_0 \tag{5}$$

where $\theta = 2\pi ft$, f is the frequency of the modulating signal or the fundamental frequency of the inverter, k is modulation index, b is magnitude of third harmonic injection and α is phase angle between third harmonic and fundamental. s_0 is arbitrary signal that is injected into the reference signal. Because the same signal is injected into the phase reference signals, the average value of phase-to-phase voltages will not be changed by this arbitrary injection signal. The arbitrary assumption of signal s_0 is valid as long as the frequency is much lower than the carrier frequency.

Based on the circuit in Fig. 1, the voltage between phases 1 and 2 can be written as:

$$v_{12} = e + Ri_{12} + L\frac{di_{12}}{dt} \tag{6}$$

If the voltage and current in the above equation are separated into the average (average over one switching cycle) and ripple components, $v_{12} = \overline{v}_{12} + \tilde{v}_{12}$ and $i_{12} = \overline{i}_{12} + \tilde{i}_{12}$, then (6) can be written as

$$\overline{v}_{12} + \tilde{v}_{12} = e + R\left(\overline{i}_{12} + \tilde{i}_{12}\right) + L\frac{d\left(\overline{i}_{12} + \tilde{i}_{12}\right)}{dt} \tag{7}$$

The average and ripple components on the left hand and right hand parts of (7) must be the same, respectively, and therefore:

$$\overline{v}_{12} = e + R\overline{i}_{12} + L\frac{d\overline{i}_{12}}{dt} \tag{8}$$

$$\tilde{v}_{12} = R\tilde{i}_{12} + L\frac{d\tilde{i}_{12}}{dt} \tag{9}$$

This analysis is concerned with high frequency ripple due to switching process. Because the ripple voltage drop across the resistance, $R\tilde{i}_{12}$, is much smaller than other component in (9), then the expression for the ripple current can be obtained as

$$\tilde{i}_{12} = \int\frac{\tilde{v}_{12}}{L}dt = \int\frac{v_{12} - \overline{v}_{12}}{L}dt \tag{10}$$

Fig. 3 shows the detailed waveforms of five-phase PWM inverter over one carrier period. In order to simplify the analysis, it is assumed that amplitude of the triangular carrier signal is unity. Because the carrier frequency is much higher than the reference ones, the reference signal over one carrier period can be assumed as constants. In Fig.3, s_1 and s_2 are the switching states of phases 1 and 2. The switching state is equal to unity (zero) when the instantaneous value of the associated reference signal is higher (lower) than the carrier signal. This figure was drawn for the case when the instantaneous value of reference signal for phase 1 is higher than the one for the phase 2.

Time interval in Fig.3 can be obtained as below:

$$\frac{T_0}{T_S} = \frac{1 - v_1^r}{4} \tag{11}$$

$$\frac{T_1}{T_S} = \frac{v_{12}^r}{4} \tag{12}$$

$$\frac{T_2}{T_S} = \frac{1 + v_2^r}{4} \tag{13}$$

where $v_{12}^r = v_1^r - v_2^r$, $T_S = 1/f_S$, and f_S is the switching of carrier frequency. Based on (10) and the output voltage waveform in Fig.3, expression of the output current ripple over one switching period can be obtained as

1268

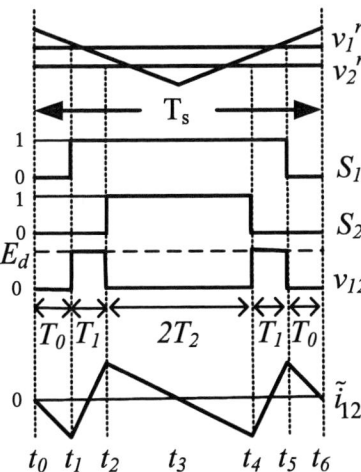

Fig. 3. Detailed output waveforms over one switching period

$$\tilde{i}_{12}=\frac{\overline{v}_{12}}{L}\begin{cases}-(t-t_0) & \text{for } t_0 \le t \le t_1 \\[2mm] \left(\dfrac{E_d}{\overline{v}_{12}}-1\right)(t-t_1)-T_0 & \text{for } t_1 \le t \le t_2 \\[2mm] -(t-t_6) & \text{for } t_2 \le t \le t_4 \\[2mm] \left(\dfrac{E_d}{\overline{v}_{12}}-1\right)(t-t_9)-\left(T_3+T_4+T_5\right) & \text{for } t_4 \le t \le t_5 \\[2mm] -\left(t-t_{12}\right) & \text{for } t_5 \le t \le t_6 \end{cases}$$

(14)

where

$$\overline{v}_{12}=E_d\frac{2T_1}{T_S}$$

(15)

is the average value of voltage between phase 1 and 2 over one carrier period. The mean square value of the output current ripple over one switching period can be obtained as

$$\tilde{I}_{12}^2=\frac{1}{T_S}\int_{t_0}^{t_0+T_S}\tilde{i}_{12}^2 dt=\frac{2}{T_S}\int_{t_0}^{t_6}\tilde{i}_{12}^2 dt$$

(16)

Substituting (13) and (14) into (16) and performing the integration, the following expression is obtained:

$$\tilde{I}_{12}^2=\frac{T_S^2 E_d^2 v_{12}^{r2}}{192L^2}\left[1+v_{12}^{r2}+3v_1^r v_2^r-v_{12}^r\right]$$

(17)

By using the same method, mean square values of current ripple for the other phases can be obtained and the results are

$$\tilde{I}_{23}^2=\frac{T_S^2 E_d^2 v_{23}^{r2}}{192L^2}\left[1+v_{23}^{r2}+3v_2^r v_3^r-v_{23}^r\right]$$

(18)

$$\tilde{I}_{34}^2=\frac{T_S^2 E_d^2 v_{34}^{r2}}{192L^2}\left[1+v_{34}^{r2}+3v_3^r v_4^r-v_{34}^r\right]$$

(19)

$$\tilde{I}_{45}^2=\frac{T_S^2 E_d^2 v_{45}^{r2}}{192L^2}\left[1+v_{45}^{r2}+3v_4^r v_5^r-v_{45}^r\right]$$

(20)

$$\tilde{I}_{51}^2=\frac{T_S^2 E_d^2 v_{51}^{r2}}{192L^2}\left[1+v_{51}^{r2}+3v_5^r v_1^r-v_{51}^r\right]$$

(21)

where

$$v_{23}^r=v_2^r-v_3^r,$$

(22)

$$v_{34}^r=v_3^r-v_4^r,$$

(23)

$$v_{45}^r=v_4^r-v_5^r,$$

(24)

$$v_{51}^r=v_5^r-v_1^r$$

(25)

Equations (17)-(25) are general expression for mean squares values of the output current ripple over one switching period. In the next section these equations will be used to minimize the output current ripple of five-phase inverter.

III. OUTPUT CURRENT RIPPLE MINIMIZATION

The total output current ripple is defined as

$$\tilde{I}_{tot}^2=\tilde{I}_{12}^2+\tilde{I}_{23}^2+\tilde{I}_{34}^2+\tilde{I}_{45}^2+\tilde{I}_{51}^2$$

(26)

The optimum injected signal can be obtained by solving the following equation

$$\frac{d\tilde{I}_{tot}^2}{ds_0}=0$$

(27)

It can be shown that

$$\frac{dv_1^r}{ds_0}=\frac{dv_2^r}{ds_0}=\frac{dv_3^r}{ds_0}=\frac{dv_4^r}{ds_0}=\frac{dv_5^r}{ds_0}=1$$

(28)

And

$$\frac{dv_{12}^r}{ds_0}=\frac{dv_{23}^r}{ds_0}=\frac{dv_{34}^r}{ds_0}=\frac{dv_{45}^r}{ds_0}=\frac{dv_{51}^r}{ds_0}=0$$

(29)

By using (28) and (29) then (27) can be solved and the result is

$$\left[\begin{matrix}v_{12}^{r2}\left(v_2^r+v_1^r\right)+v_{23}^{r2}\left(v_3^r+v_2^r\right)+v_{34}^{r2}\left(v_4^r+v_3^r\right)\\+v_{45}^{r2}\left(v_5^r+v_4^r\right)+v_{51}^{r2}\left(v_1^r+v_5^r\right)\end{matrix}\right]=0$$

(30)

Substituting (1)-(5) into (30), the optimum reference signal for minimum output current ripple is obtained as

$$s_0=k\left[c_1\sin\left(5\theta-3\alpha\right)-c_2\sin\left(5\theta-6\alpha\right)\right]$$

(31)

with

$$c_1=\frac{b\left(\cos\dfrac{3\pi}{5}\sin^2\dfrac{\pi}{5}+\sin^2\dfrac{3\pi}{5}\right)}{2\left[\sin^2\dfrac{\pi}{5}+b^2\sin^2\dfrac{3\pi}{5}\right]}$$

(32)

$$c_1=\frac{0.79775b}{\left[0.6910+1.8090b^2\right]}$$

(33)

and

$$c_2=\frac{b^2\left(\sin\dfrac{3\pi}{10}\sin^2\dfrac{3\pi}{5}+\sin^2\dfrac{\pi}{5}\right)}{2\left[\sin^2\dfrac{\pi}{5}+b^2\sin^2\dfrac{3\pi}{5}\right]}$$

(34)

$$c_2=\frac{1.0773b^2}{\left[0.6910+1.8090b^2\right]}$$

(36)

It can be seen that injection signal for reference signal to produce minimum output current ripple is fifth harmonic signal with amplitude depends on amplitude of third harmonic. Furthermore, when $b = 0$ (i.e. sinusoidal reference signal) then $c = 0$ verified that optimum reference signal for sinusoidal operation is sinusoidal reference signal [10].

The integration of (17) over one period of the fundamental frequency can be used to calculate the rms value of the total output current ripple in each phase. As the line-to-line voltages are symmetric the integration can be done only over the half-period of the fundamental frequency, that is,

$$\tilde{I}_{12rms} = \left[\frac{1}{\pi} \int_{\frac{-3\pi}{10}}^{\frac{7\pi}{10}} \tilde{I}_{12}^2 d\theta \right]^{\frac{1}{2}} \tag{34}$$

And the result is shown on equation (35) with

$$K = \frac{T_S E_d}{8\sqrt{3}L} \tag{36}$$

The results of integration in the numerical form are given in Table 1. For comparison, the results for discontinues modulation signal [8] and fifth harmonic injection modulation signal [6] are also shown.

Fig. 4 shows the calculated results of output current ripple as a function of modulation index. This figure shows that output current ripple produced by optimum reference signal is lower than other reference signal especially in high modulation index.

Fig. 5 shows the influence of third harmonic injection magnitude, b, and phase angle, α, on the output current ripple of five-phase inverter under nonsinusoidal operation with optimum reference signal. This figure shows that output current ripple vary with the value of b and α.

IV. EXPERIMENTAL RESULT

In order to verify some of the expressions derived in this paper, a five-phase inverter as shown in fig. 1 was constructed. Mosfet IRFP460 was used as power switch. A dead time of 5 μs was used to prevent a short circuit through the upper-arm and lower-arm. Load that used in these experiments is locked-rotor five-phase motor whose total resistance $R = 1.998$Ohm and total leakage inductance $L = 6.4$mH. In this experiment, the DC source voltage, $E_d = 60$ V, was obtained by using a 3-phase diode bridge rectifier. To reduce the effects of the rectifier output harmonics on the investigated inverter, a capacitor having a large capacitance (4600μF) was connected in parallel with the output terminals of rectifier. The inverter was operated at low switching frequency (2000Hz) to reduce the dead time effects in the experimental result.

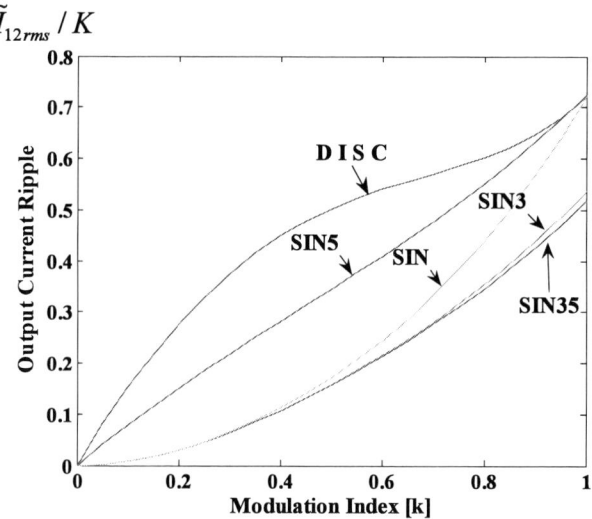

Fig.4 Output current ripple as a function of modulation index
(for SIN35, $b = 1/6$, $\alpha = 0$)

$$\tilde{I}_{12rms} = Kk \left[\begin{array}{l} 2\left[\sin^2\frac{\pi}{5} + b^2\sin^2\frac{3\pi}{5}\right] - \frac{8k}{\pi}\left[\begin{array}{l} \frac{4}{3}\sin^3\frac{\pi}{5} + \frac{4}{5}b\sin^2\frac{\pi}{5}\sin\frac{3\pi}{5}\cos(3\alpha) \\ 3b^2\sin^2\frac{3\pi}{5}\sin\frac{\pi}{5}\left(\frac{35+\cos(6\alpha)}{35}\right) + b^3\sin^3\frac{3\pi}{5}\left(\frac{1}{18}\cos(9\alpha) - \frac{1}{2}\cos(3\alpha)\right) \end{array} \right] \\[20pt] k^2\left[\begin{array}{l} \left[\frac{3}{2}\sin^2\frac{\pi}{5} + b\sin\frac{3\pi}{5}\sin\frac{\pi}{5}\left[8\sin^2\frac{\pi}{5} - \frac{9}{2}\right]\right]\cos(3\alpha) + \\ 3b^2\left[4\sin^2\frac{3\pi}{5}\sin^2\frac{\pi}{5} + \sin\frac{\pi}{10}\sin^2\frac{3\pi}{5} - \sin^2\frac{\pi}{5}\cos\left(\frac{\pi}{5}+3\alpha\right)\right] + \\ 3b^4\sin^2\frac{3\pi}{5}\left[2\sin^2\frac{3\pi}{5} - \cos\left(\frac{\pi}{5}+3\alpha\right) - \frac{1}{2}\cos(3\alpha)\right] + \\ 3b^2\left(c_1\cos(3\alpha) - c_2\right)\sin\frac{3\pi}{5}\left[\sin\frac{3\pi}{5}\sin\frac{3\pi}{10} + 2\sin\frac{\pi}{10}\sin\frac{\pi}{5}\right] + \\ 3b\left(c_1 - c_2\cos(3\alpha)\right)\sin\frac{\pi}{5}\left[\sin\frac{\pi}{5}\sin\frac{\pi}{10} - 2\sin\frac{3\pi}{5}\sin\frac{3\pi}{10}\right] + 3\left(c_2^2 - 2c_1c_2\cos(3\alpha) + c_1^2\right)\left[\sin^2\frac{\pi}{5} + b^2\sin^2\frac{3\pi}{5}\right] \end{array} \right] \end{array} \right]^{\frac{1}{2}} \tag{35}$$

TABLE I
OUTPUT CURRENT RIPPLE

Symbol	Reference Signal	Output Current Ripple ($\tilde{I}_{12rms}/K$)
SIN	Sinusodial $v_1^r = k\sin\theta$	$k\left[0.6910 - \dfrac{6.4984}{3\pi}k + 0.5182k^2\right]^{\frac{1}{2}}$
SIN3	Nonsinusoidal $v_1^r = k\left[\sin\theta + b\sin 3(\theta-\alpha)\right]$	$k\left[\begin{array}{l}\left[0.69098+1.809b^2\right]-k\left[\begin{array}{l}0.6895+0.66938b\cos(3\alpha)+4.0616b^2\left(\dfrac{35+\cos(6\alpha)}{35}\right)+\\ 2.1906b^3\left(\dfrac{1}{18}\cos(9\alpha)-\dfrac{1}{2}\cos(3\alpha)\right)\end{array}\right]\\ k^2\left[\begin{array}{l}0.51824-0.97049b\cos(3\alpha)+4.5885b^2+0.97049b^2\cos\left(\dfrac{\pi}{5}+3\alpha\right)+\\ \left[4.9088-2.7135\left(\cos\left(\dfrac{\pi}{5}+3\alpha\right)+\dfrac{1}{2}\cos(3\alpha)\right)\right]b^4\end{array}\right]\end{array}\right]^{\frac{1}{2}}$
SIN35	Optimum Nonsinusoidal $v_1^r = k\left[\sin\theta + b\sin 3(\theta-\alpha)\right]+s_0$	$k\left[\begin{array}{l}\left[0.69098+1.809b^2\right]-k\left[\begin{array}{l}0.6895+0.66938b\cos(3\alpha)+4.0616b^2\left(\dfrac{35+\cos(6\alpha)}{35}\right)+\\ 2.1906b^3\left(\dfrac{1}{18}\cos(9\alpha)-\dfrac{1}{2}\cos(3\alpha)\right)\end{array}\right]\\ k^2\left[\begin{array}{l}0.51824-0.97049b\cos(3\alpha)+4.5885b^2+0.97049b^2\cos\left(\dfrac{\pi}{5}+3\alpha\right)+\\ \left[4.9088-2.7135\left(\cos\left(\dfrac{\pi}{5}+3\alpha\right)+\dfrac{1}{2}\cos(3\alpha)\right)\right]b^4+\\ 3.2318b^2\left(c_1\cos(3\alpha)-c_2\right)-2.3932b\left(c_1-c_2\cos(3\alpha)\right)+\\ \left(c_2^2-2c_1c_2\cos(3\alpha)+c_1^2\right)\left[1.0365+2.7135b^2\right]\end{array}\right]\end{array}\right]^{\frac{1}{2}}$
DISC	Discontinues [8] (see appendix)	$k\left[2.7639 - \dfrac{44.9330}{3\pi}k + 2.5243k^2\right]^{\frac{1}{2}}$
SIN5	Fifth Harmonic Injection [5] $v_1^r = k\left[\sin\theta - 0.062\sin 5\theta\right]$	$k\left[0.6910 - \dfrac{6.4984}{3\pi}k + 0.5222k^2\right]^{\frac{1}{2}}$

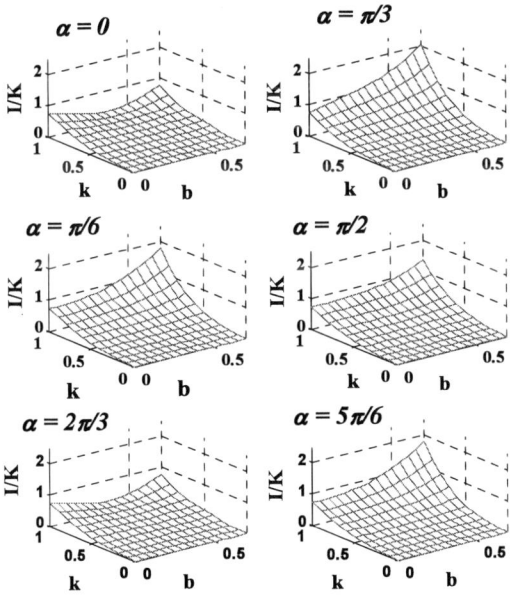

Fig. 5 Influence of b and α on output current ripple of five-phase inverter under nonsinusoidal operation

Fig 6 show output phase current under various reference signals with 0.8 modulation index. Fig.6.a. is output phase current ripple under sinusoidal reference signal, fig.6.b. is output phase current ripple under sinusoidal plus 1/6 third harmonic reference signal (i.e. b = 1/6) and fig.6.c is output phase current ripple under optimum reference signal (i.e. b = 1/6 and c = 0.139). It can be seen that output current ripple under nonsinusoidal reference signal is smaller than output current ripple under sinusoidal reference signal.

In order to measure the total harmonic the inverter output current waveform during one fundamental period is recorded by using a digital storage oscilloscope. The data in the oscilloscope is subsequently analyzed by a personal computer to determine the total harmonic.

Fig.7 shows the plots of the output current ripple of five-phase PWM inverter under various modulation signals as a function of modulation index. SIN is for pure sinusoidal reference signal (i.e. s_0 and $b = 0$). DISC is for discontinues reference signal. SIN3 is for sinusoidal reference signal with third harmonic injection (i.e. $s_0 = 0$) and SIN35 is for optimum reference signal that result in minimum output current ripple of five phase inverter in nonsinusoidal operation. Output current ripple of five-

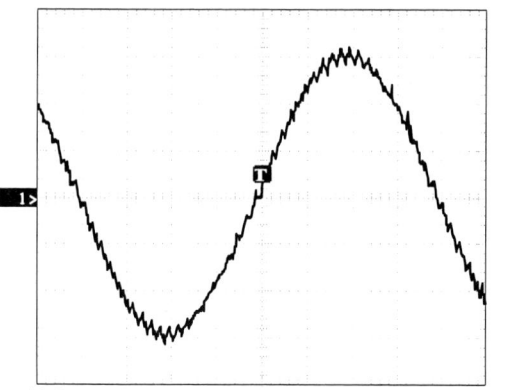

(a) Output phase current inverter under sinusoidal reference signal

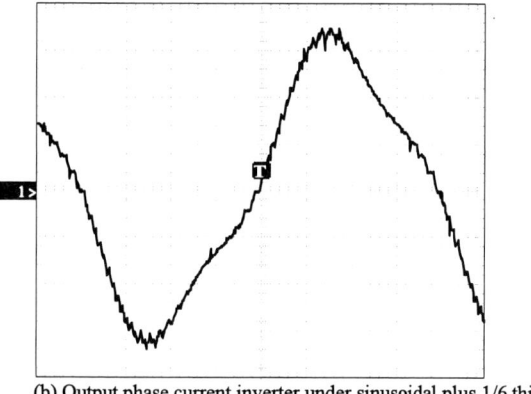

(b) Output phase current inverter under sinusoidal plus 1/6 third harmonic.

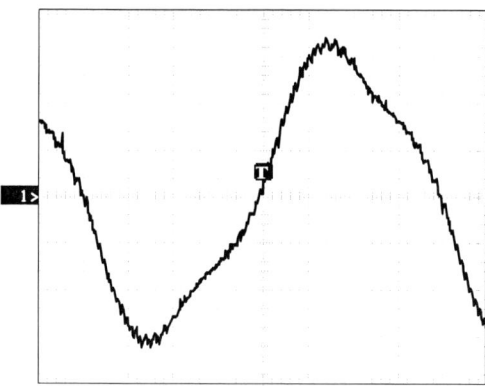

(c) Output phase current ripple inverter under optimum reference signal

Fig. 6 Output current ripple under various reference signal
($k = 0.8$, Ampere/div = 2.5A, Time/div = 2.5ms)

phase inverter with third harmonic injection plus fifth harmonic injection is lower than output current ripple of five phase inverter with third harmonic injection especially in high modulation index. This figure also shows a good agreement between the calculated and experimental results.

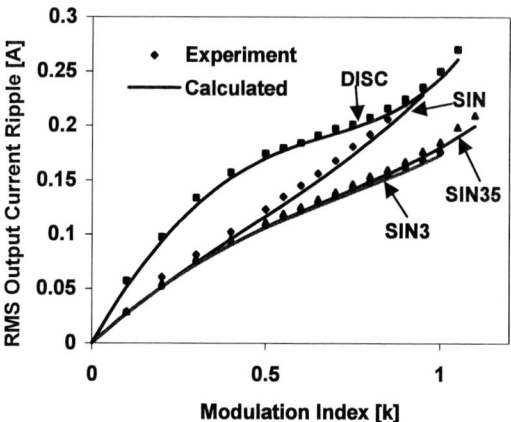

Fig.7 Experimental and calculated results of the output current ripple as a function of modulation index

V. CONCLUSION

Analysis and minimization of output current ripple of five-phase PWM inverter under non-sinusoidal emf have been presented in this paper. It shown that adding of fifth harmonic injection result in minimum output current ripple of five-phase inverter. The magnitude of fifth harmonic depends on how much third harmonic injected in five-phase inverter. Simulation and experimental result verify the analysis.

APPENDIX

Expression for reference signal of Discontinues PWM

Phase Angle	v_{1disc}^{r}
$0 - \dfrac{\pi}{5}$	$1 - v_{51}^{r}$
$\dfrac{\pi}{5} - \dfrac{2\pi}{5}$	$-1 + v_{23}^{r} + v_{12}^{r}$
$\dfrac{2\pi}{5} - \dfrac{3\pi}{5}$	1
$\dfrac{3\pi}{5} - \dfrac{4\pi}{5}$	$-1 - v_{45}^{r} - v_{51}^{r}$
$\dfrac{4\pi}{5} - \dfrac{5\pi}{5}$	$1 + v_{12}^{r}$
$\dfrac{5\pi}{5} - \dfrac{6\pi}{5}$	$-1 - v_{51}^{r}$
$\dfrac{6\pi}{5} - \dfrac{7\pi}{5}$	$1 + v_{23}^{r} + v_{12}^{r}$
$\dfrac{7\pi}{5} - \dfrac{8\pi}{5}$	-1
$\dfrac{8\pi}{5} - \dfrac{9\pi}{5}$	$1 - v_{45}^{r} - v_{51}^{r}$
$\dfrac{9\pi}{5} - \dfrac{10\pi}{5}$	$-1 + v_{12}^{r}$

REFERENCES

[1] E.A. Klingshirn, "High phase order induction motors-Part I: Description and theoretical considerations", *IEEE Trans. Power App. Syst.*, vol. PAS-102, pp.47-53, Jan. 1983.

[2] E.A.Klingshirn, "High phase order induction motors-Part II: Experimental results", *IEEE Trans. Power App.Syst.*,vol.PAS-102, pp.54-60, Jan. 1983.

[3] K.N. Pavithran, R. Parimelalagan, and M. Krishnamurthy, "Studies on Inverter-Fed Five-Phase Induction Motor Drive'', *IEEE Power Elec.*, vol. 3, No. 2, Apr. 1988, pp. 224-235.

[4] H. Xu, H.A. Toliyat, L.J. Petersen, "Five-Phase Induction Motor Drives with DSP-Based Control System", *IEEE Trans. Power. Electron.*, vol.17, no. 4, pp. 524-533, Jul. 2002.

[5] R. Shi, H.A. Toliyat, "Vector Control of Five-Phase Synchronous Reluctance Motor with Space Vector Pulse Width Modulation (SVPWM) for Minimum Switching Losses", *Proc. IEEE Applied Power Elec. Conf. APEC*, Dallas, Texas, 2002, pp. 57-63.

[6] A.Iqbal,E.Levi,M.Jones,S.N.Vukosavic, "Generalised Sinusoidal PWM with Harmonic Injection for Multi-Phase VSIs," *37th IEEE Power Electronics Specialists Conference*, June, Jeju, Korea.

[7] J.W. Kelly, E.G. Strangas, J.M. Miller, "Multiphase Space Vector Pulse Width Modulation", *IEEE Transactions on Energy Conversion*, Vol. 18, No. 2, June 2003.

[8] O. Ojo, G. Dong, "Generalized Discontinues Carrier-Based PWM Modulation Scheme for Multi-Phase Converter Machine Systems", *IEEE Industrial Application Society Annual Meeting*, 2005

[9] P.A. Dahono, Y. Sato, T. Kataoko. "Analysis and Minimization of Harmonics in the AC and DC Sides of PWM Inverter", *IEE Japan Trans. Ind. Appl.*, May 1995.

[10] P.A. Dahono, "Analysis and Minimization of Output Current Ripple of Five Phase PWM Inverters", *International Conference on Electrical Machines*, Greece, 2006.

[11] H. Xu, H.A. Toliyat, L.J Petersen, "Rotor Field Oriented Control of Five-Phase Induction Motor with the Combined Fundamental and Third harmonic Currents", *Applied Power Electronics Conference and Exposition*, Vol 1, March 2001,pp. 392 – 398

[12] H.A. Toliyat, S.P. Waikar, T.A. Lippo, "Analysis and Simulation of Five-Phase Synchronous Reluctance Machines Including Third Harmonic of Airgap MMF", *IEEE Transactions on Industry Applications*, Vol. 34, No. 2, March/April 1998.

An Improved 'DC-DC Type' High Frequency Transformer-Link Inverter by Employing Regenerative Snubber Circuit

Z. Salam, S. M. Ayob, M. Z. Ramli, and N. A. Azli

Department of Energy Conversion, Faculty of Electrical Engineering,
Universiti Teknologi Malaysia,
81310 UTM Skudai, Johor, Malaysia.
Email: zainals@fke.utm.my
Fax: +607-556 6272

Abstract–This work proposes a bidirectional high frequency (HF) link inverter using center-tapped high frequency transformer. The main advantage of this topology is the reduced size of the converter. However the utilization of the high frequency transformer results in the occurrence of high voltage spike at its secondary. To solve this problem, we incorporate a regenerative snubber to clamp the spike and subsequently feed the energy back to the power circuit. The inverter also utilizes less number of switches, which results in lower conduction and switching losses. This paper describes some of the important aspects of the snubber design. It is verified with a 1kW inverter prototype.

Index Terms—High frequency link, inverter

I. INTRODUCTION

The 'dc-dc converter type' high frequency link inverter was first described in [1]. Its basic power circuit is shown in Fig. 1. It consists of three power stages i.e. the HF PWM bridge, active rectifier and polarity-reversing bridge. The main advantage of this topology is the significant reduction in the transformer size. This is because the transformer is operated at high frequency.

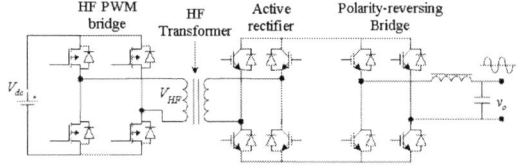

Fig. 1. The 'dc-dc converter type' high frequency link inverter.

In a previous paper [2], we have described a variation of this topology using the center-tapped transformer. We have shown that with this circuit, fewer switches are required compared to the topologies proposed in [1]. Using the center-tapped transformer however, has one major drawback; the voltage across the switch at the transformer secondary is doubled relative to the non-center tapped type. As the voltage stress across the switch is already high, additional voltage spike can make the power switch vulnerable. In this paper we proposed an improvement to our previous topology. We introduce a regenerative snubber network for the active rectifier that effectively reduces the spike to a very low value. In

addition, because of the lossless (in the ideal case) nature of the snubber, the energy contained in the voltage spike is pumped back to the main power circuit.

II. INVERTER WITH REGENERATIVE SNUBBER

A. Power Circuit

The proposed high frequency link inverter is shown in Fig. 2. The timing diagram for the key waveforms of the power stage is illustrated in Fig. 3. At the first stage, the high frequency PWM bridge converts the dc voltage to a 50Hz modulated sinusoidal PWM (SPWM) high frequency ac voltage. Then, this voltage is isolated and boosted using a center-tapped high frequency transformer. In the next stage, the high frequency SPWM waveform is rectified using a center-tapped active rectifier. The active rectifier, which consists of power switches and anti-parallel diodes, enables bidirectional power flow. For transfer of power from the source, the diodes are utilised. For reverse power flow, the power switches S3 and $\overline{S3}$ are turned-on. The rectified PWM waveform ($V_{HF\,rect}$) is then low-pass filtered to obtain the rectified fundamental component. Finally, using a polarity-reversing bridge, the second half of the rectified sinusoidal voltage waveform is unfolded at zero-crossing, and the sinusoidal output waveform is obtained. By using the center-tapped rectifier circuit, the number of bidirectional switches is reduced. Furthermore the polarity-reversing bridge utilises only supply switching frequency (50Hz) switches.

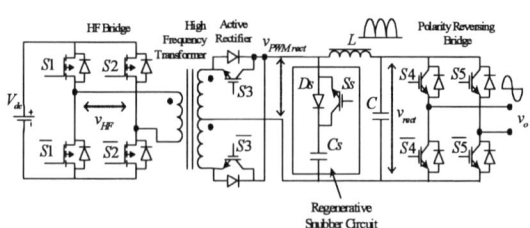

Fig. 2. Block diagram of the proposed inverter.

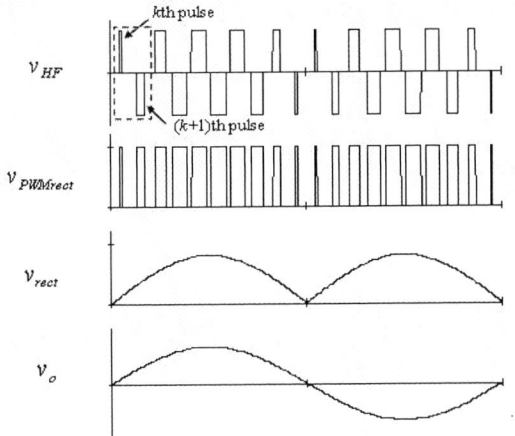

Fig. 3. Key waveforms at different stages of the dc/ac conversion.

One of the main concerns of high frequency link inverter is the unavoidable presence of leakage inductance at the transformer secondary. The energy stored in the leakage inductor is one of the primary sources of voltage spike in the power transistor. The leakage inductance carries the same current as the transistor switch when the switch is on. When the current at the transistor is turned off very quickly, the rate of change of current, i.e. di/dt is very high, resulting in high voltage spikes to appear across the switches. If not properly controlled, the spikes may result in the destruction of the switches. The normal method to dampen the spike is to use an RCD snubber network across the switch, as illustrated in Fig. 4 [2]. However, the required snubber capacitor, Cs, for adequate spike suppression can be quite large. Consequently high discharge energy will be dissipated in snubber resistor, Rs, when the switch turns back on. This mandates for the use of high power Rs, which in many cases lead to further loss in the inverter's efficiency.

In this work, we introduce a regenerative snubber at the transformer's secondary circuit. Using this snubber, the spikes across the active rectifier switches are effectively reduced to a very low value. In addition, the

energy contained in the voltage spikes is injected back to the main power stage. This will further increase the efficiency of the inverter.

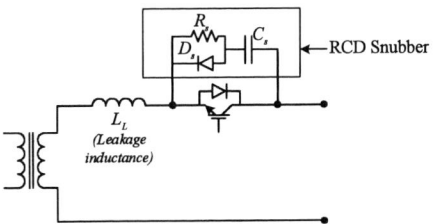

Fig. 4. RCD snubber

B. Regenerative snubber network

Fig. 5 shows the regenerative snubber network. The associated timing diagram is shown in the same figure. This snubber circuit is used to reduce the voltage spike across the active rectifier's switches (S3 and $\overline{S3}$) to a safe level. From the timing diagram in Fig. 5, it can be seen that voltage spike occurs during switching transitions. Circuit shown in Fig. 6 details the snubbering operation during this transient moment. When the gate signal (V_{PWM}) is applied to any of the rectifier's switch, that switch will turn on and the adjacent switch will turn off. The energy stored in the leakage inductance of the transformer will appear as voltage spike on the adjacent switch. Assuming the snubber diode (Ds) and switch (Ss) are ideal, the voltage level across the capacitor (Cs) without spike is v_1. When spike occurs, V_{HFrect} increases, causing the snubber diode Ds to be forward biased and consequently charges the snubber capacitor Cs. The capacitor Cs dampens the spike by reducing the di/dt. The charging process that takes place during t_1 to t_2 causes the capacitor voltage v_{cs} to rise. When v_{cs} equals v_2, i.e. when the capacitor voltage equals v_{HFrect}, the charging process stops. Snubber diode Ds is reverse biased and Cs starts to discharge its energy into the power circuit via Ss. The discharging continued until the end of the PWM pulse. When this point is reached, Ss is turned off, and the discharging process stops. Voltage vcs is maintained at its equilibrium level (v_1) until the next charging process occurs.

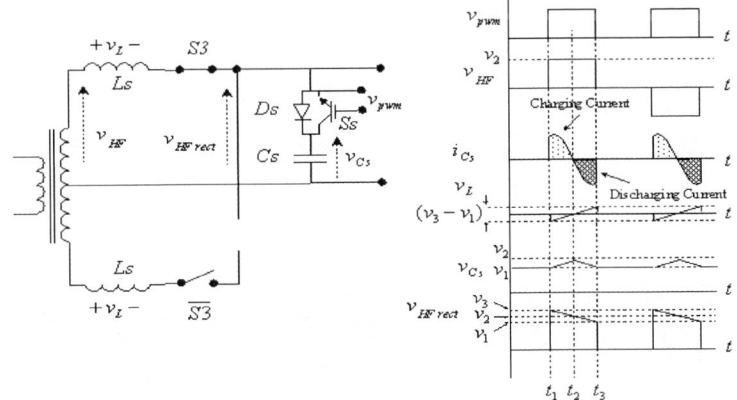

Fig. 5. Circuit of regenerative snubber network and the associated timing diagram.

1275

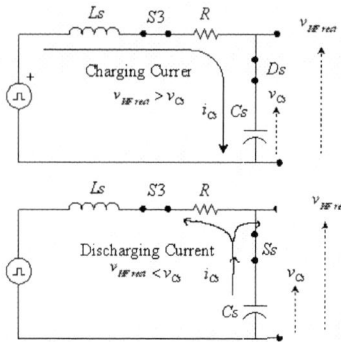

Fig. 6. Equivalent circuit of the regenerative snubber network during snubbering operation.

III. EXPERIMENTAL RESULT AND DISCUSSION

The HF bridge is constructed using the IRFP460 power MOSFET. It is a low $R_{ds(ON)}$ device with good switching capability. The active rectifier's switch is built using the IRG4PH40K IGBTs with discrete 20EFT10 fast recovery anti-parallel diodes. The rated voltage for the diode is 1200V, in precaution to the possible voltage surge that may result from the transformer leakage inductance. The polarity-reversing bridge is constructed using SK25GB065 IGBT module. Since almost all the surge voltages have been dampened before entering the polarity-reversing bridge, the chosen power switches are only rated at 600V. Using low voltage IGBT, the forward

conduction losses can be minimized. The specifications of the inverter are as follows:

- Input voltage ranged from 60V to 110V.
- Sinusoidal output voltage 220-250 Vrms, 50 Hz.
- Maximum output power of 1 kW.

The output voltage and current waveforms for resistive and inductive loads are shown in Figs.7 (a) and (b), respectively. The latter oscilogram indicates that the inverter is capable of carrying bidirectional power flow.

Fig. 9 shows the oscillograms that depicts the operation of the regenerative snubber. As can be seen, when the PWM pulse in applied to S3, at the transition point, the snubber capacitor charges and absorbs the transient energy. As a result, for the adjacent switch $\overline{S3}$, the spike is suppressed. The corresponding charging and discharging voltage at C_s is also shown in this figure. The results are identical to the theoretical timing diagram in Fig. 5. Clearly there is a close agreement between the theoretical and practical results.

Fig. 10 shows the collector emitter voltage (V_{ce}) of switch S3 before and after insertion of the regenerative snubber. It can be seen that before the regenerative snubber is inserted, voltage overshoot of about 37% of the amplitude occurs. After the insertion of the regenerative snubber, the spike has been reduced to a very low value.

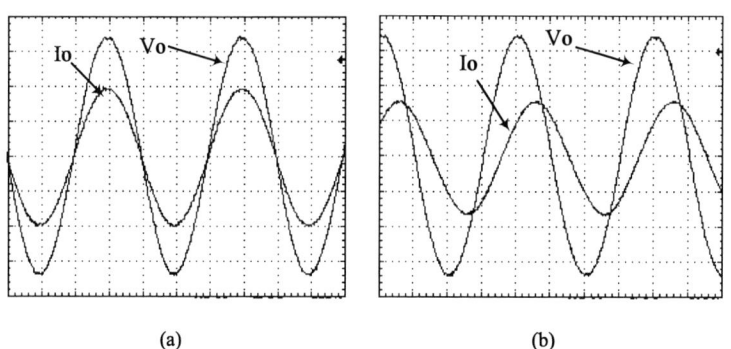

(a)   (b)

Fig. 7. Output voltage and current with (a) resistive load, (b)inductive load (power factor = 0.7)
Scales: output voltage 100V/div, output current 2A/div, time 5ms/div.

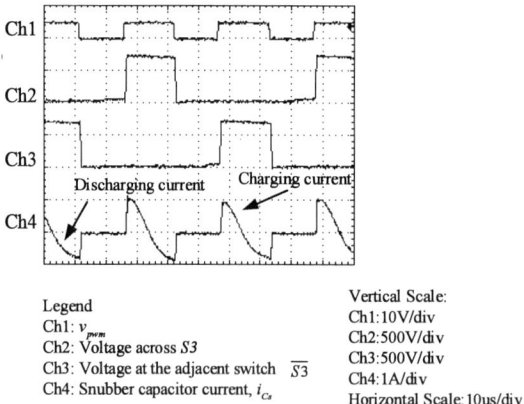

Legend
Ch1: v_{pwm}
Ch2: Voltage across $S3$
Ch3: Voltage at the adjacent switch $\overline{S3}$
Ch4: Snubber capacitor current, i_{C_s}

Vertical Scale:
Ch1:10V/div
Ch2:500V/div
Ch3:500V/div
Ch4:1A/div
Horizontal Scale:10us/div

Fig. 8. The inverter is capable of conducting bidirectional power flow.

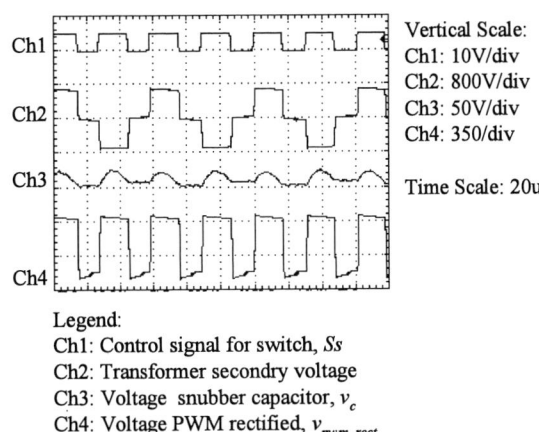

Vertical Scale:
Ch1: 10V/div
Ch2: 800V/div
Ch3: 50V/div
Ch4: 350/div

Time Scale: 20us

Legend:
Ch1: Control signal for switch, *Ss*
Ch2: Transformer secondry voltage
Ch3: Voltage snubber capacitor, v_c
Ch4: Voltage PWM rectified, v_{pwm_rect}

Fig. 9. Control signal of switch Ss and the associated output waveforms after insertion of regenerative snubber.

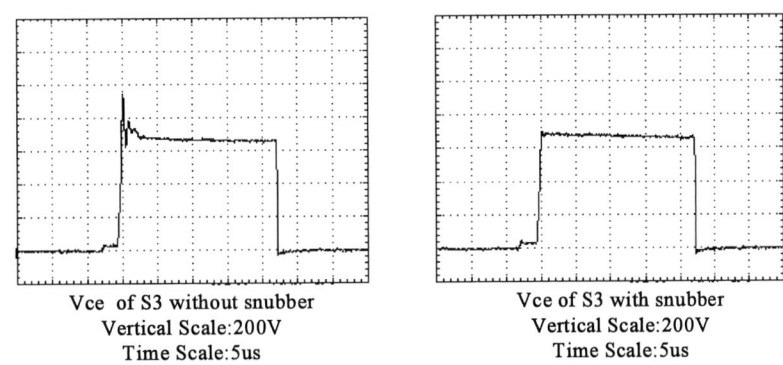

Vce of S3 without snubber
Vertical Scale:200V
Time Scale:5us

Vce of S3 with snubber
Vertical Scale:200V
Time Scale:5us

Fig. 10. Voltage Vce at switch S3 before and after the insertion of regenerative snubber.

Fig. 11 compares the inverter efficiency using regenerative snubber and the conventional RCD snubber. For the latter, the values of *Cs* and *Rs* are chosen to be 2.2 nF and 22 Ω, respectively. The chosen capacitor values for the RCD snubber components are rather conservative; typical values are much higher than this. From the figure, it can be concluded that the new snubber is able to increase the efficiency of the inverter by about 5%. A higher value for RCD components will make the difference even larger.

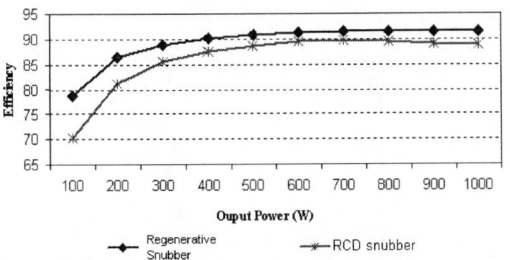

Fig. 11. Efficiency vs. output power

IV. CONCLUSION

This paper has described the operation of a snubber network that is applied to a center-tapped active rectifier high frequency transformer. To prove the concept, a 1kW prototype inverter is constructed. From the experimental results, it was found that the snubber network function as predicted. The voltage spike across the active rectifier's switch is reduced significantly. Furthermore the efficiency of the inverter is increased.

REFERENCES

[1] Koutroulis, E., and Chatzakis, J., "A bidirectional, sinusoidal, high-frequency inverter design," *IEE Proc- Electric Power App.,* vol. 148, no. 4, July 2001.

[2] Z. Salam, M. Z. Ramli, L. S. Toh and C. L. Nge, "A Bi- directional High-frequency Link Inverter Using Center- tapped Transformer" *Proc. of IEEE Power Electronics Specialist Conf. (PESC 04),* pp 3883-3888, Aachen, Germany, 20-25th June 2004.

A Novel Dimming Technique for Cold Cathode Fluorescent Lamp

K. I. Hwu, and Y. H. Chen
Center for Power Electronics Technology
National Taipei University of Technology, Taiwan
eaglehwu@ntut.edu.tw

Abstract- **In this paper, a simple dimming control is presented herein and applied to the cold cathode fluorescent lamp (CCFL). Conventionally, there are two types of dimming of CCFL, analog and digital. However, performance of analog dimming is poor due to parasitic capacitance existing between the CCFL and the ground, thereby causing a large error in the sensed current of the CCFL and hence reducing the performance of dimming, whereas the digital dimming takes back-and-fro start and cut-off of the CCFL, thereby causing surge current and voltage spike and hence reducing the life of the CCFL. Consequently, to overcome these problems, a novel dimming strategy is presented according to frequency shift. Also, a full-bridge phase-shift resonant inverter is used to power the CCFL. Some experimental results are provided to demonstrate the effectiveness of the proposed scheme.**

I. INTRODUCTION

With the fast development of the liquid crystal display (LCD) technology, the LCD possesses small size, energy-saving, low EMI, etc, thereby causing the demand in the LCD to be kept increasing greatly. However, the LCD can not emit light itself and hence needs a sufficient and well-distributed backlight light source to control the output of light via turning liquid crystal body so as to develop. Up to the present, there are two types of back light sources. One is the light-emitting diode (LED) with white light, which is costly, of small area of light and used in small-size applications, such as the mobile phone, the personal digital assistant (PDA), the global position system (GPS), the measure meter etc., and the other is the cold cathode fluorescent lamp (CCFL), which is used in large-sized applications [1]. Consequently, in order to meet requirements of size and performance of the displayer, the combination of the LCD with the CCFL is the best choice.

Besides, it is indispensable for dimming since people take life quality into great consideration. Therefore, the dimming technology is getting more and more attractive. There are two types of dimming, analog and digital. The first achieves dimming by adjusting the current flowing through the CCFL but has poor performance of dimming due to parasitic capacitance existing between the CCFL and the ground, thereby causing a large error in the sensed current of the CCFL and hence reducing the performance of dimming, whereas the digital dimming, called burst dimming, achieves dimming by driving the CCFL between no current and rated current, and hence without the problem existing in analog dimming. But, the digital dimming takes back-and-fro start and cut-off of the

CCFL, thereby causing surge current and hence reducing the life of the CCFL.

Recently, there have been three methods [2-5] widely used to dim the CCFL, such as duty-ratio control, frequency control and voltage control. The duty-ratio control is easily to realize by regulating the lamp current, but the asymmetrical lamp current and poor lamp crest factor will result in discoloration of the lamp [2]. As generally acknowledged, the frequency control [6] is the most common technique in regulating the lamp current. However, the corresponding dimming range is restricted significantly by the switching frequency. As for the voltage control, it has a good dimming performance but too complex in implementation, and hence not suitable for low-power applications. Although digital dimming is the latest technique and can enlarge the dimming range, resulting current spikes must be suppressed by appropriate control strategies [7-9] to extend the lamp life. Consequently, to overcome the disadvantages mentioned above, a novel dimming scheme is presented based on frequency shift. Also, a full-bridge phase-shift resonant inverter is employed herein to power the CCFL. Some experimental results are provided to demonstrate the effectiveness of the proposed scheme.

II. MAIN POWER STAGE FOR CCFL

Before entering into this topic, CCFL is represented as a variable resistor R_{lamp} and also omits the parasitic capacitances in the transformer and between the lamp and the ground. As shown in Fig. 1(a), a full-bridge phase-shift resonant inverter is used to power R_{lamp}, containing one main transformer MT with leakage inductances L_{r1} and L_{r2} at the primary and the secondary respectively, magnetization inductance L_m, one DC blocking capacitor C_C, one stabilization capacitor C_B, one resonance capacitor C_R, four MOSFET switches Q_1 to Q_4. Based on one appropriate control IC [10] to be mentioned below, R_{lamp} can get driven by sinusoidal voltage and current created from this full-bridge phase-shift resonant inverter. The equivalent circuit is sketched in Fig. 1(b) in the steady state, where the resonant tank [1] is composed of, C_B, C_R and L_{r2}, the primary impedance reflected to the secondary containing L_m, L_{r1} and C_C. The circuit shown in Fig. 1(b) can be redrawn to that in Fig. 1(c) and then to that in Fig. 1(d) which is a series-resonant series-parallel-loaded circuit. The relationship among parameters in Fig. 1 is described below.

978-1-4244-0644-9/07/$25.00 ©2007 IEEE

$$V_{rms} = \frac{2\sqrt{2}}{\pi} V_{dc} \qquad (1)$$

where V_{rms} is the fundamental sinusoidal wave.

$$C_s = \frac{C_C}{n^2} \qquad (2)$$

$$L_r = \frac{n^2 L_{r1} L_m}{L_{r1} + L_m} + L_{r2} \qquad (3)$$

$$C_p = C_R + \frac{C_B}{1 + \omega^2 C_B^2 R^2_{lamp}} \qquad (4)$$

$$R'_{lamp} = R_{lamp} + \frac{1}{\omega^2 C_B^2 R_{lamp}} \qquad (5)$$

where n is equal to N_2/N_1, N_1 and N_2 are turns of the primary winding and the secondary winding respectively, and ω is the switching radian frequency.

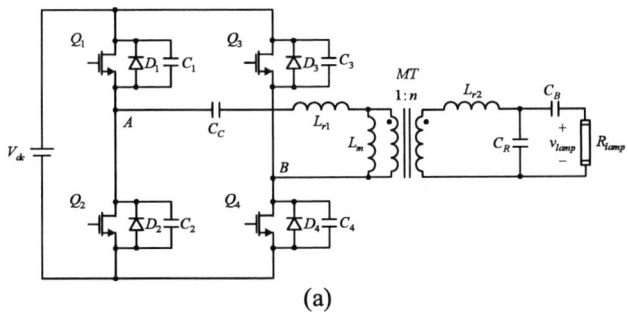

(a)

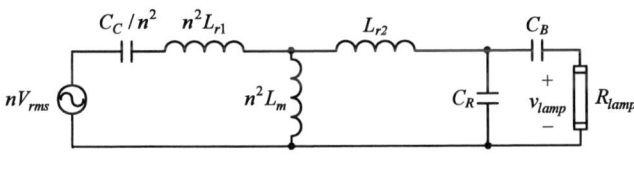

(b)

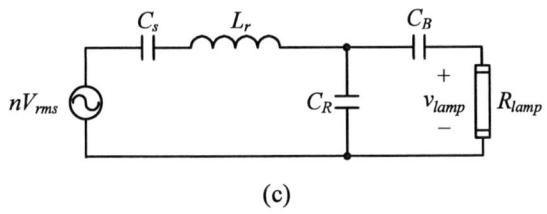

(c)

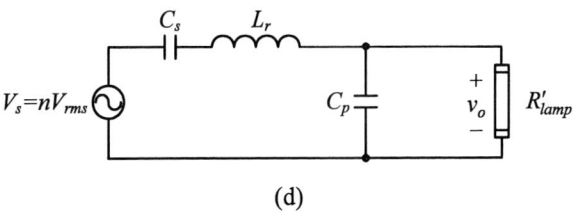

(d)

Fig. 1. (a) Full-bridge phase-shift resonant inverter powering R_{lamp}; (b) equivalent circuit simplified from (a); (c) equivalent circuit simplified from (b); (d) equivalent circuit simplified from (c).

III. ANALYSIS OF DERIVED FORMULAS

A. Behavior of CCFL during startup

The value of the equivalent resistor R_{lamp} represented for CCFL approaches to infinity during startup, so as to obtain high output voltage to drive CCFL. In this case, the operating point A shown in Fig. 2 makes the circuit operate in the series-resonant parallel-loaded characteristics with the resonant frequency

$$f_{r1} = \frac{1}{2\pi \sqrt{L_r \left(\dfrac{C_s C_p}{C_s + C_p} \right)}} \qquad (6)$$

And hence, the corresponding quality factor Q_1 is almost zero.

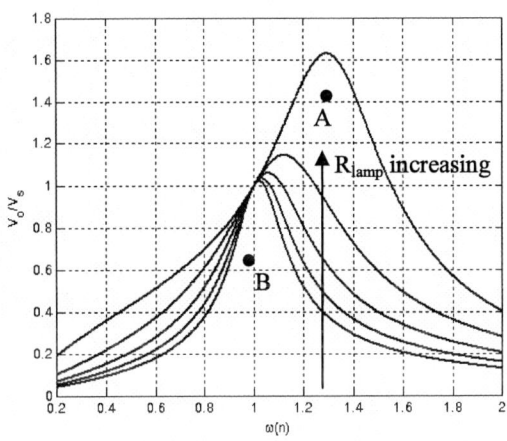

Fig. 2. Voltage gain versus frequency during startup and after startup.

B. Behavior of CCFL after startup

As the ionization phenomenon reaches stabilization, the equivalent resistance R_{lamp} is abruptly dropped and hence the effect of the parallel capacitance C_p on the circuit is decreasing. In this case, the operating point B shown in Fig. 2 makes the circuit operate in series-resonant series-loaded characteristics with the resonant frequency, to be shown as follows.

1279

$$f_{r2} = \frac{1}{2\pi\sqrt{L_r\,C_s}} \qquad (7)$$

And the corresponding quality factor Q_B is

$$Q_B = \frac{\sqrt{L_r/C_s}}{R_{lamp}} \qquad (8)$$

C. Turns ratio of the main transformer and the stabilization capacitance

As generally acknowledged, in design of the backlight module, startup should be first taken into account. Due to low voltage applied to the backlight circuit and high voltage applied to start CCFL, there is a voltage-boosting transformer between the main power stage and the load. The turns ratio n of the main transformer MT is determined as follows.

$$n \geq \frac{V_{start,rms}}{\frac{2\sqrt{2}}{\pi} \times V_{dc,min}} \qquad (9)$$

where $V_{start,rms}$ is the minimum rms value of the voltage across CCFL during startup and $V_{dc,min}$ is the minimum value of the DC input voltage.

Also, as generally acknowledged, the impedance of the stabilization capacitor C_B is about two times of the impedance of CCFL under rated conditions. That is to say, in the steady state, C_B is shown as follows.

$$C_B = \frac{I_{lamp}}{4\pi\,f_s\,V_{lamp}} \qquad (10)$$

where V_{lamp} and I_{lamp} are the rated voltage on CCFL and the rated current in CCFL respectively.

IV. PROPOSED DIMMING CONTROL

The proposed dimming control technique is based on frequency shift, as shown in Fig. 3 where irrelevant components are not shown herein. The power stage is basically controlled by one OZP6 IC [10] made by O2Micro Corp. The control of the main power stage is based on the phase shift control technique. The input voltage of the resonant circuit is controlled by the phase-shifted angle shown in Fig. 4 where v_{gs1}, v_{gs2}, v_{gs3} and v_{gs4} are gate driving signals for Q_1, Q_2, Q_3 and Q_4 respectively, thereby controlling the time of energy transfer. That's to say, the smaller the phase-shifted angle, the longer the energy transfer. The basic operation of frequency shift is based on the fact that the minimum phase shift angle is at resonant frequency f_{r2} and hence the maximum energy transfer occurs. As the frequency is shifted far from f_{r2}, the

energy transfer is reduced accordingly. The frequency after shifting is determined by the following formulas [10].

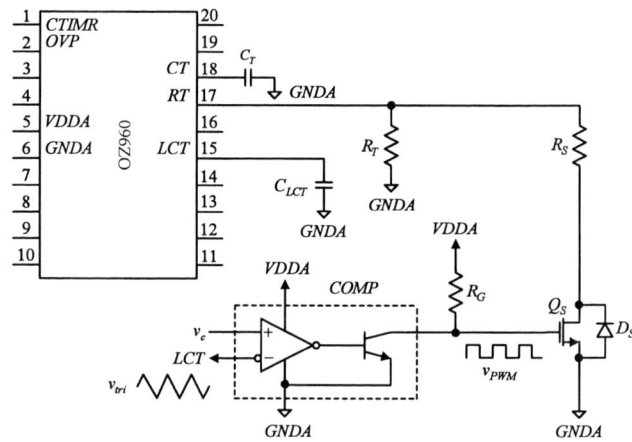

Fig. 3. Circuit for dimming control by frequency shift.

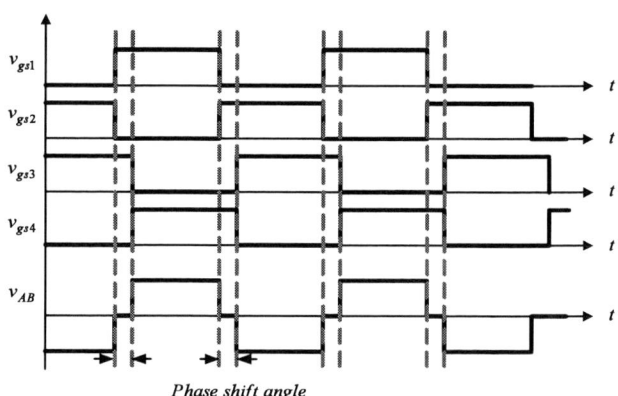

Fig. 4. Control signals for inverter switches.

$$f_s\,[kHz] = \frac{7 \times 10^4}{C_T\,(R_T\,/\!/\,R_S)} \qquad (11)$$

As shown in Fig. 3, the positive terminal of the shift resistance R_S is connected in parallel with the positive terminal of R_T which is connected to the pin RT of the OZ960 IC. Moreover, the negative terminal of R_S is connected in series with the drain of the N-MOSFET switch Q_S which is driven by the output of the comparator COMP. The inputs of COMP are the control signal v_c at the positive input terminal and the triangular wave v_{tri} at the negative input terminal, which is created from the pin LCT of the OZ960 IC by connecting one capacitor C_{LCT} between the pin LCT and the ground. Then, the duty cycle of the pulse-width-modulated (PWM) driving signal v_{PWM} from the output of COMP is used to control frequency shift based on (11). As shown in Fig. 5, the more the v_c, the larger the duty cycle of v_{PWM}, and hence the frequency of the CCFL current i_{lamp} is shifted to higher frequency than the resonant frequency f_{r2} and the corresponding value of i_{lamp} is smaller than that at f_{r2}. That's to say, the total of the energy transferred to CCFL is smaller than that at f_{r2}. And, this is why the proposed dimming technique is used without the inrush current occurring.

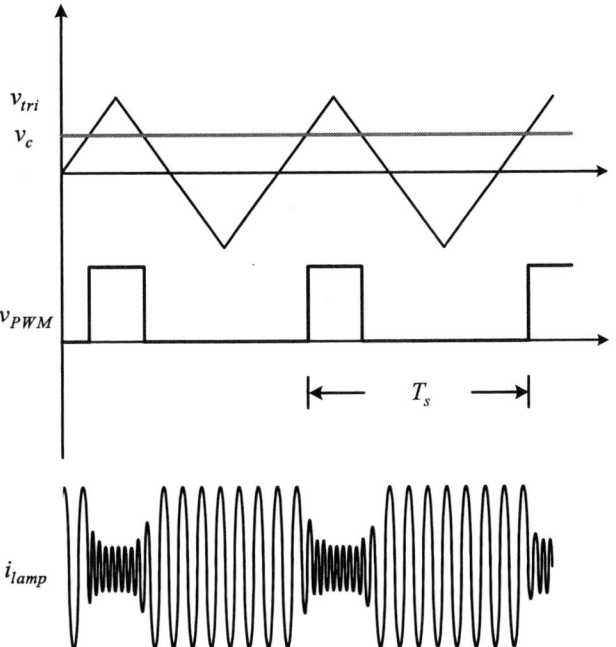

Fig. 5. Relationship between v_{PWM} and i_{lamp} due to v_c.

V. DESIGNED SPECIFICATIONS AND EXPERIMENTAL RESULTS

Before entering into this section, the specifications of CCFL are as follows. (i) rated voltage V_{lamp} and rated current I_{lamp}: $526V_{rms}$ and $5mA_{rms}$; (ii) range of start voltage V_{start}: $800\sim1200V_{rms}$; (iii) range of switching frequency f_s: $20k\sim80kHz$; and (iv) range of input voltage: $10\sim18V_{dc}$. In this case, f_s is set to 60kHz and the resonant frequency f_r is set to 50kHz. Based on the mentioned above, the key parameters such as the turns ratio n of the main transformer MT, the stabilization capacitance C_B, the blocking capacitance C_C are described as follows.

A. Turns ratio of the main transformer

Since CCFL is started by high voltage, MT needs large enough turns ratio n, which is obtained according to (9).

$$n \geq \frac{\sqrt{2}\,V_{start,rms}}{\frac{4}{\pi} \times V_{dc,min}} = \frac{\sqrt{2} \times 1200}{\frac{4}{\pi} \times 12} \cong 111 \qquad (12)$$

Finally, n is chosen to be 111.

B. Stabilization capacitance

The stabilization capacitance C_B is obtained based on (10).

$$C_B = \frac{I_{lamp}}{4\pi f_s V_{lamp}} = \frac{5 \times 10^{-3}}{4\pi \times 60k \times 526} = 12.6pF \qquad (13)$$

Eventually, C_B is selected to be 13pF.

C. Blocking capacitance

The blocking capacitor C_C is obtained based on (2) and (7).

$$C_C = \frac{n^2}{4\pi^2 L_r f_r^2} = \frac{111^2}{4\pi^2 \times 640\mu \times (50k)^2} = 1952\mu F \qquad (14)$$

where the total leakage inductance L_r is measured to be about $640\mu H$.
Finally, C_C is chosen to be $2000\mu F$.

D. Determination of parameters of the frequency-shifting circuit

In order to remove the twinkling sensitive to human eyes, the frequency of the triangular wave v_{tri} is larger than 120Hz [11], and hence, 300Hz is chosen herein. Since OZ960 IC, made by O2Micro Corp., is used as a CCFL control chip, v_{tri} is generated by connecting one capacitor C_{LCT} between the pin LCT and the ground, and the value of C_{LCT} is calculated as follows.

$$C_{LCT}[nF] = \frac{1496}{f_L[Hz]} = \frac{1496}{300} = 4.9nF \qquad (15)$$

Eventually, C_{LCT} is set to 5.6nF.

E. Experimental results

As shown in Fig. 3, C_T is set to $22\mu F$ and hence R_T is chosen to be 47kΩ based on (11) under the condition that R_T is zero. To demonstrate the proposed dimming technique, three values of R_S are chosen to be 20kΩ, 47kΩ and 150kΩ under different values of control signal v_c corresponding to various duty cycles of v_{PWM}. In Fig. 6(a) under the condition that the value of R_S is 20kΩ and the duty cycle of v_{PWM} is 50%, the related waveforms are the control signal v_c and the triangular wave v_{tri}, the PWM driving signal v_{PWM} and the CCFL current i_{lamp}. And the waveforms in Fig. 6(b) have the same conditions as the waveforms in Fig. 6(a) have except that the time scale is decreased for the convenience of observing i_{lamp} during the frequency transient. Furthermore, the waveforms shown in Figs. 7 and 8 are under the condition that the value of R_S is 47kΩ and the duty cycle of v_{PWM} is 30% and under the condition that the value of R_S is 150kΩ and the duty cycle of v_{PWM} is 20% respectively. From the experimental results mention above, it is obvious that the rms value of i_{lamp} can be adjusted by changing the amplitude of the control signal v_c or the value of R_S without any inrush current during the frequency transient, so as to dim CCFL. That is to say, the proposed control strategy can overcome the disadvantages created from analog dimming and digital dimming.

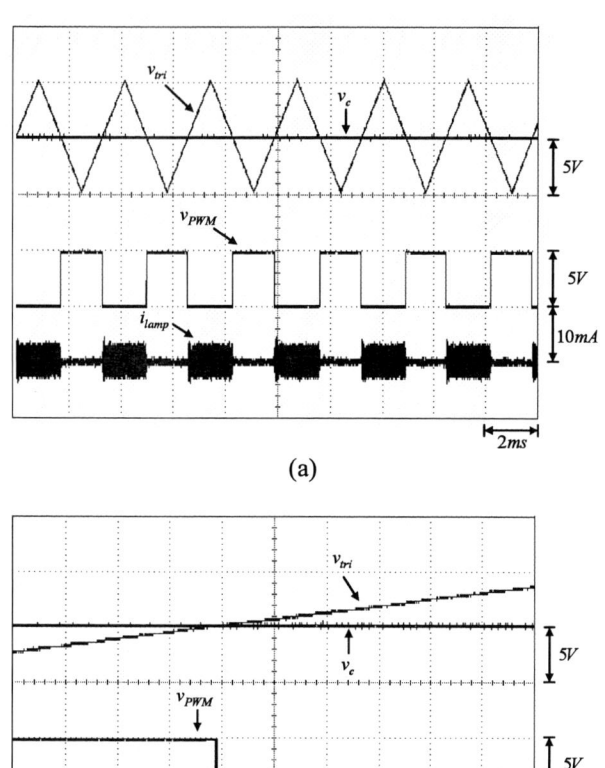

(a)

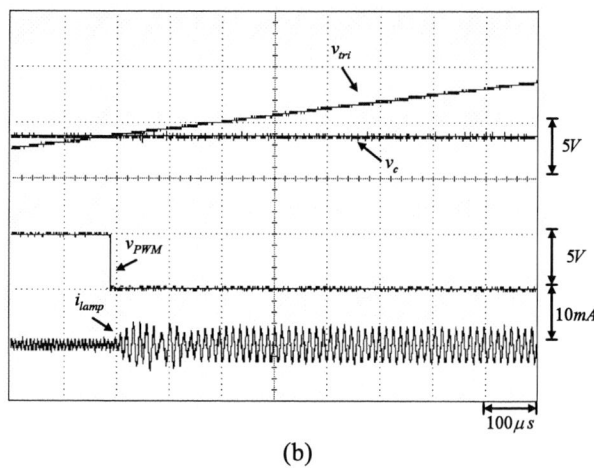

(b)

Fig. 7. (a) From the top to the bottom, waveforms for v_c and v_{tri}, v_{PWM} and i_{lamp} under $R_s = 47\text{k}\Omega$ and the duty cycle of $v_{PWM} = 0.3$; (b) waveforms enlarged from (a).

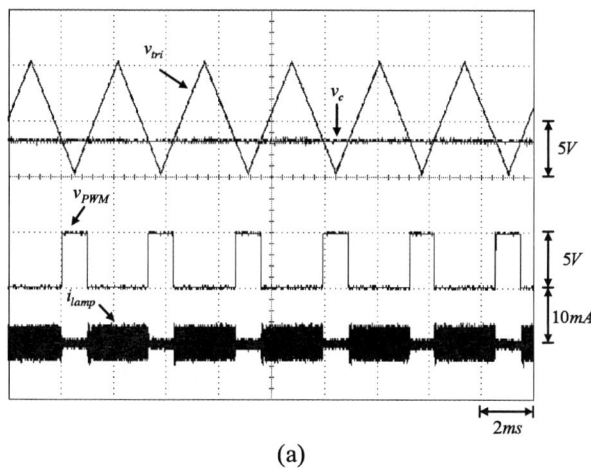

(b)

Fig. 6. (a) From the top to the bottom, waveforms for v_c and v_{tri}, v_{PWM} and i_{lamp} under $R_s = 20\text{k}\Omega$ and the duty cycle of $v_{PWM} = 0.5$; (b) waveforms enlarged from (a).

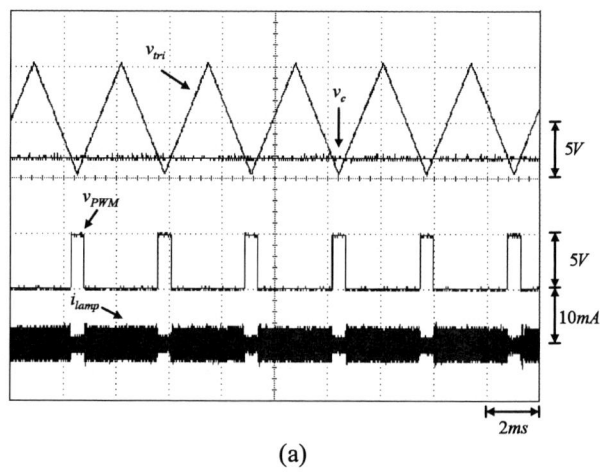

(a)

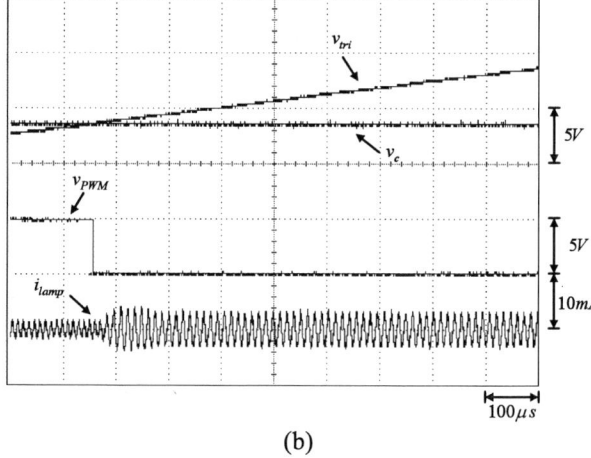

(b)

Fig. 8. (a) From the top to the bottom, waveforms for v_c and v_{tri}, v_{PWM} and i_{lamp} under $R_s = 150\text{k}\Omega$ and the duty cycle of $v_{PWM} = 0.2$; (b) waveforms enlarged from (a).

VI. CONCLUSION

In this paper, a novel dimming strategy is presented to overcome the problems existing in analog and digital dimming. Besides, this topology tends to reduce components as compared with those required in analog and digital dimming, and hence provides easy implementation and integration of the dimming circuit, thereby reducing cost significantly.

REFERENCES

[1] M. Jordan, and J. A. O' Connor, "Resonant fluorescent lamp converter provides efficient and compact solution, " *IEEE APEC93*, pp. 424-431, 1993.

[2] M. S. Lin, W. J. Ho, F. Y. Shih, D. Y. Chen and Y. P. Wu, "A cold-cathode fluorescent lamp driver circuit with synchronous primary-side dimming control," *IEEE Trans. Ind. Electron.*, vol. 45, no. 2, pp. 249-255, 1998.

[3] G. C. Hsieh, C. H. Lin, C. H. Lin and H. I. Hsieh, "Primary-side charge-pump dimming controller for the cold-cathode fluorescent lamp ballast," *IEEE TENCON'01*, vol. 2, pp. 717-723, 2001.

[4] S. W. Lee, D. Y. Ko, D. Y. Huh and Y. I. Yoo, "Simplified control technique for LCD backlight inverter systems using the mixed dimming method," *IEEE APEC'01*, pp. 447-453, 2001.

[5] F. Raiser, "Problem with lamp current using a PWM signal," *IEEE IAS'01*, vol. 1, pp. 499-503, 2001.

[6] C. H. Lin and K. J. Pai, "Differential-integral controller for the single-stage back-lighting electronic ballast," *IEEE PESC'03*, vol. 2, pp. 1000-1005, 2003.

[7] L. R. Chen, C. H. Lin, C. S. Liu, C. M. Lai and J. L. Jaw, "Phase-locked battery charge system," *IEEE INTELEC'03*, pp. 323-329, 2003.

[8] C. H. Lin and K. J. Pai, "Digital-dimming controller with current spikes elimination technique for LCD backlight electronic ballast," *IEEE APEC'04*, pp. 496-501, 2004.

[9] G. C. Hsieh and C. H. Lin," Modeling and estimation of the fluorescent lamp and its pre-heating control," *J. Light Vis. Environ.*, vol. 23, no. 1, pp. 1-9, 1999.

[10] "Data sheets of OZ960—Intelligent CCFL Inverter Controller," *O2Mirco Corp.*, 2005.

[11] Eddy Wells, "UCC3972/3BiCMOS cold cathode fluorescent lamp driver controller, evaluation board and list of materials," *Unitrode DN-102*, TI, 1999.

Time Delay Compensation For A DSP-Based Current-Source Converter Using Observer-Predictor Controller

Huu-Phuc To *, Muhammed Fazlur Rahman**, and Colin Grantham**

* Ho Chi Minh City University of Technology, Vietnam
** University of New South Wales, Sydney, Australia

Abstract– **The inevitable time delay of the digital controller can degrade the performance of a converter system. Further, this time delay may depend on both sampling and switching cycles of the converter. Therefore, with applications where the converter operates at low or medium switching frequency, the time delay can severely affect the system dynamics. Further, in controlling a current-source converter, this time delay can cause oscillation in the system. The paper proposes the use of an observer-predictor state feedback algorithm to compensate the time delay in a DSP based controller. Experimental result shows that the time delay in a DSP-based controller can have significant effect on the CSC performance. Simulation and experimental results also shows the proposed control approach can successfully compensate the time delay and provides fast dynamics to the CSC.**

Index Terms—**Current source converter, time delay, observer predictor, DSP.**

I. INTRODUCTION

Amongst two main topologies of three phase converters, voltage source converters (VSC) have been received considerably more research focus as well as practical applications than current-source converters (CSC). It may be because that the CSC is less efficient than its counterpart due to higher on-state losses in its converter bridge as well as losses in its dc-side inductor. In addition, the pulse width modulation strategies for VSC (VSC-PWM) are not only well developed, but also convenient in practice as many digital signal processors (DSP) have been equipped with the VSC-PWM generation function [1, 2].

With new advances in techniques, the above issues of CSC seems to have been solved to some extent. New reverse voltage withstanding switches such as reverse blocking IGBT [3] are now available, helping to reduce the on-state losses of the CSC. Beside, the disadvantage due to the dc-side losses in the CSC will be overcome in the near future when the superconducting coil is available with a reasonable price [4]. In terms of PWM techniques, practical approaches to convert VSC-PWM to PWM signals suitable for switching the CSC have been developed ([5, 6]). Therefore, the existing knowledge in VSC-PWM can be directly applied to the CSC. On the other hand, the current-source converter could be superior to the voltage-source converter in reliability and current waveform quality [7]. For application in power systems, the CSC is more appropriate topology to use with shunt

compensation devices, such as active power filters (APF) or Distributed Static Synchronous Compensator (DSTATCOM) as it can directly control the compensation currents.

The controller of a converter system now is often based on digital signal processor (DSP) technique as it can offer high control performance as well as flexibility in making software and design changes. In a DSP system, there is an inevitable execution time delay around one sampling cycle. An advanced DSP can use multi processors and accomplishes multiple tasks such as processing and monitoring data, executing the control algorithm and generating PWM signals to the converter. However, in this type of DSP, the time delay between the moment of sampling input data to the time of generating PWM switching signals can be longer than one sampling cycle. When the sampling process is synchronized with PWM carrier, this time delay also depends on the converter switching cycle. As a rule of thumb, the converter switching frequency should be higher than 10 to 30 times of the bandwidth of the converter current control loop so that the effect of time delay on the converter performance can be neglected [8]. However, in power application with medium switching frequency but requiring fast dynamics, such as APF or DSTACOM, this ratio may not satisfied and the system performance can be degraded.

Further, in controlling the CSC which has an oscillatory characteristics due to its resonance *LC* output filter, the time delay can result in oscillation at the converter output. Several approaches have been proposed to solve this problem. For a CSC operating as an APF in [9], the LC filter damping and the delay time compensation are combined in an open loop manner algorithm using the dynamic equation for the filter. However, the system performance with this feedforward control approach may be changed if there are some variation the actual filter parameters. In [10], the active filter CSC system is controlled with state feedback algorithm, and the time delay is modeled into state equations of the system using Pade approximation.

This paper proposes the use of an observer-predictor (OP) state feedback controller to compensate the time delay in a CSC system whose controller are fully implemented in a multi-DSP board. Simulation and experimental results on a prototype CSC operating with conventional and observer-predictor state feedback

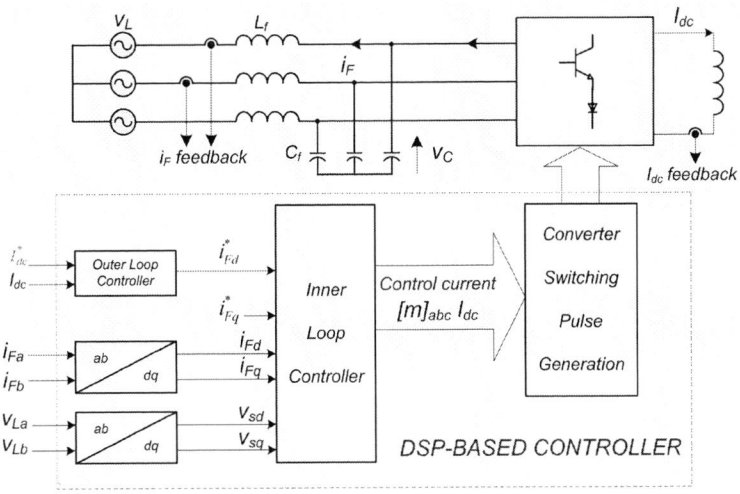

Figure 1: CSC and its controller configuration

algorithm are present to show the effect of the delay time on system response and to validate the proposed method.

II. CONFIGURATION AND CONTROL OF A CSC

A. Configuration and Model of the CSC

Fig. 1 shows the configuration of the CSC investigated in this paper. It consists of a three-phase current-source converter with an *LC* filter (L_f, C_f) at its ac-side and a dc-link inductor at its dc-side. The losses in the *LC* filter is represented by R_f. The controller of this APF is designed in cascade loop construction which has an outer loop to regulate the dc-link current I_{dc} and two inner loops to control the converter current i_F. This structure ensures fast dynamics for the converter current while maintaining I_{dc} at a desired level [11]. The inner loops operates in a *d-q* frame which synchronously rotates with the source voltage vector $\mathbf{v_L}$. The outer loop with PI controller is designed so that its response is much slower than that of the inner loops. Therefore, the coupling between the outer and inner loops is insignificant. In this case, the dc-link current can be considered as constant when investigating the inner loops.

Hereinafter, the subscripts *d* and *q* denote the *d*- and *q*-components in the synchronous *d-q* frame. Further, the superscript * represents the reference signals to the control loop. The following analysis will focus in the control of converter output current i_F which can be represented by two components: i_{Fd} and i_{Fq}.

It can be proved that the state equation for the CSC in the *d-q* frame can be given as:

$$\dot{\mathbf{x}}(t) = \mathbf{A}\mathbf{x}(t) + \mathbf{B}\mathbf{u}(t)$$
$$\mathbf{y}(t) = \mathbf{C}\mathbf{x}(t) \tag{1}$$

Where: $\mathbf{A} = \begin{bmatrix} -R_f/L_f & \omega_s & 1/L_f & 0 \\ -\omega_s & -R_f/L_f & 0 & 1/L_f \\ -1/C_f & 0 & 0 & \omega_s \\ 0 & -1/C_f & -\omega_s & 0 \end{bmatrix}$

$\mathbf{B} = \begin{bmatrix} 0 & 0 \\ 0 & 0 \\ 1/C_f & 0 \\ 0 & 1/C_f \end{bmatrix}$; $\mathbf{C} = \begin{bmatrix} 1 & 0 & 0 & 0 \\ 0 & 1 & 0 & 0 \end{bmatrix}$;

$\mathbf{x} = \begin{bmatrix} x_1 & x_2 & x_3 & x_4 \end{bmatrix}^T$: state vector (4x1),

$\mathbf{u} = \begin{bmatrix} u_1 & u_2 \end{bmatrix}^T$: control vector (2x1),

$\mathbf{y} = \begin{bmatrix} y_1 & y_2 \end{bmatrix}^T$: output vector (2x1),

And:

$x_1 = i_{Fd}$; $x_2 = i_{Fq}$; $x_3 = v_{Cd} - v_{Ld}$; $x_4 = v_{Cq} - v_{Lq}$,

$u_1 = m_d i_{dc} + C_f \omega_s v_{Lq}$; $u_2 = m_q i_{dc} - C_f \omega_s v_{Ld}$,

$y_1 = i_{Fd}$; $y_2 = i_{Fq}$,

ω_s : mains angular frequency (rad/s).

The parameters of a DSP-based controller should be designed based on the discretised model of the system. This model can be derived by integrating the continuous model in (1) over one sampling period T_s, then its form is as follows:

$$\mathbf{x}(k+1) = \mathbf{G}\mathbf{x}(k) + \mathbf{H}\mathbf{u}(k)$$
$$\mathbf{y}(k) = \mathbf{C}\mathbf{x}(k) \tag{2}$$

1285

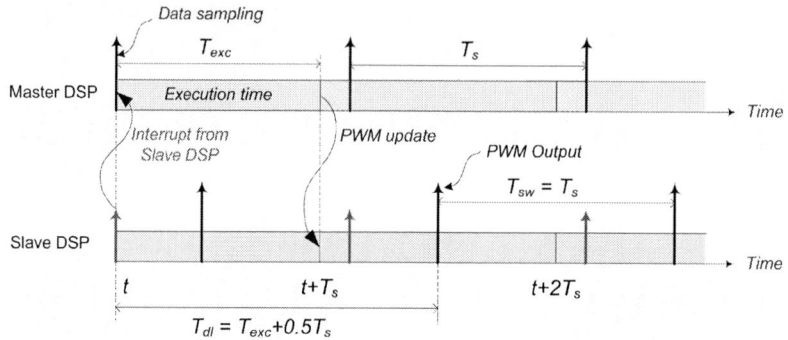

Figure 2: Time sequence of a multi DSP controller

Where: $\mathbf{G} = e^{\mathbf{A}T_s}$ and $\mathbf{H} = \left(\int_0^{T_s} e^{\mathbf{A}\lambda} d\lambda \right) \mathbf{B}$.

B. Conventional State Feedback Algorithm

For zero steady state error, the control input $\mathbf{u}$ is designed using *servo state feedback* algorithm [12] as follows:

$$\mathbf{u}(k) = -\mathbf{K_P}\mathbf{x}(k) + \mathbf{K_I}\mathbf{x_I}(k)$$
$$\mathbf{x_I}(k) = \mathbf{x_I}(k-1) + \mathbf{y}^*(k) - \mathbf{y}(k) \quad (3)$$

In (3), $\mathbf{y}^*$ is the reference input vector, $\mathbf{K_P}$ and $\mathbf{K_I}$ are the feedback gain and integrator gain, respectively, of the controller. Defining $\mathbf{x_I}$ as a new state variable results in an augmented system which is given as:

$$\begin{bmatrix} \mathbf{x}(k+1) \\ \mathbf{x_I}(k+1) \end{bmatrix} = \begin{bmatrix} \mathbf{G} & \mathbf{0} \\ -\mathbf{CG} & \mathbf{I} \end{bmatrix} \begin{bmatrix} \mathbf{x}(k) \\ \mathbf{x_I}(k) \end{bmatrix} + \begin{bmatrix} \mathbf{H} \\ -\mathbf{CH} \end{bmatrix} \mathbf{u}(k) + \begin{bmatrix} \mathbf{0} \\ \mathbf{I} \end{bmatrix} \mathbf{y}^*(k+1)$$
$$\mathbf{y}(k) = \begin{bmatrix} \mathbf{C} & \mathbf{0} \end{bmatrix} \begin{bmatrix} \mathbf{x}(k) & \mathbf{x_I}(k) \end{bmatrix}^T \quad (4)$$

The controller gains $\mathbf{K_P}$ and $\mathbf{K_I}$ can be determined using pole-placement method to allocate the desired closed-loop poles to the system in (4). These poles are selected so that the coupling between d and q current loops is eliminated as well as the required operating bandwidth of these loops can be achieved.

Further, it should be noted that at frequency higher than the resonance frequency of the LC filter ($\omega_r = 1/\sqrt{L_f C_f}$), the converter output current must be large enough to compensate the 40dB/dec. attenuation of the filter. Therefore, the designed cutoff frequency of the current loops should not exceed ω_r.

III. TIME DELAY ISSUE IN PRACTICAL CONTROLLER

A. Time Delay in a Multi-DSP Controller

In a DSP based controller, time delay is commonly estimated about one sampling period T_s. Advanced digital controller with multi-DSP board, however, can introduce higher time delay. Let us investigate the dSPACE multi-

DSP board DS1104 [13] which is used in the experimental setup of this project. This board is plugged into a personal computer (PC) and it can be programmed directly using MATLAB/SIMULINK by The MathWorks™ combined with Real-Time Interface (RTI) developed by dSPACE. The board consists of two processors: the Master DSP - a 603 PowerPC floating-point processor operating at 250MHz, and the Slave DSP - a TMS320F240 microcontroller running at 20MHz. The Master DSP executes the control algorithm, monitoring the data and interfacing with the users. The Slave DSP operates as a modulator which receives the modulation signals from the Master DSP to generate the PWM gating signals to the converter.

Fig. 2 shows timing sequence in this multi-DSP board which consists of data sampling, algorithm execution and PWM generation. As shown in this figure, the Slave DSP updates the modulation data from the Master DSP at the *middle* of its current PWM cycle, and then uses this data to generate the PWM output at the *next* switching cycle. Further, to synchronize the data sampling and processing of the Master DSP with the PWM output of the Slave DSP, an interrupt signal from the Slave DSP is used to trigger the Master DSP in every switching cycle as shown in Fig. 2. This interrupt signal is can be aligned at any position of the PWM period [13], and should be selected so that the modulation signal from the Master DSP is available before the Slave DSP can update it. Assuming that both Master and Slave DSPs are synchronized, i.e $T_s = T_{sw}$, the time delay introduced by this DSP-based controller is about:

$$T_{dl} = T_{exc} + 0.5T_s \quad (5)$$

In (5), T_{dl}, T_{exc} and T_s are the time delay, execution time of the Master DSP and sampling time, respectively. If $T_{exc} > 0.5T_s$, the time delay with this multi-DSP board is longer than one sampling cycle: $T_{dl} > T_s$.

1286

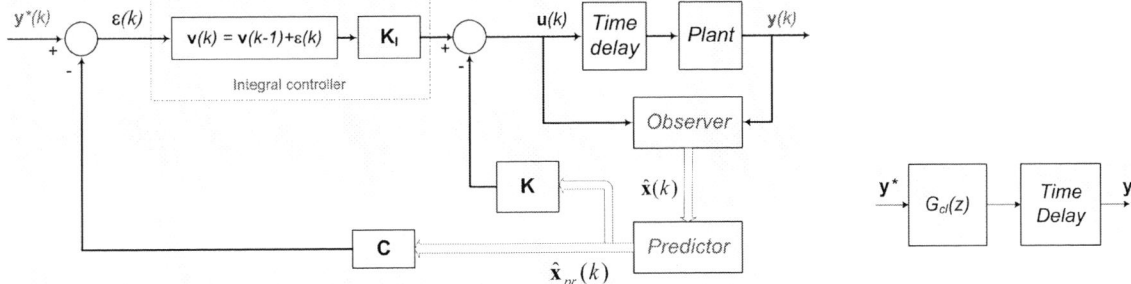

(a): Principle of the system with OP controller

(b): Equivalent block diagram
of the system with OP

Figure 3: Principle and equivalent block diagram of the system with OP controller

B. Effect of Time Delay

When there is a time delay T_{dl} in the system controller as mentioned in the above, the state equations (1) can be rewritten as follows:

$$\dot{\mathbf{x}}(t) = \mathbf{A}\mathbf{x}(t) + \mathbf{B}u(t - T_{dl})$$
$$y(t) = \mathbf{C}\mathbf{x}(t)$$
(6)

In this case, if the control input u is still generated using (3), the dynamic performance of the system can be degraded, even unstable. To overcome this problem, the desired bandwidth of the current loop must be reduced, however, this will also reduce the dynamic performance of the converter system.

IV. OBSERVER-PREDICTOR TO COMPENSATE THE TIME DELAY

An observer-predictor controller [14], which is the modification of Smith predictor, can be used to compensate the time delay in such a system. Based on this algorithm, the authors proposed the configuration of OP state feedback controller as shown in Fig. 3. With this OP controller, the control input **u** in discrete time domain is given by:

$$\mathbf{u}(k) = -\mathbf{K}_p \hat{\mathbf{x}}_{\mathbf{pr}}(k) + \mathbf{K}_I \mathbf{x}_I(k)$$
$$\mathbf{x}_I(k) = \mathbf{x}_I(k-1) + \mathbf{y}^*(k) - \mathbf{C}\hat{\mathbf{x}}_{\mathbf{pr}}(k)$$
(7)

In (6), $\hat{\mathbf{x}}_{\mathbf{pr}}(k)$ is the predicted state vector produced by the predictor, which forecast the state vector **x** one T_{dl} ahead. If $T_{dl} > T_s$, this predicted state vector is calculated as follows:

$$\hat{\mathbf{x}}_{\mathbf{pr}}(k) = \mathbf{G}_{pr}\mathbf{x}(k) + \mathbf{H}_{pr2}\mathbf{u}(k-2) + \mathbf{H}_{pr1}\mathbf{u}(k-1)$$
(8)

In (8), $\mathbf{G}_{pr}$, $\mathbf{H}_{pr2}$ and $\mathbf{H}_{pr1}$ are obtained by integrating (6) over one time delay T_{dl}, and their expressions are given as follows:

$$\mathbf{G}_{pr} = e^{\mathbf{A}T_{dl}},$$

$$\mathbf{H}_{pr1} = \left(\int_0^{T_1} e^{\mathbf{A}\lambda}d\lambda\right)\mathbf{B}, \qquad \mathbf{H}_{pr2} = \left(e^{\mathbf{A}T_1}\int_0^{T_s} e^{\mathbf{A}\lambda}d\lambda\right)\mathbf{B}.$$

Where: $T_1 = T_{dl} - T_s$.

With the OP control algorithm, the time delay is eliminated from the control loop and the equivalent of the closed loop system can be considered as shown in Fig. 3b [14]. In this case, the time delay does *not* affect the dynamic of the closed loop system G_{cl} and it just results in a time delay between the reference input y^* and the actual output y. If necessary, this delay then can be compensated by some predictions applied to y^*. Further, it should also be noted that the controller gains **K** and $\mathbf{K}_I$ in (7) are designed *as if the delay time does not exist in the system control loop*, i.e they can be determined based on the state equation in (4).

V. SIMULATION AND EXPERIMENTAL RESULTS

To validate the proposed controller, an experimental system with a prototype CSC built in our lab. The *LC* filter of this converter is: $L_f = 1\text{mH}$ and $C_f = 22\mu\text{F}$. The controller for the CSC is based on the DSP board DS1104 mentioned in section 2 above. The sampling cycle T_s of the Master DSP and the switching cycle T_{sw} of the Slave DSP are the same as they are synchronized, $T_s = T_{sw}$. The sampling cycle is selected at $T_s = 148\mu\text{s}$ so that it is higher than the execution time needed for the main program ($\approx 110\mu\text{s}$) plus the time required to transfer the data from the Master to the Slave DSP (≈ 15 to $20\mu\text{s}$). Due to the DSP operating sequence mentioned in section III, the time delay of the controller is about $1.4T_s$, thus $T_{dl} = 207\mu\text{s}$. The controller gains **K** and $\mathbf{K}_I$ are designed so that the bandwidth of the inner current loop is around LC resonance frequency $\omega_r = rad/s$.

It should be noted that the PWM signals generated by the Slave DSP of the DS1104 is only dedicated for gating VSC. However, in the experimental setup, an algorithm to convert this VSC-PWM to CSC-PWM signals has been developed and implemented in the converter gating circuits [5].

1287

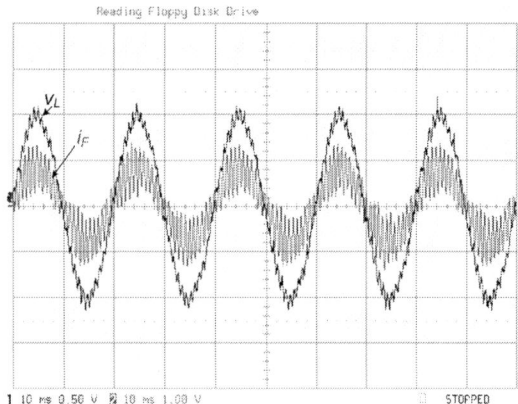

Figure 4: CSC performance with conventional algorithm
(v_L: 25V/div; i_F: 10A/div)

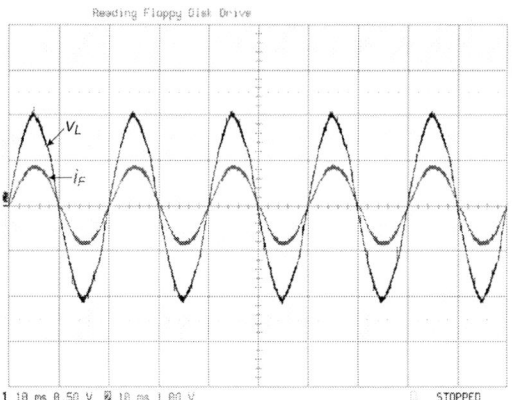

Figure 5: CSC performance with OP state feedback algorithm
(v_L: 25V/div; i_F: 10A/div)

The simulation of this CSC system is also conducted using MATLAB/SIMULINK. The discrete transfer function and appropriate delay blocks are used in the controller model to simulate the digital characteristics of the controller.

To illustrate the affect of the time delay to the CSC performance and the effectiveness of the proposed OP control algorithm, experiments are conducted on the CSC operating in rectifier mode, in which the dc-link current is maintained at 20A. and the converter current i_F is controlled to be in phase with the supply voltage v_L. The CSC is tested with both conventional and OP state feedback algorithm.

Fig. 4 shows the experimental results of converter current and supply voltage when the conventional control algorithm is applied. As the waveforms of three phases are similar, only the current and voltage of one phase are shown in this figure. It can be seen that the converter current badly oscillates because the time delay of the controller is not taken into account. This oscillation can

be eliminated if the bandwidth of current loop is reduced, however, this will also reduce the dynamics of the CSC. On the other hand, as shown in Fig. 5, when the OP control algorithm is used, the oscillation in the converter current is eliminated and the CSC operates properly.

To study dynamics of the CSC, simulation and experiment are carried out to investigate the response of converter output current to a step change of reactive input current. Fig. 6 shown the simulation of the converter current when the reactive reference current i_{Fq}^* changes from -5A to 5A and the dc-link current is maintained at 20A. As shown in this figure, the actual reactive current i_{Fq} fast tracks the reference i_{Fq}^* (Fig. 6b), resulting in the rapid change of the converter current from leading to lagging (Fig. 6a). Fig. 7 shows the experimental result of the converter current i_F due to this step change. It can be seen that this result is matched well with the simulation results in Fig. 6a.

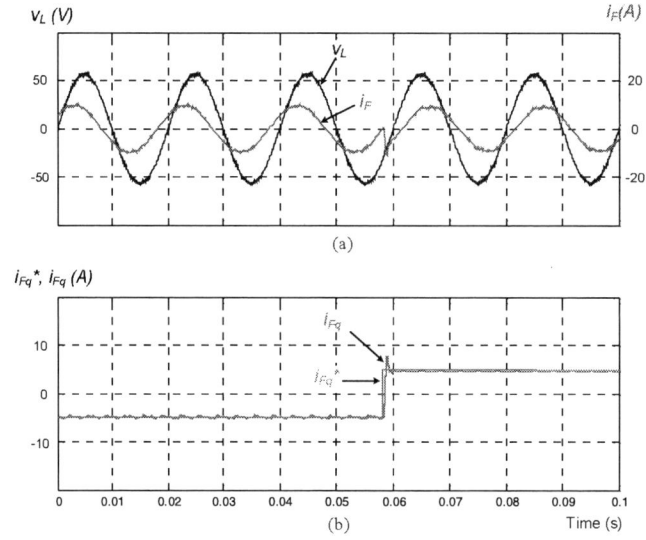

Figure 6: Simulation results of CSC response to step change of reactive current
(a): Response of CSC output current i_F, (b): Response of i_{Fq} to a step change of i_{Fq}^*

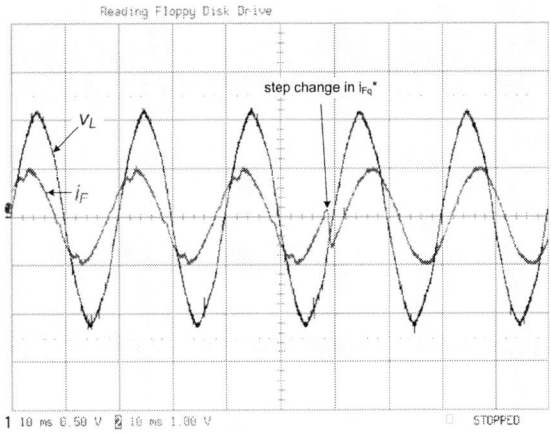

Figure 7: Experimental results of CSC response to step change of reactive current
(v_L: 25V/div; i_F: 10A/div)

VI. CONCLUSION

The paper presented an observer-predictor state feedback controller to compensate the time delay of the digital controller for a CSC. Experimental result shows that the time delay in a DSP-based controller can have significant effect on the CSC performance. Simulation and experimental results also shows the proposed control approach can successfully compensate the time delay and provides fast dynamics to the CSC.

REFERENCES

[1] J. Doval-Gandoy, A. Iglesias, C. Castro, and C. M. Penalver, "Three alternatives for implementing space vector modulation with the DSP TMS320F240," presented at IECON '99, San Jose, CA.

[2] *DS1104 R&D Controller Board - Hardware installation and configuration*: dSPACE Gmbh, 2004.

[3] T. Naito, M. Takei, M. Nemoto, T. Hayashi, and K. Ueno, "1200V reverse blocking IGBT with low loss for matrix converter," presented at ISPSD '04

[4] A. P. Malozemoff, J. Maguire, B. Gamble, and S. Kalsi, "Power applications of high-temperature superconductors: status and perspectives," *IEEE Transactions on Applied Superconductivity*, vol. 12, pp. 778 - 781, 2002.

[5] H. P To, M. F. Rahman, and C. Grantham, "Modulation of Current-Source Converters from Gating Signals for Voltage-Source Converters," presented at PESC 2006.

[6] J. Rey, J. Doval-Gandoy, F. Sanchez, C. M. Penalver, O. Lopez, and A. Nogueiras, "Converting a VSC modulation strategy for controlling a CSC used in FACTS," presented at IECON 2005.

[7] M. Routimo, M. Salo, and H. Tuusa, "Comparison of Voltage-Source and Current-Source Shunt Active Power Filters," presented at PESC 2005.

[8] V. Blasko, V. Kaura, and W. Niewiadomski, "Sampling of discontinuous voltage and current signals in electrical drives: a system approach," *IEEE Transactions on Industry Applications*, vol. 34, pp. 1123 - 1130, 1998.

[9] M. Salo and H. Tuusa, "A novel open-loop control method for a current-source active power filter," *IEEE Transactions on Industrial Electronics*, vol. 50, pp. 313-321, 2003.

[10] H. P. To, M. F. Rahman, and C. Grantham, "Time delay compensation for a current-source active power filter using state-feedback controller," presented at IEEE-IAS 2005.

[11] S. Dong and P. W. Lehn, "Modeling, analysis, and control of a current source inverter-based STATCOM," *IEEE Transactions on Power Delivery*, vol. 17, pp. 248, 2002.

[12] K. Ogata, *Discrete-time control system*: Prentice Hall International, London, 1987.

[13] *DS1104 R&D Controller Board - Feature Reference*: dSPACE GmbH, 2004.

[14] *The control handbook*: CRC Press, 1996.

Implementation of Hysteresis Current Control for Single-Phase Grid Connected Inverter

Krismadinata, Nasrudin Abd Rahim *Member IEEE*, and Jeyraj Selvaraj
Power Electronics laboratory Department of Electrical Engineering
University of Malaya Kuala Lumpur, Malaysia
krismadinata@um.edu.my

Abstract— **This paper describes a control method for single-phase grid-connected inverter system for distributed generation application. Single-band Hysteresis Current Controller is applied as the control method. The control algorithm is implemented in Digital Signal Processor (DSP) TMS320F2812. The control method provides robust current regulation and achieve unity power factor. Simulation and experimental results are provided to demonstrate the effectiveness of the design.**

Keywords: Single-Phase Grid Connected Inverter, Hysteresis Current Control and TMS320F2812

I. INTRODUCTION

Distributed energy offers solutions to many of the nation's most pressing energy and electric power problems, including blackouts and brownouts, energy security concerns, power quality issues, tighter emissions standards, transmission bottlenecks, and the desire for greater control over energy costs.

Distributed generation, also called decentralized energy or distributed energy, is a method of generating electricity from numerous small sources, such as photovoltaic, fuel cell, wind turbine etc that close to the point of use. Distributed generation reduces the amount of energy lost in transmitting electricity because the electricity is usually generated close to the place that it is used. It also reduces the number of power lines that have to be constructed.[1]

Photovoltaic as one of distributed energies need to convert to AC source in order to connect to the grid. Inverter is used to convert DC to AC. For inverter-based distributed generation systems (DG), the inverters are connected to the existing grid; therefore, the voltage cannot be controlled. The power quality is defined by the current quality. Pulse width modulation (PWM) is the most popular control technique in voltage-source inverters. As compared to the open loop voltage PWM converters, the current-controlled PWM has several advantages. One of the advantages is control of instantaneous current waveform and high accuracy. This advantage can control precisely the current injected into grid with low distortion and harmonic noise.

The strategies of current controllers can be classified as: ramp comparison controllers, predictive controllers, and hysteresis controllers. The ramp comparison controller compares the current errors to triangle wave to generate the inverter firing signals. The predictive controllers calculate the inverter voltages required to force the measured currents to follow the current reference. The hysteresis controllers utilize some type of hysteresis in the comparison of the currents to the current reference. [2]

Among the various PWM techniques, the hysteresis band current control is used very often because of its simplicity of implementation. Also, besides fast response current loop, the method does not need any knowledge of load parameters.

There are many researches in implementation hystersis current control, for example in [3]-[6], but many have yet to implement together with the grid.

This paper will present analysis and implementation of single-band hysteresis current control for single-phase grid connected inverter. The control algorithm is implemented in DSP TMS320F2812.

II. SINGLE-PHASE GRID CONNECTED INVERTER

Single-phase grid connected inverter is similar with single-phase inverters that are widely used in industrial applications such as induction heating, standby power supplies and uninterruptible supplies, however the output of the inverter is directly connected to grid. The inverter consists of four switching devices (represented as ideal switches) connected in the form of a bridge configuration.

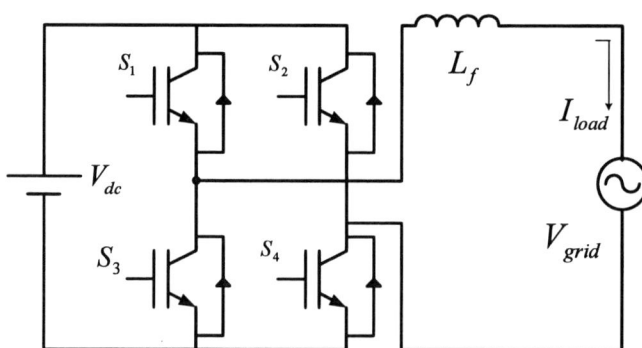

Fig. 1 Single-phase grid-connected inverter

The single-phase grid connected inverter topology is shown in Fig. 1. It is composed of a dc voltage source (V_{dc}), four power switches (S_1-S_4), a filter inductor (L_f) and utility grid (V_{grid}).

In inverter-based DG, the produced voltage from inverter must be higher than the V_{grid}. It is required to assure power flow to grid. Since V_{grid} is uncontrollable, the only way of controlling the operation of the system is by controlling the current that is following into the grid.

III. HYSTERESIS CURRENT CONTROL

Hysteresis current control is one of the easiest control methods to implement. It is simple to implement and has robust current control performance against load and source parameter changes [3]. Hysteresis current control is a method of controlling a voltage source inverter where the measured current is compared to reference current on instantaneous basis. The current error is then compared directly against a predefined band called hysteresis band to produce switching pulses for the voltage source inverter. This method controls the switches in an inverter asynchronously to ramp the current through an inductor up and down so that it tracks a reference current signal.

The implementation of the proposed single band hysteresis current control for single-phase grid connected inverter is shown in Fig. 2. In the hysteresis modulator, the current error signal δ is compared with the hysteresis band, as shown in Fig. 3. When δ crosses the upper boundary, +h, the switches S_1 and S_4 are turned on and S_2 and S_3 are turned off. When δ crosses the lower boundary, -h the switches S_1 and S_4 are turn off while S_2 and S_3 is turned on (here delays and dead times are neglected).

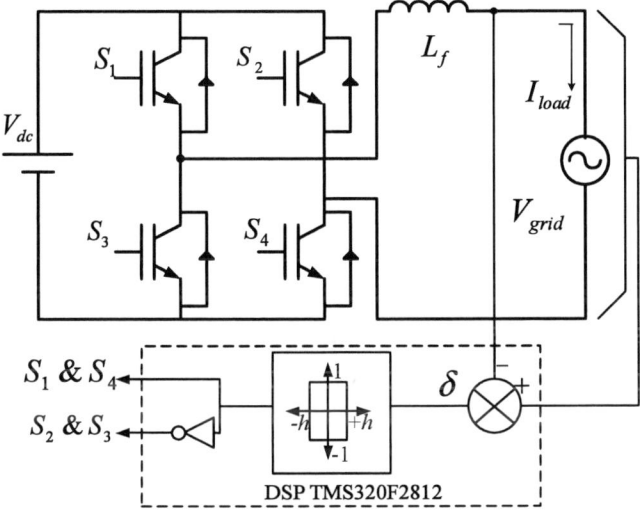

Fig.2 Proposed hystresis current control
for single-phase grid connected inverter

In this proposed hysteresis current controller, the reference signal is sensed directly from the grid. First, the grid voltage is sensed and adjusted to a desired value before converting it to current signal to become the reference current signal. This will ensure the current produced by inverter-based DG is in phase with grid voltage and also achieve unity power factor. This method is robust and effective than conventional reference signal generation by the controller and matching it with the grid voltage at later stage. This method also reduces the number of components such as Phase Lock Loop (PLL) circuits and cost significantly.

By referring to Fig. 3, when switches S_1 and S_4 receive ON signal, for instance from t_1 to t_2 and neglecting the voltage drop across internal resistance of the inductor, the following output voltage equation can be obtained:

$$V_{dc} = L_f \frac{di_0}{dt} + V_{grid} \qquad (1)$$

The output current can be considered as the reference current plus the error component, that is,

$$i_0 = i_r + \delta \qquad (2)$$

Where, δ is the error component of the output current. Substituting equation (2) in equation (1), the following can be obtained:

$$V_{dc} = L_f \frac{d}{dt}(i_r + \delta) + V_{grid} \qquad (3)$$

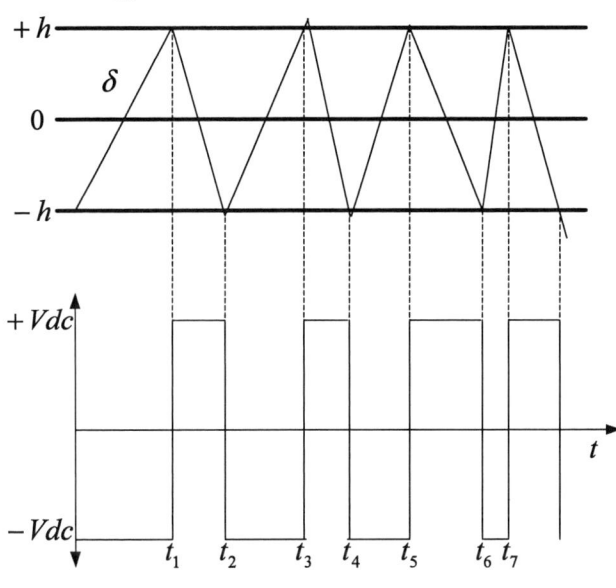

Fig.3 Hysteresis modulator

By rearranging equation (3),

$$V_{dc} - V_{grid} = L_f \frac{d}{dt}(i_r + \delta) \qquad (4)$$

For dynamic condition:

$$V_{dc} - V_{grid} = L_f \frac{d\delta}{dt} \qquad (5)$$

So perturbation error can be written as:

$$\frac{d\delta}{dt} = \frac{V_{dc} - V_{grid}}{L_f} \qquad (6)$$

During the interval when switches S_1 and S_4 is ON (T_{ON} time), the error current changes from –h to +h. thus, the ON time (t_1 to t_2) can be calculated as:

$$T_{ON} = \frac{2L_f h}{V_{dc} - V_{grid}} \qquad (7)$$

By using a similar method, the interval when switches S_2 and S_3 receive ON signals (T_{OFF} time, from t_2 to t_3) can be obtained as:

$$T_{OFF} = \frac{2L_f h}{V_{dc} + V_{grid}} \qquad (8)$$

$$T_s = T_{ON} + T_{OFF} = \frac{4V_{dc}L_f h}{(V_{dc} + V_{grid})(V_{dc} - V_{grid})} \qquad (9)$$

Thus, the switching frequency is:

$$f_s = 1/T_s = \frac{(V_{dc} + V_{grid})(V_{dc} - V_{grid})}{4V_{dc}L_f h} \qquad (10)$$

$$f_s = \frac{(V^2_{dc} - V^2_{grid})}{4V_{dc}L_f h} \qquad (11)$$

By dividing the nominator and denominator of equation (11) by V_{dc},

$$f_s = \frac{V_{dc}(1 - \frac{V^2_{grid}}{V^2_{dc}})}{4L_f h} \qquad (12)$$

Where, $V_{grid} = V_m \sin \omega_e t$, $\qquad (13)$

and $k = \dfrac{V_m}{V_{dc}}$ $\qquad (14)$

$$f_s = \frac{V_{dc}(1 - k^2 \sin^2 \omega_e t)}{4L_f h} \qquad (15)$$

k can be considered as the modulation index. The maximum switching frequency is:

$$f_{s,\max} = \frac{V_{dc}}{4L_f h} \qquad (16)$$

Thus, the maximum switching frequency varies with the dc input voltage, load inductance, and the hysteresis band. The average switching frequency can be obtained as:

$$f_{s,av} = \frac{V_{dc}}{4L_f h}\left(1 - \frac{k^2}{2}\right) \qquad (17)$$

Because the possibility of inverter to produce zero output voltage level is not used, the output voltage pattern is bipolar. Though this controller is simple, the switching devices will be switched at very high frequencies if a good output current waveform is desired. [6]

Control algorithm of single-band hysteresis current control for single-phase grid connected inverter is shown in Fig. 4 and this algorithm is applied in the DSP. Fig.5 shows block diagram of experiment setup.

V_{grid} and I_{load} were sensed and sampled with sampling frequency of 100 kHz. Sampling frequency is set at a high frequency in order to obtain good and accurate result.

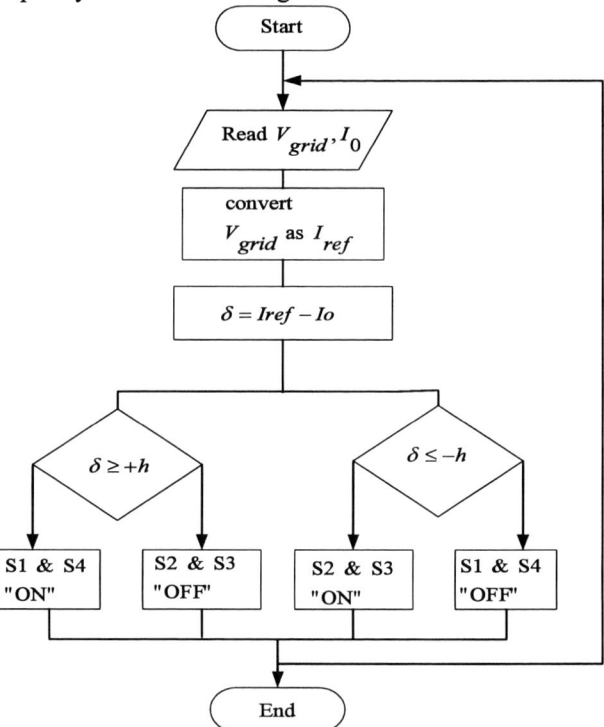

Fig.4 Flowchart single-band hysteresis current control

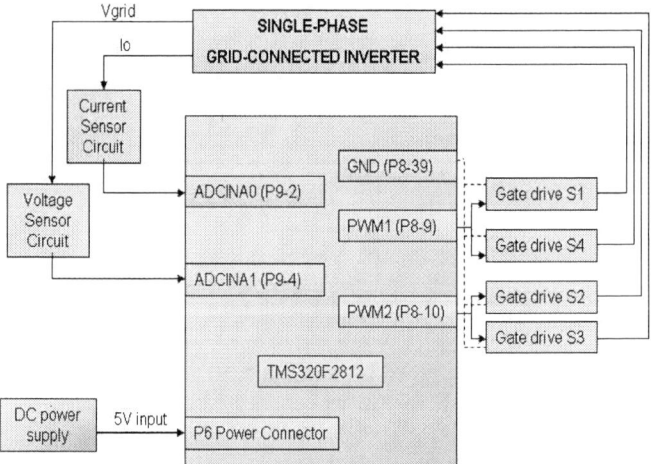

Fig. 5 Block diagram of experiment setup

VI. SIMULATION RESULT

Simulations have been conducted for a single-phase grid-connected inverter using PSIM power electronic simulation software. In the simulations, DC voltage source is 400V, the grid voltage is 240V and the grid frequency is 50Hz; the output filter inductance is 5mH. Fig.6 shows the simulation diagram of the proposed inverter with using PSIM. Error signal from comparison between reference current and actual current is shown in Fig.7. From this figure, it can be seen that the error signal is crossing upper and lower hysteresis band.

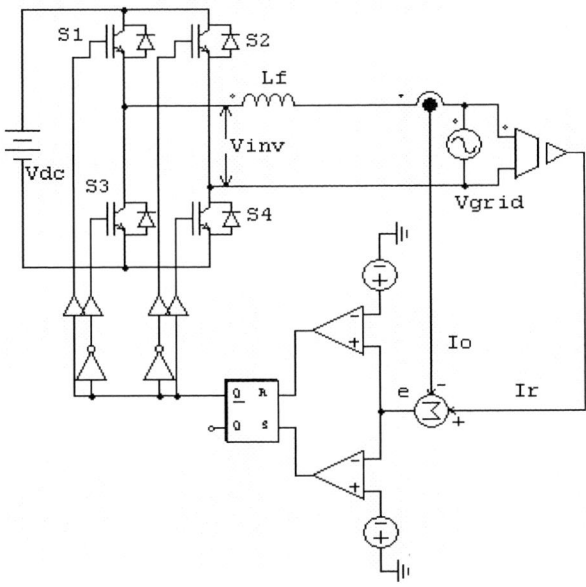

Fig. 6 Diagram simulation single-phase grid-connected inverter with PSIM

Switching frequency pattern resulting from the algorithm is shown in Fig. 8, while inverter output voltage before inductor or filter side is shown in Fig. 9. The output voltage is switched from plus to minus of DC source and vice-versa. This switching frequency pattern is called bipolar switching.

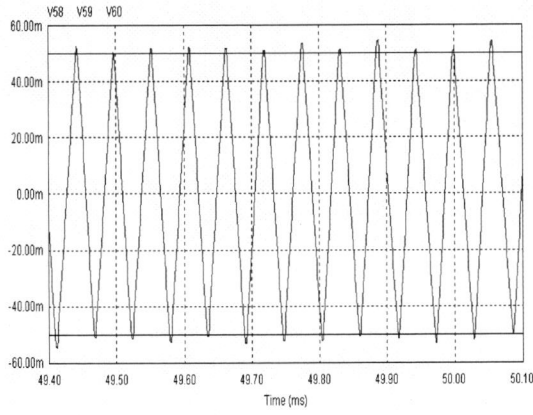

Fig. 7 Error signal and hysteresis band

This inverter is designed for 3.4kW. The injected root mean square (RMS) current to grid is 14A. The output current waveform under the hysteresis current control is shown in Fig. 10. In this simulation $+h$ and $-h$ is set as $+0.05$ and -0.05 respectively.

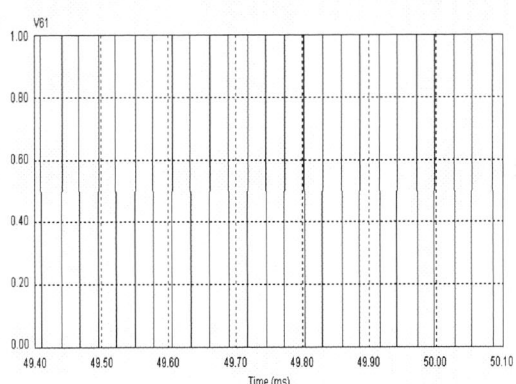

Fig.8. Switching pattern of hysteresis current control

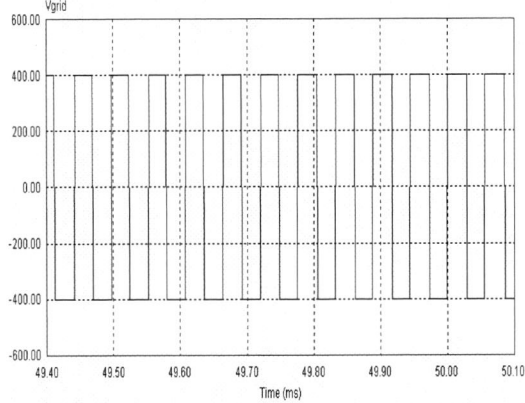

Fig.9 Inverter output voltage before inductor

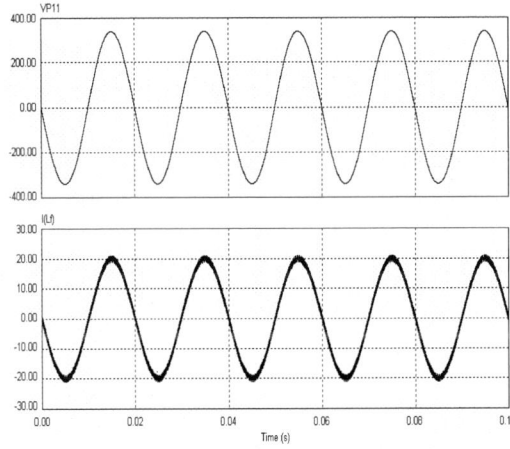

Fig.10 Simulation result Grid voltage and Grid current

Fig. 10 shows the waveform of V_{grid} and waveform of injected current to grid I_{load}. Waveform of produced current is at unity power factor with grid voltage. This can be shown in Fig. 11. The scale of current waveform is enlarged to be 5 times.

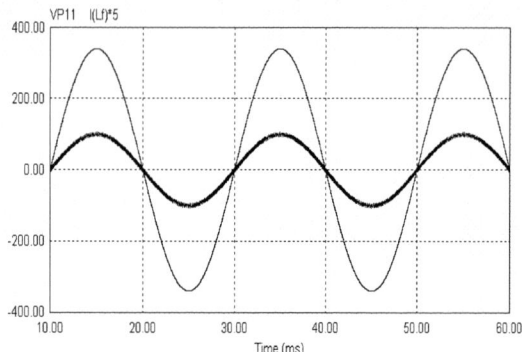

Fig. 11 Simulation result Grid voltage
and Grid current at unity power factor

VI. EXPERIMENT RESULT

The control algorithm discussed so far has been implemented and tested experimentally on a single-phase grid connected inverter. The inverter's parameters are equal to parameters in the simulation. IRG4PH50UD is used as switches and 5mH inductor as filter.

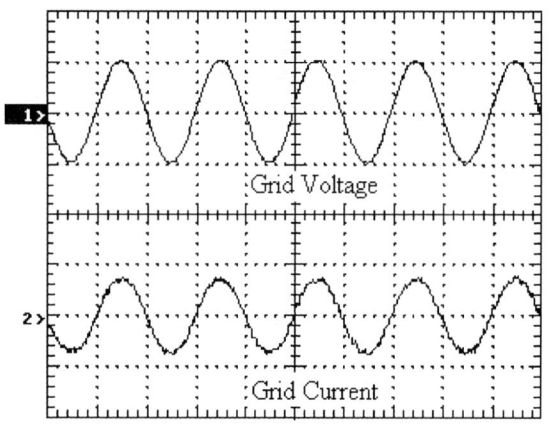

Fig. 12 Experiment result Grid voltage and Grid current

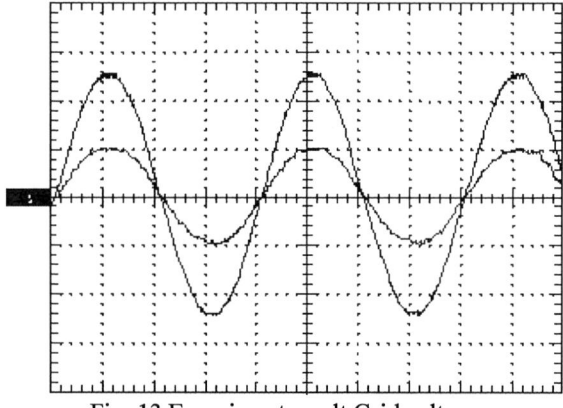

Fig. 13 Experiment result Grid voltage
and Grid current at unity power factor

As far as the controller implementation is concerned, a fixed-point 32-bit DSP (TMS320F2812) has been used. Its

peripheral units allow a straightforward implementation of the control strategy; in fact, being conceived for the control of electrical drives. The chip comprises capture and compare programmable I/O pins, counters and, on the system's board, digital-to-analog converters (DAC's).

Fig. 12 shows experiment result for produced current while Fig. 13 shows the combined waveform of the current and grid voltage to show that the system is operating at unity power factor. The experiment results are similar with the simulation result to show that the hysteresis current control can be implemented with grid connected inverter.

VII. CONCLUSION

This paper presents a single-band hysteresis current control for single-phase grid connected inverter. The effectiveness of the control scheme has been verified both by simulation and experimentally. The current produced by this inverter is in phase with grid voltage and also achieve unity power factor. This method is robust and effective than conventional reference signal generation by the controller and matching it with the grid voltage at later stage. This method also reduces the number of components such as Phase Lock Loop (PLL) circuits and cost significantly.

VIII. ACKNOWLEDGMENT

This work has been supported by United Nations Development Programme-Global Environment Facility, Pusat Tenaga Malaysia and University of Malaya, for Malaysia Building Integrated Photovoltaic Technology Application Project, The authors would like to thank all the people who are concern.

IX. REFERENCES

[1] http://en.wikipedia.org/wiki/Distributed_generation
[2] Kojabadi, et. al, "A Novel DSP-Based Current-Controlled PWM Strategy for Single Phase Grid Connected Inverters", IEEE Trans. on Power Electronics, Vol. 21, No.4, July 2000, pp. 985-993
[3] Bor-Jehng Kang and Chang-Ming Liaw, "A Robust Hysteresis Current-Controlled PWM Inverter for Linear PMSM Driven Magnetic Suspended Positioning System". IEEE Trans. on Ind. Electronics, Vol. 48, No. 5, October 2001, pp 956-967
[4] S. Buso, S. Fasolo, L. Malesani, and P. Mattavelli, "A Dead-Beat Adaptive Hysteresis Current Control" IEEE Trans. on Ind. Applications, Vol. 36, No.4 July/August. 2000, pp 1174-1180
[5] P.A. Dahono and I. Krisbiantoro, "A Hystersis Current Controller for Single-Phase Full-bridge Inverter". Int. Conf. on Power Electronics and Electric Drives 2001, Indonesia, pp 415-419
[6] P.A. Dahono, "New Current Controllers for Single-Phase Full-Bridge Inverters", 2004 int. Conf. on Power System Technology - Singapore, November 2004, pp 1757- 1762

Use of Air-Cored Axial Flux Permanent Magnet Generator in Direct Battery Charging Wind Energy Systems

F.G. Rossouw and M.J. Kamper

Department of Electrical and Electronic Engineering, University of Stellenbosch, South Africa

Abstract—**Low power, low cost wind energy systems make use of uncontrolled, direct battery charging where the generator is directly connected to the battery via a diode rectifier. Air-cored axial flux permanent magnet (AFPM) generators, although excellently suited for small wind energy systems, can not be used in direct battery charging systems as optimum turbine power matching is not obtained and non-sinusoidal currents are drawn due to the rectifier; with non-sinusoidal currents these generators are shown to become extremely noisy. In this paper a method is analysed whereby better maximum power matching and sinusoidal currents are obtained using AFPM generators in direct battery charging wind energy systems.**

Index Terms—**Axial flux permanent magnet, Direct battery charging, Wind generator.**

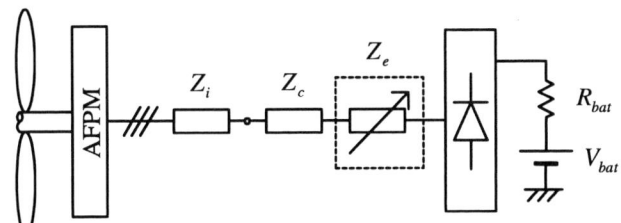

Fig. 1. Schematic diagram of the direct wind generator battery charging system.

I. Introduction

In low power stand-alone wind energy systems direct uncontrolled battery charging is often used due to its simplicity, reliability and lower cost. In these systems the wind generator is directly connected to the battery bank via a cable and diode rectifier as shown in Fig. 1. The main disadvantage of direct battery charging is that maximum power from the turbine under variable wind speed conditions is not always obtained and is largely affected by the internal impedance of the generator, Z_i, and to some extent the impedance of the cable, Z_c. The power obtained depends, thus, very much on the design and the characteristic-impedance of the generator. To compensate for the effect of the system's impedance, an external series impedance, Z_e, such as a series capacitor [1], is sometimes added as shown in Fig. 1. To obtain better power matching in small wind energy systems passive pitch-control [2], using flexible turbine blades, can be used together with controllable power converters [3]–[5] whereby the turbine-generator-cable system is voltage-decoupled from the battery-load system.

Axial flux permanent magnet (AFPM) machines showed to be ideally suited for small wind energy systems [6]. Furthermore, the use of air-cored stators in small AFPM wind generators is beneficial from the point of no iron losses, zero cogging torque, no punched laminations, easy construction and fast on-site replacement of faulty stator windings, amongst other things [7], [8]. An example of the construction of an air-cored stator AFPM machine is shown in Fig. 2; the outer PM rotor disks are interconnected and the inner stator coils are held in position by epoxy resin. The characteristic property of an air-cored AFPM machine is its large effective air gap, resulting in an extremely low internal impedance. Using the low-impedance air-cored AFPM generator in a direct battery charging system causes (i) non-optimal turbine power matching and (ii) non-sinusoidal line currents due to the diode rectifier. These harmonic line currents cause the AFPM generator to emit severe acoustic noise under load conditions, as shown in this paper. The noise can be rather annoying to persons and livestock in the vicinity of the operating wind generator. The typical non-optimal turbine power matching of an air-cored AFPM-generator in a direct battery charging system is shown by the solid line in Fig. 3. Fairly good turbine power matching is obtained at high wind speeds, but no power at all is generated at low wind speeds. The ideal would be that the generator follow the optimum power points of the wind, as indicated by the dotted peak power points.

To provide a more accurate turbine power matching and to sufficiently reduce the harmonic line current components of an air-cored AFPM-generator direct battery charging system, the effects of adding external series inductors L_e to the ac-side of the rectifier in Fig. 1 is investigated in this paper. The theoretical analysis and simulation results are presented in section II and III, followed by harmonic line current analysis and acoustic noise measurements in section IV.

978-1-4244-0644-9/07/$25.00 ©2007 IEEE

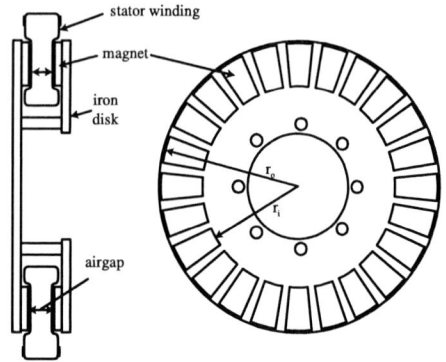

Fig. 2.   Cross-sectional schematic of an AFPM machine.

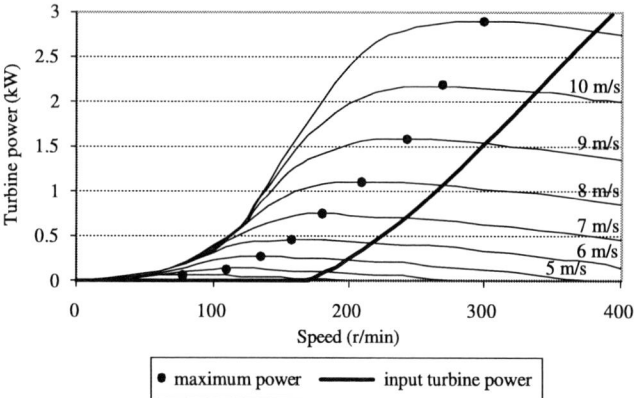

Fig. 3.   Wind turbine power versus turbine speed and power matching of a direct battery charging system.

II. Theoretical Analysis

Referring to the optimum power points in Fig. 3, two operating points, b and r, are selected on the optimum power curve. These points are shown in a simplified graph of power versus speed in Fig. 4. Operating point b is at the desired cut-in speed n_b of the system, while operating point r is at a selected rated power of the turbine, P_{tr}, developed at a rated speed n_r. Note that r is a fixed operating point that is dependent on the specific wind turbine used. Operating point b will be chosen in the region of maximum turbine power at low wind speeds on the speed-axis.

For the purpose of analysis, the synchronous generator is modelled as an ac voltage source, E_r, in series with the resistance R_s and synchronous inductance L_s, as shown in Fig. 5. The inductance L_s and resistance R_s consists of

$$L_s = L_i + L_c + L_e \tag{1}$$

and

$$R_s = R_i + R_c + R_e + R_{ac}, \tag{2}$$

where

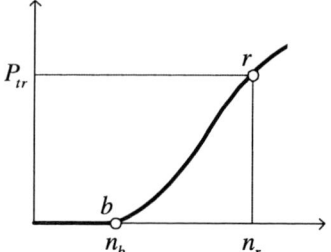

Fig. 4.   Simplified graph of power matching showing operating points b and r.

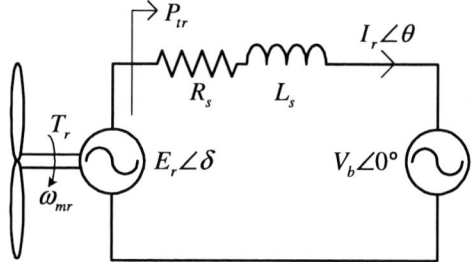

Fig. 5.   The per-phase equivalent circuit diagram.

$$R_{ac} = \frac{6R_{bat}}{\pi^2}. \tag{3}$$

R_i and L_i is the per-phase internal resistance and inductance of the generator and R_c and L_c is the per-phase resistance and inductance of the cable used to connect the generator to the rectifier-battery load. R_e and L_e is the per-phase resistance and inductance of the external inductor. R_{ac} is the equivalent per-phase ac resistance of the internal resistance of the battery, R_{bat}. The rectifier-battery load is modelled as a sinusoidal ac voltage source with constant amplitude V_b with variable frequency equal to the generator frequency f_r. The wind generator battery charging system is analysed on a per-phase basis [4], with the per-phase phasor diagram shown in Fig. 6. Note that with the diode rectifier load, the system is modelled with the phase current I_r in phase with the voltage source, thus operating at unity power factor ($\cos \theta = 1$). V_b is the fundamental phase voltage on the ac side of the rectifier and is calculated by

$$V_b = \frac{\sqrt{2}(V_{bat} + 2V_d)}{\pi} \tag{4}$$

where V_{bat} is the dc voltage value of the battery pack and V_d is the forward conducting voltage drop across a diode of the bridge rectifier [9]. From this the rated generator voltage E_r at the rated speed n_r can be calculated by

$$E_r = V_b \frac{n_r}{n_b}, \tag{5}$$

where n_b is the cut-in speed at which the generator voltage equals the battery voltage and at which speed the system starts to deliver power to the battery bank.

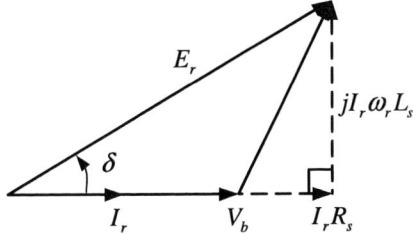

Fig. 6. Phasor diagram of the system.

The rated voltage E_r is a function of the following,

$$E_r = k_1 N B_p n_r \qquad (6)$$

where k_1 is a machine constant, N is the number of turns per coil and B_p is the peak flux density in the air gap [10]. From (5) and (6), n_b can be written as

$$n_b = \frac{V_b}{k_2 N B_p} \qquad (7)$$

where k_2 is another machine constant. To achieve the desired cut-in speed n_b, which was already decided upon at operating point b, the parameters V_b, N and/or B_p should be determined accordingly. V_b is a fixed voltage due to the battery bank voltage. The value for the air gap flux density can typically vary between from 0.5 to 0.8 T, depending on the size and thickness of the magnets and the effective air gap of the generator; hence only limited adjustments can be made in the design to B_p. The number of turns per coil, N, is therefore the most flexible parameter in the design to achieve the desired cut-in speed. It must be noted that N has an influence on R_i namely as,

$$R_i = k_3 N^2, \qquad (8)$$

where k_3 is another machine constant [10]. Thus, if N increases, R_i and the resulting copper losses will also increase. If the copper losses are deemed to be excessive, the cooling for the machine should be improved or the machine should be slightly over-sized in its design.

The turbine input power P_{tr} is calculated by

$$P_{tr} = T_r \omega_{mr} \qquad (9)$$

where

$$\omega_{mr} = \frac{\pi n_r}{30}. \qquad (10)$$

T_r is the rated torque delivered by the wind turbine at rated speed. With no core losses and wind-and-friction losses ignored, the rated turbine power P_{tr} can be calculated by,

$$P_{tr} = 3 E_r I_r \cos \delta \qquad (11)$$

where from Fig. 6

$$\cos\delta = \frac{V_b + I_r R_s}{E_r}. \qquad (12)$$

From (11) and (12), the rated current I_r can be calculated by

$$I_r = \frac{\sqrt{V_b^2 + \frac{4}{3} R_s P_{tr}} - V_b}{2 R_s}. \qquad (13)$$

From Fig. 6 we have

$$\sin\delta = \frac{I_r \omega_r L_s}{E_r} \qquad (14)$$

where

$$w_r = \frac{p \omega_{mr}}{2}, \qquad (15)$$

hence from (1), (5) and (10) the external inductance L_e can be calculated as,

$$L_e = \left(\frac{60 V_b}{\pi p n_b}\right) \frac{\sin\delta}{I_r} - L_c - L_i, \qquad (16)$$

where p is the pole number of the generator. The current I_r is calculated from (13) and δ from (12). The internal inductance of the generator L_i can be measured or calculated from finite element analysis.

As previously mentioned, from Fig. 4, the operating point b can be selected anywhere below the rated speed n_r. Together with the correctly calculated value of L_e, according to (16), the turbine power will pass through both operating points b and r of Fig. 4.

III. SIMULATION AND TESTING OF THE SYSTEM

In this section the analysis of Section II is confirmed by simulation and practical tests at a 1 - 2 kW power level. The direct battery charging system is simulated using the simulation software package Simplorer™. The equivalent circuit model used in the simulation is shown in Fig. 7. The wind generator is modelled as three single-phase voltage sources E_1, E_2 and E_3 in series with internal phase resistance and inductance R_i and L_i and cable resistance and inductance R_c and L_c. The inductance L_e in Fig. 7 represents the per-phase externally connected series inductor. The effect of L_e on the turbine power matching is investigated by varying the L_e-value in the simulation. The battery bank is modelled as a resistor R_{bat}, representing the internal resistance, in series with a voltage source V_{bat}, and both in parallel with a capacitor C_{bat} [11].

The practical test set-up is shown in Fig. 8. A three-phase induction motor controlled by a variable speed drive was used as the prime mover to drive the wind generator. The input power was measured by means of a torque transducer. A diode bridge rectifier and a lead-acid battery bank consisting of three series connected batteries acted as a load to the generator.

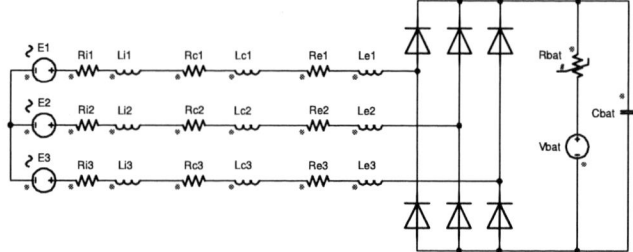

Fig. 7. Equivalent circuit model of the direct battery charging generator system.

Fig. 8. 2 kW air-cored AFPM wind generator under test.

Maximum Power Matching

To improve the power point fitting of Fig. 3, the stator winding of the generator is redesigned to obtain a much lower cut-in speed namely at $n_b = 75$ r/min as also given in Table I. P_{tr} and n_r can be chosen at at any operating point r on the power curves ot the turbine; two operating points have been selected with the data given in Table I. The values of R_i, L_i, V_{bat} were measured and are also given in Table I. The inductance L_e and the other system parameters are calculated according to the analysis of Section II; the results of these calculations are given in Table I. For the practical tests two three-phase inductors available in the laboratory were used with inductance values of $L_e = 7.5$ mH and $L_e = 10.5$ mH. These values are close to the theoretical inductance values of Table I.

The simulated and measured results of the two cases studied of the direct battery charging system are shown in Fig. 9. It shows that the generator starts producing power to the battery bank at the selected lower cut-in speed of 75 r/min. The simulated power matching through the selected operating points compares well with the measured results; the deviations between simulated and measured results at high generator speeds is attributed to the inaccurate battery model used in the simulations and the saturation of the iron core of the inductor.

In Fig. 10 the measured turbine power P_{tr} for different external inductances are plotted on the power-speed curves of

TABLE I
SIMULATION DATA

	n_b		75 r/min
n_{r1}	180 r/min	n_{r2}	225 r/min
P_{tr1}	750 W	P_{tr2}	1415 W
	R_i		620 mΩ
	L_i		650 μH
	V_{bat}		36 V
E_{r1}	40.6 V	E_{r2}	50.8 V
	V_b		16.9 V
	R_{bat}		400 mΩ
	R_{ac}		243 mΩ
I_{r1}	9.6 A	I_{r2}	14.8 A
δ_1	49.9°	δ_2	52.1°
	R_e		100 mΩ.
L_{e1}	10.1 mH	L_{e2}	6.5 mH

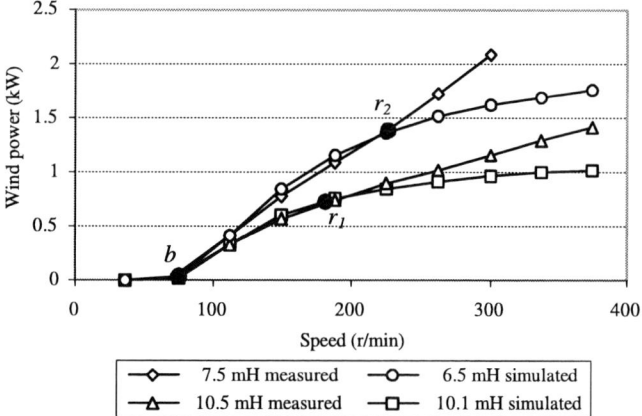

Fig. 9. Simulated and measured results of turbine power versus turbine speed with L_e a parameter.

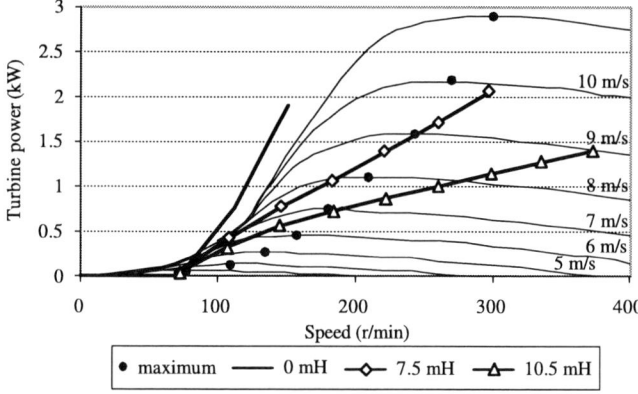

Fig. 10. Measured results of the turbine power matching with and without external inductances.

the turbine. It is shown that without any external inductance the wind generator system charges the batteries at low wind speeds but does not utilise available power at higher wind speeds. With external inductors connected a much better power matching at higher wind speeds is obtained. It is

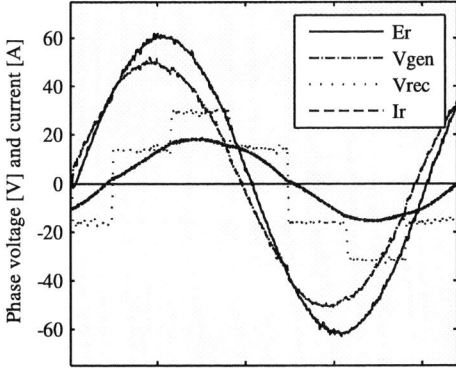

Fig. 11. Measured voltage and current waveforms of the wind generator system.

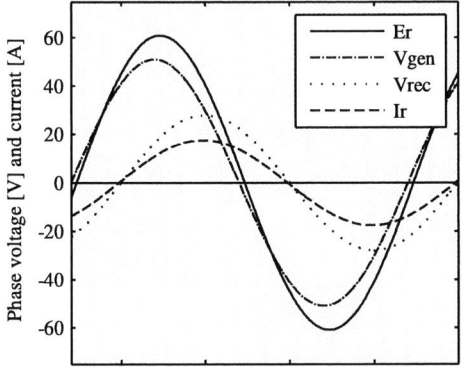

Fig. 12. Simulated single phase equivalent circuit (Fig. 5) voltage and current waveforms of the wind generator system.

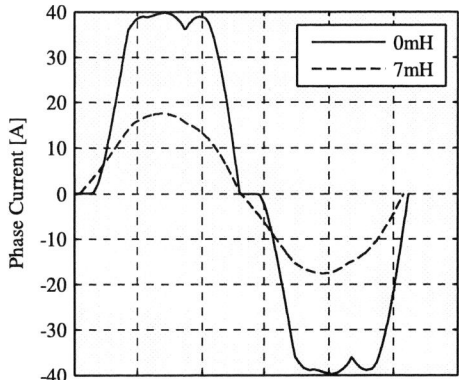

Fig. 13. Effect of series inductance on line current waveform.

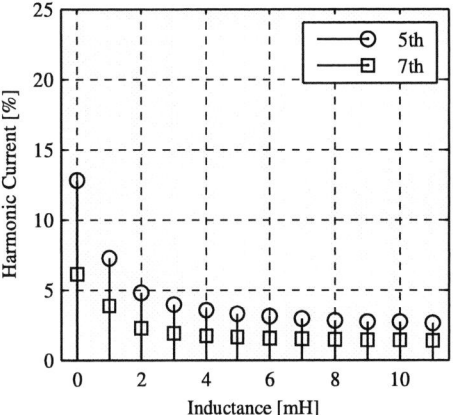

Fig. 14. Current harmonic component versus series inductance values.

furthermore shown that with a relatively large inductance value the input power delivered by the turbine becomes naturally limited at high wind speeds. Although this is not as good from a maximum power matching point of view, it is good for the protection of the system as the generator current and the resulting copper losses does not increase any further.

Fig. 11 shows the measured voltage and current waveforms of the wind generator system with the 7.5 mH external inductor connected. Note the stepped voltage, V_{rec} at the terminals of the rectifier. The measured voltage and current waveforms can be furthermore compared with the simulation results of the single phase equivalent circuit model as shown in Fig. 12; this shows excellent comparison and confirmed the analysis based on fundamental values.

IV. HARMONIC CURRENT ANALYSIS

With no external inductance in the wind generator system, the rectifier forced the line current waveform to become non-sinusoidal due to the low internal impedance of the generator. The inductance of the external inductor is much larger than the internal inductance of the AFPM generator (see Table I). The external inductor therefore filters the

higher frequency harmonic current components allowing for sinusoidal voltage and current waveforms at the terminals of the generator. This is clearly shown in Fig. 13, where the line current waveform is highly non-sinusoidal with no external inductor, but very much sinusoidal with $L_e = 7$ mH for example.

A. Harmonic Analysis

The use of a three-phase diode bridge rectifier, which act as a non-linear load with six-pulse switching characteristics, generate the following voltage/current harmonics in the wind generator system of Fig. 1,

$$h = 6k \pm 1, \qquad (17)$$

where k is any integer. From (17), it can be assumed that the 5th and 7th harmonics will be the most dominant harmonics in the system. A Fast Fourier Transform (FFT) analysis of the line current is performed to investigate the harmonic content with varying external inductance. The result of this analysis is shown in Fig. 14. It is clear that an increase in series inductance does result in a decrease of the 5th and 7th harmonic components in the line current.

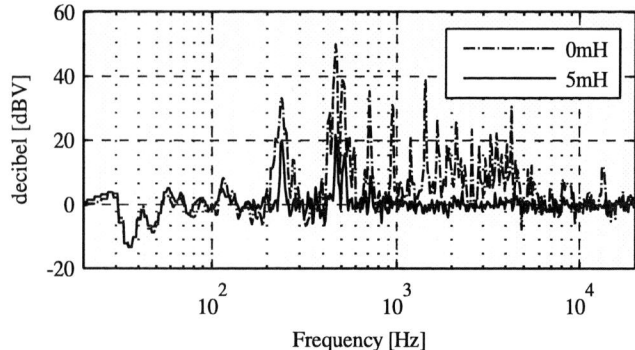

Fig. 15. Acoustic noise recorded on the audible spectrum with and without the external inductances.

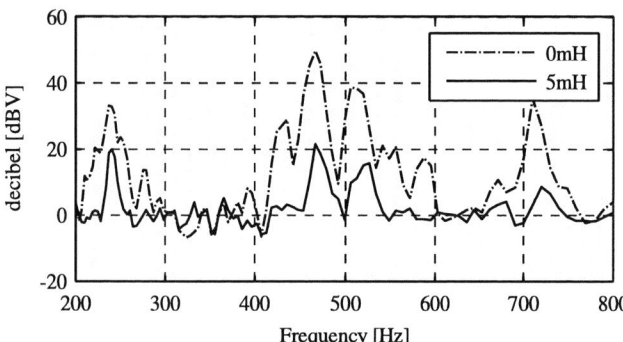

Fig. 16. Zoomed-in audible noise range with and without the external inductances.

B. Acoustic Noise Measurement

To confirm the possibility of using additional series inductors reducing the acoustic noise emitted from the wind generator, the acoustic noise of the AFPM generator was recorded under full-load test conditions. The test was conducted on a 24 pole AFPM generator together with a 5 mH series phase inductor. A sound measurement device was used to measure the emitted noise and a frequency spectrum analyser was used to record and display the data.

By keeping the speed of the generator constant, the generator's frequency and output voltage is kept constant. To cancel the effect of the switching noise of the induction motor drive and the ambient floor noise of the test laboratory, noise measurements of the system were taken with the generator open-circuited. The noise recordings were repeated with the generator feeding the rectifier-battery load with and without the proposed additional series inductors. The difference between the recorded load and no-load data is the emitted noise from the loaded generator. The acoustic noise from the loaded generator was plotted on the audible spectrum range, ranging from 20 Hz to 20 kHz, as shown in Fig. 15. The frequency components that are of the most concern are the multiples of six for the six-pulse rectifier. In all the tests the generator frequency was kept constant at 40 Hz, hence the

6th, 12th and 18th pulse frequencies of the six-pulse rectifier system would equal 240, 480 and 720 Hz respectively. The zoomed-in frequency band with peaks at these frequencies is shown in Fig. 16. From this it is clear that there is a significant reduction (20 dBV) in audible noise under loaded conditions with the addition of the 5 mH series inductor.

V. CONCLUSION

It is shown that air-cored AFPM wind generators is not suitable to be connected directly to rectifier-battery loads. This is due to (i) non-optimal turbine power matching and (ii) harmonic line currents that are causing severe acoustic vibration noise. The proposed use of series connected three-phase inductors is shown to significantly reduce the phase current harmonics. The reduction of the harmonic components in the line current of the generator reduces the acoustic noise emitted from the generator. With correctly designed series connected inductors a near-optimum turbine power matching is also obtained. The external inductance can be varied according to what takes preference in the application, because it is not a fixed part of the generator as in conventional wind generators. The disadvantage of the proposed scheme is that the wind generator is not operating any more at its highest efficiency and therefore has to be oversized in its design to still have a high efficiency.

REFERENCES

[1] S. Drouilhet, E. Muljadi, R. Holz, and V. Gevorgian, "Optimizing small wind turbine performance in battery charging applications," in *Windpower '95*, 1995.

[2] B. Borowy and Z. Salameh, "Dynamic response of a stand-alone wind energy conversion system with battery energy storage to a wind gust," *Energy Conversion, IEEE Transaction on*, vol. 12, no. 1, pp. 73–78, 1997.

[3] R. Teodorescu and F. Blaabjerg, "Flexible control of small wind turbines with grid failure detection operating in stand-alone and grid-connected mode," *Power Electronics, IEEE Transactions on*, vol. 19, no. 5, pp. 1323–1332, 2004.

[4] E. Muljadi, S. Drouilhet, and R. Holz, "Analysis of wind power for battery charging," in *ASME Wind Energy Symposium*, 1996.

[5] A. De Broe, S. Drouilhet, and V. Gevorgian, "A peak power tracker for small wind turbines in battery charging applications," *Energy Conversion, IEEE Transaction on*, vol. 14, no. 4, pp. 1630–1635, 1999.

[6] B. Chalmers and E. Spooner, "An axial-flux permanent-magnet generator for a gearless wind energy system," *Energy Conversion, IEEE Transaction on*, vol. 14, no. 2, pp. 251–257, 1999.

[7] J. Gieras, R. Wang, and M. Kamper, *Axial Flux Permanent Magnet Brushless Machines*. Kluwer, 2004.

[8] R.-J. Wang, M. Kamper, K. Van der Westhuizen, and J. Gieras, "Optimal design of a coreless stator axial flux permanent-magnet generator," *Magnetics, IEEE Transactions on*, vol. 41, no. 1, pp. 55–64, 2005.

[9] N. Mohan, T. Undeland, and W. Robbins, *Power Electronics*, B. Zobrist, Ed. Wiley, 2003.

[10] M. Kamper, R. Wang, and F. Rossouw, "Analysis and performance evaluation of axial flux air-cored stator permanent magnet machine with concentrated coils," in *International Electrical Machines and Drives*, 2007.

[11] Z. Salameh, M. Casacca, and W. Lynch, "A mathematical model for lead-acid batteries," *Energy Conversion, IEEE Transaction on*, vol. 7, no. 1, pp. 93–98, 1992.

Transverse Flux Machines for Sustainable Development – Road Transportation and Power Generation

D. Svechkarenko, A. Cosic, J. Soulard, and C. Sadarangani

Division of Electrical Machines and Power Electronics,
School of Electrical Engineering, Royal Institute of Technology
Teknikringen 33, 10044 Stockholm, Sweden, email: dmitrys@ee.kth.se

Abstract—The higher specific torque and power density of a transverse flux machine (TFM) in comparison to a conventional machine makes it a promising energy converter in various applications. In this paper, a free piston energy converter and a direct-driven wind turbine are considered. The analytical investigation of the novel TFM topology applied in these two cases is presented. The cogging torque is evaluated by use of a three-dimensional finite element method. The results of simulating generators rated in the range of 3-12 MW are discussed. Manufacturing of a linear prototype is presented.

Index Terms—Direct-driven wind turbine, free piston energy converter, sustainable development, transverse flux machine.

I. INTRODUCTION

The largest amount of man-made greenhouse gases worldwide is emitted into the atmosphere due to power generation, some industrial processes, and road transportation. A decreased emission level can be achieved by reducing consumption of fossil fuels, replacing the conventional energy sources with renewable ones, and introducing more efficient filtering systems. The work presented in this article is an attempt towards this development.

A. Road Transportation

A conventional vehicle nowadays utilizes an internal combustion engine (ICE) for the conversion of the energy stored in the fuel (normally, a petroleum-derived liquid) into mechanical energy. This process results in discharge of carbon dioxide into the atmosphere. According to the Emission Database for Global Atmospheric Research [1], as much as 14% of the man-made carbon dioxide in the atmosphere is emitted due to road transportation. A number of possible solutions to either considerably diminish or even totally exclude the usage of fossil fuels have been proposed and implemented in the past few decades.

Equipping a vehicle with fuel cells or electric batteries could achieve the desired discharge reduction. The first alternative is, however, still fairly expensive while the second one has limited application. Another possibility is to replace conventional fuels with alternative biofuels,

This work was sponsored by the Swedish Center of Excellence in Electric Power Engineering EKC2 and the European Commission.

such as bioalcohol (ethanol, biodiesel). This technology is widely investigated by a number of manufacturers worldwide and could become a commercial product in the foreseeable future, though it would require an extensive infrastructural development.

As an intermediate solution between petrol-driven vehicles nowadays and the pollution-free car of the future, the use of Free Piston Energy Converter (FPEC) depicted in Fig. 1 has been proposed. The free piston generator integrates the functions of combustion engine, crankshaft, connecting rod, and rotating electrical generator into a single unit. Thus, the system becomes more efficient, more reliable, can be multi-fuel propelled and has reduced weight. This results in reduced pollution and more work output for a given amount of fuel. Using this integration, the system, in addition, becomes self-starting and the starting motor can therefore be excluded.

B. Power Generation

To decrease the emission of greenhouse gases in power generation, the power plants whose operation is based on burning of fossil fuels should be replaced with environmentally friendly ones. Wind power is one of the fastest growing renewable energy sources over the past decade. Wind energy production is increasing by approximately 30% annually [2]. This growth is accounted for by both the amount and the size of new turbines installed. The typical size of a recently installed turbine exceeds 2 MW and in a few years turbines up to 5 MW should reasonably be expected on the market. This trend is likely to continue in the foreseeable future as more cost effective wind turbines are being developed and installed worldwide.

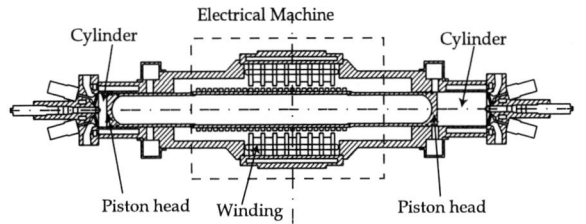

Fig. 1. Schematic of the free piston energy converter.

With the further development of wind energy and increased wind power penetration level in power systems, the issues of availability and reliability of generating units become of great importance. This particularly applies for stand-alone and offshore applications due to their often hard-to-reach locations. The overall reliability of a wind turbine is somewhat reduced by using a gearbox, which is applied to adjust a low-speed turbine shaft to a higher rotational speed of a conventional generator. In addition, a gearbox is subject to mechanical wear and vibrations, and requires lubrication and more frequent maintenance at considerable cost. As a result, a gearless wind energy system has drawn the attention of wind turbine manufacturers, such as Enercon, Made, and Harakosan. An overview of such a system, as well as the components it comprises, is presented in Fig. 2.

The gearless wind energy system requires a direct-driven low-speed generator with a large number of poles and larger than conventional generator outer diameter. Electrically excited direct-driven synchronous and induction generators are utilized by a number of wind turbine manufacturers (eg. in Enercon's *E-112* wind turbines with 4.5 MW rated power). In the last few decades, a reduced magnet price has made synchronous generators with permanent magnet excitation (PMSG) an attractive alternative. In comparison to the electrical excitation, the permanent magnet excitation favors a reduced active weight, and decreased copper losses, yet the energy yield is higher. This topology is for example utilized by Harakosan in its 2 MW wind turbine *Z72*, making it one of the largest direct-driven permanent magnet generators commercially available [3].

II. TRANSVERSE FLUX TOPOLOGY

The analyzed applications employ a transverse flux topology, shown in Fig. 3. Unlike in conventional radial-flux machines, the flux lines in this topology lie in the perpendicular or, in other words, transversal plane to the direction of movement and that of current flow.

One of the main benefits of using such a topology is the possibility to attain a high torque density [4]. By increasing the number of poles (thus decreasing the pole pitch) for given dimensions and current loading, the machine rating can be increased and, consequently, higher values of specific torque density can be achieved. Another attractive feature of a transverse flux machine (TFM) is that it allows current and magnetic loading to be set almost independently. The pole length sets the magnetic loading, whereas the machine width determines the current loading. This advantageous feature of TFM results in a

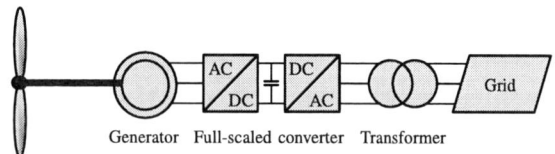

Fig. 2. A gearless wind energy system.

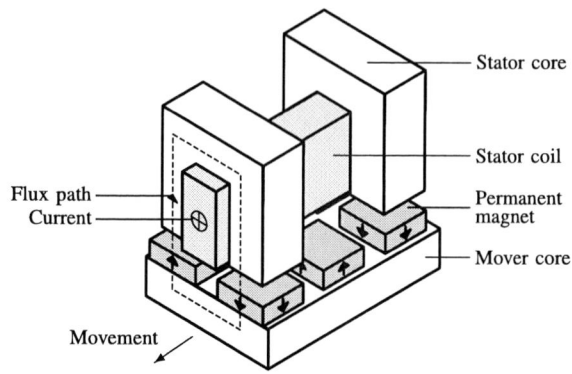

Fig. 3. Basic single-phase transverse flux topology with permanent magnet excitation.

more favorable construction as the magnetic circuit and armature winding are not competing for the same space.

One of the major drawbacks of a TFM is its high flux leakage, resulting in a poor power factor. The amount of leakage flux can be reduced to a certain extent by increasing the pole width at the cost of relinquished high torque density. Therefore, the machine designer has to consider this trade-off between machine performance and utilization of active materials and find an optimal solution. On the other hand, the large reactance would limit the short-circuit current, in case of a failure.

Another significant disadvantage of a TFM is the complicated mechanical structure of the magnetic circuit. It consists of a large number of separate small-size components, thereby resulting in a relatively weak construction and more complex manufacturing. The novel transverse flux concept analyzed in this paper can likely overcome this shortcoming, as a less complicated construction of the stator core is adopted.

The geometry depicted in Fig. 3 is a basic TFM arrangement referred to as single-sided TFM. It consists of a mover core with permanent magnets (in magnetically excited system). The stator is made of C-shaped iron cores and the winding is placed in the stator slots. The magnets polarized with alternating polarity are displaced on the mover surface, thereby producing an alternating flux in the stator iron. The winding is global as the same coil links the fluxes produced by each pole pair.

A. Novel TFM Concept

In the novel TFM concept, a single-phase structure, shown in Fig. 3 is placed along the circumference, thereby obtaining circular cross-section. It employs a single-sided topology with the inner rotor and surface-mounted magnets. The TFM consists of a hollow mover with permanent magnets, embraced by the laminated stacks with the windings placed in the slots.

The high current loading allowed in a TFM could result in considerable values of armature reaction, which cause the eddy current losses. To decrease this effect, the iron parts should be made of a laminated material. Moreover, the transverse flux concept implies that the magnetic

1302

fluxes should be carried in the plane perpendicular to the direction of movement. Therefore, the laminated material would decrease the flux in the peripheral direction and the fringing effect.

The analysis of the novel TFM topology in a linear machine with a tubular translator in the FPEC [5] and a rotational machine in a wind turbine [6] is presented in the following two sections.

III. NOVEL TFM IN FREE-PISTON ENERGY CONVERTER SYSTEM

Due to the strict requirements for the electric machine in the FPEC system, such as a low weight of the translator (movable part in the machine), the transverse flux machine can be a promising alternative [7]. Some of these requirements are listed in Table I.

TABLE I
REQUIREMENTS FOR THE ELECTRIC MACHINE IN FPEC

Property	Value
Efficiency	> 0.9
Mover mass (kg)	6
Specific power (kW/kg)	> 1
Force (kN)	4

However, the electric machine design presented in [7] was found to suffer from mechanical instability. Further investigation of the mechanical structure showed that translator should preferably have a tubular shape. In this way, the force from the pistons can be evenly transferred to the translator. Fig. 4 shows a novel transverse flux topology used in the electrical machine of a free-piston energy converter.

Two possible configurations of the TFM with the different number of slots are depicted in Fig. 5. The translator tube thickness is inversely proportional to the number of

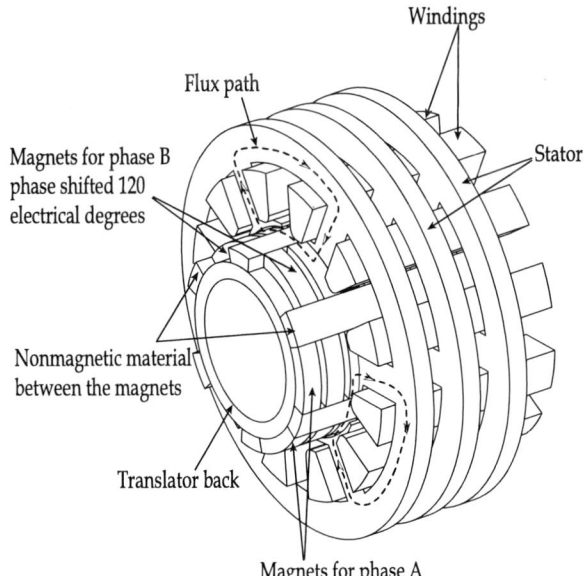

Fig. 4. Layout of the novel TFM design.

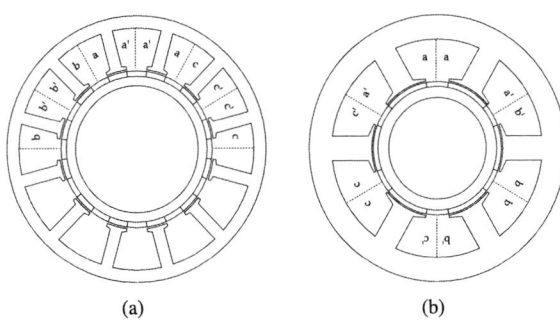

Fig. 5. Cross-section of the novel TFM topology with (a) $Q_s = 12$ and (b) $Q_s = 6$

slots. As the weight of the translator is on of the most critical requirements, the design with the higher number of slots should be preferred. However, a higher number of slots results in more magnet parts, which leads to a more complicated manufacturing process.

The force for a linear TFM according to [7] is given by

$$F = \widehat{B}_{\delta 1} \, \widehat{S} \, k_w \frac{A_{active}}{2}, \tag{1}$$

where k_w is the winding factor, $\widehat{B}_{\delta 1}$ is the peak fundamental value of the flux density in the air gap, A_{active} is the active area in the air gap, and $\widehat{S}$ is the current loading for the machine. In a TFM, the current loading is given by equation [7]

$$\widehat{S} = \frac{\sqrt{2}N_s I}{2\tau_p}, \tag{2}$$

where N_s is the number of conductors per slot per phase, I is the armature current, and τ_p is the pole pitch.

Using the analytical model developed for the surface magnet rotor design, the force of $4.1\,\text{kN}$ and the mass of the translator of $5.8\,\text{kg}$ have been obtained. In order to account for the magnetic flux leakage appeared in the third dimension, several empirical factors have been included.

A. Cogging Force

The three-dimensional magnetostatic finite element analysis have been performed to investigate the cogging force of the machine. Although the cogging force might not be crucial in the FPEC application [7], it is yet worth investigating if vibration and noise of the machine are considered.

Due to the symmetry, only one segment of the machine (two poles) has been simulated, as shown in Fig. 6. Periodic boundary conditions are set on the upper and the lower sides of the segment. These conditions require the same mesh on both sides of a segment.

Time stepping was not possible during simulations due to selected boundary conditions. Instead, a new segment for each time step has been constructed and simulated.

Fig. 7 shows the cogging force simulated only for one phase. The total cogging force is the sum of all three phases. The cogging force for the other two phases is displaced 120 electrical degrees in the axial direction.

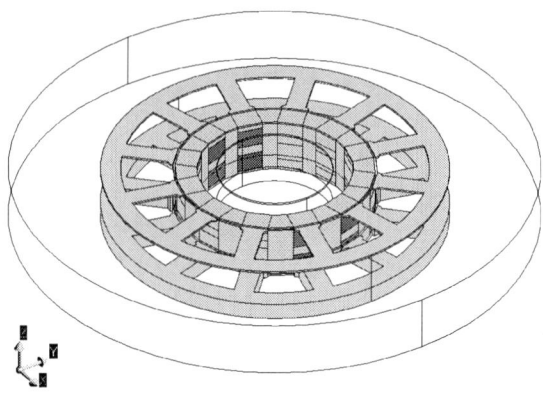

Fig. 6. 3D-FEM simulated segment of the linear machine.

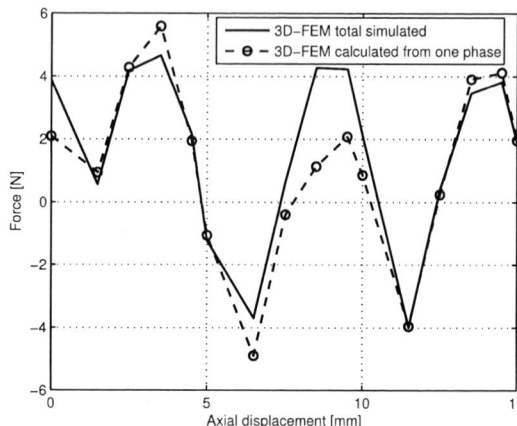

Fig. 8. Simulated and calculated cogging force in the machine for three phases.

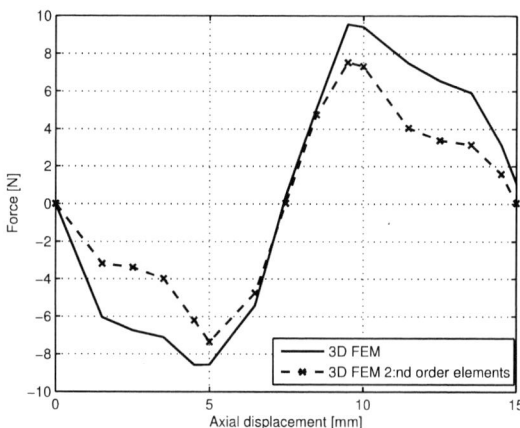

Fig. 7. Simulated cogging force in the machine for one phase.

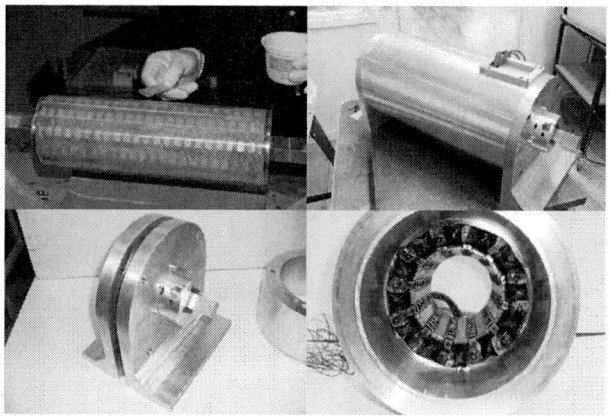

Fig. 9. Prototype machine.

The results from the one-phase simulations are used to verify the analytical results. Fig. 8 shows simulated results for all three phases and total calculated results from one phase simulations. As can be seen the agreement between the results is acceptable. As can be observed, the cogging force in the novel TFM has no significant impact on the machine performance as it is hardly exceeds 1% of the nominal force.

IV. PROTOTYPE

The manufactured prototype is downsized to develop a force of 2 kN. The pole length has been limited to a certain lowest value. Earlier experiences from the prototype production showed that the linear TFM are sensitive to the tolerance in the axial direction. A loss of about 30% of the flux due to the magnet axial displacement has been reported [7].

Some stages of the manufacturing process are presented in Fig. 9. At the time of writing, it is being prepared for the measurements.

V. NOVEL TFM IN WIND POWER GENERATION

A general view of the novel TFM direct-driven generator, as well as the radial and peripheral planes are presented in Fig. 10

The cross-section of the novel transverse flux generator is shown in Fig. 11. It employs a single-sided topology with the inner rotor and surface-mounted magnets. The generator consists of a hollow toroidal rotor with permanent magnets embraced by the laminated stacks with the windings placed in the slots.

The main machine radius R_m and the tube radius R_s are shown in Fig. 11. The cut required for the mechanical assembling of the rotor on the shaft is presented by angle 2ξ. The TFM topology allows the use of the stator winding of a simple mechanical structure, which facilitates high voltage insulation. This could be an attractive feature in the future since the voltage of wind generators has been continuously increasing and voltage levels up to 5 kV can reasonably be expected in the forthcoming generators.

A. Control Modes

The classical electric circuit model of a synchronous generator shown in Fig. 12(a) is used. The voltage drop due to winding resistance is relatively small in large synchronous machines, and therefore it is disregarded. Thus, only the phase reactance X_s, which consists of two

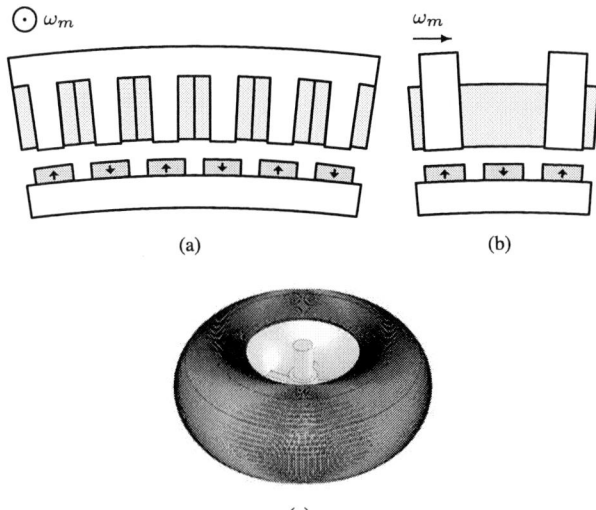

(a)   (b)

(c)

Fig. 10. Schematic representation of the novel transverse flux permanent magnet machine in stack (radial) (a) plane, rotational (peripheral) plane (b) and general view of the generator (c).

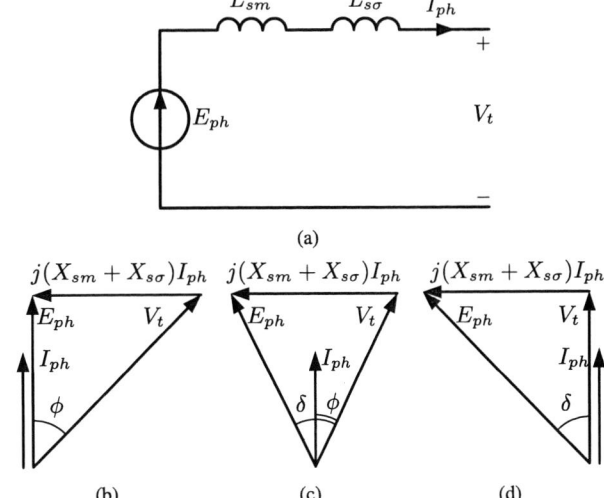

(a)

(b)   (c)   (d)

Fig. 12. The single-phase equivalent circuit of the generator (a) and the phasor diagrams for the cases when $\delta = 0$ (b), $\phi = \delta$ (c) and $\phi = 0$ (d).

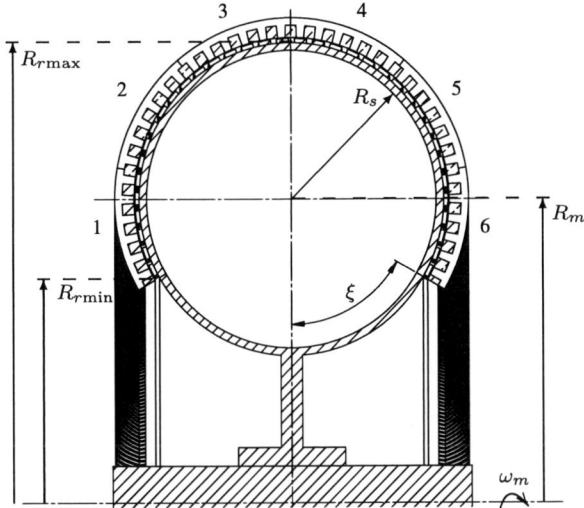

Fig. 11. Cross-section of the novel transverse flux generator in the stack plane with the main dimensions, where w_m is the direction of rotation.

components: magnetization reactance X_{sm} and leakage reactance $X_{s\sigma}$ is considered.

As the magnetization of the permanent magnet machines cannot be decreased at reduced loads (unless field weakening is applied), the induced emf E_{ph} would retain its no-load value. By selecting the magnitude of the armature current and the current angle, the terminal voltage V_t can be controlled. Three different phasor diagrams corresponding to three control strategies that could be applied for the control of the generator are presented in Figs. 12(b-d) [8], [9]. For all three modes, the forced-commutated converter is required for the grid connection of the generator [9]. The back-to-back converter has been suggested for the wind turbine applications [2]. It consists of a rectifier and an inverter with an intermediate dc-link capacitor.

The first control mode in Fig. 12(b) corresponds to the

case when the load angle $\delta = 0$, and thus terminal voltage V_t is higher than the induced emf E_{ph}. The power factor would be quite poor in this case, leading to an increasing converter rating and would contribute to the increase of the total system cost. On the other hand, in the third control mode in Fig. 12(d), the current is in phase with the terminal voltage, resulting in the unity power factor and therefore require a smaller converter.

In Fig. 12(c), a compromise control mode is presented, where the phase current is placed between the induced emf and the terminal voltage. The load angle δ and the phase angle ϕ are equal in this case and $E_{ph} \approx V_t$. This control mode could help to obtain a reasonable compromise between the generator rating and converter rating. The control mode in Fig. 12(c) has been selected in the calculations.

B. Simulation Results

The parametric study of a 5 MW wind generator has been performed, in which the influence of the design parameters on the generator performance has been investigated [10]. Based on this analysis, the main machine parameters have been selected as $\xi = 60°$ and $R_s/R_m = (0.73/1.65)$. While keeping the main parameters constant and assuming constant current loading S, the number of teeth per stator stack Q_s, rotational speed of the turbine n_m and the pole number p vary with the output power P_{out} according to Table II

The number of teeth per stator stack Q_s is subject to a slight variation in order to keep the pole pitch in the stack plane and, thus, the magnet size, nearly constant. The main machine radius R_m is selected in such a way that the current loading is approximately the same and equal to 410 A/cm (as was obtained for the 5 MW generator [10]).

The variation of the generator outer diameter with the rated power is presented in Fig. 13 and the corresponding

TABLE II
PREDEFINED PARAMETERS

Property	1	2	3	4	5
Output power P_{out} (MW)	3	5	7	10	12
Mech. speed n_m (rpm)	18.6	13.2	10.6	8.3	7.4
Pole number p	450	620	710	890	960
Number of teeth per stack Q_s	24	36	36	48	48

TABLE III
WIND GENERATORS DATA FOR VARIOUS OUTPUT POWER AND
CONSTANT CURRENT LOADING

Property	1	2	3	4	5
Output power (MW)	3	5	7	10	12
Outer diameter (m)	3.89	5.24	6.03	7.48	8.02
Axial length (m)	1.43	1.84	2.11	2.56	2.74
Current loading	419	409	409	403	413
Torque density (Nm/kg)	82	101	124	143	158
Magnet weight (kg)	1.050	2.690	4.450	8.440	11.215
Total active weight (kg)	19.400	36.775	52.400	82.750	100.400

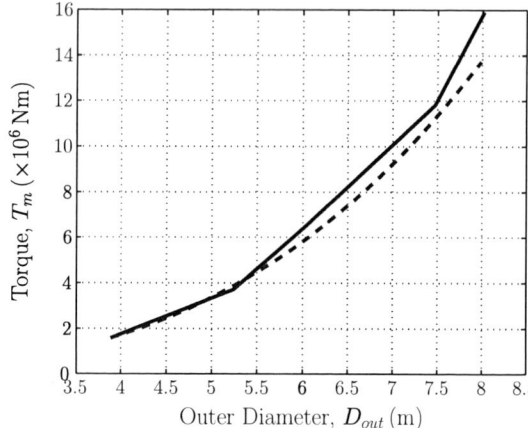

Fig. 14. Torque as a function of the outer diameter, where the solid line shows the simulated development whereas the dashed line illustrates equation $T_m \propto D_{out}^3$.

data are summarized in Table III. For an easier comparison, the scaled schematics of the generators are shown as well. An optimized 3 MW generator has an outer diameter of 3.9 m, while a 12 MW generator would require a 8 m outer diameter. The corresponding axial length increases from 1.4 m in a 3 MW generator to almost 2.7 m for the generator rated 12 MW. However, the length-to-diameter ratio is nearly the same in either case, as the axial length $l_m \approx 2R_m$ and the radii ratio R_s/R_m are constant. Advantageously, the torque density improves with the output power, eg. the 3 MW and 12 MW wind generators differ by approximately 90%. A similar trend has also been shown in [11].

The torque development with the output diameter is shown in Fig. 14 where the solid line represents the simulated results and dashed line corresponds to $T_m \propto D_{out}^3$. As can be observed, the torque varies nearly cubically with the outer diameter. That same behavior has also been described in the patent application [6].

With the developed analytical model, the generator parameters can be estimated for a specified output power. Unfortunately, the actual active weights and dimensions

of the existing wind generators are not freely available. Data for one existing permanent magnet direct-drive wind generator were found in [3]. The 2 MW generator has approximately a 4 m diameter and the rotor and stator together weigh 37 500 kg. As can be seen in Table III, a 3 MW transverse flux direct-driven generator has nearly the same outer diameter, yet lighter active weight. The investigated topology would be even more advantageous with increasing generator rating.

VI. CONCLUSIONS

A novel transverse flux topology applied for the road transportation and power generation applications has been presented in this article.

A. Free Piston Energy Converter Application

Some results of the analytical studies of a Free-Piston Energy Converter have been discussed. An analytical model for a novel TFM machine has been developed and described. The force of 4.1 kN and the mass of the translator 5.8 kg has been obtained.

In order to simplify the prototype production, no weight constraints for the translator have been considered. In addition, the pole length has been restricted to a certain minimum value. This has been done in order to account for a risk of the flux drop due to the pole shift in the axial direction.

The results from the cogging force simulations have shown that the cogging force has a little impact on the machine performance.

B. Wind Turbine Application

The novel topology has been applied in direct-driven wind generators with output power 3, 5, 7, 10, and 12 MW. Compared to the existing permanent magnet wind generator [3], the investigated TFM generator offered a more compact design and a lower active weight; however, the

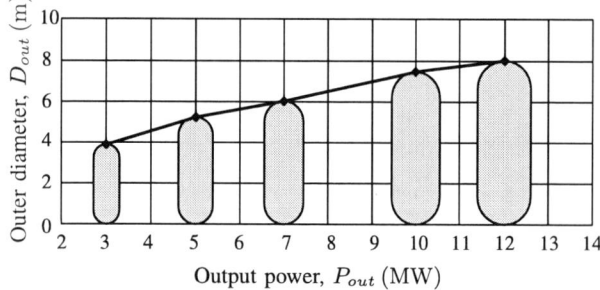

Fig. 13. The outer diameter D_{out} with the output power of wind turbines P_{out} listed in Table III.

unconventional mechanical structure would contribute to increased manufacturing cost.

The topology has been found to be promising for low-speed high-torque applications. It is also believed that the investigated TFM generator would be even more advantageous for larger ratings, thanks to the cubic relation of the torque to the output diameter.

The positive results of the analytical study encourage the performance of a more detailed investigation of the generator with three-dimensional finite element analysis.

ACKNOWLEDGMENT

The authors would like to thank CEDRAT for help with FEM simulations.

REFERENCES

[1] Emission Database for Global Atmospheric Research (EDGAR), http://www.mnp.nl/edgar/.

[2] *Wind power in power systems*, edited by T. Ackermann, John Wiley & Sons, 2005.

[3] C.J.A. Versteegh, "Design of the Zephyros Z72 wind turbine with emphasis on the direct drive PM generator", in *Proc. Nordic Workshop on Power and Industrial Electronics*, Trondheim, Norway, 2004.

[4] H. Weh, H. Hoffmann, J. Landrath, "New permanent magnet excited synchronous machine with high efficiency at low speeds", in *Proc. Int. Conf. Electrical Machines*, Vol. 3, Pisa, Italy, 1990, pp. 35-40.

[5] C. Sadarangani. Pat. nr. SE-P0401110, Transverse flux linear machine with a tubular translator construction.

[6] C. Sadarangani. Pat. nr. SE-P0401120, Hybrid transverse flux machine with high torque density for low speed applications.

[7] W. M. Arshad. "A Low-Leakage Linear Transverse-Flux Machine for a Free-Piston Generator", PhD thesis, Royal Institute of Technology, Sweden, 2003.

[8] A. Grauers, "Design of direct-driven permanent-magnet generators for wind turbines", PhD thesis, Chalmers University of Technology, Sweden, 1996.

[9] J. Hystad, "Transverse flux generators in direct-driven wind energy converters", PhD thesis, Norwegian University of Science and Technology, Norway, 2000.

[10] D. Svechkarenko, "On analytical modeling and design of a novel transverse flux generator for offshore wind turbines", Licentiate thesis, Royal Institute of Technology, Sweden, 2007.

[11] P. Anpalahan, J. Soulard and H.-P. Nee, "Design steps towards a high power factor transverse flux machine", in *Proc. European Conf. on Power Electronics and Applications*, Graz, Austria, 2001.

Low Voltage Ride-Through Capability for Wind Turbines based on Current Source Inverter Topologies

Pierluigi Tenca *, Andrew A. Rockhill †, and Thomas A. Lipo †

* Via G.Soliman 6B/32, 16154 Genova, Italy

† University of Wisconsin, 1415 Engineering Drive, Madison, WI 53706, USA

Abstract—This paper proposes a circuital solution, aimed at a class of current-source inverter topologies, that provides the ability to ride-through temporary low voltage, as well as open circuit conditions at the mains without opening the mains circuit breakers. Although the proposed solution has broader applications, it is described in the framework of a previously published current-source inverter topology for modern wind turbines. For such systems, the low voltage ride-through capability is becoming a necessary feature. The current source topology considered exploits the cable length between the nacelle and the ground to provide a significant portion of the dc-link inductance. Prior work discussed the control strategy, flexibility, design rules, protection and experimental results of this topology and acknowledged the need for low-voltage ride through capability. One possible solution, derived from an already proposed protection scheme, is presented together with selected simulation and experimental results.

Index Terms—Current-Source Topology, Low Voltage Ride-Through, Open circuit fault, Wind Turbines.

I. INTRODUCTION

Historically, wind turbine power generation was only a minute percentage of the total installed generating capacity, a situation commonly referred to as *low wind power penetration*. The somewhat sporadic nature of wind power generation is similar to the relatively sporadic nature of loads and, in the case of low wind power penetration, could be handled in a similar manner. Hence, wind turbines were allowed to disconnect from the grid whenever the situation suited the wind turbine.

However, with increasing wind power penetration, the simultaneous disconnection of a significant percentage of generating capacity can have a profound effect on the stability of the grid, especially when such an event is triggered by another grid fault, such as low-voltage due to a short-circuit or an over-load condition. In the United States, this led the Federal Energy Regulatory Commission (FERC) to propose a low voltage ride-through (LVRT) requirement in Appendix G of its Order No. 661, Large Generator Interconnection Agreement (LGIA) which was issued June 2, 2005 [1]. The new LVRT rule, applying to wind turbine facilities greater than 20 MW, specifies a depth of sag versus time envelope for which a wind turbine is expected to remain on-line and is shown as the solid line in Fig. 1. For example, an applicable wind-turbine generator would be required to remain online during a low-voltage event down to 15%

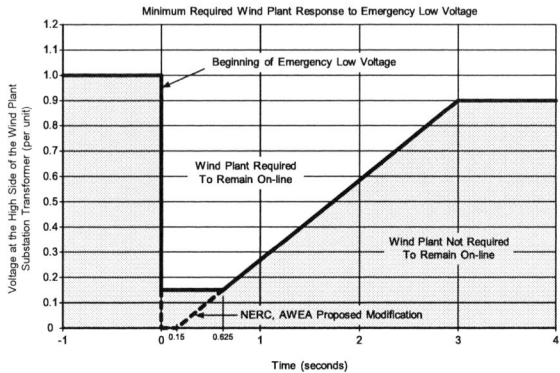

Fig. 1. Low Voltage Ride-Through (LVRT) requirement per Appendix G of FERC's Large Generator Interconnection Agreement (LGIA).

of nominal voltage for up to 0.625 seconds (37.5 cycles at 60 Hz frequency). Similar rules have been proposed or implemented in other countries with high or increasing levels of wind-power penetration as well. However, the North American Electric Reliability Council (NERC) together with the American Wind Energy Association (AWEA) have since proposed a change to the LVRT requirement of Order No. 661 to resolve differences with the existing NERC standard TPL-002-0 which dictates minimum system performance following a single grid fault. The proposed change would require generators to remain on-line for low-voltage conditions down to zero volts for a period not to exceed 9 cycles (0.15 seconds) [2]. This more severe requirement is indicated by the dashed line in Fig. 1. These grid disturbances, as well as other anomalies such as open-circuit conditions, can pose problems for many of today's variable-speed wind turbines.

II. CURRENT SOURCE TOPOLOGY FOR WIND TURBINES REQUIRING RIDE-THROUGH CAPABILITY

Fig. 2 shows the schematic of the current source topology for wind turbines previously proposed in [3]–[5]. The inverters (marked 1 and 2 in Fig. 2) employ fully-controllable switches with bidirectional voltage-blocking and unidirectional current-blocking capability. Neither inverter employs PWM modulation, but rather operates

978-1-4244-0644-9/07/$25.00 ©2007 IEEE

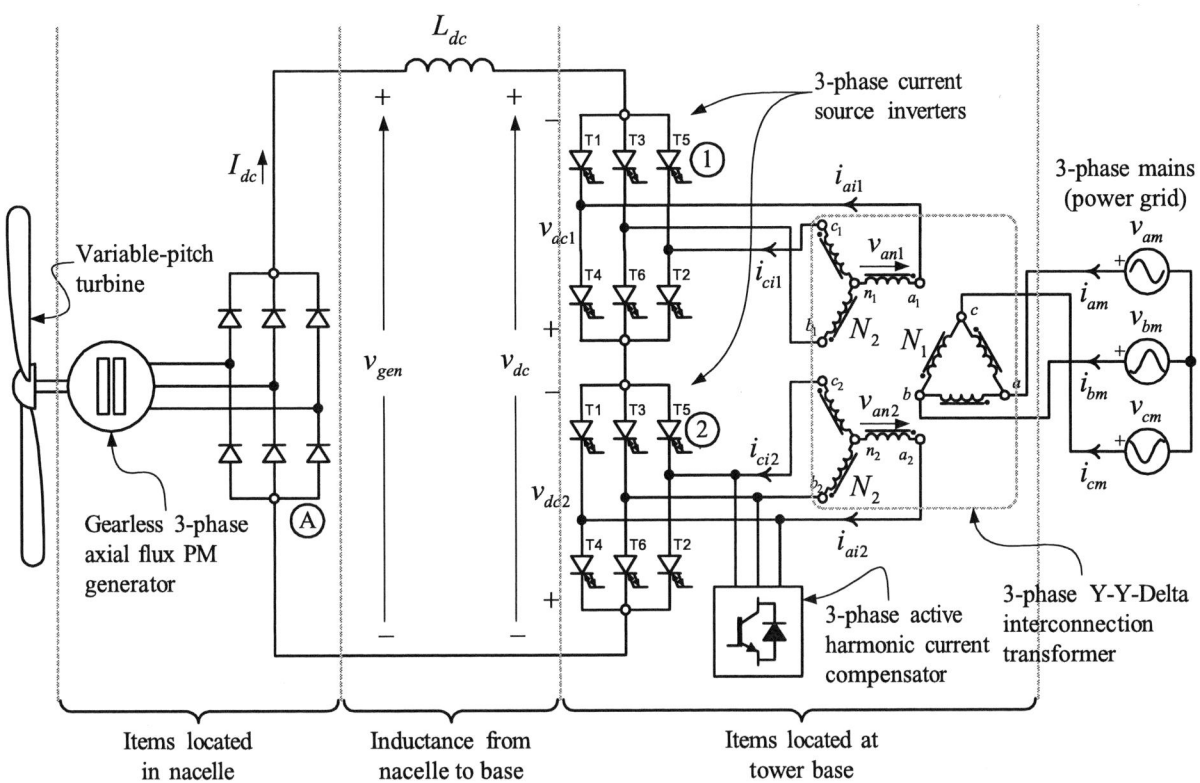

Fig. 2. Proposed Current source topology for gearless variable-speed wind turbines employing a permanent magnet synchronous generator.

under the common phase-control-angle technique, with phase control angles ϕ_1 and ϕ_2 for inverter 1 and 2, respectively. The equations describing a three-phase, phase controlled converter connected to an ideal current source I_{dc} at its two-terminal port and to a symmetrical three-phase system of ideal voltage sources at its three-terminal interconnect, under the action of phase control angle ϕ, are well known [6], [7]. The average voltage $\langle v_{dc}(t) \rangle$ measured across the two terminals of the ideal current source is a function of both the root mean square (RMS) value of the three-phase line-to-line voltage V_{LL} and the phase control angle ϕ. This relationship, in the generator convention, is given in (1). Furthermore, the n-th harmonic mains line current $i_{a(n)}(t)$, as a function of the current I_{dc} and phase control angle ϕ, is given by (2) where ω is the mains frequency expressed in radians per second.

$$\langle v_{dc}(t) \rangle = \frac{3\sqrt{2}}{\pi} V_{LL} \cos\phi \tag{1}$$

$$i_{a(n)}(t) = \frac{2I_{dc}}{\pi n} (1 - (-1)^n) \cos\left(\frac{\pi n}{6}\right) \sin\left[n\left(\phi - \omega t\right)\right] \tag{2}$$

As is evident in (2), one of the main disadvantages of the phase controlled converter is the poor fundamental displacement power factor imposed by large values of ϕ. However, the proposed topology compensates for this short-coming by employing two converters in series whose three-phase line currents are linearly combined by a proper Y-Y-Δ interconnection transformer to the

mains, which acts as an algebraic operator over such currents. This circuital structure allows to impose a desired fundamental power factor at the mains via the two independently adjustable phase control angles ϕ_1 and ϕ_2. In [4] it was shown that, through the linear mapping on the converter phase control angles as given in (3), decoupled control of the fundamental amplitude and phase of the mains line current is achievable.

$$\alpha = \frac{\phi_1 - \phi_2}{2} \qquad \beta = \frac{\phi_1 + \phi_2}{2} \tag{3}$$

For the chosen reference polarities indicated in Fig. 2, the average dc-link voltage $\langle v_{dc}(t) \rangle$ and the n-th harmonic mains line current $i_{am(n)}(t)$ are given by (4) and (5), respectively [4].

$$\langle v_{dc}(t) \rangle = -\frac{6\sqrt{6}}{\pi} \frac{N_2}{N_1} V_{LL} \cdot \cos\alpha \cdot \cos\beta \tag{4}$$

$$i_{am(n)}(t) = \frac{8I_{dc}}{\pi n} \frac{N_2}{N_1} ((-1)^n - 1) \cos\left(\frac{\pi n}{6}\right) \sin\left(\frac{\pi n}{3}\right)$$
$$\cdot \cos(n\alpha) \sin[n(\beta - \omega t)] \tag{5}$$

Hence, the dc-link voltage, and thus the dc-link current, together with the average generator torque, can be controlled via either α or β. However, for a given dc-link current, α determines the amplitude of the fundamental mains line current while β, independently determines its phase. In this way, one can control both the active and reactive power injected to the mains by the wind turbine

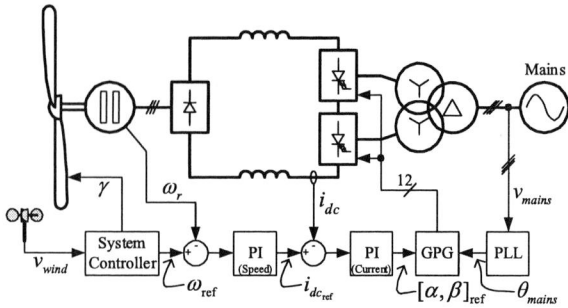

Fig. 3. Overview of the control system principle.

according to (6) and (7), respectively.

$$P_{mains} = \frac{6\sqrt{6}}{\pi} \frac{N_2}{N_1} V_{LL} \cdot I_{dc} \cos\alpha \cos\beta \quad (6)$$

$$Q_{mains} = P_{mains} \tan\beta \quad (7)$$

Fig. 3 shows a simplified control block diagram for the proposed topology. The system controller provides the high-level system commands based on various external parameters such as wind speed, VAr requirements and so on. It is responsible for the maximum power tracking capability of the wind turbine, providing the commands for both the rotor blade pitch and the generator speed. The speed control loop then develops the command for generator torque, which translates into the dc-link reference command $I_{dc_{ref}}$. The current control loop then develops the inverter voltage command, culminating in the control angles α_{ref} and β_{ref}. The control angles are then acted upon by the gate pulse generator which transforms the control angles α_{ref} and β_{ref} through an inverse mapping of (3) into the individual converter firing angles ϕ_1 and ϕ_2. It develops the inverter gate pulses with respect to the mains phase angle θ_{mains}, which is determined by a phase-lock-loop (PLL).

Selected simulation and experimental results for a 10 kW prototype of this conversion scheme were presented in [3], [4], [8] where it is also shown that specific values of α improve the spectrum of the mains currents. The harmonic content of such currents can be reduced further by providing the three-phase active harmonic compensator shown in Fig. 2. Additional simulation and experimental results (some in transient conditions), additional pictures of the laboratory prototype, as well as a more detailed description of it, are present in the technical report published by the National Renewable Energy Laboratory (NREL) [5]. It is obtainable by searching for the authors' names in the website specified in the reference.

III. FAULT PROTECTION SCHEME AT THE BASIS OF THE RIDE-THROUGH CAPABILITY

Although the ride-through requirements might seem uncorrelated with the fault protection needs, some fault scenarios do share common characteristics with a low-voltage or open circuit events at the grid. However, it will be shown that it is possible to ride-through such

events only by increasing the flexibility of the circuits originally conceived for fault protection schemes. Hence, it may be useful to revisit the fault protection circuit previously proposed by the authors in [9] in some detail before describing, in the following section, the small modifications necessary to provide the LVRT capability.

Current source topologies require protection schemes aimed at preventing faults having the "dual" character from those particular to voltage source topologies. The protection against open-circuits and the often accompanying over-voltages in current source topologies is just as important as the protection from short-circuits and the often accompanying over-currents in voltage source topologies. Nevertheless, in current source topologies it is also necessary to protect components against over-currents in order to limit electro-dynamic stresses. This is of prominent importance not only for the semiconductors, but for the passive components as well, including the distributed dc-link inductor.

In the current source based conversion circuit shown in Fig. 2 several events can occur that may result in an open circuit condition; a) a failure in the generator and/or rectifier, b) a failure in the dc-link conduction path, c) a failure in one of the inverter circuits, or d) a grid anomaly which manifests itself as an open-circuit at the mains. Without an emergency current conduction path, any one of these events could lead to destructive over-voltages. Furthermore, any possible event capable of rendering $v_{gen} > v_{dc}$ for a sufficiently long interval of time can lead to a potentially dangerous increase of the dc-link current I_{dc}. For example, inverter failures resulting in a short-circuit or a grid anomaly manifesting itself as a low voltage (or zero voltage) at the mains could lead to an uncontrollable increase in the dc-link current.

In [9] the authors introduced a decentralized scheme suitable for protection of the current source inverter topology and in particular, the dc-link inductance, against these "dual" types of dangerous events. The scheme was especially conceived to suit inductors of a distributed nature and to not incorporate any "active" components. The circuit on which the proposed protection scheme is based is shown in Fig. 4. It employs only fuses, a free wheeling diode and a thyristor triggered by a fully passive auxiliary circuit. Additionally, the circuit was conceived to avoid the need of any exchange of information between the two ports $(a1, a2)$ and $(b1, b2)$ (as shown in Fig. 4) which, in the case of a distributed inductance, may be physically located far apart.

Observing Fig. 4, it is possible to analyze briefly how the protection scheme operates in the different fault conditions. As a consequence of an open-circuit, the two following network evolutions are possible:

I) An open circuit event to the left of port $(a1, a2)$, either because of a fault in the generator/bridge or because of the intervention of fuses F_{a1} and/or F_{a2}, forces the dc-link current I_{dc} to flow in the diode D_a. During the interval of the open circuit, the voltage v_a becomes equal to the negative forward

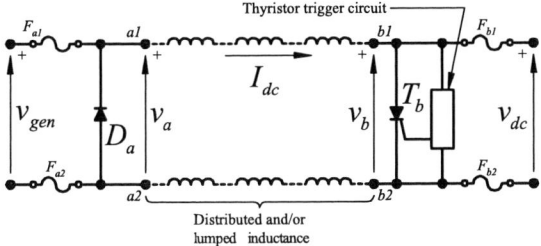

Fig. 4. Decentralized protection scheme.

voltage drop of the diode while the voltage v_b remains greater than zero. This voltage difference in conjunction with any additional dissipative phenomena forces I_{dc} to decrease, eventually, to zero.

II) An open circuit event to the right of port $(b1, b2)$, due to a grid anomaly, an inverter fault or the intervention of fuses F_{b1} and/or F_{b2}, would cause the voltage v_b to rise indefinitely unless a proper recirculation path for I_{dc} is provided. This is exactly the function of the thyristor T_b.

The thyristor trigger circuit shown in Fig. 4 is designed to switch 'on' the thyristor T_b when v_b reaches a chosen threshold value considered damaging to the converter. After its characteristic turn-on time, the thyristor T_b is in the conduction state, providing the necessary recirculation path for I_{dc}. From this instant onwards, I_{dc} could be controlled by modifying the voltage v_a in such a way that the integral function $\psi_{ab}(t)$ defined in (8) is properly limited.

$$\psi_{ab}(t) = \int_0^t [v_a(\tau) - v_b(\tau)] \, d\tau \qquad (8)$$

Conversely, if this goal is not achievable, I_{dc} will rise until at least one or both of the fuses F_{a1} and F_{a2} intervene, leading to an open circuit event belonging to scenario I described previously. In this case, the final circuit configuration of scenario II is characterized by a dc-link inductor that does not exchange energy with the rest of the network – because at least one fuse among $\{F_{a1}, F_{a2}\}$ and one among $\{F_{b1}, F_{b2}\}$ has intervened – while its current recirculates in the loop created by T_b and D_a, both in the conduction state. The current I_{dc} decreases with a rate defined by the extent the dissipative phenomena present in the loop. Further analysis, the details of the trigger circuit and experimental results are given in [9].

IV. EXTENSION OF THE PROTECTION CIRCUIT TO PROVIDE RIDE-THROUGH CAPABILITY

The analysis of the previous section showed that the evolution of the network in both scenarios I and II involve time intervals in which v_a or v_b are almost zero. This means that the basic structure of the protection network described above is already compatible with low voltage or even short-circuit scenarios at either port, but especially at the inverter port $(b1, b2)$, whose electrical quantities are immediately influenced by anomalies occurring at the point of interconnection with the mains. However, the intervention of fuses, or even the triggering of T_b, is not a suitable response to an event the system is expected to quickly recover from or to ride-through.

It will be shown in the subsequent paragraphs that the addition to the protection circuit of Fig. 4 of only one fully controllable device and associated snubber components along with an extended inverter control strategy will allow the system to cope with low-voltage and open-circuit grid anomalies in a repeatable manner without the intervention of protective devices. Fig. 5 shows the proposed modified system, complete with fault protection, capable of meeting the LVRT requirements, even down to zero-voltage, as well as automatically recovering from a grid-side open circuit.

A low-voltage condition at the mains of the proposed wind turbine topology leads to a reduction of the time-averaged quantity $\langle v_b \rangle$. A complete loss of mains voltage, due to a symmetric short-circuit at the grid, leads to $v_b \simeq 0$, and is akin to thyristor T_b conducting in scenario II, albeit the conduction path now includes the fuses F_{b1} and F_{b2}. Obviously, the intervention of the fuses is a drastic measure, that must be avoided in the LVRT strategy. Hence, as stated previously, a means of limiting I_{dc} via control of v_a is needed.

This is achieved through a sequence of repeated, and controlled, interruptions of the dc-link current exiting the diode bridge (marked A in Fig. 5). The network evolution in this case is the same as that described in scenario I because any interruption of the current in that part of the circuit is equivalent to an open-circuit event to the left of port $(a1, a2)$. When the diode D_a conducts, v_a is equal to the negative forward voltage drop of the diode D_a, providing the modulation of the voltage v_a that reduces the function $\psi_{ab}(t)$. In order to create such a controlled sequence of interruptions, the authors propose the addition of a series connected fully controllable switch T_a (for example a GTO or an IGCT) located in the nacelle, between diode bridge A and the port $(a1, a2)$.

It is necessary to provide a turn-off snubber across the power terminals of T_a as a means to absorb and then dissipate the energy stored in the generator leakage inductance as well as the stray inductance of the diode bridge and associated connections. Just after T_a is turned off, I_{dc} flows through the diode D_{sw} and the capacitor C_{sw}, charging it until the voltage v_{sw} slightly exceeds v_{gen}. When this condition is met, the current I_{dc} commutates to the diode D_a until T_a is turned on again. The capacitor C_{sw} must be selected with the aim of rendering its charging transient considerably shorter than the minimum chosen open-circuit duration, especially at low values of I_{dc}. In Fig. 5, it is straightforward to recognize the basic structure of a dc-dc buck converter whose elements are the switch T_a, the diode D_a and the dc-link inductor. The diode bridge and the dc side of the series connected CSIs constitute the two DC voltage sources between which energy conversion occurs.

1311

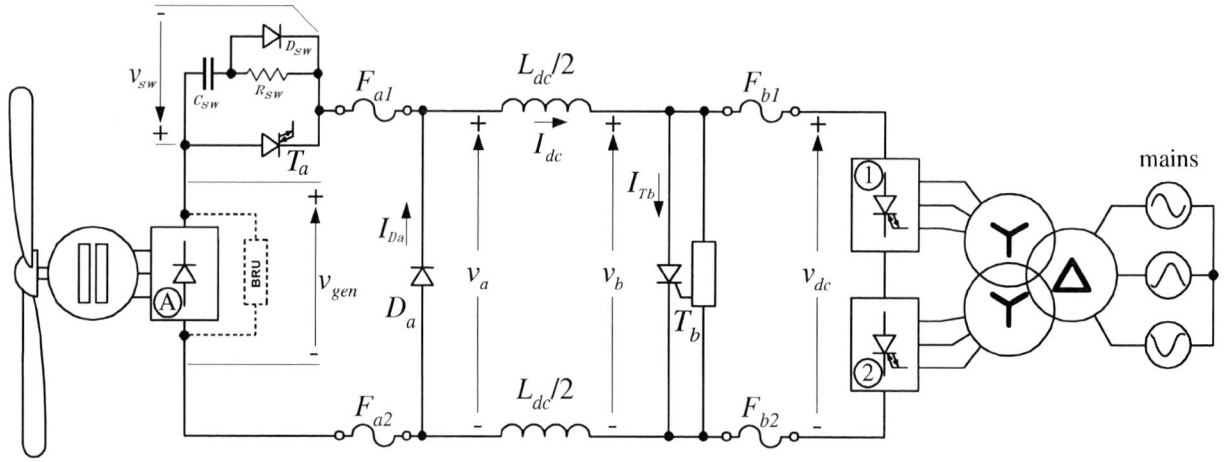

Fig. 5. Modified CSI topology to provide both fault protection and ride-through capability.

During the time interval that T_a is turned off, the current is diverted from the generator. Hence, the electromagnetic torque is zero. Without an accompanying reduction of the applied mechanical torque, the generator will accelerate. Usually the large moment of inertia of the turbine and generator prevents over-speed risks in a time interval of some seconds. In the case of an extended event, the mechanical torque applied to the generator can be decreased by means of the rotor-blade pitch controller. In extreme cases, where the previous measure is not sufficient, it would be possible to add also a braking unit (BRU) directly connected at the output of diode bridge A as is shown in Fig. 5.

The dual type of temporary grid anomaly, i.e. the open circuit condition at the mains, is similar to the open circuit scenario II. Nevertheless, riding-through this event requires that v_b does not reach the value at which the thyristor control circuit turns on T_b. This goal can be achieved by creating a recirculation path for I_{dc}, which is alternative to the one that would be provided by T_b. In order to recover from the anomalous event at any instant, it is necessary that such an alternative path can be opened or closed in a fully controllable way. The authors propose to create the alternative recirculation path by simultaneously turning on, in each of the two inverters, the two switches belonging to one, and only one, chosen phase-leg. All other switches, belonging to the other phase-legs, are simultaneously turned off. In the most general case of an inverter which uses fully bidirectional switches, this last turn off command is necessary to prevent short circuits from occurring between any possible combination of phases belonging to the same three-phase inverter.

This alternative path, obtained by voluntarily creating a short-circuit at the dc side of both inverters, satisfies the requirements since it can be created or eliminated at any moment due to the fully controllable nature of the switches employed in the inverters. This solution does not require any additional components. After the circulation path for I_{dc} is provided, as described above, the network

evolution and its related requirements are identical to the LVRT case discussed previously. Indeed, for the whole network located to the left of the port $(b1, b2)$, the artificially created short-circuit at the inverter's dc side is indistinguishable from a short-circuit in the grid.

In the extreme case where the control of I_{dc} is lost for any reason, leading to the intervention of the fuses F_{b1} and/or F_{b2} as well as F_{a1} and/or F_{a2}, the protection of the system would be ultimately provided by T_b and D_a in full accordance to the open-circuit scenario II. Further details about the implementation of the LVRT and open-circuit strategies are given in the following sections.

V. SELECTED SIMULATION RESULTS

A simulation model using PLECS, Simulink and Matlab has been constructed of the experimental setup described in [3]–[5], [8]. The parameters of the model are as follows: The nominal RMS value of the line-to-line voltage, both at the primary and the secondaries of the 60 Hz transformer is 230 V. The dc-link inductance is 100 mH and the 48-pole axial-flux permanent magnet generator is characterized by 620 V peak line-to-line voltage at no-load and 570 RPM. The generator has been designed such that, in normal operating conditions and with desired unity fundamental power factor ($\beta^*=0$), the angle α can assume the optimum value of 0.914π, leading to a significant reduction of the mains line current harmonic content [3], [4]. The chosen wind turbine average power (7 kW) leads to a dc-link reference current $I_{dc_{ref}}$ around 11.5 A. It is chosen that the overall system controller maintain $I_{dc_{ref}}$ practically constant during the LVRT and open-circuit events and that the current hysteresis band be 5 A, a value that leads to switching frequencies, as well as minimum 'on' and 'off' times, suitable for the slowest high-power semiconductor switches currently available.

The simulation results, according to each particular grid event, are described in the following subsections.

1312

A. Ride-Through of a Low Voltage event at the Mains

During normal operation, with the value of the mains line to line voltage V_{LL} inside the accepted tolerances, the series connected switch T_a is continuously gated (or turned on), therefore it produces only conduction losses.

A LVRT event begins when V_{LL} is reduced below such tolerances. Assuming a constant average rectified generator voltage $\langle v_{gen}(t) \rangle$ over the duration of the LVRT event in question and a desired fundamental power factor at the mains imposed by a specified angle β^* [4], [8], the control angle α is gradually increased, by the controller, towards π as V_{LL} continues to decrease. This occurs because the controller attempts to maintain $\langle v_{gen}(t) \rangle$ equal to $\langle v_{dc}(t) \rangle$ in order to preserve the dc-link current I_{dc} at the reference value. When α assumes the maximum value π and V_{LL} continues to decrease, β must be reduced, ultimately, to zero in order to maximize $\langle v_{dc}(t) \rangle$ according to (4). At this point, if V_{LL} continues to decrease such that $\langle v_{gen}(t) \rangle > \langle v_{dc}(t) \rangle$ for a sufficiently long interval of time, inverter control of the dc-link current is lost and I_{dc} will rise.

When I_{dc} reaches a chosen upper threshold, the switch T_a is turned off and I_{dc} then commutates from the generator to the diode D_a as previously explained. After this instant, I_{dc} begins to decrease, however the inverters will continue to synthesize the mains line current from the remaining dc-link current according to (5) to aid in grid-fault clearing. As I_{dc} decays below a specified lower threshold level T_a is turned on again. If the fault has cleared, the system will resume normal operation. Conversely, if the fault has not cleared, the process will be repeated until the LVRT event is ended. As is the case with a dc-dc buck converter, switching T_a to control I_{dc} renders $\langle v_a(t) \rangle = \langle v_{dc}(t) \rangle$.

It should be noted that special attention must be given to maintaining converter synchronization to the grid, especially during a zero-voltage event which may cause the synchronization mechanism (usually a PLL) to drift. This issue is not specific to the proposed topology, but it is shared by all types of conversion circuits, because it arises from a fundamental lack of information about the phase of the mains voltage. While this is an important practical issue, it is outside the scope of this paper. Nevertheless, assuming this information is available, the proposed LVRT technique is capable of delivering mains current even down to zero mains voltage to assist in fault clearing. In any case, it is never necessary to open the mains breakers, thus satisfying the most crucial LVRT requirement.

Figs. 6 and 7 show one selected simulation result for a LVRT event characterized by the minimum value of V_{LL} (0.15 PU) specified by the present FERC standard. Fig. 6 shows the entire event, whereas Fig. 7 shows the LVRT event on an expanded time scale in order to more closely analyze the details. The electrical quantities listed in the legend are identified by the same names used in the circuit diagrams of Figs. 2, 4 and 5.

The condition chosen to identify the beginning of a

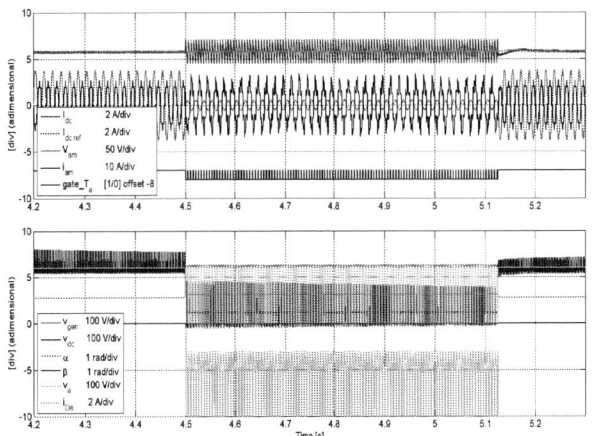

Fig. 6. Most significant quantities during a LVRT event lasting 0.625 seconds with percentage $V_{LL} = 0.15$ PU

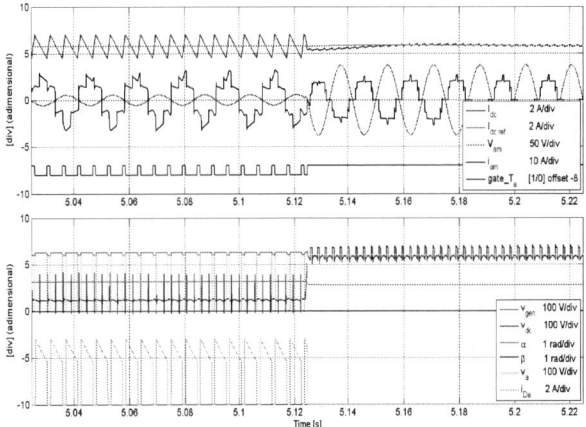

Fig. 7. Most significant quantities during the same LVRT event as Fig. 6 on an expanded time scale highlighting the recovery transient.

LVRT event and to switch to the LVRT control strategy is $(\alpha > 0.98\pi)\text{AND}(I_{dc} > I_{dc_{\text{ref}}})$. Indeed having α so close to π means that the proportional-integral (PI) dc-link current controller, whose key output is the necessary $\langle v_{dc} \rangle$, cannot increase such voltage any more to maintain the control of I_{dc}. This controller uses the measured V_{LL} and desired β^* to compute α from $\langle v_{dc} \rangle$ via (4), thereby improving the response dynamics of the converter. Entering the LVRT control mode, immediately activates the hysteretic controller of I_{dc} that commands the commutations of T_a, as is evident from Fig. 6 and 7, to achieve $\langle v_a \rangle = \langle v_{dc} \rangle$. The hysteresis thresholds are symmetrical around $I_{dc_{\text{ref}}}$. The end of the LVRT event is detected by the condition $V_{LL} > 0.9 V_{LL_{\text{nom}}}$ (where $V_{LL_{\text{nom}}} = 230$ V).

From Fig. 7, one can notice that the I_{dc} control in normal operation and, consequently, the lower harmonic content in the mains currents are recovered in less than two fundamental cycles. Because of the aforementioned use of (4) to compute α in the dc-link current PI regulator, such an angle recovers its optimum value very quickly at the end of the LVRT event. Additionally, one can observe

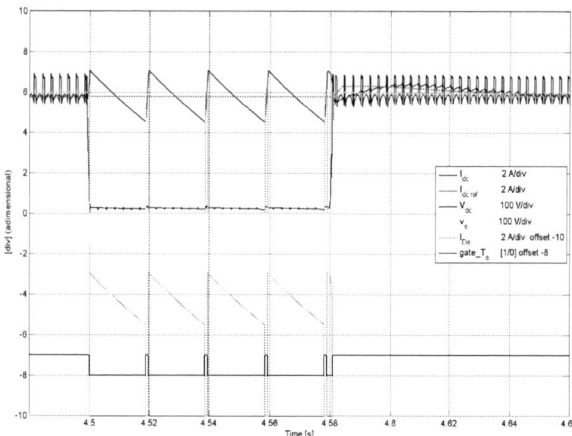

Fig. 8. Waveforms during an open-circuit event lasting 80 ms with commanded short circuit of both current source inverters

that, during the LVRT event, the pulsed voltage v_a has an average value that satisfies the necessary condition for current stability $\langle v_a \rangle = \langle v_{dc} \rangle$ and that the switching frequency of T_a is around 200 Hz with 'on' times around 1 ms. The high-frequency voltage spikes observable in v_{dc} during the LVRT event are due to the diode bridge-based snubbers, described in [4], [8], whose dc-link capacitors still possess voltage values close to the secondary line-to-line peak nominal voltage ($230\sqrt{2}$ V). This is due to the relatively short duration of the LVRT event which does not allow those capacitors to discharge significantly via their parallel connected resistors.

B. Ride-Through of an Open Circuit event at the Mains

It is important to highlight that an open-circuit event, as defined in this paper, is characterized by an open-circuit condition at the point of interconnection with the grid of one single wind turbine. This is a very different case from what is commonly called "islanding", a scenario that usually involves one or more generators and loads connected together to form a smaller grid that has become isolated from the main one. The capability to ride-through an open-circuit at the grid is addressed here because this event is particularly harmful for current source inverters.

An open-circuit event lasting 80 ms has been simulated and the most significant waveforms are shown in Fig. 8. After the detection of the open-circuit event, the two inverters are voluntarily short circuited at their dc side by gating the proper switches. As stated previously, from this instant onwards, the open-circuit event is essentially indistinguishable from a zero voltage LVRT event to the left of port $(b1, b2)$.

Observing Fig. 8, one can notice that the voltage spikes previously present in v_{dc} during the LVRT event are now absent during the open-circuit event because port $(b1, b2)$ is directly short-circuited by the inverters, whose commutations have stopped. The control of $I_{dc_{ref}}$ by the overall system controller is more difficult during a recovery from an almost zero v_{dc}. Nevertheless, the normal control of I_{dc} is resumed in about 80 ms from the end of

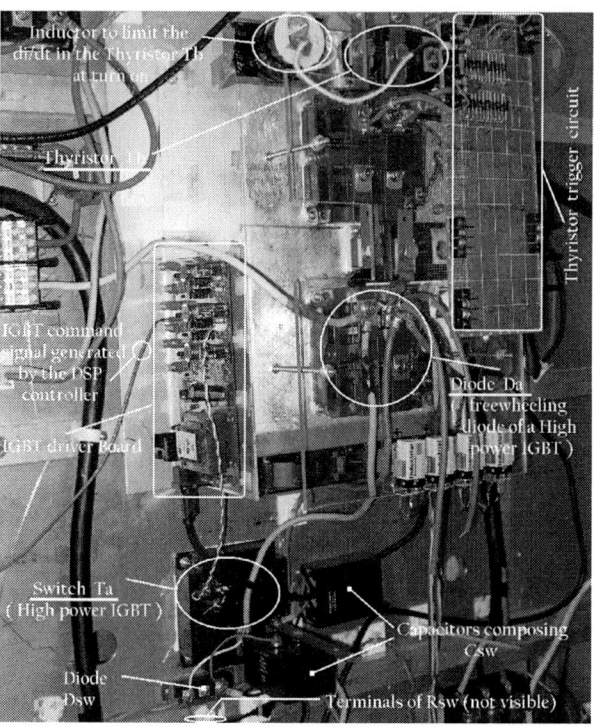

Fig. 9. Photo of the components composing the extended protection scheme to cope with the LVRT and open-circuit events

the open-circuit event. Here, the switching frequency of T_a is around 50 Hz, lower than the previously shown LVRT case, due to the fact that $v_a \approx v_{dc}$ when D_a conducts. This, in turn, leads to a much slower rate of decay of I_{dc}. As a consequence, I_{dc} requires more time to reach the low hysteresis threshold when compared to the presented LVRT case and this explains the lower switching frequency. Observing the waveforms, one can infer that the necessary condition $\langle v_a \rangle = \langle v_{dc} \rangle$ is satisfied. The switching frequency of T_a is almost constant both in the LVRT and open-circuit cases despite the strong hysteretic nature of the LVRT control mode. This occurs because the staircase-shaped voltage $v_a - v_{dc}$ applied across the inductor terminals is characterized by voltage levels having good stability to which relatively high-frequency and regular ripples are superimposed.

VI. Selected Experimental Results

Figure 10 shows a selection of waveforms extracted from the experimental test of the intervention of the thyristor trigger circuit and the subsequent turn on of the thyristor T_b. These are the most important subsystems for the ultimate fault protection of the current source inverter topology, as well as for the detection of an open-circuit event. The recorded waveforms are labeled with the names of the quantities they represent according to the names used in Figs. 2, 4 and 5. Further experimental tests and details, also concerning the internal structure and operation of the thyristor trigger circuit, are reported in [9].

An open circuit event is created to the right of port $(b1, b2)$ through a high-power IGBT connected in series

1314

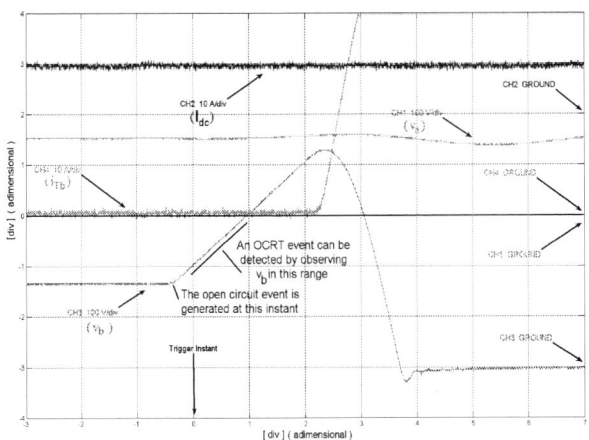

Fig. 10. Turn on of Thyristor T_b after an open-circuit event to the right of port $(b1, b2)$

and used as breaker. Immediately after having turned off the IGBT, the voltage v_b begins to rise linearly because the current I_{dc} is initially diverted into a properly designed capacitor present in the thyristor trigger circuit [9]. When v_b reaches the desired intervention threshold, T_b is turned on. It is possible to identify this instant in Fig. 10 by observing the fast rise of the current I_{T_b} (its time derivative is limited by a turn-on snubber inductance connected in series to T_b). From this point onwards I_{dc} is diverted into T_b. The current I_{T_b} temporarily exceeds the value I_{dc} because of the discharge of the aforementioned capacitor through T_b and the turn-on snubber inductance. The similar linear rise of v_b which follows an open-circuit at the grid can be used to detect the open-circuit event. The inverter bridges can be short-circuited when v_b reaches a chosen voltage threshold set well below the one which would trigger T_b.

Fig. 9 shows a picture of a specific area of the experimental setup where all key components of the extended protection scheme are located. The text superimposed on the picture identifies those components according to the names used in Figs. 2, 4 and 5. Additionally, it is possible to observe the turn-on snubber inductance, used to limit the time-derivative of the current flowing in T_b at turn on, together with the IGBT driver board used to commutate T_a which is constituted by a high-power IGBT.

VII. CONCLUSIONS

It is a fast-approaching reality that any sizable wind-turbine facility will be expected to perform in a manner similar to their more traditional brethren. As the quest for higher-performance, higher-power wind turbines continues, newly proposed topologies must also be measured with respect to their ability to react to grid anomalies. Regulatory agencies in the United States, Europe and other parts of the world have begun to set minimum standards for wind turbine performance in this regard.

With an eye towards the issues involved in scaling existing wind-turbine technology into the multi-megawatt realm, the authors have proposed a variable-speed wind-

turbine employing a gearless axial-flux permanent magnet synchronous generator coupled by a fully-rated, line-frequency switched, current source inverter, the details of which have been reported in previous works.

In this paper, the authors have explored the operation of the proposed topology with respect to both low-voltage (including short-circuit) and open-circuit grid conditions and have demonstrated similarities with fault modes existing within the converter itself.

At the system level, many different possibilities exist with respect to both the mechanical and the electrical control of generator torque during such events. The operation of the modified topology, under one possible control scenario (maintaining constant dc-link current), was successfully demonstrated through simulation. It was shown that the topology could continue to provide fault-clearing current to the mains during a low-voltage or short-circuit event and that, with an expanded inverter algorithm, the topology can repeatedly withstand, and recover from, an open-circuit event.

Experimental results of the action of the fault protection circuitry during an open circuit event were shown to demonstrate the ability of the system to withstand an open-circuit condition.

ACKNOWLEDGMENT

The authors would like to thank the National Renewable Energy Laboratory (NREL) which funded the research project under the contract number XCX-3-32227-04.

REFERENCES

[1] "Interconnection for wind energy (order no. 661 final rule, docket no. rm05-4-000)," Federal Energy Regulatory Commission (FERC), Tech. Rep., June, 2nd 2005.

[2] "On the low voltage ride-through (lvrt) provisions of the order no. 661. final rule (joint report by the north american electric reliability council (nerc) and the american wind energy associtation (awea), docket no. rm05-4-000)," Federal Energy Regulatory Commission (FERC), Tech. Rep., September, 19th 2005.

[3] P. Tenca and T. A. Lipo, "Reduced cost current source topology improving the harmonic spectrum through on-line functional minimization," in *IEEE 35th Annual Power Electronics Specialists Conference, 2004. PESC 2004.*, vol. 4, Aachen, Germany, 2004, pp. 2829–2835.

[4] P. Tenca, A. A. Rockhill, and T. A. Lipo, "Wind turbine current-source converter providing reactive power control and reduced harmonics," *IEEE Trans. Ind. Appl.*, vol. 43, no. 4, pp. 1050–1060, July/August 2007.

[5] T. A. Lipo and P. Tenca, "Design and test of a variable speed wind turbine system employing a direct drive axial flux synchronous generator," National Renewable Energy Laboratory (NREL), Available electronically at: http://www.osti.gov/bridge, Tech. Rep., July 2006.

[6] P. Wood, *Switching Power Converters*. Van Nostrand Reinhold, 1981.

[7] B. Pelly, *Thyristor Phase-Controlled Converters and Cycloconverters*. Wiley-Interscience, 1971.

[8] P. Tenca, A. A. Rockhill, and T. A. Lipo, "Current-source topology for wind turbines capable of providing leading power factor while reducing line current harmonics," in *Proceedings of the IEEE Industry Applications Conference 2006, IAS 2006*, vol. 1, Tampa, Florida, October 8-12 2006, pp. 222 – 229.

[9] P. Tenca and T. Lipo, "A decentralized protection scheme for converters utilizing a dc-link inductor," in *IEEE 31st Annual Conference of the Industrial Electronics Society. IECON 2005.*, Raleigh, North Carolina, USA, November 2005, pp. 854–859.

Optimal Control of Direct Driven Feed Axes with Flexible Structural Components

Ekkehard Batzies*, Tobias Schöller**, Volkmar Welker*** and Oliver Zirn*

*Arbeitsgruppe Mechatronik, University of Applied Sciences Giessen, 35390 Giessen, Germany,
{ekkehard.batzies, oliver.zirn}@ei.fh-giessen.de
**Rückle GmbH Werkzeugfabrik, 72587 Römerstein, Germany
***Fachbereich Mathematik und Informatik, Philipps-Universität Marburg, 35390 Marburg, Germany,
welker@mathematik.uni-marburg.de

Abstract: **In this paper we provide formulas for the controller gain for the velocity control of direct driven feed axes with flexible structural components. Our mathematical setting covers two general classes of structural flexibilities. For the first class we present a formula for the damping optimum, for the second a formula for the stability bound. Both formulas are algebraically exact. They are proved using the algebraic approach provided by Gröbner bases. In addition, the formulas are shown to fit well to experimental data from the different drives of a machine tool manipulator.**

Index Terms: **Control of servo drives, direct driven feed axes, intelligent control, structural flexibilities.**

I. INTRODUCTION

In the control of electric drives (see for example [1], [2], [3], [4]) much effort is devoted to finding optimal parameters of the controller in question. In the case of intelligent control of direct driven feed axes a central issue is the optimal choice of the velocity controller gain K_p with respect to damping. In this paper we solve this problem in the important situation when there are *internal structural flexibilities* of the drive. In addition, in the case of *elasticities of the measurement system* we also present a formula for the stability bound.

The classification of all main occurrences of *internal structural flexibilities* (i.e. *flexible load, elastic base, asymmetric drive force application* and *excitation of*

lateral flexibilities) in direct driven feed axes shows (see [5]) that mathematically, they are all equivalent. Figure 1 shows the velocity response of the C-axis of the machine tool manipulator from Figure 2. The data in Figure 1 exhibit a reasonably well damping of the *internal structural flexibility*. On the other hand, *measurement elasticity* represents the situation where the controller feedback comes from a measurement system which is mounted elastically to the drive axis. Mathematically this leads to a different situation.

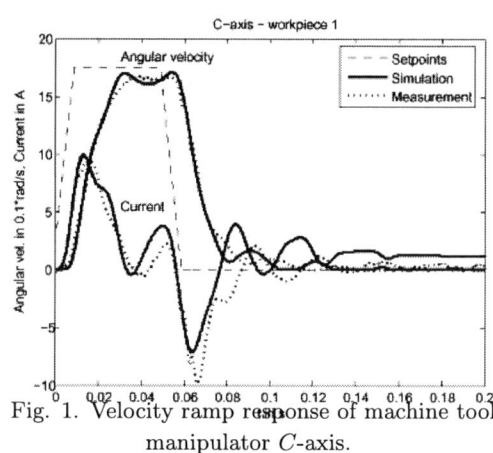

Fig. 1. Velocity ramp response of machine tool manipulator C-axis.

In [5], coarse approximation formulas for the controller gain K_p for both above mentioned general cases are presented. The objective of this paper is to derive the algebraically exact versions of these formulas. We achieve this by an algebraic approach involving the

---

978-1-4244-0644-9/07/$25.00 ©2007 IEEE   1316

Fig. 2. Machine tool manipulator.

concept of Gröbner bases. This approach allows treatment of internal damping both for *measurement elasticities* (where it constitutes the dominant effect) and also for *internal structural flexibilities* (where it provides minor correction terms).

II. STRUCTURAL FLEXIBILITIES

Both classes in the above mentioned classification (that is *internal structural flexibilities* and *measurement elasticities*) involve the following mathematical representation of a two body system with structural flexibilities:

$$\theta \ddot{\phi} = M \quad (1)$$
$$\sigma \ddot{\psi} + 2D\sigma\omega_0 \dot{\psi} + \sigma\omega_0^2 \psi = \mu M \quad (2)$$

As an example, in the case *flexible load* (see classification above and Figure 3.), these variables and parameters represent:

- ϕ: Position of center of inertia
- ψ: Difference of positions of actuator and load
- M: Input torque
- θ: Total inertia
- λ: Actuator inertia divided by total inertia
- μ: $= 1 - \lambda$: Load inertia divided by total inertia
- D: Relative internal damping according to material
- σ: $= \lambda\mu\theta$: Reduced inertia of two body system
- ω_0: Eigenfrequency of two body system

In this case, Equations (1) and (2) are equivalent to the following equations:

$$J_1 \ddot{\phi}_1 = M - d(\dot{\phi}_1 - \dot{\phi}_2) - k(\phi_1 - \phi_2) \quad (3)$$
$$J_2 \ddot{\phi}_2 = d(\dot{\phi}_1 - \dot{\phi}_2) + k(\phi_1 - \phi_2) \quad (4)$$

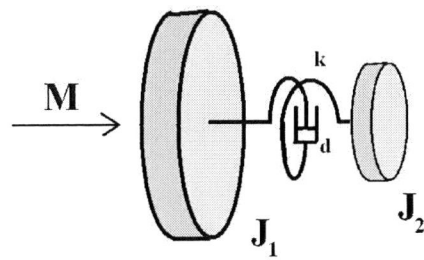

Fig. 3. Actuator and flexible load: Driving torque M, actuator inertia J_1, load inertia J_2, spring constant k, damping constant d.

where the variables and parameters represent:

- ϕ_1: Position of actuator
- ϕ_2: Position of load
- J_1: Actuator inertia
- J_2: Load inertia
- k: Torsion spring constant
- d: Torsion damping constant

The corresponding parameter transformations are given by:

$$\theta = J_1 + J_2$$
$$\sigma = \frac{J_1 J_2}{J_1 + J_2}$$
$$\lambda = \frac{J_1}{J_1 + J_2}$$
$$\omega_0 = \sqrt{k\frac{J_1 + J_2}{J_1 J_2}}$$
$$D = \frac{d}{2\sqrt{k}}\sqrt{\frac{J_1 + J_2}{J_1 J_2}}$$

The difference in the mathematical representation of the two major classes *internal structural flexibilities* and *measurement elasticities* comes from the way the above two body system is embedded into the controller loop. We treat these two classes in the following sections.

III. DAMPING OPTIMUM FOR VELOCITY CONTROL WITH INTERNAL STRUCTURAL FLEXIBILITIES

In this section, we examine the situation where a drive with *internal structural flexibilities* as described by Equations (1) and (2) is inserted into a velocity controller loop with proportional gain K_p such that the feedback measurement system is mounted non-elastically to the actuator side of the drive. We get the following result:

1317

Theorem 1 *For the proportional velocity control of a drive with internal structural flexibilities, optimal damping control is achieved by choosing the following value for the proportional velocity controller gain K_p:*

$$K_p = \frac{1}{2Z}(\sqrt{X} - Y)\theta\omega_0 \qquad (5)$$

where

$$
\begin{aligned}
X &:= \lambda(\lambda+1)^2)\dots \\
  &\quad +4D^2\lambda^3(1-\lambda)(4D^2-1)\dots \\
  &\quad -8D^2\lambda^2 + 4D^4\lambda^5 \\
Y &:= \lambda(1+3\lambda)D - 4D^3\lambda^2 \\
Z &:= 4D^4\lambda^2 - 5D^2\lambda + 1
\end{aligned}
$$

A sketch of a proof for Theorem 1 is given in Section V.

Since the relative damping value D usually is a number close to 0, linear approximation at $D = 0$ of Equation (5) yields a reasonable approximation for the value of K_p:

Corollary 2 *For $D \approx 0$, that is small internal damping in the setting of Theorem 1, optimal damping control is achieved by choosing the following value for the proportional velocity controller gain K_p:*

$$K_p := \frac{1}{2}(\sqrt{\lambda}(\lambda+1) - \lambda(1+3\lambda)D)\theta\omega_0 \qquad (6)$$

Corollary 3 *For $D = 0$, which means neglecting internal damping in the setting of Theorem 1, optimal damping control is achieved by choosing the following value for the proportional velocity controller gain K_p:*

$$K_p := \frac{1}{2}\sqrt{\lambda}(\lambda+1)\theta\omega_0 \qquad (7)$$

IV. STABILITY BOUND FOR VELOCITY CONTROL WITH MEASUREMENT ELASTICITIES

Equations (1) and (2) are also involved when describing measurement elasticities: We now interpret the whole drive (without internal structural flexibilities) as the *actuator side* and the elastically mounted feedback measurement system as the *load side*. For this setting, it is known (see [5]) that there is no optimal damping control value for the velocity controller gain K_p. Instead, increasing the value for K_p decreases damping. From the practical point of view, the stability bound is a reasonable value for K_p.

Theorem 4 *For the proportional velocity control of a drive with measurement elasticities, the stability bound for the proportional velocity controller gain K_p is given by the following value:*

$$K_p := \frac{2D}{1-2D^2}\omega_0\theta \qquad (8)$$

A sketch of the proof of Theorem 4 is given in the subsequent Section V.

V. GRÖBNER BASIS FOR POLE EQUATIONS

A. Proof of Theorem 1

When inserting a drive with internal structural flexibilities as described by Equations (1) and (2) into a velocity control loop with proportional gain K_p and with feedback measurement system mounted non-elastically to the actuator side, the resulting characteristic polynomial $\mathcal{P}$ is given by:

$$
\begin{aligned}
\mathcal{P}(s) &= s^3 + (\frac{K_p}{\lambda\theta} + 2D\omega_0)s^2 + \dots \qquad (9) \\
&\quad + \ (\omega_0^2 + 2D\omega_0\frac{K_p}{\theta})s + K_p\frac{\omega_0^2}{\theta}
\end{aligned}
$$

On the other hand, when we denote the three zeros of this characteristic polynomial by $\delta_0 \in \mathbb{R}$ and $\delta_1 \pm i\omega_1 \in \mathbb{C}$, we also get:

$$
\begin{aligned}
\mathcal{P}(s) &= s^3 + (\delta_0 + 2\delta_1)s^2 + \dots \qquad (10) \\
&\quad + \ (\delta_1^2 + \omega_1^2 + 2\delta_0\delta_1)s + \delta_0(\delta_1^2 + \omega_1^2)
\end{aligned}
$$

Comparing coefficients, we get the following three equations:

$$
\begin{aligned}
K_p + \lambda\theta(2D\omega_0 - \delta_0 - 2\delta_1) &= 0 \qquad (11) \\
\theta\omega_0^2 + 2D\omega_0 K_p \dots \\
\dots - \theta(\delta_1^2 - \omega_1^2 - 2\delta_0\delta_1) &= 0 \qquad (12) \\
K_p\omega_0^2 - \theta\delta_0(\delta_1^2 + \omega_1^2) &= 0 \qquad (13)
\end{aligned}
$$

The practical meaning of the coefficients δ_0, δ_1 and ω_1 can best explained in the example *flexible load* as shown in Figure 3. The shape of the step response of the load velocity is depicted in Figure 4. Here the coefficient δ_1 determines the exponential decrease of the envelope of the step response, ω_1 represents the step response frequency. Additionally, there is a summand $1 - \alpha\exp(\delta_0 t)$ which represents the whole-body dynamics.

Using Gröbner basis algorithms (such as provided by the Maple or Singular Computer Algebra Systems), for above equations, it is possible to obtain a

$$K_p\omega_0^2 - \theta\delta_0(\delta_1^2 + \omega_1^2) \;=\; 0 \tag{17}$$

To compute the stability bound we set $\delta_1 := 0$. This simplifies Equations (15) - (17) to:

$$2D\omega_0 - \delta_0 \;=\; 0 \tag{18}$$
$$\theta\omega_0^2 + D\omega_0 K_p - \theta\omega_1^2 \;=\; 0 \tag{19}$$
$$K_p\omega_0^2 - \theta\delta_0\omega_1^2 \;=\; 0 \tag{20}$$

This system of equations can be solved by hand and the solution is given by Equation (8).

V. Experimental Validation

The algebraic solutions given by Equations (5) - (7) for the optimum velocity controller gain fit very well to the commissioning results from many examined machine tool and robot axes.

As an example, we consider the machine tool manipulator with 3 direct driven DOF shown in Figure 2. In Table I and II we list the commissioning results using standard engineering methods (e.g. experimental step response analysis) and the gains following Equations (5), (6) and (7). As there is almost negligible damping in the plant structure, the values from the three different equations are very close to each other. (For the relative damping coefficient D, we use a value of $D = 1\%$.)

Axis	K_p foll. eq. (5)	K_p foll. eq. (6)	K_p foll. eq. (7)	exper. results for K_p
C-Axis (work-piece 1)	144.2 Nms/rad	144.1 Nms/rad	146.0 Nms/rad	130−−180 Nms/rad
C-Axis (work-piece 2)	416.2 Nms/rad	416.1 Nms/rad	423.4 Nms/rad	400−−750 Nms/rad
B-Axis	4026 Nms/rad	4024 Nms/rad	4096 Nms/rad	4000−−4500 Nms/rad
X-Axis	58146 Ns/m	58140 Ns/m	58679 Ns/m	50000−−70000 Ns/m

Table I: Experimental and algebraic controller gain values for the different axes of the machine tool manipulator

In Table II we list the system parameters of the different axes (See also [5], [7], [8]).

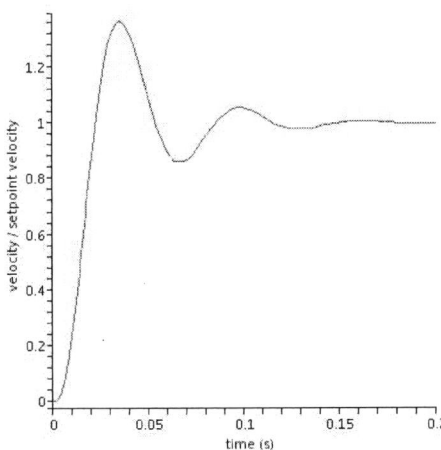

Fig. 4. Step response of load velocity.

Gröbner basis with exactly one (quite large) polynomial $f = f(\delta_1, K_p, \lambda, \theta, \omega_0, D)$, which does not contain the variables δ_0 and ω_1. Since we are interested in solutions with minimal δ_1, the theory of Lagrange multipliers gives us another equation, namely $\frac{\partial f}{\partial K_p} = 0$. Applying again Gröbner Basis algorithms to the two equations $f = 0$ and $\frac{\partial f}{\partial K_p} = 0$, we obtain a Gröbner basis with exactly one (again quite large) polynomial $g = g(K_p, \lambda, \theta, \omega_0, D)$, which does not contain the variable δ_1. g is quadratic in K_p, so it can be solved for K_p. The solution is given by Equation (5). The proof is completed by the fact that from the *Elimination Theorem* and the *Extension Theorem* (see [6, p.12]), it follows, that all solutions $(K_p, \lambda, \theta, \omega_0, D)$ of the equation $g(K_p, \lambda, \theta, \omega_0, D) = 0$ extend to solutions of the Equations (11) -(13).

B. Proof of Theorem 4

When inserting a drive (without *internal structural flexibilities*) with feedback measurement system mounted elastically to the drive into a velocity control loop with proportional gain K_p, the resulting characteristic polynomial $\mathcal{P}$ is given by:

$$\mathcal{P}(s) \;=\; s^3 + 2D\omega_0 s^2 + \ldots \tag{14}$$
$$\ldots \;+\; (\omega_0^2 + D\omega_0\frac{K_p}{\theta})s + K_p\frac{\omega_0^2}{\theta}$$

Comparing again coefficients with Equation (10) leads to the following three equations:

$$2D\omega_0 - \delta_0 - 2\delta_1 \;=\; 0 \tag{15}$$
$$\theta\omega_0^2 + D\omega_0 K_p \ldots$$
$$\ldots - \theta(\delta_1^2 - \omega_1^2 - 2\delta_0\delta_1) \;=\; 0 \tag{16}$$

1319

Axis	Eigen-frequ. ω_0	ratio λ	total inertia resp. total mass θ
C-Axis (workpiece 1)	$88\ s^{-1}$	0.57	$2.8\ kgm^2$
C-Axis (workpiece 2)	$247\ s^{-1}$	0.82	$2.08\ kgm^2$
B-Axis	$454\ s^{-1}$	0.84	$10.7\ kgm^2$
X-Axis	$204\ s^{-1}$	0.36	$705\ kg$

Table II: System parameters for the different axes of the machine tool manipulator

The experimental results also depend on additional plant parameters like encoder quantization, friction and subjective performance impressions by the commissioning engineer that are not captured by our mathematical model. These additional influences in combination with the intrinsic robustness of the cascade controller lead to a certain tolerance range for the practical commissioning.

VI. Conclusion

This paper presents an algebraic approach to the control of direct driven feed axes with structural flexibilities. Using the concept of Gröbner bases, algebraically exact optimal damping and stability bound formulas for the velocity controller gain are derived. Experimental results from a 3DOF machine tool manipulator prove their practical relevance.

Acknowledgment

The presented result has been worked out in the applied research project AiF FH3 FKZ 1733X04 founded by the German Federal Ministry of Education and Research.

References

[1] R. Crowder, "Electric Drives and Electromechanical Systems: Applications and Control", Elsevier, Amsterdam, 2006.

[2] H. Groß, J. Hamann, G. Wiegärtner, "Vorschubantriebe in der Automatisierungstechnik". Hrsg. Siemens AG, München, Publicis MCD Verlag, 2006.

[3] T. Kenjo, "Electric Motors and their Controls: An Introduction", Oxford University Press, Oxford, 1001.

[4] D. Schröder, "Elektrische Antriebe, Bd 2: Regelung von Antrieben". Springer-Verlag, Heidelberg, 1995.

[5] O. Zirn, A. Monat, T. Schöller, "Control of Direct Driven Feed Axes with Flexible Structural Components", Conference record IEEE Industrial Applications Society Annual Meeting, Hongkong, 2005.

[6] D: Cox, J. Little, D. O'Shea, "Using Algebraic Geometry". Springer-Verlag, Heidelberg, 1998.

[7] O. Zirn, S. Weikert, "Modellbildung und Simulation hochdynamischer Fertigungssysteme", Springer-Verlag, Heidelberg, 2005.

[8] O. Zirn, E. Batzies, S. Weikert, T. Schöller, "State Control of Servo Drives with Flexible Structural Components", Conference record IEEE Industrial Applications Society Annual Meeting, Tampa FL, 2006.

Leakage Energy Recovered Narrow Pulsed Voltage Generator Associated with Ultrasound Generator for Liquid Food Sterilization

S. -Y. Tseng*, Y. -D. Chang, P. -L. Huang T. -F Wu and Y. -M. Chen

*Department of Electrical Engineering
Chang Gung University
Kwei-Shan, Tao-Yuan, Taiwan
E-mail: sytseng@mail.cgu.edu.tw
Tel: 886-3-2118800; Fax: 886-3-2118700

Elegant Power Application Research Center
(EPARC)
National Chung Cheng University
Ming-Hsiung, Chia-Yi, 621 Taiwan
E-mail: d94415002@ccu.edu.tw
Tel: 886-5-2420411; Fax: 886-5-2720862

Abstract --This paper presents a narrow pulsed voltage generator along with an ultrasound generator for liquid food sterilization. The generator is derived from a bidirectional flyback converter with leakage energy recovery circuits, which can improve efficiency about 12%. In addition, the proposed generator is associated with a half-bridge inverter for generating ultrasound to clean sterilizer electrodes. In this combined generator, the switches in the converter and the inverter are integrated with the synchronous switch technique, reducing its weight, size and volume while improving efficiency significantly. Experimental results obtained from a prototype with the output voltage of $\pm 6\ kV \sim \pm 20\ kV$ and peak power of 1.2 MW have verified its feasibility.

Keywords : ultrasound generator, liquid food sterilization, bidirectional flyback converter, and half-bridge inverter

I. INTRODUCTION

Recently, pulsed electric field (PEF) used in food sterilization attracted a lot of attention [1]-[8]. In particular, for liquid food, PEFs consume power being less than 10% of that required for thermal pasteurization, and they only result in a little increase in temperature. As compared with conventional pasteurization method, the PEF one can provide consumers with safe, nutrition and fresh quality of foods. Thus, it can replace or complement conventional thermal processing methods. To further reduce power consumption during sterilization interval, PEF with narrow pulses (pulse width ≤ 1 μs) has gradually replaced that with wide pulses (pulse width > 1 μs).

To generate narrow pulses, many PEF generators use pulse forming networks (PFNs) with transmission line, or vacuum switch tubes with high voltage ratings are used to chop a dc voltage into narrow pulse voltage waveforms [9]-[11]. As a result, it is difficult to design such a narrow PEF generator and will result in high cost. A narrow PEF generator with spark gap is therefore adopted to release these drawbacks. For topology consideration, flyback converter has the merits of simple circuit structure and pulsating output current characteristic, which can charge its capacitor to any voltage levels [12]-[13]. Thus, a bidirectional flyback converter is usually cooperated with a spark gap, as shown in Fig. 1, to generate high narrow pulsed voltages. In addition, sterilizer electrodes used in liquid food sterilization could be adhered with particles

existing in the processed food, increasing contact resistance and losing its function. To avoid these problems, a bidirectional flyback converter, and a half-bridge inverter used to drive piezoelectric device for cleaning the electrodes can be adopted, as shown in Fig. 2. In Fig. 2, since leakage inductance existing in the transformer causes power loss, two active clamp circuits are introduced to the bidirectional flyback converter for reducing voltage spike across power switches and improving conversion efficiency. To simplify the circuit configuration of the proposed sterilization system, the synchronous switch technique [14] is used to integrate their active switches, which can achieve zero-voltage switching feature. Derivation of the proposed generator is illustrated in Fig. 3. The proposed narrow pulsed voltage generator can increase conversion efficiency and reduce its weight, size and volume significantly. Note that since the proposed generator are mostly operated in discontinuous conduction mode (DCM) to achieve zero-current switching (ZCS) at turn-on transition, the active clamp circuit is only used to recover the energy trapped in leakage inductor and to reduce voltage stress of switches for increasing conversion efficiency of the proposed one.

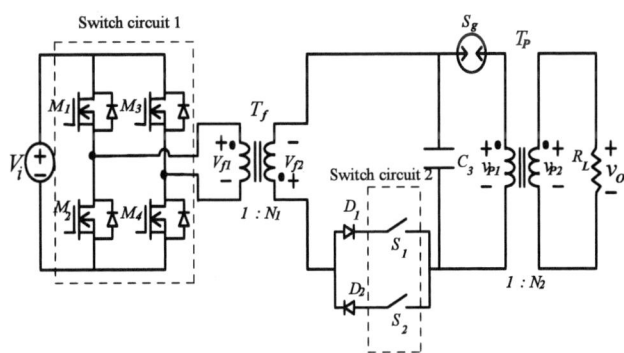

Fig. 1. Schematic diagram of a conventional bipolar hard–switching flyback converter.

978-1-4244-0644-9/07/$25.00 ©2007 IEEE

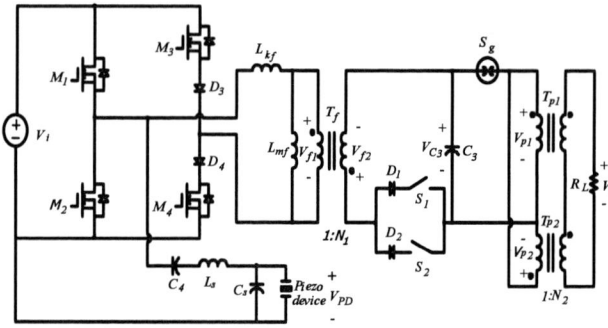

Fig. 2. A PEF generator combined with an ultrasound generator.

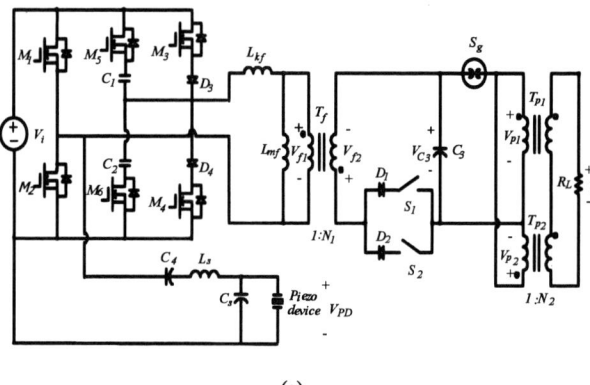

(a)

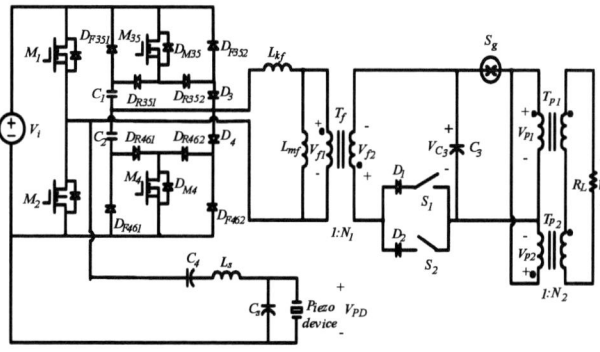

(b)

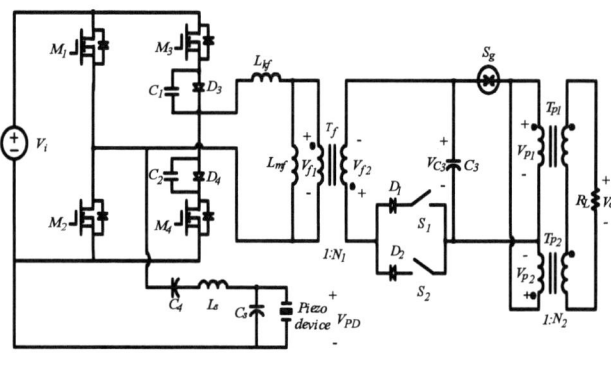

(c)

Fig. 3. Circuit derivation of the proposed PEF generator combined with ultrasound generator and energy- recovery circuits.

II. MECHANISM OF LIQUID FOOD STERILIZATION

The main task of sterilization processing is to determine sterilization parameters of narrow pulsed voltage generator. In order to determine sterilization parameters for designing NPEF generators efficiently, equivalent circuit model of cells and mechanism of sterilization processing are introduced respectively.

A. Equivalent Circuit Model of Cells:

An equivalent model of a cell is shown in Fig. 4 (a). To simplify a cell model, Fig. 4(a) is modified to be shown Fig. 4(b) [7], where R_S and C_S represent the equivalent resistor and capacitor of the suspension medium, and C_m, R_{C1} and R_{C2} represent the capacitor of outer membrane, the resistor of the outer membrane and the resistor of the cytoplasm inside the nucleus, respectively. Capacitor C_n and resistor R_n are the capacitor and resistor of the nucleus. Usually, capacitance of C_m is higher than that of C_n.

B. Mechanism of Sterilization Processing:

The sterilization mechanism of high voltage pulsed field is employed in electric field of outer membrane, actting on potential difference between outer membrane and intracellular membrane, to cause irreversible perforation on the membrane, to make the liquid flow into interior of membrane, or to affect DNA, RNA of interior nucleus and result in death of apototsis. Thus, relationship between electric field of outer membrane and its distribution of potential is very important and it is also a significant parameter to determine death of microbe.

To achieve sterilization feature, a wide pulse T_w of high narrow pulsed filed must be greater than a critical time for breaching outer membrane T_c [7], it can be represented as follows:

$$T_c = \left[\left(\frac{1+2v_c}{1-2v_c} \right) \frac{\rho 1}{2} + \rho 2 \right] C_m a_c , \qquad (1)$$

where C_m is the capacitance of the membrane per unit area, and ac is the cell diameter. The voltage across outer membrane can be expressed by

$$\Delta Vc(t) = f_a E \frac{a_c}{2} \cos\theta \left[1 - \exp(\frac{-t}{T_c}) \right], \qquad (2)$$

where E is a step electric field, and f_a is a factor of spherical cells, and θ is the angle between the cell and the direction of the electric field. When θ equals zero, equation (2) can be modified as follows:

$$\Delta Vc(t) = f_a E \frac{a_c}{2} \left[1 - \exp(\frac{-t}{T_c}) \right]. \qquad (3)$$

In (1), f_a can be expressed as

$$f_a = \frac{l_a}{l_a - a_c / 3}, \qquad (4)$$

where l_a is the length of a cell.

The proposed circuit uses high narrow pulsed electric field to sterilize liquid food. In order to get the across voltage of outer membrane, at first, we should know the across voltage of a cell, and V_{cell} can be expressed by

$$V_{cell} = f_a E a_c . \qquad (5)$$

1322

In the Fig. 1, the voltage $V_i(t)$ across the interior of the cell can be determined as

$$V_i(t) = f_a E a_c - 2\Delta V_c(t). \tag{6}$$

To derive interior voltage of the cell $V_i(t)$, we assumed the angle between the cell and the direction of the electric field θ equals zero. From (3), (6) can be modified as follows:

$$V_i(t) = f_a E \exp\left(\frac{-t}{T_c}\right). \tag{7}$$

According to (7), the electric field intensity across the cell can be expressed as

$$E_i(t) = f_a E \exp\left(\frac{-t}{T_c}\right). \tag{8}$$

From (2) and (8), the across voltage of organelle V_{org} can be expressed by

$$V_{org}(t) = f_a^2 E D \exp\left(\frac{-t}{T_c}\right), \tag{9}$$

where D is the diameter of spherical organelle, which is much smaller than that of the cell. From (9), the intracellular membrane voltage $\Delta V_i(t)$ can be expressed as

$$\Delta V_i(t) = f_a^2 E \frac{D}{2} \frac{T_c}{T_c - T_o}\left[\exp\left(\frac{-t}{T_c}\right) - \exp\left(\frac{-t}{T_o}\right)\right]. \tag{10}$$

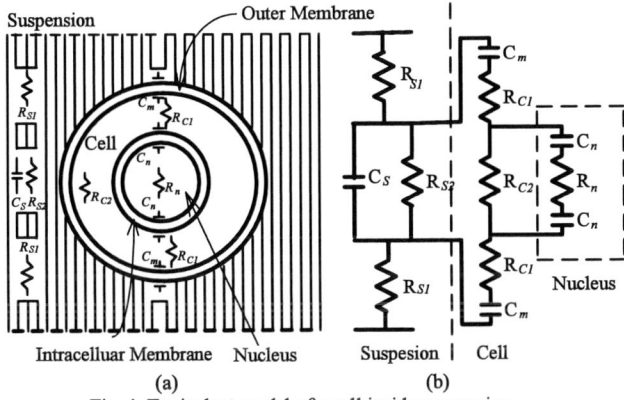

Fig. 4. Equivalent model of a cell inside suspension.

III. Derivation and Operation of the Proposed Converter

The bidirectional high narrow pulsed voltage generator is shown in Fig. 2, which is derived from a unidirectional high narrow pulsed electric field generator, and it is inserted in a full bridge inverter[11]. Sterilizer electrodes could be adhered with particles during food processing. To solve this problem, a half bridge inverter is used to drive ultrasound generator. Switches M_1 and M_2 will be integrated to the proposed circuit, and it is shown in Fig. 2. Transformer T_f with higher turns ratio can generate high leakage energy for the primary. For this reason, spike voltage would cause leakage inductance and energy-recovery problems. In order to solve this problem, the proposed circuit is associated with two active clamp circuits, as shown in Fig. 3(a). To simplify the proposed

circuit, switches M_3 and M_5 (or M_4 and M_6) are integrated into one switch set with the synchronous switch technique [12], as shown in Fig. 4(b). Since voltage stress in the M_3 and M_5 (or M_4 and M_6) are the same, therefore, the diodes D_{F351}, D_{F352}, D_{F461}, D_{F462}, D_{R351}, D_{R352}, D_{R461}, D_{R462} can be ignored. In another side, switches M35 and M46 are replaced by M_3 and M_4, respectively, and the circuit is shown you in Fig. 3(c), where the D_1 and D_2 are formed from fifteen diodes in series connection and each of which can sustain 1 kV. In addition, switches S_1 and S_2 consist of fifteen MOSFETs in series connection and each of which also can sustain 1 kV. The driving circuit is applied to a modified full bridge circuit, and its switching frequency is 1 MHz [13].

IV. Design of Proposed Circuit

The high narrow pulsed voltage generator is shown in Fig. 3(c). It consists of a DC/DC converter, a spark gap and a step-up transformer Tp. The DC/DC converter is derived from a bidirectional flyback converter with energy recovery feature. Thus, the key design issue of this research is a bidirectional flyback converter and an ultrasound generator.

A. Design Principles of Energy Recovery Capacitors C_1 and C_2

The bidirectional high narrow pulse generator is based on the switch set (M_1 and M_4 or M_2 and M_3) of flyback converter, from several ms to hundreds ms, in continuous conduction mode, while capacitor C_3 will accumulate enough energy to spark gap S_g and then it is transferred to load. When switches M_1 and M_4 turn on, as shown in Fig. 4(c), according to Kirchhoff's Voltage Law, the cross voltage of L_{mf} and L_{kf} are

$$\left(L_{kf} + L_{mf}\right)\frac{di_{Lkf}(t)}{dt} = Vi + v_{c2}(t). \tag{11}$$

The voltage of capacitor $vc_2(t)$ will be discharged to magnetizing inductance L_{mf} and leakage inductance L_{kf} of transformer T_f when switches M_1 and M_4 are turned on. Thus, during this interval, the voltage of capacitor $v_{c2}(t)$ can be expressed as follows:

$$v_{c2}(t) = V_{c2}(0) - \frac{1}{C_2}\int_0^t i_{Lkf}(\tau)d\tau, \tag{12}$$

where $Vc_2(0)$ is the initial value of $Vc_2(t)$. According to (11) and (12), we can determine leakage inductance L_{kf} as

$$i_{Lkf}(t) = \frac{V_i + V_{c2}(0)}{Z_{01}} \sin\left(\frac{t}{\sqrt{(L_{kf} + L_{mf})C_2}}\right), \tag{13}$$

where Z_{01} ($= \sqrt{(L_{kf} + L_{mf})/C_2}$) is the characteristic impedance, $vc_2(0)$ is the initial value. It will keep continuous until Lmf and Lkf release energy to zero. During this time interval, $vc_2(0)$ can be divided into three stages, as shown in Fig. 5. In the condition -I and -II, the capacitor C_2 stores energy less than that of capacitor C_3 at the output side. Thus, capacitor C_2 discharging time is less than a high narrow pulsed field period time. To analyze C_1 value, the releasing time of capacitor C_2 can be neglected. When the operation mode moves to into condition-III, according to (13), current

1323

$iL_{kf}(DT_s)(=iL_{mf}(DT_s))$ of transformer T_f can be expressed as

$$i_{Lkf}(DT_s) = i_{Lmf}(DT_s) = \frac{V_i}{L_{kf} + L_{mf}} DT_s, \qquad (14)$$

where D is duty ratio of the switch, T_S is its switching period.

When switches M_1 and M_4 are turned off, as shown in Fig. 5, the energy stored in L_{kf} is discharged to C_1. In an electric field-pulse width period, the energy stored in capacitor C_1 is gradually accumulated, and a relationship between leakage inductance L_{kf} and capacitor C_1 can be expressed as

$$C_1 v_{c1}^2(T_{sp}) = nL_{kf}i_{Lkf}^2(DT_s), \qquad (15)$$

where $n(=T_{SP}/T_S)$ is the switching cycle. From (15), capacitor C_1 can be determined as

$$C_1 = \frac{nL_{kf}i_{Lkf}^2(DT_s)}{v_{c1}^2(T_{sp})}. \qquad (16)$$

Since capacitors C_1 and C_2 play the same role, they are usually chosen with the same capacitance.

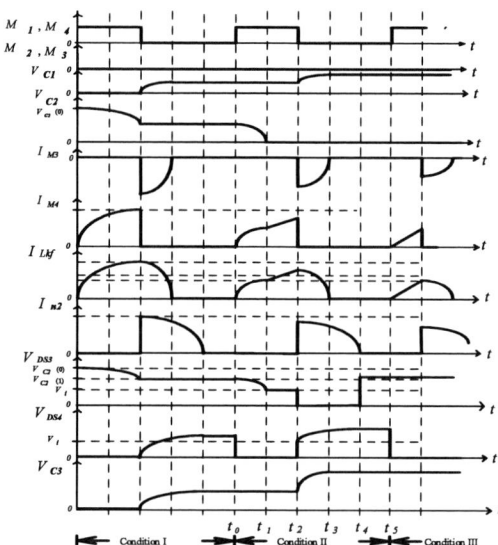

Fig. 5. Conceptual energy- recovery key waveforms of the bidirectional high narrow pulsed generator

B. Capacitor C_3

The major feature of capacitor C_3 is to store energy, and also to supply instantaneous power to the output side. However, during sterilization interval, instantaneous power of the proposed circuit can reach several thousand times of the average power, and the energy storing device must have the feature which can provide instantaneous power simultaneously. According to the above statement, the stored energy in capacitor C_3 equals to output power W_O. Hence, capacitor C_3 can be determined as

$$C_3 = \frac{2W_o}{V_{sg}^2}, \qquad (17)$$

where W_O is the maximum energy for sterilizing microbes.

C. Design of Ultrasound Resonant Network

An ultrasound generator adopts a half bridge resonant circuit to drive a piezoelectric device and produces ultrasounic wave. Although equivalent circuits of a piezoelectric device are very complex, it is usually simplified to a resistor (R), capacitor (C_a) and inductor (L) connected in series and in parallel with a capacitor (C_a), as shown in Fig. 6. In this model, C_b, R, C_a and L represent a static capacitor, equivalent load resistor, dynamic capacitor and dynamic inductor, respectively. From the equivalent model, two important parameters can be derived, which are resonant frequency and anti-resonant frequency. Their expressions can be described as follows, respectively.

$$f_r = \frac{1}{2\pi\sqrt{LC_a}}, \qquad (18)$$

and

$$f_{ar} = \frac{1}{2\pi\sqrt{L\dfrac{C_aC_b}{C_a+C_b}}}, \qquad (19)$$

where f_r is the resonant frequency and far is the anti-resonant frequency. To analyze the characteristics of a piezoelectric device, an impedance/gain-phase analyzer, Hewlett Packard 4294A, is used to measure its parameter values. A plot of its impedance and phase versus frequency is shown in Fig. 7. From Fig. 7, it can be observed that the lowest impedance occurs at the resonant frequency and the highest at the anti-resonant frequency, and its equivalent impedance is capacitive at frequencies below the resonant one or above the anti-resonant one. While, its inductive is at frequencies between the resonant and anti-resonant frequencies. When the piezoelectric device is operated at the resonant frequency, its equivalent circuit can be simplified to capacitor C_b and resistor R connected in parallel, as shown in Fig. 8, and it can attain the maximum conversion efficiency between mechanical and electrical energy. Fig. 8 shows a matching network for the piezoelectric device. From Fig. 8, the input to output voltage transfer function can be expressed as

$$\frac{V_O(j\omega_s)}{V_s(j\omega_s)} = \frac{1}{(1 - L_r C_b \omega_s^2) + j\left(\dfrac{\omega_s L_r}{R}\right)}, \qquad (20)$$

where L_r is a resonant inductor and ω_S is the resonant frequency. From (20), transfer ratio of the matching network can be determined as

$$\left|\frac{V_O(j\omega_s)}{V_s(j\omega_s)}\right| = \frac{1}{\sqrt{\left[1 - \left(\dfrac{\omega_s}{\omega_o}\right)^2\right]^2 + \left(\dfrac{\omega_s}{\omega_o Q}\right)^2}}, \qquad (21)$$

where ω_O is the natural frequency ($=1/\sqrt{L_rC_b}$) and Q is the electrical quality factor ($=R/\sqrt{L_r/C_b}$). When the operating frequency of the device is equal to the resonant one, (18) can be simplified as

$$\left|\frac{V_o(j\omega_s)}{V_s(j\omega_s)}\right| = Q. \tag{22}$$

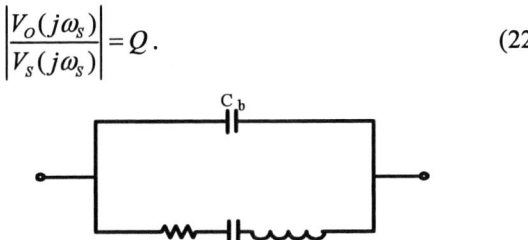

Fig.6. Equivalent circuit of the radial type of piezoelectric device.

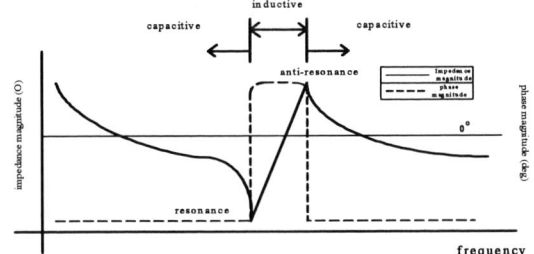

Fig.7. Impedance plot of the piezoelectric device.

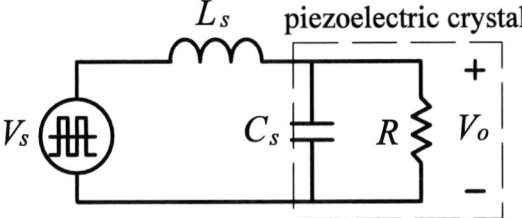

Fig.8. Equivalent circuit of the piezoelectric device operated at its resonant frequency and with a matching inductor.

V. Measured Results

To verify the performance of the proposed generator, as shown in Fig. 3(c), a prototype with the following specifications was implemented.

Measured voltage and current waveforms of switch M_1 are shown in Fig. 9, illustrating that the proposed sterilization system associated with ultrasound generator can achieve ZVS feature. Fig. 10 shows measured voltage waveforms of switches M_3 and M_4. From Fig. 10, it can be seen that the proposed generator using the energy-recovery circuit can reduce voltage stress and ringing of the switches to improve conversion efficiency. Measured voltage waveforms of diode D_3 and D_4 are shown in Fig. 11, from which it can be seen that the voltage stress of diode D_3 and D_4 can be reduced. Fig. 12 shows measured waveforms of switch voltage V_{GS1} and piezoelectric device voltage V_{PD}, from which it can be found that voltage V_{PD} is sinusoidal and it can drive the piezoelectric device. Fig. 13(a) shows measured waveforms of positive output voltage of 11 kV, while Fig. 13(b) shows that of the negative output voltage of 11 kV. We can see that the peak power output is around 1.2 MW. Experimental results have verified that the proposed converter can improve efficiency about 12% over that without energy-recovery circuit. To evaluate effectiveness of liquid food sterilization, PEFs with intensity of ±30 kV/cm, ±25 kV/cm, and ±15 kV/cm, pulse width of 500 ns and repetitive rate of 24 Hz are applied to Hylocereus Polyrhizus juice contaminated with Escherichia Coli. Its sterilization result is shown in Fig. 14, illustrating that

sterilization effectiveness is proportional to treatment time and can yield 5-order reduction at the treatment time of 4 min. These results have verified the feasibility of the proposed generator.

VI. Conclusion

A narrow pulsed voltage generator associated with ultrasound generator for liquid food sterilization is proposed in this paper. Two active clamp circuits are introduced to the bidirectional flyback converter for reducing voltage spike across power switches and recovering the energy trapped in leakage inductor. It can also produce the output voltage of ± 6 kV~± 20 kV and improve efficiency of the converter around 12%. Experimental results verify that the proposed circuit has 5D sterilization feature.

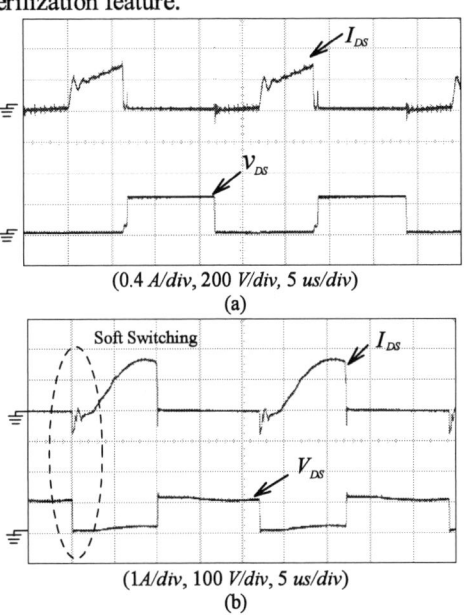

(0.4 A/div, 200 V/div, 5 us/div)
(a)

(1A/div, 100 V/div, 5 us/div)
(b)

Fig. 9. Measured voltage V_{DS} and current I_{DS} waveforms of switch M_1 (a) without ultrasound generator and (b) with the one.

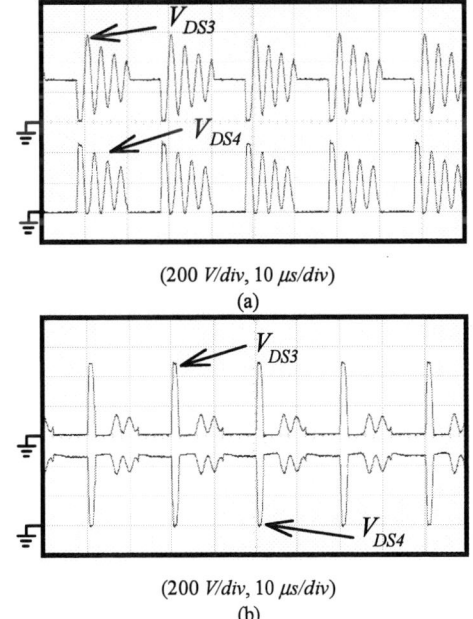

(200 V/div, 10 μs/div)
(a)

(200 V/div, 10 μs/div)
(b)

Fig. 10. Measured voltage V_{DS} waveforms of switch M_3 and M_4 (a) without the energy-recovery circuits and (b) with the one.

1325

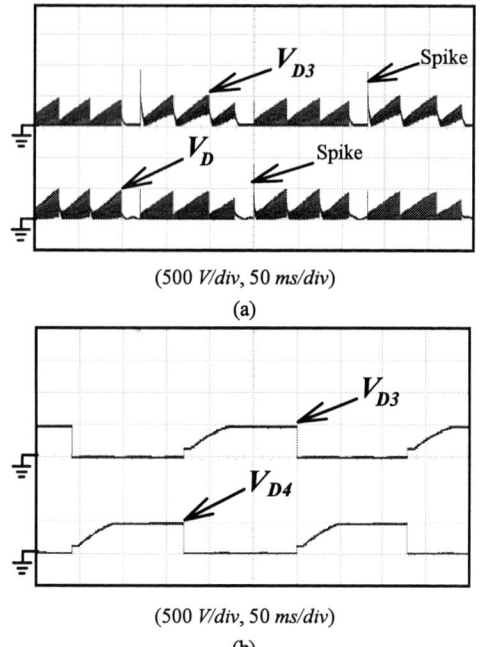

(500 *V/div*, 50 *ms/div*)

(a)

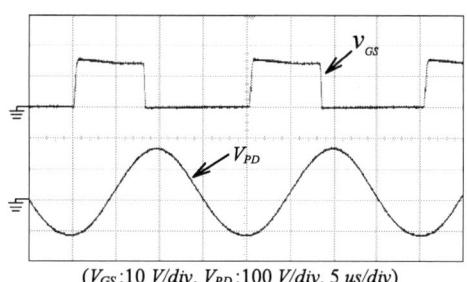

(500 *V/div*, 50 *ms/div*)

(b)

Fig. 11. Measured Voltage Waveforms of V_{D3} and V_{D4} (a) without the energy-recovery circuits and (b) with the ones.

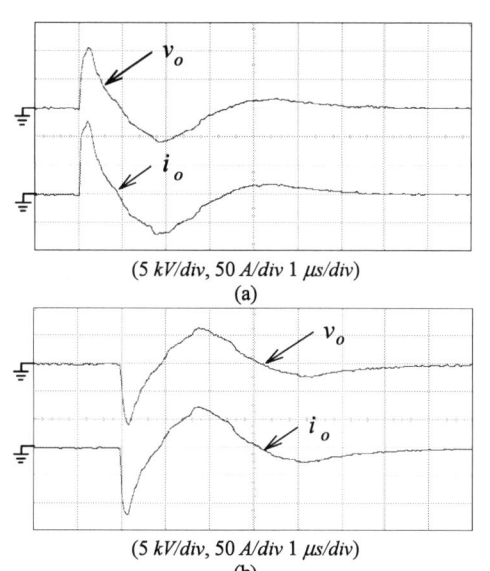

(V_{GS}:10 *V/div*, V_{PD}:100 *V/div*, 5 *us/div*)

Fig. 12. Measured waveforms of voltage V_{GS1} in the converter and voltage V_{PD} in the ultrasound generator.

(5 *kV/div*, 50 *A/div* 1 *μs/div*)

(a)

(5 *kV/div*, 50 *A/div* 1 *μs/div*)

(b)

Fig. 13. Measured waveforms of output pulse voltage vo and current io: (a) positive pulse, and (b) negative pulse.

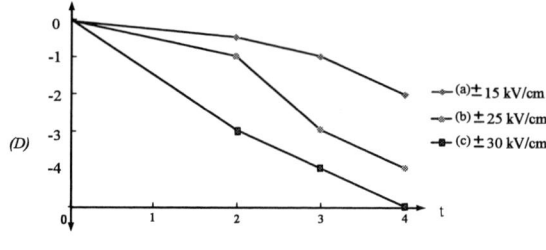

Fig. 14. Plots of survival ratio versus treatment time.

REFERENCES

[1]   X. Qiu, L. Tuhela and H. Zhang, "Application of Pulsed Power Technology in Nonthermal Food Processing and System Optimization," *Proceedings of Pulsed Power Conf.*, 1997, pp. 85-90.

[2]   J. E. dunn and J. S. Pearlman, "Methods and Apparatus of Extending the Shelf Life of Fluid Food Products," *US patent 4,695,472, 1987.*

[3]   U.-R. Pothakamury, et al., "Effect of Growth Stage and Processing Temp. on the Inactivation of E. Coli by PEF," *Journal. Food Protect.* Vol. 59, 1996, pp. 1167-1171.

[4]   Jia, M., Zhang, Q. H. and Min, D. B., "Pulsed electric field processing effects on flavor compounds and microorganisms of orange juice," *Food Chemistry 65,* 1999, pp. 445-451.

[5]   Butz, P. and Tauscher, B., "Emerging technologies: chemical aspects," *Food Research International 35,* 2002, pp.279-284.

[6]   Lado, B. H., et al., "Alternative food-preservation technologies efficacy and mechanisms," *Microbesand Infection 4,* 2002, pp.433- 440.

[7]   H. S. Karl, et al., "Bioelectrics-New Applications for Pulsed Power Technology," *IEEE Transactions on Plasma science,* Vol.30, No.1, 2002.

[8]   S. Xu, Y. C. Liang, C. Y. Lim, and D. Tien, "Investigation and Implementation of IGCT Based Solid State High Voltage Pulse," *Proceedings of the ICON,* Vol.2, 2002, pp. 966-970.

[9]   H.-Y. Lee, et al., "Sewage Sludge Treatment by Arc Discharge,"*Proceedings of the Pulsed Power Conference,* 2003 pp. 1247-1249.

[10] H.-Y. Lee, et al., "Measurement of Excellent Condition To RLC Parameter for Electrical Sterilization on Escherichia Coli," *Proceedings of the Applications of Dielectric Material,* Vol. 2, 1997, pp. 1136-1139.

[11] T. -F. Wu, S.-Y. Tseng and M.-W. Wu, "High Narrow Pulsed Electric Field Generator for Liquid Food Sterilization," *Proceedings of IEEE Applied PowerElectronics Conference, 2006,* pp. 1354-1360.

[12] T. -F. Wu, et al., "Unified Approach to Developing Single Stage Power Converters," *IEEE Trans. on Aero space and Electronic Systems,* Vol. 34, No. 1, pp. 211-223.

[13] T. -F. Wu, S.-Y. Tseng, J.-C. Guo, and M.-W. Wu, "Analysis and Design of Milti-switch Driver for High Pulsed Voltage Generators," *Proceedings of IEEE PowerElectronics Specialists Conference,* 2004, pp. 787-793.

[14] T.-F. Wu, et al., "Unified Approach to Developing Single Stage Power Converters, " *IEEE Trans. on Aerospace and Elec-tronic Systems* , Vol. 34, No. 1, pp. 211-223.

Energy Harvesting from Exercise Bicycle

Suchart Janjornmanit, Samart Yachiangkam, Aswin Kaewsingha.

Department of Electrical Engineering, Rajamangala University of Technology Lanna, Chiangmai, Thailand 50300
E-mail: jansuchart@gmail.com

Abstract-**This paper presents design of energy harvesting from exercise bicycle. The topology is based on lead-acid battery charge controller using dc-dc converter. The CUK converter is selected to deliver the energy and to regulate the charging voltage. The methodology of charging is adapted from the recommendation of IEA, which is the modified constant voltage charging with equalizing mode. The experimental exercise bicycle is constructed to verify the validity of the proposed design.**

Index Terms-Energy Harvesting, Exercise Bicycle, Renewable Energy, Alternative Energy, CUK converter, Charge control.

I. INTRODUCTION

Nowadays, the energy crisis leads the scientist to eagerly searching for alternative energy sources. There are many kinds of energy sources, such as fossil energy, solar energy, kinetic energy, wind energy, e.g. Fossil based energy generates much of the energy but pollute the environment, even though there are many ways to reduce the pollutions. Furthermore the reserves of this kind of energy are dramatically shorten, which is expected to be used up within the next few decades.

This paper proposes energy harvesting from another kind of energy source, human energy. City's lifestyle, People are worrying about their weight and trying to cut the exceeded energy stored in their body by exercise. Why don't we collect the waste energy? The energy harvesting from exercise bicycle is formulated in this research. The objective is to deliver the energy from DC generator through the converter to be kept in energy storage. We are proposing Lead-Acid battery charging control system for harvesting the energy. The battery charging system was proposed for various kinds of energy sources such as done in photovoltaic system in [1]. The recommendation of charge control in [2] and the CUK converter are selected to construct the experimental exercise bicycle in this research.

This paper is organized as follow: The constant voltage charging is firstly detailed, followed by the designing of control in proposed system section. The section of experimental exercise bicycle shows structure and performance of the proposed system. The recommendation of the design and some limitations are detailed in discussion section. Final section gives conclusion.

II. CHARGING CONTROL

The DC generator is selected to transform mechanical energy to electrical energy in this research. The nature of output voltage variation of solar panel due to the astronomy is similarly to the pattern of voltage changing in the DC generator from the bicycle. Therefore the battery charge control of the photovoltaic system is adapted to be charging system for proposed approach.

Table I and figure 1 detail some of the recommended practices of charge control in [2] for the modified constant voltage charging. Table 1 gives suggested charging setpoints for a selection of battery types. It must be noted that these should not be considered definitive, and that these may have to be adjusted depending on the type of battery, the type of controller, the system, the load and where it is install. All setpoints are given in Volts per cell (V/Cell); to find the setpoint for a normally 12 V system, for example, multiply by 6. VR setpoint is the constant voltage used to charge the battery at the intermediate stage. Prior to the VR setpoint, the lowered VR used to boost the charging and to limit the initial charging current. As soon as the minimum charging current is reached, the float voltage, which is the constant voltage lower than the VR setpoint is used. Theses setpoints are presented assuming that the controller is equipped to do equalization or an equalization charge is performed manually on a regular basis.

Figure 1 show graphs of output voltage and output current of the modified constant voltage charging. In constant voltage charging, the amount of charge current is regulated by the controller such that the battery is held at the voltage regulation setpoint. The nature of battery at initial stage of charging is that the current is drawn largely, and then is gradually dropped as the battery becomes charged. Such a large amount of current drawn is the damaging of battery and has to be avoided. Therefore at the initial stage, the charging voltage is lowered than VR setpoint, so that the charging current is regulated to the maximum allowance, which is recommended from manufacturer. The second stage, constant voltage is used when the charging current begin to lower than the maximum charging current at the VR setpoint voltage. The VR setpoint is held until the charging current is dropped down to the minimum charging current of the battery

TABLE I
Suggested Setpoints for Battery Charging Voltage.

Battery Type	Modified Constant Voltage Charging		
	VR	VR/Float	Equalize VR
Flooded/Vented Lead Antimony	2.35	2.40/2.25	2.50
Flooded/Vented Lead Calcium	2.40	2.45/2.30	2.50
Flooded/Sealed	2.35	2.45/2.30	2.50

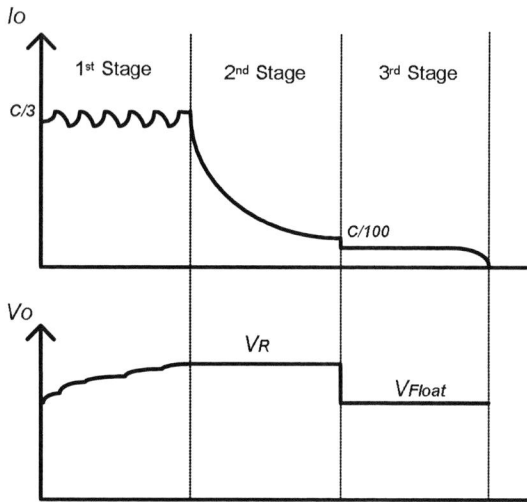

Fig. 1. Modified Constant Voltage Charging.

(C/100). At the final stage, the voltage is lowered to float voltage, which is another constant voltage used for keep charging current to the completely charge without harming the battery.

Many charge controllers shift the VR setpoint up or down depending on the condition of battery. Finding the right VR setpoint is difficult; if it is too high, the battery will be overcharge often, and if it is too low, the battery will never be fully charged. One approach is to normally apply a fairly low VR setpoint, to avoid excessive overcharge, but to occasionally raise the VR setpoint, to ensure a full charge of all cells every several weeks. This occasionally full charge is called an "equalization" charge.

III. PROPOSED SYSTEM

The goal of harvesting energy from exercise bicycle is to convert the mechanical energy to electrical energy, then deliver it to store in battery. CUK converter is used to pass the energy to battery by regulating the output voltage in according to charging current. For CUK converter, the controlling of duty cycle of PWM is the controlling of output voltage of the converter. This section gives the details of CUK converter and the proposed PWM generation.

A. CUK Converter

Figure 2 shows CUK converter for the proposed charge controller. The filter capacitor (Cf) in the circuit is used to limit voltage ripple from DC generator. The reasons of using CUK are that the output voltage from DC generator (input of the converter) may be lower or higher than voltage of battery (output of the converter) because of biking speed, and the current drawing from DC generator and the current feeding to the battery are continuous where the buck, boost and buck-boost have at least one side with pulsed current.

The CUK converter uses capacitive energy transfer and analysis is based on current balance of the capacitor. The ratio of input and output voltage of the converter is

$$\frac{V_o}{V_{in}} = -\frac{D}{(1-D)} \tag{1}$$

Since the duty ratio "D" is between 0 and 1 the output voltage can vary between lower or higher than the input voltage in magnitude. The negative sign indicates a reversal of sense of the output voltage.

B. Microprocessor Based Charging Controller

Figure 3 shows block diagram of PWM control. The PWM output is generated to achieve desired output voltage by adjusting duty cycle in according to (1). PID controller is chosen to be the regulator of the duty cycle. The measured output voltage is compared to referent voltage specified in which the output current is regulated at the proper value. The relation between output voltage and current is depicted from the relation shown in figure 1. The measured current and voltage are firstly averaged before the adjusting and comparison.

The relation between charging voltage and current is such a complex. This is hard to construct analog circuit to control all functions, but it is much easier if this is done by the fairly expensive microprocessor. The low cost digital signal processor, dsPIC is selected to control all of theses duties. The performance of the energy harvesting by using this processor is acceptable and cost effective.

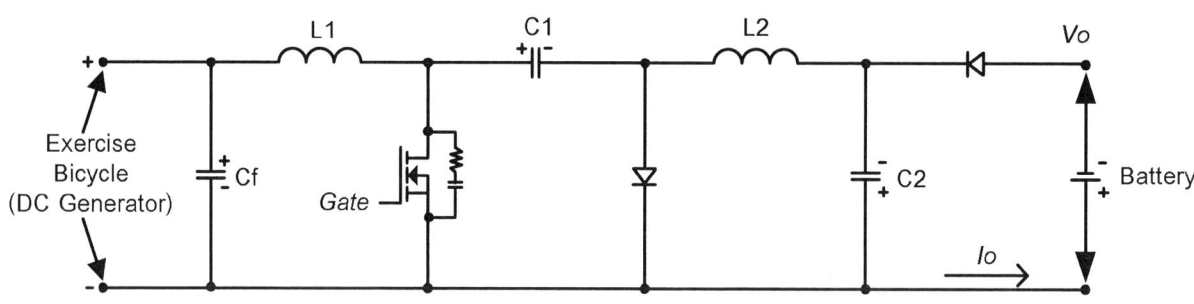

Fig. 2. CUK Converter for the charge controller.

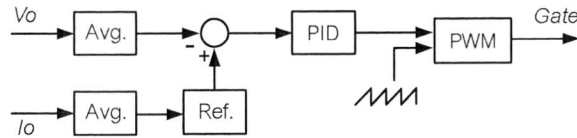

Fig 3. Diagram of the PWM Control.

IV. EXPERIMENTAL EXERCISE BICYCLE

Table II details all parameters using in the experimental module of the energy harvesting. Figure 4 shows the exercise bicycle and all control circuits connected. Figure 5 shows process values during operation. There is a little bit of voltage ripple around the target output voltage, especially when the empty battery is used. But at the middle through the end of charge stage, the charging voltage is maintained stable.

V. DISCUSSION

The output power of the exercise bicycle is limited by the size of DC generator and battery. For our module, 100W DC generator is used, 7.5AH battery is tested and the output of the bicycle is actually limited by the size of battery, which is about 30W. To gain more output power than this, the capacity of battery have to be increased and the setpoint of charging current need to be revised. There is a switch to select battery capacity in our module, which is selectable to be 7.5AH or 24AH battery. The DC generator may be another choice of increasing output power, to do this the corresponding size of battery, switching device and all set points must be considered.

TABLE II
Parameters in Experimental Exercise Bicycle.

Parameters	Value
DC Generator + Gear	100W, 24VDC
Sealed Lead-Acid Battery	12V, 7.5AH
Maximum Charging Current	C/3 (Use 2.0A)
Minimum Charging Current	C/100 (Use 75mA)
VR Setpoint	14.7 Volts
Float Voltage	13.8 Volts
Equalize Voltage (Manual Select)	15.0 Volts
Digital Signal Processor	dsPIC30F2010
Sampling Frequency	200kHz
Switching Frequency	2kHz
Switching Device	IRFZ46
Filter Capacitor	3,300uF
C1 Capacitor	3,300uF
C2 Capacitor	3,300uF
L1 Inductor	15mH
L2 Inductor	15mH

Fig. 4. Experimental Exercise Bicycle.

Fig. 5. Process values.

VI. CONCLUSION

The energy harvesting from exercise bicycle is proposed in this paper. The proposed topology is based on lead-acid battery charge controller for photovoltaic system. The modified constant voltage charging control, which is recommended by IEA, is adapted and implemented in this research. CUK converter is used to deliver the energy, from bicycle mounted DC generator, to store in the battery. The cost effective digital signal processor, dsPIC is chosen to perform PWM control in which the battery is charged properly by charging voltage and current. The experimental exercise bicycle is constructed and tested, and good performance is achieved. The design recommendation of sizing the output power is also discussed for implementation.

REFERENCES

[1] E. Koutroulis, K. Kalaitzakis. "Novel battery charging regulation system for photovoltaic applications". *Electric Power Applications, IEE Proceedings*, Volume 151, Issue 2, pp. 191 – 197, Mar. 2004.

[2] E.P. Usher, M. Ross. "Recommended practices for charge controllers". *IEA Implementing Agreement on Photovoltaic Power Systems*. Aug. 1998.

Modeling and Analysis of Igniter for HID Lamps

Weiping Zhang, and Qiang Cheng

College of Information Engineering, North China Univ. of Tech., Beijing, 100041, P.R. China
Tel. (Fax): 86-010-88802880, Email: zwp@ncut.edu.cn

Abstract—The electric properties of igniter for HID Lamps have been investigated in this paper. The main contributions are as the followings: (1) A few of models of igniter have been put forward to analyize the electric properties; (2) Based on the proposed models, an useful method has been developed; (3) Through the theoretic analyzing, some valuable conclusions for designing this Igniter have been given out, such as the both of primary and secondary of Tesla transformer should operate in resonance with its capacitor and the coupling coefficient k of Tesla transformer has a optimum value (k=0.6). A prototype has been made up. The proposed theory and approaches have been verified by the results of the experimental result and computer simulation results given in this paper.

Index Terms—Analysis , Igniter for the HID lamps, Modeling.

I. INTRODUCTION

High-Intensity-Discharge (HID) lamp has been paid more attention in residential as well as commercial lighting applications [1, 4]. The current through the lamp will tend to run away unless ballast is connected in series with the lamp, because the increment impedance of HID lamp is negative, shown in Fig.1 [2, 3]. In additional, an igniter has to be required to provide higher voltage when the HID lamps start up. Therefore, igniter plays an important role in the lighting system of HID lamps.

The basic Igniter circuit and interconnection required are illustrated in Fig.2.When the proper input power is applied to the input step-up transformer Tr2, it causes the Spark gap to break down. The spark gap breakdown allows current to flow in the resonant circuit consisting of C1and the primary of Tesla transformer Tr1. A series of damped oscillations occur (the resonant frequency is about 2 to 4 MHz) during a portion of each half cycle of the input AC line frequency (line frequency is about 50 or 60Hz). These high-frequency pulses are stepped up in voltage through Tesla transformer Tr1 and then applied to the lamp terminals. An arc is struck in the HID lamp, and when suitable voltage is available across the lamp's electrodes the lamp will ignite. Since the lamp current flows through the secondary of Tr1,the resistance—or impedance- of this winding must be small to prevent excessive heating of the igniter unit and dissipation of power, typical resistance is less than 2 milliohm for the 90 amp DC igniter and less than 0.5 milliohm for the 750 amp. Capacitor C2 provides a low impedance RF path

for the high voltage pulses through the lamp and does simultaneously a high impedance path for the AC voltage provided by the ballast. The RF trap in the secondary of transformer Tr2 is a parallel resonant circuit and its resonant frequency is exactly equal to the resonant frequency of the serous resonant circuit consisting of C1and primary of Tr1 to minimize RF losses to the power line [5].

The igniter circuit is a switch circuit because it contains a spark gap. In this paper, the electric properties of Igniter for HID Lamps have been deeply investigated. In the section 2, a few of models for Igniter has been put forward to analyize the electric properties. In the section 3, a useful analytic approach has been developed based on the proposed model and some valuable conclusions for designing this Igniter have been given out. A prototype has been made up. In the section 4, the proposed theory and approaches have been verified by the results of the experimental result and computer simulation results given in this paper.

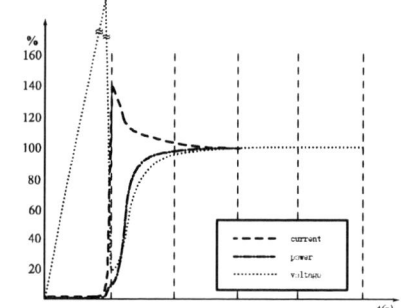

Fig.1 Electrical characteristics of HID lamp's igniting process

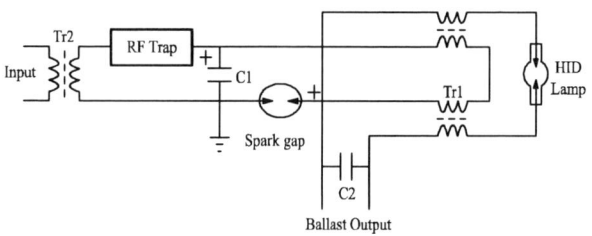

Fig.2 Igniter circuit and interconnection

II. THE ELECTRIC MODELS OF IGNITER FOR HID LAMPS

A. Model 1—an equivalent circuit for spark gap that has not broken down

Model 1, shown in Fig.3, is an equivalent circuit for Spark Gap that has not broken down. In Fig.2, Tr2 is a low frequency transformer, the output voltage of the secondary can be approximately considered as a DC

Project supported by Natural Science foundation of China(N0.50477054);Project supported by Beijing Natural Science foundation of China (No. 4052011)

978-1-4244-0644-9/07/$25.00 ©2007 IEEE 1330

voltage source, representing V1. The RF trap is almost short for line frequency signal because the line frequency is much more lower the resonant frequency. Resistor R represents the ESR of C1 and the primary loss of Tr1.Before Spark Gap is broken down, it can be equivalent as a small capacitor C(1~3nF). L1 and L2 are the primary and secondary inductor s of Tr1. C2 (10~50pF)is a capacitor of the both electrodes of the HID lamp. For the line power, the primary of Tr1 is almost short and capacitor C is open because its inductance (L1=1~5μH) and line frequency is very low (f=50~60Hz). So the voltages across C1 and C are equal to the secondary voltage V1. At $t = t_0$, the voltages across C and C1 are increased toU0, Spark Gap is broken down. $U_o \leq U_p$, Up is the peak value of V1.The energy in C1 is equal to $C_1 U_0^2 / 2$, the other component have no any storage energy. Therefore, the initial conditions are that $u_1(t_0^-) = U_o$, $i_{L1}(t_0^-) = i_{L2}(t_0^-) = 0$, $u_c(t_0^-) = u_2(t_0^-) = 0$, $u_1, u_c, u_2, i_{L1}, i_{L2}$ represent respectively the voltages across C1,C as well as C2 and the current through L1 and L2.

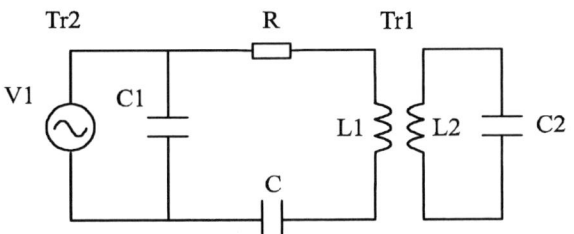

Fig.3 Model 1—an equivalent circuit for spark gap that has not broken down

B. Model 2—an equivalent circuit for Spark Gap that has broken down

Model 2, an equivalent circuit for Spark Gap that has broken down, is shown in Fig.4.In Fig.2, when the secondary voltage of Tr2 increases to the breakdown voltage U0 of Spark Gap, it is shorted and allows current to flow in the resonant circuit consisting of C1and the primary of Tr1 to generate a series-high frequency - damped pulses. In this case, RF trap shows very high impedance for the pulses and is like open. The equivalent resistance of HID lamp is very high if the lamp is not broken down. C2 (10~50pF) is a capacitor of the both electrodes of the HID lamp.

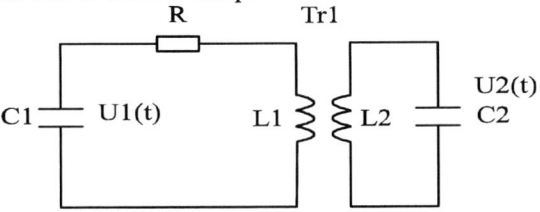

Fig.4 Model 2—an equivalent circuit for spark gap that has broken down

C. Model 3—an equivalent circuit for the lamp that has broken down

In Fig.2,when the secondary voltage of Tr1 increases to the breakdown voltage of the lamp, the lamp begins to discharge and the equivalent resistance RL is very low and the resonant circuit stops working immediately. After this monument, the ballast starts to operate and provide energy for the lamp. Therefore, an equivalent circuit for this subinterval is shown in Fig.5.

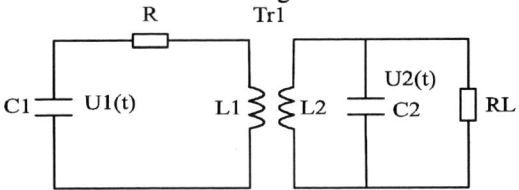

Fig.5 Model 3—an equivalent circuit for lamp that has broken down

D. Pspice model

A Pspice model for Igniter has been developed to describe overall process and shown in Fig.6. The Pspice Model consists of two parts, the principle circuit and control circuit, shown respectively Fig.6 (a) and (b). In Fig.6 (b), voltage source E1 is VCVS, and controlled by $u_2(t)$, the secondary output voltage of Tr1.If the voltage across the R1 is lower than the breakdown voltage of the lamp, the switch K, shown in Fig.6 (a), is open, or if the voltage across the R1 reaches to the breakdown voltage of the lamp, the switch K is short.

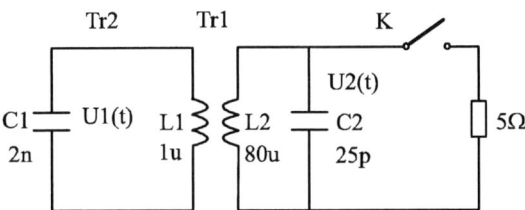

(a) The principle circuit for Spark Gap that has broken down

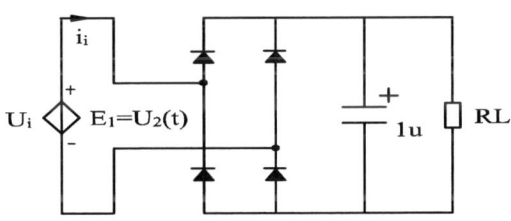

(b) Control circuit
Fig.6 Pspice Model for igniter

III. THE ANALYSIS OF ELECTRIC CHARACTERISTICS

A. No loss analysis

In order to simplify analysis, we suppose that the capacitor C1 and transformer Tr1 are ideal components and without any losses. Based on this assumption, Model 2 can be simplified as the following circuit, shown in Fig.7. The symbol M is a mutual inductance.

1331

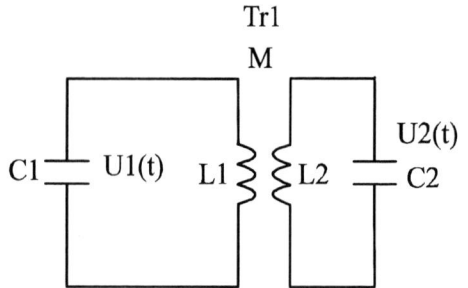

Fig 7 Simplified Model 2

If the start time point is defined as the monument of Spark Gap to be breakdown, the following equations can derived based on the simplified Model 2.

$$\begin{cases} L_1 C_1 \dfrac{d^2 u_1}{dt^2} + M C_2 \dfrac{d^2 u_2}{dt^2} + u_1(t) = 0 \\ L_2 C_2 \dfrac{d^2 u_2}{dt^2} + M C_1 \dfrac{d^2 u_1}{dt^2} + u_2(t) = 0 \end{cases} \quad (1)$$

The initial values are as the followings,

$u_1(0^+) = U_0, u_1'(0^+) = i_{L1}(0^+)/C_1 = 0,$

$u_2(0^+) = 0, u_2'(0^+) = i_{L2}(0^+)/C_2 = 0$

Through solving the equations (1), the output voltage $u_2(t)$ can be obtained,

$$u_2(t) = U_{2m} \big[\cos(s_1 t) - \cos(s_2 t) \big] / 2 \quad (2)$$

$$U_{2m} = \frac{2k\sqrt{L_2/L_1}\, U_0 \omega_2^2}{\sqrt{\omega_1^4 + \omega_2^4 - 2(1-2k^2)\omega_1^2 \omega_2^2}} \quad (3)$$

Where

$\omega_1^2 = (L_1 C_1)^{-1}, \omega_2^2 = (L_2 C_2)^{-1}, k^2 = M_2 / L_1 L_2$

$s_1^2 = (\omega_1^2 + \omega_2^2 + \sqrt{\omega_1^4 + \omega_2^4 - 2(1-2k^2)\omega_1^2 \omega_2^2})/2(1-k^2)$

$s_2^2 = \omega_1^2 + \omega_2^2 - \sqrt{\omega_1^4 + \omega_2^4 - 2(1-2k^2)\omega_1^2 \omega_2^2} / 2(1-k^2)$

Where, k is the coupling coefficient of Tr1.

The efficiency of energy transformation η is

$$\eta = \frac{1}{2} C_2 U_{2m}^2 \Big/ \frac{1}{2} C_1 U_0^2 \quad (4)$$

$$= 4k_2 / 4k^2 + (\omega_1^2/\omega_2^2 + \omega_1^2/\omega_2^2 - 2)$$

If $\omega_1 = \omega_2$, that is $L_1 C_1 = L_2 C_2$, then $\eta = 1$, the transforming efficiency reaches maximum. This maximum values is

$$U_{2m} = \frac{2k\sqrt{L_2/L_1}\, U_0 \omega^2}{\sqrt{\omega^4 + \omega^4 - 2(1-2k^2)\omega^2 \omega^2}} = \sqrt{L_2/L_1}\, U_0 = \text{constant}$$

If $\cos(s_1 t) - \cos(s_2 t) = 2$, then

$s_1 t = n\pi, s_2 t = (n-1)\pi$, $n = 1,2,3...$

$S_1 / S_2 = n/n-1$

$$\frac{s_1^2}{s_2^2} = \frac{\omega_1^2 + \omega_2^2 + \sqrt{\omega_1^4 + \omega_2^4 - 2(1-2k^2)\omega_1^2 \omega_2^2}}{\omega_1^2 + \omega_2^2 - \sqrt{\omega_1^4 + \omega_2^4 - 2(1-2k^2)\omega_1^2 \omega_2^2}}$$

$$= \frac{n^2}{n^2 - 2n + 1}$$

If $\omega_1 = \omega_2 = \omega$, then

$(2\omega^2 + 2k\omega^2)/(2\omega^2 - 2k\omega^2) = n^2/n^2 - 2n + 1$, so

$k = 2n - 1/2n^2 - 2n + 1 \quad (n = 1,2,3 \cdots)$

The curve of k vs. n is shown in Fig.8. From Fig.8, it can be known that if Tr1 is a perfect coupling transformer and its loss is neglected, then k=1 and n=1. In this case, Igniter operates in optimum condition. The efficiency and output voltage reach to maximum. However, the primary of Tr1 has to flow a large amount of current provided by ballast, a perfect coupling transformer would mean that Tr1 had a big core so that Igniter can not be installed into a lighting apparatus to connect tightly the lamp and Igniter. In general, the magnetic core of Tr1 has a big air gap to avoid magnetic saturation when the output current of the ballast flow the primary. So, let n=2, k=0.6, it is better condition for Tr1.

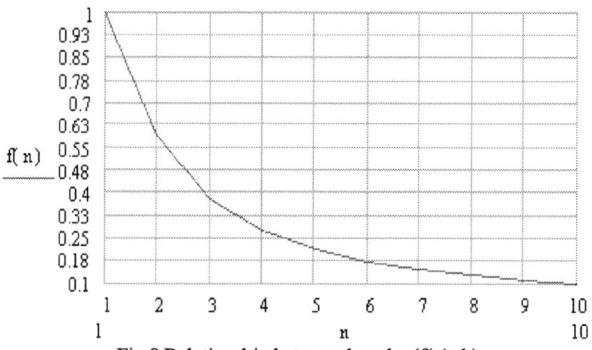

Fig.8 Relationship between k and n (f(n)=k)

B. Analysis of electric property of considering the loss

In fact, C1 has an ESR and Tr1 has winding losses. Theses losses will affect the electric property of Tr1. In general, the loss of ESR is much higher than that of the windings of Tr1; the winding losses can be neglected in the following analysis.

If the start time point is defined as the monument of Spark Gap to be breakdown, the following equations can derived based on Model 2 shown in Fig.4,

$$\begin{aligned} L_1 C_1 \frac{d^2 u_1}{dt^2} + M C_2 \frac{d^2 u_2}{dt^2} + R_1 C_1 \frac{du_1}{dt} + u_1(t) = 0 \\ L_2 C_2 \frac{d^2 u_2}{dt^2} + M C_1 \frac{d^2 u_1}{dt^2} + u_2(t) = 0 \end{aligned} \quad (5)$$

The initial values are same of equation (1).

Through solving the equations (5), the output voltage $u_2(t)$ can be obtained,

$$u_2(t) = \frac{U_{2m}}{2} (e^{-\beta t} \cos \varpi_2 t - e^{-\alpha t} \cos \varpi_1 t) \quad (6)$$

$$U_{2m} = U_0 \sqrt{L_2/L_1}\, \rho_0$$

Where, $\rho_0 = 2k / \sqrt{(1+T)^2 - 4(1-k^2)T}$

$k = M / \sqrt{L_1 L_2}$, $T = \omega_1^2 / \omega_2^2 = L_2 C_2 / L_1 C_1$

$$F_1 = \frac{1}{2(1-k^2)} \Big[1 + T + \sqrt{(1+T)^2 - 4(1-k^2)T} \Big]$$

1332

$$F_2 = \frac{1}{2(1-k^2)}\left[1+T-\sqrt{(1+T)^2-4(1-k^2)T}\right]$$

$$Q = \omega_1 L_1 / R_1 = \sqrt{L_1/C_1}/R_1$$

$$\varpi_1 = \sqrt{F_1}\omega_2 \ , \ \varpi_2 = \sqrt{F_2}\omega_2$$

$$\alpha = \sqrt{T}(F_1-1)\omega_2/2Q\sqrt{(1+T)^2-4(1-k^2)T}$$

$$\beta = \sqrt{T}(1-F_2)\omega_2/2Q\sqrt{(1+T)^2-4(1-k^2)T}$$

The efficiency of energy transformation η is

$$\eta = 4k^2/(\frac{(1+T)^2}{T}-4+4k^2) \qquad (7)$$

If $\eta = 1, T = 1$, that is $\omega_1 = \omega_2$ and $L_1C_1 = L_2C_2$. The both of primary and secondary of Tr1 should be in resonance with its capacitor for the efficiency of Tr1 want to reach to maximum. This conclusion is consistent with no loss analysis.

If $\omega_1 = \omega_2$ and the values of α and β are very small, let $\cos\varpi_2 t - \cos\varpi_1 t = 2$, then $u_2(t)$ will arrive to maximum.

$$\cos\varpi_2 t - \cos\varpi_1 t = \cos\sqrt{1/1+k}\omega - \cos\sqrt{1/1-k}\omega = 2$$

$$\sqrt{1/1+k} = n\pi, \sqrt{1/1-k} = (n+1)\pi$$

$$k = 2n+1/2n^2+2n+1(n=0,1,2,3\cdots) \qquad (8)$$

Based on equation (8), n=1 and k=0.6 are the best selection. This conclusion is consistent with no loss analysis.

IV. SIMULATION RESULTS AND EXPERIMENTAL RESULTS

A. Simulation result 1 and a simplified Model 1

Employing Model 1 do simulation to analyze the electric property of Igniter during Spark Gap that has not broken down .The main parameters are the followings: the low frequency voltage source V1 has a line frequency (f=50Hz) and 8000volts of amplitude, $C_1 = 2nF/10kv$, $R(ESR) = 0.4\Omega$ $L_1 = 1\mu H, L_2 = 80\mu H, k = 0.6$, $C_2 = 25pF$, $C = 3pF$. The simulation result shows that the current through the primary of Tr1 is less than 5μA. So, the primary of Tr1 is almost short in model 1 and a simplified Model 1, shown in Fig.9, can be obtained. This conclusion is consistent with theoretic results in section 2.1.

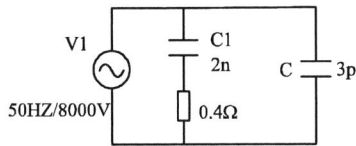

Fig.9 A Simplified Model 1

B. Simulation result 2 for Spark Gap that has broken down

Case1 R=0ohms.

Employing Model 2 do simulation to analyze the electric property of Igniter during Spark Gap that has broken down .The parameters are the same as the above section 4.2 except C1. If C1=2nF and the primary and secondary of Tr1 are all in resonance with its capacitor, the transforming efficiency is about 99.2%. However, If C1=3nF and the secondary of Tr1 is only in resonance with its capacitor, the transforming efficiency is about 89%. The simulation result shows that the maximum transforming efficiency may occur if the primary and secondary of Tr1 are all in resonance with its capacitor. This conclusion can verify the corresponding theoretic result in section 3.1.

Case2 R=0.4ohms.

Doing same simulation as case 1, the similar conclusion can be extracted as the followings. If C1=2nF, the transforming efficiency is about 94.2%. However, If C1=3nF, the transforming efficiency is about 82.2%.

Comparing with both cases, we make the following summary. (1) The maximum transforming efficiency may occur if the primary and secondary of Tr1 are all in resonance with its capacitor; (2) The loss of ESR in Capacitor C1 will hurt the efficiency very much.

C. The relationships the output voltage and efficiency with R and C1

The values of ESR and capacitor C1 are playing an important role in the output voltage and efficiency of Igniter. In this section, the relationships between the amplitude of output voltage and efficiency with the values of ESR and capacitor C1 will be discussed by applying Pspice to simulate the electric characteristics. The parameters satisfy the following conditions: the coupling coefficient k is taken as the optimal value, k=0.6; the capacitors C2 is a equivalent capacitance of 2500W HMI lamps [2], $C_2 = 25pF$; the primary and secondary inductances of Tr1 are $L_1 = 1\mu H$, $L_2 = 80\mu H$.

The curves of the normal output voltage and efficiency vs. ESR are shown in Fig.10. Theses curves explain that ESR hurts the output voltage and efficiency very much, however, if R=0~1ohms, the dropping increments of the output voltage and efficiency are in acceptance tolerance. That is ESR should be less then 1ohms.

The curves of the normal output voltage and efficiency vs. C1 are shown in Fig.11. Theses curves show that the values of C1 will affect the output voltage and efficiency. The curves explain that 2000pF is a best selection for C1.This conclusion is also consistent with the theoretic result in section 3.

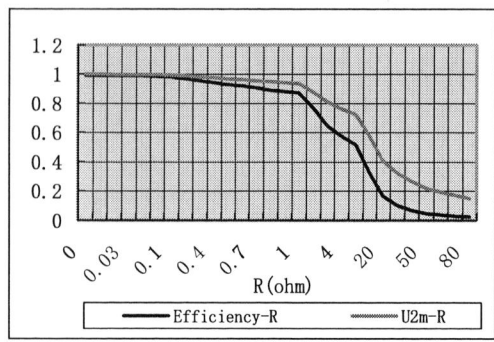

Fig.10 The normal output voltage and efficiency vs. ESR

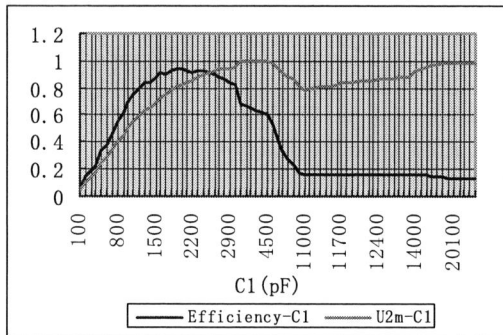

Fig.11 The normal output voltage and efficiency vs. C1

D. Overall process simulation and experimental results

In order to analysis the overall process operation of Igniter, the Pspice Model, shown in Fig.6, are used to analyze its output voltage. The simulation result is exhibited in Fig. 12.This curve shows the following conclusions: (1) the maximum amplitude occurs at the second half period; (2) the output voltage will be drop to zero if the lamp is broken down. The experimental result is given out in Fig.13.Comparing Fig.12 and Fig.13, there are the following conclusions: (1) the both waveforms are the similar shape at the first half period, but the difference of the amplitude is 3.8 times, this is because 4 times attenuator has be used; (2) the lamp is broken down at the second half period, the maximum amplitude occurs at this period.

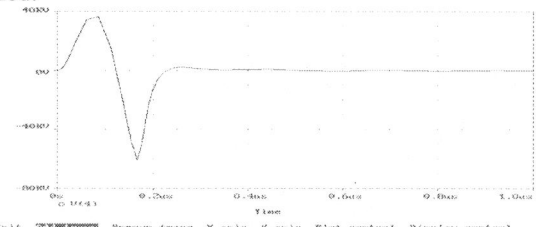

Fig.12 Simulation result for overall process operation of igniter

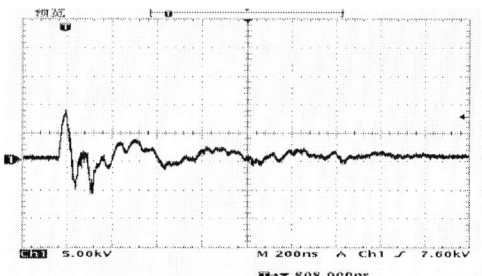

Fig.13 Experimental result

V. CONCLUSIONS

Igniter is an important component in the field of high power HID lamp lighting applications. Igniter circuit is a switch circuit because it contains a Spark Gap. Igniter has been deeply studied in this paper, a few of equivalent circuits has been proposed to analyze its electric properties and. based on the proposed model some key electric properties has been analyzed. Through the theoretic analyzing, some valuable conclusions for designing this Igniter have been given out. A prototype has been made up. The proposed theory and approaches have been verified by the results of the experimental result and computer simulation results. The main conclusions extracted in this paper can summarize as the followings: (1) the proposed models is available to studying the performances of Igniter; (2) the maximum transforming efficiency may occur if both primary and secondary of Tr1 are all in resonance with its capacitor; (3) The loss of ESR in capacitor C1 will hurt the efficiency very much, in fact, ESR should be less then 1ohms; (4) the coupling coefficient k of Tesla transformer has a optimal value, k=0.6. The above conclusions are valuable for designing and analyzing igniter.

REFERENCES

[1] E.Rasch and E. Statnic, "Behavior of Metal Halide lamps with conventional and Electronic Ballast," J.of the IES, pp.88-96, summer, 1991

[2] Weiping Zhang, et al., High Power-High Frequency Resonant Mode Electronic Ballast for Metal Halide, Proceeding of the International Power Electronics and Motion Control Conference IPEMC'04, August, 2004, Xian, China,

[3] Weiping Zhang, et al., The Controlling Strategy for Electronic Ballast of HID lamps, Proceeding of the International Power Electronics and Motion Control Conference IPEMC'06, August, 2006, Shanghai, China.

[4] OSRAM, Lighting Program Photo Optic's 05/06.

[5] L、A, Associates, Metal Halide Lamp Ballast's and Igniter 96.

Design of a Single Bi-directional DC-DC Converter for On-board Energy Improving of Zero Emission Electric Vehicles

Werachet Khan-ngern

Faculty of Engineering,
Research Center for Communications and Information Technology,
King Mongkut's Institute of Technology Ladkrabang,
Thailand, E-mail: kkveerac@kmitl.ac.th

Abstract-This paper presents the design technique of a bi-directional power flow DC-DC converter for two power sources. The power sources are composed of Proton Exchange Membrane Fuel Cells (PEMFCs) and the Nickel Metal Hydride (Ni-MH) batteries. A single DC-DC Converter is designed for both of motoring and generating modes of the DC motor load condition. Charging and discharging operating modes of NI-MH are focused. The PSpice simulated result confirms a good achievement of the energy management.

I. INTRODUCTION

Driving range per charge of electric vehicle (EV) using conventional lead-acid battery is limited due to a poor battery specific energy density. The specific energy density of a lead-acid battery varies between 15 and 45 Wh/kg [1] which is less than the 12,000 Wh/kg energy of standard automotive fuel. The Ni-MH is selected because of the higher specific energy density between 55 and 80 Wh/kg. The other source is PEMFC which powered by hydrogen and oxygen from air. The advantages of PEMFC are clean, higher specific energy, higher energy density and fast refuel [2]. These advantages result a suitable factor for EV application. The major disadvantages of the PEMFC are unable to charge the electric energy or lack of storage capacity and slow response with high cost [3].

The single DC-DC converter is designed to operate on the combination of Ni-MH charging and discharging condition. The converter simulation is done using PSpice program.

II. PROPOSED CONVERTER

The concept of single bi-directional DC-DC converter is proposed as shown in Fig. 1. Most of electric vehicles (EVs) driving system in the city are frequently operating on accelerating and braking conditions. The electric motor of the EV operates on motoring and regenerating modes. Then, bi-directional power converter is required. The bi-directional converter key performances are as follows:
1. preferable single converter
2. simplicity
3. allowing bi-directional current flows or charging and discharging modes
4. high efficiency energy management

Two energy sources are designed for DC EVs to extend the driving rage and improve the specific energy as indicated by W/kg. Ni-MH battery and PEMFC are proposed as the combination of high specific energy comparable to conventional lead-acid battery as shown in Fig. 1. It is possible to run the EV without any battery on board. This requires regenerative energy absorption part. To improve the energy management, Ni-MH battery is applied. Ni-MH battery is not only used to absorb the regenerative energy from the DC motor, but also used to supply sufficient current during accelerate or starting DC motor. However, the disadvantage of memory effect of NIMH battery should be serious considered in life cycle [4].

The DC-DC converter shown in Fig. 1 should control the energy ratio from the two sources to energize to the motor. In this case, PEMFC is dominated as the main energy supply at two times over the battery. The control technique will be described in the next section. The DC-DC converter can delivery the power to the motor and hand in the regenerative energy back to charge the energy into the Ni-MH battery.

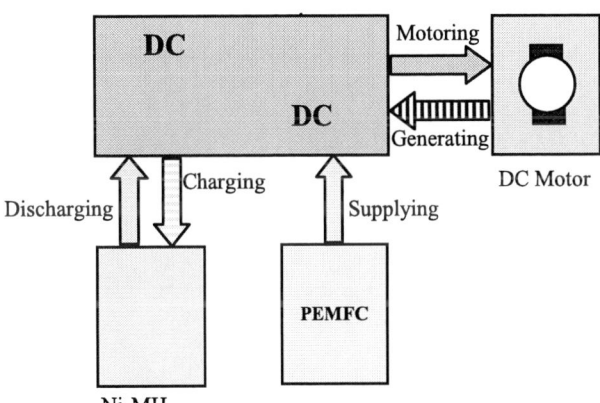

Fig 1 Block diagram of DC-DC converter operation

Fig. 2 shows the electric equivalent circuit of PEMFC and the Ni-MH. Cell voltage of the PEMFC is the result of the null voltage, activation and concentration voltage, and the series resistor. The V-I characteristic of the 500W PEMFC shows in figure 2 (c). The 36 V 8Ah, Ni-MH discharging characteristic is shown in figure 2 (d).

978-1-4244-0644-9/07/$25.00 ©2007 IEEE　　1335

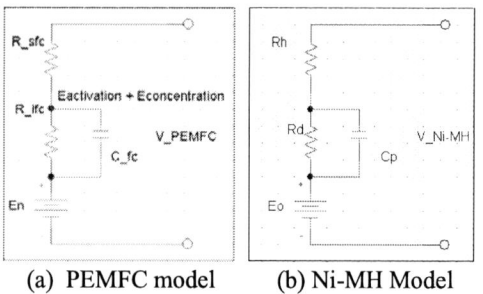

(a) PEMFC model   (b) Ni-MH Model

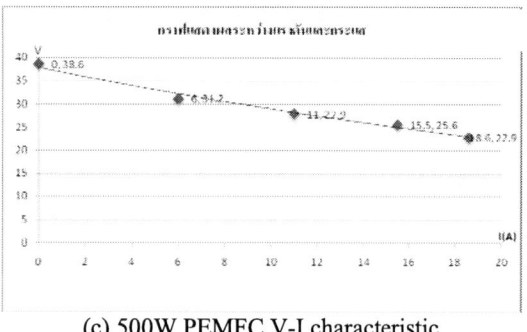

(c) 500W PEMFC V-I characteristic

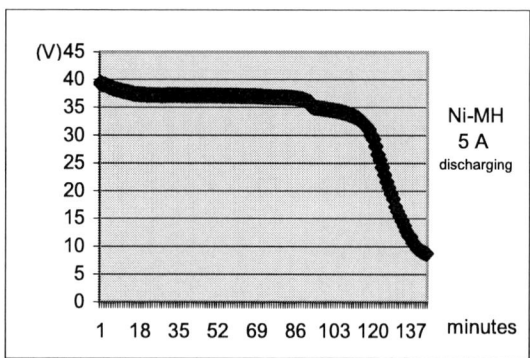

(d) Ni-MH 36V 8Ah discharging characteristic at 5 A
Fig 2 PEMFC modeling and Ni-MH modeling

Ni-MH model is the circuit as a series combination of a voltage source (Eo) a series resistance (Rh) and a parallel combination of a capacitor (Cp) and a resistor (Rd = the effective delayed resistance.

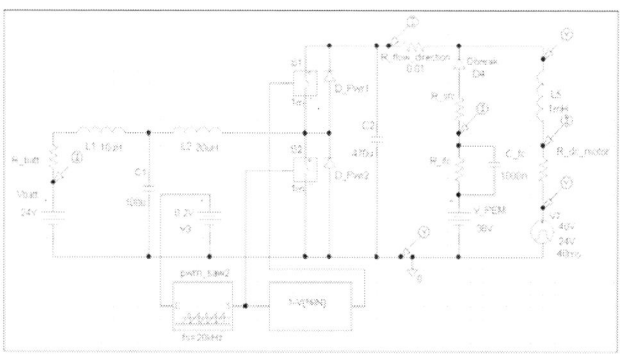

Fig 3 Single DC-DC converter schematic simulation

Converter design

The two-quadrant converter is proposed. The converter in system shown in Fig. 1 requires positive voltage for the load voltage while the load current is either positive or negative shown in Fig. 4. The load current can be negative as the effect of back emf of a dc motor. Two modes of operation can be categorized as load and the battery operation as listed in the Table 1.

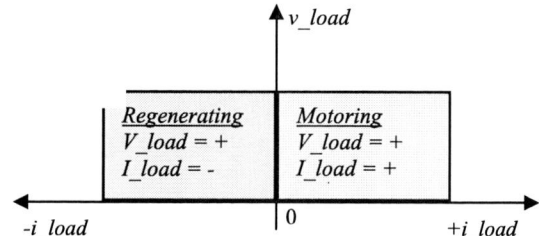

Fig 4 DC motor load operating modes

Table 1 DC motor load, battery operation and converting modes

v_load, i_load	DC motor load	Ni-MH battery	Converting modes
+, +	Motoring	Discharging	Boost
+, -	Regenerating	Charging	Buck

Converter design requirement

According to the converter key performances as above mention on a single converter, simplicity, allowing bi-directional current flows or charging and discharging modes and high efficiency energy management, the converter design requirement is proposed.

Class C converter on Buck-boost converter is designed as a single converter where two switches are needed. The two inductors, L1 and L2, can be integrated in one core inductor. The inductance current is designed to operate in continuous mode. Two capacitors, C1 and C2, are used to filter out the ripple currents on both of charging current to the Ni-MH and the dc motor load current.

Ni-MH battery discharging or Boost converting mode

S1 is the switch to step-up the voltage of the Ni-MH voltage by the function of switching duty ratio (D). The D is fixed to 0.2 to gain 1.25 time of Ni-MH terminal voltage. L2 is designed as a minimum inductance to boost or discharge which is the dominating of operating period.

Ni-MH battery charging or Buck converting mode

S2 switch is the switch to step-down the voltage of load bus which is raised by dc motor back emf during the motor braking. The Ni-MH terminal voltage is controlled by function of (1-D). L1 is series connected to the Ni-MH to filter the charging current. The deadtime between the S1 and S2 is strictly switching controlled to avoid the short through of the load bus voltage.

Energy source management

The weight limitation of on-board vehicle is the most critical of the electric vehicle. The high power energy source is carefully selected. Ni-MH is chosen for rechargeable battery due to two time of specific power density comparing to the lead-acid battery. The PEMFC is suitable for EV due to the advantage of higher specific power density with extending the driving range by carrying the hydrogen (H_2) with compact hydrogen metal hydride storage.

During the discharging or Boost converting mode, the PEMFC current is controlled to supply two time of the Ni-MH battery current. The sharing current ratio results a small energy sizing of the Ni-MH.

During the charging or Buck converting mode, the PEMFC current cannot store the energy form the dc motor back emf by the blocking diode series with the PEMFC terminal voltage. In this mode the PEMFC should supply minimum current to the Ni-MH battery to save the hydrogen consumption. The impedance and the voltage of both PEMFC and Ni-MH should be determined.

III. Simulated results

System is composed of three main parts as following:
1. Sources: PEMFC and Ni-MH battery
2. DC motor load
3. DC-DC converter

System specification:
1. 500W PEMFC, 36 Voc, 24 V at full load, air flow cooling
2. Ni-MH Battery: 24 V, 8 Ah
3. DC motor load: 24 V, 20 A
4. Buck-boost converter: 500 W

PSpice program is used to simulate the system operation as shown in Fig. 1. Figure 10 (zoom in at the last page of the paper) shows the transition of discharging or Boost converting mode and charging or Buck converting mode at 43 ms. Key waveforms, load current, inductor current, PEMFC current, load voltage and main bus voltage, show in Figures 5 and 6.

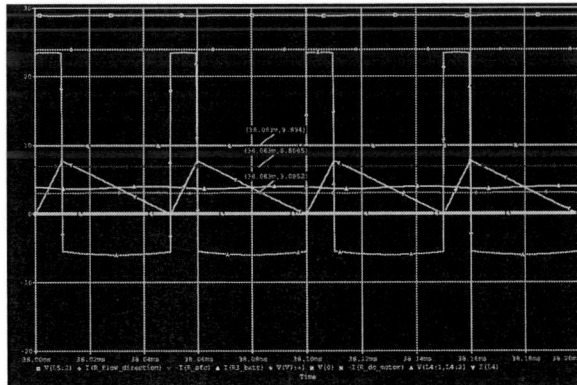

Fig. 5 Discharging or Boost converting mode

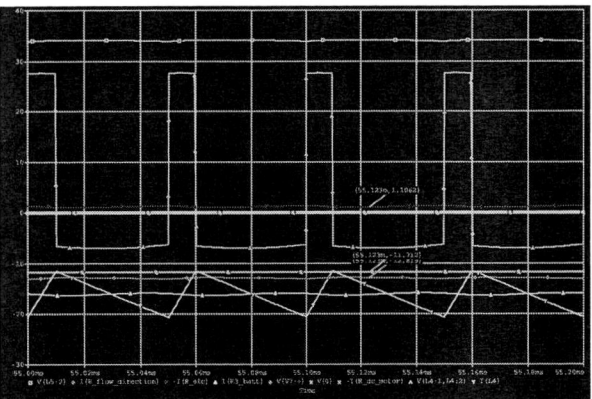

Fig. 6 Charging or Buck converting mode

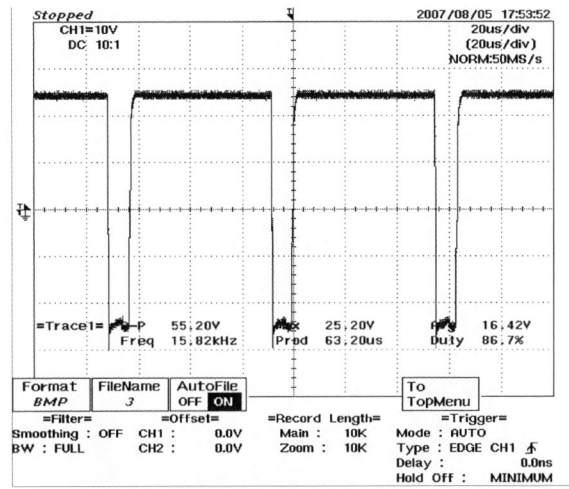

(a) Forward driving waveform

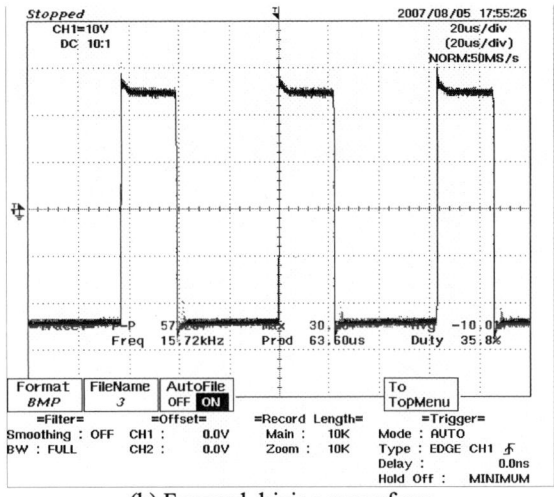

(b) Forward driving wavwform

Figure 7. Forward and reverse DC motor drive operation

Figure 7 shows the forwarded and reversed operation of the DC motor by the duty cycle variation.

1337

The achievement of power flow control can be confirmed by the simulated result as shown in Fig. 8. For example, during discharging mode, the main power supply comes from the PEMFC (196.7 W) which is about two times of the Ni-MH (93.6 W). During the charging mode when the back emf of the Dc motor occurs, the power is charged by the combination of regenerative current of the motor and the PEMFC. The power is drawn by the PEMFC (40 W) is about 10% of the power from the regenerative power of the motor (400 W). Some voltage, current and other power are listed in Table 2.

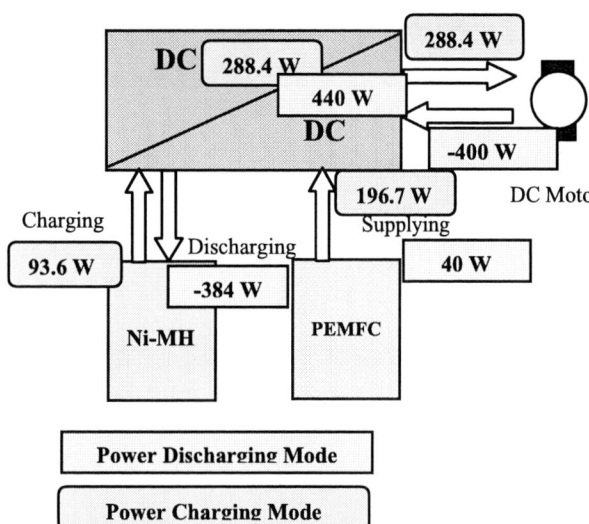

Power Discharging Mode

Power Charging Mode

Fig. 8 Power flow on discharging mode and charging mode

(a) PEMFC components

(b) PEMFC system with metal hydride hydrogen storage

(c) PEMFC electric vehicle prototype
Figure 9. PEMFC electric vehicle system

Figure 9 shows the PEMFC application in electric vehicle. Figure 9 (a) composes of the PEMFC system such as the controller, coolant system, airflow control and water output. Figure 9 (b) shows the hydrogen metal hydride storage. The prototype of electric vehicle using PEMFC based shown in figure 9 (c)

IV. Conclusion

A single bi-directional DC-DC converter is designed. The brief of converter operation is described. Advantages of the converter are as follows:

- compact,
- ease of bidirectional power flow control based on the operating of two switches
- Low ripple of input and output currents of the converter

Some of power flows are presented. Minimum of inductor is also simulated and designed for discharging mode. Ni-MH with the memory effect requires a careful charging to keep the battery life cycle. Lithium battery should be considered. Finally, the PEMFC electric vehicle prototype is introduced.

References

[1] A Jossen, J. Garche, H.Doering, M. Goetz, W, Knauapp and L. Joerissen, "*Hybrid systems with Lead-Acid Battery and PEMFC*", Journal of Power Sources, June 2005, Vol. 144, No.2, pp. 395-401.

[2] Chih-Chiang Hua, Chi-Lun Huang and Hsien-Chang Chiu, "*Research on Dynamic Response of Hybrid Power Sourec System with PEMFC and Lead-Acid batteries*", The Fourth Power Conversion Conference 2007, April 2007, pp. 739-744.

[3] Chih-Chiang Hua, Jeng-Da Lin, Jun-Wei Wu, Wei-Shi Liu, and Chi-Ruen Huang, "*Design and Implementation of a Digital Controlled Power Converter for Fuel Cell Systems*" The 2005 International Power Electronics Conference, Japan, pp. 1939-1943.

[4] Masaki Sato, Yukihiro Yanagita, Hirohito Funato, Satoshi Ogasawara, , "*New Estimation Method of State of Batteries Based on System Identification Estimation of State-of-Charge of NiMH Batteries*" The 2005 International Power Electronics Conference, Japan, pp. 1933-1938.

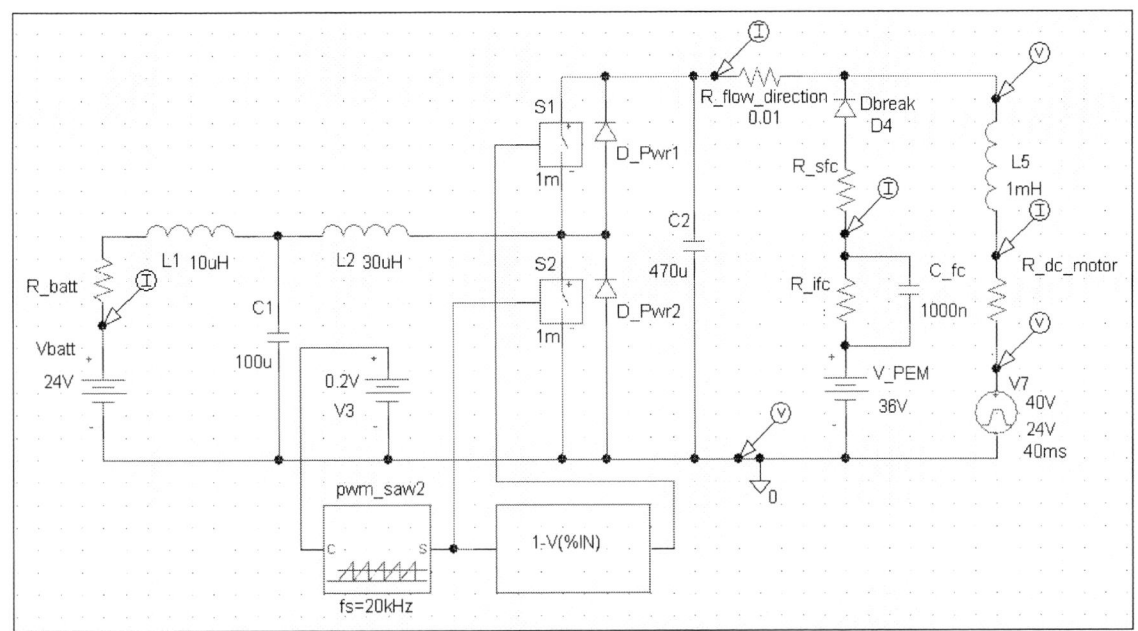

Fig 1 (Zoom in) single DC-DC converter schematic simulation

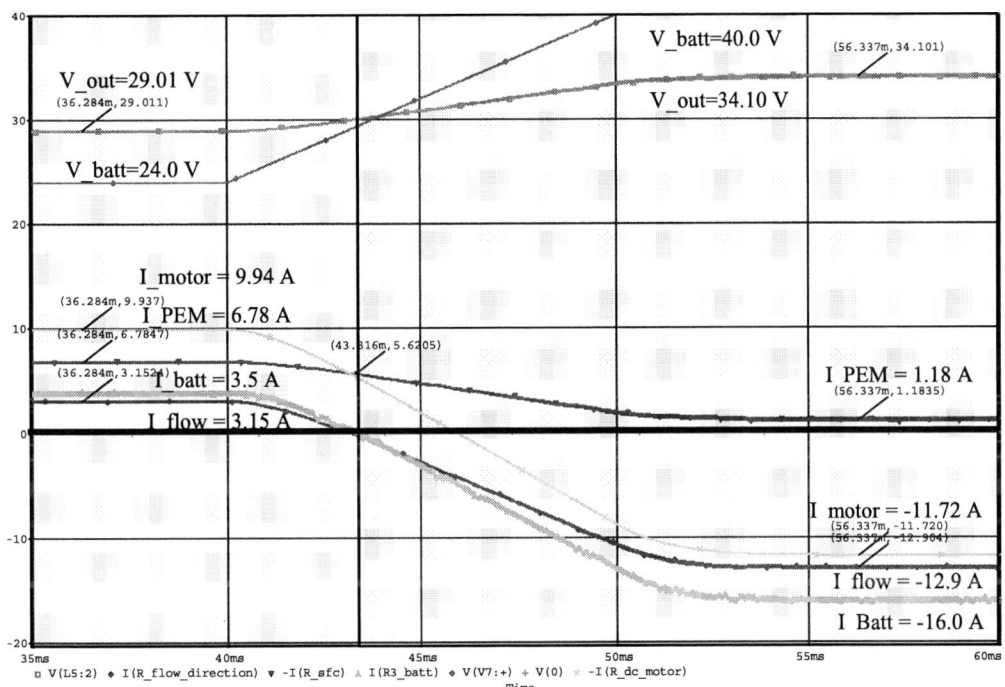

Fig 10 Battery charging and discharging transition

V_batt (A)	V_main_bus (A)	V_out (V)	I_Batt (A)	I_flow (A)	I_PEM (A)	I_motor (A)	P_batt (W)	P_PEM_t (W)	P_motor (W)	P_pem (W)	P_terminal (W)	P_flow (W)
24	29.01	24	3.9	3.15	6.78	9.94	93.6	196.68	238.56	244.08	288.35	91.38
24	34.1	40	-16	-12.9	1.18	-11.72	-384	40.238	-468.8	42.48	-399.65	-439.9

Table 2 Details of voltage, current and power flows

Speed Sensorless Control with Neuron MRAS Estimator of an Induction Machine

Dong Lei[1,2], Yang Dong[1], Liao Xiaozhong[1,2]

[1]Department of Automatic Control Engineering, Beijing Institute of Technology,

[2]Key Laboratory of Complex System Intelligent Control and Decision, Ministry of Education, Beijing, P.R. China

Abstract–In the high speed range, vector control of rotor flux orientation of an induction machine implements good performance. However, the performance in low speed rang deteriorates because of the inaccurate estimation of rotor flux and speed. In this paper, modified voltage model for rotor flux estimation and neuron model-reference adaptive system (MRAS) for speed estimation are used to improve the performance of speed sensorless vector control. To improve the accuracy of rotor flux estimation, the stator resistance is identified on-line. The experimental results show that the proposed scheme yields improved performance in low speed range.

Index Terms--induction machine, MRAS, neuron model, speed sensorless control, stator resistance identification.

I. INTRODUCTION

Vector control of induction motors has been widely used in high performance drive system. However, the performance of the vector controlled induction motors depends on the accuracy of the estimated rotor flux from the measured stator currents and the measured speed from rotor shaft encoder. Various researches have focused on speed sensorless vector control of induction motors in terms of low cost and high reliability[1]~[4].

As is well-known, the rotor flux estimation can be carried out by the so-called current model or voltage model[5][6]. The former involves rotor speed as an input variable, thus it can not be used to obtain the rotor flux in a speed sensorless rotor flux orientation system by itself. According to the latter, it achieves a good performance in high speed range. However, as the speed decreases to lower values, the rotor flux estimation deteriorates owing to the influence of inaccurate values of the stator resistance. A stator resistance error causes the estimated

rotor flux to deviate from its theoretical trajectory[7].

[8] presents a method of on-line estimation for the stator resistances of the induction motor in the vector controlled drive, using MRAS based on current model and finite compensating voltage model. [9] presents another on-line estimation for the stator resistances, using artificial neural networks. The back propagation algorithm is used for training of the neural networks. The error between the desired state variable of an induction motor and the actual state variable of a neural network model is back propagated to adjust the weights of the neural network model, so that the actual state variable tracks the desired value.

This paper presents a new method of neuron model-reference adaptive system (MRAS) for speed estimation. The current model of an induction machine is modified with neuron using as an adjustable model, and finite compensating voltage model is used as reference model. In high speed range, the back EMF is large and more accurate. The finite compensating voltage model serves as a good rotor flux estimator at high speed. However, in low speed range the back EMF is very small, and even very slight stator resistance thermal variation will corrupt the back EMF computation. The incorrect computation of back EMF, in turn, deteriorates the rotor flux estimation. To overcome these problems, the stator resistance is identified on-line to compensate the rotor flux.

II. ROTOR FLUX ESTIMATION BASED ON MODIFIED VOLTAGE MODEL

The stator and rotor voltage equations and flux equations of the induction machine in the stationary reference frame are given by

$$u_{s\alpha} = R_s i_{s\alpha} + p\psi_{s\alpha}$$
$$u_{s\beta} = R_s i_{s\beta} + p\psi_{s\beta}$$
$$0 = R_r i_{r\alpha} + p\psi_{r\alpha} + \omega_r \psi_{r\beta} \qquad (1)$$
$$0 = R_r i_{r\beta} + p\psi_{r\beta} - \omega_r \psi_{r\alpha}$$

$$\psi_{s\alpha} = L_s i_{s\alpha} + L_m i_{r\alpha}$$
$$\psi_{s\beta} = L_s i_{s\beta} + L_m i_{r\beta}$$
$$\psi_{r\alpha} = L_m i_{s\alpha} + L_r i_{r\alpha} \qquad (2)$$
$$\psi_{r\beta} = L_m i_{s\beta} + L_r i_{r\beta}$$

The rotor EMF component derived form (1)(2) is given by

$$\vec{e}_r = \frac{L_r}{L_m}\left[(\vec{u}_s - \vec{i}_s R_s) - \sigma L_s p\vec{i}_s \right] \qquad (3)$$

Where, $\sigma = 1 - \dfrac{L_m^2}{L_r L_s}$ is the total leakage factor. So voltage model of rotor flux is given by

$$\vec{\psi}_r^V = \int \vec{e}_r = \frac{L_r}{L_m}\left[\frac{1}{p}(\vec{u}_s - \vec{i}_s R_s) - \sigma L_s \vec{i}_s \right] \qquad (4)$$

where p p is d/dt. The rotor flux is the integral of the back EMF. In practice, the original voltage model of (4) is difficult to implement because the pure integration in (4) brings about the initial value and drift component problems. To avoid these problems the pure integration is replaced by the low-pass filter. However, the low-pass filter introduces new problems of amplitude damp and phase shift of rotor flux. Therefore the finite compensating voltage model will be a good solution[10]. The integrator of the finite compensating voltage model is replaced by

$$\hat{\vec{\psi}}_r^V = \frac{L_r}{L_m}[(\vec{u}_s - R_s \vec{i}_s)\frac{1}{p + (1 - k_c)\omega_c} - \sigma L_s \vec{i}_s] \qquad (5)$$

Where, $k_c = \begin{cases} 1 & |\psi_r| \le \psi_r \\ 0 & |\psi_r| > \psi_r \end{cases}$ is the compensation component.

When the magnitude of the estimated rotor flux is lower than the reference value, $k_c = 1$, the compensation component is equal to the estimated rotor flux, and the integrator becomes a pure integration. However, when the magnitude of the estimated rotor flux is higher than the reference value, the compensation component can be limited with k_c. Furthermore, if $k_c = 0$, the estimator will become to a low-pass filter, so it can effectively decrease the integration drift. Fig. 1 shows the rotor flux estimation scheme.

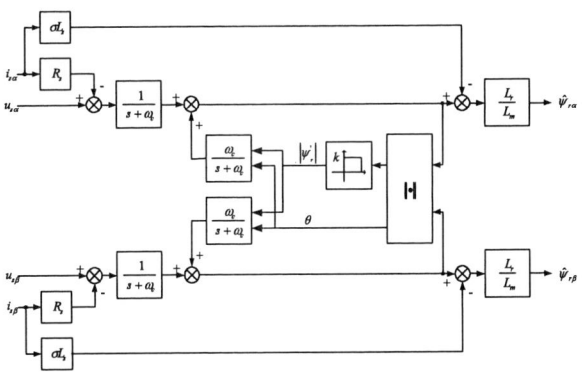

Fig. 1. Rotor flux estimation

III. SPEED ESTIMATION BASED ON NEURON MRAS SCHEME

Current model of rotor flux derived form (1)(2) is given by

$$p\vec{\psi}_r^I = (-\frac{1}{T_r}I + \omega J)\vec{\psi}_r^I + \frac{L_m}{T_r}\vec{i}_s \qquad (6)$$

where, $I = \begin{bmatrix} 1 & 0 \\ 0 & 1 \end{bmatrix}$, $J = \begin{bmatrix} 0 & -1 \\ 1 & 0 \end{bmatrix}$

The sample-data model of (6) is shown as

$$\hat{\vec{\psi}}_r^I(k) = (1 - \frac{T_s}{T_r})I\hat{\vec{\psi}}_r^I(k-1)$$
$$+ \hat{\omega}T_s J\hat{\vec{\psi}}_r^I(k-1) + \frac{T_s L_m}{T_r}\vec{i}_s(k-1) \qquad (7)$$

where, T_s is the sampling period, and T_r is the rotor time constant. Equation (7) can be rewritten as

$$\hat{\vec{\psi}}_r^N(k) = W_1 X_1 + W_2 X_2 + W_3 X_3 \qquad (8)$$

where, $W_1 = 1 - \dfrac{T_s}{T_r}$, $W_2 = \hat{\omega}T_s$, $W_3 = \dfrac{T_s L_m}{T_r}$,

$X_1 = I\vec{\psi}_r^N(k-1)$, $X_2 = J\vec{\psi}_r^N(k-1)$, $X_3 = I\vec{i}_s(k-1)$.

The single neuron current model represented by (8) is shown in Fig. 2, where W_1, W_2, W_3 represent the weights of the neuron and X_1, X_2, X_3 are the three inputs to the neuron. The single neuron model (8) involves the rotor speed in W_2, while the finite compensating voltage model (5) does not. In model reference adaptive systems, a comparison is made between the outputs of two models. The finite compensating voltage model is considered as a reference model of the induction machine. The other one may be regarded as an adjustable model, which is shown in Fig. 3. The energy function can be defined as

$$E = \frac{1}{2}[\hat{\bar{\psi}}_r^V(k) - \hat{\bar{\psi}}_r^N(k)]^2 = \frac{1}{2}\vec{\varepsilon}^2(k) \quad (9)$$

and the weight variation is given by

$$\begin{aligned}\Delta W_2(k) &= -\eta \cdot \frac{\partial E}{\partial W_2} \\ &= \eta\{\psi_{r\alpha}^N(k-1) \cdot [\psi_{r\beta}^V(k) - \psi_{r\beta}^N(k)] \\ &\quad - \psi_{r\beta}^N(k-1) \cdot [\psi_{r\alpha}^V(k) - \psi_{r\alpha}^N(k)]\}\end{aligned} \quad (10)$$

where η is the training coefficient, and the rotor speed can be calculated from W_2

$$\hat{\omega} = \frac{W_2}{T_s} \quad (11)$$

IV. ON-LINE STATOR RESISTANCE IDENTIFICATION

The performance of speed sensorless vector control of an induction machine depends on the accuracy of the estimated rotor flux and rotor speed. The accuracy of rotor flux estimation from the finite compensating voltage model is more important, because it regards as reference model of MRAS speed estimator. The accuracy of the estimated rotor flux is greatly influenced by the value of stator resistance. In low speed range, the rotor flux estimation is more sensitive to the change of stator resistance. So on-line stator resistance identification method is investigated in this paper, which is shown in Fig. 4.

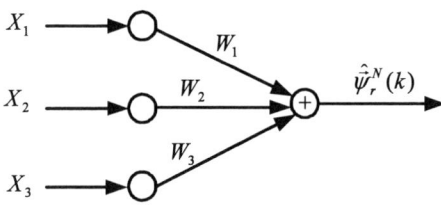

Fig. 2 . The single neuron model of rotor flux

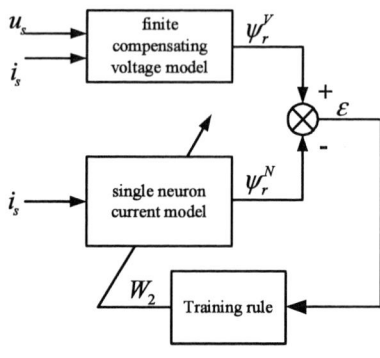

Fig. 3.   Rotor speed estimator

However, the finite compensating voltage model is considered as an adjustable model, and the current model that is in synchronously rotating reference frame is regarded as a reference model. From the state equations of the induction motor(1)(2) we can obtain the current model in synchronously rotating reference frame

$$\psi_r = \frac{L_m}{T_r p + 1} i_{sd} \quad (12)$$

As shown in Fig.4, the PI adaptation mechanism is given by

$$\Delta \hat{R}_s = (K_p + \frac{K_i}{p})\varepsilon \quad (13)$$

and the stator resistance can be calculated from (13)

$$\hat{R}_s(k) = \hat{R}_s + \Delta \hat{R}_s \quad (14)$$

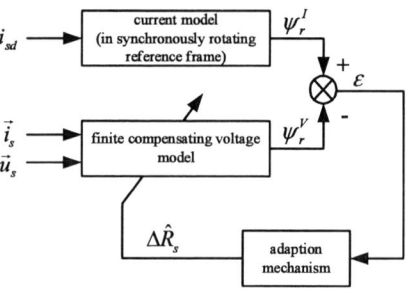

Fig. 4.   Stator resistance identification

1342

Fig. 5. Experimental set-up of induction machine drive

TABLE 1   Parameters of induction motor

Poles pairs	$n_p=2$
Rotor resistance	$R_r=1.80\,\Omega$
Stator resistance	$R_s=2.72\,\Omega$
Rotor inductance	$L_r=0.2762H$
Stator inductance	$L_s=0.2762H$
Magnetizing inductance	$L_m=0.2646H$

V. EXPERIMENTAL RESULTS

The experimental set-up for speed sensorless vector control of an induction machine implemented using a controller board based on TMS320F240 DSP. An IGBT inverter with a switching frequency of 8 kHz is used. A DC motor coupled to the induction machine is used as a load, shown in Fig. 5. A constant load torque is maintained by using the current control loop in the load circuit. A 2.2 kW induction machine has been test with this control system, and the parameters of the induction machine are shown in the table 1.

Fig. 6 shows experimental results of rotor flux estimation with the schemes according to pure integration in (4) and Fig. 1. The figure (a) in Fig. 6 shows that the estimated rotor flux drifts from real flux trajectory. However, the method of finite compensating voltage model can performs accuracy estimation of rotor flux. And this is very important for vector control of an induction machine.

Fig. 7 shows the system behavior in the case of a start course with 7Nm constant load. The reference speed is 2.5Hz. Fig. 7(a) shows the real speed and estimated speed with neuron MRAS scheme. The real speed is obtained from an encoder mounted on the rotor shaft. And the identified torque, one phase current, and rotor flux of α and β components are shown.

To check the load response, Fig. 8 and Fig. 9 show the dynamic behavior of speed sensorless control in low speed range. A load torque step up and step down are respectively set in their tests. Fig. 10(b) gives the on-line stator resistance identification behavior. The real stator resistance is 2.72Ω, and the identified stator resistance is about 2.6Ω.

VI. CONCLUSIONS

Speed sensorless vector controlled drive with an induction machine in low speed range is shown. And good dynamic performance is feasible, using finite compensating voltage model rotor flux estimator, neuron MRAS speed estimator, and stator resistance compensation algorithm. The load capacity in low speed is satisfactory for most applications. This control strategy is simplified and takes good steady and dynamic performance.

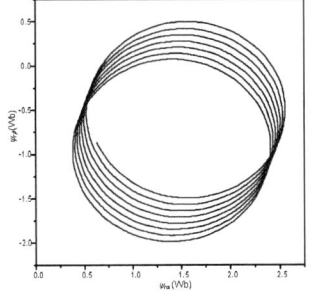

（a）Pure integration method

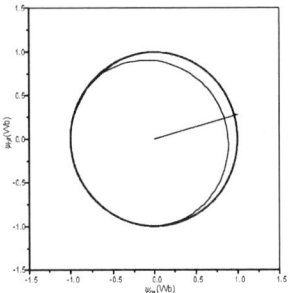

（b）finite compensating voltage model

Fig. 6.   Trajectory of rotor flux

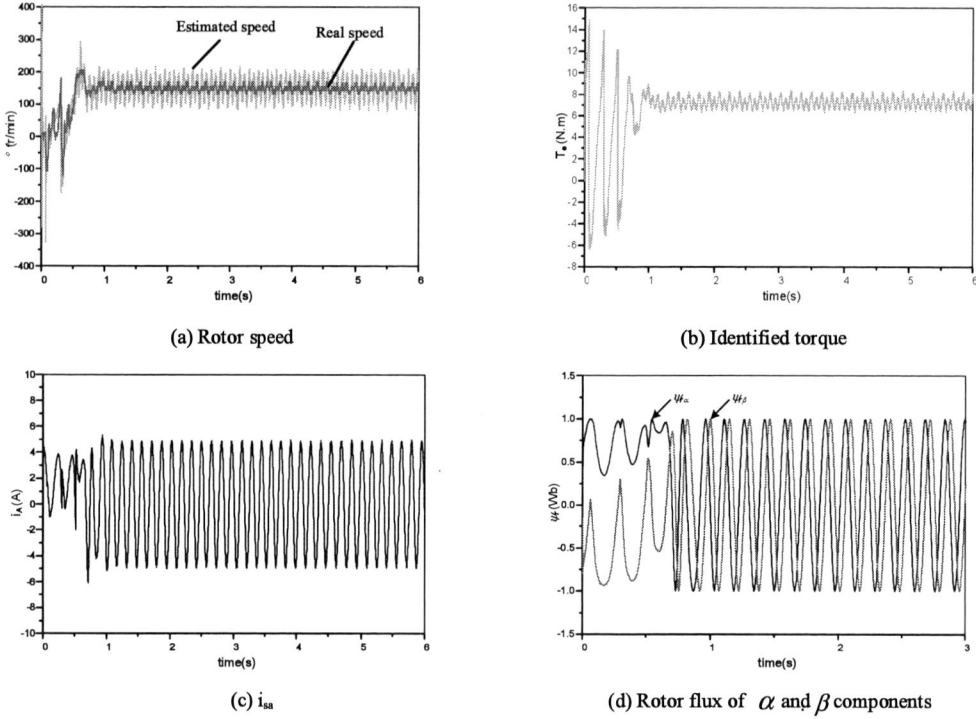

(a) Rotor speed

(b) Identified torque

(c) i_{sa}

(d) Rotor flux of α and β components

Fig. 7. Start waveform of speed sensorless control. The speed reference is set at 2.5Hz, and the load is set at 7Nm.

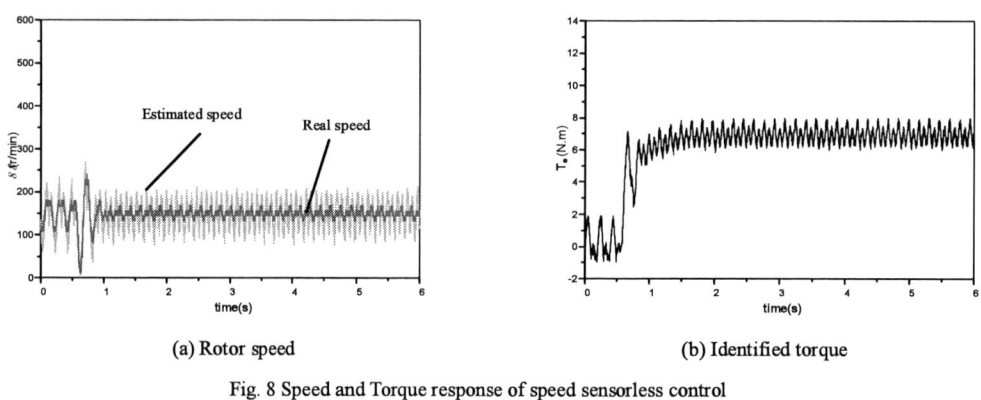

(a) Rotor speed

(b) Identified torque

Fig. 8 Speed and Torque response of speed sensorless control

for a load torque step up at 2.5Hz

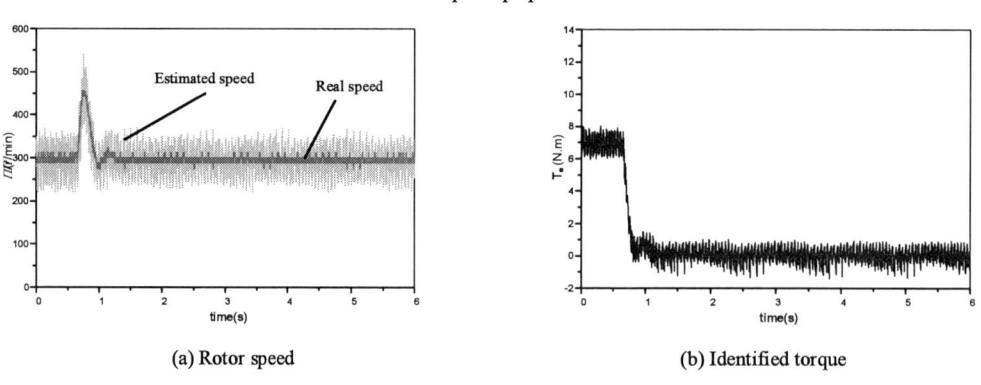

(a) Rotor speed

(b) Identified torque

Fig. 9 Speed and Torque response of speed sensorless control

for a load torque step down at 5Hz

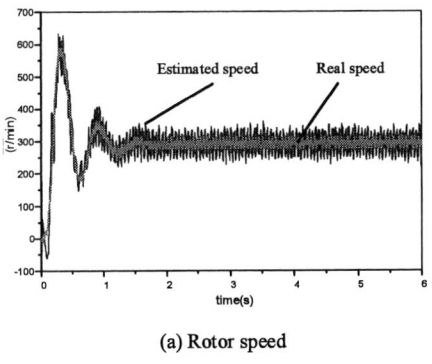
(a) Rotor speed

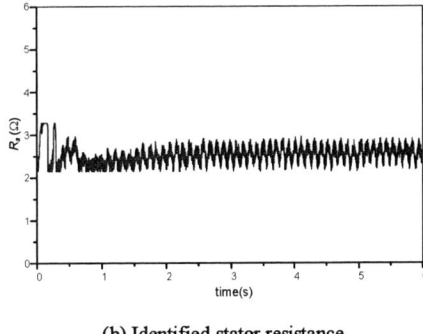
(b) Identified stator resistance

Fig. 10 Start waveform of speed sensorless control. The speed reference is set at 5Hz

REFERENCES

[1] Jiang J, Holtz J. "Speed Sensorless AC Drive for High Dynamic Performance and Steady State Accuracy." IEEE, 1995: 1029~1034.

[2] Geng Yang, Tung-hai Chin. "Adaptive-speed Identification Scheme for a Vector-contro lled Speed Sensorless Inverter-induction Motor Drive." Trans. Ind. Applicat, 1993, 29(4):820~825.

[3] P. L. Jansen and R.D. Lorenz, "A physically insightful approach to the design and accuracy assessment of flux observers for field oriented induction machine drives," IEEE Transaction on Industry Application, Vol. 30, Jan./Feb., 1994, pp. 101-110; 1696.

[4] Dong Lei, Li Yongdong , Chai Jianyun, and Wang Wensen, "Speed Sensor-less Control of An Induction Machine with Rotor Flux Tracking at Very Low Speed" ICEMS' 2001, pp. 1230-1235.

[5] Colin Schauder, "Adaptive speed identification for vector control of induction motors without rotational transducers, " in Proc. 1989 IAS Ann. Mtg. (San Diego), pp. 493-499.

[6] Hirokauz Taijima and Yoichi, "Speed sensorless field-orientation control of the induction machine," IEEE Trans. on IA, 1993, Vol. 29(1):175-180.

[7] Domenico Casadei, Giovanni Serra, and Angelo Tani, "Steady-state and transient performance evaluation of a DTC scheme in the low speed range," IEEE Transactions On Power Electronics, VOL. 16, NO. 6, Nov. 2001.

[8] Lei Dong, Yongdong Li, Xiaozhong Liao, "Novel Speed Sensorless Vector Control with Adaptive Rotor Flux Identification of Induction Motors, " IECON'03, 6 Nov, Virginia, USA.

[9] Baburaj Karanayil, Muhammed Fazlur Rahman, Colin Grantham."On-line Stator and Rotor Resistance Estimation scheme for Vector Controlled Induction motor Drive using Artificial Neural Networks." Industry Applications Conference, 2003, 38th IAS Annual Meeting. Conference Record of the Volume 1, 12-16 Oct.2003 Page(s):132-139 vol.1.

[10] Jixiong Wu, Yongdong Li, Jie Chen, Hu Hu, "Speed sensor-less direct torque control of an induction machine in low speed region," IPEMC'2000, pp. 464-468.

Adaptive Flux model for commissioning of signal injection based zero speed sensorless flux control of induction machines

T.M. Wolbank, M.A. Vogelsberger, R.H. Stumberger
Department of Electrical Drives and Machines
Vienna University of Technology
Vienna, AUSTRIA

Abstract **Speed sensorless control of ac machines at zero speed so far is only possible using signal injection methods. Especially when applied to induction machines spatial saturation leads to a heavy dependence of the control signals on the flux/load level. This dependence has to be identified on a special test stand during a commissioning procedure. To avoid the usage of a speed sensor as well as load dynamometer coupled during the commissioning an adaptive flux model is proposed that delivers an accurate reference flux angle. After the commissioning this adaptive flux model is used in combination with the signal injection method to deliver the spatial flux position.**

I. INTRODUCTION

Modern industrial ac induction machine drives are generally equipped with a speed or position sensor mounted to the rotor shaft in order to accurately determine the flux position required for controlled high dynamic operation. Especially at low fundamental frequency so called passive models based on the fundamental wave of the electrical quantities deteriorate in their performance due to the lack of feedback. At zero frequency these models fail and stable speed sensorless operation is considered only possible applying signal injection methods.

The mentioned signal injection based methods published all rely on a high frequency or transient signal impressed by the inverter on top of the fundamental wave in order to realize an additional excitation of the machine. The reaction of the machine is then influenced by inherent saliencies of the machine induced by spatial saturation, slotting of the lamination, or anisotropy. By identifying these influences in the machine reaction to the injected signal the information on the flux or rotor position necessary for control can be extracted.

Usually voltage excitation is preferred to current excitation. The excitation can be realized either harmonic as a pulsating or rotating component [6]-[9] or transient as voltage pulse sequences [10]-[12]. In this paper a transient excitation signal is used for sensorless control (INFORM method) [10]. The results however, are also applicable to the other mentioned types of signal injection.

The main problem in zero speed sensorless control currently is not the excitation or the measurement of the machine reaction but the additional signal components contained in the resulting signal that deteriorate the performance especially when the machine is loaded. These additional and disturbing signal components are caused by effects such as inverter non linearity, measurement errors, and especially the dependence of the saturation induced saliency on the magnetic point of operation. This usually most prominent saliency exhibits a heavy dependence of both magnitude as well as orientation on the flux and load level of the machine. But also the remaining signal components show a more or less distinct influence on flux/load. In order to make the resulting signal applicable for sensorless estimation of the rotor and/or flux position it is thus necessary first to identify all the operating point dependencies of the individual signal components – also denoted as the magnetic fingerprint of the machine in literature.

This identification is done during a commissioning phase for each new type of induction machine. If the goal of the commissioning is not the elimination of the saturation saliency but the exploitation of that signal component to estimate the flux position then it is necessary to ensure a reliable reference flux angle for identification. To avoid the usage of a load dynamometer as well as that of a speed/position sensor even during that phase it is necessary to establish an accurate sensorless flux model. That special model must be able both to ensure the reference angle during commissioning as well as to work in combination with the flux angle estimated using the signal injection method.

II. SIGNAL INJECTION BASED SENSORLESS CONTROL

As mentioned in the introduction, an injection method based on excitation with a voltage pulse sequence is applied in this paper. The pulse sequence is generated by the switching of the inverter and in order to minimize the influence on the current control scheme is only applied during otherwise inactive operating states. The goal is to determine the spatial position of the different inherent saliencies in order to estimate the flux or rotor position. To ensure a high sampling rate of the signal the injection of the sequence is initiated with a frequency of about 1kHz. The reaction of the machine is measured as the current slope for each single pulse. This transient current slope of an induction machine is determined by the stator equation of the machine.

978-1-4244-0644-9/07/$25.00 ©2007 IEEE

$$\underline{u}_S = r_S \cdot \underline{i}_S + l_\sigma \cdot \frac{d\underline{i}_S}{d\tau} + \frac{d\underline{\lambda}_R}{d\tau}. \qquad (1)$$

The transient change in the machine current $d\underline{i}_S/d\tau$ is thus determined by the voltage $\underline{u}_S$ applied, whose magnitude is proportional to the dc link voltage during an active inverter switching state, as well as by the value of the leakage inductance l_σ. Additional influences are the stator resistance voltage drop $r_S.\underline{i}_S$ and the back emf. The mentioned inherent saliencies have almost negligible influence on the fundamental wave behavior of the machine. They mainly affect only the transient leakage inductance and by measuring the transient current response it is thus possible to detect and locate their positions.

The influences of the stator resistance voltage drop $r_S.\underline{i}_S$ and the back emf have to be eliminated by evaluating the current change of two subsequent active voltage phasors pointing in different spatial directions. This is the main reason why a pulse sequence of at least two pulses is usually applied as excitation signal. The resulting signal contains different components related to the angular movement of the saliencies present in the machine. In Fig. 1 a typical harmonic spectrum of this signal is depicted. All signal components associated only to saturation are given in red, that mainly induced by the slotting are shown yellow colored, and the components dominantly linked to the fundamental wave are marked blue.

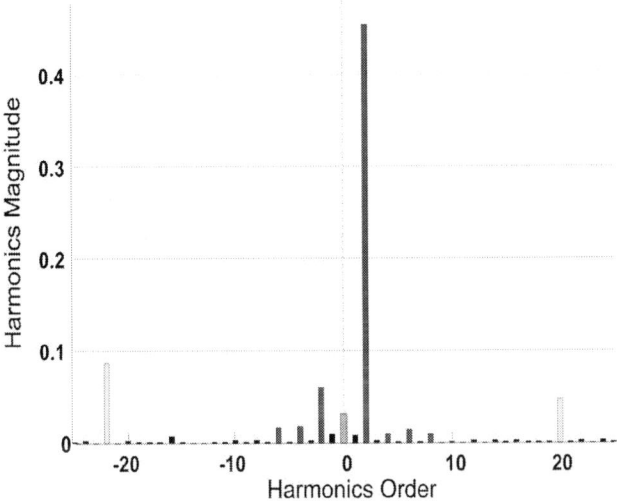

Fig. 1: Load content of current response signal after elimination of resistance and back emf. (saturation (red) slotting (yellow), fundamental wave (blue) components

The harmonic order number 1 corresponds to the fundamental wave, the dominant second harmonic is related to the saturation saliency as only the saturation level is responsible for the modulation of the leakage inductance not its direction. For the estimation of the rotor flux angle only the dominant (red) $+2^{nd}$ harmonic is useable, the remaining red-colored harmonics act as disturbance and have to be eliminated.

As the magnitudes of the harmonics all change with the operating point of the machine, there is the need for an identification of these influences during the commissioning

phase. Especially the changing magnitude and the phase shift of the dominating $+2^{nd}$ harmonic has to be accurately known for speed sensorless control.

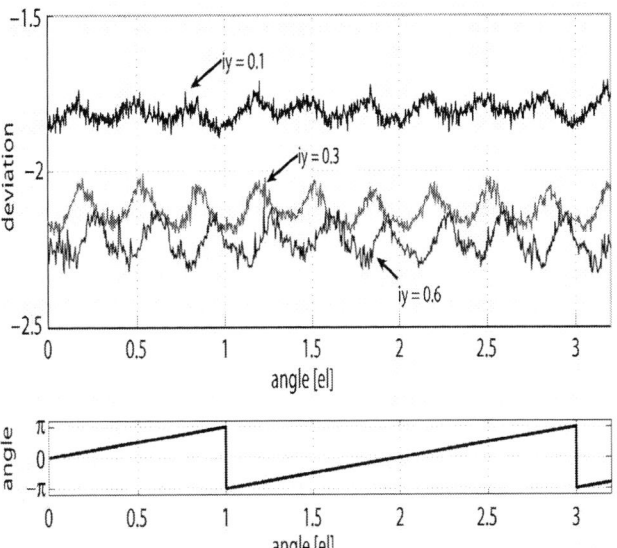

Fig. 2: Schematic angle difference between resulting signal and rotor flux for different load levels.(horizontal axes rad/π)

This effect is shown in Fig. 2 for different load levels. The horizontal axes are scaled in rad/π showing about 1,6 revolutions of the electrical angle. On the vertical axis of the upper diagram the angle difference between the flux angle and the overall angle of the resulting signal obtained from the transient current reaction is depicted in rad for three different load levels (torque current 0.1; 0.3; 0.6). The lower diagram gives the actual value of the rotor flux. It can be seen that a change of the load to higher levels is linked with an increased lagging of the angle deviation in the diagram.

In addition also the magnitude as well as orientation of a superposed higher harmonic (6th) is changed. In order to estimate the rotor flux angle using the dominant $+2^{nd}$ component the mentioned load dependence has to be identified. Such an identification is usually done on a test stand with a load dynamometer coupled and a shaft sensor mounted to the machine.

To cover the whole operating range it can be said that at low load levels the phase shift of the $+2^{nd}$ harmonic is the dominating source of disturbance, whereas for the high load levels the rise of additional disturbing harmonics are of main importance. To avoid the usage of a shaft sensor for the identification an adaptive flux model is applied.

III. ADAPTIVE FLUX MODEL

As no rotor speed or angle is used the rotor flux has to be calculated by integration of the terminal voltage using the stator equation only as shown in (2) leading to the well known voltage model.

$$\underline{\lambda}_R = \int (\underline{u}_S - r_S \cdot \underline{i}_S) d\tau - l_\sigma \cdot \underline{i}_S. \qquad (2)$$

There $\underline{u}_S$ and $\underline{i}_S$ are the stator voltage and stator current phasors, respectively. The stator resistance is denoted as r_S and the total leakage inductance is l_σ. In practical operation the integration of (2) will only be stable at higher stator frequencies when the stator voltage vector magnitude $\underline{u}_S$ is sufficiently high with respect to the stator resistance voltage drop. When approaching low speeds this condition will no longer be met and errors of estimated parameters, especially that of the stator resistance, together with measurement noise inverter non-linearity and sensor offset errors will cause the output to drift away. In industrial operation the application of voltage sensors at the inverter output is problematic and far from being standard. Thus in this investigation the stator voltage is reconstructed using the dc link voltage and the timing of the pulse width modulation (PWM). As a result the interlock dead time and the non-linear behavior of the inverter add additional errors to the input of the voltage model.

The signal injection based method is generally applied only at low speed where the performance of the flux model tends to deteriorate. For the application of the signal injection method the current reaction should be measured as near to the actual point of operation as possible. At medium and higher speed the back emf however leads to an increasing deviation of the stator current from the fundamental wave point of operation during the pulse sequence. A compensation of this effect is possible by shifting the pulse durations and sample instants accordingly. This however takes up a considerable portion of real time calculation power. Thus the whole commissioning should preferable be carried out within the low speed range of operation.

The number of papers dealing with speed sensorless flux models and performance comparison is very high. For details in the different approaches see [1]. The flux model proposed in this investigation uses modified parts of observers presented in [2], [3], [4], and [5]. The structure used in this paper is shown in Fig. 5. It consists of the stator equation (2) using additional stabilizing feedback voltages derived from flux magnitude errors as well as angle deviations with respect to the signal injection method. To improve the performance a parameter adaptation has been added as well as an additional feedback to compensate a possible offset voltage at the input of the integrator.

Based on the principle of the parameter adaptation shown in [18], the parameter tuning is done in two steps. First the deviation of the flux angle and magnitude from a reference value (starting from zero load) is calculated in a flux fixed reference frame to obtain the flux error components (Δx and Δy) expressed in equation (3) and (4).

$$\Delta x = \left| \underline{\psi}_R \right| \cdot \cos\left(\gamma_{\psi R} - \gamma_{\psi \, \mathrm{Re}f} \right) - \left| \underline{\psi}_{\mathrm{Re}f} \right| \tag{3}$$

$$\Delta y = \left| \underline{\psi}_R \right| \cdot \sin\left(\gamma_{\psi R} - \gamma_{\psi \, \mathrm{Re}f} \right) \tag{4}$$

After a transformation in a reference frame linked to the stator current the deviations can be assigned directly to errors in the parameters r_S and l_σ.

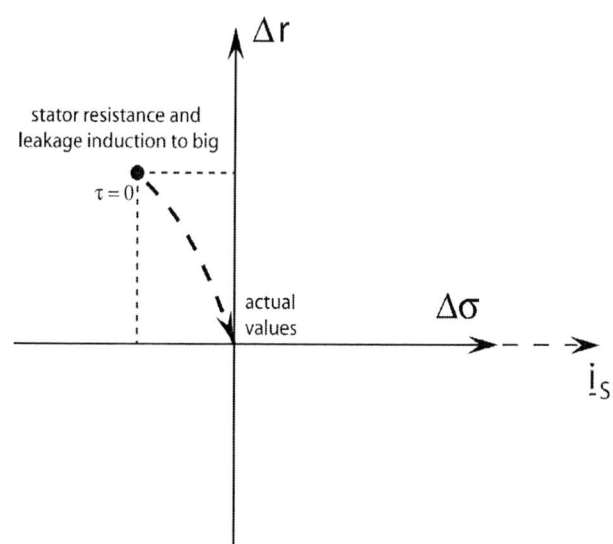

Fig. 3: Illustration of the rotor flux error represented by deviations of stator resistance and leakage inductance errors in the stator current fixed coordinate system.

In Fig. 3 the position of the rotor flux error transformed in the stator current fixed coordinate system is shown. The ordinate–axis marked the boundary for the leakage inductance error $\Delta\sigma$ (to big/to small) and the axis of the abscissa regards the border for the values of the stator resistance deviation Δr when using (2) for the flux estimation. It can be seen, that there is a fixed correlation between the quadrant of the flux failure and the following necessary tuning of the stator resistance r_S and leakage inductance l_σ. In the case, depicted in Fig. 3, the parameter adaptation is started at time $\tau=0$ with 'wrong' values (stator resistance and leakage inductance to big) to finish with the actual values.

Before the commissioning phase the parameters of the model are adjusted by activating the feed back of the block marked 'Adaptation' in Fig. 5 thus compensating also deviations in the measurement setup. During that initial phase the machine is operated at higher speed (around 0.4 p.u.) and the feedback of the flux magnitude error ($\Delta u_{\alpha,\beta}$) is deactivated.

The time trace of the parameters during the adaptation is shown in the upper diagram of Fig. 4. The lower diagram gives the trace of the flux position in the complex plane calculated with the proposed structure. The machine is operated at a speed of 0.4 p.u. and no load. As the reference angle of the flux phasor equals that of the stator current the adaptation can be performed without signal injection activated. Before the time marked $t=t_1$ the model parameters match that of the machine and no deviation in the flux is to be seen (lower diagram).

At the time instant $t=t_1$ the model parameters are changed to a multiple of the actual values. At the time instant $t=t_2$ the parameter adaptation is turned on.

As can be seen the adaptation feedback instantly changes the parameters accordingly to reduce the deviation of flux angle and magnitude. After $t=t_3$ the adaptation algorithm has reached constant parameter values and all variations have settled. It can be seen that the original values of the parameters are adjusted and no deviation is visible.

With the parameters tuned the adaptation is deactivated and the feedback of the model changed to a direct feedback of the flux magnitude error as well as the angle deviation ($\Delta u_{\alpha,\beta}$). This leads to a further stabilization of the model performance at low speed. To completely remove offset values from the input voltage an additional feedback is introduced comparable to that proposed in [4],[5].

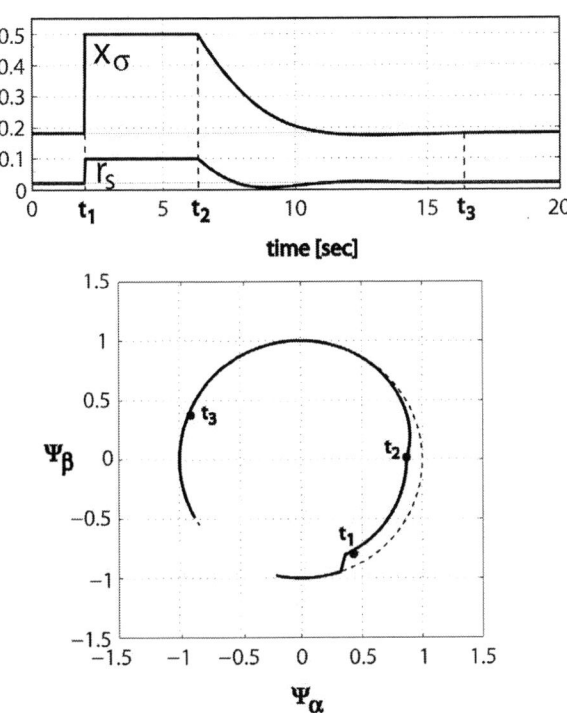

Fig. 4: Time trace of model parameters during adaptation (upper). Locus of estimated flux position with parameter adaptation active (lower). Stepwise disturbance introduced at t=t_1. adaptation activated at t=t_2.

The main difference is the calculation of the feedback value which is done in [4] using minimum maximum value comparison and the time between two zero crossings of the flux. In this investigation it was found advantageous to using a ring-buffer with a storage depth of one revolution of the flux and a summation of the past signal period. This structure offers the possibility to calculate a new feedback value every time the flux angle changes to the next position of the ring-buffer. As it actually removes the offset from the input voltage it is denoted input voltage offset elimination in Fig. 5.

One advantage of the proposed structure is that due to using pure integrators a constant phase shift of $-\pi/2$ rad is realized. In combination with the adaptation algorithm of the machine parameters a improved performance in the low speed range is reached.

Using this structure the signal injection is activated and the commissioning is started to identify, all deteriorating signal disturbances, what is also called the magnetic fingerprint in literature – as explained in the introduction. During the identification the model angle acts as the reference angle for the commissioning of the drive. As the goal of the commissioning is the sensorless estimation of the flux angle the flux angle reference is necessary.

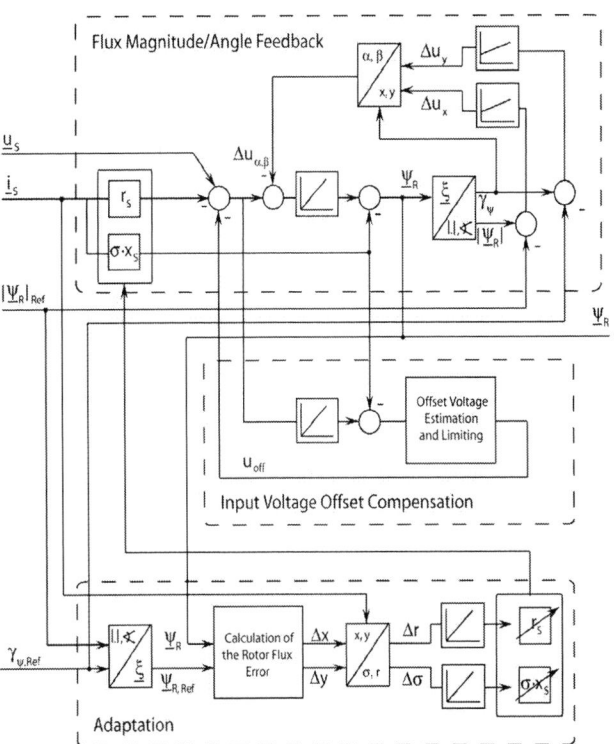

Fig. 5: Structure of the proposed fundamental wave flux model including parameter adaptation

The flux angle estimated from the transient current slope is almost independent from speed. The only significant speed influence is introduced by dead time of the excitation and signal processing. This, however can easily be removed feed forward. The whole commissioning can thus be done in the low and medium speed range where the flux model delivers accurate estimates of the flux angle.

The area of full modulation or any saturation of the current controller has however to be avoided during that phase as this would considerably change the signal contents of signal obtained from the transient current t slope.

The actual adaptation to the individual properties of the machines transient reactances is done using a neural network approach.

As type of neural network (ANN) in this paper, the structure of the well known multilayer perceptron (MLP) was applied. It consists of a input configuration with 6 input values, a single hidden layer with 25 neurons and a single neuron in the output layer. The training stage of the selected network is performed using the backpropagation algorithm. Further details in dealing with neural network in combination with the signal injection are given in [19].

After the commissioning is finished, speed sensorless control down to zero frequency is possible using the sampled transient current reaction and applying the compensation of all disturbing effects using the trained neural network. This results in an estimated flux angle that again can be combined with the fundamental wave model. This combination is possible for example by using this flux angle as reference angle ($\gamma_{\Psi,Ref}$) in Fig. 5.

In final speed sensorless operation this angle is then used

1349

to calculate the feedback voltage of what is called 'flux magnitude/angle feedback' in the figure. The parameter adaptation is then deactivated. This kind of feedback however, is only necessary at low speed. Each time the drive is again operated in medium speed range the feedback may again be switched to parameter adaptation to tune in for possible changes of the parameter values.

It has also to be stressed, that for the generation of the terminal voltages only the dc link voltage and the PWM status was used. No compensation of inverter interlock dead time or power devices voltage drops was realized.

IV.   MEASURMENTS RESULTS

The control structure described was used to carry out the commissioning of the sensorless flux control. The whole procedure including the commissioning was carried out without using a speed sensor. Using the structure depicted in Fig. 5 first the parameters of the fundamental wave model where adjusted. Then the model output was used as reference flux angle for the training of a neural network.

After the commissioning the signal obtained from the transient current response is corrected using the neural network. The resulting angle is then combined with the fundamental wave to act as stabilising reference at very low speed including zero speed.

The resulting performance of the overall structure can be seen in the following figures. The machine used for the tests has closed rotor slots.

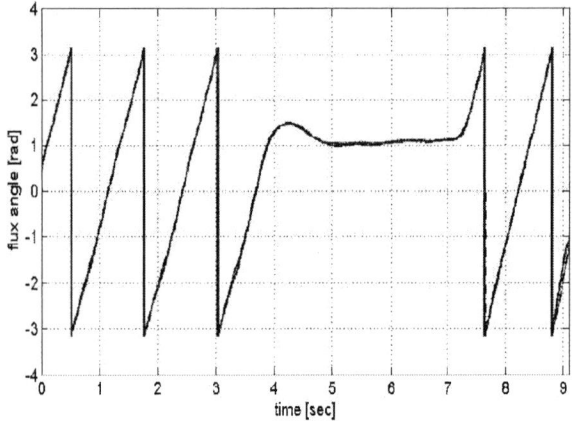

Fig. 6: Measurement results of sensorless controlled operation at 0,8 p.u. rated torque. (blue: sensorless estimated flux angle; black: reference angle calculated with sensor-based model)

The speed sensorless performance of the overall structure after commissioning is depicted in Fig. 6. There the machine was operated at 0.8 p.u. rated torque with a speed controlled load dynamometer coupled. At the left side of the figure the mechanical speed was chosen to zero. In the center of the figure the operation was changed to zero flux frequency and on the right side of the figure again zero mechanical speed is shown. The blue trace gives the speed sensorless estimated flux angle using proposed structure. It can be seen that the proposed structure enables an accurate estimate of the rotor flux angle that can effectively be used for the commissioning of the drive.

The dynamic performance of the proposed structure

during a torque step response is depicted in Fig. 7.

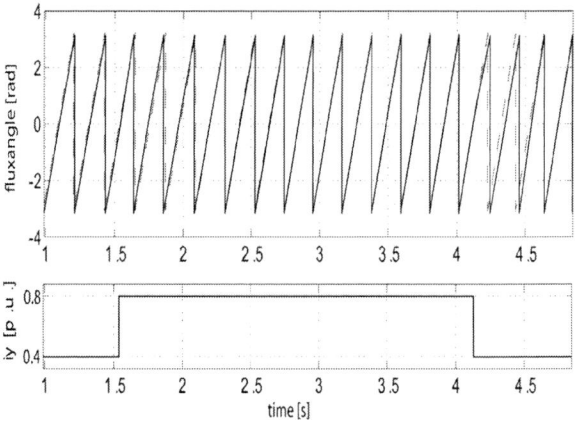

Fig. 7: Measurement results of sensorless controlled operation during step of load torque from 0.4p.u. to 0.8p.u. rated load current/torque. Upper: flux angle sensorless control structure (solid), reference angle sensor based (dashed). Lower: load current/torque step.

There a torque step is applied to the sensorless controlled test machine while it is driven by a speed controlled load machine. The mechanical speed chosen is 0.04 p.u when the torque demand is made from 0.4 p.u. to 0.8 p.u. In the upper diagram the solid trace represents the flux angle obtained from the advocated fundamental wave - sensorless-voltage model. The dashed trace represents the reference flux angle calculated with a sensor based model as a reference. In the lower diagram of this figure the load current/torque steps are shown. It can be seen that the control structure is able to guarantee stable operation also during transient operation.

V.   CONCLUSIONS

An adaptive flux model for commissioning and operation of speed sensorless control of induction machines was presented. It is based on an input voltage offset compensation and alternatively a flux magnitude/angle feedback or a parameter adaptation algorithm. The structure is able to adjust to the machine parameters and at medium speed to deliver an accurate reference angle for the commissioning of signal injection based sensorless control. After the commissioning the structure is able to guarantee stable operation at zero mechanical as well as zero flux speed during steady state as well as transient operation.

ACKNOWLEDGMENT

The authors gratefully acknowledge the financial support of the Austrian Science Foundation - "Fonds zur Förderung der wissenschaftlichen Forschung" (FWF) - under grant no. P19967-N14.

REFERENCES

[1] K. Rajashekara, A. Kawamura, and K. Matsuse, Eds., Sensorless Control of AC Motors. New York: IEEE Press, 1996.

[2] Th.M. Wolbank, H. Giuliani, R. Woehrnschimmel, J.L. Machl "Sensorless Control of Induction Machines by Combining Fundamental Wave Models with Transient Excitation Technique", Proceedings of IEEE International Electric Machines and Drives Conference, IEMDC, San Antonio, TX, USA, pp. 1379-1384, (2005)

[3] Th.M. Wolbank, "Flux Model for Field Oriented Control of an Induction Motor in the Lower Frequency Range and at Standstill Using the {INFORM} Flux Detection Method", Proceedings of International Power Electronics and Motion Control Conference, PEMC, Vol.2, pp.584-587, (1996)

[4] J. Holtz, J. Quan, "Drift- and Parameter-Compensated Flux Estimator for Persistent Zero-Stator-Frequency Operation of Sensorless-Controlled Induction Motors", IEEE Transactions on Industry Applications, Vol.39, No.4, pp.1052-1060, (2003)

[5] Q.Gao, C.S. Staines, G.M. Asher, M. Sumner, "Encoderless Operation of Induction Motor using a MRAS Observer with Zero Drift Feedback Integrator", Proceedings of IEEE EMD International Conference on Electric Machines and Drives, pp.2002-2005, (2005)

[6] Nikolas Teske, Greg M. Asher, Mark Sumner, Keith J. Bradley, "Encoderless Position Estimation for Symmetric Cage Induction Machines Under Loaded Conditions", IEEE Transactions on Industry Applications, Vol.37, No.6, pp.1793-1800, (2001)

[7] A. Consoli, "Machine Sensorless Control Techniques Based on HF Signal Injection", Proceedings of EPE-PEMC, V1,pp.98-103, (2000)

[8] M.W. Degner, R.D. Lorenz, "Using Multiple Saliencies for the Estimation of Flux, Position, and Velocity in AC Machines", Proceedings of IEEE IAS, NewOrleans, pp.760-767, (1997)

[9] J.I. Ha, S.K. Sul, "Physical Understanding of High Frequency Injection Method to Sensorless Drives of an Induction Machine", Proceedings of IEEE IAS, pp.1802-1808, (2000)

[10] M. Schroedl, "Sensorless Control of AC Machines at Low Speed and Standstill based on the Inform Method", Proceedings of IEEE IAS, Vol.1, pp.270-277, (1996)

[11] C. Caruana, G.M. Asher, J. Claire, "Sensorless Vector Control at Low and Zero Frequency Considering Zero-Sequence Current in Delta Connected Cage Induction Motors", Proceedings of IEEE IECON, pp.1460-1465, (2003)

[12] Holtz J.; "Sensorless Position Control of Induction Motors–an Emerging Technology", Proceedings of IEEE AMC, pp.1 – 14, (1998)

[13] Th.M. Wolbank, J.L. Machl, "Influence of Inverter Non-linearity and Measurement Setup on Zero Speed Sensorless Control of AC Machines Based on Voltage Pulse Injection", Proceedings of IEEE IECON, Raleigh, pp.1568-1573, (2005)

[14] F. Briz, M.W. Degner, A. Diez, R.D. Lorenz, "Measuring, Modeling and Decoupling of Saturation-Induced Saliencies in Carrier Signal Injection-Based Sensorless AC Drives", Proceedings of IEEE Industry Applications Conference, Vol. 3, pp.1842-1849, (2000)

[15] J. Holtz, H. Pan, "Elimination of Saturation Effects in Sensorless Position Controlled Induction Motors", Proceedings of IEEE Industry Applications Annual Meeting, Vol.3, pp.1695-1702, (2002)

[16] Th.M. Wolbank and J.L. Machl and Th. Jaeger, "Combination of Signal Injection and Neural Networks for Sensorless Control of Inverter Fed Induction Machines", Proceedings of IEEE Power Electronics Specialists Conference, PESC, Aachen, Germany, pp.2300-2305, (2004)

[17] P.Garcia, F.Briz, D.Raca, R.Lorenz, "Saliency tracking-based, sensorless control of ac machines using structured neural networks", Proceedings of IEEE Industry Applications Conference, pp.319-326, (2005)

[18] Th.M. Wolbank, "Analysis of a simple feed back method in field-oriented control of an induction motor without speed and positionsensor, Proceedings of IEEE IECON, Vol. 2, pp.1134-1138, (1996)

[19] T.M. Wolbank, M.A. Vogelsberger, R. Stumberger, S. Mohagheghi, T.G. Habetler, R.G. Harley, "Comparison of neural network types and learning methods for self commissioning of speed sensorless controlled induction machines" Proceedings of IEEE Power Electronics Specialists Conference (2007)

Design and Performance of a Single Stator, Dual Rotor Induction Motor

S. Sinha*, N. K. Deb**, N. Mondal** and S. K. Biswas (Sr. Member, IEEE)**

* Dept. of Electrical Engg. Central Calcutta Polytechnic, Kolkata 700014, India
** Dept. of Electrical Engg. Jadavpur University, Kolkata 700032, India

Abstract– **This paper presents the design and performance of a motor having two rotors along the same axis and under the span of a single stator. This special type of motor has two shafts projecting from the two ends, which are mechanically independent of each other and thus can meet the requirement of differential drive in an electrical vehicle. This paper presents the design, construction and performance data obtained from experimental results of the motor. An equivalent circuit of the motor has been developed and verified. The characteristics of the dual rotor motor have been obtained experimentally from a prototype and presented in this paper.**

Index Terms– **Dual Rotor Motor, Differential Drive, Electric Vehicle Drive, Induction Motor.**

I. INTRODUCTION

The concept of an induction motor having two rotors along the same axis and under the span of a single stator was presented in 1993 [1]. This special type of motor has two shafts projecting from the two ends, which are mechanically independent of each other, and thus can meet the requirement of differential drive in an electrical vehicle. Some aspects in the control methodology was published later by Kelecy & Lorenz [2], but no design or practical test data of the motor was published. The authors [2] had assumed that the new motor was equivalent to two series connected discreet motor and had thus proceeded along that linear model.

The work presented in this paper relates to the design, collection of practical data from the test motor and validation of its equivalent circuit. A motor having only one three-phase input and two output shafts in the single structure offers less space requirement as well as some unique features that will be discussed in this paper. It is also observed here that the equivalent circuit is non-linear, contrary to earlier proposals, being heavily dependent on the level of saturation of the mutual reactance, since each rotor is only half the length of the stator. However, the unique feature is that while one shaft is at standstill, the other can rotate at its normal speed with load, without significant increase in input current, permitting true differential drive for an electric vehicle.

II. CONSTRUCTION

The motor has a single stator and two rotors, as shown in Fig 1. The two rotors are identical and of squirrel cage type and are axially separated from each other inside the

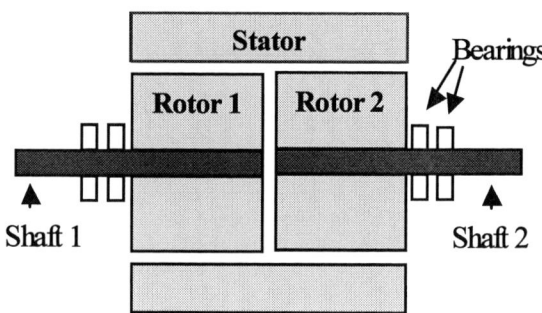

Fig. 1. Schematic diagram of motor.

stator. The rotor shafts are fixed by two bearings in each driving end only. Bearings cannot be inserted in-between the two rotors since it will be damaged due to the large eddy currents that will be generated by the leakage flux existing at that point. Thus, there is a small axial air gap of 3mm in-between the two rotors. The stator has a three phase 4 pole balanced winding distributed in its slots along the entire length. Since each rotor length does not span the full length of the stator, it is correctly expected that the leakage reactance between stator to each rotor will be higher than that between stator to rotor in normal induction motors. Thus, the rotor is designed with a double cage structure in order to partially compensate for the reduction of pull-out torque due to the higher leakage reactance.

III. DESIGN DATA

The prototype motor was designed and fabricated as per the following data :

a) *Specification of the Motor :*
7.5 HP, 3 Phase, 110 Volt, 40 Amp. 50 Hz, 4 Pole, Continuously rated, Double Cage structure, Frame = 132, Insulation = Class H, Enclosure = TEFC.

b) *Main Dimensions :*
Stator Bore D = 140 mm, Gross Iron length L = 184mm.

c) *Stator :*
Core = CRNO, Connection = Delta, Phase Voltage = 110 volt, Flux per pole = 8.3 mWb, Av. Flux Density B(av) = 0.41 T, Ampere-conductor = 19000, Turns per phase = 60, no. of slots = 36, winding factor = 0.9452, Per-phase resistance = 0.233 ohm, Length of air gap = 0.35 mm.

d) *Each Rotor (Double Cage Type) :*
Core = CRNO, Diameter = 139.3 mm, Depth of rotor core = 18 mm, Depth of slot = 23 mm, Diameter of shaft = 57.3 mm, Type of winding = squirrel cage (double cage), No. of slots = 28, Conductor = Aluminium (LMO),

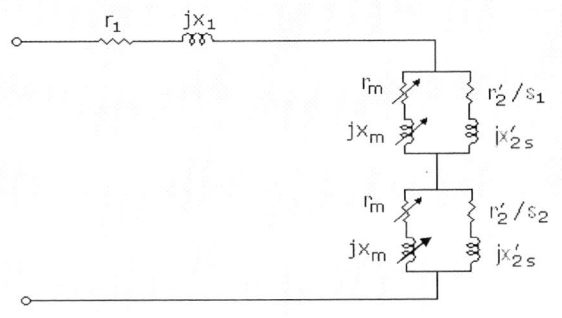

Fig. 2. Equivalent circuit (per phase).

Rotor Bar length = 92 mm.
Outer Cage each bar : Area = 18.25 sq. mm, Resistance = 0.57 ohms.
Inner Cage each bar : Area = 58.125 sq. mm, Resistance = 0.18 ohms.
End Ring Area = 140 sq. mm.

IV. EQUIVALENT CIRCUIT

The equivalent circuit of the motor is shown in Fig 2. It will be clear that it is equivalent to two rotors connected in series, but the magnetizing reactances and their core loss resistances are correctly represented by variable parameters rather than by constant parameters as described by earlier researchers [1,2]. If both the shafts are equally loaded, they run at the same speed in the same direction and share the supply voltage equally in series. However, the application of a differential load on the two shafts cause them to run at different speeds, depending on their individual slips s_1 and s_2. In the extreme situation, one shaft can be at locked condition while another is at no-load. In this case, the voltage drop across the locked rotor will be very small, resulting in an increase in the voltage drop across the no-load rotor. This causes heavy inequality of flux between the two rotors. The no-load rotor now has much higher flux than in the normal condition and goes towards saturation. This leads to reduction in the magnetizing reactance of the no-load rotor with increase in its associated losses. This situation is represented in Fig 3. In this specific situation, the input current, interestingly, is only somewhat higher that the rated current and not several times higher since the impedance of the no-load rotor is still high. Thus, the

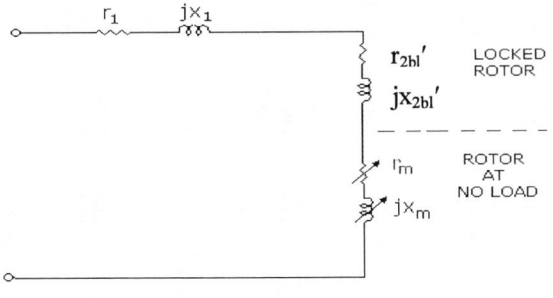

Fig. 3. Equivalent circuit with one rotor locked and other at no-load

experimental motor permits true differential movement. However, if both rotors are locked, the current becomes several times higher and the situation is the same as in normal motor.

An experimental motor with the given specification was designed and fabricated. From no load and blocked rotors (both rotors blocked) test data, assuming $x_1 : x_{21S} : x_{22S} = 1: 0.4 : 0.4$ and taking no-load rotational loss as 30 watt/phase (from experimental data), the following motor parameters are obtained experimentally :

r_1 = 0.187 ohms r_m = 0.694 ohms
r_2' = 0.172 ohms x_m = 5.06 ohms
x_{2S}' = 0.23 ohms

In the extreme case of one rotor being locked and the other rotor at no-load, the saturated values of r_m and x_m (in the locked rotor) will be different from the unsaturated values in the other (in the free running, no-load rotor). Thus, the values of r_m and x_m are correctly represented as variables in the per-phase equivalent circuit

The value of r_m and x_m as obtained from test data fed into the equivalent circuit, are :

Unsaturated : r_m = 0.694 ohms, x_m = 5.064 ohms
Saturated   : r_m = 0.311 ohms, x_m = 3.0 ohms

Thus, actual values of r_m and x_m will have to be interpolated between the two extreme sets given above, depending on the difference in the two slips. The interpolation will be non-linear as per the nature of B-H curve. The slope of the B-H curve is proportional to the inductance (and reactance for fixed frequency). Thus, as the flux in one rotor increases above the other, its x_m value falls as per the B-H curve. The increased flux increases core losses, which is accounted by the decrease of r_m while the current through it increases (still resulting in higher I^2r loss) in the series connected equivalent circuit.

In addition to the above conditions, since each rotor is double-cage, the rotor parameters will change from standstill to rated-speed condition. The experimentally obtained rotor parameters, modified according to Algers [3] are given below :

Rated-speed condition
r_2' 0.0755 ohms
x_{2S}' 0.281 ohms

Blocked rotor condition
r_{2bl}' 0.172 ohms
x_{2bl}' 0.23 ohms

V. EXPERIMENTAL DATA

A motor with the given specification was fabricated and tested in the laboratory. Two dc generators were connected to the two shafts individually so that independent loading could be carried out. Typical experimental data (per phase) are given in Table I to verify the equivalent circuit at representative conditions of slip of the two shafts (given as s_1 and s_2 respectively). Each set of data calculated from the equivalent circuit is compared with actual values obtained experimentally. The first two sets of data relate to the condition of similar loading in both shafts. The third set of data relate to the

TABLE I
TYPICAL EXPERIMENTAL DATA

Input Amperes (expt)	Input Amperes (calculated)	Power Factor (expt)	Power Factor (calculated)	Power Input Watts	Power (calculated) Watts	Rotor slip s_1 (pu)	s_2 (pu)
10.57	10.28	0.174	0.2	202	235.8	0.001	0.001
10.8	10.6	0.302	0.31	352	361.9	0.003	0.003
14.8	14.9	0.722	0.705	1118	1102	0.018	0.012
10.56	10.55	0.3	0.262	348	304	0.001	0.004

condition of different loading in both shafts. The fourth data shows the situation when one shaft is at no-load while the other is at sufficient load. It will be observed that the predicted values are close to the experimentally obtained ones, proving the validity of the proposed equivalent circuit.

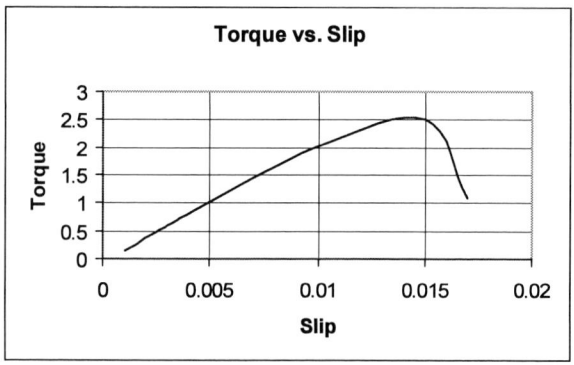

Fig. 4. Torque vs Slip characteristics with equal loading

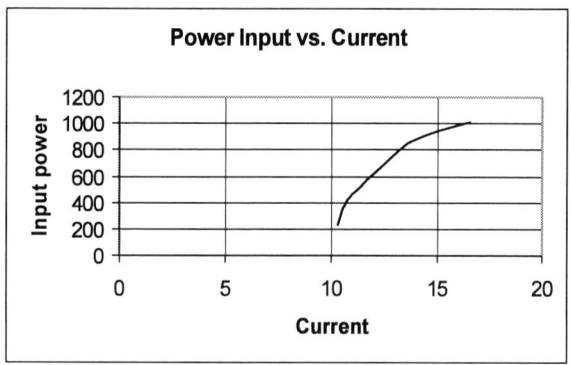

Fig. 5. Input Power vs Current characteristics with equal loading

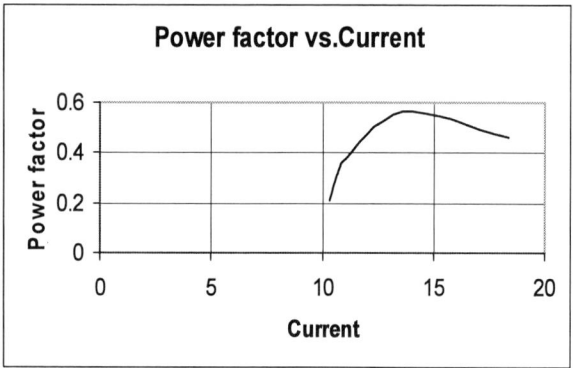

Fig. 6. Power Factor vs Current characteristics with equal loading

Performance plots for the experimental motor are given in the graphs depicted as Fig. 4 – 7, when both shafts are individually loaded to equal value. They are similar to that obtained in conventional motor with single rotor.

Detailed investigation on the motor is also presented also with unequal loading on the two shafts, resulting in their different slip conditions. These are depicted in Fig. 8 – 11. In each of the plots, slip of one shaft (say shaft 1) is held constant at a particular value to obtain a particular parameter variation data in the second shaft. For example, in Fig. 8, the Torque vs slip characteristic of shaft 2 is obtained while slip of shaft 1 is held constant at discrete values of 0.003, 0.005, 0.02, etc., resulting in a family of curves. This also shows that the torque-slip characteristic of one shaft is dependant on the slip of the other shaft. For a given torque in shaft 2, its slip reduces for an increase in slip of shaft 1. In the same way, Fig. 9 – 11 presents the dependency of Input Power, Power Factor and Efficiency when slip of shaft 1 is held constant at discrete values.

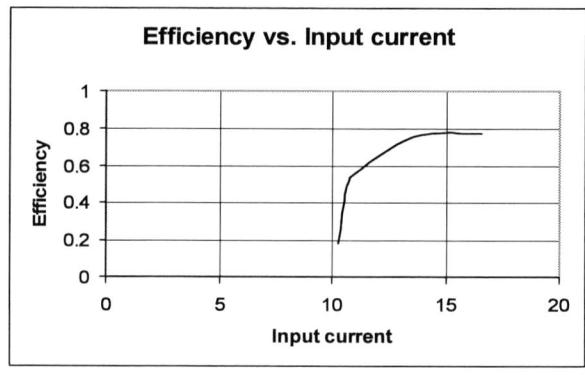

Fig. 7. Efficiency vs Current characteristics with equal loading

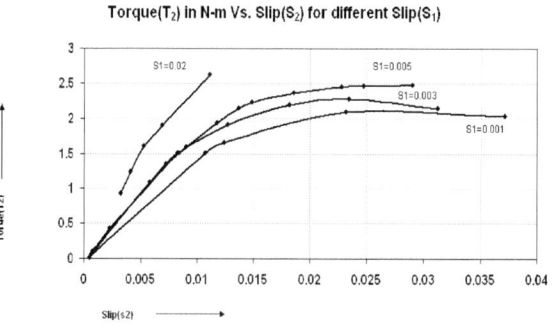

Fig. 8. Torque vs Slip characteristics with unequal loading

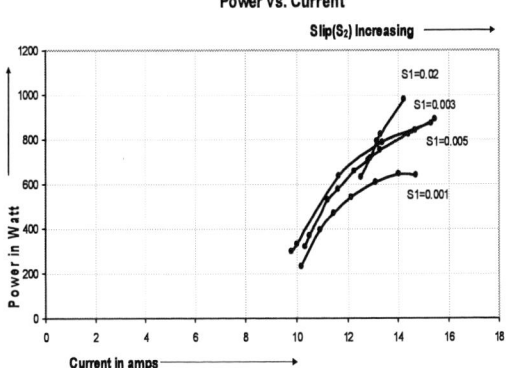

Fig. 9. Input Power vs Current characteristics with unequal loading

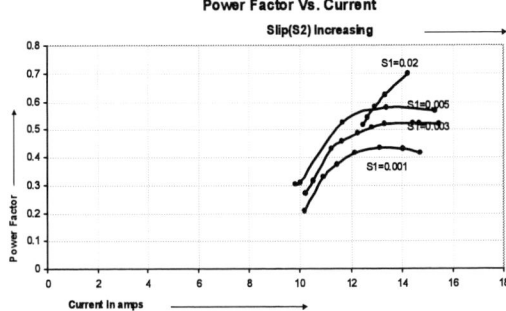

Fig. 10. Power Factor vs Current characteristics with unequal loading

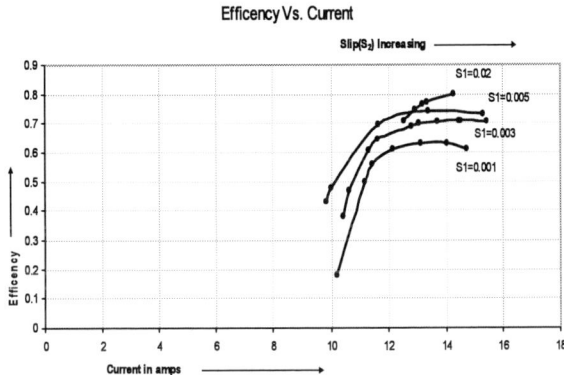

Fig. 11. Efficiency vs Current characteristics with unequal loading

Fig. 12 presents a set of bar charts of the torques in the two rotors as the slip is changed in a particular manner. The slip of rotor 1 is intentionally held constant (by suitable load variation) while the slip of rotor 2 is increased in steps. Corresponding to change in the two slips as shown in Fig. 12a, the change in the torques of the two rotors is shown in Fig.12b. From these two bar charts, it is clear that torque in rotor1 and rotor 2 both changes due to change in slip of rotor 2, even when slip of rotor 1 remains constant. This is a special feature of the single stator, dual rotor motor, making it suitable for driving an electric vehicle in differential mode.

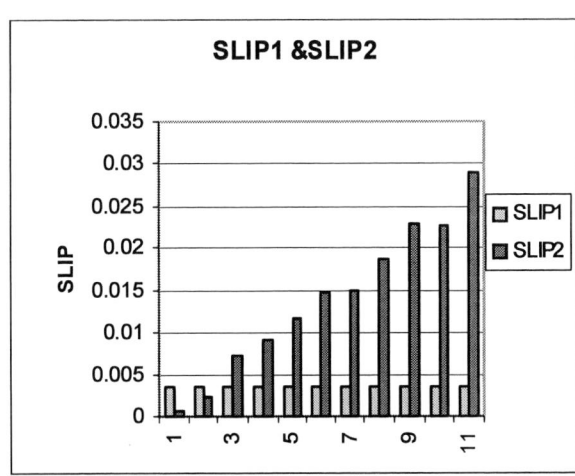

Fig. 12a. Variation in slip in rotor 1 & rotor 2

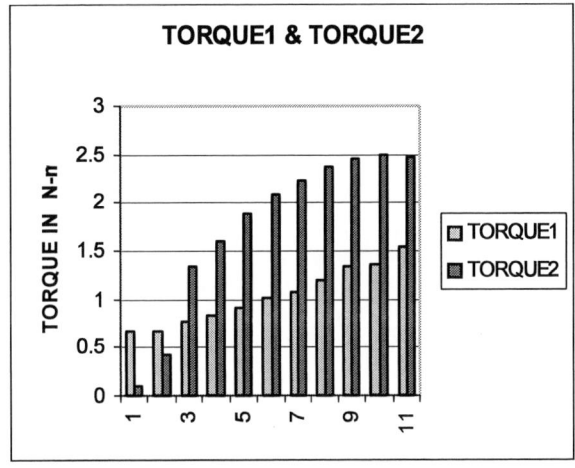

Fig. 12b. Variation in Torque in rotor 1 & rotor 2

VI. CONCLUSIONS

The actual design and its experimental verification of a new type of induction motor is presented. Its equivalent circuit is validated from experimental data. Some interesting features of the motor include true differential drive capability between its two shafts. The two shafts can run at independent speeds (depending on their individual loading) and in the extreme case, one shaft can be at locked condition, while the other is running at any load condition or no-load.

REFERENCES

[1] O. Crelerot, F. Bernot & J. F. Kauffmann, "Study of an electrical differential motor for electrical car", IEE Conference Publication no. 376, Sept 1993, pp 8-10

[2] P. M. Kelecy & R. D. Lorenz, "Control methodology for single stator, dual-rotor induction motor drives for electric vehicles", IEEE Power Electronics Specialists Conference 1995, pp 572-578.

[3] P. L. Alger, The Nature of Polyphase Induction Machine, J. Wiley & Sons, N. York, 1951.

Investigation of skew effect on the Performance of Self – Excited Induction Generators

B. Sawetsakulanond and V. Kinnares

Dept. of Electrical Engineering, King Mongkut's Institute of Technology Ladkrabang,
Bangkok 10520, Thailand
Fax 662-3269902 E-mail:Budhapon@hotmail.com E-mail: kkwijit@kmitl.ac.th

Abstract--**This paper presents the investigation of skew effects on the performance of three phase, 2.2 kW 220/380 volt 4 poles self – excited induction generators (SEIG) with skewed squirrel cage rotor. Analysis of excitation capacitor of the SEIG based on a steady – state equivalent circuit model including rotor skew effect is given. Skewed rotor with angle of 0° , 5 °, and 10 ° for the SEIG are employed. Testing and performance comparison under dynamic and steady – state operation with resistive and resistive – inductive loads have been performed. Results can be guidelines for development of effective wind induction generators.**

Index Terms--Induction generator, self - excited

I. INTRODUCTION

Owing to the continuous increase in energy need , it is difficult to meet the growing demand by exploiting energy from the limit conventional source, such as coal , oil, gas, and so on. As a consequence, a greater emphasis is now being given to harness energy from non - conventional sources   such as wind , biogas and small hydro heads[1-2,5]. A three phase induction machine can be made to work as a self - excited generator when its rotor is driven at suitable speed by wind energy and its excitation is provided by connecting a three phase capacitor bank at the stator terminals in order to build-up voltage and regulate terminal voltage. It offers various advantages over other machines such as reduced unit cost, brushless rotor (squirrel cage construction), absence of separate DC source and ease of maintenance. Numerous papers have attempted to analyze the SEIG using equivalent circuit approach [1-3,5-7]. Generally two different (but related) methods of capacitance solution for voltage build-up and terminal voltage regulation have been employed, namely, the loop impedance method and nodal admittance method. Steady state analysis of such generators seems to be more interested than dynamic analysis. Excitation capacitors affects stator current. Therefore careful selection of this capacitor is required. Controlled static var compensator in conjunction with the ac load voltage regulator based on dynamic model can be found in [8]. These techniques are based on power electronic applications which provides good performance under a wide range of operation. However few research works have paid attention on performance of SEIG with skewed slot rotor type. Therefore this paper will present the investigation of

skew effect on the performance of SEIG such as build – up voltage, relationship between terminal voltage and speed, compensated reactive power

II. EQUIVALENT CIRCUIT ANALYSIS

As shown in Fig.1, the skewed rotor slots are usually used to provide starting torque when the motors have the number of the stators slots equal to the rotor slots. It has proved that other negative influences could be reduced, such asynchronous torque harmonics, oscillating torque and stray load losses when skewed rotor slots are used. The skewed slots generate an additional leakage flux in the machines, reducing the useful flux. Therefore, a reduction of the mutual flux between stator and rotor occurs. However , rotor - bar skewing causes a decrease in the voltage induced in the rotor winding, which can be understood by considering the voltage along a given rotor bar. As the rotating air gap flux wave passes the rotor bar, the peak of the fundamental component of the flux wave will see portions of that bar at successively later times. This causes the fundamental component of the rotating flux wave to appear smaller, that is, with a reduction in the rotor voltage. As shown in equation (1), the rotor voltage is reduced by a factor equal to the skew factor (K_{sk})[3].

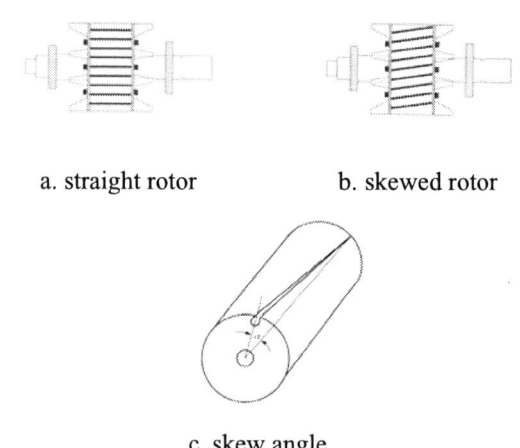

a. straight rotor   b. skewed rotor

c. skew angle

Fig.1. Slots skewing

$$K_{sk} = \frac{\sin\left(\lambda\frac{\alpha}{2}\right)}{\lambda\frac{\alpha}{2}} \qquad (1)$$

Where K_{sk} is the skew factor

$\quad\alpha$ is the skew angle

$\quad\lambda$ is the ratio between the flux for skewed slots and without skewed slots

Skewing the rotor increases the rotor resistance due to increase in bar length. The increased resistance can be written as equations (2) – (3)[10].

$$R_{b,sk} = R_b\sqrt{1+\left(\frac{\tau\alpha}{L_b\times180}\right)^2} \qquad (2)$$

$$R_{2,sk} = R_{b,sk} + R_r \qquad (3)$$

Where τ is the pole pitch

$\quad L_b$ is the bar length

$\quad R_b$ is the bar resistance without skewed slots

$\quad R_{b,sk}$ is the bar resistance with skewed slots

$\quad R_r$ is the end – ring resistance

$\quad R_{2,sk}$ is the rotor resistance with skewed slots

Apart from mutual flux reduction the skewing increases leakage reactance which is proportional to skew factor. The interest subject for this work is the study and performance analysis of SEIG with skewed slot rotor type in conjunction with an excitation and compensating capacitor bank. For the proposed capacitance consideration for the SEIG, the system can be shown in Fig.2. Capacitors are divided into two parts such as a built-up capacitor (C_b) and a compensating capacitor (C_c) for terminal voltage regulation. The C_b is responsible for no-load operation whilst both C_b and C_c are responsible for on-load operation. For no-load operation, the per phase equivalent circuit is shown in Fig.3 neglecting harmonic effect and core loss [4-5].

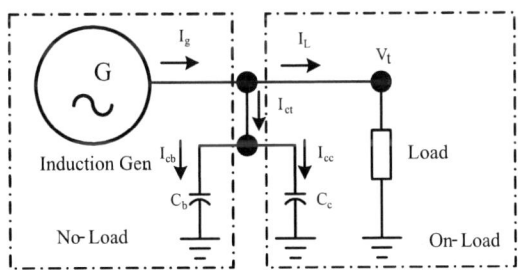

Fig.2. Single - line diagram of the SEIG

From Fig.3, impedance, Z_{CD} can be written as equations (4) - (6)

$$Z_{CD} = R_{CD} + jX_{CD} \qquad (4)$$

$$R_{CD} = \frac{(a-b)R_{2,sk}\,j(K_{sk}X_m)^2}{R_{2,sk}{}^2 + (a-b)^2\left(jK_{sk}X_m + jX_2 + j(1-K_{sk})X_m\right)^2} \qquad (5)$$

$$X_{CD} = \frac{R_{2,sk}{}^2\,jK_{sk}X_m + (a-b)^2\,jK_{sk}X_m\,jX_2\,j(1-K_{sk})X_m}{R_{2,sk}{}^2 + (a-b)^2}$$

$$\bullet\;\frac{\left(jK_{sk}X_m + (jX_2 + j(1-K_{sk})X_m)\right)}{\left(jK_{sk}X_m + jX_2 + j(1-K_{sk})X_m\right)^2} \qquad (6)$$

Where a is the per unit frequency

$\quad b$ is the per unit speed

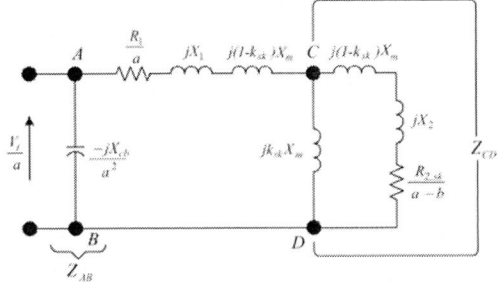

Fig.3. Per phase equivalent circuit of the SEIG with Skewing

With the reason of ease of analysis, Fig 3 can be reduced into Fig.4. impedance between nodes A and D can be expressed as following equations (5) – (8)

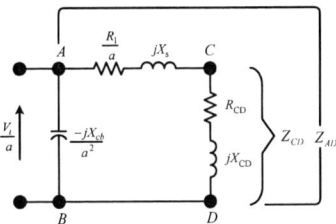

Fig.4. Simplified equivalent circuit

$$Z_{AD} = R_{AD} + jX_{AD} \qquad (7)$$

$$R_{AD} = \frac{R_1}{a} + R_{CD} \qquad (8)$$

$$jX_{AD} = jX_s + jX_{CD} \qquad (9)$$

When

$$jX_s = jX_1 + j(1-K_{sk})X_m \qquad (10)$$

Using KVL result in

$$I_g\left(\frac{-jX_{cb}}{a^2} + \left(\frac{R_1}{a} + jX_s\right) + Z_{CD}\right) = 0 \qquad (11)$$

Since I_g is not definitely zero during voltage build-up, as a consequence, equation (11) can be rewritten as

$$\left(\frac{-jX_{cb}}{a^2} + \left(\frac{R_1}{a} + jX_s\right) + Z_{CD}\right) = 0 \qquad (12)$$

From equation (12), the real part is zero. Then

$$\frac{R_1}{a} + \frac{(a-b)R_{2,sk}\,j(K_{sk}X_m)^2}{R_{2,sk}^2 + (a-b)^2(jK_{sk}X_m + jX_2 + j(1-K_{sk})X_m)^2} = 0 \quad (13)$$

According to equation (13), we can obtain a_{max} as

$$a_{max} = b - \frac{b}{2}\left[\frac{1 - \sqrt{1 - \left(\frac{b_c}{b}\right)^2}}{1 + \left(\frac{R_1}{R_{2,sk}}\right)\left(1 + \frac{jX_2 + (1-K_{sk})X_m}{jK_{sk}X_m}\right)^2}\right] \quad (14)$$

Also, b_c can be determined as

$$b_c = \frac{2R_1}{jK_{sk}X_m}\sqrt{\frac{R_2}{R_1} + \left(1 + \frac{jX_2 + j(1-K_{sk})X_m}{jK_{sk}X_m}\right)^2} \quad (15)$$

Where b_c is the critical speed

Therefore capacitance for built-up voltage during no-load can be determined as

$$C_b = \frac{1}{\left\{2\pi f_b Z_b a_{max}^2\left(jX_s + jX_{CD}\right)\right\}} \quad (16)$$

Where C_b is the per phase built – up capacitor value

Z_b is the base impedance

f_b is the base frequency

For the analysis of the on - load operation, C_c is included. According to Fig.5, when supplying the resistive – inductive load, the terminal voltage will be reduced since the load draws reactive power from the system. In order to maintain the terminal voltage constant, C_c can be determined as follows. Total reactive power of overall capacitors (i.e. C_b and C_c) is

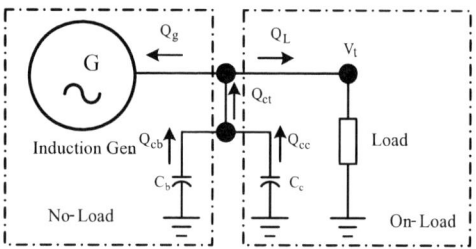

Fig.5. Power flow diagram

$$Q_{ct} = Q_g + Q_L \quad (17)$$

When

$$Q_g = \sqrt{S_g^2 - P_g^2} \quad (18)$$

$$Q_L = \frac{\left(\frac{V_t}{a}\right)^2}{X_L} \quad (19)$$

Where S_G is the per phase apparent power of the SEIG

P_G is the per phase active power of the SEIG

Q_G is the per phase reactive power of the SEIG

Q_L is the per phase reactive power of load

Q_{ct} is the per phase total reactive power of the capacitors

Compensated reactive power is

$$Q_{cc} = Q_{ct} - Q_{cb} \quad (20)$$

When

$$Q_{cb} = 2\pi\left(\frac{V_t}{a}\right)^2(fC_b) \quad (21)$$

Where Q_{cb} is the per phase reactive power of the built - up capacitor

Q_{cc} is the per phase reactive power of the compensating capacitor

Per phase current of the compensating capacitor (C_c) is determined from

$$I_{cc} = \frac{Q_{cc}}{V_t/a} \quad (22)$$

Per phase compensated capacitance is calculated as following equations

$$X_{cc} = \frac{V_t/a}{I_{cc}} \quad (23)$$

$$C_c = \frac{1}{2\pi f\left(jX_{cc}\right)} \quad (24)$$

Thus, total capacitor value for the SEIG is

$$C_t = C_b + C_c \quad (25)$$

Where C_c is the per phase value of the compensating capacitor

C_b is the per phase value of the built – up capacitor

C_t is the per phase value of the total capacitor

The procedure for determining capacitor values for the SEIG under on – load conditions when supplying the resistive load can be performed as same as for the resistive – inductive load with $Q_L = 0$

III. EXPERIMENTAL TESTS AND DISCUSSIONS

The tests have been performed in two parts. The first part is a parameter test of the SEIG and the second part is a performance test of the SEIG under load conditions.

A. Parameters Test

Three phase, 3kVA, 2.2 kW, 220/380 V, 8.7/5.0 A, 4 poles induction machines with skewed rotor types having slot angles of 0°, 5°, and 10°. Table 1 shows parameters of the under test machines obtained from the test complied with IEEE std. 112-1996 testing [9].

Table 1
Parameters of the Machines

Skew angle	R1 (Ω)	R2 (Ω)	Rc (Ω)	X1 (Ω)	X2 (Ω)	Xm (Ω)
0°	3.18	2.56	905.12	3.88	3.88	86.62
5°	3.18	2.58	1037.53	4.34	4.34	86.42
10°	3.18	2.96	1089.96	5.94	5.94	86.20

B. Operating Test

The capacitance analysis uses Maple program for determining capacitance under various load conditions and 0.85 lagging power factor Tables 2-3 show capacitance for various load power per phase and different skew angle at constant speed of 1500 rpm and regulated terminal voltage of 220 V, Y connected. Figs. 6-22 show the SEIG performance in terms of voltage build – up, power quality, frequency variation and efficiency. The next section will be detail explanation and discussion.

Table 2
Capacitance for resistive load

Skew angle	No-Load C_b (μF)	ON-Load ; C_e (μF)					C_T (μF)
		193 (W/ph)	384 (W/ph)	587 (W/ph)	768 (W/ph)	Total C_e(μF)	
0°	35	5	6	8	10	29	64
5°	35	5	7	9	12	33	68
10°	35	5	8	11	14	37	72

Table 3
Capacitance for resistive - inductive load

Skew angle	No-Load C_b (μF)	ON-Load ; C_e (μF)					C_T (μF)
		210 (W/ph)	428 (W/ph)	645 (W/ph)	868 (W/ph)	Total C_e(μF)	
0°	35	14	17	20	24	75	110
5°	35	15	18	21	26	80	115
10°	35	15	19	23	28	85	120

Fig.6. illustrates the characteristic of air – gap induced voltage (E_g) and magnetizing reactance (X_m) for various skew angles. Skew angle of 0° produces slightly higher voltage than others since effect of skewing causes slightly reduced magnetizing reactance. According to Fig.7, when increasing the prime mover of the SEIG, the SEIG with 10° skew angle has lower critical speed point

than the others. It shows the capability of better voltage build up at low speed. From Figs. 8-10 it can be seen that duration (built – up voltage time, t_b) for voltage build – up of the SEIG with 10° skew angle is longer the others.

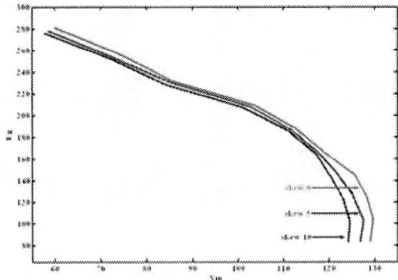

Fig.6. Characteristic of E_g and X_m

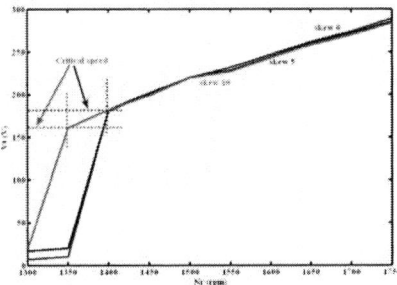

Fig.7. Variation of speed with terminal voltage for SEIG

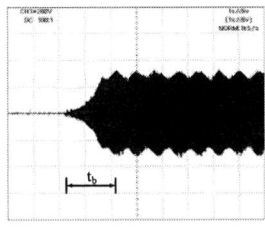

Fig.8. Built – up voltage waveform of 0° skew angle

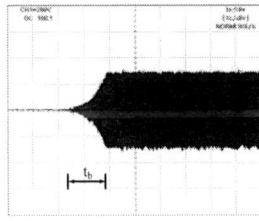

Fig.9. Built – up voltage waveform of 5° skew angle

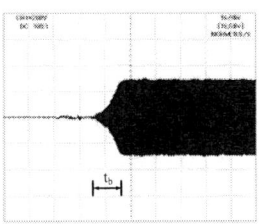

Fig.10. Built – up voltage waveform of 10° skew angle

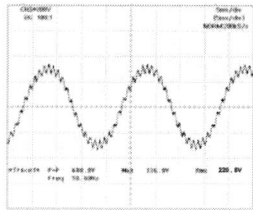

Fig.11.Steady state terminal voltage waveform at no load for 0° skew angle 0°, % THDv = 9.0 %

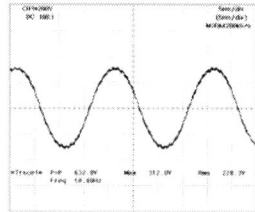

Fig.12. Steady state terminal voltage waveform at no load for 5° skew angle , % THDv = 4.7 %

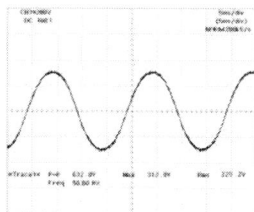

Fig.13. Steady state terminal voltage waveform at no load for 10° skew angle , % THDv = 3.9 %

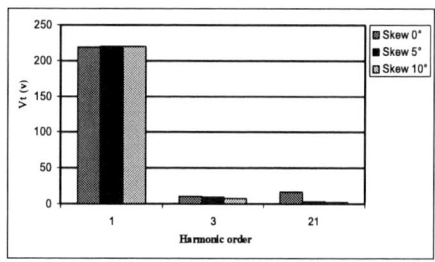

Fig.14. Harmonic voltage spectra for various skew rotor slots

As shown in Figs. 11-13, terminal voltage waveform for 10° skew angle is more nearly sinusoidal than others. The corresponding harmonic voltages are shown in Fig.14. The distortion of harmonic voltages is reduced for skewed rotor since skewing can reduce space harmonics. Figs. 15-18 illustrate reactive power supplied from the build up capacitors under various load conditions. Reactive power (Q_{cb}) for 10° skewed angle rotor is reduced more than the others when increasing load level due to the effect of frequency reduction. Compensated reactive power (Q_{cc}) for 10° skew angle is higher than for 5° skew angle and 0° skew angle respectively and is increased with a load increase. This results in using more compensated capacitance. The reason is that skew effect

causes an increase in stator leakage reactance and rotor leakage reactance. The sum of reactive power of the SEIG (Q_g) and reactive power of load (Q_L) is equal to total reactive power of the capacitors (Q_{ct}). Total reactive power of the capacitors for 10° skew angle is higher than for 5° skew angle and 0° skew angle, respectively when increasing loads. This results is using higher total capacitance due to skew effect on an increase in stator leakage and rotor leakage reactance.

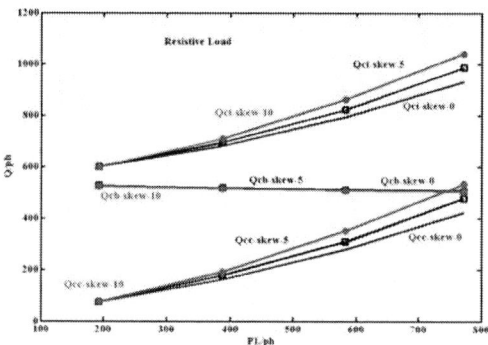

Fig.15. Variation of reactive power of capacitors with output power for resistive loads

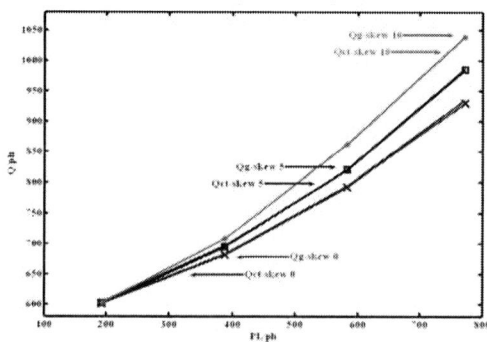

Fig.16. Variation of system reactive power with output power for resistive loads

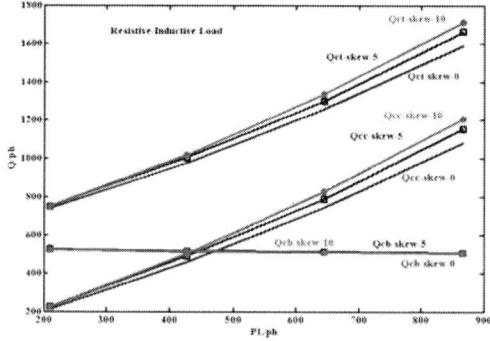

Fig.17. Variation of reactive power of capacitors with output power for resistive - inductive loads

Figs.19-20 show variation of the SEIG frequency. The change of frequency for 10° skew angle is higher than for others under resistive load and resistive-inductive loads due to skew effect on a rotor resistance increase.

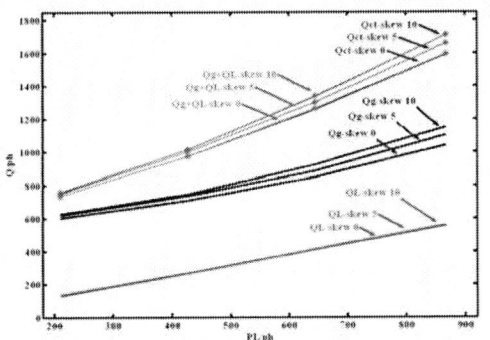

Fig.18. Variation of system reactive power with output power for resistive - inductive loads

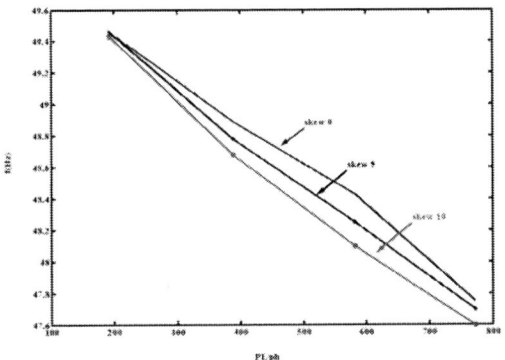

Fig.19. Variation of frequency with output power for resistive loads

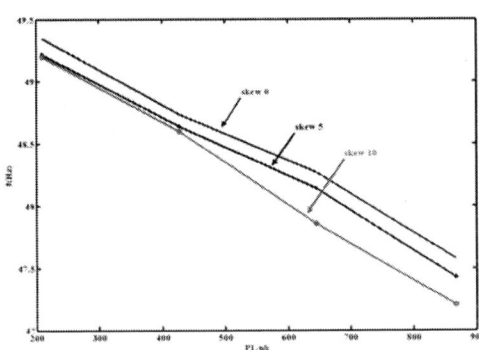

Fig.20. Variation of frequency with output power for resistive - inductive loads

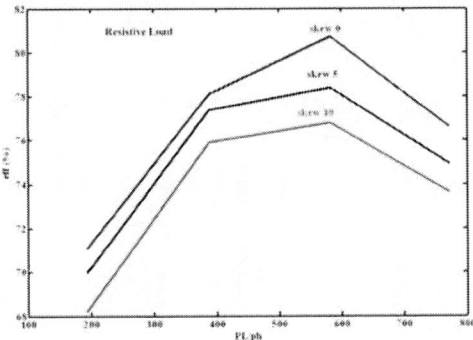

Fig.21. Variation of efficiency with output power for resistive loads

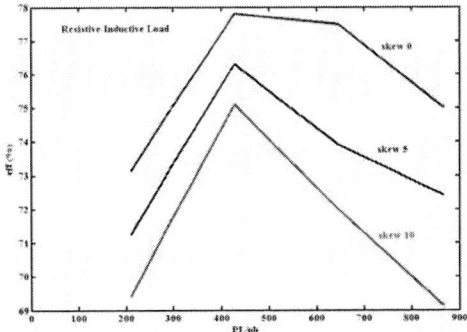

Fig.22. Variation of efficiency with output power for resistive - inductive loads

From Figs.21-22, efficiency of the SEIG with $10°$ skew angle rotor is lower than others under resistive load and resistive-inductive loads due to skew effect on an increase in rotor resistance and core loss resistance. With increased resistances, the losses increase.

IV. CONCLUSION

This paper has dealt with testing and performance analysis of the SEIG having skewed rotor with angles of $0°$, $5°$ and $10°$. It has been found that the SEIG with $10°$ skew angle of rotor is suitable for guidelines in designing of SEIG. Improvement and development of rotor of induction generator with reduced rotor resistance by increasing machine size, reduced leakage reactance for reducing winding losses and reduced compensated capacitance are required.

APPENDIX

V_b : 220 V, I_b : 5.2 A , P_b : 2.2 kW, Z_b : 42.30 Ω , Z_L : per phase resistive load 242 Ω and resistive- inductive load 166.97+j103.45 Ω

REFFERENCES

[1] S.S. Murthy , O.P. Malik , and A.K. Tandon , " Analysis of self – excited induction generators," *Proceedings of the IEE* , Vol.129, No.6, pp.260-265, November 1982.
[2] S.S. Murthy , B.P. Singh , C. Nagmani and K.V.V. Satyanarayanna , " Study on the use of conventional induction motors as self excited induction generators," *IEEE Transactions on Energy Conversion*, Vol.3, No.4, pp.842-848, December 1988.
[3] S.L Nau , "The influence of the skewed rotor slots on the magnetic noise of three phase induction motors," *Proceedings of the IEE* , Vol.140, No.444, pp.396-399, September 1997.
[4] S. Williamson and C.I.McClay ,"The Effect of Axial Variations in Saturation Due to Skew on Induction Motor Equivalent – Circuit Parameters ," *IEEE Transactions on Industry Applications*, Vol.35, No.6, pp.1323-1331, November/December 1999.

[5] T.F. Chan, "Capacitance Requirements of Self-exited Induction Generators," *IEEE Transactions on Energy Conversion*, Vol.8, No.2, pp.304-310, June 1993.

[6] S.P Singh , B.S Singh and M.P. Jain , "Comparative study on the performance of a commercially designed induction generator with induction motors operating as self excited induction generators," *Proceedings of the IEE* , Vol.140, No.5, pp.374-380, September 1993.

[7] L. Shridhar , Bhim Singh , C.S. Jha , "A step towards improvements in the characteristics of Self excited Induction Generator ," *IEEE Transactions on Energy Conversion*, Vol.8, No.1, pp.40-46, March 1993.

[8] Tarek Ahmed, Katsumi Nisshida, Mutsou Nakaoka and Hyun Woo Lee, "Self - Excited Induction generator with Simple Voltage Regulation Scheme for Wind Energy," *Proceedings of the 30 th Annual Conference of the IEEE*, Vol.140, No.5, pp.86-91, November 2004.

[9] IEEE std 112 – 1996 ,"*IEEE Standard Test Procedure for Polyphase Induction Motors and Generators* ," pp.28-53, 1996.

[10] A.F. Puchstien T.C. Lloyd and A. G.Conrad "*Alternating – Current Machines*" John Willey &Sons, Inc, pp 313-317, 1954.

Analysis of Double Loops Discrete Single Input PI Fuzzy for Single phase Inverter

S.M. Ayob, Z. Salam, and N.A. Azli

Department of Energy Conversion, Faculty of Electrical Engineering
University Teknologi Malaysia, 81310 UTM Skudai, Malaysia
Email:shahrin@fke.utm.my

Abstract–This paper presents an analysis of double loop PI-Fuzzy controller for a single phase inverter. The scheme comprises two feedback control loops arranged in a cascaded manner. The current in the filter inductor and voltage across the filter capacitor are sensed and fed back to the system to form the inner loop and the outer loop, respectively. Both loops are controlled by a discrete single input PI-Fuzzy controller. It is derived using the "sign-distance" method, which simplifies fuzzy set into a single input single output system. The work analyses the effect of fuzzy control surface for both loops in improving large-signal performance of the controller. Based on the analysis, the replacement of single input PI Fuzzy for voltage loop with a simple discrete linear PI controller will be justified.

Index Terms–PI Fuzzy, UPS inverter, Double loops controller

I. INTRODUCTION

The PI-Fuzzy and PD-Fuzzy are the general types of fuzzy logic control that have been used to control power electronic converters. Although PD type provides faster rising time and smaller overshoot compared to the PI type, the former suffered from significantly large steady state error. The existence of the error is due to the lack of integration operation in the controller itself. As a result, in many cases PD-Fuzzy controller seems to be an unpopular choice. To overcome the drawback of the PD-Fuzzy, PID-Fuzzy controller has been proposed. Theoretically such controller exhibits good dynamic performance i.e. small overshoot, fast settling time and fast rising time. However, due to the three-input based operation of fuzzy system, its 3-dimension rule table is difficult to implement. Therefore many fuzzy based controlled systems prefer to use PI type rather than PD or PID type.

Over the years, significant research has been carried out to study the fundamental mathematics of PI-Fuzzy. Compared to other nonlinear controllers, fuzzy controllers exhibit simpler mathematics and offer a higher degree of freedom in tuning its control parameters. Depending on the tuning method used, the PI-Fuzzy can be classified as a linear PI-Fuzzy and a nonlinear PI-Fuzzy controller. The linear PI-Fuzzy yields performance identically similar to its linear PI counterpart and the stability condition can be simply

assessed using conventional frequency analysis. In contrary to linear PI-Fuzzy, the nonlinear PI-Fuzzy is not supported by adequate control theory mathematics [1]. Its performance is quite unpredictable. In addition it is somewhat impossible to assess its stability due to the lack of stability theory. Most of the PI-Fuzzy designs were conducted by trial-and-error method.

However, by applying certain tuning methods, one can make the nonlinear PI-Fuzzy approximates other nonlinear controllers, such as the layered Sliding Mode controller (SMC). Previous works have shown that the nonlinear PI-Fuzzy controller is capable of performing similarly to the original SMC [2,3,4]. Furthermore, since similar mathematical principle of SMC is applied to design this nonlinear PI-Fuzzy controller, the stability assessment can simply be done using the well known Lypunov's method. This type of nonlinear PI-Fuzzy controller is also known as Fuzzy Sliding Mode controller (FSMC) [4].

Fundamentally, fuzzy control is a non-mathematical based controller. It is more of a human logic way of thinking based controller. Therefore the heuristic fuzzy design approach will always be faced and could not be totally eliminated. However, the approach can be made minimal by reducing the number of tunable parameters in the controller. A single input PI-Fuzzy employed in this paper, is a simple controller which yields an identical performance of the typical two-input PI-Fuzzy controller. The reduction in the number of inputs to a single input has significantly reduced the number of tuning parameters, thus simplifies the overall design of PI-Fuzzy control. Furthermore the input-output mapping or the control surface (Ψ) of the single input fuzzy controller can be easily constructed as a piecewise linear approximation. This provides a better insight on how exactly fuzzy controller works and makes the analysis of fuzzy controller less complex.

In this paper, an analysis on a double loop feedback single input PI-Fuzzy for inverter control is presented. For inverter systems, a single loop control structure, also referred as Voltage Mode Control (VMC) is not enough to damp the undamped poles introduced by the LC filter [10]. Hence, in order to obtain better dynamic response, a double loop consisting of current and voltage loop is preferable. Analysis of double loop with sliding mode control, deadbeat and linear PI controller for the control of single phase inverters can be found in [9,10,11],

---

This project is supported by the ScienceFund grant from the Ministry of Science, Technology and Innovation, Malaysia (MOSTI).

978-1-4244-0644-9/07/$25.00 ©2007 IEEE

respectively. However, no such concrete analysis is found for a fuzzy controller. This may be due to the complexity of analysing the behavior of Fuzzy control even for the case of the VMC.

Thus, this motivates the writers to present an analysis on the behavior of a double loop single-input PI Fuzzy controller. The controller is derived based on the signed-distance method [6]. The employment of a single-input PI-Fuzzy has made the analysis of the fuzzy control less complex as well as made the analysis of double loop possible. In the analysis, the control surface for each loop is varied and the dynamic response for the respective control surface is recorded and analysed. Then based on the simulation results conducted using MATLAB, justification on replacing the single input PI-Fuzzy for the voltage loop with a simple discrete linear PI controller is made.

II. DISCRETE SINGLE INPUT PI-FUZZY CONTROLLER

The design of the discrete single input PI Fuzzy controller is generally based on a previous work proposed by Viswanathan et. al. [7]. The conventional two input variables of PI-Fuzzy, namely the error (e) and the change of error (ce) have been replaced by a new single input variable which is the distance (d). In this method, the distance is defined as the absolute distance of any state points perpendicular to any points in the main diagonal line where by the main diagonal line is a line consisted of ZERO output membership in the rule table. The method, however can only be applied if the rule table exhibits a Toeplits structure or near Toeplits structure [7] as shown in Table I.

TABLE I
RULE TABLE WITH TOEPLITS STRUCTURE

CE\E	PB	PS	Z	NS	NB
NB	Z	PS	PB	PB	PB
NS	NS	Z	PS	PB	PB
Z	NB	NS	Z	PS	PB
PS	NB	NB	NS	Z	PS
PB	NB	NB	NB	NS	Z

The controller developed in [7] is an analog-based controller and has an identical small-signal performance of its linear PI and exhibits a superior performance over its linear controller for large load disturbance. The superiority can be achieved by varying the control surface (ψ) higher than unity gain. For digital implementation purpose, the output equation of the discrete single input PI-Fuzzy has been derived in [8]. To obtain a similar small-signal performance of its discrete linear PI controller when $\psi =1$, the controller parameters have been derived as follows,

$$\Delta u = (m+n)\psi \left[\frac{\dot{e}}{\sqrt{1+\lambda^2}} + \frac{\lambda e}{\sqrt{1+\lambda^2}} \right] \qquad (1)$$

where

$$m = K_i \left[\frac{K_p}{K_i} + \frac{T_s}{2} \right], n = K_i \left[\frac{T_s}{2} - \frac{K_p}{K_i} \right], \lambda \approx \frac{m+n}{-n}, r = m+n$$

The block diagram representation of (1) can be depicted as in Fig. 1.

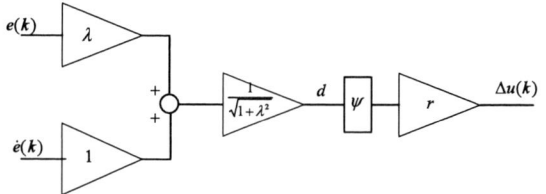

Fig. 1. Block diagram representation of (1).

In [8], analysis was conducted with both fuzzy controllers set to have identical control surfaces. This is done for simplification purposes. The control surface used in [8] and applied in this paper is as shown in Fig. 2

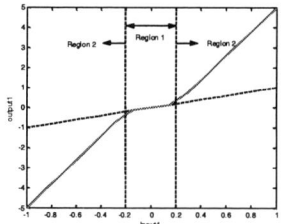

Fig. 2. Control surface used for the analysis

The control surface for the single input PI-Fuzzy controller is constructed based on two linear lines with different slopes. The first linear line is fixed at unity gain within the input range of |d|≤20 units. This is done, so that the controller can exhibit similar small-signal performance of its linear PI controller. For input range beyond |d|>20 units, the gain of the second linear line is made higher than unity so that a better performance can be obtained for large load disturbance. The control surface is set to saturate for |d|≥100 units.

III. SYSTEM DESIGN EXAMPLE

A complete single phase inverter system has been developed using MATLAB-Simulink. The system consisted of a full-bridge inverter with four switches, an L-C filter and a load. Table II summarises the parameters used in the simulations while Fig. 3 shows the block diagram of the control system. The inductor current and output voltage are sensed and fed back to the controller. Both loops are controlled by single input PI-Fuzzy controller.

TABLE II: SYSTEM PARAMETERS

Parameter	Value
V_{DC}	100V
L_{filter}	250µH
C_{filter}	33µF
Rated load	20 Ω
Reference Voltage	80V
Output Power	0.32KW
Switching frequency, f_s	20 KHz

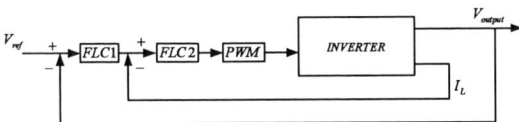

Fig. 3. Double feedback loop consisted of inductor current and output voltage

In a cascaded control scheme, the output voltage usually has slower response compared to the inductor current. Based on this, it is beneficial that the sampling time, T_s for the outer loop is set to have slower sampling rates. For the inner loop the sampling time is chosen to be similar to the switching frequency, f_s. Since, $f_s = 20$ KHz, then the sampling time is set to be equal to 50µs. Usually for the outer loop, the sampling time is set to be twice slower than the inner. Therefore the sampling time for the outer loop is set as 100µs.

To find the values of variables m and n, the linear PI controller transfer functions, C(s) for both loops are obtained using conventional frequency analysis. The controller transfer functions for the voltage loop and current loop are defined by equations (2a) and (2b), respectively. Using the Zero Order Hold (ZOH) emulation and Bilinear Transformation to transform C(s) to C(z), the parameters of m and n can be obtained for both loops as in (3a) and (3b).

$$C_v(s) = 2300 \left[\frac{0.00002174s + 1}{s} \right] \qquad (2a)$$

$$C_i(s) = 126 \left[\frac{0.00025s + 1}{s} \right] \qquad (2b)$$

$$m = K_i \left[\frac{K_p}{K_i} + \frac{T_s}{2} \right] \qquad (3a)$$

$$n = K_i \left[\frac{T_s}{2} - \frac{K_p}{K_i} \right] \qquad (3b)$$

IV. SIMULATION AND ANALYSIS

In this section, detail analysis on double-loop single input PI-Fuzzy controller with different control surfaces is presented. It should be noted that the analysis is conducted on large-signal performance of the controller.

To demonstrate that, a large load disturbance (no load to full load) is imposed on the system. To start with, let the control surface's slope for the voltage loop varies according to the values of $K_1 = 2.25$, $K_2 = 3.1875$ and $K_3 = 8.1875$. In the mean time, the control surface for the current loop is fixed to unity for the whole input range. Fig. 4 shows the result.

From Fig. 4, it can be clearly seen that there is no performance change even though the surface is varied from K_1 to K_3. This is probably because the maximum value of the distance input driven by the disturbance has not yet exceed the breakpoint, set at d=20. Applying the same procedure as above, different responses are obtained for different surface slopes when the breakpoint is set at |d|=10 as can be shown in Figure 5.

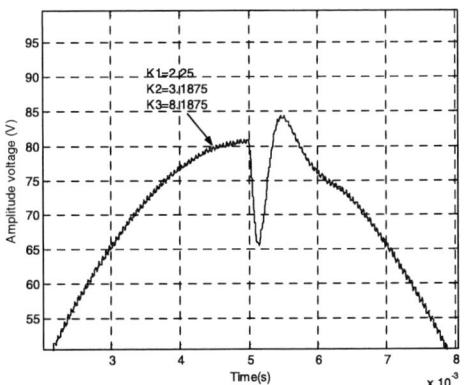

Fig. 4. Voltage output with the control surface for voltage loop is varied

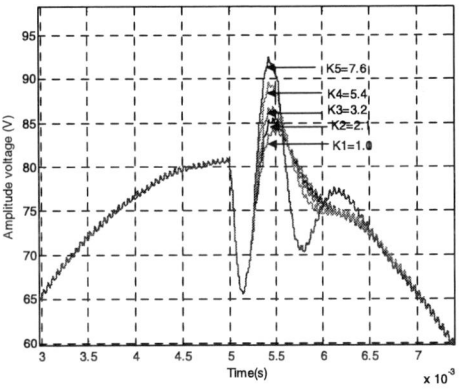

Fig. 5. Response for different slope with the breakpoint of d=10units

As can be observed in the figure, a sharp overshoot along with oscillation can be seen for a steep slope. Less overshoot and oscillation with slower rising time can be seen when the slope is set more moderate.

To analyse the effect of PI-Fuzzy for the current loop, the voltage loop control surface is set to unity for the whole range of input. Then the control surface for input |d|>20 is varied. The result for voltage response is as shown in Fig. 6. As can be observed, each slope value portrays distinct performance. Large and sharp overshoot can be examined for high gain value. A lower overshoot

1365

height is expected when the gain is reduced. Besides, an acceptable overshoot height with slower rising time can be obtained when the gain is set to unity. A better performance can be achieved by a slight increase above unity gain. A lower overshoot and faster rising time is obtained when the gain is set at 3.5. A fine-tuning process is then conducted around that value to obtain a better response. The best performance is obtained when the gain is set equal to 3.1875. As can be seen from the figure, the value gives similar rising time performance as for the case of K=3.5, but offers lower overshoot and faster settling time.

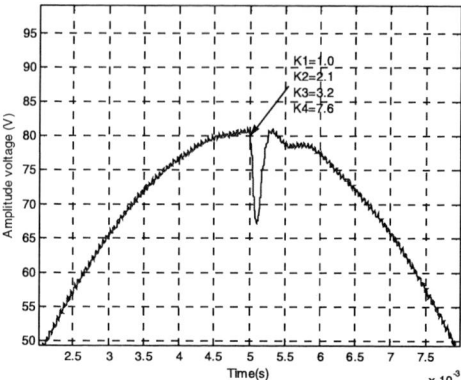

Fig. 7. Performance when both loops is PI-Fuzzy

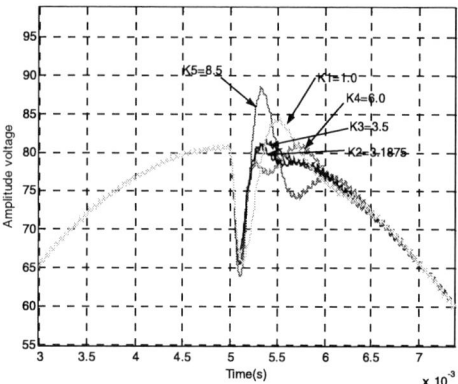

Fig. 6. Different responses are obtained when fuzzy control surface of current loop is varied

A further analysis is carried out with both loops controlled by a single input PI-Fuzzy controller with each loop having its own control surface. The analysis is carried out as follows,

- The control surface for the voltage loop has a breakpoint at $|d|=10$ and the gain for $|d|>10$ is varied.
- The control surface for the current loop has a breakpoint at $|d|=20$ and for $|d|>20$, the slope is set as 3.1875.

The result is as shown in Fig. 7. From the figure, it can be clearly seen that the performance is not affected at all even though the control surface for the voltage loop is varied.

For voltage loop, the rising time can be considered unaffected by the gain values. However, the overshoot and settling time can be controlled by varying the gain. The best performance, i.e. small overshoot and fast settling time is obtained when the gain is set equal to unity. For current loop with single input PI-Fuzzy, the rising time, overshoot and settling time is largely affected by the gain variation. Different gain values provide different performance. The best gain value that offers an excellent large-signal performance is K = 3.1875.

From the discussion, current loop seems to have a dominant impact in improving the large-signal dynamic response. The results have also shown that voltage loop has minimal effect in improving the system performance. Moreover the best performance is only achieved when the

gain is unity, which actually reflects the performance of its linear PI controller. Thus, it is sufficiently justified to replace the single input PI-Fuzzy controller with a simple linear PI controller.

V. Conclusions

In this paper, an analysis on double loop PI-Fuzzy for controlling a single phase inverter has been presented. A cascaded control structure is employed with the inductor current as the inner loop and the output voltage as the outer loop. Both loops are controlled by PI-Fuzzy controllers with each loop having a different control surface. By varying the control surface, an analysis is conducted to see the effect of each loop on the system's large-signal response. From the observations and discussions, it has been shown that fuzzy control for the voltage loop has minimal impact in improving the large-signal performance The best performance is achieved when the gain is set to unity, which is equivalent to a linear PI controller. Thus, the replacement with a linear PI controller is justified.

References

[1] B.-R. Lin, "Analysis of Neural and Fuzzy-Power Electronic Control", *IEE Proceeding of Science Measurement Technology*, Vol.144, No.1, January 1997.

[2] H. X. Li, H. B. Gatland and A. W. Green, "Fuzzy Variable Structure Control", *IEEE Transaction on System,Man and Cybernetics – Part B*, Vol.27, No.2,pp306-312, April 1997.

[3] F. Song and S. M. Smith, "A Comparison of Sliding Mode Fuzzy Controller and Fuzzy Sliding Mode Controller", *The 19th International Conference of the North American Fuzzy Information Processing Society (NAFIPS 2000)*, pp480-484, July 2000.

[4] E.-C. Chang, T.-J. Liang, J.-F. Chen and R. -L. Lin, "A Sliding-Mode Controller Based on Fuzzy Logic for PWM Inverters", *The 2004 IEEE Asia-Pacific Conference on Circuits and Systems*, pp965-968, December 2004.

[5] R. Palm, "Sliding Mode Fuzzy Control", *The 1992 IEEE International Conference on Fuzzy System*, pp.519-526, March 1992.

[6] B.J Choi, S.W. Kwak B.K.Kim, " Design and Stability Analysis of Single-Input Fuzzy Logic Controller", *IEEE Transaction on Systems, Man and Cybernetics-Part B: Cybernetics,* Vol.30, No. 2, pp303-309, April 2000.

[7] K.Viswanathan, R.Oruganti, D. Srinivasan, "Nonlinear Function Controller: A Simple Alternative to Fuzzy Logic Controller for a Power Electronic Converter", *IEEE Transaction on Industrial Electronics,* Vol. 52, pp 1439-1448, October, 2005.

[8] S. M. Ayob, Z. Salam, N. A. Azli, "Simple PI-Fuzzy Logic Controller Applied in DC-AC Converter", *The 1st International Power and Energy Conference (PECon 2006),* pp393-398, November 2006.

[9] L. Zhang, S.Qiu, "Analysis and Implementation of Sliding Mode Control for Full Bridge Inverter", *International Conference on Communications, circuits and systems,* Vol.2, 27-30, pp1380-1384, May 2005.

[10] S-L.Jung, H-S. Huang, M-Y. Chang and Y-Y. Tzou, "DSP-Based Multiple Loop Strategy for Single-Phase Inverters Used in AC Power Sources", *The 28th Annual IEEE Power Electronics Specialist Conference (PESC 1997),* Vol.1, pp706-712, June 1997

[11] P.Sanchis, A.Ursaea, E.Gubia, L. Marroyo, "Boost DC-AC Inverter: A New Control Strategy", *IEEE Transaction on Power Electronics,* Vol. 20, No. 2, pp343-353, March 2005.

A new three-phase varying-band hysteresis current controller for voltage-source inverters

Vinciane Chéreau, François Auger, Luc Loron

Institut de Recherche en Electrotechnique et Electronique de Nantes Atlantique (IREENA)
CRTT, 37 Bd de l'Université, BP 406, F-44602 Saint Nazaire cedex, France.

Abstract— **Three independent hysteresis controllers are often used to control the currents of a neutral isolated well-balanced load through a three-phase voltage-fed inverter, because of their simplicity and robustness. But this control structure also has some drawbacks, such as the destruction of the inverter if its highest switching frequency is not limited and the interdependence of the load phases. This article shows how to limit this frequency in a single-phase case and how to extend this result to the three-phase case, through a transformation avoiding the complexity of most solutions proposed until now.**

Index Terms—**adaptive hysteresis band current control, decoupling three-phase into two single-phase**

I. INTRODUCTION

Pulse Width Modulation (PWM) for three-phase voltage-fed inverters supplying three-phase isolated neutral loads has lead to a great number of current control strategies [2]. One of the simpliest is hysteresis current control. This kind of control is robust and allows to achieve a good accuracy. Nevertheless, this method presents some drawbacks compared to other strategies. Indeed, it does not take into account the interaction between the phases when used on each phase independently, and the switching frequency of the inverter can not be contained, so that it can lead to its destruction [1], [2], [3], [4]. This explains why other methods are usually preferred.

To solve the switching frequency limitation problem, some techniques are used such as using a combinatorial algorithm [5], using a quasi-sinusoidal hysteresis band [6] or a sinusoidal one [7], using feed-forward and feedback techniques [8], calculating the hysteresis band with the load parameters and the neutral potential measure [9], using phase locked loop control [10], [11] or varying the band with the system parameters (load and supply) [12].

To solve the phase dependency problem, some strategies have been developed such as using a switching table between the controllers and the inverter [13], using multi-level hysteresis controllers and a switching table [14], switching between two different control strategies (hysteresis and optimum voltage vector calculation) [15], using state-vector modulation with a region detection to choose in a switching table how to drive the inverter [16].

All these methods have their advantages and disadvantages, but a common drawback is that they complexify the

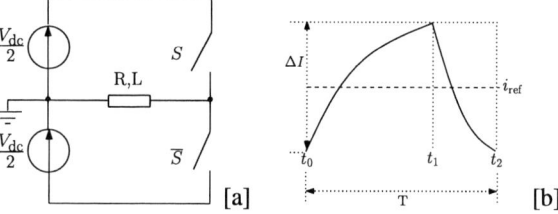

Fig. 1. [a] single-phase case, [b] Hysteresis band and switching period.

controller. This paper shows how to simply calculate the hysteresis band in the single-phase case to limit the inverter switching frequency (section II). Then it shows how to simply transform a three-phase case into a two single-phases case so that well operating single-phase strategies such as hysteresis control can be extended to three-phase applications and so that the previous calculation can be extended to the three-phase case (section III). Finally, experimental results are given (section IV).

II. SWITCHING FREQUENCY LIMITATION

In this section, we will first see how the switching frequency can be limited theoretically and then, some simulations made with Matlab/Simulink are provided to show this limitation and to compare it with two methods using a fixed-band and a sinusoidal band hysteresis controller.

A. Theoretical calculation

In this paragraph, we will design a simple method in the single-phase case to limit the inverter switching frequency, as it is one of the main drawbacks of hysteresis control. The load consists of a resistance R and an inductance L and is suplied by a voltage source $\pm V_{\mathrm{dc}}/2$ through an inverter leg (see Fig. 1.a). The current reference i_{ref} is considered as constant during each switching period. Looking at Fig. 1.b (where ΔI is the hysteresis band and T the switching period), we can define the current at each switching time by :

$$
\begin{aligned}
I_0 &= i(t_0) = i_{\mathrm{ref}} - \Delta I/2 \\
I_1 &= i(t_1) = i_{\mathrm{ref}} + \Delta I/2 \\
I_2 &= i(t_2) = i_{\mathrm{ref}} - \Delta I/2
\end{aligned}
\tag{1}
$$

Between t_0 and t_1, the current is the solution of

$$
L\frac{di(t)}{dt} + R\,i(t) = V_{\mathrm{dc}}/2
$$

and equals

$$
i(t) = I_0\, e^{-\frac{R}{L}(t-t_0)} + \frac{V_{\mathrm{dc}}}{2\,R}\left(1 - e^{-\frac{R}{L}(t-t_0)}\right)
\tag{2}
$$

---

[1]This work was supported by Electronavale Technologies 332 bd Marcel Paul, ZIL CP0604, 44806 Saint Herblain Cedex, France

Between t_1 and t_2, the current is the solution of

$$L\frac{di(t)}{dt} + R\,i(t) = -V_{\text{dc}}/2$$

and equals

$$i(t) = I_1\,e^{-\frac{R}{L}(t-t_1)} - \frac{V_{\text{dc}}}{2\,R}\left(1 - e^{-\frac{R}{L}(t-t_1)}\right) \quad (3)$$

Equations (2) and (3) clearly show that $|i(t)| \leq \frac{V_{\text{dc}}}{2\,R}$ (the maximum value is obtained when t goes to $+\infty$). Besides, Fig. 1.b, with this current limit, assumes that the following conditions are satisfied:

$$i_{\text{ref}} + \frac{\Delta I}{2} \leq \frac{V_{\text{dc}}}{2\,R} \quad (4)$$

$$i_{\text{ref}} - \frac{\Delta I}{2} \geq -\frac{V_{\text{dc}}}{2\,R} \quad (5)$$

The equations (2) to (5) also allow to compute the switching period:

$$t_1 - t_0 = \frac{L}{R}\ln\left(\frac{\frac{V_{\text{dc}}}{2\,R} + \frac{\Delta I}{2} - i_{\text{ref}}}{\frac{V_{\text{dc}}}{2\,R} - \frac{\Delta I}{2} - i_{\text{ref}}}\right) \quad (6)$$

$$t_2 - t_1 = \frac{L}{R}\ln\left(\frac{\frac{V_{\text{dc}}}{2\,R} + \frac{\Delta I}{2} + i_{\text{ref}}}{\frac{V_{\text{dc}}}{2\,R} - \frac{\Delta I}{2} + i_{\text{ref}}}\right) \quad (7)$$

$$T = t_2 - t_0 = \frac{L}{R}\ln\left(\frac{\left(\frac{V_{\text{dc}}}{2\,R} + \frac{\Delta I}{2}\right)^2 - i_{\text{ref}}^2}{\left(\frac{V_{\text{dc}}}{2\,R} - \frac{\Delta I}{2}\right)^2 - i_{\text{ref}}^2}\right) \quad (8)$$

To obtain the desired switching frequency $\frac{1}{T}$, we deduce an expression of the hysteresis band ΔI:

$$\Delta I = \frac{V_{\text{dc}}}{R}\frac{1+\delta}{1-\delta} \pm 2\sqrt{\left(\frac{V_{\text{dc}}}{R}\right)^2\frac{\delta}{(1-\delta)^2} + i_{\text{ref}}^2} \quad (9)$$

with $\delta = e^{-\frac{R}{L}T}$. Provided that $|i_{\text{ref}}| < \frac{V_{\text{dc}}}{2\,R}$, the two values of ΔI are strictly positive, but, the only one that satisfies conditions (4) and (5) is

$$\Delta I = \frac{V_{\text{dc}}}{R}\frac{1+\delta}{1-\delta} - 2\sqrt{\left(\frac{V_{\text{dc}}}{R}\right)^2\frac{\delta}{(1-\delta)^2} + i_{\text{ref}}^2} \quad (10)$$

It should be underlined that the hysteresis band does not depend on the reference sign and is lower- and upper-bounded:

$$0 < \Delta I \leq \frac{V_{\text{dc}}}{R}\frac{1-\sqrt{\delta}}{1+\sqrt{\delta}}$$

One can also notice that when R goes to $0\ \Omega$, ΔI goes to $\frac{V_{\text{dc}}T}{4\,L}$ and becomes independent of i_{ref}.

B. Simulation results

In this paragraph, some simulation results are provided to point out the effectiveness of the previous calculation (called hereafter method 3). In order to do that, we will show, first, that the switching frequency is limited and then, we will compare these results to those of a fixed band hysteresis controller (method 1) and of a sinusoidal band one [7], [6] (method 2). In our simulations, the sampling frequency is 1 MHz to avoid distorsions due to this frequency, those distorsions will be shown at the end of the paragraph. The load is a single-phase one

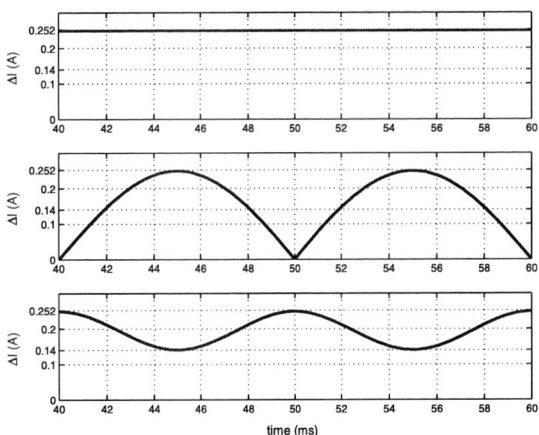

Fig. 2. Hysteresis band during one period of the reference: method 1 (top), method 2 (middle), method 3 (bottom).

with $R = 6.6\ \Omega$ and $L = 4$ mH, the DC-bus voltage is $V_{\text{dc}} = 20$ V and the desired switching frequency is $\frac{1}{T} = 5$ kHz. The reference current is a sinusoid and has an amplitude of $I_{\text{ref}} = 1$ A and a frequency of $f_{\text{ref}} = 50$ Hz. It should be noticed that the reference i_{ref} is not constant in our case but is a regularly sampled version of $i_{\text{ref}}(t) = I_{\text{ref}}\sin(2\,\pi\,f_{\text{ref}}\,t)$.

For method 3, the band is given by equation (10):

$$\Delta I(t) = \frac{V_{\text{dc}}}{R}\frac{1+\delta}{1-\delta} - 2\sqrt{\left(\frac{V_{\text{dc}}}{R}\right)^2\frac{\delta}{(1-\delta)^2} + i_{\text{ref}}^2(t)}$$

As this band should limit the switching frequency, we will use this expression for method 1 and 2. Method 1 has a fixed band and as the highest switching frequency occurs when the reference is the smallest (in our case, $i_{\text{ref}} = 0$ A), the band is

$$\Delta I(t) = \Delta I_1 = \frac{V_{\text{dc}}}{R}\frac{1-\sqrt{\delta}}{1+\sqrt{\delta}}$$

Method 2 has a sinusoidal band that follows the reference and, for the same reason as method 1, we choose

$$\Delta I(t) = \Delta I_1\,|sin(2\,\pi\,f_{\text{ref}}\,t)|$$

Fig. 2 shows how the bands evolve for each method.

On Fig. 3, the current waveforms and the spectrums are presented. Fig. 3.a shows that the switching frequency, for method 1, varies with the reference. When the reference value is the biggest ($|i_{\text{ref}}| = 1$ A), the switching frequency falls down and when this amplitude decreases to $|i_{\text{ref}}| = 0$ A, the switching frequency increases. For method 2, we have the same but as the band goes to zero, the switching frequency increases more. In fact, for this method, the switching frequency is limited by the sampling frequency ($f_{\text{switching max}} = f_{\text{sampling}}/2$) or by adding a lockout circuit [7] but the frequency of this circuit has to be higher than the desired frequency as a lockout frequency close to the desired one leads to the same results as method 1. Method 3 shows a different result as the band varies inversely with the reference amplitude, so the switching frequency stays nearly constant. Indeed, the band calculation was made

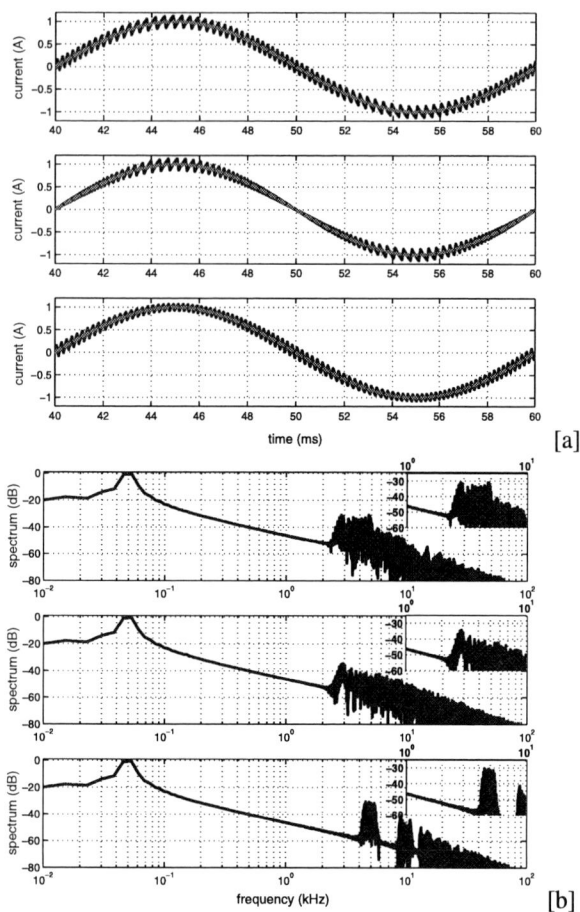

[a]

[b]

Fig. 3. Current waveform (load current (solid blue) and reference (dashed red)) [a] and load current spectrum [b]: method 1 (top), method 2 (middle), method 3 (bottom).

under the hypothesis that the currents at the beginning and at the end of the switching period are the same. Since the reference varies, this hypothesis is not satisfied and so the switching frequency is not strictly constant.

Fig. 3.b confirms these results. We can see that for method 1, the spectrum (except the fondamental) is under -30 dB and is located between 2.5 kHz and 5 kHz (see the zoom between 1 and 10 kHz). Therefore, the switching frequency is limited under the chosen frequency (5 kHz) and varies. For method 2, the spectrum is under -32 dB and is not well located, there are harmonics and sub-harmonics higher than the desired frequency and this can lead to the inverter destruction. For method 3, the spectrum is under -30 dB and is much more located than for the other methods, as it varies between 4 kHz and 5 kHz only.

Table I shows the total harmonic distorsion (THD) for each method. For a sampling frequency of 1 MHz, our method is between method 1 and 2. As we have not added a lockout circuit, the THD yield by method 2 is better than if we would have added it and the switching frequency is only limited by the sampling frequency. For a sampling frequency of 50 kHz, we have the same results but the THD is worse as sampling distorts the currents.

To conclude, we can say that our method gives better

TABLE I
SINGLE-PHASE SIMULATION THD

Sampling frequency	Method		
	1	2	3
1 MHz	10.30 %	7.31 %	8.22 %
50 kHz	12.97 %	9.54 %	10.45 %

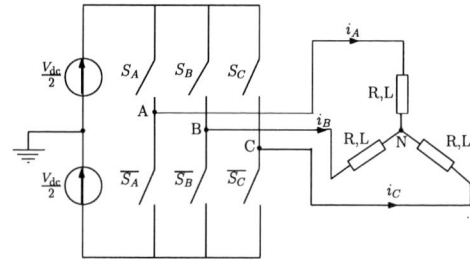

Fig. 4. Three phase load with DC-AC inverter.

results than a fixed band as it limits the upper bound of the switching frequency and its variations, and the current is less distorted. The sinusoidal band seems to give better results than our method as the THD and the spectrum amplitude are less important but the switching frequency cannot be limited below the desired frequency with this method. Besides, this method only works with a sinusoidal reference, whereas our method does not care of the reference shape. Therefore, our method seems to be more powerfull.

III. 3-TO-2 PHASE TRANSFORMATION

In this section, we will first show how to transform a three-phase case into a two single-phase case to take the dependency between the load phases into account and to extend the previous calculation to limit the inverter switching frequency. Some simulation results are also provided.

A. Theoretical calculation

In this paragraph, we will see how to decouple the three phases to obtain two single-phase systems so that the previous calculation can easily be extended.

The load is a well-balanced three-phase R-L load with isolated neutral (see Fig. 4) supplied by a DC voltage source $\pm V_{\mathrm{dc}}/2$ through a DC-AC inverter (with switches S_A, S_B and S_C). The phase currents can be written as

$$
\begin{aligned}
L\frac{di_A}{dt} + R\,i_A &= V_A - V_N \\
L\frac{di_B}{dt} + R\,i_B &= V_B - V_N \\
L\frac{di_C}{dt} + R\,i_C &= V_C - V_N,
\end{aligned}
\tag{11}
$$

where V_X is the potential at point X. V_A, V_B and V_C can only be equal to $V_{\mathrm{dc}}/2$ or $-V_{\mathrm{dc}}/2$ (depending on the inverter switches) and as the phase current sum equals zero, the neutral potential is:

$$
V_N = \frac{V_A + V_B + V_C}{3}
\tag{12}
$$

1370

TABLE II
2-PHASE CURRENTS

case	S_A	S_B	S_C	$i_\alpha(t)$	$i_\beta(t)$
1	0	0	0	$i_{\alpha_0}(t)$	$i_{\beta_0}(t)$
2	0	0	1	$i_{\alpha_0}(t) - \frac{V_{dc}}{3}\frac{1-\gamma(t)}{R}$	$i_{\beta_0}(t) - \frac{V_{dc}}{\sqrt{3}}\frac{1-\gamma(t)}{R}$
3	0	1	0	$i_{\alpha_0}(t) - \frac{V_{dc}}{3}\frac{1-\gamma(t)}{R}$	$i_{\beta_0}(t) + \frac{V_{dc}}{\sqrt{3}}\frac{1-\gamma(t)}{R}$
4	0	1	1	$i_{\alpha_0}(t) - \frac{2\,V_{dc}}{3}\frac{1-\gamma(t)}{R}$	$i_{\beta_0}(t)$
5	1	0	0	$i_{\alpha_0}(t) + \frac{2\,V_{dc}}{3}\frac{1-\gamma(t)}{R}$	$i_{\beta_0}(t)$
6	1	0	1	$i_{\alpha_0}(t) + \frac{V_{dc}}{3}\frac{1-\gamma(t)}{R}$	$i_{\beta_0}(t) - \frac{V_{dc}}{\sqrt{3}}\frac{1-\gamma(t)}{R}$
7	1	1	0	$i_{\alpha_0}(t) + \frac{V_{dc}}{3}\frac{1-\gamma(t)}{R}$	$i_{\beta_0}(t) + \frac{V_{dc}}{\sqrt{3}}\frac{1-\gamma(t)}{R}$
8	1	1	1	$i_{\alpha_0}(t)$	$i_{\beta_0}(t)$

TABLE III
2 SINGLE-PHASE CURRENTS

case	S_α	S_β	$i_\alpha(t)$	$i_\beta(t)$
2	0	0	$i_{\alpha_0}(t) - \frac{V_{dc}}{3}\frac{1-\gamma(t)}{R}$	$i_{\beta_0}(t) - \frac{V_{dc}}{\sqrt{3}}\frac{1-\gamma(t)}{R}$
3	0	1	$i_{\alpha_0}(t) - \frac{V_{dc}}{3}\frac{1-\gamma(t)}{R}$	$i_{\beta_0}(t) + \frac{V_{dc}}{\sqrt{3}}\frac{1-\gamma(t)}{R}$
6	1	0	$i_{\alpha_0}(t) + \frac{V_{dc}}{3}\frac{1-\gamma(t)}{R}$	$i_{\beta_0}(t) - \frac{V_{dc}}{\sqrt{3}}\frac{1-\gamma(t)}{R}$
7	1	1	$i_{\alpha_0}(t) + \frac{V_{dc}}{3}\frac{1-\gamma(t)}{R}$	$i_{\beta_0}(t) + \frac{V_{dc}}{\sqrt{3}}\frac{1-\gamma(t)}{R}$

TABLE IV
EQUIVALENCE 1-PHASE AND 2 VIRTUAL 1-PHASE

changing element	1-phase	α	β
DC voltage	$\frac{V_{dc}}{2}$	$\frac{V_{dc}}{3}$	$\frac{V_{dc}}{\sqrt{3}}$
reference phase	0	0	$-\frac{\pi}{2}$

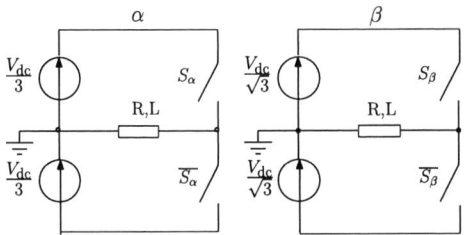

Fig. 5.   Equivalent two single-phases.

Substituting (12) in (11) leads to:

$$L\frac{di_A}{dt} + R\,i_A = \frac{2\,V_A - V_B - V_C}{3}$$
$$L\frac{di_B}{dt} + R\,i_B = \frac{2\,V_B - V_A - V_C}{3} \qquad (13)$$
$$L\frac{di_C}{dt} + R\,i_C = \frac{2\,V_C - V_A - V_B}{3}$$

Equations (13) can be solved with $(t_0, I_{A_0}, I_{B_0}, I_{C_0})$ as initial conditions:

$$i_A(t) = I_{A_0}\,\gamma(t) + \frac{2\,V_A - V_B - V_C}{3\,R}(1 - \gamma(t))$$
$$i_B(t) = I_{B_0}\,\gamma(t) + \frac{2\,V_B - V_A - V_C}{3\,R}(1 - \gamma(t)) \quad (14)$$
$$i_C(t) = I_{C_0}\,\gamma(t) + \frac{2\,V_C - V_A - V_B}{3\,R}(1 - \gamma(t))$$

with $\gamma(t) = e^{-\frac{R}{L}(t-t_0)}$. To go from 3-phase (A,B,C) to 2-phase (α, β), the Clarke transform [16] is used:

$$i_\alpha(t) = \frac{2}{3}\left(i_A(t) - \frac{i_B(t) + i_C(t)}{2}\right) = i_A(t) \quad (15)$$
$$i_\beta(t) = \frac{\sqrt{3}}{3}(i_B(t) - i_C(t)) \qquad (16)$$

As the potential at each point A, B and C can only take two different values, this leads to eight possible combinations (see Table II, where, for the switches S_A, S_B and S_C, 1 stands for on and 0 for off and where i_α and i_β are achieved by substituting (14) in (15) and (16), by replacing the potentials by their values and by setting $i_{\alpha_0}(t) = I_{A_0}\,\gamma(t)$ and $i_{\beta_0}(t) = \frac{I_{B_0} - I_{C_0}}{\sqrt{3}}\gamma(t)$).

If we extract from Table II the 4 cases (2, 3, 6, 7), it can be noticed, in Table III, that i_α and i_β have expressions

that are equivalent to single-phase current ones (they look like equations (2) and (3)). As a consequence, we can control a three-phase load as two single-phase loads (see Fig. 4 and 5). Furthermore, there is no need for switching tables, as it is easy to see in Table II and III that the virtual switch S_α of Fig.5 is directly the switch S_A and the virtual switch S_β of this figure is the switch S_B and the complementary of the switch S_C. It should be noticed that this transform can be extented to applications with back-emf such as AC motors.

Since we have shown that it is possible to control a three-phase load as two single-phase loads, we can extend the calculation previously performed for a simple single-phase load to control the currents i_α and i_β.

In order to do that, we have built table IV which gives the changing elements between the single-phase model (Fig. 1.a) and the two equivalent single-phase models in α and β (Fig. 5). The current references are chosen as sampled versions of:

$$i_{\mathrm{ref}_A}(t) = I_{\mathrm{ref}}\,\sin(2\,\pi\,f_{\mathrm{ref}}\,t) \qquad (17)$$
$$i_{\mathrm{ref}_B}(t) = I_{\mathrm{ref}}\,\sin(2\,\pi\,f_{\mathrm{ref}}\,t - \frac{2\,\pi}{3}) \qquad (18)$$
$$i_{\mathrm{ref}_C}(t) = I_{\mathrm{ref}}\,\sin(2\,\pi\,f_{\mathrm{ref}}\,t + \frac{2\,\pi}{3}) \qquad (19)$$

and so, with Clarke transform:

$$i_{\mathrm{ref}_\alpha}(t) = I_{\mathrm{ref}}\,\sin(2\,\pi\,f_{\mathrm{ref}}\,t) \qquad (20)$$
$$i_{\mathrm{ref}_\beta}(t) = I_{\mathrm{ref}}\,\sin(2\,\pi\,f_{\mathrm{ref}}\,t - \frac{\pi}{2}) \qquad (21)$$

We can achieve the hysteresis band for the α-phase and for the β-phase by substituting in equation (10) the elements of the second column of Table IV by those of the third column for ΔI_α, and by those of the fourth column for ΔI_β (with $\delta = e^{-\frac{R}{L}T}$):

$$\Delta I_\alpha(t) = \frac{2\,V_{dc}}{3\,R}\frac{1+\delta}{1-\delta}$$
$$-2\sqrt{\left(\frac{2\,V_{dc}}{3\,R}\right)^2\frac{\delta}{(1-\delta)^2} + i_{\mathrm{ref}_\alpha}^2(t)} \quad (22)$$

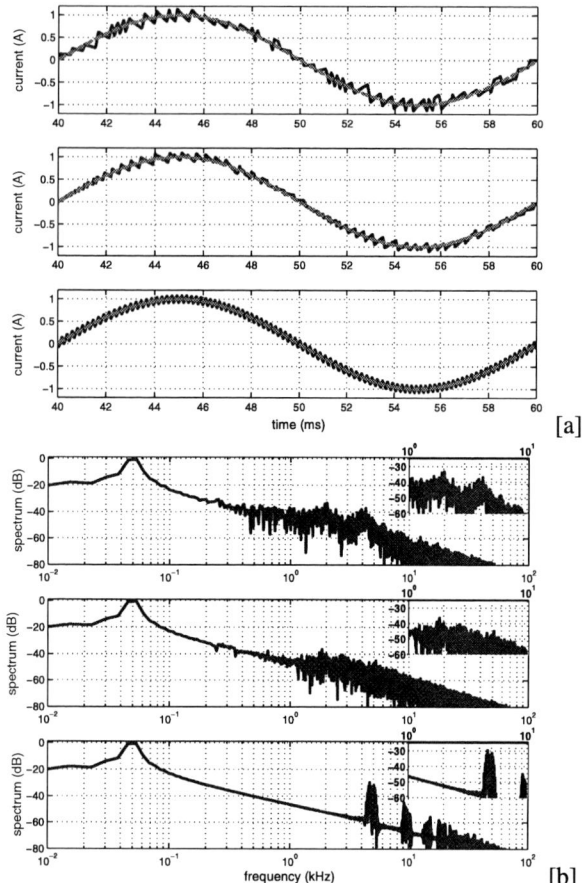

Fig. 6. Phase A current waveform (load current (solid blue) and reference (dashed red)) [a] and load current spectrum [b]: method 1 (top), method 2 (middle), method 3 (bottom).

$$\Delta I_\beta(t) = \frac{2V_{dc}}{\sqrt{3}R}\frac{1+\delta}{1-\delta}$$
$$-2\sqrt{\left(\frac{2V_{dc}}{\sqrt{3}R}\right)^2\frac{\delta}{(1-\delta)^2}+i^2_{ref_\beta}(t)} \quad (23)$$

This choice for ΔI_α and ΔI_β allows to limit the switching frequency on the virtual α and β phases. Since $S_\alpha = S_A$ and $S_\beta = S_B = \overline{S_C}$ (tables II and III), limiting switching frequency of S_α and S_β is equivalent to limiting the frequency of S_A, S_B, S_C.

B. Simulation results

In this paragraph, some simulations are presented to illustrate the previous theoretical results. We will compare our method with two others, as for the single-phase case:

- method 1: an hysteresis controller with a fixed band on each three phase,
- method 2: an hysteresis controller with a sinusoidal band on each three phase [7] without lockout circuit,
- method 3: an hysteresis controller on each two virtual phase (α, β) with a variable band as defined by equations (22) and (23).

The sampling period is 1 MHz, as in the single-phase case, to remove the current distorsion due to this frequency. We will also give the THD computed for each

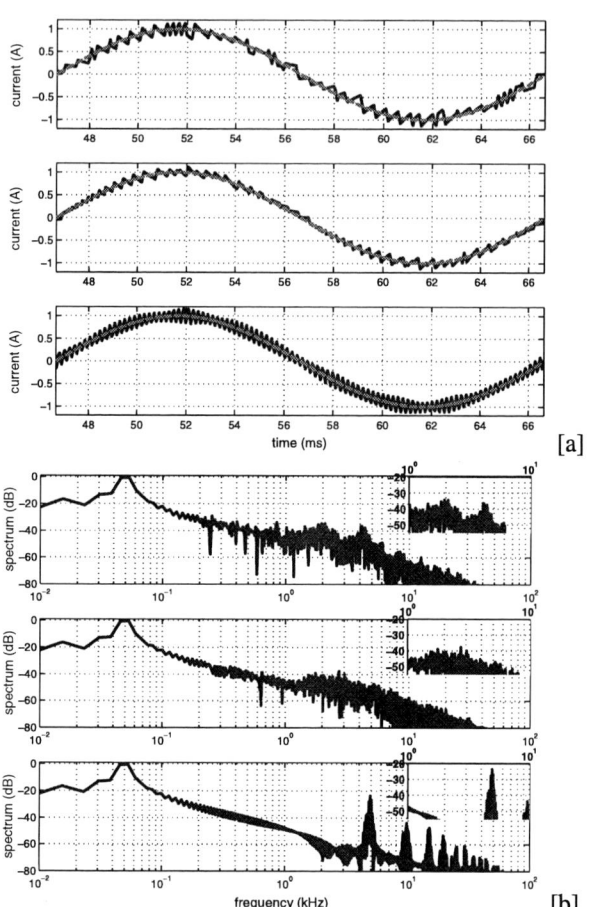

Fig. 7. Phase B current waveform (load current (solid blue) and reference (dashed red)) [a] and load current spectrum [b]: method 1 (top), method 2 (middle), method 3 (bottom).

method with a sampling frequency of 50 kHz to compare with experimental results. The load parameters are $R = 2.2\ \Omega$ and $L = 4$ mH, the DC-bus voltage is $V_{dc} = 20$ V and the desired switching frequency is $\frac{1}{T} = 5$ kHz. The reference current has an amplitude of $I_{ref} = 1$ A and a frequency of $f_{ref} = 50$ Hz.

The hysteresis bands are computed with the same reasoning as with a single-phase load. For method 1, we choose the phase bands as:

$$\Delta I_A(t) = \Delta I_B(t) = \Delta I_C(t) = \Delta I_1$$

with ΔI_1 defined in section II-B.
For method 2, the phase bands are:

$$\begin{aligned}
\Delta I_A(t) &= \Delta I_1\,|\sin(\omega t)| \\
\Delta I_B(t) &= \Delta I_1\,|\sin(\omega t - \frac{2\pi}{3})| \\
\Delta I_C(t) &= \Delta I_1\,|\sin(\omega t + \frac{2\pi}{3})|
\end{aligned}$$

For method 3, $\Delta I_\alpha(t)$ and $\Delta I_\beta(t)$ are calculated with eq. (22) and (23).

Fig. 6.a and 7.a show the current waveforms for phase A and B. The currents for methods 1 and 2 have flat parts due to the phase interferences. These phenomenons cannot be found on the currents of the third method as

TABLE V
3-PHASE SIMULATION THD

Sampling frequency	Phase	Method		
		1	2	3
1 MHz	A	10.69 %	6.66 %	6.51 %
1 MHz	B	11.39 %	7.92 %	11.34 %
50 kHz	A	12.43 %	8.08 %	8.02 %
50 kHz	B	12.92 %	9.18 %	13.87 %

the phases are decoupled thanks to our transform. For method 3, the current on phase B is distorted compared to the current on phase A. This is due to the fact that the proposed transform implies that phase A is phase α and so is directly controlled as a single-phase, whereas, phase B is controlled with phase C through phase β. For method 1 and 2, the switching frequency on each phase is varying, whereas, for method 3, it seems to stay nearly fixed.

Fig. 6.b and 7.b are the current spectrums of phase A and B for each method. First, method 1 and 2 has a phase A spectrum close to phase B one. It is not the case for method 3 for the reason explained previously. Then, we can see that the spectrum is well located for method 3 whereas it is not for the other methods. Indeed, for phase A, method 1 spectrum is greater than -40 dB until 4.5 kHz, method 2 spectrum is under this value but does not show clear limits for the switching frequency. Method 3 has its spectrum located between 4.2 kHz and 5.2 kHz, its value is higher than for the other methods but the switching frequency is nearly constant. For phase B, method 1 and 2 have the same results whereas method 3 shows a highest value of spectrum for the switching frequency than for phase A but also a well located spectrum.

The THD has been computed for each method (see table V). For a sampling frequency of 1 MHz, the THD obtained with method 3 is better on phase A. For phase B, method 3 is close to method 1. This is due to the distorsion implied by the chosen transform. It should be noticed that, as for the single-phase simulation, method 2 has no lockout circuit, so the THD is better than it should be. For a sampling frequency of 50 kHz, the THD rates are worse as a lower sampling frequency leads to more distorsions on the current. Phase A still shows better results for method 3 but phase B has its worst THD for this method. So, the sampling frequency has to be high enough to perform correctly.

To conclude, we have shown that our transform allows to decouple the phases easily and that the proposed band calculation can be extended to a three-phase load and leads to a limited switching frequency. The importance of a sufficiently high sampling frequency has been underlined to reduce the current distorsions for the proposed method.

The reduced computational cost of our algorithm should

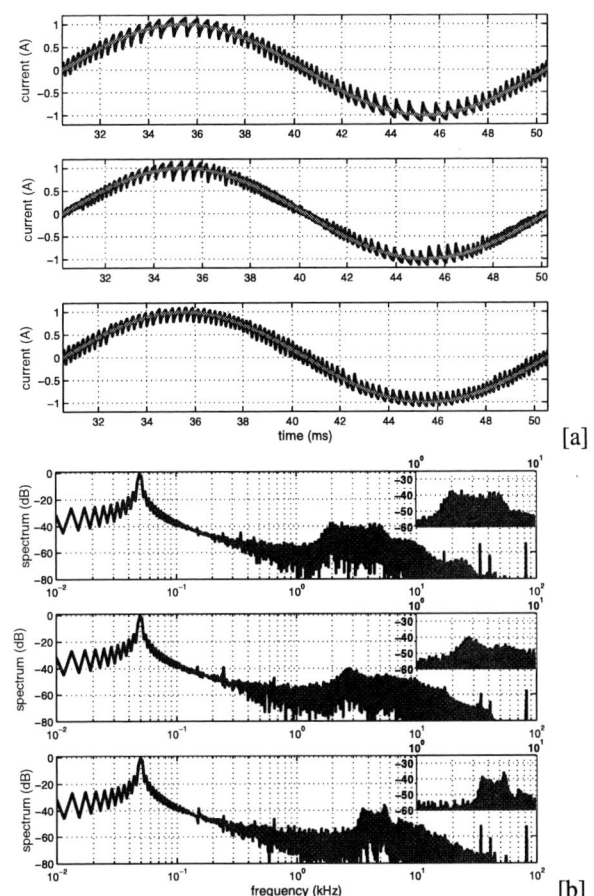

Fig. 8. Current waveform (load current (solid blue) and reference (dashed red)) [a] and load current spectrum [b]: method 1 (top), method 2 (middle), method 3 (bottom).

allow its implementation on current microcontrollers.

IV. EXPERIMENTAL RESULTS

In this section, we will compare the experimental results of the three hysteresis control strategies, with the simulation results presented in paragraphs II-B and III-B. These methods are implemented on a DS1103 dSpace environment with a sampling frequency of 50 kHz. The system's parameters are still $L = 4$ mH, $V_{\mathrm{dc}} = 20$ V, $\frac{1}{T} = 5$ kHz, $I_{\mathrm{ref}} = 1$ A and $f_{\mathrm{ref}} = 50$ Hz. The load resistance is $R = 6.6$ Ω for the single-phase load and $R = 2.2$ Ω for the three-phase one. The experimental data have been recorded with a Yokogawa ScopeCorder DL750 with a sampling rate of 500 kHz.

A. Single-phase results

Fig. 8.a shows the current waveforms for each method. We can notice that, as in simulation, the switching frequency seems to stay mostly constant for method 3 than for the others. Method 1 has its greater switching frequency when the reference amplitude decreases. Method 2 has the same behaviour but its switching frequency increases more.

Fig. 8.b shows the current spectrum of each method. All the methods give a spectrum under -40 dB but method 3 has the most located spectrum. Indeed, it is

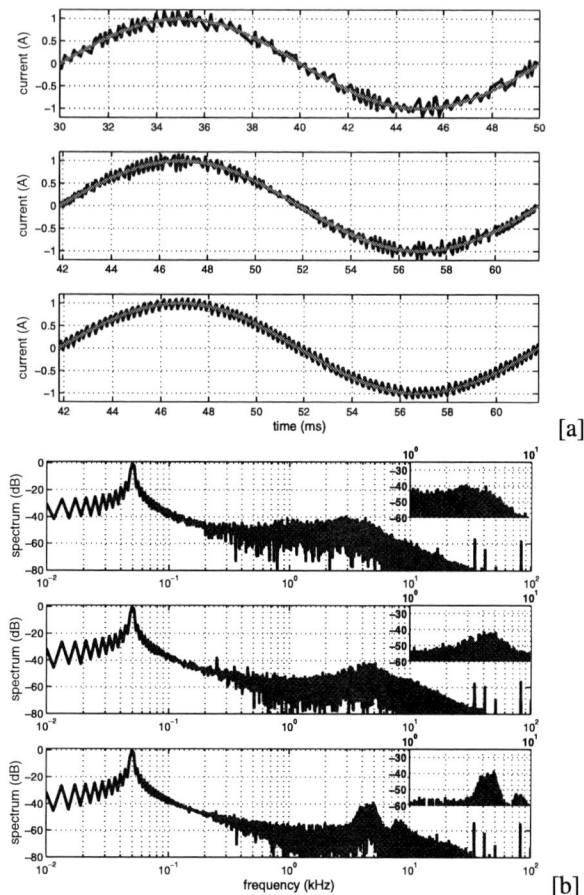

[a]

[b]

Fig. 9. Phase A current waveform (load current (solid blue) and reference (dashed red)) [a] and load current spectrum [b]: method 1 (top), method 2 (middle), method 3 (bottom).

TABLE VI
SINGLE-PHASE THD

Sampling frequency	Method		
	1	2	3
50 kHz	13.85 %	11.03 %	12.19 %

located between 3.5 kHz and 5.2 kHz whereas method 1 spectrum goes from 2 kHz to 5 kHz and method 2 one goes from 3 kHz to nearly 6 kHz.

The THD has been computed for each method (table VI). As in simulation, method 3 is between methods 1 and 2 but method 2 has no lockout circuit, so, its THD should be worse.

These results confirm that the proposed method limits the switching frequency and yields a correct THD.

B. Three-phase results

Fig. 9.a and 10.a shows the current waveforms obtained for a three-phase load for phase A and B. As in simulation, we can see that method 3 gives more regular currents than the two other methods, without interferences between phases. Phase B current is more distorted than phase A because of our transform.

Fig. 9.b and 10.b are the current spectrums for each

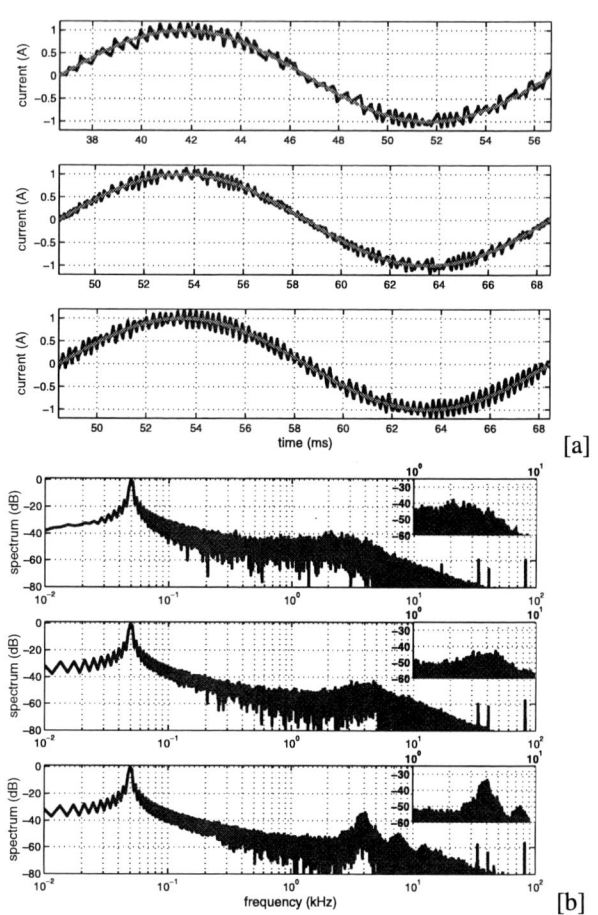

[a]

[b]

Fig. 10. Phase B current waveform (load current (solid blue) and reference (dashed red)) [a] and load current spectrum [b]: method 1 (top), method 2 (middle), method 3 (bottom).

TABLE VII
3-PHASE THD

Sampling frequency	Phase	Method		
		1	2	3
50 kHz	A	12.41 %	9.26 %	8.81 %
50 kHz	B	11.87 %	8.49 %	13.17 %

method. Method 3 has a well located spectrum on phase A (from 4 kHz to 5 kHz) and on phase B (from 3.5 kHz to 4.5 kHz) whereas methods 1 and 2 have spectrums with no distinct limits. The harmonics and sub-harmonics have values under -40 dB for method 1, under -42 dB for phase A and -45 dB for phase B for method 2 and under -40 dB for phase A and -35 dB for phase B for method 3. Therefore, there is not a great difference between the three methods.

The THD has been computed for each method (see table VII). The results are similar to those obtained in simulation for phase A. Method 3 has better results than the other methods whereas, for phase B, it has worse results than these methods. This is due to the switching frequency and to the fact that method 2 has no lockout circuit.

1374

To conclude, we have shown that the experimental results obtained leads to the same conclusion as the simulation ones: the switching frequency is limited with the proposed method and the phase currents are well decoupled thanks to the proposed transform.

V. Conclusion

In this article, we have first shown how to achieve the inverter switching frequency limitation through a simple hysteresis band calculation in a single-phase case. To this aim, we assume that the current reference is constant during each switching period. As it is not generally the case, the switching frequency slightly varies instead of remaining constant. However, it can be considered as an advantage, as a small variation on the switching frequency avoids uncomfortable noises.

Then, we have transformed, in a very simple manner, a three-phase load application with isolated neutral in a two single-phases application, so that we can easily apply single-phase control strategies to three-phase cases. Besides, it allows to extend the previous single-phase calculation to a three-phase application.
For this, we used the Clarke transform (α β plane). By choosing four of the eight possible configurations of the inverter switches, we succeeded to decouple the phases. Furthermore, these four configurations allow to control directly the switching frequency on phase A through the phase α, and to control the switching frequencies on phases B and C through the phase β. Therefore, there is no need of a switching table.

The simulation and experimental results obtained with our hysteresis band calculation and our transform have shown their interest compared to two other control strategies: a fixed-band hysteresis comparator and a sinusoidal-band hysteresis one [7] for a single-phase load and fixed-band hysteresis controllers and sinusoidal-band hysteresis controllers [7] on each phase for a three-phase load.
In the single-phase case, we have shown that the switching frequency can be controlled easily with our band calculation and in the three-phase case, the phases are decoupled, thanks to our transform, so that we can control the load current without phase interferences and, as with a single-phase load, the switching frequency is nearly constant.
It should also be noticed that the system parameters and the sampling frequency have a great importance on the results obtained. If the system parameters have values that leads to a very small variation of the band, our method is equivalent to a fixed-band one. If the sampling frequency is not high enough, the current is distorted and therefore, the THD becomes worse.

Finally, the proposed method leads to a located spectrum below the desired switching frequency. Hence, applications that need to filter the load current to extract the fundamental can use this method as the filter design will be easier.

References

[1] J. Holtz, *Pulsewidth modulation: A Survey*, IEEE Trans. on Industrial Electronics, december 1992, vol. 39, n° 5, pp. 410-420.

[2] J. Holtz, *Pulsewidth modulation for Electronic Power Conversion*, Proceedings of the IEEE, august 1994, vol. 82, n° 8, pp. 1194-1214.

[3] M.P. Kazmierkowski and L. Malesani, *Current Control Techniques for Three-Phase Voltage-Source PWM Converters: A Survey*, IEEE Trans. on Industrial Electronics, october 1998, vol. 45, n° 5, pp. 691-703.

[4] S. Buso, L. Malesani and P. Mattavelli, *Comparison of Current Control Techniques for Active Filter Applications*, IEEE Trans. on Industrial Electronics, october 1998, vol. 45, n° 5, pp. 722-729.

[5] A. Tilli and A. Tonielli, *Sequential Design of Hysteresis Current Controller for Three-Phase Inverter*, IEEE Trans. on Industrial Electronics, october 1998, vol. 45, n° 5, pp. 771-781.

[6] K.M. Rahman, M. Rezman Khan, M.A. Choudhury and M.A. Rahman, *Variable-Band Hysteresis Current Controllers for PWM Voltage-Source inverters*, IEEE Trans. on Power Electronics, november 1997, vol. 12, n° 6, pp. 964-970.

[7] A. Tripathi and P.C. Sen, *comparative Analysis of Fixed and Sinusoidal Band Hysteresis Current Controllers for Voltage Source Inverters*, IEEE Trans. on Industrial Electronics, february 1992, vol. 39, n° 1, pp. 63-73.

[8] Q. Yao and D.G. Holmes, *A Simple, Novel Method for Variable-Hysteresis-Band Current Control of a three-phase Inverter with Constant Switching Frequency*, Industry Applications Society Annual Meeting 1993, pp. 1122-1128.

[9] T.W. Chun and M.K. Choi, *Development of Adaptive Hysteresis Band Current Control Strategy of PWM Inverter with Constant Switching Frequency*, Applied Power Electronics Conference and Exposition 1996, pp. 194-199.

[10] L. Malesani and P. Tenti, *A Novel Hysteresis Control Method for Current-Controlled Voltage-Source PWM Inverters with Constant Modulation Frequency*, IEEE Trans. on Industry Applications, january/february 1990, vol. 26, n° 1, pp. 88-92.

[11] L. Malesani, P. Mattavelli and P. Tomasin, *Improved Constant-Frequency Hysteresis Current Control of VSI Inverters with Simple Feedforward Bandwidth Prediction*, IEEE Trans. on Industry Applications, october 1997, vol. 33, n° 5, pp. 1194-1202.

[12] B.K. Bose, *An Adaptive Hysteresis-Band Current Control Technique of a Voltage-Fed PWM Inverter for Machine Drive System*, IEEE Trans. on Industrial Electronics, october 1990, vol. 37, n° 5, pp. 402-408.

[13] M.P. Kazmierkowski and W. Sulkowski, *A Novel Vector Scheme for Transistor PWM Inverter-Fed Induction Motor Drive*, IEEE Trans. on Industrial Electronics, february 1991, vol. 38, n° 1, pp. 41-47.

[14] A. Ackva, H. Reinold and R. Olesinski, *A Simple and Self-Adapting High-Performance Current Control Scheme for three-phase Voltage Source Inverters*, PESC 1992, pp. 435-442.

[15] H. Le-Huy and L. Dessaint, *An Adaptive Current Control Scheme for PWM Synchronous Motor Drives: Analysis and Simulation*, IEEE Trans. on Power Electronics, october 1989, vol. 4, n° 4, pp. 486-495.

[16] B.H. Kwon, T.W. Kim and J.H. Youm, *A Novel SVM-Based Hysteresis Current Controller*, IEEE Trans. on Power Electronics, march 1998, vol. 13, n° 2, pp. 297-307.

Diode-Assisted Buck-Boost Current Source Inverters

F. Gao[1], C. Liang[1], P. C. Loh[1] and F. Blaabjerg[2]

[1]School of Electrical and Electronic Engineering
Nanyang Technological University
Nanyang Avenue, Singapore 639798
Email: gaof0001@ntu.edu.sg

[2]Institute of Energy Technology
Aalborg University
DK-9220 Aalborg East, Denmark
Email: fbl@iet.aau.dk

Abstract--This paper presents a couple of novel current source inverters (CSIs) with the enhanced current buck-boost capability. With the unique diode-inductor network added between current source inverter circuitry and current boost elements, the proposed buck-boost current source inverters demonstrate a double current boost capability when comparing with the recently reported buck-boost CSIs. For modulating the presented CSIs, two modulation schemes are proposed for achieving either optimized harmonic performance or minimal commutation count, meanwhile keeping the important current buck-boost operation uninfluenced. Lastly, all theoretical findings were verified experimentally using constructed laboratory prototypes.

Index Terms--Buck-boost; current source inverter; pulse-width modulation.

I. INTRODUCTION

To date, current source inverters (CSIs) have been well developed with some attractive advantages, for instance, the better load voltage and current shape, inherent short circuit protection and regeneration capability when the line commutated thyristor rectifier is used in front, which are the main reasons why CSIs can be good alternatives in several applications from low to medium power range. In particular, CSIs can either be used in programmable ac sources (PACSs) and uninterruptible power supplies (UPSs) or serve as ac drives and reactive power compensators. So far, the CSIs have already been appreciated in medium voltage large power induction drives [1], and are currently being investigated even in the low-power applications [2].

Undoubtedly, conventional CSIs would suffer their disadvantages with one particular constraint being its "only voltage-boost" functionality, which limit their applications not suitable for low voltage operation. Actually, an additional controlled front-end buck rectifier can be used to step down the dc link voltage in order to achieve both buck and boost conversion. But unfortunately, having a controlled rectifier would definitely increase the complication of the inverter control and synchronization needed between the added rectifier and the rear-end inverter, and might not function well under severely distorted supply conditions. In addition, this method significantly increases the total system cost and reduces the final conversion efficiency.

Alternatively, referring to [3, 4], Ćuk and SEPIC dc-dc power conversion concepts can be added into CSI topologies to realize current buck-boost operation whose front-end circuit consist of at least one inductor and one capacitor to store inductive energy. And the dc current boost ratio is firmly related to the duty ratio of additional switch located in the front-end current boost circuitry. Developing from the newly proposed dc-dc converter [5], this paper then presents an alternative solution based on the Ćuk- and SEPIC-derived theories in CSI operation to further achieve double current boost capability by using an additional X-shape diode-inductor network, in which the inherent operational principle of the unique passive network allows the inductor currents to be regulated to flow either in series or in parallel with the help of unidirectional diodes, therefore, the effectively boosted dc-link current can be enhanced with the careful control techniques using specifically designed modulation schemes. For properly modulating the proposed buck-boost CSIs, two modulation schemes are introduced with either optimized harmonic performance or reduced total commutation count achieved, respectively. Finally, all theoretical findings were verified experimentally using the constructed laboratory prototypes.

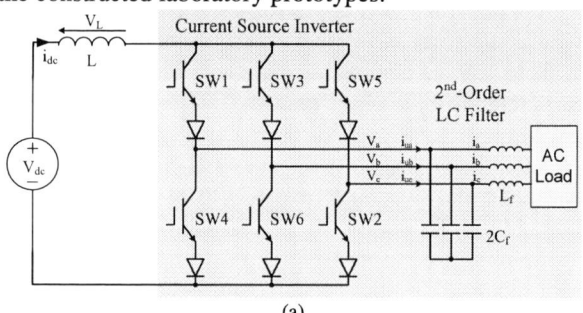

(a)

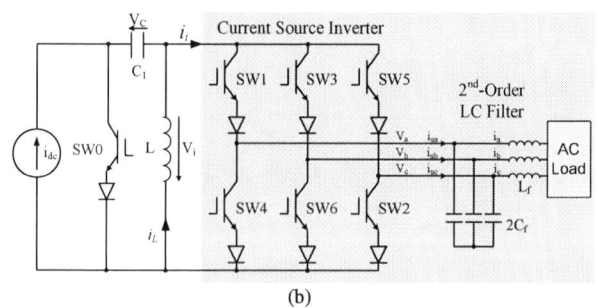

(b)

Fig. 1. Topologies of (a) conventional CSI and (b) SEPIC-derived buck-boost CSI.

II. Review of Buck-Boost Current Source Inverters

Referring to [4], the buck-boost current source inverter with extra active and passive components added has been developed from the conventional CSI to SEPIC-derived topology as illustrated in Fig.1, where besides the common circuit appeared in the shaded area which consists of a CSI circuitry, a 2nd order LC filter and an ac load, the rest front-end circuit performs the dc current boost capability. For conventional CSI shown in Fig. 1(a), the input dc current i_{dc} can be kept as constant as expected when large inductor L is used, resulting that the peak value of current i_x ($x = a$, b or c) after the LC filter cannot exceed i_{dc} theoretically. To overcome the current buck limitation, Fig. 1(b) shows a possible solution using capacitor C_1 as energy storage intermediate in the front-end circuit which can be derived from the SEPIC dc-dc converter. When SW0 is switched OFF, current i_{dc} charges capacitor C_1 and flows through CSI circuitry, meanwhile the inductor current of i_L also flows through the CSI circuitry resulting in the dc link current being the summation of i_{dc} and i_L, which of course is larger than the original input dc current i_{dc}. When SW0 is turned ON, capacitor C_1 is discharged, where the discharging current flows through SW0 and L implying that the CSI should be in the null states because the inductor currents i_L is not intentionally regulated to flow through the rear-end CSI circuitry. Assuming the conductive duty ratio of SW0 is 1-k, the peak value of dc-link current i_i and ac output current $\hat{i}$ can be derived as:

$$i_i = \frac{1}{1-k} \cdot i_{dc} \; ; \; \hat{i} = \frac{M}{(1-k)} \cdot i_{dc} \qquad (1)$$

Where, M refers to the modulation ratio normally used in the inverter control. Due to the limitation of k, M, with respect to the vertical span defined by the three-phase sinusoidal references, must be numerically equal to or less than k in practice in order to maintain the normalized volt-sec average unchanged, otherwise CSI circuitry would suffer alternating dc-link currents between i_i and zero during active intervals.

Besides the buck-boost topology shown in Fig.1 (b), another buck-boost topology derived from Ćuk converter is illustrated in Fig. 2, whose two distinct operating modes with respect to the charging and discharging process of capacitor C are also determined by the switching status of dc side switch SW0. By analyzing its operational principle like above analysis, it is noted that the boosted dc-link current and ac output current can still be expressed as (1).

It is noted that the reviewed current buck-boost topologies assume minimal passive components in the dc current boost circuitry in front of the conventional CSI bridge, mainly because the added capacitor and inductor are the basic elements for performing current boost operation. Although the buck-boost CSIs can achieve any desired value for the boosted currents theoretically as indicated in (1), where $1/(1$-$k)$ can be tuned to be infinite, their practical outputs are limited by the parasitic components and the rating of all front-end passive and

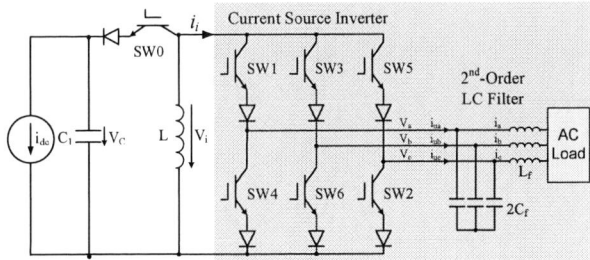

Fig. 2. Topology of Ćuk-derived buck-boost CSI.

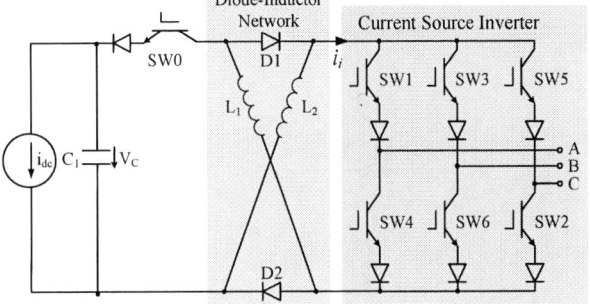

Fig. 3. Topology of Ćuk-derived buck-boost CSI with a diode-inductor network.

active components, hence the desired high current is hard to achieve, implying the need to design new buck-boost CSIs with enhanced current buck-boost capability.

III. Topologies and Operational Principles of Buck-Boost Current Source Inverters with a Diode-Inductor Network

The above reviewed buck-boost CSIs with boost ratio $1/(1$-$k)$ can theoretically boost dc input current to any desired value, however in practice, A large boost ratio would cause the inductor L suffering a huge boosted current, which definitely increases the requirement for L. In this section, a diode-inductor network is assumed to share the boosted dc-link current so that the current stress across inductor can be significantly decreased and meanwhile the dc-link current could achieve double boost capability comparatively if the same modulation condition is assumed. The detailed topologies and operational principles for the two proposed CSIs would be presented in the following subsections, respectively.

A. Ćuk-Derived Buck-Boost CSI with a Diode-Inductor Network

In [5], a Diode-Capacitor Network was introduced into a DC-DC boost converter to increase the boost ratio of output voltage by regulating the capacitor charging current properly, which to some extent, can be extended to DC-AC inversion for achieving enhanced voltage boost capability beyond the traditional voltage source inverters with only the inherent voltage-buck operation. For current source inverters, a diode-inductor network can further be introduced accordingly, as shown in Fig. 3, where an X-shape network comprised of two inductors and two diodes is inserted between the front-end circuit and inverter circuitry. Compared with the topology in Fig.

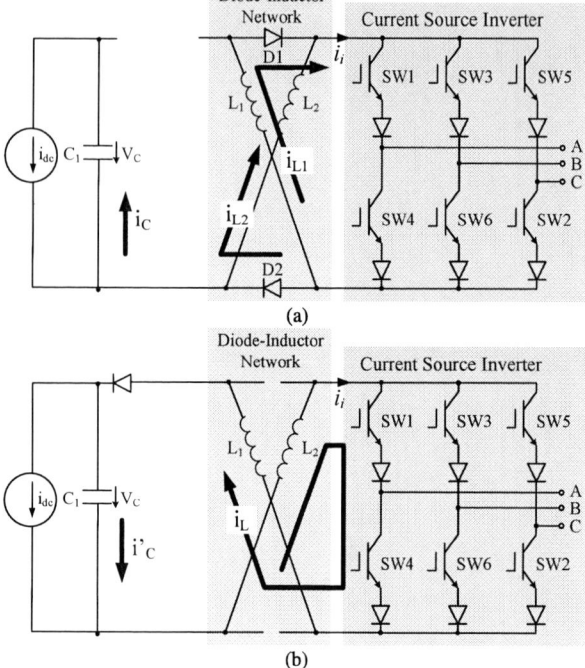

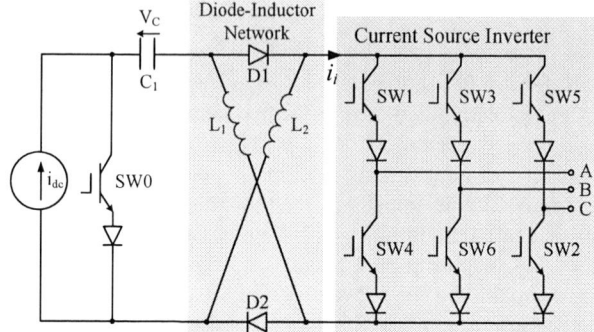

Fig. 5. Topology of SEPIC-derived buck-boost CSI with a diode-inductor network.

Fig. 4. Equivalent circuits of Ćuk-derived buck-boost inverter with a diode-inductor network when (a) SW0 is switched OFF and (b) SW0 is switched ON.

2, two diodes and one inductor are added intentionally in order to realize the expected performance. When the dc side switch SW0 is turned OFF, the dc current i_{dc} flows through the capacitor C_1 to boost electrical energy stored in capacitor C_1. At the same time, the diode D1 and D2 are now forward-biased to connect L_1 and L_2 in parallel, whose equivalent circuit is shown in Fig. 4(a), where SW0 effectively disconnects the diode-inductor network from the input dc source making the current of capacitor C_1 equal to input dc current. When SW0 is turned ON, the capacitor discharging current of i_C' with input current i_{dc} together would flow through diode-inductor network and inverter circuitry to charge the inductive energy stored in L_1 and L_2. Because diodes D1 and D2 are naturally reverse-biased in this case, the inductors L_1 and L_2 are forced to be connected in series as the current flow path indicated in Fig. 4(b). Note that the inductor currents need to flow through CSI bridge served as the free-wheeling path, hence the null state should be commanded during this period in order to not influence the normalized volt-sec average, which implies the switching state of SW0 = ON must be strictly commanded during only null intervals. Properly controlling the front-end active switch and inverter bridge, an alternative series charging and parallel discharging operation of inductor currents therefore can be achieved whose charging current is the same as illustrated in Fig. 2, but discharging current released to ac load is the summation of two inductor currents resulting in the double current boost seen from ac load when compared with the operation implied in Fig. 2.

To derive the relationship between dc input and ac output correctly, the capacitor current should be calculated first. Assuming the switching period is T and the conductive duty ratio of SW0 is $1-k$ ($0 < k < 1$). During SW0 = OFF interval, the capacitor voltage increases with a constant slope of i_{dc}/C. During SW0 = ON interval, the capacitor voltage falls with a constant slope of $(i_L-i_{dc})/C$. Because the average value of capacitor current is expected to be zero in a switching period, the inductor current $i_L = i_{L1} = i_{L2}$ can be derived as:

$$\frac{i_{dc}}{C} \cdot kT = \frac{i_L - i_{dc}}{C} \cdot (1-k)T \; ; \; i_L = i_{L1} = i_{L2} = \frac{1}{1-k} i_{dc} \quad (2)$$

To fully utilize the boosted current, it is preferred to use the summation of inductor currents as the effective dc-link current with reference to the analyzed SW0 = OFF state, therefore, the ac output peak current can be expressed as:

$$\hat{i} = M(i_{L1} + i_{L2}) = \frac{2Mi_{dc}}{1-k} \quad (3)$$

If a triplen harmonic offset is added to the modulation references, the modulation ratio M would increase with a factor of 1.15. Obviously, the output ac current is double boosted compared with (1) with the help of an X-shape diode-inductor network and can be stepped up or down by controlling M and k properly. Note that the six active intervals in CSI switching should appear in the SW0 = OFF interval if the current buck-boost operation is desired to use the doubled inductor current as dc-link current. When active intervals locate in both SW0 = ON and SW0 = OFF intervals, the dc-link current would alternate between i_L and $2i_L$ causing the distorted output because only i_L not $i_{L1} + i_{L2}$ flows through inverter circuitry during SW0 = ON interval.

B. SEPIC-Derived Buck-Boost CSI with a Diode-Inductor Network

Alternatively, the SEPIC-derived buck-boost CSI in Fig. 1(b) can also be developed to couple a diode-inductor network between inverter circuitry and the front-end boost circuit to realize the enhanced current boost capability as shown in Fig. 5. Analyzing the charging and discharging process of capacitor current, it is observed that the SEPIC-derived topology shows different performance features which would be illustrated below. Again, when the dc side switch SW0 turns OFF, the constant dc input current begins to charge capacitor C_1, meanwhile, diodes D1 and D2 are forward biased

1378

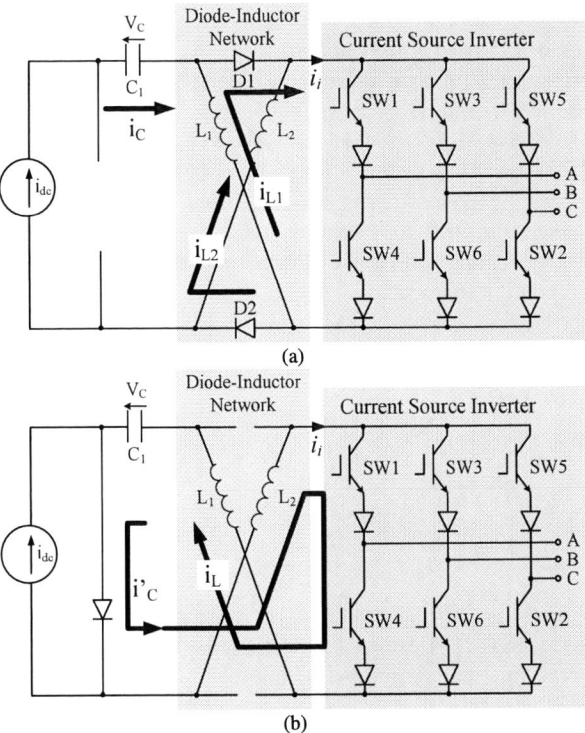

Fig. 6. Equivalent circuits of SEPIC-derived buck-boost inverter with a diode-inductor network when (a) SW0 is switched OFF and (b) SW0 is switched ON.

naturally making i_{L1} and i_{L2} flow in parallel, as shown in the equivalent circuit of Fig. 6(a). Therefore, the dc-link current can then be expressed as:

$$i_i = i_{L1} + i_{L2} + i_{dc} \qquad (4)$$

During SW0 = ON interval, dc current i_{dc} flows through SW0 and its series diode for free-wheeling without charging any passive component, and the discharging capacitor current i_C' flows through two inductors and inverter circuitry in series, as shown in Fig. 6(b), which means the inductor current equals to capacitor discharging current. Assuming the conductive duty ratio of SW0 is 1-k again, the capacitor discharging current can be expressed as:

$$k \cdot i_C = (1-k) \cdot i_C' \Rightarrow i_C' = i_L = \frac{k \cdot i_C}{1-k} = \frac{k \cdot i_{dc}}{1-k} \qquad (5)$$

Therefore, the maximum dc-link current and peak ac output current can be derived from (4) as:

$$i_i = \frac{2k \cdot i_{dc}}{1-k} + i_{dc} = \frac{1+k}{1-k} \cdot i_{dc} ; \hat{i} = \frac{(1+k)}{(1-k)} \cdot Mi_{dc} \qquad (6)$$

Compared to Ćuk-derived topology, both the inductor current i_L and the peak ac output current in SEPIC-derived topology are comparatively lesser if the same modulation conditions are assumed for equations (2), (3) and (5), (6), since $k < 1$ and consequently, $1 + k < 2$ in (6).

IV. MODULATION SCHEMES OF BUCK-BOOST CSIs WITH A DIODE-INDUCTOR NETWORK

Because the gating signals of CSI can be mapped from VSI switching states by using proper logic mapping technique [6], therefore, the intentional operation of

buck-boost CSI for unique output performance can be carried out by first controlling the corresponding VSI modulator. Depending on the different modulation targets, the gating signals of both SW0 and rear-end inverter can be generated for achieving either optimized harmonic performance or minimal commutation count, meanwhile maintaining the basic characteristic of current buck-boost operation.

A. PWM Switching Scheme with Optimized Harmonic Performance

To achieve optimized harmonic performance in two-level inverters, the null intervals in each switching sequence must be kept equal, which is easy to be implemented by adding a triplen harmonic offset ($V_{off} = -0.5(\max(V_A, V_B, V_C)+\min(V_A, V_B, V_C))$) to the sinusoidal references. Because the dc-link current is alternating between $i_{dc}/(1-k)$ and $2i_{dc}/(1-k)$ for Ćuk-derived topology and $ki_{dc}/(1-k)$ and $(1+k)i_{dc}/(1-k)$ for SEPIC-derived topology, respectively, according to the switching status of SW0, therefore, the large value of dc-link current is preferred to be used in the proposed buck-boost CSIs so as to demonstrate the advantage of enhanced current boost capability. During SW0 = OFF interval, the large dc-link current appears in both topologies, the traditional active states, therefore, should be restricted in this interval to keep normalized volt-sec average unchanged. Fig. 7 shows the exampled switching sequence designed for achieving optimized harmonic performance in buck-boost VSI [7], which can be mapped to CSI modulation easily, where SA, SB, SC represent the upper switches of full bridge VSI and SW refers to the dc side switch used in buck-boost VSI whose switching signal is the complement of SW0 in CSI modulation. Obviously, !SW0 = SW = OFF states always locate in null intervals to maintain the normalized volt-sec average unchanged. Following this modulation criteria, the modulation references suitable for buck-boost VSI operation are derived as:

$$\begin{cases} V_a = M \cos(\omega t) + V_{off} \\ V_b = M \cos(\omega t - 120°) + V_{off} \\ V_c = M \cos(\omega t + 120°) + V_{off} \\ V_u = -V_d = k \end{cases} \qquad (7)$$

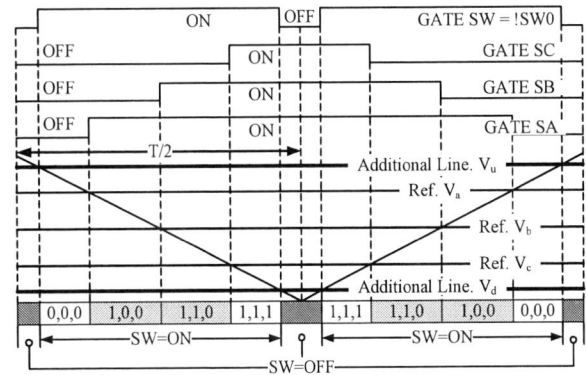

Fig. 7. Exampled switching sequence for buck-boost VSI with optimized harmonic performance.

1379

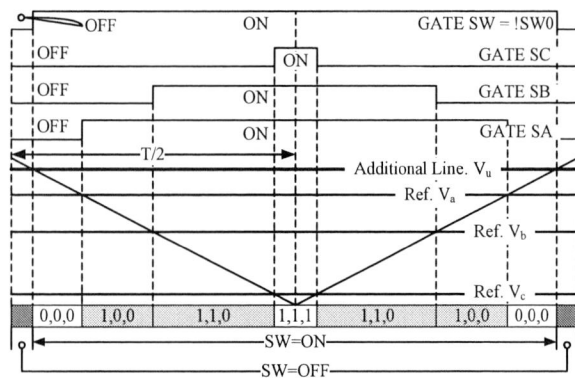

Fig. 8. Exampled switching sequence for buck-boost VSI with minimal commutation count.

Where, V_u and V_d are the designed simple linear references for defining the conductive duty ratio of both dc side switch SW for buck-boost VSIs and SW0 for buck-boost CSIs by comparing with the specific triangular carrier. In order to avoid the alternating dc-link current after the signal mapping in the proposed buck-boost CSIs, k should be restricted to be equal to or larger than the amplitude of sinusoidal references.

B. PWM Switching Scheme with Minimal Commutation Count

Observing Fig. 7, it is found that SW0 switches twice in each switching cycle, which is due to the centralized arrangement of active states forces the SW0 = ON interval distributes in two null intervals. Although SW0 = ON state can locate only in either the first null interval {0, 0, 0} or the last null interval {1, 1, 1} (e.g.) of the demonstrated switching period to reduce its commutation count from two to one, the additional current stress would impose on components because of the additionally unnecessary duty ratio of SW0, which causes the actual conductive duty ratio of SW0 being samller than 1-k. However, the modulation ratio M is still restricted by k, therefore, the modulation index is not fully utilized, especially when needing high boosted current, resulting in the unnecessary current stress crossing components. In order to reduce commutation count and current stress simultaneously, this subsection presents an improved modulation method whose exampled switching sequence is shown in Fig. 8. Being different from Fig. 7, only an additional linear reference V_u, instead of V_u and V_d, is used to compare with the specific triangular carrier. Three sinusoidal references are then consequently moved

downwards in vertical to reduce null interval {1, 1, 1} without changing the normalized volt-sec average, therefore, two null states are no longer locate symmetrically in each switching sequence resulting in the inferior harmonic performance consequently. Now the conductive duty ratio k' of SW = !SW0 is equal to $(1+k)/2$ when the same boosted dc current is desired. The inductor current i_L in equation (2) and (5) can then be changed to the following equations, respectively, for clearly distinguishing the difference between these two modulation schemes.

$$i_L = i_{L1} = i_{L2} = \frac{1}{1-k'} \cdot i_{dc}$$
$$k' = \frac{1+k}{2} \Rightarrow i_{L1} = i_{L2} = \frac{2}{1-k} \cdot i_{dc} \quad (8)$$

$$i_L = i_{L1} = i_{L2} = \frac{k'}{1-k'} \cdot i_{dc}$$
$$k' = \frac{1+k}{2} \Rightarrow i_{L1} = i_{L2} = \frac{1+k}{1-k} \cdot i_{dc} \quad (9)$$

In (8) and (9), k' refers to the conductive duty ratio of SW (=!SW0). Besides the triplen harmonic added for increasing modulation ratio, another offset for shifting sinusoidal references should also be introduced to reduce current stress. To summarize, the resulted modulation references are expressed as:

$$\begin{cases} V_a = M\cos(\omega t) + V_{off} - (1-k)/2 \\ V_b = M\cos(\omega t - 120°) + V_{off} - (1-k)/2 \\ V_c = M\cos(\omega t + 120°) + V_{off} - (1-k)/2 \\ V_u = -V_d = k \end{cases} \quad (10)$$

Since SW0 = ON states and null states are the same when seen from ac load, therefore, it's preferred to replace null interval with SW0 = ON interval as much as possible to reduce unnecessary current stress, implying the redundancy of null interval is significantly decreased.

After the generation of VSI gating signals, a digital logic mapping for CSI commands can be implemented using one-to-one mapping and many-to-many mapping as listed in Table I so that the diode-assisted buck-boost CSIs can be controlled properly.

V. EXPERIMENTAL RESULTS

For verifying the operational principles of proposed diode-assisted buck-boost CSIs and modulation schemes presented, laboratory prototypes for Ćuk- and SEPIC-

TABLE I. DIGITAL LOGIC MAPPING FOR PRESENTED BUCK-BOOST CSI GATING SIGNALS.

Bit Patterns by (7) {SA, SB, SC, SA', SB',SC'}	Required CSI Gating Signals {SW1, SW3, SW5, SW4,SW6, SW2} {SW0}	CSI Gate Type
SV1 = {1,0,0,0,1,1}	SC1 = {1,0,0,0,0,1} {0}	Active (One-to-One Mapping)
SV2 = {1,1,0,0,0,1}	SC2 = {0,1,0,0,0,1} {0}	
SV3 = {0,1,0,1,0,1}	SC3 = {0,1,0,1,0,0} {0}	
SV4 = {0,1,1,1,0,0}	SC4 = {0,0,1,1,0,0} {0}	
SV5 = {0,0,1,1,1,0}	SC5 = {0,0,1,0,1,0} {0}	
SV6 = {1,0,1,0,1,0}	SC6 = {1,0,0,0,1,0} {0}	
SV0 = {0,0,0,1,1,1} SV7 = {1,1,1,0,0,0}	SC7 = {1,0,0,1,0,0} {0}or{1} SC8 = {0,1,0,0,1,0} {0}or{1} SC9 = {0,0,1,0,0,1} {0}or{1}	Null (Many-to-Many Mapping)

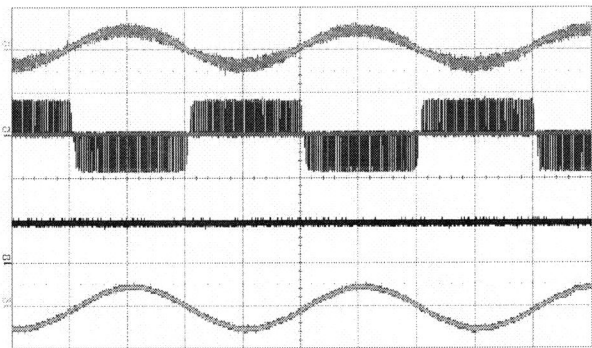

Fig. 9. Experimental waveforms of Ćuk-derived diode-assisted buck-boost CSI with optimized harmonic performance when $M=k=0.3$. TOP to BOTTOM: capacitor voltage, 20V/div; switched current, 5A/div; input current, 1.5A/div; filtered phase current, 2A/div. Time scale: 5ms/div.

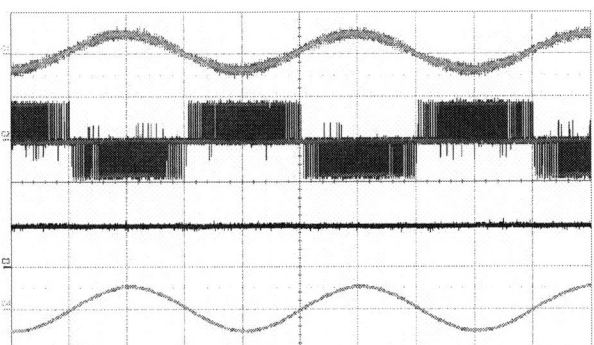

Fig. 10. Experimental waveforms of Ćuk-derived diode-assisted buck-boost CSI with optimized harmonic performance when $M=k=0.7$. TOP to BOTTOM: capacitor voltage, 100V/div; switched current, 10A/div; input current, 1.5A/div; filtered phase current, 10A/div. Time scale: 5ms/div.

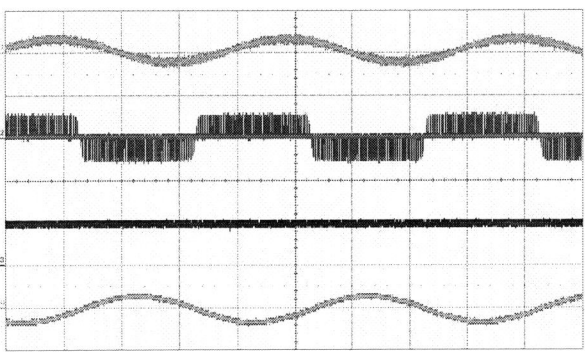

Fig. 11. Experimental waveforms of SEPIC-derived diode-assisted buck-boost CSI with optimized harmonic performance when $M=k=0.3$. TOP to BOTTOM: capacitor voltage, 20V/div; switched current, 5A/div; input current, 1.5A/div; filtered phase current, 2A/div. Time scale: 5ms/div.

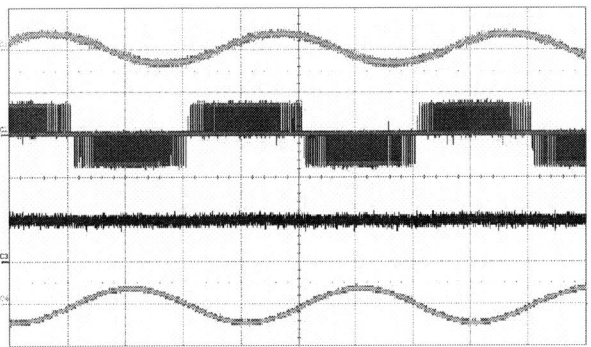

Fig. 12. Experimental waveforms of SEPIC-derived diode-assisted buck-boost CSI with optimized harmonic performance when $M=k=0.7$. TOP to BOTTOM: capacitor voltage, 100V/div; switched current, 10A/div; input current, 1.5A/div; filtered phase current, 10A/div. Time scale: 5ms/div.

derived diode-assisted buck-boost current source inverters were constructed using the passive components $L_1 = L_2 = 20$ mH and $C = 15$ μF and powered with the constant dc input current $i_{dc} = 1.5$A. When the laboratory prototype was configured as Ćuk-derived diode-assisted buck-boost CSI, the captured experimental waveforms with $1.15*M = 1.15*0.3$ and $k = 0.3$ for optimized harmonic performance are shown in Fig. 9, where the switched current was effectively boosted to $2*1.5/(1-0.3)$ ≈ 4.3A, and the peak filtered ac current was also in accordance with the theoretical calculation from equation (3). To verify its corresponding current boost capability, Fig. 10 shows the captured experimental waveforms with $1.15*M = 1.15*0.7$ and $k = 0.7$, where the switched current was boosted around $2*1.5/(1-0.7) = 10$A and the filtered phase current exhibited the enhanced current boost capability with the significant increase of current amplitude achieved. Since the resistance of inductors can perturb the steady state operation point [8], the ac output currents in both figures therefore have a slight degradation of amplitude unlike the value acquired from the theoretical calculations. Next, to illustrate the operational validity of SEPIC-derived diode-assisted buck-boost CSI, Fig. 11 and Fig. 12 show the captured waveforms under the same modulation condition used for Fig. 9 and 10, respectively. Both Fig. 11 and 12 demonstrate the little inferior current boost capability

when compared with Fig. 9 and 10, because the inherent operational principles of SEPIC-derived diode-assisted topology limit its current boost capability according to equation (5) and (6). The current stress across inductor L_1 and L_2, however, is significantly reduced, which implies that the rating of inductors of diode-inductor network in SEPIC-derived diode-assisted topology can be chosen less than those used in Ćuk-derived diode-assisted topology in practice.

Comparatively, the verifications of modulation scheme with minimal commutation count are shown in Fig. 13 and Fig. 14 for Ćuk-derived diode-assisted topology and Fig. 15 and Fig. 16 for SEPIC-derived diode-assisted topology, respectively, under the same modulation conditions as those used for the modulation scheme with optimized harmonic performance. Clearly, the similar output performances are achieved as expected, except of the inferior harmonic performance obviously observed from the waveforms of capacitor voltage (top waveform in all experimental results).

VI. CONCLUSION

This paper proposes a set of enhanced current buck-boost CSIs using a special designed diode-inductor network coupled between inverter circuitry and front-end circuit to perform series charging and parallel discharging processes between inductors, which is simply achieved

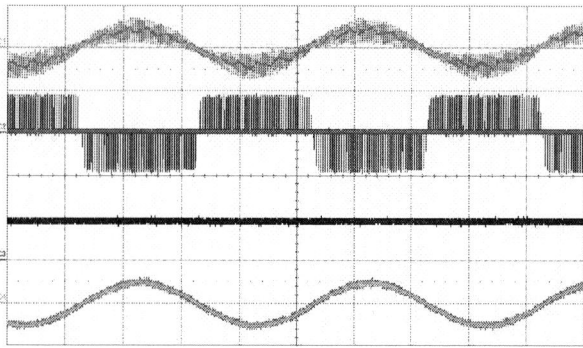

Fig. 13. Experimental waveforms of Ćuk-derived diode-assisted buck-boost CSI with minimal commutation count when $M=k'=0.3$. TOP to BOTTOM: capacitor voltage, 20V/div; switched current, 5A/div; input current, 1.5A/div; filtered phase current, 2A/div. Time scale: 5ms/div.

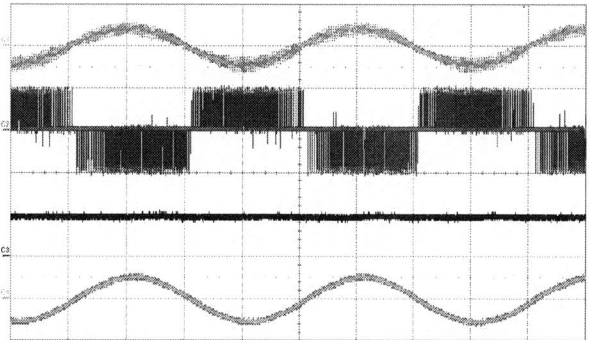

Fig. 14. Experimental waveforms of Ćuk-derived diode-assisted buck-boost CSI with minimal commutation count when $M=k'=0.7$. TOP to BOTTOM: capacitor voltage, 100V/div; switched current, 10A/div; input current, 1.5A/div; filtered phase current, 10A/div. Time scale: 5ms/div.

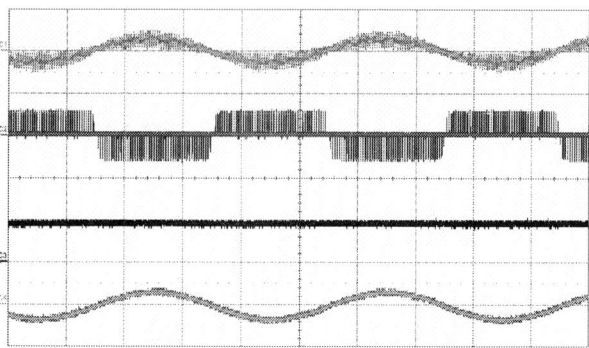

Fig. 15. Experimental waveforms of SEPIC-derived diode-assisted buck-boost CSI with minimal commutation count when $M=k'=0.3$. TOP to BOTTOM: capacitor voltage, 20V/div; switched current, 5A/div; input current, 1.5A/div; filtered phase current, 2A/div. Time scale: 5ms/div.

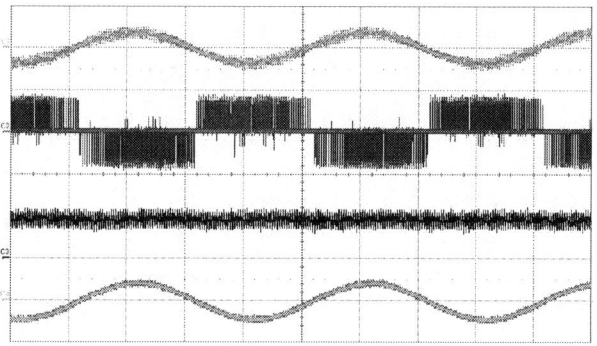

Fig. 16. Experimental waveforms of SEPIC-derived diode-assisted buck-boost CSI with minimal commutation count when $M=k'=0.7$. TOP to BOTTOM: capacitor voltage, 100V/div; switched current, 10A/div; input current, 1.5A/div; filtered phase current, 10A/div. Time scale: 5ms/div.

by the block property of unidirectional diodes in the X-shape network. Modulation wise, the proposed inverter can be controlled with the designed modulation schemes for achieving either optimized harmonic performance or minimal commutation count, meanwhile keeping the current stress across components as little as possible. Although the SEPIC-derived diode-assisted topology cannot acquire the same boosted capability as Ćuk-derived diode-assisted topology when the same modulation condition is assumed, it exhibits the requirement for the current rating of inductors is comparatively less. All theoretical findings were verified experimentally using the constructed laboratory prototypes with captured results for visual confirmation.

REFERFENCES

[1] J. R. Espinoza and G. Joos, "A current-source-inverter-fed induction motor drive system with reduced losses," IEEE Trans. Ind. Appl., vol. 34, pp. 796–805, Jul./Aug. 1998.

[2] V. D. Colli, P. Cancelliere, F. Marignetti, and R. Di Stefano, "Influence of voltage and current source inverters on low-power induction motors," *Electric Power Applications, IEE Proceedings* -, vol. 152, pp. 1311-1320, Sept. 2005.

[3] J. Kikuchi and T. A. Lipo, "Three-phase PWM boost-buck rectifiers with power-regenerating capability," *Industry Applications, IEEE Transactions on,* vol. 38, pp. 1361-1369, Sept./Oct. 2002.

[4] P. C. Loh, P. C. Tan, F. Blaabjerg, and T. K. Lee, "Topological Development and Operational Analysis of Buck-Boost Current Source Inverters for Energy Conversion Applications," in *Proc. IEEE-PESC'06*, 2006, pp. 1033-1038.

[5] Hiroshi Nomura, Kenichiro Fujiwara, Masanobu Yoshida, "A New DC-DC Converter Circuit with Larger Step-up/down Ratio," in *Proc. IEEE-PESC'06*, 2006, pp. 3006-3012.

[6] P. C. Loh, D. M. Vilathgamuwa, C. J. Gajanayake, L. T. Wong, and C. P. Ang, " Z-Source Current-Type Inverters: Digital Modulation and Logic Implementation," *Power Electronics, IEEE Transactions on*, vol. 22, pp. 169-177, Jan. 2007.

[7] F. Gao, P. C. Loh, D. M. Vilathgamuwa, F. Blaabjerg, C. K. Goh and J. Q. Zhang, "Topological and Modulation Design of a Buck-Boost Three-Level Dual Inverter," in *Proc. IEEE-IECON'06*, 2006, pp. 2408-2413.

[8] B. P. McGrath and D. G. Holmes, "Natural Current Balancing of Multicell Current Source Converters," in *Proc. IEEE-PESC'07*, 2007, pp. 968-974.

Single-Stage Fluorescent Lamps Electronic Ballast Using Class-DE Low dv/dt Rectifier for Power-Factor Correction

Chainarin Ekkaravarodome*, Adisak Nathakaranakule*, and Itsda Boonyaroonate**

*Division of Energy Technology, School of Energy, Environment and Materials
**Power Electronics & Motor Drives Laboratory (PEMD LAB)
Department of Electrical Engineering, Faculty of Engineering
King Mongkut's University of Technology Thonburi, Bangkok, Thailand 10140
Phone (66-2) 470-0039, E-mail: ekkaravarodome@hotmail.com

Abstract--A single-stage high-power-factor (HPF) electronic ballast with a Class-DE low dv/dt rectifier as a PFC has been proposed in this paper. The proposed is archived by using a bridge rectifier which acts as the Class-DE rectifier. By using this topology the conduction angle of the bridge rectifier diode current can be increased then the low line current harmonic was obtained. The Class-DE low dv/dt rectifier is driven by a high frequency sinusoidal current source, which an electronic ballast prototype operating 84 kHz has been implemented to drive 36-W fluorescent lamp. Experimental results verify the theoretical analysis. It is shown that the designed electronic ballast has 0.99 power factor, 1.3% total harmonic distortion, 1.32 lamp current crest factor and 90% efficiency at full power.

Index Terms--Electronic ballast, Power- factor correction, Class-DE low dv/dt rectifier, Class-D parallel resonant inverter.

I. INTRODUCTION

Basically, the ballast is needed for gas-discharge lamps such as fluorescent lamps because the lamp exhibits a negative dynamic resistance [1]-[2]. Therefore, they cannot be connected directly to an ac voltage source. The main functions of ballasts are to provide high striking voltage during starting and current limiting control for steady-state operation. An attractive solution of Electronic ballasts in terms of energy saving, such system reduces the consumption of electric energy compared to previous electromagnetic ballasts, as well as several other advantages. These include light weight, reduced volume, long lamp life, high-quality light due to elimination of flickering, high luminous efficacy of fluorescent lamps operating at high-frequencies, high power factor and low harmonics in utility line current [2] satisfy lighting equipment of the IEC 61000-3-2 Class-C standard. The electronic ballast for fluorescent lamps is developed by using the two-stage circuit, have been previously presented [2]-[3]. The main problem of the two-stage increases the component count and thus a high cost. To reduce cost by integrate the PFC stage with dc/ac resonant inverter into a single-stage have been proposed [4]-[7]. In this paper is to introduce a new topology of a single-stage fluorescent lamps electronic ballast using Class-DE low dv/dt rectifier for power-factor correction,

as well as to present its circuit description, principle of operation, design procedure, experimental results and conclusion.

II. CIRCUIT DESCRIPTION

Fig. 1 shows a circuit of the proposed single stage electronic ballast. It consists of an EMI filter $L_f - C_f$, a bridge rectifier $D_1 - D_4$, a high-frequency stored charge capacitor C_d connected in parallel with the diode D_3 of bridge rectifier, a Class-DE current driven low dv/dt rectifier [8] for input current shaping (ICS), a series-resonant circuit $L_p - C_p$ and a Class-D parallel resonant inverter $L_r - C_r - C_c - R_L$ [2] the power switches can exhibit either zero-voltage switching (ZVS) and bulk capacitor C_B. The series-resonant circuit $L_p - C_p$ is fed by a square-wave output voltage of the Class-D parallel resonant inverter and converts to a sinusoidal current source to drive Class-DE rectifier.

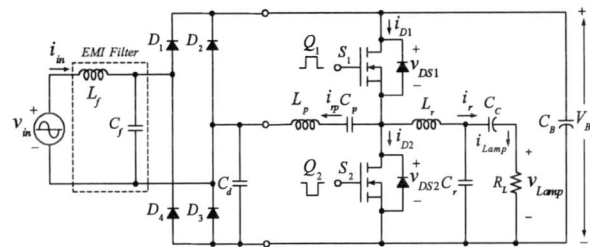

Fig. 1. Proposed electronic ballast with Class-DE low dv/dt rectifier as the ICS stage.

III. PRINCIPLE OF OPERATION

The principle of operation of the Class-DE low dv/dt rectifier in the ICS stage is explained by equivalent circuit show in Fig. 2(a). If $D_1 - D_4$ are fast recovery diode. The diode D_1 and D_3 in the diagonal of the bridge rectifier during the positive half of the cycle of the line voltage $v_{in} = V_{in} \sin \omega_L t$ and D_2 and D_4 during the negative half of the cycle. The model of the line-voltage rectifier output is a rectified full-wave sinusoidal voltage source $v_{REC} = |v_{in}| = V_{in} |\sin \omega_L t|$. The shape of the current waveform i_{rp} through the series-resonant circuit $L_p - C_p$ depends on the loaded quality factor Q_L of this circuit. If Q_L of the resonant circuit is high enough current waveform of the resonant circuit that drives the Class-DE rectifier is close to a sine wave, the D_1 and voltage source v_{REC} are connected in series can be move D_1 and voltage source v_{REC}, as show in Fig. 2(b). In this circuit the voltage source V_B and v_{REC} are connected in series and can be combined into one voltage source $V_B - v_{REC}$, as show in Fig. 2(c).

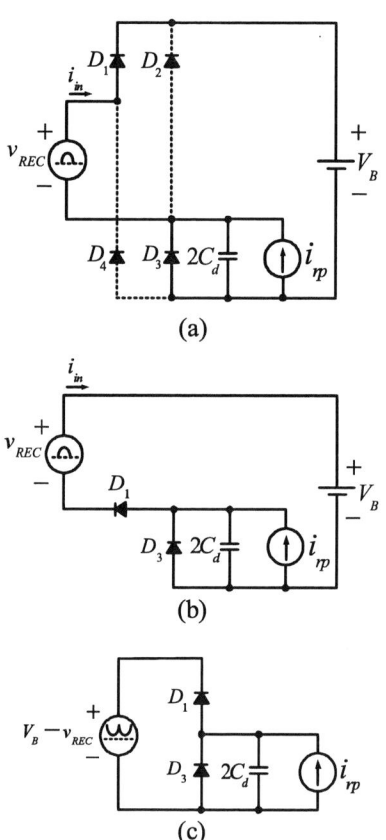

Fig. 2. Circuit derivation of the ICS with the Class-DE rectifier. (a) Equivalent circuit. (b) Equivalent circuit when move D_1 and voltage source v_{REC}. (c) Equivalent circuit with two voltages source V_B and v_{REC} combined into one voltage source $V_B - v_{REC}$.

Conceptual waveforms of the proposed ICS with the Class-DE rectifier are show in Fig. 3. Fig. 3(a) depicts a sinusoidal line-voltage waveform. Fig. 3(b) and (c) shows the rectified line voltage v_{REC} and the combined voltage waveform $V_B - v_{REC}$ respectively. If the instantaneous value of v_{in} is positive and low, the voltage $V_B - v_{REC}$ of the Class-DE rectifier is high, and the duty cycle D of the rectifier diode current is low. Therefore, the average value of the rectifier diode current over one switching cycle is low. Conversely, if the instantaneous value

of v_{in} is positive and high, the voltage $V_B - v_{REC}$ of the Class-DE rectifier is low, and the duty cycle D of the rectifier diode is high. Thus, the average value of the diode current over one switching cycle is high. For the half cycle with a negative line voltage, the bridge rectifier rectifiers the negative values of v_{in} to the positive values and causes the same effect on the diode duty cycle as the half cycle with the position line voltage. The conduction angle modulation of the rectifier diode over the line frequency f_L and the line-input current i_{in} are show in Fig. 3(d).

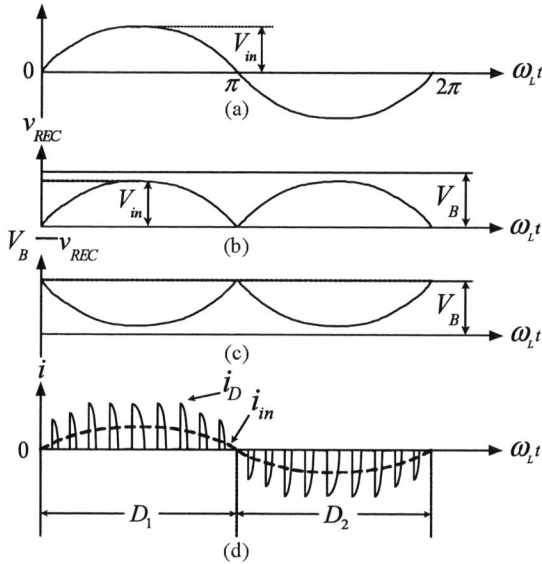

Fig. 3. Conceptual waveforms of proposed ICS. (a) Line voltage waveform v_{in}. (b) The rectified line voltage v_{REC}. (c) The combined Voltage $V_B - v_{REC}$. (d) Input current waveform i_{in} is the filtered average diode current and follows the shape of the line voltage v_{in}.

Fig.4 shows the rectifier output voltage as a function of load resistance. If the instantaneous value of rectifier output voltage V_o is low, the load resistance R_{DE} of the Class-DE rectifier is low. Therefore, the value of rectifier output current is high. Conversely, if the instantaneous value of rectifier output voltage V_o is high, the load resistance R_{DE} of the Class-DE rectifier is high. Thus, the value of rectifier output current is low.

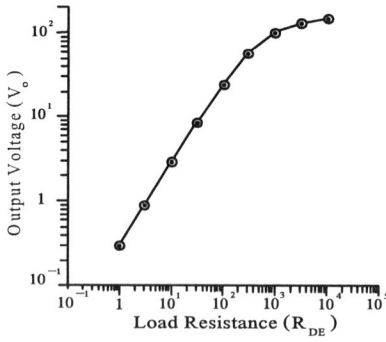

Fig. 4. Rectifier output voltage as a function of load resistance.

1384

Normalized effective input impedance of the Class-DE rectifier is show in Fig. 5 [8], the no-load condition at duty cycle $(D_d) = D_{min} = 0$ and the full-load condition at duty cycle $(D_d) = D_{max} = 0.5$, which the Normalized effective input impedance of Class-DE rectifier is $Z_i' = \omega_s 2C_d Z_i = R_i' + jX_i'$ and load resistance R_L of the Class-DE rectifier is $R_{DE} = V_o / I_o$.

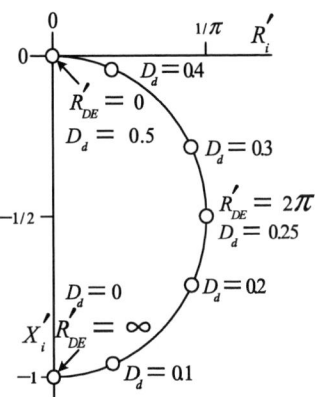

Fig. 5. Normalized effective input impedance of the Class-DE low dv/dt rectifier.

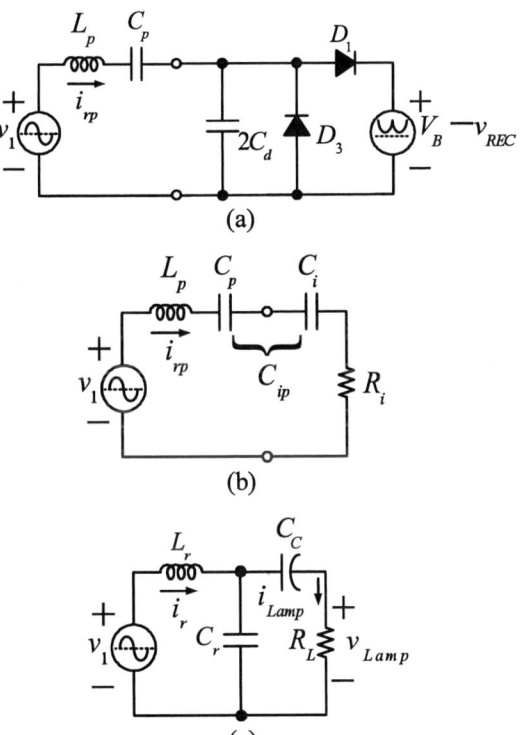

Fig. 6. Equivalent circuits of electronic ballast. (a) ICS Class-DE rectifier with equivalent sine-wave voltage source v_1. (b) Simplified equivalent circuit of (b). (c) Equivalent circuit of inverter semi-stage.

The proposed electronic ballast can be divided into two parts: an ICS semi-stage and an inverter semi-stage. Fig. 6(a) shows an equivalent circuit of the ICS semi-stage. Fig. 6(b) shows a simplified circuit of the ICS Class-DE rectifier. Fig. 6(c) shows an equivalent circuit of the inverter semi-stage.

IV. DESIGN PROCEDURE

To exemplify the design procedure of the proposed electronic ballast, the 36-W electronic ballast will be designed for a line rms voltage V_{irms} of 220V and the line frequency f_L of 50 Hz. Assume that the total ballast efficiency η is equal to 0.9 and the ballast draws a sin-wave input current. The input power is therefore obtained from $P_{in} = P_{out} / \eta \approx 40$ W. The amplitude of the ballast input current is calculated from $I_{in} = I_{omax} = \sqrt{2} P_{in} / V_{irms} \approx 0.25$ A. To design the ICS Class-DE rectifier, the no-load condition at duty cycle $(D_d) = D_{min} = 0$ and the full-load condition at duty cycle $(D_d) = D_{max}$ are considered. If a low value of D_{max} is used, a low THD of i_{in} is achieved, but the main switches have high voltage stresses. If a high value of D_{max} is chosen, the main switches have low voltage stresses, but a high THD of i_{in} occurs. The maximum duty cycle $D_{max} = D_d = (\pi - \phi)/2\pi = 0.4$ was selected because it givens a good compromise between a low distortion of i_{in} and a reasonable value of the switch voltage stress. The switching frequency f_s is 84 kHz, and, therefore, the conduction angle of diode is $\phi = 0.628$ rad/s, and high-frequency stored charge capacitor C_d is determined by [8]

$$C_d = \frac{R_{DE}'}{2\omega_s R_{DE}} \tag{1}$$

$$R_{DE} = \frac{V_B - V_{in}}{I_{omax}}, R_i' = \frac{8R_{DE}'}{(2\pi + R_{DE}')^2} \tag{2}$$

where R_{DE} is load of Class-DE rectifier, the value of C_d is obtained by solving (1) and (2), the resulting value is $C_d = 8.35$ nF. Therefore, a standard value of $C_d = 10$ nF is selected for C_d. The normalized effective input impedance of the Class-DE rectifier is given by [8]

$$Z_i' = \omega_s 2C_d Z_i = R_i' + jX_i' \tag{3}$$

$$R_i' = \frac{\sin^2 \phi}{\pi}, X_i' = \frac{\sin \phi \cos \phi - \phi}{\pi} \tag{4}$$

The value of input impedance of the Class-DE low dv/dt rectifier is obtained by solving (3) and (4). The resulting value is $Z_i = 11 + j4.8$. The amplitude of i_{rp} at full-load is determined by

$$I_{rpf} = \sqrt{\frac{2I_{omax}(V_B - V_{in})}{R_i}} \qquad (5)$$

The dc bus voltage $V_B = 330\,\text{V}$, therefore i_{rp} is $I_{rpf} = 0.94\,\text{A}$. The magnitude of the impedance of V_1 is given by

$$V_1 = I_{rpf}\,|Z_f| \qquad (6)$$

where the magnitude of the impedance of the series resonant circuit is

$$|Z_f| = \sqrt{R_i^2 + \left(\omega_s L_p - \frac{1}{\omega_s C_i}\right)^2} \qquad (7)$$

The magnitude of i_{rp} at the no-load condition is determined by

$$I_{rpn} = \frac{V_B / 2}{|-j / \omega_s 2C_d|} \qquad (8)$$

Therefore, $I_{rpn} = 1.74\,\text{A}$. and the magnitude of the equivalent voltage source V_1 is

$$V_1 = I_{rpn}\left| j\omega_s L_p - \frac{j}{\omega_s 2C_d}\right| \qquad (9)$$

For a finite value of capacitance C_p, an additional inductance L_e can be added to the inductance L_p to compensate for the reactance of the capacitance $C_p = 100\,\text{nF}$, the value of the additional inductance is given by

$$L_e = \frac{1}{\omega_s^2 C_p} \qquad (10)$$

Therefore, the total inductance is $L_p = 438.48\,\mu\text{H}$, for the bulk capacitor C_B is used for energy storage and its capacitance can be obtained by

$$C_B \geq \frac{P_{in}}{\omega_L V_B^2 V_{ripple}} \qquad (11)$$

For less than 1.5% ripple voltage, the value of the bulk capacitor is $C_B \geq 70.15\,\mu\text{F}$, Therefore a standard value of

$100\,\mu\text{F}$ is selected for C_B. The Class-D parallel resonant inverter is designed to drive a T8-36W fluorescent lamp by using the design procedure given in [2].

V. EXPERIMENTAL RESULTS

The measured input line voltage and current waveforms are shown in Fig. 7. The THD of input current was about 1.3%. The input power factor was 0.99 are shown in Fig. 8. Fig. 9 shows the experimental waveforms of the diode current and the diode voltage for the Class-DE rectifier near the peak and the zero crossing of the line voltage, respectively. The waveform of the switch voltage of and the switch current of the Class-DE low dv/dt rectifier are shown in Fig. 10.The waveform of the switch voltage of and the switch current of the Class-D parallel resonant inverter are shown in Fig. 11. The measured lamp voltage and current waveforms are shown in Fig. 12. Fig. 13 shows the measured waveform of the lamp current. The peak value of lamp-current envelope was 525 mA and the rms value of the lamp current was 398.45 mA. The lamp current crest factor was 1.32. The total measured efficiency of the ballast was 90%. The operating switching frequency was 84 kHz.

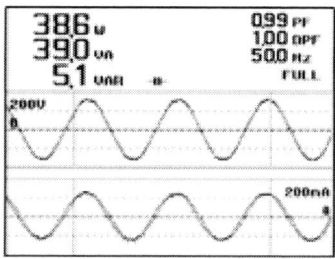

Fig. 7. The input line voltage and the current waveforms from the power analyzer

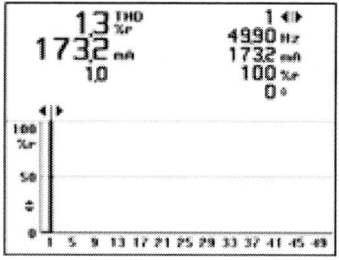

Fig. 8. The measured THD of i_{in} from the power analyzer

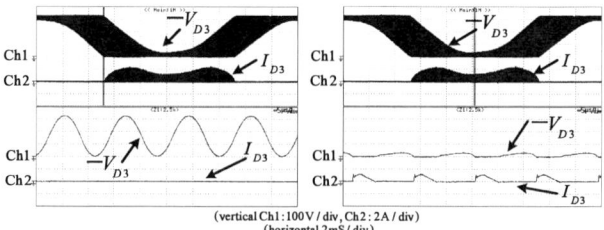

(vertical Ch1 : 100 V / div, Ch2 : 2A / div)
(horizontal 2mS / div)

Fig. 9. The measured waveforms of diode voltage and the diode current of D_3 for the Class-DE low dv/dt rectifier (a) Near the zero crossing of line voltage (b) Near the peak of line voltage

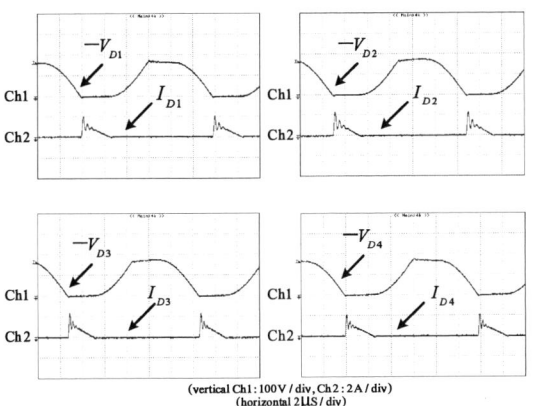

(vertical Ch1 : 100 V / div, Ch2 : 2 A / div)
(horizontal 2 μS / div)

Fig. 10.The measured waveforms of diode voltage and diode current of $D_1 - D_4$ for Class-DE rectifier

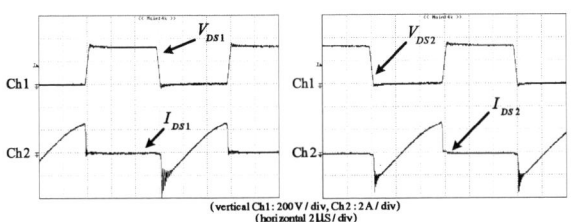

(vertical Ch1 : 200 V / div, Ch2 : 2 A / div)
(horizontal 2 μS / div)

Fig. 11. The measured waveforms of switch voltage and switch current of S_1, S_2 for Class-D parallel resonant inverter

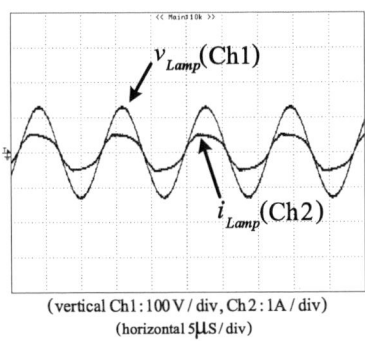

(vertical Ch1 : 100 V / div, Ch2 : 1A / div)
(horizontal 5 μS / div)

Fig. 12. The measured waveforms of lamp voltage and lamp current

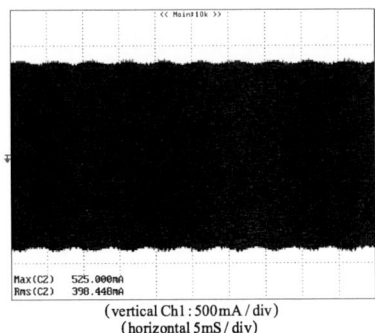

(vertical Ch1 : 500mA / div)
(horizontal 5mS / div)

Fig. 13. The experimental envelope waveform of lamp current

VI. CONCLUSION

A single-stage high-power-factor (HPF) electronic ballast with a Class-DE low dv/dt rectifier as a PFC has been proposed in this paper. The proposed is archived by using a bridge rectifier which acts as the Class-DE rectifier. By using this topology the conduction angle of the bridge rectifier diode current can be increased then the low line current harmonic was obtained and achieve nearly unity power factor. The Class-DE low dv/dt rectifier is driven by a high frequency sinusoidal current source, the power switches can exhibit either zero-voltage switching (ZVS), which an electronic ballast prototype operating 84 kHz has been implemented to drive T8-36W fluorescent lamp. Experimental results verify the theoretical analysis. It is shown that the designed electronic ballast has 0.99 power factor, 1.3% total harmonic distortion satisfy lighting equipment of the IEC 61000-3-2 Class-C standard, 1.32 lamp current crest factor and 90% efficiency at full power which is a very good result.

REFERENCES

[1] C. S. Moo *et al.*, "A fluorescent lamp model for high-frequency electronic ballasts," in *Proc. IAS'00.*,Vol. 5, pp. 3361 – 3366, Oct. 8-12, 2000.

[2] M.K. Kazimierczuk, and W. Szaraniec, "Electronic ballast for fluorescent lamps," *IEEE Trans. Power Electron.*, Vol. 8, pp. 386 – 395, Oct. 1993.

[3] E. Santi *et al.*, "High frequency electronic ballast provides line frequency lamp current," *IEEE Trans. Power Electron.*, Vol. 16, no. 5, pp. 667–675, Sep. 2003.

[4] Jinrong Qian, and F.C. Lee, "Charge pump power-factor-correction technologies Part II: Ballast applications," *IEEE Trans. Power Electron.*, Vol 15 , pp. 130 – 139, Jan. 2000.

[5] R. de Oliveira Brioschi, and J.L.F. Vieira, "High-power-factor electronic ballast with constant DC-link voltage," *IEEE Trans. Power Electron.*,Vol. 13, pp.1030 – 1037, Nov.1998.

[6] J. Calleja and J.M. Alonso *et al.*, "Design and experimental results of an input-current-shaper based electronic ballast," *IEEE Trans. Power Electron.*, Vol. 18, no. 2, pp. 547–557, Mar. 2003.

[7] K. Jirasereeamornkul and M.K. Kazimierczuk *et al.*, "Single-stage electronic ballast with Class-E rectifier as power-factor corrector," *IEEE Trans. Circuit and Systems.*, Vol. 53, no. 1, pp. 139–148, Jan. 2006.

[8] D.C. Hamill, "Class DE inverters and rectifiers for DC-DC conversion," *27th Annual IEEE Power Electronics Specialists Conference*, PESC '96 Record., Vol 1, pp. 854-860, June. 1996.

Output Impedance Design Consideration of Three Control Schemes for Bus Converter in On-Board Distributed Power System

Seiya Abe*, Masahiko Hirokawa**, and Tamotsu Ninomiya*

* Kyushu University Department of EESE, 744, Motooka, Nishi-ku, Fukuoka, 819-0395 , Japan
** TDK Corporation, 2-15-7, Higashi-Ohwada, Ichikawa, Chiba, 272-8558, Japan

Abstract-- **The power supply system which requires the low-voltage / high-current output has been changing from conventional centralized power system to distributed power system. The distributed power system consists of bus converter and POL. The most important factor is the system stability in bus architecture design. The overlap between the output impedance of bus converter and input impedance of POL causes system instability, and it has been an actual problem. Increasing the bus capacitor, system stability can be reduced easily. However, due to the limited space on the system board, increasing of bus capacitors is impractical. The urgent solution of the issue is desired strongly. This paper presents the output impedance design for on-board distributed power system by means of three control schemes of bus converter. The output impedance peak of the bus converter and the input impedance of the POL are analyzed, and it is conformed by experimentally for stability criterion. Furthermore, the design process of each control schemes for system stability is proposed.**

Index Terms—**Distributed power system, stability criterion, input and output impedance, bus converter design.**

I. INTRODUCTION

Various LSI is used in the telecommunication application equipments and the driving voltage is various. On the other hand, increase of load current is also remarkable by advanced function of LSI. Since the present LSI is designed in accordance with semiconductor manufacture technology, the tolerance level of operation voltage is very narrow. Consequently, the voltage drop by the wiring impedance of power line causes malfunction of LSI. In order to reduce the malfunction of LSI by the voltage drop, it is proposed that the converter is arranging very close to the LSI. This converter is called POL. Thus, the power supply system which requires the low-voltage / high-current output has been changing from conventional centralized power system to distributed power system. The distributed power system consists of first-stage isolated DC-DC converter as a bus converter and second-stage non-isolated DC-DC converter as a POL. However, the instability phenomenon in a distributed power system is posing a problem recently. This is instability phenomenon resulting from overlapping between the output impedance of bus converter and the input impedance of POL. Increasing the bus capacitor, system stability can be reduced easily. However, due to the limited space on the system board, increasing of bus

capacitors is impractical. The urgent solution of the issue is desired strongly, and the various discussion of system stability has been reported[1-8]. Then, we also have reported the detailed discussion of system stability by control schemes of bus converter (Un-regulated, Semi-regulated and Full-regulated)[9-14]. However, so far, the detailed discussion of practical design of bus converter about on-board distributed power system has not been reported. This paper presents the optimal design of bus converter for on-board distributed power system by means of three control schemes of bus converter.

II. STABILITY CRITERION

Figure 1 shows the distributed power system consisting of bus converter and POL. Even if each converter has stable operation, the instability phenomenon may occur by connecting two converters in series. Bus converter and POL have input-to-output voltage transfer function Gvb(s) and Gvp(s), respectively. The overall input-to-output voltage transfer function Gvv(s) is given following equation;

$$G_{vv}(s) = \frac{G_{vb}(s)G_{vp}(s)}{1 + Z_o(s)/Z_{in}(s)} \quad (1)$$

where Zo (s) is the output impedance of bus converter, and Zin(s) is the input impedance of POL. From Eq. (1), the input and output impedance is greatly concerned with the system stability. The stability of closed-loop system is decided with the characteristics equation $1+Z_o(s)/Z_{in}(s)$. This means relation between Zo(s) and Zin(s) decides the system stability. This system may become unstable when both impedances are overlapped, as shown in Fig. 2 (a). It is necessary to eliminate this impedance overlap for system stability. However, eliminating this impedance overlap for all frequency range is very difficult.

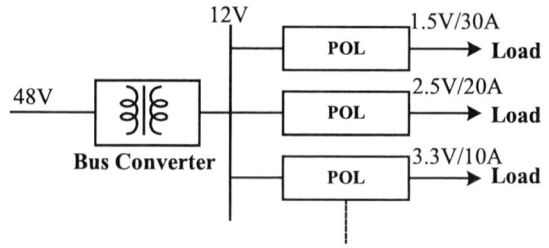

Fig. 1. Distributed power system.

On the other hand, it may have stable operation even if both impedances are overlapped. This is because the phase margin becomes large under the influence of the input impedance. When the bandwidth of POL is enough wider than the bandwidth of bus converter, if the peak value of output impedance becomes almost equal to the steady-state value of input impedance |Zin(0)| as shown in Fig. 2 (b), then this system becomes stability limit as shown in Fig. 2 (b). Moreover, this system becomes unstable if the peak value of output impedance exceeds |Zin(0)|. From mentioned above consideration, the new stability criterion can be defined as follows[14].

$$\begin{cases} \left| Z_{in}(0) \right| \geq Z_{o\_peak} & : \quad Stable \\ \left| Z_{in}(0) \right| < Z_{o\_peak} & : \quad Unstable \end{cases}$$

III. IMPEDANCE ANALISYS

The half-bridge converter with the most popular circuit of the power–stage is used as a bus converter, and the synchronous buck converter with the most popular circuit is used as POL. The output impedance of bus converter and the input impedance of POL can be derived by applying the stage space averaging method[15,16].

A. Input impedance

At first, the low-frequency value |Zin(0)| of input impedance is estimated. The input impedance of POL can be derived as following equation[17].

$$\frac{1}{Z_{in}(s)} = \frac{1}{Z_N(s)} \cdot \frac{T_p(s)}{1+T_p(s)} + \frac{1}{Z_D(s)} \cdot \frac{1}{1+T_p(s)} \quad (2)$$

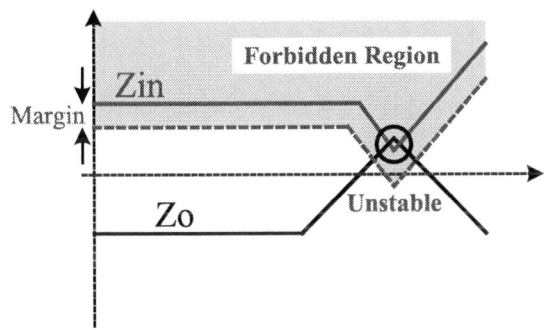

(a) General stability criterion

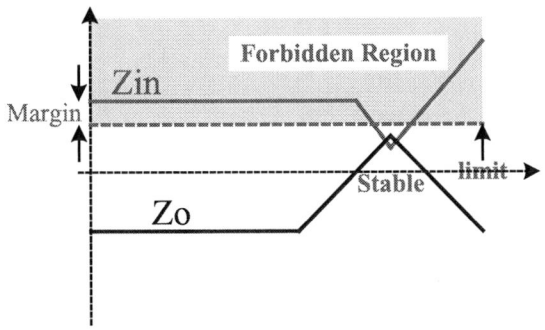

(b) New stability criterion
Fig. 2. Stability criterion.

From Eq. (2), the low-frequency value of input impedance |Zin(0)| is given by following equation.

$$\left| Z_{in}(0) \right|_{(dB\Omega)} \approx 20\log\left(\frac{R+r_L}{D^2}\right) \quad (dB\Omega) \quad (3)$$

|Zin(0)| has minimum value at rated load, so the estimation of |Zin(0)| must be at rated load. Next, the output impedance is examined.

B. Output impedance

The output impedance of bus converter can be derived as following equations.

Open loop

$$Z_o(s) = \frac{s^2 L_b C_b r_{c_b} + s\left(L_b + C_b r_{L_b} r_{c_b}\right) + r_{L_b}}{s^2 L_b C_b + s C_b\left(r_{L_b} + r_{c_b}\right) + 1} \quad (4)$$

Closed loop

$$Z_{oc}(s) = \frac{Z_o(s)}{1+T_b(s)} \quad (5)$$

where,

$$T_b(s) = k \cdot PWM \cdot G_{dv_b}(s) \quad (6)$$

$$G_{dv_b}(s) = \frac{V_s}{P_b(s)}\left(sC_b r_{c_b} + 1\right) \quad (7)$$

$$P_b(s) = s^2 L_b C_b + s C_b\left(r_{L_b} + r_{c_b}\right) + 1 \quad (8)$$

k : sense gain products error amp. gain,
PWM : gain of the comparator.

In open loop case, the peak frequency is the same resonant frequency fp of the loop gain T(s) as shown in Fig. 3, and the peak value of the output impedance can be derived from Eq. (4).

$$Z_{o\_peak} = \frac{L_b}{C_b\left(r_{L_b} + r_{c_b}\right)} \quad (9)$$

In closed loop case, the output impedance peak moves to crossover frequency fc as shown in Fig. 3. In this instant the peak value of the closed loop output impedance can be derive following equation.

$$Z_{o\_peak} = \frac{L_b}{C_b\left\{(1+\alpha)r_{c_b} + r_{L_b}\right\}} \quad (10)$$

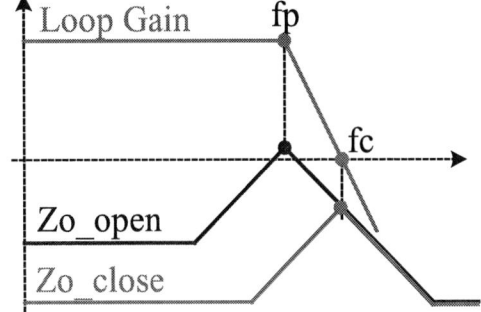

Fig. 3. Output impedance peak.

1389

whrer,

$$\alpha = |T(0)| = k \cdot PWM \cdot V_{in} \quad (11)$$

Moreover, from transfer function of loop gain, the crossover frequency fc is expressed as follows by means of peak frequency fp of loop gain.

$$f_c = \sqrt{1+\alpha} f_p \quad (12)$$

From Eq. (10) (12), the peak value of closed loop output impedance is expressed as follows.

$$Z_{oc\_peak} = \frac{L_b}{C_b \left\{ \left(\dfrac{f_c}{f_p}\right)^2 r_{C_b} + r_{L_b} \right\}} \quad (13)$$

As shown in Eq. (13), if fc is set to 1, it becomes the same as Eq(9). Therefore, the peak value of output impedance is calculable by means of Eq. (13).

IV. OUTPUT IMPEDANCE SPECIFICATION

The output impedance characteristic of each control shames is different, and each bus converter has different operation. Therefore, the output impedance design suitable for the feature of each control method is required. From now, the output impedance design for each control shames is considered.

A. Un-regulated

In un-regulated case, the output impedance is the same as open-loop output impedance because of this control method has no control loop.

In order to reduce the peak value of output impedance, it is effective to make inductance small or to enlarge capacitance. Generally, un-regulated bus converter is operated at maximum duty ratio. Therefore, the inductor of the bus converter can be reduced as small as possible to reduce the system instability. The peak value of output impedance is reducing with small inductor. Figure 4 shows the experimental result of the relation between the output impedance and inductance. Moreover, Fig. 5 shows the analytical and experimental results of the relation between the peak value of output impedance and inductance. Both results agreed well. As mentioned above, the peak of impedance is easily obtained from Eq. (13). However, this method depends on converter topology that has a double-ended circuit at secondary side such as half-bridge or full-bridge. Moreover, there are some limits such that high accurate input voltage or a POL with wide input range.

B. Semi-regulated

Semi-regulated bus converter has a control loop. However, regulation is related to variation of input voltage, therefore the output impedance is same as un-regulated case. In this case, the duty ration is changed, and the inductor of the bus converter cannot be reduced. Therefore, very large bus capacitor is needed to reduce the peak value of output impedance.

Figure 6 shows the experimental result of the relation between the output impedance and capacitance. Moreover, Fig. 7 shows the analytical and experimental results of the relation between the peak value of output impedance and capacitance. Both results agreed well.

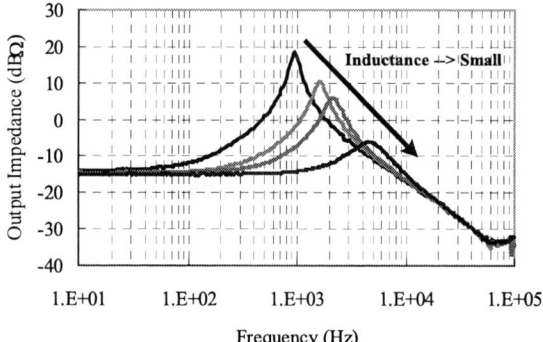

Fig. 4. Inductance and output impedance.

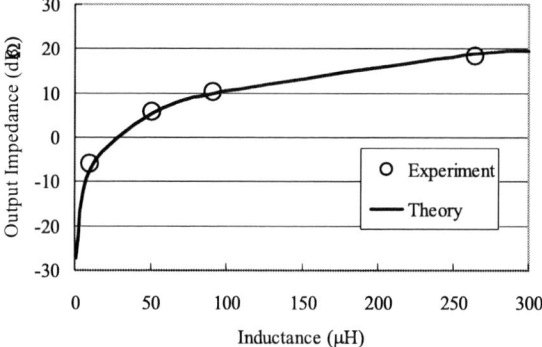

Fig. 5. Inductance and peak value of Zo.

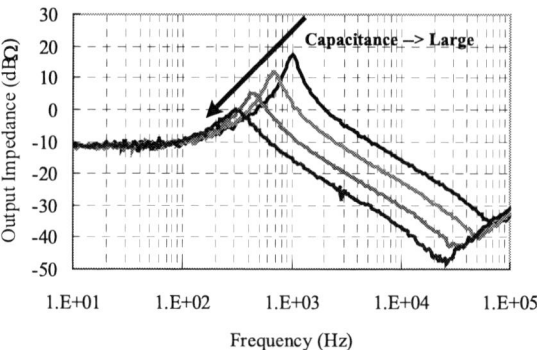

Fig. 6. Capacitance and output impedance.

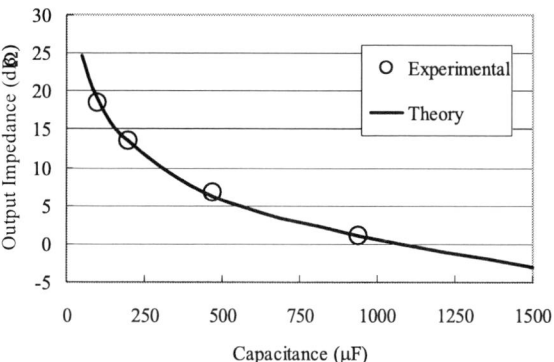

Fig. 7. Capacitance and peak value of Zo.

In semi-regulated case, essentially it becomes very unstable and we have found that the demerit is very large capacitors are needed at the intermediate bus in order to be stable. However, it can be used at limited conditions such as wide input range (36-75V) and POL with low power (in other words, POL with very high input impedance).

C. Full-regulated

Full-regulated bus converter has a feedback loop, so the output impedance characteristic is changed. Therefore, output impedance can be made small with wide bandwidth. Figure 8 shows the experimental result of the relation between the output impedance and bandwidth. Moreover, Fig. 9 shows the analytical and experimental results of the relation between the peak value of output impedance and bandwidth. Both results agreed well.

V. DESIGN EXAMPLE

In order to evaluate the performance of this system, the experiment circuits are implemented using the specifications and parameters in Table 1. Here, the case with two POLs is discussed for actual example. The practical design process is shown below.

A. Input impedance estimation

The low-frequency value $|Zin(0)|$ of input impedance is given by Eq. (3).

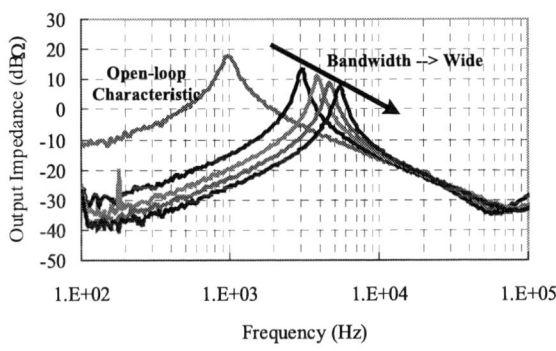

Fig. 8. Bandwidth and output impedance.

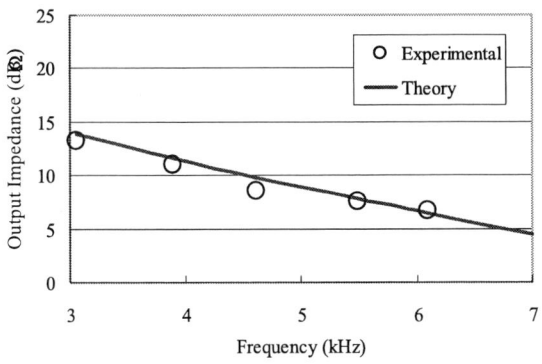

Fig. 9. Bandwidth and peak value of Zo

The duty ratio is D=0.275 and output resistance is R=0.66(Ω) from the relation input and output. In this case, the $|Zin(0)|$ is 20.6(dBΩ). When two POLs of same condition are connecting in parallel, $|Zin(0)|$ is 14.6(dBΩ). Figure 10 shows the experimental result of input impedance. The low-frequency value $|Zin(0)|$ is around 15(dBΩ) as shown in Fig. 10.

The experimental results and analytical results are agreed well. If the stability margin is set to 6(dBΩ), then the peak value of output impedance must be set to 8.6(dBΩ).

B. Output impedance design

Figure 11 shows the output impedance characteristic of the basic case using the parameter of Table 1. As shown in Fig. 11, the peak value of the output impedance is around 18(dBΩ). From mentioned above calculation, the peak value of the output impedance needs to set around 8.6(dBΩ) for sufficient system stability.

In un-regulated case, the optimal inductance value is considered because the stability is improved by small inductance. From Eq. (13), the optimal inductance value can be derived as following equation.

TABLE I
CIRCUIT PARAMETERS.

	Symbol	Description	Value
Bus Converter	Vin	Input Volotage	48V
	Vb	Bus Volotage	12V
	Lb	Output Inductor of Bus Converter	270μH
	Cb	Output Capacitor of Bus Converter	100μF
	rlb	Registance of Lb	300mΩ
	rcb	ESR of Cb	25mΩ
POL	Vo/Io	Output Condition	3.3V/5A
	Lo	Output inductor	2.8μH
	Co	Output capacitor	820μF
	rl	Registance of Lo	25mΩ
	rc	ESR of Co	10mΩ

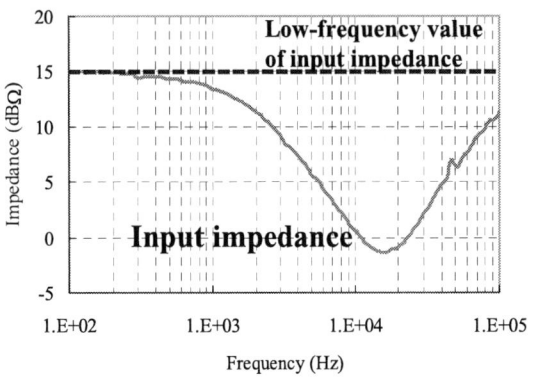

Fig. 10. Input impedance characteristic.

$$L_{b\_optimal} = C_b \left(r_{C_b} + r_{L_b} \right) Z_{o\_peak} \quad (14)$$

where, the unit of |Zo_peak| is Ω.

Since the output impedance must be set to 8.6(dBΩ), the inductance value is set to around 87(μH) from Eq. (14). Figure 12 shows the experimental result of the output impedance with small inductance. The inductance value is around 90 (μH), and the peak value of output impedance is around 8.5 (dBΩ). The experimental results and analytical results are agreed well. Moreover, in open loop case, since the rL is generally larger than rc, the output impedance does not become smaller than rL as shown in Fig. 4. Therefore, the inductance value has minimum value. From Eq. (14), the minimum value of the inductance is given by following equation.

$$L_{b\_min} = C_b \left(r_{C_b} + r_{L_b} \right) r_{L_b} \quad (15)$$

In this case, the minimum value of inductance is around 10 (μH).

In semi-regulated case, the optimal capacitance value is considered because the stability is improved by large capacitance. From Eq. (13), the optimal capacitance value can be derived as following equation.

$$C_{b\_optimal} = \frac{L_b}{\left(r_{C_b} + r_{L_b} \right) Z_{o\_peak}} \quad (16)$$

where, the unit of |Zo_peak| is Ω.

Since the output impedance must be set to 8.6(dBΩ), the capacitance value is set to around 300(μF) from Eq. (16). In this case, the influence of ESR is considered. Because the ESR becomes small when the capacitor is connected in parallel.

Figure 13 shows the experimental result of output impedance with large capacitance. The capacitance value is 300 (μF), and the peak value of output impedance is around 8 (dBΩ). The experimental results and analytical results are agreed well. Moreover, the output impedance does not become smaller than rL as shown in Fig. 6. Therefore, the capacitance value has maximum value. From Eq. (16), the maximum value of the capacitance is given by following equation.

$$C_{b\_max} = \frac{L_b}{\left(r_{C_b} + r_{L_b} \right) r_{L_b}} \quad (17)$$

In this case, the maximum value of capacitance is around 2.8 (mF).

In full-regulated case, the optimal bandwidth is considered because the stability is improved by wide bandwidth. From Eq. (13), the optimal bandwidth can be derived as following equation.

$$f_{c\_optimal} = f_p \sqrt{\frac{\dfrac{L_b}{C_b Z_{oc\_peak}} - r_{L_b}}{r_{C_b}}} \quad (18)$$

where, the unit of |Zo_peak| is Ω.

Since the output impedance must be set to 8.6(dBΩ), the bandwidth is set to around 5.1kHz from Eq. (18). Figure 14 shows the experimental result of the output impedance with wide bandwidth. The bandwidth is around4.7kHz, and the peak value of output impedance is around 8.5 (dBΩ). The experimental results and analytical results are agreed well.

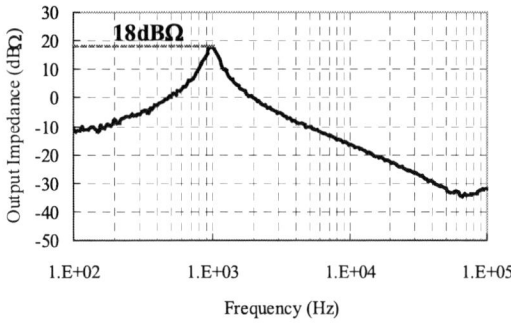

Fig. 11. Output impedance (Basic parameters).

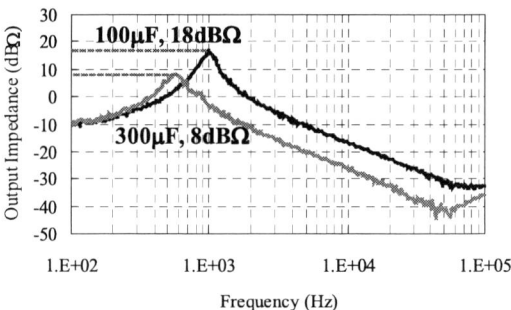

Fig. 13. Output impedance with large capacitor.

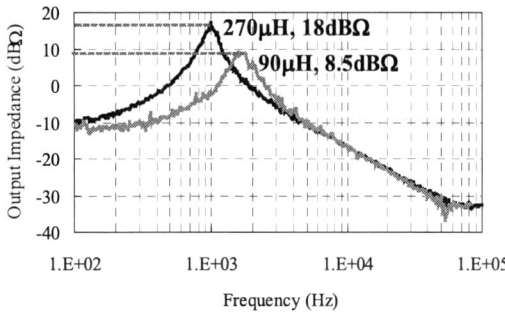

Fig. 12. Output impedance with small inductor.

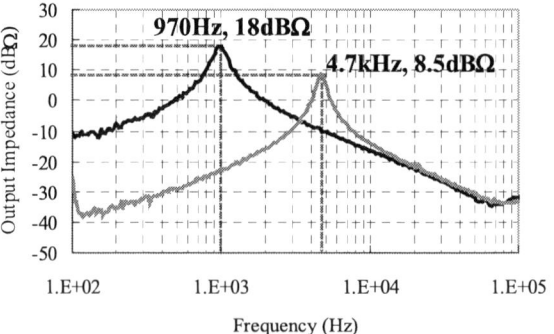

Fig. 14. Output impedance with wide bandwidth.

VI. CONCLUSIONS

This paper presents the output impedance design for on-board distributed power system by means of three control methods of bus converter. The output impedance peak of the bus converter and the input impedance of the POL were analyzed, and it was conformed by experimentally for stability criterion. As a result, the standard of the discrimination of stability on a frequency response of input and output impedance was clarified. Furthermore, the design process of each control method for system stability was proposed.

REFERENCES

[1] C. M. Wildrick, F. C. Lee, B. H. Cho, B. Choi, "A Method of Defining the Load Impedance Specification for A Stable Distributed Power System", IEEE Transactions on Power Electronics Vol. 10. No. 3. May 1995, pp. 280-284.

[2] X. Feng, Z. Ye, K. Xing, F. C. Lee, D. Borojevic, "Individual Load Impedance Specification for a Stable DC Distributed Power System", IEEE Applied Power Electronics Conference (APEC) 1999, pp. 923-929.

[3] X. Feng, F. C. Lee, "On-line Measurement on Stability Margin of DC Distributed Power System", IEEE Applied Power Electronics Conference (APEC) 2000, pp. 1190-1196.

[4] M. P. Sayani, J. Wanes, " Analyzing and Determining Optimum On-Board Power Architectures for 48V-input Systems", IEEE Applied Power Electronics Conference (APEC) 2003.

[5] K. Hisanaga, K. Harada, "Stability Analysis of the Distributed Power System with Intermediate Bus Converter", IEICE Technical Report, Vol.103, No.199, pp.19-24, Jul. 2003 (in Japanese).

[6] K. Hisanaga, K. Harada, "Stability Analysis of the Distributed Power System with Intermediate Bus Converter (2nd Report)", IEICE Technical Report, Vol.103, No.652, pp.7-12, Feb. 2004 (in Japanese).

[7] Y. Ren, M. Xu, K. Yao, Y. Meng, F. C. Lee, J. Guo, "Two-Stage Approach for 12V VR", IEEE Applied Power Electronics Conference (APEC) 2004.

[8] J. Wei, F. C. Lee, "An Output Impedance-Based Design of Voltage Regulator Output Capacitors for High Slew-Rate Load Current Transients", IEEE Applied Power Electronics Conference (APEC) 2004.

[9] S. Abe, M. Hirokawa, T. Zaitsu, T. Ninomiya, "Stability Design of Bus Converter Following by POLs in Distributed Power System", IASTED Circuits, Signals,and Systems (CSS), 2005, pp552-557.

[10] S. Abe, H. Nakagawa, M. Hirokawa, T. Zaitsu , T. Ninomiya, "Comparison of System Stability in Distributed Power System Based on Control Method of Bus Converter", IASTED Energy and Power Systems (EPS) 2005, pp109-114.

[11] S. Abe, H. Nakagawa, M. Hirokawa, T. Zaitsu , T. Ninomiya, "System Stability of Full-Regulated Bus Converter in Distributed Power System", International Telecommunications Energy Conference (INTELEC) 2005, pp563-568.

[12] S. Abe, H. Nakagawa, M. Hirokawa, T. Zaitsu , T. Ninomiya, " Stability Improvement of Distributed Power System by Using Full-Regulated Bus Converter ", Annual Conference of the IEEE Industrial Electronics Society (IECON) 2005, pp2549-2553.

[13] S. Abe, T. Ninomiya, M. Hirokawa, T. Zaitsu , "Stability Comparison of Three Control Schemes for Bus Converter in Distributed Power System ", International Conference on Power Electronics and Drive Systems (PEDS) 2005, pp1244-1249.

[14] S. Abe, M. Hirokawa, T. Zaitsu , T. Ninomiya, " Stability Design Consideration for On-Board Distributed Power System Consisting of Full-Regulated Bus Converter and POLs ", IEEE Power Electronics Specialists Conference (PESC) 2006, pp2669-2673.

[15] R.D. Middlebrook, S. Cuk, "A General Unified Approach to Modeling Switching-Converter Power Stages," IEEE Power Electronics Specialists Conference (PESC) 1976, pp. 18-34.

[16] T. Ninomiya, M. Nakahara, T. Higashi, K. Harada, "A Unified Analysis of Resonant Converters," IEEE Transactions on Power Electronics Vol. 6. No. 2. April 1991, pp. 260-270.

[17] R. D. Middlebrook, "Input Filter Considerations in Design and Application of Switching Regulators", IAS'76, 1976, pp. 91-107.

Optimal Generation Rescheduling for Security Operation of Power Systems Using Optimal Control Theory

J. Q. Sun *, K. W. Chan *, and D. Z. Fang **

* Department of Electrical Engineering, The Hong Kong Polytechnic University, Hong Kong SAR
** School of Electrical Engineering and Automation, Tianjin University, Tianjin 300072, P. R. China

Abstract--An approach of optimal generation rescheduling for security operation of power systems is proposed with a heuristic constraint of transient stability by setting the maximum generator angle difference less than a threshold value in a designated post-fault duration. With this constraint, a concept called stability performance index is introduced in the approach for developing optimal control model. Rules for evaluating the stability performance index are presented. A technique of inverse integration is utilized to evaluate the rescheduling actions by using the gradient of the index. This approach has the advantage that it is applicable to systems represented by complex detailed models. Case studies on the 10-machine New England test power system are given to show the effectiveness and efficiency of the approach. The results suggest that this approach is able to ensure the system operation security in avoiding any potential transient instability.

Index Terms--transient stability, preventive control, optimal control, trajectory sensitivity.

I. INTRODUCTION

Generation rescheduling has been carried out in the industry mostly by trial and error. That is to say, it is based mainly on the engineer's experiences for the last few decades. In recent years some progress has been made in generation rescheduling for dynamic security dispatch [1-5]. Most of these methods for determining remedial action as suggested in the literatures use first-order sensitivity of some energy function to derive the rescheduling. They are not only able to determine the stability or instability of the system, but also provide an index of the stability or instability of the system. In [1], sensitivities to certain parameters are built into the system equations before applying the BCU method. In [2-4], the sensitivities of the energy margin to system parameters are used to determine new schedules of stability limits by using some optimization method. The extended equal area criterion is used in [5] to develop another sensitivity method. In some of these methods the controlling unstable equilibrium point (UEP) is assumed not to change after rescheduling. This assumption is quite limiting because significant rescheduling is bound to change the controlling UEP, as shown in [1]. At the same time, the main drawback of using the energy margin as

the security measure is in the computation required for the calculation of the energy margin itself which is used to identify those severe cases needing remedial actions. The bulk of the computation needed is in determining the controlling UEP [1,6].

Time domain simulation is still the main method for analysis transient stability in modern power systems, especially when power systems are represented by detailed model. With the development of computer technology, it is not difficulty to construct mathematical programming or integration algorithm for evaluating the disturbance trajectory, but how to derive the quantitative information reflected stability degree is the hot topic around the world. In [7-13], the power system transient stability model is converted into algebraic set of equations for each time step of the simulation. The approach reported in [14-17] is based on trajectory sensitivity in which the gradient of stability constraints with respect to control parameters is used to indicate generation rescheduling. The initial conditions for trajectory sensitivity are set to be zeros or ones.

In this paper, we propose a new method for generation rescheduling by focusing on the development of a more systematic generation rescheduling methodology. Combined with disturbance trajectory derived by time domain simulation, nonlinear programming techniques are constructed for preventive control. Stability constraints of control model used are formed by setting generator angle to be less than a threshold value at fixed terminal time. By using principles of optimal control, the preventive control model based on differential- algebraic formulation of power systems is converted to an optimal control problem, in which the relative rotor angles of generators are controlled to be less than a threshold value through generation rescheduling when some severe contingency occurs. Case studies using the 10-machine New England test power system are given to show effectiveness of the method. Accuracy of the results suggests that this method will be potentially useful in avoiding transient instability. Hence, the technique can be used in an online preventive control scheme for secure operations of system. At the same time, the initial conditions of conjugate equation are functional of rotor angle, which has specific physical meaning.

The authors gratefully acknowledge the support of the Hong Kong Polytechnic University, the Research Grants Council of the Hong Kong Special Administrative Region (RGC No: PolyU 5200/03E).

II. OPTIMAL CONTROL

For formulating the dynamic optimization problem, let us suppose that we have a general non-linear dynamical system described by the state space equations:

$$\begin{cases} \dot{x}(t) = f_1(x,y,u) \\ 0 = g_1(x,y,u) \end{cases} \quad t \in [t_0, t_{cl}] \quad (1)$$

$$\begin{cases} \dot{x}(t) = f_2(x,y,u) \\ 0 = g_2(x,y,u) \end{cases} \quad t \in [t_{cl}, t_f] \quad (2)$$

where x and y are the state variables and the algebraic variables of power system, u is control variable, t_{cl} is the fault clearing time, t_f is the study period. The initial condition for state $x(t_0)$ is assumed known.

Equation (1) could be used to describe many physical systems. Essentially our control problem is to choose the control trajectories u ($t_0 \le t \le t_f$), so as to ensure that the system described the equation (1) has a "desirable" dynamic behavior. For example, if the system is operating normally in the steady state when it receives a disturbance which changes its state to some known (or measurable) state x_0, the 'desirable' dynamic behavior could be represented by the control vector u which minimizes some functions of the states and controls all along the trajectory; a suitable way, thus, of choosing u may be to find the which minimizes the cost function.

$$J[u(\cdot)] = \phi(x(t_f), t_f) < 0 \quad (3)$$

Lagrange multiplier function vectors $\lambda(t)$ and $\xi(t)$, are introduced to construct a new function (4).

$$\tilde{J}(x(t_f), u) = \phi(x(t_f))$$
$$+ \int_{t_0}^{t_{cl}} [\lambda^T (f_1(x,y,u) - \dot{x}) + \xi^T g_1(x,y,u)] dt \quad (4)$$
$$+ \int_{t_{cl}}^{t_f} [\lambda^T (f_2(x,y,u) - \dot{x}) + \xi^T g_2(x,y,u)] dt$$

From equation (1) and (2), it is easy to show that:

$$\tilde{J}(x(t_f), u) = J(x(t_f), u) \quad (5)$$

With the formula of integration by parts it can be derived that:

$$\int_{t_0}^{t_{cl}} \lambda^T \dot{x} dt = \int_{t_0}^{t_{cl}} \dot{\lambda}^T(t) x(t) dt + \lambda^T(t_{cl}) x(t_{cl}) - \lambda^T(t_0) x(t_0) \quad (6)$$

and

$$\int_{t_{cl}}^{t_f} \lambda^T \dot{x} dt = \int_{t_{cl}}^{t_f} \dot{\lambda}^T(t) x(t) dt + \lambda^T(t_f) x(t_f) - \lambda^T(t_{cl}) x(t_{cl}) \quad (7)$$

Substituting (6) and (7) into (4)

$$\tilde{J}(x(t_f), u) = \phi(x(t_f)) + \lambda^T(t_0) x(t_0) - \lambda^T(t_f) x(t_f)$$
$$+ \int_{t_0}^{t_{cl}} [\lambda^T f_1(x,y,u) + \dot{\lambda}^T(t) x(t) + \xi^T g_1(x,y,u)] dt \quad (8)$$
$$+ \int_{t_{cl}}^{t_f} [\lambda^T f_2(x,y,u) + \dot{\lambda}^T(t) x(t) + \xi^T g_2(x,y,u)] dt$$

In mathematics, (9) can be derived by taking variation of both sides of (8) in considering $\delta\lambda(t) = 0$ and $\delta\xi(t) = 0$.

$$\delta\tilde{J}(x(t_f), u) = (\frac{\partial\phi}{\partial x} - \lambda)^T \delta x(t_f) + \lambda^T(t_0) \delta x(t_0)$$
$$+ \int_{t_0}^{t_{cl}} (\lambda^T \frac{\partial f_1}{\partial x} + \xi^T \frac{\partial g_1}{\partial x} + \dot{\lambda}^T) \delta x dt$$
$$+ \int_{t_0}^{t_{cl}} (\lambda^T \frac{\partial f_1}{\partial y} + \xi^T \frac{\partial g_1}{\partial y}) \delta y dt + \int_{t_0}^{t_{cl}} (\lambda^T \frac{\partial f_1}{\partial u} + \xi^T \frac{\partial g_1}{\partial u}) \delta u dt$$
$$+ \int_{t_{cl}}^{t_f} (\lambda^T \frac{\partial f_2}{\partial x} + \xi^T \frac{\partial g_2}{\partial x} + \dot{\lambda}^T) \delta x dt$$
$$+ \int_{t_{cl}}^{t_f} (\lambda^T \frac{\partial f_2}{\partial y} + \xi^T \frac{\partial g_2}{\partial y}) \delta y dt + \int_{t_{cl}}^{t_f} (\lambda^T \frac{\partial f_2}{\partial u} + \xi^T \frac{\partial g_2}{\partial u}) \delta u dt \quad (9)$$

It is obvious that (9) is hold for arbitrary multiplier function vectors $\lambda(t)$ and $\xi(t)$. The basic idea is how to choose an system about $\lambda(t)$ and $\xi(t)$ to make the right hand side of (9) be independent of variations $\delta x(t)$, $\delta y(t)$ and $\delta x(t_f)$. To achieve above objective adjoint systems (10) and (11) are adopt.

$$\begin{cases} \dot{\lambda}^T = -\lambda^T \frac{\partial f_2}{\partial x} - \xi^T \frac{\partial g_2}{\partial x} \\ 0 = \lambda^T \frac{\partial f_2}{\partial y} + \xi^T \frac{\partial g_2}{\partial y} \quad t \in [t_{cl}, t_f] \quad (10) \\ \lambda(t_f) = \frac{\partial\phi(x(t_f), t_f)}{\partial x(t_f)} \end{cases}$$

$$\begin{cases} \dot{\lambda}_1^T = -\lambda^T \frac{\partial f_1}{\partial x} - \xi^T \frac{\partial g_1}{\partial x} \\ 0 = \lambda^T \frac{\partial f_1}{\partial y} + \xi^T \frac{\partial g_1}{\partial y} \end{cases} \quad t \in [t_0, t_{cl}] \quad (11)$$

Thus, (9) becomes:

$$\delta\tilde{J}(x(t_f), u) = \lambda^T(t_0) \frac{\partial x(t_0)}{\partial u} \delta u +$$
$$\int_{t_0}^{t_{cl}} (\lambda^T \frac{\partial f_1}{\partial u} + \xi^T \frac{\partial g_1}{\partial u}) \delta u dt + \int_{t_{cl}}^{t_f} (\lambda^T \frac{\partial f_2}{\partial u} + \xi^T \frac{\partial g_2}{\partial u}) \delta u dt \quad (12)$$

Therefore

$$\frac{\partial J(x(t_f), u)}{\partial u} = \lambda^T(t_0) \frac{\partial x(t_0)}{\partial u} +$$
$$\int_{t_0}^{t_{cl}} (\lambda^T \frac{\partial f_1}{\partial u} + \xi^T \frac{\partial g_1}{\partial u}) dt + \int_{t_{cl}}^{t_f} (\lambda^T \frac{\partial f_2}{\partial u} + \xi^T \frac{\partial g_2}{\partial u}) dt \quad (13)$$

III. PRINCIPLE FOR ALGORITHM

Operation based transient stability enhancements is conducted at the system control center and would include such measure of power flow control that include both preventive action before a possible contingency and corrective control as an after the fact action [18]. In the case of dynamic instability there is not much of a corrective nature that can be done under practical operation conditions, and therefore, here we focus on the preventive generation rescheduling in normal system operations instead.

For systems represented by (1) and (2), the optimal generation rescheduling model to maintain the secure operation of the system in the sense of transient stability can be formulated as a nonlinear programming problem as shown in (14).

$$\min \sum_{i=1}^{n} |\Delta P_{gi}|$$
$$s.t. \quad J(x(t_f), u) < 0$$
$$\sum_{i=1}^{n} \Delta P_{gi} = 0 \qquad (14)$$
$$P_{gi}^{\min} - P_{gi} \le \Delta P_{gi} \le P_{gi}^{\max} - P_{gi}$$

Assuming that system operation before stability control is desirable with respect to generation schedule, the objective function (14) is to find a secure operating point when the changes in generation are minimized for the given operating condition and contingencies. Similar objective functions are also used in [2,10,19-21]. In the programming model P_{gi} is taken as control vector u, and ΔP_{gi}, $P_{gi}^{\min}$ and $P_{gi}^{\max}$ are for power output change and generation limits of generator i.

IV. ALGORITHM FOR GENERATION RESCHEDULING

Assumed in the system operation, a contingency c_l is detected as harmful, i.e. constraint (3) is violated for c_l. The generation rescheduling approach is to evaluate control actions to make c_l become harmless with control model (14) developed in the previous section. For this approach, several assumptions have been made as follows:

1) The initial operation condition is given.
2) Loads remain constant in evaluation of the control actions.

Before conducting the following rescheduling algorithm, a tolerance parameter ε should be set.

Generation rescheduling algorithm:

Step 1: $k \leftarrow 0$.

Step 2: Perform load flow analysis.

Step 3: Perform trajectory simulation and then estimate $J(x(t_f), u)$ for contingency c_l.

Step 4: If $J(x(t_f), u) \ge 0$, calculate $\partial J_l(x(t_f), u)/\partial u$ according to the procedure shown in section II, solve (14) for control action ΔP_g^k and do $P_g^{k+1} \leftarrow P_g^k + \Delta P_g^k$; $k \leftarrow k+1$; $\alpha \leftarrow 1$; go to *Step 2*.

Step 5: If $J(x(t_f), u) < 0$ and $\sum |\Delta P_{gi}^{k-1}| < \varepsilon$, goto *Step 7*.

Step 6: $\alpha \leftarrow \alpha/2$; $\Delta P_g^{k-1} \leftarrow \alpha \Delta P_g^{k-1}$; $P_g^k = P_g^{k-1} + \Delta P_g^{k-1}$, goto *Step 2*.

Step 7: $P_g^* \leftarrow P_g^k$ and stop.

V. CASE STUDY

The proposed method has been tested on the New England 10-machine, 39-bus system as shown in Fig.1 [22]. Generator 1 is used as the reference generator. All synchronous machines are represented by classical model whilst all loads are represented by composite model of 40% constant impedance and 60% constant power. In the time domain simulation, the time step is 0.02s and the maximum integration time interval is 4s. 180° is selected as stability threshold which could be changed on demand for different requirements or systems. The fixed tolerance ε for optimal control is set to 0.1MW. The lower and upper generation limits for each generator are given in Table I.

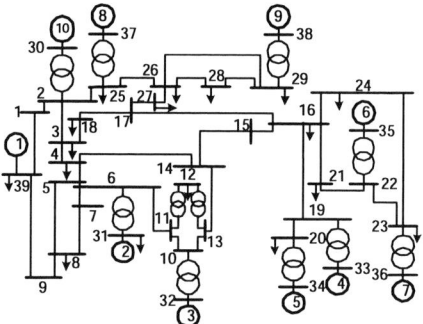

Fig.1 10-generator New England test power system

TABLE I
GENERATOR BASE OUTPUT AND LIMITS

G	Base (MW)	Upper (MW)	Lower (MW)
1	1014.60	1200	600
2	601.97	650	200
3	684.20	800	300
4	608.23	750	300
5	493.85	650	250
6	628.93	750	300
7	537.87	750	250
8	542.47	700	250
9	830.00	900	400
10	248.10	350	100

Case A: *Three-phase-to-ground-fault at bus 22:* The fault is simulated at bus 22 and cleared by tripping line 22-21 at $t_{cl} = 0.12$ s, which is greater than the critical clearing time for the base case. Appling the algorithm of section IV, the new schedule for generators is shown in Table II. The system has become transient stable in this load condition for the same fault. The relative rotor angles before and after the generator rescheduling are shown in Fig.2.

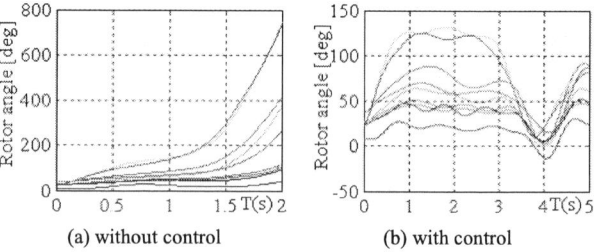

(a) without control   (b) with control

Fig.2 Relative rotor angle for contingency 22*-21, tcl=0.12

Case B: *Three-phase-to-ground-fault at bus 25:* The fault is simulated at bus 25 and cleared by tripping line 25-2 at $t_{cl} = 0.16$ s, which is again greater than the critical clearing time for the base case. Appling the algorithm of section VI, the new schedule for generators is shown in Table II. The system has become transient stable in this load condition for the same fault. The relative rotor angles before and after the generation reschedule are shown in Fig.3.

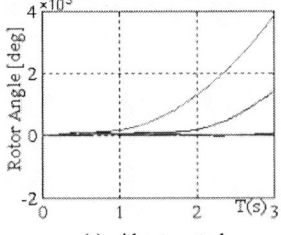 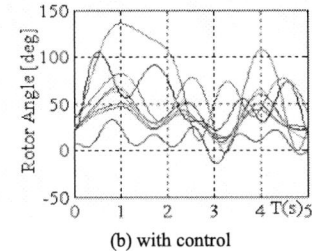

(a) without control   (b) with control

Fig.3 Relative rotor angle for contingency 22*-21, tcl=0.16

TABLE II
OPTIMAL GENERATION RESCHEDULING RESULTS

G	Case A (MW)	Case B (MW)
1	1017.9	1018.0
2	605.03	607.23
3	687.09	689.37
4	609.86	612.93
5	494.92	498.48
6	617.90	633.70
7	528.09	542.50
8	545.44	527.26
9	832.54	805.87
10	251.52	253.86

As indicated from the Fig. 2 and 3, the system stability is improved through generation rescheduling, and the proposed approach is very effective in ensuring the secure operation of the power systems.

VI. CONCLUSIONS

While computing technologies for fast stability simulation is advancing, theoretical works for computing the stability limits precisely is progressing slowly. In this paper, a novel approach for optimal generation rescheduling has been proposed to solve the challenging transient stability constrained optimization problem. Stability constraints adopted for the control model are formed by setting the generator angle to be less than a threshold value at a fixed termination time. By using the principles of optimal control, the preventive control model based on the differential-algebraic formulation of power systems is converted to an optimal control problem, in which the relative rotor angles of generators are controlled to be less than a threshold value through generation rescheduling when some severe contingency occurs. The optimization problem is to find an admissible control u that causes the dynamic system represented by the differential-algebraic equations to follow an admissible trajectory x that minimizes the performance index function J. The proposed approach is very effective in ensuring the secure operation of the power systems and inherits the advantages of time domain simulation that it has little limitations on the power system models and is robust.

REFERENCES

[1] J. Tong, H. D. Chiang and T. P. Conneen. "A sensitivity-based BCU method for fast derivation of stability limits in electric power systems," 92WM149-5 PWRS.

[2] A. A. Fouad. J. Tong, "Stability constrained Optimal rescheduling of generation, "IEEE, Trans. Power Systems. Vol. 8, No.1, pp.105-112, 1993.

[3] V. Vittal, E.Z. Zhou, C. Hwang and A. A. Fouad, "Derivation of Stability Limits Using Analytical Sensitivity of transient Energy Margin," IEEE Trans on. Power System, vol. 4, pp.363-1372, Oct. 1989

[4] P. W. Sauer, K. D. Demaree and M. A. Pai, "Stability limited load supply and interchange capability," IEEE Trans. Power Apparatus and Systems, vol.PAS-102, No.1,pp.3637-3643, Nov. 1983.

[5] Y. Xue, V. Cutsem and M. Pavella, "Real-time analytic sensitivity method for transient security assessment and preventive control," Proc. IEE, vol.135, pt. C, No. 2, pp.107-117, Mar. 1988.

[6] M. A. Pai, Energy Function Analysis for power system stability, Kluwer Academic Publishers, Boston, 1989.

[7] M. La Scala, M. Trovato, C. Antonelli, "On-line dynamic preventive control: an algorithm for transient security dispatch", IEEE Trans on Power Systems,1998, 13(2): 601 –610.

[8] E. De Tuglie, M. La Scala, P. Scarpellini, "Real-time preventive actions for the enhancement of voltage-degraded trajectories", IEEE Trans on Power Systems,1999,14(2): 561 – 568.

[9] E. De Tuglie, M. Dicorato, M. La Scala, P. Scarpellini, "A static optimization approach to assess dynamic available transfer capability", IEEE Trans on Power Systems,2000,15(3): 1069 – 1076.

[10] D. Gan, R.J. Thomas, R.D. Zimmerman, "Stability-constrained optimal power flow", IEEE Trans on Power Systems,2000,15(2): 535 – 540.

[11] D. Gan, D. Chattopadhyay, X.C. Luo, "Stability constrained OPF: new results", Proceedings of the 35th Southeastern Symposium on System Theory 2003, 16-18 March 2003: pp273 – 277.

[12] Y. Yuan, J. Kubokawa, H. Sasaki, "A solution of optimal power flow with multicontingency transient stability constraints", IEEE Trans on Power Systems, 2003,18(3):1094–1102.

[13] Y. Yuan, J. Kubokawa, H. Sasaki, et al., "Interior point method based optimal power flow with transient stability constraints", Automation of Electric Power Systems, 2002, 26(13):14-19.

[14] Chen, L.; Taka, Y, et.al., "Optimal operation solutions of power systems with transient stability constraints", IEEE Trans on Circuits and Systems,2001,48(3):327–339.

[15] Luonan Chen; Ono, A, et.al., "Optimal power flow constrained by transient stability", PowerCon 2000 - International Conference on Power System Technology, Perth, Australia, 2000, 1:1 – 6.

[16] X.L. Yang, Y.Z. Sun, H.F. Wang, "A new approach for optimal power flow with transient stability constraints", Automation of Electric Power Systems, 2003, 27(14):13-17.

[17] M.B. Liu, Y. Xia, J. Wu, "Calculation of available transfer capability with transient stability constraints", Proceedings of the CSEE, 2003, 23(9):28-33.

[18] A.K. David, J.L. Xu, "Dynamic security enhancement in power-market system", IEEE Trans on Power Systems Vol.17, No.2, pp.431-438, 2002.

[19] W. Li and A. Bose, "A coherency based rescheduling method for dynamic security," in Proceedings of PICA, 1997, pp254-259.

[20] D. Gan and Z. Qu, et.al., "Methodology and computer package for generation rescheduling", IEE Proceedings-Generation Transmission and Distribution, vol.144, no.3,pp.301-307,May,1997.

[21] Y. Kato, S. Iwamoto, "Transient stability preventive control for stable operating condition with Desired CCT", IEEE Trans on Power Systems Vol.17, No.4, pp.1154-1160, 2002.

[22] M. A. Pai, "Energy Function Analysis for Power System Stability". USA: Kluwer Academic Publisher, 1989.

Improvement of Transient Response of Thermal Power Plant Using VVVF Inverter

N. Matsui * and F. Kurokawa **

* Chouryou Control System CO., LTD, Japan
** Department of Electrical and Electronic Engineering, Nagasaki University, Japan
1-1, Akunoura machi, Nagasaki 850-8610, Japan
Fax:+81-95-828-7714
E-mail:nobumasa_matsui@ryousei.co.jp

Abstract--The construction of a thermal power plant is relatively little. However, the thermal power plant holds approximately 65% of world power source. Recently, the concern with the request of improvement for deterioration in the efficiency has been growing in the existing thermal power plant. Especially, it is required the efficiency for a small thermal power plant that is operated in the partial load. An inverter power source is applied to the field of various power electronics for the energy saving. When the inverter power source occurs abnormally, what is at stake is time delay in the transient response of air flow controller to the heating balance of the boiler and NO_X of environment factor. Therefore, the purpose of this paper is to present a new air flow control to improve of transient response in partial load for safety operation in the boiler when the inverter of forced draft fan is occurred abnormally. As a result, the combustion remained gas oxygen is back to original condition less than 1.0 second in the all case.

I. INTRODUCTION

The construction of a thermal power plant is relatively little. However, the thermal power plant holds approximately 65% of world's power source [1]. Recently, the concern with the request of improvement for deterioration in the efficiency has been growing in the existing thermal power plant. Especially, it is required the efficiency for a small thermal power plant that is operated in the partial load. An inverter power source is applied to the field of various power electronics for the energy saving. There are many results in each field, that is, a counter-measure of the abnormal occurrence time due to instantaneous power failure, an unexpected power source surge and noise and so forth are devised. In general, an inverter has two methods as counter-measure of the abnormal. After an inverter performs a self-judgment when an inverter does not have abnormality, it is a method to return to original speed. Another is a method to change to a commercial power source. it is a method to increase to the rated speed [2]. These counter-measures do not cover all the influence that machinery applying an inverter power source gives to a total system. When the inverter power source occurs abnormally, the power of equipment is lost primarily by both counter-measures. As a result, the speed of machinery falls according to the relationship between the load and inertia. This is unfavorable for a boiler.

A simulation is very effective to study the influence to the whole plant. There are many kind of machine in a thermal power plant that is nonlinear system. J.Kenndy and R.Eberheart suggested a method of solving the nonlinear problem that is PSO (Particle Swarm Optimization) as the optimization algorism [3]. Further more, K.Sagara and T.Kishikawa suggested a method of digital simulation for boiler heating surface [4]. It is not so difficult to build a thermal power plant model and to be able to emulate as same as actual plant controller by using DBSS (Data Base Simulation System) [5],[6] at present.

It has already reported that recommended time of transfer is 0.5 second from variable voltage and variable frequency (VVVF) to commercial power source when inverter of forced draft fan (FDF) occurs abnormally in the full load [7]. However, in a small existing thermal power plant, since it is not usually operated in the full load, it is important that it operated partial load so much at present. In this case, what is at stake is time delay in the transient response of air flow controller to the heating balance of the boiler and NO_X of environment factor.

This paper presents a new air flow control method to improve of transient response in partial load for safety operation in the boiler when the inverter of FDF is occurred abnormally.

After reviewing, at first, chapter II describes both then plant model and the control model. Next, chapter III describes the dynamic characteristics that compares with simulated and measured when inverter of FDF (Forced Draft Fan) is trip on abnormal test. An error is satisfied the standard of ISA [8] for the dynamic simulation of power plant. Finally, it describes a simulation at the partial load as the case study for safety operation.

II. SIMULATION MODEL

Figure 1 shows the constitution of the plant model and the control model using the DBSS (Data Base Simulation System). The equipment of the plant model and control model in Fig. 1 is shown in Table I. The plant model consists of 5 kinds of fluid that is air, flue gas, fuel, water and steam. The calculation of plant model interval time is 50 milliseconds. The control model consists of 9 kinds of controller. The calculation of control model interval time is 1 second as same as the boiler control system of actual plant. The air flow controller makes two control signals in the part of slanted line box of air flow controller using

978-1-4244-0644-9/07/$25.00 ©2007 IEEE

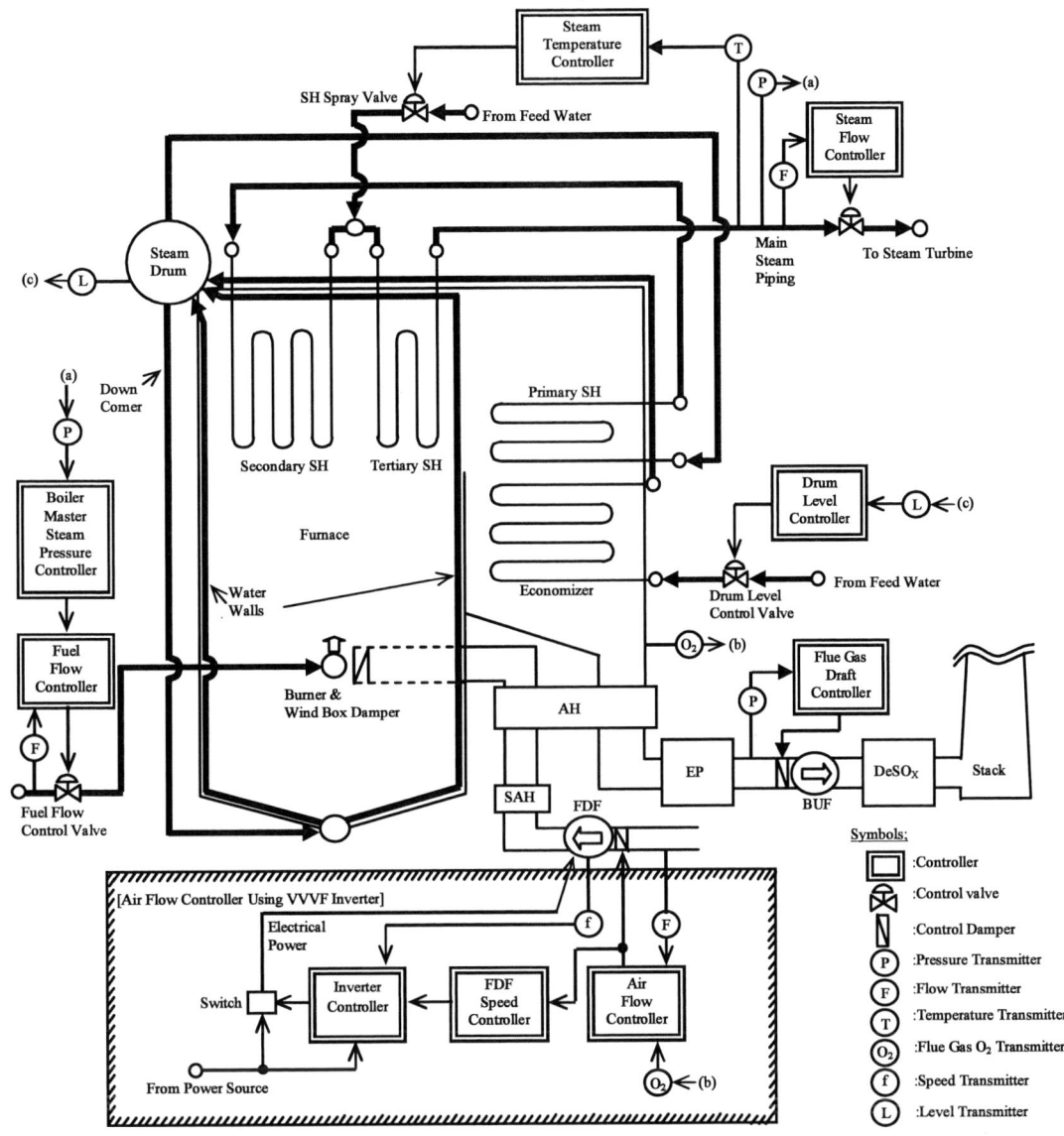

Fig. 1. Constitution of plant model and control model.

TABLE I Equipment of plant model and control model.

Equipment of Plant Model		Control Model
<Water and Steam side>	<Air, Flue Gas and Fuel>	
Economizer	Forced Draft Fan (FDF)	Boiler Master
Steam Drum	Steam Air Heater (SAH)	Fuel Flow Controller
Down Comer	Air Heater (AH)	Air Flow Controller
Water Walls	Wind Box Damper	FDF Speed Controller
Super Heaters (SH)	Furnace	Inverter Controller
Super Heater Spray Valve	Electrostatic Precipitator(EP)	Steam Flow Controller
Main steam piping	Boost Up Fan (BUF)	Draft Controller
Steam Flow Control Valve	De-Sulfurization (DeSO$_X$)	Drum Level Controller
	Fuel Flow Control Valve	Steam Temperature Controller
	Fuel Burner	
	Stack	

1399

inverter, one is the damper position demand of fan, another is the inverter speed demand. The inverter controller operates the power of fan that obeys form air flow controller demand of speed. The output of fan model is air flow to the boiler by the operated damper position, speed and mass balance of the boiler system. The output of boiler is the combustion remained gas oxygen (O_2) and main steam flow as the combustion condition. It is able to evaluate a state of the boiler with O_2 which is combustion output because the boiler heating balance is decided by amount of fuel and air. More over, it simulates to consider the steam and water condition for the total plant simulation. Since the plant model constitutes it in the nonlinearity which assumed physics basics type, a constant is given from design data in the plant. The paragraph from A to C are described the equation a part of plant model. The paragraph D is described the air flow controller.

A. Piping and Duct model

Piping and duct model is simulated by following equation;

$$P_a = P_e + \frac{L_e - L_a}{10000\,v} - \xi v W^2 \tag{1}$$

where

Pe	:Inlet Pressure	(atg)
Pa	:Outlet Pressure	(atg)
Le	:Inlet floor height	(m)
La	:Outlet floor height	(m)
v	:Specific Volume	(m³/kg)
W	:Weight flow	(kg/s)
ξ	:Coefficient of Pressure Loss	(-)

B. Fan model

Speed of fan is simulated by following mechanical balance equation;

$$J\frac{d}{dt}\omega = T \tag{2}$$

where

J	:2nd moment of Fan	(m⁴)
ω	:Corner speed	(rad/s)
T	:Torque	(kg-m)

C. Combustion model of furnace

Furnace pressure is calculated by the equation on the state of gas that is the both mass balance and energy balance of furnace. The mass balance equation is as follows;

$$\frac{d}{dt}V\gamma = W_e - W_a \tag{3}$$

The heat balance equation is as follows;

$$\frac{d}{dt}V\gamma E = H_e W_e - H_a W_a - Q \tag{4}$$

$$V\frac{d}{dt}\gamma H - VA\frac{d}{dt}P = H_e W_e - H_a W_a - Q \tag{5}$$

where E is given by the first law of thermodynamics;

$$E = H - AvP \tag{6}$$

Specific weight γ and enthalpy H is variable, the following equation is given from Eq. (5);

$$VH_0\frac{d}{dt}\gamma + V\gamma_0\frac{d}{dt}H - VA\frac{d}{dt}P = H_e W_e - H_a W_a - Q \tag{7}$$

From Eq. (3) and Eq. (7), given the following equation;

$$V\gamma_0\left(\frac{d}{dt}H - AR\frac{d}{dt}T\right) = (H_e - H_0)\cdot W_e - (H_a - H_0)\cdot W_A - Q \tag{8}$$

where

$$v\frac{d}{dt}P = R\frac{d}{dt}T \tag{9}$$

Furnace gas enthalpy balance equation is as follows;

$$\frac{d}{dt}H = \frac{\kappa}{V\gamma_0}\{(H_e - H_0)\cdot W_e - (H_a - H_0)\cdot W_a - Q\} \tag{10}$$

where

$$AR = C_p - C_v \tag{11}$$

$$\kappa = \frac{C_p}{C_v} \tag{12}$$

$$dT = \frac{dH}{C_p} \tag{13}$$

Furnace gas temperature is given from Eq. (10), function of gas contents as follows;

$$T = f(H, Cg) \tag{14}$$

where

V	: Furnace inner volume	(m³)
γ	: Specific weight	(kg/m³)
E	: Furnace inner energy	(kcal/kg)
We	: Weight flow into furnace	(kg/s)
Wa	: Weight flow out of furnace	(kg/s)
He	: Enthalpy into furnace	(kcal/kg)
Ha	: Enthalpy out of furnace	(kcal/kg)
Q	: Radiation of heat from gas	(kcal/s)
A	: Mechanical equivalent of heat	(kcal/kgf m)
R	: Gas constant	(kgf m/kg K)
H	: Furnace gas enthalpy	(kcal/kg)
P	: Furnace gas pressure	(mmAq)
Cp	: Specific heat at constant pressure	(kcal/kg K)
Cv	: Specific heat at constant volume	(kcal/kg K)
k	: adiabatic constant	(-)
Cg	: Gas contents	(-)

D. Air flow Control model

Figures 2(a) and (b) show air flow controller. The diagram of conventional air flow controller is shown in Fig. 2(a). The damper position is controlled by PI controller that input is the air flow difference between the set point and measured air flow. The air flow set point is given by the function of combustion demand of boiler master in Fig. 1. The diagram of air flow controller with inverter is shown in Fig. 2(b). Air flow is controlled by the damper position as same as Fig. 2(a). The inverter speed demand is controlled by PI controller that input is the damper position difference between setting and demand. The set point of damper position is given by the function of Fx1.

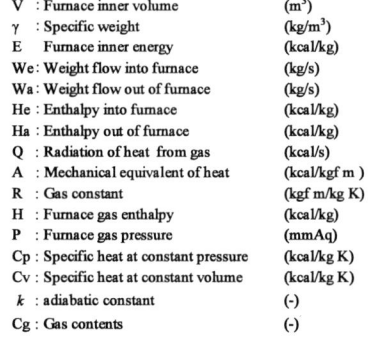

Fig. 2(a). Diagram of conventional air flow controller.

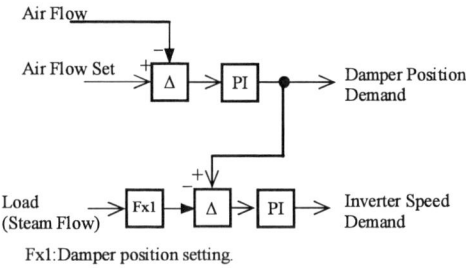

Fig. 2(b). Diagram of air flow controller with inverter.

III. DYNAMIC CHARACTERISTICS

When the inverter of FDF occurs abnormally, air flow controller has two actions at the same time. One is the speed of FDF from partial speed to the rated speed because the power source is changed to commercial power source. The other is damper position of FDF is reduced static characteristics position. These two actions are carried out at this time.

The transient response of a plant influences is not only a characteristic but also dimensions of the constitution, drive and sensor. These parameters have a design margins, however, there are some deterioration and kind of fouling in the existing plant. So, an error spreads between a design and actual plant. Therefore, some parameters have a coefficient, and plant model is carried out identification by J of 2nd moment of fan of Eq. (2) in chapter II. T of torque of Eq. (2) in chapter II is the balance between electrical power and load of fan. These powers are have a time delay, so that time delay is identified.

Figures 3(a) through (d) show the comparing the both measured of actual plant and simulated at the 70% load when the inverter of FDF occurs abnormal testing.

As the model identification result of fan speed response at inverter trip is shown in Fig. 4(a). The starting point in Fig. 4(a) is the timing of inverter trip. The changing time in the interval I is the time from inverter power source to commercial power source. An error of reducing speed is 0.3% in this period. Reaching time to rated speed in the interval II is the time until rated speed after power source is changed completely. An error of rising speed is less than 2.5% in this period. Further more, an error of reaching time is 3.4%. The comparing air flow with measured and simulated is shown in Fig 4(b). An error is less than 2.0%. FDF out pressure and furnace pressure the comparing with measured and simulated is shown in Fig 4(c) and (d). An error is less than 4.0%. The instrument standard of America (ISA) defines an error less than 10% on a dynamic simulation for power plant. Since an error is less than 4.0% in dynamics simulation is presented in the comparing, this shows that the accuracy of the proposed model is 2.5 times as much as a standard one.

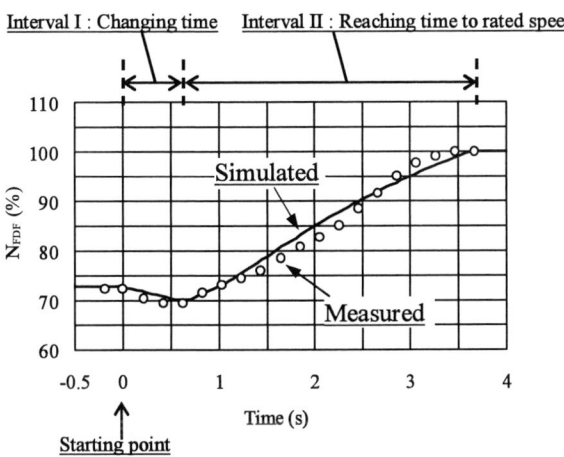

Fig. 3(a). Compare FDF speed with measured and simulated in 70% load at inverter abnormal.

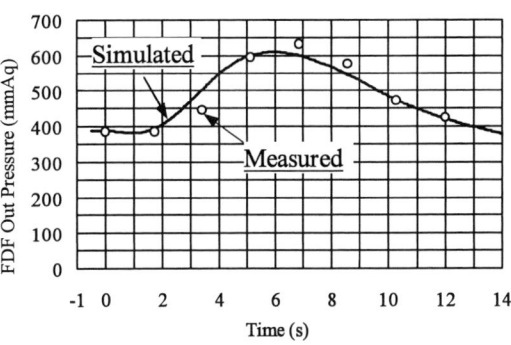

Fig. 3(c). Compare FDF out pressure with measured and simulated in 70% load at inverter abnormal.

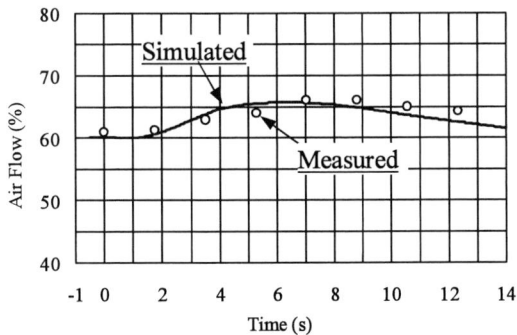

Fig. 3(b). Compare the air flow with measured and simulated in 70% load at inverter abnormal.

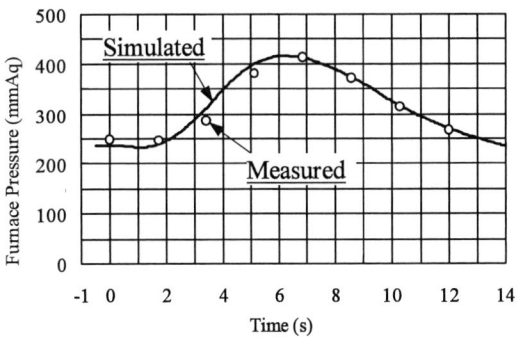

Fig. 3(d). Compare the furnace pressure with measured and simulated in 70% load at inverter abnormal.

IV. THE CONFIGURATION OF SYSTEM AND CONVENTIONAL PERFORMANCE CHARACTERISTICS

Figure 4 shows the diagram of abnormal mode of air flow controller with inverter in Fig. 3(b). When the inverter occurs abnormally, the inverter speed demand is set to 100% as the rated speed. Fx1 in Fig. 4 is the setting of damper position with inverter and static characteristics without inverter is shown in Fig. 8. The setting of damper position is designed to realize high efficiency in the useful operation load so much. Therefore, the set point of FDF's damper position is +5% opening against 78% opening of static position at 100% load and +10% opening against 50% opening of static position at 30% load. The most different is 50% load in this plant, its set point is +15% opening against static opening position.

Figure 5 shows the simulation result using the air flow controller in Fig. 4 when the inverter of FDF occurs abnormally. Case 1 is in 50% load, Case 2 is in 75% load and Case 3 is in full load, respectively, in Fig. 5. While FDF speed (N_{FDF}) increases to the rated speed, after the combustion gas O_2 increases according to air flow increasing. It takes a long time that the combustion gas O_2 becomes original condition by the damper of FDF.

Especially, in 50% load of Case 1, it increases O_2 from 3% to 7%. This result is the problem both the heating balance of the boiler and NO_X of environment factor, and so forth.

V. IMPROVEMENT OF TRANSIENT RESPONSE

Figure 6 shows the diagram of improvement controller of air flow that is added advanced part. Fx2 in Fig. 6 is the advanced damper position difference between the setting with inverter and static characteristics without inverter is shown in Fig. 9. The advanced controller is used while the inverter is in-serviced. When the inverter occurs abnormally, the advance controller is decreased to 0% through the ramp function.

Figure 7 shows the simulation result using the air flow controller in Fig. 6 when the inverter of FDF occurs abnormally. While N_{FDF} increases to the rated speed, the combustion gas O_2 is suppressed to increase. After FDF reaches the rated speed, the combustion gas O_2 becomes original condition less than 1.0 second in the all case is shown in Fig. 7. As a result, improvement of air flow controller is satisfied as the safety operation for the total plant.

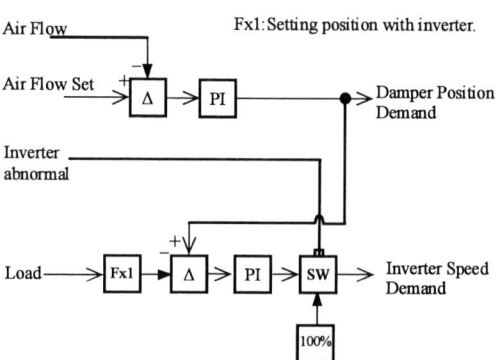

Fig. 4. Diagram of abnormal mode of air flow controller in Fig. 3(b).

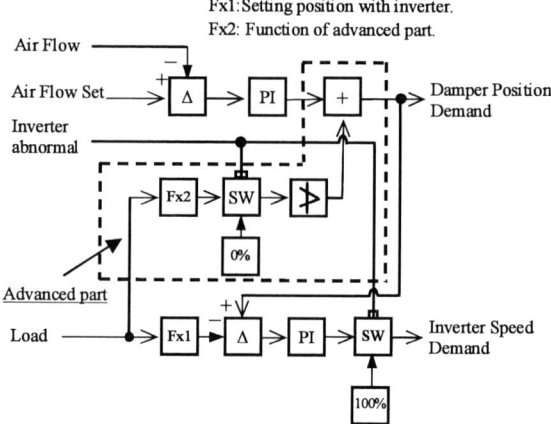

Fig. 6. Diagram of improvement of air flow controller added advanced part.

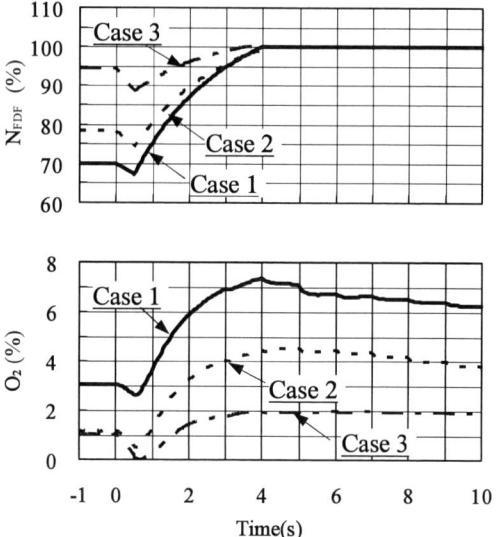

Fig. 5. Simulation result in conventional controller.

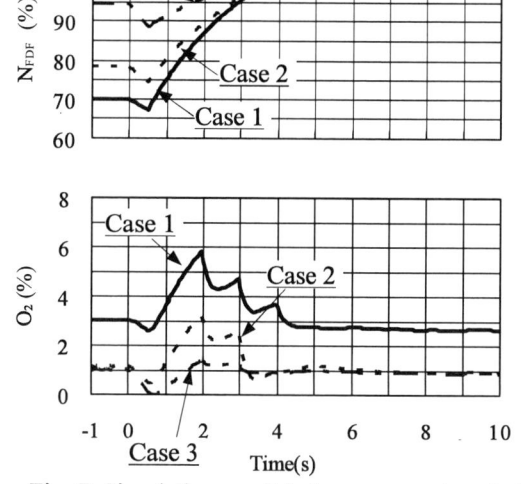

Fig. 7. Simulation result in improvement controller.

1402

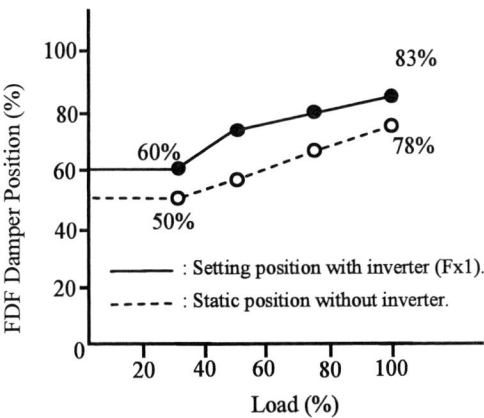

Fig. 8. Damper position.

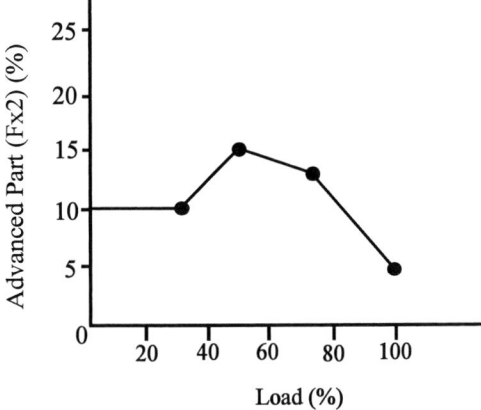

Fig. 9. Function of advanced part.

VI. CONCLUSIONS

A new air flow control method to improve of transient response in partial load for safety operation in the boiler when the inverter of FDF occurs abnormally is presented in this paper. This result is as follows.

(1) The plant model and the control model are built for the whole power plant simulation.
(2) The proposal simulation model was compared with actual plant measured data when the inverter of FDF occurs abnormal testing. An error is lass than 4.0 %. This result is satisfied less than 10% as the standard that ISA defines.
(3) Simulation case study is carried out on base of accuracy model, it clarified that the transient response of the combustion gas O_2 needs to improve. Since air flow controller was improved by the advanced part, the O_2 became original condition less than 1.0 second in the all case after FDF reached the rated speed when inverter occurs abnormally.

Moreover, this method of simulation is carried out the same as actual plant controller to use control emulator, therefore, it is easy to apply the parameter of controller for the actual plat controller according to a result of simulation.

VII. REFERENCES

[1] Energy balance of OECD countries 2002-2003.
[2] Yasukawa Electric: "Inverter Driver technique", 2nd ed, Nikkan Kougyou Inc., 1997.
[3] J. Kenndy and R. Eberheart : "Particle Swarm Optimization", Proceedings of the 1995 IEEE International Conference On Neural Networks, vol. IV, pp.1942-1948, Pert, Australia, 1995.
[4] K. Sagara and T. Kishikawa: "A New Method of Digital Simulation for Boiler Heating Surface", Simulation of System, L.Dekker, editor (c) North-Holland Publishing Company, 1976.
[5] T. Kishikawa: "Development and the present conditions of DBSS", vol. 5, Technical Report of Cyouryou Control System co., ltd., pp.53-59, 2005.
[6] N. Matsui, T. Ozaki, M. Mizoe, T.Matsuoka: "Engineering of Simulation", vol. 5, Technical Report of Cyouryou Control System co., ltd., pp.45-52, 2005.
[7] N. Matsui and F. Kurokawa: "Boiler Simulation at the Trip of Forced Draft Fan", IEEJ, The 2007 Annual meeting on Toyama, No.4, pp.290, Mar 2007.
[8] ISA: The Instrument Standard of America.

A Novel Circuit Topology for Three-Phase Four-Wire Distribution Electronic Power Transformer

*H.Mirmousa** and M.R.Zolghadri**

*Sharif University of Technology, Tehran, Iran, Email: zolghadr@sharif.edu
**FARAB Co. Main Contractor of Energy and Water Projects, Tehran, Iran, Email: hamid.mirmousa@gmail.com

Abstract--Distribution Power Electronic Transformers are introduced to improve the power quality issues as well as to reduce the size and the cost of the transformers once technological progress made it possible. In this paper first, some existing configurations are overviewed. Then a new circuit topology is introduced for input and output stages of a Power Electronic Transformer. This circuit comprises three-phase input matrix converter, three-phase four-wire output matrix converter and the high frequency transformer. Control scheme of input and output link is analyzed and simulation result of the system is presented. Reduced size, lower price, higher efficiency and better voltage regulation are some of the advantages of this approach.

Index Terms-- Electronic Power Transformer, Control Scheme, Input and Output Link

I. INTRODUCTION

Transformers are used widely in electric power distribution/conversion systems to perform functions like galvanic isolation, voltage transformation, noise decoupling, etc. Transformers are one of the heaviest and most expensive parts in an electrical distribution system. The size of the transformer is a function of the saturation flux density of the core material and maximum allowable core and winding temperature rise. Saturation flux density is inversely proportional to frequency and increasing the frequency allows higher utilization of the steel magnetic core and reduction in the size of transformer. Electronic Power Transformer (EPT) is a new type of power transformer which can benefit the increase of the frequency to reduce the size. Frequency conversion in the input and the output stage is done via Power Electronic converters. Therefore the input Power Factor Correction (PFC) and the output voltage regulation can be achieved through proper control of these converters. Technological limits in the rating and the frequency of the existing switches appear as a constraint to increase the frequency so more and achieve a better performance in EPT applications.

As noted before the primary purpose of EPT is to reduce the size of core and winding by improving the frequency. Versus of this reduction, some electronic switches, largest cooling system and some capacitors and reactors are added to the system. So overall size and cost of system may be unchanged or even more with compare of ordinary transformers. Of course further and scholastic study on this subject will conducive to achieve the adapted configuration for such system. New proposed system represented in this paper is much acceded to mentioned configuration as of economical and practical aspects.

II. LITERATURE REVIEW

Many topologies are proposed for EPT. All of them are based on converting line input voltage to higher frequency and apply it to the magnetic core. Input current PFC and output voltage regulation are some more features of them. Some configurations, which are proposed up to now, are shown in Fig. 1. to 4.

In the circuit shown in Fig. 1, 50Hz Medium Voltage (MV) or High Voltage (HV) is converted to DC voltage [1,2]. In the next stage, DC voltage converts to Medium Frequency (MF) for feeding the magnetic core. In the output, the MF voltage is converted to DC and finally it is inverted to 50Hz via three distinct single phase inverters. Far number of switches, need to huge capacitor at the DC link and capacitor in the HV side, which necessitates using higher voltage switches, are some disadvantages of this system.

The system shown in Fig. 2 is very similar to the earlier one. It does not have a 4wire output circuit which is the other disadvantage of this structure to be used as a distribution transformer [3].

System shown in Fig. 3 uses a matrix converter in the input stage to convert line frequency directly to a medium frequency. In the output stage a PWM rectifier is used to convert MF voltage to DC voltage. Then by using a three phase inverter, the 50 Hz voltage is introduced [4,5].

Fig. 4 shows a 3wire system, which is inconvenient for 4 wire distribution systems. [6].

All of these systems have added an electronic layer in the transformer structure so many proper functions such as PFC, voltage regulation, control of power system performance and other ancillary applications can be implemented anywhere needed. Meanwhile increasing the frequency, can lead to the reduction of the size of the transformer.

978-1-4244-0644-9/07/$25.00 ©2007 IEEE

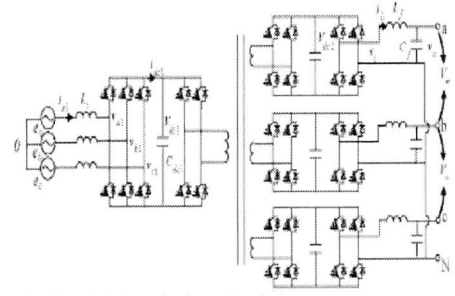

Fig. 1. One of the proposed circuits [1], [2]

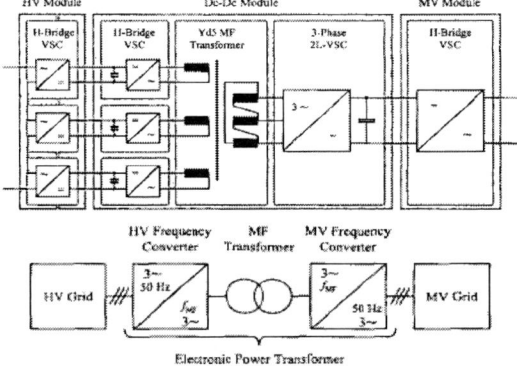

Fig. 2. Proposed circuit for use in 3wire systems [3]

of the winding is used as the fourth wire of the output link to be used as the null wire. The proposed system consists of matrix converters on both the primary and secondary side exciting the transformer synchronously. The advantages of proposed approach are:

1. Lower number of switches.
2. No need for capacitor at MV side.
3. Fewer losses due to fewer switches and consequently higher efficiency.
4. Endurance of heavily load unbalancing without any shock to core or windings.
5. Ability of bilateral power transition.

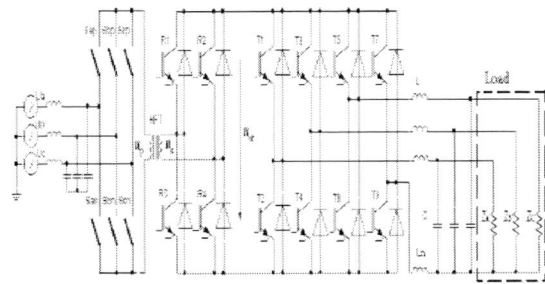

Fig. 3. One other proposed circuits for use in 4wire systems [4], [5]

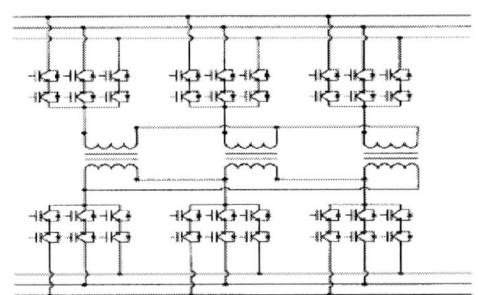

Fig. 4. Other proposed circuit for use in 3wire systems [6]

III. NEW PROPOSED SYSTEM

A. Power circuit

New system which is shown in Fig. 5 is suitably configured for application in the distribution system. Input link is a matrix converter with six bilateral switches. It converts directly the 50Hz sinusoidal main voltages to the medium frequency voltage. The inverse function is carried out in the output link in order to convert medium frequency to 50Hz sinusoidal voltage. The switching scheme of the output is similar to a PWM inverter. It needs the zero crossing detection of medium frequency voltage for control scheme implementation. In order to reduce output voltage and current THD, an LC filter is needed. For a better clarification of the conversion operation, simulation result is presented.

It should be noted that existing semiconductor switches do not permit to implement this system with such a high frequency (2kHz to 10kHz) in power ratings as high as 2000kVA.

The primary purpose of the size, weight, and price reduction is achieved through using circuit topology that employs the least number of switches, using a standard grain oriented silicon steel core transformer with an operating frequency of 2.5 kHz to 5 kHz. The transformer middle end

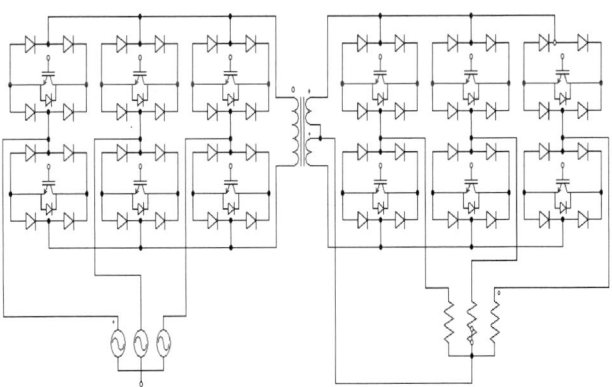

Fig. 5. New proposed circuit

1405

B. Control scheme

Control system of the input and the output links are isolated. This means that power factor correction of the input is independent from output voltage regulation.

In the output link the 50Hz voltage is produced directly from medium frequency voltage delivered to the LV winding of the transformer. As shown in Fig. 6, switching signals for generating directly the 50Hz voltage from medium frequency is produced similar to a simple sinusoidal PWM generator by comparing the required sinusoidal waveform with the triangular carrier waveform.

Fig. 7 shows a typical control loop for the output voltage regulation. In this method the control system is dependent on circuit elements such as the size of the output filter capacitor and reactor. Any variation in the circuit element will affect the control system. Meanwhile using this scheme is a usual practice in power electronics and drive systems [7].

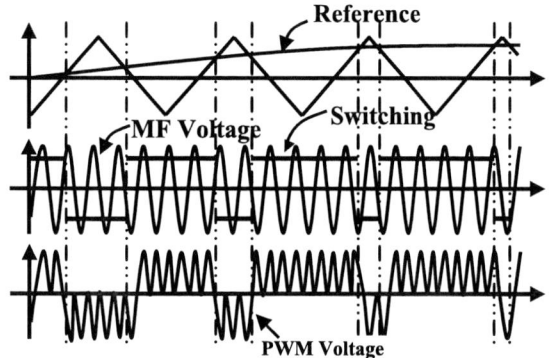

Fig. 6. Switching scheme of output link

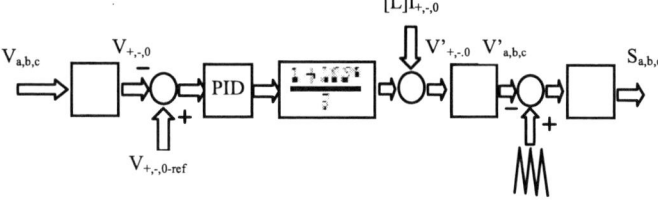

Fig. 7. Control loop of output link

Fig. 8 shows the block diagram of the new method proposed for the control of the output three phase voltages. In this scheme, regulation of voltage amplitude and voltage phase is isolated. In the first loop the amplitude of each phase is regulated and by second loop the phase angle of each voltage is regulated and finally by merging the two generated values for amplitude and phase angle the balance three phase voltages will be introduced.

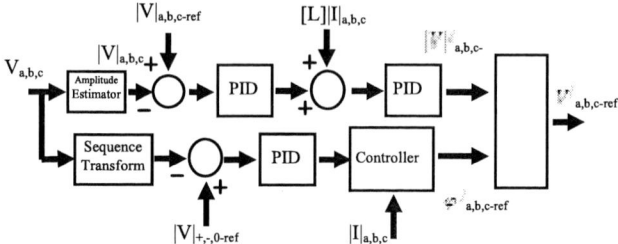

Fig. 8. Modified control loop of output link

Generally, the three-phase voltages and currents can be transformed to positive, negative, and zero-sequence components that are [6]:

$$\begin{pmatrix} \vec{U}^+ \\ \vec{U}^- \\ \vec{U}^0 \end{pmatrix} = \begin{pmatrix} U^+ e^{j(\alpha t + \psi_U^+)} \\ U^- e^{j(\alpha t + \psi_U^-)} \\ U^0 e^{j(\alpha t + \psi_U^0)} \end{pmatrix} = 1/3 \begin{pmatrix} 1 & \alpha & \alpha^2 \\ 1 & \alpha^2 & \alpha \\ 1 & 1 & 1 \end{pmatrix} \begin{pmatrix} \vec{U}_a \\ \vec{U}_b \\ \vec{U}_c \end{pmatrix} \tag{1}$$

Where α is equal to $\exp(j(2/3)\pi)$ and instantaneous power is as follow:

$$
\begin{aligned}
P(t) &= 3U^+ I^+ Cos(\psi_U^+ - \psi_I^+) + 3U^- I^- Cos(\psi_U^- - \psi_I^-) \\
&+ 3U^0 I^0 (\psi_U^0 - \psi_I^0) + 3U^+ I^- Cos(2\alpha t + \psi_U^+ + \psi_i^-) \\
&+ 3U^- I^+ Cos(2\alpha t + \psi_U^- + \psi_i^+) + 3U^0 I^0 Cos(2\alpha t + \psi_U^0 + \psi_i^0)
\end{aligned}
\tag{2}
$$

Where +, - and 0 superscripts stand for positive, negative and zero sequence values correspondingly. ψ_I is the current phase angle of each sequence and ψ_U is the voltage phase angle of each sequence. In the input link PFC under unbalanced mains voltage is based on power factor correction of positive sequence of voltage and current, so the input power factor can be controlled by positive current phase angle ψ_I^+.

Fig. 7 shows the block diagram of the control system used for the output stage. The output reference voltage is calculated using the measured out put voltage and current as shown in equations (3) to (9).

$$V_a^{/} = V_a + LSi_a + LCS^2 V_a \tag{3}$$

$$V_a^{/} = (1 + LCS^2)V_a + LSi_a \tag{4}$$

Small signal analysis of equation (4) is as below:

$$V_a^{/} + v_a^{/} = (V_a + v_a)(1 + LCS^2) + LS(I_a + i_a) \tag{5}$$

So, equation (5) is divided into two parts as small signal and steady state mode. For the sake of brevity, only the equation of small signal is presented now. In fact, behavior of system is related to small signal characteristics.

1406

$$v_a^/ = V_{a-ref}^/ - V_{a-exist}^/ \quad , \quad v_a = V_{a-ref} - V_a \tag{6}$$

$$(V_{a-ref}^/ - V_{a-exist}^/) = (V_{a-ref} - V_a)(1 + LCS^2) + LSi_a \tag{7}$$

$$\int(V_{a-ref}^/ - V_{a-exist}^/) = \int((V_{a-ref} - V_a)(1 + LCS^2) + LSi_a) \tag{8}$$

Let $V_{a-exist}^/ = 0$ so equation (9) is achieved for reference voltage before filter. This reference compared with the ramp carrier signal to generate the switching command of each phase.

$$V_{a-ref}^/ = (V_{a-ref} - V_a)(\frac{1 + LCS^2}{S}) + Li_a \tag{9}$$

$V_a^/$ is the voltage of the output before filter and V_a is the load voltage. L and C are the output filter inductor and capacitor inductance and capacitance.

In the input link the three equal square signals are applied to each phase switches with equal phase difference. To achieve the MF low harmonics voltage and for achieving the unity power factor in the input the switching so as [5] is applied. The size of the input reactor and the phase difference between each phase switching signal are effective on the input PFC control. Pre regulation of voltage in the output of LV winding of transformer is based on voltage drop regulation on the transformer propagating reactor. So the size of this reactor should be noticed whenever transformer is designed. Voltage drop on the reactor is controlled by changing the frequency of the generated voltage.

IV. SIMULATION RESULTS

In this section the simulation results of circuit shown in Fig. 5 is presented. All simulations are performed using MATLAB SIMULINK. IGBT switches and all other elements are of MATLAB standard library of power circuit. It should be noted that switching loss has ignored in the simulation.

A. Simulation of the output link

To select the output stage of this topology, the ability to supply single phase and three phase unbalance loads is the major criteria. Fig. 9 shows one phase of the output stage of the circuit. The load shown in Fig. 9 is resistive but in the simulation it substituted by reactive load.

In this circuit L is equal to 2mH and C is equal to 300uF. Input voltage frequency is 2.5kH. Output voltage and current are shown in Fig. 10.

Transient behavior of system at starting and load change conditions is shown in Fig. 11 to 17. When an unbalance load is carried out by this system, the rate of voltage growth for each phase is different according to its load. For example Fig. 11 shows the current of an unbalance load. In this state phase a is full load and phases b and c have 20% and 50% of full load respectively.

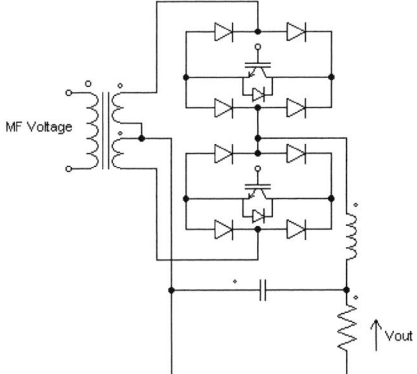

Fig. 9. Single phase circuit of output

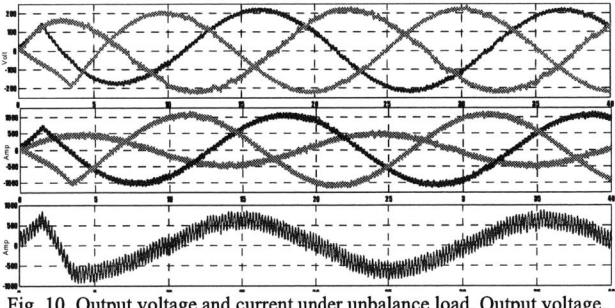

Fig. 10. Output voltage and current under unbalance load. Output voltage, output current and current of null wire respectively from top

Starting of system or any change in loads will lead to variation in voltages of each phase temporarily until reaching the regulated conditions. Fig. 12 shows the profile of each phase voltage at starting mode with load shown in Fig. 11. It is observable that rate of grow up and the maximum over shoot of voltages are dependent on amount of load of its corresponding phase.

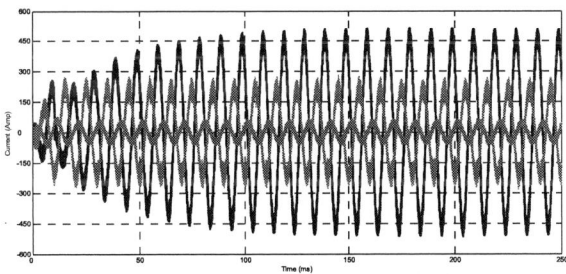

Fig. 11. Output current at unbalance load

Sinusoidal behavior of the output voltages at starting under unbalance load is shown in Fig. 13. As it can be seen the sinusoidal shape of voltage is retained while its amplitude changes. Simulations results show that the harmonic contents of the voltage increases with decreasing load.

1407

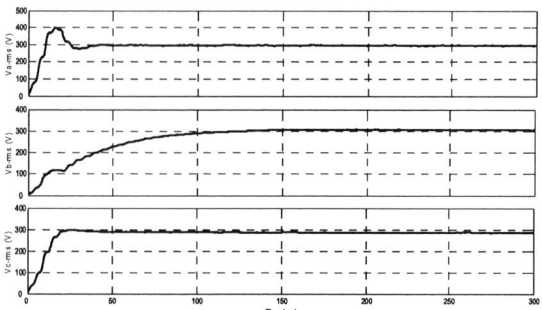

Fig. 12. Variation of output voltage at starting time under unbalance load. Phase *c* (20% load), phase *a* (full load) and phase *b* (50% load) respectively from top.

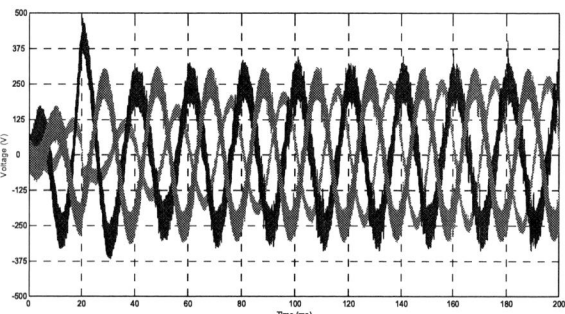

Fig. 13. Variation of output sinusoidal voltage at starting time

Fig. 14 shows the 50% enhancement of load and Fig. 15 shows the variation of voltage of this phase. It is visible that a load increase will lead to a voltage drop temporarily. Bearing of control system in order to revive the voltage in this status is shown in Fig. 15. Also a reduction of load will lead to temporary voltage swell according to Fig. 16.

Fig. 17. Shows the voltage variation of all three phases when one of the phases has load change. It is visible that phases with no load variation don't have any changes in voltage.

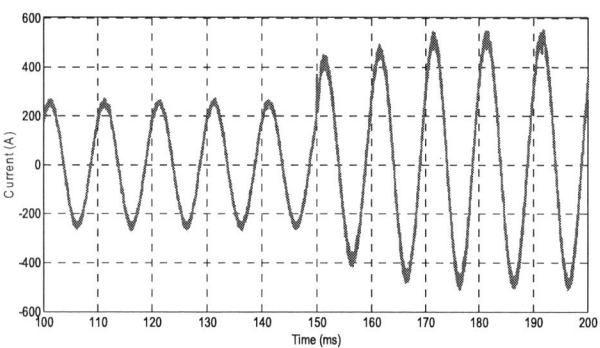

Fig. 14. 50% Load enhancement

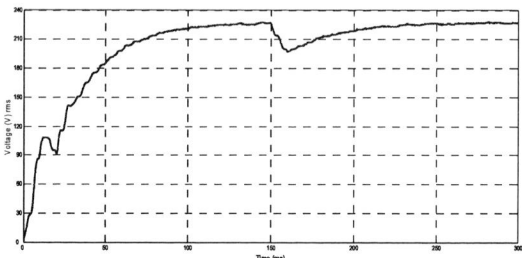

Fig. 15. Voltage variation under load enhancement

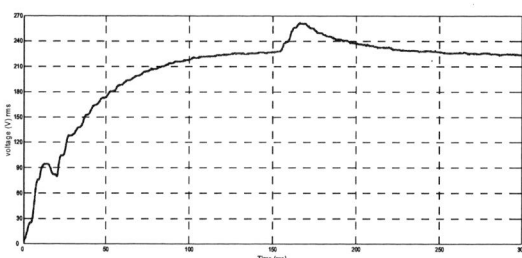

Fig. 16. Voltage variation under load reduction

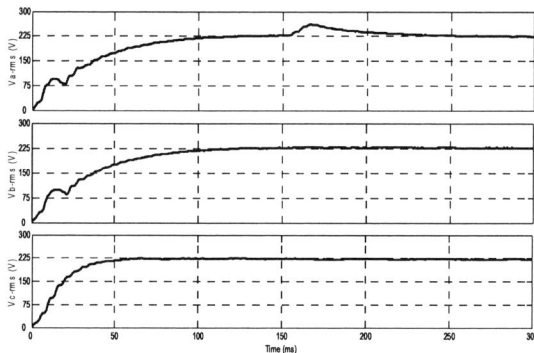

Fig. 17. Three phase voltage variation under load reduction

Current grow at start up is shown in Fig. 18. In this Fig. the loads are 100%, 60% and 0% for each phase.

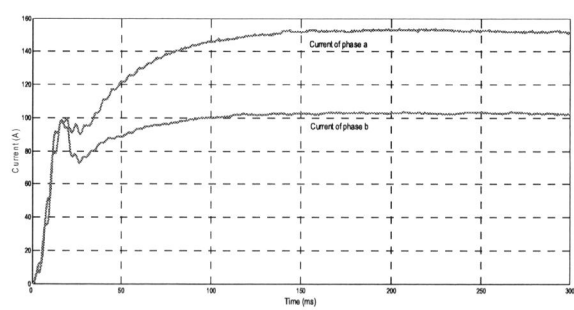

Fig. 18. Current grow at start up mode

Analysis of output voltage with positive, negative and zero sequence will be advantage to clarify the balancing of produced voltages. Fig. 19 shows the sequence voltages of output synchronized with starting time. It is visible that the positive sequence is regulated to expected value and negative and zero sequences tend to zero.

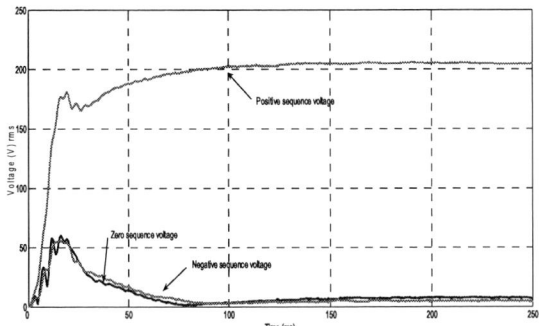

Fig. 19. Positive, negative and zero sequence components of the output voltage.

Design of the output filter is important. Achieving to standard rate of THD in voltage and current will be the criteria of filter designing. Optimum design of filter in this system is based on achieving the maximum THD of 3% in voltage at full load and 5% at no load. These conditions lead to having an LC filter with values of 0.8mH for reactor and 150uF for capacitor. Fig. 21 to 23 show the THD of voltage at full, 50% and no load conditions. The voltage and current in aforementioned condition are presented in Fig. 20. According to Fig. 21 to 23 the THD of voltage at no load is equal to 4.77%, at 50% load is equal to 3.61% and at full load is equal to 1.8%.

Other important analysis of output link is the performance of system in bilateral transition mode. This has studied in section IV-C.

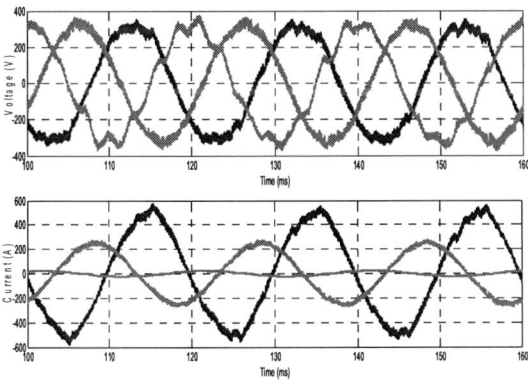

Fig. 20. Output voltages and currents for study of THD

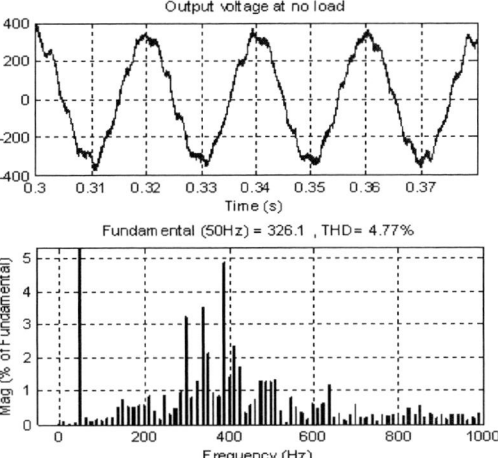

Fig. 21. Output voltage and its related THD at no load

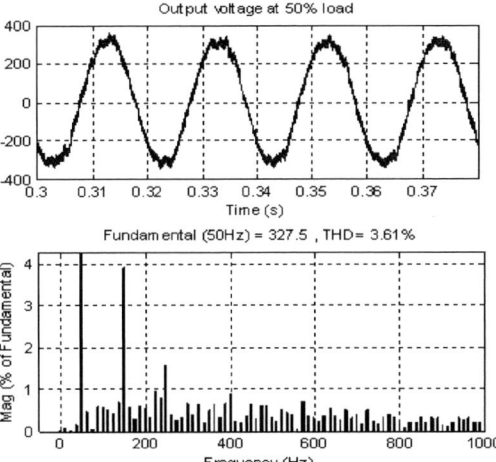

Fig. 22. Output voltage and its related THD at 50% load

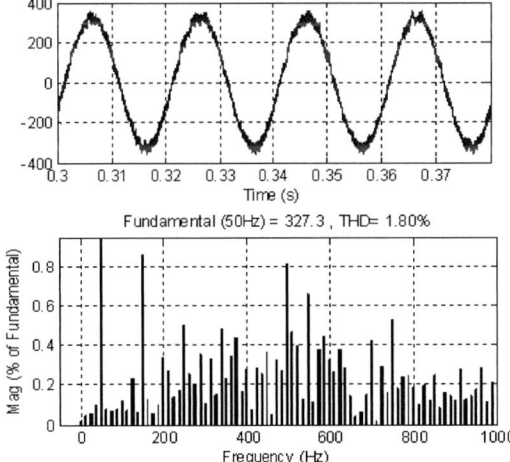

Fig. 23. Output voltage and its related THD at full load

B. Simulation of input link

Ideal performance of the input is to have fully power factor correction and generate the medium frequency voltage with low THD. In addition, having low harmonics in the input current and balanced current are important issues too. In this system by regulating the switching frequency and phase difference between each phase switching signals, the ability to achieving PFC and harmonic control is realized. So the simulation result which is shown in Fig. 24 represents the performance of the input link.

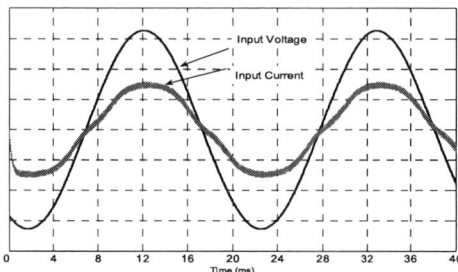

Fig. 24. Input voltage and current

Fig. 25 shows the three phase input currents. It is visible that all three phase current are balance

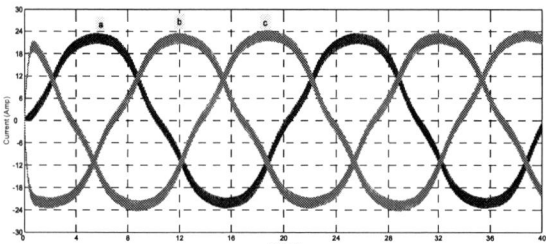

Fig. 25. Input three phase current

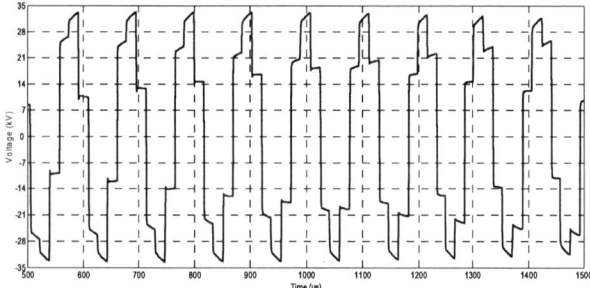

Fig. 26. Medium frequency voltage in the primary

Medium frequency generated voltage which is applied to winding is shown in Fig. 26. Due to limited switching frequecy, no PWM signal can apply for generating this voltage. Therfore the voltage in the primary winding have some more harmonics. Fig. 27 shows the voltage appeared in the secondary winding. It is visible that the harmonic content

of voltage at the secondary winding is decreased because of reactive properties of transformer winding.

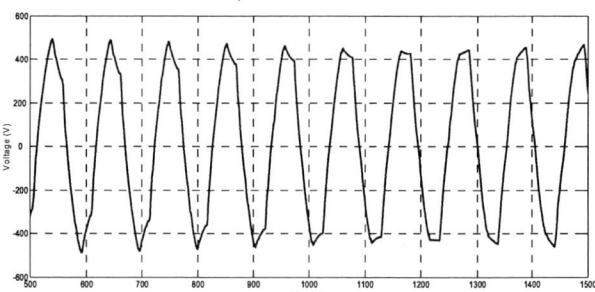

Fig. 27. Medium frequency voltage in the secondary

C. Bilateral power transition

Other important feature which is mentioned as the advantage of this system is its ability for bilateral power transition. In order to show this concept, a synchronous machine has applied as regenerative load. Firstly, by regulating the exciting current and mechanical torque, the state of consumer load is achieved. Then according to Fig. 28 at 70ms a change in exciting current and torque is applied so that the load has become regenerative. Ability of power flow in the inverse direction is shown clearly in Fig. 28. So without any change in switching algorithm in the output link, achieving of bilateral power transition is realized.

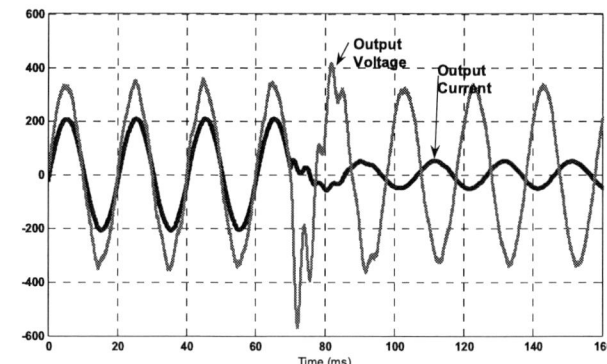

Fig. 28. Bilateral power transition in the output link

V. CONCLUSION

EPT is a new solution for enhancement of power distribution systems performance. It can be controlled to achieve PFC in its input stage, and a high voltage regulation even for unbalance loads in the output stage. In the proposed configuration, the input capacitor is eliminated and consequently the voltage level of input switches has been reduced. In addition, using midpoint of the LV winding of the transformer as the neutral point of the four wire three phase output, number of switched in the output stage converter as well as their voltage rating are reduced. Simulation results for a simple control scheme is presented to show the ability of the system to achieve PFC in the input stage and supplying single phase loads as well as three phase unbalance loads with an acceptable voltage regulation. The proposed system is a bilateral EPT which makes it suitable for using in any transformer application in power systems.

ACKNOWLEDGMENT

This work is partially supported by the Centre of Excellence in Power System Management and Control (CEPSMC)-Sharif University of Technology.

REFERENCES

[1] D.Wang, C.Mao, J.Lu, S.Fan, and L.Chen, "The Research on Characteristics of Electronic Power Transformer for Distribution System" *2005 IEEE/PES Transmission and Distribution Conference & Exhibition, Asia and Pacific Dalian, China*

[2] D.Wang, C.Mao, J.Lu, S.Fan, F.Peng, "Theory and application of distribution electronic power transformer" *Elsevier, Electric Power Systems Research*, 2006

[3] H.Wrede, V.Staudt, A.Steimel, "Design of an Electronic Power Transformer" *IEEE*, 2002

[4] J.Aijuan, L.Hangtian, L.Shaolong, "A New High-Frequency AC Link Three-Phase Four-Wire Power Electronic Transformer" *IEEE*, 2006

[5] J.Aijuan, L.Hangtian, L.Shaolong, "A New Matrix Type Three-Phase Four-Wire Power Electronic Transformer" This work is sponsored by Shanghai Municipal Education Commission development foundation program *(No: 05EZ06)*

[6] M.Kang, P.N.Enjeti, I.J.Pitel, "Analysis and Design of Electronic Transformer for Electric Power Distribution System" *IEEE*, 1997

[7] J.W.Kolar, U.Drofenik, F.C.Zach, "VIENNA Rectifier II—A Novel Single-Stage High-Frequency Isolated Three-Phase PWM Rectifier System" *IEEE Transactions on Industrial Electronics, Vol. 46, No. 4*, August 1999

Hamid Mirmousa was born in Tehran, Iran in 1983. He received his B.Sc in Electrical Engineering from University of Science and Technology of Iran in 2005 and currently is M.Sc student of Sharif University of Technology. His research interests are power electronic systems and its application in the power systems. His recent work is simulation and design of power circuit and control strategy of electronic power transformer which is represented in this paper. He is working as an electrical engineer in the FARAB Company which is the main contractor of energy and water as well as oil and petrochemical projects located in Tehran, Iran.

Dr MohammadReza Zolghadri was born in Tehran, Iran in 1963.He received his B.Sc. and M.Sc. on 1989 and 1992 from Sharif University of Technology, Tehran, Iran and his Ph.D. on 1997 from Institute Polytechnic of Grenoble (INPG) France all in Electrical Engineering. Since 1992 he is with the Electrical Engineering Dept of Sharif University of Technology. His research interests are Power Electronics and its application in Power Systems and Variable Speed Drives.

A Half-Bridge DC/DC Converter for Plasma Cutting Machine

N. Sanajit and A. Jangwanitlert

Faculty of Engineering , King Mongkut' s Institute Technology Ladkrabang
Bangkok Thailand , Tel./ Fax. (662) 326-4550
Email: kjanuwat@kmitl.ac.th

Abstract-**This paper presents the design and construction of a plasma cutting machine by using a half-bridge dc/dc converter for negative direct current power supply and initial exciting arc by a high voltage high frequency circuit. When the gas passes a torch electrode, it will explode to be plasma arc and transfer to the work piece. This process has been designed for an output power of plasma cutting machine at 2 kW by employing Isolate Gate Bipolar Transistors (IGBTs) as switches. The switching frequency is 20 kHz. The plasma cutting machine can cut 1 mm of steel. When compared with a linear transformer plasma cutting machine, the designed plasma cutting machine has an efficiency quite similar. In addition, the plasma cutting machine is small, light, and compatible.**
Index Terms- **Plasma cutting machine, Half-bridge**

I. INTRODUCTION

Equipment for cutting metals is developed continuously. From Saw, gas to electric arc, technicians widely use electric cutting machine more than others. Because it is light, fast, and smooth for cutting metals, electric cutting machine can divide into two types. One is linear power supply electric cutting machine, the other one is switching power supply electric cutting machine. Linear power supply and switching power supply electric cutting machine, operating to cut metals, can achieve similarly good cutting quality. The arc jet between torch and workpiece is called " plasma jet". Plasma jet is very hot and can be utilized for cutting metals [1]. In the past, there are a few paper [2-3] mentions the construction of plasma cutting machine. Therefore, this paper presents the design and construction of dc switching power supply using a half-bridge inverter to provide ac high frequency voltage. The ac high frequency voltage is transferred to the high frequency center-tap transformer and then to the ultra-fast recovery diodes in order to rectify the voltage. The rectified voltage provides the negative dc voltage for main arc. The other part is connected with the coupling filter inductor. This part, creating high voltage high frequency (HV-HF), comes from Flyback converter of Television set in order to energize the molecules of gases at the torch electrode. Then, the molecules are separated in order to get plasma jet as shown in Fig. 1 [4]. The air flow rate is

controlled appropriately in the designed plasma cutting machine. This machine can be a prototype for high power applications in the future.

II. PLASMA ARC CUTTING

Plasma arc cutting is plasma occurred by heat together with electric arc. In this paper, the electric arc, occurred by providing an HV-HF, helps the discharge easily. In case of anode and cathode being not contacted for arc, the dc generator is provided as shown in Fig. 2. Arc, occurred by HV-HF, has two types: transferred and non transferred method as shown in Fig. 3. Transferred arc uses workpiece at anode electrode in order to provide plasma jet. The non-transferred arc does not use workpiece as anode electrode like transferred arc [5]. In this paper, non-transferred arc is used.

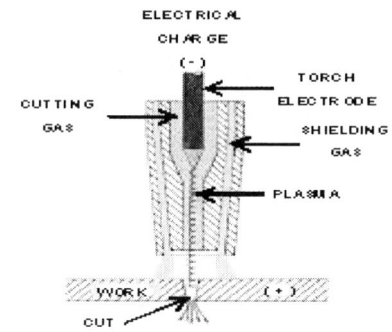

Fig. 1 Principle of torch electrode.

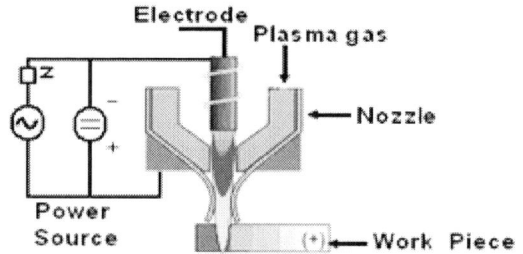

Fig. 2 Main Arc occurred by HV-HF.

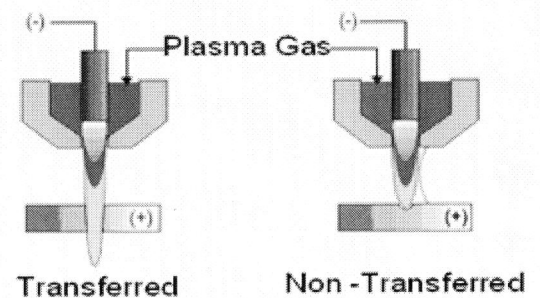

Fig. 3 Types of plasma arc using HV-HF.

III. PROTOTYPE DESIGN

From the above principle, the design of dc/dc half-bridge converter used as a switching power supply for plasma cutting machine [6] as shown in Fig. 4.

The output voltage (Vo) of dc/dc half-bridge converter is given by (1)

$$V_O = (\frac{N_S}{N_P})(V_{in}/2 - V_{GE(sat)})\frac{t_{ON}}{(T/2)} - V_D \qquad (1)$$

Where:
Np is the turn of primary winding of transformer,
Ns is the turn of secondary winding of transformer,
V_{GE} is the voltage across collector and emitter,
T_{on}/T is the duty cycle of IGBT,
V_D is the voltage across diode D_1 or D_2,
V_{in} is the dc input voltage.

The center-tap transformer reduces the ultra-fast diodes for rectifying. The specifications are given as follows:

Output power (P_o)   2000 W

Primary voltage (V_{pri})   150 ± 5% V

Output voltage (V_o)   150 V

Switching frequency (f_s)   20 kHz

Turn ratio of transformer ($Ratio$) 1:1

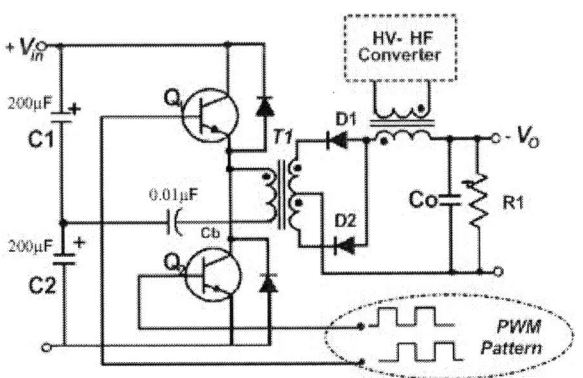

Fig. 4 DC/DC half-bridge converter.

Table I
Air Classifier [7]

Air	Nitrogen (N₂)	Oxygen (O₂)	Arkon (Ar)	Carbon dioxide (CO₂)
% by volume	78.08	20.95	0.93	0.03
% by weight	75.53	23.14	1.28	0.05

In this research, air is used as plasma gases. Gases are composed of nitrogen, oxygen, and inert gas. In this case, it is suitable for plasma jet.

Control Circuit for Plasma cutting Machine

Control circuit is defined for operating in the dc/dc half-bridge converter. In this case, the timing diagram shown in Fig. 5 is used to control air control valve for activating arc by an HV-HF method. Furthermore, the overall control circuitry is shown in Fig. 6. This figure includes the dc/dc half-bridge converter. The HV-HF circuit in dash line provides the arc from the beginning to the end cutting process. The HV-HF is built from flyback converter of Television set. The HV-HF signal is transferred by the coupling transformer used as a coupling inductor as well.

IV. TEST AND EVALUATION

The test is divided into two parts. One is testing for half-bridge dc/dc switching converter at various loads. The other test is for cutting metals at various parameters.

• Half-bridge dc/dc switching converter performance
The output power is tested from 200 W at light output power to 2 kW at rated output power by adjusting duty cycle from 34 % to 44 %. The efficiency and output current are shown in Figs. 7 and 8, where switching frequency is fixed at 20 kHz.

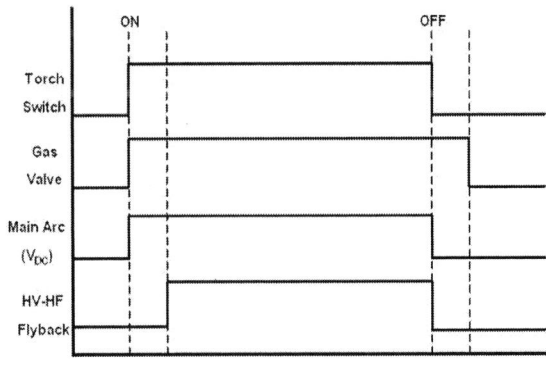

Fig. 5 Timing diagram of plasma cutting process.

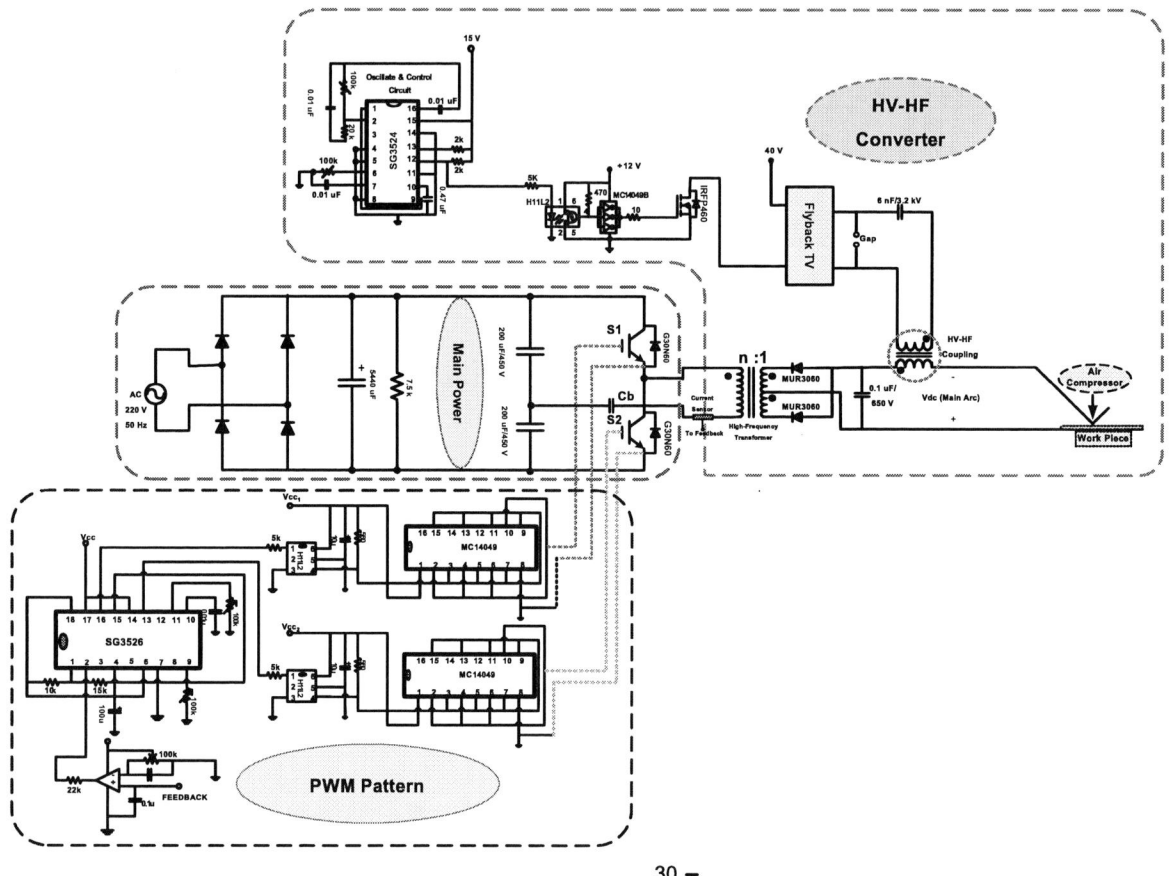

Fig. 6 Overall of plasma cutting machine.

Fig. 6 shows the principle of a half-bridge dc-dc converter for plasma cutting machine. It can be briefly described as follows: when the control circuit drives a half-bridge dc-dc converter, power flows from the center-tap transformer passing to the bridge rectifiers. In order to provide the negative polarity for plasma jet, the power supply is from the HV-HF circuit using Flyback converter. This converter is connected with the coupling with output inductor in order to activate the initial plasma jet.

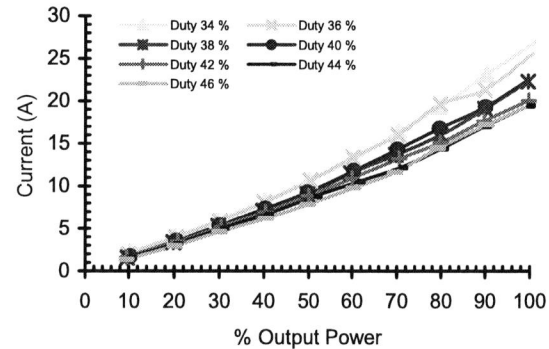

Fig. 8 Current vs. output power while adjusting duty cycle at frequency of 20 kHz.

From Figs. 7 and 8, when the duty cycle is increased, the efficiency of system is also increased. However, the output current is decreased due to the fact that time interval of conduction is increased. While the average voltage is increased, the same load current is decreased and dropped voltage is decreased as well.

Fig. 9 shows the efficiency vs. output power while adjusting switching frequency at the duty cycle of 35%. The trend of all curves is quite similar. When the output power is high, the efficiency is low, and vice versa.

Test for cutting metals

The test of the plasma cutting machine, which effects to cutting metals, is evaluated. The parameters for cutting are changed. The analyses are considered on the workpiece.

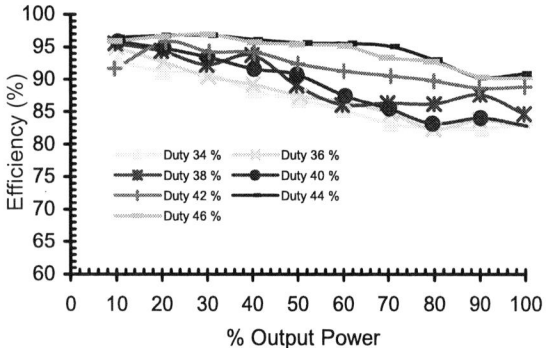

Fig. 7 Efficiency vs. output power while adjusting duty cycle at frequency of 20 kHz

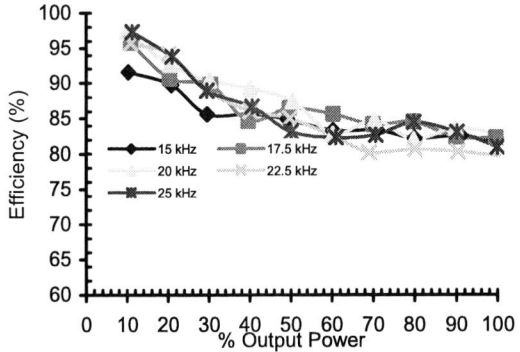

Fig. 9 Efficiency vs. output power while adjusting
switching frequencies at duty cycle of 35%.

Fig. 10 shows the experimental result that mentions the output inverter voltage and primary current as cutting carbon steel at duty cycle of 35% and switching frequency of 20 kHz. The result provides the Zero-Voltage Switching (ZVS) condition as a soft switching technique due to the fact that the primary current lags the output inverter voltage. That means the voltage across switch goes to be zero before turning on switch.

1) Test of cutting carbon steel : Consider duty cycle

The results from workpieces are analyzed. The air pressure is kept at 1.5 kg/cm^2 and the frequency is 20 kHz. However, the duty cycle is adjusted. It is found that at the duty cycle of 40%, the indentation or trail from cutting is continuous due to the period of switch conduction is longer than using the duty cycle of 20%. Also, it is found that the carbon steel cannot be cut at the duty cycle of 20% because the ion plasma is not completely occurred and period of the switch conduction is not enough. It makes the trail discontinuously as shown in Fig. 11.

2) Test of cutting carbon steel : Consider at air pressure

The results from workpieces are analyzed. The duty cycle is kept at 35% and the frequency is fixed at 20 kHz. However, the air pressure is adjusted. It is found that air pressure at 1 kg/cm^2 have the indentation from cutting better than using air pressure at 3 kg/cm^2 because the air

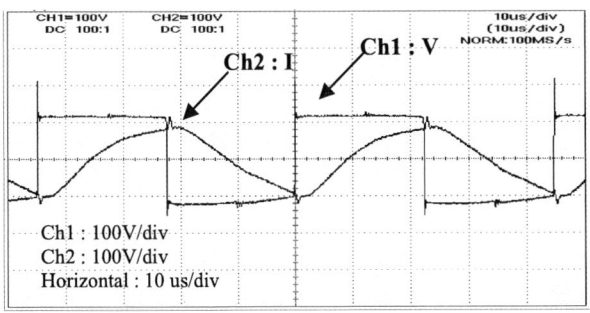

Fig. 10 Output inverter voltage and primary current as
cutting carbon steel at duty cycle of 35%,
frequency of 20 kHz.

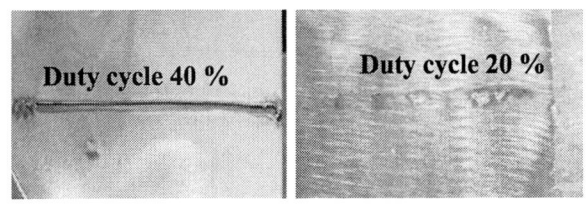

Fig. 11 Comparison of the trail on cutting carbon steel
between duty cycle at 20 % and 40%.

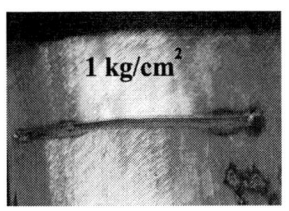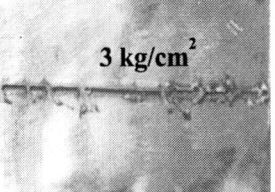

Fig. 12 Comparison of the trail on cutting carbon steel
between air pressure at 1 kg/cm^2 and 3 kg/cm^2.

pressure at 1 kg/cm^2 has the absolute ion at the electrode that can provide heat higher.

Therefore, the period of switch conduction and air pressure has the effect on cutting the metals. In order to achieve high efficiency, the parameters should be selected appropriately. The results are shown in Fig. 12.

3) Consider on cutting stainless steel plate

The results from the cutting stainless steel plate at duty cycle of 35 % and switching frequency of 20 kHz and air pressure of 1.5 kg/cm^2 are analyzed. It is observed that the indentation is quite well as shown in Fig. 13. When compared between the carbon steel and stainless steel at the same as switching frequency of 20 kHz and duty cycle of 35%, the indentation are quite similar; where cutting the carbon steel uses the air pressure at 1 kg/cm^2, whereas cutting the stainless steel uses the air pressure on 1.5 kg/cm^2. The reason is that the structure of stainless steel has a toughness more than that of carbon steel.

4) Comparison of the advantage between constructed prototype and the market type.

The physical structure and trail between constructed prototype and market type are compared.

The comparison results indicate that the plasma cutting prototype, which is built, is smaller and lighter than the market type because of the plasma cutting prototype applied from switching power supply as shown in Fig. 14. That can help the technician use and move this prototype easily. In addition, the trails between both types are not quite different significantly because the trail is based on air pressure.

1415

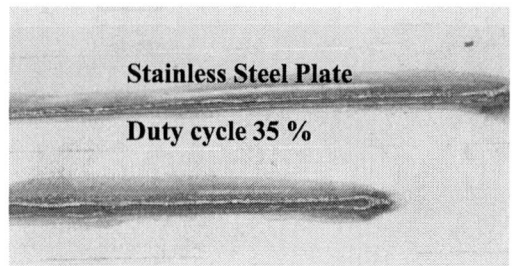

Fig. 13 The trail on stainless steel plate at 1.5 kg/cm^2, duty cycle of 35%, and frequency of 20 kHz.

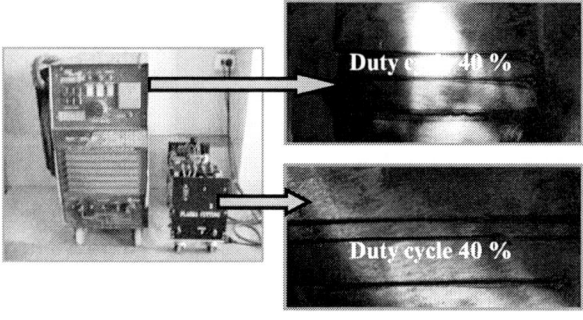

Fig. 14 Comparison of the structure and trail between constructed prototype and the market type. : market type (right Top), Plasma cutting prototype (right bottom).

V. CONCLUSION

From the experimental results, the plasma cutting machine is based on the dc/dc half-bridge converter. The rated power is 2 kW. It can cut metals following the design and construction. The plasma cutting machine is developed for the education or commercial system. It is made from switching power supply which is light, small. Also, the workpiece quality after cutting is better than the one from gas cutting process. The indentation or trail of cutting is small. It does not waste steel very much and the sparkle is less.

ACKNOWLEDGMENT

The authors wish to thank Weewan Engineering Co. Ltd., where suggest us to construct of the prototype.

REFERENCES

[1] S. Ramakrinshnan, M. Gershenzon, and F. Polivka, "Plasma generation for the plasma cutting process," *IEEE Trans. on Plasma Science*, Vol 25, No. 5, 1997, pp. 937-946.

[2] G. R. Kamath, "A passive coupled-inductor flying capacitor lossless snubber circuit for plasma cutting power supply," *Proc. in IEEE APEC Conf Rec.*, 2005, pp. 237-243.

[3] _____, "A passive reduced rating output rectifier snubber for plasma cutting power supply," *Proc. in IEEE APEC Conf Rec.*, 2006, pp. 85-91.

[4] B. Lucas and D. Hilton., "Cutting processes - plasma arc cutting - process and equipment considerations,"*http://www.twi.co.uk/professional/ protected/band_3/jk51.html*, May 1 ,2007, p.1.

[5] B. L. Bemis and G. S. Settles, "Ultraviolet Imaging of the Anode Attachment in Transferred-Arc Plasma Cutting," *IEEE Trans. on Plasma Science*, Vol. 27,No. 1,February 1999,pp.44-45.

[6] M. H. Rashid, *Power Electronics*, Prentice Hall, 2004.

[7] C. Landry, "Improving plasma cutting in sheet Metal applications," *http://archive.metalforming magazine .com/1997/09/plasma/997plasma.htm*, May 5, 2006, pp.1-5.

Narongrit Sanajit received B.Eng. in Electrical Engineering from Mahanakorn Univercity of Technology, Thailand, in 2003. He is currently a M.Eng. student at King Mongkut's Institute Technology Ladkrabang. His research interests include power electronics and power converter.

Anuwat Jangwanitlert received B.Eng. and M.Eng. in Electrical Engineering from King Mongku's Instutute of Technology Ladkrabang and Chulalongkorn University, Thailand, in 1991,1995, respectively. He graduated Ph.D. in Electrical Engineering from the University of Arkansas, USA in 2004. He is currently an assistant professor at King Mongkut's Institute Technology Ladkrabang. His research works are focused on power switching converter and power electronic applications.

Ripple Estimation for Paralleled Converter System with Automatic Interleaving Function

Teruhiko Kohama*, Ryota Tsunesada*, and Tamotsu Ninomiya**

*Department of Electrical Engineering, Fukuoka University,8-19-1 Nanakuma, Jonan-ku, Fukuoka 814-0180, JAPAN

*Department of EESE, Kyushu University, 744 Motooka, Nishi-ku, Fukuoka 819-0395, JAPAN

Abstract—Current ripple for any paralleled converter system under ideal out-of-phase operation is estimated with simplified circuit model. The ideal out-of-phase operation is achieved by introducing automatic interleaving function. Relationships between the ripple and circuit parameters such as duty ratio and number of modules are revealed which provide simple design guideline for ripple-minimized paralleled converter system.

Index Terms — interleaving operation, paralleled converter system, ripple cancellation, zero ripple

I. INTRODUCCTION

Parallel operation of DC-DC converter module is a key issue to increase current capacity of power supply. Fast dynamic response is also achieved by reducing inductance of smoothing filters in converter modules. However, small inductance causes large current ripple which results

in large voltage ripple in the output. Interleaving technique is applied to reduce the voltage ripple by cancelling the current ripple in the output capacitor [1-3]. As an example a two-paralleled converter system with interleaving control is shown in Fig.1. Current ripples in modules are cancelled each other in output capacitor under out-of-phase operation. It is known that the ripple is minimized when the phases between converter modules are equal to $2\pi/N$[rad], where N is the number of converter modules. Normally, special design consideration or pre-adjustment is necessary to achieve the ideal out-of-phase operation for individual paralleled system.

We proposed an automatic interleaving control for any paralleled converter system[4,5]. It reduces voltage and current ripples in output stage. However, the effectiveness of the ripple cancellation depends on the number of converter modules and their duty ratios. For instance Fig.2 shows the current waveforms for two-paralleled converter system operating at different duty cycle. Even though Figs.2(a) and (b) are operating under ideal out-of-phase condition, the effectiveness of ripple cancellation is different. In this paper simple ripple estimation for any paralleled converter system under ideal out-of-phase operation is performed with simplified circuit model. Experimental results are shown to confirm the effectiveness of the model.

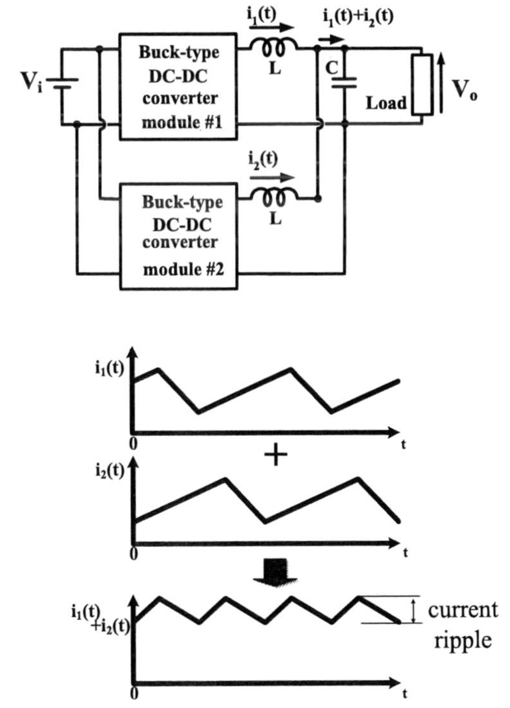

Fig.1 Current waveforms for two-paralleled converter system.

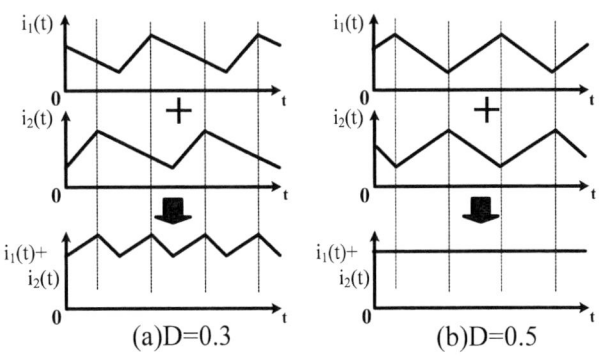

Fig.2 Current waveforms for two-paralleled converter system.

978-1-4244-0644-9/07/$25.00 ©2007 IEEE

II. AUTOMATIC INTERLEAVING TECHNIQUE

A. Principle of automatic interleaving technique

Automatic interleaving function proposed in [5] is shown in Fig.3. In Fig.3 module #1(master) sends a synchronizing signal to the next module #2(slave) with time delay of τ. The module #2 restarts the own switching operation by the synchronizing signal and sends a synchronizing signal to the next module #3 with the same delay τ. The delay τ is determined by V_R which depends on the total number of modules N including master. The master has a current source I_m in series with resistance R. Other resistances R in slave modules are connected in parallel with the master. Therefore, V_R is determined by the following equation.

$$V_R = V_m/N \qquad (1)$$

where, V_m is an amplitude of sawtooth-wave in the oscillator which is given by

$$V_m = RI_m. \qquad (2)$$

From Fig.3 there is no difference between master and slave module except for "Sync in" signal and current source I_m. Detailed circuit diagram for slave module is shown in Fig.4(a). Each module has a sawtooth oscillator, synchronizing circuit and delay circuit. Sawtooth oscillator[6,7] consists of a capacitor C, two comparators(CP_1 and CP_2) and a charging/discharging circuit. In the normal operation, capacitor C is charged and discharged by constant current source I_{ref1} and I_{ref2}. Toggle switch SW is controlled by RS-flipflop whose state is determined by CP_1 and CP_2 with the threshold of V_{cmax} and V_{cmin} respectively. Once synchronizing signal with narrow pulse is fed into the base terminal of the transistor next to capacitor C, V_c dropped into V_{cmin} immediately and restarts the charging operation. Synchronizing signal with time delay τ is obtained by using comparator CP with the threshold of V_{TH}, which is given by

$$V_{TH} = V_R + V_{cmin}. \qquad (3)$$

In the case of N=2, for example, $V_R = V_m/2$, and CP changes the output state when $V_c = V_m/2 + V_{cmin}$ as shown in Fig.4(b). As a result, the next synchronizing signal is generated with time delay of $\tau = T/2$, where T is a switching period of master module. In the case of N=3, V_R decreases

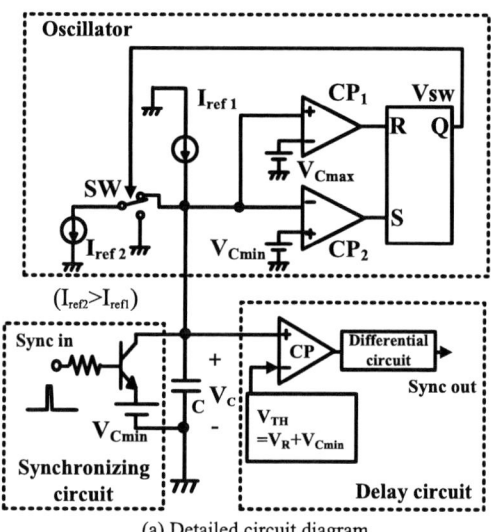

(a) Detailed circuit diagram

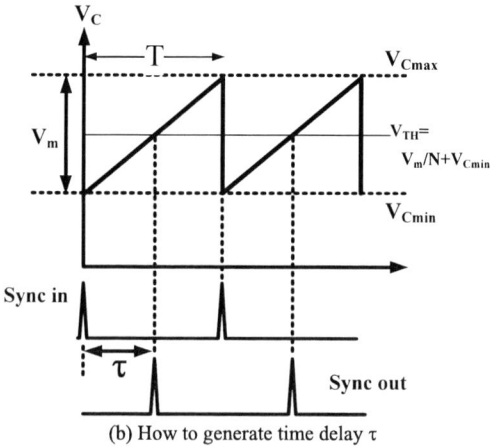

(b) How to generate time delay τ

Fig.4 Detailed circuit diagram and its operation in slave module.

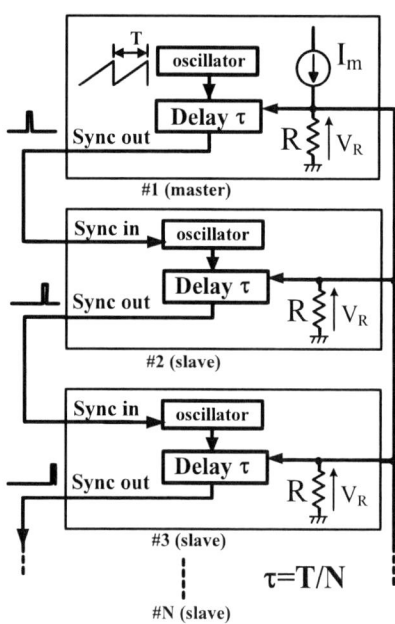

Fig.3 Principle of automatic interleaving technique.

to$V_m/3$ and V_{TH} is set to $V_m/3+V_{cmin}$ which makes delay of T/3. Similar discussion is given for any number of N. From (1) and sawtooth-waveform in Fig.4(b), τ is determined by following equation.

$$\tau=(V_R/V_m)T=T/N \qquad (4)$$

Therefore no pre-adjustment or individual circuit consideration is necessary for any paralleled system. Delay τ is automatically determined instantly by connecting the converter modules. This control circuit is easily implemented into conventional PWM controller because of simple analog circuit compared with other interleaving technique.

B. Experimental results

Fig.5(a) shows experimental result of sawtooth oscillators for two paralleled converter system(N=2), where $\tau=T/2$ and ideal out-of-phase operation is achieved. Similar results are obtained for N=3, N=4 in Figs.5(b) and 5(c) respectively.

III. RIPPLE ESTIMATION

A. Simplified circuit model for ripple estimation

A module of buck type DC-DC converter in Fig.6(a) is simply represented by a pulse voltage source V_p and a smoothing LC filter as shown in Fig.6(b). The shape of V_p is determined by input voltage V_i and duty ratio D of the module. In order to simplify the circuit model all modules in paralleled converter system are assumed to be on the same condition except for phase delay. During interleaving operation, voltage ripple is relatively small compared to DC level of V_o. Therefore the shape of module current $i_1(t)$ in Fig.6 is assumed saw-tooth waveform. From the above discussion the shape of current ripple in module #1 is shown in Fig.7(a), where the slope m_1 and m_2 are given as follows.

$$m_1 = \frac{V_i - V_o}{L} \qquad (5)$$

$$m_2 = \frac{V_o}{L} \qquad (6)$$

From Fig.7(a) amplitude ΔI of the ripple $\Delta i_1(t)$ is determined by

$$\Delta I = m_1 DT_S = m_2(1-D)T_S \qquad (7)$$

If Vo is nearly constant during a switching period of T_s, Fig.6(b) is represented by triangular current source as shown in Fig.7(b).

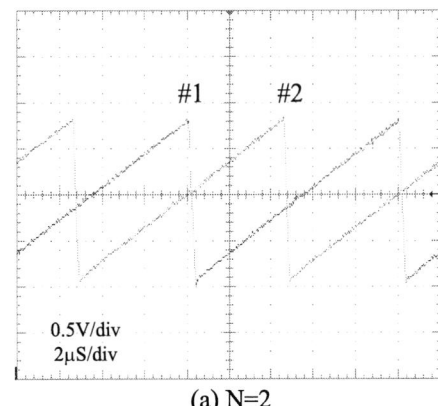

(a) N=2

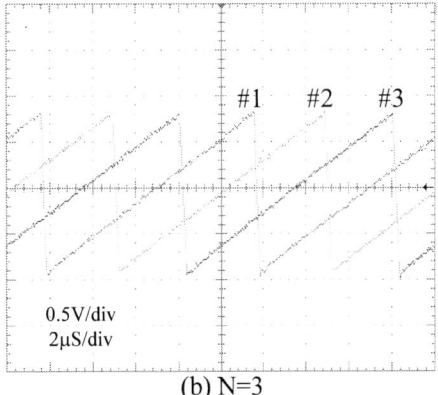

(b) N=3

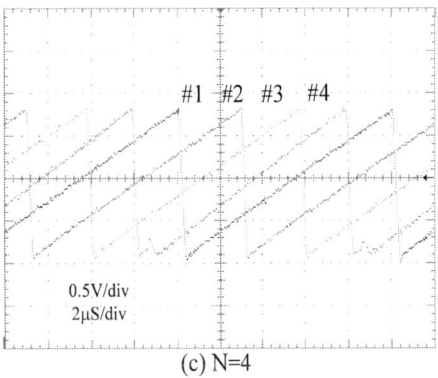

(c) N=4

Fig.5 Waveforms of oscillator with automatic interleaving control.

The ripple $\Delta i_1(t)$ in module #1 is derived from Fig.7(a).

$$\Delta i_1(t) = \begin{cases} m_1 t - \Delta I/2 & (0 < t < DT_S) \\ \Delta I/2 - m_2 t & (DT_S < t < T_S) \end{cases} \qquad (8)$$

Current ripple in module #k under ideal interleaving control is also given by

$$\Delta i_k(t) = i_1\left(t - (k-1)\frac{T_S}{N}\right) \qquad (9)$$

1419

where, N is the number of converter modules.

B. Ripple in output stage

From Fig.7(b), a circuit model focused on current ripple for N-paralleled converter system is shown in Fig.8. From this figure the current $i_c(t)$ in output capacitor C is given by

$$i_C(t) = \sum_{k=1}^{N} \Delta i_k(t) . \tag{10}$$

Amplitude ΔI_C of the current ripple $i_c(t)$ is given by

$$\Delta I_C = Max(i_c(t)) - Min(i_c(t)) . \tag{11}$$

Where, Max(x) is a function for maximum value of x(t) in one switching period.

Min(x) is a minimum value of x(t) for one switching period.

Voltage ripple $\Delta v_c(t)$ across the output capacitor is easily obtained by integrating Eq.(10) as follows.

$$\Delta v_C(t) = \frac{1}{C} \int i_C(t)dt . \tag{12}$$

Therefore the amplitude ΔV_C of $\Delta v_c(t)$ is given by

$$\Delta V_C = Max(v_c(t)) - Min(v_c(t)) . \tag{13}$$

IV. DISCUSSION

A. Ripple estimation and simple design guideline

The ripples ΔI_c is calculated with Mathcad. Figure 9

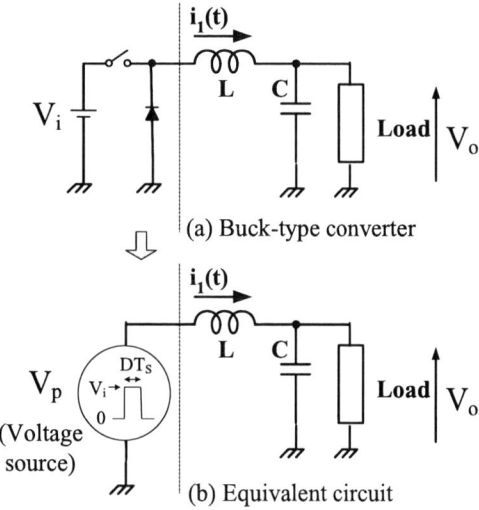

(a) Buck-type converter

(b) Equivalent circuit

Fig.6 Simplified circuit for a buck-type converter module.

shows the relationships between the current ripples and number of modules N in the paralleled system as a parameter of duty ratio D. Current ripple ΔI for one module is assumed 5[A] as an example. In Fig.9(a) ΔI decreases to zero as N increases to 10 for D=0.1. Figs.9(b), 9(d) and 9(f) show that the ripple becomes zero at N=5 and 10 for D=0.2, 0.4 and 0.6. In case of D=0.5, zero current ripple is achieved at N=2,4,6,8,10. On the other hand the ripple for D=0.3 becomes zero only at N=10.

From Fig.9, it is clear that zero ripple operation is achieved by the condition of

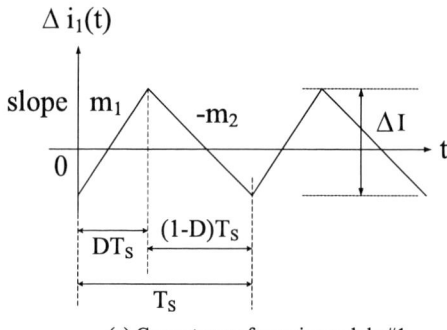

(a) Current waveforms in module #1

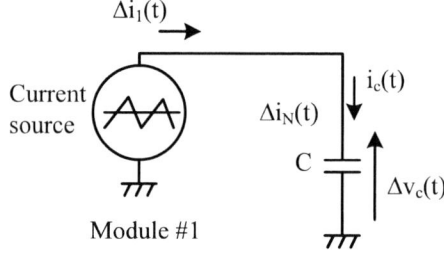

(b) Equivalent circuit model for module #1

Fig.7 Ripple estimation model for module #1

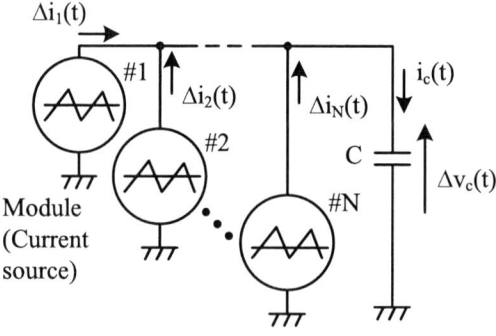

Fig.8 Simplified circuit model for N-paralleled converter system

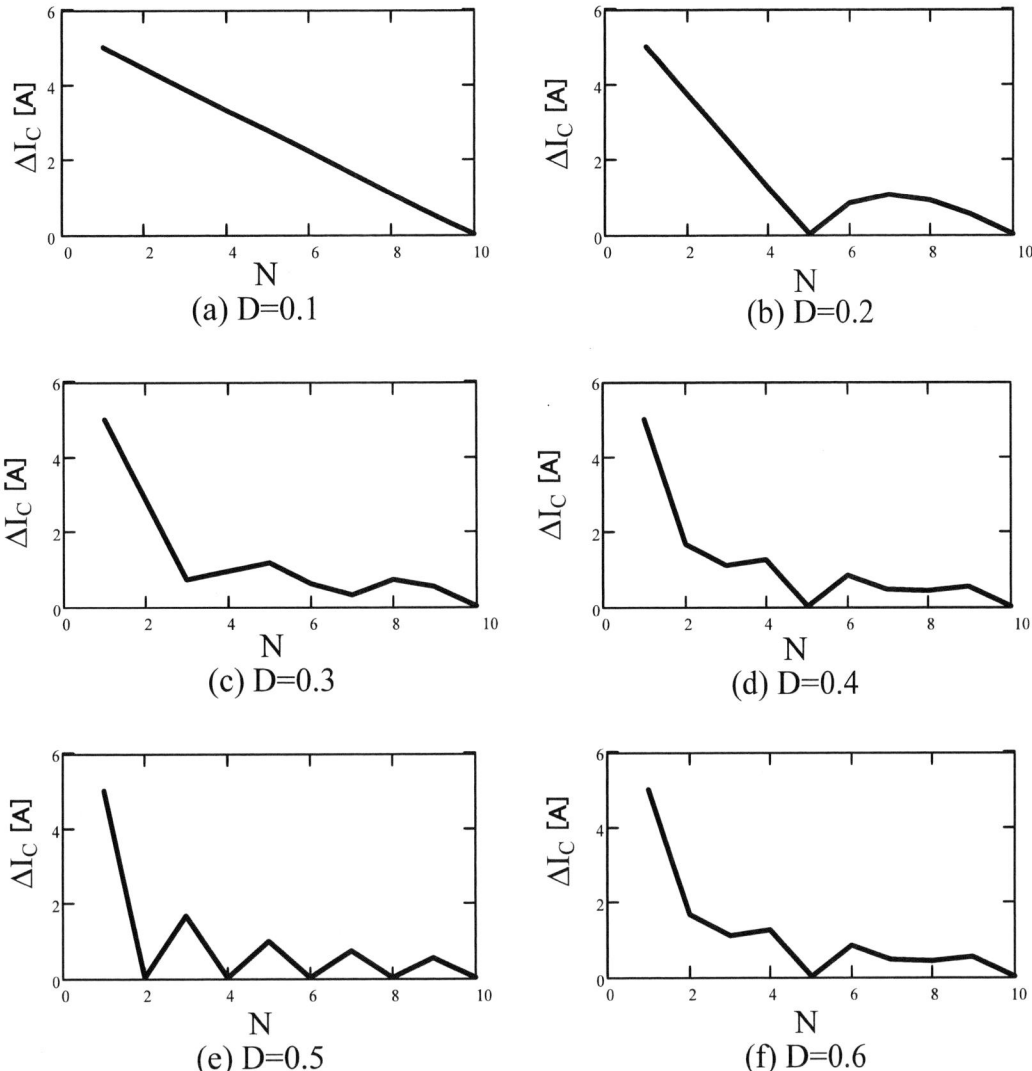

Fig.9 Current ripple estimation for N-paralleled converter system.

$$DN = \mathrm{int}\, eger \,. \qquad (14)$$

Equation (14) shows that zero ripple converter system can be realized by selecting DN =1,2,3.... Although D depends on Vi, Vo, load, and circuit topology of the module, the principle of zero ripples is essential to decrease LC filter dramatically and to achieve fast dynamic response of output voltage. The above ripple estimation is based on ideal interleaving operation. Proposed automatic interleaving control is suitable for that operation.

B. Experimental results

Figs.10(a) and 10(b) show relationships between current ripple and duty ratio for N=2, N=3 respectively. Current ripple ΔI_c is normalized by the ripple in single module.

In Fig.10(a), the ripple is minimized at D=0.5 that satisfies Eq.(14)(DN=1).

ΔI_c for N=3 is also shown in Fig.10(b). Minimum current ripple is achieved at D=0.33(DN=1) and D=0.66(DN=2). Both experimental and theoretical results are agree well that shows the effectiveness of proposed circuit model.

V. CONCLUSIONS

Simple ripple estimation for any paralleled DC-DC converter system with automatic interleaving control is performed through simplified circuit model. Zero ripple converter system can be achieved under specific condition of duty ratio and number of modules.

Experimental results show the effectiveness of proposed ripple estimation with simplified circuit model.

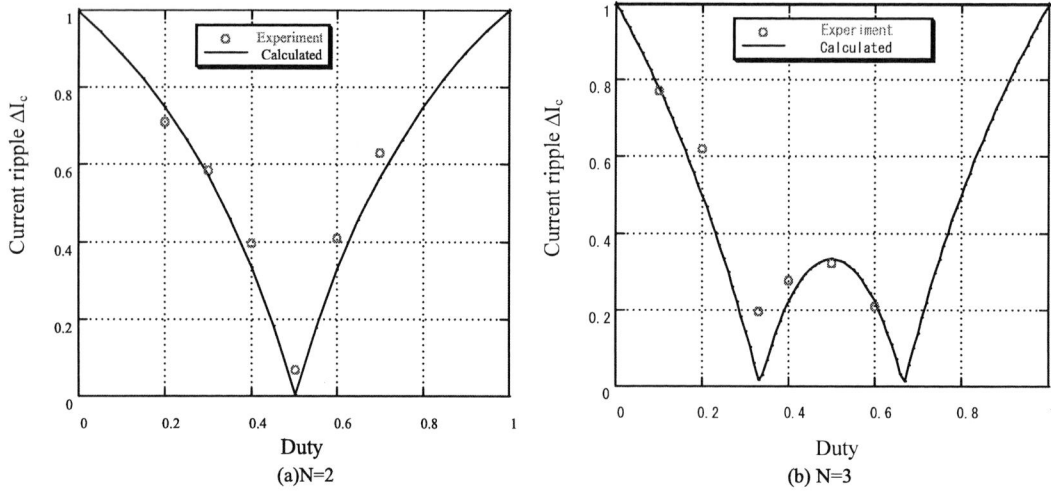

Fig.10 Current ripple vs. duty ratio of converter module.

REFERENCES

[1] B.A.Miwa, D.M.Otten, and M.F.Schlecht, "High efficiency power factor correction using interleaving techniques," IEEE Applied power Electronics Conference, pp.557-568, 1992.

[2] D.J.Perreault and J.G.Kassakian,"Distributed Interleaving of Paralleled Power Converters," IEEE Trans. on Circuit and Systems, Vol.44, No.8, pp.728-734, 1997.

[3] J.Wei and F.C.Lee, "A Novel Soft-Switched, High-Frequency, High-Efficiency, High-Current 12V Voltage Regulator – The Phase-Shift Buck Converter," IEEE 18th Applied Power Electronics Conference, pp.724-730, 2003.

[4] T.Kohama, G.Endo, H.Shimamori, T.Ninomiya,"New Synchronizing Circuit Suitable for Paralleled Converter System with Automatic Interleaving Operation," Proceedings of IEEE 19th Applied Power Electronics Conference and Exposition, pp.647-653, 2004.

[5] T.Kohama, G.Endo, H.Shimamori, T.Ninomiya, "Simple Multiphase Control for Paralleled Converter System", IEICE Trans. on communication, Vol.E88-B, No.12, pp.4636-4642, 2005.

[6] "High Speed PWM Controller UC3823A,B/3825A,B", Unitrode Product & Applications Handbook, pp.6-158~6-165 1995-1996.

[7] "High Speed PWM Controller UC3824", Unitrode Product & Applications Handbook, pp.6-166~6-172, 1995-1996.

Author Index

A

Abdi, Ehsan 1096
Abe, Seiya 1388
Abjadi, N. R. 1442
Achara, P. 394
Adélaide, L. 569
Adya, A. 731
Afjei, E. 722
Agarwal, Pramod 1810
Agarwal, Vineeta 1891
Ahmadian, H. Molla 1147
Ahn, Jin-Woo 1857
Almardy, M. 677
Alonge, F. 959
Amaral, Acácio M. R. 587, 643
Amirudin, Dessy 534
Amrane, F. 569
An, Young-Joo 1527
Andersen, P. Scavenius 886
Ang, Y. 382
Ang, Yong-Ann 376
Aodsup, K. 937
Arab, G.R. 1449
Aree, P. 703
Arvindan, A. N. 480
Ashaibi, Ahmed Ali 1242
Ataei, S. 722
Athab, Hussain S. 869, 874
Attaviriyanupap, Pathom 1102
Auger, Francois 1368
Ayob, S. M. 1274, 1363, 1682
Azli, N. A. 475, 1041, 1274, 1363, 1682

B

Bac, Nguyen Xuan 1501
Baharom, R. 1626
Baiju, M.R. 1047
Banerjee, Subrata 812
Bartholet, M.T. 257
Batzies, Ekkehard 1316
Bhat, A.K.S. 677
Bhuvaneswari, G. 310
Bi, C. 1082
Bina, M. Tavakoli 465, 1060, 1065, 1799
Binder, A. 249
Bingham, C. M. 382
Bingham, Chris 376
Binh, Tran Cong 1195
Biswas, S. K. 1352, 1605
Blaabjerg, Frede 226, 541, 1247, 1376
Boonchiam, P. N. 937, 1851
Boonyaroonate, Itsda 1078, 1383
Bosing, M. 1160
Branco, P.J. Costa 917
Brauer, Helge J. 716
Buatti, Gustavo M. 643

Bunlaksananusorn, C. 977, 1575

C

Cangemi, T. 959
Cardoso, A. J. Marques 587, 643
Carstensen, Christian 912
Chalermyanont, Kusumal 295
Champa, P. 703
Chan, K. W. 1394, 1727, 1804
Chan, Shun-Yu 305
Chang, David 305
Chang, Tsin-Yuan 581
Chang, Y. D. 1321
Chao, Ma Xian 665
Chatratana, S. 1495
Chaudhari, M. A. 1708
Chen, Jiaxin 510
Chen, L. 1538
Chen, Sufen 1262
Chen, W. C. 440
Chen, Wei 1636
Chen, Y. H. 456, 1278
Chen, Y. M. 1321
Chenfeng, Yang 745
Cheng, Chien-Lung 749, 1703
Cheng, K. W. Eric 1691, 1697
Cheng, K.W.E. 1727
Cheng, Qiang 1330
Chengfeng, Yang 270, 427
Chereau, Vinciane 1368
Chern, Shyi-Ching 749, 1703
Cheung, N.C. 1727
Cheung, Norbert C 1691, 1697
Chi, Chien-An 388
Chiang, Wen-Jung 824
Chien, F.T. 660
Chin, Li-Yuan 305
Chivite-Zabalza, F. Javier 788, 796, 804
Cho, B. H. 401
Cho, Kyu Min 665, 1142
Choi, S. J. 401
Choudhary, Sonika 1757
Chrin, P. 1575
Chudamani, R. 1827
Chudoung, Nakharet 1213
Chun, Tae-Won 1857
Chunkag, V. 863
Ciobotaru, M. 226
Colak, Baris 1507
Corradini, L. 600
Cosic, A. 1301
Cruden, A. 1182

D

Dahlan, N.Y. 527
Dahono, P. A. 1267
Dahono, Pekik Argo 534

Author Index

Dai, Z. 1885
Dananjayan, P. 626
Davat, Bernard 1
Deb, N. K. 1352, 1605
Dehbonei, H. 1657
Deleroi, W. 1495
Deng-Em, S. 1851
Deni, ... 534, 1267
Densei-Lambda, K.K. 280
Desai, Hardik P. 829
Dhomane, G. A. 1590
Dick, Christian P. 448
D'ippolito, F. 959
Ditmanson, C. 556
Doki, Shinji 999
Doncker, R. W. De 907, 912, 1160
Doncker, Rik W. De 213, 327, 333, 448, 710, 716
Dong, Lei 1691
Dong, Ming-Chui 607, 614
Dong, Yang 1340
Dorkmai, Pramoch 697
Dorrell, David G 886, 922, 1167, 1174
Duan, S.X. 551
Duan, Shanxu 836, 842
Dwivedi, Avneesh 1757
Dzung, Phan Quoc 1195, 1202, 1501

E

Ekkaravarodome, Chainarin 1383
Ertan, H. Bulent 1507
Eskandari, B. 1060, 1065

F

Fang, D. Z. 1394, 1804
Fang, Kuo-Lun 1610, 1712
Fang, Tzu-Hsuan 1717
Fei, Wanmin 350, 354, 1672
Ferraz, Antonio 817
Ferreira, O.C. 1017
Fidler, Peter 327
Fingerhuth, S. 1160
Finney, S.J. 299, 1242
Finney, Steve J. 1255
Foroosh, S. Chini 1465
Forsyth, Andrew J. 788, 796, 804
Foster, M. P. 382
Foster, Martin 376
Fuengwarodsakul, Nisai H. 710
Fukuda, Shoji 1070
Fukushima, K. 1885

G

Gairola, Sanjay 738, 899
Gao, F. 1247, 1376
Garg, Vipin 310
Geethalakshmi, B. 626

Goel, P.K. 941
Gonthier, L. 322
Gopinath, Anish 1047
Goyal, Devendra 1520
Grant, D M 368
Grantham, Colin 1284
Gruber , W. 574
Guan, Xiaohan 994, 1752
Gueldner, H. 1006
Guldner, H. 556
Guo, Youguang 275, 510, 1662
Gupta, H.O. 1810
Gupta, J.R.P. 731

H

Hai, Quach Thanh 1033
Hajian, M. 1449
Hamzah, M.K. 527, 1626
Hamzah, N.R. 527, 1626
Han, Ying-Duo 607, 614
Hansen, P. E. 886
Haque, M. Tarafdar 620
Harada, Y. 1885
Hasegawa, Masaru 1543
Hellinger, R. 1006
Hennen, Martin D. 716, 907
Hew, W.P. 1514
Heyun, Lin 270, 427, 745
Higuchi, Kohji 280
Hinkkanen, Marko 406
Hirokawa, Masahiko 1388
Hirota, Atsushi 1740
Ho, Shine-Tzong 498, 1788
Hoang, Nguyen Minh 1195, 1501
Hofmann, W. 781, 1538
Hotait, Hadi A 299, 1255
Hothongkham, Prasopchok 1236
Hsieh, C. T. 1567
Hsu, Chih-Jen 286
Huang, P. L. 1321
Hung, Tsung-You 1762, 1767
Hwu, K. I. 338, 456, 692, 1278

I

Idris, Z. 527
Iov, F. 226
Ishitobi, Manabu 504
Islam, S. 1834
Iso, Osamu 1102

J

Jang, B.H. 1657
Jangjaempradit, Saksit 1641
Jangwanitlert, A. 989, 1412
Janjornmanit, Suchart 1327
Jayashree, E. 1555

Author Index

Jeevananthan, S. 1221
Jegathesan, V. 1677
Jerome, Jovitha 1677
Jeung, Giwoo 1092
Jian, Guo 270, 427, 745
Jou, Hurng-Liahng 493, 824
Jovanovic, Milutin G 922
Junge, Christian 1533
Jwo, W. S. 1560

K

Kadir, M. N. Abdul 1514
Kaewsingha, Aswin 1327
Kamnarn, U. 863
Kamper, M.J. 420, 1017, 1295
Kando, M. 488
Kang, Yong 551, 836
Kano, Masaru 414
Kanthaphayao, Y. 863
Kanzi, K. 465
Karunakar, K 1620
Karutz, P. 574
Kasal, Gaurav Kumar 357
Kasper, K. A. 1160
Kavitha, A. 595
Kazimierczuk, Marian K 1136
Kennel, R.M. 1017
Kerz, O. 363
Khaehintung, Noppadol 847, 1429
Khajeh, A. 1455
Khalil, Ahmed G. Abo- 1471
Khan, P. K. Shadhu 869, 874
Khan-Ngern, Werachet 460, 1335
Khomfoi, Surin 1055, 1228
Khun, C. 488
Kim, Dong-Hun 1092
Kim, H. S. 1846
Kim, Hee Jun 665, 1142
Kim, Heung-Geun 1092
Kim, I. C. 1846
Kim, In Dong 1092
Kinnaraes, V. 1483
Kinnares, V. 1356, 1489
Kinnares, Vijit 1236
Kittiratsatcha, S. 977
Ko, S. H. 1657, 1846, 1846
Ko, T.K. 1657
Ko, Yi-Pin 388
Kobayashi, Takayuki 265
Kock, H.W. De 1017
Koenig, Andreas 327
Kohama, Teruhiko 1417
Kok, W. Sae- 368
Kolar, J.W. 257, 574
Kongsuk, P. 937
Kongthawornwattana, P. 977
Konig, Andreas 448

Krein, Philip T. 221
Krismadinata, ... 1290
Kubota, Hisao 265
Kulvitit, Youthana 342, 697
Kumar, S. Ganesh 1632
Kumar, S. Krishna 1632
Kumchaiyo, Ruthapong 1078
Kunakorn, Anantawat 1429
Kuo, J. S. 440
Kuo, Jian-Long 1717, 1722
Kurokawa, F. 968, 1398
Kusuhara, Yoshito 954
Kwok, K. W. 1727
Kwok, Y. L. 1727
Kwon, Soon Kurl 504

L

Laczynski, T. 1645
Lafzi, A. 620
Lai, Y. M. 1262
Lai, Yen-Shin 1586
Lakhdari, Z. 569
Lan, Yi-Hung 749
Lee, Chien-Min 1586
Lee, Dong-Choon 1471
Lee, Dong-Hee 1527, 1857
Lee, Hong Hee 1027, 1033
Lee, S. R. 1657, 1846
Lee, S. W. 1657, 1846
Lee, Yuang-Shung 286, 388
Lei, Dong 949, 1340, 1697
Lei, Yuzhou 291, 1777
Leibfried, T. 363, 726
Lenke, Robert U. 213
Lenwari, W. 470
Leou, Rong Ceng 546
Lerdudomsak, Smith 999
Li, X. 551
Li, Y. J. 440
Liang, C. 1376
Liao, C.N. 660
Liao, Xiaozhong 1691
Lijie, Wang 949
Lim, P. Y. 475, 1041
Lim, S. H. 1846
Lim, T.C. 299
Lin, Chang-Hua 1610, 1712, 1762, 1767
Lin, Chih-Hong 1549
Lin, H. C. 1567
Lin, Hung-Chih 581
Lin, Min 1423
Lipo, Thomas A. 1308
Liu, B.Y. 551
Liu, Bangyin 836, 842, 1636
Liu, Dikai 275
Liu, Fei 842
Liu, Maw-Yang 1610, 1712

Author Index

Liu, Xian-Lin 1722
Liu, Yi-Hwa 546
Liu, Yuanchao 291, 1615, 1752, 1777
Liu, Z. 551
Loh, P. C. 1247, 1376, 1620
Loh, Poh Chiang 541
Loron, Luc 1368
Lu, Haiyan 275
Lu, Y. 1727
Lu, Zhengyu 354, 1088
Luomi, Jorma 406

M

Ma, Yu 632, 1088, 1601
Macheiner, P. 880
Madawala, U. K. 648, 654
Makany, Ph. 569
Makino, Tomoaki 1740
Manmek, Thip 773
Mao, Peng 1615
Markadeh, Gh. R. Arab 1442
Marques, Gil D. 636
Martin, F. 363
Martins, J. F. 894, 1875
Masoum, Amir S. 767
Masoum, M.A.S. 767, 1834
Massoud, A.M. 299
Massoud, Ahmed M. 1255
Massoud, Ahmed 1242
Matsui, Keiju 1543
Matsui, N. 1398
Matsui, Y. 1109
Matsui, Yasuaki 1102
Matsuo, K. 1686
Matsuse, K. 394, 685
Matsuse, Kouki 521, 1460
Mattavelli, P. 600
Mattavelli, Paolo 760
Mcmahon, Richard 1096
Medagam, Peda V. 1477
Mekhilef, S. 1514
Meng, Peipei 632
Mertens, A. 1645
Meyer, Christoph 213
Milani, A. Roshan 620
Miri, A. M. 726
Mirmousa, H. 1404
Mishima, Tomokazu 563
Mithulananthan, N. 937
Mittal, A.P. 731
Mittal, Raghu K. 1757
Miura, T. 1686
Miyamoto, Hiroyuki 1773
Moallem, Ali 983, 1147
Modak, J. P. 1708
Moghani, J. S. 1455
Mondal, N. 1352, 1605

Moon, Y.H. 1657
Morimoto, Masayuki 1641, 1773
Morita, Katsuaki 1773
Moses, Paul S. 767
Mossner, K. 363
Mudannayake, Chathura P. 773
Mun, Sang Pil 504, 563
Mura, Florian 213
Muraoka, Hidekazu 563
Murthy, S.S. 941, 1123, 1757

N

Nabeshima, Takashi 858, 1423, 1734
Naetiladdanon, Sumate 755
Nagai, Satoshi 1740
Nakagawa, Shin 954
Nakanishi, Hirotaka 858, 1734
Nakano, Kazushi 280
Nakano, Tadao 858, 1734
Nakaoka, Mutsuo 504, 563
Nakayama, Asahi 954
Nandhakumar, R. 1221
Nathakaranakule, Adisak 1383
Navi, K. 722
Nazarzadeh, Jalal 1782
Neammanee, B. 1129
Neuhaus, Christoph R. 710
Ngern, W. Khan- 488, 515, 1667, 1794
Ngoc, Ha Pham 1102
Nguyen, Binhminh 1434
Nho, Eui-Chel 1527
Nho, Nguyen Van 1027, 1033
Nia, S.Hosein 1449
Ninomiya, T. 1885
Ninomiya, Tamotsu 954, 1388, 1417
Nishijima, Kimihiro 858, 1423, 1734
Nishimura, Jun 1460
Noguchi, Toshihiko 414, 1595, 1651
Noor, S.Z. Mohammad 1626
Norigoe, I. 1885
Nussbaumer, T. 257, 574

O

Obata, S. 671
Ogura, K. 1109
Oh, Won Seok 1142
Oka, Kazuo 521, 1460
Okuma, Shigeru 999
Omori, Hideki 563
Opanuruk, Puckapon 342
Oranpiroj, Kosol 318
Owatchaiphong, Satit 912
Ozdemir, Engin 1055
Ozdemir, Sule 1055

Author Index

P

Pai, Kai-Jun 1762, 1767
Pal, Jayanta 812
Palandurkar, M.V. 1708
Panda, Sanjib K. 852
Park, Hong-Geuk 1471
Park, J. H. 401
Pashajavid, E. 465
Passal, A. 322
Patel, H. K. 829
Pavitra, G. 1757
Peng, S.T. 930
Phuong, Le Minh 1195, 1202, 1501
Piboonwattanakit, K. 1667
Piippo, Antti 406
Pinto, A.J.P. 1123
Pires, A. J. 894, 917
Pires, V. Fernao 894, 1875
Plum, Thomas 327, 333
Pothana, Aravind 1152
Pothi, N. 1208
Pourboghrat, Farzad 1477
Prasad, Dinkar 812
Prasertsit, Anuwat 295
Premrudeepreechacharn, Suttichai 318, 1208
Pusorn, W. 1851

Q

Qian, Zhaoming 632, 1088, 1601
Qu, Yilong 1822, 1880

R

Rafael, Silviano 917
Rahim, Nasrudin Abd 1290
Rahimzadeh, S. 1799
Rahman, Muhammed Fazlur 1284
Rakpenthai, C. 1208
Ramalingam, C.S. 1827
Ramli, M. Z. 1274
Randewijk, P.J. 420, 1744
Rentzsch, M. 556
Ribeiro, Antonio C. 636
Ribeiro, Hugo 643
Ritchie, E. 1167
Rizqiawan, Arwindra 534
Rockhill, Andrew A. 1308
Rong, Runjie 541
Rossouw, F.G. 1295
Rost, J. 1006

S

Sadarangani, Chandur 1012, 1301
Saggini, S. 600
Saha, Bishwajit 504, 563
Saito, Y. 671

Sakulhirirak, D. 515
Salam, Z. 1274, 1363, 1682
Sanajit, N. 1412
Sangampai, Pairote 295
Sangwongwanich, Somboon 1213
Sankar, S. Siva 1632
Sano, Kohji 1595
Saparon, A. 527, 1626
Saritsiri, Kritsada 460
Sato, Akira 1651
Sato, S. 1109
Sato, Terukazu 858, 1423, 1734
Sawatpipat, P. 1840
Sawetsakulanond, B. 1356, 1483, 1489
Schmidt, I. 1115
Schneider, T. 249
Scholler, Tobias 1316
Schroder, D. 1187
Schuster, H. 1187
Sebastiao, Pedro J. 636
Sekine, T. 671
Sekiya, Hiroo 1136
Selvaraj, Jeyraj 1290
Senicar, Florian 1533
Sera, D. 226
Sezgin, Volkan 1507
Shah, Laxman 1182
Shahbazi, M. 1455
Shao, Shiyi 1096
Shariatmadar, S. Mohammad 1782
Sharma, Deepen 973
Sharma, V. K. 480
Shen, C.L. 930
Shi, Hu 949
Shiang, J. Z. 433, 1560
Shibano, Yusuke 265
Shisha, Samer 1012
Shuang, Gao 949, 1697
Shuhua, Fang 270, 427, 745
Silber, S. 257
Silva, J. Fernando 1875
Sim, J. M. 401
Sing, Bhim 941
Singer, A. 781
Singh, Bhim 58, 310, 357, 731, 738, 899, 1520, 1816
Singhal, Varun 1816
Sinha, S. 1352, 1605
Sirisuk, Phaophak 847, 1429
Sirisumrannukul, S. 1495
Skorokhod, Y.Y. 1868
Sode-Yome, A. 937
Soh, C.S. 1082
Soltani, J. 1442, 1449
Somsiri, P. 703
Son, Kwang-Myoung 1471
Songboonkaew, J. 989
Soter, Stefan 1533
Soulard, J. 1301

Author Index

Sousa, Duarte M. 636, 817
Srisongkram, W. 1851
Stone, D. A. 382
Stone, David 376
Stumberger, R.H. 1346
Su, Ching-Hung 1610, 1712
Su, Y.-H. 433
Subsingha, W. 1851
Sudhakar, S. Bala 1581
Sudmee, W. 1129
Suetsugu, Tadashi 1136
Sugawara, A. 1109
Sugimura, Hisayuki 504, 563
Sukita, S. 968
Sumner, M. 470
Sun, J. Q. 1394
Sun, Yu-Hua 493
Supriatna, E. G. 1267
Suryawanshi, H. M. 1590
Svechkarenko, D. 1301

T

Ta, Minh C. 1434
Tahami, F. 1147, 1465
Tai, Sio-Un 607, 614
Takeda, T. 1109
Takegami, Eiji 280
Tan, K. 1834
Tan, Siew-Chong 1262
Tan, Weipu 1822, 1880
Taniguchi, T. 1686
Tansatit, Tanvaa 342, 697
Tarateeraseth, V. 515
Tarnekar, S. G. 1708
Tayjasanant, T. 1840, 1862
Tedeschi, Elisabetta 760
Tenca, Pierluigi 1308
Teng, Jen-Hao 305
Teng, L. Y. 1041
Tenti, P. 600
Tenti, Paolo 760
Teo, K.K. 1082
Teodorescu, R. 226, 1247
Teshnizi, Hesameddin Mirzaee 983
Theinmontri, Surapon 295
Thrimawithana, D. J. 648, 654
Tiwari, S.K. 941
To, Huu-Phuc 1284
Tolbert, Leon M. 1055, 1228
Tomihisa, Yoshihiro 858, 1734
Tomioka, Satoshi 280
Tomita, H. 671
Trevisan, D. 600
Tsai, Y.T. 660
Tse, Chi K. 1262
Tseng, S. Y. 440, 433, 1321, 1560, 1567
Tsukakoshi, K. 1885

Tsunesada, Ryota 1417
Tungpimonrut, K. 703

U

Ueda, Shigeta 1070
Ulinuha, A. 1834
Uma, G. 595, 1555, 1632
Uyaisom, C. 1794

V

Vadirajacharya, K. 1810
Vaigundamoorthi, M. 1555
Vargas, Ismael Araujo- 788, 796
Vasudevan, Krishna 1152, 1827
Veerachary, M. 973, 1581
Veszpremi, K. 1115
Vilathgamuwa, D M 1247, 1620
Vinh, Pham Quang 1195, 1202, 1501
Viriya, P. 394, 685
Vishwakarma, Alok 1891
Vogelsberger, M.A. 1346
Volskiy, S.I. 1868
Vorlander, M. 1160

W

Walker, J. A. 1167
Wang, Chengzhi 1636
Wang, Chien-Ming 1610, 1712, 1762, 1767
Wang, Hua 1662
Wang, Peng 541
Wang, Qi 350
Wang, R-J. 420
Wang, Shoufang 1672
Wang, Shuhong 275
Wang, Shun-Chung 546
Wang, Xixi. 1691
Wangsathitwong, S. 1495
Watanabe, Kazushi 280
Watanabe, Takayuki 1136
Wegener, Ralf 1533
Weihrauch, N. C. 886
Welker, Volkmar 1316
Weller, A. 1006
Westermaier, C. 1187
Williams, Barry W 299, 1182, 1242, 1255
Wipasuramonton, P. 703
Wolbank, T.M. 880, 1346
Wong, Man-Chung 607, 614
Wu, Jinn-Chang 493, 824
Wu, Ming-Yi 1703
Wu, T. F 1321
Wu, Xinhui 852

X

Xiaozhong, Liao 949, 1340, 1697

Author Index

Xie, Xiaogao 1601
Xiping, Liu...270, 427, 745
Xu, Hai...1142
Xu, Jianxin..852
Xu, Pengwei 842
Xu, Yun 1636

Y

Yachiangkam, Samart.......................................1327
Yang, C. M. 1560
Yang, Yihan...1822, 1880
Yau, Y. T. 338, 692
Yeh, Jim-Chwen 749, 1703
Yeon, Jae Eul 665
Yingkayun, Krisda 318
Yongyuth, N. 685
Yoothanom, N. 515
Yoshida, Takatsugu 1070
Yoshimura, S. 671
Yoshioka, Satoshi 1543
Yossombut, K. 1840
Yun, S. T...401

Z

Zanchetta, P. 470
Zhan, Yuedong 1662
Zhang, Dongyan 994
Zhang, H.B. 299
Zhang, Junming 632
Zhang, Weiping 291, 994, 1330, 1615, 1752, 1777
Zhang, Xiaofeng 1088
Zhang, Xiaoqiang 291, 1777
Zhang, Yanli 350, 354, 1672
Zhao, Xusen 1752
Zhijun, E. 1804
Zhu, G.R. 551
Zhu, Jianguo 275, 510, 1662
Zirn, Oliver 1316
Zolghadri, M.R. 1404
Zolghadri, Mohammadreza....................................983
Zoller, T. 726
Zou, Yunping..1636